DNA REPLICATION IN EUKARYOTIC CELLS

COLD SPRING HARBOR MONOGRAPH SERIES

The Lactose Operon
The Bacteriophage Lambda
The Molecular Biology of Tumour Viruses
Ribosomes
RNA Phages
RNA Polymerase
The Operon
The Single-Stranded DNA Phages
Transfer RNA:
  Structure, Properties, and Recognition
  Biological Aspects
Molecular Biology of Tumor Viruses, Second Edition:
  DNA Tumor Viruses
  RNA Tumor Viruses
The Molecular Biology of the Yeast *Saccharomyces:*
  Life Cycle and Inheritance
  Metabolism and Gene Expression
Mitochondrial Genes
Lambda II
Nucleases
Gene Function in Prokaryotes
Microbial Development
The Nematode *Caenorhabditis elegans*
Oncogenes and the Molecular Origins of Cancer
Stress Proteins in Biology and Medicine
DNA Topology and Its Biological Implications
The Molecular and Cellular Biology of the Yeast *Saccharomyces:*
  Genome Dynamics, Protein Synthesis, and Energetics
  Gene Expression
Transcriptional Regulation
Reverse Transcriptase
The RNA World
Nucleases, Second Edition
The Biology of Heat Shock Proteins and Molecular Chaperones
Arabidopsis
Cellular Receptors for Animal Viruses
Telomeres
Translational Control
DNA Replication in Eukaryotic Cells

DNA REPLICATION IN EUKARYOTIC CELLS

Edited by

Melvin L. DePamphilis
National Institute of Child Health
and Human Development
National Institutes of Health

COLD SPRING HARBOR LABORATORY PRESS
1996

DNA REPLICATION
IN EUKARYOTIC
CELLS

Monograph 31

Printed in the United States of America
Book design by Emily Harste

ISBN 0-87969-459-9
ISSN 0270-1847
LC 95-71969

All Cold Spring Harbor Laboratory Press publications may be ordered directly from Cold Spring Harbor Laboratory Press, 10 Skyline Drive, Plainview, New York 11803-2500. Phone: 1-800-843-4388 in Continental U.S. and Canada. All other locations: (516) 349-1930. FAX: (516) 349-1946. E-mail: cshpress@cshl.org. For a complete catalog of all Cold Spring Harbor Laboratory Press publications, visit our World Wide Web Site http://www.cshl.org/

Contents

Preface

Science should be made as simple as possible, but not simpler.
Albert Einstein (1879–1955)

In 1959, Arthur Kornberg received the Nobel Prize in medicine and physiology for his pioneering work on the biological synthesis of DNA, an event that marked the beginning of a tremendous effort to understand how cells and viruses replicate their genetic information. This effort produced a new field of scientific research known as DNA replication, accompanied by an extensive literature that has accumulated over the past 35 years. In 1991, Arthur Kornberg and Tania Baker dealt with this vast and complex literature by publishing a wonderful synopsis of DNA replication that covers all aspects of the subject, but whose particular strength lies in its clearly articulated description of the many manifestations of DNA synthesis observed in prokaryotic cells, their bacteriophages, and their plasmids. The goal of this book is to provide a more detailed treatment of the subject of DNA replication in eukaryotic cells.

During the past two decades, advances in technology have permitted analyses of viral, mitochondrial, and cellular genomes that previously had not been possible. Interest in this field accelerated as it became clear that regulation of DNA replication was central to understanding regulation of cell and viral proliferation, events that have a direct impact on our understanding of human diseases. Given the complexity of eukaryotic cells and their need to coordinate proliferation with differentiation during animal development, it is not surprising that the subject of DNA replication in eukaryotic cells is more complex than may have been anticipated. On the order of 30 to 40 proteins are involved directly in the process of DNA replication, as are components of the transcription machinery, chromosome structure, and nuclear structure. DNA replication in eukaryotes involves nuclear and mitochondrial genomes as well as viral genomes containing single-stranded or double-stranded DNA or RNA. It involves specific DNA sequences that determine where replication begins (replication origins) and that mark the ends of linear chromosomes (telomeres). It involves decisions of when, where, and how to initiate DNA replication, repair DNA damage that would otherwise interfere with replication, and ensure that one, and only one, complete and accurate copy of its genome is made before a cell attempts to divide. It utilizes many of the same principles found in prokaryotic cells but offers the possibility that new principles, uniquely suited to problems encountered in eukaryotes, still await discovery. Nevertheless, progress in this field has been impressive.

Our knowledge of events at DNA replication forks in eukaryotic DNA now rivals that in prokaryotic systems: All of the DNA and RNA replication intermediates have been identified, and the proteins that synthesize and repair DNA, as well as the genes that produce these proteins, are well on the way to full disclosure. Even the more subtle processes that initiate and regulate replication are yielding to investigation. Several viral origins have been characterized and shown to contain transcription elements and to function in soluble systems with purified components, behaving in many ways like prokaryotic replication origins. The genomes of mitochondria and single-cell organisms also contain well-characterized replication origins, and many of the proteins that interact with them have been identified. However, initiation of replication at these origins has not yet been made to occur outside of the cell. Intriguingly, the genomes of metazoan organisms contain more complex replication origins whose function depends on nuclear structure as well as soluble proteins. Many of the proteins that regulate DNA replication have been identified, and their chemical modifications and associations with other proteins have been monitored throughout the cell division cycle. Thus, Bruce Stillman, Director of the Cold Spring Harbor Laboratory, and John Inglis, Executive Director of the Cold Spring Harbor Laboratory Press, believed that the time had come to collate this information and summarize our current understanding of DNA replication in eukaryotic cells. I was invited to orchestrate the effort.

The purpose of the book is threefold. First, it is for teachers who would appreciate a current survey of the concepts involved, but who do not have the time to sort through the vast amount of information on which these concepts are based. Second, it is for scientists already working in this and other fields who would appreciate a synopsis of the facts and concepts that have emerged so far. Finally, it is for those who would like to relate DNA replication to other biological problems in eukaryotes, but who may feel intimidated by the sheer size and complexity of the subject. Accordingly, we have organized the book into three parts: concepts, proteins, and systems.

The first part presents the major concepts involved in DNA replication as well as correlative subjects such as DNA repair, chromatin structure, protein phosphorylation, and cell cycle control. Each chapter collates results from many different experimental systems, distills the essential ideas from these results, searches for common underlying themes, and presents to the reader as coherent a view of the subject as possible. The second part describes each class of proteins that are directly involved in DNA replication. These chapters summarize the current information concerning the structure of these proteins, their catalytic activities, their mechanism of action, and useful inhibitors and genetic mutations. The third group of chapters describes what is known about how a particular genome replicates. Each chapter identifies the DNA replication intermediates, the specific DNA or RNA sequences that are required, and the replication proteins that are involved. Although this approach results in a certain amount of redundancy, it also allows the same material to be presented by different authors with different points of view.

To ensure that each aspect of the subject was accurately, completely, and clearly presented, 92 scientists in the field of DNA replication participated in assembling this book. Each of the 40 chapters was written independently by its respective author(s), then reviewed by the editor as well as one or more experts in that aspect of the subject. These included authors of other chapters as well as the reviewers acknowledged below. In addition, this work would not have been possible without the patient, cheerful, and professional editorial staff at the Cold Spring Harbor Laboratory Press. In particular, we are indebted to Patricia Barker and Joan Ebert for bringing this book to fruition. I also thank Dr. Ernst Winnacker, Director of the Gene Center at the University of Munich in Germany, and the Humboldt Society for enabling me to take a brief sabbatical in an excellent scientific environment that contributed to the success of this project.

On behalf of the authors, the external reviewers, and the members of the Cold Spring Harbor Laboratory Press, I express the hope that those who read this book will enjoy it as much as did those who wrote it.

March, 1996 **Melvin L. DePamphilis**

Acknowledgments

The following scientists generously donated their time and effort by reviewing individual chapters. We are indebted to them for sharing the burden of writing this book.

Avril Arthur, *Institute for Biochemistry, Munich, Germany*
Kenneth Berns, *Cornell University Medical College, New York, New York*
Craig Chinault, *Baylor College of Medicine, Houston, Texas*
Louise Chow, *University of Alabama, Birmingham, Alabama*
Donald Ganem, *University of California Medical Center, San Francisco, California*
David Gilbert, *SUNY Health Science Center, Syracuse, New York*
David Glover, *University of Dundee, Dundee, Scotland*
Linda Hanley-Bowdoin, *North Carolina State University, Raleigh, North Carolina*
Chris Hutchison, *University of Dundee, Dundee, Scotland*
Laurie Kaguni, *Michigan State University, East Lansing, Michigan*
Takehiko Kobayashi, *National Institute for Basic Biology, Okazaki, Japan*
Hans Krokan, *University of Trondheim, Trondheim, Norway*
Robert Lehman, *Stanford University Medical School, Stanford, California*
Lawrence Loeb, *University of Washington, Seattle, Washington*
Grant McFadden, *University of Alberta, Edmonton, Alberta, Canada*
Neil Osheroff, *Vanderbilt University Medical School, Nashville, Tennessee*
Dan Ray, *University of California, Los Angeles, California*
J.B. Schvartzman, *Centro de Investigaciones Biológicas, Madrid, Spain*
José Sogo, *ETH, Zürich, Switzerland*
Allan Spradling, *Carnegie Institution of Washington, Baltimore, Maryland*
William Sugden, *University of Wisconsin, Madison, Wisconsin*
David Weaver, *Dana Farber Cancer Institute, Boston, Massachusetts*
Sandra Weller, *University of Connecticut Health Center, Farmington, Connecticut*

DNA REPLICATION IN EUKARYOTIC CELLS

1

Mechanisms for Replicating DNA

George S. Brush and Thomas J. Kelly
Department of Molecular Biology and Genetics
The Johns Hopkins University School of Medicine
Baltimore, Maryland 21205

COMMON STEPS IN DNA REPLICATION

In the most fundamental sense, the general mechanism of DNA replication was first suggested by Watson and Crick as an immediate and obvious consequence of the complementarity of the two strands of the DNA structure. Thus, all replication processes simply involve the melting apart of the two strands followed by the polymerization of each complementary strand on the resulting single-stranded templates. However, when one looks a bit closer at the details of the process of genome duplication, one finds that cells, plasmids, and viruses have evolved a bewildering variety of particular solutions to the problem (Kornberg and Baker 1992). In many cases, the level of complexity of the enzymatic machinery for DNA replication is considerably greater than might have been expected, given that the information required to generate two daughter genomes is encoded in the structure of the parental genome in such a simple way. Such complexity presumably evolved to increase the efficiency and fidelity of DNA replication and to ensure that the duplication of the genome is coordinated with other events in the life of a cell. Below we attempt to distill the observed complexity down to the few basic processes that are common to most DNA replication pathways.

Initial Opening of the Duplex at Origins of Replication

DNA replication usually begins at one or more specific sites within the genome, referred to as origins of DNA replication. The first essential event in the initiation of DNA synthesis is the local opening of the duplex to provide access to the template strands. Origins of replication serve to increase the efficiency of initiation of DNA replication by providing loci for the assembly of multiprotein complexes that mediate DNA synthesis. If the individual components required for DNA synthesis were capable of interacting with random sites along the DNA, a sufficient local concentration of all of the essential factors might be achieved infrequently. Thus, the required components are usually brought together in one place by specific protein-DNA and protein-protein interactions. Origins of replication also provide specific points for the control of cellular DNA replication, ensuring that replication occurs at the right point in the cell cycle and that each segment of DNA is replicated precisely once. The initial opening of the DNA duplex at origins is generally mediated by specific initiator proteins and can be facilitated by certain structural features of the DNA (e.g., negative supercoiling, easily unwound sequences) and by certain accessory proteins (e.g., single-stranded DNA binding proteins [SSBs]).

Duplex Unwinding at Replication Forks

The initial opening of the duplex allows the establishment of a replication fork(s). The essence of this process is the loading of a DNA helicase on one or both of the exposed single strands. DNA helicases are enzymes that utilize the energy of ATP hydrolysis to translocate unidirectionally along a DNA strand, melting the duplex. At some origins, helicases are loaded onto both DNA strands, resulting in the establishment of two active replication forks (bidirectional DNA replication). In other cases, only a single fork is established (unidirectional DNA replication). An important characteristic of a given helicase activity is its polarity of translocation. Some helicases track along in the 3′ to 5′ direction of the so-called "leading strand" template of the replication fork, and others move in the 5′ to 3′ direction of the "lagging strand" template.

Priming of DNA Synthesis

In most DNA replication systems, the process of starting new DNA chains is distinct from the process of elongating established chains. Thus, all of the known DNA polymerases are incapable of starting chains de novo and require a primer to begin DNA synthesis. In eukaryotic cells, DNA synthesis is generally primed by short RNA chains. The separation of initiation from elongation and the use of RNA, rather than DNA, to initiate DNA synthesis are probably consequences of the requirement for extremely high fidelity in DNA replication. One major mechanism for achieving high fidelity involves proofreading the products of a given polymerization step before proceeding to the next polymerization step. A proofreading exonuclease built into most DNA polymerases recognizes and efficiently excises mismatched nucleotides at the primer terminus. It is likely that this type of proofreading mechanism would be rather inefficient in detecting errors at or near the beginning of a new DNA chain. This problem is apparently solved by "marking" sites of initiation with a chemically distinct RNA chain. The replication machinery can later remove the RNA and fill the resulting gap by extension of an upstream DNA chain by DNA polymerase with proofreading function. This is a relatively costly solution, but one that maintains the accuracy of the genome.

Although both nuclear and mitochondrial DNA replication make use of RNA priming, the enzymatic mechanism of primer synthesis is different in the two cases. The synthesis of RNA primers for nuclear DNA synthesis is accomplished by a primase enzyme that is a component of DNA polymerase-α (pol-α:primase). This primase appears to be the only enzyme that is capable of priming chromosomal DNA synthesis in eukaryotes and is distinct from the RNA polymerases involved in the transcription of nuclear genes. A few eukaryotic viruses, such as SV40, utilize the cellular priming apparatus for the replication of their genomes. There is no evidence for the presence of pol-α:primase in mitochondria, and it is believed that the synthesis of RNA primers for mitochondrial DNA replication is carried out by uniquely mitochondrial enzymes. Interestingly, initiation of DNA synthesis at the primary origin of DNA replication in the mitochondrial genome is mediated by mitochondrial RNA polymerase, the same enzyme that is responsible for mitochondrial transcription.

Some eukaryotic viruses have evolved mechanisms for priming DNA synthesis that do not involve oligoribonucleotide synthesis. Two examples are discussed below. The parvovirus genomes have self-complementary hairpin termini that allow the 3′ termini of the genomic single strands to prime DNA synthesis. This example represents a very simple case, since the primers for DNA replication are incorporated into the viral genome itself. Adenoviruses make use of a protein to prime DNA replication. In this interesting case, the first phosphodiester bond is formed between the terminal nucleotide and a serine residue of a virus-encoded protein.

The distribution and frequency of priming events on the two parental strands determine the general pattern of DNA replication in a given system (Fig. 1). In continuous DNA replication there is only a single priming event per template strand, so each progeny DNA strand is synthesized continuously from one end to the other. Examples include mitochondrial DNA replication and the replication of the parvoviruses and adenoviruses. In semidiscontinuous DNA replication one progeny

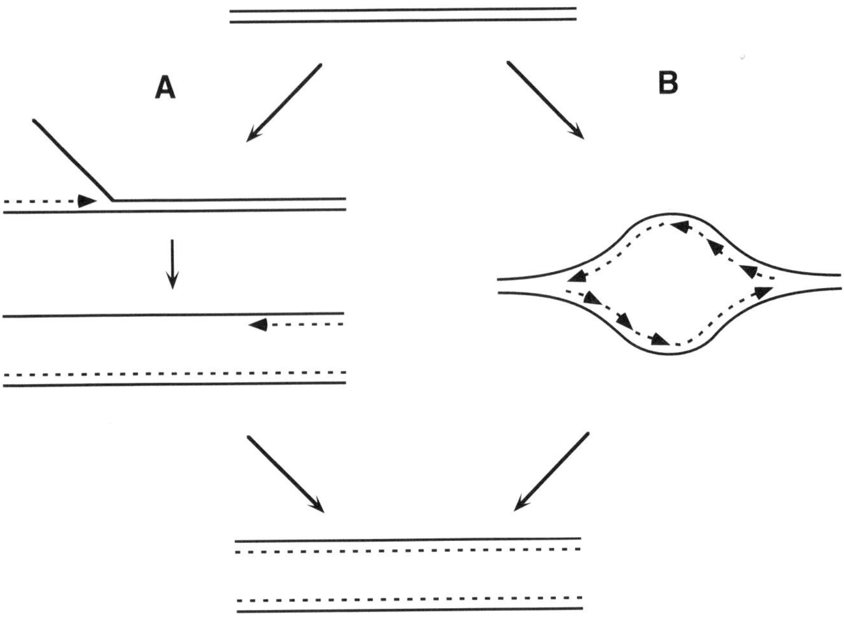

Figure 1 Continuous (*A*) versus semidiscontinuous (*B*) DNA replication. The two basic mechanisms for replicating DNA are shown, with nascent DNA (*broken lines*) synthesized from primers in the direction of the arrows.

strand (leading strand) is elongated continuously from a single primer, whereas the other progeny strand (lagging strand) is constructed from many short DNA chains elongated from multiple primers. Polymerization of the leading strand occurs in the same direction as replication fork movement, and polymerization of the lagging strand occurs in the opposite direction. Nuclear DNA replication and the replication of some viruses (e.g., SV40, herpesviruses) proceed by a semidiscontinuous mechanism. In semidiscontinuous DNA synthesis, completion of the lagging strand requires a repair system to remove the primers, fill in the resulting gaps, and join together the short nascent DNA strands.

Elongation of Nascent DNA Strands

Eukaryotic cells contain several DNA polymerase activities. The nuclear pol-α:primase complex mentioned above appears to be largely concerned with the synthesis of primers during semidiscontinuous DNA synthesis. The DNA polymerase activity of the enzyme is capable of extending the oligoribonucleotides synthesized by the intrinsic primase activity. However, pol-α:primase is relatively nonprocessive, so that only a short DNA chain is synthesized before the enzyme dissociates from the template. The relatively rapid turnover of the enzyme is consistent with its primary role in the synthesis of multiple primers on the lagging strand as the replication fork advances. The RNA-DNA primers synthesized by pol-α:primase can be extended by the highly processive DNA polymerases δ (pol-δ) and ε (pol-ε). Both of these enzymes can be assembled into complexes with the eukaryotic processivity factor proliferating cell nuclear antigen (PCNA) at primer termini. PCNA serves as a topological clamp, tethering pol-δ or pol-ε to the template, so that many thousands of nucleotides can be polymerized before the enzyme dissociates. It is likely that the bulk of chromosomal DNA synthesis is mediated by either pol-δ or pol-ε or both. In addition to their high processivity, which contributes to the efficiency of DNA replication, both enzymes have active proofreading exonuclease activities that enhance the fidelity of DNA replication.

As described in greater detail below, mitochondria possess a DNA polymerase activity distinct from the enzymes involved in nuclear DNA replication. Although some eukaryotic viruses, like SV40, utilize the resident cellular DNA polymerases, others, like the adenoviruses, encode DNA polymerases of their own.

Maturation of Nascent DNA Strands

A consequence of the widespread use of RNA priming mechanisms to initiate DNA synthesis is the requirement for an enzymatic machinery to remove the primers and replace them with DNA. In the case of nuclear DNA replication, this process is rapid and efficient, occurring soon after Okazaki fragment synthesis. Recent work suggests that primers are removed through the combined action of ribonuclease H (RNase H) and a 5′ to 3′ exonuclease. The resulting gaps are filled by DNA polymerase, probably pol-δ or pol-ε, and the final nick is sealed by DNA ligase I. Since this maturation process may result in the nearly complete removal of both RNA and DNA portions of the primers originally generated by the pol-α:primase complex, nearly all of the final DNA product is the result of polymerization by the more processive (and more accurate) DNA polymerases δ and ε. The process of primer replacement in mitochondria is probably mediated by a different set of enzymes, but little work has been done on the problem. A significant fraction of mitochondrial genomes contain residual ribonucleotides at the initiation sites of DNA synthesis, suggesting that removal of RNA primers may be relatively slow and/or inefficient within the organelle.

The DNA-priming mechanism utilized by the parvoviruses requires a maturation mechanism that is quite different from that needed for RNA priming. Since DNA synthesis is primed by a duplex hairpin at the terminus of the genome (see below), there remains the problem of replicating the hairpin primer sequence itself. This is accomplished by a novel and interesting mechanism in which the hairpin sequence first serves as a primer and then is transferred to the newly synthesized DNA strand where it can serve as a template for DNA synthesis (see below). This maturation scheme regenerates the full-length parvovirus genome. Although a similar scheme could, in principle, be used for the maturation of the ends (telomeres) of linear cellular chromosomes, it is now clear that telomeres are synthesized by a special enzyme, telomerase, which polymerizes short DNA repeats at chromosome termini using an RNA template that is intrinsic to the enzyme. Further discussion of telomerase is beyond the scope of this chapter.

CONTINUOUS DNA REPLICATION

As indicated above, mitochondrial DNA and a number of viruses that replicate in eukaryotic cells have evolved replication mechanisms in which both progeny strands are synthesized continuously from one end to the other. To illustrate this general mode of DNA replication, we summarize the replication pathways of two viral systems, the parvoviruses and the adenoviruses, and then discuss mammalian mitochondrial DNA replication. The examples chosen also illustrate the major modes of priming DNA synthesis: DNA self-priming, protein priming, and RNA priming.

Mechanism of Parvovirus (AAV) DNA Replication

The parvoviruses are the simplest of the viruses that infect animal cells (for reviews, see Berns 1990a; Muzyczka 1992; Berns and Linden 1995). They contain linear single-stranded DNA genomes that are converted to duplex replicating intermediates within infected cells. Progeny single strands are generated from the duplex intermediates by a self-priming, strand-displacement mechanism. Viral DNA replication is largely dependent on host enzymes, probably the factors normally involved in leading-strand synthesis at chromosomal replication forks. Only one viral gene, *rep*, is required for DNA replication, and its product(s) is mainly concerned with processing of the termini of the genome.

Two general classes of parvoviruses have been identified (Berns 1990a). Adeno-associated viruses (AAV) generally require coinfection with a helper virus for efficient replication, although the basis for this requirement is not understood. Other parvoviruses are capable of autonomous replication in the absence of any helper virus (Cotmore and Tattersall 1987). All members of the parvovirus group have a similar genetic organization and follow a roughly similar replication pathway. The replication mechanism of AAV is discussed here as prototypical of the group.

AAV Genome

The AAV genome contains two large open reading frames (ORFs) (Srivastava et al. 1983). The 5′ ORF, referred to as the *rep* gene, encodes four distinct, but overlapping, polypeptide chains generated by alternate promoter utilization and alternate splicing. The two largest gene products, Rep78 and Rep68, have been implicated in AAV DNA replication by genetic studies (Hermonat et al. 1984; Tratschin et al. 1984; Yang et al. 1992). It is not known whether they play different roles in DNA replication in vivo, but either protein appears to be capable of supporting AAV DNA replication in vitro. The Rep proteins have site-specific endonuclease and helicase activities that mediate the novel processing reaction responsible for regenerating AAV termini (Ashktorab and Srivastava 1989; Im and Muzyczka 1990; Snyder et al. 1990a,b; McCarty et al. 1994; Ni et al. 1994).

Replication Pathway

Current evidence indicates that AAV DNA replication is completely continuous (Berns and Hauswirth 1979; Challberg and Kelly 1989; Muzyczka 1992; Berns and Linden 1995). The termini of the viral genome contain 145-bp palindromic sequences, referred to as inverted terminal repeats (ITRs) (Lusby et al. 1980). When folded in such a way as to maximize base-pairing, the ITRs are capable of forming T-shaped hairpin structures containing only seven unpaired bases (Fig. 2). The 3′ end of the genome represents a primer terminus that is used initially to convert the infecting single-stranded genome to a linear duplex form whose strands are covalently linked at one end via the terminal hairpin (Fig. 2) (Hauswirth and Berns 1979). The hairpin terminus is converted to an open duplex through the action of the Rep protein(s) and host enzymes (Tattersall and Ward 1976; Hauswirth and Berns 1979; Snyder et al. 1990b). This processing event, called terminal resolution or hairpin

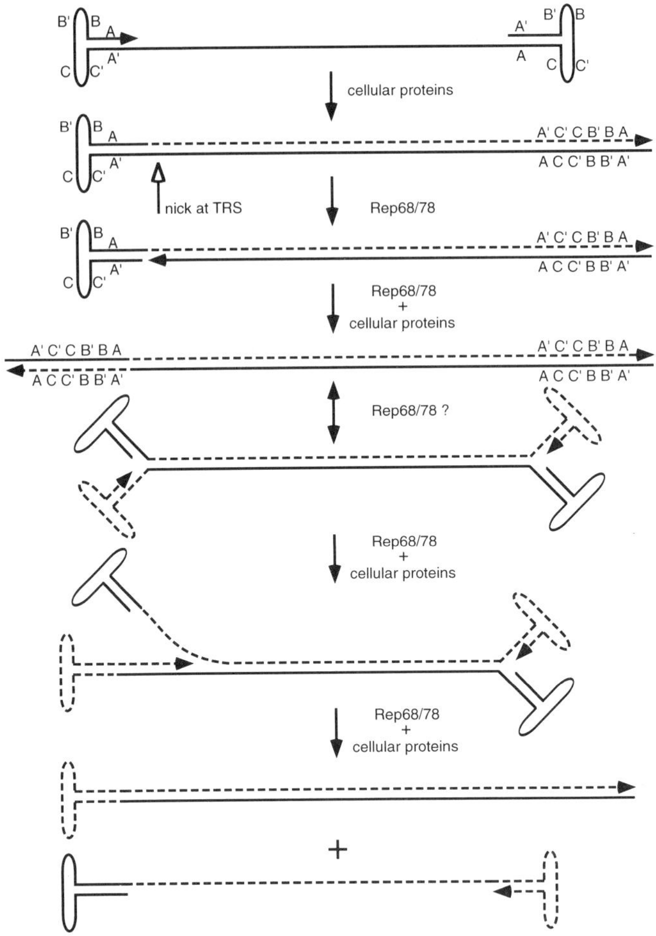

Figure 2 Model for AAV DNA replication. The single-stranded AAV genome contains self-complementary sequences at the termini (indicated by the letters A,A′,B,B′,C,C′) and folds to form the T-shaped hairpin structures shown. The 3′-OH end of one of the two hairpins serves to prime AAV DNA synthesis by host enzymes. Newly synthesized DNA is indicated by broken lines. The terminal resolution site (*trs*) is nicked by the AAV Rep protein to begin the terminal resolution process. DNA synthesis proceeds from the newly created 3′-OH terminus at the *trs* and continues to the end of the genome, resulting in the formation of an open duplex replication intermediate. The two strands of this intermediate are identical to the infecting AAV genomes except for inversion of the terminal palindromic sequence. After rearrangement of the termini to reform the hairpin structures, DNA synthesis proceeds by a displacement mechanism. Each round of DNA replication produces a progeny single strand and a duplex that can undergo terminal resolution to produce the open duplex replication intermediate.

transfer, is initiated by a single endonucleolytic cleavage at a site opposite the original 3′ terminus of the genome. Cleavage at the terminal resolution site (TRS) is carried out by the Rep protein, which becomes covalently linked to the 5′-phosphoryl terminus at the TRS cleavage site (Im and Muzyczka 1990; Snyder et al. 1990a). The 3′-OH terminus is then extended by a cellular polymerase(s) to replicate the terminal hairpin. It is likely that the helicase activity of the covalently bound Rep protein facilitates this process by unwinding the hairpin. The result of the terminal resolution reaction is the generation of a linear duplex replication intermediate whose strands are identical to those of the infecting AAV genome except for inversion of the terminal palindromic sequence (Lusby et al. 1981). The intermediate is then replicated via a self-priming, strand-displacement mechanism (Fig. 2). Priming of this reaction probably involves a rearrangement of the termini reforming the hairpin structures. The resulting structure resembles a replication fork and is presumably a competent substrate for the cellular elongation machinery. It is not clear whether rearrangement of the termini is catalyzed by Rep or a cellular protein or whether it occurs spontaneously. The products of the second stage of the AAV replication reaction are a displaced single strand and a linear duplex with covalently joined termini. The latter can be resolved to an open duplex by Rep and the process can be repeated. Since the two termini of AAV are essentially equivalent, equal numbers of the two complementary strands are synthesized. Both strands are packaged into progeny virus particles (Berns 1990b).

Enzymology of AAV DNA Replication

Several in vitro systems that carry out various aspects of the AAV DNA replication reaction have been developed (Hong et al. 1992; Ni et al. 1994; Ward et al. 1994). When duplex DNA molecules containing functional AAV origins are used as templates, either Rep68 or Rep78 is required for extensive DNA synthesis (Ni et al. 1994). It has also been reported that DNA synthesis in vitro can be stimulated significantly when the extracts are prepared from adenovirus-infected cells, suggesting that a protein(s) encoded or induced by the helper may play some role in AAV DNA replication (Ni et al. 1994). However, it seems likely that most of the proteins involved in viral DNA replication are derived from the host cell. Recent work indicates involvement of the cellular replication proteins PCNA, replication factor C (RF-C), and replication protein A (RP-A) (N. Muzyczka, pers. comm.). The DNA polymerase responsible for AAV DNA chain elongation has not yet been identified with certainty, but, given the involvement of the processivity factor PCNA, pol-δ and pol-ε are likely candidates. Thus, AAV is probably replicated mainly by the apparatus responsible for leading-strand synthesis at cellular replication forks (see below). It is not known whether movement of the AAV replication fork is catalyzed by the helicase activity of Rep68/78 or whether a cellular helicase mediates this function. Since the AAV genome is linear and rather short, there is no apparent need for a DNA topoisomerase, consistent with the observation that antibodies against mammalian topoisomerase I and II do not inhibit DNA replication in vitro (N. Muzyczka, pers. comm.).

Mechanism of Adenovirus DNA Replication

Adenovirus Genome

The mechanism of adenovirus DNA replication has been well characterized because of the early development of a cell-free replication system (Challberg and Kelly 1979). The most extensively studied viral serotypes are the closely related Ad2 and Ad5 (for reviews, see Challberg and Kelly 1989; Hay and Russell 1989; Stillman 1989; Salas 1991; Van der Vliet 1991; Kornberg and Baker 1992). The Ad2/5 genome is a linear double-stranded DNA molecule of 36 kb with two novel structural features: (1) The nucleotide sequences at the ends of the genome are identical for the first 103 nucleotides and (2) the 5′end of each strand is covalently linked to the virus-encoded, 55-kD terminal protein (TP). Viral DNA replication requires three viral proteins: an 80-kD precursor to the terminal protein (pTP), a 140-kD DNA polymerase (Ad pol), and a single-stranded DNA-binding protein (Ad DBP). The efficiency of DNA replication is increased significantly by several cellular proteins, including the transcription factors nuclear factor I (NFI) and octamer-binding protein 1 (Oct-1) and a DNA topoisomerase (NFII) (Nagata et al. 1982; Pruijn et al. 1986; Rosenfeld and Kelly 1986; Rosenfeld et al. 1987). Initiation of adenovirus DNA replication takes place by a protein-priming mechanism at the termini of the genome, and each daughter strand is synthesized continuously from one end to the other (Lechner and Kelly 1977; Rekosh et al. 1977; Challberg et al. 1980; Ikeda et al. 1982).

Initiation of Adenovirus DNA Replication

The adenovirus origin of replication, encompassing roughly the first 50 bp of the viral genome, is the locus of binding of several proteins. NFI and Oct-1 bind to specific recognition sites within the region between nucleotides 19 and 51 (Nagata et al. 1983b; Rawlins et al. 1984; Pruijn et al. 1986; Rosenfeld and Kelly 1986; Rosenfeld et al. 1987; Wides et al. 1987). A complex of the two viral proteins, pTP and Ad pol (pTP-pol), appears to recognize sequence elements within the first 18 nucleotides that constitute the minimal essential origin (Ikeda et al. 1982; Mul and Van der Vliet 1992; Temperley and Hay 1992). Binding of pTP-pol to the viral origin of replication is the critical first step in the initiation reaction (Fig. 3). This step is greatly facilitated by protein-protein interactions between pTP-pol and the bound cellular proteins, NFI and Oct-1 (Bosher et al. 1990; Chen et al. 1990; Mul et al. 1990). The binding reaction may also be stimulated by the presence of the 55-kD terminal protein at the 5′ end of one of the parental strands (see Fig. 3). Once the pTP-pol is bound at the terminus of the genome, DNA synthesis is initiated by the formation of a phosphodiester bond between the first nucleotide in the new DNA chain, dCMP, and the β-hydroxyl group of a serine residue in the pTP (Challberg et al. 1980). This novel protein-priming reaction is presumably catalyzed by the adenovirus DNA polymerase, which then functions to extend the nascent chain (Challberg et al. 1980; Ikeda et al. 1982). An early intermediate in the elongation reaction of Ad5 is a trinucleotide covalently attached to the pTP (pTP-CAT) (King and Van der Vliet 1994). Interestingly, it has recently been demonstrated that the template for synthesis of pTP-CAT is not the first three nucleotides in the parental strand, but nucleotides 4–6. The sequence at the terminus of the template strand (3′ GTAGTAGTTA...5′) is repetitive, so it is thought that the pTP-CAT, synthesized opposite residues 4–6, translocates to positions 1–3 to start elongation. The function of this rather baroque mechanism is not yet clear, but it has been suggested that it may serve to protect the integrity of the terminal sequences of the viral genome during DNA replication (King and Van der Vliet 1994).

Elongation of Adenovirus DNA Strands

Following initiation at one of the termini, DNA synthesis proceeds by a displacement mechanism, producing a daughter duplex and a free single strand (Fig. 3) (Lechner and Kelly 1977). The latter can cyclize via self-complementary termini to form a panhandle structure. Since the duplex panhandle is identical to the termini of the original genome, initiation of DNA synthesis can presumably occur by the same mechanism, leading to the completion of a second daughter duplex. In vitro studies have demonstrated that the elongation of nascent adenovirus DNA strands re-

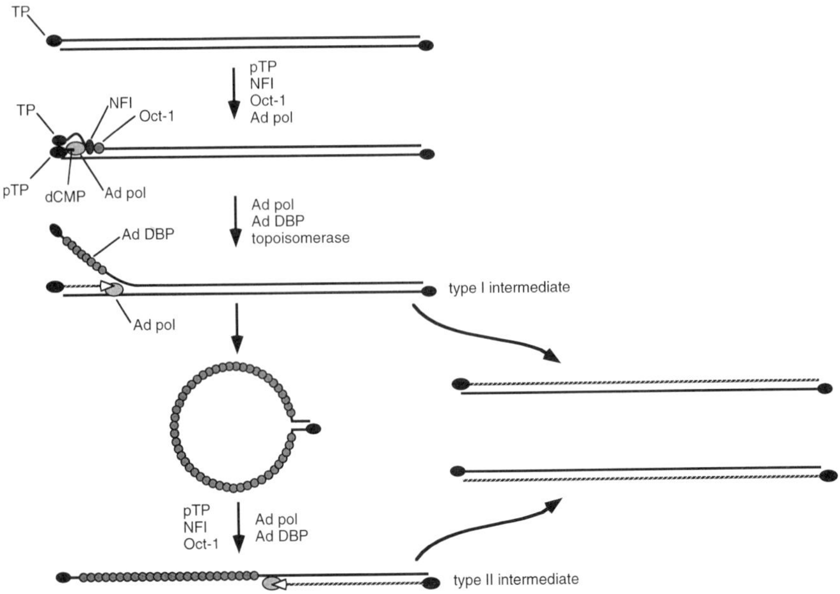

Figure 3 Model for adenovirus DNA replication. The 5′ ends of the adenovirus genome are covalently linked to the 55-kD terminal protein (TP). Initiation of DNA synthesis requires the 80-kD preterminal protein (pTP), the 140-kD adenovirus DNA polymerase (Ad pol), and the cellular proteins nuclear factor I (NFI) and octamer-binding protein 1 (Oct-1). Initiation is facilitated by the adenovirus DNA-binding protein (Ad DBP), which enhances NFI binding. Initiation involves the formation of a phosphodiester bond between the β-OH of a serine residue in pTP and the 5′ phosphoryl group of dCMP (see text for details). Elongation of nascent adenovirus DNA chains (*dotted lines*) requires the Ad pol, Ad DBP, and a cellular topoisomerase (NFII). Displacement synthesis on type I intermediates leads to the generation of a progeny duplex and a displaced single strand. The displaced single strand circularizes via self-complementary termini and serves as a substrate for a second initiation event, identical to the first. Completion of DNA synthesis on type II intermediates results in the formation of a second progeny duplex.

quires only two proteins, Ad pol and Ad DBP (Lindenbaum et al. 1986). The adenovirus DNA polymerase is a highly processive enzyme that extends DNA primers much more efficiently than RNA primers (Field et al. 1984; Lindenbaum et al. 1986). Its activity is specifically stimulated by the Ad DBP; prokaryotic or other eukaryotic DBPs cannot replace Ad DBP in the replication reaction (Lindenbaum et al. 1986). Movement of the displacement fork during adenovirus DNA replication does not appear to require a separate helicase activity. The Ad DNA pol, acting in concert with the Ad DBP, is sufficient to unwind the parental duplex. The energy required for unidirectional fork movement is derived exclusively from hydrolysis of the deoxyribonucleoside triphosphate precursors for DNA synthesis (Pronk et al. 1994). Thus, adenovirus appears to have evolved a highly efficient two-protein engine for DNA synthesis that combines the functions of helicase and polymerase. Interestingly, in vitro studies have suggested that the replication of the adenovirus genome is facilitated by a cellular DNA topoisomerase (NFII) even though the genome is linear (Nagata et al. 1983a). Replication proceeds efficiently in the presence of Ad DNA pol and Ad DBP until replication is about 25% complete, after which fork movement is slow unless topoisomerase activity is present. The basis for this requirement is not completely clear, but presumably it reflects some hindrance to the free rotation of the unreplicated parental duplex in adenovirus replication intermediates.

Mechanism of Mitochondrial DNA Replication

General Features of Mitochondrial DNA Replication

Much of the early work on the mechanism of mitochondrial DNA replication was done in mammalian systems, but studies in other organisms, particularly yeast, have been increasingly important for identifying and characterizing required replication proteins (for reviews, see Clayton 1991, 1992; Kornberg and Baker 1992; Schmitt and Clayton 1993). The available evidence suggests that the basic features of mitochondrial DNA replication have been conserved from yeast to man, so the following description focuses primarily on mammalian cells (Clayton 1982, 1991). The mammalian mitochondrial genome is a closed duplex circle of about 16 kb. Cells contain on the order of 10^3–10^4 mitochondrial genomes with an average of 5–10 per organelle. Unlike the nuclear DNA, mitochondrial genomes are replicated throughout the cell cycle, and templates appear to be drawn at random from the pool of genomes. Thus, in a given cell cycle, some mitochondrial genomes are replicated more than once and some are not replicated at all. The regulatory mechanisms that determine the total number of mitochondrial genomes per cell are not understood.

Studies of the structures of replication intermediates have provided strong evidence that the replication of the mitochondrial genome occurs by a completely continuous mechanism (Fig. 4). In mammalian cells there are two origins of DNA replication, one for each strand, which are located about 11 kb apart (Clayton 1982). At each of these origins, DNA synthesis is primed by specific RNA molecules (see below). A single priming event apparently suffices for the complete synthesis of each strand. The origin for heavy (H) strand synthesis (O_H) is activated first, resulting in the establishment of a replication fork that moves unidirectionally (Gillum and Clayton 1979; Chang and Clayton 1985). At this fork a new H strand is continuously elongated, and the parental H strand is displaced. The origin for light (L) strand synthesis (O_L) is activated only when it has been rendered single-stranded by the passage of the H-strand replication fork (Wong and Clayton 1985a,b). It is likely that a specific secondary structure that forms in the displaced strand is recognized by the protein(s) responsible for RNA primer synthesis. The L strand is continuously elongated from the primer in what is essentially a large gap-filling reaction. Since the two origins of DNA replication are separated by approximately two-thirds of the genome, the replication of the two strands of mitochondrial DNA is quite asynchronous (see Fig. 4). When the synthesis of the H strand is completed, a process that takes about an hour, two products are generated. One product is a duplex circle with a full-length progeny H strand, and the other is a gapped circle with an incomplete progeny L strand, which is subsequently extended to full length.

The primary event in mitochondrial DNA replication is the activation of the H-strand origin, since the activation of the L-strand origin is secondary to the establishment of the H-strand replication fork. Thus, one likely point of regulation of mitochondrial DNA synthesis is the generation of the initial RNA primer. There may also be a second control point that operates at the level of elongation of the H strand. A majority of mitochondrial genomes in mammalian cells contain a so-called D loop in the vicinity of O_H (Gillum and Clayton 1979). This structure consists of a short nascent H DNA whose 5′ end is located at the origin for H-strand synthesis. The nascent H strands in D loops appear to be metabolically labile and are turned over at a rate that significantly exceeds the rate of replication of mitochondrial genomes. The existence of the short nascent H strands suggests that there may be a barrier to chain elongation just downstream from O_H (Clayton 1982; Madsen et al. 1993). It is possible that mechanisms exist that regulate whether or not the H-strand replication fork can pass this barrier, but at present there is no evidence on this point.

Less work has been done on the replication of the mitochondrial DNA of yeast, but it is likely that DNA synthesis is largely continuous, although there may be more than one priming event per strand (Blanc and Dujon 1980; Baldacci et al. 1984; de Zamaroczy et al. 1984; Schmitt and

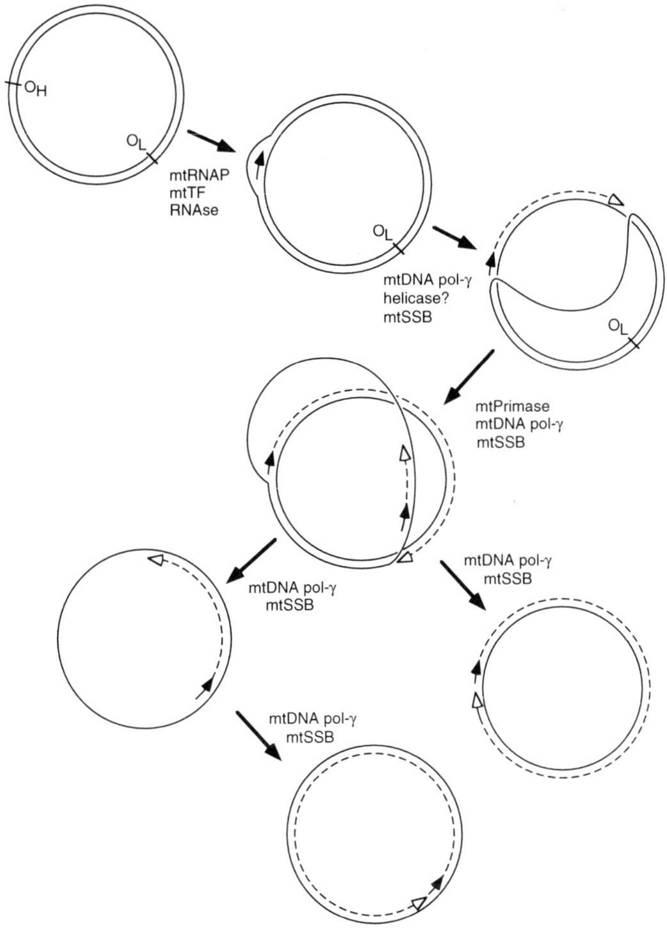

Figure 4 Model for mammalian mitochondrial DNA replication. The two mitochondrial origins O_H and O_L are located about two-thirds of the genome apart. Initiation of DNA replication begins with the synthesis of a transcript by the mitochondrial RNA polymerase (mtRNAP) and an essential transcription factor (mtTF). The transcripts are either processed by an endoribonuclease (RNase) to form a primer for mitochondrial DNA replication or continue to be extended to form mRNA. If cleavage occurs, a nascent H strand (*broken line*) is synthesized by the mitochondrial DNA polymerase (mtDNA pol-γ). Other proteins, including a putative mitochondrial helicase and SSB (mtSSB) probably participate in the chain elongation process. Synthesis of the L strand is initiated by a mitochondrial primase (mtPrimase) after the displacement fork passes O_L. Completion of H-strand synthesis and separation of the two progeny molecules occurs before completion of L-strand synthesis.

Clayton 1993). At least four putative origins of replication have been located in the yeast mitochondrial genome. It is likely that more origins are required because of the large size of the yeast mitochondrial genome. The structural features of yeast origins are similar to those of higher eukaryotes (Schmitt and Clayton 1993). However, unlike mammalian origins, the sites of initiation of synthesis of the two strands are in close proximity rather than being separated by many kilobases.

Priming of Mitochondrial DNA Synthesis

Transcription of the mammalian mitochondrial genome is directed by two divergent promoters located within the D-loop region (Chang and Clayton 1984). Transcripts initiated at one of these promoters (LSP) also serve to prime H-strand synthesis during mitochondrial DNA replication (Chang and Clayton 1985; Chang et al. 1985). A fraction of the LSP transcripts are cleaved by an endoribonuclease at one of several discrete sites at O_H, approximately 100 nucleotides from the promoter. The 3′-OH termini of the transcripts are then elongated by the mitochondrial DNA replication apparatus. Transcription from the LSP requires the mitochondrial RNA polymerase and a specific transcription factor, mtTFA (Fisher and Clayton 1985; Parisi et al. 1993). An RNase capable of processing nascent transcripts at sites near the transition from RNA synthesis to DNA synthesis has been identified in mammalian cells and yeast (Chang and Clayton 1987; Schmitt and Clayton 1992; Stohl and Clayton 1992). The enzyme, called RNase MRP, contains a required RNA moiety that may be involved in recognition of the specific cleavage sites. Recently, it has been suggested that a second mitochondrial nuclease, endonuclease G, may also play a role in processing nascent transcripts to generate primers for DNA replication (Cote and Ruiz-Carrillo 1993).

Priming of L-strand synthesis at O_L appears to be mediated by a mitochondrial primase that recognizes a specific stem-loop structure which forms in the H-strand template (Wong and Clayton 1985a,b; Hixson et al. 1986). The enzyme initiates the synthesis of a short oligoribonucleotide at a run of T residues in this loop. The transition from RNA synthesis to DNA synthesis occurs immediately adjacent to the base of the stem. The mitochondrial primase appears to require an RNA component for activity, and the most likely candidate is cellular 5.8S RNA (Wong and Clayton 1986). Thus, the priming mechanisms for the synthesis of both strands of the mitochondrial genome require ribonucleoprotein enzymes.

Mechanism of Chain Elongation

Although some of the required enzymes have been identified, the detailed mechanism of continuous DNA synthesis in mitochondria is not well understood. This is mainly due to the absence of an efficient cell-free replication system. All of the components involved in mitochondrial DNA replication, including both the priming and elongation stages, are encoded by nuclear genes and imported into the organelle. A highly processive mitochondrial DNA polymerase, referred to as DNA polymerase-γ, has been purified from several sources (Wernette and Kaguni 1986; Wernette et al. 1988; Insdorf and Bogenhagen 1989a,b; Gray and Wong 1992). The human enzyme consists of subunits of 140 kD and 54 kD. The larger subunit contains the polymerase active site as well as a potent 3′ to 5′ exonuclease activity. The latter shows strong preference for unpaired primer termini and presumably plays a proofreading role. The yeast gene encoding the catalytic subunit of DNA polymerase-γ (*MIP1*) has been cloned (Foury 1989). The amino acid sequence of the Mip1 protein is similar to both eukaryotic nuclear DNA polymerases and reverse transcriptases, but has no discernible resemblance to prokaryotic DNA polymerases.

Mitochondrial DBPs have also been identifed from several organisms (Pavco and Van Tuyle 1985; Mignotte et al. 1988; Van Dyck et al. 1992; Tiranti et al. 1993; Stroumbakis et al. 1994). Interestingly, these proteins are homologous to *Escherichia coli* SSB, and their physicochemical properties are quite similar to the prokaryotic enzyme as well. Yeast mutants lacking the *RIM1* gene, which encodes the mitochondrial SSB, are completely devoid of mitochondrial DNA, consistent with an essential role for the protein in mitochondrial DNA replication (Van Dyck et al. 1992). Presumably, the mitochondrial SSB facilitates the elongation of nascent DNA chains by the mitochondrial DNA polymerase. However, the functional interactions of the two proteins have not been extensively analyzed to date.

Another likely accessory protein for mitochondrial DNA synthesis is a DNA helicase, since H-strand synthesis requires melting of the parental strands at the replication fork. A helicase activity has been identified in highly purified mitochondria from bovine brain (Hehman and Hauswirth 1992). This activity, which translocates in the 3′ to 5′ direction on the single-stranded portion of partially duplex substrates, is a likely candidate for a replicative helicase in mitochondria, but this role has not been verified directly. The Pif1 helicase of budding yeast has been implicated in the repair, recombination, and replication of the mitochondrial genome (Lahaye et al. 1991). Genetic interactions have been observed between the *PIF1* gene and the *RIM1* gene, encoding the mitochondrial SSB (Van Dyck et al. 1992). Moreover, a yeast strain deficient in the Pif1 enzyme loses mitochondrial DNA at elevated temperatures.

Mitochondrial DNA replication remains the best example of a naturally occurring, completely continuous DNA replication system in eukaryotic cells. Although a number of the components required for DNA replication have been identified, and we have a general picture of the overall replication pathway, much remains to be learned about the biochemical mechanisms involved. Future in vitro studies of mitochondrial DNA synthesis with purified components will likely fill in many of the gaps in our knowledge.

SEMIDISCONTINUOUS DNA SYNTHESIS

Replication of Chromosomal DNA

In the continuous DNA replication systems that we have discussed so far, the two complementary DNA strands are synthesized relatively independently and asynchronously. In contrast, the replication of eukaryotic chromosomal DNA occurs by a semidiscontinuous mechanism in which the synthesis of the two strands is strongly coupled in space and time. One advantage of semidiscontinuous DNA replication over completely continuous DNA replication is that the generation of single-stranded DNA is much more localized and transient. This may help preserve the integrity of the genome, since breaks and other lesions in single-stranded DNA are difficult to repair.

Semidiscontinuous DNA replication is mediated by a complex protein machine assembled at each replication fork. The proteins comprising this replication machine act in concert to unwind the parental strands and carry out the simultaneous synthesis of the two progeny strands. Both progeny strands are synthesized in the 5′ to 3′ direction, but since the parental DNA strands are antiparallel, two distinct mechanisms of DNA synthesis are required. One of the two progeny strands (leading strand) is synthesized continuously in the direction of fork movement. The other strand (lagging strand) is synthesized discontinuously in the direction opposite to fork movement. Discontinuous DNA synthesis on the lagging-strand template involves the repeated synthesis of oligoribonucleotide primers, which are then elongated into short DNA chains (Okazaki fragments). Following their synthesis, Okazaki fragments are processed to remove the RNA primers and joined together to form an uninterrupted progeny strand.

The semidiscontinuous DNA replication mechanism appears to be quite ancient in origin, as its essential features have been conserved from bacteria to man. Although the basic biochemical processes that occur at eukaryotic and prokaryotic replication forks are similar, there are many differences in detail. For example, in *E. coli*, a single DNA polymerase is thought to form a dimeric complex that mediates DNA synthesis on both the leading and lagging strands (Maki et al. 1988; McHenry 1988), whereas eukaryotes contain at least three distinct DNA polymerase activities, all of which participate in DNA replication. The critical protein-protein interactions that are required for formation of an efficient replica tion machine also appear to differ between prokaryotes and eukaryotes. In *E. coli*, the primase activity responsible for initiating new DNA chains forms a specific complex with the helicase activity responsible for unwinding the parental strands as the fork moves (Kornberg and Baker 1992) whereas in eukaryotes, the primase is a subunit of one of the DNA polymerase molecules (Lehman and Kaguni 1989; Wang 1991). Some of the accessory proteins involved in DNA synthesis have more complex structures in eukaryotes than in prokaryotes. One example is the single-stranded DNA-binding protein, which is a single polypeptide chain in *E. coli* (Meyer and Laine 1990) but a large heterotrimeric protein in eukaryotes (Fairman and Stillman 1988; Wold and Kelly 1988). The greater complexity may be related to the more complex mechanisms required to regulate DNA replication in eukaryotes. Despite these apparent differences, there are many similarities between prokaryotic and eukaryotic DNA replication that reflect a common evolutionary origin (see Stillman, this volume). One particularly striking example is the remarkable structural similarity of the β subunit of *E. coli* DNA polymerase III and the eukaryotic protein PCNA, both of which function as "sliding clamps" to increase the processivity of DNA synthesis (Kong et al. 1992; Krishna et al. 1994).

A novel feature of eukaryotic chromosomes is the packaging of DNA into chromatin. This fact may account for the surprising disparity in the rate of fork movement in prokaryotes and eukaryotes. Whereas the *E. coli* replication machinery moves at the prodigious rate of 100 kb/minute, unwinding DNA at some 10,000 rpm, eukaryotic fork movement is much slower (0.5–5 kb/min) (Kornberg and Baker 1992). The presence of histones may limit the maximal rate at which DNA polymerization can occur. Such a constraint might have contributed to the evolution of intrinsically slower polymerases in eukaryotes. In addition, special mechanisms may have evolved to allow fork movement without disruption of either the replication complex or the nucleosomes. In this manner, the DNA remains packaged even while DNA synthesis is taking place. However, the ability of large protein assemblies to negotiate one another is not unique to eukaryotes, as in vitro experiments have shown that the phage T4 replication apparatus can bypass an RNA polymerase complex moving in either direction without complete dissociation of either set of proteins (Liu et al. 1993; Liu and Alberts 1995).

Viral Model Systems

Direct biochemical investigation into the mechanisms underlying cellular DNA replication has proven difficult. With the exception of special cases such as *Xenopus* eggs, it has not yet been possible to develop an in vitro cellular DNA replication system. In the absence of such a system, investigators have turned to viral models to study the mechanism of cellular DNA replication. Viruses present several advantages, including small genomes and well-defined origins of replication. The Papovaviridae, which include the SV40, polyoma, and papilloma viruses, have been particularly important to the study of chromosomal replication. Upon infection of suitable host cells, the genomes of these viruses are transported to the nucleus, where the double-stranded viral minichromosomes are replicated by mechanisms that closely resemble the cellular process. Initiation occurs at a single origin within the viral genome and proceeds bidirectionally in a semidiscontinuous manner. The great advantage of SV40 and other papovaviruses is that many of the important steps in DNA replication are performed by host proteins.

A significant advance in the study of eukaryotic DNA replication came with the development of a cell-free SV40 system in 1984 (Li and Kelly 1984). Primate cytoplasmic extract and a single viral protein, the large T antigen, carry out all the functions required for complete replication of SV40-origin-containing plasmid DNA. T antigen recognizes the origin and unwinds the duplex, providing access for the numerous host replication proteins that function coordinately to synthesize the daughter DNA strands (for review, see Stillman 1989; Borowiec et al. 1990; Kelly 1991). Fractionation of human extract and reconstitution of replication activity using an in vitro complementation assay have led to the identification of those cellular proteins necessary and sufficient for viral DNA replication (Ishimi et al. 1988; Wold et al. 1989; Tsurimoto et al. 1990; Weinberg et al. 1990; Eki et al. 1992; Waga et al. 1994). Given that the mechanisms of SV40 and cellular DNA replication appear to be very similar, there is considerable confidence that the same proteins have identical functions in the replication of the chromosomes. It should be noted that analysis of the viral system has not identified all of the proteins involved in eukaryotic DNA replication. For example, genetic studies in yeast strongly suggest that pol-ε is necessary for DNA replication in vivo (Araki et al. 1992; Budd and Campbell 1993), but this enzyme was not originally identified as necessary for SV40 DNA replication.

As discussed previously, the natural template for eukaryotic DNA replication is chromatin. Whereas the cell-free system was originally developed with naked DNA templates, a number of studies have attempted to more accurately reflect the in vivo replicative process by employing chromatin templates. Although in vitro replication of such minichromosomes is very inefficient relative to that of naked DNA, the repression of initiation due to tightly bound histones can be relieved by the presence of transcription factors bound near the origin (Cheng and Kelly 1989; Cheng et al. 1992). A similar stimulatory effect of transcription factors on replication efficiency has been observed in vivo. Several possible mechanisms could account for this derepression, including increased accessibility of the origin to the initiator protein or activation of the replication apparatus by contact with factors bound near the origin (see DePamphilis, this volume). Further studies have focused on the fate of parental nucleosomes during DNA replication and indicate that nucleosomes remain associated with the replicating DNA during DNA synthesis (Randall and Kelly 1992). In addition, the assembly of new chromatin appears to be coupled to SV40 DNA replication in vitro (Stillman 1986). Experiments such as these have added to our understanding of the mechanism by which the replication machinery negotiates other DNA-bound structures, a process that must occur continuously in the nucleus.

Enzymology of the Replication Fork

Replication Proteins

Studies employing the SV40 model system have resulted in the identification of many cellular replication proteins. Recent work employing both genetic and biochemical techniques has added to this list. A summary of known replication proteins is presented in Table 1 and in some detail below.

Helicases

The focal point of all replication forks is the helicase, which catalyzes the transition from double- to single-stranded DNA. In *E. coli,* the DnaB

Table 1 Cellular replication proteins

Protein	Subunit (kD)[a]	Replicative function
DNA polymerases		
pol-α:primase	180, 70, 58, 48	DNA polymerase, primase
pol-δ	125, 48	DNA polymerase, 3′ to 5′ exonuclease
pol-ε	258, 55	DNA polymerase, 3′ to 5′ exonuclease
Accessory proteins		
RP-A	70, 32, 14	single-stranded DNA binding
PCNA	36	pol-δ/ε processivity factor
RF-C	145, 40, 38, 37, 36.5	loads PCNA onto template
Nucleases		
ribonuclease H1	89	Okazaki fragment maturation
FEN-1 (MF-1)	44	Okazaki fragment maturation
Others		
DNA ligase I	102	joins Okazaki fragments
topoisomerase I	100	unlinks parental strands
topoisomerase II	172	unlinks parental strands and progeny duplexes

References include those cited within the text as well as Kesti et al. (1993), Syvaoja and Linn (1989), Eder and Walder (1991), and Miller et al. (1981).

[a]Approximiate molecular weights of human polypeptides.

helicase translocates in a 5′ to 3′ direction while unwinding DNA (LeBowitz and McMacken 1986) and therefore is bound to the lagging-strand template during DNA replication. The same mechanism is employed by the phages T4 and T7 (Matson et al. 1983; Richardson and Nossal 1989), which induce their own helicases to engage in semi-discontinuous replication. In direct contrast, SV40 T antigen is bound to the leading-strand template and moves in the 3′ to 5′ direction advancing the fork (Goetz et al. 1988; Wiekowski et al. 1988). Although the polarity of translocation is not conserved, the replicative helicases of *E. coli*, T4, T7, and SV40 are all hexamers, suggesting a common quaternary structure for enzymes with this function. To date, several eukaryotic helicases have been identified, but conclusive identification of the enzyme that acts at chromosomal replication forks must await further biochemical and genetic investigation.

Single-stranded DNA-binding Protein

As the replication fork advances, a helix-destabilizing protein is required to maintain the single-stranded DNA structure that serves as a template for RNA priming and DNA synthesis. In eukaryotic cells, replication protein A (RP-A; RF-A; HSSB) performs this function (Wobbe et al. 1987; Fairman and Stillman 1988; Wold and Kelly 1988). This phosphoprotein has three subunits with molecular weights of 70,000, 32,000, and 14,000. The large subunit contains the DNA-binding activity but cannot support SV40 DNA replication in vitro by itself (Kenny et al. 1990; Erdile et al. 1991); therefore, at least one of the two smaller subunits is likely to be required for replication. In support of this hypothesis, all three yeast genes encoding RP-A are essential for viability (Heyer et al. 1990; Brill and Stillman 1991), and antibodies directed against any of the three subunits inhibit SV40 DNA replication in vitro (Erdile et al. 1990, 1991; Kenny et al. 1990; Umbricht et al. 1993). Although the roles of the two smaller RP-A subunits remain unknown, the middle subunit is phosphorylated during S phase of the cell cycle (Din et al. 1990), suggesting that this subunit may play some role in regulating DNA replication.

The DNA-binding properties of RP-A have been investigated extensively, but there remains some disagreement about fundamental aspects of the protein-DNA interaction. RP-A binds relatively nonspecifically to single-stranded DNA but exhibits a modest preference for DNA sequences rich in pyrimidines (Kim et al. 1992). Several different binding site sizes have been reported, ranging from 8 to 30 nucleotides for human RP-A and up to 100 nucleotides for the yeast protein (Alani et al. 1992; Kim et al. 1992; Blackwell and Borowiec 1994). Some evidence has been presented that RP-A forms two different types of complexes with DNA, which may account for the variability in site size estimates. Further disagreement centers on the degree of cooperativity of the DNA-binding reaction. Most prokaryotic single-stranded DNA-binding proteins display a high level of cooperativity, allowing the rapid and complete binding of any exposed single-stranded regions in the genome. Steady-state fluorescence experiments indicate that this is also the case with yeast RP-A (Alani et al. 1992). However, a series of studies employing electrophoretic mobility shift of oligonucleotides in the presence of human RP-A have demonstrated little cooperativity (Kim et al. 1992; Kim and Wold 1995). Other studies employing a similar technique suggest that only one of two proposed binding modes is highly cooperative (Blackwell and Borowiec 1994). The lack of consensus with regard to the DNA-binding properties may result from differences in experimental technique, or possibly differences in the state of the protein itself, such as the extent of phosphorylation.

DNA Pol-α:Primase Complex

DNA synthesis is initiated by the bifunctional pol-α:primase complex, a heterotetrameric phosphoprotein (Wang 1991). The primase activity resides in the 48-kD D subunit and is tightly associated with the 58-kD C subunit, which is thought to tether the primase to the 180-kD polymerase A subunit (Copeland and Wang 1993; Santocanale et al. 1993; Bakkenist and Cotterill 1994; Stadlbauer et al. 1994). With the exception of the *Drosophila* enzyme (Cotterill et al. 1987), there is no proofreading exonuclease activity associated with the A subunit. The remaining 70-kD B subunit has no known catalytic function, but it may contribute to recruitment of pol-α:primase to the replication fork (Collins et al. 1993) (see below).

The main function of pol-α:primase is to serve as a priming enzyme. The primase catalyzes the synthesis of complementary oligoribonucleotides, which are then extended a short distance by the A subunit DNA polymerase activity. Although high concentrations of pol-α:primase can support the complete replication of SV40 origin-containing plasmid DNA in vitro (Ishimi et al. 1988), it is unlikely that cellular replication relies heavily on this polymerase during elongation. The low processivity of the enzyme and the lack of an associated proofreading exonuclease suggest that pol-α:primase serves exclusively to initiate DNA synthesis on the lagging strand. Dissociation of pol-α:primase from the DNA provides a primer terminus for the assembly of the PCNA/pol-δ or /pol-ε complexes, highly processive polymerases that can efficiently extend the RNA/DNA primers originally synthesized by pol-α:primase.

DNA Polymerases δ and ε

The heterodimeric DNA polymerases δ and ε are involved in the elongation stage of DNA replication. Unlike pol-α:primase, polymerases δ and ε do not act alone but require the action of two auxiliary factors. The multisubunit replication factor C (RF-C) (Tsurimoto and Stillman 1989; Lee et al. 1991a) binds to the primer terminus immediately after RNA/DNA primer synthesis has been completed by pol-α:primase, allowing the subsequent assembly of a functional pol-δ or pol-ε complex. Once bound to the primer-template junction, RF-C loads PCNA onto the DNA in an energy-dependent reaction. PCNA then functions as a processivity factor, binding to pol-δ or pol-ε and maintaining a stable interaction between polymerase and template (Tan et al. 1986; Prelich et al. 1987; Burgers 1991; Lee et al. 1991a; Tsurimoto and Stillman 1991; Podust and Hubscher 1993). Ultimately, one or both of these processive polymerases, with their intrinsic proofreading activities, probably synthesize all of the cellular DNA, thereby ensuring faithful duplication of the genome (see below).

DNA polymerases δ and ε have some similarities in catalytic function and are both essential for viability in yeast (for reviews, see Wang 1991; So and Downey 1992). However, it is not yet clear whether they have specialized roles in replication. pol-ε does not substitute well for pol-δ in the cell-free SV40 DNA replication system with purified proteins (Lee et al. 1991b), so it may contribute to DNA synthesis only when certain additional factors are present. It is also possible that these enzymes have important roles that are not limited to DNA replication. For example, recent experiments indicate that pol-ε is one member of the S-phase checkpoint pathway (Navas et al. 1995), a biochemical feedback mechanism that delays cell-cycle progression upon damage to the DNA or inhibition of DNA synthesis during the replicative phase. Thus, in addition to synthesizing DNA, pol-ε may be involved in monitoring the status of DNA replication.

Nucleases

Synthesis of the lagging strand results in the generation of DNA fragments with RNA primers at the 5′ ends. For lagging-strand synthesis to be completed, these primers must be removed and the resulting gap must be filled. In *E. coli,* DNA polymerase I, with its intrinsic 5′ to 3′ exonuclease activity, mediates both functions (Kornberg and Baker 1992). None of the replicative eukaryotic polymerases contains a 5′ to 3′ exonuclease, indicating that other factors are involved in the initial processing of Okazaki fragments. A 44-kD 5′ to 3′ exonuclease (FEN-1; MF-1) that is required for the formation of covalently closed circular DNA during in vitro SV40 DNA replication has been identified in human cells (Ishimi et al. 1988; Harrington and Lieber 1994; Waga et al. 1994). Recently, studies employing purified calf proteins and a model lagging-strand template have shown that FEN-1 and RNase H1 act together to process Okazaki fragments. RNase H1 nicks the primer on the 5′ side of the 3′ ribonucleotide, providing a suitable substrate for FEN-1, which removes the 3′-terminal ribonucleotide of the RNA primer. The oligoribonucleotide is displaced and the gap is filled by DNA polymerase, resulting in nicked double-stranded DNA (Turchi and Bambara 1993; Huang et al. 1994; Murante et al. 1994; Turchi et al. 1994).

DNA Ligase

After removal of the RNA primers and extension of the DNA chains through the resulting gaps, the nascent DNA fragments must be joined to complete the synthesis of the lagging strand. Of the three DNA ligases that have been identified in mammalian cells (Lindahl and Barnes 1992), DNA ligase I is the most likely candidate to carry out this function in vivo. In a cell-free SV40 system containing highly purified proteins, DNA ligase I, but not DNA ligase III, catalyzes the formation of covalently closed daughter molecules (Waga et al. 1994). Although DNA ligase II has not been tested, this enzyme is present at low abundance and is likely to be involved in DNA repair. Genetic studies support the conclusion that DNA ligase I is involved in cellular replication in vivo. It has been shown that the human DNA ligase I gene can complement a *Saccharomyces cerevisiae* cdc9 mutant, which is defective in DNA replication due to DNA ligase deficiency (Barnes et al. 1990).

Topoisomerases

As replication proceeds and the parental strands are unwound, positive supercoils are potentially introduced ahead of the replication fork. The resulting accumulation of torsional strain could lead to inhibition of fork movement if not relieved by a DNA topoisomerase. In eukaryotic cells, two types of topoisomerases have been discovered (Wang 1985). The type I enzyme introduces a transient single-strand break in the DNA, thereby relaxing either negatively or positively supercoiled DNA. Type II topoisomerase introduces a transient double-strand DNA break through which duplex DNA is passed. In addition to relaxing DNA, this enzyme can decatenate intertwined molecules. Studies with the SV40 model system employing naked DNA have revealed that either type of topoisomerase is capable of removing the positive supercoils ahead of the fork, allowing rapid and efficient DNA synthesis (Yang et al. 1987). However, the progeny DNA molecules that are formed remain multiply intertwined because of failure to remove all of the links between the parental strands during DNA synthesis. Topoisomerase II is required to resolve this tangled structure into two separate progeny genomes, allowing subsequent segregation (Yang et al. 1987).

The topological problems accompanying SV40 DNA replication in vitro are likely to arise during cellular replication as well. The length of chromosomal DNA and its association with nuclear proteins to form chromatin probably precludes free rotation of the DNA during replication. Therefore, the two eukaryotic topoisomerases are likely to function in vivo as they do in the SV40 system. It has been suggested that topoisomerase II may have a special role in the completion of DNA synthesis where two adjacent forks converge (Ishimi et al. 1992). However, this conclusion is based on studies of a cell-free SV40 system containing only one DNA polymerase (pol-α:primase) and may not accurately reflect the normal cellular process.

Organization of the Replication Fork

Most of the studies aimed at defining the organization of the cellular replication fork have revolved around the SV40 system. Although the identity and mechanism of the cellular helicase remain unknown, it is likely that the known cellular proteins play similar roles in chromosomal replication as in the viral model. The current picture of the eukaryotic replication fork is summarized in Figure 5. Leading-strand synthesis is performed by a processive polymerase complex (PCNA/pol-δ or PCNA/pol-ε) and a single auxiliary protein (RP-A). Lagging-strand synthesis requires additional proteins. pol-α:primase synthesizes RNA/DNA primers that are extended by complexes similar to those operating on the leading strand. Primer removal and completion of DNA synthesis require FEN-1 nuclease, RNase H1, and DNA ligase I as described above.

Studies with both crude and purified systems have demonstrated that assembly of replication proteins at the replication fork is mediated by specific protein-protein contacts. These interactions are critical for the coordinated synthesis of both leading and lagging strands, and are likely to be important at all stages of the replicative process. Specific mechanisms involving multiple proteins at the replication fork are discussed below.

Protein-protein Interactions

In the SV40 system, the viral T antigen and the cellular proteins pol-α:primase and RP-A are required for initiation of DNA replication, and a variety of studies have provided evidence for significant molecular interactions among the three proteins. For example, biochemical analysis of model reactions has uncovered a number of instances in which one protein affects the activity of another (Kenny et al. 1989; Collins and Kelly 1991; Collins et al. 1993; Melendy and Stillman 1993; Murakami

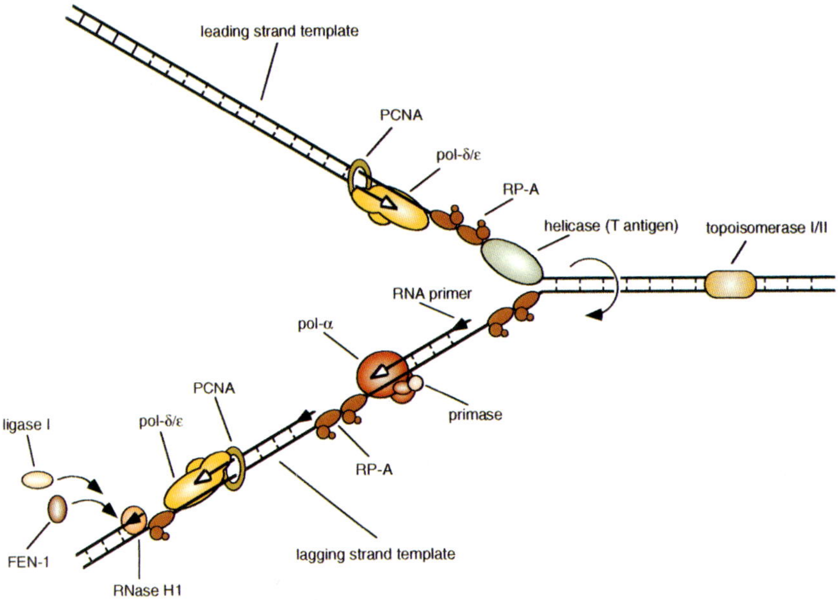

Figure 5 Semidiscontinuous DNA synthesis. This figure shows a diagrammatic view of the organization of the eukaryotic replication fork. The pol-δ/ε auxiliary factor RF-C is not included in this diagram because it is not known whether this protein stays associated with the replication apparatus after loading PCNA. See text for further details.

and Hurwitz 1993a). These biochemical data are supported by physical studies demonstrating direct association of each combination of two proteins. In particular, much information has been gathered on the pol-α/T-antigen interaction. This association appears complex, possibly involving three of the four pol-α subunits (A, B, and D) (Dornreiter et al. 1990, 1992; Gannon and Lane 1990; Collins et al. 1993; Bruckner et al. 1995).

It is very likely that the molecular interactions among initiation proteins are important determinants of efficiency and specificity in the initiation reaction. Unwinding of the SV40 origin requires T antigen and RP-A, and contact between these two proteins is likely to be important at this early stage of replication. Although the isolated large subunit of RP-A does not interact with T antigen, intact RP-A does (Dornreiter et al. 1992), suggesting that one or both of the smaller subunits could play a role in this interaction. Once the duplex is unwound, interaction between T antigen and pol-α may contribute to efficient priming on the RP-A-coated, single-stranded DNA template. It is possible that the different interaction domains on pol-α have different functional roles during replication. For example, the D subunit may help to coordinate initiation at the origin (Schneider et al. 1994), and the A and B subunits may be important for tethering pol-α:primase to the advancing helicase (see below). Priming is also likely to be influenced by an RP-A/primase interaction, which has been demonstrated to occur in vitro through the large subunit of RP-A (Dornreiter et al. 1992). The close association of the three proteins involved in initiation of SV40 DNA replication is quite striking and suggests that similar interactions may occur at cellular origins of replication. Confirmation of this possibility awaits identification of the cellular counterpart(s) of T antigen.

Protein-protein interactions are also critically important during elongation of the nascent DNA chains. As discussed above, efficient elongation requires a highly processive polymerase, and it is likely that pol-δ (or pol-ε), with its auxiliary factors RF-C and PCNA, carries out this function on both leading and lagging strands. The processivity of pol-δ is absolutely dependent on interaction with PCNA, which is topologically linked to the template DNA. There is no evidence for interaction between the processive polymerases and other replication proteins, although RP-A does stimulate pol-δ activity to some extent (Kenny et al. 1989).

Recent data suggest the possibility that protein-protein interactions may even be important during the final stage of DNA replication, when the Okazaki fragments are joined to form a continuous DNA chain. Studies with purified proteins have shown that DNA ligase I, but not DNA ligase III, can catalyze the formation of covalently closed daughter DNA molecules in the SV40 model system (Waga et al. 1994). Since both ligases are capable of sealing nicks in double-stranded DNA, the strict requirement for DNA ligase I may reflect specific contacts between the enzyme and other proteins involved in the maturation of nascent strands.

A number of interesting interactions between replication proteins and other cellular factors have been uncovered. Although the functions of these interactions are not entirely clear, some of these may be involved in regulating replication. RP-A interacts with several proteins, including the transcription factors p53, GAL4, and VP16 (He et al. 1993; Li and Botchan 1993) and the repair proteins XPA and XPG (He et al. 1995; Matsuda et al. 1995). Another example, discovered with the SV40 model system, is the coupling of chromatin assembly to DNA replication (Stillman 1986). This coupling is strongly suggestive of a direct interaction between the chromatin assembly apparatus and the replication machinery and probably evolved to ensure rapid and efficient packaging of the progeny DNA following synthesis.

Synthesis of Okazaki Fragments and Cycling of Pol-α:Primase

Because synthesis of the lagging strand proceeds in the direction opposite to fork movement, repeated priming events by the pol-α:primase are required. There are two general mechanisms by which this priming could occur. One possibility is that pol-α:primase completely dissociates from the template following completion of each RNA/DNA primer (distributive mechanism). In this scenario, synthesis of each RNA/DNA primer would involve association of a different molecule of pol-α:primase with the template. Alternatively, pol-α:primase might be tethered to the replication complex at the fork via specific protein-protein interactions. In this case, the enzyme would not leave the domain of the template following the completion of an RNA/DNA primer but would immediately reassociate with the template at a new site to initiate primer synthesis (processive mechanism).

At this point, the available data are not sufficient to choose between the two mechanisms. Evidence supporting a distributive model has come from dilution experiments in the cell-free SV40 DNA replication system (Murakami and Hurwitz 1993b). When the pol-α:primase concentration is decreased by dilution, the rate of DNA replication decreases proportionately. Although these data clearly indicate that pol-α:primase dissociates from the template at a measurable rate during DNA synthesis, it has not yet been possible to directly measure the number of priming events mediated by a given pol-α:primase prior to dissociation. Evidence for processive priming has been provided by the strong physical interaction between pol-α and T antigen (Collins et al. 1993). It is known that T antigen is capable of unwinding long segments of DNA without dissociation. Binding of pol-α:primase to T antigen could significantly increase the lifetime of pol-α:primase at the fork. Thus, a single pol-α:primase molecule could proceed through several priming cycles, allowing efficient synthesis of the lagging strand. In agreement with this model, the presence of T antigen increases the apparent processivity of pol-α:primase on model templates (Collins and Kelly 1991).

In addition to possibly increasing processivity of priming, interaction between pol-α:primase and T antigen may facilitate access of pol-α:primase to the template during replication. Experiments with model templates have shown that RP-A-coated single-stranded DNA is resistant to priming. In the presence of T antigen, this inhibition is relieved (Collins et al. 1993; Melendy and Stillman 1993). Therefore, interaction of pol-α:primase with T antigen may be required for primase to productively interact with the template DNA. During cellular DNA replication, a similar mechanism must be employed to allow efficient priming of the RP-A-coated DNA. It is possible that the cellular process is also mediated by a direct physical interaction between pol-α:primase and the replicative helicase.

Proofreading Mechanisms

The catalytic subunits of polymerases δ and ε contain 3′ to 5′ exonuclease activities, allowing high-fidelity DNA synthesis. Therefore, the majority of nuclear DNA synthesis employs the conventional proofreading mechanism first outlined in prokaryotic systems. However, pol-α: primase contains no obvious proofreading activity. It has been reported that a cryptic 3′ to 5′ exonuclease is uncovered when the *Drosophila* pol-α:primase complex is dissociated (Cotterill et al. 1987), but this phenomenon has not been observed with pol-α:primase complexes from other species. It is not clear how errors are corrected in DNA synthesized by pol-α:primase. One possibility is that a separate exonuclease proofreading activity exists. However, it is more likely that maturation of lagging-strand DNA fragments leads to removal of mismatches downstream from the RNA primer in pol-α-catalyzed regions. Prior to ligation of the nicked strand, FEN-1 nuclease and a proofreading polymerase could act in conjunction to accurately replace the DNA originally polymerized by pol-α. Since FEN-1 does not function on a gapped template but requires a nick, removal of DNA and resynthesis would have to occur concurrently. Alternating nucleolytic and synthetic steps by these two enzymes would closely resemble the "nick translation" reaction of *E. coli* DNA polymerase I (Murante et al. 1994).

Replication Centers

An interesting recent development has been the discovery that DNA replication may occur in relatively discrete foci in the nuclei of mammalian cells. When nascent DNA is labeled by exposure of cells to a short pulse of BrdU, the label is localized to a relatively small number of intranuclear sites. Enumeration of these sites or "replication centers" suggests that each may contain as many as 100 replication forks. Thus, DNA replication may be highly compartmentalized in eukaryotic cells (for review, see Cook 1991). Both PCNA and RP-A have been detected at these replication centers by immunofluorescence (Bravo and Macdonald-

Bravo 1987; Cardoso et al. 1993; Brenot-Bosc et al. 1995). Interestingly, the large subunit of RP-A has been detected at foci prior to the onset of replication and is a possible component of a prereplication complex at the origin (Adachi and Laemmli 1992).

The organization of replication centers is not well understood. One hypothesis invokes the nuclear matrix as an insoluble support that serves as a foundation for DNA replication and organizes the replication centers. However, a clear understanding of the structure and function of foci will have to await further biochemical investigation.

Control of Eukaryotic DNA Replication—Unanswered Questions

The basic enzymology of the eukaryotic replication fork has been uncovered through the identification and characterization of the replication proteins required for DNA synthesis in vitro. However, many fundamental questions remain regarding the control of replication in vivo. These problems focus on the coordination of DNA replication with other events in the cell cycle, including the timing of initiation, the prevention of multiple rounds of replication during a single replicative phase of the cell cycle, and the response to insults that may compromise the faithful duplication of the genetic material. Much of the current research on eukaryotic DNA replication is directed at characterizing these mechanisms.

Initiator proteins, which recognize origins of replication and unwind the double-stranded DNA, are likely to be central to the regulation of cellular DNA replication. The presence or the activation of these proteins may be responsible for controlling the timing of initiation during the cell cycle, and their subsequent removal or inactivation may be necessary to prevent re-replication during a single S phase. Identification of eukaryotic initiator proteins has proven to be quite difficult. However, a complex of six polypeptides that specifically recognizes yeast origin sequences has been purified recently (Bell and Stillman 1992). This origin recognition complex (ORC) is likely to function as an initiator, and thorough characterization of the protein should provide insight into the mechanisms that operate to regulate initiation. Studies have already shown that ORC protein level is constant through the cell cycle (Diffley and Cocker 1992). This observation suggests that the initiator protein is activated at the G_1/S transition, possibly by protein phosphorylation or by specific interactions with other cellular proteins.

Studies with the cell-free SV40 system have provided some evidence that protein phosphorylation may be involved in replication control.

SV40 DNA replication does not occur until the G_1/S transition has been reached in the host cell. Interestingly, extracts prepared from cells in G_1 are incompetent for in vitro SV40 DNA replication unless supplemented with protein phosphatase 2Ac (PP2Ac) or cdc2 kinase (Virshup et al. 1989; D'Urso et al. 1990). Although the key target of these G_1-activating enzymes is not yet clear, T antigen is one possible candidate. Both PP2Ac and cdc2 kinase can directly modify the phosphorylation state of T antigen in vitro, thereby regulating its ability to unwind the SV40 origin. It is possible that cellular initiator proteins may be activated at the G_1/S boundary by similar phosphorylation and/or dephosphorylation events.

Two cellular proteins involved in the initiation of SV40 DNA replication in vitro, pol-α:primase and RP-A, are phosphorylated in cell-cycle-dependent manners (Din et al. 1990; Nasheuer et al. 1991). There has been no clear demonstration that phosphorylation affects the replicative function of either protein, but this remains a reasonable possibility. It has also been suggested that RP-A phosphorylation might be involved in a signaling pathway that coordinates DNA replication with the cell cycle (Brush et al. 1994). In this case, RP-A would have both a replicative and a regulatory role. Such bifunctionality has recently been demonstrated with pol-ε, which appears to act as a replicative polymerase and as a member of the S-phase checkpoint pathway (Navas et al. 1995). This finding may help to explain the need for two rather redundant DNA polymerase activities in eukaryotic cells.

Observations such as those described above have provided some clues into the mechanism of DNA replication control, but it is clear that the majority of the process is not well understood. The characterization of the pathways and their components that allow communication between the replication machine and other cellular apparati will rely on both the biochemistry of the viral model systems and the genetics afforded by yeast. Since the basic mechanism of DNA replication has been conserved from yeast to humans, many of the regulatory mechanisms are likely to be conserved as well. As a result, these systems will continue to be used interchangeably to provide us with an understanding of how the cell efficiently and faithfully replicates its DNA in preparation for cell division.

REFERENCES

Adachi, Y. and U.K. Laemmli. 1992. Identification of nuclear pre-replication centers poised for DNA synthesis in *Xenopus* egg extracts: Immunolocalization study of replication protein A. *J. Cell Biol* **119:** 1–15.

Alani, E., R. Thresher, J.D. Griffith, and R.D. Kolodner. 1992. Characterization of DNA-binding and strand-exchange stimulation properties of y-RPA, a yeast single-strand-DNA-binding protein. *J. Mol. Biol.* **227:** 54–71.

Araki, H., P.A. Ropp, A.L. Johnson, L.H. Johnston, A. Morrison, and A. Sugino. 1992. DNA polymerase II, the probable homolog of mammalian DNA polymerase epsilon, replicates chromosomal DNA in the yeast *Saccharomyces cerevisiae. EMBO J.* **11:** 733–740.

Ashktorab, H. and A. Srivastava. 1989. Identification of nuclear proteins that specifically interact with adeno-associated virus type 2 inverted terminal repeat hairpin DNA. *J. Virol.* **63:** 3034–3039.

Bakkenist, C.J. and S. Cotterill. 1994. The 50-kDa primase subunit of *Drosophila melanogaster DNA* polymerase alpha: Molecular characterization of the gene and functional analysis of the overexpressed protein. *J. Biol. Chem.* **269:** 26759–26766.

Baldacci, G., B. Cherif-Zahar, and G. Bernardi. 1984. The initiation of DNA replication in the mitochondrial genome of yeast. *EMBO J.* **3:** 2115–2120.

Barnes, D.E., L.H. Johnston, K. Kodama, A.E. Tomkinson, D.D. Lasko, and T. Lindahl. 1990. Human DNA ligase I cDNA: Cloning and functional expression in *Saccharomyces cerevisiae. Proc. Natl. Acad. Sci.* **87:** 6679–6683.

Bell, S.P. and B. Stillman. 1992. ATP-dependent recognition of eukaryotic origins of DNA replication by a multiprotein complex [see comments]. *Nature* **357:** 128–134.

Berns, K.I. 1990a. Parvoviridae and their replication. In *Virology* (ed. B.N. Fields and D.M. Knipe), pp. 1743–1763. Raven Press, New York.

———. 1990b. Parvovirus replication. *Microbiol. Rev.* **54:** 316–329.

Berns, K.I. and W.W. Hauswirth. 1979. Adeno-associated viruses. *Adv. Virus Res.* **25:** 407–449.

Berns, K.I. and R.M. Linden. 1995. The cryptic life style of adeno-associated virus. *BioEssays* **17:** 237–245.

Blackwell, L.J. and J.A. Borowiec. 1994. Human replication protein A binds single-stranded DNA in two distinct complexes. *Mol. Cell. Biol.* **14:** 3993–4001.

Blanc, H. and B. Dujon. 1980. Replicator regions of the yeast mitochondrial DNA responsible for suppressiveness. *Proc. Natl. Acad. Sci.* **77:** 3942–3946.

Borowiec, J.A., F.B. Dean, P.A. Bullock, and J. Hurwitz. 1990. Binding and unwinding—How T antigen engages the SV40 origin of DNA replication. *Cell* **60:** 181–184.

Bosher, J., E.C. Robinson, and R.T. Hay. 1990. Interactions between the adenovirus type 2 DNA polymerase and the DNA binding domain of nuclear factor I. *New Biol.* **2:** 1083–1090.

Bravo, R. and H. Macdonald-Bravo. 1987. Existence of two populations of cyclin/proliferating cell nuclear antigen during the cell cycle: Association with DNA replication sites. *J. Cell Biol.* **105:** 1549–1554.

Brenot-Bosc, F., S. Gupta, R.L. Margolis, and R. Fotedar. 1995. Changes in the subcellular localization of replication initiation proteins and cell cycle proteins during G1- to S-phase transition in mammalian cells. *Chromosoma* **103:** 517–527.

Brill, S.J. and B. Stillman. 1991. Replication factor-A from *Saccharomyces cerevisiae* is encoded by three essential genes coordinately expressed at S phase. *Genes Dev.* **5:** 1589–1600.

Bruckner, A., F. Stadlbauer, L.A. Guarino, A. Brunahl, C. Schneider, C. Rehfuess, C. Prives, E. Fanning, and H.P. Nasheuer. 1995. The mouse DNA polymerase alpha-primase subunit p48 mediates species-specific replication of polyomavirus DNA in vitro. *Mol. Cell. Biol.* **15:** 1716–1724.

Brush, G.S., C.W. Anderson, and T.J. Kelly. 1994. The DNA-activated protein kinase is required for the phosphorylation of replication protein A during simian virus 40 DNA replication. *Proc. Natl. Acad. Sci.* **91:** 12520–12524.

Budd, M.E. and J.L. Campbell. 1993. DNA polymerases delta and epsilon are required for chromosomal replication in *Saccharomyces cerevisiae. Mol. Cell. Biol.* **13:** 496–505.

Burgers, P.M. 1991. *Saccharomyces cerevisiae* replication factor C. II. Formation and activity of complexes with the proliferating cell nuclear antigen and with DNA polymerases delta and epsilon. *J. Biol. Chem.* **266:** 22698–22706.

Cardoso, M.C., H. Leonhardt, and G.B. Nadal. 1993. Reversal of terminal differentiation and control of DNA replication: Cyclin A and Cdk2 specifically localize at subnuclear sites of DNA replication. *Cell* **74:** 979–992.

Challberg, M.D. and T.J. Kelly. 1979. Adenovirus DNA replication in vitro. *Proc. Natl. Acad. Sci.* **76:** 655–659.

———. 1989. Animal virus DNA replication. *Annu. Rev. Biochem.* **58:** 671–717.

Challberg, M.D., S.V. Desiderio, and T.J. Kelly. 1980. Adenovirus DNA replication in vitro: Characterization of a protein covalently linked to nascent DNA strands. *Proc. Natl. Acad. Sci.* **77:** 5105–5109.

Chang, D.D. and D.A. Clayton. 1984. Precise identification of individual promoters for transcription of each strand of human mitochondrial DNA. *Cell* **36:** 635–643.

———. 1985. Priming of human mitochondrial DNA replication occurs at the light-strand promoter. *Proc. Natl. Acad. Sci.* **82:** 351–355.

———. 1987. A novel endoribonuclease cleaves at a priming site of mouse mitochondrial DNA replication. *EMBO J.* **6:** 409–417.

Chang, D.D., W.W. Hauswirth, and D.A. Clayton. 1985. Replication priming and transcription initiate from precisely the same site in mouse mitochondrial DNA. *EMBO J.* **4:** 1559–1567.

Chen, M., N. Mermod, and M.S. Horwitz. 1990. Protein-protein interactions between adenovirus DNA polymerase and nuclear factor I mediate formation of the DNA replication preinitiation complex. *J. Biol. Chem.* **265:** 18634–18642.

Cheng, L. and T.J. Kelly. 1989. Transcriptional activator nuclear factor I stimulates the replication of SV40 minichromosomes in vivo and in vitro. *Cell* **59:** 541–551.

Cheng, L.Z., J.L. Workman, R.E. Kingston, and T.J. Kelly. 1992. Regulation of DNA replication in vitro by the transcriptional activation domain of GAL4-VP16. *Proc. Natl. Acad. Sci.* **89:** 589–593.

Clayton, D.A. 1982. Replication of animal mitochondrial DNA. *Cell* **28:** 693–705.

———. 1991. Replication and transcription of vertebrate mitochondrial DNA. *Annu. Rev. Cell Biol.* **7:** 453–478.

———. 1992. Transcription and replication of animal mitochondrial DNAs. *Int. Rev. Cytol.* **141:** 217–232.

Collins, K.L. and T.J. Kelly. 1991. Effects of T antigen and replication protein A on the initiation of DNA synthesis by DNA polymerase alpha-primase. *Mol. Cell. Biol.* **11:** 2108–2115.

Collins, K.L., A.A. Russo, B.Y. Tseng, and T.J. Kelly. 1993. The role of the 70 kDa subunit of human DNA polymerase alpha in DNA replication. *EMBO J.* **12:** 4555–4566.

Cook, P.R. 1991. The nucleoskeleton and the topology of replication (Review). *Cell* **66:** 627–635.

Copeland, W.C. and T.S. Wang. 1993. Enzymatic characterization of the individual mammalian primase subunits reveals a biphasic mechanism for initiation of DNA replication. *J. Biol. Chem.* **268:** 26179–26189.

Cote, J. and A. Ruiz-Carrillo. 1993. Primers for mitochondrial DNA replication generated by endonuclease G. *Science* **261:** 765–769.

Cotmore, S.F. and P. Tattersall. 1987. The autonomously replicating parvoviruses of vertebrates. *Adv. Virus Res.* **33:** 91–174.

Cotterill, S.M., M.E. Reyland, L.A. Loeb, and I.R. Lehman. 1987. A cryptic proofreading 3′→5′ exonuclease associated with the polymerase subunit of the DNA polymerase-primase from *Drosophila melanogaster. Proc. Natl. Acad. Sci.* **84:** 5635–5639.

de Zamaroczy, M., G. Faugeron-Fonty, G. Baldacci, R. Goursot, and G. Bernardi. 1984. The ori sequences of the mitochondrial genome of a wild-type yeast strain: Number, location, orientation and structure. *Gene* **32:** 439–457.

Diffley, J.F.X. and J.H. Cocker. 1992. Protein-DNA interactions at a yeast replication origin. *Nature* **357:** 169–172.

Din, S., S.J. Brill, M.P. Fairman, and B. Stillman. 1990. Cell-cycle-regulated phosphorylation of DNA replication factor A from human and yeast cells. *Genes Dev.* **4:** 968–977.

Dornreiter, I., A. Hoss, A.K. Arthur, and E. Fanning. 1990. SV40 T antigen binds directly to the large subunit of DNA polymerase α. *EMBO J.* **9:** 3329–3336.

Dornreiter, I., L.F. Erdile, I.U. Gilbert, W.D. von Winkler, T.J. Kelly, and E. Fanning. 1992. Interaction of DNA polymerase alpha-primase with cellular replication protein A and SV40 T antigen. *EMBO J.* **11:** 769–776.

D'Urso, G., R.L. Marraccino, D.R. Marshak, and J.M. Roberts. 1990. Cell cycle control of DNA replication by a homologue from human cells of the p34cdc2 protein kinase. *Science* **250:** 786–791.

Eder, P.S. and J.A. Walder. 1991. Ribonuclease H from K562 human erythroleukemia cells: Purification, characterization, and substrate specificity. *J. Biol. Chem.* **266:** 6472–6479.

Eki, T., T. Matsumoto, Y. Murakami, and J. Hurwitz. 1992. The replication of DNA containing the simian virus 40 origin by the monopolymerase and dipolymerase systems. *J. Biol. Chem.* **267:** 7284–7294.

Erdile, L.F., M.S. Wold, and T.J. Kelly. 1990. The primary structure of the 32-kDa subunit of human replication protein A. *J. Biol. Chem.* **265:** 3177–3182.

Erdile, L.F., W.-D. Heyer, R. Kolodner, and T.J. Kelly. 1991. Characterization of cDNA encoding the 70-kDa subunit of human replication protein A (RP-A), a single-stranded DNA binding protein involved in DNA replication and recombination. *J. Biol. Chem.* **266:** 12090–12098.

Fairman, M.P. and B. Stillman. 1988. Cellular factors required for multiple stages of SV40 DNA replication in vitro. *EMBO J.* **7:** 1211–1218.

Field, J., R.M. Gronostajski, and J. Hurwitz. 1984. Properties of the adenovirus DNA polymerase. *J. Biol. Chem.* **259:** 9487–9495.

Fisher, R.P. and D.A. Clayton. 1985. A transcription factor required for promoter recognition by human mitochondrial RNA polymerase. Accurate initiation at the heavy- and light-strand promoters dissected and reconstituted in vitro. *J. Biol. Chem.* **260:** 11330–11338.

Foury, F. 1989. Cloning and sequencing of the nuclear gene MIP1 encoding the catalytic subunit of the yeast mitochondrial DNA polymerase. *J. Biol. Chem.* **264:** 20552–20560.

Gannon, J.V. and D.P. Lane. 1990. Interactions between SV40 T antigen and DNA polymerase α. *New Biol.* **2:** 84–92.

Gillum, A.M. and D.A. Clayton. 1979. Mechanism of mitochondrial DNA replication in mouse L-cells: RNA priming during the initiation of heavy-strand synthesis. *J. Mol. Biol.* **135:** 353–368.

Goetz, G.S., F.B. Dean, J. Hurwitz, and S.W. Matson. 1988. The unwinding of duplex regions in DNA by the simian virus 40 large tumor antigen-asssociated DNA helicase activity. *J. Biol. Chem.* **263:** 383–392.

Gray, H. and T.W. Wong. 1992. Purification and identification of subunit structure of the human mitochondrial DNA polymerase. *J. Biol. Chem.* **267:** 5835–5841.

Harrington, J.J. and M.R. Lieber. 1994. The characterization of a mammalian DNA structure-specific endonuclease. *EMBO J.* **13:** 1235–1246.

Hauswirth, W.W. and K.I. Berns. 1979. Adeno-associated virus DNA replication: Nonunit-length molecules. *Virology* **93:** 57–68.

Hay, R.T. and W.C. Russell. 1989. Recognition mechanisms in the synthesis of animal virus DNA. *Biochem J.* **258:** 3–16.

He, Z., L.A. Henricksen, M.S. Wold, and C.J. Ingles. 1995. RPA involvement in the damage-recognition and incision steps of nucleotide excision repair. *Nature* **374:** 566–569.

He, Z., B.T. Brinton, J. Greenblatt, J.A. Hassell, and C.J. Ingles. 1993. The transactivator proteins VP16 and GAL4 bind replication factor A. *Cell* **73:** 1223–1232.

Hehman, G.L. and W.W. Hauswirth. 1992. DNA helicase from mammalian mitochondria. *Proc. Natl. Acad. Sci.* **89:** 8562–8566.

Hermonat, P.L., M.A. Labow, R. Wright, K.I. Berns, and N. Muzyczka. 1984. Genetics of adeno-associated virus: Isolation and preliminary characterization of adeno-associated virus type 2 mutants. *J. Virol.* **51:** 329–339.

Heyer, W.-D., M.R.F. Rao, L.F. Erdile, T.J. Kelly, and R.D. Kolodner. 1990. An essential *Saccharomyces cerevisiae* single-stranded DNA binding protein is homologous to the large subunit of human RP-A. *EMBO J.* **9:** 2321–2329.

Hixson, J.E., T.W. Wong, and D.A. Clayton. 1986. Both the conserved stem-loop and divergent 5′-flanking sequences are required for initiation at the human mitochondrial origin of light-strand DNA replication. *J. Biol. Chem.* **261:** 2384–2390.

Hong, G., P. Ward, and K.I. Berns. 1992. In vitro replication of adeno-associated virus DNA. *Proc. Natl. Acad. Sci.* **89:** 4673–4677.

Huang, L., Y. Kim, J.J. Turchi, and R.A. Bambara. 1994. Structure-specific cleavage of the RNA primer from Okazaki fragments by calf thymus RNAse H1. *J. Biol. Chem.* **269:** 25922–25927.

Ikeda, J.E., T. Enomoto, and J. Hurwitz. 1982. Adenoviral protein-primed initiation of DNA chains in vitro. *Proc. Natl. Acad. Sci.* **79:** 2442–2446.

Im, D.S. and N. Muzyczka. 1990. The AAV origin binding protein Rep68 is an ATP-dependent site-specific endonuclease with DNA helicase activity. *Cell* **61:** 447–457.

Insdorf, N.F. and D.F. Bogenhagen. 1989a. DNA polymerase γ from *Xenopus laevis*. I. The identification of a high molecular weight catalytic subunit by a novel DNA polymerase photolabeling procedure. *J. Biol. Chem.* **264:** 21491–21497.

———. 1989b. DNA polymerase gamma from *Xenopus laevis*. II. A 3′→5′ exonuclease is tightly associated with the DNA polymerase activity. *J. Biol. Chem.* **264:** 21498–21503.

Ishimi, Y., A. Claude, P. Bullock, and J. Hurwitz. 1988. Complete enzymatic synthesis of

DNA containing the SV40 origin of replication. *J. Biol. Chem.* **263:** 19723–19733.

Ishimi, Y., K. Sugasawa, F. Hanaoka, T. Eki, and J. Hurwitz. 1992. Topoisomerase II plays an essential role as a swivelase in the late stage of SV40 chromosome replication in vitro. *J. Biol. Chem.* **267:** 462–466.

Kelly, T.J. 1991. DNA replication in mammalian cells: Insights from the SV40 model system. *Harvey Lect.* **85:** 173–188.

Kenny, M.K., S.H. Lee, and J. Hurwitz. 1989. Multiple functions of human single-stranded-DNA binding protein in simian virus 40 DNA replication: Single-strand stabilization and stimulation of DNA polymerases alpha and delta. *Proc. Natl. Acad. Sci.* **86:** 9757–9761.

Kenny, M.K., U. Schlegel, H. Furneaux, and J. Hurwitz. 1990. The role of human single-stranded DNA binding protein and its individual subunits in simian virus 40 DNA replication. *J. Biol. Chem.* **265:** 7693–7700.

Kesti, T., H. Frantti, and J.E. Syvoaja. 1993. Molecular cloning of the cDNA for the catalytic subunit of human DNA polymerase ε. *J. Biol. Chem.* **268:** 10238–10245.

Kim, C. and M.S. Wold. 1995. Recombinant human replication protein A binds to polynucleotides with low cooperativity. *Biochemistry* **34:** 2058–2064.

Kim, C., R.O. Snyder, and M.S. Wold. 1992. Binding properties of replication protein A from human and yeast cells. *Mol. Cell. Biol.* **12:** 3050–3059.

King, A.J. and P.C. Van der Vliet. 1994. A precursor terminal protein-trinucleotide intermediate during initiation of adenovirus DNA replication: Regeneration of molecular ends in vitro by a jumping back mechanism. *EMBO J.* **13:** 5786–5792.

Kong, X.P., R. Onrust, M. O'Donnell, and J. Kuriyan. 1992. Three-dimensional structure of the beta subunit of *E. coli* DNA polymerase III holoenzyme: A sliding DNA clamp. *Cell* **69:** 425–437.

Kornberg, A. and T.A. Baker. 1992. *DNA replication*, 2nd edition. Freeman, New York.

Krishna, T.S., X.P. Kong, S. Gary, P.M. Burgers, and J. Kuriyan. 1994. Crystal structure of the eukaryotic DNA polymerase processivity factor PCNA. *Cell* **79:** 1233–1243.

Lahaye, A., H. Stahl, S.D. Thines, and F. Foury. 1991. PIF1: A DNA helicase in yeast mitochondria. *EMBO J.* **10:** 997–1007.

LeBowitz, J.H. and R. McMacken. 1986. The *Escherichia coli* dnaB replication protein is a DNA helicase. *J. Biol. Chem.* **261:** 4738–4748.

Lechner, R.L. and T.J. Kelly. 1977. The structure of replicating adenovirus 2 DNA molecules. *Cell* **12:** 1007–1020.

Lee, S.H., A.D. Kwong, Z.Q. Pan, and J. Hurwitz. 1991a. Studies on the activator 1 protein complex, an accessory factor for proliferating cell nuclear antigen-dependent DNA polymerase delta. *J. Biol. Chem.* **266:** 594–602.

Lee, S.H., Z.Q. Pan, A.D. Kwong, P.M. Burgers, and J. Hurwitz. 1991b. Synthesis of DNA by DNA polymerase epsilon in vitro. *J. Biol. Chem.* **266:** 22707–22717.

Lehman, I.R. and L.S. Kaguni. 1989. DNA polymerase alpha. *J. Biol. Chem.* **264:** 4265–4268.

Li, J.J. and T.J. Kelly. 1984. Simian virus 40 DNA replication in vitro. *Proc. Natl. Acad. Sci.* **81:** 6973–6977.

Li, R. and M.R. Botchan. 1993. The acidic transcriptional activation domains of VP16 and p53 bind the cellular replication protein A and stimulate in vitro BPV-1 DNA replication. *Cell* **73:** 1207–1221.

Lindahl, T. and D.E. Barnes. 1992. Mammalian DNA ligases. *Annu. Rev. Biochem.* **61:** 251–281.

Lindenbaum, J.O., J. Field, and J. Hurwitz. 1986. The adenovirus DNA binding protein and adenovirus DNA polymerase interact to catalyze elongation of primed DNA templates. *J. Biol. Chem.* **261:** 10218–10227.

Liu, B. and B.M. Alberts. 1995. Head-on collision between a DNA replication apparatus and RNA polymerase transcription complex. *Science* **267:** 1131–1137.

Liu, B., M.L. Wong, R.L. Tinker, E.P. Geiduschek, and B.M. Alberts. 1993. The DNA replication fork can pass RNA polymerase without displacing the nascent transcript. *Nature* **366:** 33–39.

Lusby, E., R. Bohenzky, and K.I. Berns. 1981. Inverted terminal repetition in adeno-associated virus DNA: Independence of the orientation at either end of the genome. *J. Virol* **37:** 1083–1086.

Lusby, E., K.H. Fife, and K.I. Berns. 1980. Nucleotide sequence of the inverted terminal repetition in adeno-associated virus DNA. *J. Virol.* **34:** 402–409.

Madsen, C.S., S.C. Ghivizzani, and W.W. Hauswirth. 1993. Protein binding to a single termination-associated sequence in the mitochondrial DNA D-loop region. *Mol. Cell. Biol.* **13:** 2162–71.

Maki, H., S. Maki, and A. Kornberg. 1988. DNA polymerase III holoenzyme of *Escherichia coli.* IV. The holoenzyme is an asymmetric dimer with twin active sites. *J. Biol. Chem.* **263:** 6570–6578.

Matson, S.W., S. Tabor, and C.C. Richardson. 1983. The gene 4 protein of bacteriophage T7. Characterization of helicase activity. *J. Biol. Chem.* **258:** 14017–14024.

Matsuda, T., M. Saijo, I. Kuraoka, T. Kobayashi, Y. Nakatsu, A. Nagai, T. Enjoji, C. Masutani, K. Sugasawa, F. Hanaoka, et al. 1995. DNA repair protein XPA binds replication protein A (RPA). *J. Biol. Chem.* **270:** 4152–4157.

McCarty, D.M., J.H. Ryan, S. Zolotukhin, X. Zhou, and N. Muzyczka. 1994. Interaction of the adeno-associated virus Rep protein with a sequence within the A palindrome of the viral terminal repeat. *J. Virol.* **68:** 4998–5006.

McHenry, C.S. 1988. DNA polymerase III holoenzyme of *Escherichia coli. Annu. Rev. Biochem.* **57:** 519–550.

Melendy, T. and B. Stillman. 1993. An interaction between replication protein A and SV40 T antigen appears essential for primosome assembly during SV40 DNA replication. *J. Biol. Chem.* **268:** 3389–3395.

Meyer, R.R. and P.S. Laine. 1990. The single-stranded DNA-binding protein of *Escherichia coli. Microbiol. Rev.* **54:** 342–380.

Mignotte, B., J. Marsault, and M. Barat-Gueride. 1988. Effects of the *Xenopus laevis* mitochondrial single-stranded DNA-binding protein on the activity of DNA polymerase gamma. *Eur. J. Biochem.* **174:** 479–484.

Miller, K.G., L.F. Liu, and P.T. Englund. 1981. A homogeneous type II DNA topoisomerase from HeLa cell nuclei. *J. Biol. Chem.* **256:** 9334–9339.

Mul, Y.M. and P.C. Van der Vliet. 1992. Nuclear factor I enhances adenovirus DNA replication by increasing the stability of a preinitiation complex. *EMBO J.* **11:** 751–760.

Mul, Y.M., C.P. Verrijzer, and P.C. van der Vliet. 1990. Transcription factors NFI and NFIII/oct-1 function independently, employing different mechanisms to enhance adenovirus DNA replication. *J. Virol.* **64:** 5510–5518.

Murakami, Y. and J. Hurwitz. 1993a. DNA polymerase alpha stimulates the ATP-dependent binding of simian virus tumor T antigen to the SV40 origin of replication. *J. Biol. Chem.* **268:** 11018–11027.

———. 1993b. Functional interactions between SV40 T antigen and other replication proteins at the replication fork. *J. Biol. Chem.* **268:** 11008–11017.

Murante, R.S., L. Huang, J.J. Turchi, and R.A. Bambara. 1994. The calf 5′- to 3′-exonuclease is also an endonuclease with both activities dependent on primers annealed upstream of the point of cleavage. *J. Biol. Chem.* **269:** 1191–1196.

Muzyczka, N. 1992. Use of adeno-associated virus as a general transduction vector for mammalian cells. *Curr. Top. Microbiol. Immunol.* **158:** 97–129.

Nagata, K., R.A. Guggenheimer, and J. Hurwitz. 1983a. Adenovirus DNA replication in vitro: Synthesis of full-length DNA with purified proteins. *Proc. Natl. Acad. Sci.* **88:** 4266–4270.

———. 1983b. Specific binding of a cellular DNA replication protein to the origin of replication of adenovirus DNA. *Proc. Natl. Acad. Sci.* **80:** 6177–6181.

Nagata, K., R.A. Guggenheimer, T. Enomoto, J.H. Lichy, and J. Hurwitz. 1982. Adenovirus DNA replication in vitro: Identification of a host factor that stimulates synthesis of the preterminal protein-dCMP complex. *Proc. Natl. Acad. Sci.* **79:** 6438–6442.

Nasheuer, H.P., A. Moore, A.F. Wahl, and T.S. Wang. 1991. Cell cycle-dependent phosphorylation of human DNA polymerase alpha. *J. Biol. Chem.* **266:** 7893–7903.

Navas, T.A., Z. Zhou, and S.J. Elledge. 1995. DNA polymerase epsilon links the DNA replication machinery to the S phase checkpoint. *Cell* **80:** 29–39.

Ni, T.H., X. Zhou, D.M. McCarty, I. Zolotukhin, and N. Muzyczka. 1994. In vitro replication of adeno-associated virus DNA. *J. Virol.* **68:** 1128–1138.

Parisi, M.A., B. Xu, and D.A. Clayton. 1993. A human mitochondrial transcriptional activator can functionally replace a yeast mitochondrial HMG-box protein both in vivo and in vitro. *Mol. Cell. Biol.* **13:** 1951–1961.

Pavco, P.A. and G.C. Van Tuyle. 1985. Purification and general properties of the DNA-binding protein (P16) from rat liver mitochondria. *J. Cell. Biol.* **100:** 258–264.

Podust, V.N. and U. Hubscher. 1993. Lagging strand DNA synthesis by calf thymus DNA polymerases alpha, beta, delta and epsilon in the presence of auxiliary proteins. *Nucleic Acids Res.* **21:** 841–846.

Prelich, G., C.K. Tan, M. Kostura, M.B. Mathews, A.G. So, K.M. Downey, and B. Stillman. 1987. Functional identity of proliferating cell nuclear antigen and a DNA polymerase-delta auxiliary protein. *Nature* **326:** 517–520.

Pronk, R., W. Van Driel, and P.C. Van der Vliet. 1994. Replication of adenovirus DNA in vitro is ATP-independent. *FEBS Lett.* **337:** 33–38.

Pruijn, G.J., W. van Driel, and P.C. van der Vliet. 1986. Nuclear factor III, a novel sequence-specific DNA-binding protein from HeLa cells stimulating adenovirus DNA replication. *Nature* **322:** 656–659.

Randall, S.K. and T.J. Kelly. 1992. The fate of parental nucleosomes during SV40 DNA replication. *J. Biol. Chem.* **267:** 14259–14265.

Rawlins, D.R., P.J. Rosenfeld, R.J. Wides, M.D. Challberg, and T.J. Kelly. 1984. Structure and function of the adenovirus origin of replication. *Cell* **37:** 309–319.

Rekosh, D.M., W.C. Russell, A.J. Bellet, and A.J. Robinson. 1977. Identification of a protein linked to the ends of adenovirus DNA. *Cell* **11:** 283–295.

Richardson, R.W. and N.G. Nossal. 1989. Characterization of the bacteriophage T4 gene 41 DNA helicase. *J. Biol. Chem.* **264:** 4725–4731.

Rosenfeld, P.J. and T.J. Kelly. 1986. Purification of nuclear factor I by DNA recognition site affinity chromatography (published erratum appears in *J. Biol. Chem.* [1986] **261:** 10015) *J. Biol. Chem.* **261:** 1398–1408.

Rosenfeld, P.J., E.A. O'Neill, R.J. Wides, and T.J. Kelly. 1987. Sequence-specific interactions between cellular DNA-binding proteins and the adenovirus origin of DNA replication. *Mol. Cell. Biol.* **7:** 875–886.

Salas, M. 1991. Protein-priming of DNA replication. *Annu. Rev. Biochem.* **60:** 39–71.

Santocanale, C., M. Foiani, G. Lucchini, and P. Plevani. 1993. The isolated 48,000-dalton subunit of yeast DNA primase is sufficient for RNA primer synthesis. *J. Biol. Chem.* **268:** 1343–1348.

Schmitt, M.E. and D.A. Clayton. 1992. Yeast site-specific ribonucleoprotein endoribonuclease MRP contains an RNA component homologous to mammalian RNase MRP RNA and essential for cell viability. *Genes Dev.* **6:** 1975–1985.

———. 1993. Conserved features of yeast and mammalian mitochondrial DNA replication. *Curr. Opin. Genet. Dev.* **3:** 769–774.

Schneider, C., K. Weisshart, L.A. Guarino, I. Dornreiter, and E. Fanning. 1994. Species-specific functional interactions of DNA polymerase alpha-primase with SV40 T antigen require SV40 origin DNA. *Mol. Cell. Biol.* **14:** 3176–3185.

Snyder, R.O., D.S. Im, and N. Muzyczka. 1990a. Evidence for covalent attachment of the adeno-associated virus (AAV) rep protein to the ends of the AAV genome. *J. Virol.* **64:** 6204–6213.

Snyder, R.O., R.J. Samulski, and N. Muzyczka. 1990b. In vitro resolution of covalently joined AAV chromosome ends. *Cell* **60:** 105–113.

So, A.G. and K.M. Downey. 1992. Eukaryotic DNA replication. *Crit. Rev. Biochem. Mol. Biol.* **27:** 129–155.

Srivastava, A., E.W. Lusby, and K.I. Berns. 1983. Nucleotide sequence and organization of the adeno-associated virus 2 genome. *J. Virol.* **45:** 555–564.

Stadlbauer, F., A. Brueckner, C. Rehfuess, C. Eckerskorn, F. Lottspeich, V. Forster, B.Y. Tseng, and H.P. Nasheur. 1994. DNA replication in vitro by recombinant DNA-polymerase-alpha-primase. *Eur. J. Biochem.* **222:** 781–793.

Stillman, B. 1986. Chromatin assembly during SV40 DNA replication in vitro. *Cell* **45:** 555–565.

———. 1989. Initiation of eukaryotic DNA replication in vitro. *Annu. Rev. Cell. Biol.* **5:** 197–245.

Stohl, L.L. and D.A. Clayton. 1992. *Saccharomyces cerevisiae* contains an RNase MRP that cleaves at a conserved mitochondrial RNA sequence implicated in replication priming. *Mol. Cell. Biol.* **12:** 2561–2569.

Stroumbakis, N.D., Z. Li, and P.P. Tolias. 1994. RNA- and single-stranded DNA-binding (SSB) proteins expressed during *Drosophila melanogaster* oogenesis: A homolog of bacterial and eukaryotic mitochondrial SSBs. *Gene* **143:** 171–177.

Syvaoja, J. and S. Linn. 1989. Characterization of a large form of DNA polymerase δ from HeLa cells that is insensitive to proliferating cell nuclear antigen. *J. Biol. Chem.* **264:** 2489–2497.

Tan, C.K., C. Castillo, A.G. So, and K.M. Downey. 1986. An auxiliary protein for DNA polymerase-delta from fetal calf thymus. *J. Biol. Chem.* **261:** 12310–12316.

Tattersall, P. and D.C. Ward. 1976. Rolling hairpin model for replication of parvovirus and linear chromosomal DNA. *Nature* **263:** 106–109.

Temperley, S.M. and R.T. Hay. 1992. Recognition of the adenovirus type 2 origin of DNA replication by the virally encoded DNA polymerase and preterminal proteins. *EMBO J.* **11:** 761–768.

Tiranti, V., M. Rocchi, S. DiDonato, and M. Zeviani. 1993. Cloning of human and rat cDNAs encoding the mitochondrial single-stranded DNA-binding protein (SSB). *Gene* **126:** 219–225.

Tratschin, J.D., I.L. Miller, and B.J. Carter. 1984. Genetic analysis of adeno-associated virus: Properties of deletion mutants constructed in vitro and evidence for an adeno-associated virus replication function. *J. Virol.* **51:** 611–619.

Tsurimoto, T. and B. Stillman. 1989. Purification of a cellular replication factor, RF-C, that is required for coordinated synthesis of leading and lagging strands during simian virus 40 DNA replication in vitro. *Mol. Cell. Biol.* **9:** 609–619.

———. 1991. Replication factors required for SV40 DNA replication in vitro. I. DNA structure-specific recognition of a primer-template junction by eukaryotic DNA polymerases and their accessory proteins. *J. Biol. Chem.* **266:** 1950–1960.

Tsurimoto, T., T. Melendy, and B. Stillman. 1990. Sequential initiation of lagging and leading strand synthesis by two different polymerase complexes at the SV40 DNA replication origin. *Nature* **346:** 534–539.

Turchi, J.J. and R.A. Bambara. 1993. Completion of mammalian lagging strand DNA replication using purified proteins. *J. Biol. Chem.* **268:** 15136–15141.

Turchi, J.J., L. Huang, R.S. Murante, Y. Kim, and R.A. Bambara. 1994. Enzymatic completion of mammalian lagging-strand DNA replication. *Proc. Natl. Acad. Sci.* **91:** 9803–9807.

Umbricht, C.B., L.F. Erdile, E.W. Jabs, and T.J. Kelly. 1993. Cloning, overexpression, and genomic mapping of the 14-kDa subunit of human replication protein A. *J. Biol. Chem.* **268:** 6131–6138.

Van der Vliet, P.C. 1991. The role of cellular transcription factors in the enhancement of adenovirus DNA replication. *Semin. Virol.* **2:** 271–280.

Van Dyck, E., F. Foury, B. Stillman, and S.J. Brill. 1992. A single-stranded DNA binding protein required for mitochondrial DNA replication in *S. cerevisiae* is homologous to *E. coli* SSB. *EMBO J.* **11:** 3421–3430.

Virshup, D.M., M.G. Kauffman, and T.J. Kelly. 1989. Activation of SV40 DNA replication in vitro by cellular protein phosphatase 2A. *EMBO J.* **8:** 3891–3898.

Waga, S., G. Bauer, and B. Stillman. 1994. Reconstitution of complete SV40 DNA replication with purified replication factors. *J. Biol. Chem.* **269:** 10923–10934.

Wang, J.C. 1985. DNA topoisomerases. *Annu. Rev. Biochem.* **54:** 665–697.

Wang, T.S. 1991. Eukaryotic DNA polymerases. *Annu. Rev. Biochem.* **60:** 513–552.

Ward, P., E. Urcelay, R. Kotin, B. Safer, and K.I. Berns. 1994. Adeno-associated virus DNA replication in vitro: Activation by a maltose binding protein/Rep 68 fusion protein. *J. Virol.* **68:** 6029–6037.

Weinberg, D.H., K.L. Collins, P. Simancek, A. Russo, M.S. Wold, D.M. Virshup, and T.J. Kelly. 1990. Reconstitution of simian virus 40 DNA replication with purified proteins. *Proc. Natl. Acad. Sci.* **87:** 8692–8696.

Wernette, C.M. and L.S. Kaguni. 1986. A mitochondrial DNA polymerase from embryos of *Drosophila melanogaster*. Purification, subunit structure, and partial characterization. *J. Biol. Chem.* **261:** 14764–14770.

Wernette, C.M., M.C. Conway, and L.S. Kaguni. 1988. Mitochondrial DNA polymerase from *Drosophila melanogaster* embryos: Kinetics, processivity, and fidelity of DNA polymerization. *Biochemistry* **27:** 6046–6054.

Wides, R.J., M.D. Challberg, D.R. Rawlins, and T.J. Kelly. 1987. Adenovirus origin of DNA replication: Sequence requirements for replication in vitro. *Mol. Cell. Biol.* **7:** 864–874.

Wiekowski, M., M.W. Schwartz, and H. Stahl. 1988. Simian virus 40 large T antigen helicase: Characterization of the ATPase-dependent DNA unwinding activity and its substrate requirements. *J. Biol. Chem.* **263:** 436–442.

Wobbe, C.R., L. Weissbach, J.A. Borowiec, F.B. Dean, Y. Murakami, P. Bullock, and J. Hurwitz. 1987. Replication of SV40-origin-containing DNA with purified proteins. *Proc. Natl. Acad. Sci.* **84:** 1834–1838.

Wold, M.S. and T.J. Kelly. 1988. Purification and characterization of replication protein A, a cellular protein required for in vitro replication of simian virus 40 DNA. *Proc. Natl. Acad. Sci.* **85:** 2523–2527.

Wold, M.S., D.H. Weinberg, D.M. Virshup, J.J. Li, and T.J. Kelly. 1989. Identification of cellular proteins required for simian virus 40 DNA replication. *J. Biol. Chem.* **264:** 2801–2809.

Wong, T.W. and D.A. Clayton. 1985a. In vitro replication of human mitochondrial DNA: Accurate initiation at the origin of light-strand synthesis. *Cell* **42:** 951–958.

———. 1985b. Isolation and characterization of a DNA primase from human mitochondria. *J. Biol. Chem.* **260:** 11530–11535.

———. 1986. DNA primase of human mitochondria is associated with structural RNA that is essential for enzymatic activity. *Cell* **45:** 817–825.

Yang, L., M.S. Wold, J.J. Li, T.J. Kelly, and L.F. Liu. 1987. Roles of DNA topoisomerases in SV40 DNA replication in vitro. *Proc. Natl. Acad. Sci.* **84:** 950–954.

Yang, Q., A. Kadam, and J.P. Trempe. 1992. Mutational analysis of the adeno-associated virus *rep* gene. *J. Virol.* **66:** 6058–6069.

2

Origins of DNA Replication

M.L. DePamphilis
National Institute of Child Health and Human Development
National Institutes of Health
Bethesda, Maryland 20892-2753

Initiation of DNA replication involves three sequential steps. First, one or more specific *trans*-acting proteins bind to specific *cis*-acting DNA sequences referred to as origins of replication (DePamphilis 1993a,b; Kornberg and Baker 1992). Second, DNA unwinding of the two complementary templates begins. This step is usually carried out by a DNA helicase but can also be done by some DNA polymerases (e.g., adenovirus [Ad]). DNA unwinding is facilitated by single-strand-specific DNA-binding proteins such as replication protein A (RP-A), which coat the templates, and by topoisomerase I, which releases torsional stress generated by unwinding DNA. Third, DNA synthesis is initiated on one or both templates. In cellular chromosomes and DNA viruses that do not encode their own DNA polymerase (e.g., SV40, PyV, and PV), DNA polymerase-α:DNA primase complex synthesizes a short RNA-primed nascent DNA chain referred to as an Okazaki fragment. The first Okazaki fragment initiated on each template is extended continuously by DNA polymerase-δ and its accessory proteins to become the long nascent DNA strand on the forward arm of each of the two replication forks. The net result of these steps is bidirectional DNA replication employing bubble and fork structures such as those found in the chromosomes of prokaryotic and eukaryotic cells (Fig. 1). DNA replication is coupled to chromatin assembly, resulting in the random distribution of pre-fork histone octamers to both arms of the fork and rapid assembly of new histone octamers in the intervening regions of newly replicated DNA. Initiation of DNA replication can also occur in only one direction instead of both directions (geminiviruses, parvovirus, Ad, mtDNA) and can utilize preexisting DNA primers (parvovirus), RNA primers (retroviruses), and protein-nucleotide primers (Ad), instead of de novo synthesis of RNA primers. The sequences, proteins, and mechanisms referred to in each step are discussed in detail in various chapters of this book.

DNA Replication in Eukaryotic Cells
 0-87969-459-9/96 $5 + .00

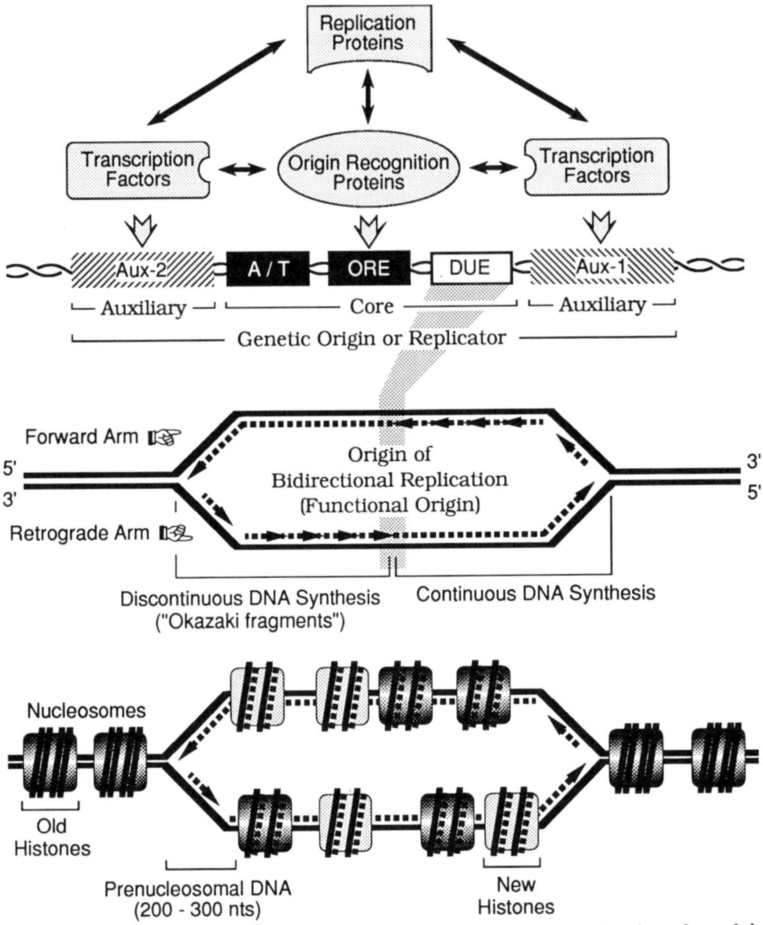

Figure 1 The modular concept of origins of DNA replication found in simple genomes (DePamphilis 1993a). Replication is initiated by a combination of replication proteins and transcription factors interacting with specific DNA sequences and with each other. Bidirectional DNA replication originates at the DUE, resulting in a replication bubble with two replication forks traveling in opposite directions. The structure of replication forks (DePamphilis and Wassarman 1980; Brush and Kelly; Stillman; both this volume), their organization into chromatin (Cusick et al. 1989; Wolffe, this volume), and the process of chromatin assembly (Gruss and Sogo 1992; Wolffe, this volume) have been described in detail previously.

Central to understanding how animal cells regulate DNA replication is understanding the nature of DNA sites where replication begins. Every genome analyzed so far contains at least one replication origin per chromosome, and the genomes of eukaryotic cells contain about one origin every 10–330 kb (Hand 1978). Replication origins play two important roles in DNA replication. First, they ensure that each time a cell divides, the entire genome is replicated efficiently. For example, a single mammalian cell contains about 1.8 meters of DNA that must be replicated in 6–8 hours, and early embryos of frogs, flies, and sea urchins replicate comparable amounts of DNA in 10–40 minutes. Second, replication origins provide a way in which to regulate when and where initiation events occur (Table 1). Replication of cellular chromosomes is restricted to one phase of the cell proliferation cycle (S phase), and initiation at each of several thousand replication origins is restricted to once per S phase (Blumenthal et al. 1974). Multiple initiation events at the same locus (gene amplification) can occur in tumors and transformed cell lines, but only rarely are genes amplified during normal animal development (Tlsty 1990). In contrast, replication of mtDNA and large viral DNA genomes such as herpes and vaccinia is not dependent on the cell entering S phase. mtDNA replicates randomly throughout the cell division cycle, and large viral genomes usurp the cell's machinery to provide their own replication components. Replication of small viral DNA genomes such as papovaviruses is restricted to S phase, but each genome copy may undergo two or more rounds of replication during a single S phase. Thus, it appears that mitochondrial and viral genomes in mammalian cells were designed to escape the very controls required for cellular DNA replication. This may not be true for simpler organisms such as flagellated protozoa, where replication of mtDNA (kinetoplast DNA) appears to follow the same rules as nuclear DNA.

Origins of DNA replication that function in eukaryotic cells can be divided into two groups: those found in "simple genomes" such as animal viruses (SV40, polyomavirus [PyV], PV, Ad, herpes simplex virus [HSV], Epstein-Barr virus [EBV]), mitochondria (human, mouse), protozoa (*Tetrahymena*), yeast (*Saccharomyces cerevisiae, Schizosaccharomyces pombe*), and slime mold (*Physarum*); and those found in "complex genomes" of metazoa such as flies (*Drosophila*, *Sciara*), frogs (*Xenopus*), and mammals (rodents, human). Simple origins that function in eukaryotes are similar to those that function in prokaryotes (Kornberg and Baker 1992). They have a modular anatomy composed of unique DNA sequence motifs and interactions with soluble proteins (DePamphilis 1993a,b). Whether or not metazoan chromosomes contain sequence-specific replication origins analogous to those found in simpler genomes is controversial, and therefore is discussed in more detail. What is clear is that replication begins at specific sites in metazoan chromosomes, that initiation of DNA replication requires nuclear structure, and that nuclear structure imposes site specificity.

Table 1 Characteristics of replication origins that function in eukaryotic cells

	Initiation of DNA replication					
Genome	maps to a specific site	requires specific DNA sequences	is restricted to S phase	occurs ?/origin/ cell cycle	exhibits ARS activity	requires nuclear structure
Mitochondria	~300 bp	30 and ~500 bp	no	>1	no	no
Animal viruses (nuclear)	1 to ~200 bp	18 to ~1000[b] bp	yes	>1	yes	no
Yeast	~300 bp	100 to 200 bp	yes	1	yes	
Protozoa (*Tetrahymena* macro nucleus)	<1 kb	<2 kb	yes			
Slime mold	<1 kb		yes	1		
Flies (DNA amplification)	yes[a]	1 to 8 kb[c]	no	>1	no	
(S phase, cultured cells)	yes[a]		yes	1		
Mammals	yes[a]	1 to 8 kb[d]	yes	1	yes/no	yes

[a]See Table 4.
[b]EBV *oriP*, the largest origin, includes nonrequired sequences between *ori*-core and its enhancer.
[c]Estimated from genetic analyses of *Drosophila* chorion gene amplification locus.
[d]Estimated from four reported ARS activities and one natural deletion mutant (see Table 3).

CHARACTERISTICS OF SIMPLE ORIGINS

Replication origins in simple genomes are composed of modular units acting in concert to determine where and when DNA replication will occur (Table 1). In this sense, replication origins are equivalent to transcription promoters, but whereas each promoter is uniquely responsible for transcription of its associated gene, large chromosomes like yeast often contain more origins than are needed for their replication (Newlon et al. 1993). This flexibility in origin number is allowed because DNA regions that lack a replication origin can be replicated by replication forks that originate from origins located many kilobases away.

Simple origins exhibit four characteristics.

1. Simple origins are composed of unique, genetically required, sequences. These sequences occupy from 50 bp to about 1000 bp and are defined by *cis*-acting mutations that prevent DNA replication. They are anatomically similar to origins that function in prokaryotic cells and consist of at least four elements, three of which (origin recognition element [ORE], DNA unwinding element [DUE], A/T-rich element) are essential and therefore referred to as the origin core (*ori*-core) (Fig. 1) (DePamphilis 1993a,b). *ori*-cores occupy from 18 bp to about 120 bp. *ori*-core function is critically dependent on the spacing, orientation, and arrangement of these three elements (DePamphilis 1993a,b; Parsons and Tegtmeyer 1992; Harrison et al. 1994). In general, there is a lack of symmetry in both the organization of sequence elements and their specific functions in replication origins. As a result, DNA synthesis initiation events within *ori*-core may occur only on one DNA strand (e.g., Ad, mtDNA, SV40, PyV). Therefore, initiation of bidirectional replication (e.g., SV40, PyV) is unlikely to be a symmetrical event with two forks moving simultaneously out of the ORE. More likely is that forward-arm synthesis begins on one strand of *ori*-core and then progresses beyond *ori*-core before DNA synthesis is initiated in the opposite direction on the complementary strand (see "Initiation zone model" in DePamphilis et al. 1988).

   The fourth element consists of transcription-factor-binding sites that flank one or both sides of *ori*-core (DePamphilis 1993c): *aux-1* is proximal to the DUE element, and *aux-2* is proximal to the A/T element (Fig. 1). These transcription-factor-binding sites are referred to as *ori*-auxiliary components (*aux*), because they facilitate *ori*-core activity from 2- to 1000-fold, depending both on the origin and on experimental conditions, but are not required for replication under all conditions and do not affect the mechanism by which replication occurs (DePamphilis 1993a,b). Some yeast origins, for example, contain a binding site (element B3) for transcription factor ABF1, whereas others do not. In this sense, auxiliary components are analogous to transcription enhancers. However, although enhancers that stimulate transcription are generally independent of their distance and orientation relative to the promoter, orientation and spacing between auxiliary sequences and *ori*-core are critical in some origins (Ad, SV40, PyV), whereas in other origins (EBV, yeast), these parameters are flexible. These differences presumably reflect differences in the specificity and strength of interactions between transcription factors and origin recognition proteins, as well as differences in the mechanism by which *ori*-auxiliary components function.

2. The genetic origin is coincident with the functional origin. The functional origin is the site where DNA synthesis actually begins. It has been mapped to within ±300 bp of the genetic origin in yeast chromosomes and plasmids, and the transition from discontinuous to continuous DNA synthesis on each template (Fig. 1) mapped with nucleotide resolution to the genetic origins in SV40, PyV, Ad, mtDNA, *Escherichia coli* (Kohara et al. 1985; Rokeach and Zyskind 1986; Seufert and Messer 1987), and bacteriophage λ (Yoda et al. 1988). However, these transitions can lie outside *ori* when SV40 DNA replication is initiated in vitro (Bullock et al. 1991), suggesting that extensive DNA unwinding can precede initiation of DNA synthesis, thereby providing DNA polymerase-α:DNA primase with the opportunity to begin synthesis outside of *ori* (see "Initiation zone model" in DePamphilis et al. 1988). In vivo, the rate of DNA unwinding may be retarded by chromatin structure.
3. Simple origins can act as autonomously replicating sequences (ARSs). ARS elements confer on other DNA molecules the ability to replicate when transferred to either cells or cell extracts containing the required replication proteins. So far, ARS elements have been demonstrated only with viral and yeast origins. Functional origins in the chromosomes of *S. cerevisiae* have been shown to correspond to individual ARS elements that are genetically required for origin activity (Newlon, this volume). The same appears true for *S. pombe*, although sequence requirements for origins in *S. pombe* appear more diffuse and origin function less efficient than in *S. cerevisiae* (Caddle and Calos 1994; Dubey et al. 1994; Wohlgemuth et al. 1994).
4. Simple origins can function in a soluble cell-free DNA replication system. So far, this has been demonstrated only with viral origins, a

problem that may be overcome when large amounts of purified replication proteins become available for other systems.

ANATOMY OF SIMPLE REPLICATION ORIGINS

Origin Recognition Element

An ORE is the DNA-binding site for one or more origin recognition proteins (Table 2) that are required for initiation of DNA replication. This DNA-protein interaction can be regulated by posttranslational modifications, such as phosphorylation of specific amino acid residues in SV40 T antigen (see Weisshart and Fanning, this volume). Sequences within or flanking the ORE often exhibit the characteristics of bent DNA (Zahn and Blattner 1985; Deb et al. 1986; Williams et al. 1988), which may facilitate protein binding (Ryder et al. 1986). Moreover, although an ORE may exhibit a twofold symmetry, interaction with its recognition protein can be asymmetrical, leading to an asymmetrical opening of *ori*-core (SenGupta and Borowiec 1994).

Origin recognition proteins serve at least two functions. The first is to initiate DNA unwinding using either their own helicase activity (e.g., parvovirus, SV40, PyV, PV, HSV) or one that associates with it (EBV may be an example). The second function is to guide other replication proteins to the origin. For example, SV40 and PyV T antigen associate with DNA polymerase-α:DNA primase, the enzyme responsible for initiation of the first RNA-primed DNA chain (Okazaki fragment) at their respective origins, and with RP-A, a single-stranded DNA-binding protein that stimulates both helicase and polymerase action (Melendy and Stillman 1993; Schneider et al. 1994 and references therein). Similarly, Ad preterminal protein, the protein-dCTP primer that initiates Ad DNA synthesis, is complexed with Ad DNA polymerase, the enzyme responsible for Ad DNA synthesis (see Hay, this volume). In yeast, a six-protein origin recognition complex binds to an approximately 50-bp sequence that includes elements A and B1. All yeast origins require element A (containing the 11-bp conserved ARS consequence sequence) and at least two of the three B elements that lie to one side of element A (see Newlon, this volume). Origin recognition proteins such as EBV EBNA1 that do not have their own helicase activity (Frappier and O'Donnell 1991) presumably associate with a cellular helicase. These interactions between origin recognition proteins and replication proteins can determine the host-cell specificity of a replication origin (Melendy and Stillman 1993; Schneider et al. 1994).

Table 2 Proteins that activate replication origins

	Origin recognition proteins[a]			Transcription factors[b]			
Origin	protein	enzymatic activities	associated rep. proteins	Aux-2	associated rep. proteins	Aux-1	associated rep. proteins
Gemini	Rep (AL1)	endonuclease, ATPase					
Parvo							
MVM	NS1	endonuclease, helicase		ATF, NS1	NS1		
AAV	Rep 68 and 78	endonuclease, helicase		Rep	Rep		
Papova							
SV40	T antigen	helicase	pol-α:primase RP-A	AP1, Sp1, NF1>T antigen (Gal4, VP16, c-Jun, GR)	RP-A	T antigen	T antigen
polyoma	T antigen	helicase	pol-α:primase RP-A	AP1>>Gal4, VP16, PEA3 c-Jun, v-Jun, E1A, E2, Sp1, Rel, E2F (CREB, pRB)	RP-A	T antigen	T antigen
Papilloma	E1	helicase		E2, VP16	E1, RP-A	E2, VP16	E1, RP-A
Herpes simplex	UL9	helicase[c]		several candidates			
Ad2	preterminal prot. Ad DNA pol	prime DNA synthesis DNA synthesis	Ad DNA pol preterminal prot.	NF1, Oct-1	Ad DNA pol		
Ad4	(same as Ad2)			(no Aux-2 factors)			
Epstein-Barr							
oriP	EBNA1	(*no* helicase)		EBNA1			
oriLyt	BZLF-1 ?			BZLF-1 (Jun, E2, Myc, VP16)			
Mitochondria							
oriH	CSB1 RNase MRP	 endoribonuclease		mtTFA, mtRNA pol			
oriL	mtDNA primase	RNA primer synthesis					
Kinetoplast	p13.5	binds ssDNA at ori					
S. cerevisiae	6-protein complex				ABF1, RAP1 Gal4		

[a]Adenovirus origin recognition protein is a complex of preterminal protein and Ad DNA polymerase. Pol-α:primase is DNA polymerase-α:DNA primase.
[b]Transcription factors in parentheses did not stimulate origin activity in vivo.
[c]Helicases encoded by papova and papilloma viruses can unwind DNA at the origin, but HSV helicase cannot. References are found in the appropriate chapters.

DNA Unwinding Element

DNA unwinding appears to begin at an easily unwound DNA sequence referred to as the DNA unwinding element (DUE, Fig. 1). A DUE is identified by *cis*-acting mutations that both increase the stability of the double helix (i.e., make DNA unwinding more difficult) and reduce DNA replication (Umek and Kowalski 1990b). Computer programs are available that determine DNA helical stability in known sequences (Rychlik and Rhoads 1989; Natale et al. 1992), and the effects of mutations on helical stability can be assessed within the context of a supercoiled plasmid using a single-strand-specific endonuclease, or two-dimensional gel electrophoresis of plasmid topoisomers. These approaches have identified DUEs in *E. coli oriC* (Kowalski and Eddy 1989), yeast ARS elements in plasmids (Natale et al. 1992; Huang and Kowalski 1993; Miller and Kowalski 1993), and replication origins in yeast chromosomes (Huang and Kowalski 1993) and SV40 (Lin and Kowalski 1994). In yeast replication origins the genetic and physical properties of element B2 are consistent with those of a DUE (Rao et al. 1994; Theis and Newlon 1994).

Although primary sequence is an important determinant of the energy required for DNA unwinding (because of the importance of base-stacking interactions), there is no unique consensus sequence for a DUE. DUE sequences in yeast are not conserved, and easily unwound DNA sequences that are not components of origins can substitute for the DUEs in both yeast and *E. coli* (Umek and Kowalski 1988; Kowalski and Eddy 1989). Therefore, DUEs are unlikely to be binding sites for specific replication proteins. Instead, it appears that one of the proteins binding to the ORE must interact with the DUE because the spatial relationship between these two core elements is critical. In yeast, the DUE always is located 3′ to the T-rich strand of the ARS consensus sequence (Natale et al. 1993), and reversal of the orientation of the ARS consensus sequence with respect to the DUE abolishes DNA replication (Holmes and Smith 1989). Binding of proteins to ORE results in DNA unwinding in the DUEs of *E. coli oriC* (Kowalski and Eddy 1989), SV40, and PyV. In SV40 and PyV, the DUE, the site where T antigen begins unwinding DNA, and the origin of bidirectional replication (OBR; defined by the transition between continuous and discontinuous DNA synthesis on each template [Fig. 1]) are all coincident (DePamphilis 1993a; Lin and Kowalski 1994). Similarly, in yeast ARS307, a majority of leading strands emanate from the proposed DUE (unpublished data cited in Theis and Newlon 1994). Therefore, the DUE appears to be the entry site for the replication machinery.

Not all easily unwound DNA sequences are part of replication origins, but the fact that such sequences can substitute for known DUEs qualifies them as "potential DUEs." Since the energetic, length, and spacing requirements of true DUEs remain to be defined, it is difficult to estimate the frequency at which potential DUEs occur in natural DNA. For purposes of comparison, a potential DUE can be defined as a 100-bp sequence whose helical stability is 20 or more kcal/mole below the average for the sequence analyzed. This definition is stringent enough to exclude some real DUEs. Such a potential DUE would be expected once every 3.2 kb in a random sequence of 60% A+T content (D. Natale, pers. comm.), suggesting that potential DUEs occur much more frequently than replication origins in yeast (1 origin/36 kb) and mammalian (1 origin/100 kb) DNA. Thus, the ability of an easily unwound sequence to function as a DUE likely depends on conditions such as its proximity to other origin elements (Fig. 1), the concentration of initiation factors, the influence of chromatin structure, and the amount of negative superhelical energy available (Umek and Kowalski 1990a).

A/T-rich Element

Most, but not all (e.g., EBV), replication origins contain an A/T-rich sequence consisting of a T-rich and an A-rich strand. The length of the A/T-rich element is critical for origin function (Gerard and Gluzman 1986; Koff et al. 1991), a fact that may be related to its bent character (Deb et al. 1986). Bent DNA can interact more easily with proteins, which may account for the fact that binding of origin recognition proteins to OREs frequently distorts the A/T-rich element (Koff et al. 1991; Gillette et al. 1994; SenGupta and Borowiec 1994). Distortion of the A/T-rich element may facilitate either binding to the ORE or melting of the DUE.

Auxiliary Components

ori-auxiliary components stimulate replication only when they bind one or more transcription factors, and only when the transcription factor contains an activation domain that specifically interacts with the replication machinery (Table 2) (DePamphilis 1993c; van der Vliet, this volume). In some genomes (SV40, PyV, PV, EBV), the same sequence elements that function as promoters or enhancers in transcription also function as *aux* components in replication; *cis*-acting mutations that affect one process also affect the other. Auxiliary components could be used to regulate origin activity in two ways. First, the ability of a particular transcription factor to stimulate a particular origin may be limited to specific members of a transcription factor family, and to the availability of specific coactivator proteins (Guo and DePamphilis 1992). Thus, auxiliary components can stimulate the same origin to different extents in different cell types as the composition of available transcription factors changes during animal development (Rochford and Villarreal 1991). Second, just as transcription factors can initiate transcription of different genes at different times during the cell division cycle, they could regulate the temporal order of DNA replication during S phase, accounting for the observation that active genes are replicated early and inactive genes are replicated late (see Simon and Cedar, this volume).

There are four basic mechanisms by which transcription factors can stimulate *ori*-core (DePamphilis 1993c):

1. An upstream promoter can direct transcription through *ori*-core. The resulting mRNA is then cut by an endonuclease at specific sites to generate RNA primers for initiation of DNA synthesis. This mechanism occurs at mtDNA *oriH, E. coli* filamentous and T-odd bacteriophage, and *E. coli* plasmid ColE1 (Kornberg and Baker 1992).
2. Transcription factors can facilitate binding of origin recognition proteins to *ori*-core. For example, NFI binding to *aux-2* facilitates binding of subsaturating concentrations of the Ad2 preterminal protein/Ad DNA polymerase complex (pTP-pol) (see Hay, this volume). Binding of PV-encoded enhancer-specific activation protein, E2, to *aux-1* and possibly *aux-2* facilitates binding of E1 to *ori*-core (see Stenlund, this volume). Binding of EBV-encoded EBNA1 protein to the EBV enhancer (*aux-2*) stabilizes interaction of EBNA1 to the EBV *ori*-core (Frappier et al. 1994).
3. Transcription factors can facilitate the activity of an initiation complex after it has formed. For example, SV40 *ori*-auxiliary components stimulate SV40 *ori*-core by facilitating T-antigen-dependent DNA unwinding at *ori*-core (Gutierrez et al. 1990). This may occur by stabilizing the interaction of T antigen with ssDNA as it disrupts its own binding site by unwinding it (Gutierrez et al. 1990), and by recruiting RP-A, a single-strand DNA-binding protein, to stabilize ssDNA (He et al. 1993; Li and Botchan 1993). The T-antigen dimer that binds to *aux-1* appears to interact with the T-antigen hexamer bound to ORE (Guo et al. 1991), thus accounting for the observation that the need for *aux-1* to stimulate papovavirus origin activity is inversely related to the ability of T antigen to activate *ori*-core (Sock et al. 1993). In the chromosomes of *E. coli*, bacteriophage λ, and plasmid R6K, tran-

scription or association with RNA polymerase near *ori* can stimulate *ori* activity by removing torsional stress in DNA and thus facilitating DNA unwinding (Kornberg and Baker 1992).

4. Transcription factors can prevent chromatin structure from interfering with binding of replication factors to origins. Nucleosomes can repress replication origins in yeast (Simpson 1990), *Drosophila* (Karpen and Spradling 1990), and mammalian chromosomes (Forrester et al. 1990). Therefore, since transcriptionally active DNA sequences are not incorporated into nucleosomes (Morse et al. 1992), transcription through a replication origin may provide access to replication factors. Alternatively, binding of transcription factors to enhancers may allow interactions between the enhancer and *ori*-core, analogous to those between enhancer and promoter (Majumder and DePamphilis 1994), that prevent nucleosomes from repressing origins in the same way that it prevents nucleosomes from repressing promoters (Paranjape et al. 1994). This mechanism may apply to PyV, where the PyV enhancer (*aux-2*) is dispensable under conditions where a repressive chromatin structure appears to be absent (Prives et al. 1987; Martínez-Salas et al. 1988; Majumder et al. 1993), and to SV40 *aux-2* where prebinding some transcription factors (e.g., NFI, Gal4:VP16) can prevent chromatin assembly from interfering with SV40 DNA replication in vitro (see Hassell and Brinton, this volume). However, the facts that Gal4:VP16 does not stimulate the SV40 origin in vivo, and that prebinding T antigen alone can also prevent nucleosome repression, suggest that other mechanisms should be considered (DePamphilis 1993c).

VIRAL ORIGINS AS MODELS FOR CELLULAR ORIGINS

PV and EBV have been considered models for cellular DNA replication because their genomes replicate in the nucleus, their DNA replication is restricted to S phase, and they maintain a low number of genome copies per cell. Moreover, early studies on PV DNA replication concluded that a complex interaction between positive and negative controls restricted initiation of replication to once per origin per S phase (Roberts and Weintraub 1988). However, later studies (see Stenlund, this volume) revealed that PV origins are remarkably similar to those in papovaviruses, and that although *cis*-acting PV sequences can suppress the activity of lytic origins such as SV40 and PyV, they do not restrict them to one initiation event per S phase (Nallaseth and DePamphilis 1994). In fact, initiation of PV DNA replication is not restricted to once per S phase (Gilbert and Cohen 1987; Ravnan et al. 1992). EBV remains a candidate because its DNA replicates at the same rate as cellular chromosomes (Haase and Calos 1991; Yates and Guan 1991), but it remains to be determined whether or not EBV DNA replication, like cellular DNA replication, does not reinitiate when cells are limited to a single S phase in the presence of a mitotic inhibitor such as nocodazole (Nallaseth and DePamphilis 1994). This test should be applied to all putative mammalian ARS elements (Table 3) (Krysan et al. 1993; Masukata et al. 1993).

COMPLEX (METAZOAN) ORIGINS

In comparison with simple genomes, origins of replication in multicellular animals (the metazoa) often appear paradoxical. Early attempts to identify *ori* sequences in mammalian chromosomes by their ability to function as ARS elements were difficult to reproduce and therefore controversial (Gutierrez et al. 1988; Burhans and Huberman 1994), although some recent reports (Table 3) look promising. Nevertheless, most large (>10 kb) DNA fragments from mammalian chromosomes can provide some ARS activity in mammalian cells (Krysan et al. 1993; Masukata et al. 1993), suggesting that DNA length is more critical than DNA sequence. The same conclusion is reached when DNA is injected into the eggs of frogs, sea urchins, or fish, or when DNA is added to extracts of *Xenopus* eggs or *Drosophila* embryos (Coverley and Laskey 1994; see Laskey and Madine; Blow and Chong; both this volume). DNA replication is initiated at a single randomly chosen site within virtually any DNA molecule (Hines and Benbow 1982; Méchali and Kearsey 1984; Hyrien and Méchali 1992; Mahbubani et al. 1992). This lack of site-specific initiation also appears during chromosome replication in *Xenopus* (Hyrien and Méchali 1993) and *Drosophila* (Shinomiya and Ina 1991) embryos, suggesting that the lack of sequence requirements and site specificity observed when DNA is introduced into cultured cells, eggs, or egg extracts accurately reflects chromosome replication in situ.

Another approach to identify replication origins in metazoan chromosomes is in situ mapping of initiation sites for DNA replication in the hope that subsequent physical and genetic analysis of those loci will reveal the nature of a metazoan origin of DNA replication. This approach has revealed that DNA replication in metazoan chromosomes occurs at specific sites.

Table 3 Metazoan origins of replication are genetically determined

Organism	Location	Same origin in single copy and multicopy genomes[a]	Translocated to other sites	Deletion in OBR	ARS activity	APE activity
Hamster	DHFR gene (oriβ)	chromosomal (1)	active (2)			yes (3)
Mouse	rRNA gene					yes (4)
Drosophila	chorion gene		active (5)			
Human	hsp70 gene				yes (6)	
Mouse	IgH gene enhancer				yes (7)	
Mouse	ADA gene (late S phase)	episomal (8)			yes (8)	
Human	c-*myc* gene				yes (9)	
Human	cDNA 343				yes (10)	
Hamster	CAD gene group	episomal (11)				
Human	β-globin gene			inactive (12)		
Mouse	ADA gene (early S phase)	episomal (13)				

[a]Amplified gene copies are found in either chromosomal or episomal locations.

References (in parentheses): (1) Handeli et al. 1989; Burhans et al. 1990, 1991; Vassilev et al. 1990; Dijkwel and Hamlin 1992, 1995; DePamphilis 1993d; Tasheva and Roufa 1994a; Gilbert et al. 1995; M. Giacca, unpubl.; (2) Handeli et al. 1989; (3) Stolzenburg et al. 1994; (4) Hermann et al. 1994; (5) Orr-Weaver 1991; (6) Taira et al. 1994; (7) Ariizumi et al. 1993; Iguchi-Ariga et al. 1993; (8) Virta-Pearlman et al. 1993; (9) Berberich et al. 1995; (10) Wu et al. 1995; (11) Kelly et al. 1995; (12) Kitsberg et al. 1993; (13) Pearson et al. 1994.

MAPPING ORIGINS OF REPLICATION IN METAZOAN CHROMOSOMES

Methods for mapping DNA replication initiation sites fall into two categories: those that begin by labeling nascent DNA chains, and those that begin by fractionating DNA structures. The first category analyzes nascent DNA strands labeled by incorporation of radioactive or density-substituted deoxyribonucleotides during DNA replication, annealing them to sequence-specific probes in order to determine the amount and direction of synthesis that occurs at specific DNA sites (for discussion, see Vassilev and DePamphilis 1992). There are four basic methods.

The first method identifies the earliest labeled DNA fragments in cells that have been synchronized at their G_1/S border, permeabilized, and then released into S phase in the presence of a labeled deoxyribonucleotide. Labeled DNA should appear first at origins of DNA replication. Identification of the earliest labeled DNA fragment can be facilitated by cross-linking the DNA templates with psoralen prior to initiation of replication to prevent migration of replication forks away from the origin region. These methods are generally applied to cells containing amplified DNA sequences. Thus, replication sites identified under these conditions could either be unique to amplified DNA sequences or an artifact of the cell-synchronization conditions.

Examination of single-copy sequences in exponentially proliferating cells in the absence of metabolic inhibitors is possible with the help of the DNA polymerase chain reaction (PCR). Labeled nascent DNA strands are separated from unreplicated DNA and then fractionated according to their length. Replication origins can be localized either from the length (Vassilev et al. 1990) or abundance (Giacca et al. 1994; Yoon et al. 1995) of nascent DNA strands passing through a specific genomic sequence. An origin of bidirectional replication (OBR) (Fig. 1) lies halfway along the shortest nascent DNA strand passing through a specific sequence. In addition, the closer a specific sequence lies relative to an OBR, the greater its abundance in shorter nascent DNA chains relative to longer nascent DNA chains. By examining several specific sequences on either side of a putative OBR, bidirectional replication events can be distinguished from unidirectional events, and the resolution of an OBR increases. Quantitation of the number of nascent DNA chains that contain a specific sequence (and therefore OBR resolution) can be improved by competition between hybridization of the PCR primer to nascent DNA chains and a competitor DNA standard (Giacca et al. 1994).

The third group measures the distribution of Okazaki fragments between the two arms of a replication fork. If DNA replication occurs by the replication fork mechanism (Fig. 1), then DNA synthesis on the forward arm (leading strand) is a continuous process, but DNA synthesis on the retrograde arm (lagging strand) is a discontinuous process in which short (40–300 nucleotides) RNA-primed, nascent DNA chains (Okazaki fragments) are repeatedly formed and joined into longer chains (see Brush and Kelly, this volume). Therefore, the ratio of Okazaki fragments that anneal to unique DNA sequences representing the retrograde-arm template versus those representing the forward-arm template provides a minimum estimate of the fraction of replication forks traveling in a specified direction. By measuring the distribution of Okazaki fragments between the templates at several different genomic locations, the transition from continuous to discontinuous DNA synthesis that defines an OBR can be mapped (Fig. 1). With small circular genomes such as SV40 and PyV, sufficient Okazaki fragments can be isolated from infected cells to localize the SV40 and PyV OBRs to 2 bp and 20 bp, respectively, by digesting the DNA hybrids at a unique restriction site and sequencing the nascent DNA strands (DePamphilis et al. 1988). With metazoan chromosomes, it is necessary to synchronize cells at their G_1/S border in order to collect Okazaki fragments only from newly initiated replication origins and not from replication forks traveling through from upstream or downstream origins. This limited availability of labeled Okazaki fragments restricts resolution of OBRs in mammalian cells to the sizes of the sequence-specific probes used in the hybridization reaction.

A fourth method measures replication fork polarity in exponentially proliferating cells by inhibiting protein synthesis in vivo. Under these conditions, Okazaki fragment synthesis is preferentially inhibited, allowing accumulation of labeled long nascent DNA strands synthesized on the forward arms of replication forks. Again, the fraction of labeled strands that anneal to each of the two strands of a specific sequence identifies fork polarity. In this procedure, however, nascent DNA should preferentially anneal to the template complementary to that recognized by Okazaki fragments. Initial studies using this mapping protocol assumed that it depended on preferential segregation of pre-fork histone octamers to the forward arm of replication forks in the absence of histone synthesis, and therefore employed micrococcal nuclease to digest nascent DNA on retrograde arms (Handeli et al. 1989). Subsequent studies demonstrated that this mapping protocol did not depend on chromatin structure and therefore did not require micrococcal nuclease (Burhans et al. 1991; Kitsberg et al. 1993).

The second category of methods for identifying replication origins is based on fractionating DNA according to its size and shape using two-dimensional (2-D) gel electrophoresis (Fangman and Brewer 1991; Huberman 1994). Structures specifically associated with DNA replication, such as bubbles and forks (depicted in Fig. 1), can be released by digestion with restriction endonucleases. They are then recognized by their mobility patterns during gel electrophoresis, and their genomic locations are identified by blotting-hybridization with sequence-specific radiolabeled probes. This approach is particularly useful in systems such as yeast, where incorporation of labeled nucleotides into nascent DNA is difficult. The direction in which replication forks travel can be determined by first fractionating double-stranded DNA according to its size at neutral pH, and then fractionating it in a second dimension at alkaline pH to observe the lengths of single-stranded (nascent) DNA released from replication forks. The closer to an OBR, the greater the abundance of short nascent DNA strands, analogous to measuring the lengths of nascent (radiolabeled) DNA strands described above. 2-D neutral/neutral and 2-D neutral/alkaline gel electrophoresis fractionations can be run sequentially to determine more precisely the sizes of forks and bubbles at specific genomic locations (Liang and Gerbi 1994). These methods have localized replication origins in PV, yeast, and slime mold to specific sites of 0.2 to 1 kb (see Stenlund; Newlon; Pierron and Bénard; all this volume).

METAZOAN CHROMOSOMES INITIATE DNA REPLICATION AT SPECIFIC SITES

Mapping initiation sites for DNA replication at 18 different genomic locations (Table 4) has yielded the following characteristics of metazoan DNA replication origins:

1. DNA synthesis does not initiate randomly throughout cellular chromosomes, but at specific DNA sites. Therefore, at some point during animal development, specific replication origins are formed. These sites, however, appear to consist of a primary origin (OBR) flanked by many secondary origins (initiation zone).
2. From 80% to 95% of DNA synthesis occurs bidirectionally from specific genomic loci referred to as OBRs. This conclusion is based on the fraction of replication forks traveling in the same direction, as determined by 2-D neutral/alkaline gels (Liang et al. 1993; Liang and Gerbi 1994; Shinomiya and Ina 1994), the ratio of Okazaki fragments that hybridize to the two strands of a unique DNA probe (Burhans et

Table 4 Metazoan chromosomes initiate DNA replication at specific sites

Organism	Location	OBR[a] (kb)	Initiation zone[b] (kb)	Reference
Hamster	DHFR gene (*ori*β)	0.5–3	55	1
Human	rRNA genes	8	31	2
Mouse	rRNA genes	3		3
Drosophila	DNA polymerase α gene	5	10	4
Drosophila	chorion gene	1[c]	12	5
Sciara	chromosome-2, locus 9	1[c]	6	6
Human	Hsp70 gene	0.4		7
Human	lamin B2 gene	0.5		8
Mouse	IgH gene enhancer	0.6		9
Mouse	ADA gene (late S phase)	2		10
Human	c-*myc* gene	2		11
Hamster	ribosomal protein S14 gene	2		12
Human	cDNA 343 (early S phase)	2		13
Human	β-globin gene	2		14
Mouse	CAD gene group	5		15
Hamster	rhodopsin gene	5		16
Hamster	DHFR gene (*ori*γ)	8		17
Mouse	ADA gene (early S phase)	11		18

[a]Mapped by labeling of nascent DNA strands.

[b] Mapped by 2-D gel analyses of replication bubbles.

[c]Origins of gene amplification that were mapped by 2-D gel analyses of replication fork direction.

References: (1) Handeli et al. 1989; Burhans et al. 1990, 1991; Vassilev et al. 1990; Dijkwel and Hamlin 1992, 1995; DePamphilis 1993d; Tasheva and Roufa 1994a; Gilbert et al. 1995; M. Giacca, unpubl.; (2) E. Gogel and F. Grummt, unpubl.; (3) Little et al. 1993; Yoon et al. 1995; (4) Shinomiya and Ina 1994; (5) Delidakis and Kafatos 1989; Heck and Spradling 1990; (6) Liang et al. 1993; Liang and Gerbi 1994; (7) Taira et al. 1994; (8) Giacca et al. 1994; (9) Ariizumi et al. 1993; Iguchi-Ariga et al. 1993; (10) Virta-Pearlman et al. 1993; (11) Vassilev and Johnson 1990; Berberich et al. 1995; (12) Tasheva and Roufa 1994a; (13) Wu et al. 1993; Pearson et al. 1994; (14) Kitsberg et al. 1993; (15) Kelly et al. 1995; (16) Gale et al. 1992; (17) Handeli et al. 1989; (18) Carroll et al. 1993.

al. 1990; Carroll et al. 1993; Tasheva and Roufa 1994a; Berberich et al. 1995; Kelly et al. 1995; Gilbert et al. 1995), and the ratio of long nascent DNA strands from forward arms of replication forks that hybridize to the two strands of a unique DNA probe (Handeli et al. 1989; Burhans et al. 1991; Kitsberg et al. 1993; Kelly et al. 1995). In addition, quantitative analysis of specific sequences (Gilbert et al. 1995), particularly those within long nascent DNA strands (Vassilev and Johnson 1990; Vassilev et al. 1990; Giacca et al. 1994; Yoon et al. 1995), reveals that most of them originate bidirectionally from a small chromosomal locus that resides within an initiation zone (see point 4).

3. An OBR is contained within 0.5 kb to 2 kb. This conclusion is based on 11 different OBRs that appear to lie within a 2-kb region and on the DHFR *ori*β locus where five different nascent DNA strand methods have been applied with remarkable agreement (Table 4). Some OBRs lie within larger regions of 5–11 kb. All of these estimates are for the maximum size of an OBR; resolution is limited by the difficulty in preparing probes large enough to give a strong hybridization signal with radiolabeled nascent DNA chains, and devoid of repetitive sequences. Future refinements in origin-mapping techniques will likely resolve these OBRs to a smaller locus. Thus, metazoan replication origins appear to be 3–10 times larger than replication origins in simple genomes (0.05–1 kb). The fact that 18 OBRs have been identified by independent investigators using a variety of different methods gives confidence that site-specific initiation is not an artifact of the experimental conditions used to map them. Similar results were obtained with synchronized and unsynchronized cells, with cells containing single-copy sequences and with cells containing amplified multicopy sequences, with untreated cells and with cells treated with metabolic inhibitors, and with different methods for detecting specific DNA sequences.

   Neutral/alkaline 2-D gel electrophoresis has been used to map an OBR to 1 kb at an amplification locus in *Sciara* (Liang et al. 1993; Liang and Gerbi 1994), giving credence to an earlier interpretation of 2-D gel electrophoresis mapping data that 80% of replication forks at the chorion gene amplification locus in *Drosophila* originate from a specific 1-kb site (Heck and Spradling 1990). Neutral/alkaline 2-D gel electrophoresis also identified an S-phase OBR 15–20 kb downstream from the *Drosophila* DNA polymerase α gene (Shinomiya and Ina 1994).
4. Replication bubbles are detected throughout a larger "initiation zone" of 6–55 kb that includes the OBR. This conclusion is based on analyses of DNA structures by neutral/neutral 2-D gel electrophoresis at five different genomic loci (Table 3) (Delidakis and Kafatos 1989; Heck and Spradling 1990; Dijkwel and Hamlin 1992, 1995; Liang et al. 1993; Little et al. 1993; Liang and Gerbi 1994), two of which (*Drosophila* chorion gene, *Sciara* locus 9) are developmentally programmed amplification origins. Four of these "initiation zones" encompass an OBR that was detected either by nascent strand analyses (rRNA genes, DHFR *ori*β) or by measuring the direction of fork movement using neutral/alkaline 2-D gel electrophoresis (*Sciara* locus 9, *Drosophila* pol α gene). One study on the *Drosophila*

chorion gene locus (Heck and Spradling 1990) concluded that although multiple initiation sites may exist within a 12-kb locus, a model in which a single origin is preferred 70–80% of the time could explain their neutral/neutral 2-D gel electrophoresis data.

Results of neutral/neutral 2-D gel analyses are consistent with newly synthesized DNA analyses and most neutral/alkaline gel analyses if one assumes that the frequency of initiation events at the OBR is much greater than the frequency of initiation events outside the OBR. In fact, replication bubbles detected by neutral/neutral 2-D gel analyses appear more abundant in the 12-kb region containing the DHFR *ori*β OBR (Dijkwel and Hamlin 1992), and in the 8-kb region at the 5′-end of the rRNA transcription unit (Little et al. 1993) where nascent DNA strand analyses revealed a >10-fold excess of newly synthesized DNA relative to other sites within the initiation zone (DePamphilis 1993d; Gilbert et al. 1995; Yoon et al. 1995). In practice, the relative number of initiation events in different DNA segments is difficult to quantify by 2-D gel analysis because of concerns over variable loss of replication bubbles and other technical problems (Dijkwel and Hamlin 1992; Krysan et al. 1993; Little et al. 1993), whereas analysis of labeled nascent DNA chains lends itself readily to quantification and thus reveals the preference for one site relative to another. For example, the ratio of DNA synthesis between the two templates of a specific DNA fragment automatically provides the minimum fraction of replication forks moving in the same direction through this region. Initiation events distributed randomly outside the OBR simply contribute to the background level in these mapping protocols.

METAZOAN ORIGINS OF REPLICATION ARE GENETICALLY DETERMINED

The simple fact that metazoan origins map to specific sites that replicate at specific times during S phase (see Simon and Cedar, this volume) demonstrates that origins of replication are inherited from one cell division to the next. This conclusion is reinforced by reports that the same OBR identified in cells containing two copies per diploid genome is also identified in cells containing 1,000 (hamster DHFR gene) to 30,000 (mouse ADA gene) tandem copies of either chromosomal or extrachromosomal (episomal) sequences (Table 3). Therefore, each copy of the amplified region that initiates replication must use the same OBR; otherwise, initiation would appear to occur at many different sites within the same DNA locus.

Direct evidence that metazoan replication origins are genetically determined comes from reports that DHFR *ori*β (Handeli et al. 1989) and the chorion gene amplification origin (Orr-Weaver 1991) retain their activity when translocated to other chromosomal sites. Conversely, an 8-kb deletion between the human δ-globin and β-globin genes that includes the only OBR found within a 135-kb region eliminates bidirectional replication from this site; all replication forks now move in one direction through this 135-kb region (Kitsberg et al. 1993). These data demonstrate that metazoan origins of replication are determined by as yet undefined DNA sequences. Nevertheless, identification of genetically required DNA sequences that function as ARS elements has been difficult.

To date, five reports of ARS elements that function in mammalian cells and cell extracts have been documented in detail and shown to correspond to sites where replication occurs in mammalian chromosomes (Table 3). In other plasmid assays, replication appears to depend on the distribution of as yet undefined sequence signals over a large area (>10 kb), signals that are more prevalent in human DNA than in bacterial DNA (Krysan et al. 1993). Sequences have been identified in human DNA that stimulate plasmid replication about 3-fold and are present at a 2-kb OBR mapped in chromosomal DNA (Masukata et al. 1993). These results suggest that replication is stimulated by simple sequence features that occur frequently in mammalian DNA and therefore may promote initiation events throughout the initiation zone.

ARS activity in mammalian cells may depend on several variables. For example, some OBR regions may exhibit stronger ARS activity than others. Incubation of negatively supercoiled plasmid DNA with DNA polymerase-α:DNA primase, RP-A, T-antigen helicase, and DNA gyrase resulted in site-specific initiation of DNA replication at the strong yeast origin, ARS1, and at the c-*myc* OBR (Ishimi et al. 1994). These conditions employ the energy derived from negative superhelical turns to initiate DNA replication at DUEs that can be unwound by T antigen in the presence of RP-A. However, in the DHFR *ori*β region where ARS activity has not been detected (Burhans et al. 1990), preference for the OBR region was observed, but it was less pronounced than with the other two origins.

Other studies on plasmid DNA replication in human cells (Caddle and Calos 1992) or in *Xenopus* eggs and egg extracts (Gilbert et al. 1995) failed to observe either preferential replication of plasmids containing the DHFR *ori*β region or site-specific initiation within *ori*β in those plasmids that contained this sequence. However, when nuclei from G_1-phase hamster cells were incubated in *Xenopus* egg extract, DNA replication was initiated specifically at or near the same *ori*β OBR utilized by hamster cells (Gilbert et al. 1995). Therefore, site-specific initiation of DNA replication in metazoan chromosomes involves nuclear structure, a requirement that may be difficult to fulfill with plasmid DNA. For example, matrix (scaffold) attachment regions (Schlake et al. 1994) or locus control regions (Bonifer et al. 1994) can increase transcription rates for integrated but not episomal templates, demonstrating that some potential components of replication origins function only in the context of cell chromosomes. Conversely, many sequences that can function as ARS elements in plasmids do not function as replication origins in chromosomes (Kipling and Kearsey 1990; Newlon et al. 1993). Therefore, plasmid DNA replication may not be an appropriate model for metazoan cellular DNA replication, because metazoan origins may function efficiently only within the context of a real chromosome.

Finally, the sequence context of an origin can strongly affect its activity. When two or more yeast ARS elements are in close proximity (~6 kb), the efficiency of each is reduced, and only one is activated in each cell cycle (Brewer and Fangman 1993; Marahrens and Stillman 1994). This phenomenon has been demonstrated in *S. pombe* chromosomes, where initiation zones have been shown to be composed of two or three independent origins (Dubey et al. 1994; Wohlgemuth et al. 1994). Other sequences in the neighborhood also can affect origin activity, making one yeast ARS element preferred over its neighbor (Newlon et al. 1993; Brewer and Fangman 1994). Thus, one could imagine that a metazoan initiation zone is composed of many "simple origins" of the type found in yeast, for example, and that the resulting interference patterns from neighboring origins and extraneous sequences would impose an OBR at one particular site. Moreover, the anatomical complexity observed for metazoan initiation zones could vary considerably as a function of the number and arrangement of the simple origins that comprise them.

The ability to detect ARS activity in mammalian cells thus may depend on a number of factors, among which are negative superhelical density in the extrachromosomal DNA, sequence context of the cellular OBR, number and proximity of initiation signals that comprise a replication origin, size of the extrachromosomal DNA, and the relative strengths of various OBRs. In addition, detection of ARS activity may require stringent selection conditions (Virta-Pearlman et al. 1993). Detection of ARS in *S. pombe*, for example, is more difficult than in *S. cerevisiae*, because virtually every DNA sequence, even vector DNA, is capable of replicating to a limited extent and therefore requires stringent selection conditions to identify "true" ARS elements (Caddle and Calos 1994;

Dubey et al. 1994; Wohlgemuth et al. 1994). Furthermore, if replication sites in nuclei are limited, only a small number of extrachromosomal origins will be accommodated, and detection may require a sensitive PCR-based assay (Taira et al. 1994).

An alternative assay for *cis*-acting sequences that initiate DNA replication is to look for an amplification promoting element (APE) that promotes formation of large numbers of integrated copies of a DNA sequence, rather than replication of extrachromosomal DNA sequences. A 370-bp APE has been identified in the nontranscribed spacer region of mouse rRNA gene (Hermann et al. 1994) and mediates a 40- to 800-fold amplification of the vector DNA in transformed cells. This DNA segment also contains an OBR that maps from 0.5 to −3.5 kb upstream of the transcription initiation site for mouse rRNA gene (E. Gogel and F. Grummt, unpubl.), in agreement with the OBR at the 5′end of human rRNA genes (Little et al. 1993; Yoon et al. 1995). The 4.5-kb OBR region in the hamster DHFR *ori*β also acts as an APE and contains homologies with the APE found in rRNA genes (Stolzenburg et al. 1994). APE activity may provide a more reproducible assay for metazoan replication origin sequences, if these sequences function efficiently only in the context of a large chromosome.

DNA FEATURES OF A METAZOAN REPLICATION ORIGIN

Metazoan replication origins contain a number of structural features that may be related to their role in DNA replication, although none have so far been shown to be required for initiation of chromosomal DNA replication (see Heintz, this volume). Among these are binding sites for transcription factors, and those for c-Myc protein (Taira et al. 1994) and octamer-binding transcription factors (Iguchi-Ariga et al. 1993) have been reported to contribute to ARS activity in mammalian cells. Replication and transcription sites are colocalized in mammalian nuclei (Hassan et al. 1994), but transcription through cellular replication origins does not appear to be required for replication, since only 3 of the 18 OBRs in Table 4 are located within a transcribed region (cDNA 343, RPS14 gene, CAD gene). More likely, both processes are facilitated by an open chromatin structure that provides access to initiation factors and negative superhelical energy that facilitates DNA unwinding. These features may be provided by a particular nuclear structure.

One feature likely to be shared by all OBRs is the presence of one or more DUEs, since DNA unwinding must precede DNA synthesis and simple OBRs appear to initiate DNA unwinding at a DUE (see above).

Potential DUEs exist at or close to several OBRs, including DHFR *ori*β (Dobbs et al. 1994).

A second feature is a densely methylated island (DMI) that consists of 127 bp (RPS14 gene OBR) to 512 bp (DHFR *ori*β) of DNA in which all dC residues are methylated on both strands, regardless of the adjacent nucleotide (Tasheva and Roufa 1994b). This unusual methylation pattern has been observed only in association with replication origins, and then only in proliferating cells. Intriguingly, DNA methyltransferase, the enzyme responsible for converting hemimethylated sites to methylated sites in nascent DNA, becomes associated with replication foci during S phase (Leonhardt et al. 1992). DMIs might act in a positive way by providing a binding site for a replication-specific factor, by altering DNA structure to promote unwinding at a neighboring DUE, or by altering chromatin structure to increase accessibility to initiation factors. In fact, the DMI overlapping the DHFR *ori*β OBR is flanked by binding sites for RIP60, a protein that can link the two sites to form a 736-bp DNA loop that encompasses the DMI and flanks a potential DUE (Mastrangelo et al. 1993). By analogy to *E. coli* DnaA, bacteriophage λ O protein (Kornberg and Baker 1992), and SV40 T-antigen (see Borowiec, this volume), binding of origin recognition proteins can impose superhelical tension that causes untwisting of DNA in nearby DUEs. Alternatively, by analogy to *E. coli oriC* (Herrick et al. 1994), the DMI may help to limit initiation to once per S phase. *oriC* is methylated on both strands at 11 sites. When these sites become hemimethylated as a result of replication, *oriC* associates with an outer membrane component, delaying rebinding of its origin recognition protein, DnaA, and thus delaying reinitiation.

A third feature is attachment sites for nuclear matrix or nuclear scaffold (Nakayasu and Berezny 1989; Hozák et al. 1993), components of nuclear structure that are commonly associated with newly replicated DNA and, in some cases, origins of cellular replication (see Laskey and Madine, this volume).

A fourth feature is palindromic sequences that can collapse into cruciform structures when sufficient negative superhelical energy is provided. These structures may promote initiation of DNA replication at specific sites. Cruciform extrusion at the origin of *E. coli* plasmid pT181 is promoted by the plasmid-encoded initiator protein RepC (Noirot et al. 1990), and antibodies directed against DNA cruciforms can stimulate overall DNA synthesis and copy number of specific genes in permeabilized mammalian cells (Zannis-Hadjopoulos et al. 1988). Moreover, staining mammalian cells with anti-cruciform antibodies suggests that cruciform structures accumulate as cells prepare to enter S phase (Ward et al. 1991). A role for cruciforms in replication origins must be considered cautiously, however, since cruciform structures can be a direct result of aberrant DNA replication involved in gene amplification (Cohen et al. 1994).

THE ROLE OF NUCLEAR STRUCTURE IN METAZOAN DNA REPLICATION

Prokaryotic genomes and animal virus genomes can all replicate in the presence of purified soluble proteins and cofactors; no requirement for a cellular structure has been observed, although there exists a transient interaction between *E. coli oriC* and the outer membrane that regulates the rate at which reinitiation can occur at *oriC* (Herrick et al. 1994). Whether or not the same is true for replication origins in simple eukaryotic organisms remains to be seen. However, one of the most striking requirements for initiation of DNA replication in metazoan chromosomes is that of nuclear structure (see Laskey and Madine, this volume).

DNA replication in metazoan chromosomes occurs at discrete nuclear foci. Clusters of replication origins initiate replication synchronously (Hand 1978), giving rise to discrete "replication complexes" that contain from 100 to 300 replication forks (see Laskey and Madine, this volume). Formation of these replication complexes accounts for the many observations that newly synthesized DNA is preferentially bound to components of nuclear structure generally referred to as nuclear matrix or nuclear scaffold (Nakayasu and Berezney 1989; Hozák et al. 1993). These replication complexes appear to be assembled in an energy-dependent process prior to S phase at the sites where replication begins (Fig. 2). RP-A, a heterotrimeric single-strand DNA-binding protein that is required for replication of metazoan chromosomes (Fang and Newport 1993), is bound at discrete foci in nuclei prior to DNA unwinding and DNA synthesis (Adachi and Laemmli 1994). High levels of cyclin B/ cdc2 protein kinase, an enzyme that is required for entrance into mitosis, prevents the appearance of these RP-A foci, consistent with their absence in mitotic chromosomes (Adachi and Laemmli 1994). Cyclin-A-dependent cdk2 protein kinase, an enzyme that is required for entrance into S phase (Fang and Newport 1991), colocalizes with RP-A (Cardoso et al. 1993). Proliferating cell nuclear antigen (PCNA), a cofactor for DNA polymerase-δ, and DNA polymerase-α, an enzyme required for synthesis of Okazaki fragments, are also found at replication foci in S-phase nuclei (Kill et al. 1991). Whether or not they are prebound to these foci before replication begins is not clear. Presumably, licensing factor (see Laskey and Madine; Blow and Chong; both this volume), a cytoplasmic initiation factor that gains access to replication origins only when the nuclei become permeable during mitosis, also binds to preinitiation complexes.

Initiation of DNA replication in metazoan chromosomes requires an intact nuclear structure. Replication of DNA introduced into either

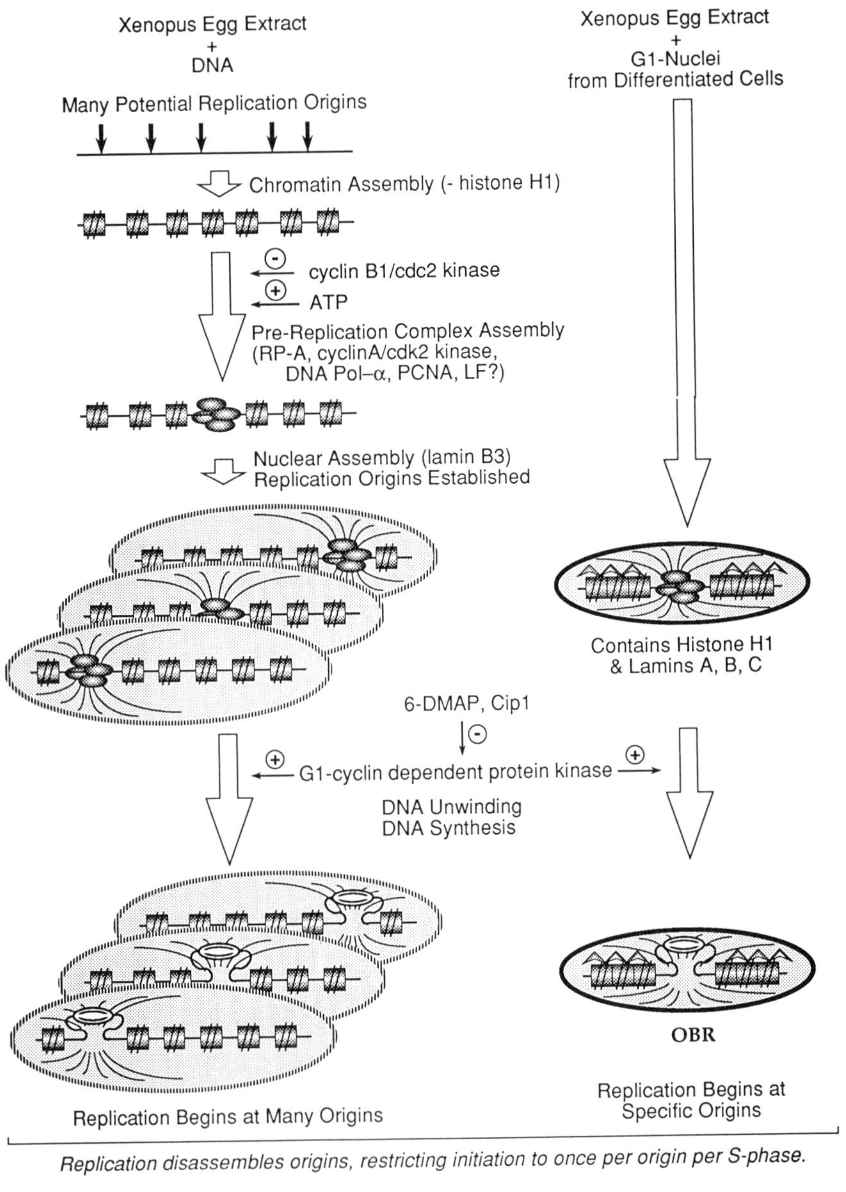

Figure 2 (See facing page for legend.)

Xenopus eggs or egg extracts does not occur unless DNA is first assembled into chromatin and then organized into a nuclear structure that includes lamin B3 and functional nuclear pores (see Laskey and Madine; Blow; both this volume). In addition, the nuclear envelope is instrumental in regulating the onset of S phase, apparently by regulating access of chromosomal DNA to one or more initiation factors (licensing factor) present in the cytoplasm (Coverley and Laskey 1994).

Site-specific initiation requires an intact nuclear structure. *Xenopus* egg extract can initiate DNA replication in purified DNA molecules only after the DNA is organized into a pseudo-nucleus, but under these conditions, DNA replication is independent of DNA sequence and begins at many sites distributed throughout the molecules. However, *Xenopus* egg extract can initiate DNA replication at specific sites in mammalian chromosomes, but only when the DNA is presented in the form of an intact nucleus from differentiated cells (Gilbert et al. 1995). Initiation of DNA synthesis in nuclei isolated from G_1-phase hamster cells is distinguished from continuation of DNA synthesis at preformed replication forks in S-phase nuclei by a delay that precedes DNA synthesis, a dependence on soluble *Xenopus* egg factors, sensitivity to the protein kinase inhibitor 6-dimethylaminopurine (DMAP), and complete labeling of nascent DNA chains. Initiation sites for DNA replication were mapped downstream from the amplified DHFR gene region by (1) identification of the earliest labeled DNA fragments (Gilbert et al. 1993), (2) quantitative hybridization of newly synthesized DNA to double-stranded DNA probes to reveal genomic loci where DNA synthesis began, and (3)

Figure 2 Acquisition of site-specific DNA replication. *Xenopus* eggs or egg extracts assemble bare DNA or sperm chromatin into a relaxed nuclear structure that permits initiation of DNA replication within many sequences, allowing the early-cleavage-stage amphibian embryo to rapidly replicate its genome. In contrast, preformed nuclei from G_1-phase differentiated mammalian cells initiate DNA replication under the same conditions at or near a site-specific OBR that was selected by the mammalian cell to be used as a replication origin during its subsequent S phase. Selection of initiation sites may be restricted by chromatin structure (nucleosome , histone H1) masking some potential origins, and nuclear matrix (scaffold) associated regions ( ) stabilizing DNA unwinding at other potential origins. At some point during animal development, changes occur in nuclear organization that restrict the number of sites that can be used as origins of replication. In *Xenopus* (and presumably other animals whose embryos undergo rapid cell cleavages) this transition appears to occur after the mid-blastula transition (see text). (LF) Licensing factor.

quantitative hybridization of Okazaki fragments to single-stranded DNA probes to reveal the transition between continuous and discontinuous DNA synthesis on each template within this initiation locus. When bare DNA substrates are used, then *Xenopus* eggs or egg extracts do not distinguish between prokaryotic DNA, hamster DNA that does not contain a replication origin, and hamster DNA that does contain a replication origin. Moreover, initiation events were distributed equally throughout a 30-kb cosmid containing the DHFR *ori*β region. When nuclei are used, *Xenopus* egg extract continues DNA synthesis in S-phase nuclei at sites that had been initiated in hamster cells (e.g, DHFR *ori*β). When the integrity of the nuclear membrane is preserved, *Xenopus* egg extract initiates DNA replication in G_1-phase nuclei specifically at or near the OBR (*ori*β) utilized by hamster cells. When nuclear integrity is damaged, preference for initiation at *ori*β is significantly reduced or eliminated. Therefore, initiation sites for DNA replication in mammalian cells are established prior to S phase by some component of differentiated nuclear structure, and this replication origin can be recognized by soluble initiation factors present in *Xenopus* eggs.

Subsequent studies (Wu and Gilbert 1996) have revealed that *Xenopus* egg extract initiates replication at many sites throughout the DHFR gene region in nuclei isolated from early G_1-phase hamster cells, whereas the same extract initiates specifically at *ori*β in nuclei isolated from late G_1-phase hamster cells. Therefore, specific origins of replication in mammalian chromosomes are reestablished during each cell division cycle several hours after nuclear assembly occurs.

MODEL FOR METAZOAN ORIGINS OF DNA REPLICATION

There are three basic models for understanding the cumulative data on metazoan replication origins. The first is the "strand separation model" in which extensive DNA unwinding precedes initiation of RNA-primed DNA synthesis at many sites concurrently on both templates (Benbow et al. 1992). This model was based largely on reports of single-stranded DNA in *Xenopus* and *Drosophila* embryos, but the single-stranded DNA bubbles that one would expect as replication intermediates have not been detected by 2-D gel analyses of DNA replication in *Xenopus* eggs (Hyrien and Méchali 1992; Mahbubani et al. 1992) or *Drosophila* embryos (Liang and Gerbi 1994), or by electron microscopy of DNA replicating in mammalian cells (Hamlin et al. 1992; Gruss et al. 1994). On the contrary, the fact that replication forks and bubbles are detected by these methods, and the fact that replication fork polarity is observed by hybridization of either Okazaki fragments or long nascent DNA chains to complementary DNA templates (see above, Mapping Origins of Replication in Metazoan Chromosomes), provide compelling evidence that DNA synthesis occurs at replication forks in both simple and complex genomes.

A second model is one in which a replication complex assembles at the OBR, but then migrates to other positions on either side of the OBR before initiating DNA unwinding and DNA synthesis. This would generate a Gaussian distribution of initiation events with the OBR at its apex. However, one would expect the OBR sequence to strongly stimulate DNA replication under all conditions, a prediction that is difficult to reconcile with the lack of DNA sequence preference observed in the eggs and cleavage-stage embryos of frogs, flies, fish, and sea urchins.

A third model is suggested by the Jesuit dictum that "many are called, but few are chosen" and perhaps offers the simplest way of interpreting all the data (Fig. 3) (DePamphilis 1993a,b,d; Burhans and Huberman 1994). Whereas naked DNA contains many possible sites where replication can begin, assembly of DNA into chromatin can suppress initiation at some of these sites, and organization of chromatin into a nuclear structure can activate DNA replication at other sites by promoting DNA unwinding. Thus, initiation sites for DNA replication in metazoan chromosomes would consist of three parts:

1. *OBR.* Most initiation events occur bidirectionally within a 0.5- to 2-kb locus. In differentiated cells, where most potential replication origins are suppressed, favored sites may consist of an easily unwound DNA sequence in combination with other origin components such as DMIs, OREs, and transcription-factor-binding sites. These features, in addition to sequences that may attach to nuclear matrix (scaffold) in order to stabilize unwound DNA at the OBR (Bode et al. 1992), would comprise the site-specific, heritable replication origins that have been mapped in metazoan chromosomes (Table 1).
2. *Initiation zone.* Some initiation events are detected randomly distributed throughout a 6-kb to 55-kb region that encompasses the OBR. These sequences may remain accessible to replication factors so that the same cell can occasionally initiate replication at a nearby DUE. Alternatively, some cells within a population simply may have selected a different OBR in the same region, or are capable of utilizing more than one OBR in the same region, the choice of which can change during successive S phases. Secondary initiation sites also occur in simple genomes, although usually at a lower frequency. For example, papovavirus replication origins are at least 100-fold more effi-

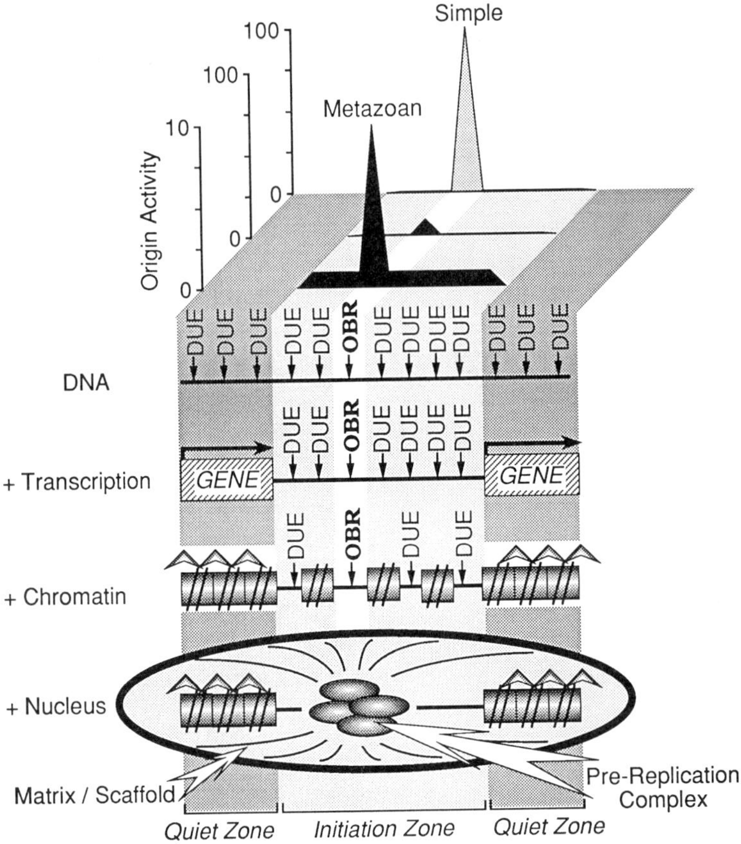

Figure 3 The "Jesuit Model" for metazoan replication origins and their relationship to simple replication origins. An OBR in simple genomes appears much more efficient than an OBR in metazoan genomes, thus reducing the relative intensity of secondary initiation events observed in the initiation zone surrounding the OBR. "Quiet zones" are sequences where no initiation events are detected. Eukaryotic DNA contains many potential origins; some may be simply DUEs whereas other, more efficient replication origins may be analogous to those in simple genomes (Fig. 1). The number of potential origins that can become active origins is restricted by actively transcribed DNA regions ("Genes") and chromatin structure (nucleosome , histone H1 ), whereas nuclear structure, which is required for initiation of metazoan DNA replication, activates selected origins.

cient at initiating replication than neighboring DNA sequences, but initiation events still occur at other sites in the DNA molecule, although they are difficult to detect (Martin and Setlow 1980; Tack and Proctor 1987). Similarly, bacteriophages T4 and T7 contain both a primary replication origin and several secondary origins (Kornberg and Baker 1992). If we compare site specificity in a metazoan genome with that in a simple genome, initiation events outside the metazoan OBR would go undetected unless the scale is expanded to accommodate the lower activity of a metazoan replication origin (Fig. 3).

3. *Quiet zone.* In contrast to DNA replication in rapidly cleaving embryos where initiation events are detected throughout the genome, initiation events in differentiated cells are restricted to specific sites (OBR + initiation zone). Initiation events outside the initiation zone may be further suppressed by active transcription (Haase et al. 1994) and higher-order chromatin structure (Fig. 3) (Hand 1978; Forrester et al. 1990; Simpson 1990; Ferguson and Fangman 1992; Karpen and Spradling 1992), two features that are generally absent in rapidly cleaving embryos (Fig. 2) (Wolffe 1994).

DNA replication in *Xenopus* egg extracts provides a simple paradigm for the selection and assembly of initiation sites for DNA replication (Fig. 2). Naked DNA contains many potential sites where DNA replication can begin. These sites most likely correspond to easily unwound DNA sequences that are components of replication origins in simple (Natale et al. 1993) as well as metazoan (Dobbs et al. 1994) genomes, and that can promote site-specific replication in plasmid DNA (Ishimi et al. 1994). When naked DNA is introduced into *Xenopus* eggs or egg extract, chromatin is assembled in the absence of histone H1, which is required for compaction of DNA into 30-nm fibers (Wolffe 1994). In addition, nuclei assembled in *Xenopus* eggs are less compact than nuclei in differentiated cells. This may be due partially to the fact that the composition of nuclear lamin proteins, a component of nuclear structure required for DNA replication (Newport et al. 1990; Jenkins et al. 1993), is simpler in *Xenopus* embryos than in *Xenopus* differentiated cells (Benavante et al. 1985; Stick and Hausen 1985). This relaxed environment may allow assembly of replication complexes at any one of the many potential origins located within a large DNA region, thus facilitating rapid replication of the *Xenopus* embryonic genome. The fact that replication does not initiate at more than one site within a single plasmid molecule (Hines and Benbow 1982; Méchali and Kearsey 1984; Mahbubani et al. 1992; Hyrien and Méchali 1992) suggests that replication in embryos undergoing rapid nuclear division cycles (e.g., frogs, flies, fish, sea urchins) occurs at only one of many potential sites within a large DNA region. Formation of replication complexes does not require nuclear membrane formation (Adachi and Laemmli 1994), but initiation of DNA replication does, suggesting that some component of nuclear structure interacts directly with each replication origin. Moreover, activation of replication complexes appears to require the action of G_1-cyclin-dependent protein kinase cdk2. General inhibitors of protein kinases (e.g., DMAP) prevent initiation of DNA replication in either sperm chromatin or G_1 nuclei, but do not inhibit active replication forks (Blow 1993; Kubota and Takisawa 1993; Gilbert et al 1995). Cip1 protein, a specific inhibitor of cdk2 kinase, exhibits a similar effect, and this inhibition can be overcome with cyclin E (Strausfeld et al. 1994). Cip1 inhibition occurs after formation of replication complexes but before DNA unwinding has begun (Adachi and Laemmli 1994).

It has long been observed that the density of initiation sites (origin-to-origin distance) in metazoan chromosomes varies in different cell types and under different experimental conditions, leading to the conclusion that many more replication sites are available than are used during a normal S phase, and that the number of active replication origins depends on the ratio of initiation factors to DNA and accessibility of origins as determined by higher-order chromatin structure (Hand 1978). As animal development progresses and cells begin to undergo differentiation, a 5- to 10-fold reduction can occur in the frequency of initiation sites (Blumenthal et al. 1974; McKnight and Miller 1977; Buongiorno-Nardelli et al. 1982; Shinomiya and Ina 1991). In fact, although initiation events in *Xenopus* rRNA gene region are distributed randomly throughout these sequences during early development (prior to gastrulation), initiation events become confined to the intergenic regions as development progresses (Hyrien et al. 1995). Changes in chromatin and nuclear structure that occur at the mid-blastula transition in *Xenopus*, for example, may repress initiation of DNA replication at some loci while facilitating it at others (Fig. 3) (Gilbert et al. 1995). In yeast, for example, at least half of the ARS elements on chromosome III are not active as chromosomal replication origins (Newlon et al. 1993). Factors that likely contribute to these changes in metazoa include the onset of transcription and cell differentiation, and the appearance of histone H1 and lamins A and C. Thus, some of the initiation sites chosen in fertilized *Xenopus* eggs may remain as origins throughout development, whereas other sites may be eliminated.

SUMMARY AND PERSPECTIVE

DNA replication in differentiated eukaryotic cells is initiated at specific DNA sites, regardless of whether the genome is viral or mitochondrial, or whether the organism consists of one cell type or many cell types. Perhaps the most basic difference among origins is whether they are designed to initiate once per cell division cycle or more than once. However, whereas the anatomy of initiation sites in most simple genomes is fairly clear and their mechanism of action is understood in principle, if not in detail, initiation sites in metazoan chromosomes remain elusive: Their requirement for nuclear structure makes them more complex than those in simple genomes, and the sequences that determine them have not yet been identified. Metazoan replication origins may follow the same principles revealed by replication origins in simple genomes but solve the problems of determining where, when, and how to initiate DNA synthesis in different ways. Why such complexity appears in the chromosomes of metazoa but not single-cell organisms may involve differences in chromatin structure and nuclear organization that accompany the appearance of histone H1, nuclear lamins, and nuclear envelope breakdown during mitosis in metazoans.

The concept of a metazoan origin as a specific nucleoprotein complex that interacts with soluble replication factors, rather than simply a specific sequence that is recognized by soluble factors, can account for three important features of metazoan DNA replication. First, it provides a way to change the number of origins per chromosomal locus at different stages in animal development, thus accommodating the need for shorter S phases in the early developmental stages of some animals. Second, it provides a way to limit the number of origins per genome without simultaneously limiting the cell's capacity to sustain genetic alterations. A large number of less efficient origins that demand less sequence specificity allow more flexibility for rearranging genetic information than a small number of highly efficient origins that require a great deal of sequence specificity. Finally, it provides a way to restrict initiation at each origin to once per S phase. Initiation at each replication origin would be restricted to once per cell cycle if the nucleoprotein complex that defines a replication origin in G_1 nuclei was disrupted by the act of replication. In that event, reinitiation at the same site would be prevented, even in the presence of excess soluble initiation factors.

Metazoan replication origins in G_1 nuclei of differentiated mammalian cells may be analogous to the six-protein origin replication complex assembled at yeast ARS sites (see Newlon, this volume) or the EBNA1 protein/origin core DNA complex in EBV episomes (see Yates, this volume), both of which appear to be stable DNA-protein complexes that exist throughout the cell cycle. Thus, a licensing factor (see Laskey and Madine, this volume) may recognize this DNA/protein platform, permitting it to be activated by a G_1-cyclin-dependent protein kinase. Licensing factor is inactivated by the replication process and then replaced when the nuclear membrane becomes permeable to cytoplasmic factors during mitosis.

ACKNOWLEDGMENTS

I am indebted to Darren Natale, David Gilbert, William Burhans, and many other friends and colleagues for their intellectual contributions to this paper.

REFERENCES

Adachi. Y. and U.K. Laemmli. 1994. Study of the cell cycle-dependent assembly of the DNA pre-replication centres in *Xenopus* egg extract. *EMBO J.* **13:** 4153–4164.

Ariizumi, K., Z. Wang, and P.W. Tucker. 1993. Immunoglobulin heavy chain enhancer is located near or in an initiation zone of chromosomal DNA replication. *Proc. Natl. Acad. Sci.* **90:** 3695–3699.

Benavante, R., G. Krohne, and W.W. Franke. 1985. Cell type-specific expression of nuclear lamina proteins during development of *Xenopus laevis*. *Cell* **41:** 177–190.

Benbow, R.M., J. Zhao, and D.D. Larson. 1992. On the nature of origins of DNA replication in eukaryotes. *BioEssays* **14:** 661–670.

Berberich, S., A. Trivedi, D.C. Daniel, E.M. Johnson, and M. Leffak. 1995. In vitro replication of plasmids containing human c-*myc* DNA. *J. Mol. Biol.* **245:** 92–109.

Blow, J.J. 1993. Preventing re-replication of DNA in a single cell cycle: Evidence for a replication licensing factor. *J. Cell Biol.* **122:** 993–1002.

Blumenthal, A.B., H.J. Kriegstein, and D.S. Hogness. 1974. The units of DNA replication in *Drosophila melanogaster* chromosomes. *Cold Spring Harbor Symp. Quant. Biol.* **38:** 205–223.

Bode, J., Y. Kohwi, L. Dickinson, T. Joh, D. Klehr, C. Mielke, and T. Kohwi-Shigematsu. 1992. Biological significance of unwinding capability of nuclear matrix-associating DNAs. *Science* **55:** 195–197.

Bonifer, C., N. Yannoutsos, G. Krüger, F. Grosveld, and A.E. Sippel. 1994. Dissection of the locus control function located on the chicken lysozyme gene domain in transgenic mice. *Nucleic Acids Res.* **22:** 4202–4210.

Brewer, B.J. and W.L. Fangman. 1993. Initiation at closely spaced replication origins in a yeast chromosome. *Science* **262:** 1728–1731.

———. 1994. Initiation preference at a yeast origin of replication. *Proc. Natl. Acad. Sci.* **91:** 3418–3422.

Bullock, P.A., Y.S. Seo, and J. Hurwitz. 1991. Initiation of SV40 DNA synthesis in vitro. *Mol. Cell. Biol.* **11:** 2350–2361.

Buongiorno-Nardelli, M., G. Micheli, M.T. Carri, and M. Marilley. 1982. A relationship between replicon size and supercoiled loop domains in the eukaryotic genome. *Nature* **298:** 100–102.

Burhans, W.C. and J.A. Huberman. 1994. DNA replication origins in animal cells: A question of context? *Science* **263:** 639–640.

Burhans, W.C., L.T. Vassilev, M.S. Caddle, N.H. Heintz, and M.L. DePamphilis. 1990. Identification of an origin of bidirectional DNA replication in mammalian chromosomes. *Cell* **62:** 955–965.

Burhans, W.C., L.T. Vassilev, J. Wu, J.M. Sogo, F.N. Nallaseth, and M.L. DePamphilis. 1991. Emetine allows identification of origins of mammalian DNA replication by imbalanced DNA synthesis, not through conservative nucleosome segregation. *EMBO J.* **10:** 4351–4360.

Caddle, M.S. and M.P. Calos. 1992. Analysis of the autonomous replication behavior in human cells of the DHFR putative origin of replication. *Nucleic Acids Res.* **20:** 5971–5978.

———. 1994. Specific initiation at an origin of replication from *Schizosaccharomyces pombe*. *Mol. Cell. Biol.* **14:** 1796–1805.

Cardoso, M.C., H. Leonhardt, and B. Nadal-Ginard. 1993. Reversal of terminal differentiation and control of DNA replication: Cyclin A and cdk2 specifically localize at subnuclear sites of DNA replication. *Cell* **74:** 979–992.

Carroll, S.M., M.L. DeRose, J.L. Kolman, G.H. Nonet, R.E. Kelly, and G.M. Wahl. 1993. Localization of a bidirectional DNA replication origin in the wild type and in episomally amplified murine ADA loci. *Mol. Cell. Biol.* **13:** 2971–2981.

Cohen, S., D. Hassin, S. Karby, and S. Lavi. 1994. Hairpin structures are the primary amplification products: A novel mechanism for generation of inverted repeats during gene amplification. *Mol. Cell. Biol.* **14:** 7782–7791.

Coverley, D. and R.A. Laskey. 1994. Regulation of eukaryotic DNA replication. *Annu. Rev. Biochem.* **63:** 745–776.

Cusick, M.E., P.M. Wassarman, and M.L. DePamphilis. 1989. Application of nucleases to visualizing chromatin organization at replication forks. *Methods Enzymol.* **170:** 290–316.

Deb, S.P., A.L. DeLucia, C. Bauer, A. Koff, and P. Tegtmeyer. 1986. Domain and structure of the SV40 core origin of replication. *Mol Cell. Biol.* **6:** 1663–1670.

Delidakis, C. and F.C. Kafatos. 1989. Amplification enhancers and replication origins in the autosomal chorion gene cluster of *Drosophila*. *EMBO J.* **8:** 891–901.

DePamphilis, M.L. 1993a. Eukaryotic DNA replication: Anatomy of an origin. *Annu. Rev. Biochem.* **62:** 29–63.

———. 1993b. Origins of DNA replication that function in eukaryotic cells. *Curr. Opin. Cell Biol.* **5:** 434–441.

———. 1993c. How transcription factors regulate origins of DNA replication in eukaryotic cells. *Trends Cell Biol.* **3:** 161–167.

———. 1993d. Origins of DNA replication in metazoan chromosomes. *J. Biol. Chem.* **268:** 1–4.

DePamphilis, M.L. and P.M. Wassarman. 1980. Replication of eukaryotic chromosomes: A close-up of the replication fork. *Annu. Rev. Biochem.* **49:** 627–666.

DePamphilis, M.L., E. Martínez-Salas, D.Y. Cupo, E.A. Hendrickson, C.E. Fritze, W.R. Folk, and U. Heine. 1988. Initiation of polyomavirus and SV40 DNA replication, and the requirements for DNA replication during mammalian development. *Cancer Cells* **6:** 165–175.

Dijkwel, P.A. and J.L. Hamlin. 1992. Initiation of DNA replication in the DHFR locus is confined to the early S-period in CHO cells synchronized with the plant amino acid mimosine. *Mol. Cell. Biol.* **12:** 3715–3722.

———. 1995. The Chinese hamster DHFR origin consists of multiple potential nascent strand start sites. *Mol. Cell. Biol.* **15:** 3023–3031.

Dobbs, D.L., W.-L. Shaiu, and R.M. Benbow. 1994. Modular sequence elements associated with origin regions in eukaryotic chromosomal DNA. *Nucleic Acids Res.* **22:** 2479–2489.

Dubey, D.D., J. Zhu, D.L. Carlson, K. Sharma, and J.A. Huberman. 1994. Three ARS elements contribute to the ura4 replication origin region in the fission yeast, *Schizosaccharomyces pombe. EMBO J.* **13:** 3638–3647.

Fang, F. and J. Newport. 1991. Evidence that the G1-S and G2-M transitions are controlled by different cdc2 proteins in higher eukaryotes. *Cell* **66:** 731–742.

———. 1993. Distinct roles of cdk2 and cdc2 in RP-A phosphorylation during the cell cycle. *J. Cell Sci.* **106:** 983–994.

Fangman, W.L. and B.J. Brewer. 1991. Activation of replication origins within yeast chromosomes. *Annu. Rev. Cell Biol.* **7:** 375–402.

Ferguson, B.M. and W.L. Fangman. 1992. A position effect on the time of replication origin activation in yeast. *Cell* **68:** 333–339.

Forrester, W.C., E. Epner, M.C. Driscoll, T. Enver, M. Brice, T. Papayannopoulou, and M. Groudine. 1990. A deletion of the human β-globin locus activation region causes a major alteration in chromatin structure and replication across the entire β-globin region. *Genes Dev.* **4:** 1637–1649.

Frappier, L. and M. O'Donnell. 1991. Overproduction, purification and characterization of EBNA1, the origin binding protein of Epstein-Barr virus. *J. Biol. Chem.* **266:** 7819–7826.

Frappier, L., K. Goldsmith, and L. Bendell. 1994. Stabilization of the EBNA1 protein on the EBV latent origin of DNA replication by a DNA looping mechanism. *J. Biol. Chem.* **269:** 1057–1062.

Gale, J.M., R.A. Tobey, and J.A. D'Anna. 1992. Localization and DNA sequence of a replication origin in the rhodopsin gene locus of Chinese hamster cells. *J. Mol. Biol.* **224:** 343–352.

Gerard, R. and Y. Gluzman. 1986. Functional analysis of the role of the A/T rich region and upstream flanking sequences in SV40 DNA replication. *Mol. Cell. Biol.* **6:** 4570–4577.

Giacca, M., L. Zentilin, P. Norio, S. Diviacco, D. Dimitrova, G. Contreas, G. Biamonti, G. Perini, F. Weighardt, S. Riva, and A. Falaschi. 1994. Fine mapping of a replication origin of human DNA. *Proc. Natl. Acad. Sci.* **91:** 7119–7124.

Gilbert, D.M. and S.N. Cohen. 1987. BPV plasmids replicate randomly in mouse fibroblasts throughout S-phase of the cell cycle. *Cell* **50:** 59–68.

Gilbert, D.M., H. Miyazawa, and M.L. DePamphilis. 1995. Site-specific initiation of DNA replication in *Xenopus* egg extract requires nuclear structure. *Mol. Cell. Biol..* **15:** 2942–2954.

Gilbert, D.M., H. Miyazawa, F.S. Nallaseth, J.M. Ortega, J.J. Blow, and M.L. DePamphilis. 1993. Site-specific initiation of DNA replication in metazoan chromosomes and the role of nuclear organization. *Cold Spring Harbor Symp. Quant. Biol.* **58:** 475–485.

Gillette, T.G., M. Lusky, and J. Borowiec. 1994. Induction of structural changes in the BPV-1 origin of replication by the viral E1 and E2 proteins. *Proc. Natl. Acad. Sci.* **91:** 8846–8850.

Gruss, C. and J.M. Sogo. 1992. Chromatin replication. *BioEssays* **14:** 1–8.

Gruss, C., J. Wu, and JM. Sogo. 1994. Disruption of nucleosomes at the replication fork.

EMBO J. **12:** 4533–4545.

Guo, Z.-S. and M.L. DePamphilis. 1992. Specific transcription factors stimulate both simian virus 40 and polyomavirus origins of DNA replication. *Mol. Cell. Biol.* **12:** 2514–2524.

Guo, Z.-S., U. Heine, and M.L. DePamphilis. 1991. T-antigen binding to site I facilitates initiation of SV40 DNA replication but does not affect bidirectionality. *Nucleic Acids Res.* **19:** 7081–7088.

Gutierrez, C., Z-S. Guo, J. Roberts, M.L. DePamphilis. 1990. Simian virus origin auxiliary sequences weakly facilitate binding of T-antigen, but strongly facilitate initiation of DNA unwinding. *Mol. Cell. Biol.* **10:** 1719–1728.

Gutierrez, C., Z.-S. Guo, W. Burhans, M.L. DePamphilis, J. Farrell-Towt, and G. Ju. 1988. Is c-myc protein directly involved in DNA replication? *Science* **240:** 1202–1203.

Haase, S.B. and M.P. Calos. 1991. Replication control of autonomously replicating human sequences. *Nucleic Acids Res.* **19:** 5053–5058.

Haase, S.B., S.S. Heinzel, and M.P. Calos. 1994. Transcription inhibits the replication of autonomously replicating plasmids in human cells. *Mol. Cell. Biol.* **14:** 2516–2526.

Hamlin, J.L., P.A. Dijkwel, and J.P. Vaughn. 1992. Initiation of replication in the Chinese hamster DHFR domain. *Chromosoma* (suppl. 1) **102:** 17–23.

Hand, R. 1978. Eucaryotic DNA: Organization of the genome for replication. *Cell* **15:** 317–325.

Handeli, S., A. Klar, M. Meuth, and H. Cedar. 1989. Mapping replication units in animal cells. *Cell* **57:** 909–920.

Harrison, S., K. Fisenne, and J. Hearing. 1994. Sequence requirements of the EBV latent origin of DNA replication. *J. Virol.* **68:** 1913–1925.

Hassan, A.B., R.J. Errington, N.S. White, D.A. Jackson, and P.R. Cook. 1994. Replication and transcription sites are colocalized in human cells. *J. Cell Sci.* **107:** 425–434.

He, Z., B.T. Brinton, J. Greenblatt, J.A. Hassell, and C.J. Ingles. 1993. Transactivator proteins VP16 and GAL4 bind replication factor A. *Cell* **73:** 1223–1232.

Heck, M.M.S. and A.C. Spradling. 1990. Multiple replication origins are used during *Drosophila* chorion gene amplification. *J. Cell Biol.* **4:** 903–914.

Hermann, C., E. Gartner, U.H. Weidle, and F. Grummt. 1994. High copy expression vector based on amplification promoting sequences. *DNA Cell Biol.* **13:** 437–445.

Herrick, J., R. Kern, S. Guha, A. Landulsi, O. Fayet, A. Malki, and M. Kohiyama. 1994. Parental strand recognition of the DNA replication origin by the outer membrane in *E. coli*. *EMBO J.* **13:** 4695–4703.

Hines, P.J. and R.M. Benbow. 1982. Initiation of replication at specific origins in DNA molecules microinjected into unfertilized of the frog *Xenopus laevis*. *Cell* **30:** 459–468.

Holmes, S.G. and M.M. Smith. 1989. Interaction of the H4 autonomously replicating sequence core consensus sequence and its 3′-flanking domain. *Mol. Cell. Biol.* **9:** 5464–5472.

Hozák P., A.B. Hassan, D.A. Jackson, and P.R. Cook. 1993. Visualization of replication factories attached to nucleoskeleton. *Cell* **73:** 361–373.

Huang, R.-Y. and D. Kowalski. 1993. A DNA unwinding element and an ARS consensus comprise a replication origin within a yeast chromosome. *EMBO J.* **12:** 4521–4531.

Huberman, J.A. 1994. Analysis of DNA replication origins and directions by two-dimensional gel electrophoresis. In *The cell cycle: A practical approach* (ed. P. Fantes and R.F. Brooks), pp. 213–234. Oxford University Press, United Kingdom.

Hyrien, O. and M. Méchali. 1992. Plasmid replication in *Xenopus* eggs and egg extracts:

A 2D gel electrophoretic analysis. *Nucleic Acids Res.* **20:** 1463–1469.

———. 1993. Chromosomal replication initiates and terminates at random sequences but at regular intervals in the ribosomal DNA of *Xenopus* early embryos. *EMBO J.* **12:** 4511–4520.

Hyrien, O., C. Maric, and M. Méchali. 1995. Transition in specification of embryonic metazoan DNA replication origins. *Science* **270:** 994–997.

Iguchi-Ariga, S.M.M., N. Ogawa, and H. Ariga. 1993. Identification of the initiation region of DNA replication in the murine immunoglobulin heavy chain gene and possible function of the octamer motif as a putative DNA replication origin in mammalian cells. *Biochim. Biophys. Acta* **1172:** 73–81.

Ishimi, Y., K. Matsumoto, and R. Ohba. 1994. DNA replication from initiation zones of mammalian cells in a model system. *Mol. Cell. Biol.* **14:** 6489–6496.

Jenkins, H., T. Holman, C. Lyon, B. Lane, R. Stick, and C. Hutchison. 1993. Nuclei that lack a lamina accumulate karyophilic proteins and assemble a nuclear matrix. *J. Cell Sci.* **106:** 275–285.

Karpen, G.H. and A.C. Spradling. 1992. Reduced DNA polytenization of a minichromosome region undergoing position effect variegation in *Drosophila. Cell* **63:** 97–107.

Kelly, R.E., M.L. DeRose, B.W. Draper, and G.M. Wahl. 1995. Identification of an origin of bidirectional replication within the coding region of the ubiquitously expressed CAD gene. *Mol. Cell. Biol.* **15:** 4136–4148.

Kill, I.R., J.M. Bridges, K.H.S. Campbell, G. Maldonado-Codina, and C.J. Hutchison. 1991. The timing of the formation and usage of replicase clusters in S-nuclei of human diploid fibroblasts. *J. Cell Sci.* **100:** 869–876.

Kipling, D. and S.E. Kearsey. 1990. Reversion of autonomously replicating sequence mutations in *S. cerevisiae*: Creation of a eucaryotic replication origin within procaryotic vector DNA. *Mol. Cell. Biol.* **10:** 265–272.

Kitsberg, D., S. Selig, I. Keshet, and H. Cedar. 1993. Replication structure of the human β-globin gene domain. *Nature* **366:** 588–590.

Koff, A., J.F. Schwedes, and P. Tegtmeyer. 1991. HSV origin-binding protein (UL9) loops and distorts the viral replication origin. *J. Virol.* **65:** 3284–3292.

Kohara, Y., N. Tohdoh, X.-W. Jiang, and T. Okazaki. 1985. The distribution and properties of RNA primed initiation sites of DNA synthesis at the replication origin of *Escherichia coli* chromosome. *Nucleic Acids Res.* **13:** 6847–6866.

Kornberg, A. and T. Baker. 1992. *DNA replication*. W.H. Freeman, New York.

Kowalski, D. and M.J. Eddy. 1989. The DNA unwinding element: A novel, *cis*-acting component that facilitates opening of the *Escherichia coli* replication origin. *EMBO J.* **8:** 4335–4344.

Krysan, P.J., J.G. Smith, and M.P. Calos. 1993. Autonomous replication in human cells of multimers of specific human and bacterial DNA sequences. *Mol. Cell. Biol.* **13:** 2688–2696.

Kubota, Y. and H. Takisawa. 1993. Determination of initiation of DNA replication before and after nuclear formation in *Xenopus* egg cell free extracts. *J. Cell Biol.* **123:** 1321–1331.

Leonhardt, H, A.W. Page, H.U. Weier, and T.H. Bestor. 1992. A targeting sequence directs DNA methyltransferase to sites of DNA replication in mammalian nuclei. *Cell* **71:** 865–873.

Li, R. and M.R. Botchan. 1993. Acidic transcription activation domains of VP16 and p53 bind cellular replication protein A and stimulate in vitro BPV-1 DNA replication. *Cell* **73:** 1207–1221.

Liang, C. and S.A. Gerbi. 1994. Analysis of an origin of DNA amplification in *Sciara coprophila* by a novel three dimensional gel method. *Mol. Cell. Biol.* **14:** 1520–1530.

Liang, C., J.D. Spitzer, H.S. Smith, and S.A. Gerbi. 1993. Replication initiates at a confined region during DNA amplification in *Sciara* DNA puff II/9A. *Genes Dev.* **7:** 1072–1084.

Lin, S. and D. Kowalski. 1994. DNA helical instability facilitates initiation at the SV40 replication origin. *J. Mol. Biol.* **235:** 496–507.

Little, R.D., T.H.K. Platt, and C.L. Schildkraut. 1993. Initiation and termination of DNA replication in human rRNA genes. *Mol. Cell. Biol.* **13:** 6600–6610.

Mahbubani, H.M., T. Paull, J.K. Elder, and J.J. Blow. 1992. DNA replication initiates at multiple sites on plasmid DNA in *Xenopus* egg extracts. *Nucleic Acids Res.* **22:** 1457–1462.

Majumder, S. and M.L. DePamphilis. 1994. TATA-dependent enhancer stimulation of promoter activity in mice is developmentally acquired. *Mol. Cell. Biol.* **14:** 4258–4268.

Majumder, S., M. Miranda, and M.L. DePamphilis. 1993. Analysis of gene expression in mouse preimplantation embryos demonstrates that the primary role of enhancers is to relieve repression of promoters. *EMBO J.* **12:** 1131–1140.

Marahrens, Y. and B. Stillman. 1994. Replicator dominance in a eukaryotic chromosome. *EMBO J.* **13:** 3395–3400.

Martin, R.G. and V.P. Setlow. 1980. Initiation of SV40 DNA synthesis is not unique to the replication origin. *Cell* **20:** 381–391.

Martínez-Salas, E., D.Y. Cupo, and M.L. DePamphilis. 1988. The need for enhancers is acquired upon formation of a diploid nucleus during early mouse development. *Genes Dev.* **2:** 1115–1126.

Mastrangelo, I.A., P.G. Held, L. Dailey, J.S. Wall, P.V.C. Hough, N. Heintz, and N.H. Heintz. 1993. RIP60 dimers assemble link structures at an origin of bidirectional replication in the DHFR amplicon of CHO cells. *J. Mol. Biol.* **232:** 766–778.

Masukata, H., H. Satoh, C. Obuse, and T. Okazaki. 1993. Autonomous replication of human chromosomal DNA fragments in human cells. *Mol. Biol. Cell* **4:** 1121–1132.

McKnight, S.L. and O.L. Miller, Jr. 1977. Electron microscopic analysis of chromatin replication in the cellular blastoderm *Drosophila melanogaster* embryo. *Cell* **12:** 795–804.

Méchali, M. and S. Kearsey. 1984. Lack of specific sequence requirement for DNA replication in *Xenopus* eggs compared with high sequence specificity in yeast. *Cell* **38:** 55–64.

Melendy, T. and B. Stillman. 1993. An interaction between replication protein A and T-antigen appears essential for primosome assembly during SV40 DNA replication. *J. Biol. Chem.* **268:** 3389–3395.

Miller, C.A. and D. Kowalski. 1993. *cis*-Acting components in the replication origin from ribosomal DNA of *Saccharomyces cerevisiae. Mol. Cell. Biol.* **13:** 5360–5369.

Morse, R.H., S.Y. Roth, and R.T. Simpson. 1992. A transcriptionally active tRNA gene interferes with nucleosome positioning in vivo. *Mol. Cell. Biol.* **12:** 4015–4025.

Nakayasu, H. and R. Berezney. 1989. Mapping replication sites in the eukaryotic cell nucleus. *J. Cell Biol.* **108:** 1–11.

Nallaseth, F.S. and M.L. DePamphilis. 1994. Papillomavirus contains *cis*-acting sequences that can suppress but not regulate origins of DNA replication. *J. Virol.* **68:**

3051–3064.
Natale, D.A., A.E. Schubert, and D. Kowalski. 1992. DNA helical stability accounts for mutational defects in a yeast replication origin. *Proc. Natl. Acad. Sci.* **89:** 2654–2658.
Natale, D.A., R.M. Umek, and D. Kowalski. 1993. Ease of DNA unwinding is a conserved property of yeast replication origins. *Nucleic Acids Res.* **21:** 555–560.
Newlon, C., I. Collins, A. Dershowitz, A.M. Deshpande, S.A. Greenfeder, L.Y. Ong, and J.F. Theis. 1993. Analysis of replication origin function on chromosome III of *Saccharomyces cerevisiae. Cold Spring Harbor Symp. Quant. Biol.* **58:** 415–423.
Newport, J.W, K.L. Wilson, and W.G. Dunphy. 1990. A lamin-independent pathway for nuclear envelope assembly. *J. Cell Biol.* **111:** 2247–2259.
Noirot, P., J. Bargonetti, and R.P. Novick. 1990. Initiation of rolling circle replication in pT181 plasmid: Initiator protein enhances cruciform extrusion at the origin. *Proc. Natl. Acad. Sci.* **87:** 8560–8564.
Orr-Weaver, T.L. 1991. *Drosophila* chorion genes: Cracking the eggshell's secrets. *BioEssays* **13:** 97–105.
Paranjape, S.M., R.T. Kamakaka, and J.T. Kadonaga. 1994. Role of chromatin structure in the regulation of transcription by RNA polymerase II. *Annu Rev. Biochem.* **63:** 265–297.
Parsons, R. and P. Tegtmeyer. 1992. Spacing is crucial for coordination of domain functions within the SV40 core origin. *J. Virol.* **66:** 1933–1942.
Pearson, C.E., A. Shihab-el-Deen, G.B. Price, and M. Zannis-Hadjopoulos. 1994. Electron microscopic analysis of *in vitro* replication products of ors 8, a mammalian origin enriched sequence. *Somatic Cell Mol. Genet.* **20:** 147–152.
Prives, C., Y. Murakami, F.G. Kern, W. Folk, C. Basilico, and J. Hurwitz. 1987. DNA sequence requirements for replication of polyomavirus DNA in vivo and in vitro. *Mol. Cell. Biol.* **7:** 3694–3704.
Rao, H., Y. Marahrens, and B. Stillman. 1994. Functional conservation of multiple elements in yeast chromosomal replicators. *Mol. Cell. Biol.* **14:** 7643–7651.
Ravnan, J.-B., D.M. Gilbert, G. Kelly, T. Hagen, and S.N. Cohen. 1992. Random-choice replication of extrachromosomal BPV molecules in heterogeneous, clonally derived BPV-infected cell lines. *J. Virol.* **66:** 6946–6952.
Roberts, J.M. and H. Weintraub. 1988. *cis*-Acting negative control of DNA replication in eukaryotic cells. *Cell* **52:** 397–404.
Rochford, R. and L.P. Villarreal. 1991. Polyomavirus DNA replication in the pancreas and in a transformed pancreas cell line has distinct enhancer requirements. *J. Virol.* **65:** 2108–2112.
Rokeach, L.A. and J.W. Zyskind. 1986. RNA termination within the *E. coli* origin of replication: Stringent regulation and control by DnaA protein. *Cell* **46:** 763–771.
Rychlik, W. and R.E. Rhoads. 1989. A computer program for choosing optimal oligonucleotides for filter hybridization, sequencing and in vitro amplification of DNA. *Nucleic Acids Res.* **17:** 8543–8551.
Ryder, K., S. Silver, A.L. DeLucia, E. Fanning, and P. Tegtmeyer. 1986. An altered DNA conformation in origin region I is a determinant for the binding of SV40 large T-antigen. *Cell* **44:** 719–725.
Schlake, T., D. Klehr-Wirth, M. Yoshida, T. Beppu, and J. Bode. 1994. Gene expression within a chromatin domain: The role of core histone hyperacetylation. *Biochemistry* **33:** 4197–4206.
Schneider, C., K. Weisshart, L.A. Guarino, I. Dornreiter, and E. Fanning. 1994. Species

specific functional interactions of DNA polymerase-α-primase with SV40 T-antigen require SV40 origin DNA. *Mol. Cell. Biol.* **14:** 3176–3185.

SenGupta, D.J. and J.A. Borowiec. 1994. Strand and face: Topography of interactions between the SV40 origin of replication and T-antigen during initiation of replication. *EMBO J.* **13:** 982–992.

Seufert, W., and W. Messer. 1987. Start sites for bidirectional in vitro DNA replication inside the replication origin, oriC, of *E. coli*. *EMBO J.* **6:** 2469–2472.

Shinomiya, T. and S. Ina. 1991. Analysis of chromosomal replicons in early embryos of *Drosophila melanogaster* by two-dimensional gel electrophoresis. *Nucleic Acids Res.* **19:** 3935–3941.

———. 1994. Mapping an initiation region of DNA replication at a single-copy chromosomal locus in *Drosophila melanogaster* cells by two-dimensional gel methods and PCR-mediated nascent-strand analysis: Multiple replication origins in a broad zone. *Mol. Cell. Biol.* **14:** 7394–7403.

Simpson, R.T. 1990. Nucleosome positioning can affect the function of a *cis*-acting DNA element in vivo. *Nature* **343:** 387–389.

Sock, E., M. Wegner, E.A. Fortunato, and F. Grummt. 1993. Large T-antigen and sequences within the regulatory region of JC virus both contribute to the features of JC virus DNA replication. *Virology* **197:** 537–548.

Stick, R. and P. Hausen. 1985. Changes in the nuclear lamina composition during early developmnet of *Xenopus laevis*. *Cell* **42:** 191–200.

Stolzenburg, F., R. Gerwig., E. Dinkl, and F. Grummt. 1994. Structural homologies and functional similarities between mammalian origins of replication and amplification promoting sequences. *Chromosoma* **103:** 209–214.

Strausfeld, U.P., M. Howell, R. Rempel, J.L. Maller, T. Hunt, and J.J. Blow. 1994. Cip1 blocks the initiation of DNA replication in *Xenopus* extracts by inhibition of cyclin-dependent kinases. *Curr. Biol.* **4:** 876–883.

Tack, L.C. and G.N. Proctor. 1987. Two major replicating SV40 chromosome classes: Synchronous replication fork movement is associated with bound large T antigen during elongation. *J. Biol. Chem.* **262:** 6339–6349.

Taira, T., S.M.M. Iguchi-Ariga, and H. Ariga. 1994. A novel DNA replication origin identified in the human heat shock protein 70 gene promoter. *Mol. Cell. Biol.* **14:** 6386–6397.

Tasheva, E.S. and D.J. Roufa. 1994a. A mammalian origin of bidirectional DNA replication within the Chinese hamster RPS14 locus. *Mol. Cell. Biol.* **14:** 5628–5635.

———. 1994b. Densely methylated DNA islands in mammalian chromosomal replication origins. *Mol. Cell. Biol.* **14:** 5636–5644.

Theis, J.F. and C.S. Newlon. 1994. Domain B of ARS307 contains two functional elements and contributes to chromosomal replication origin function. *Mol. Cell. Biol.* **14:** 7652–7659.

Tlsty, T.D. 1990. Normal diploid human and rodent cells lack a detectable frequency of gene amplification. *Proc. Natl. Acad. Sci.* **87:** 3132–3136.

Umek, R.M. and D. Kowalski. 1988. The ease of DNA unwinding as a determinant of initiation at yeast replication origins. *Cell* **52:** 559–567.

———. 1990a. The DNA unwinding element in a yeast replication origin functions independently of easily unwound sequences present elsewhere on a plasmid. *Nucleic Acids Res.* **18:** 6601–6605.

———. 1990b. Thermal energy suppresses mutational defects in DNA unwinding at a yeast replication origin. *Proc. Natl. Acad. Sci.* **87:** 2486–2490.

Vassilev, L.T. and M.L. DePamphilis. 1992. Guide to identification of origins of DNA replication in eukaryotic cell chromosomes. *Crit. Rev. Biochem. Mol. Biol.* **27:** 445–472.

Vassilev, L.T. and E.M. Johnson. 1990. An initiation zone of chromosomal DNA replication located upstream of the c-*myc* gene in proliferating HeLa cells. *Mol. Cell Biol.* **10:** 4899–4904.

Vassilev, L.T., W.C. Burhans, and M.L. DePamphilis. 1990. Mapping an origin of DNA replication at a single copy locus in exponentially proliferating mammalian cells. *Mol. Cell. Biol.* **10:** 4685–4689.

Virta-Pearlman, V.J., P.H. Gunaratne, and A.C. Chinault. 1993. Analysis of a replication initiation sequence from the adenosine deaminase region of the mouse genome. *Mol. Cell. Biol.* **13:** 5931–5942.

Ward, G.K., A. Shihab-el-Deen, and M. Zannis-Hadjopoulos. 1991. DNA cruciforms and the nuclear supporting stucture. *Exp. Cell. Res.* **195:** 92–98.

Williams, J.S., T.T. Eckdahl, and J.N. Anderson. 1988. Bent DNA as a replication enhancer in *S. cerevisiae. Mol. Cell. Biol.* **8:** 2763–2769.

Wohlgemuth, J.G., G.H. Bulboaca, M. Moghadam, M.S. Caddle, and M.P. Calos. 1994. Physical mapping of origins of replication in the fission yeast *Schizosaccharomyces pombe. Mol. Biol. Cell.* **5:** 839–849.

Wolffe, A.P. 1994. The role of transcription factors, chromatin structure and DNA replication in 5S RNA gene regulation. *J. Cell Sci.* **107:** 2055–2063.

Wu, C., M. Zannis-Hadjopoulos, and G.B. Price. 1993. In vivo activity for initiation of DNA replication resides in a transcribed region of the human genome. *Biochim. Biophys. Acta* **1174:** 258–268.

Wu, J.-R. and D.M. Gilbert. 1996. A distinct G1-phase step required to specify a mammalian replication origin. *Science* (in press).

Yates, J.L. and N. Guan. 1991. Epstein-Barr virus-derived plasmids replicate only once per cell cycle and are not amplified after entry into cells. *J. Virol.* **65:** 483–488.

Yoda, K.-Y., H. Yasuda, X.-W. Jiang, and T. Okazaki. 1988. *Nucleic Acids Res.* **16:** 6531–6546.

Yoon, Y., J.A. Sanchez, C. Brun, and J.A. Huberman. 1995. Mapping of replication initiation sites in human ribosomal DNA by nascent strand abundance analysis. *Mol. Cell. Biol.* **15:** 2482–2489.

Zahn, K. and F.R. Blattner. 1987. Direct evidence for DNA bending at the λ-replication origin. *Science* **236:** 416–422.

Zannis-Hadjopoulos, M., L. Frappier, M. Khoury, and G.B.Price. 1988. Effect of anti-cruciform DNA monoclonal antibodies on DNA replication. *EMBO J.* **7:** 1837–1844.

3

Roles of Transcription Factors in DNA Replication

Peter C. van der Vliet
Laboratory for Physiological Chemistry
Utrecht University
3508 TA Utrecht
The Netherlands

Sequence-specific recognition of DNA by proteins is essential for both transcription and DNA replication. Transcription factors form a growing family of regulatory proteins that can positively or negatively influence transcription by binding to regulatory elements in DNA contacting components of the basal transcription machinery. In recent years, it has become clear that several transcription factors are multifunctional and also directly influence initiation of DNA replication. This was first detected during studies of the replication of eukaryotic viruses, such as adenovirus and papovaviruses, but it may well be a more general phenomenon. Many viral origins consist of a core origin and auxiliary regions that contribute to the initiation of replication and are required for optimal viral growth (for review, see DePamphilis 1988, 1993a; Challberg and Kelly 1989; Stillman 1989). These auxiliary regions contain transcription factor-binding sites. In polyomavirus, presence of the transcriptional enhancer stimulated DNA replication up to 1000-fold, and presence of the enhancer increased SV40 replication approximately 100-fold. The auxiliary region of adenovirus stimulates initiation of replication up to 200-fold, although this region is not directly involved in transcription. The transcription factors used for enhancement of replication can be cellular as well as viral. Cellular transcription factors are involved in adenovirus, SV40, and polyomavirus replication. For papillomavirus and Epstein-Barr virus, the viral proteins (BPV-E2, EBNA1) combine functions in transcription and replication, as is the case for T antigen of SV40 and polyomavirus.

The availability of initiation systems that can be reconstituted with purified proteins for adenovirus and papovaviruses, combined with efficient transfection assays with origin-containing plasmids, has facilitated studies of the mechanism by which transcription factors enhance initiation. A number of distinct modes of action have been discovered. As a general rule, transcription factors enhance replication by facilitating rate-limiting steps in the initiation process. As summarized in Table 1 and Figure 1, these steps include:

1. Recruiting initiation proteins such as adenovirus DNA polymerase, the single-strand binding protein RP-A, or BPV-E1 helicase to the origin. By binding these proteins and targeting them to the origin employing their DNA-binding domains, transcription factors can facilitate the assembly of a multiprotein preinitiation complex and also stabilize such a complex. This method of action requires, understandably, a strict positioning of the transcription factors relative to the core origin.
2. Changing the activity of initiation proteins. Several examples exist in which the helicase activities of SV40 T antigen or BPV-E1 are influenced by transcription factors.
3. Induction of structural changes in origin DNA. This could lead to facilitated formation of a preinitiation complex by bringing initiation proteins closer together or by winding DNA on the surface of a multiprotein complex. Many transcription factors such as ABF-1, NFI, and Oct-1 can bend DNA. Bending could also be instrumental in origin opening.
4. Global changes in the chromatin structure surrounding the origin. Opening up the nucleosome structure could prevent the inhibitory effect of histones and nucleosomes on initiation (antirepressor effect) and could lead to a destabilization of the origin region, which could facilitate binding of initiation proteins. For this mode of action, the position of the transcription factor-binding site relative to the core origin is less fixed.

Of course, these various mechanisms are not mutually exclusive and may operate together to optimize the initiation process. The contribution of each individual mechanism depends on the system and the conditions. No evidence has been obtained so far that transcription factors also are involved in elongation or even origin clearance. On the contrary, several transcription factors seem to contact their targets only during a limited time span, being released soon after initiation or even before the actual initiation reaction has taken place.

Whether transcription factors employ the same mechanisms to enhance both processes, DNA replication and transcription, is not yet clear.

Table 1 How can transcription factors enhance replication?

	System	Transcription factor	Target protein
Recruitment of replication proteins (leading to assembly and stabilization of a multimeric initiation complex)	Ad	NFI	pTP-polymerase
		Oct-1	pTP-polymerase
	polyoma	T-ag	T-ag, pol-α, RP-A
		AP1	RP-A
	SV40	T-ag	T-ag, pol-α, RP-A
		SP1	?
	BPV	E2	E1
	EBV oriP	EBNA1	EBNA1
	oriLyt	BLTF-1	?
Changing the activity of an initiator protein	SV40	T-ag	T-ag helicase
	BPV	E2	E2 helicase
Changing the global architecture or nucleosome structure of the origin, e.g., by bending or looping	Ad	Oct-1	
	BPV	E2	
	S. cerevisiae	ABF-1	
	EBV	EBNA1	

Only naturally bound transcription factors are mentioned. For substitution of binding sites by synthetic ones, see Table 2. (T-ag) Large T antigen.

Global changes of chromatin structure could be instrumental in both processes. Such changes require, in general, the transcription activation domains of the transcription factors. On the contrary, interactions with initiation proteins are specific and use protein interfaces that are different from the ones involved in contacting the transcription machinery.

In most cases, the role of transcription factors in DNA replication is independent of the transcription process itself. Bound transcription factors do not appear to activate DNA replication by increasing transcription from, or through, the origin of DNA replication. An exception is mitochondrial DNA replication (Clayton 1991). Similar to prokaryotic systems (Baker and Kornberg 1992), the role of transcription factors is to regulate the synthesis of transcripts that act as a primer or a regulator. This is beyond the scope of this chapter; mitochondrial DNA replication is discussed in detail in Clayton (this volume). Below, I focus on recent results obtained in several viral systems as well as in yeast. Several reviews on the role of transcription factors in DNA replication have appeared previously (DePamphilis 1988, 1993b; Van der Vliet 1991; Heintz 1992).

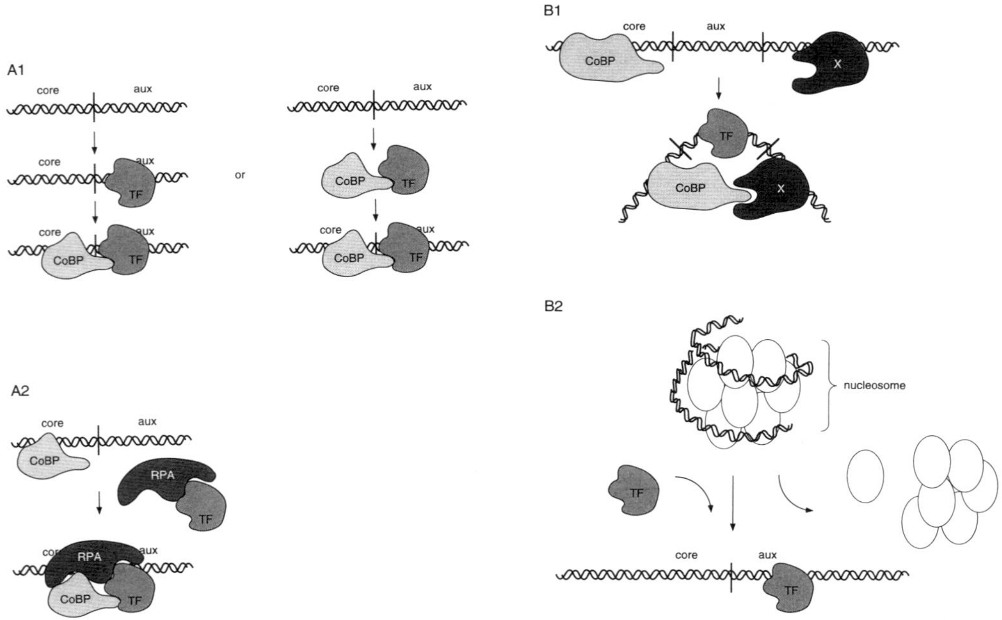

Figure 1 Ways in which transcription factors can stimulate initiation of DNA replication. (*A*) Direct contacts: (1) A core origin binding protein (CoBP) is recruited by a transcription factor (TF) binding to the auxiliary origin region (aux). Formation of a CoBP-TF complex may occur in solution or on the DNA. In this way a transcription factor stabilizes binding of CoBP to the core origin and may also influence the activity of CoBP. (2) A transcription factor may also recruit auxiliary initiation proteins, e.g., RP-A, that could help unwind the origin. (*B*) Structural changes in DNA or chromatin. (1) Bending of DNA by a transcription factor could bring two essential components, CoBP and X, another protein of the initiation complex, in close proximity. (2) Transcription factors disrupt the nucleosome structure, thereby enabling the initiation proteins to bind to the core origin (antirepressor effect).

VIRAL SYSTEMS

Adenovirus

The 36-kbp linear genome of human adenoviruses replicates via protein-primed initiation followed by elongation via a displacement mechanism (see Hay, this volume). The development of an in vitro initiation system has led to the purification of novel viral and cellular proteins that could be used to reconstitute efficient replication in vitro. Genetic and biochemical analysis indicates that three viral proteins are required for replication. These are the precursor terminal protein (pTP) and the viral DNA polymerase (pol), forming a pTP-pol heterodimer, and the DNA-binding protein (DBP).

The viral proteins can only sustain a very limited level of initiation, and two nuclear transcription factors stimulate initiation considerably. These are nuclear factor 1 (NFI, Nagata et al. 1982; Leegwater et al. 1985; Rosenfeld and Kelly 1986) and nuclear factor III (Pruijn et al. 1986; O'Neill and Kelly 1988). Nuclear factor III is identical to the octamer-binding transcription factor Oct-1 (O'Neill et al. 1988; Pruijn et al. 1989). NFI and Oct-1 function by binding to the origins of replication located at each molecular end. These consist, for Ad2 and Ad5, of a conserved core region, located at nucleotides 9–18 from the termini, two less well conserved space regions (1–9 and 18–24), and an auxiliary region (25–55). The core region binds pTP-pol, and the auxiliary region represents the binding site for NFI and Oct-1. The various domains within the origin are closely spaced, and enhancement of replication by NFI and Oct-1 depends critically on the spatial arrangement of their binding sites (Adhya et al. 1986; Wides et al. 1987; Bosher et al. 1990; Coenjaerts et al. 1991), suggesting essential protein-protein interactions.

How Does NFI Stimulate Initiation?

Human NFI is the collective name of a family of transcription factors (52–66 kD) generated by differential splicing from a single gene. They contain a highly conserved DNA-binding and dimerization domain (NFI-BD) located within the 220–240 amino-terminal amino acids (Paonessa et al. 1988; Santoro et al. 1988; Meisterernst et al. 1989). This region suffices for stimulation of replication (Mermod et al. 1989; Gounari et al. 1990). The NFI-BD dimer binds to positions 25–39 by contacting two successive regions in the major groove of the DNA, with a consensus TGGCA-N5-GCCAA (Nagata et al. 1983; Gronostajski 1986; De Vries et al. 1987). Interestingly, binding of NFI or NFI-BD to the origin is enhanced by the viral DBP (Cleat and Hay 1989; Stuiver and Van der Vliet 1990; Bosher et al. 1991), presumably through structural changes that DBP induces in origin DNA (Stuiver et al. 1992).

The level of stimulation by NFI is strongly dependent on the pTP-pol concentration and drops from 60-fold at low concentrations to approximately 2-fold at high concentrations of pTP-pol (Mul et al. 1990). This suggests that NFI recruits pTP-pol to the origin. In agreement with this, a direct DNA-independent interaction between NFI and pTP-pol has been reported (Bosher et al. 1990; Chen et al. 1990; Mul et al. 1990). The domain interacting with polymerase is located in NFI-BD between amino acids 68 and 150 (Chen et al. 1990) and is genetically distinguishable from the DNA-binding domain. Mutations in NFI that fail to bind pTP-pol are also defective for replication stimulation (Armentero et al. 1994). NFI not only recruits pTP-pol, but also stabilizes its interaction with the origin, as shown by template competition experiments and gel retardation (Mul and Van der Vliet 1992). Kinetic considerations are in accord with an increase in the amount of active preinitiation complex rather than an effect on the activity of pTP-pol, since NFI increases the V_{max} without changing the K_m (Mul and Van der Vliet 1993). The stabilizing effect of NFI, as measured by the half-life of a pTP-pol-DNA complex, was only approximately 10-fold compared to a 60-fold enhancement of initiation. This may indicate that NFI acts in multiple ways, possibly involving also structural changes within the origin (Zorbas et al. 1989). A role for NFI in origin unwinding is less likely, however, since NFI is still required when the template strand in the core origin is already unwound (Kenny and Hurwitz 1988).

Since NFI does not bind efficiently to single-stranded DNA, it can be anticipated that unwinding of the NFI site by the passing DNA polymerase leads to dissociation of NFI from DNA. Unexpectedly, dissociation of NFI occurs very early in initiation as soon as the DNA polymerase encounters dNTPs (Coenjaerts and Van der Vliet 1994). This result may be explained by conformational changes in the DNA polymerase and in origin DNA during initiation that could lead to disruption of both the NFI-pTP-pol interaction and the interaction between NFI and DNA. Apparently NFI delivers pTP-pol to the template but is released and recycled after initiation. This is reminiscent of similar effects observed in the BPV E1-E2 complex, where E2 dissociates as soon as E1 starts its helicase action (Lusky et al. 1994) and to the recruitment of DnaB helicase by DnaC or by λP protein (Baker and Kornberg 1992). In the latter case, λP protein must be removed by heat shock proteins before replication can start.

How Does Oct-1 Stimulate Initiation?

Oct-1 is a 91-kD ubiquitous transcription factor belonging to the POU-protein family (Herr et al. 1988; Verrijzer and Van der Vliet 1993). It binds to the sequence ^{40}ATGATAATGA^{49} in the Ad2 origin and stimulates initiation approximately 7-fold, independently of NFI (Mul et al. 1990; Coenjaerts et al. 1994). Like the NFI site (Hay 1985), the Oct-1 site is important for optimal replication in vivo. If other compensating DNA sequences are present, removal of the Oct-1 site is without consequence (Hatfield and Hearing 1993).

Deletion analysis indicated that the DNA-binding POU domain suffices for stimulation of replication (Verrijzer et al. 1990). This domain is highly conserved, and POU domain transcription factors from different subclasses all enhance replication, albeit to slightly different extents (Verrijzer et al. 1992).

In several aspects, similarities between enhancement of replication by Oct-1 and NFI exist. Like NFI, Oct-1 enhances the V_{max} of the initiation reaction and still stimulates initiation when the origin is partially unwound. Stimulation by Oct-1 is also dependent on the pTP-pol concentration, although less sharply than observed for NFI (Mul and Van der Vliet 1992, 1993; Coenjaerts et al. 1994). In contrast to NFI, direct contacts between pTP-pol and Oct-1 or the POU domain were not observed in solution, but recently an interaction could be found employing a GST-POU fusion protein immobilized on glutathione-agarose beads (Coenjaerts et al. 1994). Previous attempts to detect this interaction in solution might have been hampered by a low affinity. Binding was also observed with the POU_{HD}, although more weakly than with the intact POU domain, indicating that additional contacts with POUs cannot be excluded. Interestingly, the POU-pTP-pol interaction could not be competed away by NFI, suggesting that the regions contacted on pTP-pol are different for NFI and Oct-1 (Coenjaerts et al. 1994). This also explains the independent stimulatory effects observed with these two transcription factors.

DNA bending by the POU domain (Verrijzer et al. 1991; Johansen et al. 1993) may contribute to the stimulation, but the way in which this occurs is not well understood. Bending may facilitate the interaction between the various proteins and thus have an architectural role in the formation and assembly of a multiprotein initiation complex. It is not likely that bending per se is sufficient for enhancement, since substitution of the Oct-1-binding site by an AP1 site did not lead to stimulation in the presence of Fos-Jun combinations that are able to bend the DNA to a similar extent as Oct-1 (Coenjaerts et al. 1994). This stresses the importance of specific protein-protein interactions between Oct-1 and pTP-pol and explains the strict positioning of its binding site.

In conclusion, NFI and Oct-1 stimulate initiation mainly by a direct interaction with different, so far undefined, domains in pTP-pol. Small distortions of the origin structure induced by these transcription factors may further facilitate the assembly of a preinitiation complex. POU_{HD} and the NFI dimer bind to the same side of the DNA helix and, together with pTP-pol, the preinitiation complex spans five helical turns requiring considerable bending of the DNA to allow the necessary surface contacts. In addition to NFI and Oct-1, the terminal protein and DBP are required for optimal initiation. The presence of the terminal protein in the viral DNA leads to a modest increase in binding of pTP-pol to the origin (Pronk and Van der Vliet 1993), whereas bound DBP enhances binding of NFI. Together, these five proteins optimize formation and stability of a preinitiation complex. NFI and Oct-1 seem to function only in assembly of the preinitiation complex. Evidence for a function of NFI or Oct-1 in elongation or any other event during initiation is lacking. Unwinding of the origin or changing the kinetics of initiation appears to be mainly a function of the viral DBP (Mul and Van der Vliet 1993; Zijderveld and Van der Vliet 1994; Monaghan et al. 1994). A nucleosome antirepression effect, as suggested for papovaviruses, is unlikely since optimal stimulation is detected in the absence of nucleosomes, and adenovirus DNA is not constrained in a regular nucleosome structure in infected cells. Apparently, the viral origin and pTP-pol complex have evolved to allow interactions with conserved DNA-binding domains of transcription factors. Why these particular transcription factors are employed is not clear, and the adaptation process varies between serotypes, since Ad4 does not have an NFI-binding site. Presumably, the Ad4 pTP-pol complex has a stronger interaction with the core origin, thereby bypassing the need for NFI recruitment (Harris and Hay 1988). This is reminiscent of the T antigens of polyomavirus and SV40, which also differ in their affinity for the core origin, leading to an increased need for transcription factors in polyomavirus, which encodes a T antigen that binds more weakly to the origin than SV40 T antigen.

SV40

The 5243-bp circular SV40 DNA replicates bidirectionally from a fixed origin composed of a 64-bp core and adjacent auxiliary regions, Aux-1 and Aux-2, which each span approximately 40 bp. The core origin contains a 24-bp central region that forms the binding site for the SV40 T-antigen double hexamer flanked on one side by a 17-bp AT-rich element and on the other by a 10-bp DNA unwinding element in which the origin of bidirectional replication is located (see also Hassell and Brinton, this volume).

Although the core origin suffices for initiation, the flanking auxiliary sequences function synergistically to increase the frequency of initiation up to 100-fold, depending on the conditions and methods used to measure DNA replication (Guo et al. 1989; Gutierrez et al. 1990). Aux-1 contains binding sites for T antigen, and precise positioning of Aux-1 relative to the core origin is required for its enhancing effect. Deletion of Aux-1 had only a very slight (1.6-fold) effect on binding of T antigen to the core origin or on the bidirectionality of initiation, even under conditions of limiting T antigen. In contrast, a strong effect on T-antigen-dependent DNA unwinding was observed (Guo et al. 1991). This suggests that any interaction between T-antigen molecules bound at Aux-1 and at the core origin influences the helicase activity rather than increasing the binding of T antigen. The level of stimulation by Aux-1 depends on the type of papovavirus. JC virus relies on Aux-1 to the greatest extent, then SV40, and BK virus the least (Sock et al. 1993). This may be related to small differences in the interaction between the various T antigens, leading to different strengths of the intramolecular interactions.

Aux-2 also contains binding sites for T antigen, albeit weak ones, and is recognized by several cellular transcription factors, including SP1, LSF, AP1, AP2, AP4, GT-1B, and p53 (Table 2). Due to the complexity of this region, it has been difficult to establish, by mutational analysis, which transcription factor-binding sites are essential for stimulation of initiation. Therefore, another approach was taken. Aux-2 was replaced by multimers of one site or by nonnatural binding sites for various transcription factors, and their effects were studied by transfection (see Table 2). Substitution of Aux-2 by two synthetic SP1-binding sites restored replication to wild-type levels (Guo and DePamphilis 1992; Lednicky and Folk 1992), whereas one copy was less effective. Such a copy-number-dependent stimulation was also found upon substitution with NFI or ATF sites (Hoang et al. 1992) and is reminiscent of the results obtained with transcriptional enhancers.

In a more extensive study using a number of different transcription factors, AP1 and, to a lesser extent, T antigen, were shown to be active. However, other DNA-binding proteins like Gal4 could not substitute, even when fused to the strong *trans*-activation domains of VP16 or c-Jun (Guo and DePamphilis 1992). The same was true for binding of *Escherichia coli lac* repressor adjacent to the origin (Cheng et al. 1992). This shows that binding of proteins to Aux-2 is insufficient and indicates

Table 2 Natural auxiliary regions can sometimes be substituted by synthetic auxiliary regions

System	Natural sites	Substituting sites and proteins	Activity in vivo	Activity in vitro	References
Adenovirus	Oct-1	AP1		–	1
SV40 Aux-2	SP1, T-ag, LSF,	(SP1)3	+	–	2,3
	p53, AP1, AP2,	(AP1)3	+	–	2
	AP4, GI-IB	(Tag)3	+	+	2
		VP16	–	+	2,4,5
		c-Jun	–		2
		Gal4	–		2
		NFI	+		4,6
Polyoma Aux-2	AP1 (PEA1),	(SP)3	+		2
	AP3 (EBP-20),	(AP1)3	+		2
	T-ag, cETS	(Tag)3	+		2
	(PEA3), PEA2	VP16	+		2,8
		c-Jun	+		2,8,10,11
		v-Jun	+		8,10,11
		Gal4	+		2,7,8
		CREB	–		10
		Rel	+		12
		E1A	+		8
		Fos	+		8
S. cerevisiae	ABF-1	RAP-1	+		13
ARS-1 B3		Gal4	+		13

Transcription activation domains were bound to different DNA-binding domains, with comparable results. The positions of the binding sites vary. The levels of activity differ considerably, but this is not indicated in this table. For details, see the references. Since the conditions for testing are different, a direct comparison of the results is not always possible. PEA3, PEA1, and EBP-20 are the murine homologs of human c-ETS, AP1, and AP3, respectively. References: (1) Coenjaerts et al. 1994; (2) Guo and DePamphilis 1992; (3) Lednicky and Folk 1992; (4) Hoang et al. 1992; (5) Cheng et al. 1992; (6) Cheng and Kelly 1989; (7) Bennett Cook and Hassell 1991; (8) Baru et al. 1991; (9) Morgan et al. 1993; (10) Murakami et al. 1991; (11) Wasylyk et al. 1990; (12) Ishikawa et al. 1993; (13) Marahrens and Stillman 1992.

specificity, presumably related to the possibility to interact with other replication proteins such as T antigen or RP-A.

Binding of the active transcription factors to T antigen has not been reported, but a direct interaction was observed between VP16 and the large subunit of RP-A, RPA-1 (He et al. 1993; Li and Botchan 1993). This has led to the interesting hypothesis that recruitment of RP-A could stabilize single-stranded DNA in the origin or assist in RP-A-dependent unwinding by the T-antigen helicase. RP-A can also specifically interact with DNA polymerase-α and SV40 T antigen (Dornreiter et al. 1992; Melendy and Stillman 1993), and these protein-protein interactions may facilitate primosome assembly. Therefore, RP-A may function to influence two rate-limiting steps, the formation of an initiation complex and the subsequent helicase activity. For recruitment of RP-A to be important, the intracellular or local concentrations of RP-A should be limiting. RP-A is a rather abundant protein involved in many different processes; within the nucleus, competition for the protein may be strong. Thus, it is of advantage for the virus to recruit as much RP-A as possible for its own replication. Nevertheless, some caution in the interpretation is needed. The VP16 *trans*-activation domain is notoriously promiscuous in its interactions and, in addition to RP-A, has been reported to bind to many other proteins, including TBP, TFIIB, TFIIH, TAF40, and Oct-1. Moreover, Gal4-VP16 does not stimulate initiation in vivo (Guo and DePamphilis 1992; Hoang et al. 1992) despite its interaction with RP-A. This may be caused by sufficiently high concentrations of RP-A present in the cells used, or it may indicate that in vivo this mechanism is not important.

Another explanation of the effect of transcription factors on Aux-2 is based on replacement of Aux-2 by NFI-binding sites. In vivo, strong stimulation was observed, but the effect of adding NFI in a standard SV40 cell-free system was negligible. However, NFI specifically prevented the repression of DNA replication occurring when the template was preassembled into chromatin, suggesting a role as antirepressor (Cheng and Kelly 1989). A similar antirepressing effect was observed for Gal4-VP16 (Cheng et al. 1992). This agrees with results obtained with the naturally bound SP1, which can counteract histone H1-mediated inhibition of RNA pol II transcription (Croston et al. 1991; Laybourn and Kadonaga 1991). Several observations, however, argue against a model in which perturbation of the local nucleosome structure is the primary mechanism by which transcription factors stimulate SV40 replication. As mentioned above, Gal4-VP16 does not stimulate in vivo. Furthermore, a number of transcription factors, including T antigen itself, can inhibit chromatin assembly and prevent nucleosomes from repressing initiation (Ishimi 1992; Gruss et al. 1993). Finally, it has been reported that enhancement by SP1 is maintained when nucleosomes have been stripped (Lednicky and Folk 1992). Thus, although antirepression can be observed, it is not necessarily the primary reason for enhancement but, rather, could be a consequence of transcription factor binding. This explanation has also been put forward for similar effects observed upon binding of the BPV E1-E2 complex (see below).

Although binding of transcription factors to Aux-2 is the most likely way in which these proteins function, it may not be the only way. Recently, Oct-1 was shown to bind to the AT-rich element of the core origin, and binding inhibits the T-antigen-dependent helicase activity (Kilwinski et al. 1995). This may provide another level of control by transcription factors.

Polyomavirus

The polyomavirus (Py) genome is closely related to SV40 but replicates only in murine cells. The origin consists also of a core region flanked by two auxiliary regions. Like SV40, Py encodes a large T antigen essential for replication that binds to the core origin as a hexamer, has DNA-dependent ATPase and helicase activities, and binds DNA polymerase-α:DNA primase. However, Py T antigen binds more weakly to DNA and thus may rely more heavily on auxiliary factors (see Hassell and Brinton, this volume).

Binding of T antigen to Aux-1 activates replication 5- to 10-fold and presumably serves the same role as in SV40. The 200-bp Aux-2 region, coinciding with the enhancer element (de Villiers et al. 1984), stimulates 200- to 1000-fold, which is more than in SV40. Within Aux-2, two elements have been recognized (α and β) that are functionally redundant. Proteins binding to these sites are known as polyoma enhancer binding proteins (PEBP) or polyoma enhancer activators (PEA) and are homologs of the AP1 and ETS families of transcription factors, respectively. Several experiments indicate that binding of these transcription factors is important for replication and that overexpression of AP1 leads to stimulation of DNA replication (Murakami et al. 1991). Fusion proteins of various DNA-binding domains and the *trans*-activation domains of Fos-Jun were active in a DNA-binding-dependent manner, indicating the need for *trans*-activation domains (Wasylyk et al. 1990; Morgan et al. 1993). Interestingly, stimulation of replication was under control of growth-promoting agents such as the phorbol ester TPA. This suggests that protein kinase C-dependent signal transduction pathways are coupled directly to DNA replication, possibly through phosphorylation of the *trans*-activation domain. This effect seemed to be specific for AP1, since overexpression of CREB, also recognizing the AP1 site, stimulated only transcription but not replication (Murakami et al. 1991). So far, specific interactions between AP1 and other replication proteins have not been reported.

To uncouple effects on transcription and replication and to determine if other transcription factors could substitute for AP1 or ETS, a similar approach as for SV40 was employed. Natural Aux-2 sequences were replaced by synthetic binding sites capable of binding just one of the transcription factors, either individually or in tandem. Also tested were sites for well-characterized transcription factors such as AP1 and NF-κB (Guo and DePamphilis 1992; Ishikawa et al. 1993), and factors without a mammalian counterpart, such as Gal4 or fusion proteins between the Gal4 DNA-binding domain and the *trans*-activation domains of VP16, E1A, c-Jun, or BPV-E2 (Table 2) (Wasylyk et al. 1990; Baru et al. 1991; Murakami et al. 1991; Guo and DePamphilis 1992; Morgan et al. 1991). These binding sites, in the presence of appropriate proteins, could all substitute for Aux-2, but to a different extent and dependent on the positions of the sites.

In contrast to SV40, stimulation by VP16 was observed, although natural AP1 stimulates better than Gal4-VP16, Gal4, and c-Jun. Except for Rel oncoproteins (Ishikawa et al. 1993), *trans*-activation domains were essential. Synergistic activation was achieved by multimerization of the Gal4-binding sites (Baru et al. 1991; Bennett Cook and Hassell 1991), which should be juxtaposed to the core origin for optimal functioning. Gal4-VP16 stimulated replication of polyomavirus but not SV40 DNA in vivo, which may be related to the weak binding of Py T antigen and correspondingly higher requirement for auxiliary factors. Alternatively, there may be different levels of coactivators in the cells used for transfection.

The models put forward to explain the mechanisms by which these *trans*-activation domains work are the same as described above for SV40. A chromatin interference model is not excluded but is difficult to reconcile with the finding that one cluster of five identical Gal4-binding sites stimulates replication, whereas two such clusters are inhibitory (Baru et al. 1991). A strong argument for the importance of an interaction with RP-A comes from mutagenesis studies. Mutations in VP16 that reduce the ability of Gal4-VP16 to stimulate replication are also defective in RP-A binding (He et al. 1993). An extensive analysis (B.T. Brinton et al., in prep.; Hassell and Brinton, this volume) revealed that mutation of 6 of the 46 amino acids that make up the amino-terminal domain of VP16 coordinately reduced DNA replication, as well as binding to RPA-1. This provides strong evidence in favor of RP-A recruitment as the main mechanism for stimulation of DNA replication. Whether RP-A is a limiting factor is not known, but even if RP-A is present in excess, binding of VP16 could be essential to activate RP-A or to induce structural changes.

Papillomaviruses

The papillomavirus genome is slightly larger than that of papovaviruses and encodes two, rather than one, proteins required for replication (E1 and E2). E2 is a viral transcription factor. Much of the information on viral DNA replication has been derived from bovine papillomavirus type-1 (BPV-1) and has been extended recently for several human serotypes (HPV) (see Stenlund, this volume).

BPV-1 contains an origin of bidirectional replication spanning approximately 60 bp. A core origin consisting of an AT-rich region and an imperfect inverted repeat is flanked by two auxiliary regions (BS11 and BS12) that stimulate replication considerably (Ustav et al. 1991). The 68-kD E1 phosphoprotein (Sun et al. 1990) binds specifically to the core origin (Ustav et al. 1991). It possesses DNA-dependent ATPase and helicase activities (Seo et al. 1993a; Thorner et al. 1993; Yang et al. 1993) and distorts the origin structure severely upon binding in an ATP-dependent fashion, as shown by $KMnO_4$ sensitivity assays (Gillette et al. 1994). Interestingly, E2 can also bind the p180 subunit of DNA polymerase-α:DNA primase (Park et al. 1994). Thus, in many respects E1 resembles T antigen and is the essential initiator protein, although only limited sequence homology with T antigen exists. Under in vitro conditions, E1 is sufficient for initiation if added in high amounts to the reaction (Li and Botchan 1993). In vivo the minimal origin includes an E2-binding site (Ustav et al. 1993).

E2 binds as a dimer to a 12-bp palindromic sequence, ACC N6 GGT, present in BS11 and BS12, and binding leads to severe bending of the DNA. The E2 open reading frame can encode three proteins by differential splicing: E2, E2-TR, and E8/E2. All three contain the 85-amino-acid carboxy-terminal DNA-binding domain, but only the full-length 48-kD E2 enhances replication and transcription. The other two proteins function as repressors of transcription and possibly also DNA replication. The regions of E2 required for enhancement of replication and transcription are not identical (Winokur and McBride 1992).

Enhancement of initiation by E2 is directly related to its capacity to bind the E1 helicase. The interaction between E1 and E2 is cooperative and leads to an increase in the affinity of E1 for its binding site (Mohr et al. 1990; Blitz and Laimins 1991; Lusky et al. 1993; Seo et al. 1993b). E2 can even rescue an E1 mutant defective in DNA binding (Spalholz et al. 1993; Thorner et al. 1993). Moreover, the amount of E1 required to generate structural changes in the origin is lowered by the presence of E2 (Gillette et al. 1994). E2 also enhances origin-dependent unwinding by E1 (Seo et al. 1993b). Thus, E2 stimulates replication by enhancing several properties of E1; in agreement with this, the highest level of stimulation by E2 in vitro occurs at low E1 concentrations (Müller et al. 1994). Interestingly, an intact DNA-binding domain is not a strict requirement for E2 to enhance replication (Winokur and McBride 1992). This has led to the hypothesis that the cooperative nature of the interaction, accompanied by structural changes in E1 upon complex formation with E2, suffices for enhancement of E1 binding to DNA. This targeting role of E2 may be short-lived, as indicated by band-shift experiments (Lusky et al. 1994). The E1-E2 complex binds to the origin but is still replication-inactive. It may function as a focal point for attracting other replication proteins, leading to a replication-active complex that can start unwinding. As soon as this occurs, E2 is released, possibly through structural changes induced in E1 and the origin. This is reminiscent of the early release of NFI during initiation of adenovirus DNA replication (Coenjaerts and Van der Vliet 1994) and of the recruitment of DnaB helicase by DnaC or λP protein (Baker and Kornberg 1992).

The mechanism of action by E2 might well involve more than just interacting with E1. E2 can also recruit RP-A, and mutants defective in RP-A binding are less effective in enhancing BPV replication in vitro (Li and Botchan 1993). Moreover, E2 interferes with nucleosome binding, but this antirepressor action is not specific for E1 and E2 and may be a consequence of E1/E2 binding rather than a primary effect (Li and Botchan 1994).

The results obtained with BPV were largely confirmed for several HPV serotypes, although some differences were observed (Frattini and Laimins 1994; Gopalakrishnan and Khan 1994). Interestingly, mutational analysis of HPV-11 showed that the presence of an E1-binding site is less important than the presence of an E2-binding site (Russell and Botchan 1995), stressing the function of E2 as the major targeting protein for the E1 helicase, which binds with lower specificity. Moreover, E2 also suppressed nonspecific initiation by E1 in vitro (Kuo et al. 1994).

In conclusion, the combination of E1 and E2 has many similarities to T antigen. Not only does it have similar enzymatic activities, but it also attracts both RP-A and DNA polymerase-α:DNA primase. However, the binding of E1 to the core origin is not always strong enough, and the most important role of E2 appears to be to facilitate binding of limiting amounts of E1. The intracellular ratio of E1 to E2 may be an important factor in determining the replication efficiency and thus the copy number. Once they have been delivered by E2, possibly assisted by E2-induced DNA bending and by the cooperative nature of the E1-E2 interaction, E1 multimers may attract other replication proteins, and a preinitiation complex will be formed. The cooperative binding of E1 and E2 leads to ATP-induced structural changes that may release E2 and activate the helicase activity.

Herpesvirus

DNA replication has been most extensively studied in the α-herpesvirus HSV and the γ-herpesvirus Epstein-Barr virus (EBV). In the 150-kbp linear HSV-DNA, three distinct origins have been described, ori_L and two copies of ori_S, located in the inverted repeats of the S component of the DNA. Seven viral proteins required for replication have been described (see Challberg, this volume).

As in other DNA viruses, both ori_S and ori_L are flanked by *cis*-acting sequences containing transcriptional regulatory elements that have been postulated to function in DNA replication as shown by deletion analysis (Stow and McMonagle 1983; Wong and Schaffer 1991). Whether binding of transcription factors to these elements is essential and how these transcription factors contribute to replication efficiency will be difficult to establish in view of the lack of an origin-dependent in vitro system. Besides the mechanisms proposed for papovaviruses, even a role of transcription itself is not excluded since the elements are still active in transcription during replication. This transcription process could influence ori function by modifying the local DNA structure in a way similar to that described for *E. coli oriC* (Baker and Kornberg 1988).

EBV oriP Is Activated by Transcription Factor-induced DNA Looping

The 192-kbp linear genome of EBV is maintained as multiple copies of a plasmid in latently infected B lymphocytes. Latent replication originates from *oriP*, the plasmid origin of replication, and requires only one viral protein, EBNA1, which also functions as a transcription factor.

oriP contains two noncontiguous elements, DS (dyad symmetry) and FR (family of repeats), separated by approximately 1 kbp. DS forms the functional origin of replication and contains four imperfect binding sites for EBNA1, two within a 65-bp region of dyad symmetry and two flanking this sequence. FR contains 20 tandemly repeated copies of a 30-bp sequence, each containing a 12-bp palindrome core that forms the EBNA1-binding site. This region functions as an enhancer both for transcription and DNA replication and also governs stable segregation of *oriP*-containing plasmids.

Upon initiation, the EBNA1 dimers bind first to the sites in FR, then to sites in DS. EBNA1 complexes formed on the two elements interact, and this leads to looping-out of the DNA (Frappier and Donnell 1991; Su et al. 1991). Binding of EBNA1 to FR or DS alone is much weaker than to both elements, indicating that the intramolecular interactions between

EBNA1 bound to FR and DS stabilize the DNA/protein complex, as shown by competition with nonspecific DNA (Frappier and Donnell 1991; Su et al. 1991). Thus, looping enables site saturation at lower concentrations via cooperativity. These interactions take place in a part of the EBNA1 molecule separate from the normal dimerization domain or the DNA-binding domain (Chen et al. 1993) and are located in a small region between amino acids 350 and 361 (Frappier et al. 1994). Contact with FR-bound EBNA1 leads to stabilization of EBNA1 at the DS element and could well be a trigger to activate initiation of DNA synthesis, either by recruitment of other replication factors or by a change in DNA structure. In particular, the protein-mediated looping might facilitate unwinding by introducing topological stress. Just DS-DS-bound EBNA1 does not stabilize binding. Interestingly, looping is also an important mechanism for bacteriophage R6k replication, where origin-bound Rep protein stabilizes the other Rep proteins bound weakly to *oriα*. This cooperativity leads also to efficient replication. In these cases, apparently specific interacting surfaces can only make optimal contacts when positioned on the DNA such that a loop is occurring. Only this position leads to optimal protein-protein interactions, thus imposing torsional strain on the DNA that can be used for origin opening.

Lytic Origin of Replication

EBV has both a latent state and a lytic replication cycle. The lytic phase can be induced by treating cells with phorbol esters or by introducing a vector expressing the *BZLF1* gene encoding a viral transcription factor. Two copies of *oriLyt* separated from *oriP* have been found, one in DS-L and one in DS-R (Hammerschmidt and Sugden 1988). These origins contain multiple regions required for replication and additional sequences that increase replication and can be functionally substituted by a transcriptional enhancer. Similar to HSV-1, at least seven viral genes required for replication have been found (Fixman et al. 1992). The DS-L origin is quite large and complex and consists of a 225-bp AT-rich region that is presumed to be the site of initiation, flanked by two elements, 320 and 370 bp long, respectively (Schepers et al. 1993b). These flanking elements contain promoters of two divergently transcribed genes, *BHLF1* and *BHRF1*, and also act as enhancers for replication. The BHLF1 promoter/replication enhancer is controlled by the viral transcription factor BZLF1 (synonyms: EBI, zta, ZEBRA, z) which belongs to the b-Zip class and is related to the AP1 family. Four binding sites are found in the promoter of *BHLF1* and three in *BHRF1*, to which BZLF1 binds with different affinities. Another *trans*-activator that binds to the *BHRF1* promoter is Rta, which may also be involved in enhancement of replication.

Employing a transient replication and cotransfection protocol, it could be shown that for replication enhancement, the amino-terminal transcription activation domains of *BZLF1* are required (Schepers et al. 1993a). In contrast to results in the papovavirus system, other transcription factors (c-Jun, BPV-E2, Myc, or VP16) could not substitute for BZLF1, suggesting the requirement for specific interactions, although the target proteins have not been identified.

Other Viral Systems

The need for transcription factors to optimize DNA replication is likely not confined to the systems described above. Evidence is also present for Geminivirus (see Bisaro, this volume) and parvoviruses such as minute virus of mice (MVM). Here the active origin (50 bp) contains an ATF-binding site (Cotmore and Tattersall 1994). However, in these cases, the evidence for a function of these transcription factors is still circumstantial.

Cellular Origins

The use of transcription factors by viruses for the enhancement of their replication seems to be a rule rather than an exception. This is likely caused by adaptation of the viral genome to the presence of these proteins in their host cells. The need for economic use of genetic information due to packaging constraints may have forced the use of sequence information in *cis*-acting elements such as enhancers for dual purposes. This may also hold for the employment of viral transcription factors such as BPV-E2 and EBNA1 for multiple functions. Such sequence constraints do not exist in eukaryotic cells, and therefore an extrapolation of the need for transcription factors to cellular DNA synthesis is not a matter of fact. Nevertheless, recent results have indicated a strong link between transcription and replication also in cellular origins, and at least two proteins required for the initiation of DNA replication in *Saccharomyces cerevisiae* are also essential for control of transcription.

Origins in Yeast

In *S. cerevisiae*, replication origins have been well characterized, and although an in vitro system for initiation is not yet established, information on the replication proteins has recently been obtained both by biochemical and by genetic means. Yeast origins coincide with ARS elements and are located in AT-rich regions of the genome with an average spacing of approximately 100 kbp (see Newlon, this volume). ARS elements are approximately 100 bp in length and contain two essential elements, A and B. The A element contains an 11-bp consensus sequence (ACS), which is the only region common to all *S. cerevisiae* origins. The B element is composed of three functional elements, B1, B2, and B3, which are also important for origin functions but are less conserved. The A element is bound in vitro by the multiprotein origin recognition complex (ORC) (Bell and Stillman 1992), which can also be observed in vivo (Diffley and Cocker 1992). ORC binds the ACS in an ATP-dependent manner and also covers part of the B1 element. The complex is composed of six different polypeptides, ORC1–ORC6, which are essential for yeast viability. Interestingly, the 72-kD ORC2 protein is also implicated in transcriptional silencing of the HMR-E locus (Rivier and Rine 1992; Bell et al. 1993; Foss et al. 1993; Micklem et al. 1993). In agreement with this, the *cis*-acting sequences required for silencing can function as ARS elements. These results provide strong evidence for a role in vivo of ORC in replication as well as in transcriptional control, in particular in silencing. Possibly the ORC2 protein can be present in different complexes, dependent on its function. Alternatively, DNA replication could be involved directly in transcription silencing by some form of mechanistic coupling in which ORC is involved (Rivier and Rine 1992). Due to the lack of an in vitro system, the exact role of ORC2 in replication remains obscure.

The ORC complex remains bound to the ACS throughout the cell cycle but can occur in two states differing in footprint pattern at different parts of the cycle (Diffley et al. 1994). Near the end of mitosis and through G_1, an additional region of protection overlapping the ORC footprint is observed that is lost after entry into S phase. Diffley and coworkers suggest that this postreplicative state, which resembles the footprint pattern obtained in vitro with ORC and ABF-1, represents a ground state laid down immediately after S phase and remaining throughout the cell cycle even into G_0. Thus, ORC and ABF-1, a transcription factor described below, may have to be activated to create a prereplicative state.

The B3 element contains the binding site for the multifunctional transcription factor ABF1 (Buchman et al. 1988; Diffley and Stillman 1988). ABF1-binding sites have been identified in several ARSs at variable positions up to 1.2 kbp from the A element, indicating that they can function in a position- and orientation-independent manner to stimulate replication, thereby resembling transcriptional enhancers. ABF-1 is an abundant 81-kD phosphoprotein. It binds in a sequence-specific manner to a consensus sequence $GCAN_4Y_2RCTR$ (R=purine, Y=pyrimidine). A detailed contact point analysis has indicated that ABF-1 is mainly oriented along one face of the helix (McBroom and Sadowski 1994a). Considerable bending (120°) of the DNA has been inferred from circular permutation and phasing analysis (McBroom and Sadowski 1994b). ABF-1 is involved in positive control of a number of genes, in silencing of the mating-type loci HML and HMR, and in DNA replication.

To understand its role in DNA replication in more detail, the B3 element from ARS-1 has been substituted by the binding sites of several other transcription factors including RAP-1, Gal4, and a LexA-Gal4 fusion protein (Marahrens and Stillman 1992). These binding sites substituted efficiently, in both orientations and in a DNA-binding-dependent manner in the presence of the appropriate transcription factors, indicating that the B3 element may be a rather nonspecific enhancer.

Two hypotheses, not mutually exclusive, have been put forward to explain the role of ABF-1. One is a direct contact between the ABF-1 activation domain and replication proteins, present at the core origin. A candidate is RP-A, which can be bound directly by Gal4 (He et al. 1993). Recruitment of RP-A followed by binding to the B1-B2 unwinding element and stabilizing single-stranded DNA or assisting in unwinding is an attractive hypothesis. However, a direct interaction between ABF-1 and RP-A has not been described, and the ABF-1 site can vary in position and orientation without loss of activity. Moreover, such a recruitment model is difficult to reconcile with the continuous presence of ABF-1 in vivo at the core origin (Diffley et al. 1994). As shown by in vivo footprinting of an ARS element in which the ABF-1 site is mutated, ABF-1 is not involved in recruitment of ORC (J. Cocker and J. Diffley, pers. comm.) or the ORC-bound factor Dbf4 (Dowell et al. 1994). An alternative is that the *trans*-activation domain of ABF-1 is an architectural component facilitating the interaction between multiprotein complexes and DNA or influencing the local chromatin structure, employing its capacity to bend DNA. Interestingly, RAP-1, which can substitute for ABF-1, can also strongly bend DNA (Gilson et al. 1993).

Another transcription factor that has been implicated in DNA replication is the TATA-box-binding protein TBP (Lue and Kornberg 1993). TBP can bind to the ACS, and TBP-binding sites can activate replication origins in vivo. Moreover, the ARS-1 B domain can act as a weak upstream activating sequence in transcription. However, this could just be fortuitous; no example of an origin within a transcription unit in yeast chromosomes has been found. There is also no evidence that ARS elements have promoter activity (Fangman and Brewer 1991).

On the basis of genetic evidence, another class of proteins, Mcm (*m*ini*c*hromosome *m*aintenance), have been suggested to influence DNA replication (Christ and Tye 1991). One of these, Mcm-1, is a transcription factor that bridges the α1 and α2 regulatory homeobox proteins involved in mating-type control (Smith and Jóhnson 1992) and also interacts with other transcription factors. Its potential role in DNA replication is based on a plasmid maintenance defect in Mcm-1 mutants. As yet there is no evidence that Mcm-1 interacts directly with ARSs, and its role may be limited to transcriptional enhancement of genes coding for essential replication proteins.

In summary, the evidence for a link between DNA replication enhancement and transcriptional control, in particular silencing of HML-E, is compelling, but the mechanisms by which ABF-1 and ORC2 exert their multiple functions remain unclear. Most evidence points to an architectural role of ABF-1 rather than a direct interaction with other replication proteins.

Origins in Higher Eukaryotes

A number of potential eukaryotic origins in metazoa have been described recently (see DePamphilis, this volume). In those cases where sequence analysis of the potential origins is available, a large number of transcription factor-binding sites have been found. Since no mutational analysis has been performed, a functional analysis is not possible, and the significance of these sites is not clear. Although it is tempting to assume that they may act as enhancers of replication in a similar fashion as in yeast and the viral systems, they may just be localized close to origins by chance and represent promoter regions without any functional relationship to initiation of DNA replication. (Details of the binding sites in cellular origins can be found in DePamphilis, this volume.)

SUMMARIZING REMARKS

Why Are Transcription Factors Used for Replication?

Although at first sight the lack of a separate set of replication-enhancing proteins seems strange, it is understandable from an evolutionary point of view. Only a limited number of basic principles to recognize DNA have been found to date. This number may expand upon further structural analysis, but nevertheless, it seems that Nature uses variations on successful binding modes such as helix-turn-helix, zinc fingers, and helix-loop-helix above development of completely novel modes. These conserved modes of DNA recognition can apparently be used for multiple purposes, in both DNA replication and transcription. A similar situation is found for RP-A, also multifunctional, and for TFIIH, active in transcription and repair (Schaeffer et al. 1993; Drapkin et al. 1994). An advantage of the use of transcription factors is that, in general, they open up chromatin structure. This may explain why transcriptionally active regions of chromosomal DNA replicate early in S phase. But why should transcription activation domains not have been adapted specifically to interact with replication proteins? Presumably, the flexible nature of activation domains increases the potential for different, specific interactions. Moreover, the specificity might be increased by the use of different coactivators. Transcription activation domains can be mimicked by reiteration of short peptide segments, revealing their modular organization (Tanaka et al. 1994). Their flexibility has so far hampered detailed structural analysis, but they may adopt a more fixed structure upon binding their target proteins functioning in the transcriptional machinery or in initiation of DNA replication. Such a flexible interaction exists also for major histocompatibility complex molecules in binding peptides (Young et al. 1994) and in the interaction between molecular chaperones and unfolded proteins (Braig et al. 1994). Thus, transcriptional activation domains may be very suitable to permit interactions with different molecules of the replication machinery.

A further advantage of the use of the transcription factors is that modification, in particular phosphorylation, could enable a subtle regulation of initiation. This has been described for transcription control by Fos-Jun (AP1), and in that respect it is interesting that a natural AP1 site in polyomaviruses plays such a prominent role and may be involved in replication timing.

Why would a virus not circumvent the need for transcription factors by enhancing the levels of rate-limiting factors directly? One reason may be that, when viral replication proteins and DNA are in low concentration early in infection, cellular proteins could enhance replication, thus enabling a quick start. However, this would just give a small advantage in time. More likely, for limiting factors such as E1 and pTP-pol, the ability to bind in a sequence-specific manner would be counterproductive. This holds in particular for DNA polymerase, which must clear the origin to enable elongation. Transcription factor guiding can also prevent nonproductive action outside the origin, which might occur if high levels of the initiator are present (Kuo et al. 1994). Consistent with this notion, the BPV-E2 protein stimulated the origin-specific unwinding reaction catalyzed by E1, but had no stimulatory effect on nonspecific helicase activity (Seo et al. 1993b).

Are Similar Mechanisms Used to Activate Transcription and Replication?

As summarized in Table 3, the various mechanisms proposed for enhancement of replication apply to a great extent also for transcription. For the latter, various target proteins have been detected, like TBP, TAFs, or TFIIB, and even TFIIH or RNA polymerase itself. In one instance (TFIIB), a change in the structure is observed upon interaction with VP16 (Roberts and Green 1994). Transcription factors are involved in modulation of the global chromatin structure required for transcription, as exemplified by the MMTV-LTR promoter and NFI. These effects on chromatin structure might be similar in transcription and replication and therefore require the same domains of the transcription factors.

In most cases where direct protein-protein interactions have been observed, however, a high specificity is found, and the regions required for transcription activation and replication enhancement are different. This is most conspicuous when just a DNA-binding domain is required for repli-

Table 3 Transcription factor action in replication and transcription compared

Action	Transcription	Replication
Recruitment of basal proteins leading to stable preinitiation complex	TBP	Ad pTP-pol
	TFIIB	BPV-E1
	TFIIF	RP-A
	TFIIH	pol-α:primase
Influencing helicase activity	TFIIH?	BPV-E1
		T-ag (SV40)
Interaction with coactivators	TAF 40, 110, 150	?
	CREB-binding protein	
Multimerization	many	EBNA1
		BPV-E1
		T-ag
Influencing target conformation	TFIIB	?
Effect on chromatin structure	SWI/SNF	yes
Promoter/origin clearance	TFIIE	?
	TFIIH	

cation (adenovirus: NFI, Oct-1; polyoma: Rel oncoprotein). Despite their name, these DNA-binding domains can interact not only with DNA, but with other proteins as well. For instance, the Oct-1 POU homeodomain can interact with several other proteins besides the Ad DNA polymerase and employs different amino acids to contact these different targets (Coenjaerts et al. 1994).

Future experiments will doubtless reveal the details of interacting domains and the consequences for these interactions, employing reconstituted replication systems as well as structural studies. A major question will be whether these transcription factors only optimize viral replication systems, or whether they are also involved in cellular replication, in particular, timing and control of initiation of cell division. Potential candidates are transcription factors binding to the locus control region of the β-globin gene, which influences the time in S phase when the locus is replicated (Forrester et al. 1990; Kitsberg et al. 1993).

ACKNOWLEDGMENTS

I thank Brad Brinton and Jackie Russell for sending submitted manuscripts or preprints. Experiments performed in the author's laboratory were supported by the Netherlands Foundation for Chemical Research (SON) with financial support from the Netherlands Organization for Scientific Research (NWO). Critical reading of the manuscript by H.T.M. Timmers and J. Dekker is gratefully acknowledged.

REFERENCES

Adhya, S., P.S. Shneidman, and J. Hurwitz. 1986. Reconstruction of adenovirus replication origins with a human nuclear factor I binding site. *J. Biol. Chem.* **261:** 3339–3346.

Armentero, M.-T., M. Horwitz, and N. Mermod. 1994. Targeting of DNA polymerase to the adenovirus origin of DNA replication by interaction with nuclear factor I. *Proc. Natl. Acad. Sci.* **91:** 11537–11541.

Baker, T.A. and A. Kornberg. 1988. Transcriptional activation of initiation of replication from the *E. coli* chromosomal origin: An RNA-DNA hybrid near oriC. *Cell* **55:** 113–123.

———. 1992. *DNA replication*. W.H. Freeman, San Francisco.

Baru, M., M. Shlissel, and H. Manor. 1991. The yeast GAL4 protein transactivates the polyomavirus origin of DNA replication in mouse cells. *J. Virol.* **65:** 3496–3503.

Bell, S.P. and B. Stillman. 1992. ATP-dependent recognition of eukaryotic origins of DNA replication by a multiprotein complex. *Nature* **357:** 128–134.

Bell, S.P., R. Kobayashi, and B. Stillman. 1993. Yeast origin recognition complex functions in transcription silencing and DNA replication. *Science* **262:** 1844–1849.

Bennett Cook, E.R. and J.A. Hassell. 1991. Activation of polyomavirus DNA replication by yeast GAL4 is dependent on its transcriptional activation domains. *EMBO. J.* **10:** 959–969.

Blitz, I.L. and L.A. Laimins. 1991. The 68-kilodalton E1 protein of bovine papillomavirus is a DNA binding phosphoprotein which associates with the E2 transcriptional activator in vitro. *J. Virol.* **65:** 649–656.

Bosher, J., E.C. Robinson, and R.T. Hay. 1990. Interactions between the adenovirus type 2 DNA polymerase and the DNA binding domain of nuclear factor I. *New Biol.* **2:** 1083–1090.

Bosher, J., I.R. Leith, S.M. Temperley, M. Wells, and R.T. Hay. 1991. The DNA-binding domain of nuclear factor I is sufficient to cooperate with the adenovirus type 2 DNA binding protein in viral DNA replication. *J. Gen. Virol.* **72:** 2975–2980.

Braig, K., Z. Otwinowski, R. Hedge, D.C. Boisvert, A. Joachimiak, A.L. Horwich, and P.B. Sigler. 1994. The crystal structure of the bacterial chaperonin GroEL at 2.8 Å. *Nature* **371:** 578–580.

Buchman, A.R., N.F. Lue, and R.D. Kornberg. 1988. Connections between transcriptional activators upstream activation sequences, autonomously replicating sequences, and telomeres in *Saccharomyces cerevisiae. Mol. Cell. Biol.* **8:** 210–225.

Challberg, M.D. and T.J. Kelly. 1989. Animal virus DNA replication. *Annu. Rev. Biochem.* **58:** 671–717.

Chen, M., N. Mermod, and M.S. Horwitz. 1990. Protein-protein interactions between adenovirus DNA polymerase and nuclear factor 1 mediate formation of the DNA replication preinitiation complex. *J. Biol. Chem.* **265:** 18634–18642.

Chen, M.R., J.M. Middeldorp, and S.D. Hayward. 1993. Separation of the complex DNA binding domain of EBNA-1 into DNA recognition and dimerization subdomains of novel structure. *J. Virol.* **67:** 4875–4885.

Cheng, L. and T.J. Kelly. 1989. Transcriptional activator nuclear factor 1 stimulates the replication of SV40 minichromosomes in vivo and in vitro. *Cell* **59:** 541–551.

Cheng, L.Z., J.L. Workman, R.E. Kingston, and T.J. Kelly. 1992. Regulation of DNA replication in vitro by the transcriptional activation domain of GAL4-VP16. *Proc. Natl. Acad. Sci.* **89:** 589–593.

Christ, C. and B.-K. Tye. 1991. Functional domains of the yeast transcription/replication factor MCM1. *Genes Dev.* **5:** 751–764.

Clayton, D.A. 1991. Replication and transcription of vertebrate mitochondrial DNA. *Annu. Rev. Cell Biol.* **7:** 453–478.

Cleat, P.H. and R.T. Hay. 1989. Co-operative interactions between NFI and the adenovirus DNA binding protein at the adenovirus origin of replication. *EMBO J.* **8:** 1841–1848.

Coenjaerts, F.E.J. and P.C. Van der Vliet. 1994. Early dissociation of nuclear factor I from the origin during initiation of adenovirus DNA replication studied by origin immobilization. *Nucleic Acids Res.* **22:** 5235–5240.

Coenjaerts, F.E.J., J.A.W.M. Van Oosterhout, and P.C. Van der Vliet. 1994. The Oct-1 POU domain stimulates adenovirus DNA replication by a direct interaction between the viral precursor terminal protein-DNA polymerase complex and the POU homeodomain. *EMBO J.* **13:** 5401–5409.

Coenjaerts, F.E.J., E. De Vries, G.J.M. Pruijn, W. Van Driel, S.M. Bloemers, N.M.T. van der Lugt, and P.C. Van der Vliet. 1991. Enhancement of DNA replication by transcription factors NFI and NFIII/Oct-1 depends critically on the positions of their binding sites in the adenovirus origin of replication. *Biochim. Biophys. Acta* **1090:** 61–69.

Cotmore, S.F. and P. Tattersall. 1994. An asymmetric nucleotide in the parvoviral 3′ hairpin directs segregation of a single active origin of DNA replication. *EMBO J.* **13:** 4145–4152.

Croston, G.E., L.A. Kerrigan, L.M. Lira, D.R. Marshak, and J.T. Kadonaga. 1991. Sequence-specific antirepression of histone H1-mediated inhibition of basal RNA polymerase II transcription. *Science* **251:** 643–649.

DePamphilis, M.L. 1988. Transcriptional elements as components of eukaryotic origins of DNA replication. *Cell* **52:** 635–638.

———. 1993a. Eukaryotic DNA replication: Anatomy of an origin. *Annu. Rev. Biochem.* **62:** 29–62.

———. 1993b. How transcription factors regulate origins of DNA replication in eukaryotic cells. *Trends Cell Biol.* **3:** 161–167.

de Villiers, J., W. Schaffner, C. Tyndall, S. Lupton, and R. Kamen. 1984. Polyoma virus DNA replication requires an enhancer. *Nature* **312:** 242–246.

De Vries, E., W. Van Driel, S.J. van den Heuvel, and P.C. Van der Vliet. 1987. Contact point analysis of the nuclear factor I recognition site reveals symmetric binding at one side of the DNA helix. *EMBO J.* **6:** 161–168.

Diffley, J.F.X. and J.H. Cocker. 1992. Protein-DNA interactions at a yeast replication origin. *Nature* **357:** 169–172.

Diffley, J.F.X. and B. Stillman. 1988. Purification of a yeast protein that binds to origins of DNA replication and a transcriptional silencer. *Proc. Natl. Acad. Sci.* **85:** 2120–2124.

Diffley, J.F.X., J.H. Cocker, S.J. Dowell, and A. Rowley. 1994. Two steps in the assembly of complexes at yeast replication origins in vivo. *Cell* **78:** 303–316.

Dornreiter, I., L.F. Erdile, I.U. Gilbert, D. Von Winkler, T.J. Kelly, and E. Fanning. 1992. Interaction of DNA polymerase α-primase with cellular replication protein A and SV40 T antigen. *EMBO J.* **11:** 769–776.

Dowell, S.J., P. Romanowski, and J.F.X. Diffley. 1994. Interaction of Dbf4, the Cdc7 protein kinase regulatory subunit, with yeast replication origins in vivo. *Science* **265:** 1243–1246.

Drapkin, R., J.T. Reardon, A. Ansari, J.-C. Huang, L. Zawel, K. Ahn, A. Sancar, and D. Reinberg. 1994. Dual role of TFIIH in DNA excision repair and in transcription by RNA polymerase II. *Nature* **368:** 769–771.

Fangman, W.L. and B.J. Brewer. 1991. Activation of replication origins within yeast chromosomes. *Annu. Rev. Cell Biol.* **7:** 375–402.

Fixman, E.D., G.S. Hayward, and S.D. Hayward. 1992. *trans*-Acting requirements for replication of Epstein-Barr virus ori-Lyt. *J. Virol.* **66:** 5030–5039.

Forrester, W.C., E. Epner, M.C. Driscoll, T. Enver, M. Brice, and M. Papaya Groudine. 1990. A deletion of the human β-globin locus activation region causes a major alteration in chromatin structure and replication across the entire β-globin locus. *Genes Dev.* **4:** 1637–1649.

Foss, M., F.J. McNally, P. Laurenson, and J. Rine. 1993. Origin recognition complex (ORC) in transcriptional silencing and DNA replication in *S. cerevisiae. Science* **262:** 1838–1845.

Frappier, L. and M. Donnell. 1991. Epstein-Barr virus nuclear antigen 1 mediates a DNA loop within the latent replication origin of Epstein-Barr virus. *Proc. Natl. Acad. Sci.* **88:** 10875–10879.

Frappier, L., K. Goldsmith, and L. Bendell. 1994. Stabilization of the EBNA1 protein on the Epstein-Barr virus latent origin of DNA replication by a DNA looping mechanism. *J. Biol. Chem.* **269:** 1057–1062.

Frattini, M.G. and L.A. Laimins. 1994. The role of the E1 and E2 proteins in the replication of human papillomavirus type 31b. *Virology* **204:** 799–804.

Gillette, T.G., M. Lusky, and J.A. Borowiec. 1994. Induction of structural changes in the bovine papillomavirus type 1 origin of replication by the viral E1 and E2 proteins. *Proc. Natl. Acad. Sci.* **91:** 8846–8850.

Gilson, E., M. Roberge, R. Giraldo, D. Rhodes, and S.M. Gasser. 1993. Distortion of the DNA double helix by RAP1 at silencers and multiple telomeric binding sites. *J. Mol. Biol.* **231:** 293–310.

Gopalakrishnan, V. and S.A. Khan. 1994. E1 protein of human papillomavirus type 1α is sufficient for initiation of viral DNA replication. *Proc. Natl. Acad. Sci.* **91:** 9597–9601.

Gounari, F., R. De Francesco, J. Schmidt, P.C. Van der Vliet, R. Cortese, and H.G. Stunnenberg. 1990. Amino terminal domain of NFI binds to DNA as a dimer and activates adenovirus DNA replication. *EMBO J.* **9:** 559–566.

Gronostajski, R.M. 1986. Analysis of nuclear factor I binding to DNA using degenerate oligonucleotides. *Nucleic Acids Res.* **14:** 9117–9132.

Gruss, C., J. Wu, T. Koller, and J.M. Sogo. 1993. Disruption of the nucleosomes at the replicational fork. *EMBO J.* **12:** 4533–4545.

Guo, Z.S. and M.L. DePamphilis. 1992. Specific transcription factors stimulate simian virus 40 and polyomavirus origins of DNA replication. *Mol. Cell. Biol.* **12:** 2514–2524.

Guo, Z.S., U. Heine, and M.L. DePamphilis. 1991. T antigen binding to site I facilitates initiation of SV40 DNA replication but does not affect bidirectionality. *Nucleic Acids Res.* **19:** 7081–7088.

Guo, Z.S., C. Gutierrez, U. Heine, J.M. Sogo, and M.L. DePamphilis. 1989. Origin auxiliary sequences can facilitate initiation of simian virus 40 DNA replication in vitro as they do in vivo. *Mol. Cell. Biol.* **9:** 3593–3602.

Gutierrez, C., Z.S. Guo, J. Roberts, and M.L. DePamphilis. 1990. Simian virus 40 origin auxiliary sequences weakly facilitate T antigen binding but strongly facilitate DNA unwinding. *Mol. Cell. Biol.* **10:** 1719–1728.

Hammerschmidt, W. and B. Sugden. 1988. Identification and characterization of oriLyt, a lytic origin of DNA replication of Epstein-Barr virus. *Cell* **55:** 427–433.

Harris, M.P.G. and R.T. Hay. 1988. DNA sequences required for the initiation of adenovirus type 4 DNA replication in vitro. *J. Mol. Biol.* **201:** 57–68.

Hatfield, L. and P. Hearing. 1993. The NFIII/Oct-1 binding site stimulates adenovirus DNA replication in vivo and is functionally redundant with adjacent sequences. *J. Virol.* **67:** 3931–3939.

Hay, R.T. 1985. Origin of adenovirus DNA replication. Role of nuclear factor I binding site in vivo. *J. Mol. Biol.* **186:** 129–136.

He, Z., B.T. Brinton, J. Greenblatt, J.A. Hassell, and C.J. Ingles. 1993. The transactivator proteins VP16 and GAL4 bind replication factor A. *Cell* **73:** 1223–1232.

Heintz, N.H. 1992. Transcription factors and the control of DNA replication. *Curr. Opin. Cell Biol.* **4:** 459–467.

Herr, W., R.A. Sturm, R.G. Clerc, L.M. Corcoran, D. Baltimore, P.A. Sharp, H.A. Ingraham, M.G. Rosenfeld, M. Finney, G. Ruvkun, and H.R. Horvitz. 1988. The POU domain: A large conserved region in the mammalian pit-1, oct-1, oct-2 and *Caenorhabditis elegans* unc-86 gene products. *Genes Dev.* **2:** 1513–1516.

Hoang, A.T., W. Wang, and J.D. Gralla. 1992. The replication activation potential of selected RNA polymerase II promoter elements at the simian virus 40 origin. *Mol. Cell. Biol.* **12:** 3087–3093.

Ishikawa, H., M. Asano, T. Kanda, S. Kumar, C. Gelinas, and Y. Ito. 1993. Two novel functions associated with the Rel oncoproteins: DNA replication and cell-specific transcriptional activation. *Oncogene* **8:** 2889–2896.

Ishimi, Y. 1992. Preincubation of T antigen with DNA overcomes repression of SV40 DNA replication by nucleosome assembly. *J. Biol. Chem.* **267:** 10910–10913.

Johansen, T., U. Moens, T. Holm, A. Fjose, and S. Krauss. 1993. Zebrafish pou[c]: A divergent POU family gene ubiquitously expressed during embryogenesis. *Nucleic Acids Res.* **21:** 475–483.

Kenny, M. and J. Hurwitz. 1988. Initiation of adenovirus DNA replication. II. Structural requirements using synthetic oligonucleotide adenovirus templates. *J. Biol. Chem.* **263:** 9809–9817.

Kilwinski, J., M. Baack, S. Heiland, and R. Knippers. 1995. Transcription factor Oct1 binds to the AT-rich segment of the simian virus 40 replication origin. *J. Virol.* **69:** 575–578.

Kitsberg, D., S. Selig, I. Keshet, and H. Cedar. 1993. Replication structure of the human β-globin gene domain. *Nature* **366:** 588–590.

Kuo, S.R., J.S. Liu, T.R. Broker, and L.T. Chow. 1994. Cell-free replication of the human papillomavirus DNA with homologous viral E1 and E2 proteins and human cell extracts. *J. Biol. Chem.* **269:** 24058–24065.

Laybourn, P.J. and J.T. Kadonaga. 1991. Role of nucleosomal cores and histone H1 in regulation of transcription by RNA polymerase II. *Science* **254:** 238–245.

Lednicky, J. and W.R. Folk. 1992. Two synthetic Sp1-binding sites functionally substitute for the 21-base-pair repeat region to activate simian virus 40 growth in CV-1 cells. *J. Virol.* **66:** 6379–6390.

Leegwater, P.A.J., W. Van Driel, and P.C. Van der Vliet. 1985. Recognition site of nuclear factor I, a sequence specific DNA-binding protein from HeLa cells that stimulates adenovirus DNA replication. *EMBO J.* **4:** 1515–1521.

Li, R. and M.R. Botchan. 1993. The acidic transcriptional activation domains of VP16 and p53 bind the cellular replication protein A and stimulate in vitro BPV-1 DNA replication. *Cell* **73:** 1207–1221.

———. 1994. Acidic transcription factors alleviate nucleosome-mediated repression of DNA replication of bovine papillomavirus type 1. *Proc. Natl. Acad. Sci.* **91:** 7051–7055.

Lue, N.F. and R.D. Kornberg. 1993. A possible role for the yeast TATA-element-binding protein in DNA replication. *Proc. Natl. Acad. Sci.* **90:** 8018–8022.

Lusky, M., J. Hurwitz, and Y.S. Seo. 1993. Cooperative assembly of the bovine papilloma virus E1 and E2 proteins on the replication origin requires an intact E2 binding site. *J. Biol. Chem.* **268:** 15795–15803.

———. 1994. The bovine papillomavirus E2 protein modulates the assembly of but is not stably maintained in a replication-competent multimeric E1-replication origin complex. *Proc. Natl. Acad. Sci.* **91:** 8895–8899.

Marahrens, Y. and B. Stillman. 1992. A yeast chromosomal origin of DNA replication defined by multiple functional elements. *Science* **255:** 817–823.

McBroom, L.D. and P.D. Sadowski. 1994a. Contacts of the ABF1 protein of *Saccharomyces cerevisiae* with a DNA binding site at MATα. *J. Biol. Chem.* **269:** 16455–16460.

———. 1994b. DNA bending by *Saccharomyces cerevisiae* ABF1 and its proteolytic fragments. *J. Biol. Chem.* **269:** 16461–16468.

Meisterernst, M., L. Rogge, R. Foechler, M. Karaghiosoff, and E.L. Winnacker. 1989. Structural and functional organization of a porcine gene coding for nuclear factor I. *Biochemistry* **28:** 8191–8200.

Melendy, T. and B. Stillman. 1993. An interaction between replication protein A and SV40 T antigen appears essential for primosome assembly during SV40 DNA replication. *J. Biol. Chem.* **268:** 3389–3395.

Mermod, N., E.A. O'Neill, T.J. Kelly, and R. Tjian. 1989. The proline-rich transcriptional activator of CTF/NFI is distinct from the replication and DNA binding domain. *Cell* **58:** 741–753.

Micklem, G., A. Rowley, J. Harwood, K. Nasmyth, and J.F.X. Diffley. 1993. Yeast origin recognition complex is involved in DNA replication and transcriptional silencing. *Nature* **366:** 87–89.

Mohr, I.J., R. Clark, S. Sun, E.J. Androphy, P. MacPherson, and M.R. Botchan. 1990. Targeting the E1 replication protein to the papillomavirus origin of replication by complex formation with the E2 transactivator. *Science* **250:** 1694–1699.

Monaghan, A., A. Webster, and R.T. Hay. 1994. Adenovirus DNA binding protein: Helix destabilising properties. *Nucleic Acids Res.* **22:** 742–748.

Morgan, I.M., M. Asano, L.S. Havarstein, H. Ishikawa, T. Hiiragi, Y. Ito, and P.K. Vogt. 1993. Amino acid substitutions modulate the effect of Jun on transformation, transcriptional activation and DNA replication. *Oncogene* **8:** 1135–1140.

Mul, Y.M. and P.C. Van der Vliet. 1992. Nuclear factor I enhances adenovirus DNA replication by increasing the stability of a preinitiation complex. *EMBO J.* **11:** 751–760.

———. 1993. The adenovirus DNA binding protein effects the kinetics of DNA replication by a mechanism distinct from NFI or Oct-1. *Nucleic Acids Res.* **21:** 641–647.

Mul, Y.M., C.P. Verrijzer, and P.C. Van der Vliet. 1990. Transcription factors NFI and NFIII/Oct-1 function independently, employing different mechanisms to enhance adenovirus DNA replication. *J. Virol.* **64:** 5510–5518.

Müller, F., Y.-S. Seo, and J. Hurwitz. 1994. Replication of bovine papillomavirus type 1 origin-containing DNA in crude extracts and with purified proteins. *J. Biol. Chem.* **269:** 17086–17094.

Murakami, Y., M. Satake, Y. Yamaguchi Iwai, M. Sakai, M. Muramatsu, and Y. Ito. 1991. The nuclear protooncogenes c-*jun* and c-*fos* as regulators of DNA replication. *Proc. Natl. Acad. Sci.* **88:** 3947–3951.

Nagata, K., R.A. Guggenheimer, and J. Hurwitz. 1983. Specific binding of a cellular DNA replication protein to the origin of replication of adenovirus DNA. *Proc. Natl. Acad. Sci.* **80:** 6177–6181.

Nagata, K., R.A. Guggenheimer, T. Enomoto, J.H. Lichy, and J. Hurwitz. 1982. Adenovirus DNA replication in vitro: Identification of a host factor that stimulates synthesis of the preterminal protein-dCMP complex. *Proc. Natl. Acad. Sci.* **79:** 6438–6442.

O'Neill, E.A. and T.J. Kelly. 1988. Purification and characterization of nuclear factor III (origin recognition protein C), a sequence-specific DNA binding protein required for efficient initiation of adenovirus DNA replication. *J. Biol. Chem.* **263:** 931–937.

O'Neill, E.A., C. Fletcher, C.R. Burrow, N. Heintz, R.G. Roeder, and T.J. Kelly. 1988. Transcriptional factor OTF-1 is functionally identical to the DNA replication factor NFIII. *Science* **241**: 1210-1213.

Paonessa, G., F. Gounari, R. Frank, and R. Cortese. 1988. Purification of a NFI-like DNA-binding protein from rat liver and cloning of the corresponding cDNA. *EMBO J.* **7:** 3115–3123.

Park, P., W. Copeland, L. Yang, T. Wang, M.R. Botchan, and I.J. Mohr. 1994. The cellular DNA polymerase α-primase is required for papillomavirus DNA replication and associates with the viral E1 helicase. *Proc. Natl. Acad. Sci.* **91:** 8700–8704.

Pronk, R. and P.C. Van der Vliet. 1993. The adenovirus terminal protein influences binding of replication proteins and changes the origin structure. *Nucleic Acids Res.* **21:** 2293–2300.

Pruijn, G.J.M., W. Van Driel, and P.C. Van der Vliet. 1986. Nuclear factor III, a novel sequence-specific DNA-binding protein from HeLa cells stimulating adenovirus DNA replication. *Nature* **322:** 656–659.

Pruijn, G.J.M., P.C. Van der Vliet, N.A. Dathan, and I.W. Mattaj. 1989. Anti-OTF-1 antibodies inhibit NFIII stimulation of in vitro adenovirus DNA replication. *Nucleic Acids Res.* **17:** 1845–1863.

Rivier, D.H. and J. Rine. 1992. An origin of DNA replication and a transcription silencer require a common element. *Science* **256:** 659–663.

Roberts, S.G.E. and M.R. Green. 1994. Activator-induced conformational change in general transcription factor TFIIB. *Nature* **371:** 717–720.

Rosenfeld, P.J. and T.J. Kelly. 1986. Purification of nuclear factor I by DNA recognition site affinity chromatography. *J. Biol. Chem.* **261:** 1398–1408.

Russell, J. and M.R. Botchan. 1995. *cis*-Acting components of human papillomavirus (HPV) DNA replication: Linker substitution analysis of the HPV type 11 origin. *J. Virol.* **69:** 651–660.

Santoro, C., N. Mermod, P.C. Andrews, and R. Tjian. 1988. A family of human CAAT-box-binding proteins active in transcription and DNA replication: Cloning and expression of multiple cDNAs. *Nature* **334:** 218–224.

Schaeffer, L., R. Roy, S. Humbert, V. Moncollin, W. Vermeulen, J.H.J. Hoeijmakers, P. Chambon, and J.-M. Egly. 1993. DNA repair helicase: A component of BTF2 (TFIIH) basic transcription factor. *Science* **260:** 58–63.

Schepers, A., D. Pich, and W. Hammerschmidt. 1993a. A transcription factor with homology to the AP-1 family links RNA transcription and DNA replication in the lytic cycle of Epstein-Barr virus. *EMBO. J.* **12:** 3921–3929.

Schepers, A., D. Pich, J. Mankertz, and W. Hammerschmidt. 1993b. *cis*-Acting elements in the lytic origin of DNA replication of Epstein-Barr virus. *J. Virol.* **67:** 4237–4245.

Seo, Y.S., F. Muller, M. Lusky, and J. Hurwitz. 1993a. Bovine papilloma virus (BPV)-encoded E1 protein contains multiple activities required for BPV DNA replication. *Proc. Natl. Acad. Sci.* **90:** 702–706.

Seo, Y.S., F. Muller, M. Lusky, E. Gibbs, H.Y. Kim, B. Phillips, and J. Hurwitz. 1993b. Bovine papilloma virus (BPV)-encoded E2 protein enhances binding of E1 protein to the BPV replication origin. *Proc. Natl. Acad. Sci.* **90:** 2865–2869.

Smith, D.L. and A.D. Johnson. 1992. A molecular mechanism for combinatorial control in yeast: MCM1 protein sets the spacing and orientation of the homeodomains of an α2 dimer. *Cell* **68:** 133–142.

Sock, E., M. Wegner, E.A. Fortunato, and F. Grummt. 1993. Large T antigen and sequences within the regulatory region of JC virus both contribute to the features of JC virus DNA replication. *Virology* **197:** 537–548.

Spalholz, B.A., A.A. McBride, T. Sarafi, and J. Quintero. 1993. Binding of bovine papillomavirus E1 to the origin is not sufficient for DNA replication. *Virology* **193:** 201–212.

Stillman, B.M. 1989. Initiation of eukaryotic DNA replication in vitro. *Annu. Rev. Cell Biol.* **5:** 197–245.

Stow, N.D. and E.C. McMonagle. 1983. Characterization of the TRs/IRs origin of DNA replication of herpes simplex virus type 1. *Virology* **130:** 427–438.

Stuiver, M.H. and P.C. Van der Vliet. 1990. The adenovirus DNA binding protein forms a multimeric protein complex with double-stranded DNA and enhances binding of nuclear factor I. *J. Virol.* **64:** 379–386.

Stuiver, M.H., W.G. Bergsma, A.C. Arnberg, H. Van Amerongen, R. Van Grondelle, and P.C. Van der Vliet. 1992. Structural alterations of double-stranded DNA in complex with the adenovirus DNA-binding protein. Implications for its function in DNA replication. *J. Mol. Biol.* **225:** 999–1011.

Su, W., T. Middleton, B. Sugden, and H. Echols. 1991. DNA looping between the origin of replication of Epstein-Barr virus and its enhancer site: Stabilization of an origin complex with Epstein-Barr nuclear antigen 1. *Proc. Natl. Acad. Sci.* **88:** 10870–10874.

Sun, S., L. Thorner, M. Lentz, P. McPherson, and M. Botchan. 1990. Identification of a 68-kilodalton nuclear ATP-binding phosphoprotein encoded by bovine papillomavirus type 1. *J. Virol.* **64:** 5093–5105.

Tanaka, M., W.M. Clouston, and W. Herr. 1994. The Oct-2 glutamine-rich and proline-rich activation domains can synergize with each other or duplicates of themselves to activate transcription. *Mol. Cell. Biol.* **14:** 6046–6055.

Thorner, L.K., D.A. Lim, and M.R. Botchan. 1993. DNA-binding domain of bovine papillomavirus type 1 E1 helicase: Structural and functional aspects. *J. Virol.* **67:** 6000–6014.

Ustav, M., E. Ustav, P. Szymanski, and A. Stenlund. 1991. Identification of the origin of replication of bovine papillomavirus and characterization of the viral origin recognition factor E1. *EMBO J.* **10:** 4321–4329.

———. 1993. The bovine papillomavirus origin of replication requires a binding site for the E2 transcriptional activator. *Proc. Natl. Acad. Sci.* **90:** 898–902.

Van der Vliet, P.C. 1991. The role of cellular transcription factors in the enhancement of adenovirus DNA replication. *Semin. Virol.* **2:** 271–280.

Verrijzer, C.P. and P.C. Van der Vliet. 1993. POU domain transcription factors. *Biochim. Biophys. Acta* **1173:** 1–21.

Verrijzer, C.P., A.J. Kal, and P.C. Van der Vliet. 1990. The DNA binding domain (POU domain) of transcription factor Oct-1 suffices for stimulation of DNA replication. *EMBO J.* **9:** 1883–1888.

Verrijzer, C.P., M. Strating, Y.M. Mul, and P.C. Van der Vliet. 1992. POU domain transcription factors from different subclasses stimulate adenovirus DNA replication. *Nucleic Acids Res.* **20:** 6369–6375.

Verrijzer, C.P., J.A.W.M. Van Oosterhout, W. Van Weperen, and P.C. Van der Vliet. 1991. POU proteins bend DNA via the POU-specific domain. *EMBO J.* **10:** 3007–3014.

Wasylyk, C., J. Schneikert, and B. Wasylyk. 1990. Oncogene v-*jun* modulates DNA replication. *Oncogene* **5:** 1055–1058.

Wides, R.J., M.D. Challberg, D.R. Rawlins, and T.J. Kelly. 1987. Adenovirus origin of DNA replication: Sequence requirements for replication in vitro. *Mol. Cell. Biol.* **7:** 864–874.

Winokur, P.L. and A.A. McBride. 1992. Separation of the transcriptional activation and replication functions of the bovine papillomavirus-1 E2 protein. *EMBO. J.* **11:** 4111–4118.

Wong, S.W. and P.A. Schaffer. 1991. Elements in the transcriptional regulatory region flanking herpes simplex virus type 1 oriS stimulate origin function. *J. Virol.* **65:** 2601–2611.

Yang, L., I. Mohr, E. Fouts, D.A. Lim, M. Nohaile, and M.R. Botchan. 1993. The E1 protein of bovine papilloma virus 1 is an ATP-dependent DNA helicase. *Proc. Natl. Acad. Sci.* **90:** 5086–5090.

Young, A.C.M., W. Zhang, J.C. Sacchettini, and S.G. Nathenson. 1994. The three-dimensional structure of H-2Db at 2.4 Å resolution: Implications for antigen-determinant selection. *Cell* **76:** 39–50.

Zijderveld, D.C. and P.C. Van der Vliet. 1994. Helix-destabilizing properties of the adenovirus DNA binding protein. *J. Virol.* **68:** 1158–1164.

Zorbas, H., L. Rogge, M. Meisterernst, and E.L. Winnacker. 1989. Hydroxyl radical footprints reveal novel structural features around the NFI binding site in adenovirus DNA. *Nucleic Acids Res.* **17:** 7735–7748.

4

Roles of Nuclear Structure in DNA Replication

Ronald Laskey and Mark Madine
Wellcome/CRC Institute
Cambridge CB2 1QR
United Kingdom

Department of Zoology
University of Cambridge
Cambridge CB2 3EJ
United Kingdom

The cell nucleus is the defining feature of eukaryotes. It is bounded by a nuclear envelope consisting of two concentric layers of membrane perforated by nuclear pores that serve as channels of communication between the nucleus and the cytoplasm. DNA is not randomly packed into the nucleus but packaged precisely in such a way that all regions are accessible for replication each cell cycle. Partial access and therefore partial replication would result in chromosome breakage or nondisjunction at mitosis, with disastrous consequences. The packing hierarchy involves radial loop organization from an axial scaffold, as well as the compaction resulting from coiling DNA twice around the nucleosome subunits of chromatin (Schedl and Grosveld 1995; Van Holde et al. 1995).

A crucial feature of eukaryotic chromosomal DNA replication is that it always occurs within a nucleus. In lower eukaryotes such as fungi or *Physarum*, the nuclear membrane remains intact throughout the cell cycle, whereas it breaks down during mitosis of higher eukaryotes. Nevertheless, replication is constrained to interphase when the nuclear membrane is intact. We argue that this constraint has important regulatory consequences.

Further key features of eukaryotic DNA replication are that multiple initiations occur within a single chromosome and that these initiations are coordinated so that each region of the chromosome replicates, but replicates only once in any cell cycle. We argue that nuclear structure has essential roles to play in coordinating multiple initiations to replicate the chromosome exactly once.

NUCLEAR STRUCTURE IS REQUIRED FOR CELLULAR DNA REPLICATION

Studies of eukaryotic DNA replication have been held back by a shortage of cell-free systems that initiate chromosomal DNA replication efficiently in vitro. Efficient replication systems have been developed for several viruses (see other chapters in this volume), but the only systems that clearly initiate efficiently on cellular DNA are derived from animal eggs. Initiating cell-free replication systems have been developed from unfertilized eggs of *Xenopus* and *Drosophila* (Blow and Laskey 1986; Newport 1987; Crevel and Cotterill 1991). In both cases, the egg contains a prefabricated stockpile of materials for exceptionally rapid DNA replication and nuclear assembly. *Xenopus* and *Drosophila* embryos reach 10,000 cells faster than a proliferating mammalian cell divides once.

These features make animal eggs attractive systems to study cell proliferation (see Blow and Chong, this volume), but they have some compensating disadvantages. For example, they lack G_0, G_1, and G_2 phases of the cell cycle, making them unsuitable for studies of transition between normal cell-cycle phases. A cell-free system that initiates cellular chromosomal replication efficiently in extracts of mammalian cells or yeast would have important applications.

A conspicuous feature of both the replication systems from *Xenopus* eggs is their dependence on nuclear structure to initiate DNA replication. Initiation is only observed when the template DNA is enclosed within a complete nuclear membrane (Newport 1987; Sheehan et al. 1988; Blow and Sleeman 1990). Nuclei whose membranes have been permeabilized by nonionic detergents or lysolecithin must be repaired by membrane vesicles before initiation is observed (Leno et al. 1992; Coverley et al. 1993).

Two further lines of evidence emphasize the importance of nuclear structure in initiation of DNA replication. First, purified DNA is replicated reasonably efficiently when added to *Xenopus* egg extracts (Blow and Laskey 1986; Newport 1987; Blow and Sleeman 1990; Newport et al. 1990; Cox and Laskey 1991), but replication is observed only when the DNA is assembled into pseudonuclei. Furthermore, the efficiency of replication depends on the efficiency of nuclear assembly, and only the DNA that is enclosed within the pseudonuclei replicates (Blow and Sleeman 1990).

The second line of evidence that nuclear structure is crucial for DNA replication comes from studies of nuclear membrane assembly in vitro. These indicate that the nuclear membrane defines the DNA it encloses as an integral unit of replication. Thus, Blow and Watson (1987) found that individual sperm nuclei in a common egg-extract environment act as integrated and independent units of replication. All the DNA in a nucleus replicates roughly synchronously, but not synchronously with its neighbors. This conclusion was strengthened by Leno and Laskey (1991), who exploited an unusual behavior of chicken erythrocyte nuclei in egg extract. These nuclei tend to clump to form aggregates. The response of the extract is to assemble a single nuclear membrane around the entire clump, not around the individual nuclei it contains. In this circumstance, all the nuclei enclosed within a single nuclear membrane replicated in precise synchrony, even though different clumps replicated at different times, and even though individual nuclei outside the clumps replicated asynchronously. Therefore, the nuclear membrane defines the DNA it encloses as a unit for initiation of replication.

One conspicuous way in which the nuclear membrane could regulate initiation of DNA replication is by concentrating nuclear proteins in the nucleus from the cytoplasm. The importance of this process for initiation of replication has been demonstrated by use of the inhibitor wheat germ agglutinin (WGA). This inhibits protein import and also prevents initiation of replication (Cox 1992).

A second way in which replication depends on an aspect of nuclear structure concerns the nuclear lamina. When lamins are removed from egg extract by immunodepletion, nuclear assembly and protein import are not inhibited, but the nuclei formed are unable to initiate replication (Newport et al. 1990; Meier et al. 1991; Jenkins et al. 1993).

The conclusion we offer from this section is that nuclear structure is essential for initiation of DNA replication in eukaryotes. This in turn suggests that attempts to obtain cell-free initiation of DNA replication from mammalian cells should focus on using nuclei as the template.

REPLICATION OCCURS AT DISCRETE SITES WITHIN THE NUCLEUS

Replication forks are not distributed diffusely throughout the nucleus. Instead, they are clustered in replication foci or factories (Fig. 1) (Nakamura et al. 1986; Mills et al. 1989; Nakayasu and Berezney 1989; Cox and Laskey 1991; Fox et al. 1991; O'Keefe et al. 1992; Hozak et al. 1993). These may contain tens to several hundreds of replication forks, depending on the type of cell. They are revealed by pulse-labeling sites of DNA replication with precursors that can be detected fluorescently.

The significance of replication fork clustering is not clear. It might facilitate coordination between polymerases and accessory proteins on the

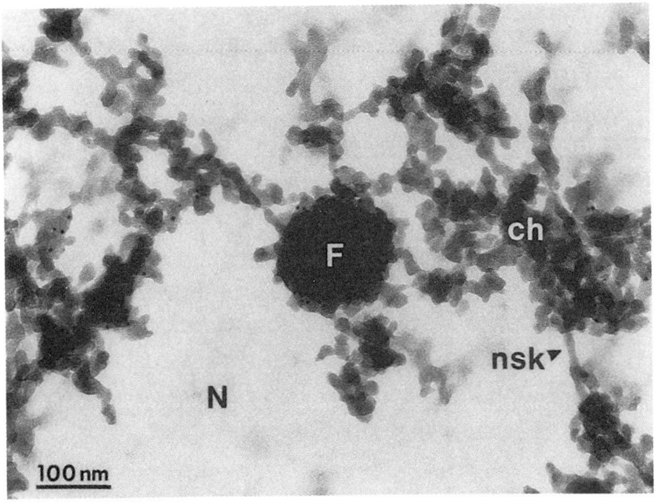

Figure 1 A replication factory. Resinless electron microscopy of HeLa nuclei following chromatin digestion reveals residual chromatin (ch) attached to an underlying nucleoskeleton (nsk). In S-phase cells, dense "factories" (F) are sites of DNA synthesis. Bar, 100 nm. (By courtesy of P. Hozak and P. Cook.)

two sides of each fork and on the two forks resulting from each initiation. More interestingly, it might provide a structural framework for the task of ensuring that all DNA is completely replicated.

The clustering of replication forks has added fuel to the debate of whether mobile DNA polymerases move along a fixed DNA track, or alternatively, whether mobile DNA is spooled through immobilized replication machinery (Hozak et al. 1993; Jackson 1995). Opinions remain divided on this issue, although it appears to us that evidence for spooling through fixed replication sites is slowly growing stronger. Electron microscopy of the initial DNA unwinding reaction for SV40 DNA replication clearly illustrates two T-antigen complexes at the replication forks which remain together while loops of single-stranded DNA are extruded (Wessel et al. 1992). If spooling is the correct interpretation, then one particular problem arises from the need to replicate the last regions of DNA between two replication foci or factories. Are they pulled in two directions? Are they completed by mobile polymerases detaching from the cluster? Alternatively, does DNA move in the same direction through consecutive clusters? Answers to these questions and the underlying issue of spooling would help us to understand the structural basics of replication.

Patterns of replication fork clusters remain the same throughout S phase of *Xenopus* sperm nuclei replicating in *Xenopus* egg extracts (Mills et al. 1989), but in somatic mammalian cells they change as S phase progresses (Nakayasu and Berezney 1989; Fox et al. 1991; O'Keefe et al. 1992). Some foci appear late, and further clustering together of entire foci occurs late in replication. Perhaps this aggregation of foci contributes to solving the problem referred to above of how the last stretch of DNA between adjacent foci is replicated.

What specifies the pattern of foci? This question has been addressed by two extreme examples seen in Figure 2. At one extreme, purified DNA from bacteriophage has been used as a template for reassembly into pseudonuclei and replication in egg extracts. Even phage DNA is replicated under strict cell-cycle control in *Xenopus* eggs (Harland and Laskey 1980; Méchali et al. 1983; Newport 1987). As shown in Figure 2, phage DNA is also replicated under the egg's spatial control. Even though it lacks eukaryotic sequences, the pattern of clustered replication forks strikingly resembles that of *Xenopus* sperm nuclei, indicating that specific eukaryotic DNA sequences are not required for this level of spatial regulation.

At the opposite extreme, polytene nuclei from salivary glands of *Drosophila* larvae also have a similar pattern of clustered replication forks superimposed on them by *Xenopus* egg extract (Fig. 2). The polytene chromatin is remodeled and decondensed by the extract during this process, losing its banded appearance. From these examples, we conclude that the pattern of clustered replication forks can arise de novo even on prokaryotic DNA that has never been subjected to such patterns, or on highly organized polytene chromosomes. The pattern is specified by the egg extract, not by the incoming nuclei, and it must be imposed by a structural measuring mechanism that is independent of DNA sequence.

One series of experiments demonstrates that *Xenopus* eggs can recognize and use preexisting nuclear structure, at least transitionally, for initiation of replication. Gilbert et al. (1995) showed that nuclei from Chinese hamster ovary cells with highly amplified genes for dihydrofolate reductase initiate replication nonrandomly at preferred sites. These sites are the same as those used by cells in vivo. However, when the nuclear membrane was disrupted or naked DNA was used as the substrate, specificity of initiation was lost.

REPLICATION LICENSING AND THE NUCLEAR MEMBRANE

Above, we argued that the nuclear membrane plays a regulatory role by coordinating the replication of all the DNA it contains. Here we argue

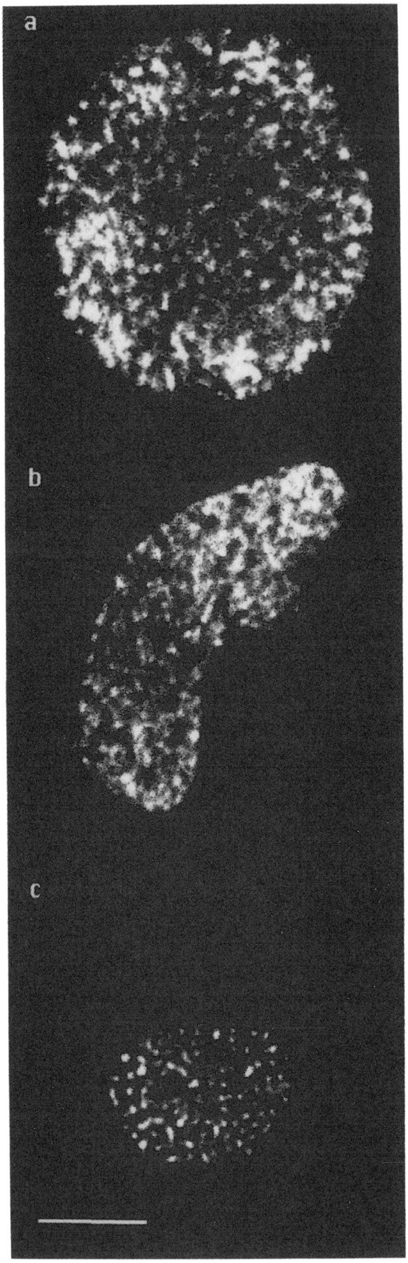

Figure 2 Discrete sites of replication seen as fluorescent foci in three types of nuclei replicating in *Xenopus* egg extracts: (*a*) *Drosophila* polytene nuclei, (*b*) *Xenopus* sperm nuclei, and (*c*) bacteriophage λ DNA pseudonuclei. Bar, 5 μm. (Original photographs provided by A.D. Mills and L.S. Cox.)

that it plays a further regulatory role by preventing reinitiation of DNA replication within a single cell cycle. The evidence for this assertion comes from experiments in which replicated, G_2 nuclei are transferred to fresh egg extract. They do not reinitiate replication if their nuclear membranes are intact. If, however, their nuclear membranes are permeabilized by nonionic detergents or lysolecithin before they are transferred, they reinitiate replication efficiently (Fig. 3). These experiments have been performed using *Xenopus* sperm nuclei as the templates or using synchronized HeLa nuclei (Blow and Laskey 1988; Leno et al. 1992; Coverley et al. 1993; Madine et al. 1995). These observations can be explained by a licensing model (Fig. 4A) (Blow and Laskey 1988). This model postulates an essential initiation factor (originally called "licensing factor," but see below) that is necessary, but not sufficient, for initiation of replication and that is unable to cross the nuclear envelope because it lacks a nuclear localization signal. Therefore, it could only bind to chromatin when the nuclear membrane breaks down in mitosis.

The model postulates that the factor would be inactivated by the act of replication so, in this way, each round of replication would need to be individually licensed by nuclear envelope breakdown at mitosis. Permeabilizing the nuclear membrane artificially would simply mimic this effect of mitosis, allowing the factor to act on chromatin to generate

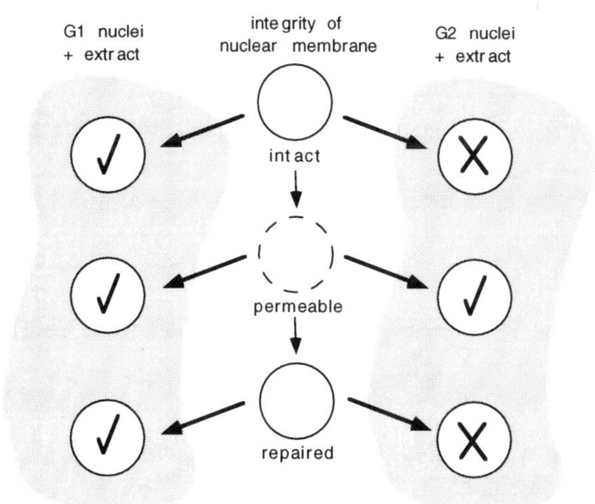

Figure 3 Summary of replication capacity of synchronized HeLa nuclei in *Xenopus* egg extract. G_1 nuclei replicate whether or not their nuclear membranes are intact, whereas G_2 nuclei are only able to replicate when their nuclear membrane is permeable (for references, see text).

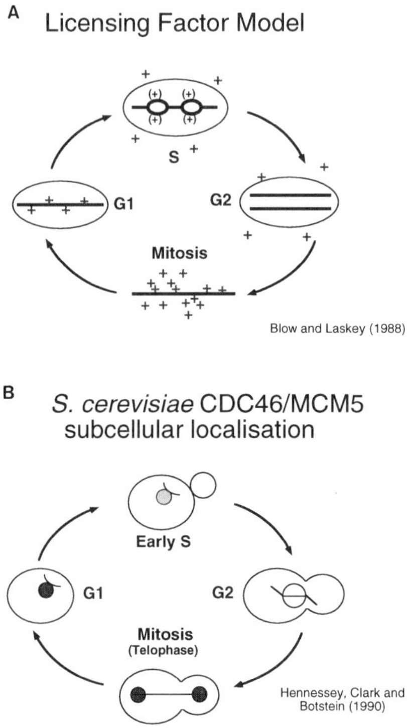

Figure 4 (*A*) "Licensing factor" model for mechanism limiting DNA replication to once per cell cycle. Licensing factor binds to DNA during mitosis but is unable to enter the nuclear envelope during interphase. It is inactivated by initiation or passage of a replication fork so that replication cannot reinitiate until the nuclear envelope breaks down again in the next mitosis. (Modified from Blow and Laskey 1988.) (*B*) Cartoon depicting nuclear localization of CDC46/MCM5 during the cell cycle of *S. cerevisiae* (Hennessey et al. 1990). Dark shading represents nuclear localization.

a new license. Evidence supporting a positive factor of this type has come from three laboratories (Blow 1993; Coverley et al. 1993; Kubota and Takisawa 1993) and has been extended recently (Chong et al. 1995; Kubota et al. 1995; Madine et al. 1995). These studies suggest, however, that the activity can be resolved into at least two components.

A family of proteins that were discovered in yeast have become excellent candidates for one of the components required for replication licensing. These proteins are becoming a growth industry. They are called MCM or *m*ini *c*hromosome *m*aintenance proteins. They have been discovered independently in *Saccharomyces cerevisiae, Schizosaccharomyces pombe*, mouse, *Pleurodeles*, and *Notophthalmus*, and have also been cloned from several other species, including human and *Xenopus*. In yeast, they are required for DNA replication, and their localization is strikingly similar to that predicted for a licensing factor. Not only are they required for replication, but they are cytoplasmic until mitosis, when they enter the nucleus at anaphase and remain there until S phase, when they are removed (Fig. 4B).

This coincidence was investigated by cloning, expressing, and raising antibodies against *Xenopus* MCM3 (Madine et al. 1995). Affinity-purified antibodies precipitated a complex containing several MCM proteins. Removal of this complex by immunodepletion inhibits replication of *Xenopus* sperm nuclei (Fig. 5). Replication is restored by re-addition of the MCM complex. When synchronized HeLa cells were used as templates, permeable G_2 nuclei replicated in mock-depleted extract as expected, but not in extract depleted of the MCM complex. In contrast, G_1 HeLa nuclei replicated whether or not the *Xenopus* MCM complex was present, implicating the MCM complex in the mechanism that distinguishes G_1 from G_2 nuclei and preventing further replication of G_2 nuclei (Fig. 6). Similar conclusions were reached independently by Kubota et al. (1995) and Chong et al. (1995) using different approaches but also implicating the MCM complexes.

An obvious question is, Can the MCM proteins cross the nuclear membrane? Experiments with an MCM3 fusion protein indicated that it cannot cross an interphase nuclear membrane (Kubota et al. 1995). How-

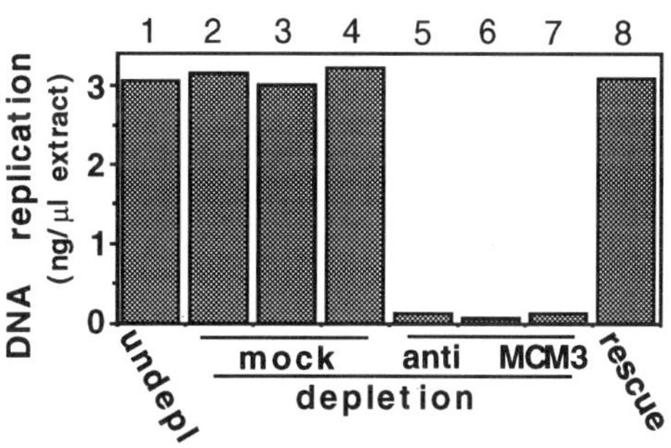

Figure 5 Immunodepletion by anti-XMCM3 antibodies inhibits DNA replication of *Xenopus* sperm nuclei in *Xenopus* egg extracts. Sperm nuclei replicate efficiently in undepleted (track *1*) extract or mock immunodepleted extracts (tracks *2–4*). In contrast, immunodepletion of the MCM protein complex with any one of three anti-XMCM3 antibodies abolishes replication (tracks *5–7*). Readdition of MCM complex purified from a fourth antibody restores replication (for details, see Madine et al. 1995). (Reprinted, with permission, from Madine et al. 1995 [copyright Macmillan].)

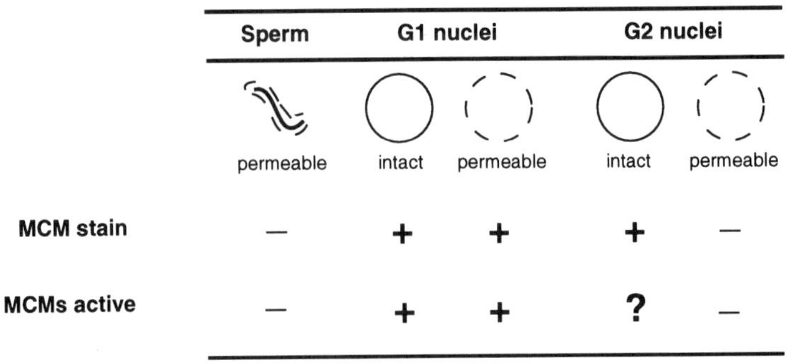

Figure 6 Summary of the MCM status of nuclei from *Xenopus* sperm or synchronized G_1 and G_2 HeLa cells. "MCM stain" refers to immunofluorescence with antibodies against MCM3. "MCMs active" refers to the ability of nuclei to replicate in an egg extract from which the MCM complex has been depleted (for details, see Madine et al. 1995).

ever, when transport of the native MCM complex is examined, MCM3 does cross an intact nuclear membrane (Madine et al. 1995). One possible explanation of this difference is a folding problem of the recombinant fusion protein, but a more interesting possibility is that another member of the MCM family is responsible for targeting the whole MCM complex to the nucleus.

If the MCM complex can cross an intact nuclear membrane, then how does it relate to the requirement for nuclear membrane breakdown or permeability to reinitiate DNA replication? Clues may come from two directions. First, M.A. Madine et al. (in prep.) have shown that nuclear envelope permeability is needed to allow binding of MCM3 to chromatin, rather than for entry into the nucleus. Second, Chong et al. (1995) have shown by fractionation that a second activity is required for replication licensing. Thus, it appears that two separate activities are required for replication licensing. It is possible to argue semantically about which of these should be called "licensing factor." We suggest that this argument is unhelpful and that the term licensing factor should be replaced by referring instead to "replication licensing," since at least two classes of factors are involved.

ACKNOWLEDGMENTS

We thank A.D. Mills and L.S. Cox for providing original photographs for Figure 2; special thanks to A.D. Mills for help with preparation of figures and to Christine Fox for preparation of the manuscript. The work in this laboratory was supported by the Cancer Research Campaign.

REFERENCES

Blow, J.J. 1993. Preventing re-replication of DNA in a single cell cycle: Evidence for a replication licensing factor. *J. Cell Biol.* **122:** 993–1002.

Blow, J.J. and R.A. Laskey. 1986. Initiation of DNA replication in nuclei and purified DNA by a cell-free extract of *Xenopus* eggs. *Cell* **47:** 577–587.

———. 1988. A role for the nuclear envelope in controlling DNA replication within the cell cycle. *Nature* **332:** 706–709.

Blow, J.J. and A.M. Sleeman. 1990. Replication of purified DNA in *Xenopus* egg extracts is dependent on nuclear assembly. *J. Cell Sci.* **95:** 383–391.

Blow, J.J. and J.V. Watson. 1987. Nuclei act as independent and integrated units of replication in a *Xenopus* cell-free system. *EMBO J.* **6:** 1997–2002.

Chong, J.P.J., H.M. Mahbubani, C.-Y. Khoo, and J.J. Blow. 1995. Purification of an Mcm-containing complex as a component of the DNA replication licensing factor. *Nature* **374:** 418–421.

Coverley, D., C.S. Downes, P. Romanowski, and R.A. Laskey. 1993. Reversible effects of nuclear membrane permeabilisation on DNA replication: Evidence for a positive licensing factor. *J. Cell Biol.* **122:** 985–992.

Cox, L.S. 1992. DNA replication in cell-free extracts from *Xenopus* eggs is prevented by disrupting nuclear envelope function. *J. Cell Sci.* **101:** 43–53.

Cox, L.S. and R.A. Laskey. 1991. DNA replication occurs at discrete sites in pseudonuclei assembled from purified DNA in vitro. *Cell* **66:** 271–275.

Crevel, G. and S. Cotterill. 1991. DNA replication in cell-free extracts from *Drosophila melanogaster. EMBO J.* **10:** 4361–4369.

Fox, M.H., D.J. Arndt-Jovin, T.M. Jovin, P.H. Baumann, and M. Robert-Nicoud. 1991. Spatial and temporal distribution of DNA replication sites localised by immunofluorescence and confocal microscopy in mouse fibroblasts. *J. Cell Sci.* **99:** 247–53.

Gilbert, D.M., H. Miyazawa, and M. DePamphilis. 1995. Site-specific initiation of DNA replication in *Xenopus* egg extracts requires nuclear structure. *Mol. Cell. Biol.* **15:** 2942–2954.

Harland, R. and R.A. Laskey. 1980. Regulated replication of DNA microinjected into eggs of *Xenopus laevis. Cell* **21:** 761–771.

Hennessy, K.M., C.D. Clark, and D. Botstein. 1990. Subcellular localization of yeast CDC46 varies with the cell cycle. *Genes Dev.* **4:** 2252–2263.

Hozak, P., A.B. Hassan, D.A. Jackson, and P.R. Cook. 1993. Visualization of replication factories attached to a nucleoskeleton. *Cell* **73:** 361–373.

Jackson, D.A. 1995. Nuclear organization: Uniting replication foci, chromatin domains and chromosome structure. *BioEssays* **17:** 587–591.

Jenkins, H., T. Holman, C. Lyon, B. Lane, R. Stick, and C. Hutchison. 1993. Nuclei that lack a lamina accumulate karyophilic proteins and assemble a nuclear matrix. *J. Cell Sci.* **106:** 275–285.

Kubota, Y. and H. Takisawa. 1993. Determination of initiation of DNA replication before and after nuclear formation in *Xenopus* egg cell free extracts. *J. Cell Biol.* **123:** 1321–1331.

Kubota, Y., S. Mimura, S. Nishimoto, H. Takisawa, and H. Nojima. 1995. Identification of the yeast MCM3 related protein as a component of *Xenopus* DNA replication licensing factor. *Cell* **81:** 601–609.

Leno, G.H. and R.A. Laskey. 1991. The nuclear membrane determines the timing of

DNA replication in *Xenopus* egg extracts. *J. Cell Biol.* **112:** 557–566.

Leno, G.H., C.S. Downes, and R.A. Laskey. 1992. The nuclear membrane prevents replication of human G2 nuclei but not G1 nuclei in *Xenopus* egg extract. *Cell* **69:** 151–158.

Madine, M.A., C.-Y. Khoo, A.D. Mills, and R.A. Laskey. 1995. An MCM3 complex is required for cell cycle regulation of DNA replication in vertebrate cells. *Nature* **375:** 421–424.

Méchali, M., F. Méchali, and R.A. Laskey. 1983. Tumour promoter TPA increases initiation of replication on DNA injected into *Xenopus* eggs. *Cell* **35:** 63–69.

Meier, J., K.H.S. Campbell, C.C. Ford, R. Stick, and C.J. Hutchison. 1991. The role of lamin L substructure in nuclear assembly and DNA replication, in cell free extracts of *Xenopus* eggs. *J. Cell Sci.* **98:** 271–278.

Mills, A.D., J.J. Blow, J.G. White, W.B. Amos, D. Wilcock, and R.A. Laskey. 1989. Replication occurs at discrete foci spaced throughout nuclei replicating in vitro. *J. Cell Sci.* **94:** 471–477.

Nakamura, H., T. Morita, and C. Sato. 1986. Structural organisations of replicon domains during DNA synthetic phase in the mammalian nucleus. *Exp. Cell Res.* **165:** 291–297.

Nakayasu, H. and R. Berezney. 1989. Mapping replicational sites in the eukaryotic cell nucleus *J. Cell Biol.* **108:** 1–11.

Newport, J. 1987. Nuclear reconstitution in vitro: Stages of assembly around protein-free DNA. *Cell* **48:** 205–217.

Newport, J., K.L. Wilson, and W.G. Dunphy. 1990. A lamin independent pathway for nuclear envelope assembly. *J. Cell Biol.* **111:** 2247–2259.

O' Keefe, R.T., S.C. Henderson, and D.L. Spector. 1992. Dynamic organisation of DNA replication in mammalian cell nuclei: Spatially and temporally defined replication of chromosome-specific α-satellite DNA sequences. *J. Cell Biol.* **116:** 1095–1110.

Schedl, P. and F. Grosveld. 1995. Domains and boundaries. In *Chromatin structure and gene expression* (ed. S.C.R. Elgin), pp. 172–191. IRL/Oxford University Press, Oxford, United Kingdom.

Sheehan, M.A., A.D. Mills, A.M. Sleeman, R.A. Laskey, and J.J. Blow. 1988. Steps in the assembly of replication-competent nuclei in a cell-free system from *Xenopus* eggs. *J. Cell Biol.* **106:** 1–12.

Van Holde, K., J. Zlatanova, G. Arents, and E. Moudrianakis. 1995. Elements of chromatin structure: Histones, nucleosomes and fibres. In *Chromatin structure and gene expression* (ed. S.C.R. Elgin), pp. 1–21. IRL/Oxford University Press, Oxford, United Kingdom.

Wessel, R., J. Schweizer, and H. Stahl. 1992. Simian virus 40 T-antigen DNA helicase is a hexamer which forms a binary complex during bidirectional unwinding from the viral origin of DNA replication. *J. Virol.* **66:** 804–815.

5

Mechanisms for Priming DNA Synthesis

Margarita Salas
Centro de Biología Molecular "Severo Ochoa" (CSIC-UAM)
Universidad Autónoma, Canto Blanco
28049 Madrid, Spain

Jennifer T. Miller and Jonathan Leis
Department of Biochemistry
Case Western Reserve University School of Medicine
Cleveland, Ohio 44106-4935

Melvin L. DePamphilis
National Institute of Child Health and Human Development
National Institutes of Health
Bethesda, Maryland 20892-2753

DNA replication is a semiconservative process in which a DNA polymerase uses one DNA strand as a template for the synthesis of a second, complementary, DNA strand. However, in contrast to RNA polymerases, which can initiate RNA synthesis on a DNA template de novo, all DNA polymerases require a preexisting primer on which to initiate DNA synthesis (Kornberg and Baker 1992). One apparent exception to this rule is a mitochondrial DNA (mtDNA)-encoded reverse transcriptase (RT) in *Neurospora* (Wang and Lambowitz 1993). Preexisting primers can be classified into four groups. The simplest primer consists of the 3′-hydroxyl (3′-OH) termini of DNA chains that are complementary to the DNA template and thereby form a stable duplex structure at the site where DNA synthesis begins. This primer is used for DNA repair (Friedberg and Wood, this volume), parvovirus DNA replication (Brush and Kelly; Cotmore and Tattersall; both this volume), some RTs. The second type of primer consists of a deoxyribonucleoside monophosphate that is covalently attached to a specific serine, threonine, or tyrosine residue of a protein. Examples are bacteriophage, plasmids, and animal viruses that replicate as a linear DNA genome, and animal viruses such as hepadnaviruses whose genome is partially double-stranded and partially single-stranded. The third type of primer consists of tRNA molecules that anneal to specific sequences in the RNA genomes of retroviruses where their 3′-OH termini are utilized by RT. The fourth class of primers consists of nascent RNA chains. These comprise nascent RNA transcripts that are processed to create a primer at a specific site in the template and short nascent oligoribonucleotides (initiator RNA) that are synthesized at many sites in the template and rapidly extended into short RNA-DNA primers by DNA polymerase-α:DNA primase (pol-α:primase). Nascent RNA transcripts are used during initiation of mtDNA replication, whereas replication forks in cellular chromosomes and double-stranded DNA (dsDNA) viral genomes that replicate within the nucleus use the initiator RNA mechanism.

INITIATION OF DNA SYNTHESIS ON PROTEIN PRIMERS

From prokaryotes to eukaryotes, linear DNAs exist in nature that cannot form circular structures through cohesive ends, hairpin structures at their termini, or concatemers during replication. In such cases, initiation of replication cannot take place by either RNA or DNA priming. However, most of these linear DNAs contain a protein covalently linked to their 5′ ends that acts as primer during initiation of DNA replication. This protein is called terminal protein (TP), and its role in DNA replication is summarized in Figure 1 and has been previously reviewed by Salas (1991).

The first evidence for the existence of protein attached at the ends of a linear dsDNA was the finding that the 19.3-kb virion DNA of *Bacillus subtilis* phage ϕ29 could be isolated as circular molecules and concatemers that were converted into unit-length linear DNA by treatment with proteolytic enzymes (Ortín et al. 1971). Such a treatment greatly reduced the capacity of ϕ29 DNA to transfect competent *B. subtilis* cells (Hirokawa 1972), and DNA isolated from a ϕ29 temperature-sensitive mutant in gene 3 was thermolabile for transfection (Yanofsky et al. 1976). Later it was shown that the 28-kD protein product of viral gene 3 was covalently linked at each 5′ end of the viral DNA (Salas et al. 1978). Other *B. subtilis* phages that contain linear dsDNA and TP of similar size fall into three serological classes: (1) ϕ15, PZA, and PZE in the group of ϕ29; (2) Nf, M2, and B103; and (3) GA-1. All phages of the ϕ29 family have a short inverted terminal repeat (ITR, Fig. 2), which is six nucleotides long for ϕ29, PZA, ϕ15, and B103 DNAs; eight nucleotides long for Nf and M2 DNAs; and seven nucleotides long for GA-1 DNA (Salas 1991). The linkage between ϕ29 TP and DNA is a phosphoester bond between the OH group of Ser-232 in TP and 5′-

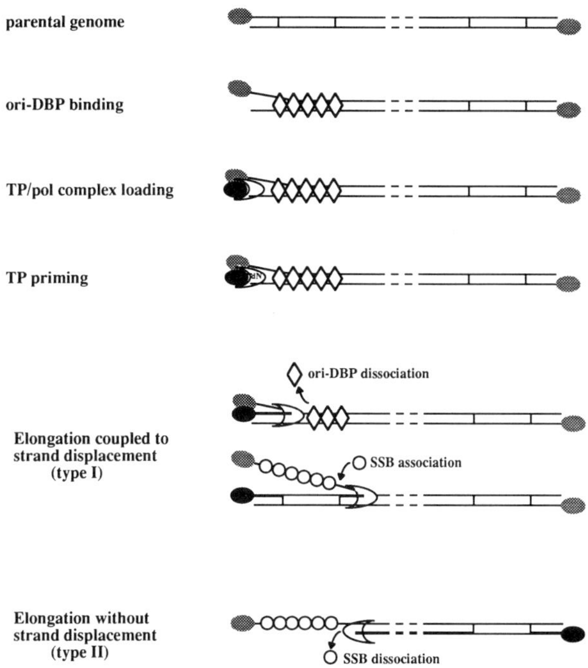

Figure 1 Model for protein priming of DNA replication. Only initiation starting at one DNA end is shown.

dAMP. Prediction of secondary structure in this region suggests that Ser-232 is located in a β turn, probably in the external part of the molecule, preceded by an α helix (Hermoso et al. 1985).

Replication of linear genomes starts at either DNA end, non-simultaneously, and proceeds by strand displacement toward the other end (Fig. 1). A dsDNA-binding protein (ori-DBP) binds to sequences close to the DNA ends, forming a nucleoprotein complex that helps to open the DNA ends to facilitate the interaction of a heterodimer that forms between TP and the DNA polymerase encoded by the linear genome. Initiation of replication occurs at either DNA terminus by the covalent linkage of a specific dNMP to the OH group of a specific serine, threonine, or tyrosine residue on TP, in a reaction catalyzed by the DNA polymerase and directed by an internal nucleotide at the 3′ end of the template. After sliding-back (or jumping-back) (see below), elongation occurs coupled to strand displacement giving rise to type I molecules, with the likely dissociation of ori-DBP and the association of single-stranded DNA-binding protein (SSB) to the displaced ssDNA. Type I molecules are unit-length linear dsDNA with one or more single-

B. subtilis phages	φ29	3' TTTCATCCCATG
	φ15	3' TTTCATCCCATG
	PZA	3' TTTCATCCCATG
	Nf	3' TTTCATTCCAAG
	M2	3' TTTCATTCCAAG
	B103	3' TTTCATCCCATG
	GA1	3' TTTATCTAAGGG
E. coli phage	PRD1	3' CCCCTATGCACG
S. pneumoniae phage	Cp-1	3' TTTCGTACATGA
Eukaryotic virus	Adeno 2	3' GTAGTAGTTATT
Linear plasmids	S1	3' TTTTCATATGTT
	S2	3' TTTTCATATGTT
	pSKL	3' TTTTCCATATCT
	pGKL1	3' TGTGTATTGTAT
	pGKL2	3' TTTTCCATATAT
	pMC3-2	3' CAACACATATGG
	Kalilo	3' CACATTCCCCGT
	Maranhar	3' XXCCCCGTAGCC
	pSCP1	3' (CC)CCCCGCCTCT
	pSCL	3' GGGCGCCTCGCC
	pSLP2	3' GGGCGCCTCGCC
	pSLA2	3' GGGCGCCTCGCC
Bacterial chromosome	*S. lividans*	3' GGGCGCCTCGCC

Figure 2 The 3′-terminal nucleotide sequences of TP-containing DNA genomes.

stranded tails from the same or a different DNA end. Type II molecules are then formed in which elongation occurs without strand displacement, with the concomitant dissociation of the SSB protein. Type II molecules are unit-length linear molecules partially double-stranded and partially single-stranded. Thus, two fully replicated molecules result from initiation at each DNA end.

Requirements for Initiation of DNA Replication in the Phage φ29 Genome

To test whether a fully displaced parental DNA strand can be replicated in vivo, recombinant φ29 DNA molecules containing parental TP at only one DNA end were constructed. No replication in *B. subtilis* protoplasts was obtained, suggesting that the fully displaced DNA strand is not a template for replication (Escarmís et al. 1989). When the above recombinant φ29 DNA molecules were used as templates in the in vitro system (see below), no type II molecules were found (Gutiérrez et al. 1991a). Moreover, when replication of φ29 TP-DNA was studied in the in vitro system, a significant amount of type II replicative intermediates was found at an incubation time at which no synthesis of full-length φ29 DNA was detected. These results indicate that the appearance of type II replicative intermediates does not require synthesis of full-length DNA and full displacement of the parental strand; rather, the results support a model in which initiation of replication can occur from both DNA ends, and type II molecules are produced by separation of the two displacement forks when they meet.

When extracts from φ29-infected *B. subtilis* were incubated with [α-^{32}P]dATP in the presence of φ29 TP-DNA, a ^{32}P-labeled protein with the electrophoretic mobility of the TP, product of the viral gene 3, was found (Peñalva and Salas 1982; Shih et al. 1982). Incubation of the labeled protein with piperidine released 5′-dAMP, indicating the formation of a TP-dAMP covalent complex (Peñalva and Salas 1982). A covalent complex between the M2 TP and 5′-dAMP was also found when extracts from phage M2-infected *B. subtilis* were used (Matsumoto et al. 1983).

Phage φ29 genes 2 and 3 and phage M2 genes G and E were shown to be essential for the in vitro initiation reaction, that is, the formation of the TP-dAMP covalent complex using the corresponding TP-DNA as template (Blanco et al. 1983; Matsumoto et al. 1983). Phage φ29 genes 2 and 3 were cloned, and both proteins, p2 (Blanco and Salas 1984) and p3 (TP) (Prieto et al. 1984; Watabe et al. 1984a), were overproduced and purified in a functional form. Purified p2, in addition to catalyzing the initiation reaction, has DNA polymerase activity (Blanco and Salas 1984; Watabe et al. 1984a) and 3′→5′ exonuclease activity, believed to be involved in proofreading (Watabe et al. 1984b; Blanco and Salas 1985a), which is about 10-fold more active on single- than on double-stranded DNA (Garmendia et al. 1992). φ29 DNA polymerase can also catalyze TP-deoxynucleotidylation in the absence of DNA template (Blanco et al. 1992).

The in vitro initiation reaction is greatly stimulated by NH_4^+ ions (Blanco and Salas 1985b), which stabilize the formation of a heterodimer between DNA polymerase and TP (Blanco et al. 1987), required for the initiation of replication. In fact, the two proteins are purified as a complex when extracts from φ29-infected cells are used (Watabe et al. 1983; Matsumoto et al. 1984). In addition, Mn^{++} ions are better activators of the initiation reaction than Mg^{++} due to a 50-fold decrease in the K_m for dATP (Esteban et al. 1992).

Once the viral DNA polymerase catalyzes the covalent linkage of dAMP to the OH group of Ser-232 in the TP, the same DNA polymerase processively elongates the nascent DNA strand in vitro to give full-length ϕ29 DNA (Blanco and Salas 1985b; Blanco et al. 1989). Further studies using primed M13 DNA as template indicated that the ϕ29 DNA polymerase is a highly processive enzyme (over 70 kb) that can catalyze strand displacement without need of accessory proteins (Blanco et al. 1989). Other viral proteins required for ϕ29 DNA replication are the origin-binding protein p6 and SSB p5 (see below).

Using the four ϕ29 replication proteins: TP, DNA polymerase, origin-binding protein p6, and SSB p5, and limiting amounts of ϕ29 TP-DNA, it is possible to amplify in vitro ϕ29 DNA by a factor of 1000-fold (Blanco et al. 1994).

Interaction of ϕ29 DNA Polymerase and Terminal Protein

Mutants obtained in each putative active-site residue at the Exo I, Exo II, or Exo III motifs of ϕ29 DNA polymerase, located at the amino end, did not impair protein-primed initiation and DNA polymerization; however, exonuclease activity was strongly reduced in all cases (Bernad et al. 1989; Soengas et al. 1992; Esteban et al. 1994). Unexpectedly, these mutant proteins were almost inactive when assayed for ϕ29 TP-DNA replication. This defect was shown to be due mainly to a 10- to 50-fold decrease in the rate of DNA synthesis coupled to strand displacement. Therefore, the strand-displacement activity of the ϕ29 DNA polymerase resides in the amino-terminal domain, probably overlapping with the 3′→5′ exonuclease active site (Soengas et al. 1992). Site-directed mutagenesis of the conserved motifs in the carboxy-terminal portion of ϕ29 DNA polymerase indicated that this domain of the ϕ29 DNA polymerase contains the protein-primed initiation and DNA-polymerization activities of this enzyme (Blanco and Salas 1995). Assuming that the ϕ29 DNA polymerase structure is similar to that of the Klenow fragment of *Escherichia coli* DNA polymerase I (Ollis et al. 1985; Blanco et al. 1991), the polymerase DNA-binding cleft is proposed to be also the TP-binding site (Salas et al. 1993). Thus, the TP molecule bound to the DNA polymerase cleft could place the specific priming residue (Ser-232), acting as OH donor, next to the dNTP-binding site.

Deletion mutagenesis studies in the ϕ29 TP indicated the existence of two DNA polymerase-binding regions, located at positions 72–80 and 241–261, and three DNA-binding regions, at positions 13–18, 30–51, and 56–71 (Salas 1991). Site-directed mutagenesis showed that changing Ser-232 to a threonine inactivated the protein, whereas changing it to cysteine reduced the priming activity to about 0.7% of wild-type TP.

Changing Leu-220, Ser-223, and Ser-226 independently into proline resulted in mutant proteins with 3%, 140%, and 1%, respectively, of wild-type TP priming activity. All of the mutant TPs could interact with DNA polymerase and DNA, suggesting that Leu-220 and Ser-226, in addition to Ser-232, form part of a functional domain involved in initiation of DNA replication (Salas 1991).

The amino acid motif RGD, characteristic of cell adhesion proteins, is present in the TP of phage ϕ29, and it was proposed to be involved in the interaction of the TP/DNA polymerase heterodimer with the parental TP bound to the DNA (Kobayashi et al. 1991a,b). In addition, a sequence very similar to KKGCPPDD, found in the β subunit of the fibronectin receptor protein, is also present in ϕ29 TP. Analysis of synthetic peptides suggested that this sequence acts as a receptor in the parental TP, whereas the RGD sequence acts as an effector in the new TP primer.

ϕ29 Origin of DNA Replication

Treatment of ϕ29 TP-DNA with proteinase K inactivates it as a template for DNA replication. However, when the residual peptide that remains attached to the 5′ end is removed with piperidine, template activity is restored, although it is 5- to 10-fold less than that obtained with ϕ29 TP-DNA. The minimal ϕ29 replication origins are located within the terminal 12 bp at each DNA end (Salas 1991). Template activity is also obtained with single-stranded oligonucleotides 12 bases long corresponding to the 3′-terminal sequence at the right or left ϕ29 DNA ends, giving rise to formation of a TP-dAMP initiation complex that can be fully elongated (Méndez et al. 1992). Interestingly, deoxynucleotidylation of the TP can also be obtained in the absence of any template, although in this case, any of the four dNTPs work, and the affinity for the nucleotide is greatly decreased (Blanco et al. 1992). Therefore, the template provides nucleotide specificity and increases the affinity of TP for its DNA polymerase.

ϕ29 Protein p6 and Formation of a Nucleoprotein Complex

Protein p6 is a 123-amino-acid protein that stimulates initiation of ϕ29 DNA replication by reducing the K_m for dATP and facilitating the transition from initiation to elongation (Blanco et al. 1986, 1988). Stimulation by p6 is due to formation of a nucleoprotein complex that spans 200–300 bp from each DNA end (Prieto et al. 1988; Serrano et al. 1989). p6 binds preferentially the ϕ29 replication origins at nucleotides 46–68 at the left end of ϕ29 DNA and nucleotides 62–125 at the right end (Serrano et al. 1989). These regions do not show any sequence similarity, but they contain DNA sequences predicted to be bendable every 12 bp, suggesting that bendability may be the major determinant for protein p6 recognition (Serrano et al. 1989). In agreement with this, tandem repetitions of a 24-bp bendable sequence, present in one of the main p6 recognition regions, bind protein p6 in the same positions and with higher affinity than the ϕ29 DNA replication origins (Serrano et al. 1993).

An α-helical structure located in the amino-terminal region of protein p6 is involved in binding DNA through its minor groove (Freire et al. 1994). A p6 dimer binds 24 bp, bending or kinking the DNA every 12 bp. Binding is highly cooperative, giving rise to a large multimeric complex in which a right-handed superhelix wraps around a protein core (Serrano et al. 1990), restraining positive supercoils when formed on a covalently closed plasmid (Prieto et al. 1988). p6 binding results in a 4.2-fold compaction of the DNA in which one superhelical turn has 63 bp (2.6 protein p6 dimers) with a pitch of 5.1 nm and a diameter of 6.6 nm. Therefore, the DNA should be bent 66° every 12 bp and underwound 11.5 bp per turn (Serrano et al. 1993). These features could facilitate the initial unwinding of DNA required to start replication by the DNA polymerase. In agreement with this, activation by p6 is greatest at lower temperatures (Salas et al. 1993).

ϕ29 Protein p5

Protein p5 is a 13-kD SSB that stimulates ϕ29 DNA replication (Salas 1991). Stimulation does not result from an increase either in formation of the TP-dAMP initiation complex (Martín et al. 1989) or in DNA elongation rate (Gutiérrez et al. 1991b). p5 binds nonspecifically to ssDNAs, including the ssDNA portions of replicative intermediates produced during ϕ29 DNA replication in vitro (Gutiérrez et al. 1991b). Each p5 monomer covers 3–4 nucleotides with a binding constant of 10^5 M^{-1} and an unlimited cooperativity parameter of 50–80 (Soengas et al. 1994). p5 can facilitate removal of secondary structure in the displaced ssDNA in replicative intermediates and displace oligonucleotides annealed to ssDNA (Soengas et al. 1995). Therefore, in addition to protecting the ssDNA produced during ϕ29 DNA replication from nuclease degradation (Martín et al. 1989) and preventing unproductive binding of ϕ29 DNA polymerase to ssDNA (Gutiérrez et al. 1991b), ϕ29 SSB could help to unwind secondary structure that may form in the displaced ssDNA during ϕ29 DNA replication. Consistent with this hypothesis, the rate of ϕ29

DNA elongation by φ29 DNA polymerase mutants defective in strand displacement is stimulated about 5-fold by addition of φ29 SSB (Soengas et al. 1995).

The interaction of the proteins and replication origin described above for φ29 DNA replication is summarized in Figure 1. In many respects, φ29 DNA replication is a paradigm for replication of other linear DNA genomes.

Protein-primed DNA Replication in Other Genomes

Phage PRD1

Phage PRD1 infects gram-negative bacteria including *E. coli* and *Salmonella typhimurium*. The 5′ termini of the 14.9-kb PRD1 linear genome are linked to a 28-kd TP by a phosphoester bond between dGMP and tyrosine residue 190 (Bamford et al. 1983; Bamford and Mindich 1984; Shiue et al. 1991). DNA of PRD1 and closely related phages PR3, PR4, PR5, PR722, and L17 all have a 110-bp-long ITR (Mindich and Bamford 1988). TPs from PRD1, φ29, PZA, Nf, adenovirus, and the core antigen of duck hepatitis B virus all contain the conserved motif Tyr-Ser-Arg-Leu-Arg-Thr (Hsieh et al. 1990).

PRD1 DNA replication appears quite similar to that of φ29. It requires viral genes I (DNA polymerase) and VIII (TP) (Hsieh et al. 1987; Jung et al. 1987; Savilahti and Bamford 1987) in addition to some host components. Purified DNA polymerase catalyzes the formation of TP-dGMP covalent complex, has processive DNA chain-elongation activity, can catalyze strand displacement, and contains 3′→5′ exonuclease activity (Savilahti et al. 1991; Zhu and Ito 1994). As with φ29 DNA polymerase, initiation is strongly activated by Mn^{++} (Caldentey et al. 1992). PRD1 DNA polymerase can also catalyze TP-deoxynucleotidylation in the absence of DNA template (Caldentey et al. 1992). Mutagenesis of conserved Lys-340 in the Kx_3NSxYG motif of PRD1 DNA polymerase generated a protein that had lost protein-priming and polymerization activities without affecting the 3′→5′ exonuclease activity (Zhu et al. 1994). Mutagenesis in residue Arg-174 of the PRD1 TP, corresponding to the conserved motif YSRLRT, resulted in an inactive TP that was unable to form an initiation complex (Hsieh et al. 1990). As with φ29, linear duplex, protein-free, DNA molecules containing the 20 bp from the PRD1 DNA ends can undergo replication by protein priming in vitro. Similarly, a 27-mer single-stranded oligonucleotide containing the 20 bases of the 3′ end of the PRD1 genome supported the formation of the TP-dGMP complex (Yoo and Ito 1991).

Phage Cp-1

The 5′ termini of the 19.3-kb DNA of *Streptococcus pneumoniae* phages Cp-1, Cp-5, and Cp-7 are covalently linked to a 28-kD TP and contain ITRs of 236 bp, 343 bp, and 347 bp, respectively (García et al. 1983; Salas 1991). The amino acid sequence of Cp-1 TP is 71% homologous to that of ϕ29 (Martín et al. 1995) and contains a phosphoester bond between threonine and 5′-dAMP that can be elongated in vitro (García et al. 1986). Phage Cp-1 DNA polymerase is 96% homologous to that of phage ϕ29 (Martín et al. 1995). The 40-kb DNA of HB3 and related phages HB-623 and HB-746 (Romero et al. 1990) carries a 23-kD TP whose role in DNA replication remains to be determined, although the invertron model has also proposed a role in integration (Sakaguchi 1990).

Linear Plasmids

A variety of linear plasmids have been isolated from bacteria, yeast, fungi, and higher plants. In most cases, long ITRs have been characterized, and in many of them, evidence for the existence of a TP linked at the 5′ ends of the DNA has been reported (Meinhardt et al. 1990; Salas 1991; Rohe et al. 1992). Linear DNA of yeast killer plasmids pGKL1 and pGKL2 replicates by a strand-displacement mechanism similar to that described for phage ϕ29 (Fujimura et al. 1988). Extracts from cells carrying pGKL1 contain an activity called terminal region recognition factor 1 (TRFI) that recognizes the termini of both pGKL1 (bp 107–183 within the ITR) and pGKL2 (bp 126–179 within the ITR) (McNeel and Tamanoi 1991). This binding protein could be similar in function to protein p6 and factors NFI and NFIII that bind to the replication origins of phage ϕ29 and Ad, respectively, and stimulate the initiation of replication. Furthermore, ORF1 in plasmid pGKL1 and ORF2 in pGKL2 contain regions of amino acid homology located at the carboxyl region of eukaryotic-type DNA polymerases, including the protein-primed DNA polymerases. By analogy with linear viral genomes, it is likely that the TPs of pGKL1 and pGKL2 are plasmid-encoded (Salas 1991).

Replication of plasmid pAI2 from *Ascobolus immersus* also starts at the DNA ends. In addition, a large open reading frame (ORF) spanning 1202 amino acids shares homology with eukaryotic-type DNA polymerases and with putative DNA polymerases from linear plasmids (Kempken et al. 1989). ORF1 from the *Claviceps purpurea* plasmid pClK1 encodes a protein of 1097 amino acids, which is also likely to be a DNA polymerase according to the above criteria (Oeser and Tudzynski 1989).

The gene encoding TP is unknown. However, taking into account that the amino acid involved in the linkage of ϕ29 and Ad TPs to DNA is located in a β turn preceded by an α helix (Hermoso et al. 1985), it is possible that the TP of pClK1 is encoded by 400 amino acids at the amino-terminal part of ORF1. In fact, serine residue 327 of this ORF was reported to meet the criteria for a nucleotide-linking site (Oeser and Tudzynski 1989). ORF3 of plasmid S1 also codes for a protein with amino acid homology with eukaryotic-type DNA polymerases (Paillard et al. 1985). A similar case is that of ORFs of plasmids pEM from *Agaricus bitorquis*, pMC3-2 from *Morchella conica*, and the *kalilo* and *maranhar* plasmids from *Neurospora* (for review, see Rohe et al. 1992). All the above data suggest that the linear plasmids replicate by a protein-priming mechanism.

Linear Chromosomes

Two copies of a DNA sequence similar or identical to the right end of the linear plasmid pSLP2 were found at the ends of the 8-Mb *Streptomyces lividans* linear chromosome. The telomeres contain a 25-kb ITR and carry covalently bound protein. The TP is removed by piperidine treatment, suggesting that, like phage ϕ29, the linkage to DNA occurs through a serine residue in the *S. lividans* TP (Lin et al. 1993). Chromosomal DNA of six other *Streptomyces* species also behave as linear molecules of about 8 Mb, suggesting that chromosomal linearity may be common among the streptomycetes (Lin et al. 1993).

A functional *oriC* also has been located at the center of the *S. lividans* chromosome (Zakrzewska-Czerwinska and Schrempf 1992). Since this chromosome can be circularized by joining the two ends by artificial targeted recombination or by spontaneous deletions spanning both telomeres, it appears to exist as either a linear or a circular molecule (Lin et al. 1993). The linear chromosome appears to replicate bidirectionally from its center. Presumably, replication of the ends is completed by protein-primed DNA replication. It has been postulated that the more primitive bacterial chromosomes replicated from the telomeres, and that the complex machinery for internal initiation came later. It is an open question whether bacterial chromosomes that replicate exclusively from the telomeres still exist in bacteria (Chen et al. 1995).

Adenovirus

The human adenovirus (Ad) genome consists of linear, dsDNA of 36 kb with ITRs of about 100 bp (Steenbergh et al. 1977) and a protein of 55 kD covalently linked at the 5′ ends of the DNA (Carusi 1977; Rekosh et al. 1977). The Ad TP is synthesized as a precursor, the preTP (pTP) (Challberg et al. 1980), and pTP forms a phosphoester bond between serine residue 580 and 5′-dCMP (Challberg et al. 1980; Desiderio and Kelly 1981; Smart and Stillman 1982).

Initiation of Ad DNA replication occurs at either DNA end and proceeds by a strand-displacement mechanism giving rise to type I and type II molecules, as described above for phage φ29 DNA replication (Brush and Kelly; Hay; both this volume). A model first proposed by Rekosh et al. (1977) suggested that a free molecule of the TP could act as a primer for the initiation of replication by formation of a covalent linkage with dCMP, the 5′-terminal nucleotide, that would provide the 3′-OH group needed for elongation by the DNA polymerase. The 5′ end of each nascent daughter strand is covalently linked to a protein of 80 kD (Challberg et al. 1980; Lichy et al. 1981), structurally related to the 55-kD protein covalently linked to the 5′ ends of Ad DNA. The 80-kD pTP is processed to the 55-kD TP during virus maturation (Challberg and Kelly 1981). As with φ29 and PRD1 DNA replication, pTP and a 140-kD Ad DNA polymerase are required for Ad DNA replication. pTP and Ad DNA polymerase form a heterodimer that binds to the core origin comprising the 20 terminal nucleotides (Temperley and Hay 1992). Initiation of replication occurs by DNA polymerase-catalyzed transfer of dCMP onto the OH group of Ser-580 in pTP. Ad DNA polymerase, like φ29 and PRD1 DNA polymerases, has, in addition to its initiation and polymerization activities, 3′→5′ exonuclease activity presumably involved in proofreading (Lindenbaum et al. 1986). In most of the mutants obtained, both initiation and polymerization activities were lost, indicating that these functions might be closely linked, as is the case with the φ29 DNA polymerase. The amino terminus of pTP is essential for priming activity and DNA binding (Pettit et al. 1989).

Another viral protein, the DNA-binding protein (DBP), binds ssDNA and dsDNA and stimulates the initiation of replication. Elongation of Ad DNA replication catalyzed by Ad DNA polymerase proceeds by strand displacement and also requires DBP. Ad DBP has properties that it shares with both p6 (dsDNA-binding protein) and p5 (SSB) of phage φ29. Like p6, DBP binds to the replication origins forming a nucleoprotein complex and stimulates the initiation of replication by decreasing the K_m for the initiating dNTP. Like p5, DBP binds to ssDNA and has unwinding activity. NFI and NFIII/Oct-1 are transcription factors that bind to the auxiliary region of the Ad replication origin, where they enhance formation of a stabilized nucleoprotein structure with the three viral proteins (van der Vliet, this volume).

Hepadnaviruses: Reverse Transcription

Hepadnaviruses contain a 3-kb circular, dsDNA genome in which the minus (uncoding) strand is complete with unique 5′ and 3′ termini. The plus strand is incomplete with a unique 5′ end and a heterogeneous 3′ end (Seeger and Mason, this volume). Three members of this virus family (human hepatitis B, ground squirrel hepatitis, and duck hepatitis B) contain a protein covalently linked to the 5′ end of the complete minus strand (Gerlich and Robinson 1980; Molnar-Kimber et al. 1983; Weiser et al. 1983).

The mechanism by which hepadnavirus DNA replicates involves reverse transcription (Seeger and Mason, this volume). The RNA template for reverse transcription, the pre-genome, is produced by copying the minus-strand DNA in the closed circular dsDNA molecule formed from the virion DNA (Summers and Mason 1982; Ganem and Varmus 1987; Seeger et al. 1991). The polymerase gene product of hepadnaviruses encodes the TP at the amino-terminal quarter of this 785-amino-acid polypeptide, the DNA polymerase/RT in the central part, and the RNase H at the carboxy-terminal region. Furthermore, TP and RT domains are linked by a tether region that tolerates amino acid substitutions and deletions (Bartenschlager and Schaller 1988; Chang et al. 1990; Radziwill et al. 1990). Tyr-96 donates the OH group for the formation of the covalent bond between the polymerase gene product and viral DNA (Zoulim and Seeger 1994). Interestingly, Tyr-96 is part of a Gly-x-Tyr motif present in the TP of picornavirus at the site where the protein is linked to the RNA (Khudyakov and Makhov 1989). Thus, unlike other RTs that use tRNA as primers, hepatitis B virus RT uses a protein as primer. However, in contrast to other genomes that use proteins to prime DNA synthesis, DNA priming and polymerase activities in hepadnaviruses reside on the same polypeptide.

Sliding-back (Jumping-back) Mechanism for the Transition from Initiation to Elongation and for the Maintenance of the DNA Ends

Use of single-stranded homopolymers in the in vitro ϕ29 replication system led to the conclusion that TP priming is a template-directed event (Méndez et al. 1992). Therefore, a mutational analysis of the ϕ29 DNA right replication origin was carried out using as templates 12-mer single-stranded oligonucleotides containing the 3′end of the natural ϕ29 DNA sequence (TTTCAT.......) or mutant derivatives with single changes in the first, second, or third T, to determine which T residue directs the formation of the TP-dAMP initiation complex. The results obtained clearly indicate that the second nucleotide of the template directs the linkage of dAMP to the primer protein (Méndez et al. 1992). This unexpected result was confirmed using a variety of single-stranded oligonucleotides and also dsDNA fragments containing the natural ϕ29 DNA left replication origin, or with a mutation at the second nucleotide from the end. Addition of the origin-binding protein p6 to the dsDNA fragments gives similar results, indicating that the initiation site of ϕ29 DNA replication is the second 3′-terminal nucleotide. The physiological role of this internal initiation event is supported by the fact that all the nucleotides in the template, including the 3′-terminal one, are replicated. Moreover, a terminal repetition of at least two nucleotides is required for efficient elongation of the initiation complex.

A sliding-back mechanism was proposed for the transition from initiation to elongation (Fig. 3) (Méndez et al. 1992). Once the TP-dAMP initiation complex has been formed, directed by the second nucleotide (T) at the 3′ end, the TP-dAMP complex slides backward, locating the dAMP in front of the first nucleotide of the template (asymmetric translocation). Then the next nucleotide (A) is incorporated to the TP-dAMP complex, using again the second T of the template as a director. Further nucleotide addition involves normal translocation of both template and DNA primer terminus (normal elongation).

This strategy for maintaining the integrity of the ϕ29 DNA ends could also increase the fidelity of the TP-primed initiation reaction (Esteban et al. 1993). If a mismatched initiation event occurs, transition to elongation would not be efficient, and the incorrect initiation complex could dissociate from the DNA. Concomitant removal of both TP and the mismatched nucleotide may be the only possibility for editing, because the exonucleolytic activity of ϕ29 DNA polymerase, which acts as a proofreading enzyme during elongation (Garmendia et al. 1992), cannot excise the first dNMP linked to the TP (Esteban et al. 1993). Nonetheless, if a mutation is established in the first nucleotide of the DNA, the second 3′ nucleotide of the template would be used again, restoring the terminal repetition at the end of the molecule. Therefore, according to the sliding-back model, errors in the first replication event would be lost because the 3′-terminal nucleotide is not used as the initiation site.

A similar strategy was employed to show that in PRD1 DNA, the fourth base from the 3′ end of the template directs, by base complementarity, the dNMP to be linked to the TP in the initiation reaction (Fig. 3) (Caldentey et al. 1993). Thus, phage PRD1 maintains its 3′-end DNA sequences via a sliding-back mechanism. Unlike TP-DNA, the

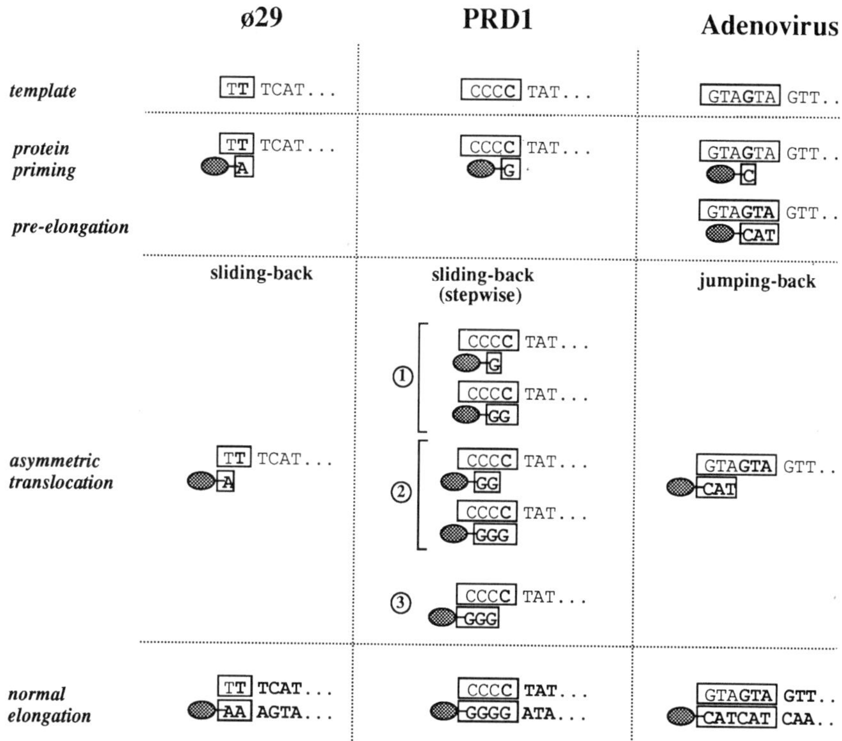

Figure 3 Sliding-back (jumping-back) model for the transition from initiation to elongation.

ssDNA templates could not be elongated by TP and DNA polymerase in vitro. Nevertheless, analysis of the transition products obtained with TP-DNA and origin-containing oligonucleotides suggest that sliding-back occurs stepwise, the fourth base being the directing position during the entire process.

In the Ad genome, the terminal sequence of the template strand has the more complex reiteration 3′-GTAGTA. Changes in the first G residue do not affect the formation of the pTP-dCMP complex, whereas additional substitution of the G residue at position four drastically impairs this reaction (Dobbs et al. 1990). Therefore, the G residue at position four could act as the template nucleotide during formation of the pTP-dCMP complex (Méndez et al. 1992). Recently, it was reported that a kinetic barrier to further elongation of the Ad pTP-CAT product is relieved by a high dCTP concentration (Mul and van der Vliet 1993). This is consistent with a rate-limiting, sliding-back step in which the pTP-CAT product, initiated from the fourth template residue, slides back in a single jump to be paired with the first three bases of the template, thus regenerating the Ad DNA ends (Caldentey et al. 1993). Recent results (King and van der Vliet 1994) support the model in which the pTP-CAT intermediate, synthesized opposite to positions 4–6, jumps back to positions 1–3 of the template to start elongation (Fig. 3). This jumping-back mechanism ensures the integrity of terminal sequences during replication of the linear genome.

Other linear genomes that contain TP have some sequence repetition at the DNA ends (Fig. 2), suggesting that internal initiation followed by sliding-back either in one step (phage ϕ29), stepwise (phage PRD1), or in a single jump (Ad) could be applicable to other systems that use a protein primer. This mechanism may also apply to RNA replication. It has been suggested that initiation of Tacaribe arenavirus RNA replication occurs at the second 3′-terminal nucleotide, and that the initiation complex slips backward before elongation can continue (Garcin and Kolakofsky 1992). Another example may be hepatitis B virus, where the polymerase binds to an RNA hairpin that serves as a template for formation of a short DNA primer covalently linked to protein. Following its synthesis, the nascent DNA strand apparently dissociates from its template and reanneals with complementary sequences at the 3′ end of the RNA genome where DNA synthesis continues (Wang and Seeger 1993).

INITIATION OF DNA SYNTHESIS ON PREFORMED RNA PRIMERS

Retroviruses, as well as retroelements, recruit preexisting host-encoded tRNAs to prime DNA synthesis by RT (Weiss et al. 1985). Various tRNA species are utilized, depending on the retrovirus (Leis et al. 1993); $tRNA^{Trp}$ is used by avian sarcoma leukosis viruses and $tRNA^{Pro}$, $tRNA^{Lys,1,2}$, and $tRNA^{Lys,3}$ are used by a variety of murine, feline, simian, or human viruses.

Encapsidation of tRNA Primers by Retroviruses

Specific tRNA primers are encapsidated into virions through interactions with RT (Panet et al. 1975; Levin and Siedman 1979, 1981; Peters and Hu 1980). These interactions involve the TψC and the DHU arms of the tRNA (Hu and Dahlberg 1983; Barat et al. 1991). Furthermore, the recognition of the primer occurs at the level of the virus-encoded *gag-pol* precursor rather than that of free RT, since viral protease activation is not necessary for selective encapsulation of primers into virions (Crawford and Goff 1985; Stewart et al. 1990; Mak et al. 1994). The presence of modified bases in the tRNA also does not appear to be important either for binding or functionality as a primer (Aiyar et al. 1992; Wohrl et al. 1993; Huang et al. 1994).

Minus-strand Priming during Retrovirus Replication

Annealing of the primer to viral RNA is catalyzed by RT (Barat et al. 1989; Aiyar et al. 1992). The viral nucleocapsid protein may stimulate this process, but it is not necessary (Aiyar et al. 1994). The site to which the primer binds is in the 5′-untranslated region of viral RNA referred to as the primer binding site (PBS) (Weiss et al. 1985). The acceptor stem of the tRNA is unwound, and between 14 and 22 nucleotides (Weiss et al. 1985) form a base-paired duplex with the PBS (Fig. 4). The viral RNA 5′ of the PBS is termed U5, and the RNA 3′ of the PBS is termed leader. Complete base-pairing between the PBS and the primer tRNA is not necessary for priming of DNA synthesis. As many as 9 of the 18 base pairs between the human immunodeficiency virus (HIV-1) PBS and tRNA ^{Lys3} (Wakefield et al. 1994) and 3 of the 18 base pairs between the Rous sarcoma virus (RSV) PBS and tRNA^{Trp} (Aiyar et al. 1994) can be mismatched without abrogating primer function, provided that the mismatches do not interfere with the base-pairing at the 3′ end of the primer. Li et al. (1994) have shown that mutations that replace the HIV-1 PBS with sequences complementary to the closely related tRNA^{Lys1,2} or tRNA^{Phe} are biologically functional, although they produce viruses with delayed growth properties. When one examines the complement of tRNAs packaged into these mutant virions, one finds that they do not differ from wild type. This indicates that the PBS is not involved in selective packaging of tRNAs into virions. Subsequent rounds of infection by these mutant viruses result in reversion of the PBS to wild-type sequences.

RNA Secondary Structure Influences Initiation of Minus-strand DNA Synthesis during Reverse Transcription

Although there is significant base-pairing between the primer tRNA and the PBS sequence in viral RNA, the ability to utilize the primer efficiently involves an additional set of viral RNA secondary structures. Such structures are best defined in avian retroviruses (Cobrinik et al. 1988, 1991; Aiyar et al. 1992, 1994). The potential secondary structures of nucleotides 56–130 around the PBS of avian sarcoma/leukosis virus RNA is depicted in Figure 4. As shown, there is a potential to form base

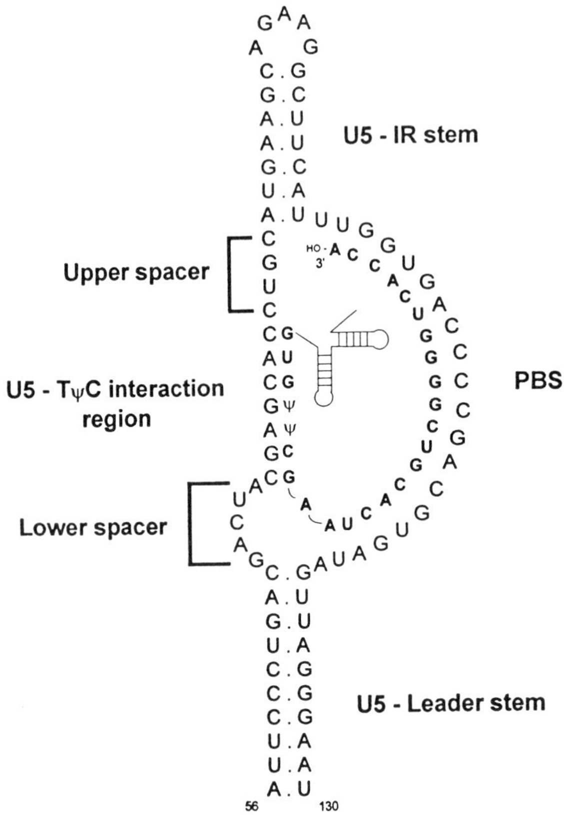

Figure 4 A linear representation of the 5′ terminus of avian sarcoma/leukosis virus RNA. The U5 and leader regions are as indicated. A $tRNA^{Trp}$ primer is shown schematically annealed to the primer binding site (PBS). Two sets of inverted repeat sequences that flank the PBS are represented by base-paired regions termed the U5-IR stem and the U5-leader stem. An interaction with the TψC arm of the primer and U5 RNA is also shown. Non-base-paired regions between the various duplex elements are shown by the square brackets.

pairs in the viral RNA between inverted repeat sequences near the PBS called the U5-leader and U5-IR stems. If either duplex is disrupted by deletion or base substitutions, partial defects to initiation of reverse transcription result and viruses containing these mutations grow more slowly (Cobrinik et al. 1988, 1991; Aiyar et al. 1992, 1994). Mutations to the U5-IR stem also result in viral integration defects, since these sequences at the level of DNA form part of the 5′ long terminal repeat (LTR) integrase recognition site (Cobrinik et al. 1991). In addition to the U5-IR and U5-leader stems, an interaction between 7 nucleotides of the TψC arm of the $tRNA^{Trp}$ primer and sequences in U5 lying between the stem structures is also necessary for efficient initiation of reverse transcription (Fig. 4) (Aiyar et al. 1992). Interactions of this type were originally proposed by Haseltine et al. (1977) and suggested by the data of Cordell et al. (1979).

Each of the above RSV duplex elements is separated by short non-base-paired spacers (Fig. 4). Small insertions or deletions, but not base substitutions, within these regions cause decreases in initiation of reverse transcription in vitro (Aiyar et al. 1994). These data suggest a need to maintain a specific orientation of these RNA structures with respect to one another, and this is further supported by the studies of Olson et al. (1992), who placed a 4-nucleotide insertion into U5 RNA between the U5-leader stem and the U5-TψC interaction site in a reticuloendotheliosis virus-based plasmid vector. After one round of replication, the majority of replication-competent viruses contained viral RNA with wild-type U5 sequences. The biological reason for these structures and their specific spatial orientation is not fully understood, but they may be required by RT to properly recognize and form the initiation complex. For a more detailed review of these secondary structure interactions and the effect of mutations that disrupt their potential base-pairing, the reader is referred to Leis et al. (1993).

Several other retroviruses are proposed by sequence analysis to form similar secondary structures in viral RNA. In HIV-1 (Baùdin et al. 1993), HIV-2 (Berkhout and Schoneveld 1993), and Moloney murine leukemia virus (Mo-MLV) (Mougel et al. 1993), the existence of the U5-IR and U5-leader stems has been substantiated by direct chemical and enzymatic probing. Additionally, Murphy and Goff (1989) demonstrated for Mo-MLV that a deletion mutant in the region corresponding to the U5-IR stem was defective in initiation of cDNA synthesis. For HIV-1 replication, RNA structural interactions with sequences downstream from the PBS (Kohlstaedt and Steitz 1992) or further upstream of the U5-leader stem (Isel et al. 1993) have been reported to be required for initiation of reverse transcription. Bacterial retrons also require specific secondary structures at the 5′ end of the template RNA for the cDNA-priming reaction (Shimamoto et al. 1993). However, the priming in this case probably involves a 2′ rather than a 3′-hydroxyl group (Shimada et al. 1994).

Plus-strand Priming during Retrovirus Replication

The presence of the PBS near the 5′ terminus of viral RNA was originally unexpected, since RT extends DNA synthesis from the 3′-OH end of the primer in the 5′ to 3′ direction. Thus, to synthesize full-length minus-strand DNA, the primer and RT must be repositioned at the 3′ end of the viral RNA. A mechanism to accomplish this has been described (Fig. 5) and involves multiple steps dependent on the various enzymatic activities associated with RT (RNA- and DNA-dependent DNA polymerase, RNase H, unwinding activity) (Weiss et al. 1985; Coffin 1990). In contrast to the recruitment of cellular tRNAs as primers for cDNA synthesis, primers for plus-strand DNA synthesis are enzymatically produced through the action of RT-associated RNase H on the viral RNA. The initial plus-strand primer(s) is created in a polypurine tract at the 3′ end of viral RNA after reverse transcription has produced an RNA-DNA hybrid substrate (Fig. 5) (Smith et al. 1984; Rattray and Champoux 1987; Charneau and Clavel 1991). An analysis of viral DNA in cells shortly after infection indicates that, in contrast to minus-strand cDNA synthesis, plus-strand DNA synthesis is discontinuous. This implies that there must be multiple plus-strand initiation events. Consistent with this is the finding of a second plus-strand initiation site in the HIV-1 *pol* gene by Charneau et al. (1992). Full-length plus-strand DNA is sub sequently assembled by the action of DNA ligase. The final product of reverse transcription is a linear viral DNA flanked by two LTRs, which then integrates into the host chromosome in a reaction that is dependent on the viral integrase (Brown et al. 1987; Craigie et al. 1990; Katz et al. 1990).

Retrotransposons: Initiation of Reverse Transcription

Transposable genetic elements of yeast and higher eukaryotes replicate via a mechanism closely related to that of retroviruses. Like retroviruses, these elements possess a PBS sequence near their 5′ ends and recruit a host tRNA to prime cDNA synthesis. For the yeast (*S. cerevisiae*) transposable elements TY1, TY2, and TY3, the cellular initiator $tRNA_i^{Met}$ is used as primer (Warmington et al. 1985; Hansen and Sandmeyer 1990; Chapman et al. 1992; Von Pawel-Rammingen et al. 1992). For TY4, $tRNA^{Asn}$ may be used (Janetzky and Lehle 1992; Stucka et al. 1992). Several higher eukaryotic elements such as gypsy, 297, and 412 contain PBS sequences complementary to tRNAs other than $tRNA_i^{Met}$ (Britten and Springer 1993). As with retroviruses, the primer is selectively packaged into virus-like particles (Pochart et al. 1993). However, unlike retroviruses, the base-pairing between the PBS and the tRNA is not tolerant of mismatching. Transposition in *S. cerevisiae* cannot occur if heterologous tRNAs are provided from the *Schizosaccharomyces pombe* strain of yeast (Voytas and Boeke 1993). More recently, Lauermann and

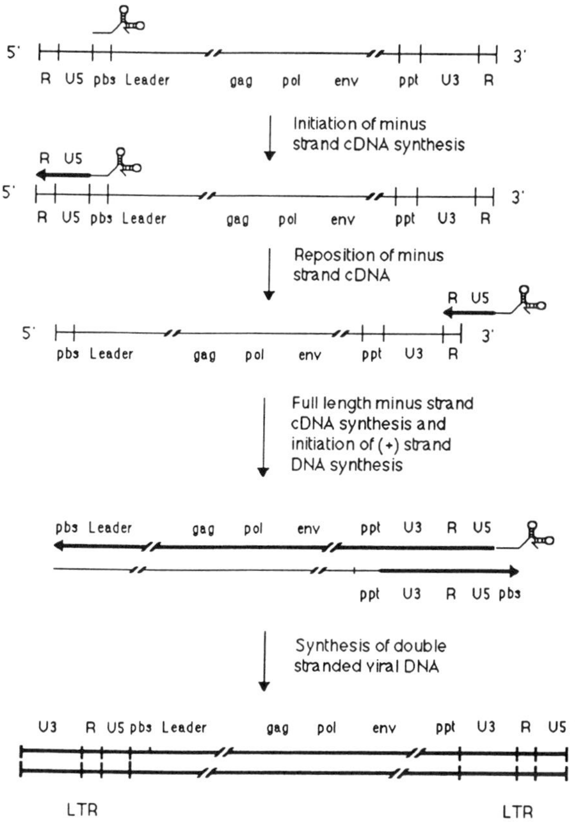

Figure 5 A diagrammatic representation of early reverse transcription of retroviral RNA. Major steps in this process are outlined to indicate key features. Thick lines with arrows indicate the direction of DNA synthesis. Features of the viral RNA (*thin line*) are as follows: (R) repeat sequence; (U5) unique 5′ RNA; (pbs) primer binding site; (Leader) 5′-untranslated RNA; (*gag*, *pol*, *env*) coding sequences; (ppt) polypurine tract; (U3) unique 3′-RNA. The primer tRNA is shown annealed to the PBS.

Boeke (1994) reported that a single base change introduced into the TY1 PBS is not "healed" during the reverse transcription process. This latter result suggests that the primer sequence is not copied into the PBS during reverse transcription and thus may represent a fundamental difference between reverse transcription catalyzed by retroviruses and retrotransposons.

Several retrotransposable elements, including copia from *Drosophila melanogaster*, TY5 of *S. cerevisiae*, and elements in *Physarum polycephalum* and *Volvox carteri* (Kikuchi et al. 1986; Rothnie et al.

1991; Voytas and Boeke 1992; Lindauer et al. 1993), contain PBS sequences that are not homologous to the 3′ end of any natural tRNA molecule. Instead, the PBS is complementary to 15 nucleotides of the 3′ end of a 39-nucleotide fragment derived from the 5′ end of $tRNA_i^{Met}$. The nuclease that is responsible for the cleavage of the $tRNA_i^{Met}$ is not known. However, it has been shown that the *E. coli* RNA-processing enzyme, RNase P, is capable of cleaving $tRNA_i^{Met}$ at several sites, including one that would produce the above primer fragment (Kikuchi and Sasaki 1992). More recently, Hayashi and Stark (1994) have identified an endoribonuclease (RNase Zma) from *Zea mays* that may be the eukaryotic equivalent of the *E. coli* RNase P.

INITIATION OF DNA SYNTHESIS ON NASCENT RNA PRIMERS

Initiation of mtDNA Synthesis

Initiation of mtDNA replication in mammalian systems requires an RNA molecule to prime DNA synthesis but differs from the retroelements in several aspects. The most salient difference is that the primer is not a host cellular tRNA, but a nascent RNA transcript derived from its own genome which is then cleaved to create the primer. Replication of mtDNA initiates in a 150-bp region that controls both transcription and DNA replication (Clayton, this volume). A site-specific endoribonuclease, termed RNase mitochondrial RNA processing (RNase MRP), has been described that cleaves RNA polymerase transcripts near the site of transition from RNA to DNA synthesis (Chang and Clayton 1987). This enzyme is related to *E. coli* RNase P in that it shares a common secondary structure for its associated RNA (Forster and Altman 1990; Schmitt et al. 1993) and an antigenic determinant (Liu et al. 1994).

The Mauriceville and closely related Varkud mitochondrial plasmids of *Neurospora crassa* probably initiate DNA synthesis via multiple mechanisms, one of which may involve self-priming utilizing the 3′-OH end of a 3′-terminal tRNA-like structure as the primer (Wang and Lambowitz 1993; Kennell et al. 1994).

Initiation of Okazaki Fragment Synthesis

In those genomes in which DNA replication initiates internally and DNA synthesis occurs concomitantly on both templates, DNA synthesis is continuous on one arm but discontinuous on the other (Brush and Kelly, this volume). DNA synthesis in the direction of fork movement (forward arm of fork or leading-strand template) involves the continuous incorporation of dNTPs by DNA polymerase-δ to form long nascent DNA strands, whereas DNA synthesis in the direction opposite fork movement (retrograde arm of fork or lagging-strand template) is carried out discontinuously by repeated initiation, elongation, and, finally, joining of short nascent DNA chains (Okazaki fragments) to the 5′ end of the long nascent DNA strand. Examples of this mechanism are found in prokaryotic and eukaryotic cell chromosomes, plasmids that can replicate in prokaryotic cells or in the nuclei of eukaryotic cells, and bacteriophage and nuclear animal viruses with dsDNA genomes (Kornberg and Baker 1992).

The first Okazaki fragment initiated on each template of the replication origin becomes the continuously synthesized nascent DNA strand on the forward arm of the newly created replication fork (DePamphilis et al. 1988). Once replication forks are established, Okazaki fragments originate predominantly, perhaps exclusively, from the retrograde arm. This has been shown for Okazaki fragments synthesized during DNA replication in vivo (SV40, Perlman and Huberman 1977; Kaufmann et al. 1978; Kaufmann 1981; Hay and DePamphilis 1982; Hay et al. 1984; polyomavirus, Hendrickson et al. 1987a,b; mammalian chromosomes, Burhans et al. 1990; Carroll et al. 1993; Tasheva and Roufa 1994; Berberich et al. 1995; Kelly et al. 1995), as well as in vitro in the absence of DNA ligase (Ishimi et al. 1988). Initiation of DNA synthesis on either arm of replication forks in eukaryotic cells involves a series of intermediates (Fig. 6). Okazaki fragments are transient intermediates in DNA replication ($t_{1/2}$ ~ 1 min at 30°C in mammalian cells) that consist of nascent DNA chains from 40 to 300 nucleotides long with an oligoribonucleotide covalently linked to their 5′end (DePamphilis and Wassarman 1980). Synthesis of this oligoribonucleotide is the first step in Okazaki fragment metabolism,

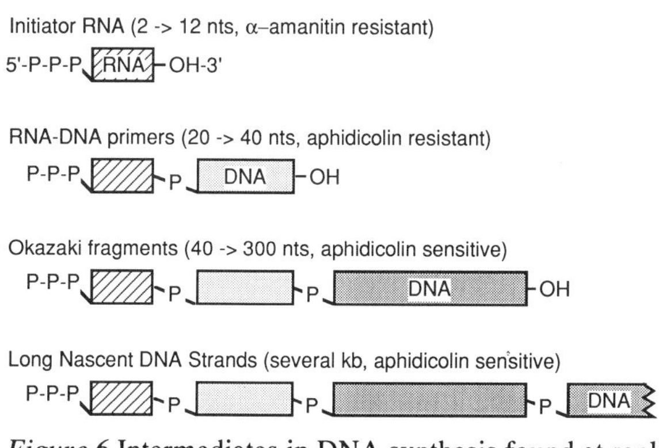

Figure 6 Intermediates in DNA synthesis found at replication forks.

providing an RNA primer on which to initiate DNA synthesis. This RNA primer has been designated initiator RNA (iRNA), because of its unique role in DNA replication (Reichard et al. 1974).

iRNA

Characteristics of iRNA on Newly Replicated DNA

The data documenting the structure, size, and sequence composition of iRNA (Fig. 7) synthesized at the sites where DNA synthesis begins come from extensive studies on SV40, polyomavirus, and mammalian nuclear DNA replication in whole cells or in isolated nuclei incubated with cytoplasm. These results have been reviewed previously in detail (DePamphilis and Wassarman 1980, 1982; DePamphilis and Bradley 1986; DePamphilis 1987). The following is an updated summary of the key points.

Transient, covalent phosphodiester linkages between RNA and DNA have been demonstrated by radiolabeling nascent DNA chains during their synthesis with a specific [α-^{32}P]dNTP substrate and then incubating them in alkali in order to transfer $^{32}PO_4$ from the 5′-terminal deoxyribonucleotide to its neighboring ribonucleotide. The frequency of each of the four possible rN-p-dN linkages is revealed by the amount of each of the four [2′(3′)-^{32}P]rNMPs produced. With replicating SV40 (Anderson et al. 1977), polyomavirus (Reichard et al. 1974; Hunter and Francke 1974), and mammalian cell (Waqar and Huberman 1975; Tseng and Goulian 1975) DNA, all 16 possible rN-p-dN linkages are present at frequencies consistent with a random distribution along the DNA template, and rN-p-dN linkages are excised at the same rate that Okazaki fragments are joined to nascent DNA (Anderson et al. 1977).

These rN-p-dN linkages result from iRNA that can be radiolabeled internally by incorporation of [α-^{32}P]rNTPs during DNA replication (Eliasson and Reichard 1978; Tseng and Goulian 1980; Kaufmann 1981). iRNA synthesis is insensitive to α-amanitin, a specific inhibitor of RNA polymerases II and III, revealing that these enzymes are not required for iRNA synthesis. The enzyme that is required (DNA primase) can incorporate dNTPs in place of their corresponding rNTPs when low concentrations of rNTPs are used (Eliasson and Reichard 1979; Tseng and Goulian 1980). Nascent DNA can be removed from the [^{32}P]iRNA by digesting with DNase I, leaving 1–3 dNMPs at the 3′ end of the iRNA. This material migrated during gel electrophoresis as oligonucleotides of 10 ± 2 residues. iRNA present on isolated replicating intermediates of SV40 and polyomavirus DNA also has been radiolabeled at its 5′

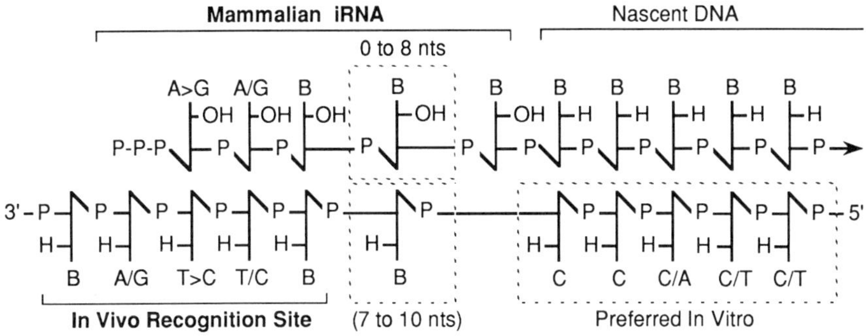

Enzyme	In Vitro Recognition Site
Mammalian DNA Primase:DNA Polymerase–α	PPPPu Pu 3'-PuPy(Py)$_5$
HSV Helicase-Primase Complex (UL5/8/52)	PPPPu Pu GPy(Py)$_4$
E. coli DNA Primase (dnaG)	PPPAG GTC
Phage T4 DNA Helicase-Primase complex (genes 41/61)	PPPPuC TPyG
Phage T7 DNA Primase (gene 4)	PPPACC(A/C) CTGG(T/G) PPPACAC CTGTG

Figure 7 DNA primase recognition sequences. (*Top* panel) Typical initiation site for RNA-primed DNA nascent DNA synthesized in whole cells or in isolated nuclei in cell lysates is diagrammed with specific bases indicated by A, G, C, or T. B indicates any one of the four bases. Purified pol-α:primase prefers pyrimidine-rich sites in vitro, sometimes upstream of a CC(C/A) consensus sequence. (*Bottom* panel) Typical initiation sites for purified DNA primase alone or in the company of other proteins. Template sequences are 3′-5′.

terminus using [γ-^{32}P]ATP and T4 polynucleotide kinase and then digested with T4 exonuclease to remove all but one 3′-deoxyribonucleotide (Hay et al. 1984; Hendrickson et al. 1987b). This iRNA contained 6–9 rNMPs that began primarily with A (60%) or G (25%).

To distinguish full-length iRNA from degraded RNA primers or RNA fragments from other sources, the 5′-terminal ribonucleotide has been radiolabeled exclusively by incorporation of [β-^{32}P]rATP or [β-^{32}P] rGTP (Eliasson and Reichard 1978; Tseng et al. 1979; Kaufmann 1981). The length of this iRNA was the same as described above. Newly replicated DNA has also been isolated first and then radiolabeled with

[α-^{32}P]rGTP using vaccinia virus guanylyltransferase, the enzyme that "caps" mRNA (Hendrickson et al. 1987b). This procedure labeled only iRNA with a 5′-end terminal ribonucleoside di- or triphosphate. Following digestion of nascent DNA, most of this iRNA contained 6–9 rNMPs, although iRNA as short as 2 and as long as 12 residues was detected. Release of the GTP-capped terminal nucleotide with P1 nuclease revealed that 70% of the primers began with A and 30% with G.

The presence of a short oligoribonucleotide at one end of nascent DNA chains has been confirmed by degrading RNA with either alkali or ribonuclease, and then measuring the change in the size of nascent DNA chains. This strategy revealed that 50% of the Okazaki fragments synthesized in vivo by SV40 contained iRNA 6–10 residues long (Kaufmann et al. 1977; Hay et al. 1984), and 90% of the Okazaki fragments synthesized by mammalian chromosomes contained iRNA 8–12 residues long (Burhans et al. 1991).

Site Specificity of iRNA Synthesis In Vivo

Evolution would be at a distinct disadvantage if DNA replication required specific DNA sequence signals distributed throughout the genome every 100 or so base pairs. Thus, it is not surprising that initiation of DNA synthesis at replication forks does not require specific template sequences, although it does show preference for certain sequence motifs. Nascent DNA chains in replicating SV40 (Hay and DePamphilis 1982; Hay et al. 1984) and polyomavirus (Hendrickson et al. 1987a; DePamphilis et al. 1988) DNA isolated from virus-infected cells have been selectively radiolabeled at their iRNA-p-DNA junction by removing the iRNA with alkali to unmask a 5′-terminal OH group on the nascent DNA chain. These 5′ termini were then labeled using [γ-^{32}P]rATP and T4 polynucleotide kinase. The resulting ^{32}P-labeled DNA chains were annealed to their template, cut at a unique restriction site, and then fractionated by gel electrophoresis alongside their own DNA sequence ladder in order to identify the position of their 5′-terminal nucleotide (i.e., the iRNA-p-DNA junction). The 5′ terminus of iRNA-DNA chains (i.e., the starting nucleotide location for iRNA) was determined by mapping the locations of end-labeled nascent DNA chains before and after removing their iRNA.

Results from these experiments revealed that >99% of initiation events occurred on the retrograde template and that at least 88% of these initiation events occurred at either 3′-purine-T-pyrimidine-5′ (PuTPy) or 3′-purine-C-pyrimidine-5′ (PuCPy) sites in the template. iRNA synthesis began at the underlined nucleotide with either ATP or GTP. PuTPy sites comprised 78% of initiation sites, consistent with a strong preference for initiation with ATP, and iRNA-p-DNA linkages revealed no sequence preferences, as expected from the nearest-neighbor analyses described above. Moreover, iRNA initiation sites encoded 8 ± 2 ribonucleotides (range = 2–12). This mapping strategy also revealed the transition from discontinuous (presence of initiation sites) to continuous (absence of initiation sites) that marks the origin of bidirectional replication (DePamphilis, this volume).

DNA Polymerase-α:DNA Primase

Pol-α:primase consists of four subunits whose structure is well conserved from yeast to mammals (Wang, this volume). Judged by its sensitivities to a variety of DNA replication inhibitors and its requirement for SV40, polyomavirus, and cellular DNA replication in vitro, pol-α:primase is the enzyme responsible for initiation of Okazaki fragment synthesis (DePamphilis and Bradley 1986; Brush and Kelly; Wang; Hassell and Brinton; all this volume). This enzyme contains a 49-kD and a 58-kD polypeptide, both of which are involved in iRNA synthesis in higher eukaryotes (Lehman and Kaguni 1989), and both of which are essential for DNA replication in yeast (Foiani et al. 1989a). The 49-kD subunit contains the rNTP-binding site and DNA primase catalytic activity (Nasheuer and Grosse 1988; Foiani et al. 1989b; Bakkenist and Cotterill 1994), whereas the 58-kD subunit may help to stabilize NTP binding to the 49-kD subunit (Santocanale et al. 1993).

The properties of DNA primase are consistent with the characteristics of iRNA synthesis. DNA primase is insensitive to α-amanitin and template-dependent, although it readily substitutes dNTPs in place of the corresponding rNTP (Sheaff and Kuchta 1994). Initiation of RNA primer synthesis is the rate-limiting step in DNA synthesis in vitro; RNA synthesis is about 100-fold slower than DNA synthesis (Grosse and Krauss 1985; Sheaff and Kuchta 1993). Pol-α:primase appears to slide along its template looking for primase recognition sites whose K_m values range from 7 pM to 100 pM (Davey and Faust 1990). Initiation depends on formation of an enzyme-DNA-NTP_1-NTP_2 complex complementary to the first two nucleotides in the template initiation site (Sheaff and Kuchta 1993), emphasizing the importance of the 3′-PuTPy-5′ template sequence as a recognition signal. This importance of the second or third nucleotide in primase site selection is reminiscent of site selection by protein primers (Fig. 3).

Pol-α:primase does not synthesize an RNA primer of unique length, but a family of oligoribonucleotides ranging from 2 to 14 residues, most of which are 7 to 13 residues long. Their size is strongly influenced by template sequence, the ratio of ATP to GTP (Yamaguchi et al. 1985a,b; Suzuki et al. 1993), and their proximity to secondary DNA structures such as cruciforms (Tseng and Prussak 1989). In the absence of polymerase, multimeric RNA primers are made, demonstrating that primase is responsible for determining RNA primer length (Tseng and Ahlem 1983; Singh et al. 1986; Cotterill et al. 1987). Pol-α prefers elongating primers at least 7 residues long (Kuchta et al. 1990). However, two accessory proteins (C_1C_2) are frequently associated with pol-α and stimulate its activity on ssDNA templates from 180- to 1800-fold by reducing the K_m of its primer (Pritchard et al. 1983; Kawasaki et al. 1986). C_1C_2 appear to eliminate nonproductive binding of this enzyme to ssDNA, allowing the polymerase to slide along the template until it recognizes a primer. C_1C_2 also decrease the average length of RNA primers synthesized by pol-α:primase by an average of 3–4 bases (Viswanatha et al. 1986). Since DNA replication appears dependent on these proteins (Kumble et al. 1992), they may facilitate initiation of DNA synthesis at replication forks.

Site Specificity of Pol-α:Primase In Vitro

As with iRNA initiation sites selected on viral chromosomes replicating in nuclei, no specific sequences emerge as pol-α:primase start sites on ssDNA templates in vitro. However, in vitro, this enzyme prefers to initiate RNA-primed DNA synthesis at or near pyrimidine-rich sequences (Fig. 7). The minimum size of the recognition site for mammalian pol-α:primase is 3′-purine-$[pyrimidine]_6$-5′ (Davey and Faust 1990; Suzuki et al. 1993). Selection of RNA primer initiation sites is strongly affected by the ratio of ATP to GTP and therefore affects the nucleotide composition of RNA primers and the frequency with which they initiate with either A or G (Yamaguchi et al. 1985b; Suzuki et al. 1993). The K_m for ATP is 100 μM (Kuchta et al. 1990). Mapping of initiation sites for purified pol-α:primase from monkey cells on unique segments of SV40 ssDNA revealed a preference for 3′-$(Py)_nC\underline{T}TT(Py)_n$ (80%) and 3′-$(Py)_nC\underline{C}C(Py)_n$ (20%), where the underlined nucleotide is complementary to the first nucleotide in the RNA primer (Yamaguchi et al. 1985a; Vishwanatha et al. 1986). These sites were less frequent (1/16 bases) than those for iRNA synthesis in vivo (1/7 bases), the RNA primers were shorter (7 ± 1 bases in vitro, 10 ± 1 bases in vivo), and the sites chosen in vitro were not the same as those chosen in vivo. Similar results were obtained from mapping initiation sites for RNA-primed DNA synthesis on plasmid DNA undergoing replication in a HeLa cell extract (Bullock et al. 1994). RNA primers most frequently began at 3′-NTT sites located, on average, once every 19 bases. However, the authors' conclusion that RNA primers had no preferred 5′-terminal nucleotide is contrary to all other studies and probably resulted from a 1-base mapping error. Other sequence preferences have also been reported. RNA primers that begin with pppAG or pppGG are favored, and 3′-dCT is strongly favored over 3′-dAC in the template (Sheaff and Kuchta 1993). Initiation sites in which 3′-CC(C/A)-5′ occurs 10–13 nucleotides downstream from the RNA primer start site are strongly preferred (Davey and Faust 1990), since point mutations that disrupt this motif decrease the K_m for DNA approximately 7-fold. One explanation for the variation observed in identifying initiation sites in vitro could be the protein composition of the enyzme preparation tested. For example, DNA primase alone (Tseng and Prussak 1989) appears to select far fewer sites than when it is associated with DNA polymerase-α. Phage T4 DNA-binding protein confines synthesis of RNA primers by the phage T4 DNA primase/DNA helicase complex to those sites where Okazaki fragment synthesis begins (Cha and Alberts 1990).

RNA-DNA Primers

iRNA is first extended into pppRNA-p-DNA chains of ≤40 nucleotides with a mean length of about 35 nucleotides. These RNA-DNA chains were originally referred to as "DNA primers" (Nethanel et al. 1988), but are referred to here as RNA-DNA primers to distinguish them from DNA primers that appear in parvovirus replication and DNA repair. RNA-DNA primers can be labeled during SV40-driven DNA replication in isolated nuclei or cell extracts by briefly incubating them with [α-^{32}P]NTPs or [α-^{32}P]dNTPs and then fractionating the labeled DNA by gel electrophoresis (Nethanel et al. 1988, 1992; Nethanel and Kaufmann 1990; Bullock et al. 1994). Like Okazaki fragments (Anderson and DePamphilis 1979), DNA primers contain iRNA, originate predominantly from the lagging-strand template, are transient intermediates in DNA replication, and are separated from the 5′ends of nascent downstream DNA chain by a short gap. Synthesis of RNA-DNA primers is insensitive to aphidicolin, a specific inhibitor of DNA pols α, δ, and ε (Nethanel et al. 1988), and so is the first 30–40 nucleotides polymerized by pol-α:primase (Decker et al. 1986), suggesting that RNA-DNA primers are synthesized by this enzyme. Moreover, butylphenyl-dGTP, a specific inhibitor of pol-α, and neutralizing antibodies against pol-α, inhibit synthesis of RNA-DNA primers. Assembly of RNA-DNA primers into longer DNA chains requires proliferating cell nuclear antigen (PCNA) (Bullock et al. 1991) and ATP (Nethanel et al. 1992), suggesting that this step requires pol-δ or -ε and an ATP-requiring protein such as DNA ligase. The limited processivity of pol-α:primase in DNA synthesis (10–15 dNMPs [Yamaguchi et al. 1985b]) should facilitate the switch to another, more processive enzyme, such as pol-δ.

So far, RNA-DNA primers have been reported only in SV40 replicating intermediates. However, with the exception of T antigen, SV40 DNA replication depends entirely on cellular proteins. Therefore, it is likely that replication of cellular chromosomes, as well as host-dependent viral chromosomes such as other papovaviruses and papillomaviruses, will also involve RNA-DNA primers.

Summary of Priming by Nascent RNA

Nascent iRNA usually begins with pppA and sometimes pppG, but not with pppU or pppC. Most iRNA consists of 6–9 ribonucleotides, although iRNA as short as 2 and as long as 12 ribonucleotides has been detected. iRNA initiation sites show a preference for PuTPy and PuCPy where T and C, respectively, are complementary to the first ribonucleotide. Other than this, neither the RNA moeity nor the RNA-p-DNA junction exhibits any sequence specificity. Therefore, the transition from RNA to DNA synthesis depends on primer length rather than on template sequence signal. However, since primer length appears to vary from one initiation site to another, template sequence must indirectly determine the nucleotide site where DNA synthesis begins. For example, the transition from iRNA to DNA synthesis, as well as from RNA-DNA primers to Okazaki fragment synthesis, may result from changes in secondary structure that result from the fact that RNA:DNA hybrids adopt the A form whereas DNA:DNA hybrids adopt the B form (Selsing and Wells 1979; Arnott et al. 1986).

Initiation of RNA-primed DNA synthesis in eukaryotes and prokaryotes shares certain features, although differences exist in their preferred template initiation sites (Fig. 7). All DNA primases initiate synthesis only with a purine ribonucleoside triphosphate (ATP>GTP) and can substitute dNTPs for rNTPs at a low frequency during iRNA synthesis. All DNA primase template recognition sequences include a preferred nucleotide at position −1 relative to the start site. Herpes simplex virus (HSV) DNA primase, like mammalian primase, prefers a pyrimidine-rich template sequence (Tenney et al. 1995; Challberg, this volume). *E. coli* DNA primase most frequently begins with pppAG at the template sequence 3′-GTC-5′ (Swart and Griep 1993), and the consensus recognition sequence for *E. coli* DNA primase-DnaB helicase is 3′-PuPyPy-5′ (Yoda and Okazaki 1991). This is strikingly similar to iRNA synthesis in mammalian nuclei. DNA primases from bacteriophages T4 and T7 prefer to begin with pppAC at a template sequence that contains a pyrimidine at position −1 (Cha and Alberts 1990; Mendelman and Richardson 1991). On the whole, these results suggest a conservation of function between prokaryotic and eukaryotic DNA primases.

Models for Okazaki Fragment Synthesis

There are two models for synthesis of Okazaki fragments (Fig. 8). The first is the "initiation zone" model, which proposes that one iRNA initiation site is selected stochastically among the many potential sites located within a defined template region (initiation zone). Formation of an Okazaki fragment would then follow an ordered sequential pathway: (1) synthesis of iRNA by DNA primase, (2) extension of iRNA into a DNA primer by pol-α, (3) synthesis of Okazaki fragment by pol-α or -δ, (4) excision of the iRNA in front of the Okazaki fragment (a two-step process [Anderson and DePamphilis 1979] involving two different nucleases [Brush and Kelly, this volume]), (5) filling in the resulting gap by pol-δ or -ε, and (6) ligation of the 3′ end of the Okazaki fragment to the 5′ end of the long nascent DNA strand. Thus, the size of an Okazaki fragment would vary from the length of a single RNA-DNA primer (~40 bases) to the maximum length of the initiation zone (~300 bases), and only a small subset of potential iRNA initiation sites would be used in a single DNA molecule during each round of DNA replication. In order for Okazaki fragments of different sizes to accumulate waiting for iRNA excision, gap filling, and ligation to occur (Anderson and DePamphilis 1979), initiation of RNA-primed DNA synthesis must be slow relative to the rate of DNA synthesis that follows. This is consistent with the properties of pol-α:primase (Sheaff and Kuchta 1993) and pol-δ (Wang, this volume).

What determines the size of the initiation zone? One possibility is the periodic structure of chromatin in front of the replication fork (DePamphilis and Wassarman 1980; DePamphilis et al. 1988). If the rate-limiting step in replication is the ability of the fork to move from

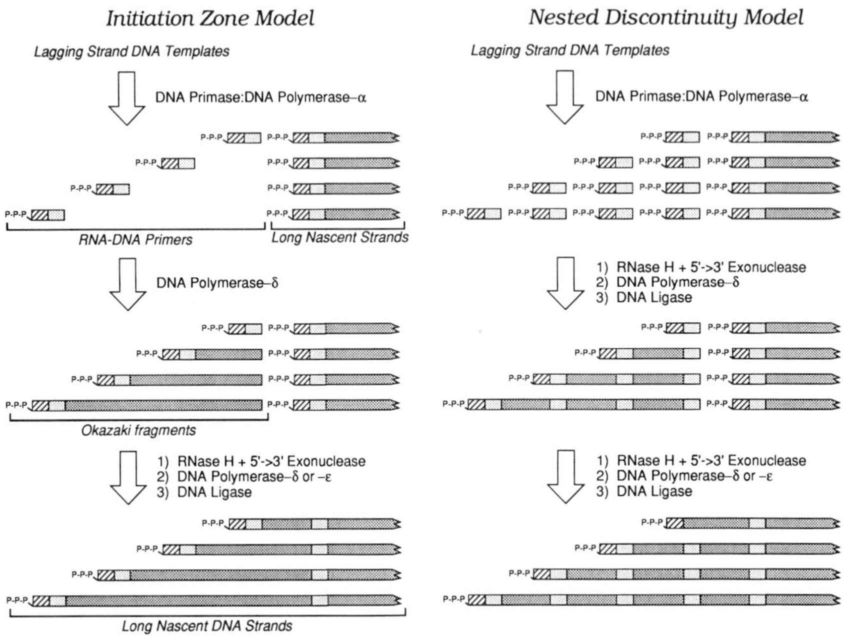

Figure 8 Two models for Okazaki fragment synthesis at DNA replication forks. Newly synthesized DNA is represented on four individual lagging-strand templates. Symbols are the same as in Fig. 6. See text for description.

one nucleosome to the next (i.e., nucleosome disassembly), then the amount of template DNA exposed on the retrograde arm of the fork (i.e., lagging-strand DNA template) will be the same as the average length of internucleosomal DNA. This would produce lagging-strand template on the retrograde arm with the mean length and size range of Okazaki fragments. In prokaryotic cells where nucleosomes are absent, Okazaki fragments average about 1500 nucleotides in length (Kornberg and Baker 1992). A second possibility is the organization of a replication fork. The average length of an Okazaki fragment may be determined by the length of retrograde DNA arm that is wrapped around the DNA replication complex (Stillman, this volume). This model is supported by the fact that Okazaki fragments are produced on newly initiated DNA molecules in vitro with a mean size and length distribution that is similar to those made in vivo (Ishimi et al. 1988). However, the added DNA substrate is rapidly assembled into nucleosomes under these conditions, resulting in the appearance of nucleosomes both in front of and behind replication forks (Gruss et al. 1990).

The second model proposes that synthesis of an Okazaki fragment itself is a discontinuous process in which Okazaki fragments result from assembly of a nested series of RNA-DNA primers (nested discontinuity model, Fig. 8; Nethanel et al. 1988, 1992). Instead of a stochastic selection of iRNA initiation sites, the first RNA-DNA primer is synthesized close to the 5′ end of the long nascent DNA strand. This is followed in rapid sucession by synthesis of additional RNA-DNA primers as the pol-α:primase complex travels in a 5′→ 3′ direction on the lagging-strand template. iRNA excision occurs concomitantly, and the resulting gaps between RNA-DNA primers are filled in by pol-δ or -ε to form an Okazaki fragment. The final iRNA excision/gap-filling event then occurs in front of the Okazaki fragment, and it is ligated to the long nascent DNA strand. What delays the first RNA-DNA primer from joining to the long nascent DNA strand is not clear, although it could be an inherent property of the replication complex.

ACKNOWLEDGMENTS

M.S. thanks her colleagues for many helpful discussions, L. Blanco for his help in designing figures, those who sent preprints of their work, and Ms. M. Angeles Martínez for typing the manuscript. Work in the laboratory of M.S. was supported by grants from the National Institutes of Health (5 RO1 GM-27242-15), Dirección General de Investigación Científica y Técnica (PB93-0173), European Economic Community (CHRX-CT93-0248), and Fundación Ramón Areces.

REFERENCES

Aiyar, A., Z. Ge, and J. Leis. 1994. A specific orientation of RNA secondary structures is required for initiation of reverse transcription. *J. Virol.* **68:** 611–618.

Aiyar, A., D. Cobrinik, Z. Ge, H.-J. Kung, and J. Leis. 1992. Interaction between retroviral U5 RNA and the TΨC loop of the $tRNA^{Trp}$ primer is required for efficient initiation of reverse transcription. *J. Virol.* **66:** 2464–2472.

Anderson, S. and M.L. DePamphilis. 1979. Metabolism of Okazaki fragments during simian virus 40 DNA replication. *J. Biol. Chem.* **254:** 11495–11504.

Anderson, S., G. Kaufmann, and M.L. DePamphilis. 1977. RNA primers in simian virus 40 DNA replication: Identification of transient RNA-DNA covalent linkages in replication DNA. *Biochemistry* **16:** 4990–4998.

Arnott, S., R. Chandrasekaran, R.P. Millane, and H-S Park. 1986. DNA-RNA hybrid secondary structures. *J. Mol. Biol.* **188:** 631–640.

Bakkenist, C.J. and S. Cotterill. 1994. The 50 kd primase subunit of *Drosophila melanogaster* DNA polymerase-α. *J. Biol. Chem.* **269:** 26759–26766.

Bamford, D.H. and L. Mindich. 1984. Characterization of the DNA-protein complex at the termini of the bacteriophage PRD1 genome. *J. Virol.* **50:** 309–315.

Bamford, D.H., T. McGraw, G. MacKenzie, and L. Mindich. 1983. Identification of a protein bound to the termini of bacteriophage PRD1 DNA. *J. Virol.* **47:** 311–316.

Barat, C., S. LeGrice, and J.-L. Darlix. 1991. Interaction of the HIV-1 reverse transcriptase with a synthetic form of its replication primer, $tRNA^{Lys3}$. *Nucleic Acids Res.* **19:** 751–757.

Barat, C., V. Lullien, O. Shatz, G. Keith, M.T. Nugeyre, F. Gruninger-Leitch, F. Barre-Sinoussi, S.F.J. LeGrice, and J.L. Darlix. 1989. HIV-1 reverse transcriptase specifically interacts with the anticodon domain of its cognate primer tRNA. *EMBO J.* **8:** 3279–3285.

Bartenschlager, R. and H. Schaller. 1988. The amino-terminal domain of the hepadnaviral P-gene encodes the terminal protein (genome-linked protein) believed to prime reverse transcription. *EMBO J.* **7:** 4185–4192.

Baudin, F., R. Marquet, C. Isel, J.-L. Darlix, B. Ehresmann, and C. Ehresmann. 1993. Functional sites in the 5′ region of human immunodeficiency virus type 1 RNA form defined structural domains. *J. Mol. Biol.* **229:** 382–397.

Berberich, S., A. Trivedi, D.C. Daniel, E.M. Johnson, and M. Leffak. 1995. In vitro replication of plasmids containing human c-*myc* DNA. *J. Mol. Biol.* **245:** 92–109.

Berkhout, B. and I. Schoneveld. 1993. Secondary structure of the HIV-2 leader RNA comprising the tRNA-primer binding site. *Nucleic Acids Res.* **21:** 1171–1178.

Bernad, A., L. Blanco, J.M. Lázaro, G. Martín, and M. Salas. 1989. A conserved 3′→5′ exonuclease active site in prokaryotic and eukaryotic DNA polymerases. *Cell* **59:** 219–228.

Blanco, L. and M. Salas. 1984. Characterization and purification of a phage ϕ29 coded DNA polymerase required for the initiation of replication. *Proc. Natl. Acad. Sci.* **81:** 5325-5329.

———. 1985a. Characterization of a 3′→ 5′ exonuclease activity in the phage ϕ29-encoded DNA polymerase. *Nucleic Acids Res.* **13:** 1239–1249.

———. 1985b. Replication of ϕ29 DNA with purified terminal protein and DNA polymerase: Synthesis of full-length ϕ29 DNA. *Proc. Natl. Acad. Sci.* **82:** 6404–6408.

———. 1995. Mutational analysis of bacteriophage ϕ29 DNA polymerase. *Methods Enzymol.* **262:** 283–294.

Blanco, L., A. Bernad, and M. Salas. 1988. Transition from initiation to elongation in the protein-primed ϕ29 DNA replication. Salt-dependent stimulation by the viral protein p6. *J. Virol.* **62:** 4167–4172.

Blanco, L., A. Bernad, M.A. Blasco, and M. Salas. 1991. A general structure for DNA-dependent DNA polymerases. *Gene* **100:** 27–38.

Blanco, L., A. Bernad, J.A. Esteban, and M. Salas. 1992. DNA-independent deoxynucleotidylation of the ϕ29 terminal protein by the ϕ29 DNA polymerase. *J. Biol. Chem.* **267:** 1225–1230.

Blanco, L., J.A. García, M.A. Peñalva, and M. Salas. 1983. Factors involved in the initiation of phage ϕ29 DNA replication *in vitro*: Requirement of the gene 2 product for the formation of the protein p3-dAMP complex. *Nucleic Acids Res.* **11:** 1309–1323.

Blanco, L., J. Gutiérrez, J.M. Lázaro, A. Bernad, and M. Salas. 1986. Replication of phage ϕ29 DNA *in vitro*: Role of the viral protein p6 in initiation and elongation. *Nucleic Acids Res.* **14:** 4923–4937.

Blanco, L., J.M. Lázaro, M. de Vega, A. Bonnin, and M. Salas. 1994. Terminal protein-primed DNA amplification. *Proc. Natl. Acad. Sci.* **91:** 12198–12202.

Blanco, L., A. Bernad, J.M. Lázaro, G. Martín, C. Garmendia, and M. Salas. 1989. Highly efficient DNA synthesis by the phage ϕ29 DNA polymerase. Symmetrical mode of DNA replication. *J. Biol. Chem.* **264:** 8935–8940.

Blanco, L., I. Prieto, J. Gutiérrez, A. Bernad, J.M. Lázaro, J.M. Hermoso, and M. Salas. 1987. Effect of NH_4^+ ions on ϕ29 DNA-protein p3 replication: Formation of a complex between the terminal protein and the DNA polymerase. *J. Virol.* **61:** 3983–3991.

Britten, R.J. and M.S. Springer. 1993. Phylogenetic relationships of reverse transcriptase and RNase H sequences and aspects of genome structure in the gypsy group of retrotransposons. *Mol. Biol. Evol.* **10:** 1370–1379.

Brown, B., B. Bowerman, H.E. Varmus, and J.M. Bishop. 1987. Correct integration of retroviral DNA *in vitro*. *Cell* **49:** 347–356.

Bullock, P.A., Y.S. Seo, and J. Hurwitz. 1991. Initiation of SV40 DNA synthesis in vitro. *Mol. Cell. Biol.* **11:** 2350–2361.

Bullock, P.A., S. Tevosian, C. Jones, and D. Denis. 1994. Mapping initiation sites for SV40 DNA synthsis events in vitro. *Mol. Cell. Biol.* **14:** 5043–5055.

Burhans, W.C., L.T. Vassilev, M.S. Caddle, N.H. Heintz, and M.L. DePamphilis. 1990. Identification of an origin of bidirectional DNA replication in mammalian chromosomes. *Cell* **62:** 955–965.

Burhans, W.C., L.T. Vassilev, J. Wu, J.M. Sogo, F. Nallaseth, and M.L. DePamphilis. 1991. Emetine allows identification of origins of mammalian DNA replication by imbalanced DNA synthesis, not through conservative nucleosome segregation. *EMBO J.* **10:** 4351–4360.

Caldentey, J., L. Blanco, D.H. Bamford, and M. Salas. 1993. *In vitro* replication of bacteriophage PRD1 DNA. Characterization of the protein-primed initiation site. *Nucleic Acids Res.* **21:** 3725–3730.

Caldentey, J., L. Blanco, H. Savilahti, D.H. Bamford, and M. Salas. 1992. *In vitro* replication of bacteriophage PRD1 DNA. Metal activation of protein-primed initiation and DNA elongation. *Nucleic Acids Res.* **20:** 3971–3976.

Carroll, S.M., M.L. DeRose, J.L. Kolman, G.H. Nonet, R.E. Kelly, and G.M. Wahl. 1993. Localization of a bidirectional DNA replication origin in the wild type and in episomally amplified murine ADA loci. *Mol. Cell. Biol.* **13:** 2971.

Carusi, E.A. 1977. Evidence for blocked 5′ termini in human adenovirus. *Virology* **76:** 389–394.

Cha, T.-A. and B.M. Alberts. 1990. Effects of the bacteriophage T4 gene 41 and gene 32 proteins on RNA primer synthesis: Coupling of leading and lagging strand DNA synthesis at a replication fork. *Biochemistry* **29:** 1791–1798.

Challberg, M.D. and T.J. Kelly. 1981. Processing of the adenovirus terminal protein. *J. Virol.* **38:** 272–277.

Challberg, M.D., S.V. Desiderio, and T.J. Kelly. 1980. Adenovirus DNA replication *in vitro*: Characterization of a protein covalently linked to nascent DNA strands. *Proc. Natl. Acad. Sci.* **77:** 5105–5109.

Chang, D.D. and D.A. Clayton. 1987. A novel endoribonuclease cleaves at a priming site of mouse mitochondrial DNA replication. *EMBO J.* **6:** 409–417.

Chang, L.-J., R.C. Hirsch, D. Ganem, and H.E. Varmus. 1990. Effects of insertional and point mutations on the functions of the duck hepatitis B virus polymerase. *J. Virol.* **64:** 5553–5558.

Chapman, K.B., A.S. Bystrom, and J.D. Boeke. 1992. Initiator methionine tRNA is essential for Ty1 transposition in yeast. *Proc. Natl. Acad. Sci.* **89:** 3236–3240.

Charneau, P. and F. Clavel. 1991. A single-stranded gap in human immunodeficiency virus unintegrated linear DNA defined by a central copy of the polypurine tract. *J. Virol.* **65:** 2415–2421.

Charneau, P., M. Alizon, and F. Clavel. 1992. Second origin of DNA plus-strand synthesis is required for optimal human immunodeficiency virus replication. *J. Virol.* **66:** 2814–2820.

Chen, C.W., Y.S. Lin, Y.L. Yang, M.F. Tsou, H.M. Chang, H.M. Kieser, and D.A. Hopwood. 1995. The linear chromosomes of *Streptomyces*: Structure and dynamics. *Actinomycetologia* (in press).

Cobrinik, D., L. Soskey, and J. Leis. 1988. A retroviral RNA secondary structure required for efficient initiation of reverse transcription. *J. Virol.* **62:** 3622–3630.

Cobrinik, D., A. Aiyar, Z. Ge, M. Katzman, H. Huang, and J. Leis. 1991. Overlapping retrovirus U5 sequence elements are required for efficient integration and initiation of reverse transcription. *J. Virol.* **65:** 3864–3872.

Coffin, J.M. 1990. Retroviridae and their replication. In *Virology*, 2nd edition (ed. B.N. Fields et al.), pp. 1437–1500. Raven Press, New York.

Cordell, B., R. Swanstrom, H.M. Goodman, and J.M. Bishop. 1979. $tRNA^{Trp}$ as primer for RNA-directed DNA polymerase: Structural determinants of function. *J. Biol. Chem.* **254:** 1866–1874.

Cotterill, S., G. Chui, and I.R. Lehman. 1987. DNA polymerase-primase from embryos of *Drosophila melanogaster* DNA primase subunits. *J. Biol. Chem.* **262:** 16105–16108.

Craigie, R., T. Fujiwara, and F. Bushman. 1990. The IN protein of Moloney murine leukemia virus processes the viral DNA ends and accomplishes their integration in vitro. *Cell* **62:** 829–837.

Crawford, S. and S.P. Goff. 1985. A deletion mutation in the 5′ part of the *pol* gene of Moloney murine leukemia virus blocks proteolytic processing of the Gag and Pol proteins. *J. Virol.* **53:** 899–907.

Davey, S.K. and E.A. Faust. 1990. Murine DNA polymerase-α:primase initiates RNA primed DNA synthesis preferentially upstream of a 3′-CC(C/A)-5′ motif. *J. Biol. Chem.* **265:** 3611–3614.

Decker, R.S., M. Yamaguchi, R. Possenti, and M.L. DePamphilis. 1986. Initiation of SV40 DNA replication *in vitro:* Aphidicolin causes accumulation of early replicating intermediates and allows determination of the initial direction of DNA synthesis. *Mol. Cell. Biol.* **6:** 3815–3825.

DePamphilis, M.L. 1987. Replication of simian virus 40 and polyoma virus chromosomes. In *Molecular aspects of papovaviruses* (ed. Y. Aloni), pp. 1–40. Martinus Nijhoff, Boston.

DePamphilis, M.L. and M.K. Bradley. 1986. Replication of simian virus 40 and polyoma virus chromosomes. In *The Papovaviridae* (ed. N.P. Salzman), vol. 1, pp. 99–246. Plenum Press, New York.

DePamphilis, M.L. and P.M. Wassarman. 1980. Replication of eukaryotic chromosomes: A close-up of the replication fork. *Annu. Rev. Biochem.* **49:** 627–666.

———. 1982. Organization and replication of papovavirus DNA. In *Organization and replication of viral DNA* (ed. A.S. Kaplan), pp. 37–114, CRC Press, Boca Raton, Florida.

DePamphilis, M.L., E. Martínez-Salas, D.Y. Cupo, E.A. Hendrickson, C.E. Fritze, W.R. Folk, and U. Heine. 1988. Initiation of polyomavirus and SV40 DNA replication, and the requirements for DNA replication during mammalian development. *Cancer Cells* **6:** 165–175.

Desiderio, S.V. and T.J. Kelly. 1981. Structure of the linkage between adenovirus DNA and the 55,000 molecular weight terminal protein. *J. Mol. Biol.* **145:** 319–337.

Dobbs, L., L.J. Zhao, G. Sripad, and R. Padmanabhan. 1990. Mutational analysis of single-stranded DNA templates active in the *in vitro* initiation assay for adenovirus DNA replication. *Virology* **178:** 43–51.

Eliasson, R. and P. Reichard. 1978. Replication of polyoma DNA in isolated nuclei. Synthesis and distribution of initator RNA. *J. Biol. Chem.* **253:** 7469–7475.

———. 1979. Replication of polyoma DNA in isolated nuclei. VII. Initiator RNA synthesis during nucleotide depletion. *J. Mol. Biol.* **129:** 393–409.

Escarmís, C., D. Guirao, and M. Salas. 1989. Replication of recombinant ϕ29 DNA molecules in *Bacillus subtilis* protoplasts. *Virology* **169:** 152–160.

Esteban, J.A., M. Salas, and L. Blanco. 1993. Fidelity of ϕ29 DNA polymerase. Comparison between protein-primed initiation and DNA polymerization. *J. Biol. Chem.* **268:** 2719–2726.

Esteban, J.A., A. Bernad, M. Salas, and L. Blanco. 1992. Metal activation of synthetic and degradative activities of ϕ29 DNA polymerase, a model enzyme for protein-primed DNA replication. *Biochemistry* **31:** 350–359.

Esteban, J.A., M.S. Soengas, M. Salas, and L. Blanco. 1994. 3′→5′ exonuclease active site of ϕ29 DNA polymerase. Evidence favoring a metal-ion-assisted reaction mechanism. *J. Biol. Chem.* **269:** 31946–31954.

Foiani, M., C. Santocanale, P. Plevani, and G. Lucchini. 1989a. A single essential gene, PRI2, encodes the large subunit of DNA primase in *Saccharomyces cerevisiae. Mol. Cell. Biol.* **9:** 3081–3087.

Foiani, M., A.G. Linder, G.R. Hartmann, G. Lucchini, and P. Plevani. 1989b. Affinity labeling of the active center and ribonucleoside triphosphate binding site of yeast DNA primase. *J. Biol. Chem.* **264:** 2189–2194.

Forster, A.C. and S. Altman. 1990. Similar cage-shaped structures for the RNA components of all ribonuclease P and ribonuclease MRP enzymes. *Cell* **62:** 407–409.

Freire, R., M. Salas, and J.M. Hermoso. 1994. A new protein domain for binding to DNA through the minor groove. *EMBO J.* **13:** 4353–4360.

Fujimura, H., T. Yamada, F. Hishinuma, and N. Gunge. 1988. DNA replication *in vivo* of linear DNA killer plasmids pGKL1 and pGKL2 in *Saccharomyces cerevisiae. FEMS Microbiol. Lett.* **49:** 441–444.

Ganem, D. and H.E. Varmus. 1987. The molecular biology of hepatitis B viruses. *Annu. Rev. Biochem.* **67:** 651–693.

García, E., A. Gómez, C. Ronda, C. Escarmís, and R. López. 1983. Pneumococcal bacteriophage Cp-1 contains a protein tightly bound to the 5′ termini of its DNA. *Virology* **128:** 92–104.

García, P., J.M. Hermoso, J.A. García, E. García, R. López, and M. Salas. 1986. Formation of a covalent complex between the terminal protein of pneumococcal bacteriophage Cp-1 and 5′-dAMP. *J. Virol.* **58:** 31–35.

Garcin, A. and D. Kolakofsky. 1992. Tacaribe arenavirus RNA synthesis *in vitro* is primer dependent and suggests an unusual model for the initiation of genome replication. *J. Virol.* **66:** 1370–1376.

Garmendia, C., A. Bernad, J.A. Esteban, L. Blanco, and M. Salas. 1992. The bacteriophage ϕ29 DNA polymerase, a proofreading enzyme. *J. Biol. Chem.* **267:** 2594–2599.

Gerlich, W.H. and W.S. Robinson. 1980. Hepatitis B virus contains protein attached to the 5′ terminus of its complete DNA strand. *Cell* **21:** 801–809.

Grosse, F. and G. Krauss. 1985. The primase activity of DNA polymerase-α from calf thymus. *J. Biol. Chem.* **260:** 1881–1888.

Gruss, C., C. Gutiérrez, W.C. Burhans, M.L. DePamphilis, T. Koller, and J.M. Sogo. 1990. Nucleosome assembly in mammalian cell extracts before and after DNA replication. *EMBO J.* **9:** 2911–2922.

Gutiérrez, C., J.M. Sogo, and M. Salas. 1991a. Analysis of replicative intermediates produced during bacteriophage ϕ29 DNA replication *in vitro*. *J. Mol. Biol.* **222:** 983–994.

Gutiérrez, C., G. Martín, J.M. Sogo, and M. Salas. 1991b. Mechanism of stimulation of DNA replication by bacteriophage ϕ29 SSB protein p5. *J. Biol. Chem.* **266:** 2104–2111.

Hansen, L.J. and S.B. Sandmeyer. 1990. Characterization of a transpositionally active Ty3 element and identification of the Ty3 integrase protein. *J. Virol.* **64:** 2599–2607.

Haseltine, W.A., A.M. Maxam, and W. Gilbert. 1977. Rous sarcoma virus genome is terminally redundant: The 5′ sequence. *Proc. Natl. Acad. Sci.* **74:** 989–993.

Hay, R.T. and M.L. DePamphilis. 1982. Initiation of simian virus 40 DNA replication in vivo: Location and structure of 5′-ends of DNA synthesized in the *Ori* region. *Cell* **28:** 767–779.

Hay, R.T., E.A. Hendrickson, and M.L. DePamphilis. 1984. Sequence specificity for the initiation of RNA primed-SV40 DNA synthesis in vivo. *J. Mol. Biol.* **175:** 131–157.

Hayashi, D.K. and B.C. Stark. 1994. A novel tRNA precursor cleaving endoribonuclease from *Zea mays*. *Arch. Biochem. Biophys.* **309:** 123–128.

Hendrickson, E.A., C.E. Fritze, W.R. Folk, and M.L. DePamphilis. 1987a. The origin of bidirectional DNA replication in polyoma virus. *EMBO J.* **6:** 2011–2018.

———.1987b. Polyoma virus DNA replication is semi-discontinuous. *Nucleic Acids. Res.* **15:** 6369–6385.

Hermoso, J.M., E. Méndez, F. Soriano, and M. Salas. 1985. Location of the serine residue involved in the linkage between the terminal protein and the DNA of phage ϕ29. *Nucleic Acids Res.* **13:** 7715–7728.

Hirokawa, H. 1972. Transfecting deoxyribonucleic acid of *Bacillus* bacteriophage ϕ29 that is protease sensitive. *Proc. Natl. Acad. Sci.* **69:** 1555–1559.

Hsieh, J.-C., S.K. Yoo, and J. Ito. 1990. An essential arginine residue for initiation of protein-primed DNA replication. *Proc. Natl. Acad. Sci.* **87:** 8665–8669.

Hsieh, J.-C., G. Jung, M.C. Leavitt, and J. Ito. 1987. Primary structure of the DNA terminal protein of bacteriophage PRD1. *Nucleic Acids Res.* **15:** 8999–9009.

Hu, J.C., and J.E. Dahlberg. 1983. Structural features required for the binding of $tRNA^{Trp}$ to avian myeloblastosis virus reverse transcriptase. *Nucleic Acids Res.* **11:** 4823–4833.

Huang, Y., J. Mak, Q. Cao, Z. Li, M.A. Wainberg, and L. Kleiman. 1994. Incorporation of excess wild-type and mutant $tRNA_3^{Lys}$ into human immunodeficiency virus type 1. *J. Virol.* **68:** 7676–7683.

Hunter, T. and B. Francke. 1974. In vitro polyoma DNA synthesis: Involvement of RNA in discontinuous chain growth. *J. Mol. Biol.* **83:** 123–130.

Isel, C., R. Marquet, G. Keith, C. Ehresmann, and B. Ehresmann. 1993. Modified nucleotides of $tRNA_3^{Lys}$ modulate primer/template loop-loop interaction in the initiation complex of HIV-1 reverse transcription. *J. Biol. Chem.* **268:** 25269–25272.

Ishimi, Y., A. Claude, P. Bullock, and J. Hurwitz. 1988. Complete enzymatic synthesis of DNA containing the SV40 origin of replication. *J. Biol. Chem.* **263:** 19723–19733.

Janetzky, B. and Lehle, L. 1992. Ty4, a new retrotransposon from *Saccharomyces cerevisiae*, flanked by *tau*-elements. *J. Biol. Chem.* **267:** 19798–19805.

Jung, G., M.C. Leavitt, J.-C. Hsieh, and J. Ito. 1987. Bacteriophage PRD1 DNA polymerase: Evolution of DNA polymerases. *Proc. Natl. Acad. Sci.* **84:** 8287–8291.

Katz, R.A., G. Merkel, J. Kulkosky, J. Leis, and A.M. Skalka. 1990. The avian retroviral IN protein is both necessary and sufficient for integrative recombination *in vitro*. *Cell* **63:** 87–95.

Kaufmann, G. 1981. Characterization of initiator RNA from replicating SV40 DNA synthesized in isolated nuclei. *J. Mol. Biol.* **147:** 25–39.

Kaufmann, G., S. Anderson, and M.L. DePamphilis. 1977. RNA primers in simian virus 40 DNA replication: Distribution of 5′-terminal oligoribonucleotides in nascent DNA. *J. Mol. Biol.* **116:** 549–567.

Kaufmann, G., R. Bar-Shavit, and M.L. DePamphilis. 1978. Okazaki pieces grow opposite to the replication fork direction during simian virus 40 replication. *Nucleic Acids Res.* **5:** 2535–2545.

Kawasaki K., T. Enomoto, M. Suzuki, M. Seki, F. Hanaoka, and M. Yamada. 1986. Detection and characterization of a novel factor that stimulates DNA polymerase-α. *Biochemistry* **25:** 3044–3050.

Kelly, R.E., M.L. DeRose, B.W. Draper, and G.M. Wahl. 1995. Identification of an origin of bidirectional replication within the coding region of the ubiquitously expressed *CAD* gene. *Mol. Cell. Biol.* **15:** 4136–4148.

Kempken, F., F. Meinhardt, and K. Esser. 1989. In *organello* replication and viral affinity of linear, extrachromosonal DNA of the ascomycete *Ascobolus immersus*. *Mol. Gen. Genet.* **218:** 523–530.

Kennell, J.C., H. Wang, and A. Lambowitz, 1994. The Mauriceville plasmid of *Neurospora* spp. uses novel mechanisms for initiating reverse transcription in vivo. *Mol. Cell. Biol.* **14:** 3094–3107.

Khudyakov, Y.E. and A.M. Makhov. 1989. Prediction of terminal protein and ribonuclease H domains in the gene P product of hepadnaviruses. *FEBS Lett.* **243:** 115–118.

Kikuchi, Y. and N. Sasaki. 1992. Hyperprocessing of tRNA by the catalytic RNA of RNase P. *J. Biol. Chem.* **267:** 11972–11976.

Kikuchi, Y., Y. Ando, and T. Shiba. 1986. Unusual priming mechanism of RNA-directed DNA synthesis in copia retrovirus-like particles of *Drosophila*. *Nature* **323:** 824–826.

King, A.J. and P.C. van der Vliet. 1994. A precursor terminal protein-trinucleotide intermediate during initiation of adenovirus DNA replication: Regeneration of molecular ends in vitro by a jumping-back mechanism. *EMBO J.* **13:** 5786–5792.

Kobayashi, H., K. Kitabayaski, K. Matsumoto, and H. Hirokawa. 1991a. Primer protein of bacteriophage M2 exposes the RGD receptor site upon linking the first deoxynucleotide. *Mol. Gen. Genet.* **226:** 65–69.

———. 1991b. Receptor sequence of the terminal protein of bacteriophage M2 that interacts with an RGD (Arg-Gly-Asp) sequence of the primer protein. *Virology* **185:** 901–903.

Kohlstaedt, L.A. and T.A. Steitz. 1992. Reverse transcriptase of human immunodeficiency virus can use either $tRNA_3^{Lys}$ or *Escherichia coli* $tRNA_2^{Gln}$ as a primer in an *in vitro* primer-utilization assay. *Proc. Natl. Acad. Sci.* **89:** 9652–9656.

Kornberg, A. and T. Baker. 1992. *DNA replication*. W.H. Freeman, New York.

Kuchta, R.D., B. Reid, and L.M. Chang. 1990. DNA primase processivity and the primase to polymerase-α activity switch. *J. Biol. Chem.* **265:** 16158–16165.

Kumble, K.D., P.L. Iversen, and J.K. Vishwanatha. 1992. The role of primer recognition proteins in DNA replication: Inhibition of cellular proliferation by antisense oligodeoxyribonucleotides. *J. Cell Sci.* **101:** 35–41.

Lauermann, V. and J.D. Boeke. 1994. The primer tRNA sequence is not inherited during Ty1 retrotransposition. *Proc. Natl. Acad. Sci.* **91:** 9847–9851.

Lehman, I.R. and L.S. Kaguni. 1989. DNA polymerase α. *J. Biol. Chem.* **264:** 4265–4268.

Leis, J., A. Aiyar, and D. Cobrinik. 1993. Regulation of initiation of reverse transcription of retroviruses. In *Reverse transcriptase* (ed. A.M. Skalka and S.P. Goff), pp. 33–47. Cold Spring Harbor Laboratory Press, Cold Spring Harbor, New York.

Levin, J.G. and J.G. Seidman. 1979. Selective packaging of host tRNA's by murine leukemia virus particles does not require genomic RNA. *J. Virol.* **29:** 328–335.

———. 1981. Effect of polymerase mutations on packaging of primer $tRNA^{Pro}$ during murine leukemia virus assembly. *J. Virol.* **38:** 403–408.

Li, X., J. Mak, E.J. Arts, Z. Gu, L. Kleiman, M.A. Wainberg, and M.A. Parniak. 1994. Effects of alterations of primer-binding site sequences on human immunodeficiency virus type 1 replication. *J. Virol.* **68:** 6198–6206.

Lichy, J.H., M.S. Horwitz, and J. Hurwitz. 1981. Formation of a covalent complex between the 80,000-dalton adenovirus terminal protein and 5′-dCMP *in vitro*. *Proc. Natl. Acad. Sci.* **78:** 2678–2682.

Lin, Y.S., H.M. Kieser, D.A. Hopwood, and C.W. Chen. 1993. The chromosomal DNA of *Streptomyces lividans* 66 is linear. *Mol. Microbiol.* **10:** 923–933.

Lindauer, A., D. Fraser, M. Bruderlein, and R. Schmitt. 1993. Reverse transcriptase families and a copia-like retrotransposon, *osser*, in the green alga *Volvox carteri*. *FEBS Lett.* **319:** 261–266.

Lindenbaum, J.D., J. Field, and J. Hurwitz. 1986. The adenovirus DNA-binding protein and adenovirus DNA polymerase interact to catalyze elongation of primed DNA templates. *J. Biol. Chem.* **261:** 10218–10227.

Liu, M.-H., Y. Yuan, and R. Reddy. 1994. Human RNase P RNA and nucleolar 7-2 RNA share conserved 'T_o' antigen-binding domains. *Mol. Cell. Biochem.* **130:** 75–82.

Mak, J., M. Jiang, M.A. Wainberg, M.-L. Hammarskjold, D. Rekosh, and L. Kleiman. 1994. Role of $Pr160^{gag\text{-}pol}$ in mediating the selective incorporation of $tRNA^{Lys}$ into human immunodeficiency virus type 1 particles. *J. Virol.* **68:** 2065–2072.

Martín, A.C., R. López, and P. García. 1995. Nucleotide sequence and transcription of the left early region of *Streptococcus pneumoniae* bacteriophage Cp-1 coding for the terminal protein and the DNA polymerase. *Virology* **211:** 21--32.

Martín, G., J.M. Lázaro, E. Méndez, and M. Salas. 1989. Characterization of the phage ϕ29 protein p5 as a single-stranded DNA binding protein. Function in ϕ29 DNA-protein p3 replication. *Nucleic Acids Res.* **17:** 3663–3672.

Matsumoto, K., T. Saito, and H. Kirokawa. 1983. *In vitro* initiation of bacteriophage ϕ29 and M2 DNA replication: Genes required for formation of a complex between the terminal protein and 5′ dAMP. *Mol. Gen. Genet.* **191:** 26–30.

Matsumoto, K., T. Saito, C.I. Kim, T. Ando, and H. Hirokawa. 1984. Bacteriophage ϕ29 DNA replication in vitro: Participation of the terminal protein and the gene 2 product on elongation. *Mol. Gen. Genet.* **196:** 381–386.

McNeel, D.G. and F. Tamanoi. 1991. Terminal region recognition factor 1, a DNA-binding protein recognizing the inverted terminal repeats of the pGKL1 linear DNA plasmids. *Proc. Natl. Acad. Sci.* **88:** 11398–11402.

Meinhardt, F., F. Kempken, J. Kämper, and K. Esser. 1990. Linear plasmids among eukaryotes: Fundamentals and application. *Curr. Genet.* **17:** 89–95.

Mendelman, L.V. and C.C. Richardson. 1991. Requirements for primer synthesis by bacteriophage T7 63-kDa gene 4 protein. *J. Biol. Chem.* **266:** 23240–23250.

Méndez, E., L. Blanco, J.A. Esteban, A. Bernad, and M. Salas. 1992. Initiation of ϕ29 DNA replication occurs at the second 3′ nucleotide of the linear template: A sliding-back mechanism for protein-primed DNA replication. *Proc. Natl. Acad Sci.* **89:** 9579–9583.

Mindich, L. and D.H. Bamford. 1988. Lipid-containing bacteriophages. In *The bacteriophages* (ed. R. Calendar), vol. 2, pp. 475–520. Plenum Press, New York.

Molnar-Kimber, K.L., J.W. Summers, J.M. Taylor, and W.S. Mason. 1983. Protein covalently bound to minus strand DNA intermediates of duck hepatitis B virus. *J. Virol.* **45:** 165–172.

Mougel, M., N. Tounekti, J.-L. Darlix, J. Paoletti, B. Ehresmann, and C. Ehresmann. 1993. Conformational analysis of the 5′ leader and the *gag* initiation site of Mo-MuLV RNA and allosteric transitions induced by dimerization. *Nucleic Acids Res.* **21:** 4677–4684.

Mul, Y.M. and P.C. van der Vliet. 1993. The adenovirus DNA binding protein effects the kinetics of DNA replication by a mechanism distinct from NFI or Oct-1. *Nucleic Acids Res.* **21:** 641–647.

Murphy, J.E. and S.P. Goff. 1989. Construction and analysis of deletion mutations in the U5 region of Moloney murine leukemia virus: Effects on RNA packaging and reverse transcription. *J. Virol.* **63:** 319–327.

Nasheuer, H.P. and F. Grosse. 1988. DNA polymerase α-primase from calf thymus. Determination of the polypeptide responsible for primase activity. *J. Biol. Chem.* **263:** 8981–8988.

Nethanel, T. and G. Kaufmann. 1990. Two DNA polymerases may be required for synthesis of the lagging strand of SV40. *J. Virol.* **64:** 5912–5918.

Nethanel, T., T. Zlotkin, and G. Kaufmann. 1992. Assembly of SV40 Okazaki pieces from DNA primers is reversibly arrested by ATP depletion. *J. Virol.* **66:** 6634–6640.

Nethanel, T., S. Reisfeld, G. Dinter-Gottlieb, and G. Kaufmann. 1988. An Okazaki piece of SV40 may be synthesized by ligation of shorter precursor chains. *J. Virol.* **62:** 2867–2873.

Oeser, B. and P. Tudzynski. 1989. The linear mitochondrial plasmid pClK1 of the phytopathogenic fungus *Claviceps purpurea* may code for a DNA polymerase and an RNA polymerase. *Mol. Gen. Genet.* **217:** 132–140.

Ollis, D.L., R. Brick, R. Hamlin, N.G. Xuong, and T.A. Steitz. 1985. Structure of the large fragment of *Escherichia coli* DNA polymerase I complexed with dTMP. *Nature* **313:** 762–766.

Olson, P., H.M. Temin, and R. Dornburg. 1992. Unusually high frequency of reconstitution of long terminal repeats in U3-minus retrovirus vectors by DNA recombination or gene conversion. *J. Virol.* **66:** 1336–1343.

Ortín, J., E. Viñucla, M. Salas, and C. Vásquez. 1971. DNA-protein complex in circular DNA from phage ϕ29. *Nat. New Biol.* **234:** 275–277.

Paillard, M., R.A. Sederoff, and C.S. Levings III. 1985. Nucleotide sequence of the S-1 mitochondrial DNA from the S cytoplasm of maize. *EMBO J.* **4:** 1125–1128.

Panet, A., W.A. Haseltine, D. Baltimore, G. Peters, F. Harada, and J.E. Dahlberg. 1975. Specific binding of tryptophan transfer RNA to avian myeloblastosis virus RNA-dependent DNA polymerase (reverse transcriptase). *Proc. Natl. Acad. Sci.* **72:** 2535–2539.

Peñalva, M.A. and M. Salas. 1982. Initiation of phage ϕ29 DNA replication *in vitro*: Formation of a covalent complex between the terminal protein, p3, and 5′-dAMP. *Proc. Natl. Acad. Sci.* **79:** 5522–5526.

Perlman, D. and J.A. Huberman. 1977. Asymmetric Okazaki piece synthesis during replication of SV40 DNA in vivo. *Cell* **12:** 1029–1043.

Peters, G.G. and J. Hu. 1980. Reverse transcriptase as the major determinant for selective packaging of tRNA's into avian sarcoma virus particles. *J. Virol.* **36:** 692–700.

Pettit, S.C., M.S. Horwitz, and J.A. Engler. 1989. Mutations of the precursor to the terminal protein of adenovirus serotype 2 and 5. *J. Virol.* **63:** 5244–5250.

Pochart, P., B. Agoutin, C. Fix, G. Keith, and T. Heyman. 1993. A very poorly expressed $tRNA^{Ser}$ is highly concentrated together with replication primer initiator $tRNA^{Met}$ in the yeast Ty1 virus-like particles. *Nucleic Acids Res.* **21:** 1517–1521.

Prieto, I., J.M. Lázaro, J.A. García, J.M. Hermoso, and M. Salas. 1984. Purification in a functional form of the terminal protein of *Bacillus subtilis* phage ϕ29. *Proc. Natl. Acad. Sci.* **81:** 1639–1643.

Prieto, I., M. Serrano, J.M. Lázaro, M. Salas, and J.M. Hermoso. 1988. Interaction of the bacteriophage ϕ29 protein p6 with double-stranded DNA. *Proc. Natl. Acad. Sci.* **85:** 314–318.

Pritchard, C.G., D.T. Weaver, E.F. Baril, and M.L. DePamphilis. 1983. DNA polymerase-α cofactors C_1C_2 function as primer recognition proteins. *J. Biol. Chem.* **258:** 9810–9819.

Radziwill, G., W. Tuckek, and H. Schaller. 1990. Mutational analysis of the hepatitis B virus P gene product: Domain structure and RNAse H activity. *J. Virol.* **64:** 613–620.

Rattray, A.J. and J.J. Champoux. 1987. The role of Moloney murine leukemia virus RNase H activity in the formation of plus-strand primers. *J. Virol.* **61:** 2843–2851.

Reichard, P., R. Eliasson, and G. Soderman. 1974. Initiator RNA in discontinuous polyoma DNA synthesis. *Proc. Natl. Acad. Sci.* **71:** 4901–4905.

Rekosh, D.M.K., W.C. Russell, A.J.D. Bellet, and A.J. Robinson. 1977. Identification of a protein linked to the ends of adenovirus DNA. *Cell* **11:** 283–295.

Rohe, M., J. Schründer, P. Tudzynski, and F. Meinhardt. 1992. Phylogenetic relationship of linear, protein-primed replicating genomes. *Curr. Genet.* **21:** 173–176.

Romero, A., R. López, R. Lurz, and P. García. 1990. Temperate bacteriophages of *Streptococus pneumoniae* that contain protein covalently linked to the 5′ ends of their DNA. *J. Virol.* **64:** 5149–5155.

Rothnie, H.M., K.J. McCurrah, L.A. Glover, and N. Hardman. 1991. Retrotransposon-like nature of Tp1 elements: Implications for the organization of highly repetitive, hypermethylated DNA in the genome of *Physarum polycephalum. Nucleic Acids Res.* **19:** 279–286.

Sakaguchi, K. 1990. Invertrons, a class of structurally and functionally related genetic elements that includes linear DNA plasmids, transposable elements, and genomes of adeno-type viruses. *Microbiol. Rev.* **54:** 66–74.

Salas, M. 1991. Protein-priming of DNA replication. *Annu. Rev. Biochem.* **60:** 39–71.

Salas, M., R.P. Mellado, E. Viñuela, and J.M. Sogo. 1978. Characterization of a protein covalently linked to the 5′ termini of the DNA of *Bacillus subtilis* phage ϕ29. *J. Mol. Biol.* **119:** 269–291.

Salas, M., J. Méndez, J.A. Esteban, M. Serrano, C. Gutierrez, J.M. Hermoso, A. Bravo, M.S. Soengas, J.M. Lázaro, M.A. Blasco, R. Freire, A. Bernad, J.M. Sogo, and L. Blanco. 1993. Terminal protein priming of DNA replication: Bacteriophage ϕ29 as a model system. In *Virus strategies, molecular biology and pathogenesis* (ed. W. Doerfler and P. Böhm), pp. 3–19. Verlag Chemie. Weinheim, Germany.

Santocanale, C., M. Foiani, G. Lucchini, P. Plevani. 1993. The isolated 48,000-dalton subunit of yeast DNA primase is sufficient for RNA primer synthesis. *J. Biol. Chem.* **268:** 1343–1348.

Savilahti, H. and D. Bamford. 1987. The complete nucleotide sequence of the left very early region of *Escherichia coli* bacteriophage PRD1 coding for the terminal protein and the DNA polymerase. *Gene* **57:** 121–130.

Savilahti, H., J. Caldentey, K. Lundström, J.E. Syväoja, and D.H. Bamford. 1991. Overexpression, purification and characterization of *Escherichia coli* bacteriophage PRD1 DNA polymerase. *J. Biol. Chem.* **266:** 18737–18744.

Schmitt, M.E., J.L. Bennett, D.J. Dairaghi, and D.A. Clayton. 1993. Secondary structure of RNase MRP RNA predicted by phylogenetic comparison. *FASEB J.* **7:** 208–213.

Seeger, C., J. Summers, and W.S. Mason. 1991. Viral DNA synthesis. *Curr. Top. Microbiol. Immunol.* **168:** 41–59.

Selsing, E. and R.D. Wells. 1979. Polynucleotide block polymers consisting of a DNA•RNA hybrid joined to a DNA•DNA duplex. *J. Biol. Chem.* **254:** 5410–5416.

Serrano, M., M. Salas, and J.M. Hermoso. 1990. A novel nucleoprotein complex at a replication origin. *Science* **248:** 1012–1016.

Serrano, M., C. Gutiérrez, M. Salas, and J.M. Hermoso. 1993. Superhelical path of the DNA in the nucleoprotein complex that activates the initiation of phage ϕ29 DNA replication. *J. Mol. Biol.* **230:** 248–259.

Serrano, M., J. Gutiérrez, I. Prieto, J.M. Hermoso, and M. Salas. 1989. Signals at the bacteriophage ϕ29 DNA replication origins required for protein p6 binding and activity. *EMBO J.* **8:** 1879–1885.

Sheaff, R.J. and R.D. Kuchta. 1993. Mechanism of calf thymus DNA primase: Slow initiation, rapid polymerization, and intelligent termination. *Biochemistry* **32:** 3027–3037.

———. 1994. Misincorporation of nucleotides by calf thymus DNA primase and elongation of primers containing multiple noncognate nucleotides by DNA polymerase-α. *J. Biol. Chem.* **269:** 19225–19231.

Shih, M.F., K. Watabe, and J. Ito. 1982. *In vitro* complex formation between bacteriophage ϕ29 terminal protein and deoxynucleotide. *Biochem. Biophys. Res. Commun.* **105:** 1031–1036.

Shimada, M., H. Hosaka, H. Takaku, J.S. Smith, M.J. Roth, S. Inouye, and M. Inouye. 1994. Specificity of priming reaction of HIV-1 reverse transcriptase, 2-OH or 3′ OH. *J. Biol. Chem.* **269:** 3925–3927.

Shimamoto, T., M.-Y. Hsu, S. Inouye, and M. Inouye. 1993. Reverse transcriptases from bacterial retrons require specific secondary structures at the 5′ end of the template for the cDNA priming reaction. *J. Biol. Chem.* **268:** 2684–2692.

Shiue, S.Y., J.C. Hsieh, and J. Ito. 1991. Mapping of the linking tyrosine residue of the PRD1 terminal protein. *Nucleic Acids Res.* **19:** 3805–3810.

Singh, H., R.G. Brooke, M.H. Pausch, G.T. Williams, C. Trainor, and L.B. Dumas. 1986. Yeast DNA primase and DNA polymerase activities. An analysis of RNA priming and its coupling to DNA synthesis. *J. Biol. Chem.* **261:** 8564–8569.

Smart, J.E. and B.W. Stillman. 1982. Adenovirus terminal protein precursor. *J. Biol. Chem.* **257:** 13499–13506.

Smith, J.K., A. Cywinski, and J.M. Taylor. 1984. Specificity of initiation of plus-strand DNA by Rous sarcoma virus. *J. Virol.* **52:** 314–319.

Soengas, M.S., C. Gutiérrez, and M. Salas. 1995. Helix-destabilizing activity of ϕ29 single-stranded DNA binding protein: Effect on the elongation rate during strand displacement DNA replication. *J. Mol. Biol.* **253:** 517–529.

Soengas, M.S., J.A. Esteban, M. Salas, and C. Gutiérrez. 1994. Complex formation between phage ϕ29 single-stranded DNA binding protein and DNA. *J. Mol. Biol.* **239:** 213–226.

Soengas, M.S., J.A. Esteban, J.M. Lázaro, A. Bernad, M.A. Blasco, M. Salas, and L. Blanco. 1992. Site-directed mutagenesis at the Exo III motif of ϕ29 DNA polymerase. Overlapping structural domains for the 3′ → 5′ exonuclease and strand-displacement activities. *EMBO J.* **11:** 4227–4237.

Steenbergh, P.H., J. Maat, H. van Ormondt, and J.S. Sussenbach. 1977. The nucleotide sequence at the termini of adenovirus type 5 DNA. *Nucleic Acids Res.* **4:** 4371–4389.

Stewart. L., G. Schatz, and V.M. Vogt. 1990. Properties of avian retrovirus particles defective in viral protease. *J. Virol.* **64:** 5076–5092.

Stucka, R., C. Schwarzlose, H. Lochmuller, U. Hacker, and H. Feldmann. 1992. Molecular analysis of the yeast Ty4 element: Homology with Ty1, copia, and plant retrotransposons. *Gene* **122:** 119–128.

Summers, J. and W.S. Mason. 1982. Replication of the genome of a hepatitis B-like virus by reverse transcription of an RNA intermediate. *Cell* **29:** 403–415.

Suzuki, M., E. Savoysky, S. Izuta, M. Tatebe, T. Okajima, and S. Yoshida. 1993. RNA priming coupled with DNA synthesis on natural template by calf thymus DNA polymerase-α-primase. *Biochemistry* **32:** 12782–12792.

Swart, J.R. and M.A. Griep. 1993. Primase from *E. coli* primes single stranded templates in the absence of single stranded DNA binding protein or other auxiliary proteins. *J. Biol. Chem.* **268:** 12970–12976.

Tasheva, E.S. and D.J. Roufa. 1994. A mammalian origin of bidirectional DNA replication within the Chinese hamster RPS14 locus. *Mol. Cell. Biol.* **14:** 5628–5635.

Temperley, S.M. and R.T. Hay. 1992. Recognition of the adenovirus type 2 origin of DNA replication by the virally encoded DNA polymerase and preterminal proteins. *EMBO J.* **11:** 761–768.

Tenney, D.J., A.K. Sheaffer, W.W. Hurlburt, M. Bifano, and R.K. Hamatake. 1995. Sequence-dependent primer synthesis by the herpes simplex virus helicase primase complex. *J. Biol. Chem.* **270:** 9129–9136.

Tseng, B.Y. and C.N. Ahlem. 1983. A DNA primase from mouse cells: Purification and partial characterization. *J. Biol. Chem* **258:** 9845–9849.

Tseng, B.Y. and M. Goulian. 1975. Evidence for covalent association of RNA with nascent DNA in human lymphocytes. *J. Mol. Biol.* **99:** 339–346.

———. 1980. Initiator RNA synthesis upon ribonucleotide depletion: Evidence for base substitutions. *J. Biol. Chem.* **255:** 2062–2066.

Tseng, B.Y. and C.E. Prussak. 1989. Sequence and structural requirements for primase initiation in the SV40 origin of replication. *Nucleic Acids Res.* **17:** 1953–1963.

Tseng, B., J.M. Erickson, and M. Goulian. 1979. Initiator RNA of nascent DNA from animal cells. *J. Mol. Biol.* **129:** 531.

Vishwanatha, J.K., M. Yamaguchi, M.L. DePamphilis, and E.F. Baril. 1986. Selection of template initiation sites and lengths of RNA primers synthesized by DNA primase are strongly affected by its organization in a multiprotein DNA polymerase-α complex. *Nucleic Acids Res.* **14:** 7305–7323.

Von Pawel-Rammingen, U., S. Astrom, and A.S. Bystrom. 1992. Mutational analysis of conserved positions potentially important for initiator tRNA function in *Saccharomyces cerevisiae*. *Mol. Cell. Biol.* **12:** 1432–1442.

Voytas, D.F. and J.D. Boeke. 1992. Yeast retrotransposon revealed (letter). *Nature* **358:** 717.

———. 1993. Yeast retrotransposons and tRNAs. *Trends Genet.* **9:** 421–427.

Wakefield, J.K., H. Rhim, and C.D. Morrow. 1994. Minimal sequence requirements of a functional human immunodeficiency virus 1 primer binding site. *J. Virol.* **68:** 1605–1614.

Wang, G.H. and C. Seeger. 1993. Novel mechanism for reverse transcription in hepatitis B viruses. *J. Virol.* **67:** 6507–6512.

Wang, H. and A. Lambowitz. 1993. The Mauriceville plasmid reverse transcriptase can initiate cDNA synthesis de novo and may be related to reverse transcriptase and DNA polymerase progenitor. *Cell* **75:** 1071–1081.

Waqar, M.A., and J.A. Huberman. 1975. Covalent attachment of RNA to nascent DNA in mammalian cells. *Cell* **6:** 551–557.

Warmington, J.R., R.B. Waring, C.S. Newlon, K.J. Indge, and S.G. Oliver. 1985. Nucleotide sequence characterization of Ty1-17, a class II transposon from yeast. *Nucleic Acids Res.* **13:** 6679–6693.

Watabe, K., M. Leusch, and J. Ito. 1984a. Replication of bacteriophage ϕ29 DNA in vitro: The roles of terminal protein and DNA polymerase. *Proc. Natl. Acad. Sci.* **81:** 5374–5378.

———. 1984b. A 3′ to 5′ exonuclease activity is associated with phage ϕ29 DNA polymerase. *Biochem. Biophys. Res. Commun.* **123:** 1019–1026.

Watabe, K., M.F. Shih, and J. Ito. 1983. Protein-primed initiation of phage ϕ29 DNA replication. *Proc. Natl. Acad. Sci.* **80:** 4248–4252.

Weiser, B., D. Ganem, C. Seeger, and H.E. Varmus. 1983. Closed circular viral DNA and asymmetrical heterogeneous forms in livers from animals infected with ground squirrel hepatitis virus. *J. Virol.* **48:** 1–19.

Weiss, R., N. Teich, H. Varmus, and J. Coffin, eds. 1985. *RNA tumor viruses,* 2nd edition, parts 1 and 2. Cold Spring Harbor Laboratory, Cold Spring Harbor, New York.

Wohrl, B.M., B. Ehresmann, G. Keith, S.F.J. LeGrice. 1993. Nuclease footprinting of human immunodeficiency virus reverse transcriptase/$tRNA^{Lys\text{-}3}$ complexes. *J. Biol. Chem.* **268:** 13617–13624.

Yamaguchi, M., E.A. Hendrickson, and M.L. DePamphilis. 1985a. DNA primase-DNA polymerase-α from simian cells: Sequence specificity of initiation sites on simian virus 40 DNA. *Mol. Cell. Biol.* **5:** 1170–1183.

———. 1985b. DNA primase-DNA polymerase-α from simian cells: Modulation of RNA primer synthesis by ribonucleoside triphosphates. *J. Biol. Chem.* **260:** 6254–6263.

Yanofsky, S., F. Kawamura, and J. Ito. 1976. Thermolabile transfecting DNA from temperature-sensitive mutant of phage ϕ29. *Nature* **259:** 60–63.

Yoda, K. and T. Okazaki. 1991. Specificity of the recognition sequence for *E. coli* primase. *Mol. Gen. Genet.* **227:** 1–8.

Yoo, S.K. and J. Ito. 1991. Sequence requirements for protein-primed DNA replication of bacteriophage PRD1. *J. Mol. Biol.* **218:** 779–789.

Zakrzewska-Czekwinska, J. and H. Schrempf. 1992. Characterization of an autonomously replicating region from the *Streptomyces lividans* chromosome. *J. Bacteriol.* **147:** 2688–2693.

Zhu, W. and J. Ito. 1994. Purification and characterization of PRD1 DNA polymerase. *Biochim. Biophys. Acta* **1219:** 267–276.

Zhu, W., M.C. Leavitt, G. Jung, and J. Ito. 1994. Mutagenesis of a highly conserved lysine 340 of the PRD1 DNA polymerase. *Biochim. Biophys. Acta* **1219:** 260–266.

Zoulim, F. and C. Seeger. 1994. Reverse transcription in hepatitis B viruses is primed by a tyrosine residue of the polymerase. *J. Virol.* **68:** 6–13.

6

Mechanisms for Completing DNA Replication

Deepak Bastia and Bidyut K. Mohanty
Department of Microbiology
Duke University Medical Center
Durham, North Carolina 27710

DNA replication consists of three steps: initiation, ongoing replication, and termination. The termination of replication is important because the synchrony of the termination process with subsequent cell division should be a significant factor in the equal and orderly distribution of genetic material to the daughter cells of both prokaryotes and eukaryotes.

Termination in a circular chromosome in prokaryotes involves arrest of replication forks at specific sequences called replication arrest sites (for reviews of this process, see Hill 1992; Baker 1995). There are some exceptions to the phenomenon of sequence-specific replication arrest in both prokaryotic and eukaryotic chromosomes; e.g., the early stage of bacteriophage λ DNA replication (Valenzuela et al. 1976) and SV40 DNA replication (Lai and Nathans 1975). Replication fork arrest is the first step in the termination process. Termination also involves, in *Escherichia coli*, decatenation of the arrested, catenated daughter molecules of DNA by a special topoisomerase called topo IV (Kato et al. 1988, 1990; Schmid 1990; Adams et al. 1992; Hiasa and Marians 1994b). Mutants that are defective in topo IV, a heterodimeric enzyme, are also defective in nucleoid segregation (Kato et al. 1988; Schmid 1990; Adams et al. 1992). More information on topoisomerases can be found in the chapter by Hangaard Andersen et al. (this volume). A third step in termination and proper segregation of the newly completed daughter molecules involves a site-specific recombination system. The system resolves multimers generated by an odd number of recombinations which may occur between the two separating daughter, circular DNA molecules (Blakely et al. 1991; Kuempel et al. 1991).

Although a great deal is known about the termination of replication in prokaryotes, very little is known about this process in eukaryotes. In eukaryotic chromosomes, the linear DNA molecule faces two major problems for completing a round of replication. First, the newly synthesized strand at its 5′ end would have a gap created by the removal of the last primer RNA. The filling-in of the gap to complete a round of replication requires special structures at the ends of the DNA molecules, such as terminal redundancy, inverted repeats, or telomeres (for a discussion of the problem, see Kornberg and Baker 1992). Telomeres have been reviewed in a separate chapter in this volume (see Greider et al.). Second, since initiation of replication occurs at many origins in linear eukaryotic DNA, each replication unit must terminate or merge with the preceding and the succeeding units at internal termini. Whether these internal termini are sequence-specific remains unknown.

Since the movement of the replication fork is driven by two key enzymes, namely DNA helicase and DNA polymerase, interference with the activities of either of the two enzymes would cause pausing or arrest of replication forks. Interference with the movement of the replication fork may be due to specific DNA sequences, lesions on DNA, or specific DNA/protein complexes; interference may be transient (pausing) or may lead to arrest of the replisome complex for a longer period.

We first discuss pausing or arrest of replication forks at certain DNA sequences and the nature of these sequences. These pause or arrest sites do not necessarily define authentic replication termination sites but have interesting biological implications and, therefore, are discussed. A major part of this chapter focuses on our present knowledge of sequence-specific replication arrest systems in the prokaryotic organisms *E. coli* and *Bacillus subtilis*. Replication arrest at these sequences to which specific proteins bind has been referred to as termination process in the past. In fact, the fork arrest constitutes only the first step of replication termination. The few examples of known sequence-specific internal arrest sites in linear eukaryotic chromosomes are discussed at the end of the chapter.

PAUSING OF REPLICATION FORKS CAUSED BY DNA SEQUENCE/DNA POLYMERASE INTERACTION

Replication Pause Sites

Both prokaryotic and eukaryotic DNA polymerases pause or are arrested at certain DNA sequences (LaDuca et al. 1983; for review, see Bierne and Michel 1994). The arresting sequences are hairpin loops, polypurine stretches, or sequences that could adopt triple helix structures when subjected to negative supercoiling (Wells et al. 1988; Lindsey and Leach 1989; Dayn et al. 1992).

Weaver and DePamphilis (1984) have systematically analyzed the role of palindromic and nonpalindromic sequences in arresting DNA synthesis in vivo and in vitro. They observed that DNA polymerase-α was arrested before entering a palindromic sequence (class I sites). The arrest was nonpolar. There were also nonpalindromic (class II) sites that arrested DNA polymerase-α, and the sequences in front of and behind the point of arrest were necessary for the arrest to occur. Interestingly, the complementary sequence did not arrest polymerase-α-catalyzed DNA synthesis. The physiological role of these pause sites is mostly unknown at this time. An interesting sequence that causes fork arrest was discovered by the investigation of mitomycin-C-induced "onion skin" replication of mammalian chromosomal DNA initiated from the origin of replication of an integrated polyomavirus DNA. The pause-inducing site present in the adjacent host DNA contained the sequence $(dG\text{-}dA)_{27}$-$(dT\text{-}dC)_{27}$ (Baran et al. 1983, 1987). This sequence, when cloned into SV40 DNA, caused altered plaque morphology and arrested SV40 replication in vivo for approximately a minute. It is conceivable that such sequences are positioned to limit onion skin replication to certain regions of the chromosome (Rao et al. 1988).

Tapper and DePamphilis (1980) examined the pausing of the two replication forks approaching the replication terminus of SV40 and discovered that replication pause sites were spread over several hundred base pairs. In the absence of a specific termination site in SV40 (Lai and Nathans 1975), it is conceivable that the pause sites promote confluence of the two forks within a given region of SV40 DNA.

G-rich polypurine tracts have been shown to arrest DNA synthesis in vitro, and the arrest of DNA polymerase at the homopurine tracts was not relieved by the addition of single-stranded DNA-binding proteins or other accessory proteins (d'Ambrosio and Furano 1987). Near the left end of the linear chromosome of bacteriophage ϕ29, a strong pause site has been observed in vitro (M. Salas, pers. comm.) and, interestingly, this pause site arrests the fork moving left to right to prevent it from moving in a direction opposite to the direction of transcription of strongly transcribed early genes. Thus, the pause site could serve to prevent DNA replication from entering actively transcribed regions.

The mechanistic aspects of DNA polymerase pausing caused by DNA sequences are not entirely clear and thus remain as a potentially interesting topic for future investigation.

Possible Biological Roles of Pause Sites

Analysis of the available data provides very few clues to the possible physiological roles of replication pause sites. The pausing of replication forks in the regions near the replication termini could be a mechanism to terminate replication preferentially in that segment of the chromosome, in the absence of a sequence-specific replication terminus. The early and late transcription of SV40 DNA is controlled by promoters located near the origin, and the early and late transcription proceeds counterclockwise and clockwise, respectively, toward the terminus. It is conceivable that the pause sites fine-tune the bidirectional fork movement to proceed in the same directions as that of early and late transcripts and, thus, prevent the replication forks from entering actively transcribed regions from the opposite direction and from colliding with the transcriptional apparatus.

In summary, several types of DNA sequences can cause pausing of the movement of DNA polymerases in vivo and in vitro. In some cases, the pausing could serve physiological functions such as limiting DNA amplification by onion skin replication to a certain region of the chromosome or preventing replication forks from entering actively transcribed regions from a direction opposite to that of the transcribing RNA polymerase (Brewer 1988). The polypurine pause site that was discovered at the termini of adjacent chromosomal DNA that was replicated by polyomavirus-initiated onion skin replication of transformed rat cells could be acting to limit gene amplification to a limited region of the chromosome. However, since polyoma DNA can integrate into many chromosomal sites, it is likely that the replication is arrested wherever the forks find a polypurine stretch by chance. It is not known what function the polypurine stretches perform in host replication or recombination. Since pause sites are potentially recombinogenic, these sites could have evolved to make certain DNA regions available for the initiation of the recombinogenic process.

REPLICATION FORK ARREST AT THE TERMINI OF PROKARYOTES

Methods for Detection of Replication Termini

Most prokaryotic chromosomes initiate replication from one origin at a time, and the replication forks usually move bidirectionally until they meet each other at the terminus to generate two daughter molecules. The terminus region in most of the bacterial chromosomes and plasmids contains sequence-specific replication arrest sites that limit the end of the replication cycle to this region by blocking the fork progression (for review, see Hill 1992). To locate the origin and the terminus and to determine the direction of the fork movement, the frequencies of genetic markers located near the origin, termini, and the region in between are measured, during either exponential or synchronized growth. More than three decades ago, the idea of marker frequency analysis was elegantly used by Yoshikawa and Sueoka (1963) to localize the replication origin and the terminus and to determine the direction of fork progression in *B. subtilis*. The principle behind the work is shown in Figure 1A. In a linear chromosome of unit length, when replication is proceeding unidirectionally from an origin located at 0 toward the terminus located at 1, the frequency $g(x)$ of a marker x is obtained by solving the integral shown in Figure 1A. The solution of the integral yields 2^{1-x}. When $x = 0$, $g(x) = 2$; when $x = 1$, $g(x) = 1$. Thus, the frequency of markers located close to the origin will be two times that of the markers located at or near the terminus. The frequencies of the markers located between the *ori* and the terminus follow a gradient of values between 1 and 2 shown in the curve of $g(x)$ plotted as a function of the map location. In practice, Yoshikawa and Sueoka (1963) determined the marker frequencies by extracting the sheared DNA from an exponentially growing prototrophic strain of *B. subtilis*, transforming a multiply marked auxotroph and measuring the number of transformants for each marker. The marker frequencies were normalized using a marker located in the middle of the chromosome, such as leucine (Fig. 1A). The same idea can be extended to investigate the movement of a bidirectionally replicating chromosome.

If the replication terminus is located at a position that is 180° from the origin, the symmetry of the replication fork movement will normally make it difficult to determine if the replication fork arrest is sequence-specific. If the movement of the forks is synchronous, the two forks meet each other at a location diametrically opposite to the origin even in the absence of an arresting sequence at the terminus. Thus, an asymmetry has to be generated in the fork progression to detect a sequence-specific terminus. The asymmetry can be created by moving the origin with respect to the terminus, either by introducing a replication origin at a new location or by introducing deletions or duplications into one arm of the chromosome between the *ori* and the terminus. A replication terminus that arrests forks can be detected by the arrest of one fork before the arrival of the second fork at an asymmetric location with respect to *ori*.

In *B. subtilis*, asymmetry was generated by making one arm from the *ori* to the terminus longer (than the other arm) by 25% by a chromosomal duplication (see Schnieder et al. 1983). Using this strain, O'Sullivan and Anagnostopoulus (1982) performed marker frequency analysis to localize an arresting terminus. Similarly, asymmetry in the *E. coli* chromosome was created by initiating replication from an integrated copy of P2 phage origin and by introducing deletions (Kuempel et al. 1977; Louarn et al. 1977; Kuempel and Duerr 1979; Francois et al. 1989, 1990). His-

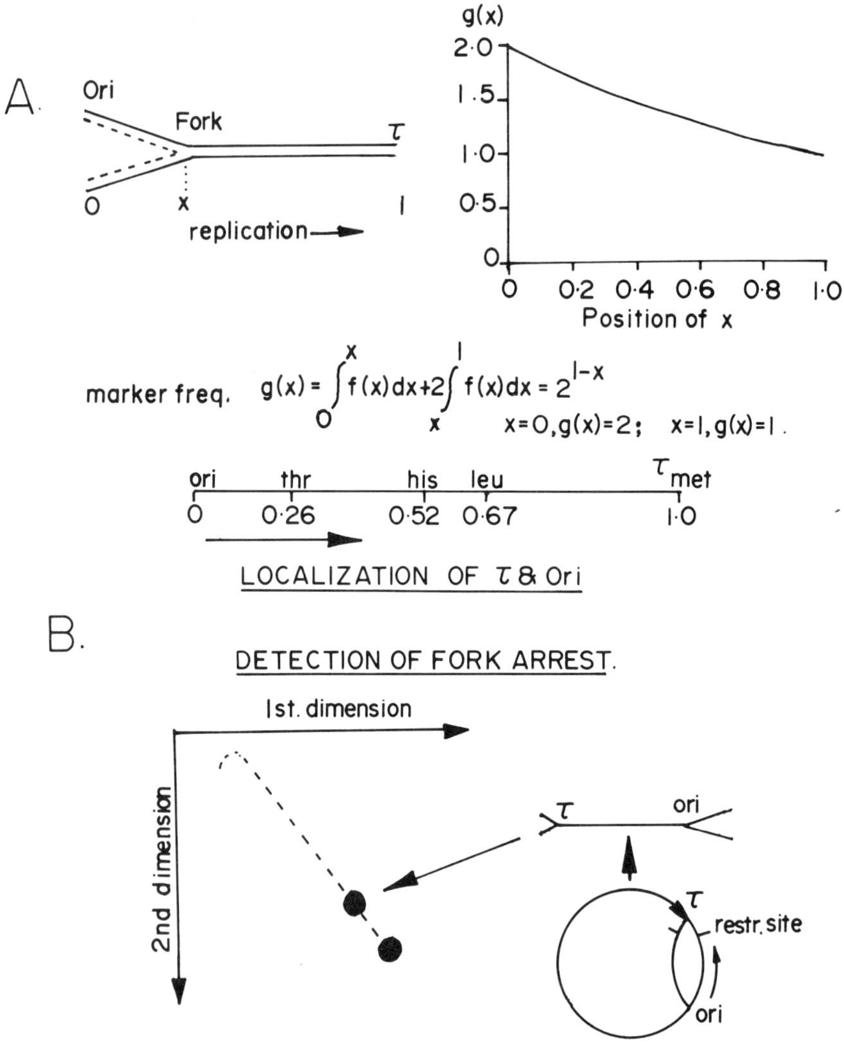

Figure 1 Methods for detection of replication origins and termini. (*A*) Marker frequency analysis developed by Yoshikawa and Sueoka (1963). The chromosome is linear with a replication origin at 0 and the terminus at 1.0. The fork movement is unidirectional. The equation for marker frequency $g(x)$, the plot of $g(x)$ as a function of map location x, and the chromosomal map of a segment of *B. subtilis* chromosome are shown. The equation predicts that the frequency of the markers at *ori*, in a log-phase cell population, would be twice that of the markers at the terminus. (*B*) Two-dimensional Brewer-Fangman gel electrophoresis of a unidirectional replicon. The *ori* and terminus (τ) are shown. The circular replicon is linearized at the unique restriction site, and the DNA is fractionated in the first dimension by molecular mass. The second dimension includes ethidium bromide, resulting in shape discrimination.

torically, the discovery of the specific terminus in *E. coli* was made by Kuempel et al. (1977) and Louarn et al. (1977). The replication terminus was identified in plasmid R6K by electron microscopy of the forks impeded by an asymmetrically located terminus (Lovett et al. 1975; Crosa et al. 1976). The bacterial termini were localized initially by measuring marker frequency by DNA-DNA hybridizations and comparing the frequency of markers located immediately before and after the terminus.

The termini of R6K were cloned into a unidirectionally replicating vector, and the replication was visualized by electron microscopy of replication intermediates (Kolter and Helinski 1978). The results showed that the replication termini transiently arrested the replication fork, and then the forks were released to finish replication near the *ori*. Using smaller and smaller pieces of the terminal DNA, the termini were localized to a 216-bp piece of DNA (Bastia et al. 1981a). Subsequent work revealed that the DNA fragment contained two sites of opposite polarity (Horiuchi and Hidaka 1988), each of which could block the replication fork approaching from one direction.

Another technique developed to map an arrested fork in larger bacterial chromosomes is one-dimensional gel electrophoresis designed to detect a Y-shaped stalled fork (Weiss and Wake 1984; Horiuchi and Hidaka 1988; Pelletier et al. 1988).

A technique that has greatly aided the investigation of replication origins, fork movement analysis, localization of replication arrest sites, and the termini is the two-dimensional neutral agarose gel electrophoresis developed by Brewer and Fangman (1987). A variant procedure using electrophoresis in the first dimension in a neutral agarose gel followed by a second dimension in alkaline condition has been developed by Huberman and colleagues (Linskens and Huberman 1988). For detailed discussion of the procedures, the reader is referred to the original papers. The Brewer-Fangman technique, as it applies to detecting replication termini in a unidirectionally replicating plasmid, is shown in Figure 1B. A replication terminus (τ) was cloned into a pUC19 vector approximately 450 bp away from the unidirectional replication origin. The replication intermediates were isolated from cells growing in the exponential phase, and the plasmid DNA was restricted at a unique restriction site between the *ori* and the terminus. Note that the arrested intermediate should have a double Y structure. The DNA was resolved by electrophoresis in the first dimension in an agarose gel at neutral pH and was run on the second dimension at the same pH in a gel containing ethidium bromide. The dye intercalates between DNA base pairs to different extents, depending on the shape of the DNA, and helps discriminate between linear, Y-shaped, or X-shaped DNA molecules. The expected distribution is a hook-shaped pattern of DNA that is visualized by Southern blotting and hybridization by a labeled DNA probe. The monomeric linear DNA forms the prominent lower spot, whereas the termination intermediate is marked by the spot just above and to the left of the monomer spot. By cutting the DNA at other unique restriction sites, one can localize the terminator site (in this case already predetermined) on the plasmid DNA (Fig. 1B). The neutral-alkaline two-dimensional gel procedure is equally effective in detecting and localizing termination sites (Linskens and Huberman 1988).

Structure and Chromosomal Location of Replication Arrest Sequences of Prokaryotes

Definitive information on replication fork arrest sequences has come mostly from the studies of prokaryotic chromosomes. The relatively small amount of available information from eukaryotes such as ribosomal DNA of yeast, human, and plants, and from mammalian mitochondrial DNA, is discussed in a later section.

The replication arrest sites of chromosomes of *E. coli* and *B. subtilis* and of plasmid R6K are shown in Figure 2,A–C. The information on the replication fork arrest in *E. coli* comes from the work of Hill, Kuempel, and their associates (Hill et al. 1988) and from the work of Horiuchi and his associates (Hidaka et al. 1988, 1991). Most of the information on the structure and in vivo analysis of the replication arrest system of *B. subtilis* has come from the work done in the laboratory of Wake and coworkers (for review, see Lewis and Wake 1991 and many other papers by Wake and colleagues cited in this chapter). The replication arrest sites of plasmids of *E. coli* are almost indentical to those of the *E. coli* host chromosomes (Bastia et al. 1981b; Hidaka et al. 1988; Sista et al. 1989).

There are six known replication arrest sites in both *E. coli* (Hidaka et al. 1991; Hill 1992) and *B. subtilis* (Franks et al. 1995). In both *E. coli* and *B. subtilis* the sites are arranged in two sets of three, and in each set all three sites have the same polarity (Fig. 2). The polarity is defined as the orientation of the site for fork arrest with respect to the origin of replication (Hidaka et al. 1988; Hill et al. 1988; Sista et al. 1989; Smith and Wake 1992; Sahoo et al. 1995a). The location and orientation of the arrest sites are such that a replication fork approaching from left to right passes through the first set of three sites in both *E. coli* and *B. subtilis* and through one of the two sites of R6K and is arrested at the first site of

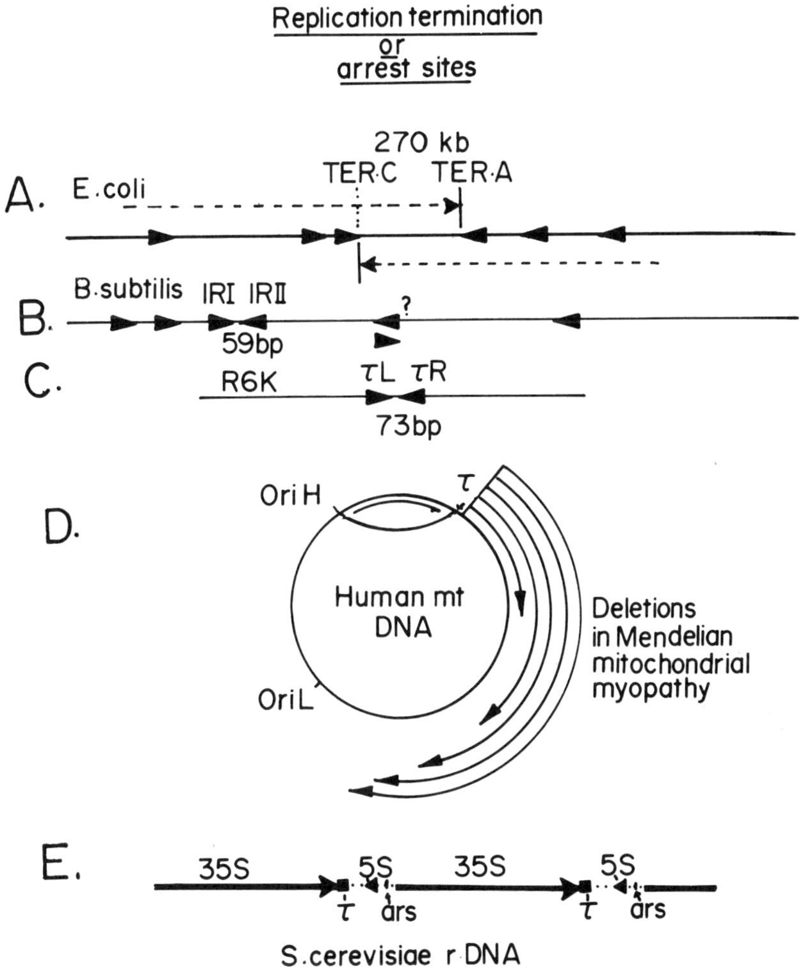

Figure 2 Replication termini in various systems. (*A*) Replication termini of *E. coli*. The dotted lines with arrows show the points of arrest of the two replication forks of the bidirectional replicon. (*B, C*) Replication termini of *B. subtilis* and the plasmid R6K. (*D*) D-loop of the human mitochondrial DNA. τ indicates the point of arrest of the ~600-nt-long newly synthesized H strand. In Mendelian mitochondrial myopathy, deletions extend from near τ, clockwise toward the origin of light-strand synthesis *oriL*. (*E*) Replication fork arrest sites (called replication fork barrier or RFB) τ in yeast rDNA. Replication forks moving in a direction opposite to the transcription of 35S rRNA are arrested at τ (RFB).

the second set that the fork encounters. The converse is true for the fork approaching from right to left. Once the first fork is arrested at a given site, the second fork approaching from the opposite direction stops at the first site, thus preventing the reduplication of the region between two sets of arrest sites. The sites TerA, TerB, and TerC (also called τ1, τ2, and τ3) have been reported to be the most frequently used sites in *E. coli* under laboratory conditions (Pelletier et al. 1988; Louarn et al. 1991). TerI (also called IRI) of *B. subtilis* is reported to be the site that arrests the replication fork moving right to left, which usually arrives first at the terminus (Carrigan et al. 1991; Smith and Wake 1992; Franks et al. 1995).

The replication arrest sites of *B. subtilis* are related to one another by a consensus sequence, as are the sites of *E. coli* and plasmid R6K (Table 1). There is no sequence homology between the sites of gram-positive *B. subtilis* and gram-negative *E. coli*. Unlike the sites of *E. coli* and R6K, the sites of *B. subtilis* have two overlapping sequences called the core (or B site) and the auxiliary sequence (or A site) (Lewis et al. 1990; Sahoo et al. 1995a). The sequences of the fork arrest sites are binding sites for the replication terminator proteins, Ter (Tus) and RTP, encoded by *E. coli* and *B. subtilis*, respectively.

Replication Terminator Protein of *E. coli*

The existence of a terminator protein that causes replication fork arrest was first suggested by in vitro replication experiments carried out by Germino and Bastia (1981). A hybrid replicon containing the plasmid ColE1 origin of replication and the replication terminus of R6K was replicated in cell extracts of an *E. coli* strain that did not harbor a resident R6K plasmid. Replication fork initiated from the ColE1 origin in vitro was arrested at the R6K terminus. Considering that the plasmid did not encode a terminator protein and that the terminus sequence did not have features (e.g., hairpins, polypurine stretches) that are known to impede polymerase movement, the work suggested that the host cell extract probably contained a terminator protein that recognized the R6K terminus.

The terminator protein was first purified from cell extracts of *E. coli*, and the protein bound to two sites present in a 216-bp DNA fragment of R6K (Hidaka et al. 1989; Sista et al. 1989). The sites corresponded to the left and the right arrest sites of R6K (Horiuchi and Hidaka 1988). The *tus* gene encoding the Ter (Tus) protein was discovered by Hill et al. (1989) and Hidaka et al. (1989) and the protein, by direct analysis and as deduced from the DNA sequence, was found to be 36 kD in molecular mass. Cross-linking studies (Sista et al. 1991) and sedimentation equilibrium studies (Gottlieb et al. 1992) both showed a monomeric protein in solution.

Table 1 Sequences of replication fork arrest sites of different organisms

E. coli chromosome	
TerA	5′-AATTAGTATGTTGTAACTAAAGT-3′
TerB	5′-AATAAGTATGTTGTAACTAAAGT-3′
TerC	5′-ATATAGGATGTTGTAACTAATAT-3′
TerD	5′-CATTAGTATGTTGTAACTAAATG-3′
TerE	5′-TTAAAGTATGTTGTAACTAAGNN-3′
TerF	5′-CCTTCGTATGTTGTAACGACGAT-3′
R6K plasmid	
TerR1	5′-CTCTTGTGTGTTGTAACTAAATC-3′
TerR2	5′-CTATTGAGTGTTGTAACTACTAG-3′
Consensus	AAA AA A
	5′-NN G TGTTGTAACTA NNN-3′
	TTT TG C
B. subtilis chromosome (strain 168)	
TerI (IRI)	5′-ACTAAGAAAACTATGTACCAAATGTTCAGT-3′
TerII (IRII)	5′-ACTGACAACACTAGTTACTAAATATTCAAT-3′
TerIII	5′-ACTAATTGATCTATGTACTAAATATTCATA-3′
TerIV	5′-ACTAACTAAACTATGTACTAAATATTCACT-3′
TerV	5′-ACTAAATAAATAATGTACTAAATATTCAAC-3′
TerVI	5′-ACTAAATAATCTATGTACCAAATGTTCAAT-3′
B. subtilis chromosome (strain W23)	
IRI	5′-ACTAAGTGAACTGTGAACCAAATGTTCACT-3′
IRII	5′-ACTGAGAACACTATGTACTAAATATTCAAT-3′
Consensus	A TAAACTA T T A
	5′-ACT AN TG AC AAAT TTCANN-3′
	G AGCTTAG A C G
Saccharomyces cerevisiae rRNA	
	5′-TTGCCCGGACAGTTTGCTTCATGGAGCAGTTTTTTCCGCACCATC AGAGCGGCAAACATGAGTGCTTGTATAAGTTTAGAGAATTGAGA-3′
Mitochondrial D-loop sequence that may be a termination sequence	
	5′-TTGACTGTACATAGTACATTATGTCAAATTC-3′

The contact points of Ter protein with the replication arrest sites of R6K (Sista et al. 1991) and of *E. coli* (Gottlieb et al. 1992) have been analyzed showing a pattern of asymmetric contacts involving both strands of the cognate sites. The equilibrium dissociation constant varied between the Ter sites. Sista et al. (1991) have found the K_d for the R6K Ter site to be 5 x 10^{-9} moles/liter, whereas Gottlieb et al. (1992) found the K_d for the *E. coli* Ter site to be 10^{-11} moles/liter. For the more efficient TerB sites of *E. coli*, Gottlieb et al. (1992) found the K_d value to be 3.4 x 10^{-13} moles/liter. The differences between two different Ter sites may be due to differences in their contact points, and the differences in the values for the same site in different observations may be due to buffer conditions and methods used in the different experiments. Table 2 shows characteristics of different Ter sites. Gottlieb et al. (1992) have compared K_d, dissociation rate constant, and half-life values of Ter-Tus complex with that of the *lac* repressor-operator complex. Although the K_d values for both the complexes were found to be similar, the half-life of the Ter-Tus complex was very high in comparison to that of *lac* repressor-operator complex.

The Ter protein, added either to cell extracts of *tus*⁻ *E. coli* (Khatri et al. 1989; MacAllister et al. 1990) or to a defined replication system for ColE1 type origins (Hill and Marians 1990) or for the *oriC* system of *E. coli* (Lee et al. 1989; Lee and Kornberg 1992), caused orientation-specific arrest of replication forks. Although both the leading and the lagging strands were arrested at the terminus, a transient intermediate, consistent with a D-loop structure, could be detected in the in vitro reaction (MacAllister et al. 1990). Both Hill and Marians (1990) and Lee and Kornberg (1992) mapped the point of arrest in vitro of the newly synthesized DNA at the terminator sites. The site was located just within the region protected in hydroxyradical and DNase I footprinting experiments.

A significant develoment in understanding the biochemistry of fork arrest emerged from the discovery by Khatri et al. (1989) and Lee et al. (1989) that the Ter (Tus) protein of *E. coli* was able to impede, in one orientation of the terminus, the activity of the main replicative helicase, DnaB. The DnaB helicase translocates in the $5' \rightarrow 3'$ direction (LeBowitz and McMacken 1986), whereas the helicase PriA (factor Y) of *E. coli* translocates in the $3' \rightarrow 5'$ direction (Lee and Marians 1987). The Ter (Tus) protein of *E. coli* was also able to impede PriA helicase in an orientation-specific manner (Hiasa and Marians 1992; Lee and Kornberg 1992).

Interestingly, the Ter protein was able to impede the helicase activity of SV40 large T antigen in an orientation-dependent mode in vitro (Bedrosion and Bastia 1991; Amin and Hurwitz 1992; Hidaka et al. 1992). A hybrid replicon with SV40 origin and the R6K terminus, when

Table 2 Characteristics of replication fork arrest sites and the terminator proteins

Sequence	Name	Source	Protein	Mol. mass (kD)	Gene	K_d (moles/l)
● ● 5′-TGAGTGTTGTAACTACTA-3′ 3′-ACTCACAACATTGATGAT-5′ →| ● ●	TerR	R6K	Ter (Tus)	36	*Tus*	1×10^{-11} 5×10^{-9}
● ● ● 5′-AATAAGTATGTTGTAACTAAAGT-3′ 3′-TTATTCATACAACATTGATTTCA-5′ →| ●	TerB	*E. coli*	Ter (Tus)	36	*Tus*	3.4×10^{-13}
● ● 5′-ACTAAGAAAACTATGTACCAAATGTTCAGT-3′ 3′-TGATTCTTTTGATACATGGTTTACAAGTCA-5′ ● ● ● ● |←	TerI (IRI)	*B. subtilis*	RTP	14	*rtp*	1.2×10^{-11} 6×10^{-11}

Closed circles show methylation protection of G residues. Arrows and vertical lines indicate the site and polarity of replication blockage.

replicated in HeLa cell extract containing the SV40 T antigen and Ter protein, showed polar fork arrest at the R6K terminus (Bedrosian and Bastia 1991; Amin and Hurwitz 1992).

What is the mechanism of orientation-specific fork arrest? Does it involve strictly Ter protein/terminus DNA interaction that imposes a polar roadblock to most helicases? Alternatively, are both terminator protein/DNA interaction and helicase/terminator protein interaction involved in polar fork arrest?

The issue of helicase specificity of the Ter protein has been debated (Khatri et al. 1989; Lee et al. 1989; Hiasa and Marians 1992; Lee and Kornberg 1992). Some laboratories have reported that Ter protein blocks all helicases, including DnaB and SV40 Tag, that promote Cairns-type replication to generate θ-shaped replication intermediates and also the helicases involved in rolling-circle replication, e.g., Rep helicase; in DNA repair, e.g., helicase II; and in conjugative DNA transfer, e.g., helicase I (Lee et al. 1989; Hidaka et al. 1992; Lee and Kornberg 1992).

Khatri et al. (1989) and Hiasa and Marians (1992) have reported that Ter protein does not block helicase II or Rep helicase. The inhibition of helicase II reported by Bedrosian and Bastia (1991) was due to high concentrations of Ter protein used in the work. In a detailed study of the helicase specificity of the contrahelicase (antihelicase) activity of Ter protein, Sahoo et al. (1995b) discovered that the Ter protein did not impede helicase I or Rep helicase over a wide range of enzyme (helicase)-to-substrate and terminator protein-to-substrate ratios. Under the same conditions, the Ter protein readily impeded activities of both DnaB and PriA helicases in a polar fashion.

Lee and Kornberg (1992) have reported that Ter protein impedes the activities of the Klenow fragment of DNA polymerase I and the DNA polymerases of T7 and T5 phages. The blocks showed minimal orientation dependence of the terminus sequence (a twofold difference). It is likely that the polymerase block is mainly due to a nonspecific roadblock created on DNA by Ter protein and is not considered to be of any major physiological significance.

Recently, Mohanty et al. (1996) have discovered that the terminator proteins of both *E. coli* and *B. subtilis* block the elongation of RNA chains by several prokaryotic RNA polymerases in a completely polar fashion. The antihelicase activity and the RNA polymerase anti-elongation activities were isopolar. Mohanty et al. (1996) have investigated the possible biological significance of the RNA polymerase impedance by the terminator proteins. Using model substrates, they discovered that passage of an RNA transcript through the non-blocking end of the terminus abrogates the contrahelicase activity of both Ter and RTP. Transcriptional passage also releases the replication forks arrested at the τ2 (TerB) terminus of *E. coli*. The in vivo significance of the results can be understood by considering the finding along with the observations of Roecklein and Kuempel (1992) that transcription elongation initiated from an upstream promoter (P1) and directed toward the TerB (τ2) sequence in vivo in *tus*$^+$ cells was arrested at the τ2 site but invaded the τ2 terminus in *tus*$^-$ cells. Thus, the anti-transcriptase activity of the Ter protein and RTP most likely has evolved to protect the functional integrity of the replication arrest sites from transcriptional invasion.

Consistent with this notion, Hidaka et al. (1988) had observed that the non-blocking ends of each of the three frequently used replication termini τ1, τ2, and τ3 (Fig. 2A) were flanked by sequences with a GC-rich hairpin loop and an AT-rich tail characteristic of ρ-independent terminator sites. Considered together, these results support the idea that the ability of the terminator protein to block transcriptional elongation is a mechanism to protect the replication termini from functional inactivation by invading transcripts.

Replication Terminator Protein of *B. subtilis*

Unlike the Ter protein of *E. coli*, RTP of *B. subtilis* is a dimer with subunit molecular mass of 14.5 kD (Lewis et al. 1989, 1990). Two dimers of RTP bind to each arrest site of *B. subtilis* in a stepwise manner. A single dimer first binds to the core site (B site) and then, by apparent cooperative protein-protein interaction, promotes the binding of a second dimer to the auxiliary (A) site (Fig. 3). The core site by itself, binding to a single monomer, is incapable of arresting replication forks (Smith and Wake 1992; Sahoo et al. 1995a).

The mode of interaction of RTP with the cognate binding site has been studied both by missing nucleoside hydroxyradical footprinting (Langley et al. 1993) and by methylation protection and interference studies (Sahoo et al. 1995a). The results show that the protein dimer contacts the core site (B site) more frequently than the auxiliary site (A site), and also that the contacts at the core site are on both strands of DNA whereas the contacts at the auxiliary site were mostly one-stranded. Sahoo et al. (1995a) reported that RTP showed purine contacts on both strands of the core but with only one strand of the auxiliary site (Table 2).

There have been two major breakthroughs in the biochemical and structure-function analysis of RTP. Although there is very little primary

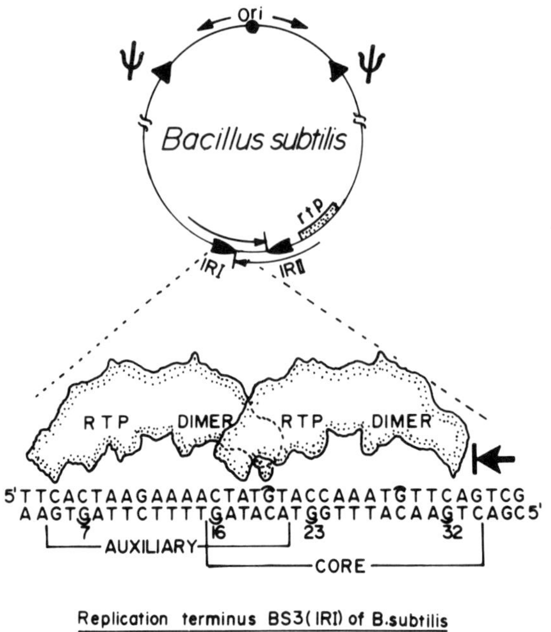

Figure 3 Replication of a *B. subtilis* chromosome showing the termini ψ that are active during stringent response. The IRI and IRII termini contain overlapping core and auxiliary sequences. A single dimer of RTP binds to the core and, by cooperative protein-protein interaction, promotes the binding of a second dimer to the auxiliary site. RTP is encoded by the *rtp* gene. (Reprinted, with permission, from Manna et al. 1996.)

structural homology between Ter of *E. coli* and RTP of *B. subtilis*, Kaul et al. (1994) discovered that RTP functions both in vivo and in vitro in *E. coli.* It blocks the activity of both DnaB and PriA helicases of *E. coli* (Kaul et al. 1994; Sahoo et al. 1995b). Independently, Wake and colleagues have also reported that RTP functions in vivo in *E. coli* (Young and Wake 1994). Since there is no currently available in vitro replication system for *B. subtilis*, it is now possible to analyze the biochemical functions of RTP using the well-defined surrogate *E. coli* in vitro replication system. A second major breakthrough has been the solving of the crystal structure of RTP apoprotein at 2.6 Å (Bussiere et al. 1995).

Crystal Structure of RTP

The crystal structure of the RTP apoprotein has the following features: (1) an amino-terminal disordered region; (2) four α helices with the carboxy-terminal, longest α helix (α4) forming an antiparallel coiled-coil dimerization domain; (3) one short and two long β sheets, the β2 and β3 being connected by an extended loop (Fig. 4).

On the basis of the known activities of RTP, one would expect the protein to have the following domains: (1) a DNA-binding domain, (2) a dimerization domain, (3) a dimer-dimer interaction surface, (4) a helicase-blocking surface, and (5) a region that blocks elongation of RNA chains by prokaryotic RNA polymerases unless, of course, the postulated helicase-blocking domain is also involved in blocking chain extension by RNA polymerase. An interesting question to consider is how a symmetric dimer of RTP is able to act in an asymmetric fashion by blocking helicase activity in a polar mode.

The "Winged Helix" DNA-binding Domain of RTP

The DNA-binding domain of RTP is of interest for at least two reasons: First, the DNA/protein interaction at the terminus presents the terminator protein to the approaching helicase; second, the DNA/protein interaction is likely to generate functional asymmetry from two interacting, symmetric dimers, since DNA/protein contacts are different at the core (B site) and the auxiliary site (A site). Figure 3 shows diagrammatically the interaction between two dimers of RTP on the IRI (BS3) terminus. IRI stands for inverted repeat I and is synonymous with binding site 3, BS3 (see Sahoo et al. 1995a).

Pai et al. (1996) have localized the DNA-binding domain by systematic saturation mutagenesis of DNA encoding RTP and then by screening of the mutants for any defect in DNA binding by employing the genetic selection scheme of Elledge et al. (1990). The mutants that showed defects in DNA binding by the in vivo genetic assay were further characterized by biochemical analysis. By site-directed mutagenesis, mutants of RTP were isolated with cysteine residues substituted at selected sites, derivatized with azido-phenacyl bromide, and cross-linked to the DNA. The cross-linking experiments provided direct evidence for either the contact of α3, β2, and the amino-terminal arm with DNA or the very close proximity of these regions to the DNA. One mutant protein showing a severe DNA-binding defect was crystallized, and the crystal structure was solved and compared with the wild-type RTP structure. The structures were almost identical, thereby showing no misfolding caused by the mutation. The combination of these approaches showed that the amino-terminal unstructured arm, the β2 sheet, and the α3 helix make contacts with the terminus DNA (see Fig. 5). The structure has been referred to as a "winged helix." The β2-β3 sheets and the extended loops

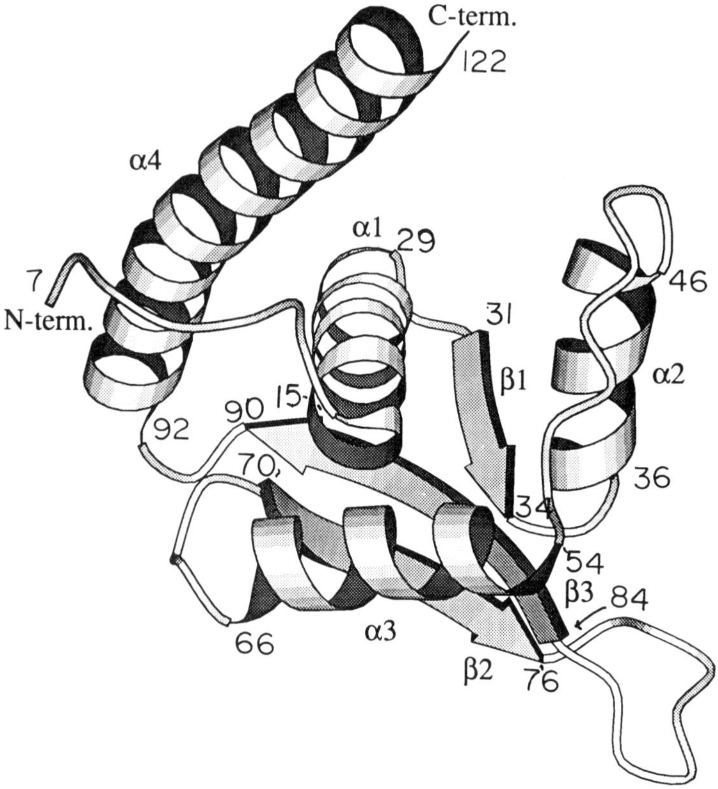

Figure 4 Ribbon diagram of an RTP monomer. The protein contains an unstructured amino-terminal region, four α helices, three β strands, and an extended loop connecting β2 with β3.

connecting the two form the wings (Figs. 4 and 5). Swindels (1995) has compared the structure of RTP with that of the eukaryotic fork head transcription factor, histone H5, and LexA, and has noted the remarkable similarity in the tertiary structures of these, otherwise unrelated, proteins.

Dimerization Domain

The carboxy-terminal α helices form the antiparallel, coiled-coil dimerization domain of RTP (Bussiere et al. 1995).

Dimer-dimer Interaction Domain

Wake and his colleagues (Lewis et al. 1990; Carrigan et al. 1991; Langley et al. 1993) have observed that in vivo, a single dimer of RTP

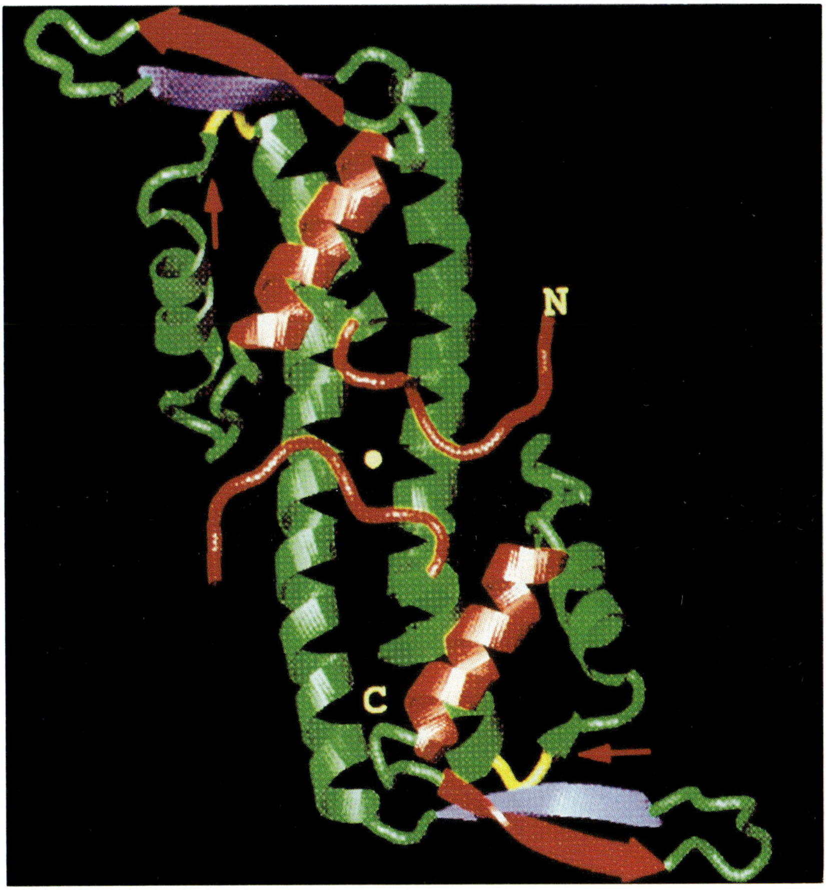

Figure 5 Ribbon diagram of an RTP dimer. (*Red*) DNA-binding regions; (*blue*) dimer-dimer interaction region; (*yellow*) helicase-blocking region. Mutations at the regions marked by red arrows result in loss of RNA polymerase block with no change in DNA binding. The carboxy-terminal α4 helices form an antiparallel, coiled-coil dimerization domain.

binding to a core site fails to impede replication forks. Sahoo et al. (1995a) have extended these results in an in vitro system and have shown that a single dimer of RTP, binding to a core site of a terminator DNA, cannot impede DnaB helicase or replication forks. Thus, two dimers of RTP binding to the core and the auxiliary sites (see Fig. 3) are needed to form a functional replication terminus.

The dimer-dimer interaction domain is therefore an important region of RTP structure. Since RTP crystallizes as a dimer, crystallography did not provide any clue as to the location of the dimer-dimer interaction domain. We resorted to mutagenesis and biochemical analysis of the mutant forms of the protein to localize the dimer-dimer interaction site. The site seems to be located in the β3 strand of RTP (blue region, Fig. 5). A tyrosine residue, located at the coordinate 88 on the β3 strand, upon mutation to a phenylalanine, yielded a protein that binds to the core site but fails to bind to the auxiliary site. The tyrosine at the position 88 is within hydrogen-bond-forming distance of a glycine residue at position 34. Mutation of the glycine at 34 to an arginine also abolishes dimer-dimer interaction. It appears that β3-β3 stacking between two dimers may be the basis of dimer-dimer interaction (Manna et al. 1996). The mutant is completely defective in impeding helicases, replication forks, and RNA polymerases. In addition, a valine to glycine substitution at position 85 and leucine to serine substitution at 82, at the loop connecting β2 to β3, also abolishes dimer-dimer interaction.

Helicase-blocking Surface

Mechanistically, the helicase-blocking surface is of considerable importance, because the existence of such a surface would strongly support a mechanism of fork arrest that involves specific blocking of the helicase activity by protein-protein interaction with RTP. The following criteria were used to look for such a mutant. The mutations affecting the putative helicase-blocking surface of RTP should not affect DNA binding. If the impedance of RNA polymerase-catalyzed chain elongation is controlled by a separate region of RTP, the mutants impairing helicase block should not affect the ability to block RNA polymerase. Using the above criteria, we examined mutants of the region of an exposed hydrophobic patch with a few charged residues located between the α1 helix and the β1 sheet (Figs. 4 and 5). We discovered that a mutation in the region colored yellow in Figure 5 met the criteria stated above. The mutant bound to DNA almost normally, blocked RNA chain extension normally, but failed to block DnaB helicase, PriA helicase, and replication forks in vitro (A.C. Manna et al., in prep.). This result is significant in two ways. First, the mutant RTP marks the helicase-blocking domain of RTP and supports a mechanism of contrahelicase activity that involves RTP/helicase interaction. Second, the results indicate that different regions of the protein are probably involved in the antihelicase and RNA chain anti-elongation activities. A third mutation at the site marked by red arrows in Figure 5 greatly reduced RNA polymerase block without detectably affecting DNA binding. Thus, this region marked by the mutation identifies the RNA polymerase-blocking surface of RTP (Fig. 5). The mutant defective in RNA polymerase block was also defective in helicase block.

The structure-function analysis of RTP, guided by the crystal structure, has been very informative with regard to the mechanism of replication fork impedance. Future work will be directed toward mapping of the sites of DnaB helicase, of SV40 T antigen, and of T7 RNA polymerase that interact with RTP.

It should be noted that the replication terminator proteins of *E. coli* and *B. subtilis*, without the aid of any other protein, block not only the translocation of DnaB and PriA helicases, but also the authentic unwinding of DNA, regardless of the length of the duplex region in helicase substrates (Sahoo et al. 1995a,b). These findings disagree with an earlier report that the replication terminator protein of *E. coli* by itself can inhibit helicase translocation on DNA, but for blockage of DNA unwinding, needed the participation of other replisomal proteins (Hiasa and Marians 1992). It would be surprising, however, if the fork arrest is not more efficient in the presence of a full complement of proteins that drive the replication fork.

Replication Arrest Sites of *B. subtilis* Active under Stringent Conditions

Simone Seror and coworkers (Henckes et al. 1989) have shown that conditions that generate high levels of the alarmone ppGpp in the cell, either by treating *B. subtilis* with hydroxamate or by shifting a temperature-sensitive mutant of the DnaB gene to the nonpermissive temperature (note the DnaB of *B. subtilis*, unlike the protein with the same designation in *E. coli*, is not a helicase but is an initiator protein), cause the replication forks initiated at the *ori* to be arrested approximately 200 kb away on either side at the so-called stringent termini (marked ψ in Fig. 3). Seror's group has shown further that the arrest of replication forks at the stringent termini ψ requires RTP (Levine et al. 1995). Although the marker frequency analysis by DNA-DNA hybridization used to map the arrest site did not allow a precise localization of the ψ sites, the requirement for RTP might indicate the presence of arrest sites such as those present at the normal replication terminus (Fig. 2) that is used under relaxed conditions. How does RTP arrest forks at ψ sites under high ppGpp but not under relaxed conditions? It is known that many promoters such as that for rRNA are shut off during stringent conditions (Cashel and Rudd 1987). Combining this observation with the finding of Mohanty et al. (1996) that replication arrest sites of *B. subtilis* can be rendered ineffective in blocking replication forks by passage of an RNA transcript, a possible mechanism of conditional usage of a replication arrest site can be hypothesized. It is tempting to suggest that the ψ sites are normally kept in a nonfunctional state by transcripts directed through the sequences by promoters that are sensitive to high ppGpp. Under stringent conditions, we suggest that these promoters are turned off, thus allowing RTP bound to the ψ sites to block replication forks. Under low ppGpp, we suggest that the promoters are turned on, and the passage of RNA transcript through the ψ sequences keeps them inactive. In this scheme, ψ sites may be simply regular terminator sequences that bind RTP. Alternative models would involve modification of RTP under stringent conditions. The scheme proposed can be experimentally tested once the ψ sites have been more precisely localized. Levine et al. (1995) have pointed out that the ψ sites serve as checkpoints of replication under stringent conditions. Interestingly, in *E. coli*, ppGpp causes replication arrest at the origin of the *E. coli* chromosome and not at other sites (Levine et al. 1991).

Regulation of Synthesis of Ter Protein

Both in vitro studies by Natarajan et al. (1991) and in vivo studies by Roecklein et al. (1991) showed that Ter protein is a transcriptional repressor of its own synthesis and prevents RNA polymerase from binding to a promoter that is present immediately upstream of the sequence encoding the Ter (Tus) protein. Thus, Ter (Tus) autoregulates its synthesis by binding to the τ2 site that acts as an operator of the autoregulated promoter. Interestingly, Roecklein and Kuempel (1992) also discovered that transcription from an upstream promoter is impeded at the elongation stage by the τ2 sequence in a tus^+ but not in a tus^- host. Natarajan et al. (1993) have reported that a partially purified protein fraction of *E. coli* was able to abrogate both the contrahelicase activity and the fork arrest by the Ter protein in vitro. This opens the possibility of regulation of replication arrest function at a different level by protein-protein interaction.

Possible Physiological Roles of Sequence-specific Replication Arrest System

Since the *tus* gene of *E. coli* and the *rtp* gene of *B. subtilis* can be deleted without causing lethality, specific fork arrest and termination of replication do not appear to be essential for cell viability. Yet the fact that multiple replication arrest sites have been maintained against mutational drift suggests that there is a physiological need for the sites. At this time, there are no definitive clues as to why specific arrest sites have been maintained in the chromosomes. Several suggestions have been made regarding the possible physiological role of these sites. Brewer (1988) has suggested that the sites prevent replication forks from entering the chromosome from a direction opposite to that of transcription, in order to avoid collision between RNA polymerase and the replisomal machinery. Liu and Alberts (1995) have shown that such collision can cause pausing of replication forks. The pausing may invite recombination involving the free DNA ends exposed by falling off of the replication proteins, thus exposing free DNA ends that may lead to genomic instability. Lee et al. (1989) have suggested that the replication arrest sites prevent the θ-type replication from turning into a σ-type or rolling-circle replication. Lewis and Wake (1991) have speculated that the arrest sites might provide the proper structure for more efficient decatenation of daughter molecules. Dasgupta et al. (1991) integrated a copy of the R1 plasmid into the *E. coli* chromosome and caused chromosomal replication to be initiated from the R1 origin in the *tus*$^+$ cell of *E. coli*. The location of the R1 *ori* was such that the almost unidirectional initiation caused the replication fork to travel over a much greater length of DNA to reach the terminus. Under this condition, cell division was perturbed, causing a lack of proper septation and filament formation. Deletion of the *tus* gene greatly reduced filament formation by eliminating fork arrest at the terminus, allowing the forks to pass through and finish replication. Dasgupta et al. (1991) have speculated that the Ter sites restrict each fork to travel one-half the length of the chromosome before meeting each other at a site located diametrically opposite to the *ori* and thus maintain symmetry of replication. Perhaps this also maintains proper gene dosage of markers on both arms of the chromosome. Thus, the Ter system ensures the symmetry of replication. Do termination events signal subsequent cell division at the end of a replication cycle? Although the existence of such a mechanism has been proposed, recent work using synchronized initiation of a single cycle of replication and a DNA synthesis inhibitor shows that septa can form on partially replicated chromosomes that have not reached the terminus. Thus, under these experimental conditions, no linkage of cell septation to completion of the termination of replication was observed. Nordstrom et al. (1991) have also argued for separate mechanisms controlling termination from that of cell division. Recently Hiasa and Marians (1994a), using minichromosome templates containing *oriC* and Ter-binding sites, have shown that the Tus-Ter (Ter-τ) complex prevents overreplication of bidirectionally replicating template. In summary, information on the physiological role of the replication arrest system remains largely obscure.

Decatenation of Circular Daughter Chromosomes

The separation of the daughter molecules after replication requires decatenation; i.e., separation of the two intertwined DNA molecules. The separation at the termini could occur in two steps: In the first, melting of the hydrogen bonds between the remaining parental duplex by a helix-destabilizing protein or a helicase is followed by repair of the gaps of the two catenated rings; in the second step, a topoisomerase decatenates the two daughter molecules (Adams et al. 1992). Topo IV has been identified as the enzyme responsible for decatenation (Kato et al. 1988, 1990; Peng and Marians 1993). Topo IV mutants are defective in partition and accumulate nucleoids in the middle of the cell (Kato et al. 1988; Schmid 1990). Inhibition of topo IV in temperature-sensitive mutants of the two structural genes *par*C and *par*E of *Salmonella* causes the accumulation of catenated dimers. The catenanes were right-handed and parallel, consistent with a melting step followed by a decatenation step of chromosome separation at the terminus (see Fig. 6) (Adams et al. 1992; Zechiedrich and Cozzarelli 1995). A more detailed discussion of topoisomerases can be found in the chapter by Hangaard Andersen et al. (this volume).

The movement of the fork generates positive superhelical stress that can be removed by gyrase, which is uniquely capable of putting in negative superhelicity. However, gyrase can not substitute for topo IV in decatenation (Zechiedrich and Cozzarelli 1995). It is conceivable that topo IV is compartmentalized in a termination complex that gyrase cannot penetrate (Adams et al. 1992). Under certain conditions, topo IV can carry out *oriC* replication in vitro, replacing the need for gyrase, but only under substoichiometric ratios of enzymes to substrate (Hiasa and Marians 1994b). At stoichiometric ratios, the enzyme has been found to relax the substrate.

Pulse-chase experiments with bacterial plasmids under conditions where topo IV is inhibited yielded only about 10% accumulation of catenated dimers, raising questions as to whether topo IV was the true decatenating enzyme at the terminal stages of replication. Recent work in which such pulse-chase experiments were carried out using topo IV ts-gyrA^R (quinoline resistant) and topo IV ts-gyrA^+ double mutants showed that substantial amounts of catenated dimers accumulated when gyrase activity was blocked by quinolines under conditions that also blocked topo IV. Thus, catenated dimers are authentic and major kinetic intermediates of termination of replication. These intermediates are subject to decay by gyrase at a rate that is one one-hundredth of that of the decatenating activity of topo IV (Zechiedrich and Cozzarelli 1995).

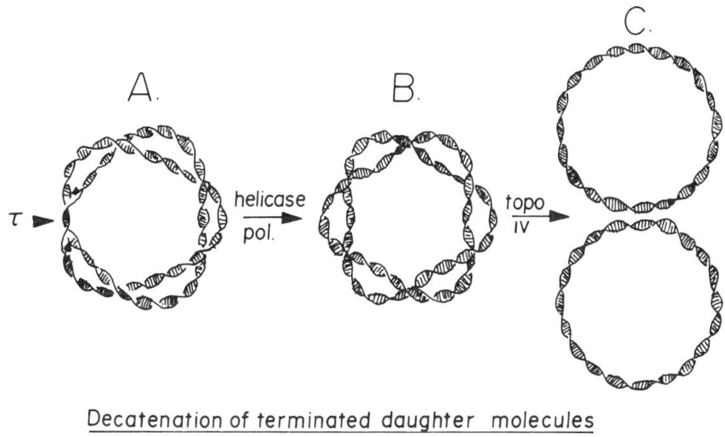

Figure 6 Schematic representation of the decatenation and segregation of two terminated daughter molecules. The scheme is redrawn from Adams et al. (1992) and Zechiedrich and Cozzarelli (1995). (*A*) Two daughter molecules with the replication forks arrested at the terminus τ. The molecules are still held together by base-pairing of a single turn of the unreplicated parental DNA. (*B*) Unreplicated parental DNA is separated by melting of the base pairs by a helicase, followed by synthesis to fill in the gap, thus generating two fully replicated catenated daughter molecules. (*C*) Topo IV acts on the catenated daughter molecules, thus separating the two. The two-step process of decatenation is supported by topological evidence (Adams et al. 1992).

In plasmid DNA replication in eukaryotic cells (driven by the SV40 origin of replication), the sequence at the termination site strongly affects the fraction of catenated dimers that form (Weaver et al. 1985; Fields-Berry and DePamphilis 1989). For example, the normal termination site for SV40 DNA replication and the yeast CEN3 sequence both promote formation of catenated intertwines when replication terminates in these sequences, but not when these sequences were placed elsewhere on the plasmid and termination occurred outside of them. Therefore, topo II must act behind the replication forks as they enter the termination region, and these two sequences must impede the ability of topo II to resolve catenated intertwines. If the action of topo II did not occur until replication was completed and catenated dimers were formed, then the ability of topo II to resolve catenation should not be affected by the sequence at the termination region, because the catenated intertwines would be distributed throughout the two interlocked sibling molecules.

Multimer Resolution

Because of an odd number of random crossovers between the catenated daughter molecules, multimers are generated. The terminal region of the *E. coli* chromosome has a 33-bp-long site called *dif* (*d*eletion *i*nduced *f*ilamentation) that interacts with the host-encoded recombinases *XerC* and *XerD* to resolve the multimers into monomers. Deletions of the *dif* site or mutations in *XerC* or *XerD* cause filamentation of cells (Blakely et al. 1991; Kuempel et al. 1991; Leslie and Sherratt 1995). The 33-bp *dif* site can substitute for the natural *dif* region, but only in a relatively location-specific manner (Tecklenberg et al. 1995).

The Replication Terminus Is a Recombinogenic Hot Spot

Several groups of workers have reported that the replication terminus of *E. coli* is a hot spot for recombination (Bierne et al. 1991; Louarn et al. 1991; Horiuchi et al. 1994; Horiuchi and Fujimura 1995). Using a λ cI 857 temperature-sensitive repressor-encoding prophage integrated at several sites on the *E. coli* chromosome, Louarn et al. (1991) discovered that the prophage inserted at the terminus was excised at a significantly higher frequency than phage insertions elsewhere on the chromosome. The excision was compatible with recombination events occurring at each replication cycle. The authors have proposed a model which postulates that RecA-mediated recombination plays a critical role in the resolution of catenated dimers. The model is an interesting one but does not seem to take into account the known requirement for topo IV in catenane resolution (Kato et al. 1990; Adams et al. 1992). A hybrid replicon of pBR322 and M13 origins containing the replication arrest site with blocking end toward both origins yielded, in a tus^{+} cell, deletions of which the majority mapped to within 5–6 nucleotides of the replication arrest site of leading-strand synthesis. The authors pointed out the recombinogenic potential of the replication arrest sites and the consequent propensity to cause genome instability (Bierne et al. 1991).

Horiuchi and Fujimura (1995) constructed *E. coli* strains with two replication arrest sites that blocked both forks of the bidirectionally expanding replication bubble. These strains were hyperrecombinogenic and were dependent on $recA^{+}$ and $recB^{+}(C^{+})$ genes for growth and induced SOS response constitutively. The authors have postulated a recombinogenic event caused by a double strand break at the stalled fork and reconstruction of the fork by recBCD-dependent recombination events. In summary, stalled recombination forks at the replication arrest sites generate a hyperrecombinogenic state that would cause genome instability and genome rearrangement.

REPLICATION ARREST SITES IN EUKARYOTES

Arrest of DNA Replication Forks at Specific Sites

Replication forks do not travel at a continuous rate along the genome, but pause at various sequences (Tapper and DePamphilis 1980). Some sequences have been identified at specific sites in the genomes of animal viruses, eukaryotic cells, and mitochondria that can arrest progress of DNA replication forks. Brewer (1988) has suggested that the primary function of arrest sites for DNA replication forks in the rRNA gene repeats is to prevent collision between replication forks and the RNA polymerase I transcription apparatus. Whether or not this hypothesis can be generally applied to other genes awaits further investigation.

The Epstein-Barr virus (EBV) genome contains a tandem repeat of 20 copies of a 30-bp sequence that binds tightly the virally encoded EBNA1 protein and functions both as an enhancer for viral promoters and as a component of the EBV origin of DNA replication (*oriP*). Replication forks originating at *oriP* do not traverse this EBV enhancer region (Gahn and Schildkraut 1989). Termination of DNA replication in circular plasmids driven by the EBV *oriP* and EBNA1 terminate replication within the enhancer. The EBV enhancer, in the presence of EBNA1 protein, also can arrest replication forks originating from the SV40 replication origin (Dhar and Schildkraut 1991). Arrest of DNA replication forks required at least two, but not more than six, tandem repeats. More recent studies show that this EBNA1-binding sequence can prevent SV40 T-antigen helicase from unwinding DNA (C. Schildkraut, pers. comm.). Whether or not the ability of EBNA1 DNA-binding sites to arrest replication forks exhibits polarity remains to be determined.

The best-studied replication fork barriers in eukaryotes are found in the nontranscribed spacer regions of yeast rRNA genes (Brewer and Fangman 1988; Linskens and Huberman 1988; Brewer et al. 1992). Actively transcribed rRNA gene tandem repeats exhibit a barrier to replication forks traveling upstream (i.e., opposite to the direction of transcription), as diagrammed in Figure 2E. Replication forks seldom originate from origins of replication located immediately downstream from inactive genes (Lucchini and Sogo 1994). A site in the intergenic spacer sequence exhibits polarity by blocking forks coming from one direction but not from the other (Brewer et al. 1992; Kobayashi et al. 1992). Transcription by yeast RNA polymerase I of a DNA fragment containing a transcription terminator near the putative replication arrest site revealed three sites that could arrest transcription (Lang and Reeder 1993). The first site bound the yeast Reb1 protein and arrested RNA polymerase I arriving from one direction, but not the other. The second site did not exhibit polarity. It may bind as-yet-unidentified proteins. Interestingly, the third site, which also lacked polarity, overlapped the putative replication arrest site. Recently, Horiuchi and Kobayashi (cited in Horiuchi and Fujimura 1995) have confirmed the earlier observations of Voelkel-Meimon et al. (1987) that the putative replication arrest site in yeast rRNA gene repeats is associated with a recombination hot spot (HOT 1 sequence). One of the mutants with reduced recombination activity at this HOT spot also showed reduced arrest of DNA replication forks. These results suggest that DNA replication forks in yeast may be subject to the same type of orientation-specific replication arrest sites that have been described in bacteria. DNA replication arrest sites analogous to those described in budding yeast have been mapped in the tandem repeats of rRNA genes in human cells (Little et al. 1993), plant cells (Hernandez et al. 1988, 1993), mouse cells and fission yeast (Lopez-Estrano et al. 1996), and the mouse rDNA replication fork arrest sites exhibit polarity. Thus, DNA replication fork arrest sites appear to be a common feature of rRNA gene repeats in nature.

Another example of a site-specific DNA replication arrest site is found in mitochondrial DNA. Mammalian (and other animal) mitochondrial DNA carry out replication in two stages. The parental light strand serves as the template for initiation from Ori_H, and the fork, after copying about 600 nucleotides, is arrested at a specific site forming a D-loop (Fig. 2E) (for reviews, see Clayton 1991a,b). Eventually, the daughter H strand elongates to expose Ori_L, which initiates the daughter light strand. The arrest of the heavy strand occurs near the terminus-associated sequences (TAS), which are evolutionarily conserved in vertebrates. A protein isolated from bovine cells binds to the mitochondrial TAS (Madsen et al. 1993). No further information is presently available on the physiological role of the putative protein in D-loop biogenesis.

The region of the D-loop, as in cases of replication termination sites, appears to be recombinogenic. In a Mendelian, autosomal dominant disorder called mitochondrial myopathy, the carriers develop an accumulation of deletions in the mitochondrial DNA, starting from a unique location to within a few nucleotides of the 3′ end of the arrested nascent H-strand DNA (Zeviani et al. 1989). In a more recent report, Zeviani (1992) has shown that some of the deletions in the mitochondrial DNA of myopathic patients are found downstream from the D-loop.

The origins of deletions in mitochondrial DNA triggered by a mutant nuclear gene are interesting to contemplate. First, the deletions could be caused by a mutant replication terminator protein, encoded by a nuclear gene, but acting to cause H-strand arrest. If the mutant protein has a propensity to allow formation of more free ends or gaps or a hyper-recombinogenic substrate, deletions could be generated by non-homologous recombination. The lesions found downstream from the 3′ end of the arrested H strand (Zeviani 1992) could also be potentiated by the extension and downstream pausing of the daughter strand. The isolation and analysis of the probable human mitochondrial replication arrest protein and the nuclear gene encoding the protein are interesting problems to tackle in the future.

Sequence-specific arrest sites for DNA replication forks may also exist in regions where DNA amplification events occur. For example, cells in culture and cancer cells in animals undergo constant genomic rearrangements and gene amplification events that can be demonstrated by selecting for cells that continue to proliferate in the presence of various metabolic inhibitors (for review, see Schimke 1984; Stark et al. 1989; Di Leonardo et al. 1993; Stark 1993; Wintersberger 1994). Developmentally orchestrated gene amplification occurs in the chorion genes of *Drosophila* (Heck and Spradling 1990) and the polytene chromosomes of *Chironomous* and *Sciara* (Gerbi et al. 1993). A mechanism that can account for selective gene amplification is the "onion skin model" first proposed to explain the excision of integrated SV40 genomes from the chromosomes of virus-transformed cell lines (Botchan et al. 1979). Direct evidence for this model comes from analysis of chorion gene amplification in *Drosophila* embryos (Kafatos et al. 1985).

The onion skin mechanism for gene amplification requires that multiple initiation events occur at the same origin of replication before the completion of replication of the replication unit, and that replication forks emanating from this origin are arrested at sites on either side of the sequence that is amplified. Thus, the second set of forks resulting from reinitiation at the origin eventually catches up with the first set of forks, creating multiple, concentric replication bubbles (like the layers of an onion). At this time, there is no direct evidence for specific sequences that arrest replication forks flanking known amplification loci, but one could anticipate finding them. In addition, there may be specific terminator proteins produced by the specialized cells in which programmed gene amplification occurs. Such systems should provide fertile ground for future experiments.

Two Pathways for Separation of Sibling Molecules at Termination Sites

Termination of DNA replication occurs whenever two oncoming replication forks collide. Since DNA replication begins at thousands of sites distributed throughout the chromosomes of eukaryotic cells, termination of DNA replication must also occur at thousands of sites during each S phase. However, in contrast to initiation of DNA replication where specific sequences determine where replication begins in the genomes of eukaryotic cells, animal viruses, and mitochondria (see DePamphilis, this volume), termination of DNA replication can occur at virtually any sequence. However, it is possible, and perhaps likely, that some replication units (e.g., rDNA) terminate replication at specific sequences. In SV40, although termination of replication does not require specific sequences (Lai and Nathans 1975), the sequence at the termination site does strongly influence the pathway by which the two sibling molecules are separated (Weaver et al. 1985; Fields-Berry and DePamphilis 1989).

As replication forks advance, DNA unwinding introduces positive superhelical turns in front of the fork that are relaxed by topoisomerase I (Minden and Marians 1986). As two oncoming replication forks approach each other, the length of unreplicated DNA between them grows shorter, and therefore the target for topo I grows smaller. Eventually, topo I can no longer act in front of the forks. When this happens, DNA unwinding at the replication fork produces one catenated intertwine in the two sibling molecules behind the replication fork for each 10 bp of DNA unwound in front of the replication fork. In eukaryotes, topo II is required to remove these catenated intertwines. Mutations in yeast topo II that inactivate its decatenation activity result in interlocked DNA cellular chromosomes during S phase and in catenated dimers in plasmids that replicate in yeast cells (DiNardo et al. 1984; Uemura et al. 1987). Inhibition of topo II in mammalian cells either by indirect means (Sundin and Varshavsky 1981; Weaver et al. 1985; Fields-Berry and DePamphilis 1989) or by the use of specific inhibitors (Richter et al. 1987; Snapka et al. 1988; Ishimi et al. 1992) results in accumulation of late-replicating intermediates (θ structures that have completed >90% of their replication) of circular DNA molecules as well as catenated dimers containing multiple intertwines. Depending on experimental conditions, the primary product can be almost all late-replicating intermediates (Ishimi et al. 1992) or almost entirely catenated dimers (Sundin and Varshavsky 1981). The fact that late-replicating intermediates accumulate when topo II is inhibited demonstrates that catenated intertwines are normally removed prior to separation of sibling DNA molecules (Weaver et al. 1985).

Neither the SV40 termination region nor the yeast CEN3 sequence promotes formation of catenated intertwines under normal physiological conditions (Koshland and Hartwell 1987; Fields-Berry and DePamphilis

1989). Therefore, catenated intertwines are normally removed prior to or concomitant with termination of DNA replication when two replication forks collide. The presence of sequence-specific termination sites in bacterial chromosomes may ensure that this occurs in bacterial chromosomes prior to cell division.

CONCLUSION

The two known replication terminator proteins, namely Ter (Tus) of *E. coli* and RTP of *B. subtilis*, arrest replication forks by their polar contrahelicase activity. These proteins also block RNA chain elongation in a polar fashion. It is interesting to note that programmed cell death of bacterial cells that lose a resident plasmid also involves a plasmid-coded contrahelicase and an antidote protein. The contrahelicase is long-lived, but the antidote (contra-contrahelicase) protein is short-lived. Thus, in the plasmid-free segregants, the antidote decays and the longer-lived contrahelicase neutralizes the helicase, thereby causing replication arrest and cell death (Ruiz-Echevarria et al. 1995).

A polar contrahelicase from a eukaryote has not yet been discovered. We strongly suspect that such proteins exist in eukaryotes and are likely to be involved in fork arrest at specific sequences. The search for such protein(s) should prove to be interesting and rewarding.

It has been noted that fork arrest sites are very recombinogenic and therefore are regions of potential chromosome instability. That certain chromosome instabilities lead to the induction of cancer has been recognized (Hartwell 1992). This fact should provide additional incentive to search for eukaryotic replication terminator proteins.

ACKNOWLEDGMENTS

We thank Drs. T. Sahoo, A. Manna, and S. Pai for permission to quote unpublished data. We thank Mrs. Waltraud Bastia for editorial work and Drs. S. White and D.E. Bussiere for many useful discussions. Preparation of this chapter became a much easier task because of the interest, encouragement, and advice given by Dr. Mel DePamphilis. Work in our laboratory was supported by a grant (GM-49264) and a Merit Award (R37 AI-19881) from the National Institutes of Health.

REFERENCES

Adams, D.E., E.M. Schekhtman, E.L. Zechiedrich, M.B. Schmid, and N. Cozzarelli. 1992. The role of topoisomerase IV in partitioning bacterial replicons and the structure of catenated intermediates in DNA replication. *Cell* **71:** 277–288.

Amin, A.A. and J. Hurwitz. 1992. Polar arrest of the simian virus 40 tumor antigen-mediated replication fork movement *in vitro* by the tus protein-terB complex of *E. coli. J. Biol. Chem.* **267:** 18612–18622.

Baker, T. 1995. Replication arrest. *Cell* **80:** 521–524.

Baran, N., A. Lapidot, and H. Manor. 1983. Onion skin replication of integrated polyoma-transformed rat cells: Termination within a specific cellular DNA segment. *Proc. Natl. Acad. Sci.* **80:** 105–109.

———. 1987. Unusual sequence element found at the end of an amplicon. *Mol. Cell. Biol.* **7:** 2636–2640.

Bastia, D., J. Germino, J.H. Crossa, and P. Hale. 1981a. Molecular cloning of the replication terminus of the plasmid R6K. *Gene* **14:** 81–89.

Bastia, D., J. Germino, J.H. Crossa, and J. Ram. 1981b. The nucleotide sequence surrounding the replication terminus of R6K. *Proc. Natl. Acad. Sci.* **78:** 2095–2099.

Bedrosian, C.L. and D. Bastia. 1991. *E coli* replication terminator protein impedes simian virus 40 SV40 replication fork movement and SV40 large tumor antigen helicase activity *in vitro* at a prokaryotic terminus sequence. *Proc. Natl. Acad. Sci.* **88:** 2618–2622.

Bierne, H. and B. Michel. 1994. When replication forks stop. *Mol. Microbiol.* **13:** 17–23.

Bierne, H., S.D. Ehrlich, and B. Michel. 1991. The replication termination signal terB of the *Escherichia coli* chromosome is a deletion hot spot. *EMBO J.* **10:** 2699–2705.

Blakely, G., S.D. Colloms, G. May, M. Burk, and D.J.C. Sherratt. 1991. *E. coli Xer*C recombinase is required for chromosomal segregation at cell division. *New Biol.* **3:** 789–798.

Botchan, M., W. Topp, and J. Sambrook. 1979. Studies on simian virus 40 excision from cellular chromosomes. *Cold Spring Harbor Symp. Quant. Biol.* **43:** 709–719.

Brewer, B. 1988. When polymerases collide: Replication and transcriptional organization of the *Escherichia coli* chromosome. *Cell* **53:** 679–686.

Brewer, B. and W.L. Fangman. 1987. The localization of replication origins on ARS plasmids in *Saccharomyces cerevisiae. Cell* **51:** 463–471.

———. 1988. A replication fork barrier at the 3′ end of yeast ribosomal RNA genes. *Cell* **55:** 637–743.

Brewer, B.J., D. Lockshon, and W.L. Fangman. 1992. The arrest of replication forks in the rDNA of yeast occurs independently of transcription. *Cell* **71:** 267–276.

Bussiere, D.E., D. Bastia, and S. White. 1995. Crystal structure of the replication terminator protein from *Bacillus subtilis* at 2.6 Å. *Cell* **80:** 651–660.

Carrigan, C.M., R.A. Peck, M.T. Smith, and R.G. Wake. 1991. Normal *ter*C region of the *Bacillus subtilis* chromosome acts in a polar manner to arrest the clockwise replication fork. *J. Mol. Biol.* **222:** 197–207.

Cashel, M. and K. Rudd. 1987. The stringent response. In Escherichia coli *and* Salmonella typhimurium: *Cellular and molecular biology* (ed. F.C. Neidhardt et al.), pp. 1410–1438. American Society for Microbiology, Washington, D.C.

Clayton, D. 1991a. Nuclear gadgets in mitochondrial DNA replication and transcription. *Trends Biochem. Sci.* **16:** 107–111.

———. 1991b. Replication and transcription of vertebrate mitochondrial DNA. *Annu. Rev. Cell Biol.* **7:** 453–478.

Crosa, J.H., L. Luttropp, and S. Falkow. 1976. Mode of replication of the conjugative R-plasmid RSF1040 in *Escherichia coli. J. Bacteriol.* **26:** 454–466.

d'Ambrosio, E. and A.V. Furano. 1987. DNA synthesis arrest sites at the right terminus of rat long interspersed repeated [LINE or L1Rn] DNA family members. *Nucleic Acids Res.* **15:** 3155–3175.

Dasgupta, S., R. Bernander, and K. Nordstrom. 1991. *In vivo* effect of the *tus* mutation on cell division in an *Escherichia coli* strain where chromosome replication is under control of plasmid R1. *Res. Microbiol.* **142:** 177–180.

Dayn, A., G.M. Samadashwily, and S. Mirkin. 1992. Intramolecular DNA triplexes: Unusual sequence requirements and influence on DNA polymerization. *Proc. Natl. Acad. Sci.* **89:** 11406–11410.

Dhar, V. and C.L. Schildkraut. 1991. Role of EBNA-1 in arresting replication forks at the Epstein-Barr virus oriP family of tandem repeats. *Mol. Cell. Biol.* **11:** 6268–6278.

Di Leonardo, A., S.P. Linke, Y. Yin, and G.M. Wahl. 1993. Cell cycle regulation of gene amplification. *Cold Spring Harbor Symp. Quant. Biol.* **58:** 655–667.

DiNardo, S., K. Voelkel, and R. Sternglanz. 1984. DNA topoisomerase mutant of *Saccharomyces cerevisiae:* Topoisomerase II is required for segregation of daughter molecules at the termination of DNA replication. *Proc. Natl. Acad. Sci.* **81:** 2616–2620.

Elledge, S.J., P. Sugino, L. Guarente, and R.W. Davis. 1990. Genetic selection for genes encoding sequence-specific DNA-binding proteins. *Proc. Natl. Acad. Sci.* **86:** 3689–3693.

Fields-Berry, S.C. and M.L. DePamphilis. 1989. Sequences that promote formation of catenated intertwines during termination of DNA replication. *Nucleic Acids Res.* **17:** 3261–3273.

Francois, V., J. Louarn, and J.-M. Louarn. 1989. The terminus region of *E. coli* chromosome is flanked by several polar replication pause sites. *Mol. Microbiol.* **3:** 995–1002.

Francois, V., J. Louarn, J.-E. Rebollo, and J.-M. Louarn. 1990. Replication termination, nondivisible zones and structure of the *Escherichia coli* chromosome. In *The bacterial chromosome* (ed. K. Drilca and M. Riley), pp. 351–359. American Society for Microbiology, Washington, D.C.

Franks, A.H., A.A. Griffith, and R.G. Wake. 1995. Identification and characterization of new DNA replication terminators in *Bacillus subtilis. Mol. Microbiol.* **17:** 13–23.

Gahn, T.A. and C.L. Schildkraut. 1989. The Epstein-Barr virus origin of replication, *ori*P, contains both the initiation and termination sites of DNA replication. *Cell* **58:** 527–535.

Gerbi, S.A., C. Liang, N. Wu, S.M. DiBartolomeis, B. Bienz-Tadmor, H.S. Smith, and F.D. Urnov. 1993. DNA amplification in DNA puff II/9A of *Sciaria coprophila. Cold Spring Harbor Symp. Quant. Biol.* **58:** 487–494.

Germino, J. and D. Bastia. 1981. Termination of DNA replication *in vitro* at a sequence-specific replication terminus. *Cell* **23:** 681–687.

Gottlieb, P.A., S. Wu, X. Zhang, M. Tecklenberg, P.L. Kuempel, and T.M. Hill. 1992. Equilibrium, kinetic, and footprinting studies of the Tus-Ter protein DNA interaction. *J. Biol. Chem.* **267:** 7434–7443.

Hartwell, L. 1992. Defects in a cell cycle checkpoint may be responsible for the genomic instability of cancer cells. *Cell* **71:** 543–546.

Heck, M.M.S. and A.C. Spradling. 1990. Multiple replication origins are used during *Drosophila* chorion gene amplification. *J. Cell Biol.* **110:** 903–914.

Henckes, G., F. Harper, A. Levine, F. Vannier, and S.J. Seror. 1989. Overreplication of the origin region in the dnaB37 mutant of *Bacillus subtilis*: Postinitiation control of chromosomal replication. *Proc. Natl. Acad. Sci.* **86:** 8660–8664.

Hernandez, P., S.S. Lamm, C.A. Bjerknes, and J. Van't Hof. 1988. Replication termini in the rDNA of the synchronized pea root cells (*Pisum sativum*). *EMBO J.* **7:** 303–308.

Hernandez, P., L. Martin-Parras, M.L. Martinez-Robles, and J.B. Schvartzman. 1993. Conserved features in the mode of replication of eukaryotic ribosomal RNA genes. *EMBO J.* **12:** 1475–1485.

Hiasa, H. and K.J. Marians. 1992. Differential inhibition of the DNA translocation and DNA unwinding activities of DNA helicases by the *E. coli* tus protein. *J. Biol. Chem.* **267:** 11379–11385.

———. 1994a. Tus prevents overreplication of *ori*C plasmid DNA. *J. Biol. Chem.* **269:** 26959–26968.

———. 1994b. Topoisomerase IV can support *ori*C DNA replication *in vitro*. *J. Biol. Chem.* **269:** 16371–16375.

Hidaka, M., M. Akiyama, and T. Horiuchi. 1988. A consensus sequence of three DNA replication terminus sites on the *E. coli* chromosome is highly homologous to the terR sites of the R6K plasmid. *Cell* **55:** 467–475.

Hidaka, M., T. Kobayashi, and T. Horiuchi. 1991. A newly identified DNA replication terminus site, TerE, on the *Escherichia coli* chromosome. *J. Bacteriol.* **173:** 391–393.

Hidaka, M., T. Kobayashi, S. Takenaka, H. Takeya, and T. Horiuchi. 1989. Purification of a DNA replication terminus (ter) site-binding protein in *Escherichia coli* and identification of the structural gene. *J. Biol. Chem.* **264:** 21031–21037.

Hidaka, M., T. Kobayashi, Y. Ishimi, M. Seki, T. Enomoto, M. Abdel-Monem, and T. Horiuchi. 1992. Termination complex in *Escherichia coli* inhibits SV40 DNA replication *in vitro* by impeding the action of T antigen helicase. *J. Biol. Chem.* **267:** 5361–5365.

Hill, T.M. 1992. Arrest of bacterial DNA replication. *Annu. Rev. Microbiol.* **46:** 603–633.

Hill, T.M. and K.J. Marians. 1990. *Escherichia coli* tus protein acts to arrest the progression of DNA replication forks *in vitro*. *Proc. Natl. Acad. Sci.* **87:** 2481–2485.

Hill, T.M., A.J. Pelletier, M.L. Tecklenberg, and P.L. Kuempel. 1988. Identification of DNA sequence from the *E. coli* terminus region that halts replication forks. *Cell* **55:** 459–466.

Hill, T.M., M. Tecklenberg, A.J. Pelletier, and P.L. Kuempel. 1989. *tus*, the *trans*-acting gene required for termination of DNA replication in *Escherichia coli*, encodes a DNA-binding protein. *Proc. Natl. Acad. Sci.* **86:** 1593–1597.

Horiuchi, T. and Y. Fujimura. 1995. Recombinational rescue of the stalled fork: A model based on analysis of an *Escherichia coli* strain with a chromosome region difficult to replicate. *J. Bacteriol.* **177:** 783–791.

Horiuchi, T. and M. Hidaka. 1988. Core sequence of two separable terminus sites of the R6K plasmid that exhibit polar inhibition of replication is a 20 bp inverted repeat. *Cell* **54:** 515–523.

Horiuchi, T., Y. Fujimura, H. Nishitani, T. Kobayashi, and M. Hidaka. 1994. The DNA replication fork blocked at the Ter site may be an entrance for the RecBCD enzyme into duplex DNA. *J. Bacteriol.* **176:** 4656–4663.

Ishimi, Y., R. Ishida, and T. Andoh. 1992. Effect of ICRF-193, a novel DNA topoisomerase II inhibitor, on simian virus 40 DNA and chromosome replication *in vitro*. *Mol. Cell. Biol.* **12:** 4007–4012.

Kafatos, F.C., W. Orr, and C. Delidakis. 1985. Developmentally regulated gene amplification. *Trends Genet.* **1:** 301–306.

Kato, J., Y. Nishimura, M. Yamada, and H. Suzuki. 1990. A new topoisomerase essential for chromosome segregation in *E. coli. Cell* **63:** 393–404.

Kato, J., Y. Nishimura, R. Imamura, H. Niki, S. Hiraga, and H. Suzuki. 1988. Gene organization in a region containing a new gene involved in chromosome partitioning in *Escherichia coli. J. Bacteriol.* **170:** 3967–3977.

Kaul, S., B.K. Mohanty, T. Sahoo, I. Patel, S. Khan, and D. Bastia. 1994. The replication terminator protein of the gram-positive bacterium *B. subtilis* functions as a polar contrahelicase in gram-negative *E. coli. Proc. Natl. Acad. Sci.* **91:** 11143–11147.

Khatri, G.S., T. MacAllister, P.R. Sista, and D. Bastia. 1989. The replication termination protein of *E. coli* is a DNA sequence specific contra-helicase. *Cell* **59:** 667–674.

Kobayashi, T., M. Hidaka, M. Nishizawa, and T. Horiuchi. 1992. Identification of a site required for DNA replication fork blocking activity in the rRNA gene cluster in *Saccharomyces cerevisiae. Mol. Gen. Genet.* **233:** 355–362.

Kolter, R. and D.R. Helinski. 1978. Activity of the replication terminus of plasmid in hybrid replicons in *Escherichia coli. J. Mol. Biol.* **124:** 425–441.

Kornberg, A. and T.A. Baker. 1992. *DNA replication*, 2nd edition. W.H. Freeman, New York.

Koshland, D. and L.H. Hartwell. 1987. The structure of sister minichromosome DNA before anaphase in *Saccharomyces cerevisiae. Science* **238:** 1713–1716.

Kuempel, P.L. and S.A. Duerr. 1979. Chromosome replication in *Escherichia coli* is inhibited in the terminus region near the *rac* locus. *Cold Spring Harbor Symp. Quant. Biol.* **43:** 563–567.

Kuempel, P.L., S.A. Duerr, and N.R. Seely. 1977. Terminus region of the chromosome of *Escherichia coli* inhibits replication forks. *Proc. Natl. Acad. Sci.* **74:** 3927–3931.

Kuempel, P.L., J.M. Henson, L. Dircks, M. Tecklenberg, and D.F. Lim. 1991. *dif*, a *rec*A-independent recombination site in the terminus region of the chromosome of *Escherichia coli. New Biol.* **3:** 799–811.

Lai, C.S. and D. Nathans. 1975. Non-specific termination of SV40 DNA replication. *J. Mol. Biol.* **97:** 113–118.

LaDuca, R.J., P.J. Fay, C. Chuang, C.S. McHenry, and R.A. Bambara. 1983. Site-specific pausing of deoxyribonucleic acid synthesis catalyzed by four forms of *Escherichia coli* DNA polymerase III. *Biochemistry* **22:** 5177–5188.

Lang, W.H. and R.H. Reeder. 1993. The Reb1 site is an essential component of a terminator for RNA polymerase I in *Saccharomyces cerevisiae. Mol. Cell. Biol.* **13:** 649–658.

Langley, D.B., M.T. Smith, P.J. Lewis, and R.G. Wake. 1993. Protein-nucleoside contacts in the interaction between the replication terminator protein of *Bacillus subtilis* and the DNA terminator. *Mol. Microbiol.* **10:** 771–779.

LeBowitz, J.O. and R. McMacken. 1986. The *Escherichia coli* DnaB protein is a DNA helicase. *J. Biol. Chem.* **261:** 4738–4748.

Lee, E.H. and A. Kornberg. 1992. Features of replication fork blockage by the *Escherichia coli* terminus binding protein. *J. Biol. Chem.* **267:** 8778–8784.

Lee, E.H., A. Kornberg, M. Hidaka, T. Kobayashi, and T. Horiuchi. 1989. *Escherichia coli* replication termination protein impedes the action of helicases. *Proc. Natl. Acad. Sci.* **86:** 9104–9108.

Lee, M.S. and K.J. Marians. 1987. *Escherichia coli* replication factor Y, a component of the primosome, can act as a DNA helicase. *Proc. Natl. Acad. Sci.* **84:** 8345–8349.

Leslie, N.R. and D.J. Sherratt. 1995. Site-specific recombination in the replication terminus region of *Escherichia coli*: Functional replacement of *dif*. *EMBO J.* **14:** 1561–1570.

Levine, A., S. Autret, and S.J. Seror. 1995. A checkpoint involving RTP, the replication terminator protein, arrests replication downstream of the origin during the stringent response in *Bacillus subtilis*. *Mol. Microbiol.* **15:** 287–295.

Levine, A., F. Vannier, M. Dehbi, G. Henckes, and S.J. Seror. 1991. The stringent reponse blocks DNA replication outside the *ori* region in *Bacillus subtilis* and at the origin in *Escherichia coli*. *Mol. Microbiol.* **219:** 605–613.

Lewis, P.J. and R.G. Wake. 1991. Termination of chromosome replication in *Bacillus subtilis*. *Res. Microbiol.* **142:** 893–900.

Lewis, P.J., M.T. Smith, and R.G. Wake. 1989. A protein involved in termination of chromosome replication in *Bacillus subtilis* binds specifically to the *ter*C site. *J. Bacteriol.* **171:** 3564–3567.

Lewis, P.J., G.B. Ralston, R.I. Christopherson, and R.G. Wake. 1990. Identification of the replication terminator protein binding sites in the terminus region of the *Bacillus subtilis* chromosome and stoichiometry of the binding. *J. Mol. Biol.* **214:** 73–84.

Lindsey, J.C. and J.R.F.C. Leach. 1989. Slow replication of palindrome-containing DNA. *J. Mol. Biol.* **206:** 779–782.

Linskens, M.H.K. and J.A. Huberman. 1988. Organization of origin of replication of rDNA in *Saccharomyces cerevisiae*. *Mol. Cell. Biol.* **8:** 4927–4935.

Little, R.D., T.H.K. Platt, and C.L. Schildkraut. 1993. Initiation and termination of DNA replication in human rRNA genes. *Mol. Cell. Biol.* **13:** 6600–6613.

Liu, B. and B. Alberts. 1995. Head-on collision between a DNA replication apparatus and RNA polymerase transcription complex. *Science* **267:** 1131–1137.

Lopez-Estrano, C., J. Schvartzman, and P. Hernandez. 1996. The replication of ribosomal RNA genes in eukaryotes. In *Chromosomes today* (ed. M. Puertas et al.). Chapman and Hall, London. (In press.)

Louarn, J., J. Patte, and J.-M. Louarn. 1977. Evidence for a fixed termination site of chromosome replication in *Escherichia coli* K12. *J. Mol. Biol.* **115:** 295–314.

Louarn, J.-M., J. Louarn, V. Francois, and J. Patte. 1991. Analysis and possible role of hyperrecombination in the termination region of the *Escherichia coli* chromosome. *J. Bacteriol.* **173:** 5097–5104.

Lovett, M.A., R.A. Sparks, and D.R. Helinski. 1975. Bidirectional replication of plasmid R6K DNA in *Escherichia coli*: Correspondence between origin of replication and a single strand break in relaxed complex. *Proc. Natl. Acad. Sci.* **72:** 2905–2909.

Lucchini, R. and J.M. Sogo. 1994. Chromatin structure and transcriptional activity around the replication forks at the 3′ end of the yeast rRNA genes. *Mol. Cell. Biol.* **14:** 318–326.

MacAllister, T., G.S. Khatri, and D. Bastia. 1990. Sequence-specific and polarized replication termination *in vitro*: Complementation of extracts of tus^- *Escherichia coli* by purified Ter protein and analysis of termination intermediates. *Proc. Natl. Acad. Sci.* **87:** 2828–2832.

Madsen, C.S., S.C. Ghivizzani, and W.W. Hauswirth. 1993. Protein binding to a single termination–associated sequence in the mitochondrial DNA D-loop region. *Mol. Cell. Biol.* **13:** 2162–2171.

Manna, A.C., K.S. Pai, D.E. Bussiere, S.W. White, and D. Bastia. 1996. The dimer-dimer interaction surface of the replication terminator protein of *Bacillus subtilis* and termination of DNA replication. *Proc. Natl. Acad. Sci.* (in press).

Minden, J.S. and K.J. Marians. 1986. *Escherichia coli* topoisomerase I can segregate replicating pBR322 daughter DNA molecules *in vitro*. *J. Biol. Chem.* **261:** 11906–11917.

Mohanty, B.K., T. Sahoo, and D. Bastia. 1996. Relationship between sequence-specific termination of DNA replication and transcription. *EMBO J.* (in press).

Natarajan, S., W.I. Kelly, and D. Bastia. 1991. Replication terminator protein of *Escherichia coli* is a transcriptional repressor of its own synthesis. *Proc. Natl. Acad. Sci.* **88:** 3867–3871.

Natarajan, S., S. Kaul, A. Miron, and D. Bastia. 1993. A 27 kd protein of *E. coli* promotes antitermination of replication *in vitro* at a sequence-specific replication terminus. *Cell* **72:** 113–120.

Nordstrom, K., R. Bernander, and S. Dasgupta. 1991. Analysis of the bacterial cell cycle using strains in which chromosome replication is controlled by plasmid R1. *Res. Microbiol.* **142:** 181–188.

O'Sullivan, M.A. and C. Anagnostopoulos. 1982. Replication terminus of the *Bacillus subtilis* chromosome. *J. Bacteriol.* **151:** 135–143.

Pai, K.S., D.E. Bussiere, F. Wang, C. Hutchison, S. White, and D. Bastia. 1996. The structure and function of the replication terminator protein of *Bacillus subtilis:* Identification of the "winged-helix" DNA binding domain. *EMBO J.* (in press).

Pelletier, A.J., T.M. Hill, and P.L. Kuempel. 1988. Location of sites that inhibit progression of replication forks in the terminus region of *Escherichia coli. J. Bacteriol.* **170:** 4295–4298.

Peng, H. and K.J. Marians. 1993. Decatenation activity of topoisomerase IV during *ori*C and pBR322 DNA replication in vitro. *Proc. Natl. Acad. Sci.* **90:** 8571–8575.

Rao, P.S., H. Manor, and R.G. Martin. 1988. Pausing in SV40 DNA replication by a sequence containing $(dG\text{-}dA)_{27}(dT\text{-}dc)_{27}$. *Nucleic Acids Res.* **16:** 8077–8094.

Richter, A., U. Strausfeld, and R. Knippers. 1987. Effects of VM26 (teniposide), a specific inhibitor of type II DNA topoisomerase, on SV40 DNA replication *in vivo. Nucleic Acids Res.* **15:** 3455–3468.

Roecklein, B.A. and P.L. Kuempel. 1992. *In vivo* characterization of *tus* gene expression in *Escherichia coli. Mol. Microbiol.* **6:** 1655–1661.

Roecklein, B., A. Pelletier, and P.L. Kuempel. 1991. The *tus* gene of *Escherichia coli*: Autoregulation, analysis of flanking sequences and identification of a complementary system in *Salmonella typhimurium. Res. Microbiol.* **142:** 169–175.

Ruiz-Echevarria, M.J., G. Gimenez-Gallego, R. Sabariegos-Jareno, and R. Diaz-Orejas. 1995. Kid, a small protein of the parD stability system of plasmid R1, is an inhibitor of DNA replication acting at the initiation of DNA synthesis. *J. Mol. Biol.* **247:** 568–577.

Sahoo, T., B.K. Mohanty, I. Patel, and D. Bastia. 1995a. Termination of DNA replication *in vitro*: Requirement for stereospecific interaction between two dimers of the replication termination protein of *Bacillus subtilis* and with the terminator site to elicit polar contrahelicase and fork impedance. *EMBO J.* **14:** 619–628.

Sahoo, T., B.K. Mohanty, A. Manna, M. Lobert, and D. Bastia. 1995b. The contrahelicase activities of the replication terminator proteins of *Escherichia coli* and *Bacillus subtilis* are helicase-specific and impede both helicase translocation and authentic DNA unwinding. *J. Biol. Chem.* **270:** 29138–29144.

Schimke, R.T. 1984. Gene amplification in cultured cells. *Cell* **37:** 705–713.

Schmid, M.B. 1990. A locus affecting nucleoid segregation in *Salmonella typhimurium.*

J. Bacteriol. **172:** 5416–5424.
Schneider, A.M., M. Gaisne, and C. Anagnostopoulos. 1983. Genetic structure and internal rearrangements of stable merodiploids from *Bacillus subtilis* strains carrying the trpE26 mutation. *Genetics* **101:** 189–210.
Sista, P.R., C.A. Hutchison, and D. Bastia. 1991. DNA-protein interaction at the replication termini of plasmid R6K. *Genes Dev.* **5:** 74–82.
Sista, P.R., S. Mukherjee, P. Patel, G.S. Khatri, and D. Bastia. 1989. A host-encoded DNA-binding protein promotes termination of plasmid replication at a sequence-specific replication terminus. *Proc. Natl. Acad. Sci.* **86:** 3026–3030.
Smith, M.T. and R.G. Wake. 1992. Definition and polarity of action of DNA replication terminators in *Bacillus subtilis. J. Mol. Biol.* **227:** 648–657.
Snapka, R.M., M.A. Powelson, and J.M. Strayer. 1988. Swivelling and decatenation of replicating simian virus 40 genomes *in vivo. Mol. Cell. Biol.* **8:** 515–521.
Stark, G.R. 1993. Regulation and mechanisms of mammalian gene amplification. *Adv. Cancer Res.* **61:** 87–113.
Stark, G.R., M. Debatisse, E. Giulotto, and G.M. Wahl. 1989. Recent progress in understanding mechanisms of mammalian DNA amplification. *Cell* **57:** 901–908.
Sundin, O. and A. Varshavsky. 1981. Terminal stages of SV40 DNA replication proceed via multiply intertwined catenated dimers. *Cell* **21:** 103–114.
Swindels, M.B. 1995. Identification of a common fold in the replication terminator protein suggests a possible mode for DNA binding. *Trends Biochem. Sci.* **20:** 300–302.
Tapper, D.P. and M.L. DePamphilis. 1980. Preferred sites are involved in the arrest and initiation of DNA synthesis during the replication of SV40 DNA. *Cell* **22:** 97–108.
Tecklenberg, M., A. Naumer, O. Nagappan, and P. Kuempel. 1995. The *dif* resolvase locus of the *Escherichia coli* chromosome can be replaced by a 33-bp sequence, but function depends on location. *Proc. Natl. Acad. Sci.* **92:** 1352–1356.
Uemura, T., H. Ohkura, Y. Adachi, K. Morino, K. Shiozaki, and M. Yanagida. 1987. DNA topoisomerase II is required for condensation and separation of mitotic chromosomes in *S. pombe. Cell* **50:** 917–925.
Valenzuela, M.S., D. Freifelder, and R. Inman. 1976. Lack of a unique terminator site for the first round of bacteriophage DNA replication. *J. Mol. Biol.* **102:** 569–589.
Voelkel-Meimon, K., R.L. Keil, and G.S. Roeder. 1987. Recombination-stimulating sequences in yeast ribosomal DNA correspond to sequences regulating transcription by RNA polymerase I. *Cell* **48:** 1071–1079.
Weaver, D.T. and M.L. DePamphilis. 1984. The role of palindromic and non-palindromic sequences in arresting DNA synthesis *in vitro* and *in vivo. J. Mol. Biol.* **180:** 961–986.
Weaver, D.T., S.C. Fields-Berry, and M.L. DePamphilis. 1985. The termination region for SV40 DNA replication directs the mode of separation for the two sibling molecules. *Cell* **41:** 565–575.
Weiss, A.S. and R.G. Wake. 1984. A unique DNA intermediate associated with termination of chromosome replication in *Bacillus subtilis. Cell* **39:** 683–689.
Wells, R.D., D.A. Collier, J.C. Hanvey, M. Shimizu, and F. Wohlrab. 1988. The chemistry and biology of unusual DNA structures adopted by oligopurine.oligopyrimidine sequences. *FASEB J.* **2:** 2939–2949.
Wintersberger, E. 1994. DNA amplification: New insights into its mechanism. *Chromosoma* **103:** 73–81.
Yoshikawa, H. and N. Sueoka. 1963. Sequential replication of *Bacillus subtilis* chromosome I: Comparison of marker frequencies in exponential and stationary growth

phases. *Proc. Natl. Acad. Sci.* **49:** 559–566.

Young, P.A. and R.G. Wake. 1994. The *Bacillus subtilis* replication terminator system functions in *Escherichia coli*. *J. Mol. Biol.* **240:** 275–280.

Zechiedrich, E.L. and N.R. Cozzarelli. 1995. Roles of topoisomerase IV and DNA gyrase in DNA unlinking during replication in *Escherichia coli*. *Genes Dev.* **9:** 2859–2869.

Zeviani, M. 1992. Nucleus-driven mutations of human mitochondrial DNA. *J. Inherited Metab. Dis.* **15:** 456–471.

Zeviani, M., S. Servidei, C. Gellera, E. Bertini, S. DiMauro, and S. DiDonato. 1989. An autosomal dominant disorder with multiple deletions of mitochondrial DNA starting at the D-loop region. *Nature* **339:** 309–311.

7

Fidelity of DNA Replication

John D. Roberts[1] and Thomas A. Kunkel[2]
[1]Laboratory of Molecular Carcinogenesis
[2]Laboratory of Molecular Genetics
National Institute of Environmental Health Sciences
Research Triangle Park, North Carolina 27709

The six billion nucleotides of the diploid human genome are replicated in only a few hours while generating so few errors that the spontaneous mutation rate may be less than 1 mutation per genome per cell division (Loeb 1991). This incredible accuracy results from three major error-avoidance processes: the high selectivity of DNA polymerases, exonucleolytic proofreading, and postreplication mismatch repair. In this chapter, we review our current understanding of the first two of these processes. Readers interested in eukaryotic mismatch repair are referred to a recent review (Modrich 1994).

We begin by describing the steps in the polymerization reaction cycle that discriminate against base substitution errors, then review studies of the substitution fidelity of the five classes of eukaryotic DNA polymerases. We then consider several ways to make errors by template-primer slippage and review what is known about eukaryotic DNA polymerase frameshift error rates. Finally, we present information on the fidelity with which the multiprotein replication machinery replicates undamaged DNA and DNA containing adducts of known carcinogens.

BASE SUBSTITUTION FIDELITY

Discrimination Steps in a Polymerization Cycle

The error discrimination steps that operate during incorporation of a single nucleotide have been worked out primarily with prokaryotic and viral DNA polymerases. The steps (Fig. 1) include binding of the polymerase to the DNA, formation of a ternary complex with the incoming deoxyribonucleoside triphosphate (dNTP), a conformational change in this complex to position the substrates for phosphodiester bond formation, the chemical reaction step to form the bond, a second conformational change following the chemical reaction, and release of pyrophos-

DNA Replication in Eukaryotic Cells

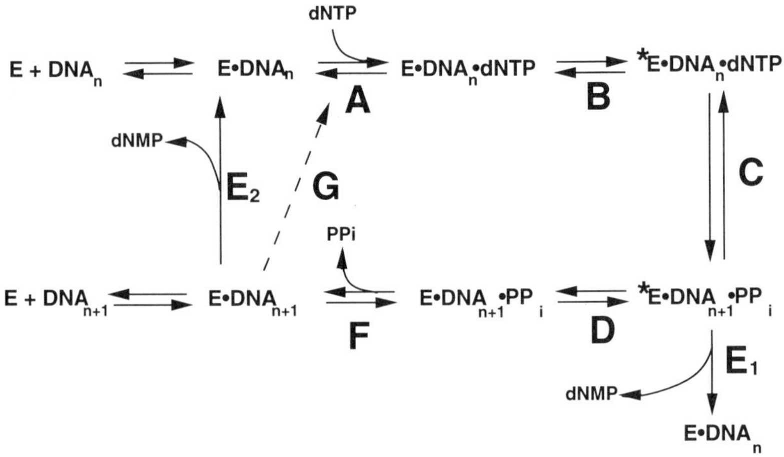

Figure 1 Reaction pathway for exonuclease-proficient DNA polymerases. Asterisks represent enzyme (E) in a different conformation. The enzyme conformations in the ternary E•DNA_n•dNTP and E•DNA_{n+1}•PPi complexes are unknown and not necessarily the same. Entry into the next cycle of polymerization is indicated by G. Translocation has not been assigned to a particular step in the reaction. Steps A–F are discussed in the text. (Reprinted, with permission, from Kunkel 1992.)

phate (PPi). After the chemical step, the polymerase translocates to commence the next cycle of polymerization, extending the nascent strand in the 5′ to 3′ direction.

DNA Polymerase Selectivity

There are several points in this cycle where discrimination against errors by the DNA polymerase occurs. The first is during binding of the dNTP to the complex formed between the polymerase and a correctly paired template primer (Fig. 1, step A). Incorrect dNTPs bind much less avidly than do correct dNTPs. At this step, discrimination against several mispairs, quantitated by kinetic analyses includes: Klenow polymerase, the large fragment of *Escherichia coli* DNA polymerase I, 0 to 23-fold (Eger and Benkovic 1992); T7 DNA polymerase, 200- to 400-fold (Wong et al. 1991; for review, see Johnson 1993); T4 DNA polymerase, ≥250-fold (Capson et al. 1992); and human immunodeficiency virus type 1 reverse transcriptase (HIV-1 RT), 25- to 400-fold (Johnson 1993; Zinnen et al. 1994). The amount of discrimination varies over a considerable range, depending on the composition of the mispair and the surrounding sequence context. This reflects differences in hydrogen-bonding potentials for various combinations of dNTPs and template bases, sequence-dependent differences in base stacking, and a demand for equivalent base-pair geometry (for review, see Echols and Goodman 1991).

For the three viral DNA polymerases mentioned above, the amount of binding discrimination exceeds that predicted from the free-energy differences between correct and incorrect base pairs in aqueous solution. Within a DNA polymerase active site, hydrogen-bonded water molecules are displaced from the transition state for complementary base pairs (Fersht 1985), and this exclusion of water from the active site may amplify base-pair free-energy differences, thus enhancing fidelity (Petruska et al. 1986; Abbotts et al. 1991; Johnson 1993). A differential ability to exclude water from the active site might partly explain DNA polymerase-dependent differences in dNTP-binding discrimination. Klenow polymerase utilizes dNTP-binding discrimination to a lesser extent than do the other three polymerases examined. This has led to the suggestion that in the "ground state" (i.e., prior to the inferred conformational change at step B), initial binding of the dNTP to Klenow polymerase may not involve base-pairing (Johnson 1993). It has also been suggested (Capson et al. 1992; Johnson 1993) that any reduction in enzymatic efficiency of *E. coli* pol I resulting from competitive inhibition by the three incorrect dNTPs might be tolerated by the cell because the primary role for this polymerase in vivo is in DNA repair or replacement of RNA primers during replication. In contrast, reduced efficiency might be less acceptable for polymerases that replicate entire genomes. All three viral polymerases mentioned above, whose role is genomic replication, discriminate more strongly against incorrect dNTP binding than does Klenow polymerase.

The next selectivity step in the polymerization cycle, inferred from several lines of evidence (Johnson 1993; Zinnen et al. 1994), involves a conformational change in the ternary complex to position the dNTP for subsequent phosphodiester bond formation (Fig. 1, step B). For T7 DNA polymerase, this change to a "closed" structure in which the polymerase is thought to lock onto the template-primer•dNTP complex (Johnson 1993) is much more rapid for correct base pairs that can adopt Watson-Crick geometry than for incorrect base pairs that cannot. This leads to a 2000- to 4000-fold faster rate of incorporation of correct nucleotides (Johnson 1993). Similarly, a difference in the rate of change in protein conformation with correctly versus incorrectly bound dNTPs has been suggested to enhance the selectivity of HIV-1 RT by 7- to 17,000-fold (Johnson 1993; Zinnen et al. 1994), depending on the mispair considered.

The situation is somewhat different with Klenow polymerase. Although a change in polymerase conformation has been inferred to limit the catalytic rate for correct incorporation by this enzyme, this step is not rate-limiting for incorrect incorporation. Rather, discrimination against incorrect incorporation is mostly due to a strong reduction in the rate of phosphodiester bond formation (Fig. 1, step C; Kuchta et al. 1988).

Steps A, B, and C in Figure 1 all contribute to polymerase selectivity against insertion of incorrect nucleotides, with the relative importance of each step dependent on the polymerase examined. The contributions of these individual steps to the fidelity of eukaryotic DNA polymerase remain to be established. However, steady-state kinetic analyses have established misinsertion and mispair extension rates of some eukaryotic DNA polymerases (see below).

Exonucleolytic Proofreading

With the Klenow polymerase, a slow step has been detected after chemistry and prior to pyrophosphate release (Carroll and Benkovic 1990; Polesky et al. 1992). This delay, suggested to result from a second conformational change in the enzyme-template•primer complex, provides an opportunity for its intrinsic 3′→5′ exonuclease to remove misinserted nucleotides (step E_1). Once a nucleotide is incorporated, the $E{\bullet}DNA_{n+1}$ complex can enter the next cycle of polymerization (Fig. 1, step G). However, the rate of correct incorporation onto a terminal mispair is much slower than onto a correctly paired terminus. This slow step provides another opportunity for excision (Fig. 1, step E_2), after transferring the misinserted nucleotide from the polymerase active site to the exonuclease active site. These sites are on separate domains of Klenow polymerase and are spatially separated in other DNA polymerases. Movement into the exonuclease active site can occur with or without enzyme dissociation from the template primer, depending on the polymerase (Joyce 1989; Donlin et al. 1991; Capson et al. 1992; Reddy et al. 1992). If the enzyme does dissociate, it may rebind to the exonuclease active site and edit the misinsertion. It is also formally possible that a terminal misinsertion may be bound and proofread by another exonuclease, one associated with a different polymerase or one not associated with a polymerase at all.

Prokaryotic polymerases containing associated 3′→5′ exonuclease activities have average base substitution error rates of about 10^{-6} to 10^{-7}. The proofreading contribution to these rates has been assessed by selectively stimulating the polymerase activity relative to the exonuclease activity by increasing the concentration of the next correct nucleotide to be incorporated after a misinsertion. Alternatively, exonuclease activity can be inhibited by adding to the reaction a nucleotide monophosphate (dNMP), the end product of exonuclease action. Proofreading can also be eliminated by changing amino acid residues essential for exonuclease activity. These residues are found in three conserved DNA sequence motifs common to the coding sequences of DNA polymerases containing intrinsic 3′→5′ exonucleases (Ito and Braithwaite 1990; Blanco et al. 1991; Chung et al. 1991; Morrison et al. 1991; Simon et al. 1991; Zhang et al. 1991). The results obtained from these three approaches suggest that proofreading contributes on average about 100-fold to fidelity. This value is consistent with estimates from in vivo studies using *E. coli* strains deficient in proofreading activity (Schaaper 1993) and with calculations suggesting that the energetic cost of improving fidelity by more than this amount using exonucleolytic activity could be unacceptably high for an organism due to too much excision of correctly paired bases (Fersht et al. 1982).

The contribution of exonucleolytic proofreading to base substitution fidelity can vary over a wide range, from only a few-fold (Bebenek et al. 1990) to almost 1000-fold (West Frey et al. 1993). This results partly from the different rate constants for polymerization from the 12 possible mispairs and also reflects enzyme- and sequence-specific influences. These are expected on the basis of the idea originally proposed by Brutlag and Kornberg (1972) and subsequently supported by extensive data (see, e.g., Bloom et al. 1994; Carver et al. 1994 and references therein) that a terminus containing a terminal mispair has a higher probability of being single-stranded ("frayed") than does a correctly paired terminus. Such a frayed end will preferentially bind to the exonuclease active site, which prefers single-stranded DNA. Similarly, a matched and, therefore, double-stranded terminus will preferentially bind to the polymerase active site, which prefers double-stranded DNA (for review, see Joyce and Steitz 1994). Because the stability of the duplex region of the template-primer will depend on its DNA sequence, proofreading efficiency is expected to differ in different sequence contexts having differing stabilities. Moreover, the degree of fraying needed to allow single-stranded DNA to bind to the exonuclease active site may vary, depending on the distance between the polymerase and exonuclease active sites. This distance, estimated to be 25–30 Å for the Klenow polymerase, could be greater for some enzymes than others (e.g., compare data in Cowart et al. [1989] to data in Capson et al. [1992]; for review, see Joyce and Steitz 1994), leading to enzyme-mediated differences in proofreading efficiency. It has also been proposed that the Klenow polymerase active site promotes movement of DNA into the exonuclease active site by rejecting aberrant primer termini (Carver et al. 1994).

Base Substitution Error Rates of Eukaryotic DNA Polymerases

The above concepts provide a framework for considering what is known about the substitution fidelity of eukaryotic DNA polymerases. Error rates during catalysis in vitro by the five template-dependent eukaryotic DNA polymerases (see Wang, this volume) are shown in Table 1. These were obtained using fidelity assays that require both misinsertion and mispair extension to score an error. The "average" values are from copying a 250-base single-stranded template sequence of the *lacZ* α-complementation gene in bacteriophage M13mp2 DNA (Kunkel 1985a). This assay scores all stable misincorporations that yield an M13 plaque

Table 1 Error rates of eukaryotic DNA polymerases

DNA polymerase	Average error rate (x 10^{-6})	Refs.	Range of error rates (x 10^{-6})	Refs.
Substitution errors				
pol-α	160[a]	1–3	1.2–380	1–13
pol-β	670	2	45–1000	2, 9
pol-δ	~10[b]	4	≤2.3–29	4
pol-ε	≤6.7	4	≤1–19[c]	4, 14
pol-γ	1.8[d]	15, 16	1.8–1200[e]	2, 9, 15, 16
One-base frameshift errors				
pol-α	50[a]	1, 4, 17		
pol-β	900	17		
pol-δ	18[b]	4		
pol-ε	5[c]	4		
pol-γ	2.4	24		

References: (1) Kunkel et al. 1989; (2) Kunkel and Alexander 1986; (3) Roberts and Kunkel 1988; (4) Thomas et al. 1991b; (5) Copeland and Wang 1991; (6) Reyland and Loeb 1987; (7) Grosse et al. 1983; (8) Kaguni et al. 1984; (9) Kunkel and Loeb 1981; (10) Brooke et al. 1991; (11) Copeland et al. 1993; (12) Dong et al. 1993a; (13) Perrino and Loeb 1989b; (14) Kunkel et al. 1987; (15) Wernette et al. 1988; (16) Kunkel and Mosbaugh 1989; (17) Kunkel 1986.

[a]Measurements have been made with pol-α preparations from several sources; the values given are averages.

[b]Reactions contained PCNA to stimulate gap-filling synthesis by pol-δ.

[c]Higher values obtained with reactions in which proofreading was compromised; thus, they probably represent a minimal estimate of the accuracy of the enzyme.

[d]Data are from reversion assays.

[e]See comment in Reference 16 regarding the higher error rate values.

with reduced blue-color intensity. The type of polymerase error is defined by sequencing the identified M13 mutants. The substitution error rates are thus average values per detectable nucleotide polymerized, for all 12 possible single-base mispairs in a variety of sequence contexts (Bebenek and Kunkel 1995). The wide range of error rates reported in the literature (Table 1) illustrates that substitution fidelity depends on the DNA polymerase, the composition of the mispair, and the local template-primer sequence. The influence of these parameters can be examined at specific template positions using steady-state kinetic approaches (Mendelman et al. 1989, 1990). This allows the amount of discrimination at both the misinsertion and mispair extension steps to be estimated separately. The fidelity of the eukaryotic DNA polymerases is considered in more detail below.

DNA Polymerase-α

This is the most extensively studied eukaryotic polymerase. DNA polymerase-α (pol-α) isolated from several sources has an average base substitution error rate of 160×10^{-6} when a variety of errors in numerous sequence contexts are considered collectively (Table 1). Similar values have been obtained when the fidelity of the yeast or human p180 catalytic subunit alone is compared to that of the four-subunit DNA polymerase-α:RNA primase complex (Kunkel et al. 1989; Copeland et al. 1993). Similar values for misinsertion fidelity, representing the product of steps A, B, and C in Figure 1, have also been observed by kinetic analyses of the *Drosophila melanogaster* DNA polymerase-α:DNA primase (Mendelman et al. 1989). Thus, highly purified DNA pol-α is not particularly accurate relative to the high fidelity required to replicate eukaryotic genomes (see below). However, a 10^{-4} error rate may more than suffice if pol-α is only responsible for synthesis of a small number of nucleotides from an RNA primer. Mistakes made here could also be removed during the RNA primer excision-replacement synthesis reaction. Moreover, pol-α fidelity estimates are thus far limited to synthesis initiated from exogenously supplied DNA primers. Since several recent observations suggest that pol-α synthesis coupled to primase activity may differ in some respects (Sheaff et al. 1994), it is possible that the fidelity of RNA-primed DNA synthesis could be higher (or lower) than present data suggest.

An average error rate of about 10^{-4} is consistent with the fact that many preparations of purified pol-α lack 3′→5′ exonuclease activity and that pol-α genes lack the three conserved sequence motifs characteristic of such exonucleases. Nonetheless, evidence exists for pol-α-associated proofreading. A 3′→5′ exonuclease activity is present in preparations of *D. melanogaster* pol-α from which the associated 70-kD subunit has been removed (Cotterill et al. 1987), and the resulting polymerase is more accurate than the pol-α:primase complex (Cotterill et al. 1987; Reyland et al. 1988). Proofreading activity may be removed during some purification schemes but retained during others. For example, mouse and human pol-α preparations have been reported that contain 3′→5′ exonuclease activity (Chen et al. 1979; Bialek et al. 1989), and the latter catalyzes high-fidelity synthesis. Precedent for proofreading by exonuclease activity of a separate gene product comes from studies in *E. coli*, where the polymerase (α subunit) and exonuclease (ε subunit) activities of the replicative Pol III holoenzyme are encoded by two different genes, designated *dnaE* and *dnaQ*, respectively (for review, see Echols and Goodman 1991). Thus, the difficulty that pol-α has in extending certain mispairs (Perrino and Loeb 1989a; Mendelman et al. 1990) could provide the opportunity for a separate exonuclease activity to proofread misinserted nucleotides, especially if pol-α dissociates from the template primer. This possibility is supported by the observation of high-fidelity DNA synthesis by pol-α in the presence of the *E. coli* ε subunit (Perrino and Loeb 1989b) or pol-δ (Perrino and Loeb 1990), which contains an intrinsic 3′→5′ exonuclease activity. The extraordinary range of error rates reported in the literature for synthesis by pol-α (Table 1) thus reflects the large number of studies performed and the use of different assay methods and pol-α preparations. It also clearly reflects differences in discrimination for mispairs of varying composition and in different local sequence contexts.

There has been much recent progress in defining the structure of DNA polymerases (for review, see Joyce and Steitz 1994). Moreover, several conserved amino acid motifs have been identified by sequence alignments of DNA polymerase genes (Delarue et al. 1990; Braithwaite and Ito 1993 and references therein). These data are the starting points to identify amino acid residues that, when changed, alter replication fidelity. Among the eukaryotic polymerases, this approach has been applied first to human pol-α. Several mutant enzymes have been constructed containing single amino acid differences in the most conserved motifs, designated I and II (for review, see Wang 1991). The proteins were overproduced, and their biochemical properties were examined. Two of the motif I mutants, having single-amino-acid differences in residues important for binding the divalent metal ion required for catalysis, have higher discrimination than the wild-type pol-α for nucleotide misinsertion and for mispair extension in Mn^{++}-activated kinetic assays (Copeland et al. 1993). Likewise, two of the motif II mutants, having single-amino-acid differences in residues suggested to be important for binding the incoming dNTP and for interacting with the primer, have improved insertion fidelity (Dong et al. 1993a), whereas another has reduced discrimination against mispair extension (Dong et al. 1993b). A continuation of this approach with pol-α should increase our understanding of the fidelity of this polymerase. Polymerases with reduced or enhanced fidelity should also be useful "biomarkers" for defining their roles in replication and repair.

DNA Polymerase-β

This smallest of the eukaryotic DNA polymerases is also the least accurate. The single subunit polymerases purified from rat hepatoma cells or chicken embryos have average substitution error rates of 670×10^{-6} (Kunkel and Alexander 1986). For individual mispairs, error rate values range from 45×10^{-6} to 1000×10^{-6}. Similar, and in a few instances even higher, error rates have been obtained for direct misinsertion by the rat enzyme, using steady-state kinetic analyses of rat pol-β (Boosalis et al. 1989). These rates are consistent with the fact that purified pol-β lacks associated 3′→5′ exonuclease activity (and the three exonuclease motifs conserved in proofreading polymerases). However, although pol-β shares this property with pol-α, it is even less accurate, demonstrating that, independent of proofreading, selectivity against substitution errors depends on the DNA polymerase.

Low-fidelity synthesis by pol-β may be consistent with a modest catalytic role in vivo, i.e., filling gaps of one or a few nucleotides during base excision repair. Alternatively, pol-β may have higher accuracy than current estimates suggest. Thus far, pol-β fidelity has been measured using template primers containing long single-stranded template regions, where synthesis is distributive rather than processive. Recently, pol-β has been shown to catalyze processive synthesis on a template adjacent to a 5′-phosphoryl end (Singhal and Wilson 1993). As this type of synthesis may more closely resemble that occurring during base excision repair, it will be interesting to determine pol-β fidelity using substrates containing single-stranded gaps of one or a few nucleotides. It is also possible that pol-β fidelity is influenced by accessory proteins. For example, mammalian DNase V, a 12-kD protein having both 3′→5′ and 5′→3′ exonuclease activity (Mosbaugh and Meyer 1980), associates with pol-β in vitro.

X-ray crystallographic structural information is now available for the ternary complex of pol-β-template•primer-dNTP (Pelletier et al. 1994). This structure provides an excellent opportunity for polymerase engineering to understand fidelity.

DNA Polymerase-δ

The fidelity of DNA synthesis in vitro by this polymerase is remarkably understudied, given its central role in eukaryotic replication. In one study (Thomas et al. 1990), the average substitution error rate in vitro of pol-δ plus proliferating cell nuclear antigen (PCNA) is about 10×10^{-6} (Table 1). The error rate of pol-δ alone was not determined because pol-δ would not complete gap-filling synthesis in the absence of PCNA. The fact that pol-δ (plus PCNA) is more accurate than pol-α or pol-β is consistent with proofreading of misinsertions by its associated 3′→5′ exonuclease activity, which has properties expected of an editing exonuclease (for review, see Bambara and Jessee 1991). The coding sequences of the yeast and human genes have the three conserved exonuclease motifs characteristic of proofreading enzymes (Chung et al. 1991; Zhang et al. 1991).

DNA Polymerase-ε

This polymerase also has an associated exonuclease activity, and the coding sequence of the gene has the three conserved exonuclease motifs characteristic of proofreading enzymes (Morrison et al. 1991). Pol-ε is highly accurate, having a very low rate for substitutions that revert a termination codon in the *lacZ* gene (Kunkel et al. 1987)[1] and an average error rate of $\leq 6.7 \times 10^{-6}$ in the forward mutation assay. In fact, substitution errors by pol-ε are only detected when dGMP, a putative inhibitor of exonucleolytic proofreading, is included in the synthesis reaction. This is consistent with the idea that the 3′→5′ exonuclease of pol-ε removes misinsertions during synthesis in vitro.

DNA Polymerase-γ

The mitochondrial replicative polymerase from several sources has high base-substitution fidelity (Table 1). The polymerase also has an associated 3′→5′ exonuclease activity (Kunkel and Soni 1988a; Insdorf and Bogenhagen 1989; Kaguni and Olsen 1989; Kunkel and Mosbaugh 1989;

[1]At the time this study was performed, what is now known to be DNA polymerase-ε was then designated DNA polymerase-δII, hence the title of the article.

Foury and Vanderstraeten 1992; Gray and Wong 1992), and the coding sequence of the yeast gene has the three conserved exonuclease motifs characteristic of proofreading enzymes (Foury and Vanderstraeten 1992). The catalytic properties of the exonuclease and the fact that fidelity is reduced in reactions containing a high dNTP concentration or dGMP (Kunkel and Soni 1988a; Kunkel and Mosbaugh 1989) are consistent with a proofreading role in vitro.

ERRORS INVOLVING TEMPLATE-PRIMER MISALIGNMENT

In addition to direct misincorporation of noncomplementary nucleotides, base-addition, -deletion, and even -substitution errors can be generated by processes involving template-primer misalignments during replication.

Frameshifts[2] Initiated by Template-primer Slippage

Strand slippage during replication of iterated[3] sequences results in misaligned intermediates stabilized by correct base pairs (Fig. 2A). Subsequent polymerization from the misaligned intermediate leads to deletion if the unpaired nucleotide(s) is in the template strand (Fig. 2A) or to addition if the unpaired nucleotide(s) is in the primer strand (not shown). This mechanism predicts (Streisinger et al. 1967) that the error rate should increase as the length of the run increases, because the potential number of correct base pairs that could stabilize the misaligned intermediates increases, as does the number of potential misaligned intermediates that can form (Fig. 2). Furthermore, the longer the run, the greater the distance between the extra nucleotide and the 3′-OH primer terminus, potentially reducing interference by the extra base during phosphodiester bond formation within the enzyme active site.

In support of this logic, frameshift error rates, expressed per nucleotide polymerized to correct for differences in the number of nucleotides in runs of different lengths, do indeed increase as the length of a homopolymeric run increases, for DNA pol-α (Kunkel 1990) and T7 DNA polymerase (Kunkel et al. 1994). In addition, error rates for one-base deletions in homopolymeric runs by pol-β (Kunkel 1986) and HIV-

[2]Although the term frameshift mutation usually refers to changes in the number of base pairs in a protein-coding sequence that are not multiples of three, for convenience it refers here to mutations resulting from any difference in the number of base pairs, regardless of location.

[3]A non-iterated nucleotide is one having 5′ and 3′ neighbors that are not identical to the nucleotide considered. An iterated nucleotide has at least one identical neighbor. Iterated means repeated; reiterated means re-repeated.

1 RT (Bebenek et al. 1993) decrease when the template sequence is altered to either shorten or eliminate the repetitive sequence. That the slippage mechanism operates in vivo is suggested by the instability in repetitive sequences associated with cancer (for review, see Loeb 1994) and neurodegenerative diseases (see, e.g., Willems 1994).

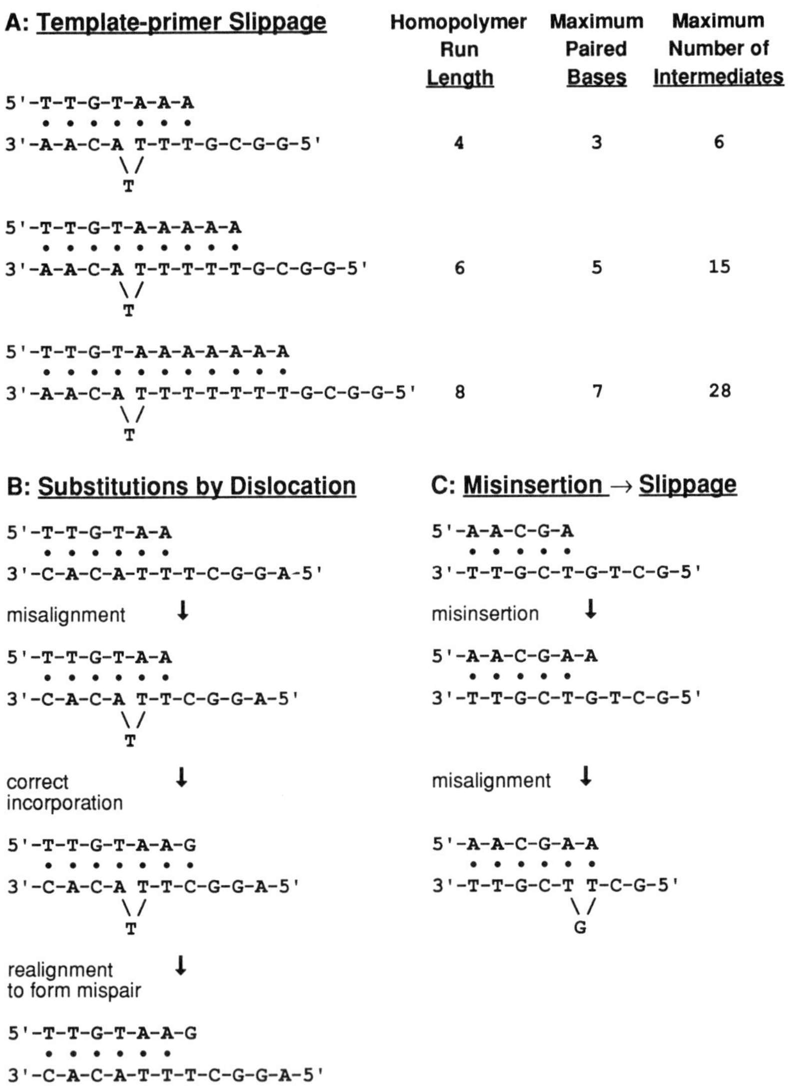

Figure 2 Pathways for errors involving misaligned template primers. See text for description.

Base Substitutions Initiated by Strand Slippage

Following slippage, correct incorporation of another nucleotide followed by realignment before continued incorporation generates a terminal mispair (Kunkel 1985a). This can yield a base substitution, but in this case, initiated by slippage rather than by misinsertion. This process has been termed dislocation mutagenesis (Kunkel and Alexander 1986) by analogy with a dislocated shoulder joint that pops out of alignment but ultimately resumes a normal position. Strong support for the model comes from fidelity studies with pol-β (Kunkel and Soni 1988b; Boosalis et al. 1989) and HIV-1 RT (Bebenek et al. 1993) in vitro. With both polymerases, strong base substitution hot spots are observed at the ends of several different homopolymeric runs, and the substitution specificity depends on the immediate template neighbor. The dislocation concept is not limited to the situation shown in Figure 2, but can involve an extra nucleotide in the primer strand (Fig. 5A in Bebenek et al. 1993), two unpaired template nucleotides (Fig. 5C in Bebenek et al. 1993), or many nucleotides (Fig. 3C in Kunkel and Soni 1988a; for review, see also Ripley 1990).

Frameshifts Initiated by Misinsertion

A distinctly different way to generate a misaligned substrate is misincorporation followed by template-primer rearrangement to provide a correct terminal base pair for continued polymerization (Fig. 2C). The resulting misalignment ultimately leads to a frameshift error, but in this case it is initiated by misinsertion rather than strand slippage. This model was suggested by several observations (for review, see Kunkel 1990) indicating that "difficult-to-extend" mispairs may realign such that synthesis proceeds from a substrate containing an extra nucleotide in the template strand but a correct base pair at the terminus.

In principle, this mechanism is possible at any template position and is not limited to the production of minus-one-base errors. Thus, frameshift errors at template runs, as well as plus and minus errors of varying numbers of nucleotides, may initiate by misincorporation. The model has been supported by fidelity studies with yeast pol-α (Kunkel et al. 1989), Klenow DNA polymerase (Bebenek et al. 1990), and HIV-1 RT (Bebenek et al. 1992) in vitro. The concept of difficult-to-extend termini led to the suggestion (Kunkel and Soni 1988b) that incorporation opposite damaged templates might also yield frameshifts by this mechanism. Studies involving replication of DNA containing several different lesions (Wang and Taylor 1992; Shibutani and Grollman 1993; Lindsley and Fuchs 1994; Napolitano et al. 1994) strongly support this suggestion.

Frameshift Error Rates of Eukaryotic DNA Polymerases

The average frameshift error rates of the five template-dependent eukaryotic DNA polymerases are shown in Table 1. These rates are errors per detectable nucleotide polymerized with the *lacZ* α-complementation gene target and are for single-base frameshift errors. Similar to the situation with substitution errors, DNA pol-α and pol-β, which lack associated proofreading activity, are less accurate than are pol-δ, pol-ε, and pol-γ, which have intrinsic exonucleases. Direct comparison of the frameshift fidelity of wild-type versus exonuclease-deficient derivatives of Klenow polymerase (Bebenek et al. 1990) and T7 DNA polymerase (Kunkel et al. 1994) strongly suggests that frameshift intermediates at both non-run sequences and in homopolymeric runs of up to 5 bp are subject to exonucleolytic proofreading. Moreover, the frameshift fidelity of pol-ε is reduced in reactions containing a high concentration of dNTPs and dGMP (Thomas et al. 1991b), suggesting that frameshift errors by this enzyme are also proofread.

Polymerase rates for minus-one-base errors are higher than for plus-one-base errors. Just as for substitution errors, frameshift error rates are sequence-dependent and polymerase-dependent, with error rates for the same mistake varying over 1000-fold (for detailed discussion, see Kunkel 1990). Polymerases can also delete or add more than a single nucleotide. Some of these errors can be explained by strand slippage between perfectly repeated DNA sequences separated by a variable number of intervening nucleotides (for review, see Kunkel 1990; Ripley 1990). More complicated models involving strand-switching, primer loop-back, and palindromic DNA sequences have also been invoked to explain complex frameshift mutations generated by DNA polymerases.

Processivity and Frameshift Fidelity

One property of polymerization relevant to polymerase frameshift fidelity at homopolymeric runs is processivity, the number of nucleotides incorporated per polymerase association/dissociation with the template primer. This was suggested by the observation that pol-α is both more accurate and more processive than pol-β (Kunkel 1985b). This correlation has been examined in greater detail with HIV-1 RT. This polymerase is inaccurate for one-base frameshifts within some but not all template runs. These hot spots for frameshift errors are also template positions where the probability of termination of processive synthesis is high (Bebenek et al. 1989). Moreover, when changes were introduced into the sequences flanking the runs, increases or decreases in frameshift error rates were observed, and these often correlated with concomitant increases or decreases in termination of processive synthesis within the run (Bebenek et al. 1993). The data thus reveal a consistent pattern wherein low processivity correlates with low frameshift fidelity, consistent with the idea that the formation and/or utilization of misaligned template primers is increased during the dissociation-reinitiation phase of a polymerization reaction.

Effect of Accessory Proteins on DNA Polymerase Fidelity

Further support for a relationship between processivity and fidelity comes from a study of the frameshift fidelity of T7 DNA polymerase with and without its processivity protein, thioredoxin (Kunkel et al. 1994). T7 DNA polymerase alone has low processivity, adding only 1–50 nucleotides before dissociating. However, when it is complexed with its accessory subunit thioredoxin, polymerization proceeds for thousands of nucleotides without dissociation. Fidelity measurements with an exonuclease-deficient mutant of T7 pol showed that the rate for one-base addition frameshifts at homopolymeric runs was 46-fold higher in the absence of thioredoxin than in its presence. This may have general significance, given that accessory proteins that enhance processivity are a general feature of multiprotein replication complexes (see Stillman, this volume). Frameshift fidelity conferred by the polymerase clamp protein is particularly interesting inasmuch as replication infidelity is one possible explanation for the instability of repetitive genomic sequences reported for several diseases (Loeb 1994; Willems 1994).

In the absence of thioredoxin, the exonuclease-deficient T7 DNA polymerase was found to be more accurate during DNA synthesis in vitro for substitutions and 1- and 2-nucleotide deletions (Kunkel et al. 1994). One possible explanation is that the premutational intermediates formed during polymerization are not successfully extended unless the polymerase is complexed with thioredoxin. The biological implication is that, under some circumstances, an accessory protein-mediated alteration in the extension rate from an unusual template primer, e.g., following incorporation opposite a damaged base or slippage at a damaged site, could actually serve a mutator function by enhancing extension synthesis to seal the error before transfer to the exonuclease active site for removal.

Another accessory protein that logically could influence fidelity is single-stranded DNA-binding protein (SSB). The substitution fidelity of synthesis by several polymerases is increased a few fold in reactions containing *E. coli* SSB (Kunkel et al. 1979, 1983). Similarly, the rate of deletions between direct repeats generated by yeast pol-α is reduced in reactions containing a yeast SSB (Roberts et al. 1990). Human replication protein A (RP-A) is a 3-subunit SSB required for replication of SV40-origin-containing DNA in HeLa cell extracts. This protein increased the fidelity of pol-α-catalyzed gap-filling synthesis by 4-fold in one study (Carty et al. 1992) but had little effect on the frameshift fidelity of yeast pol-α:primase complex in a different study (Roberts et al. 1990). Results to date thus suggest that SSB does affect fidelity, but the effects are small relative to the high degree of discrimination imposed by the polymerase and exonuclease.

FIDELITY OF MULTIPROTEIN REPLICATION COMPLEXES

A full appreciation of how genomes are stably replicated, and how instability may arise to generate disease, requires a better understanding of the fidelity of the multiprotein replication machinery. An important step toward achieving this understanding has been the development of systems that replicate double-stranded DNA in vitro (numerous chapters in this volume). One system for studying human genomic replication depends on the SV40 origin of replication (Hassell and Brinton, this volume). Circular, double-stranded DNA substrates containing the SV40 origin can be fully replicated by the proteins present in primate cells, with only the addition of SV40 T antigen needed to initiate replication at the origin. For studies in vitro, replication can be performed in extracts of cells grown in culture or by reconstitution with purified proteins (see, e.g., Waga and Stillman 1994). The latter approach has defined two polymerases (pol-α and pol-δ) and several additional proteins required for complete replication.

High Replication Fidelity with Undamaged DNA

Using forward mutation assays with either the *lacZ* (Roberts and Kunkel 1988, 1993) or *supF* (Hauser et al. 1988) reporter gene, DNA replicated in human HeLa and simian CV-1 cell extracts was found to have a mutant frequency that was not increased significantly above the background mutant frequency of unreplicated DNA. Sequence analysis of *lacZ* mutants recovered from the unreplicated as well as replicated DNA showed no significant differences, yielding error rates varying from $\leq 6.2 \times 10^{-6}$ to $\leq 0.1 \times 10^{-6}$, depending on the substitution or frameshift error considered (Thomas et al. 1991b). Inasmuch as these HeLa cell extracts have mismatch repair activity (Thomas et al. 1991a), these rates represent the sum of both replication fidelity and any heteroduplex repair occurring in the extract. Two studies (see below) suggest that mismatch repair in the extract only affects error rate determinations by 2- to 3-fold, suggesting that replication fidelity itself is high. Because no replication errors are detected with undamaged DNA and equimolar dNTP concentrations, further understanding of how high replication fidelity is achieved and how it can be perturbed requires manipulation of reaction components in order to obtain replication errors. Several approaches have been used to address specific questions.

Proofreading and the Fidelity of Leading- and Lagging-strand Replication

One obvious question is whether proofreading is partly responsible for high replication fidelity. To answer this question, reactions were performed with unequal dNTP concentrations to force specific misinsertions. For example, misincorporation of dGTP to revert a TGA opal codon can be obtained by increasing the concentration of dGTP relative to dATP (Fig. 3). The contribution of proofreading to replication fidelity can then be examined either by increasing absolute dNTP concentrations to stimulate polymerization at the expense of proofreading, or by inhibiting proofreading by adding deoxynucleoside monophosphate to the replication reaction. Results from both approaches suggest that proofreading

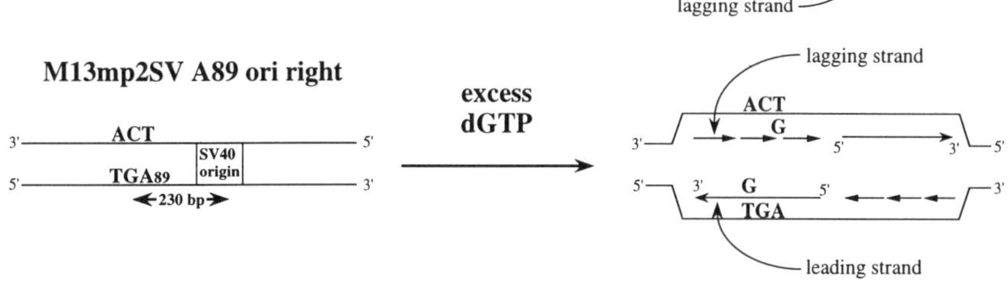

Figure 3 Representation of leading- and lagging-strand synthesis across an opal codon. See text for description.

contributes substantially to replication fidelity for base substitution (Roberts et al. 1991) and frameshift errors (Roberts et al. 1993).

The approach has been extended to examine the fidelity of the leading- and lagging-strand replication machinery. This requires comparing results with two vectors. The first (ori left, Fig. 3) contains the origin a few hundred nucleotides to the left of the reporter gene. This distance is small relative to the size of the vector (7398 bp). Previous studies (Edenberg and Huberman 1975; Li and Kelly 1985) have indicated that the rate of replication fork movement is similar in both directions from the origin. Thus, with this vector the (+) viral strand within the *lacZ* target is likely to be replicated by the lagging-strand replication apparatus. The second vector contains the origin to the right of the target, again only a few hundred nucleotides from the reporter gene (Fig. 3). In this vector, the (+) viral strand in the *lacZ* target is assumed to be replicated by leading-strand replication proteins. Fidelity measurements with these two vectors allow determination of the fidelity of replication of the same sequence by either the leading- or lagging-strand apparatus, provided that the strand on which the error was made can be assigned. This is done with dNTP substrate pool imbalances (Fig. 3), where the dNTP provided in excess is assumed to be responsible for the substitution error observed, or with template DNA damage (see below).

When this approach was combined with the reaction conditions that diminish proofreading, the data suggested that proofreading contributes to substitution fidelity during both leading- and lagging-strand replication (Roberts et al. 1991). The strategy has also been used to describe average leading- and lagging-strand replication rates for errors resulting from misinsertion of either TTP (Roberts et al. 1994) or dGTP (Izuta et al. 1995). Average error rates for replication of the *lacZ* template in a HeLa cell extract suggest that leading- and lagging-strand replication fidelity is similar for several errors but different for others (Table 2) (for additional data, see Izuta et al. 1995). There are several possible explanations for leading- and lagging-strand error rate differences. One is that mismatch repair in the extract is responsible for the asymmetry. A second is that replication of the two strands is highly asymmetric, providing unequal opportunities to make mistakes. Replication of the two strands may be performed by different DNA polymerases or perhaps the same polymerase but with a different complement of accessory proteins (for models, see Waga and Stillman 1994; Stillman; Hassell and Brinton; both this volume). This could yield differences in misinsertion rates and/or ability to extend rather than proofread mispaired or misaligned template primers. Replication on the leading strand is highly processive

Table 2 Comparative leading- and lagging-strand replication error rates

Error considered	Strand	Mispair frequency x 10^{-6} leading	lagging	Ratio lagging:leading
Average from extract reaction				
G•dTTP	–	18	39	2:1
C•dTTP	–	28	32	1:1
G•dTTP	+	2.5	82	33:1
C•dTTP	+	2.5	21	8:1
Average from reconstituted reaction				
G•dTTP	–	43	60	1:1
C•dTTP	–	≤7.1	43	≥6:1
G•dTTP	+	≤5.4	200	≥37:1
C•dTTP	+	11	43	4:1
Error rates in extract at specific sites				
T•dGTP	+ (121)	1.6	19	12:1
G•dTTP	+ (145)	≤4.8	71	≥15:1

as compared to discontinuous synthesis of Okazaki fragments on the lagging strand, which involves more than one DNA polymerase and/or one or more switches between enzymes as well as the synthesis and eventual replacement of RNA primers.

The possible influence of mismatch repair on replication fidelity in extracts has been examined by separating a HeLa cell extract into two fractions, neither of which has replication activity. When combined, these fractions reconstitute replication activity that is devoid of mismatch repair activity (Roberts et al. 1994). When the fidelity of this reaction was examined using excess dTTP (Roberts et al. 1994) or excess dGTP (Izuta et al. 1995), error rates were increased by only 2- to 3-fold over those observed in extracts having mismatch repair activity, confirming that replication is indeed highly accurate. Moreover, the error specificity was similar to that in extracts (Table 2) (Roberts et al. 1994; Izuta et al. 1995), including unequal leading- and lagging-strand rates for G•dTTP and C•dTTP errors on the plus strand (Table 2). Thus, at least these error rate asymmetries are not due to differential mismatch repair. The availability of human cell extracts having high replication activity but defective in mismatch repair due to mutant mismatch repair genes (Umar et al. 1994) will facilitate future studies of the fidelity of the human replication machinery in the absence of mismatch repair.

In another study of replication reconstituted from individual components, reactions were performed with pol-α as the only DNA polymerase (Carty et al. 1990). Fidelity was found to be intermediate between that of the unfractionated replication system and gap-filling synthesis by purified HeLa cell pol-α. This suggests that pol-α can replicate DNA with high fidelity when carrying out semiconservative DNA replication, but that additional cellular factors not present in the reconstituted system may be contributing to the higher replication fidelity of the unfractionated system. Even unfractionated extracts may be missing fidelity components that are active in vivo or simply not functioning in the extract.

Leading- and lagging-strand error rates differ at template positions 121 and 145 on the plus strand of the *lacZ* gene in M13mp2SV (Table 2). Two observations suggest that these differences may be due to differential proofreading. First, at both positions, the next correct dNTP to be incorporated following the misinsertion is the nucleotide present in excess (and, therefore, at high concentration) during replication. As explained above (Fig. 1), this situation would favor polymerization at the expense of excision of the terminal mispair. Thus, both hot spots for lagging-strand replication errors are sites where proofreading might be partially suppressed. Second, when replication reactions were repeated with added dGMP to inhibit proofreading, the error rates at both positions increased with both ori left and ori right substrates (Izuta et al. 1995). If one assumes that the addition of dGMP does not affect the inherent base selectivity of the insertion step, then the dGMP-dependent increase in error rate suggests that misinsertions are indeed occurring that, in the absence of monophosphate, are removed by the exonuclease. Inasmuch as fidelity in the absence of dGMP is higher during leading-strand replication, this suggests that leading-strand misinsertions are more effectively excised than are the lagging-strand errors that are readily detected even when dGMP is absent. Differential proofreading thus provides one mechanism to explain differences in leading- and lagging-strand replication fidelity. Since the assignment of the leading- and lagging-strand DNA polymerases during eukaryotic replication is not yet definitive, proofreading on the two strands could be carried out by any of several exonucleases.

Replication Fidelity with Damaged Substrates

A large number of genes in eukaryotic cells either control or catalyze the repair of a wide variety of physical and chemical insults, some resulting from normal cellular processes (e.g., deamination, depurination, oxidative stress, alkylation) and others from exposure to the external environment (for recent reviews, see Hanawalt 1994; Hartwell and Kastan 1994; Sancar 1994 and numerous references therein). When these repair systems fail, lesions may persist in DNA or in dNTP precursor pools. A plethora of lesions have been described over the years that have different structures and thus potentially affect replication fidelity by altering different discrimination steps. For example, an alkylated base may have altered hydrogen-bonding potential and lead to direct misinsertion errors, whereas an abasic site has lost base hydrogen-bonding potential altogether. Bulky lesions may perturb base-pair geometry or base-stacking interactions leading to misinsertion, or their structures may be inconsistent with continued replication, leading to termination or template-primer rearrangement and frameshift errors. The presence of lesions in the template strand could also affect communication between the polymerase and exonuclease active sites that is critical for proofreading. Lesion-induced replication infidelity has been examined in a number of studies with purified DNA polymerases (for review, see Echols and Goodman 1991). Discussing all this information is beyond the scope of this chapter; we briefly review only a few recent studies of the fidelity of SV40 origin-dependent replication using damaged substrates.

Replication Infidelity with 8-O-dGTP

Oxidative metabolism is known to generate mutagenic compounds within cells, one of which is 8-oxo-deoxyguanosine. The presence of several lines of defense against mutations resulting from this base analog (for review, see Michaels and Miller 1992; Grollman and Moriya 1993) suggests that it is biologically important. A variety of DNA polymerases, as well as the replication complex in HeLa cell extracts, misincorporate the triphosphate form of this base analog, 8-O-dGTP, opposite template adenines, yielding A→C transversions (Cheng et al. 1992; Pavlov et al. 1994; Minnick et al. 1995). The data suggest that 8-O-dGTP could be highly mutagenic during genomic replication in eukaryotes. This may be the case for other modified dNTPs as well (see, e.g., Feig et al. 1994). The amount of 8-O-dGTP in human cells may be modulated by hydrolysis by the human homolog (Sakumi et al. 1993) of the *E. coli mutT* protein. If so, inactivating mutations in this gene (or other enzymes that sanitize dNTP pools) might result in a mutator phenotype in human cells.

Mutagenic Translesion Replication of DNA Containing Cyclobutane Pyrimidine Dimers

Among the insults that generate lesions in DNA, ultraviolet radiation has received perhaps the greatest attention, partly due to its established role as a skin carcinogen (Brash et al. 1991). The mutagenic potential of UV photoproducts during replication by eukaryotic proteins has been examined in several studies. Pol-α is unable to synthesize past UV lesions (Moore et al. 1981), whereas DNA pol-δ in the presence of PCNA is able to replicate past *cis*-syn and *trans*-syn TT dimers (O'Day et al. 1992). Three studies have demonstrated translesion replication of cyclobutane pyrimidine dimers in cell extracts using SV40 origin-containing vectors (Carty et al. 1993; Thomas and Kunkel 1993; Thomas et al. 1993). This replication is mutagenic, with an error specificity similar to that observed in vivo (Keyse et al. 1988; Armstrong and Kunz 1990).

Mutagenic Translesion Replication of DNA Containing AAF Adducts

The ability of the replication complex in a HeLa cell extract to bypass site-specific N-2-acetylaminofluorene (AAF) adducts has also been examined (Thomas et al. 1994). The major effect was inhibition of replication, with termination occurring immediately before incorporation opposite the adduct. Among the replicated products was a higher proportion of those representing replication of the undamaged strand, leaving open the possibility that the first fork to encounter the lesion became uncoupled, i.e., replication of the damaged strand ceased while replication of the undamaged strand continued. Product analysis suggested that translesion bypass had occurred and that frameshift errors had been generated by the mechanism involving correct incorporation opposite the lesion followed by slippage (Fig. 2C).

FIDELITY OF DNA REPLICATION IN VIVO

How accurate is replication in vivo? This question has been elegantly addressed in *E. coli* by measuring mutation rates for a variety of sequence changes, using strains selectively disabled in key fidelity processes. The wild-type spontaneous mutation rate in the *lacI* reporter gene is 10^{-10} mutations per base pair replicated per generation (Table 3) (see, e.g., Schaaper 1993). A *mutL* mutant lacking methyl-directed postreplication mismatch repair has a 20- to 400-fold higher spontaneous mutation rate, depending on the type of mutation considered. With some simplifying assumptions (Schaaper 1993), the resulting mutation rate of about 10^{-7} can be considered as the fidelity of chromosomal replication. Analysis of a double mutant lacking mismatch repair and defective in proofreading by the ε subunit of the replicative DNA polymerase III holoenzyme (which replicates both the leading and lagging strands) suggests that

Table 3 Estimated contributions of the three major discrimination steps to replication fidelity

	Contribution		
	E. coli	eukaryotic	
Discrimination step	in vivo	in vitro	in vivo
Nucleotide selectivity	2×10^{-5}–2×10^{-6}	10^{-3}–10^{-6}	?
Exonucleolytic proofreading	40–200	0–200	10–≥200
Mismatch repair	20–400	–	10–≥700
Mutation rate	10^{-10}		$\leq 10^{-10}$

The values listed depend on simplifying assumptions and often involve caveats that are discussed in the text and in the references cited therein.

proofreading contributes between 40- and 200-fold to this rate, with the balance (factors of 200,000 to 2,000,000) representing the base selectivity of the replication machinery (Table 3).

Estimating replication fidelity in eukaryotes is complicated by the possibility of several types of mismatch repair, the influence of spontaneous damage, and the likelihood of multiple damage-repair pathways. Moreover, estimates are based on a few reporter genes, providing a limited view of replication fidelity for large eukaryotic genomes. Despite these qualifications, existing mutation rate data in mutant cells reveal a similar picture to that seen in *E. coli.* Mutation rates in eukaryotic cells are generally $\leq 10^{-10}$ mutations per base pair replicated per generation (see Loeb 1991 and references therein). The mutation rate in mismatch-repair-defective yeast (see, e.g., Strand et al. 1993; Prolla et al. 1994 and references therein) and human tumor cells (Kat et al. 1993; Bhattacharyya et al. 1994; Farber et al. 1994; Eshleman et al. 1995) are elevated up to several hundred-fold. Moreover, yeast mutants of pol-δ, pol-ε, and pol-γ containing substitutions for conserved exonuclease residues also have spontaneous mutation rates that are increased by up to several hundred-fold (Morrison et al. 1991, 1993; Foury and Vanderstraeten 1992), emphasizing the importance of proofreading in vivo during both nuclear and mitochondrial DNA replication.

CONCLUDING REMARKS

In many ways, our current view of eukaryotic replication fidelity is quite limited. Eukaryotic genomes are huge compared to the few hundred nucleotides scanned by current reporter genes. For example, the human genome contains a wide variety of repetitive sequence elements (see, e.g., Beckman and Weber 1992) whose instability is associated with cancer (for review, see Loeb 1994) and neurodegenerative diseases (see, e.g., Willems 1994). Despite this association, we know very little about replication error rates in repetitive sequences except for short homopolymeric runs. The fact that eukaryotic cells devote an enormous amount of energy to checkpoints in the cell cycle and a multiplicity of DNA repair processes clearly indicates the biological risk associated with unrepaired lesions. Appropriately, a great deal of attention has been paid to studies of these processes, as exemplified by the fact that DNA repair was the "molecule of the year" in 1994 (Culotta and Koshland 1994). To fully appreciate the effects of unrepaired lesions, more studies are needed to define the consequences of an encounter between an unrepaired lesion and a eukaryotic replication fork, and, possibly, the fidelity of DNA synthesis associated with the repair processes themselves. Our understanding should increase as studies are performed using a variety of systems, eventually including those that mimic replication of highly organized nuclear and mitochondrial DNA.

ACKNOWLEDGMENTS

We thank William C. Copeland and Rémi Palmantier for helpful comments on the manuscript.

REFERENCES

Abbotts, J., M. Jaju, and S.H. Wilson. 1991. Thermodynamics of A:G mismatch poly(dG) synthesis by human immunodeficiency virus 1 reverse transcriptase. *J. Biol. Chem.* **266:** 3937–3943.

Armstrong, J.D. and B.A. Kunz. 1990. Site and strand specificity of UVB mutagenesis in the *SUP-o* gene of yeast. *Proc. Natl. Acad. Sci.* **87:** 9005–9009.

Bambara, R.A. and C.B. Jessee. 1991. Properties of DNA polymerases δ and ε, and their roles in eukaryotic DNA replication. *Biochim. Biophys. Acta* **1088:** 11–24.

Bebenek, K. and T.A. Kunkel. 1995. Analyzing the fidelity of DNA polymerases. *Methods Enzymol.* (in press).

Bebenek, K., J.D. Roberts, and T.A. Kunkel. 1992. The effects of dNTP pool imbalances on frameshift fidelity during DNA replication. *J. Biol. Chem.* **267:** 3589–3596.

Bebenek, K., J. Abbotts, S.H. Wilson, and T.A. Kunkel. 1993. Error-prone polymerization by HIV-1 reverse transcriptase. Contribution of template-primer misalignment, miscoding, and termination probability to mutational hot spots. *J. Biol. Chem.* **268:** 10324–10334.

Bebenek, K., C.M. Joyce, M.P. Fitzgerald, and T.A. Kunkel. 1990. The fidelity of DNA synthesis catalyzed by derivatives of *Escherichia coli* DNA polymerase I. *J. Biol. Chem.* **265:** 13878–13887.

Bebenek, K., J. Abbotts, J.D. Roberts, S.H. Wilson, and T.A. Kunkel. 1989. Specificity and mechanism of error-prone replication by human immunodeficiency virus 1 reverse transcriptase. *J. Biol. Chem.* **264:** 16948–16956.

Beckman, J.S. and J.L. Weber. 1992. Survey of human and rat microsatellites. *Genomics* **12:** 627–631.

Bhattacharyya, N.P., A. Skandalis, A. Ganesh, J. Groden, and M. Meuth. 1994. Mutator phenotypes in human colorectal carcinoma cell lines. *Proc. Natl. Acad. Sci.* **91:** 6319–6323.

Bialek, G., H.-P. Nasheuer, H. Goetz, and F. Grosse. 1989. Exonucleolytic proofreading increases the accuracy of DNA synthesis by human lymphocyte DNA polymerase α-DNA primase. *EMBO J.* **8:** 1833–1839.

Blanco, L., A. Bernad, and M. Salas. 1991. MIP1 DNA polymerase of *S. cerevisiae*: Structural similarity with the *E. coli* DNA polymerase I-type enzymes. *Nucleic Acids Res.* **19:** 955.

Bloom, L.B., M.R. Otto, R. Eritja, L.J. Reha-Krantz, M.F. Goodman, and J.M. Beechem. 1994. Pre-steady-state kinetic analysis of sequence-dependent nucleotide excision by the 3′-exonuclease activity of bacteriophage T4 DNA polymerase. *Biochemistry* **33:** 7576–7586.

Boosalis, M.S., D.W. Mosbaugh, R. Hamatake, A. Sugino, T.A. Kunkel, and M.F. Goodman. 1989. Kinetic analysis of base substitution mutagenesis by transient misalignment of DNA and by miscoding. *J. Biol. Chem.* **264:** 11360–11366.

Braithwaite, D.K. and J. Ito. 1993. Compilation, alignment, and phylogenetic relationships of DNA polymerases. *Nucleic Acids Res.* **21:** 787–802.

Brash, D.E., J.A. Rudolph, J.A. Simon, A. Lin, G.J. McKenna, H.P. Baden, A.J. Halperin, and J. Pontén. 1991. A role for sunlight in skin cancer: UV-induced p53 mutations in squamous cell carcinoma. *Proc. Natl. Acad. Sci.* **88:** 10124–10128.

Brooke, R.G., R. Singhal, D.C. Hinkle, and L.B. Dumas. 1991. Purification and characterization of the 180 kDa and 86 kDa subunits of the *Saccharomyces cerevisiae* DNA primase-DNA polymerase protein complex. The 180 kDa subunit has both DNA polymerase and 3′→5′-exonuclease activities. *J. Biol. Chem.* **266:** 3005–3015.

Brutlag, D. and A. Kornberg. 1972. Enzymatic synthesis of deoxyribonucleic acid. XXXVI. A proofreading function for the 3′→5′ exonuclease activity in deoxyribonucleic acid polymerases. *J. Biol. Chem.* **247:** 241–248.

Capson, T.L., J.A. Peliska, B. Fenn Kaboord, M. West Frey, C. Lively, M. Dahlberg, and S.J. Benkovic. 1992. Kinetic characterization of the polymerase and exonuclease activities of the gene 43 protein of bacteriophage T4. *Biochemistry* **31:** 10984–10994.

Carroll, S.S. and S.J. Benkovic. 1990. Mechanistic aspects of DNA polymerases: *Escherichia coli* DNA polymerase I (Klenow fragment) as a paradigm. *Chem. Rev.* **90:** 1291–1307.

Carty, M.P., A.S. Levine, and K. Dixon. 1992. HeLa cell single-stranded DNA-binding protein increases the accuracy of DNA synthesis by DNA polymerase α. *Mutat. Res.* **274:** 29–43.

Carty, M.P., J. Hauser, A.S. Levine, and K. Dixon. 1993. Replication and mutagenesis of UV-damaged DNA templates in human and monkey cell extracts. *Mol. Cell. Biol.* **13:** 533–542.

Carty, M.P., Y. Ishimi, A.S. Levine, and K. Dixon. 1990. DNA polymerase α from HeLa cells synthesizes DNA with high fidelity in a reconstituted replication system. *Mutat. Res.* **232:** 141–153.

Carver, T.E., Jr., R.A. Hochstrasser, and D.P. Millar. 1994. Proofreading DNA: Recognition of aberrant DNA termini by the Klenow fragment of DNA polymerase I. *Proc. Natl. Acad. Sci.* **91:** 10670–10674.

Chen, Y.-C., E.W. Bohn, S.R. Planck, and S.H. Wilson. 1979. Mouse DNA polymerase α. Subunit structure and identification of a species with associated exonuclease. *J. Biol. Chem.* **254:** 11678–11687.

Cheng, K.C., D.S. Cahill, H. Kasai, S. Nishimura, and L.A. Loeb. 1992. 8-Hydroxyguanine, an abundant form of oxidative DNA damage, causes G→T and A→C substitutions. *J. Biol. Chem.* **267:** 166–172.

Chung, D.W., J. Zhang, C.K. Tan, E.W. Davie, A.G. So, and K.M. Downey. 1991. Primary structure of the catalytic subunit of human DNA polymerase δ and chromosomal location of the gene. *Proc. Natl. Acad. Sci.* **88:** 11197–11201.

Copeland, W.C. and T.S.-F. Wang. 1991. Catalytic subunit of human DNA polymerase α overproduced from Baculovirus-infected insect cells. *J. Biol. Chem.* **266:** 22739–22748.

Copeland, W.C., N.K. Lam, and T.S.-F. Wang. 1993. Fidelity studies of the human DNA polymerase α. The most conserved region among α-like DNA polymerases is responsible for metal-induced infidelity in DNA synthesis. *J. Biol. Chem.* **268:** 11041–11049.

Cotterill, S.M., M.E. Reyland, L.A. Loeb, and I.R. Lehman. 1987. A cryptic proofreading 3′→5′ exonuclease associated with the polymerase subunit of the DNA polymerase-primase from *Drosophila melanogaster*. *Proc. Natl. Acad. Sci.* **84:** 5635–5639.

Cowart, M., K.J. Gibson, D.J. Allen, and S.J. Benkovic. 1989. DNA substrate structural requirements for the exonuclease and polymerase activities of procaryotic and phage DNA polymerases. *Biochemistry* **28:** 1975–1983.

Culotta, E. and D.E. Koshland Jr. 1994. DNA repair works its way to the top. *Science* **266:** 1926–1929.

Delarue, M., O. Poch, N. Tordo, D. Moras, and P. Argos. 1990. An attempt to unify the structure of polymerases. *Protein Eng.* **3:** 461–467.

Dong, Q., W.C. Copeland, and T.S.-F. Wang. 1993a. Mutational studies of human DNA polymerase α. Identification of residues critical for deoxynucleotide binding and misinsertion fidelity of DNA synthesis. *J. Biol. Chem.* **268:** 24163–24174.

———. 1993b. Mutational studies of human DNA polymerase α. Serine 867 in the second most conserved region among α-like DNA polymerases is involved in primer binding and mispair primer extension. *J. Biol. Chem.* **268:** 24175–24182.

Donlin, M.J., S.S. Patel, and K.A. Johnson. 1991. Kinetic partitioning between the exonuclease and polymerase sites in DNA error correction. *Biochemistry* **30:** 538–546.

Echols, H. and M.F. Goodman. 1991. Fidelity mechanisms in DNA replication. *Annu. Rev. Biochem.* **60:** 477–511.

Edenberg, H.J. and J.A. Huberman. 1975. Eukaryotic chromosome replication. *Annu. Rev. Genet.* **9:** 245–284.

Eger, B.T. and S.J. Benkovic. 1992. Minimal kinetic mechanism for misincorporation by DNA polymerase I (Klenow fragment). *Biochemistry* **31:** 9227–9236.

Eshleman, J.R., E.Z. Lang, G.K. Bowerfind, R. Parsons, B. Vogelstein, J.K.V. Wilson, M.L. Veigl, W.D. Sedwick, and S.D. Markowitz. 1995. Increased mutation rate at the *hprt* locus accompanies microsatellite instability in colon cancer. *Oncogene* **10:** 33–37.

Farber, R.A., T.D. Petes, M. Dominska, S.S. Hudgens, and R.M. Liskay. 1994. Instability of simple sequence repeats in a mammalian cell line. *Hum. Mol. Genet.* **3:** 253–256.

Feig, D.I., L.C. Sowers, and L.A. Loeb. 1994. Reverse chemical mutagenesis: Identification of the mutagenic lesions resulting from reactive oxygen species-mediated damage to DNA. *Proc. Natl. Acad. Sci.* **91:** 6609–6613.

Fersht, A.R. 1985. *Enzyme structure and mechanism*. W.J. Freeman, New York.

Fersht, A.R., J.W. Knill-Jones, and W.-C. Tsui. 1982. Kinetic basis of spontaneous mutation. Misinsertion frequencies, proofreading specificities and cost of proofreading by DNA polymerases of *Escherichia coli*. *J. Mol. Biol.* **156:** 37–51.

Foury, F. and S. Vanderstraeten. 1992. Yeast mitochondrial DNA mutators with deficient proofreading exonucleolytic activity. *EMBO J.* **11:** 2717–2726.

Gray, H. and T.W. Wong. 1992. Purification and identification of subunit structure of the human mitochondrial DNA polymerase. *J. Biol. Chem.* **267:** 5835–5841.

Grollman, A.P. and M. Moriya. 1993. Mutagenesis by 8-oxoguanine: An enemy within. *Trends Genet.* **9:** 246–249.

Grosse, F., G. Krauss, J.W. Knill-Jones, and A.R. Fersht. 1983. Accuracy of DNA polymerase-α in copying natural DNA. *EMBO J.* **2:** 1515–1519.

Hanawalt, P.C. 1994. Transcription-coupled repair and human disease. *Science* **266:** 1957–1958.

Hartwell, L.H. and M.B. Kastan. 1994. Cell cycle control and cancer. *Science* **266:** 1821–1828.

Hauser, J., A.S. Levine, and K. Dixon. 1988. Fidelity of DNA synthesis in a mammalian *in vitro* replication system. *Mol. Cell. Biol.* **8:** 3267–3271.

Insdorf, N.F. and D.F. Bogenhagen. 1989. DNA polymerase ψ from *Xenopus laevis*. 2. A 3′→5′ exonuclease is tightly associated with the DNA polymerase activity. *J. Biol. Chem.* **264:** 21498–21503.

Ito, J. and D.K. Braithwaite. 1990. Yeast mitochondrial DNA polymerase is related to the family A DNA polymerases. *Nucleic Acids Res.* **18:** 6716.

Izuta, S., J.D. Roberts, and T.A. Kunkel. 1995. Replication error rates for G•dGTP, T•dGTP and A•dGTP mispairs: Evidence for differential proofreading by leading and lagging strand DNA replication complexes. *J. Biol. Chem.* **270:** 2595–2600.

Johnson, K.A. 1993. Conformational coupling in DNA polymerase fidelity. *Annu. Rev. Biochem.* **62:** 685–713.

Joyce, C.M. 1989. How DNA travels between the separate polymerase and 3′-5′-exonuclease sites of DNA polymerase I (Klenow fragment). *J. Biol. Chem.* **264:** 10858–10866.

Joyce, C.M. and T.A. Steitz. 1994. Function and structure relationships in DNA polymerases. *Annu. Rev. Biochem.* **63:** 777–822.

Kaguni, L.S. and M.W. Olsen. 1989. Mismatch-specific 3′→5′ exonuclease associated with the mitochondrial DNA polymerase from *Drosophila* embryos. *Proc. Natl. Acad. Sci.* **86:** 6469–6473.

Kaguni, L.S., R.A. DiFrancesco, and I.R. Lehman. 1984. The DNA polymerase-primase from *Drosophila melanogaster*. Rate and fidelity of polymerization on single-stranded DNA templates. *J. Biol. Chem.* **259:** 9314–9319.

Kat, A., W.G. Thilly, W.-H. Fang, M.J. Longley, G.-M. Li, and P. Modrich. 1993. An alkylation-tolerant, mutator human cell line is deficient in strand-specific mismatch repair. *Proc. Natl. Acad. Sci.* **90:** 6424–6428.

Keyse, S.M., F. Amaudruz, and R.M. Tyrrell. 1988. Determination of the spectrum of mutations induced by defined-wavelength solar UVB (313-nm) radiation in mammalian cells by use of a shuttle vector. *Mol. Cell. Biol.* **8:** 5425–5431.

Kuchta, R., P. Benkovic, and S.J. Benkovic. 1988. Kinetic mechanism whereby DNA polymerase I (Klenow) replicates DNA with high fidelity. *Biochemistry* **27:** 2716–2725.

Kunkel, T.A. 1985a. The mutational specificity of DNA polymerase-β during *in vitro* DNA synthesis. Production of frameshift, base substitution, and deletion mutants. *J. Biol. Chem.* **260:** 5787–5796.

———. 1985b. The mutational specificity of DNA polymerases-α and -γ during *in vitro* DNA synthesis. *J. Biol. Chem.* **260:** 12866–12874.

———. 1986. Frameshift mutagenesis by eukaryotic DNA polymerases *in vitro*. *J. Biol. Chem.* **261:** 13581–13587.

———. 1990. Misalignment-mediated DNA synthesis errors. *Biochemistry* **29:** 8003–8011.

———. 1992. DNA replication fidelity. *J. Biol. Chem* **267:** 18251–18254.

Kunkel, T.A. and P.S. Alexander. 1986. The base substitution fidelity of eucaryotic DNA polymerases. Mispairing frequencies, site preferences, insertion preferences, and base substitution by dislocation. *J. Biol. Chem.* **261:** 160–166.

Kunkel, T.A. and L.A. Loeb. 1981. Fidelity of mammalian DNA polymerases. *Science* **213:** 765–767.

Kunkel, T.A. and D.W. Mosbaugh. 1989. Exonucleolytic proofreading by a mammalian DNA polymerase γ. *Biochemistry* **28:** 988–995.

Kunkel, T.A. and A. Soni. 1988a. Exonucleolytic proofreading enhances the fidelity of DNA synthesis by chick embryo DNA polymerase-γ. *J. Biol. Chem.* **263:** 4450–4459.

———. 1988b. Mutagenesis by transient misalignment. *J. Biol. Chem.* **263:** 14784–14789.

Kunkel, T.A., R.R. Meyer, and L.A. Loeb. 1979. Single-strand binding protein enhances fidelity of DNA synthesis *in vitro*. *Proc. Natl. Acad. Sci.* **76:** 6331–6335.

Kunkel, T.A., S.S. Patel, and K.A. Johnson. 1994. Error-prone replication of repeated DNA sequences by T7 DNA polymerase in the absence of its processivity subunit. *Proc. Natl. Acad. Sci.* **91:** 6830–6834.

Kunkel, T.A., R.D. Sabatino, and R.A. Bambara. 1987. Exonucleolytic proofreading by calf thymus DNA polymerase δ. *Proc. Natl. Acad. Sci.* **84:** 4865–4869.

Kunkel, T.A., R.M. Schaaper, and L.A. Loeb. 1983. Depurination-induced infidelity of deoxyribonucleic acid synthesis with purified deoxyribonucleic acid replication proteins *in vitro*. *Biochemistry* **22:** 2378–2384.

Kunkel, T.A., R.K. Hamatake, J. Motto-Fox, M.P. Fitzgerald, and A. Sugino. 1989. Fidelity of DNA polymerase I and the DNA polymerase I-DNA primase complex from *Saccharomyces cerevisiae*. *Mol. Cell. Biol.* **9:** 4447–4458.

Li, J.J. and T.J. Kelly. 1985. Simian virus 40 DNA replication *in vitro*: Specificity of initiation and evidence for bidirectional replication. *Mol. Cell. Biol.* **5:** 1238–1246.

Lindsley, J.E. and R.P.P. Fuchs. 1994. Use of single-turnover kinetics to study bulky adduct bypass by T7 DNA polymerase. *Biochemistry* **33:** 764–772.

Loeb, L.A. 1991. Mutator phenotype may be required for multistage carcinogenesis. *Cancer Res.* **51:** 1–5.

———. 1994. Microsatellite instability: Marker of a mutator phenotype in cancer. *Cancer Res.* **54:** 5059–5063.

Mendelman, L.V., J. Petruska, and M.F. Goodman. 1990. Base mispair extension kinetics: Comparison of DNA polymerase α and reverse transcriptase. *J. Biol. Chem.* **265:** 2338–2346.

Mendelman, L.V., M.S. Boosalis, J. Petruska, and M.F. Goodman. 1989. Nearest neighbor influences on DNA polymerase insertion fidelity. *J. Biol. Chem.* **264:** 14415–14423.

Michaels, M.L. and J.H. Miller. 1992. The GO system protects organisms from the mutagenic effect of the spontaneous lesion 8-hydroxyguanine (7,8-dihydro-8-oxoguanine). *J. Bacteriol.* **174:** 6321–6325.

Minnick, D.T., Y.I. Pavlov, and T.A. Kunkel. 1995. The fidelity of the human leading and lagging strand DNA replication apparatus with 8-oxodeoxyguanosine triphosphate. *Nucleic Acids Res.* **22:** 5658–5664.

Modrich, P. 1994. Mismatch repair, genetic stability, and cancer. *Science* **266:** 1959–1960.

Moore, P.D., K.K. Bose, S.D. Rabkin, and B.S. Strauss. 1981. Sites of termination of *in vitro* DNA synthesis on ultraviolet- and N-acetylaminofluorene-treated ϕX174 templates by prokaryotic and eukaryotic DNA polymerases. *Proc. Natl. Acad. Sci.* **78:** 110–114.

Morrison, A., J.B. Bell, T.A. Kunkel, and A. Sugino. 1991. Eukaryotic DNA polymerase amino acid sequence required for 3′→5′ exonuclease activity. *Proc. Natl. Acad. Sci.* **88:** 9473–9477.

Morrison, A., A.L. Johnson, L.H. Johnston, and A. Sugino. 1993. Pathway correcting DNA replication errors in *Saccharomyces cerevisiae*. *EMBO J.* **12:** 1467–1473.

Mosbaugh, D.W. and R.R. Meyer. 1980. Interaction of mammalian deoxyribonuclease V, a double-strand 3′ to 5′ and 5′ to 3′ exonuclease with deoxyribonucleic acid polymerase-β from Novikoff hepatoma. *J. Biol. Chem.* **255:** 10239–10247.

Napolitano, R.L., I.B. Lambert, and R.P.P. Fuchs. 1994. DNA sequence determinants of carcinogen-induced frameshift mutagenesis. *Biochemistry* **33:** 1311–1315.

O'Day, C.L., P.M.J. Burgers, and J.-S. Taylor. 1992. PCNA-induced synthesis past *cis*-syn and *trans*-syn-1 thymine dimers by calf thymus DNA polymerase δ *in vitro*. *Nucleic Acids Res.* **20:** 5403–5406.

Pavlov, Y.I., D.T. Minnick, S. Izuta, and T.A. Kunkel. 1994. DNA replication fidelity with 8-oxodeoxyguanosine triphosphate. *Biochemistry* **33:** 4695–4701.

Pelletier, H., M.R. Sawaya, A. Kumar, S.H. Wilson, and J. Kraut. 1994. Structures of ternary complexes of rat DNA polymerase β, a DNA template-primer, and ddCTP. *Science* **264:** 1891–1903.

Perrino, F.W. and L.A. Loeb. 1989a. Differential extension of 3′ mispairs is a major contribution to the high fidelity of calf thymus DNA polymerase-α. *J. Biol. Chem.* **264:** 2898–2905.

———. 1989b. Proofreading by the ε subunit of *Escherichia coli* DNA polymerase III increases the fidelity of calf thymus DNA polymerase α. *Proc. Natl. Acad. Sci.* **86:** 3085–3088.

———. 1990. Hydrolysis of 3′-terminal mispairs *in vitro* by the 3′→5′ exonuclease of DNA polymerase δ permits subsequent extension by DNA polymerase α. *Biochemistry* **29:** 5226–5231.

Petruska, J., L.C. Sowers, and M.F. Goodman. 1986. Comparison of nucleotide interactions in water, proteins, and vacuum: Model for DNA polymerase fidelity. *Proc. Natl. Acad. Sci.* **83:** 1559–1562.

Polesky, A.H., M.D. Dahlberg, S.J. Benkovic, N.D.F. Grindley, and C.M. Joyce. 1992. Side chains involved in catalysis of the polymerase reaction of DNA polymerase I from *Escherichia coli*. *J. Biol. Chem.* **267:** 8417–8428.

Prolla, T.A., D.-M. Christie, and R.M. Liskay. 1994. Dual requirement in yeast DNA

mismatch repair for *MLH1* and *PMS1*, two homologs of the bacterial *mutL* gene. *Mol. Cell. Biol.* **14:** 407–415.
Reddy, M.K., S.E. Weitzel, and P.H. von Hippel. 1992. Processive proofreading is intrinsic to T4 DNA polymerase. *J. Biol. Chem.* **267:** 14157–14166.
Reyland, M.E. and L.A. Loeb. 1987. On the fidelity of DNA replication. Isolation of high fidelity DNA polymerase-primase complexes by immunoaffinity chromatography. *J. Biol. Chem.* **262:** 10824–10830.
Reyland, M.E., I.R. Lehman, and L.A. Loeb. 1988. Specificity of proofreading by the 3′→5′ exonuclease of the DNA polymerase-primase of *Drosophila melanogaster*. *J. Biol. Chem.* **263:** 6518–6524.
Ripley, L.S. 1990. Frameshift mutation: Determinants of specificity. *Annu. Rev. Genet.* **24:** 189–213.
Roberts, J.D. and T.A. Kunkel. 1988. Fidelity of a human cell DNA replication complex. *Proc. Natl. Acad. Sci.* **85:** 7064–7068.
———. 1993. Fidelity of DNA replication in human cells. In *Gene and chromosome analysis*, Part B, (ed. K.W. Adolph), pp. 295–313. Academic Press, San Diego.
Roberts, J.D., D. Nguyen, and T.A. Kunkel. 1993. Frameshift fidelity during replication of double-stranded DNA in HeLa cell extracts. *Biochemistry* **32:** 4083–4089.
Roberts, J.D., D.C. Thomas, and T.A. Kunkel. 1991. Exonucleolytic proofreading of leading and lagging strand DNA replication errors. *Proc. Natl. Acad. Sci.* **88:** 3465–3469.
Roberts, J.D., S. Izuta, D.C. Thomas, and T.A. Kunkel. 1994. Mispair-, site-, and strand-specific error rates during simian virus 40 origin-dependent replication *in vitro* with excess deoxythymidine triphosphate. *J. Biol. Chem.* **269:** 1711–1717.
Roberts, J.D., R.K. Hamatake, M.S. Fitzgerald, A. Sugino, and T.A. Kunkel. 1990. Effect of accessory proteins on the fidelity of DNA synthesis by eukaryotic replicative polymerases. In *Mutation and the environment,* part A: *Basic mechanisms* (ed. M.L. Mendelsohn and R.J. Albertini), pp. 91–100. Wiley-Liss, New York.
Sakumi, K., M. Furuichi, T. Tsuzuki, T. Kakuma, S. Kawabata, H. Maki, and M. Sekiguchi. 1993. Cloning and expression of cDNA for a human enzyme that hydrolyzes 8-oxo-dGTP, a mutagenic substrate for DNA synthesis. *J. Biol. Chem.* **268:** 23524–23530.
Sancar, A. 1994. Mechanisms of DNA excision repair. *Science* **266:** 1954–1956.
Schaaper, R.M. 1993. Base selection, proofreading, and mismatch repair during DNA replication in *Escherichia coli*. *J. Biol. Chem.* **268:** 23762–23765.
Sheaff, R.J., R.D. Kuchta, and D. Ilsley. 1994. Calf thymus DNA polymerase α-primase: "Communication" and primer•template movement between the two active sites. *Biochemistry* **33:** 2247–2254.
Shibutani, S. and A.P. Grollman. 1993. On the mechanism of frameshift (deletion) mutagenesis *in vitro*. *J. Biol. Chem.* **268:** 11703–11710.
Simon, M., L. Giot, and G. Faye. 1991. The 3′ to 5′ exonuclease activity located in the DNA polymerase δ subunit of *Saccharomyces cerevisiae* is required for accurate replication. *EMBO J.* **10:** 2165–2170.
Singhal, R.K. and S.H. Wilson. 1993. Short gap-filling synthesis by DNA polymerase β is processive. *J. Biol. Chem.* **268:** 15906–15911.
Strand, M., T.A. Prolla, R.M. Liskay, and T.D. Petes. 1993. Destabilization of tracts of simple repetitive DNA in yeast by mutations affecting DNA mismatch repair. *Nature* **365:** 274–276.

Streisinger, G., Y. Okada, J. Emrich, J. Newton, A. Tsugita, E. Terzaghi, and M. Inouye. 1967. Frameshift mutations and the genetic code. *Cold Spring Harbor Symp. Quant. Biol.* **31:** 77–84.

Thomas, D.C. and T.A. Kunkel. 1993. Replication of UV-irradiated DNA in human cell extracts: Evidence for mutagenic bypass of pyrimidine dimers. *Proc. Natl. Acad. Sci.* **90:** 7744–7748.

Thomas, D.C., J.D. Roberts, and T.A. Kunkel. 1991a. Heteroduplex repair in extracts of human HeLa cells. *J. Biol. Chem.* **266:** 3744–3751.

Thomas, D.C., D.C. Nguyen, W.W. Piegorsch, and T.A. Kunkel. 1993. Relative probability of mutagenic translesion synthesis on the leading and lagging strands during replication of UV-irradiated DNA in a human cell extract. *Biochemistry* **32:** 11476–11482.

Thomas, D.C., J.D. Roberts, M.P. Fitzgerald, and T.A. Kunkel. 1990. Fidelity of animal cell DNA polymerases α and δ and of a human DNA replication complex. In *Antimutagenesis and anticarcinogenesis mechanisms*, vol. 2 (ed. K. Kuroda et al.), pp. 289–297. Plenum Press, New York.

Thomas, D.C., X. Veaute, T.A. Kunkel, and R.P.P. Fuchs. 1994. Mutagenic replication in human cell extracts of DNA containing site specific N-2-acetylaminofluorene adducts. *Proc. Natl. Acad. Sci.* **91:** 7752–7756.

Thomas, D.C., J.D. Roberts, R.D. Sabatino, T.W. Myers, C.K. Tan, K.M. Downey, A.G. So, R.A. Bambara, and T.A. Kunkel. 1991b. Fidelity of mammalian DNA replication and replicative DNA polymerases. *Biochemistry* **30:** 11751–11759.

Umar, A., J.C. Boyer, D.C. Thomas, D.C. Nguyen, J.I. Risinger, J. Boyd, Y. Ionov, M. Perucho, and T.A. Kunkel. 1994. Defective mismatch repair in extracts of colorectal and endometrial cancer cell lines exhibiting microsatellite instability. *J. Biol. Chem* **269:** 14367–14370.

Waga, S. and B. Stillman. 1994. Anatomy of a DNA replication fork revealed by reconstitution of SV40 DNA replication in vitro. *Nature* **369:** 207–212.

Wang, C.-I. and J.-S. Taylor. 1992. *In vitro* evidence that UV-induced frameshift and substitution mutations at T tracts are the result of misalignment-mediated replication past a specific thymine dimer. *Biochemistry* **31:** 3671–3681.

Wang, T.S.-F. 1991. Eukaryotic DNA polymerases. *Annu. Rev. Biochem.* **60:** 513–552.

Wernette, C.M., M.C. Conway, and L.S. Kaguni. 1988. Mitochondrial DNA polymerase from *Drosophila melanogaster* embryos: Kinetics, processivity, and fidelity of DNA polymerization. *Biochemistry* **27:** 6046–6054.

West Frey, M., N.G. Nossal, T.L. Capson, and S.J. Benkovic. 1993. Construction and characterization of a bacteriophage T4 DNA polymerase deficient in 3′→5′ exonuclease activity. *Proc. Natl. Acad. Sci.* **90:** 2579–2583.

Willems, P.J. 1994. Dynamic mutations hit double figures. *Nat. Genet.* **8:** 213–215.

Wong, I., S.S. Patel, and K.A. Johnson. 1991. An induced-fit kinetic mechanism for DNA replication fidelity: Direct measurement by single-turnover kinetics. *Biochemistry* **30:** 526–537.

Zhang, J., D.W. Chung, C.-K. Tan, K.M. Downey, and A.G. So. 1991. Primary structure of the catalytic subunit of calf thymus DNA polymerase δ: Sequence similarities with other DNA polymerases. *Biochemistry* **30:** 11742–11750.

Zinnen, S., J.-C. Hsieh, and P. Modrich. 1994. Misincorporation and mispaired primer extension by human immunodeficiency virus reverse transcriptase. *J. Biol. Chem* **269:** 24195–24202.

8

DNA Excision Repair Pathways

Errol C. Friedberg
Laboratory of Molecular Pathology
Department of Pathology
University of Texas
Southwestern Medical Center
Dallas, Texas 75235

Richard D. Wood
Imperial Cancer Research Fund
Clare Hall Laboratories
South Mimms, Herts, EN6 3LD
United Kingdom

Most life forms have the ability to respond to alterations in genomic DNA that occur spontaneously or are caused by environmental agents. Generally, these responses take one of two forms. Cells can either repair the damage and restore the genome to its normal physical and functional state, or they can tolerate lesions in a way that reduces their lethal effects (Friedberg et al. 1995). This brief overview exclusively considers the former cellular response to DNA damage, which represents true DNA repair. However, the tolerance of base damage, typically by replicative bypass, sets the stage for permanent mutations in DNA. In fact, a major function of DNA repair is the prevention of mutations, which can have significant phenotypic consequences, including neoplastic transformation in mammalian cells (Friedberg et al. 1995).

The repair of altered bases in DNA is frequently classified into two major categories that have important mechanistic distinctions. A relatively limited group of lesions in DNA can be repaired in single-step reactions that directly reverse the damage. The light-dependent monomerization of cyclobutane pyrimidine dimers by DNA photolyase is a well-characterized example (Kim and Sancar 1993). DNA photolyases have been extensively characterized from many prokaryotes and fungi, and from vertebrates, including fish (Yasuhira and Yasui 1992) and marsupials (Yasui et al. 1994). However, this mode for the repair of the quantitatively major form of DNA damage induced by ultraviolet (UV) radiation seems to have been lost in placental mammals (Li et al. 1993).

 0-87969-459-9/96 $5 + .00

The direct removal of small alkyl groups (such as methyl groups) specifically from the O^6 position of guanine and the O^4 position of thymine in DNA is another notable example of repair by the reversal of base damage. The enzyme that removes alkyl groups from these specific sites in guanine and thymine is designated O^6-methylguanine-DNA methyltransferase and is ubiquitous in nature.

A more general mode of DNA repair, which is the primary topic of this review, is the physical excision of damaged or inappropriate bases from the genome by multistep biochemical reactions. At present, three specific modes of such repair have been identified and characterized in eukaryotes. These are referred to as base excision repair, nucleotide excision repair, and mismatch repair.

BASE EXCISION REPAIR OF DNA IN EUKARYOTES

The term base excision repair (BER) was coined to emphasize that this DNA repair mechanism is characterized by the excision of nucleic acid base residues in the free form (Friedberg et al. 1995). In contrast, nucleotide excision repair (NER) removes damaged nucleotides as part of fragments which are about 30 nucleotides long. The primary and initiating event of BER is the hydrolysis of the N-glycosyl bond linking a nitrogenous base to the deoxyribose-phosphate chain, thereby releasing the free base (Fig. 1). The hydrolysis of N-glycosyl bonds in DNA is catalyzed by a class of enzymes called DNA glycosylases. Multiple DNA glycosylases have been identified in eukaryotic cells (Table 1). Each enzyme removes a limited spectrum of base alterations. Uracil-DNA glycosylase is particularly specific, as it catalyzes the excision exclusively of the base uracil (and 5-fluorouracil) from DNA. Uracil in DNA usually results from the spontaneous or chemically induced deamination of cytosine, although it can occasionally arise by incorporation from small intracellular pools of dUTP or from the deamination of dCTP (Friedberg et al. 1995). Since deamination of cytosine in DNA generates a U•G mispair, excision of uracil is important to avoid G•C→A•T transition mutations during subsequent semiconservative DNA replication. The enzyme from human cells is encoded by the *UNG* gene, which maps to chromosome 12 (Aasland et al. 1990). The amino acid sequence of human uracil-DNA glycosylase shares extensive amino acid sequence identity with the *ung* gene product from a variety of other organisms.

Another highly specific DNA glycosylase is the thymine mismatch-DNA glycosylase, which to date has been identified only in extracts of human cells (Neddermann and Jiricny 1993). This enzyme specifically

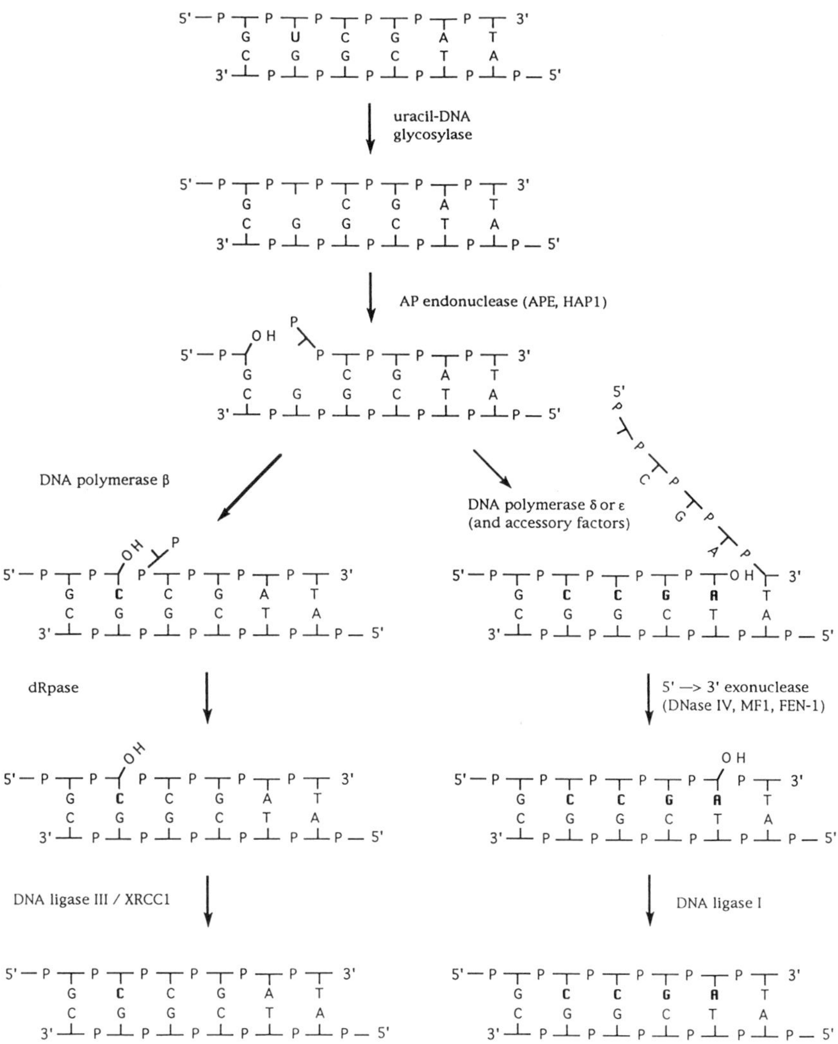

Figure 1 Base excision repair (BER) of DNA in eukaryotes. Repair of a uracil residue initiated by uracil-DNA glycosylase is shown as an example. A branched pathway of repair synthesis is depicted, resulting in either single nucleotide repair patches (*left*) or longer patches of variable size (*right*), depending on which enzymes are utilized. Enzymes implicated in the longer patch repair mode are tentatively assigned. (Adapted, with permission, from Lindahl et al. 1995.)

catalyzes the excision of T when it is mispaired with G, a mispairing that results from the deamination of 5-methylcytosine in DNA. Hence, this form of BER represents one of the several biochemical strategies that hu-

Table 1 DNA glycosylases for base excision repair in eukaryotes

DNA glycosylase	*E. coli* homolog	Substrates
Uracil-DNA glycosylase	*ung*	uracil or fluorouracil in DNA
3-MeA-DNA glycosylase	*alkA*	3-methyladenine hypoxanthine O^2-methylthymine (minor groove alterations)
FaPy-DNA glycosylase	*fpg* (*mutM*)	formamidopyrimidines 8-hydroxyguanine (oxidized and ring-opened purines)
Pyrimidine hydrate-DNA glycosylase	*nth (endoIII)*	thymine glycol cytosine hydrate urea (oxidized and ring-opened pyrimidines)
G:T mismatch glycosylase	–	thymine when paired to guanine, O^6-methyl-guanine, or 6-thioguanine

man cells possess for the repair of mismatched bases (see below).

The excision of bases from duplex DNA generates apurinic or apyrimidinic (AP) sites (Fig. 1). The repair of these sites of base loss utilizes a specific class of endonucleases designated AP endonucleases (Friedberg et al. 1995). Prokaryotes such as *Escherichia coli* have at least two such enzymes. However, studies of yeast and mammalian cells have thus far resulted in the purification and characterization of a single major endonuclease that catalyzes the incision of phosphodiester linkages exclusively 5′ to AP sites, generating 3′-OH and 5′-deoxyribose-phosphate residues (Fig. 1). This enzyme also has 3′-phosphatase activity, and, in some organisms, a weak 3′→5′ exonuclease activity. In mammalian cells, the gene that encodes this AP endonuclease is variously called *BAP1* (bovine AP endonuclease), *APEX* (AP endonuclease/exonuclease), *HAPI* (human AP endonuclease), or *APE* (AP endonuclease) (Demple et al. 1991; Robson et al. 1991, 1992; Seki et al. 1991). Intriguingly, the HAPI/APE protein was also independently isolated as REF-1, a redox factor that can regulate the Fos-Jun transcriptional activation proteins (Walker et al. 1993; Xanthoudakis et al. 1994). Thus, in addition to its role in BER, this protein may play a role in transducing signals associated with oxidative stress to a regulatory network which involves genes associated with the metabolism of reactive oxygen species. Like human uracil-DNA glycosylase, the HAP1 protein shows extensive evolutionary conservation at the amino acid sequence level, and the protein can correct some of the mutant phenotypes of *E. coli* cells defective in Xth protein, the major AP endonuclease in that organism.

The completion of BER requires the removal of the 5′-terminal deoxyribose-phosphate residue generated by the AP endonuclease, followed by repair synthesis and DNA ligation. A 47-kD enzyme activity designated DNA deoxyribophosphodiesterase (dRpase) has been identified in human cells and can remove such sugar-phosphate residues from duplex DNA (Price and Lindahl 1991). If this enzyme is indeed utilized in vivo, a single nucleotide gap would be generated that can be filled in by a DNA polymerase (Fig. 1). In vitro, such repair synthesis is efficiently catalyzed by DNA polymerase-β (Dianov et al. 1992). It has been suggested that because of its limited fidelity, polymerase-β may be utilized in this particular very short patch mode of repair synthesis during BER and not in repair synthesis associated with longer repair patches.

BER is sometimes associated with the generation of longer repair patches. There are experimental indications that DNA polymerases δ and ε are primarily involved in this second BER synthesis mode in both yeast and human cells. These longer repair tracts are thought to result from a nick translation reaction accompanied by strand displacement in the 5′→3′ direction, thereby generating a flap type of structure (Fig. 1). Removal of the overhanging 5′-terminal single-stranded region of DNA is believed to be effected by a 5′ single-strand/duplex junction-specific nuclease originally designated DNase IV (Lindahl et al. 1969; Robins et al. 1994) and more recently as FEN-1 (Harrington and Lieber 1994). This nuclease, which is conserved in the yeasts *Saccharomyces cerevisiae* as Rad27 protein and in *Schizosaccharomyces pombe* as rad2 protein (Murray et al. 1994), has also been implicated in lagging-strand DNA synthesis during semiconservative replication (Ishimi et al. 1988; Turchi et al. 1994; Waga et al. 1994). Inspection of the predicted amino acid sequences of DNase IV, Rad27, *S. pombe* rad2 protein, and various prokaryotic DNA polymerases endowed with 5′→3′ exonuclease activity (such as *E. coli* DNA polymerase I), shows limited regions of amino acid sequence homology (Table 2). Interestingly, this homology is shared with junction-specific nucleases that are involved in NER in *S. cerevisiae* and in mammalian cells (Table 2; see later discussion).

The final biochemical event during BER is DNA ligation. Mammalian cells (and possibly the yeast *S. cerevisiae*) contain several DNA ligases. These are fully discussed in chapter 20. A recently recognized

Table 2 Amino acid sequence homology between various nuclease families

E. coli pol I	(101)	-MGLPLL--SGV**EAD**D-IG-LAR-A---G
Bacillus caldotenax pol I	(97)	-Y-IP-Y----Y**EAD**D-IG-LA--A---G
Streptococcus pneumoniae pol I	(102)	-MGI--Y--A-Y**EAD**D-IG-L-K-A---G
Thermus flavus pol I	(104)	LLGL--L--PGF**EAD**D-LA-LAK-A---G
T4 rnh (orfA)	(119)	YMPY-VM----Y**EAD**D-IAVL-K---L-G
T5 D15	(116)	---FP-F---GV**EAD**D--AYI-K----L-
T7 6	(121)	---F-CI--P-L**EGD**D-MGVIA--P--FG
S. cerevisiae Rad27	(145)	LMGIPYI--AP-**EAE**AQCA-LAK-GKVYA
S. pombe Rad2	(147)	LMGIPFV--APC**EAE**AQCA-LARSGKVYA
Mus musculus FEN-1	(145)	LMGIPYL--AP-**EAE**A-CA-LAKAGKVYA
Homo sapiens DNase IV	(147)	LMGIPYL--AP-**EAE**A-CA-L-KAGKVYA
S. cerevisiae Rad2	(781)	-FGIPYI--APM**EAE**AQCA-L-----V-G
S. pombe Rad13	(767)	LFGLPYI--AP-**EAE**AQCS-L-----V-G
Xenopus laevis XPG	(811)	LFGIPYI--APM**EAE**AQCAIL--T----G
H. sapiens XPG	(778)	LFGIPYI--APM**EAE**AQCAIL--T----G

The various nucleases shown are grouped according to known functions. The top group are DNA polymerases with associated nuclease activities. The next group are members of the eukaryotic DNase IV family, followed by the *S. cerevisiae* Rad2 nuclease family. The numbers in parentheses refer to amino acid position. Only identical or related amino acids are shown. The dashes indicate nonconserved positions. The EA(G)E(D) motif that is conserved in all 15 sequences is shown in bold. (Adapted, with thanks, from Stuart G. Clarkson.)

aspect of DNA metabolism that can affect the kinetics of BER is the potential competition for strand breaks in DNA between the enzymatic machinery required for the completion of this repair process and the enzyme poly(ADP-ribose) polymerase. This enzyme normally has a high affinity for strand breaks in DNA. However, in the presence of NAD^+, the bound enzyme undergoes extensive autoribosylation, which results in decreased binding affinity for DNA and its eventual dissociation (Molinete et al. 1993; Satoh et al. 1993). It has been suggested that the binding of non-ribosylated poly(ADP ribose) polymerase to strand breaks introduced by the sequential action of a DNA glycosylase and AP endonuclease during BER may signal a slowing or cessation of DNA replication while BER takes place. Alternatively, or additionally, the binding of poly(ADP-ribose) polymerase to DNA may serve to reduce the potential for the initiation of recombination at sites of strand breakage (Lindahl et al. 1995).

NUCLEOTIDE EXCISION REPAIR OF DNA IN EUKARYOTES

NER is characterized by the excision of damaged bases in oligonucleotide fragments. In contrast to the limited substrate specificity of most DNA glycosylases, NER operates on a large spectrum of base damage, particularly that produced by environmental mutagenic and carcinogenic agents which produce bulky, helix-distorting perturbations in DNA structure. In human cells, NER is the principal mechanism by which base damage produced by UV radiation is removed from DNA. Individuals with the rare inherited disorder xeroderma pigmentosum (XP) have defects in NER genes (Table 3), generally leading to a greatly increased risk of sunlight-induced skin cancer (Cleaver and Kraemer 1989).

There is now compelling evidence that in both prokaryotes and eukaryotes, the oligonucleotides excised during NER are generated by dual incisions which flank sites of base damage. Unlike BER, which is believed to require no more than 5 proteins to complete the entire process, there is good evidence that in eukaryotes the events that precede repair synthesis and DNA ligation during NER require the participation of between 15 and 20 gene products (Table 3). This degree of biochemical complexity is reminiscent of that associated with the initiation of basal transcription by RNA polymerase II. Indeed, it has been suggested that the generation of a transcription "bubble" as part of the process of promoter clearance during RNA polymerase II transcription, and the generation of a bubble that defines the single-strand/duplex DNA junctions required for bimodal incision during NER (Fig. 2), may be mechanistically related (Hoeijmakers and Bootsma 1994). This suggestion is to a large extent prompted by the recent discovery that both in the yeast *S. cerevisiae* and in human cells the processes of NER and RNA polymerase II-mediated transcription share multiple proteins in common (Table 3) (Bootsma and Hoeijmakers 1993; Chalut et al. 1994; Drapkin and Reinberg 1994; Friedberg et al. 1994).

Among the many proteins required for the initiation of RNA polymerase II transcription in *S. cerevisiae* (all of which are encoded by essential genes) are a complex of six polypeptides designated core TFIIH (Table 3) (Svejstrup et al. 1995). This core complex is believed to assemble with three other proteins (TFIIK) endowed with kinase activity for the carboxy-terminal domain of the largest subunit of RNA polymerase II, to yield a holoTFIIH supercomplex, the form of TFIIH that is functional in transcription initiation (Svejstrup et al. 1995). Four of the six subunits of core TFIIH have been directly shown to be indispensable for NER in this yeast (Table 3) (Svejstrup et al. 1995). In crude extracts of *S. cerevisiae* that are competent for RNA polymerase II basal transcription,

Table 3 Eukaryotic nucleotide excision repair genes and proteins

Human gene	Human map position	*S. cerevisiae* homolog	*S. pombe* homolog	M_r (human gene product unless indicated otherwise)	Comments
XPA	9q34.1	*RAD14*		31 kD (40/42 kD on gels)	binds damaged DNA
XPB/ERCC3	2q21	*SSL2* (*RAD25*)	*ercc3*$^{sp+}$	89 kD (89 kD on gels)	3′→5′ DNA helicase; in TFIIH
XPC	3p25	*RAD4*		106 kD (125 kD on gels)	binds ssDNA
HHR23B	3p25	*RAD23*		43 kD (58 kD on gels)	associated with XPC
XPD/ERCC2	19q13.2	*RAD3*	*rad15*$^+$	87 kD (80 kD on gels)	5′→3′ DNA helicase; in TFIIH
XPG/ERCC5	13q33	*RAD2*	*rad13*$^+$	133 kD (180–200 kD on gels)	CS in some affected individuals; DNA nuclease
XPF/ERCC4? (ERCC4)	16p13.13	*RAD1*	*rad16*$^+$	126 (*S. cerevisiae*)	component of DNA nuclease
ERCC1	19q13.2	*RAD10*	*swi10*$^+$	31 (human) (39 kD on gels)	component of DNA nuclease
p44	5q13	*SSL1*		44 kD (yeast 50 kD)	in TFIIH
p62	11p14-15.1	*TFB1*		70–73 kD in yeast	in TFIIH
p52	6p21.3-22.2	*TFB2*		55 kD in yeast	in TFIIH
CSA/ERCC8	5	*RAD28*		44 kD	WD-repeat protein
CSB/ERCC6	10q11.2	*RAD26*		168 kD	DNA helicase? transcription coupling?
RPAp70	17p13	*RFA1*		68 kD (70 kD on gels)	binding to
RPAp32	1p35-36.1	*RFA2*		29 kD (34 kD on gels)	single-stranded
RPAp14	7p21-22	*RFA3*		13.6 kD	DNA
LIG1	19q13.2-3	*CDC9*	*cdc17*$^+$	102 kD (120 kD on gels)	DNA ligase I
RFC1		*CDC44*		140 kD	RF-C large subunit
PCNA	20	*POL30*	*pcn1*$^+$	29 kD (36 kD on gels)	toroidal sliding clamp

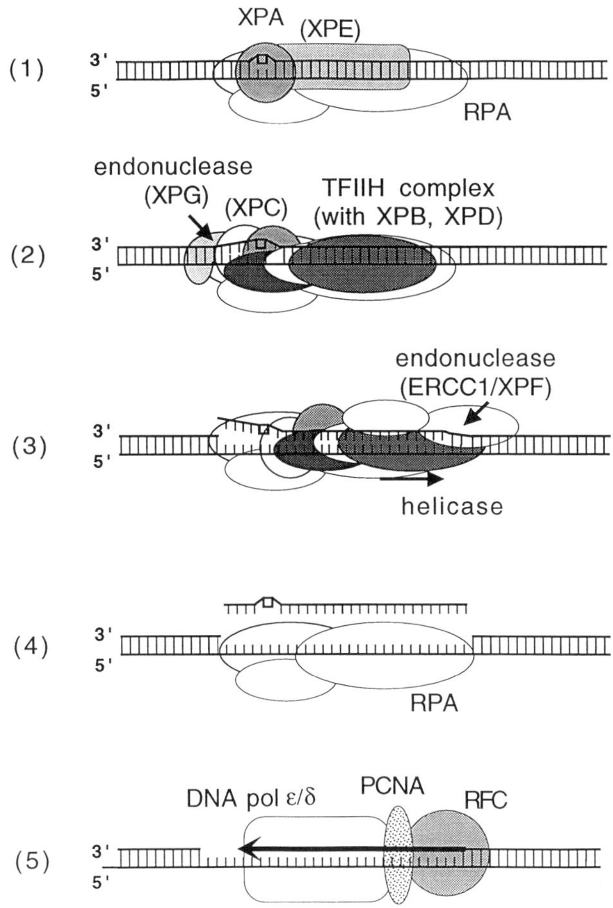

Figure 2 Nucleotide excision repair (NER) in mammalian cells. The steps shown are (1) DNA damage recognition; (2) incision on the 3′ side of the damage; (3) the generation of a further opened structure by the helicase function(s) of TFIIH and incision on the 5′ side of the damage; (4) release of a damage-containing oligonucleotide; (5) repair synthesis.

a different supercomplex can be identified that includes all six subunits of core TFIIH plus at least five other proteins which are indispensable for NER but are not known to be required for transcription, namely, Rad1, Rad2, Rad4, Rad10, and Rad14. This supercomplex comprising at least these 11 proteins is referred to as the nucleotide excision repairosome (Svejstrup et al. 1995). There are indications of comparable protein complexes for transcription and NER in human cells (Roy et al. 1994).

The observation that core TFIIH is common to both repair and transcription may help explain the fact that in both yeast and mammalian cells NER takes place significantly faster in the template strand of transcriptionally active genes than in the nontranscribed strand (Bohr 1992; Hanawalt 1992). However, it remains to be determined whether the particular form of TFIIH that participates in transcriptionally coupled NER is the same as that loaded onto the DNA during transcription initiation as part of the TFIIH holocomplex. It is intuitively compelling to consider that when the TFIIH holocomplex is loaded onto promoter sites, the NER proteins are retained in the transcription elongation complex. Thus, if transcription is arrested at sites of base damage in the template strand, the core TFIIH complex would constitute a strategically positioned nucleation site for the assembly of a functionally active repairosome. However, there is as yet no direct experimental evidence for this mechanism. An alternative possibility is that TFIIH does not participate in the process of transcription elongation. There is indeed some evidence for this view (Drapkin and Reinberg 1994; Goodrich and Tjian 1994). In this event, repair proteins would be recruited to sites of arrested transcription at base damage by a mechanism(s) that is yet to be specifically determined. Additionally, in mammalian cells, most of the genome is transcriptionally silent, so NER frequently occurs in the absence of a coupling to transcription. Yet components of TFIIH are still required, and it remains to be discovered precisely how these are delivered to damaged sites in transcriptionally silent regions of the genome.

Biochemical functions have been identified for several NER proteins (Table 3). The bimodal incision mechanism involves single-strand/duplex junction-specific nucleases. In *S. cerevisiae* the nuclease that is believed to incise DNA 5′ to sites of base damage is carried in the Rad1/Rad10 protein complex (Tomkinson et al. 1993). The mammalian homologs of Rad1/Rad10 are ERCC4 and ERCC1, respectively (Table 3). In vitro, the yeast Rad1/Rad10 endonuclease (and presumably the ERCC1/XPF complex in mammalian cells) cuts splayed-arm substrates specifically at 3′ single-strand/duplex junctions where the single strand has a 3′ end (Bardwell et al. 1994). Reciprocally, purified human XPG protein cuts splayed-arm substrates specifically at single-strand/duplex junctions where the single strand has a 5′ end, and additionally cuts bubble substrates with a consistent polarity (O'Donovan et al. 1994). Thus, XPG protein is believed to incise DNA on the 3′ side of damage during NER. Presumably the homologous Rad2 protein acts similarly (Harrington and Lieber 1994). Rad2 protein (Habraken et al. 1993), Rad1/Rad10 complex (Tomkinson et al. 1993; Sung et al. 1993), and XPG protein (O'Donovan et al. 1994; Habraken et al. 1994) also can cut bacteriophage M13 DNA, presumably at single-strand/duplex junctions at hairpin loops in such DNA. Thus, it seems reasonable to conclude that bimodal damage-specific incision during NER in eukaryotes is achieved by the concomitant or sequential actions of the Rad1/Rad10 and Rad2 nucleases in yeast, and by the ERCC1/XPF and XPG nucleases in mammalian cells (Fig. 2). There are indications that the XPC protein may also participate in the incision process (Shivji et al. 1994).

Studies on NER suggest that the excised oligonucleotide fragments have a precise size of approximately 30 ± 2 nucleotides in human (and presumably in yeast) cells. Based on the established specificity of the endonucleases just discussed for single-strand/duplex junctions, we are led to the model that an open structure of about 30 nucleotides is somehow generated in damaged DNA during NER in eukaryotes. Regardless of precisely how the TFIIH core complex is loaded onto DNA, the known biochemical properties of two components of this complex may help explain how such an open structure might arise. Both the yeast Rad3 (human XPD) and Ssl2 (human XPB) proteins are DNA helicases with opposite directionality. It is therefore possible that one or both of these helicases unwind limited regions of duplex DNA on either side of a damaged site during NER (Fig. 2). There is indeed extensive evidence that the $5'\rightarrow3'$ helicase function of Rad3 protein is specifically required for NER but not for transcription. The $3'\rightarrow5'$ helicase function of yeast Ssl2 (XPB) protein is essential for the viability of yeast cells and is therefore presumably indispensable for transcription as well.

It is not yet established how endonucleolytic cleavage is directed specifically to the DNA strand containing a lesion. The yeast Rad14 and homologous human XPA proteins are DNA-binding proteins with preferential affinity for certain types of DNA damage (Robins et al. 1991; Guzder et al. 1993; Jones and Wood 1993; Asahina et al. 1994). Presumably, these proteins play some role in the recognition of base damage. The XPE protein may also participate in this process (Chu and Chang 1988; Hirschfeld et al. 1990; Keeney et al. 1993; Takao et al. 1993; Payne and Chu 1994). It has also been shown that the helicase function of purified Rad3 (human XPD) protein is arrested by the presence of many types of base damage, specifically in the strand on which the protein translocates (Naegeli et al. 1992; Sung et al. 1994). Hence, Rad3 (XPD) protein may also be an important player in damage recognition, both in the nontranscribed bulk of the genome and in transcribed regions of DNA. Specific protein-protein interactions, such as found between XPA and ERCC1, may help direct endonucleolytic cleavage to the correct strand (Li et al. 1994; Park and Sancar 1994).

In addition to the proteins discussed above, there are indications that other polypeptides are involved in NER. In yeast, these include the *RAD7*, *RAD16*, and *RAD23* gene products, but the precise functional role(s) of these proteins remains to be determined. Recent studies using a cell-free system for NER in yeast indicate an absolute requirement for Rad7 protein (Z. Wang and E.C. Friedberg, unpubl.). The single-stranded DNA-binding replication protein A (RP-A) is necessary for NER supported by mammalian cell extracts in vitro, where it participates during DNA repair synthesis. Additionally, RP-A is required for damage-specific incision (Coverley et al. 1992; Shivji et al. 1992). The homolog of this heterotrimeric protein in *S. cerevisiae* is encoded by the *RFA* genes, and yeast strains carrying mutations in the *RFA1* gene are abnormally sensitive to UV radiation, suggesting a role for RP-A in NER in vivo (Longhese et al. 1994).

The repair synthesis step of NER in mammalian cells requires the DNA polymerase accessory factor proliferating cell nuclear antigen (PCNA) in vitro (Shivji et al. 1992), and there is also evidence that PCNA is involved in vivo (Celis and Madsen 1986; Toschi and Bravo 1988; Miura et al. 1992; Hall et al. 1993; Jackson et al. 1994). These observations, together with experiments with chemical inhibitors, implicate the PCNA-dependent DNA polymerases δ or ε in the repair synthesis step. Perhaps either enzyme works in this function, but DNA polymerase-ε appears to be the most suitable candidate in vivo (Nishida et al. 1988; Syväoja et al. 1990). Recent studies have shown that DNA polymerase-ε is also functionally well suited for NER in vitro, and in the presence of RP-A, replication protein C (RP-C) is also required (R.D. Wood et al., unpubl.). RP-C functions to load PCNA onto a DNA template in order to initiate DNA synthesis (Podust et al. 1994). The repair synthesis patch in vitro (Hansson et al. 1989; Shivji et al. 1992) and in vivo (Cleaver et al. 1991) is about 30 nucleotides long, reflecting precise filling of the gap created by the excision of oligonucleotides.

MISMATCH REPAIR IN EUKARYOTES

Mismatches can arise in DNA by two primary mechanisms. Replicative DNA polymerases do not copy templates with complete accuracy, so mismatches can arise because of replicative errors. Additionally, heteroduplexes formed as recombination intermediates between two homologous pieces of DNA (such as two alleles of a gene) can contain mismatches arising from polymorphisms. The former mechanism is the most important source of mismatches in somatic cells.

General (so-called long-patch) mismatch repair is best understood in *E. coli*, where the core enzymes of the system are the products of the *mutH*, *mutL*, and *mutS* genes (Modrich 1991; Friedberg et al. 1995). Mismatch repair can only protect cells from permanent mutations if the parental strand (containing the correct information) can be accurately distinguished from the daughter strand. In *E. coli*, the strand discrimination signal is provided by adenine methylation in GATC sequences; newly replicated DNA is not yet methylated on the daughter strand (Modrich 1991; Friedberg et al. 1995). The MutH protein binds to DNA at hemimethylated GATC sequences and effects incision on the unmethylated strand. MutS protein recognizes and binds to the mismatch, and the intervening region (often hundreds of nucleotides long) is excised and recopied by a DNA polymerase (Modrich 1991; Friedberg et al. 1995). The MutL protein mediates communication between the distantly bound MutH and MutS products, bringing them together by looping out the intervening region of DNA (Fig. 3) (Modrich 1991).

Long-patch mismatch excision repair is a highly conserved process that appears to work in a similar way in eukaryotes as in *E. coli*, except that strand discrimination does not appear to be methyl-directed in eukaryotes (Friedberg et al. 1995). Thus, there may be no eukaryotic MutH homolog. Instead, the strand-discrimination signal is thought to be provided by single-stranded nicks or gaps in newly replicated DNA, which have not yet been joined by a DNA ligase. A protein that recognizes these nicks or gaps would replace the function of MutH. However, several structural homologs of MutL and MutS have been isolated from yeast and from mammalian cells. Of the two known MutS homologs, one (hMSH2) is a nuclear protein and the other (hMSH1) is mitochondrial. Interestingly, there are at least three MutL homologs in humans (hMLH1, hPMS1, and hPMS2). hMSH2 has been demonstrated to be a DNA mismatch-binding protein in vitro (Palombo et al. 1994). It seems probable that the MutS and MutL proteins form complexes with one another in different combinations to facilitate recognition of a wide range of different types of mismatches, ranging from common single base-pair mismatches such as G•T, through loop-outs of one, two, or more nucleotides.

Inactivation of mismatch repair genes in humans is clearly implicated in the pathogenesis of hereditary non-polyposis colon cancer (HNPCC). Individuals with this condition inherit an inactivated allele of a mismatch repair gene, and colorectal carcinomas in these individuals (as well as sporadic tumors from non-HNPCC patients) have two inactivated alleles. The mismatch repair defect increases the spontaneous mutation rate in

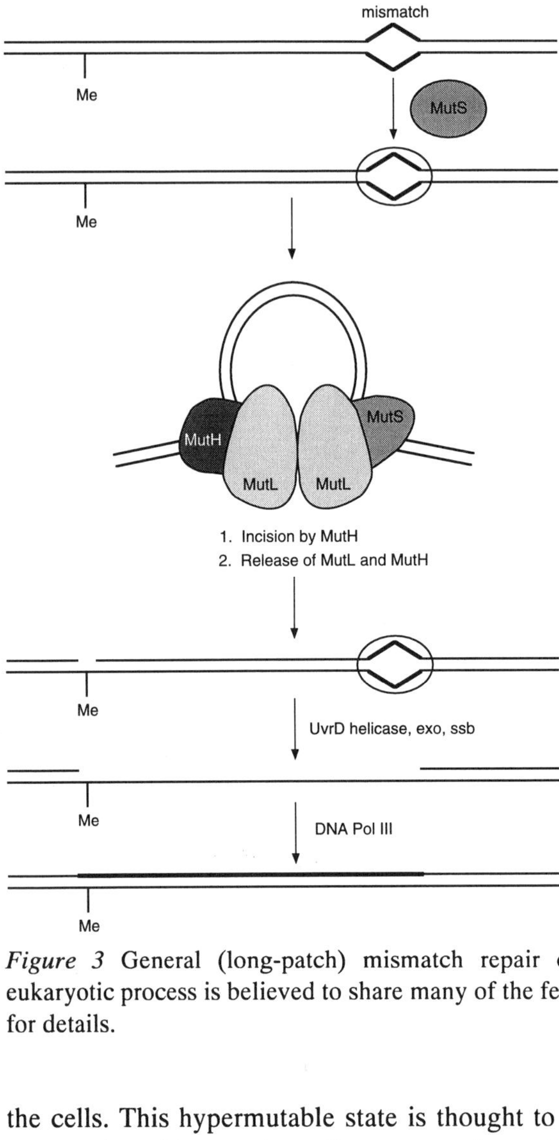

Figure 3 General (long-patch) mismatch repair of DNA in *E. coli*. The eukaryotic process is believed to share many of the features shown here. See text for details.

the cells. This hypermutable state is thought to be an early event in the progression of tumors to malignancy, as it greatly accelerates the acquisition of mutations in other tumor suppressor genes and oncogenes. Thus far, HNPCC families have been found with mutations in the hMSH2, hMLH1, hPMS1, and hPMS2 genes (Fishel et al. 1993; Leach et al. 1993; Bronner et al. 1994; Nicolaides et al. 1994; Papadopoulos et al. 1994). A notable characteristic of colorectal carcinoma cell lines is their high rate of polymorphism in microsatellite repeat sequences. During DNA replication, di- or trinucleotide repeat units in such sequences can be accidentally lost or gained by replication fork slippage. In the absence of mismatch repair, the lengths of the microsatellite sequences change more rapidly than in normal cells, reflecting the hypermutable state (Aaltonen et al. 1993).

An interesting feature of mismatch repair-defective cells is an association with tolerance to simple DNA N-nitroso-methylating agents such as MNNG and MNU (Karran and Bignami 1994). This tolerance results because the most toxic DNA adduct produced by such agents, O^6-methylguanine, can pair with thymine and the resulting $G^{O6Me}{\cdot}T$ base pair is recognized as a mismatch. However, G•T mismatches are nearly always repaired by removal of the T residue, so futile cycles of mismatch repair are initiated. The major alternative pathway for the removal of O^6-methylguanine is by the O^6-methylguanine-DNA methyltransferase referred to earlier, but many cells, particularly those that are transformed, spontaneously lose expression of this enzyme. In such cases, the repeated excision cycles of futile mismatch repair eventually lead to lethal strand breaks (Karran and Bignami 1994). Thus, there may be selective pressure for loss of mismatch repair in cells that are frequently exposed to methylating agents. Colon cells are exposed to bile acids that can be converted into methylating compounds (Karran and Bignami 1994). Cells with defective O^6-methylguanine-DNA methyltransferase and defective mismatch binding proteins become more resistant to DNA-methylating agents, but are hypermutable and show microsatellite instability (Branch et al. 1993; Kat et al. 1993; Aquilina et al. 1994).

Mismatch repair can be studied in vitro and the entire repair process can be carried out in mammalian cell extracts, so details of the biochemistry of this excision repair mode are emerging rapidly.

CONCLUDING REMARKS

The general topic of excision repair of DNA in eukaryotes has undergone many exciting developments in recent years. The biochemistry of BER is essentially fully defined in vitro and is consonant with the genetics of this process. It has additionally been firmly established that NER is a complex biochemical process involving a large number of gene products. The findings in a number of laboratories that some components of the NER machinery are also components of the RNA polymerase II basal transcription apparatus have added new and exciting dimensions to this DNA repair mode, including new insights into the possible molecular pathogenesis of human hereditary diseases associated with defective NER (Friedberg et al. 1994; Vermeulen et al. 1994). Equally dramatic strides have been made in deciphering the mechanism of strand-directed mismatch repair in eukaryotes. The association of defective NER and mismatch repair with a variety of human cancers provides convincing support for the somatic mutation hypothesis of neoplastic transformation and the crucial role of DNA repair in protecting against this consequence of DNA damage.

ACKNOWLEDGMENTS

We apologize to many of our colleagues for the restrictive reference list necessitated by space limitations. We thank our laboratory colleagues for critical review of the manuscript. E.C.F. and R.D.W. acknowledge grant support from the U.S. Public Health Service and the Imperial Cancer Research Fund, respectively.

REFERENCES

Aaltonen, L.A., P. Peltomäki, F.S. Leach, P. Sistonen, L. Pylkkänen, J.-P. Mecklin, H. Järvinen, S.M. Powell, J. Jen, S.R. Hamilton, G.M. Petersen, K.W. Kinzler, B. Vogelstein, and A. de la Chapelle. 1993. Clues to the pathogenesis of familial colorectal cancer. *Science* **260:** 812–816.

Aasland, R., L.C. Olsen, N.K. Spurr, H.E. Krokan, and D.E. Helland. 1990. Chromosomal assignment of human uracil-DNA glycosylase to chromosome 12. *Genomics* **7:** 139–141.

Aquilina, G., P. Hess, P. Branch, C. MacGeoch, I. Casciano, P. Karran, and M. Bignami. 1994. A mismatch recognition defect in colon-carcinoma confers DNA microsatellite instability and a mutator phenotype. *Proc. Natl. Acad. Sci.* **91:** 8905–8909.

Asahina, H., I. Kuraoka, M. Shirakawa, E.H. Morita, N. Miura, I. Miyamoto, E. Ohtsuka, Y. Okada, and K. Tanaka. 1994. The XPA protein is a zinc metalloprotein with an ability to recognize various kinds of DNA damage. *Mutat. Res.* **315:** 229–237.

Bardwell, A.J., L. Bardwell, A.E. Tomkinson, and E.C. Friedberg. 1994. Specific cleavage of model recombination and repair intermediates by the yeast Rad1-Rad10 DNA endonuclease. *Science* **265:** 2082–2085.

Bohr, V.A. 1992. Gene-specific DNA repair: Characteristics and relations to genomic instability. In *DNA repair mechanisms* (ed. V.A. Bohr et al.), pp. 217–227. Munksgaard, Copenhagen.

Bootsma, D. and J.H.J. Hoeijmakers. 1993. Engagement with transcription. *Nature* **363:** 114–115.

Branch, P., G. Aquilina, M. Bignami, and P. Karran. 1993. Defective mismatch binding and a mutator phenotype in cells tolerant to DNA damage. *Nature* **362:** 652–654.

Bronner, C.E., S.M. Baker, P.T. Morrison, G. Warren, L.G. Smith, M.K. Lescoe, M. Kane, C. Earabine, J. Lipford, A. Lindblom, P. Tannergård, R.J. Bollag, A.R. Godwin, D.C. Ward, M. Nordenskjøld, R. Fishel, R. Kolodner, and R.M. Liskay. 1994. Mutation in the DNA mismatch repair gene homologue *hMLH1* is associated with hereditary non-polyposis colon cancer. *Nature* **368:** 258–261.

Celis, J.E., and P. Madsen. 1986. Increased nuclear cyclin/PCNA antigen staining of non S-phase transformed human amnion cells engaged in nucleotide excision DNA repair. *FEBS Lett.* **209:** 277–283.

Chalut, C., V. Moncollin, and J.M. Egly. 1994. Transcription by RNA polymerase II: A process linked to DNA repair. *BioEssays* **16:** 651–655.

Chu, G. and E. Chang. 1988. Xeroderma pigmentosum group E cells lack a nuclear factor that binds to damaged DNA. *Science* **242:** 564–567.

Cleaver, J.E. and K.H. Kraemer. 1989. Xeroderma pigmentosum. In *The metabolic basis of inherited disease,* 6th edition (ed. C.R. Scriver et al.), pp. 2949–2971. McGraw-Hill, New York.

Cleaver, J.E., J. Jen, W.C. Charles, and D.L. Mitchell. 1991. Cyclobutane dimers and (6-4)photoproducts in human-cells are mended with the same patch sizes. *Photochem. Photobiol.* **54:** 393–402.

Coverley, D., M.K. Kenny, D.P. Lane, and R.D. Wood. 1992. A role for the human single-stranded DNA binding protein HSSB/RPA in an early stage of nucleotide excision repair. *Nucleic Acids Res.* **20:** 3873–3880.

Demple, B., T. Herman, and D.S. Chen. 1991. Cloning and expression of ape, the cDNA-encoding the major human apurinic endonuclease-definition of a family of DNA-repair enzymes. *Proc. Natl. Acad. Sci.* **88:** 11450–11454.

Dianov, G., A. Price, and T. Lindahl. 1992. Generation of single-nucleotide repair patches following excision of uracil residues from DNA. *Mol. Cell. Biol.* **12:** 1605–1612.

Drapkin, R. and D. Reinberg. 1994. The multifunctional TFIIH complex and transcriptional control. *Trends Biochem. Sci.* **19:** 504–508.

Fishel, R., M.K. Lescoe, M.R.S. Rao, N.G. Copeland, N.A. Jenkins, J. Garber, M. Kane, and R. Kolodner. 1993. The human mutator gene homolog *MSH2* and its association with hereditary nonpolyposis cancer. *Cell* **75:** 1027–1038.

Friedberg, E.C., G.C. Walker, and W. Siede. 1995. *DNA repair and mutagenesis.* ASM Press, Washington, D.C.

Friedberg, E.C., A.J. Bardwell, L. Bardwell, Z. Wang, and G. Dianov. 1994. Transcription and nucleotide excision repair-reflections, considerations and recent biochemical insights. *Mutat. Res.* **307:** 5–14.

Goodrich, J.A. and R. Tjian. 1994. Transcription factors IIE and IIH and ATP hydrolysis direct promoter clearance by RNA polymerase II. *Cell* **77:** 145–156.

Guzder, S.N., P. Sung, L. Prakash, and S. Prakash. 1993. Yeast DNA-repair gene *RAD14* encodes a zinc metalloprotein with affinity for ultraviolet-damaged DNA. *Proc. Natl. Acad. Sci.* **90:** 5433–5437.

Habraken, Y., P. Sung, L. Prakash, and S. Prakash. 1993. Yeast excision repair gene *RAD2* encodes a single-stranded DNA endonuclease. *Nature* **366:** 365–368.

———. 1994. Human xeroderma-pigmentosum group-G gene encodes a DNA endonuclease. *Nucleic Acids Re.s* **22:** 3312–3316.

Hall, P.A., P.H. McKee, H. Menage, R. Dover, and D.P. Lane. 1993. High levels of p53 protein in UV-irradiated normal human skin. *Oncogene* **8:** 203–207.

Hanawalt, P.C. 1992. Transcription-dependent and transcription-coupled DNA repair responses. In *DNA repair mechanisms* (ed. V.A. Bohre et al.), pp. 231–242. Munksgaard, Copenhagen.

Hansson, J., M. Munn, W.D. Rupp, R. Kahn, and R.D. Wood. 1989. Localization of DNA repair synthesis by human cell extracts to a short region at the site of a lesion. *J.*

Biol. Chem. **264:** 21788–21792.

Harrington, J.J. and M.R. Lieber. 1994. Functional domains within FEN-1 and Rad2 define a family of structure-specific endonucleases—Implications for nucleotide excision-repair. *Genes Dev.* **8:** 1344–1355.

Hirschfeld, S., A.S. Levine, K. Ozato, and M. Protic. 1990. A constitutive damage-specific DNA-binding protein is synthesized at higher levels in UV-irradiated primate cells. *Mol. Cell. Biol.* **10:** 2041–2048.

Hoeijmakers, J.H.J. and D. Bootsma. 1994. Incisions for excision. *Nature* **371:** 654–655.

Ishimi, Y., A. Claude, P. Bullock, and J. Hurwitz. 1988. Complete enzymatic synthesis of DNA containing the SV40 origin of replication. *J. Biol. Chem.* **263:** 19723–19733.

Jackson, D.A., A.B. Hassan, R.J. Errington, and P.R. Cook. 1994. Sites in human nuclei where damage induced by ultraviolet-light is repaired—Localization relative to transcription sites and concentrations of proliferating cell nuclear antigen and the tumor-suppressor protein, p53. *J. Cell Sci.* **107:** 1753–1760.

Jones, C.J. and R.D. Wood. 1993. Preferential binding of the xeroderma pigmentosum group A complementing protein to damaged DNA. *Biochemistry* **32:** 12096–12104.

Karran, P. and M. Bignami. 1994. DNA damage tolerance, mismatch repair and genome instability. *BioEssays* **16:** 833–839.

Kat, A., W.G. Thilly, W.H. Fang, M.J. Longley, G.M. Li, and P. Modrich. 1993. An alkylation-tolerant, mutator human cell line is deficient in strand-specific mismatch repair. *Proc. Natl. Acad. Sci.* **90:** 6424–6428.

Keeney, S., G.J. Chang, and S. Linn. 1993. Characterization of a human DNA-damage binding protein implicated in xeroderma pigmentosum E. *J. Biol. Chem.* **268:** 21293–21300.

Kim, S.T. and A. Sancar. 1993. Photochemistry, photophysics, and mechanism of pyrimidine dimer repair by DNA photolyase. *Photochem. Photobiol.* **57:** 895–904.

Leach, F.S., N.C. Nicolaides, N. Papadopoulos, B. Liu, J. Jen, R. Parsons, P. Peltomäki, P. Sistonen, L.A. Aaltonen, M. Nyström-Lahti, X.-Y. Guan, J. Zhang, P.S. Meltzer, J.-W. Yu, F.-T. Kao, D.J. Chen, K.M. Cerosaletti, R.E.K. Fournier, S. Todd, T. Lewis, R.J. Leach, S.L. Naylor, J. Weissenbach, J.-P. Mecklin, H. Järvinen, G.M. Petersen, S.R. Hamilton, J. Green, J. Jass, P. Watson, H.T. Lynch, J.M. Trent, A. de la Chappelle, K.W. Kinzler, and B. Vogelstein. 1993. Mutations of a *mutS* homolog in hereditary nonpolyposis colorectal cancer. *Cell* **75:** 1215–1225.

Li, L., S.J. Elledge, C.A. Peterson, E.S. Bales, and R.J. Legerski. 1994. Specific association between the human DNA repair proteins XPA and ERCC1. *Proc. Natl. Acad. Sci.* **91:** 5012–5016.

Li, Y.F., S.T. Kim, and A. Sancar. 1993. Evidence for lack of DNA photoreactivating enzyme in humans. *Proc. Natl. Acad. Sci.* **90:** 4389–4393 .

Lindahl, T., J.A. Gally, and G.M. Edelman. 1969. Deoxyribonuclease IV: A new exonuclease from mammalian tissues. *Proc. Natl. Acad. Sci.* **62:** 597–603.

Lindahl, T., M.S. Satoh, and G. Dianov. 1995. Enzymes acting at strand interruptions in DNA. *Philos. Trans. R. Soc. Lond. Biol. Sci.* **347:** 57–62.

Longhese, M.P., P. Plevani, and G. Lucchini. 1994. Replication factor A is required in vivo for DNA replication, repair, and recombination. *Mol. Cell. Biol.* **14:** 7884–7890.

Miura, M., M. Domon, T. Sasaki, and Y. Takasaki. 1992. Induction of proliferating cell nuclear antigen (PCNA) complex-formation in quiescent fibroblasts from a xeroderma pigmentosum patient. *J. Cell. Physiol.* **150:** 370–376.

Modrich, P. 1991. Mechanisms and biological effects of mismatch repair. *Annu. Rev.*

Genet. **25:** 229–253.

Molinete, M., W. Vermeulen, A. Burkle, J. Menissier-De Murcia, J.H. Kupper, J.H.J. Hoeijmakers, and G. de Murcia. 1993. Overproduction of the poly(ADP-ribose) polymerase DNA-binding domain blocks alkylation-induced DNA-repair synthesis in mammalian cells. *EMBO J.* **12:** 2109–2117.

Murray, J.M., M. Tavassoli, R. Al-Harithy, K.S. Sheldrick, A.R. Lehmann, A.M. Carr, and F.Z. Watts. 1994. Structural and functional conservation of the human homologue of the *Schizosaccharomyces pombe rad2* gene, which is required for DNA repair and chromosome segregation. *Mol. Cell. Biol.* **14:** 4878–4888.

Naegeli, H., L. Bardwell, and E.C. Friedberg. 1992. The DNA helicase and adenosine-triphosphatase activities of yeast rad3 protein are inhibited by DNA damage—A potential mechanism for damage-specific recognition. *J. Biol. Chem.* **267:** 392–398.

Neddermann, P. and J. Jiricny. 1993. The purification of a mismatch-specific thymine-DNA glycosylase from HeLa cells. *J. Biol. Chem.* **268:** 21218–21224.

Nicolaides, N.C., N. Papadopolous, B. Liu, Y.F. Wei, K.C. Carter, S.M. Ruben, C.A. Rosen, W.A. Haseltine, R.D. Fleischmann, and C.M. Fraser. 1994. Mutations of 2 pms homologs in hereditary nonpolyposis colon cancer. *Nature* **371:** 75–80.

Nishida, C., P. Reinhard, and S. Linn. 1988. DNA repair synthesis in human fibroblasts requires DNA polymerase δ. *J. Biol. Chem.* **263:** 501–510.

O'Donovan, A., A.A. Davies, J.G. Moggs, S.C. West, and R.D. Wood. 1994. XPG endonuclease makes the 3′ incision in human DNA nucleotide excision repair. *Nature* **371:** 432–435.

Palombo, F., M. Hughes, J. Jiricny, O. Truong, and J. Hsuan. 1994. Mismatch repair and cancer. *Nature* **367:** 417.

Papadopoulos, N., N.C. Nicolaides, Y.-F. Wei, S.M. Ruben, K.C. Carter, C.A. Rosen, W.A. Haseltine, R.D. Fleischmann, C.M. Fraser, M.D. Adams, J.C. Venter, S.R. Hamilton, G.M. Petersen, P. Watson, H.T. Lynch, P. Peltomaki, J.-P. Mecklin, A. de la Chapelle, K.W. Kinzier, and B. Vogelstein. 1994. Mutation of a *mutL* homolog in hereditary colon cancer. *Science* **263:** 1625–1629.

Park, C.H. and A. Sancar. 1994. Formation of a ternary complex by human XPA, ERCC1, and ERCC4(XPF) excision-repair proteins. *Proc. Natl. Acad. Sci.* **91:** 5017–5021.

Payne, A. and G. Chu. 1994. Xeroderma-pigmentosum group-e binding factor recognizes a broad spectrum of DNA damage. *Mutat. Res.* **310:** 89–102.

Podust, L.M., V.N. Podust, C. Floth, and U. Hübscher. 1994. Assembly of DNA polymerase δ and ε holoenzymes depends on the geometry of the DNA template. *Nucleic Acids Res.* **22:** 2970–2975.

Price, A. and T. Lindahl. 1991. Enzymatic release of 5′-terminal deoxyribose phosphate residues from damaged DNA in human cells. *Biochemistry* **30:** 8631–8637.

Robins, P., D.J.C. Pappin, R.D. Wood, and T. Lindahl. 1994. Structural and functional homology between mammalian DNase IV and the 5′ nuclease domain of *Escherichia coli* DNA polymerase I. *J. Biol. Chem.* **269:** 28535–28538.

Robins, P., C.J. Jones, M. Biggerstaff, T. Lindahl, and R.D. Wood. 1991. Complementation of DNA repair in xeroderma pigmentosum group A cell extracts by a protein with affinity for damaged DNA. *EMBO J.* **10:** 3913–3921.

Robson, C.N., A.M. Milne, D.J.C. Pappin, and I.D. Hickson. 1991. Isolation of cDNA clones encoding an enzyme from bovine cells that repairs oxidative DNA damage *in vitro*—Homology with bacterial repair enzymes. *Nucleic Acids Res.* **19:** 1087–1092.

Robson, C.N., D. Hochhauser, R. Craig, K. Rack, V.J. Buckle, and I.D. Hickson. 1992. Structure of the human DNA-repair gene *HAP1* and its localization to chromosome 14q 11 2-12. *Nucleic Acids Res.* **20:** 4417–4421.

Roy, R., J.P. Adamczewski, T. Seroz, W. Vermeulen, J.-P. Tassan, L. Schaeffer, E.A. Nigg, J.H.J. Hoeijmakers, and J.-M. Egly. 1994. The MO15 cell cycle kinase is associated with the TFIIH transcription-DNA repair factor. *Cell* **79:** 1093–1101.

Satoh, M.S., G.G. Poirier, and T. Lindahl. 1993. NAD+-dependent repair of damaged DNA by human cell extracts. *J. Biol. Chem.* **268:** 5480–5487.

Seki, S., S. Ikeda, S. Watanabe, M. Hatsushika, K. Tsutsui, K. Akiyama, and B. Zhang. 1991. A mouse DNA-repair enzyme (APEX nuclease) having exonuclease and apurinic apyrimidinic endonuclease activities—Purification and characterization. *Biochim. Biophys. Acta* **1079:** 57–64.

Shivji, M.K.K., A.P.M. Eker, and R.D. Wood. 1994. DNA repair defect in xeroderma pigmentosum group C and complementing factor from HeLa cells. *J. Biol. Chem.* **269:** 22749–22757.

Shivji, M.K.K., M.K. Kenny, and R.D. Wood. 1992. Proliferating cell nuclear antigen is required for DNA excision repair. *Cell* **69:** 367–374.

Sung, P., P. Reynolds, L. Prakash, and S. Prakash. 1993. Purification and characterization of the *Saccharomyces cerevisiae RAD1-RAD10* endonuclease. *J. Biol. Chem.* **268:** 26391–26399.

Sung, P., J.F. Watkins, L. Prakash, and S. Prakash. 1994. Negative superhelicity promotes ATP-dependent binding of yeast Rad3 protein to ultraviolet-damaged DNA. *J. Biol. Chem.* **269:** 8303–8308.

Svejstrup, J.Q., Z. Wang, W.J. Feaver, X. Wu, T.F. Donahue, E.C. Friedberg, and R.D. Kornberg. 1995. Different forms of TFIIH for transcription and DNA repair: Holo-TFIIH and a nucleotide excision repairosome. *Cell* **80:** 21–28.

Syväoja, J., S. Suomensaari, C. Nishida, J.S. Goldsmith, G.S.J. Chui, S. Jain, and S. Linn. 1990. DNA polymerases α, δ, and ε: Three distinct enzymes from HeLa cells. *Proc. Natl. Acad. Sci.* **87:** 6664–6668.

Takao, M., M. Abramic, M. Moos, V.R. Otrin, J.C. Wootton, M. McLenigan, A.S. Levine, and M. Protic. 1993. A 127 kDa component of a UV-damaged DNA-binding complex, which is defective in some xeroderma pigmentosum group E patients, is homologous to a slime-mold protein. *Nucleic Acids Res.* **21:** 4111–4118.

Tomkinson, A.E., A.J. Bardwell, L. Bardwell, N.J. Tappe, and E.C. Friedberg. 1993. Yeast DNA repair and recombination proteins Rad1 and Rad10 constitute a single-stranded-DNA endonuclease. *Nature* **362:** 860–862.

Toschi, L. and R. Bravo. 1988. Changes in cyclin/proliferating cell nuclear antigen distribution during DNA repair synthesis. *J. Cell Biol.* **107:** 1623–1628.

Turchi, J.J., L. Huang, R.S. Murante, Y. Kim, and R.A. Bambara. 1994. Enzymatic completion of mammalian lagging-strand DNA replication. *Proc. Natl. Acad. Sci.* **91:** 9803–9807.

Vermeulen, W., A.J. van Vuuren, M. Chipoulet, L. Schaeffer, E. Appeldoorn, G. Weeda, N.G.J. Jaspers, A. Priestly, C.F. Arlett, A.R. Lehmann, M. Stefanini, M. Mezzina, A. Sarasin, D. Bootsma, J.-M. Egly, and J.H.J. Hoeijmakers. 1994. Three unusual repair deficiencies associated with transcription factor BTF2(TFIIH). Evidence for the existence of a transcription syndrome. *Cold Spring Harbor Symp. Quant. Biol.* **59:** 317–329.

Waga, S., G. Bauer, and B. Stillman. 1994. Reconstitution of complete SV40 DNA-replication with purified replication factors. *J. Biol. Chem.* **269**: 10923–10934.

Walker, L.J., C.N. Robson, E. Black, D. Gillespie, and I.D. Hickson. 1993. Identification of residues in the human DNA-repair enzyme HAP1 (REF-1) that are essential for redox regulation of jun DNA-binding. *Mol. Cell. Biol.* **13:** 5370–5376.

Xanthoudakis, S., G.G. Miao, and T. Curran. 1994. The redox and DNA-repair activities of REF-1 are encoded by nonoverlapping domains. *Proc. Natl. Acad. Sci.* **91:** 23–27.

Yasuhira, S. and A. Yasui. 1992. Visible light-inducible photolyase gene from the goldfish *Carassius auratus. J. Biol. Chem.* **267:** 25644–25647.

Yasui, A., A.P.M. Eker, S. Yasuhira, H. Yajima, T. Kobayashi, M. Takao, and A. Oikawa. 1994. A new class of DNA photolyases present in various organisms including aplacental mammals. *EMBO J.* **13:** 6143–6151.

9

Chromatin Structure and DNA Replication: Implications for Transcriptional Activity

Alan P. Wolffe
Laboratory of Molecular Embryology
National Institute of Child Health and Human Development
National Institutes of Health
Bethesda, Maryland 20892-2710

The period in the cell cycle when the genome is replicated (S phase) is crucially important for the establishment and maintenance of programs of differential gene activity. Not only must DNA be replicated, but the chromosome itself must be duplicated. The majority of genes in the proliferating cell of a defined type retain the same states of transcriptional activity through cell division. This requires the duplication of the precise nucleoprotein complexes directing gene transcription or repression on the nascent DNA templates. The maintenance of these specific regulatory complexes through replication reflects the commitment of a defined cell type or line to a particular state of determination. Preexisting chromosomal structures are transiently disrupted by transit through the replication elongation complex. Most of these structures are faithfully reassembled following replication through mechanisms discussed in this chapter. However, the transient disruption of these structures also offers a window of opportunity for modifying regulatory nucleoprotein complexes. These alterations can either activate genes through the disruption of repressed states, or direct the repression of previously active genes. Thus, cell division offers a molecular mechanism to redirect the commitment of a cell toward a particular determined state. A consideration of the processes occurring at the eukaryotic replication fork suggests how this important development process might be accomplished.

DNA Replication in Eukaryotic Cells
 0-87969-459-9/96 $5 + .00

IMPLICATIONS OF DNA REPLICATION FOR STABLE STATES OF TRANSCRIPTIONAL ACTIVITY

Active and Repressed States of Eukaryotic Genes

The local nucleoprotein complexes required to maintain a eukaryotic gene in an active or repressed state have been defined in some detail (Tjian and Maniatis 1994; Wolffe 1994a). Transcriptional activity for a given gene depends on a number of sequence-specific transcription factors (e.g., SP1), structural proteins (e.g., HMGI/Y), and non-DNA-binding proteins associated with the promoter interacting to recruit general transcription factors (e.g., TFIIA, TFIIB) together with the TFIID complex (containing TBP [*T*ATA *b*inding *p*rotein] and the TAFs [*T*ATA *a*ssociated *f*actors]). The assembly of this large nucleoprotein complex is initiated through the association of DNA-binding proteins and requires many intermediate steps leading to the recruitment of RNA polymerase II and, eventually, to transcription itself. Conversely, several features may determine a gene to be transcriptionally inactive. A common mechanism appears to be a deficiency in an essential component required for the assembly of the active complex. If this component is a DNA-binding protein, the cognate DNA sequence might become associated with the histone proteins. Specific nucleosomal structures assembled by the core histones (H2A, H2B, H3, and H4) might restrict the subsequent association of either sequence-specific DNA-binding proteins or the basal transcriptional machinery (Simpson 1991; Wolffe 1994b). Other proteins may stabilize repressive higher-order chromatin structures dependent on prior association of the core histones (Hansen and Wolffe 1994); these include linker histone variants (Khochbin and Wolffe 1994) or the chromodomain (*chro*matin *m*odification *o*rganizer) proteins such as HP1 and Polycomb (James and Elgin 1986; Paro and Hogness 1991).

Generally, the assemblies of nucleosomes or transcription complexes on the promoter of a eukaryotic gene are mutually exclusive. The prior assembly of nucleosomes can prevent transcription factors from binding to DNA and, conversely, the prior assembly of a transcription complex prevents nucleosome formation from repressing transcription (Fig. 1). Although these results provide an excellent molecular basis for the maintenance of stable states of gene expression in a terminally differentiated nondividing cell, they do not explain why either transcriptionally active or inactive states are assembled onto DNA in the first place, nor do they explain how such states can be propagated through cell division. Clearly, because both nucleoprotein structures can incorporate the same DNA molecule, the possibility exists of a competition occurring between the assembly of the two structures. This competition, in fact, occurs dur-

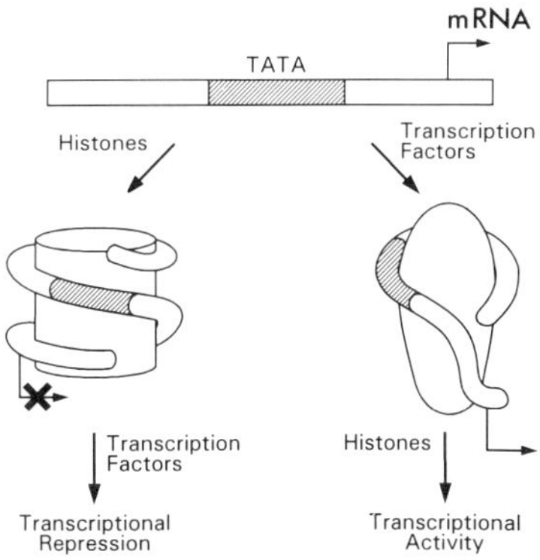

Figure 1 Nucleosome assembly and transcription complex assembly are often mutually exclusive. Two alternate pathways are shown for the association of DNA-binding proteins with a promoter containing a TATA homology. The start site of transcription of mRNA is indicated by the bent arrow.

ing the staged assembly of either active or repressed states following replication (see Almouzni et al. 1990a; Aparicio and Gottschling 1994). Molecular mechanisms that influence the outcome of this competition direct the commitment of a cell to a particular state of determination or facilitate developmentally regulated switches in cell fate. However, to appreciate how this competition occurs, we must first discuss the consequences of DNA replication for preexisting chromatin structures.

Impact of DNA Replication on Preexisting Chromatin Structures

Chromatin consists of long arrays of nucleosomal DNA interspersed with specific regulatory nucleoprotein complexes. The replication fork moves through chromatin without apparent impediment. Replication fork progression disrupts preexisting nucleosomes; however, the fate of regulatory nucleoprotein complexes depends on the particular structure examined.

Nucleosomes

Major considerations for preexisting nucleosomes during the replication process are whether the histones present in the nucleosome stay together on nascent DNA, and whether nucleosomes are randomly or conservatively segregated to daughter DNA strands. DNA replication requires the transient unwinding of duplex parental DNA into two single-stranded regions. Although histones associate with single-stranded DNA (Almouzni et al. 1990b), they do not assemble nucleosomes. This property, coupled with the competing protein-DNA interactions involved in DNA synthesis at the replication fork, probably accounts for nucleosome disruption. Histones released from the parental chromatin during replication in vitro can be easily sequestered onto competitor DNA (Gruss et al. 1993). However, in vivo these histones are sequestered onto daughter DNA molecules close to the replication fork (see Fig. 4) (Sogo et al. 1986; Perry et al. 1993).

A nucleosome contains an octamer consisting of two molecules each of the four core histones (H2A, H2B, H3, and H4) and a single molecule of a fifth linker histone (H1). The four core histones and the linker histone have very selective interactions with each other (Fig. 2A) (Arents et al. 1991; Arents and Moudrianakis 1993). Our most detailed understanding of nucleosomal architecture and construction has relied on in vitro experiments that have attempted to reconstruct nucleosomes with purified histones. These experiments have been informative, although they involve dialysis from high salt to low salt concentrations and do not employ the molecular chaperones used in vivo (see below). The central "kernel" of the nucleosome is made up of two heterodimers of histones H3 and H4. Only when this "tetramer" is bound to DNA can two heterodimers of H2A and H2B bind to complete assembly of the histone octamer (Hayes et al. 1991). One heterodimer of H2A and H2B binds to either side of the histone tetramer in an interaction dependent on both protein-protein and protein-DNA contacts. Only when the complete octamer of core histones has assembled on DNA can a single molecule of linker histone be stably bound (Fig. 2B) (Hayes and Wolffe 1993). The exact position of the linker histone within the nucleosome is currently the subject of controversy (Pruss et al. 1995); however, in the one case in which it has been mapped within a specific nucleosome, it occupies an asymmetric position within the nucleosome (Fig. 2C) (Hayes et al. 1994).

In vivo a comparable assembly of the nucleosome occurs. A tetramer $(H3, H4)_2$ and a dimer (H2A, H2B) are stable at physiological ionic strengths (Eickbusch and Mondrianakis 1978). However, they will not associate together in the absence of DNA under physiological conditions. The tetramer $(H3, H4)_2$ must again associate with DNA on newly replicated DNA before H2A and H2B can complete the nucleosome core (Almouzni et al. 1990a). Histone H1 is the last protein to be stably sequestered, completing the nucleosome (Fig. 3) (Worcel et al. 1978). I expand on the details of replication-coupled chromatin assembly later in this chapter.

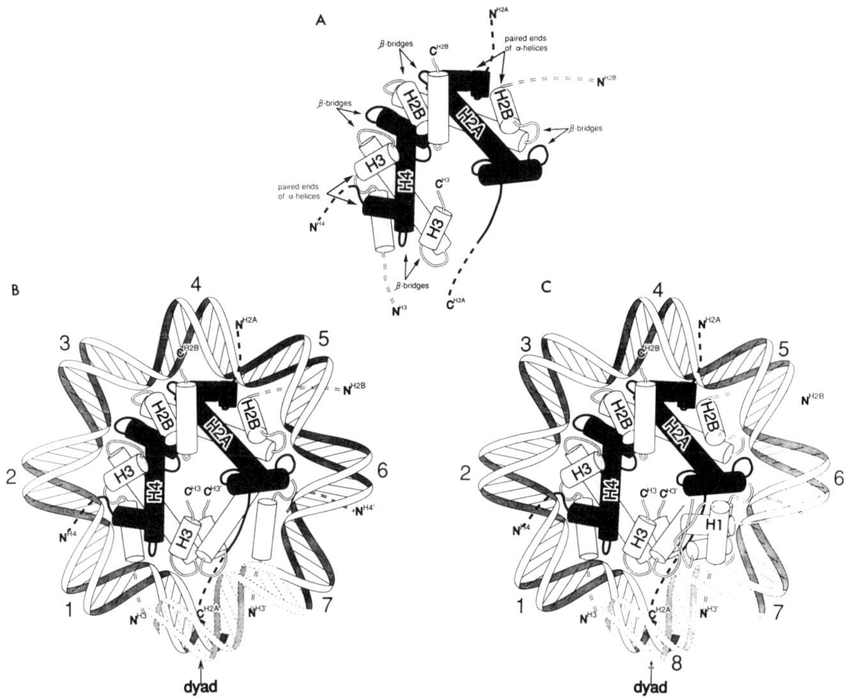

Figure 2 Nucleosomal architecture. (*A*) The histone fold and DNA-binding motifs. The relative juxtaposition of the two histone heterodimers as viewed "from the top" (i.e., along the superhelical axis of the DNA) is shown. The approximate positions of the flexible histone tails are shown by broken lines. Note the six regularly spaced domains (*double arrows*) predicted to be involved in DNA binding. (*B*) A view down the superhelical axis of the nucleosome core. The helical turns of DNA are numbered relative to the dyad axis (0). (*C*) One potential position for the linker histone H1 globular domain within the nucleosome (Pruss et al. 1995).

Jackson (1987, 1990) determined that a substantial fraction of the preexisting octamers associated with DNA within the chromosome in vivo fell apart following replication into dimers (H2A, H2B) and tetramers $(H3, H4)_2$. Tetramers from these preexisting nucleosomes rapidly reassociate with daughter DNA duplexes (Gruss et al. 1993). Newly synthesized dimers (H2A, H2B) can then be sequestered to complete the octamer, mixing old and new histones into a single structure (see Fig. 4). The disruption of preexisting nucleosomal structure at the replication fork, coupled to dissociation of the histones from DNA,

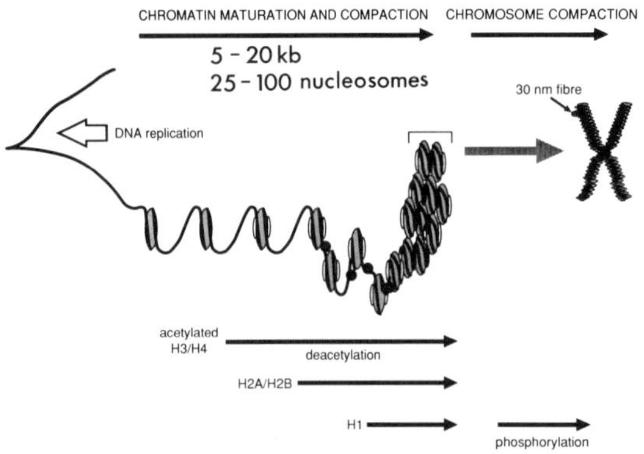

Figure 3 An in vivo pathway for de novo chromatin assembly coupled to replication. Chromatin assembly on nascent DNA. Acetylated histones H3 and H4 (*stippled ellipsoids*) are sequestered by the DNA first, histones H2A and H2B (*open ellipsoids*) follow, and, finally, histone H1 (*dark circle*) binds, stabilizing chromatin folding within the irregular 30-nm fiber. During this progressive assembly of chromatin, DNA is compacted, nucleosome formation leads to a sevenfold compaction of DNA, and the subsequent formation of the 30-nm fiber contributes a further sevenfold compaction. These compactions represent the major topological constraints of DNA in a eukaryotic nucleus. Mature chromatin is predominantly deacetylated. During mitosis, histone H1 is phosphorylated and 30-nm fibers pack together to assemble the mitotic chromosome. As much as 5–20 kb of nascent DNA, including 25–100 nucleosomes, may be present as "immature" chromatin associated with the replication fork at various levels of compaction (see text for details).

strongly suggests that the dispersive segregation of these histones to both daughter DNA duplexes occurs during replication (Cusick et al. 1984; Sogo et al. 1986; Burhans et al. 1991; Krude and Knippers 1991; Randall and Kelly 1992; Sugasawa et al. 1992). Importantly, the incorporation of preexisting histone tetramers $(H3, H4)_2$ into nascent chromatin provides a means of maintaining and propagating a stable state of gene activity. The old H3 and H4 present in the nascent chromatin retain their preexisting posttranslational modification state (Perry et al. 1993). This differs from that of newly synthesized H3 and H4 and can potentially influence subsequent transcription of the associated DNA (see below). The dispersive segregation of "old" histones coupled to maintenance of their preexisting states of modification provides a molecular mechanism whereby an epigenetic imprint might be propagated through replication (Fig. 4, see below).

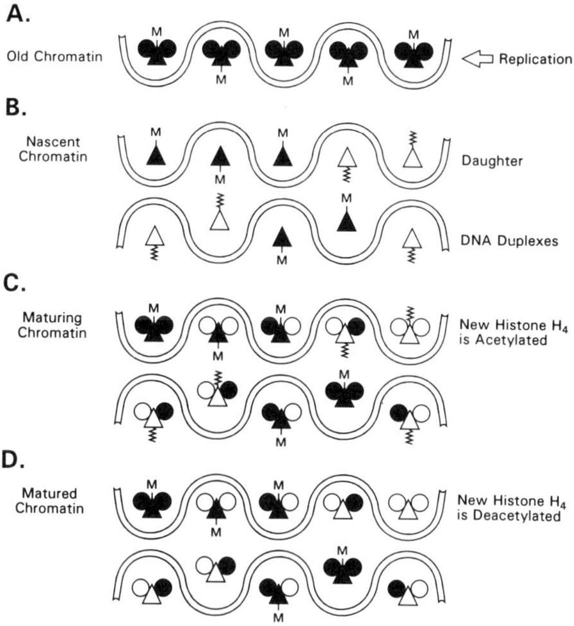

Figure 4 Nucleosome disruption during replication and reassembly following replication. (*A*) Old chromatin consisting of preexisting nucleosomes (histone octamer plus DNA) containing a tetramer $(H3, H4)_2$ (*filled triangle*) and two dimers (H2A, H2B) (*filled circles*). The histones in the tetramer are modified (M). Replication displaces these histones from DNA; the octamer can fall apart into tetramers and dimers. (*B*) Nascent chromatin. Old tetramers associate with both daughter DNA duplexes. Newly synthesized tetramers (*open triangles*) containing diacetylated histone H4 (*zigzag line*) also associate with daughter DNA in a process facilitated by CAF-1. (*C*) Maturing chromatin. Old and new dimers (*open circles*) bind to the tetramers. (*D*) Matured chromatin. New tetramers are deacetylated.

Regulatory Complexes

A special case for a regulatory nucleoprotein complex maintaining association with DNA throughout the cell cycle is the protein assembly that regulates use of an origin of replication itself. Stable association of proteins with an origin through the cell cycle has been established through in vivo footprinting methodologies on the replication origin of Epstein-Barr virus (Hsieh et al. 1993) and on a yeast chromosomal ARS element (Diffley and Cocker 1992). Implicit in the maintenance of these regulatory complexes through S phase is the concept that they duplicate themselves. An attractive mechanism for the maintenance of regulatory complexes through replication requires multiple copies of a given *trans-*

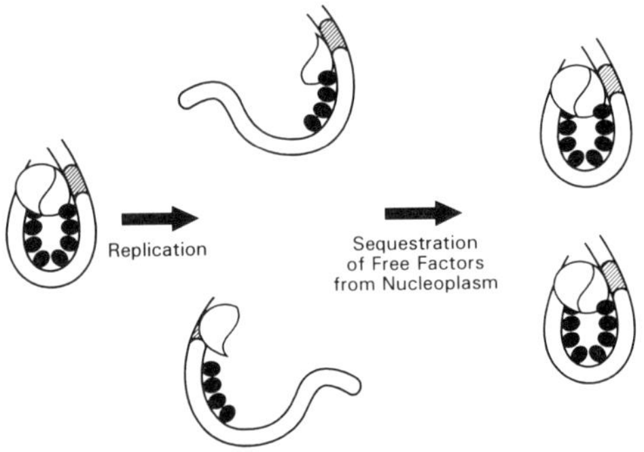

Figure 5 A regulatory nucleoprotein complex could make use of multiple protein-DNA interactions to maintain integrity through replication. Following replication, proteins partition to daughter DNA duplexes. Free factors are then sequestered from the nucleoplasm to reassemble two daughter complexes.

acting factor to bind to the regulatory DNA sequences (Fig. 5) (Brown 1984). This could be determined by sequence or structural selectivity. If the preexisting multimeric protein complex is split during replication, copies of the *trans*-acting factors could be segregated to both daughter DNA duplexes. These *trans*-acting factors could then either directly sequester other factors from the nucleoplasm making use of protein-protein interactions, or they could maintain the regulatory DNA sequences accessible in the face of ongoing chromatin assembly, such that when other factors became available they could bind to DNA. Structurally driven protein association is consistent with the maintenance of DNA distortion throughout the cell cycle at the Epstein-Barr viral origin (Hsieh et al. 1993).

In contrast to origin complexes, the basal transcriptional machinery appears to be removed from promoter elements by the passage of a replication fork (Wolffe and Brown 1986). Replication is found to be dominant to the transcription process, and a direct consequence of replication fork progression through an active 5S rRNA gene is the displacement of transcription factors. Several correlations from in vivo work support the generality of this observation. There is a clear antagonism between transcription and replication on efficiently replicating SV40 DNA molecules (Lebkowski et al. 1985; Lewis and Manley 1985). Replication forks invade the transcriptionally active ribosomal RNA genes in yeast (Saffer and Miller 1986; Lucchini and Sogo 1994). Thus, replication apparently resets the transcriptional status of a chromosome to "ground zero." The component protein molecules that determine transcriptional activity have to reassemble regulatory complexes de novo on the daughter DNA duplexes. This reassembly occurs not on naked DNA, but on a nascent chromatin template.

Chromatin Assembly Has Replication-dependent and -independent Pathways

Replication-independent Pathways

Early work on physiological chromatin assembly pathways made use of cell-free preparations from *Xenopus* oocytes and eggs (Laskey et al. 1978; Glikin et al. 1984). More recently, extracts of *Drosophila* embryos have been used with similar results (Becker and Wu 1992). For both systems, chromatin assembly on nonreplicating DNA is relatively slow, taking several hours to assemble nucleosomes to a physiological density (one nucleosome per 180–200 bp). This contrasts with the rapid assembly of chromatin in vivo during early embryogenesis in *Xenopus* and *Drosophila*, where entire cell cycles take only 30 minutes and 10 minutes, respectively. Thus, the molecular mechanisms that mediate chromatin assembly in the absence of DNA replication have questionable physiological relevance. Nevertheless, these systems have provided useful information on the biochemistry of the assembly process.

In *Xenopus* oocytes, histones are synthesized under the control of distinct regulatory mechanisms that operate outside of S phase. Tetramers $(H3, H4)_2$ are stored in a complex with the molecular chaperone N1/N2 (Kleinschmidt et al. 1986). Dimers (H2A, H2B) are stored in a complex with the chaperone nucleoplasmin (Dilworth et al. 1987). Both chaperones exchange histones onto DNA at physiological ionic strength. N1/N2 must function before nucleoplasmin to assemble a nucleosome (Kleinschmidt et al. 1990). During normal development, nucleoplasmin has a specialized role in the remodeling of *Xenopus* sperm chromatin, where it facilitates the exchange of sperm-specific basic proteins for histones H2A and H2B (Philpott and Leno 1992). Nucleoplasmin and N1/N2 allow large amounts of histones to be stably sequestered in the *Xenopus* oocyte and egg; however, a role for these proteins in directly mediating chromatin assembly during early embryogenesis remains to be established.

Replication-dependent Pathways

In vivo in normal somatic cells, the vast bulk of the histone proteins are synthesized during S phase. These histones are immediately assembled onto nascent DNA at the replication fork (Ruiz-Carrillo et al. 1975; Jackson et al. 1976). Stillman (1986) discovered that the chromatin assembly process is coupled to replication. The molecular chaperone mediating the process is chromatin assembly factor 1 (CAF-1), which requires ongoing DNA replication to function (Smith and Stillman 1989). CAF-1 directs the association of the tetramer (H3, H4)$_2$ with replicating DNA. Dimers (H2A/H2B) then bind in a CAF-1-independent process to complete the histone octamer (Smith and Stillman 1991; see also Fotedar and Roberts 1989). CAF-1 requires a modified tetramer (H3, H4)$_2$ from the cytosol of human cells in order to function (Kaufman and Botchan 1994). This is potentially a key regulatory event in distinguishing the biochemistry of replication-dependent and -independent chromatin assembly pathways. It is possible that the phosphorylation and diacetylation of histone H4 coupled to its synthesis (Ruiz-Carrillo et al. 1975; Jackson et al. 1976; Dimitrov et al. 1994) may be necessary for chromatin assembly. Whether CAF-1 has specific interactions either with highly modified H4 and/or with the replication machinery itself are important questions yet to be resolved.

Almouzni and colleagues (Almouzni and Méchali 1988a,b; Almouzni et al. 1990b, 1991) established that replication-coupled pathways of chromatin assembly also exist in *Xenopus*. However, the molecular chaperones that couple replication to chromatin assembly, such as the CAF-1 found in somatic cells, remain to be defined. These replication-dependent pathways direct the efficient assembly of nucleosomes both in vitro and in vivo with kinetics that could easily accommodate a cell cycle duration of 30 minutes (Almouzni and Méchali 1988a; Almouzni et al. 1990b; Almouzni and Wolffe 1993). The mechanism of enhanced assembly involves both the rapid deposition of the histone tetramer (H3, H4)$_2$ and facilitation of the subsequent deposition of dimers (H2A, H2B) (Almouzni et al. 1990b). Similar results consistent with a facilitated two-step assembly of chromatin have been obtained in mammalian systems (Gruss et al. 1990).

The de novo assembly of chromatin on replicating templates in vitro provides a useful independent confirmation of earlier work on the staged assembly of chromatin during S phase in vivo. As discussed earlier, DNA replication disrupts preexisting nucleoprotein structures within the chromosome. Histones that are displaced during replication reassociate with newly synthesized DNA, but do so randomly on both daughter DNA duplexes. A consequence of this segregation is that nascent chromatin has a 50% enrichment of preexisting histones. The remainder of the histones incorporated into chromatin are newly synthesized. Radiolabeling of these newly synthesized histones has allowed the kinetics of their incorporation into chromatin and subsequent modification to be determined.

Newly synthesized and preexisting histone tetramers (H3, $H4)_2$ associate with nascent DNA (Worcel et al. 1978; Jackson 1987, 1990); this is followed over the space of several minutes by the sequestration of both preexisting and newly synthesized histone dimers (H2A, H2B). Thus, the majority of nucleosomes behind a replication fork are hybrids of both old and new core histones. Finally, a mixture of newly synthesized and preexisting histone H1 stably associates with the nascent chromatin.

The overall process of chromatin maturation as assayed by nuclease sensitivity requires as long as 10–20 minutes in a rapidly proliferating mammalian cell (Cusick et al. 1983). Assuming a rate of replication fork movement of 0.5–1 kb of DNA per minute, this implies that 25–100 nucleosomes are present on both of the nascent DNA duplexes as "immature" chromatin during S phase (see Fig. 3). The initial rapid deposition of old and new histones H3 and H4 on newly synthesized DNA reflects the nuclease-sensitive stage, whereas the subsequent deposition of histone dimers (H2A, H2B) and histone H1 correlates with the appearance of regular nucleosomal arrays and nuclease resistance (Smith et al. 1984). The sequential sequestration of histones is clearly once again related to the structure of the nucleosome, since the tetramer (H3, $H4)_2$ forms the core of the structure, whereas histones H2A and H2B bind at the periphery of the nucleosome, and histone H1 can only associate in its proper place after two turns of DNA are wrapped around the core histones (Fig. 2) (Hayes et al. 1991; Hayes and Wolffe 1993).

Newly synthesized histone H4 is phosphorylated and acetylated in the amino-terminal tail domain (Ruiz-Carillo et al. 1975; Jackson et al. 1976). Approximately 30 minutes after deposition during chromatin assembly, the diacetylated H4 is deacetylated to its mature form. If H4 deacetylation is inhibited, chromatin never achieves the nuclease resistance of bulk chromatin, indicative of the formation of stable higher-order structures. Histone H1 may be less efficiently incorporated into chromatin containing acetylated H4 (Perry and Annunziato 1989; but see Ura et al. 1994). Thus, histone diacetylation is likely to maintain nascent chromatin in a structure that is more accessible to other DNA-binding proteins.

In summary, chromatin assembly in vivo is coupled to replication, most probably through the activity of specific molecular chaperones such as CAF-1. Nucleosome assembly occurs in stages and involves transient posttranslational modifications of core histones synthesized during S phase (Figs. 3, 4).

Epigenetic Mechanisms: The Assembly of Active and Repressed Transcriptional States

In vivo experiments using *Saccharomyces cerevisiae* suggest that replication disassembles repressed chromatin states and facilitates the access of *trans*-acting factors to DNA (Aparicio and Gottschling 1994). Other experiments using yeast suggest that replication has an essential role in facilitating the repression of specific genes (Miller and Nasmyth 1984). We have discussed how biochemical experiments indicate that replication introduces a dynamic aspect to chromosomal structure, both directing the disassembly of preexisting structures and facilitating the assembly of nucleosomes. A central issue in gene regulation is how the assembly of nucleoprotein structures following replication can maintain or alter states of potential transcriptional activity.

Repression

Replication and transcription are most clearly seen to be linked in yeast. Components of the yeast origin recognition complex (ORC) regulate both the initiation of replication within the chromosome and the repression of transcription within the same chromosomal domain (Bell et al. 1993). The molecular mechanisms responsible for the repression of transcription directed by ORC are unknown. Two possible explanations are (1) that the ORC compartmentalizes adjacent chromatin into a transcriptionally incompetent environment within the nucleus, or (2) that the ORC influences the type of chromatin assembled adjacent to it. The ORC complex may be a greatly streamlined version of the replication factories of larger eukaryotes (Cook 1991). These replication factories represent special nuclear compartments at which proteins involved in the replication process are sequestered. It is possible that a gene adjacent to the origin is directed by the ORC to reside in a replication-competent but transcriptionally incompetent environment. Alternatively, if replication itself is essential for transcriptional repression (Miller and Nasmyth 1984; Laurensen and Rine 1992), then the coupling of chromatin assembly to the replication process could contribute to repression. Preexisting transcriptionally active complexes would be displaced by the replication fork. *trans*-Acting factors would then have to compete for assembly against the deposition of histones. In vivo and in vitro experiments in *Xenopus* demonstrate that the coupling of nucleosome assembly to replication can very effectively repress basal transcription (Almouzni et al. 1990a; Almouzni and Wolffe 1993). As discussed earlier, the ORC complex provides one biological example of the maintenance of sequence-specific or structure-dependent protein-DNA interactions through the replication process. However, since the ORC also serves to initiate the replication process, maintenance of the ORC may occur under circumstances distinct from the transcription complexes or chromatin structures that are exposed to the fully assembled replication-elongation complex.

We have discussed how the histones already on the template during replication are segregated randomly to the daughter DNA duplexes, but within the context of small groups of nucleosomes. This maintenance of histone modification states potentially influences *trans*-acting factor access to DNA. Moreover, if proteins that modify the subsequent folding of nucleosomal arrays or that modify histones themselves, for example, by acetylation or deacetylation, are also partitioned in this way, the properties of a chromatin domain might be stably propagated. For example, histone H4 acetylation may interfere with the association of histone H1 with chromatin (Perry and Annunziato 1989). Histone H1 is known to repress specific genes in vivo (Bouvet et al. 1994). Other proteins that might recognize properties of the "old" histones within nascent chromatin include the chromodomain proteins that initiate the formation of heterochromatin. These are also good candidates for propagating preexisting states of chromatin-mediated transcriptional repression (Fig. 6).

Activation

In vitro experiments using cell-free preparations of *Xenopus* eggs indicate that stable states of gene activity can be propagated in a nuclear environment (Wolffe 1993; Barton and Emerson 1994). How might this occur?

The simplest situation leading to continued gene activity would be the case in which a superabundance of transcription factors specific for a given gene was available within the nucleus throughout the cell cycle, including S phase. The factors would always be able to bind to their regulatory elements should they become accessible, recruiting the basal transcriptional machinery to the nascent promoter DNA, and thereby preventing histones or other proteins from binding to the TATA homology. Several features of nascent chromatin facilitate the association of transcription factors (Wolffe 1991). For example, the complex of the 5S rRNA gene with the tetramer $(H3, H4)_2$ is not repressive to transcription (Wolffe 1989; Tremethick et al. 1990; Almouzni et al. 1991), whereas the complete octamer of core histones $(H2A, H2B, H3, H4)_2$ is repressive at high densities of octamers bound to DNA (Clark and Wolffe 1991; Hayes and Wolffe 1992). Moreover, acetylation of the core

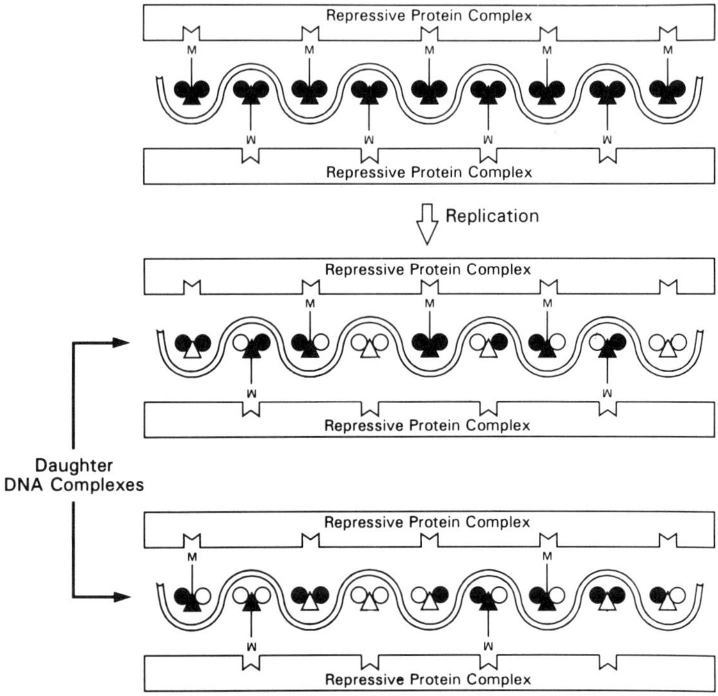

Figure 6 Preexisting histone modifications could provide an epigenetic imprint. A repressive protein complex (e.g., containing chromodomain proteins) recognizes a histone modification (M). Following replication, modified histones are segregated to both daughter DNA duplexes sufficient to sustain interaction with the repressive protein complex.

histones facilitates transcription factor access to DNA even when the complete octamer is bound (Lee et al. 1993). The histone tetramer (H3, $H4)_2$ recognizes the DNA sequences that position the nucleosome containing the 5S rRNA gene (Hayes et al. 1991), hence it is probable that the formation of a specific chromatin structure also has a role in allowing transcription factors access to the template. Thus, following replication, it is probable that sequence-specific DNA-binding proteins have an opportunity to reassociate with daughter DNA molecules (Fig. 7). Replication might under certain circumstances facilitate gene activation (Enver et al. 1988; Wilson and Patient 1993; Aparicio and Gottschling 1994).

The regulated activity of a transcription factor, such that it becomes able to bind to DNA or to function during S phase, could lead to transcription activation in a way that is replication-dependent. In a developmental context, this event might be coupled to a particular embryonic

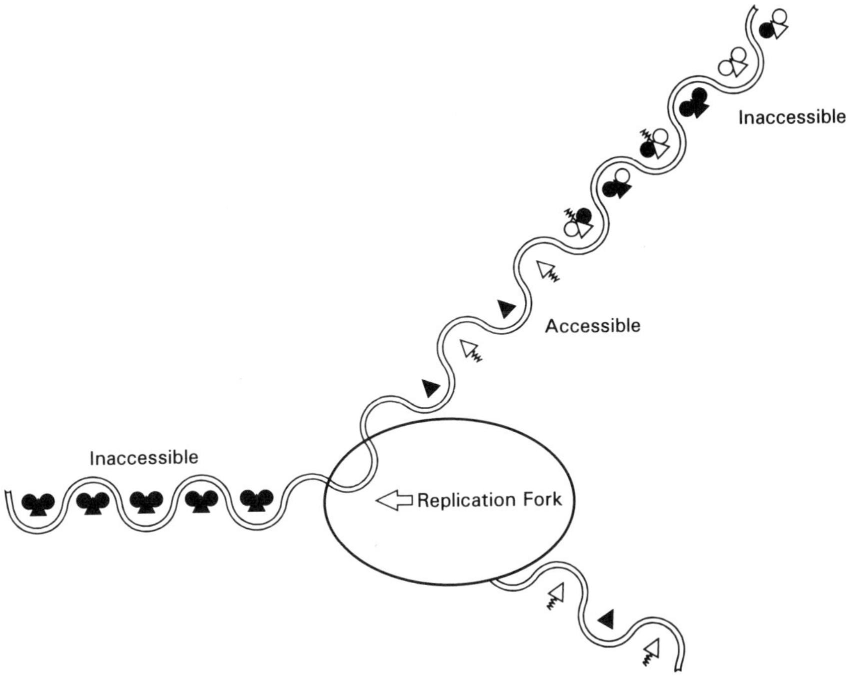

Figure 7 Replicative disruption of preexisting chromatin structures provides a window of opportunity for transcription factors to program genes. Nucleosome assembly is represented as in Fig. 4. The accessibility of nascent, maturing, and mature chromatin to *trans*-acting factors is indicated.

cleavage cycle or to a regulated period of cell division. For example, in *Caenorhabditis elegans* and the sea urchin, replication events are correlated with changes in the commitment of cells to a particular developmental fate (Mita-Miyazawa et al. 1985; Edgar and McGhee 1988). Similar changes can occur in differentiated cells that express one set of specialized genes and that can switch to another program of gene expression only after one or more cell divisions (e.g., Wolffian regeneration of the lens; Takata et al. 1964). However, replication events are not necessarily essential for changing gene expression within a particular cell (Chiu and Blau 1984; Blau et al. 1985). This is not surprising, since chromatin structure is not completely inert in vivo. Histones H2A, H2B, and H1 are known to exchange with a pool of free histones in a cell (Louters and Chalkley 1985). Complexes of DNA with only histones H3 and H4 therefore exist for a limited amount of time. However, a comparison of the rate and efficiency of gene activation in the presence or absence of cell proliferation has not yet been made.

The maintenance of specific transcription factor-DNA interactions through replication, as discussed earlier for the ORC, might be facilitated by considering the promoter, the enhancer, and locus control regions not as separate entities, but as contributory components to a single structure (Wolffe 1990). This could be achieved through protein-protein interactions between the distinct nucleoprotein complexes assembled at each regulatory element. One reason for the separation of these regulatory elements over extensive distances may be that any single structure might be independently disrupted by DNA replication, while the other would remain intact. If protein binding to one sequence element influences the binding of proteins to the other, then the intact nucleoprotein complex might facilitate the re-formation of the disrupted one.

Replication Timing

Chromatin organization outside the ORC may also have significance for the initiation of replication and the timing of this initiation in S phase. If replication disrupts both active and repressed chromatin structures, then the entire nucleus has to be remodeled after each replication event. I have suggested a means of accomplishing this remodeling; however, the re-formation of nuclear structures has other implications. If there are limiting transcription factors available in a cell, then a gene that is replicated early in S phase has more opportunity for the assembly of an active transcription complex than a gene that replicates late. This is simply because the gene that replicates early is available for transcription factors to bind before all of the early-replicating portion of the genome has sequestered these factors. A late-replicating gene therefore experiences a relative deficiency in factor availability (Gottesfeld and Bloomer 1982; Wormington et al. 1982). Conversely, it is also possible that the type of chromatin assembled early in S phase is more accessible to transcription factors than chromatin assembled late in S phase. For example, early-replicating chromatin may sequester histones that are more highly acetylated and, consequently, more accessible to the transcription factors that maintain continued transcription activity. Transcriptionally active genes replicate early in S phase (Goldman et al. 1984; Gilbert 1986; Wolffe 1993). The reason for this early replication is unknown, but possibilities include the local disruption of chromatin structure by transcription complexes, such that the DNA within those chromatin domains becomes more accessible to the replication machinery (Wolffe and Brown 1988). Many transcription factors may also be replication factors (DePamphilis 1988, 1993); consequently, local concentrations of transcription factors may favor the assembly of replication initiation complexes.

The issue very much is one of which came first: the chicken or the egg, or both? It is possible to argue that active transcription complexes open chromatin to admit replication factors, or, alternatively, that these sites are replicated first and are thus more accessible to transcription factors. Whether either or both of the much discussed mechanisms operate in vivo remains to be established.

CURRENT PROBLEMS AND FUTURE PROSPECTS

Chromatin structure is now realized to reflect a dynamic interaction between the many protein complexes that both organize DNA and fulfill regulatory roles. A much simplified picture suggests that replication disrupts local chromatin structures that preexist on the chromosome before replication. The subsequent reassembly of the nucleosome necessitates a staged process using modified histones that might be more accessible to transcription factors. This would provide a window of opportunity for reestablishing particular states of transcriptional activity. On a more global scale (>1–2 kb), chromatin proteins that retain a particular modification (e.g., acetylation) or that cooperatively influence chromosome structure toward activation or repression could provide an imprint on chromatin activity through DNA replication and chromosome duplication. Replication is established as having a major impact on preexisting nucleoprotein structures and a major role in their reassembly. Although significant attention has been given to the enzymology of the duplication of DNA, relatively little progress has been made concerning the enzymology of chromosomal duplication. The molecular mechanisms of chromatin assembly are not defined in any detail. The definition of molecular chaperones such as CAF-1 is a major advance; however, how CAF-1 functions is unknown. Does CAF-1 have a catalytic or structural role? Does it interact with the elongation complex? What are the special features of the histones that allow CAF-1 to utilize them for nucleosome assembly? On a more mundane level, we do not know the precise sequences or structures of the histone proteins necessary for chromatin assembly. The enzymes that transform nascent chromatin into a mature structure are yet to be defined at the molecular level. How mature chromatin is recognized by other proteins that influence states of gene repression, such as the chromodomain proteins, is unknown.

At this time, only the simple 5S rRNA genes of *Xenopus* have been extensively analyzed with respect to the significance of intermediates in chromatin assembly for the capacity to program genes for future transcription. This work is greatly in need of extension to a broader spectrum of eukaryotic genes. Preliminary work in *Drosophila* is consistent with a transition from programmable to stable repressed states as chromatin matures (Kamakaka et al. 1993). Caution must be used with many of the interpretations concerning the general impact of chromatin structure on transcription, since in vitro chromatin assembly systems currently make use of oocyte or embryonic extracts. These systems contain histone variants or modifications not found in normal somatic cells (Dimitrov et al. 1994). It is to be hoped that chromatin assembly systems coupled to replication that employ biochemically defined histones from normal somatic cells will be developed.

The impact of DNA replication on gene expression is readily analyzed through yeast genetics (Miller and Nasmyth 1984; Aparicio and Gottschling 1994; Laurenson and Rine 1992); however, homologous biochemical systems are currently lacking to test the many hypotheses proposed to explain the phenomena observed. Much progress in the biochemistry of yeast replication can be anticipated. The further reconstruction of determinative events in development will require continued consideration of the fate of regulatory nucleoprotein complexes during replication (Diffley and Cocker 1992). This is an important focus for future research. At a biochemical level, in vitro systems capable of maintaining states of gene expression through replication offer considerable promise (Wolffe 1993; Barton and Emerson 1994). The future clearly has the exciting prospect of understanding and thus reconstructing chromosomal duplication in all its complexity at a molecular level.

REFERENCES

Almouzni, G. and M. Méchali. 1988a. Assembly of spaced chromatin promoted by DNA synthesis in extracts from *Xenopus* eggs. *EMBO J.* **7:** 664–672.

———. 1988b. Assembly of spaced chromatin: Involvement of ATP and DNA topoisomerase activity. *EMBO J.* **7:** 4355–4365.

Almouzni, G. and A.P. Wolffe. 1993. Replication coupled chromatin assembly is required for the repression of basal transcription *in vivo*. *Genes Dev.* **7:** 2033–2047.

Almouzni, G., M. Méchali, and A.P. Wolffe. 1990a. Competition between transcription complex assembly and chromatin assembly on replicating DNA. *EMBO J.* **9:** 573–582.

———. 1991. Transcription complex disruption caused by a transition in chromatin structure. *Mol. Cell. Biol.* **11:** 655–665.

Almouzni, G., D.J. Clark, M. Méchali, and A.P. Wolffe. 1990b. Chromatin assembly on replicating DNA *in vitro*. *Nucleic Acids Res.* **18:** 5767–5774.

Aparicio, O.M. and D.E. Gottschling. 1994. Overcoming telomeric silencing: A *trans*-activator competes to establish gene expression in a cell cycle dependent way. *Genes Dev.* **8:** 1133–1146.

Arents, G. and E.N. Moudrianakis. 1993. Topography of the histone octamer surface:

Repeating structural motifs utilized in the docking of nucleosomal DNA. *Proc. Natl. Acad. Sci.* **90:** 10489–10493.

Arents, G., R.W. Burlingame, B.W. Wang, W.E. Love, and E.N. Moudrianakis. 1991. The nucleosomal core histone octamer at 3.1Å resolution: A tripartite protein assembly and a left-handed superhelix. *Proc. Natl. Acad. Sci.* **88:** 10148–10152.

Barton, M.C. and B.M. Emerson. 1994. Regulated expression of the β-globin gene locus in synthetic nuclei. *Genes Dev.* **8:** 2453–2465.

Becker, P.B. and C. Wu. 1992. Cell-free system for assembly of transcriptionally repressed chromatin from *Drosophila* embryos. *Mol. Cell. Biol.* **12:** 2241–2249.

Bell, S.P., R. Kobayashi, and B. Stillman. 1993. Yeast origin recognition complex functions in transcription silencing and DNA replication. *Science* **262:** 1844–1849.

Blau, H.M., G.K. Pavlath, E.C. Hardeman, C.P. Chiu, L. Silberstein, S.G. Webster, S.C. Miller, and C. Webster. 1985. Plasticity of the differentiated state. *Science* **230:** 758–766.

Bouvet, P., S. Dimitrov, and A.P. Wolffe. 1994. Specific regulation of chromosomal 5S rRNA gene transcription *in vivo* by histone H1. *Genes Dev.* **8:** 1147–1159.

Brown, D.D. 1984. The role of stable complexes that repress and activate eukaryotic genes. *Cell* **37:** 359–365.

Burhans, W.C., L.T. Vassilev, J. Wu, J.M. Sogo, F. Nallaseth, and M.L. DePamphilis. 1991. Emetine allows identification of origins of mammalian DNA replication by imbalanced DNA synthesis, not through conservative nucleosome segregation. *EMBO J.* **10:** 4351–4360.

Chiu, C.P. and H.M. Blau. 1984. Reprogramming cell differentiation in the absence of DNA synthesis. *Cell* **37:** 879–887.

Clark, D.J. and A.P. Wolffe. 1991. Superhelical stress and nucleosome mediated repression of 5S RNA gene transcription *in vitro*. *EMBO J.* **10:** 3419–3428.

Cook, P.R. 1991. The nucleoskeleton and the topology of replication. *Cell* **66:** 627–635.

Cusick, M.E., M.L. DePamphilis, and P.M. Wasserman. 1984. Dispersive segregation of nucleosomes during replication of simian virus 40 chromosomes. *J. Mol. Biol.* **178:** 249–271.

Cusick, M.E., K.S. Lee, M.L. DePamphilis, and P.M. Wasserman. 1983. Structure of chromatin at deoxyribonucleic acid replication forks: Nuclease hypersensitivity results from both prenucleosomal deoxyribonucleic acid and an immature chromatin structure. *Biochemistry* **22:** 3873–3884.

DePamphilis, M.L. 1988. Transcriptional elements as components of eukaryotic origins of DNA replication. *Cell* **52:** 635–638.

———. 1993. How transcription factors regulate origins of DNA replication in eukaryotic cells. *Trends Cell Biol.* **3:** 161–167.

Diffley, J.F.X. and J.H. Cocker. 1992. Protein DNA interactions at a yeast replication origin. *Nature* **357:** 169–172.

Dilworth, S.M., S.J. Black, and R.A. Laskey. 1987. Two complexes that contain histones are required for nucleosome assembly *in vitro*: Role of nucleoplasmin and N1 in *Xenopus* egg extracts. *Cell* **51:** 1009–1018.

Dimitrov, S., M.C. Dasso, and A.P. Wolffe. 1994. Remodeling sperm chromatin in *Xenopus laevis* egg extracts: The role of core histone phosphorylation and linker histone B4 in chromatin assembly. *J. Cell Biol.* **126:** 591–601.

Edgar, L.G. and J.D. McGhee. 1988. DNA synthesis and the control of embryonic gene expression in *C. elegans*. *Cell* **53:** 589–599.

Eickbusch, T.H. and E.N. Moudrianakis. 1978. The histone core complex: An octamer assembled by two sets of protein-protein interactions. *Biochemistry* **17:** 4955–4965.

Enver, T., A.C. Brewer, and R.K. Patient. 1988. Role of DNA replication in β-globin gene activation. *Mol. Cell. Biol.* **8:** 1301–1308.

Fotedar, R. and J.M. Roberts. 1989. Multistep pathway for replication dependent nucleosome assembly. *Proc. Natl. Acad. Sci.* **86:** 6459–6463.

Gilbert, D.M. 1986. Temporal order of replication of *Xenopus laevis* 5S ribosomal RNA genes in somatic cells. *Proc. Natl. Acad. Sci.* **83:** 2924–2928.

Glikin, G.C., I. Ruberti, and A. Worcel. 1984. Chromatin assembly in *Xenopus* oocytes: *In vitro* studies. *Cell* **37:** 33–41.

Goldman, M.A., G.P. Holmquist, M.C. Gray, L.A. Caston, and A. Nag. 1984. Replication timing of genes and middle repetitive sequences. *Science* **224:** 686–692.

Gottesfeld, J.M. and L.S. Bloomer. 1982. Assembly of transcriptionally active 5S RNA gene chromatin *in vitro*. *Cell* **28:** 781–791.

Gruss, C., J. Wu, T. Koller, and J.M. Sogo. 1993. Disruption of nucleosomes at replication forks. *EMBO J.* **12:** 4533–4545.

Gruss, C., C. Gutierrez, W.C. Burhans, M.L. De Pamphilis, T. Koller, and J.M. Sogo. 1990. Nucleosome assembly in mammalian cell extracts before and after DNA replication. *EMBO J.* **9:** 2911–2922.

Hansen, J.C. and A.P. Wolffe. 1994. A role for histones H2A/H2B in chromatin folding and transcriptional repression. *Proc. Natl. Acad. Sci.* **91:** 2339–2343.

Hayes, J.J. and A.P. Wolffe. 1992. Histones H2A/H2B inhibit the interactions of transcription factor IIIA with the *Xenopus borealis* somatic 5S RNA gene in a nucleosome. *Proc. Natl. Acad. Sci.* **89:** 1229–1233.

———. 1993. Preferential and asymmetric interaction of linker histones with 5S DNA in the nucleosome. *Proc. Natl. Acad. Sci.* **90:** 6415–6419.

Hayes, J.J., D.J. Clark, and A.P. Wolffe. 1991. Histone contributions to the structure of DNA in the nucleosome. *Proc. Natl. Acad. Sci.* **88:** 6829–6833.

Hayes, J.J., D. Pruss, and A.P. Wolffe. 1994. Contacts of the globular domain of histone H5 and core histones with DNA in a chromatosome. *Proc. Natl. Acad. Sci.* **91:** 7817–7821.

Hsieh, D.-J., S.M. Camiolo, and Y.L. Yates. 1993. Constitutive binding of EBNA1 protein to the Epstein-Barr virus replication origin, oriP, with distortion of DNA structure during latent infection. *EMBO J.* **12:** 4933–4944.

Jackson, V., A. Shires, N. Tanphaichitr, and R. Chalkley. 1976. Modification of histones immediately after synthesis. *J. Mol. Biol.* **104:** 471–483.

Jackson, V. 1987. Deposition of newly synthesized histones: New histones H2A and H2B do not deposit in the same nucleosome with new histones H3 and H4. *Biochemistry* **26:** 2315–2325.

———. 1990. *In vivo* studies on the dynamics of histone-DNA interaction: Evidence for nucleosome dissolution during replication and transcription and a low level of dissolution independent of both. *Biochemistry* **29:** 719–731.

James, T.C. and S.C. Elgin. 1986. Identification of a non histone chromosomal protein associated with heterochromatin in *Drosophila melanogaster* in its gene. *Mol. Cell. Biol.* **6:** 3862–3872.

Kamakaka, R.T., M. Bulger, and J.T. Kadonaga. 1993. Potentiation of RNA polymerase II transcription by Gal4-VP16 during but not after DNA replication and chromatin assembly. *Genes Dev.* **7:** 1779–1795.

Kaufman, P.D. and M.R. Botchan. 1994. Assembly of nucleosomes: Do multiple assembly factors mean multiple mechanisms? *Curr. Opin. Genet. Dev.* **4:** 229–235.

Khochbin, S. and A.P. Wolffe. 1994. Developmentally regulated expression of linker-histone variants in vertebrates. *Eur. J. Biochem.* **225:** 501–510.

Kleinschmidt, J.A., A. Seiter, and H. Zentgraf. 1990. Nucleosome assembly *in vitro*: Separate histone transfer and synergistic interaction of native histone complexes purified from nuclei of *Xenopus laevis* oocytes. *EMBO J.* **9:** 1309–1318.

Kleinschmidt, J.A., C. Dingwall, G. Maier, and W.W. Franke. 1986. Molecular characterization of a karyophilic histone-binding protein: cDNA cloning amino acid sequence and expression of nuclear protein N1/N2 of *Xenopus laevis*. *EMBO J.* **5:** 3547–3552.

Krude, T. and R. Knippers. 1991. Transfer of nucleosomes from parental to replicated chromatin. *Mol. Cell. Biol.* **11:** 6257–6267.

Laskey, R.A., B.M. Honda, A.D. Mills, and J.T. Finch. 1978. Nucleosomes are assembled by an acidic protein which binds histones and transfers them to DNA. *Nature* **275:** 416–420.

Laurensen, P. and J. Rine. 1992. Silencers, silencing and heritable transcriptional states. *Microbiol. Rev.* **56:** 543–592.

Lebkowski, J.S., S. Clancy, and M.P. Calos. 1985. Simian virus 40 replication in adenovirus-transformed human cells antagonizes gene expression. *Nature* **317:** 169–171.

Lee, D.Y., J.J. Hayes, D. Pruss, and A.P. Wolffe. 1993. A positive role for histone acetylation in transcription factor binding to nucleosomal DNA. *Cell* **72:** 73–84.

Lewis, E.D. and J.L. Manley. 1985. Repression of simian virus 40 early transcription by viral DNA replication in human 293 cells. *Nature* **317:** 172–175.

Louters, K. and R. Chalkley. 1985. Exchange of histones H1, H2A, and H2B *in vivo*. *Biochemistry* **24:** 3080–3085.

Lucchini, R. and J.M. Sogo. 1994. Chromatin structure and transcriptional activity around the replication forks arrested at the 3′ end of the yeast rRNA genes. *Mol. Cell. Biol.* **14:** 318–326.

Miller, A.M. and K.A. Nasmyth. 1984. Role of DNA replication in the repression of silent mating type loci in yeast. *Nature* **312:** 247–251.

Mita-Miyazawa, I., S. Ikegami, and N. Saitoh. 1985. Histospecific acetylcholinesterase development in the presumptive muscle cells isolated from 16 cell stage ascidian embryos with respect to the number of DNA replications. *J. Embryol. Exp. Morphol.* **87:** 1–12.

Paro, R. and D.S. Hogness. 1991. The Polycomb protein shares a homologous domain with a heterochomatin associated protein of *Drosophila*. *Proc. Natl. Acad. Sci.* **88:** 263–267.

Perry, C.A. and A.T. Annunziato. 1989. Influence of histone acetylation on the solubility, H1 content and DNaseI sensitivity of newly replicated chromatin. *Nucleic Acids Res.* **17:** 4275–4291.

Perry, C.A., C.D. Allis, and A.T. Annunziato. 1993. Parental nucleosomes segregated to newly replicated chromatin are underacetylated relative to those assembled *de novo*. *Biochemistry* **32:** 13615–13623.

Philpott, A. and G.H. Leno. 1992. Nucleoplasmin remodels sperm chromatin in *Xenopus* egg extracts. *Cell* **69:** 759–767.

Pruss, D., J.J. Hayes, and A.P. Wolffe. 1995. Nucleosomal anatomy—Where are the histones? *BioEssays* **17:** 161–170.

Randall, S.K. and T.J. Kelly. 1992. The fate of parental nucleosomes during SV40 DNA replication. *J. Biol. Chem.* **267:** 14259–14265.

Ruiz-Carrillo, A., L.J. Wangh, and V.G. Allfrey. 1975. Processing of newly synthesized histone molecules. *Science* **190:** 117–128.

Saffer, L.D. and O.L. Miller, Jr. 1986. Electron microscopic study of *Saccharomyces cerevisiae* rDNA chromatin replication. *Mol. Cell. Biol.* **6:** 1147–1157.

Simpson, R.T. 1991. Nucleosome positioning: Occurrence, mechanisms and functional consequences. *Prog. Nucleic Acids Res. Mol. Biol.* **40:** 143–184.

Smith, P.A., V. Jackson, and R. Chalkley. 1984. Two stage maturation process for newly replicated chromatin. *Biochemistry* **23:** 1576–1581.

Smith, S. and B.W. Stillman. 1989. Purificiation and characterization of CAF1, a human cell factor required for chromatin assembly during DNA replication in vitro. *Cell* **58:** 15–25.

———. 1991. Stepwise assembly of chromatin during DNA replication *in vitro. EMBO J.* **10:** 971–980.

Sogo, J.M., H. Stahl, T. Koller, and R. Knippers. 1986. Structure of the replicating SV40 minichromosomes: The replication fork, core histone segregation and terminal structures. *J. Mol. Biol.* **189:** 189–204.

Stillman, B.W. 1986. Chromatin assembly during SV40 DNA replication in vitro. *Cell* **45:** 555–565.

Sugasawa, K., Y. Ishimi, T. Eki, J. Hurwitz, A. Kikuchi, and F. Hanaoka. 1992. Nonconservative segregation of parental nucleosomes during SV40 chromosome replication in vitro. *Proc. Natl. Acad. Sci.* **89:** 1055–1059.

Takata, C., J.F. Albright, and T. Yomada. 1964. Lens antigens in a lens regenerating system studied by the immunofluorescent technique. *Dev. Biol.* **9:** 385–397.

Tjian, R. and T. Maniatis. 1994. Transcriptional activation: A complex puzzle with few easy pieces. *Cell* **77:** 5–8.

Tremethick, D., K. Zucker, and A. Worcel. 1990. The transcription complex of the 5S RNA gene, but not transcription factor TFIIIA alone, prevents nucleosomal repression of transcription. *J. Biol. Chem.* **265:** 5014–5023.

Ura, K., A.P. Wolffe, and J.J. Hayes. 1994. Core histone acetylation does not block linker histone binding to a nucleosome including a *Xenopus borealis* 5S rRNA gene. *J. Biol. Chem.* **269:** 27171–27174.

Wilson, A.C. and R.K. Patient. 1993. DNA replication facilitates the action of transcriptional enhancers in transient expression assays. *Nucleic Acids Res.* **21:** 4296–4304.

Wolffe, A.P. 1989. Transcriptional activation of *Xenopus* class III genes in chromatin isolated from sperm and somatic nuclei. *Nucleic Acids Res.* **17:** 767–780.

———. 1990. Transcription complexes. *Prog. Clin. Biol. Res.* **322:** 171–186.

———. 1991. Implications of DNA replication for eukaryotic gene expression. *J. Cell Sci.* **99:** 201–206.

———. 1993. Replication timing and *Xenopus* 5S RNA gene transcription *in vitro. Dev. Biol.* **157:** 224–231.

———. 1994a. Transcription: In tune with the histone. *Cell* **77:** 13–16.

———. 1994b. Nucleosome positioning and modification: Chromatin structures that potentiate transcription. *Trends Biochem. Sci.* **19:** 240–244.

Wolffe, A.P. and D.D. Brown. 1986. DNA replication *in vitro* erases a *Xenopus* 5S RNA gene transcription complex. *Cell* **47:** 217–227.

———. 1988. Developmental regulation of two 5S ribosomal RNA genes. *Science* **241:**

1626–1632.
Worcel, A., S. Han, and M.L. Wong. 1978. Assembly of newly replicated chromatin. *Cell* **15:** 969–977.
Wormington, W.M., M. Schlissel, and D.D. Brown. 1982. Developmental regulation of *Xenopus* 5S RNA genes. *Cold Spring Harbor Symp. Quant. Biol.* **47:** 879–884.

10

Roles of Phosphorylation in DNA Replication

Klaus Weisshart[1] and Ellen Fanning[1,2]
[1]Institute for Biochemistry
81375 Munich, Germany

[2]Department of Molecular Biology
Vanderbilt University
Nashville, Tennesee 37235

Phosphorylation of proteins by protein kinases and dephosphorylation by protein phosphatases represents one of the most common, versatile, and perhaps confusing regulatory mechanisms in eukaryotic cells. Many, perhaps most, proteins in the eukaryotic nucleus are phosphoproteins, among them the proteins involved in DNA replication and its control. The importance of protein phosphorylation as a regulatory mechanism lies in its ready response to intracellular or extracellular signaling, its reversibility, and its ability to act as a measuring device to translate gradual changes into a molecular switch thrown at a threshold level of phosphorylation of a key target protein. The net level of protein phosphorylation is determined by the balance between the activities of protein kinases and those of protein phosphatases (for review, see Cohen 1989; Hubbard and Cohen 1993). The activities of kinases and phosphatases themselves are regulated by their own phosphorylation state. To complicate matters further, the number of known protein kinases, phosphatases, and regulatory subunits is growing rapidly, and their specificity is not always predictable from the primary sequence of the substrate protein (see Moreno and Nurse 1990; Cegielska et al. 1994a). Moreover, the effect of phosphorylation on the activity of a given target protein usually depends on the exact sites that are modified and unmodified. Thus, to understand how protein phosphorylation regulates the activity of replication proteins, one must know which sites in a target protein are phosphorylated, how physiological signals affect phosphorylation at each site, and how this phosphorylation influences the protein's activity. Knowledge of protein kinases or phosphatases that act at each phosphorylation site is also useful in elucidating the role of protein phosphorylation in regulating the target protein.

DNA Replication in Eukaryotic Cells

The role of protein phosphorylation in control of the timing of DNA replication in the cell cycle has been studied extensively (for review, see Norbury and Nurse 1992; Coverley and Laskey 1994), and these phosphorylation events, as well as the cyclins, cyclin-dependent kinases, and their effects on substrates such as the retinoblastoma tumor suppressor protein are discussed by Nasmyth (this volume). Here we discuss specific examples of DNA replication proteins whose regulation by phosphorylation has been described in detail, summarize our current understanding of the regulation of several cellular replication proteins by phosphorylation, and, finally, speculate how phosphorylation could control the activity of some DNA replication proteins in the cell cycle.

PROTEINS THAT RECOGNIZE AND ACTIVATE REPLICATION ORIGINS

Simian Virus 40 Large Tumor Antigen

The replication of SV40 DNA is a useful model for eukaryotic DNA replication (see Brush and Kelly; Hassell and Brinton; both this volume). SV40 DNA is double-stranded, circular, and supercoiled. In infected monkey cells and in the virus particle, viral DNA is complexed with cellular histones, forming nucleosomes that resemble those in cellular chromatin (for review, see Fanning and Knippers 1992). A set of ten human replication proteins that is sufficient to replicate SV40 DNA in vitro has been purified and characterized (Waga et al. 1994; see Brush and Kelly; Hassell and Brinton; both this volume). Many of these mammalian proteins are functionally conserved in yeast, and the genes encoding many of them have been shown to be essential for viability in yeast (for discussion of these proteins, see Stillman; Wang; Coen; Hübscher et al.; Borowiec; Nash and Lindahl; all this volume), providing strong support for the validity of the SV40 model system and arguing that mammalian chromosomes are probably replicated by the same set of cellular proteins. Like mammalian chromosomal DNA, SV40 DNA replicates in the nucleus during the S phase in the cell cycle. It is not clear how SV40 DNA replication is limited to the S phase, but it is evident that cell-cycle-dependent protein kinases, phosphatases, and their regulatory subunits play an important role. However, unlike mammalian chromosomes, SV40 chromatin undergoes multiple rounds of replication during S phase, indicating that SV40 somehow circumvents the cellular mechanisms that limit chromosomal replication to a single round per S phase. Thus, viral replication is subject to some of the cellular controls on DNA replication, but not all of them.

The only viral protein needed for SV40 DNA replication in vitro and in cultured monkey cells is the large tumor antigen (T antigen). T antigen is a multifunctional phosphoprotein whose intrinsic biochemical activities include sequence-specific binding to the SV40 origin of DNA replication, ATP binding and hydrolysis, and 3′–5′ DNA helicase activity (for review, see Fanning 1992; Fanning and Knippers 1992). T antigen resides primarily in the nucleus of infected cells, where it serves to stimulate the cell to progress to the S phase prior to viral DNA replication and controls expression of viral and cellular genes and virion assembly, in addition to participating directly in viral DNA replication.

A basic understanding of the process of SV40 DNA replication and the action of the proteins involved has been developed through work performed using cell-free systems (for review, see Kelly 1988; Stillman 1989; Borowiec et al. 1990; Hurwitz et al. 1990). In the current model, SV40 T antigen binds specifically to sequences in duplex origin DNA and assembles in the presence of ATP into a double hexamer on the origin, causing local distortion or melting in the origin DNA sequences flanking the T-antigen-binding site (see Borowiec, this volume). In the next step of the reaction, the DNA helicase activity of the double hexamers, driven by ATP hydrolysis, bidirectionally unwinds the two parental strands by reeling the DNA through the complex (Dean et al. 1992; Wessel et al. 1992). The single-stranded regions are stabilized by the single-stranded DNA-binding protein RP-A, and the enzyme responsible for initiation of synthesis, DNA polymerase-α:primase (pol-α:primase), joins the protein-DNA complex. The assembly of this complex is directed by specific protein-protein interactions that occur among T antigen, RP-A, and pol-α:primase (Dornreiter et al. 1990, 1992, 1993; Melendy and Stillman 1993; Murakami and Hurwitz 1993; for review, see Diffley 1992). After synthesis of the first primer and Okazaki fragment, replication factor C (RF-C) displaces pol-α:primase from the primer-template junction and recruits proliferating cell nuclear antigen (PCNA). The RF-C/PCNA complex then targets DNA polymerase-δ to the junction, where it takes over leading-strand DNA synthesis (Lee et al. 1991a; Tsurimoto and Stillman 1991a, b, c). On the lagging strand, pol-α:primase initiates a new RNA primer, extends it into a short DNA primer, dissociates and then repeats this process to generate the next DNA primer (see Salas et al., this volume). To extend the DNA primers on the lagging strand into Okazaki fragments, polymerase δ, again with the help of RF-C and PCNA, binds to the termini and extends the fragments until it encounters the previous Okazaki fragment. RNase H, together with maturation factor I (MFI), removes the RNA primer, and ligase I joins the two fragments (Turchi et al. 1994; Waga and Stillman 1994; Waga et al. 1994; see Brush and Kelly, this volume). Topoisomerases I and II release the torsional stress caused by the progression of the replication fork, and topoisomerase II also functions in the resolution of catenated replication products (Yang et al. 1987; Ishimi et al. 1988).

The replication functions of SV40 T antigen are tightly controlled by its phosphorylation state (for review, see Prives 1990; Fanning 1992, 1994). T antigen expressed in mammalian cells is phosphorylated on multiple serine (S) and threonine (T) residues clustered at the carboxyl and amino termini of the protein (Fig. 1A) (Scheidtmann et al. 1991). Mutational analysis of the T antigen phosphorylation sites in cell culture demonstrated that phosphorylation at six of these sites (S106, S112, S639, S677, S679, and T701) was dispensable, whereas T-antigen mutants bearing alanine in place of S120, S123, or T124 were defective

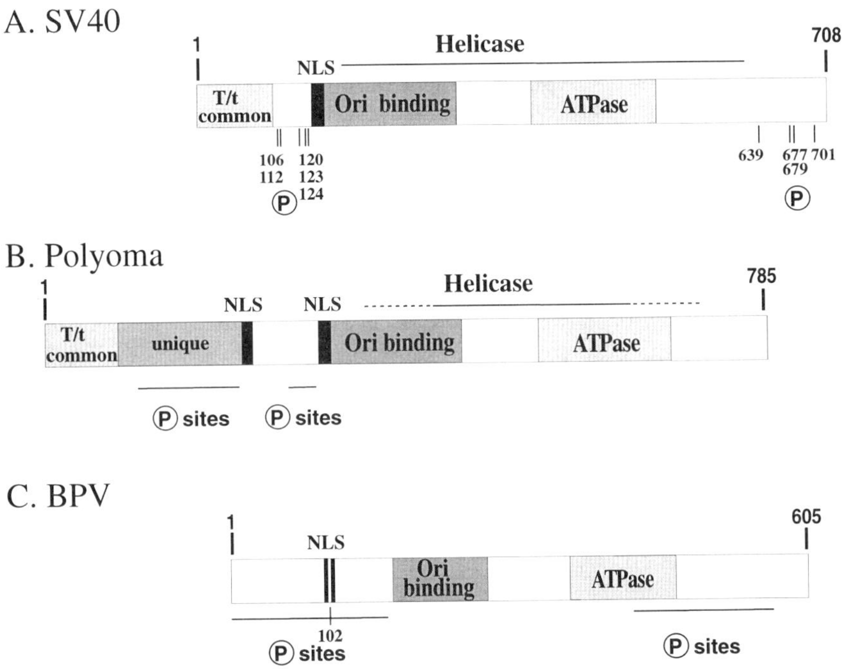

Figure 1 Functional domains and known phosphorylation sites of viral origin recognition proteins. (*A*) SV40 T antigen; (*B*) polyoma T antigen; (*C*) bovine papillomavirus E1 protein. The data were compiled from Fanning (1994); E.H. Wang et al. (1993); Lambert (1991); and unpublished data kindly provided by Mike Lentz. Phosphorylation sites (P) are indicated below each polypeptide by residue numbers (*A*), and regions containing phosphorylation sites by lines (*B*, *C*).

in viral replication in vivo (Schneider and Fanning 1988), suggesting that modification of these sites is required for SV40 DNA replication in monkey cells.

The role of phosphorylation of T antigen at these three sites in the control of viral DNA replication has been further investigated in cell-free reactions using purified proteins. These studies confirm the requirement for phosphorylation of T124 that was observed in vivo but suggest a quite different role for modification of S120 and S123, namely, that it inhibits viral DNA replication in vitro. For example, the bulk of the T antigen isolated from SV40-infected mammalian cells is hyperphosphorylated and inactive in SV40 DNA replication in vitro (Fig. 2) (Virshup et al. 1989). T antigen expressed in bacteria is not modified at these sites and is also unable to initiate SV40 DNA replication in vitro (Fig. 2) (McVey et al. 1989). However, T antigen expressed in recombinant baculovirus systems has an intermediate level of phosphorylation, with T124 modified in nearly all molecules, but with most serine sites remaining unmodified (Höss et al. 1990); this T antigen is highly active in SV40 replication in vitro (Fig. 2).

In agreement with these findings, treatment of T antigen with protein kinases and phosphatases that alter the phosphorylation state of T124, S120, or S123 activates or inhibits its replication activity in vitro. Phosphorylation of T124 by cdc2-cyclin B activated the replication activity of unphosphorylated bacterially expressed T antigen (McVey et al. 1989). The positive effect of phosphorylation of T124 on the replication activity of T antigen was reversed by treating it with isoforms of protein phosphatase PP2A that specifically targeted T124, i.e., PP2A-D and PP2A-T55 (Cegielska et al. 1994b). Treatment of hyperphosphorylated T antigen from mammalian cells with the catalytic subunit of PP2A (C36, $PP2A_c$) or with an isoform of PP2A (PP2A-T72) stimulated SV40 replication by dephosphorylating S120 and S123 (Virshup et al. 1992; Cegielska et al. 1994b). More recently, an isoform of casein kinase I was shown to phosphorylate these serines in vitro (Cegielska and Virshup 1993; Cegielska et al. 1994a). Kinase treatment of hypophosphorylated replication-active T antigen isolated from baculovirus-infected insect cells led to inactivation of SV40 replication activity. Taken together, these results indicate that SV40 DNA replication in vitro is regulated positively and negatively by the phosphorylation state of T antigen (Fig. 2).

The biochemical mechanism by which phosphorylation of T antigen regulated its replication activity was elucidated by comparing the activities of baculovirus-expressed wild-type and mutant T antigens.

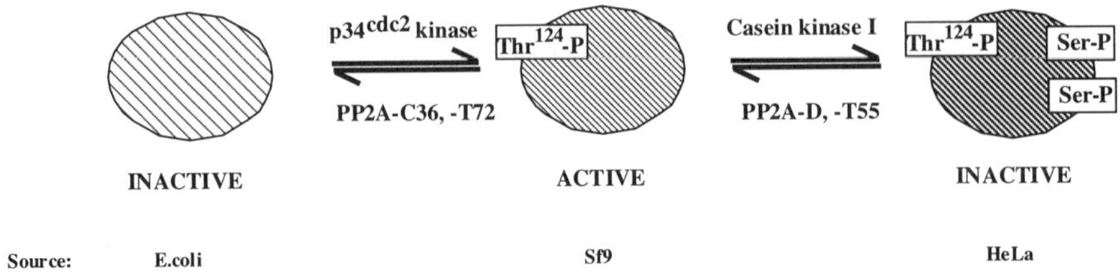

Figure 2 Phosphorylation controls SV40 T-antigen replication activity in vitro. Unphosphorylated T antigen synthesized in bacteria is inactive in replication unless it is first phosphorylated on Thr-124. Hypophosphorylated T antigen expressed in insect cells using baculovirus vectors is active, whereas hyperphosphorylated T antigen expressed in mammalian cells is inactive. Protein kinases and phosphatases (Cegielska et al. 1994a,b) that act on the crucial Thr-124, Ser-120, and Ser-123 sites are indicated. (Reprinted, with permission, from Fanning 1994.)

Phosphorylation at T124 is necessary for the bidirectional unwinding of duplex SV40 origin DNA by T antigen in the initiation reaction (McVey et al. 1993; Moarefi et al. 1993b). In contrast, other enzymatic activities like hexamer assembly, unidirectional DNA helicase, DNA binding, and origin DNA distortion were not markedly influenced by phosphorylation of T124. The specific effect of phosphorylation on bidirectional unwinding appears to result from its ability to stabilize hexamer-hexamer contacts that are necessary for bidirectional, but not unidirectional, DNA unwinding (K. Weisshart and E. Fanning, unpubl.; T. Kelly, pers. comm.). Phosphorylation of T124 also results in cooperative assembly of the two hexamers on the SV40 origin (K. Weisshart and E. Fanning, unpubl.; T. Kelly, pers. comm.).

Interestingly, phosphorylation of T antigen on the inhibitory serine residues also decreased the cooperative assembly of T antigen double hexamers on SV40 origin DNA, blocking bidirectional origin-unwinding activity and viral DNA replication in vitro (Virshup et al. 1992; Cegielska and Virshup 1993; Cegielska et al. 1994a). Thus, T antigen unphosphorylated on T124 or phosphorylated on S120 and S123 could potentially bind to the origin as a double hexamer and distort it locally without initiating bidirectional unwinding and replication until it became phosphorylated on T124 and dephosphorylated on S120 or 123, thereby facilitating interactions between the two hexamers that are required for bidirectional unwinding (Fig. 3). We suggest that the interactions between the hexamers are dynamic, undergoing cycles of association and dissociation as unwinding proceeds (Fig. 3). Although this mechanism

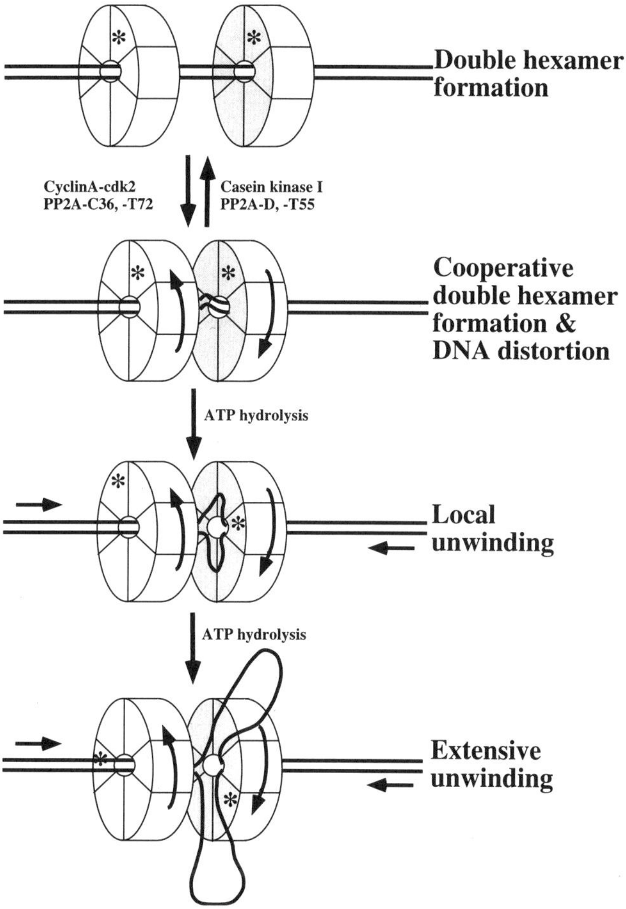

Figure 3 The phosphorylation state of SV40 T antigen governs hexamer-hexamer interactions in bidirectional origin unwinding. Prior to viral DNA replication in infected mammalian cells, T antigen accumulates primarily in hyperphosphorylated forms that can assemble as a double hexamer on the SV40 origin, but the hexamers cannot interact properly with each other to unwind the origin and permit initiation. Replication-active hexamers can be formed upon dephosphorylation of Ser-120 or Ser-123, or by phosphorylation of newly synthesized unphosphorylated T antigen on Thr-124. When a sufficient concentration of this hypophosphorylated form of T antigen is reached in the infected cell, ATP hydrolysis and dynamic hexamer-hexamer interactions lead to bidirectional unwinding at the origin and initiation of replication. Subsequent phosphorylation of Ser-120 and Ser-123 or dephosphorylation of Thr-124 would block reinitiation of replication. (Reprinted, with permission, from Fanning 1994.)

remains speculative, the data demonstrate clearly that the phosphorylation state of T antigen controls SV40 DNA replication both negatively and positively at the level of unwinding of origin DNA. It will be interesting to see whether phosphorylation of cellular origin recognition or unwinding proteins regulates their function in a similar manner.

The significance of T antigen phosphorylation in viral DNA replication in vivo remains for the most part unresolved. Although it is well established that phosphorylation of T124 is essential for SV40 DNA replication in vivo (Schneider and Fanning 1988), this modification may play additional roles in regulation of replication. One attractive model proposes that phosphorylation of T antigen at T124 could link the replication activity of T antigen to the cell cycle (McVey et al. 1989, 1993). T antigen is associated with cdk2-cyclin A in the cell (Adamczewski et al. 1993) and is a much better substrate for this kinase than for cdc2-cyclin B in vitro (C. Voitenleitner et al., unpubl.). cdk2-cyclin A is activated at the G_1-S transition and is found associated with replicating SV40 DNA (Fotedar and Roberts 1991). Together these data suggest a role for cdk2-cyclin A as the best candidate for phosphorylation of T124 in vivo and support the notion that this modification event could link SV40 DNA replication to S phase. However, the phosphate turnover on this site is relatively slow in SV40-infected cells, and most T-antigen molecules carry the modification (Scheidtmann 1986; for review, see Fanning and Knippers 1992). Furthermore, there is no direct evidence to suggest that modification of T124 in SV40-infected or -transformed cells is cell-cycle-dependent. Thus, it remains an open question whether the cell-cycle-dependence of SV40 DNA replication is controlled by this phosphorylation event.

The physiological role of phosphorylation of T-antigen residues S120 and S123 in vivo is even more puzzling. Whereas phosphorylation of S120 and S123 inhibits viral DNA replication in vitro, phosphorylation at these sites is required in vivo (Schneider and Fanning 1988). Reexamination of the kinetics of SV40 DNA replication in cultured cells with the S120A and S123A mutants yielded preliminary evidence that the mutant T antigens did begin viral DNA replication, but only a small amount of DNA accumulated (Moarefi et al. 1993a). To resolve the apparent discrepancy between the in vivo and in vitro results, it has been suggested that phosphorylation of these serine residues is necessary for regulatory functions not detectable in the in vitro replication system (Fanning 1994). Support for this hypothesis comes from the finding that phosphate turnover on these serines is rapid (Scheidtmann 1986), so that the balance between phosphorylation and dephosphorylation on these two serines could determine the fraction of T antigen active in initiation of DNA replication (Fanning 1992). Phosphorylation at these serines may be required in the termination of replication, or it could influence the association of viral DNA with replication foci (Hozak and Cook 1994). These or other mechanisms might reduce the ability of T antigen to sustain multiple rounds of viral DNA replication in vivo, resulting in the defect observed with the S120 and S123 mutants in cell culture but not in vitro (Schneider and Fanning 1988; Moarefi et al. 1993a). Further work is required to resolve this paradox.

The physiological relevance of PP2A and casein kinase I action on T antigen in vivo also remains obscure. Casein kinase I is ubiquitously expressed and not regulated in the cell cycle. In addition, PP2A is found primarily in the cytoplasm rather than in the nucleus. However, one cannot exclude the possibility that the different isoforms of kinase and phosphatase that affect T antigen have different subcellular localizations. Indeed, there is evidence that the activity of PP2A on T antigen is restricted to S phase (Ludlow 1992). Other protein phosphatases that act at the G_1-S border may also target T antigen, such as a mammalian homolog of SIT4 from budding yeast and *Drosophila* (Fernandez-Sarabia et al. 1992; Mann et al. 1993) or a mammalian homolog of the yeast cdc25A phosphatase (Hoffmann et al. 1994; Jinno et al. 1994).

Polyomavirus Large Tumor Antigen

The initiator protein for polyomavirus (PyV) replication is the polyoma large T antigen, Py T, which interacts with specific sequences within the PyV origin of replication (Cowie and Kamen 1984; Hassell and Brinton, this volume). Like its SV40 counterpart, Py T antigen unwinds duplex DNA (Wang and Prives 1991) using its ATPase/helicase function (Seki et al. 1990). Py T antigen is a phosphoprotein with two clusters of phosphorylation sites located at the amino terminus (Fig. 1B) (Hassauer et al. 1986; Bockus and Schaffhausen 1987). No assessment of the importance of single phosphorylation sites has been undertaken, but treatment with calf intestinal phosphatase had differential effects on Py T antigen, depending on the degree of phosphate removal (E.H. Wang et al. 1993). Treatment with limiting amounts of phosphatase stimulated origin binding, whereas incubation with high concentrations of phosphatase had the opposite effect. These data are reminiscent of early results obtained with partially dephosphorylated SV40 T antigen (Grässer et al. 1987) and suggest that the replication activity of PyV T antigen might be regulated through differential phosphorylation. It will be interesting to test whether phosphorylation of Py T regulates initiation of viral DNA replication by controlling the ability of Py T to catalyze bidirectional unwinding of the origin DNA as in the SV40 system.

Papillomavirus E1 Protein

Two bovine papillomavirus (BPV)-encoded proteins are necessary and sufficient for BPV DNA replication in vivo, the 68-kD E1 protein and the 48-kD E2 protein (for review, see Lambert 1991; Ustav and Stenlund 1991; Ustav et al. 1993; Stenlund, this volume). The E2 protein can be dispensable in vitro when high levels of E1 are used (Seo et al. 1993b; Spalholz et al. 1993). All other replication proteins are supplied by the host cell. E1, like SV40 and Py T antigens, is a phosphoprotein that possesses several activities necessary for replication function (Ustav and Stenlund 1991; Seo et al. 1993a; Thorner et al. 1993; Yang et al. 1993; Park et al. 1994; Bonne-Andrea et al. 1995): origin binding activity, polymerase-α binding, ATPase activity, and 3′–5′ helicase activity (Fig. 1C). E2 protein is a transcriptional activator (Spalholz et al. 1993) and also a phosphoprotein (Bream et al. 1993) that stimulates the replication activity of E1 protein in vitro, at least in part by binding to it (Seo et al. 1993b; Spalholz et al. 1993; Benson and Howley 1995). It has been proposed that the role of E2 is to tether E1 more tightly to the origin and to recruit cellular replication factors, e.g., RP-A and pol-α:primase (Li and Botchan 1993), to the BPV origin.

The facts that BPV DNA is maintained as an episome at a constant copy number in latently transformed cells (Law et al. 1981) and that it replicates during S phase along with the host chromosomal DNA (Gilbert and Cohen 1987) might suggest that the activities of E1 and E2 are regulated in the cell cycle. Since both proteins are phosphoproteins, this regulation could depend on phosphorylation events. The E1 residue T102 near the nuclear localization signal is a substrate in vitro for cdc2-cyclin kinase (Fig. 1C). However, replacing that threonine by isoleucine showed no observable effect on either the nuclear localization of E1 or its efficiency in DNA replication (Lentz et al. 1993). Thus, the functional role, if any, of phosphorylation of the BPV replication proteins remains unknown.

Yeast Origin Recognition Complex

The most likely candidate for the replication initiator in yeast is the origin recognition complex (ORC) (for review, see Bell and Stillman 1992; Diffley 1992; Huberman 1992; Li and Alberts 1992; see Newlon;

Nasmyth; Stillman; all this volume). It consists of six polypeptides with molecular masses of 50, 53, 56, 62, 72, and 120 kD, respectively, that have been termed ORC-1 to ORC-6 according to the decrease in molecular weight. The ORC complex binds specifically to a bipartite site in yeast origins of replication in an ATP-dependent manner (Rao and Stillman 1995; Rowley et al. 1995). Mutations in the gene coding for ORC-2 have shown that ORC is necessary for the initiation of DNA replication and transcriptional silencing at the mating-type locus (Bell et al. 1992; Foss et al. 1993; Micklem et al. 1993; for review, see Newlon 1993; Huberman 1994; Kelly et al. 1994). ORC-6 has been shown to be important for the initiation of DNA replication (Li and Herskowitz 1993).

Evidence has been obtained by in vivo footprinting that ORC is bound to ARS-1 sequences in yeast cells in the same manner as it is in vitro (Diffley and Cocker 1992). Interestingly, ORC remains bound to the origin throughout the cell cycle. These data suggested that ORC could repress replication except in S phase, when it could be activated for replication. ORC-2 and ORC-6 contain multiple potential phosphorylation sites for cyclin-dependent kinases and other kinases (Bell et al. 1992; Li and Herskowitz 1993; Micklem et al. 1993), so that phosphorylation by cell-cycle-regulated kinases may influence ORC activity. However, mutations in the potential sites of phosphorylation by cyclin-dependent kinases have not yet revealed any loss of replication function (J. Diffley, pers. comm.). Interestingly, DBF4, the CDC7 protein kinase regulatory subunit, interacts with the yeast replication origin (Dowell et al. 1994), and activation of CDC7 by DBF4 at the G_1/S-phase transition is important for the initiation of S phase (Jackson et al. 1993; for review, see Newlon 1993; Nasmyth, this volume). Thus, the interaction of DBF4 with ORC-2 may target the CDC7 catalytic subunit to the ORC complex, enabling it to phosphorylate the ORC-2 polypeptide (for review, see Barinaga 1994). However, the functional effects of this phosphorylation have not yet been described.

Although the footprint of ARS sequences remains detectable throughout the cell cycle, there is an increase in the protected area during G_1 phase, which is indicative of prereplicative complexes binding in the vicinity of ORC (Diffley et al. 1994). The enlarged footprint may arise through the association of additional proteins with the ORC complex. There is evidence that ORC-6 interacts with CDC6 and CDC46/MCM5 proteins (Li and Herskowitz 1993), which are good candidates for components of the prereplicative complex. CDC6 of budding yeast and its counterpart Cdc18 from fission yeast are important for entry into S phase

(Kelly et al. 1993). The CDC6 gene product shows ATPase activity, but no detectable DNA unwinding activity (Zwerschke et al. 1994). CDC46 also acts at an early stage in DNA replication. The subcellular localization of CDC46 and other members of the MCM 2-3-5 family is regulated in the cell cycle (Hennesey et al. 1991; Y. Chen et al. 1992; Yan et al. 1993; for review, see Tye 1994). These proteins are nuclear only between M and S phase, consistent with the notion that they could participate in the G_1-specific enlargement of the footprint on the origin of replication, but no data are available on this point. One might expect that cell-cycle-dependent phosphorylation events would regulate the functions of the MCM proteins, but again, mutations in the potential sites of phosphorylation of MCM proteins by cyclin-dependent kinases have not affected their function (B. Tye, pers. comm.) (but see section below). It has been speculated based on their amino acid sequences that the MCM proteins may function as ATP-dependent helicases (Koonin 1993).

Xenopus and Mammalian Initiation Proteins

The MCM proteins are found not only in yeast, but also in frogs, mice, and humans (for review, see Tye 1994). Their importance in control of DNA replication is underscored by their recent identification (Chong et al. 1995; Kubota et al. 1995; Madine et al. 1995) as components of a replication licensing factor that was postulated by Blow and Laskey (1988) to limit eukaryotic chromosomal DNA replication to one round per cell cycle (for review, see Coverley and Laskey 1994). Taken together, the data strongly support the notion that a complex of different MCM proteins is required for chromatin replication in the *Xenopus* in vitro system, that their presence in the nucleus before replication is necessary, and that upon replication, they are lost or become easily extractable from the nucleus. Interestingly, replication licensing factor activity is blocked by treatment of *Xenopus* extracts with protein kinase inhibitors, suggesting that phosphorylation is directly or indirectly involved in their function (for review, see Coverley and Laskey 1994; Tye 1994).

A mammalian homolog of MCM3, the P1 protein, was found associated with DNA pol-α:primase (Thömmes et al. 1992b). The nuclear localization of P1 is regulated during the cell cycle like that of its yeast and frog counterparts (Kimura et al. 1994). Nuclear P1 appears to exist in distinct subpopulations: one that is hyperphosphorylated and loosely bound in the nucleus, and another that is underphosphorylated and tightly bound in the nucleus. In early S phase, the tightly bound form of P1 was predominant, but as S phase progressed, the loosely bound P1 gradually increased and became the major form by the end of S phase (Kimura et al. 1994). This behavior would be consistent with a role for P1 in prereplicative chromatin complexes and disassembly of these complexes after cell-cycle-dependent phosphorylation.

OTHER PROTEINS THAT ARE REQUIRED FOR DNA REPLICATION

Pol-α:Primase

Pol-α:primase is composed of four subunits with molecular masses of 180, 68, 58, and 48 kD (p180, p68, p58, and p48) (Thömmes and Hübscher 1990; Wang 1991 and this volume). All the cDNAs encoding the human, mouse, and yeast subunits have been cloned (Lucchini et al. 1987; Wong et al. 1988; Collins et al. 1993; Miyazawa et al. 1993; Stadlbauer et al. 1994). The structure and function of pol-α:primase are highly conserved in eukaryotic organisms (Wang 1991). Pol-α:primase is an essential component of the cellular replication machinery (Pizzagalli et al. 1988; Francesconi et al. 1991), where it functions both in initiation at origins of replication and in the synthesis of Okazaki fragments on the lagging strand of the replication fork (Focher et al. 1988; Prelich and Stillman 1988; Thömmes and Hübscher 1990; Wang 1991). Pol-α:primase also initiates the leading strand at the origin of SV40 replication (Tsurimoto et al. 1990). The p180 subunit harbors the DNA polymerase activity (Pizzagalli et al. 1988; Wong et al. 1988; Nasheuer et al. 1991), the p68 subunit is thought to play a role in the initiation process (Collins et al. 1993; Foiani et al. 1994), and the p58 and p48 subunits comprise the primase activity (Santocanale et al. 1993).

During activation of quiescent mammalian cells, the expression of all subunits is up-regulated. However, in actively cycling cells, gene expression is constitutive and mRNA and protein levels are only marginally enhanced prior to S phase (Wahl et al. 1988; Miyazawa et al. 1993). Thus, if pol-α:primase function is regulated during the cell cycle, it is likely to involve a posttranslational mechanism. Indeed, both the p180 and the p68 subunits are phosphorylated on serine and threonine residues in a cell-cycle-dependent manner (Wong et al. 1986). The p180 subunit is phosphorylated throughout the cell cycle but becomes hyperphosphorylated in the G_2-M transition, whereas the p68 subunit is phosphorylated beginning in S and more heavily at the G_2-M transition (Nasheuer et al. 1991).

Phosphorylation of DNA polymerase-α:primase at different points in the cell cycle has also been studied in budding yeast (Foiani et al. 1995). Both the p180 and p86 (B) subunits are phosphorylated in yeast, but only the B subunit showed a reduced electrophoretic migration upon modification. The kinetics of phosphorylation during the cell cycle revealed that newly synthesized B subunit occurred in an underphosphorylated form during S phase and became modified during G_2. Maternal B subunit was stable and became phosphorylated in early S phase. Both populations of B subunit were dephosphorylated during exit from mitosis (Foiani et al. 1995). The observation that the B subunit is required very early in S phase for initiation, but has no effect on the enzymatic functions of yeast pol-α:primase, suggested that phosphorylation may regulate the activity of the B subunit in initiation of replication (Fig. 4) (Foiani et al. 1995).

The functional relevance of phosphorylation of pol-α:primase has been investigated in vitro. Treatment of pol-α:primase with casein kinase II in vitro phosphorylated the p180 and p58 subunits, but had no influence on either polymerase or primase activity (Podust et al. 1990). Some of the in vivo cell-cycle-dependent phosphorylation sites on the p180 and p68 subunits of mammalian pol-α:primase were shown to be phosphorylated by $p34^{cdc2}$ kinases in vitro, implying these kinases as possible regulators. However, these kinases also had little effect on the enzymatic activities of pol-α:primase; only a slight reduction in affinity for single-stranded DNA was found (Nasheuer et al. 1991).

The functional effect of phosphorylation of pol-α:primase by cyclin-dependent kinases has recently been reexamined using purified proteins. Pol-α:primase expressed in the baculovirus system was modified on the p180 and p68 subunits by cyclin E-, A-, and B-dependent kinases (C. Voitenleitner et al., unpubl.). All of these kinases utilized pol-α:primase approximately equally well as a substrate. Phosphorylation of the enzyme did not affect its polymerase activity, in agreement with published data, although it stimulated primase activity severalfold. In contrast, the ability of pol-α:primase to initiate SV40 DNA replication in vitro in a reaction performed with purified proteins was markedly inhibited after phosphorylation by cyclin-A-dependent kinases. Treatment with cyclin-B-dependent kinase reduced initiation activity more modestly. Cyclin-E-dependent kinase caused no reduction in initiation activity; indeed, a small stimulation was observed in some experiments (C. Voitenleitner et al., unpubl.).

These results raise the question whether phosphorylation of pol-α:primase may influence its interaction with other initiation proteins. Both the p180 and the p68 subunits interact with SV40 T antigen (Dornreiter et al. 1990, 1992, 1993; Collins and Kelly 1991; Collins et al. 1993; Murakami and Hurwitz 1993), and these interactions play a sub-

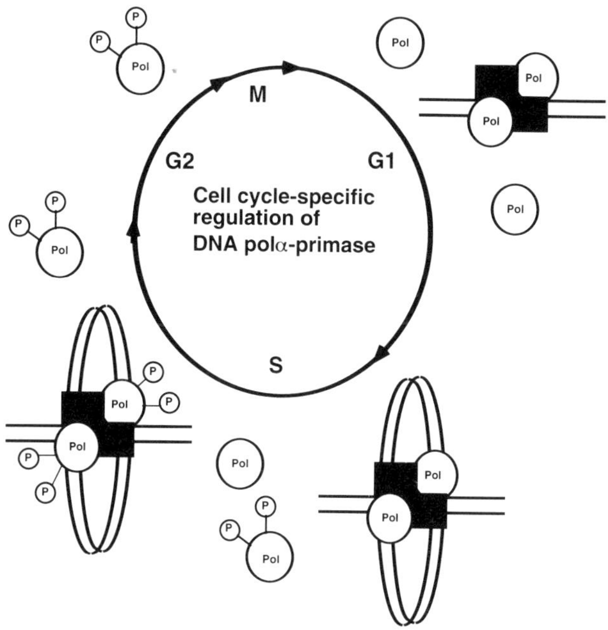

Figure 4 Model for regulation of origin-specific initiation activity of pol-α:primase by cell-cycle-dependent phosphorylation. Hypophosphorylated pol-α:primase that appears in late M or early G_1 (Foiani et al. 1995) is suggested to assemble in a prereplicative complex together with cellular initiation proteins on the cognate origin of DNA replication (Diffley et al. 1994). The complex would remain dormant until activation at the G_1/S border, an event that probably requires cyclin-E-dependent kinase directly or indirectly (Knoblich et al.1994; Jackson et al. 1995; Sauer et al. 1995). Upon initiation, phosphorylation of pol-α:primase is postulated to inhibit its ability to initiate new rounds of replication at an origin. Perhaps phosphorylation of pol-α:primase at the origin is even involved in the transition from initiation to elongation. If cyclin A-dependent kinases, which accumulate as the cell progresses from S through M phase, were responsible for these modifications, the fraction of initiation-active pol-α:primase would be predicted to decline from S to M phase without a concomitant reduction in lagging-strand DNA replication activity. A similar mechanism may regulate the SV40 initiation activity of pol-α:primase in virus-infected cells, thereby facilitating assembly of new virus particles in late S phase and G_2.

stantial role in the initiation of SV40 replication. The ability of phosphorylated pol-α:primase to undergo these interactions has not yet been assayed directly. Intriguingly, however, only a subpopulation of mouse pol-α:primase was associated with Py T antigen, and only this population was able to stimulate viral DNA replication (Moses and

Prives 1994). It will be interesting to investigate the phosphorylation state of these different fractions of pol-α:primase.

Taken together, the results of these studies in different systems with different sets of reagents argue that cell-cycle-dependent phosphorylation of pol-α:primase regulates its function in initiation of DNA replication. Figure 4 depicts a speculative model for this regulation. Since the model makes several predictions that can be tested experimentally, it may be useful in gaining a better understanding of the role of protein phosphorylation in replication.

Replication Protein A

Replication protein A (RP-A), also known as replication factor A (RF-A) and human single-stranded DNA-binding protein (HSSB), was identified as an essential factor for SV40 DNA replication in vitro (Wobbe et al. 1987; Fairman and Stillman 1988; Wold and Kelly 1988). It is the major single-stranded DNA-binding protein in mammalian cells (Seroussi and Lavi 1993) and is highly conserved in all eukaryotes (Wold and Kelly 1988; Brill and Stillman 1989; Mitsis et al. 1993). RP-A is a heterotrimeric protein with subunits of 14, 32, and 70 kD (p14, p32, and p70) (Fairman and Stillman 1988; Wold and Kelly 1988). The cDNAs encoding all three subunits have been cloned from human (Erdile et al. 1990, 1991; Umbricht et al. 1993) and yeast (Heyer et al. 1990; Brill and Stillman 1991) and are essential for cell viability in yeast (Heyer et al. 1990; Brill and Stillman 1991; Longhese et al. 1994).

Human RP-A binds tightly to single-stranded DNA (Wobbe et al. 1987; Fairman and Stillman 1988; Wold and Kelly 1988) with low cooperativity (Kim and Wold 1995) in two distinct complexes, one occupying 8–10 nucleotides, the other 20–30 nucleotides (Blackwell and Borowiec 1994; Kim et al. 1994). Binding affinity to RNA and double-stranded DNA is much lower (Brill and Stillman 1989; Wold et al. 1989; Kim et al. 1992). RP-A has been shown to be associated with an intrinsic DNA-unwinding activity (Georgaki et al. 1992). The nucleic acid-binding properties of the homolog from *Drosophila melanogaster* (dRPA) were reported to be quite similar to those of human RP-A (Mitsis et al. 1993). However, RP-A from *Saccharomyces cerevisiae* (yRPA) was found to be quite different, binding to a site of about 90 nucleotides with high cooperativity (Alani et al. 1992). Yeast RP-A, however, cannot substitute for its human counterpart in SV40 DNA replication in vitro (Brill and Stillman 1991).

The p70 subunit of RP-A contains the intrinsic DNA-binding activity (Wold et al. 1989; Kenny et al. 1990). The p32 and p14 subunits form a stable subcomplex (Henricksen et al. 1994; Stigger et al. 1994) that is important for proper formation of the heterotrimeric complex (Henricksen et al. 1994). The biological functions of the 32-kD and 14-kD subunits are not known. However, monoclonal antibodies directed against the 70-kD, 32-kD, or 14-kD subunits block SV40 DNA replication, suggesting a role for all three subunits in this process (Erdile et al. 1990; Kenny et al. 1990; Umbricht et al. 1993).

RP-A has multiple functions in the replication process (Kenny et al. 1989). During bidirectional unwinding of the SV40 origin of replication by T antigen, it stabilizes the created single-stranded regions. Moreover, it is essential during primosome assembly (Melendy and Stillman 1993) and primer synthesis on the SV40 origin (Schneider et al. 1994), reflecting the specific protein-protein interactions that occur between RP-A, T antigen, and pol-α:primase (Dornreiter et al. 1992, 1993; Nasheuer et al. 1992; Murakami and Hurwitz 1993; K. Weisshart and E. Fanning, unpubl.). RP-A also plays a role in the elongation reaction, where it has been shown to stimulate DNA polymerases α (Kenny et al. 1989; Erdile et al. 1991; Brown et al. 1992), δ (Kenny et al. 1989; Tsurimoto and Stillman 1989; Lee et al. 1991a), and ε (Lee et al. 1991b). In addition, it increases the fidelity of DNA synthesis by pol-α:primase (Carty et al. 1992; Roberts et al. 1993; Suzuki et al. 1994) and stimulates copurifying helicases (Seo et al. 1991; Thömmes et al. 1992a; Georgaki et al. 1994).

Besides its essential functions in DNA replication, RP-A is also required for DNA repair and recombination in vivo (Longhese et al. 1994) and nucleotide excision repair of UV-damaged DNA in vitro (Coverley et al. 1991). It also stimulates in vitro strand-exchange proteins involved in homologous recombination (Heyer et al. 1990; Moore et al. 1991; Alani et al. 1992). RP-A has also been shown to interact with the acidic domains of transcription factors VP16, GAL4, and p53 (Dutta et al. 1993; He et al. 1993; Li and Botchan 1993). The significance of these interactions is still unknown. On the one hand, these factors stimulated BPV replication (Li and Botchan 1993); on the other hand, p53 inhibited SV40 replication (Dutta et al. 1993). It was found recently that yRP-A binds specifically to sequence elements that are involved in the regulation of transcription (Luche et al. 1993; Singh and Samson 1995), so that the interaction of RP-A with transcription factors could play a role in transcriptional regulation.

Several posttranslational modifications of the RP-A subunits have been observed. Both the p70 and p32 subunits are pol-ADP-ribosylated

(Eki and Hurwitz 1991) and phosphorylated (Din et al. 1990; Fang and Newport 1993). The abundance of RP-A remains constant throughout the cell cycle, but the p32 subunit is phosphorylated in a cell-cycle-specific manner on multiple serines (Din et al. 1990). In cycling cells, p32 becomes extensively phosphorylated at the G_1-S transition and remains phosphorylated until G_2-M, when dephosphorylation occurs (Din et al. 1990). The hyperphosphorylated form of p32 shows decreased mobility during gel electrophoresis and runs as a 36-kD species.

On the basis of amino acid sequence, human RP-A p32 has two potential sites for cyclin-dependent kinases, five for DNA-PK, and five for casein kinase II. There is good evidence that cyclin-dependent kinases and DNA-PK phosphorylate RP-A p32 in vitro and in vivo. In vitro, cdc2-cyclin B kinase phosphorylates p32 at Ser-23 and Ser-29, sites that are also phosphorylated in vivo (Dutta and Stillman 1992), concomitant with a modest stimulation of SV40 DNA replication and origin unwinding in G_1 extracts. The same sites were also targeted specifically within the replication initiation complex by cdk2-cyclin A kinase (Fotedar and Roberts 1991, 1992; Dutta and Stillman 1992; Elledge et al. 1992; Pan et al. 1993a). However, mutant RP-A lacking the two cyclin-dependent kinase sites in the p32 subunit supported SV40 DNA replication in vitro as well as wild-type RP-A, demonstrating that phosphorylation of RP-A by cyclin-dependent kinases is dispensable in this system (Henricksen and Wold 1994). Consistent with these results, phosphorylated and unphosphorylated RP-A were equally active in SV40 DNA replication in vitro (Pan et al. 1995).

Under the conditions of the in vitro SV40 replication system, underphosphorylated RP-A becomes hyperphosphorylated (Fotedar and Roberts 1992; Henricksen and Wold 1994). The kinase involved appears to be DNA-PK, but cyclin A-cdk2 may also play a role. RP-A p32 is predominantly phosphorylated when bound to DNA, and the kinase that phosphorylates it is itself bound to single-stranded DNA (Fotedar and Roberts 1992). The p32 subunit in RP-A from cycling embryonic *Xenopus* extracts is also phosphorylated in S phase only in nuclear-localized RP-A bound to single-stranded DNA (Fang and Newport 1993). Hyperphosphorylation of RP-A p32 by DNA-PK in a cyclin-A-activated G_1 extract was reported to require prephosphorylation by cyclin A-cdk2 (Pan et al. 1994). However, mutant RP-A lacking the two potential cyclin-dependent kinase sites in p32 still becomes hyperphosphorylated (Henricksen and Wold 1994; Lee and Kim 1995). Moreover, replication-competent extracts depleted of cyclin-dependent kinases still hyperphosphorylate DNA-bound RP-A in vitro (Dutta and Stillman

1992), whereas extracts depleted of DNA-PK do not (Brush et al. 1994).

Human RP-A is also found hyperphosphorylated in G_1 cells that have been exposed to ionizing radiation (Liu and Weaver 1993) or UV light (Carty et al. 1994). A good candidate for the kinase that phosphorylates RP-A during the damage response is the DNA-PK (Lees-Miller and Anderson 1991), a kinase that is known to be involved in DNA repair and is stimulated by double-strand breaks (for review, see Gottlieb and Jackson 1994). Purified DNA-PK phosphorylates RP-A p32 in vitro (Brush et al. 1994; Pan et al. 1994), and cells that are deficient in DNA-PK activity fail to hyperphosphorylate RP-A after irradiation (Boubnov and Weaver 1995). Hyperphosphorylated RP-A in extracts from irradiated cells appeared to be inactive in SV40 DNA replication in vitro, and additional RP-A purified from unirradiated cells relieved that block (Carty et al. 1994), suggesting that phosphorylation on the potential DNA-PK sites might activate RP-A for its tasks in repair while blocking its function in the replication process. However, other workers have found that phosphorylation of RP-A by DNA-PK did not affect its activity in SV40 DNA replication in vitro (Brush et al. 1994).

Based on the evidence from in vitro SV40 DNA replication, phosphorylation of RP-A appears to have little or no functional importance. However, it may play a role in vivo, such as correct localization to replication centers (for review, see Hozak and Cook 1994). Chromatin replication in *Xenopus* egg extracts is dependent on RP-A (Adachi and Laemmli 1994). Before being assembled into replication centers, RP-A is targeted to the prereplicative sites that form after mitosis and prior to initiation (Adachi and Laemmli 1994). This association is loose due to the absence of single-stranded DNA and is likely to be mediated by protein-protein interactions. Assembly of RP-A into prereplicative complexes does not occur in the presence of active cyclin B-cdc2 kinase, suggesting that only the underphosphorylated form of RP-A found after mitosis can be targeted to the complexes. Active cyclin A-cdk2 kinase did not facilitate or inhibit the assembly of prereplicative complexes (Adachi and Laemmli 1994). However, it colocalized with RP-A, and presumably many other replication proteins, to replication foci (Adachi and Laemmli 1992, 1994; Cardoso et al. 1993), suggesting a possible function in this process.

In summary, it seems likely that phosphorylation of RP-A regulates its activity in vivo, but the mechanism of this regulation remains poorly understood. Much of the confusion may stem from the multiple functions of RP-A in the cell, but it should be noted that the assay system used most frequently to assess the effects of phosphorylation, in vitro SV40

DNA replication, probably does not reveal all of the relevant steps in in vivo regulation of DNA replication.

DNA Polymerases δ and ε

Polymerases δ and/or ε are thought to play a pivotal role in leading-strand DNA synthesis as well as lagging-strand synthesis (Burgers 1991; Nethanel et al. 1992; Podust and Hübscher 1993; see Wang, this volume). Polymerase-δ is a two-subunit complex with polypeptides of approximately 124 kD and 55 kD (Chung et al. 1991) whose activity and processivity are stimulated by complexing with PCNA (Bauer and Burgers 1988). Polymerase-ε consists of a catalytic subunit of 258 kD (Kesti et al. 1993) that copurifies with a smaller 55-kD polypeptide in human (Syväoja and Linn 1989; Lee et al 1991b) and with polypeptides of 80, 34, 30, and 29 kD in yeast extracts (Hamatake et al. 1990). RF-C and PCNA also function to stimulate polymerase-ε (Burgers 1991; Lee et al. 1991b). Besides their involvement in DNA replication, both polymerase-δ and polymerase-ε are implicated in DNA repair (Z. Wang et al. 1993; Blank et al. 1994; Zeng et al. 1994b; Budd and Campbell 1995).

Polymerase-δ mRNA and protein expression is somewhat increased at the G_1-S border, but more significantly, the protein is most actively phosphorylated during S phase (Zeng et al. 1994a). It will be important to assess the enzymatic activities of hyper- and hypophosphorylated polymerase-δ and to look for differences in its interactions with other replication proteins like PCNA (Bauer and Burgers 1988) or RP-A (Longhese et al. 1994).

Polymerase-ε mRNA expression and protein levels are strongly dependent on cell proliferation (Kesti et al. 1993; Tuusa et al. 1995), but their abundance fluctuated less dramatically in actively cycling cells. The phosphorylation state of polymerase-ε has not been investigated.

PCNA and RF-C

PCNA is a phosphoprotein and its phosphorylation state appears to vary in a cell cycle-dependent fashion (Prosperi et al. 1993, 1994). Interestingly, highly phosphorylated PCNA associated with insoluble nuclear structures primarily during S phase, whereas soluble PCNA was found in early G_1 and G_2/M in a weakly phosphorylated form. Expression of PCNA appears to be independent of the cell cycle in actively growing cells. On the other hand, its expression declines in quiescent cells and is strongly induced by mitogens (Mathews et al. 1984). It is also subject to regulation by binding to cdk-cyclin kinases and their inhibitors (for review, see Heichman and Roberts 1994; Pines 1994).

RF-C is a multisubunit protein consisting of five polypeptides (p36.5, p37, p38, p40, and p140; also designated RFC1–5) (Lee et al. 1991b; Tsurimoto and Stillman 1991c). The mammalian and yeast genes encoding the RF-C subunits have been cloned (M. Chen et al. 1992a,b; Bunz et al. 1993; Pan et al. 1993b; Li and Burgers 1994a,b; Luckow et al. 1994; Noskov et al. 1994; Cullman et al. 1995; Hübscher et al., this volume). The p140 subunit has an ATP-binding domain and binds unspecifically to DNA. Its transcripts are most abundant in strongly proliferative tissues but are also relatively high in brain (Luckow et al. 1994). The p40 polypeptide has an ATP-binding motif and shows ATPase activity. It seems to interact with PCNA, polymerase-δ, and the p37 subunit of RF-C (Pan et al. 1993b). The steady-state levels of the mRNA fluctuate only slightly during the cell cycle (Noskov et al. 1994). The p38 subunit binds ATP, and its ATPase activity is stimulated by single-stranded DNA (Li and Burgers 1994a). The p37 subunit shows specific affinity to polymerase-ε and binds specifically to primer-template termini (Pan et al. 1993b). The p37 mRNA levels do not change significantly during the cell cycle (Li and Burgers 1994b). Since RNA levels of two of the subunits are constant during the cell cycle, protein modifications like phosphorylation might contribute to the regulation of RF-C function. A thorough investigation of RF-C expression and modification, however, must await the availability of suitable immunological reagents.

Topoisomerases

Topoisomerases are involved in the maintenance of chromatin and DNA structure (see Hangaard-Andersen et al., this volume). Topoisomerase I relaxes supercoiled DNA, whereas topoisomerase II catalyzes decatenation and unknotting of topologically linked DNA circles, as well as the relaxation of supercoiled DNA (Wang 1985). Both topoisomerases have been shown to be phosphorylated.

Dephosphorylation of serine/threonine residues in topoisomerase I isolated from different sources reduced or abolished its enzymatic activity (Durban et al. 1983; Kaiserman et al. 1988; Pommier et al. 1990; Tournier et al. 1992). Phosphorylation of the protein on tyrosine decreased activity as well (Tse-Dinh et al. 1984). Conversely, phosphorylation of topoisomerase I by casein kinase II and related kinases (Durban et al. 1983; Kaiserman et al. 1988) or protein kinase C (Pommier et al. 1990; Samuels and Shimizu 1992) stimulated activity.

The decatenation activity of topoisomerase II was shown to be stimulated by phosphorylation by casein kinase II, protein kinase C, cdc2-cyclin kinases, and Ca^{++}/calmodulin-dependent protein kinase (Sahyoun et al. 1986; Rottman et al. 1987; Ackerman et al. 1988; Cardenas and Gasser 1993; for review, see Poljak and Käs 1995). The major in vivo kinase acting on topoisomerase II in *S. cerevisiae* has been identified as casein kinase II (Cardenas et al. 1992). Topoisomerase II is hyperphosphorylated during G_2/M phase (Cardenas et al. 1992), consistent with the finding that this phosphorylation is required for activity of topoisomerase II during chromosome segregation. The stimulation of decatenation induced by phosphorylation was found to be at least partly due to an increase in the DNA-binding affinity of topoisomerase II (Dang et al. 1994) and an increased rate of ATP hydrolysis (Corbett et al. 1993).

SUMMARY AND OUTLOOK

Our understanding of the role of protein phosphorylation in regulation of DNA replication remains fragmentary. All of the proteins that recognize and activate eukaryotic origins of replication are phosphorylated, but with the exception of SV40 T antigen, there is no clear picture of how these modifications affect the function of the proteins. Similarly, the key replication enzymes in eukaryotic cells are phosphoproteins, but the functional importance of phosphorylation is established only for topoisomerase II, and in preliminary fashion for pol-α:primase. The lack of data to support a role for phosphorylation in regulation of DNA replication is not due to lack of effort. The mutational analysis of potential phosphorylation sites that was successful in investigating SV40 T antigen has now been extended to many of the cellular origin recognition proteins and several replication enzymes, including pol-α:primase and RP-A, without conspicuous success. Does this mean that these phosphorylation events are fortuitous? The cell-cycle-dependence of phosphorylation of replication proteins, the correlation between phosphorylation of MCM proteins with changes in their subcellular localization, and the importance of the cyclin-dependent kinases, other cell-cycle-dependent kinases, and the DNA-dependent protein kinase suggest that phosphorylation of most replication proteins is probably physiologically relevant. How then can one demonstrate this importance more directly? It should be noted that the current understanding of phosphorylation in control of SV40 T antigen's replication activity was not achieved solely through mutagenesis. Not only were the in vivo phosphorylation sites mapped, but several overexpression systems and a wide variety of biochemical assays were available in many laboratories to probe the functions of T antigen and its mutants before and after treatment with purified kinases and phosphatases over the course of the past decade. As outlined in the introduction, such a broad approach will likely be required to understand the roles of phosphorylation in the control of DNA replication.

ACKNOWLEDGMENTS

We thank Andreas Zeitvogel and Achim Dickmanns for assistance with the figures; Mel DePamphilis, Vladimir Podust, and Christoph Rehfuess for helpful criticism of the manuscript; and M. Lentz, M. Foiani, C. Lehner, T. Kelly, B. Tye, and J. Diffley for communication of results before publication. Work in the authors' laboratories was supported by the Deutsche Forschungsgemeinschaft, Fonds der Chemischen Industrie, BMFT-Genzentrum, the European Community (CHRX-CT93-0248 DG12), and the National Institutes of Health (1 RO1 GM-52948-01).

REFERENCES

Ackerman, P., C.V. Glover, and N. Osherhoff. 1988. Phosphorylation of DNA topoisomerase *in vivo* and in total homogenates of *Drosophila* Kc cells. *J. Biol. Chem.* **263:** 12653–12660.

Adachi Y. and U. Laemmli. 1992. Identification of nuclear pre-replication centers poised for DNA synthesis in *Xenopus* egg extracts: Immunolocalization study of replication protein A. *J. Cell Biol.* **119:** 1–15.

———. 1994. Study of the cell cycle dependent assembly of DNA pre-replication centers in *Xenopus* egg extracts. *EMBO J.* **13:** 4153–4164.

Adamczewski, J.P., J.V. Gannon, and T. Hunt. 1993. Simian virus 40 large T-antigen associates with cyclin A and $p33^{cdk2}$. *J. Virol.* **67:** 6551–6557.

Alani, E., R. Thresher, J.D. Griffith, and R.D. Kolodner. 1992. DNA binding properties of yRP-A and interactions with yeast and *E. coli* strand-exchange proteins. *J. Mol. Biol.* **227:** 54–71.

Barinaga, M. 1994. Yeast enzyme finds fame in link to DNA replication. *Science* **265:** 1175–1176.

Bauer, G.A. and P.M.J. Burgers. 1988. The yeast analog of mammalian cyclin/proliferating cell nuclear antigen interacts with mammalian DNA polymerase δ. *Proc. Natl. Acad. Sci.* **85:** 7506–7510.

Bell, S.P. and B. Stillman. 1992. ATP-dependent recognition of eukaryotic origins of DNA replication by a multiprotein complex. *Nature* **357:** 128–134.

Bell, S.P., R. Kobayashi, and B. Stillman. 1993. Yeast origin recognition complex functions in transcription silencing and DNA replication. *Science* **262:** 1844–1849.

Benson, J.D. and P.M. Howley. 1995. Amino-terminal domains of the bovine papillomavirus type 1 E1 and E2 proteins participate in complex formation. *J. Virol.* **69:** 4364–4372.

Blackwell, L.J. and J.A. Borowiec. 1994. Human replication protein A binds single-stranded DNA in two distinct complexes. *Mol. Cell. Biol.* **14:** 3993–4001.

Blank, A., B. Kim, and L.A. Loeb. 1994. Polymerase δ is required for base excision repair of DNA methylation damage in *Saccharomyces cerevisiae. Proc. Natl. Acad. Sci.* **91:** 9047–9051.

Blow, J.J. and R.A. Laskey. 1988. A role for the nuclear envelope in controlling DNA replication within the cell cycle. *Nature* **332:** 546–548.

Bockus, B.J. and B. Schaffhausen. 1987. Localization of the phosphorylations of polyomavirus large T antigen. *J. Virol.* **61:** 1155–1163.

Bonne-Andrea, C., S. Santuccii, P. Clertant, and F. Tillier. 1995. Bovine papillomavirus E1 protein binds specifically DNA polymerase α but not replication protein A. *J. Virol.* **69:** 2341–2350.

Borowiec, J.A., F.B. Dean, P.A. Bullock, and J. Hurwitz. 1990. Binding and unwinding—How T antigen engages the SV40 origin of DNA replication. *Cell* **60:** 181–184.

Boubnov, N.V. and D.T. Weaver. 1995. *scid* cells are deficient in Ku and replication protein A phosphorylation by the DNA-dependent protein kinase. *Mol. Cell. Biol.* **15:** 5700–5706.

Bream, G.L., C.-A. Ohmstede, and W.C. Phelps. 1993. Characterization of human papillomavirus type 11 E1 and E2 proteins expressed in insect cells. *J. Virol.* **67:** 2655–2663.

Brill, S.J. and B. Stillman. 1989. Replication factor-A from *Saccharomyces cerevisiae* functions in the unwinding of SV40 origin of replication. *Nature* **342:** 92–95.

———. 1991. Replication factor-A from *Saccharomyces cerevisiae* is encoded by three essential genes coordinately expressed at S-phase. *Genes Dev.* **5:** 1589–1600.

Brown, G.W., T.E. Melendy, and D.S. Ray. 1992. Conservation of structure and function of DNA replication protein A in the trypanosomatid *Crithidia fasciculata. Proc. Natl. Acad. Sci.* **89:** 10227–10231.

Brush, G.S., C.W. Anderson, and T.J. Kelly. 1994. The DNA-activated protein kinase is required for the phosphorylation of replication protein A during simian virus 40 DNA replication. *Proc. Natl. Acad. Sci.* **91:** 12520–12524.

Budd, M.E. and J.L. Campbell. 1995. DNA polymerases required for repair of UV-induced damage in *Saccharomyces cerevisiae. Mol. Cell. Biol.* **15:** 2173–2179.

Bunz, F., K. Kobayashi, and B. Stillman. 1993. cDNAs encoding the large subunit of human replication factor C. *Proc. Natl. Acad. Sci.* **90:** 11014–11018.

Burgers, P. 1991. *Saccharomyces cerevisiae* replication factor C. II. Formation and activity of complexes with the proliferating cell nuclear antigen and with DNA polymerases δ and ε. *J. Biol. Chem.* **266:** 22698–22706.

Cardenas, M.E. and S.M. Gasser. 1993. Casein kinase II copurifies with yeast topoisomerase II and reactivates the dephosphorylated enzyme. *J. Cell Sci.* **104:** 533–543.

Cardenas, M.E., Q. Dang, C.V.C. Glover, and S.M. Gasser. 1992. Casein kinase II phosphorylates the eukaryote-specific C-terminal domain of topoisomerase II *in vivo. EMBO J.* **11:** 1785–1795.

Cardoso, M.C., H. Leonhardt, and B. Nadal-Ginard. 1993. Reversal of terminal differentiation and control of DNA replication: Cyclin A and cdk2 specifically localize at subnuclear sites of DNA replication. *Cell* **74:** 979–992.

Carty, M.P., A.S. Levine, and K. Dixon. 1992. HeLa cell single-stranded DNA-binding protein increases the accuracy of DNA synthesis by DNA polymerase α in vitro. *Mutat. Res.* **274:** 29–43.

Carty, M.P., M. Zernik-Kobak, S. McGrath, and K. Dixon. 1994. UV light-induced DNA synthesis arrest in HeLa cells is associated with changes in phosphorylation of human single-stranded DNA-binding protein. *EMBO J.* **13:** 2114–2123.

Cegielska, A. and D.M. Virshup. 1993. Control of simian virus 40 replication by the HeLa nuclear kinase casein kinase I. *Mol. Cell. Biol.* **13:** 1202–1211.

Cegielska, A., I. Moarefi, E. Fanning, and D.M. Virshup. 1994a. T-antigen kinase inhibits simian virus 40 DNA replication by phosphorylation of intact T antigen on serines 120 and 123. *J. Virol.* **68:** 269–275.

Cegielska, A., S. Shaffer, R. Derua, J. Goris, and D.M. Virshup. 1994b. Different oligomeric forms of protein phosphatase 2A activate and inhibit simian virus 40 DNA replication. *Mol. Cell. Biol.* **14:** 4616–4623.

Chen, M., Z.-Q. Pan, and J. Hurwitz. 1992a. Sequence and expression in *Escherichia coli* of the 40-kDa subunit of activator 1 (replication factor C) of HeLa cells. *Proc. Natl. Acad. Sci.* **89:** 2516–2552.

———. 1992b. Studies of the cloned 37-kDa subunit of activator 1 (replication factor C) of HeLa cells. *Proc. Natl. Acad. Sci.* **89:** 5211–5215.

Chen, Y., K. Hennessy, D. Botstein, and B.-K. Tye. 1992. CDC46/MCM5, a yeast protein whose subcellular localization is cell-cycle regulated, is involved in DNA replication at autonomously replicating sequences. *Proc. Natl. Acad. Sci.* **89:** 10459–10463.

Chong, J.P.J., H.M. Mahbubani, C.-Y. Khoo, and J.J. Blow. 1995. Purification of an MCM-containing complex as a component of the DNA replication licensing system. *Nature* **375:** 418–421.

Chung, D.W., J. Zhang, C.-K. Tan, E.W. Davie, A.G. So, and K.M. Downey. 1991. Primary structure of the catalytic subunit of human DNA polymerase δ and chromosomal location of the gene. *Proc. Natl. Acad. Sci.* **88:** 11197–11201.

Cohen, P. 1989. The structure and regulation of protein phosphatases. *Annu. Rev. Biochem.* **58:** 453–508.

Collins, K.L. and T.J. Kelly. 1991. The effects of T antigen and replication protein A on the initiation of DNA synthesis by polymerase α:primase. *Mol. Cell. Biol.* **11:** 2108–2115.

Collins, K.L., A.A.R. Russo, B.Y. Tseng, and T.J. Kelly. 1993. The role of the 70 kDa subunit of human DNA polymerase α in DNA replication. *EMBO J.* **12:** 4555–4566.

Corbett, A.H., A.W. Fernald, and N. Osterhoff. 1993. Protein kinase C modulates the catalytic activity of topoisomerase II by enhancing the rate of ATP hydrolysis: Evidence for a common mechanism of regulation by phosphorylation. *Biochemistry* **32:** 2090–2097.

Coverley, D. and R.A. Laskey. 1994. Regulation of eukaryotic DNA replication. *Annu. Rev. Biochem.* **63:** 745–776.

Coverley, D., M.K. Kenny, M. Munn, W.D. Rupp, D.P. Lane, and R.D. Wood. 1991. Requirement for the replication protein SSB in human excision repair. *Nature* **349:** 538–541.

Cowie, A. and R. Kamen. 1984. Multiple binding sites for polyomavirus large T antigen within regulatory sequences of polyomavirus DNA. *J. Virol.* **52:** 750–760.

Cullman, G., K. Fien, R. Kobayashi, and B. Stillman. 1995. Characterization of the five replication factor C genes of *Saccharomyces cerevisiae. Mol. Cell. Biol.* **15:** 4661–4671.

Dang, Q., G.-C. Alghisi, and S.M. Gasser. 1994. Phosphorylation of the C-terminal domain of yeast topoisomerase II by casein kinase II affects DNA-protein interaction. *J. Mol. Biol.* **243:** 10–24.

Dean, F.B., J.A. Borowiec, T. Eki, and J. Hurwitz. 1992. The simian virus 40 T antigen double hexamer assembles around the DNA at the replication origin. *J. Biol. Chem.* **267:** 14129–14137.

Diffley, J.F.X. 1992. Early events in eukaryotic DNA replication. *Trends Cell Biol.* **2:** 298–303.

Diffley, J.F.X. and J.H. Cocker. 1992. Protein-DNA interactions at a yeast replication origin. *Nature* **357:** 169–172.

Diffley, J.F.X., J.H. Cocker, J. Dowell, and A. Rowley. 1994. Two steps in the assembly of complexes at yeast replication origins *in vivo. Cell* **78:** 303–316.

Din, S.U., S.J. Brill, M.P. Fairman, and B. Stillman. 1990. Cell cycle regulated phosphorylation of DNA replication factor A from human and yeast cells. *Genes Dev.* **4:** 968–977.

Dornreiter, I., W.C. Copeland, and T.S.F. Wang. 1993. Initiation of simian virus 40 DNA replication requires the interaction of a specific domain of human DNA polymerase α with large T antigen. *Mol. Cell. Biol.* **13:** 809–820.

Dornreiter, I., A. Höss, A.K. Arthur, and E. Fanning. 1990. SV40 T antigen binds directly to the catalytic subunit of DNA polymerase α. *EMBO J.* **9:** 3329–3336.

Dornreiter, I., L.F. Erdile, U. Gilbert, D. von Winkler, T.J. Kelly, and E. Fanning. 1992. Interaction of DNA polymerase α-primase with replication protein A and SV40 T antigen. *EMBO J.* **11:** 769–776.

Dowell, S.J., P. Romanowski, and J.F.X. Diffley. 1994. Interaction of Dbf4, the Cdc7 protein kinase regulatory subunit, with yeast replication origins *in vivo. Science* **265:** 1243–1246.

Durban, E., J.S. Mills, D. Roll, and H. Busch. 1983. Phosphorylation of purified Novikoff hepatoma topoisomerase I. *Biochem. Biophys. Res. Commun.* **111:** 897–905.

Dutta, A. and B. Stillman. 1992. cdc2 family kinases phosphorylate a human cell DNA replication factor, RPA, and activate DNA replication. *EMBO J.* **11:** 2189–2199.

Dutta, A., J.M. Ruppert, J.C. Aster, and E. Winchester. 1993. Inhibition of DNA replication by p53. *Nature* **365:** 79–82.

Eki, T. and J. Hurwitz. 1991. Influence of poly (ADP-ribose) polymerase on the enzymatic synthesis of SV40 DNA. *J. Biol. Chem.* **266:** 3087–3100.

Elledge, S.J., R. Richman, F.L. Hall, R.T. Williams, N. Lodgson, and J.W. Harper. 1992. *CDK2* encodes a 33-kDa cyclin A-associated protein kinase and is expressed before *CDC2* in the cell cycle. *Proc. Natl. Acad. Sci.* **89:** 2907–2911.

Erdile, L.F., M.S. Wold, and T.J. Kelly. 1990. The primary structure of the 32-kDa subunit of human replication protein A. *J. Biol. Chem.* **265:** 3177–3182.

Erdile, L.F., W.-D. Heyer, R.D. Kolodner, and T.J. Kelly. 1991. Characterization of a cDNA encoding the 70-kDa single-stranded DNA-binding subunit of human replication protein A and its role in DNA replication. *J. Biol. Chem.* **266:** 12090–12098.

Fairman M.P. and B. Stillman. 1988. Cellular factors required for multiple stages of SV40 DNA replication *in vitro. EMBO J.* **7:** 1211–1218.

Fang, F. and J.W. Newport. 1993. Distinct roles of cdk2 and cdc2 in RP-A phosphorylation during the cell-cycle. *J. Cell Sci.* **106:** 983–994.

Fanning, E. 1992. Simian virus 40 large T antigen: The puzzle, the pieces, and the emerging picture. *J. Virol.* **66:** 1289–1293.

———. 1994. Control of SV40 DNA replication by protein phosphorylation: A model for cellular DNA replication. *Trends Cell. Biol.* **4:** 250–255.

Fanning, E. and R. Knippers. 1992. Structure and function of simian virus 40 large T antigen. *Annu. Rev. Biochem.* **61:** 55–85.

Fernandez-Sarabia, M.J., A. Sutton, T. Zhong, and K.T. Arndt. 1992. SIT4 protein phosphatase is required for the normal accumulation of SWI4, CLN1, CLN2, and HCS26 RNAs during late G1. *Genes Dev.* **2:** 2417–2428.

Focher, F., E. Ferrari, S. Spadari, and U. Hübscher. 1988. Do DNA polymerases δ and α act coordinately as leading and lagging strand replicases. *FEBS Lett.* **229:** 6–10.

Foiani, M., G. Liberi, G. Lucchini, and P. Plevani. 1995. Cell cycle-dependent phosphorylation and dephosphorylation of the yeast DNA polymerase α-primase B subunit. *Mol. Cell. Biol.* **15:** 883–891.

Foiani, M., F. Marini, D. Gamba, G. Lucchini, and P. Plevani. 1994. The B subunit of the DNA polymerase α-primase complex in *Saccharomyces cerevisiae* executes an essential function at the initial stage of DNA replication. *Mol. Cell. Biol.* **14:** 923–933.

Foss, M., F.J. McNally, P. Laurenson, and J. Rine. 1993. Origin recognition complex (ORC) in transcriptional silencing and DNA replication in *S. cerevisiae*. *Science* **262:** 1838–1844.

Fotedar, R. and J.M. Roberts. 1991. Association of $p34^{cdc2}$ with replicating DNA. *Cold Spring Harbor Symp. Quant. Biol.* **56:** 325–333.

———. 1992. Cell cycle regulated phosphorylation of RPA-32 occurs within the replication initiation complex. *EMBO J.* **11:** 2177–2187.

Francesconi, S., M.P. Longhese, A. Piseri, C. Santocanale, G. Lucchini, and P. Plevani. 1991. Mutations in the conserved yeast DNA primase domains impair DNA replication in vivo. *Proc. Natl. Acad. Sci.* **88:** 3877–3881.

Georgaki, A., B. Strack, V. Podust, and U. Hübscher. 1992. DNA unwinding activity of replication protein A. *FEBS Lett.* **308:** 240–244.

Georgaki, A., N. Tuteja, B. Sturzenegger, and U. Hübscher. 1994. Calf thymus DNA helicase F, a replication protein A copurifying enzyme. *Nucleic Acid Res.* **22:** 1128–1134.

Gilbert, D.M. and S.N. Cohen. 1987. Bovine papillomavirus plasmids replicate randomly in mouse fibroblasts throughout S-phase of the cell cycle. *Cell* **50:** 59–68.

Gottlieb, T.M. and S.P. Jackson. 1994. Protein kinases and DNA damage. *Trends Biochem. Sci.* **19:** 500–503.

Grässer, F.A., K. Mann, and G. Walter. 1987. Removal of serine phosphates from simian virus 40 T antigen increases its ability to stimulate DNA replication *in vitro* but has no effect on ATPase and DNA binding. *J. Virol.* **61:** 3373–3380.

Hamatake, R.K., H. Hasegawa, A.B. Clark, K. Bebenek, T. Kunkel, and A. Sugino. 1990. Purification and characterization of DNA polymerase II from the yeast *Saccharomyces cerevisiae*. *J. Biol. Chem* **265:** 4072–4083.

Hassauer, M., K.H. Scheidtman, and G. Walter. 1996. Mapping of phosphorylation sites in polyomavirus large T antigen. *J. Virol.* **58:** 805–816.

He, Z., B.T. Brinton, J. Greenblatt, J.A. Hassell, and C.J. Ingles. 1993. The transactivator proteins VP16 and GAL4 bind replication factor A. *Cell* **73:** 1223–1232.

Heichman, K.A. and J.M. Roberts. 1994. Rules to replicate by. *Cell* **79:** 557–562.

Hennesey, K., C.D. Clarek, and D. Botstein. 1991. Subcellular localisation of yeast *CDC*

46 varies with the cell cycle. *Genes Dev.* **4:** 2252–2263.

Henricksen, L.A. and M.S. Wold. 1994. Replication protein A mutants lacking phosphorylation sites for p34cdc2 kinase support DNA replication. *J. Biol. Chem.* **269:** 24203–24208.

Henricksen, L.A., C.B. Umbricht, and M.S. Wold. 1994. Recombinant replication protein A: Expression, complex formation, and functional characterization. *J. Biol. Chem.* **269:** 11121–11132.

Heyer, W.-D., M.R.S. Rao, L.F. Erdile, T.J. Kelly, and R.D. Kolodner. 1990. An essential *Saccharomyces cerevisiae* single-stranded DNA binding protein is homologous to the large subunit of RP-A. *EMBO J.* **9:** 2321–2329.

Hoffmann, I., G. Draetta, and E. Karsenti. 1994. Activation of the phosphatase activity of human cdc25A by a cdk2-cyclin E dependent phosphorylation at the G1/S transition. *EMBO J.* **13:** 4302–4310.

Höss, A., I. Moarefi, K.-H. Scheidtmann, L.J. Cisek, J.L. Corden, I. Dornreiter, A.K. Arthur, and E. Fanning. 1990. Altered phosphorylation pattern of simian virus 40 T antigen expressed in insect cells by using a baculovirus vector. *J. Virol.* **64:** 4799–4807.

Hozak, P. and P.R. Cook. 1994. Replication factories. *Trends Cell Biol.* **4:** 48–52.

Hubbard, M.J. and P. Cohen. 1993. On target with a new mechanism for the regulation of protein phosphorylation. *Trends Biochem Sci.* **18:** 172–177.

Huberman, J.A. 1992. Quest's end and question's begin. *Curr. Biol.* **2:** 351–352.

———. 1994. A tale of two functions. *Nature* **367:** 20–21.

Hurwitz, J., F.B. Dean, A.D. Kwong, and S.H. Lee. 1990. The *in vitro* replication of DNA containing the SV40 origin. *J. Biol. Chem.* **265:** 18043–18046.

Ishimi, Y., A. Claude, P. Bullock, and J. Hurwitz. 1988. Complete enzymatic synthesis of DNA containing the SV40 origin of replication. *J. Biol. Chem.* **263:** 19723–19733.

Jackson, A.L., P.M.B. Pahl, K. Harrison, J. Rosamond, and R.A. Sclafani. 1993. Cell cycle regulation of the yeast Cdc7 protein kinase by association with Dbf4 protein. *Mol. Cell. Biol.* **13:** 2899–2908.

Jackson, P.K., S. Chevalier, M. Philippe, and M.W. Kirschner. 1995. Early events in DNA require cyclin E and are blocked by $p21^{CIP1}$. *J. Cell Biol.* **130:** 755–769.

Jinno, S., K. Suto, A. Nagata, M. Igarashi, Y. Kanaoka, H. Nojima, and H. Okayama. 1994. Cdc25A is a novel phosphatase functioning early in the cell cycle. *EMBO J.* **13:** 1549–1556.

Kaiserman, H.B., T.S. Ingebritsen, and R.M. Benbow. 1988. Regulation of *Xenopus laevis* DNA topoisomerase I activity by phosphorylation *in vitro*. *Biochemistry* **27:** 3216–3222.

Kelly, T.J. 1988. SV40 DNA replication. *J. Biol. Chem.* **263:** 17889–17892.

Kelly, T.J., P.V. Jallepalli, and R.K. Clyne. 1994. Silence of the ORCs. *Curr. Biol.* **4:** 238–241.

Kelly, T.J., G.S. Martin, S.L. Forsburg, R.J. Stephen, A. Russo, and P. Nurse. 1993. The fission yeast $cdc18^+$ gene product couples S phase to START and mitosis. *Cell* **74:** 371–382.

Kenny, M.K., S.-H. Lee, and J. Hurwitz. 1989. Multiple functions of human single-stranded-DNA binding protein in simian virus 40 DNA replication: Single-strand stabilization and stimulation of DNA polymerases α and δ. *Proc. Natl. Acad. Sci.* **86:** 9757–9761.

Kenny, M., U. Schlegel, H. Furneaux, and J. Hurwitz. 1990. The role of human single-stranded DNA binding protein and its individual subunits in simian virus 40 DNA replication. *J. Biol. Chem.* **265:** 7693–7700.

Kesti, T., H. Frantti, and J.E. Syväoja. 1993. Molecular cloning of the cDNA for the catalytic subunit of human DNA polymerase ε. *J. Biol. Chem.* **268:** 10238–10245.

Kim, C. and M.S. Wold. 1995. Recombinant human replication protein A binds to polynucleotides with low cooperativity. *Biochemistry* **34:** 2058–2064.

Kim, C., B.F. Paulus, and M.S. Wold. 1994. Interactions of human replication protein A with oligonucleotides. *Biochemistry* **33:** 14197–14206.

Kim, C., R.O. Snyder, and M.S. Wold. 1992. Binding properties of replication protein A from human and yeast cells. *Mol. Cell. Biol.* **12:** 3050–3059.

Kimura, H., N. Nozaki, and K. Sugimoto. 1994. DNA polymerase α associated protein P1, a murine homolog of yeast MCM3, changes its intranuclear distribution during the DNA synthetic period. *EMBO J.* **13:** 4311–4320.

Knoblich, J.A., K. Sauer, L. Jones, H. Richardson, R. Saint, and C.F. Lehner. 1994. Cyclin E controls S phase progression and its down-regulation during *Drosophila* embryogenesis is required for the arrest of cell proliferation. *Cell* **77:** 107–120.

Koonin, E.V. 1993. A common set of conserved motifs in a vast variety of putative nucleic acid-dependent ATPases including MCM proteins involved in the initiation of eukaryotic DNA replication. *Nucleic Acid. Res.* **21:** 2541–2547.

Kubota, Y., S. Mimura, S.-I. Nishimoto, H. Takisawa, and H. Nojima. 1995. Identification of the yeast MCM3-related protein as a component of *Xenopus* DNA replication licensing factor. *Cell* **81:** 601–609.

Lambert, P.F. 1991. Papillomavirus DNA replication. *J. Virol.* **65:** 341–342.

Law, M.-F., D.R. Lowy, I. Dvoretzky, and P.M. Howley. 1981. Mouse cell transformed by bovine papillomavirus contain only extrachromosomal viral DNA sequences. *Proc. Natl. Acad. Sci.* **78:** 2727–2731.

Lee, S.-H. and D.K. Kim. 1995. The role of the 34-kDa subunit of human replication protein A in simian virus 40 DNA replication *in vitro*. *J. Biol. Chem.* **270:** 12801–12807.

Lee, S.-H., A.D. Kwong, Z.Q. Pan, and J. Hurwitz. 1991a. Studies on the activator 1 protein complex, an accessory factor for proliferating cell nuclear antigen-dependent DNA polymerase δ. *J. Biol. Chem.* **266:** 594–602.

Lee, S.-H., Z.-Q. Pan, A.D. Kwong, P.M.J. Burgers, and J. Hurwitz. 1991b. Synthesis of DNA by DNA polymerase ε *in vitro*. *J. Biol. Chem.* **266:** 22707–22717.

Lees-Miller, S.P. and C.W. Anderson. 1991. The DNA-activated protein kinase DNA-PK: A potential coordinator of nuclear events. *Cancer Cells* **3:** 341–346.

Lentz, M.R., D. Pak, I. Mohr, and M.R. Botchan. 1993. The E1 replication protein of bovine papillomavirus type 1 contains an extended nuclear localization signal that includes a $p34^{cdc2}$ phosphorylation site. *J. Virol.* **67:** 1414–1423.

Li, J.J. and B.M. Alberts. 1992. Eukaryotic initiation rites. *Nature* **357:** 114–115.

Li, J.J. and I. Herskowitz. 1993. Isolation of ORC6, a component of the yeast origin recognition complex by a one-hybrid system. *Science* **262:** 1870–1874.

Li, R. and M.R. Botchan. 1993. The acidic transcriptional activation domains of VP16 and p53 bind the cellular replication protein A and stimulate *in vitro* BPV-1 DNA replication. *Cell* **73:** 1207–1221.

Li, X. and P.M.J. Burgers. 1994a. Molecular cloning and expression of the *Saccharomyces cerevisiae RFC3* gene, an essential component of replication factor C. *Proc. Natl. Acad. Sci.* **91:** 868–872.

———. 1994b. Cloning and characterization of the essential *Saccharomyces cerevisiae RFC4* gene encoding the 37-kDa subunit of replication factor C. *J. Biol. Chem.* **269:** 21880–21884.

Liu, V.F. and D.T. Weaver. 1993. The ionizing-radiation induced replication protein A phosphorylation response differs between ataxia telangiectasia and normal human cells. *Mol. Cell. Biol.* **13:** 7222–7231.

Longhese, M.P., P. Plevani, and G. Lucchini. 1994. Replication factor A is required *in vivo* for DNA replication, repair and recombination. *Mol. Cell. Biol.* **14:** 7884–7890.

Lucchini, G., S. Francesconi, M. Foiani, G. Badaracco, and P. Plevani. 1987. Yeast DNA polymerase-primase complex: Cloning of *PRI1*, a single essential gene related to DNA primase activity. *EMBO J.* **6:** 737–742.

Luche, R.M., W.C. Smart, T. Marion, M. Tillman, R.A. Sumrada, and T.G. Cooper. 1993. *Saccharomyces cerevisiae* BUF protein binds to sequences participating in DNA replication in addition to those mediating transcriptional repressing (URS1) and activation. *Mol. Cell. Biol.* **13:** 5749–5761.

Luckow, B., F. Bunz, B. Stillman, P. Lichter, and G. Schütz. 1994. Cloning, expression and chromosomal localization of the 140-kDa subunit of replication factor C from mice and humans. *Mol. Cell. Biol.* **14:** 1626–1634.

Ludlow, J.W. 1992. Selective ability of S-phase cell extracts to dephosphorylate large T antigen in vitro. *Oncogene* **7:** 1011–1014.

Madine, M.A., C.-Y. Khoo, A.D. Mills, and R.A. Laskey. 1995. MCM3 complex required for cell cycle regulation of DNA replication in vertebrate cells. *Nature* **375:** 421–424.

Mann, D.J., V. Dombradi, and P.-T.W. Cohen. 1993. *Drosophila* protein phosphatase V functionally complements a SIT4 mutant in *Saccharomyces cerevisiae* and its amino-terminal region can confer this complementation to a heterologous phosphatase catalytic domain. *EMBO J.* **12:** 4833–4842.

Mathews, M.B., R.M. Bernstein, B.R. Franza, and J.T. Garrels. 1984. Identity of the proliferating cell nuclear antigen and cyclin. *Nature* **309:** 374–376.

McVey, D., L. Brizuela, I. Mohr, D.R. Marshak, Y. Gluzman, and D. Beach. 1989. Phosphorylation of large tumor antigen by cdc2 stimulates SV40 DNA replication. *Nature* **341:** 503–507.

McVey, D., S. Ray, Y. Gluzman, L. Berger, A.G. Wildeman, D.J. Marshak, and P. Tegtmeyer. 1993. Cdc2 phosphorylation of threonine 124 activates the origin-unwinding functions of simian virus 40 T antigen. *J. Virol.* **67:** 5206–5215.

Melendy, T. and B. Stillman. 1993. An interaction between replication protein A and SV40 T antigen appears essential for primosome assembly during SV40 DNA replication. *J. Biol. Chem.* **268:** 3389–3395.

Micklem, G., A. Rowley, J. Harwood, K. Nasmyth, and J.F.X. Diffley. 1993. Yeast origin recognition complex is involved in DNA replication and transcriptional silencing. *Nature* **366:** 87–89.

Mitsis, P.G., S.C. Kowalczykowski, and I.R. Lehman. 1993. A single-stranded DNA binding protein from *Drosophila melanogaster*: Characterization of the heterotrimeric protein and its interaction with single-stranded DNA. *Biochemistry* **32:** 5257–5266.

Miyazawa, H., M. Izumi, S. Tada, R. Takada, M. Masutani, M. Ui, and F. Hanaoka. 1993. Molecular cloning of the cDNAs for the four subunits of mouse DNA polymerase α-primase complex and their gene expression during cell proliferation and the cell cycle. *J. Biol. Chem.* **268:** 8111–8122.

Moarefi, I., C. Schneider, K. van Zee, A. Höss, A.K. Arthur, and E. Fanning. 1993a. Control of SV40 DNA replication by protein phosphorylation. In *DNA replication and the cell cycle* (ed. E. Fanning et al.), pp. 157–169. Springer Verlag, Berlin.

Moarefi, I., D. Small, I. Gilbert, M. Höffner, S.K. Randall, C. Schneider, A.A.R. Russo, U. Ramsperger, A.K. Arthur, H. Stahl, T.J. Kelly, and E. Fanning. 1993b. Mutation of cyclin-dependent kinase phosphorylation site in simian virus 40 (SV40) large T antigen specifically blocks SV40 origin DNA unwinding. *J. Virol.* **67:** 4992–5002.

Moore, S.P., L. Erdile, T. Kelly, and R. Fishel. 1991. The human homologous pairing protein HPP-1 is specifically stimulated by the cognate single-stranded binding protein hRP-A. *Proc. Natl. Acad. Sci.* **88:** 9067–9071.

Moreno, S. and P. Nurse. 1990. Substrates for $p34^{cdc2}$: In vivo veritas? *Cell* **61:** 549–551.

Moses, K. and C. Prives. 1994. A unique subpopulation of murine DNA polymerase α/primase specifically interacts with polyomavirus T antigen and stimulates DNA replication. *Mol. Cell. Biol.* **14:** 2767–2776.

Murakami, Y. and J. Hurwitz. 1993. DNA polymerase α stimulates the ATP-dependent binding of simian virus tumor T antigen to the SV40 origin of replication. *J. Biol. Chem.* **268:** 11018–11027.

Nasheuer, H.-P., A. Moore, A.F. Wahl, and T.S.-F. Wang. 1991. Cell cycle-dependent phosphorylation of human DNA polymerase α. *J. Biol. Chem.* **266:** 7893–7903.

Nasheuer, H.-P., D. von Winkler, C. Schneider, I. Dornreiter, I. Gilbert, and E. Fanning. 1992. Purification and functional characterization of bovine RP-A in an *in vitro* replication system. *Chromosoma* **102:** S52–S59.

Nethanel, T., T. Zlotkin, and G. Kaufmann. 1992. Assembly of simian virus 40 Okazaki pieces from DNA primers is reversibly arrested by ATP depletion. *J. Virol.* **66:** 6634–6640.

Newlon, C.S. 1993. Two jobs for the origin replication complex. *Science* **262:** 1830–1831.

Norbury, C. and P. Nurse. 1992. Animal cell cycles and their control. *Annu. Rev. Biochem.* **61:** 441–470.

Noskov, V., S. Maki, Y. Kawasaki, S.-H. Leem, B.-I. Ono, H. Araki, Y. Pavlov, and A. Sugino. 1994. The *RFC2* gene encoding a subunit of replication factor C of *Saccharomyces cerevisiae*. *Nucleic Acids Res.* **22:** 1527–1535.

Pan, Z.-Q., A. Amin, and J. Hurwitz. 1993a. Characterization of the *in vitro* reconstituted cyclin A or B1 dependent cdk2 and cdc2 kinase activities. *J. Biol. Chem.* **268:** 20443–20451.

Pan, Z.-Q., M. Chen, and J. Hurwitz. 1993b. The subunits of activator 1 (replication factor C) carry out multiple functions for proliferating-cell nuclear antigen-dependent DNA synthesis. *Proc. Natl. Acad. Sci.* **90:** 6–10.

Pan, Z.-Q., A.A. Amin, E. Gibbs, H. Niu, and J. Hurwitz. 1994. Phosphorylation of p34 subunit of human single-stranded-DNA-binding protein in cyclin A-activated G_1 extracts is catalyzed by cdk-cyclin A complex and DNA-dependent protein kinase. *Proc. Natl. Acad. Sci.* **91:** 8343–8347.

Pan, Z.-Q., C.-H. Park, A.A. Amin, J. Hurwitz, and A. Sancar. 1995. Phosphorylated and unphosphorylated forms of human single-stranded DNA-bindng protein are equally active in simian virus 40 DNA replication and in nucleotide excision repair. *Proc. Natl. Acad. Sci.* **92:** 4636–4640.

Park, P., W. Copeland, L. Yang, T. Wang, M.R. Botchan. and I.J. Mohr. 1994. The cellular DNA polymerase α-primase is required for papillomavirus DNA replication and

associates with the viral E1 helicase. *Proc. Natl. Acad. Sci.* **91:** 8700–8704.
Pines, J. 1994. p21 inhibits cyclin shock. *Nature* **369:** 520–521.
Pizzagalli, A., P. Valsasnini, P. Plevan, and G. Lucchini. 1988. DNA polymerase I gene of *Saccharomyces cerevisiae*: Nucleotide sequence, mapping of temperature sensitive mutation, and protein homology with other DNA polymerases. *Proc. Natl. Acad. Sci.* **85:** 3772–3776.
Podust, V. and U. Hübscher. 1993. Lagging strand DNA synthesis by calf thymus DNA polymerases α, β, δ, and ε in the presence of auxiliary proteins. *Nucleic Acids Res.* **21:** 841–846.
Podust, V., G. Bialek, H. Sternbach, and F. Grosse. 1990. Casein kinase II phosphorylates DNA-polymerase-α-DNA-primase without affecting its basic enzymatic properties. *Eur. J. Biochem.* **193:** 189–193.
Poljak, L. and E. Käs. 1995. Resolving the role of topoisomerase II in chromatin structure and function. *Trends Cell Biol.* **5:** 348–354.
Pommier, Y., D. Kerrigan, K.J.D. Hartman, and R.I. Glatzer. 1990. Phosphorylation of mammalian DNA topoisomerase I and activation by protein kinase C. *J. Biol. Chem.* **265:** 9418–9422.
Prelich, G. and B. Stillman. 1988. Coordinated leading and lagging strand synthesis duing SV40 DNA replication *in vitro* requires PCNA. *Cell* **53:** 117–126.
Prives, C. 1990. The replication functions of SV40 T antigen are regulated by phosphorylation. *Cell* **61:** 735–738.
Prosperi, E., A.I. Scovassi, L.A. Stivala, and L. Bianchi. 1994. Proliferating cell nuclear antigen bound to DNA synthesis sites: Phosphorylation and association with cyclin D1 and cyclin A. *Exp. Cell Res.* **215:** 257–262.
Prosperi, E., L.A. Stivala, E. Sala, I.A. Scovassi, and L. Bianchi. 1993. Proliferating cell nuclear antigen complex formation induced by ultraviolet irradiation in human quiescent fibroblasts as detected by immunostaining and flow cytometry. *Exp. Cell Res.* **205:** 320–325.
Rao, H. and B. Stillman. 1995. The origin recognition complex interacts with a bipartite DNA binding site within yeast replicators. *Proc. Natl. Acad. Sci.* **92:** 2224–2228.
Roberts, J.D., D. Nguyen, and T.A. Kunkel. 1993. Frameshift fidelity during replication of double-stranded DNA in HeLa cell extracts. *Biochemistry* **32:** 4083–4089.
Rottman, M., H.C. Schroder, M. Gramzow, K. Reinneisen, B. Kurelec, A. Dorn, U. Friese, and W.E. Muller. 1987. Specific phosphorylation of proteins in pore complex-laminae from sponge *Geodia cydonium* by the homologous aggregation factor and phorbol ester. Role of protein kinase C in phosphorylation of DNA topoisomerase II. *EMBO J.* **6:** 3939–3944.
Rowley, A., J.H. Cocker, J. Harwood, and J.F.X. Diffley. 1995. Initiation complex assembly at budding yeast replication origins begins with the recognition of a bipartite sequence by limiting amounts of the initiator, ORC. *EMBO J.* **14:** 2631–2641.
Sahyoun, N., M. Wolf, J. Besterman, T. Hsieh, M. Sander, H. LeVine, K.J. Chang, and P. Cuatrecasas. 1986. Protein kinase C phosphorylates topoisomerase II: Topoisomerase activation and its possible role in phorbol ester-induced differentiation of HL-60 cells. *Proc. Natl. Acad. Sci.* **83:** 1603–1607.
Samuels, D.S. and N. Shimizu. 1992. DNA topoisomerase I phosphorylation in mouse fibroblasts treated with 12-O-tetradecanoylphorbol-13-acetate *in vitro* by protein kinase C. *J. Biol. Chem.* **267:** 11156–11162.
Santocanale, C., M. Foiani, G. Lucchini, and P. Plevani. 1993. The isolated 48,000-dalton

subunit of yeast DNA primase is sufficient for RNA primer synthesis. *J. Biol. Chem.* **268:** 1343–1348.

Sauer, K., J.A. Knoblich, H. Richardson, and C.F. Lehner. 1995. Distinct modes of cyclin E/cdc2c kinase regulation and S-phase control in mitotic and endoreduplication cycles of *Drosophila* embryogenesis. *Genes Dev.* **9:** 1327–1339.

Scheidtmann, K.-H. 1986. Phosphorylation of simian virus 40 large T antigen: Cytoplasmic and nuclear phosphorylation sites differ in their metabolic stability. *Virology* **150:** 85–95.

Scheidtmann, K.-H., M. Buck, J. Schneider, D. Kalderon, E. Fanning, and A.E. Smith. 1991. Biochemical characterization of phosphorylation site mutants of simian virus 40 large T antigen: Evidence for interactions between amino- and carboxy-terminal domains. *J. Virol.* **65:** 1479–1490.

Schneider, C., K. Weisshart, L.A. Guarino, I. Dornreiter, and E. Fanning. 1994. Species-specific functional interactions of DNA polymerase α-primase with simian virus 40 (SV 40) T antigen require SV 40 origin DNA. *Mol. Cell. Biol.* **14:** 3176–3185.

Schneider, J. and E. Fanning. 1988. Mutations in the phosphorylation sites of simian virus 40 (SV40) T antigen alter its origin DNA-binding specificity for site I or II and affect SV40 DNA replication activity. *J. Virol.* **62:** 1598–1605.

Seki, M., T. Enomoto, T. Eki, A. Miyahima, Y. Murakami, R. Hanaoka, and M. Ui. 1990. DNA helicase and nucleotide 5′ triphosphate activities of polyoma virus large tumor antigen. *Biochemistry* **29:** 1003–1009.

Seo, Y.-S., S.-H. Lee, and J. Hurwitz. 1991. Isolation of a DNA helicase from HeLa cells requiring the multisubunit human single-stranded DNA-binding protein for activity. *J. Biol. Chem.* **266:** 13161–13170.

Seo, Y.-S., F. Müller, M. Lusky, and J. Hurwitz. 1993a. Bovine papillomavirus (BPV)-encoded E1 protein contains multiple activities required for BPV DNA replication. *Proc. Natl. Acad. Sci.* **90:** 702–706.

Seo, Y.-S., F. Müller, M. Lusky, E. Gibbs, H.-Y. Kim, B. Phillips, and J. Hurwitz. 1993b. Bovine papillomavirus (BPV)-encoded E2 protein enhances binding of E1 protein to the BPV replication origin. *Proc. Natl. Acad. Sci.* **90:** 2865–2869.

Seroussi, E. and S. Lavi. 1993. Replication protein A is the major single-stranded DNA binding protein detected in mammalian cell extracts by gel retardation assays and UV-crosslinking of long and short single-stranded DNA molecules. *J. Biol. Chem.* **268:** 7147–7154.

Singh, K.K. and L. Samson. 1995. Replication protein A binds to regulatory elements in yeast DNA repair and DNA metabolism genes. *Proc. Natl. Acad. Sci.* **92:** 4907–4911.

Spalholz, B.A., A.A. McBride, T. Sarafi, and J. Quintero. 1993. Binding of bovine papillomavirus E1 to the origin is not sufficient for DNA replication. *Virology* **193:** 201–212.

Stadlbauer, F., A. Brueckner, C. Rehfuess, C. Eckerskorn, F. Lottspeich, V. Förster, B.Y. Tseng, and H.-P. Nasheuer. 1994. DNA replication *in vitro* by recombinant DNA-polymerase α-primase. *Eur. J. Biochem.* **222:** 781–793.

Stigger, E., F.B. Dean, J. Hurwitz, and S.-H. Lee. 1994. Reconstitution of functional single-stranded DNA-binding protein from individual subunits expressed by recombinant baculoviruses. *Proc. Natl. Acad. Sci.* **91:** 579–583.

Stillman, B. 1989. Initiation of eukaryotic DNA replication *in vitro*. *Annu. Rev. Cell Biol.* **5:** 197–245.

Suzuki, M., S. Izuta, and S. Yoshida. 1994. DNA polymerase α overcomes an error-prone pause site in the presence of replication protein-A. *J. Biol. Chem.* **269:** 10225–10228.

Syväoja, J. and S. Linn. 1989. Characterization of a large form of DNA polymerase δ from HeLa cells that is insensitive to proliferating cell nuclear antigen. *J. Biol. Chem.* **264:** 2489–2497.

Thömmes P. and Hübscher, U. 1990. Eukaryotic DNA replication: Enzymes and proteins acting at the fork. *Eur. J. Biochem.* **194:** 699–712.

Thömmes, P., E. Ferrari, R. Jessberger, and U. Hübscher. 1992a. Four different DNA helicases from calf thymus. *J. Biol. Chem.* **267:** 6063–6073.

Thömmes, P., R. Fett, B. Schray, R. Burkhart, M. Barnes, C. Kennedy, N.C. Brown, and R. Knippers. 1992b. Properties of the nuclear P1 protein, a mammalian homolog of the yeast Mcm3 replication protein. *Nucleic Acids Res.* **20:** 1069–1074.

Thorner, L.K., D.A. Lim, and M.R. Botchan. 1993. DNA-binding domain of papillomavirus type 1 E1 helicase: Structural and functional aspects. *J. Virol.* **67:** 6000–6014.

Tournier, M.-F., J. Sobczak, B. Nechaud, and M. Duguet. 1992. Comparison of biochemical properties of DNA-topoisomerase I from normal and proliferating liver. *Eur. J. Biochem.* **210:** 359–364.

Tse-Dinh, W.-C., T.W. Wong, and A.R. Goldberg. 1984. Virus and cell-encoded protein tyrosine kinases inactivate DNA topoisomerases *in vitro*. *Nature* **312:** 785–786.

Tsurimoto, T. and B. Stillman. 1989. Multiple replication factors augment DNA synthesis by two eukaryotic DNA polymerases, α and δ. *EMBO J.* **8:** 3883–3889.

———. 1991a. Replication factors required for SV40 DNA replication *in vitro*. I. DNA structure specific recognition of a primer-template junction by eukaryotic DNA polymerases and their accessory proteins. *J. Biol. Chem.* **266:** 1950–1960.

———. 1991b. Replication factors required for SV40 DNA replication *in vitro*. II. Switching of DNA polymerase α and δ during initiation of leading and lagging strand synthesis. *J. Biol. Chem.* **266:** 1961–1968.

———. 1991c. Purification of cellular replication factor, RF-C, that is required for coordinated leading and lagging strands during simian virus 40 DNA replication *in vitro*. *Mol. Cell. Biol.* **9:** 609–619.

Tsurimoto, T., T. Melendy, and B. Stillman. 1990. Sequential initiation of lagging and leading strand synthesis by two different polymerase complexes at the SV40 DNA replication origin. *Nature* **346:** 534–539.

Turchi, J.J., L. Huang, R.S. Murante, Y. Kim, and R.A. Bambara. 1994. Enzymatic completion of mammalian lagging-strand DNA replication. *Proc. Natl. Acad. Sci.* **91:** 9803–9807.

Tuusa, J., L. Uitto, and J.E. Syväoja. 1995. Human DNA polymerase ε is expressed during cell proliferation in a manner characteristic of replicative DNA polymerases. *Nucleic Acids Res.* **23:** 2178–2183.

Tye, B.-K. 1994. The MCM2-3-5 proteins: Are they replication licensing factors? *Trends Cell Biol.* **4:** 160–166.

Umbricht, C.B., L.F. Erdile, E.W. Jabs, and T.J. Kelly. 1993. Cloning, overexpression, and genomic mapping of the 14-kDa subunit of replication protein A. *J. Biol. Chem.* **268:** 6131–6138.

Ustav, E. and A. Stenlund. 1991. Transient replication of BPV-1 requires two viral polypeptides encoded by the E1 and E2 open reading frames. *EMBO J.* **10:** 449–457.

Ustav, E., M. Ustav, P. Szymanski, and A. Stenlund. 1993. The bovine papillomavirus origin of replication requires a binding site for the E2 transcriptional activator. *Proc. Natl. Acad. Sci.* **90:** 898–902.

Virshup, D.M., M.G. Kaufmann, and T.J. Kelly. 1989. Activation of SV40 DNA replication *in vitro* by cellular protein phosphatase 2A. *EMBO J.* **8:** 3891–3898.

Virshup, D.M., A.A.R. Russo, and T.J. Kelly. 1992. Mechanism of activation of simian virus 40 DNA replication by protein phosphatase 2A. *Mol. Cell. Biol.* **12:** 4883–4895.

Waga, S. and B. Stillman. 1994. Anatomy of a DNA replication fork revealed by reconstitution of SV40 DNA replication *in vitro*. *Nature* **369:** 207–212.

Waga, S., G. Bauer, and B. Stillman. 1994. Reconstitution of complete SV40 DNA replication with purified DNA replication factors. *J. Biol. Chem.* **269:** 10923–10934.

Wahl, A.F., A.M. Geis, B.H. Spain, S.W. Wong, D. Korn, and T.S.-F. Wang. 1988. Gene expression of human DNA polymerase α during cell proliferation and the cell cycle. *Mol. Cell. Biol.* **8:** 5016–5025.

Wang, E.H. and C. Prives. 1991. DNA helicase and duplex DNA unwinding activities of polyoma and SV40 large T antigen display similarities and differences. *J. Biol. Chem.* **266:** 12668–12675.

Wang, E.H., S. Bhattacharyya, and C. Prives. 1993. The replication functions of polyomavirus large tumor antigen are regulated by phosphorylation. *J. Virol.* **67:** 6788–6796.

Wang, J.C. 1985. DNA topoisomerases. *Annu. Rev. Biochem.* **54:** 665–697.

Wang, T.S.-F. 1991. Eukaryotic DNA polymerases. *Annu. Rev. Biochem.* **60:** 513–552.

Wang, Z., X. Wu, and E.C. Friedberg. 1993. DNA repair synthesis during base excision repair *in vitro* is catalyzed by polymerase ε and is influenced by polymerase α and δ in *Saccharomyces cerevisiae*. *Mol. Cell. Biol.* **13:** 1051–1058.

Wessel, R., J. Schweizer, and H. Stahl. 1992. Simian virus 40 T-antigen DNA helicase is a hexamer which forms a binary complex during bidirectional unwinding from the viral origin of DNA replication. *J. Virol.* **66:** 804–815.

Wobbe, C.R., L. Weissbach, J.A. Borowiec, F.B. Dean, Y. Murakami, P. Bullock, and J. Hurwitz. 1987. Replication of simian virus 40 origin-containing DNA with purified components. *Proc. Natl. Acad. Sci.* **84:** 1834–1838.

Wold, M.S. and T. Kelly. 1988. Purification and characterization of replication protein A, a cellular protein required for *in vitro* replication of simian virus 40 DNA. *Proc. Natl. Acad. Sci.* **85:** 2523–2527.

Wold, M.S., D.H. Weinberg, D.M. Virshup, J.J. Li, and T.J. Kelly. 1989. Identification of cellular proteins required for simian virus 40 DNA replication. *J. Biol. Chem.* **264:** 2801–2809.

Wong, S.W., L.R. Paborski, P.A. Fisher, T.S.-F. Wang, and D. Korn. 1986. Structural and enzymological characterization of immunoaffinity-purified DNA polymerase α-primase complex from KB cells. *J. Biol. Chem.* **261:** 7958–7968.

Wong, S.W., A.F. Wahl, P.-M. Yuan, N. Arai, B.E. Pearson, K.-I. Arai, D. Korn, M.W. Hunkapiller, and T.S-F. Wang. 1988. Human DNA polymerase α gene expression is cell proliferation dependent and its primary structure is similar to both prokaryotic and eukaryotic replicative DNA polymerases. *EMBO J.* **7:** 37–47.

Yan, H., A.M. Merchant, and B.K. Tye. 1993. Cell cycle-regulated nuclear localization of MCM2 and MCM3, which are required for the initiation of DNA synthesis at chromosomal replication origins in yeast. *Genes Dev.* **7:** 2149–2160.

Yang, L., M.S. Wold, J.J. Li, T.J. Kelly, and L.F. Liu. 1987. Roles of DNA topoisomerases in simian virus 40 DNA replication *in vitro*. *Proc. Natl. Acad. Sci.* **84:** 950–954.

Yang, L., I. Mohr, E. Fouts, D.A. Lim, M. Nohaile, and M. Botchan. 1993. The E1 protein of bovine papillomavirus 1 is an ATP-dependent DNA helicase. *Proc. Natl. Acad. Sci.* **90:** 5086–5090.

Zeng, X.-R., H. Hao, Y. Jiang, and M.Y.W.T. Lee. 1994a. Regulation of human polymerase δ during the cell cycle. *J. Biol. Chem.* **269:** 24027–24033.

Zeng, X.-R., Y. Jiang, S.J. Zhang, H. Hao, and M.Y.W.T. Lee. 1994b. Polymerase δ is involved in the cellular response to UV damage in human cells. *J. Biol. Chem.* **269:** 1348–1351.

Zwerschke, W., H.-W. Rottjakob, and H. Küntzel. 1994. The *Saccharomyces cerevisiae* CDC6 gene is transcribed at late mitosis and encodes an ATP/GTPase controlling S phase initiation. *J. Biol. Chem.* **269:** 23351–23356.

11

Control of S Phase

Kim Nasmyth
Institute for Molecular Pathology
A1030, Vienna, Austria

Sustained cell proliferation requires the duplication and segregation between daughter cells of every cell constituent. Because chromosomes are present in only one or two copies per cell, their duplication and segregation are particularly tightly regulated. To ensure that daughter cells inherit at least one copy of each of their genes and to maintain an appropriate balance in their relative numbers, the vast majority of eukaryotic cells replicate each sequence in their genomes once and only once per cell cycle. To ensure that daughter cells inherit sufficient but not excessive cytoplasm for the execution of their genetic programs, i.e., for protein synthesis, the duplication and segregation of chromosomes must occur no more or less frequently than cells duplicate the rest of their constituents. Finally, to ensure that cells maintain their ploidy, which is essential if they are to contribute to the germ line, reduplication must not recur until sister chromatids have been segregated at anaphase.

STARTING DNA REPLICATION IN G_1

Coordinating DNA Replication and Cell Growth

Pulse labeling with radioactive tracers showed that DNA replication occupies a defined window within the interdivision period, which is separated from the previous and preceding M phases by two gap periods, called G_1 and G_2, respectively (Howard and Pelc 1951). Chromosome duplication and segregation are therefore periodic processes. Unlike chromosomes, most cellular constituents like ribosomes and enzymes are present in large numbers and are synthesized fairly continuously during the cell cycle (Mitchison 1970). During sustained cell proliferation, successive rounds of DNA replication and chromosome segregation (known as the chromosome cycle) must occur with the same frequency as cells double the number of their ribosomes and other cell constituents, known as cell growth (see Fig. 1). Coordination between these two processes is

DNA Replication in Eukaryotic Cells

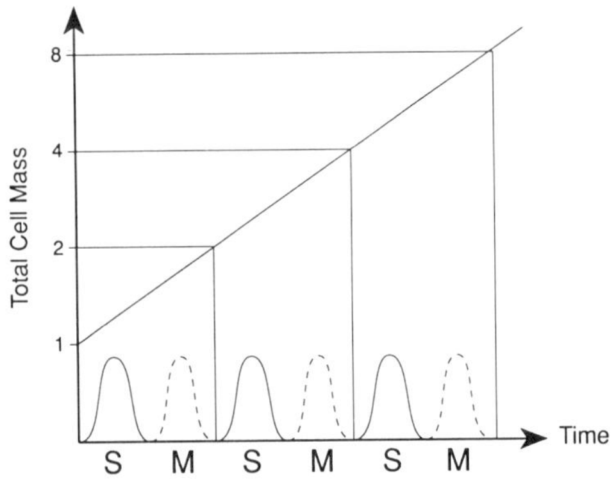

Figure 1 Coordinating the chromosome cycle and cell growth. Total cell mass increases exponentially (for a short while even in the absence of progression through the chromosome cycle), and the signals that trigger S and M phases (shown as oscillations on the abcissa) must occur with the same frequency as cytoplasmic mass doubling if nuclear/cytoplasmic ratios are to be kept within narrow bounds.

needed to maintain the ratio between nucleus and cytoplasm within certain bounds. It may be achieved either by a dependence of the chromosome cycle on the growth cycle or vice versa.

Independence of Cell Growth from the Chromosome Cycle

Dependence of cell growth on the chromosome cycle has been most carefully analyzed in the the fission yeast *Schizosaccharomyces pombe*, where increases in the rates of total protein or RNA synthesis are largely unaffected for several hours after inactivation of the Cdk1 (cdc2) protein kinase, which specifically blocks the chromosome cycle (Creanor and Mitchison 1984). To a first approximation, therefore, DNA replication does not limit the rate of cell growth. A similar conclusion can be drawn from the behavior of chromosome cycle mutants of the budding yeast *Saccharomyces cerevisiae* (Hartwell et al. 1974). A lack of DNA replication does eventually cause cell growth to slow down, after two to three generation times. Presumably, templates for mRNA synthesis eventually become limiting, at least for a few key growth genes.

Mutants in key cell cycle regulators do not exist to test the dependence of growth on the chromosome cycle in mammalian cells, so inhibitors have instead been employed. Some cell lines, such as HeLa, appear to shut down protein synthesis in the presence of DNA synthesis inhibitors, whereas others, such as CHO, continue cell growth like yeast *cdk1* mutants (Kung et al. 1993). Whether the growth cycle is largely independent of the chromosome cycle in animal or plant cells is therefore an unresolved but important issue that would best be investigated using primary cell lines, or better still, analyzing cells in their natural environment in vivo. Arrest in late G_1 using newly identified cyclin kinase inhibitors would be preferable to replication inhibitors, because the shutdown in protein synthesis seen in HeLa cells could be a specific response to arrest in S phase itself. Anecdotal evidence on the behavior of senescent cells, whose cell cycles are blocked in G_1, suggests that they become very large, indicating that their cell cycle arrest does not interfere with their growth (Lucibello et al. 1993). Many differentiated cells, for example, neurons, must continue growth long after they have ceased proliferating. An intrinsic independence of growth from the chromosome cycle would therefore not only favor microorganisms whose growth rates would be restricted by a tight linkage, but would also facilitate the formation of very differently sized differentiated cells in metazoa.

Dependence of S Phase on Growth

In the absence of any tight dependence of growth on the chromosome cycle, nuclear/cytoplasmic ratios are largely maintained by a dependence of the chromosome cycle on growth. The chromosome cycles of most eukaryotic cells are, with few exceptions, tightly dependent on RNA and protein synthesis. The dependence of DNA replication on cell growth was first noticed by Killander and Zetterberg, who observed that there was much greater variation in the size of fibroblast cells at birth than at the onset of S phase and proposed that DNA replication depends on growth to a critical cell size (Killander and Zetterberg 1965). A particularly clear example of this size control is found in the yeast *S. cerevisiae*, whose asymmetric division by budding gives rise to mother and daughter cells with very different sizes. Mother cells are born equal to or larger than the critical size and have very short G_1 periods, whereas the daughter cells are born well below the critical size and must spend long periods growing before they can enter S phase (Hartwell and Unger 1977). Many eukaryotic cells like amoeba, *Physarum*, and the fission yeast *S. pombe* have very short G_1 periods and appear to break this rule. Size controls over S-phase onset nevertheless exist, even in these organisms. In *S. pombe*, for example, cells are usually born larger than the size needed for S-phase onset due to size controls over entry into mitosis. Abolition of this mitotic control creates cells smaller than the critical size, which, like fibroblasts or budding yeast daughter cells, must grow considerably before they can undergo DNA replication (Nurse and Thuriaux 1977). Thus, size controls over the onset of S phase can be obscured by controls that restrict passage through other cell cycle transitions. An extreme example is the repeated rounds of replication without any growth during embryonic cleavage divisions, which are possible only because prior growth of the egg ensures that it can undergo many divisions before cells below a critical size are produced.

One consequence of the asymmetric interdependence of cell growth and the chromosome cycle is that sustained increases in cellular proliferation cannot be driven simply by accelerating the chromosome cycle but must instead be driven by the forces that cause cells to grow. This is well established for microorganisms, where accelerating the chromosome cycle merely leads to reductions in cell size (Nurse 1975), but it is not widely appreciated that it might be valid also for metazoan cell proliferation. It is often assumed that deregulation of chromosome cycle activators, for example Cdk1, -2, and -4 cyclin-dependent kinases, should accelerate cell division (Quelle et al. 1993) and thereby contribute to transformation. This can only be true if the chromosome cycle or regulators concerned with it also affect the rate of a cell's growth, which does not occur in microorganisms and remains to be established for mammalian cells.

Growth-related Protein Synthesis Needed for S-phase Onset

How cell growth to a critical size triggers S phase is an important but still largely unsolved problem. Size per se is not the critical variable but rather the nuclear/cytoplasmic ratio, because cell size is directly proportional to ploidy. One corollary of the observation that increases in the rate of protein synthesis are largely independent of the chromosome cycle is that a cell's rate of protein synthesis is largely proportional to cell size in growing G_1 cells. Thus, the total rate of protein synthesis in a cell may be a good measure of its size, and a critical level may be required for S-phase onset. The widespread observations that passage from G_1 to S phase is extremely sensitive to inhibitors of protein synthesis are consistent with this notion (Cross et al. 1989).

In yeast, there are at least two broadly different programs of protein synthesis, one occurring in quiescent stationary-phase cells, which includes synthesis of several heat shock and polyubiquitin proteins (Iida and Yahara 1984), and another occurring in growing cells, which includes synthesis of growth-related proteins like ribosomal proteins and presumably those needed specifically for the chromosome cycle. These two programs of protein synthesis are distinguished by their different dependence on the initiation factor eIF-4e, which binds to mRNA caps and is encoded by *CDC33* (Brenner et al. 1988). Growth-related but not stationary-phase-related protein synthesis depends on eIF-4e. The G_1 arrest of temperature-sensitive *cdc33* mutations at the restrictive temperature implies, not surprisingly, that S-phase onset depends on proteins synthesized as part of the growth-related program. How cells turn on the eIF-4e dependent and growth-related program of protein synthesis in response to nutrients or growth factors is therefore a very important aspect of S-phase control but is outside the realm of this review. Growth-factor- or nutrient-induced transcription of growth-related genes clearly also has an important part in the stimulation of growth-related protein synthesis, but changes in the translational apparatus that allow these mRNAs to be translated may be equally, if not more, important.

Two Parables about Growth and Division

Certain phenomena concerning the control of S phase cannot be understood without appreciating the distinction between cell growth and the chromosome cycle. A good example is the part played by cAMP in *S. cerevisiae*. Mutants defective in adenyl cyclase or cAMP-dependent protein kinases cannot sustain growth-related protein synthesis, and they arrest in G_1 (Matsumoto et al. 1982), from which it has been concluded that cAMP must be a positive effector of S phase. More recently, it has been found that cAMP is involved in modulating cell size in response to nutrients. Cells growing in poor media are smaller than those growing in rich media, principally because of a reduction in the critical size needed for entry into S phase and budding (Lorincz and Carter 1979). Mutants whose cAMP-dependent protein kinases cannot respond to cAMP cannot alter their size in response to nutrients (Tokiwa et al. 1994). Furthermore, addition of cAMP alone to the medium mimics the effects of transferring cells to a richer medium; that is, it increases the cell size needed for S phase. It is thought to do this by inhibiting the accumulation of G_1 cyclins needed for S phase (Baroni et al. 1994; Tokiwa et al. 1994). cAMP must therefore be considered as an inhibitor of S phase; but how can it be both an inhibitor and a promoter of S phase? This paradox is resolved only if one appreciates that cAMP is regulating two very different types of physiological processes, both of which are needed for S phase. On the one hand, it is involved (directly or indirectly) in signaling activation of growth-related protein synthesis, whereas, on the other hand, it inhibits the accumulation of mRNAs for G_1 cyclins, whose translation could be part of the growth-related protein synthesis program and which have highly specific roles in triggering S phase. In having these two effects, cAMP helps to increase the rate of protein synthesis in cells transferred to rich media while simultaneously increasing their cell size. The function of this size control is not clear, but it could be beneficial for the rapid execution of the chromosome cycle due to a cell size-related increase in protein synthetic capacity.

A factor that reduces growth-related protein synthesis while accelerating the onset of S phase (i.e., the opposite effect to cAMP) may have an important role in the differentiation of erythrocytes (Dolznig et al. 1995). Undifferentiated precursor cells whose proliferation is supported by mixture of growth factors can be induced to differentiate synchronously into mature nondividing erythrocytes by transfer to medium containing erythropoetin. The rate of protein synthesis remains fairly constant during the first 72 hours but gradually switches from growth-related proteins to proteins specific to erythrocytes like hemoglobin. During much of this period, cell division is temporarily accelerated due to a dramatic reduction in the length of G_1, which is all the more remarkable because the end product is a cell that cannot proliferate at all. Thus, like cAMP, erythropoetin has at least two effects: It promotes the chromosome cycle, presumably by increasing the activity of cyclin-dependent protein kinases (see later), but it also shuts down growth-related protein synthesis. The consequence of accelerating division but not cell growth is the production of erythrocytes that are many times smaller than their precursor cells. A moral to be drawn from these two stories is that changes in cell size are sometimes a better indicator of changes in the regulation of the chromosome cycle than are changes in mass doubling times.

S- and M-phase Promoting Factors

What is synthesized by growing cells that triggers DNA replication and then mitosis? What do oocytes store up that allows fertilized eggs to undergo successive cleavages in the absence of any growth? What do differentiating erythrocytes transiently make more of that reduces their G_1 period? Do cells by mysterious means measure all their constituents, for example, the number of ribosomes or the total rate of growth-related protein synthesis, or do they instead simply measure the quantity of one or a small number of components that are synthesized as part of the general growth program of the cell and whose abundance is proportional to cell size? Evidence for such a "sampling" mechanism has come from two approaches: the isolation of yeast mutants with an altered cell size and the identification of factors and genes needed for the onset of S phase.

Lessons from Cell Fusion Studies

Due to their large size, many multinucleate cells can be readily coalesced by cutting and grafting. This enabled early cell cycle researchers to address what would happen to the chromosome cycle when cells at different stages were combined into one (Johnson and Rao 1971). Experiments with ciliates, amoebae, and, in particular, the myxomycete plasmodia like *Physarum polycephalum* showed that nuclear synchrony was achieved very rapidly. Nuclei derived from "early" cells were accelerated, whereas those from "late" cells were retarded in their progression toward mitosis. The larger the plasmodium from late cells, the more rapidly were the early nuclei accelerated. These studies suggested that a substance capable of triggering mitosis gradually accumulated during G_2.

Most of the cells on which early coalescence experiments were performed have very short G_1 periods, and it was not possible to address the nature of S-phase inducers. These were eventually addressed in experiments with HeLa cells (Rao and Johnson 1970), where it was found that fusion between G_1 and S-phase cells advanced the entry of G_1 nuclei into S phase, but fusion of G_1 with G_2 cells did not. Significantly, DNA replication of S-phase nuclei was barely, if at all, retarded by the addition of G_1 cytoplasm and nuclei. Fusion between S and G_2 cells failed to induce S phase in G_2 nuclei, but again the addition of G_2 cytoplasm and nuclei did not retard DNA replication within S-phase nuclei. These results indicated that S but neither G_1 nor G_2 cells contained a substance (*S*-phase *p*romoting *f*actor or SPF) capable of inducing DNA replication in G_1 but not in G_2 nuclei. Similar experiments suggested that mitotic HeLa cells, like those from *Physarum* (Rusch et al. 1966), contain a substance (now refered to as MPF) that is capable of inducing premature M phase in nuclei at all other stages of the cell cycle (Johnson and Rao 1970). The picture that emerged from these cell fusion experiments is one in which the abundance of S- and M-phase inducers oscillates during the cell cycle with different phases. The former appear in late G_1 and disappear as cells enter G_2, and the latter appear in late G_2 and disappear at the end of mitosis.

An M-phase-promoting factor was eventually purified from mature *Xenopus* eggs (Masui and Markert 1971), which are arrested in metaphase of meiosis II, using an assay that measured its ability to induce chromosome condensation and nuclear membrane breakdown (Lohka et al. 1988). It proved to be identical to the cdc2 protein kinase, which had by this time been implicated in the control of mitosis in *S. pombe* due to the existence of rare dominant alleles that caused cells to divide at half the normal cell size (Nurse and Thuriaux 1980; Nurse 1990). No convenient biochemical assay was ever developed for S-phase-promoting factors, and the present candidates for these factors were identified by genetic studies of the yeast cell cycle.

S-phase-promoting Factors in *S. cerevisiae*

Genes Required for the Onset of S Phase

S. cerevisiae mutants defective in the chromosome cycle fall into two classes: those with primary defects in nuclear division and those with primary defects in entering S phase (Pringle and Hartwell 1981). However, all mutants with S-phase defects eventually also ceased to undergo nuclear division, whereas most mutants with M-phase defects also ceased eventually to replicate DNA, confirming earlier conclusions, drawn for the use of inhibitors, that S and M phases were interdependent. Why cells must normally complete M phase before they can rereplicate their chromosomes is an important question that we address below.

When *S. cerevisiae* cells reach a critical size, they initiate not only DNA replication but also budding, the first step toward cytokinesis, and spindle pole body duplication, the first step toward building a bipolar mitotic spindle. This point in the yeast cell cycle, which is called Start, is also the point at which haploid cells become refractory to cell cycle inhibition by sexual pheromones and the point at which protein synthesis and nutrients become less critical for completing the cell cycle (Pringle and Hartwell 1981). Mutants with primary defects in entering S phase fell into three groups according to their degrees of pleiotropy. The least pleiotropic class included, initially, mutants in only a single gene, *CDC7*, which encodes a protein kinase that is activated in late G_1 (Hollingsworth and Sclafani 1990). *cdc7* mutants fail to initiate DNA replication. Nevertheless, they bud normally, duplicate and separate their spindle pole bodies, form mitotic spindles, and switch off the budding process, with the result that they arrest as large single-budded cells containing an undivided nucleus with a fully formed mitotic spindle (Pringle and Hartwell 1981). Subsequent studies have shown that alleles of *DBF4* (Chapman and Johnston 1989), *CDC6* (Bueno and Russell 1992), *CDC46*, *CDC47*, *CDC54* (Hennessy et al. 1991), *MCM2*, and *MCM3* (Gibson et al. 1990) arrest with similar phenotypes. *DBF4* encodes a regulatory subunit of the Cdc7 protein kinase (Jackson et al. 1993), whereas *CDC46*, *MCM3*, and *MCM5* encode members of a family of related proteins with ATPase motifs characteristic of DNA helicases (Tye 1994), homologs of which are needed for the initiation of DNA replication also in fission yeast (Coxon et al. 1992) and in vertebrates (Thommes et al. 1992). The initiation of DNA replication might be the exclusive function of this group of genes.

The next class of mutations mapped to three genes: *CDC4*, *CDC34*, and *CDC53*. These mutants are more pleiotropic than the *CDC7* class, because they fail not only to enter S phase but also to form mitotic spindles (Pringle and Hartwell 1981). However, they duplicate their spindle pole bodies and form buds. Another of their characteristics is that they fail to shut off the budding process after the formation of the first bud, with the result that the mutant cells accumulate with multiple buds (Hartwell et al. 1974). The discovery that *CDC34* encodes an E2 ubiquitin-conjugating enzyme implicated in proteolysis suggests that protein degradation as well as synthesis is needed for S-phase entry and the formation of mitotic spindles in yeast (Goebl et al. 1988).

Cyclin-dependent Kinases

The last, most pleiotropic, class of mutants, represented initially by only a single allele of *CDC28* (Hartwell 1993), failed to undergo any of the four events associated with Start; they failed to enter S phase, to duplicate spindle pole bodies, to form buds, and (in haploids) to shut off sensitivity to cell cycle arrest by mating pheromones (Hartwell et al. 1974; Reid and Hartwell 1977; Pringle and Hartwell 1981). *CDC28* encodes a type of protein kinase (Lorincz and Reed 1984) whose activity is now known to require regulatory subunits called cyclins (Nurse 1990), which had been discovered as proteins whose abundance fluctuated during sea urchin cleavage divisions (Evans et al. 1983). How crucial a role the Cdc28 protein kinase might have in cell cycle control was only fully appreciated upon the subsequent discovery that *cdc2* from *S. pombe*, which had been shown to regulate mitosis, encoded a homologous protein kinase (Beach et al. 1982). Cdc2 from *S. pombe* and Cdc28 from *S. cerevisiae* are the founding members of a large family of similar cyclin-dependent kinases now known as Cdks; they have therefore a second name, Cdk1. Through their association with different cyclin subunits, the two yeast Cdk1 kinases (Forsburg and Nurse 1991a; Nasmyth 1993) have many different cell cycle functions: starting the cell cycle, initiating S phase, orchestrating nuclear division, and ensuring the dependence of S upon M phase, to name but a few. In mammalian cells, the Cdks that participate in these different events differ in both kinase and cyclin subunits (see below). Cyclin-dependent kinases are not exclusively concerned with cell cycle control. For example, distant relatives, cyclin H/Cdk7 from mammals (Fisher and Morgan 1994) and Ccl1/kin28 (Valay et al. 1993) from yeast have been implicated in the phosphorylation of the CTD tail of RNA polymerase II (Kim et al. 1994), whereas the yeast Cdk Pho85 regulates genes involved in phosphate metabolism through its phosphorylation of the Pho4 transcription factor (Kaffman et al. 1994).

Cdks as S-phase-promoting Factors

The first real indication that the onset of S phase might be triggered by cyclin-dependent kinases like Cdk1 (Cdc28) stemmed from the isolation of yeast mutants that started their cell cycles with a smaller than normal cell size. Mutants with a reduced cell size are particularly informative about cell cycle control, because this phenotype can only arise due to hyperactivation of the chromosome cycle (Nurse 1975), whereas mutants with an increased cell size can arise due to defects in any process needed for chromosome duplication or segregation. Dominant alleles of a gene now known as *CLN3* cause yeast to start the cell cycle at half the normal cell size (Sudbery et al. 1980; Cross 1988). *CLN3* encodes a protein distantly related to cyclins A and B, which are subunits of the Cdk1 kinase that promotes M phase in *Xenopus*. The mutant alleles encoded stable variants of Cln3 due to truncation of PEST sequences at its carboxyl terminus (Nash et al. 1988). Drawing on the analogy with mitotic cyclins and the knowledge that Cdk1 is needed for starting the yeast cell cycle, it was proposed that Cln3 might be a partner of Cdk1 and that a Cln3/Cdk1 kinase might trigger early yeast cell cycle events including S phase; i.e., it might correspond to an S-phase-promoting factor.

Deletion of *CLN3* greatly increases the size needed to start the yeast cell cycle, but it is not lethal, suggesting that other cyclins might exist with overlapping functions. Many other genes encoding cyclin partners of Cdk1 have subsequently been isolated, frequently as genes whose overexpression suppresses the temperature-sensitive growth of *cdk1* (*cdc28*) mutants: *CLN1* and *CLN2*, which encode a related pair of cyclins with only distant similarity to mitotic cyclins or to CLN3 (Hadwiger et al. 1989), and *CLB1*, *CLB2*, *CLB3*, *CLB4*, *CLB5*, and *CLB6*, which encode cyclins more homologous to the mitotic B-type cyclins from animals (Ghiara et al. 1991; Surana et al. 1991; Epstein and Cross 1992;

Schwob and Nasmyth 1993). None of these genes had previously been identified by conventional genetic analyses of the cell cycle because none of them are essential genes.

Do Clns and Clbs Correspond to SPFs and MPFs?

Because of the viability of single mutants, much of the phenotypic analysis of cyclin mutants has been performed on strains carrying multiple mutations. The finding that certain combinations of cyclin mutations cause arrest specifically in G_1 or G_2 has usually been taken to mean that the cyclin genes affected had similar functions. For example, inactivation of all three Cln cyclins causes cells to arrest indefinitely as unbudded G_1 cells, whereas all combinations of double mutants are viable, although their cells are much larger than normal (Richardson et al. 1989). The conclusion was that the three Cln cyclins had equivalent functions in promoting budding and S phase. Similar studies indicated, before the discovery of Clb5 and Clb6, that the four B-type cyclins Clb1–Clb4 all participated to a greater or lesser extent in the formation of mitotic spindles but were not needed for the initiation of S phase (Fitch et al. 1992; Richardson et al. 1992).

These two sets of G_1- and G_2-specific cyclins (i.e., Clns and Clbs) are all unstable proteins whose abundance, with the exception of Cln3, fluctuates during the cell cycle. Their periodic accumulation is partly due to transcriptional controls and partly due to fluctuations in their rates of proteolysis. All genes but *CLN3* are transcribed transiently during the cell cycle: *CLN1* and *CLN2* from late G_1 to the middle of S phase (Wittenberg et al. 1990), *CLB3* and *CLB4* from S phase till the end of mitosis (Fitch et al. 1992; Grandin and Reed 1993), and *CLB1* and *CLB2* during G_2 and M phases (Ghiara et al. 1991; Surana et al. 1991). Newly synthesized Cln1, Cln2, and Cln3 proteins have half-lives between 3 and 5 minutes (Deshaies et al. 1995; Yaglom et al. 1995), which have not been reported to vary during the cell cycle. The half-life of Clb2, in contrast, varies dramatically during the cell cycle, switching from 1 minute or less during the interval between the end of mitosis and Start to 60 minutes during the interval between Start and the onset of anaphase (Amon et al. 1994). The periodic accumulation of G_1- and G_2-specific cyclins is consistent with the notion that S phase is triggered by the accumulation or activation of Cln/Cdk1 complexes in late G_1, whereas M phase is triggered by the accumulation or activation of Clb/Cdk1 complexes in G_2. The case for Cln1 and Cln2/Cdk1 kinases being S-phase triggers is particularly strong, not only because these kinases are activated shortly before S phase, but also because the onset of S phase is advanced by premature activation of *CLN2* transcription in early G_1 (see below).

Clns Promote S Phase by Activating Clbs

The simple picture of Clns and Clbs corresponding to S- and M-phase-promoting factors, respectively, was dealt a mortal blow by the discovery of two further members of the B-type class of cyclins, Clb5 and Clb6. *CLB5* and *CLB6* mRNAs are of low abundance in early G_1 cells and, like those for *CLN1* and *CLN2*, accumulate to high levels shortly before S phase (Epstein and Cross 1992; Schwob and Nasmyth 1993). Clb5 associates with Cdk1, and the kinase activity of these complexes fluctuates during the cell cycle, being absent in early G_1 and appearing shortly before S phase, only slowly declining during G_2 and M phases, and finally disappearing as cells enter anaphase (Schwob et al. 1994). Like other cyclins, neither Clb5 nor Clb6 is essential, but the onset of S phase is greatly delayed in their absence (Schwob and Nasmyth 1993). S phase is normally coincident with budding, but in *clb5 clb6* double mutants it starts at least 30 minutes later.

New data show that B-type cyclins do not merely support Cln/Cdk1 kinases in triggering S phase but have a central role in this process. There is an important distinction between the S-phase delay of *cln1 cln2* double mutants and that of *clb5 clb6* double mutants. Both budding and S phase are delayed in the former but only S phase in the latter. The lesser pleiotropy of the Clb defect indicated that Clb5 and Clb6 might have a more direct role in the initiation of S phase than Cln1 and Cln2. However, if activation of the Clb5 or Clb6/Cdk1 kinases were "the" S-phase trigger, why is S phase merely delayed in the *clb5 clb6* double mutant? The surprising answer is that the supposedly "mitotic" B-type cyclins Clb1–Clb4 and not the G_1-specific Cln cyclins assume the S-phase-promoting functions of Clb5 and Clb6 in their absence. S phase is delayed yet longer in *clb3 clb4 clb5 clb6* quadruple mutants and never occurs in sextuple mutants lacking all six Clb cyclins, despite the activity of Cln1 and Cln2/Cdk1 kinases (Schwob et al. 1994). Activation of Cln/Cdk1 kinases is therefore not sufficient for initiating S phase; Clb/Cdk1 kinases are essential. In some circumstances, Clns are not even necessary. Cells arrested in G_1 by being deprived of all three Cln cyclins can be triggered to enter S phase highly synchronously by inducing *CLB5* or even *CLB2* expression from the *GAL1-10* promoter (Schwob and Nasmyth 1993; Amon et al. 1994). None of the *CLB* cyclin genes is efficiently transcribed in the absence of Cln cyclins, and *GAL*-driven expression circumvents this. These data indicate that it is through their activation of Clb cyclins that Cln/Cdk1 kinases promote S phase. An equally important conclusion is that the type of Cdk1 kinase activated in G_1 cells does not determine whether they undergo S phase or M phase (see below).

Regulation of S-phase-promoting Cyclins

Most growth-related mRNAs are absent in quiescent yeast cells and appear rapidly upon the addition of fresh nutrients. The same is true for *CLN3*. Its mRNAs are absent in stationary-phase cells, are induced within minutes of nutrient addition, but do not thereafter fluctuate during the cell cycle (Nash et al. 1988; Hubler et al. 1993). Cln3 protein levels behave similarly; in particular, their relative abundance does not greatly vary between the birth of daughter cells born below the size critical for Start and the onset of S phase (Tyers et al. 1993). Despite this lack of control, it is clear that the amount of Cln3 protein per cell is an important determinant of when yeast cells start the cell cycle; two- or more fold increases in *CLN3* gene dosage or mutations that stabilize Cln3 protein reduce the cell size at which cells bud and enter S phase (Cross 1988; Nash et al. 1988). Because Cln3 protein has a short half-life and its synthesis is not cell-cycle-regulated, its abundance per cell should reflect the general rate of protein synthesis. It is therefore an ideal candidate for a "sample" protein whose abundance is used by the cell to measure its rate of growth. Cln3 associates with Cdk1 (Tyers et al. 1992), and genetic data indicate that its main function is to activate Cdk1 (Cross and Blake 1993). The specific activity values of Cln3/Cdk1 complexes using histone H1 as substrate are, however, so low that it has not been possible to measure them as wild-type cells progress through G_1. It is not therefore clear whether the Cln3/Cdk1 kinase is in any way regulated.

In contrast to *CLN3*, the mRNAs from *CLN1*, *CLN2*, *CLB5*, and *CLB6* are all tightly cell-cycle-regulated, being low or absent in early G_1, accumulating as cells reach a critical size in late G_1, and declining as cells progress through S and G_2 phases (Koch and Nasmyth 1994). This pattern is part of a larger program of transcriptional control exerted by a pair of related transcription factors called SBF and MBF. The mRNAs of most genes involved in DNA replication and many involved in spindle pole body duplication have a similar pattern of accumulation, but the vast majority of these genes encode stable proteins, like DNA polymerase-α, whose de novo synthesis in the preceding G_1 is not needed for S-phase entry (Falconi et al. 1993). In contrast, the short half-lives of Cln1 and Cln2 and Clb5 and Clb6 mean that de novo synthesis of these proteins in late G_1 is an important if not essential step in triggering each and every S phase.

Late G_1-specific Transcription Factors

SBF and MBF bind to SCB and MCB elements found in the promoters of late G_1-specific genes. Both factors are composed of two subunits: a common regulatory subunit Swi6 that is complexed with either Swi4 or Mbp1 DNA-binding proteins (Koch and Nasmyth 1994). Swi4 and Mbp1 have a common architecture, both containing related site-specific DNA-binding domains at their amino termini (Primig et al. 1992; Koch et al. 1993), four related ankyrin motifs within their central regions (Breeden and Nasmyth 1987), and common sequences at their carboxyl termini through which they bind to their common partner Swi6 (Andrews and Moore 1992). Swi6, which lacks a DNA-binding domain, nevertheless has a similar structure, and all three proteins are presumably derived from a common ancestor. Although SBF (Swi4/Swi6) and MBF (Mbp1/Swi6) factors bind preferentially to SCB and MCB elements, respectively, polypeptides containing only their DNA-binding domains bind to both types of sequence. G_1 cyclin genes, such as *CLN1*, *CLN2*, and *HCS26*, are mainly regulated by SBF, whereas most genes encoding DNA replication enzymes are regulated by MBF. In some cases, genes regulated by one factor become in its absence regulated by the other (C. Koch, pers. comm.).

The Yeast Cell Cycle Is Started by the SBF/MBF Transcriptional Program

Several observations suggest that the onset of all events that occur at Start is triggered by the activation of transcription due to SBF and MBF. *CLN1* and *CLN2* mRNAs are lacking in small daughter cells and only appear along with other SBF/MBF-regulated mRNAs shortly before S phase when cells attain a critical size (Koch et al. 1993). Similar levels of a *CLN2* mRNA, albeit with a *MET3* leader sequence, can be induced in early G_1 cells from a *MET3-CLN2* promoter gene fusion, and this advances the onset of S phase and budding (L. Dirick et al., in prep.). More convincing, overexpression of *SWI4* from the *GAL* promoter in cells lacking Mbp1 causes mRNAs from most, if not all, genes normally regulated by SBF and MBF to appear in very small cells; that is, as soon as daughter cells are born (M. Neuberg et al., in prep.). The mRNAs are not overproduced but merely accumulate prematurely to levels found in wild-type cells shortly before S phase. This causes DNA replication and budding in cells whose size is one-third or less than the size needed for these two Start events in wild-type cells. The proliferation of very small cells by virtue of SBF hyperactivation is dependent on the presence of *CLN1* or *CLN2* but is totally independent of *CLN3*, whose activity normally controls the size at which yeast cells start the cell cycle. This suggests that Cln3's main, if not sole, function is to facilitate activation of the SBF/MBF transcriptional program (Tyers et al. 1993) and that the appearance of Cln1 and/or Cln2/Cdk1 kinases due to this program is crucial for the induction of S phase and budding.

Activation of SBF and MBF

How is transcription due to SBF and MBF normally activated only when cells reach a critical size? The transcriptional program fails to be activated or is greatly delayed in *cdk1* (*cdc28*) mutants (Dirick and Nasmyth 1991). It also fails to occur in mutants lacking all three Cln cyclins but is immediately restored by the reactivation of any one of them (Cross and Tinkelenberg 1991). The ability of Cln1 or Cln2/Cdk1 kinases to trigger the transcription of SBF/MBF-regulated genes suggested that the sudden activation of *CLN1* and *CLN2* transcription only when cells reach a certain size could be facilitated by a positive feedback loop by which Cln1 and Cln2 promote their own transcription along with other late G_1-specific genes. This hypothesis predicts that full activation of the transcriptional program should depend on the activities of Cln1 and Cln2, which is not the case. mRNAs known to be regulated by SBF or MBF accumulate with similar kinetics and at similar cell sizes in wild type and in *cln1 cln2* double mutants, but accumulation of the same mRNAs is delayed by several hours in *cln3* mutants (Dirick et al. 1995). These data emphasize that the three Clns do not have equivalent functions. Cln3 is essential for activating SBF/MBF-mediated transcription at normal cell sizes, whereas Cln1 and Cln2 are not.

The mechanisms by which Cln3/Cdk1 activates the SBF and MBF transcription factors only when cells reach a certain size are not understood. Genomic footprinting has shown that functionally important SCBs in the *CLN2* promoter are fully occupied by SBF in cells arrested in G_1 due to a temperature-sensitive *cdk1* mutation. Lowering the temperature of these mutant cells restores Cdk1 kinase and leads to the immediate activation of *CLN2* transcription, suggesting that Cdk1 alters the activity of complexes that are already bound to SCBs (Koch et al. 1996).

Regulation of Clb/Cdk1 Kinases during G_1

Despite their lack of involvement in the activation of gene expression, Cln1 and Cln2 have a crucial role in triggering the onset of both budding and S phase (Hadwiger et al. 1989), which are advanced by premature expression of *CLN2* and which are delayed in *cln1 cln2* double mutants until cells have grown at least three times the normal size (Dirick et al. 1995). Current data are consistent with the notion that Cln1 and Cln2 promote S phase by stimulating activation of Clb/Cdk1 kinases. They do this by facilitating the accumulation of Clb proteins (Amon et al. 1994) and the disappearance of a Cdk1 inhibitory protein p40Sic1 (Schwob et al. 1994).

Rapid proteolysis of Clb2, which is largely shut off during S and G_2 phases, commences during anaphase and persists until activation of Cln1 and Cln2/Cdk1 kinases during the subsequent G_1 period. The extremely short half-life of Clb2 during early G_1 (1 minute or less) ensures that it cannot accumulate in G_1 cells lacking Cln cyclins, even when highly expressed from the *GAL* promoter (Amon et al. 1994). Clb5 protein, with a half-life of 5 minutes (Seufert et al. 1994), is more stable during this period. Cells arrested in G_1, due to deprivation or physiological inactivation of all three Clns, undergo S phase (S. Irniger, pers. comm.) upon inactivation of proteins such as Cdc16, Cdc23, and Cse1, which are required for Clb2 proteolysis (Irniger et al. 1995). This implies that the persistence of rapid Clb proteolysis until the activation of Cln1 and Cln2 could have an important role in preventing premature accumulation of Clb/Cdk1 kinase activity.

Role of a Cdk Inhibitor, p40Sic1

Probably the most important mechanism regulating the onset of S phase in yeast is the late G_1-specific destruction of a Cdk1 inhibitory protein p40Sic1 (Schwob et al. 1994). It had been found that transcription of *CLB5* in small early G_1 cells failed to advance S phase. This was surprising because premature Cln2 expression did advance S phase and yet it was thought that Clb5 had a more direct role than Cln2 in the initiation of DNA replication. The ectopic expression caused Clb5 protein accumulation, but the resulting Clb5/Cdk1 complexes were inactive due to the presence in early G_1 cells of a factor capable of inhibiting the Clb5/Cdk1 kinase. The inhibitory factor is a 40-kD protein encoded by the *SIC1* gene (Nugroho and Mendenhall 1994), which had previously been purified as a potential substrate of Cdk1 and found to be an inhibitor (Mendenhall 1993). p40Sic1 is a potent inhibitor of Clb/Cdk1 but not of Cln1 or Cln2/Cdk1 kinase activities (M. Mendenhall, pers. comm.).

Early G_1 cells contain high levels of p40Sic1 protein, but it disappears shortly before the onset of S phase and does not reappear until cells complete mitosis. This pattern of accumulation is partly due to the periodic accumulation of *SIC1* mRNAs, whose abundance increases transiently from a basal level as cells exit from mitosis (Donovan et al. 1994; Schwob et al. 1994). However, cell-cycle-dependent proteolysis mediated by the Cdc34 ubiquitin-conjugating enzyme is implicated in the sudden disappearance of p40Sic1 in late G_1. The inhibitor accumulates at the end of mitosis to similar levels in wild type and *cdc34* mutants but fails to disappear in the latter when cells ought to be entering S phase (Schwob et al. 1994). This defect is highly specific, because the SBF/MBF transcriptional program, activation of Cln1 and Cln2/Cdk1 kinases, and budding all take place in *cdc34* mutants. The accumulation of p40Sic1 probably inhibits activation of all Clb/Cdk1 kinases in *cdc34* mutants, and this must be responsible for their G_1 arrest because deletion of the *SIC1* gene permits them to undergo DNA replication. The G_1 arrest of *cdc4* and *cdc53* mutants is also due to a failure to destroy p40Sic1.

These data suggest that proteolysis of p40Sic1 mediated by Cdc4, Cdc34, and Cdc53 proteins is an essential step toward the initiation of S phase in *S. cerevisiae*. How then is this process regulated? p40sic1 protein persists in cells arrested in G_1 due to triple *cln* or *cdk1* mutations but not in sextuple *clb1-6* mutants (T. Bohm, pers. comm.), suggesting that Cln but not Clb Cdk1 kinases play a part in its destruction. Unlike SBF/MBF-mediated transcription, the disappearance of p40Sic1 is delayed in *cln1 cln2* double mutants, and this is largely responsible for their delayed S-phase entry (Dirick et al. 1995). Thus, Cln3 is insufficient and Cln1 and Cln2 are necessary to trigger disappearance of p40Sic1 on schedule. p40Sic1 is not an effective inhibitor of the Cln2/Cdk1 kinase either in vivo or in vitro but is an excellent substrate, at least in vitro (M. Mendenhall, pers. comm.). This raises the possibility that phosphorylation of p40Sic1 by Cln1 or Cln2/Cdk1 kinases might be the trigger for its proteolysis via Cdc34. Consistent with this notion, *sic1* mutations that alter potential Cln/Cdk1 phosphorylation sites delay the disappearance of p40Sic1 protein and the onset of S phase (T. Bohm, pers. comm.).

Cdc7

A piece of the S-phase regulatory jigsaw that has not yet been satisfactorily fitted into the above picture is the role and regulation of the Cdc7 kinase. Kinase activity associated with Cdc7 rises in late G_1 along with that of Clb/Cdk1 kinases (Jackson et al. 1993; Yoon et al. 1993). *cdc7* mutants fail to initiate S phase despite possessing active Clb/Cdk1 kinases. Cdc7 is therefore needed in addition to Clb/Cdk1 kinases. Cdc7 associates with the Dbf4 protein, which is also needed for replication and could be a regulatory subunit with properties (but not sequences) similar to cyclins (Jackson et al. 1993). Dbf4 associates in vivo with proteins that bind to replication origins, suggesting that some of these are substrates for the Cdc7 kinase (Dowell et al. 1994). One scenario is that Clbs activate the replication apparatus via their activation of Cdc7; i.e., Clb/Cdk1 kinases are upstream of Cdc7 on a linear pathway leading to replication (Jackson et al. 1993). It is more likely, however, that Clb/Cdk1 kinases have several roles in activating replication, only one of which might be to promote activation of the Cdc7 kinase. The need for Cdc7 and Dbf4 can be bypassed by *bob1* mutations, suggesting that Cdc7's normal function is to relieve a negative control.

Summary: Promoting S Phase in S. cerevisiae

There now exists a detailed, but by no means complete, hypothesis for how S-phase-promoting factors are produced in *S. cerevisiae* (see Fig. 2). Let us start with a small quiescent cell that has just been inoculated into fresh medium. The presence of nutrients induces the appearance, largely due to transcriptional controls, of most, if not all, mRNAs needed for cell growth, whose translation is facilitated by the simultaneous activation of translational initiation factors like eIF-4e. Neither of these events is dependent on the forthcoming chromosome cycle, but both are essential for the eventual production of specific S-phase-promoting factors. One of the many proteins whose synthesis is stimulated in this manner is Cln3, a Cdk1 cyclin, whose abundance per cell, due to its short half-life, reflects the overall rate of translation of "growth" mRNAs and thereby the size and growth rate of the cell. Few, if any, other S-phase-promoting factors are made in these small growing cells until they reach a threshold size, at which Cln3/Cdk1 suddenly activates transcription of a large battery of genes, some of which encode unstable replication proteins like Cdc6 (Piatti et al. 1995), whereas others encode further S-phase-promoting cyclins like Cln1 and Cln2 and Clb5 and Clb6. The size at which this transcriptional explosion occurs determines the size at which cells enter S phase and is regulated by external factors like nutrients, which exert their effect via cAMP; an increase in the activity of cAMP-dependent protein kinases in rich medium increases the size needed to trigger the appearance of *CLN1* and *CLN2* mRNAs. With the activation of the late

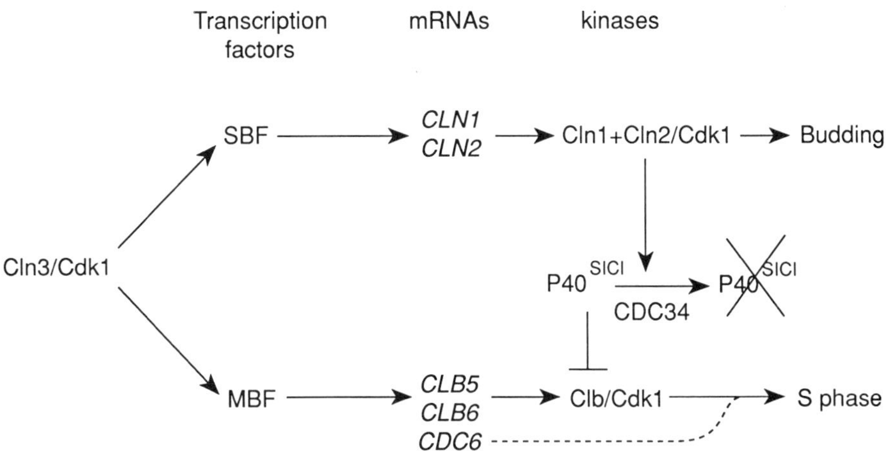

Figure 2 The production of S-phase-promoting factors in *S. cerevisiae*. The process starts with activation of transcription due to SBF and MBF transcription factors when cells reach a critical size. This leads to the synthesis of Cln1/2 and Clb5/6 cyclins and replication proteins like Cdc6 which are needed for the formation of pre-RCs at future replication origins (Cdc6 is also synthesized at the end of mitosis). Clb5 and Clb6/Cdk1 complexes remain inactive until the p40Sic1 inhibitor is destroyed via ubiquitin-mediated proteolysis involving Cdc4, Cdc34, and Cdc53, an event that is thought to be triggered by Cln1 and Cln2/Cdk1 kinases.

G_1-specific transcription program, the "baton," as it were, passes from Cln3 to Cln1 and Cln2. Accumulation of Cln1 or Cln2/Cdk1 kinases, due largely to activation of the SBF transcription factor by Cln3/Cdk1, leads to the next crucial step: activation of Clb/Cdk1 kinases, which are needed to activate the Cdc7 kinase and, presumably, many other replication proteins.

The absence of Clb/Cdk1 kinases in early G_1 cells is due partly to a lack of transcription of *CLB* genes, partly to rapid proteolysis which is started during anaphase but persists during early G_1, and partly to the accumulation at the end of mitosis or in stationary phase of the p40Sic1 protein, which is a potent inhibitor of Clb/Cdk1 kinases. Cln3/Cdk1 induces synthesis of Clbs 5 and 6, whereas Cln1 and Cln2/Cdk1 kinases promote accumulation of active Clb/Cdk1 kinases by drastically reducing proteolysis of cyclins like Clb2 and, more important still, by triggering proteolysis of p40Sic1. How activation of Clb/Cdk1 and Cdc7 kinases triggers the formation of replication forks and why they do this in G_1 but not in G_2 cells will require consideration of the substrates of these kinases.

S-phase-promoting Factors in Other Fungi

A single Cdk kinase related to Cdk1 (known as Cdc2) is needed for the onset of both S and M phases in two other ascomycetes, the fission yeast *S. pombe* (Nurse and Bissett 1981) and the filamentous fungus *Aspergillus nidulans* (Osmani et al. 1994). Less is known, however, about the cyclin partners of Cdk1 (Cdc2) that promote S phase in these two organisms. The *puc1* gene in *S. pombe* encodes a cyclin related in sequence to Cln3 (Forsburg and Nurse 1991b), but there is little or no evidence that *puc1* helps to promote the onset of S phase in vegetative cells (Forsburg and Nurse 1994). This may be less surprising now that we realize that Cln cyclins have only an indirect role in promoting S phase, even in *S. cerevisiae*, and that B-type cyclins are more intimately involved. *S. pombe* contains at least three B-type cyclins encoded by the *cig1* (Bueno et al. 1991), *cig2* (Bueno and Russell 1993; Connolly and Beach 1994; Obara Ishihara and Okayama 1994), and *cdc13* genes (Booher and Beach 1988; Hagan et al. 1988). None is necessary for the G_1-to-S-phase transition in vegetative cells, but *cdc13* is essential for M phase. There are indications that these B-type cyclins are nevertheless important for promoting S phase. For example, *cig2* is important for the terminal rounds of DNA replication that occur as *S. pombe* cells enter stationary phase upon nitrogen starvation (Obara Ishihara and Okayama 1994), and deletion of both *cig1* and *cig2* causes nuclei in vegetatively growing cells to spend longer in G_1 (Connolly and Beach 1994). *S. pombe* might therefore resemble *S. cerevisiae*, in that one of several B-type cyclins is sufficient for triggering S phase.

S phase in *S. pombe* depends on the activation of transcription in late G_1 of key replication genes by a pair of transcription factors Res1/Cdc10 (Tanaka et al. 1992; Caligiuri and Beach 1993) and Res2/Cdc10 (Tanaka et al. 1992; Zhu et al. 1994), which are related to Swi4/Swi6 and Mbp1/Swi6 in *S. cerevisiae*. Cdc10, which resembles Swi6 (Breeden and Nasmyth 1987), is necessary for both factors and is essential for the G_1-to-S-phase transition in *S. pombe* (Nurse et al. 1976). Res1/Cdc10 complexes are more important for S phase in vegetatively growing cells, whereas Res2/Cdc10 complexes are more important for premeiotic DNA replication. Fewer genes involved in DNA replication are regulated by these transcription factors than are regulated by their equivalent factors in *S. cerevisiae*. Nevertheless, transcriptional activation in late G_1 of at least two genes, *cdc18* (a homlog of *CDC6* from *S. cerevisiae* [Kelly et al. 1993]) and *cdt1*, is essential for initiation of the ensuing S phase (Hofmann and Beach 1994). Neither *cig1* nor *cdc13* mRNA are cell-cycle-regulated, but *cig2* mRNAs are periodic (Connolly and Beach 1994), accumulating to maximal levels in late G_1, and could therefore be regulated by one of the two Res/cdc10 factors. As in *S. cerevisiae*, Cdk1 is needed for late G_1-specific gene activation; in this case, there is evidence that Cdk1 is needed for the DNA-binding activity of Res/cdc10 factors (Reymond et al. 1993). A zinc-finger-containing protein, called Rep2, is needed for the activity of Res2/cdc10 but not Res1/cdc10 complexes in mitotic *S. pombe* cells (Nakashima et al. 1995).

Whether changes in the proteolysis of B-type cyclins in *S. pombe* has any role in regulating their activity during G_1 is not known. There exists, however, a Cdk inhibitory protein, encoded by the *rum1* gene, which delays the onset of S phase under certain circumstances (Moreno and Nurse 1994). In fast-growing cells, daughter cells are born larger than the threshold size needed to start the cell cycle, G_1 is extremely short, and deletion of *rum1* has little or no effect. Inactivation of the Wee1 protein kinase eliminates size control over M phase, and daughter cells are now born below the critical size and must spend one-third of their next cycle growing large enough to enter S phase. This longer G_1 period of *wee1* mutants is largely eliminated by deletion of the *rum1* gene. Rum1 is also implicated in delaying or even preventing activation of B-type cyclin/Cdk1 kinases in nitrogen-starved cells and in cells arrested in G_1 due to *cdc10* mutations. The Rum1 protein is a potent inhibitor of Cdc13/Cdk1 kinase (Correa-Bordes and Nurse 1995), but it has little or no sequence similarity to p40Sic1 from *S. cerevisiae* or Cdk inhibitors from mammalian cells, nor is it known whether or how Rum1 abundance is regulated during the cell cycle. There have been no measurements of specific B-type cyclin/Cdk1 kinase activities during G_1, so it is difficult at this stage to paint even a crude picture of how these kinases are regulated during this stage of the *S. pombe* cell cycle.

Although the details differ considerably, the broad outlines of S-phase control seem conserved between *S. pombe* and *S. cerevisiae*; this includes the synthesis of replication proteins like spCdc18 or scCdc6 due to Cdc10/Swi6-like transcription factors, the role of B-type cyclin Cdk kinases, and the control of these kinases by inhibitory proteins.

S-phase-promoting Factors in Animal Cells

It is not possible to assemble as detailed a picture of the events leading to S phase in animal cells as in yeast, not because fewer components have been identified (the advent of two-hybrid screening has largely solved this problem), but because their functional analysis is still much harder (function cannot be as reliably inactivated by antisense RNA or even antibody injections as by gene mutation, which although possible in mice, is still laborious and difficult to make conditional) and because studies are performed on many different types of cells whose behavior can be very different. Nevertheless, one can detect several parallels between animal cells and yeast, which are worth pointing out since they may represent the more highly conserved aspects of S-phase control.

The Restriction Point

It has long been known that de novo protein synthesis during G_1 is needed for S phase in tissue-culture cells. However, entry into S phase becomes less sensitive to inhibitors of protein synthesis about 2 hours before DNA replication commences. It is around this point in late G_1, called the restriction point, that many mammalian tissue-culture cells become capable of entering S phase upon withdrawal of growth factors, which are potent effectors of the cell's rate of protein synthesis (Pardee 1989). G_1-specific cyclins and their catalytic subunits may be some of the proteins whose synthesis preceding the restriction point is stimulated by growth factors and is needed for the initiation of DNA replication.

G_1-specific Cdks

Whereas ascomycetes use the same Cdk subunit to regulate S and M phases, animal cells use at least three different kinase subunits. Cdk1 (cdc2) for M phase (Riabowol et al. 1989; Hamaguchi et al. 1992), Cdk2 for S phase (Fang and Newport 1991; Paris et al. 1991), and Cdk4 or Cdk6 for progression from early to late G_1 (Sherr 1994a). As in yeast, multiple cyclins associate with a given Cdk. Cdk4 and Cdk6 associate with D-type cyclins (Matsushime et al. 1992). Cyclin D1 is absent in quiescent cells and is induced by mitogenic growth factors (Matsushime et al. 1991), reaching maximal levels several hours after their addition but long before the onset of S phase. Cyclin D1 has a half-life of around 30 minutes, which, as a percentage of the doubling time, is similar to that of yeast G_1 cyclins. It accumulates in the nucleus during G_1, but its abundance varies little if at all during the cell cycle of exponentially growing cells (Sewing et al. 1993). Antibody injection experiments had suggested that cyclin D1 is essential for progression from G_1 to S phase in several different types of cells (Baldin et al. 1993; Lukas et al. 1994). However, mouse embryos lacking any cyclin D1 gene can undergo all prenatal development (Sicinski et al. 1995). Whether this means that the effects of antibody injection were less specific than anticipated or that the cell lines used in such studies are more dependent on cyclin D1 than most embryonic cells remains to be determined. It is worth noting that even in yeast few if any cyclin gene deletions are lethal, despite having drastic effects on the kinetics of specific cell-cycle transitions, and the same could be true for mice.

Another means of testing the role of D-type cyclins in promoting S phase has been to analyze the consequences of inducing high levels of synthesis during early G_1. Hyperaccumulation of cyclin D1 caused a modest reduction in G_1 length in cycling cells and a more persuasive reduction during the outgrowth of previously quiescent cells (Quelle et al. 1993; Resnitzky et al. 1994). In a study using regulated cyclin D expression, hyperactivation did not, however, increase the rate of cell proliferation (although one suspects that it reduced cell size), which is consistent with similar findings in yeast and suggests that the G_1-to-S-phase transition is not a rate-determining step for proliferation, at least for the tissue-culture cells used in these studies. Similar experiments, performed with cyclin D transgenes in mice, have little immediate effect on cell proliferation but do eventually give rise to tumors, especially in conjunction with other oncogenes (Bodrug et al. 1994; Lovec et al. 1994).

Cdk2 is found mainly associated with cyclins E and A in mammalian cells. Both proteins oscillate in phase with the chromosome cycle, with cyclin E accumulating to maximal levels in late G_1 (Dulic et al. 1992; Koff et al. 1992), possibly coincident with the restriction point, and cyclin A somewhat later during S and G_2 phases (Pines and Hunter 1990; Tsai et al. 1991). Antibody injection experiments suggest that Cdk2 and cyclin A (Girard et al. 1991; Pagano et al. 1992; Zindy et al. 1992; Tsai et al. 1993) are essential for S phase in mammalian tissue-culture cells. Cyclin A, however, has been found unnecessary for DNA replication in extracts from *Xenopus* eggs and for at least certain divisions during *Drosophila* embryogenesis (Lehner and O'Farrell 1990; Knoblich and Lehner 1993). Cyclin E, in contrast, seems essential for replication in *Drosophila* and *Xenopus*. Maternal cyclin E transcripts disappear suddenly as *Drosophila* cells complete the last rapid mitotic division during embryogenesis (cycle 16) and start to accumulate for the first time in G_1, but the mRNAs reappear due to zygotic gene expression shortly before S phase in the minority of cells that continue cell division at this point; they also reappear during late embryogenesis in tissues that undergo endoduplication cycles. More convincing still, cyclin E mutants cease all DNA replication after cycle 16, and ectopic expression of cyclin E from a heat shock promoter induces S phase in cells that would otherwise have arrested in G_1; however, this only occurs if expression is induced soon after their exit from the previous mitosis (Knoblich et al. 1994). *Drosophila* cyclin E protein associates with a Cdk that resembles the mammalian Cdk2. In mammalian tissue culture, the artificial induction of high levels of cyclin E is insufficient to cause quiescent cells to enter S phase, but it does cause a modest reduction in the length of their G_1 period and thereby enables cells to divide at a smaller cell size (Ohtsubo and Roberts 1993; Resnitzky et al. 1994). It does not, however, alter their rate of proliferation. Thus, the accumulation of cyclin E in late G_1 may have an important role in the onset of S phase in both arthropod and vertebrate cells. Neither mammalian nor *Drosophila* cells seem to require Cdk1 (cdc2) or B-type cyclins for S phase.

Cdk Inhibitors

As in yeast, the activities of D and E type cyclin/Cdk complexes are not just regulated by the pattern of cyclin accumulation. There is a family of small proteins (p15, p16, and p18) composed of ankyrin motifs analogous to those first found in Swi6 (Breeden and Nasmyth 1987), which, it is thought, inactivate Cdk4 and Cdk6/cyclin D complexes by binding to the Cdk subunits and displacing cyclin D (Serrano et al. 1993; Guan et al. 1994; Hannon and Beach 1994). Expression of p15 is induced in keratinocytes by antimitogenic factors like transforming growth factor β (TGF-β). p16INK4, on the other hand, increases in abundance following the stimulation of fibroblasts with serum (Tam et al. 1994); its accumulation to maximal levels during S phase might play a role in inhibiting Cdk4 and Cdk6 kinases as cells enter G_2. Whether any of these inhibitors has a role in preventing premature activation of cyclin D/Cdk kinases during early G_1 is unclear. Transient overproduction of p16INK4 causes cells to arrest in G_1, and mutations in the INK4/MST1 gene for p16 are found in most tumor cell lines and in some primary tumors (Hunter and Pines 1994), suggesting that p16 could have an important role in arresting the proliferation of potential tumor cells.

Other cyclin kinase inhibitors, p21CIP1/WAF1/SDI1 (El-Deiry et al. 1993; Harper et al. 1993) and p27KIP1 (Polyak et al. 1994), bind to and inhibit a wide variety of Cdk kinases including cyclin D/Cdk4, cyclin E/Cdk2, and cyclin A/Cdk2. p27 levels decline modestly as quiescent macrophage cells are stimulated to enter S phase by growth factors (Kato et al. 1994). A proposal that this decline is important for entry into S phase remains to be tested rigorously. p27 abundance declines rapidly in T cells in response to interleukin 2 (IL-2), which could be an important event in the induction of S phase by this factor (Firpo et al. 1994). p21 levels, in contrast, rise upon growth factor stimulation, which might have a role in preventing activation of S-phase-promoting Cdks before cells have made other preparations for S phase. It is still unclear, however, whether inhibitors like p21 and p27 exert a strong effect on the kinetics of S-phase entry in cycling cells. Both certainly have the potential to block cell cycle progression. A good example is the increase in abundance of p21 in response to DNA damage, which is due to activation of p21WAF1 transcription by p53 and has a role in preventing the entry of irradiated cells into S phase (Deng et al. 1995). Other modes of p21 induction, for example, its induction by growth factors, do not involve p53 (Parker et al. 1995), and it is unknown whether p21 levels are sufficient to delay activation of S-phase-promoting Cdks in cells lacking DNA damage and, if so, how cells surmount this barrier.

Phosphorylation

The activities of all Cdks are dependent on phosphorylation of a highly conserved residue (Thr-161 [Solomon et al. 1992]), but the kinase responsible, which is another cyclin-dependent kinase composed of Cdk7(MO15) and cyclin H (Fesquet et al. 1993; Fisher and Morgan 1994; Solomon et al. 1993), and which seems identical to the kinase that phosphorylates the CTD of RNA polymerase II (Roy et al. 1994), is not apparently cell-cycle-regulated (Tassan et al. 1994). Phosphorylation at other conserved sites on Cdk1 (Tyr-15), which inhibits its kinase activity during G_2 in yeast and in animal cells, is reversed by the Cdc25 (*S. pombe*) family of phosphatases. Vertebrates possess at least three members of Cdc25 phosphatases; Cdc25A undergoes phosphorylation and activation in late G_1 and could have a role in the activation of Cdk2 kinases shortly before S phase (Hoffmann et al. 1994).

Late G_1-specific Transcription

As found in yeast, transcripts from many genes implicated in DNA replication or its control oscillate in abundance during the cell cycle, reaching maximal levels in late G_1; examples include mRNAs for thymidine kinase (Dou et al. 1992), dihydrofolate reductase (Bjorklund et al. 1990), cdc2 (Dalton 1992), DNA polymerase-α (Pearson et al. 1991), and cyclin E (Reed et al. 1992). A family of transcription factors collectively called E2F are implicated in this transcriptional program (La Thangue 1994). Most of these factors are heterodimers containing an

E2F and a DP1 subunit. There are several subtypes of each of these subunits, creating many different potential heterodimers, whose regulatory properties almost certainly differ. E2F binds to sequences that resemble the SCBs and MCBs bound by SBF and MBF in yeast, but E2F and DP1 have no sequence similarities with Swi4 or Swi6.

Analyzing patterns of mRNA accumulation in cells lacking particular genes will ultimately be necessary to establish which E2F factors are responsible for regulating which late-G_1-specific genes. Mutation of an E2F gene in *Drosophila* abolishes both S phase and developmentally regulated pulses of DNA replication enzyme mRNAs (Duronio and O'Farrell 1994; Duronio et al. 1995). Analyzing similar mutants in the mouse will undoubtedly be complicated by potential overlaps in the function of different E2F and DP-1 genes, as has been found for SBF and MBF in yeast. Another complication is that genes thought to be activated or controlled by E2F-binding sites do not have a common mode of regulation. The c-*myc* gene, for example, whose promoter contains an E2F site implicated in its transcriptional regulation, is regulated in a very different manner from many other supposedly E2F-regulated genes.

Activation of E2F in Late G_1

At least three different regulatory proteins related to the product of the retinoblastoma gene (Rb itself, p107, and p130) are found associated with different E2Fs during G_1 (Hinds and Weinberg 1994). Rb associates with E2F-1, E2F-2, and E2F-3 and can inhibit their ability to activate transcription in vitro (Dynlacht et al. 1994). Rb is phosphorylated at many residues throughout the cell cycle, but two or three specific residues become phosphorylated only in late G_1 and remain so until the end of mitosis (Sherr 1994b). This hyperphosphorylation prevents Rb from binding to E2F. It is therefore proposed that Rb hyperphosphorylation caused by cyclin D- or cyclin E-associated Cdks that become active in late G_1 enables accumulation of uncomplexed E2F, which is free to activate transcription. Although attractive as a regulatory mechanism, this theory has never been rigorously tested; for example, by characterizing the properties of Rb proteins in which residues whose phosphorylation is cell-cycle-regulated (and not others) have been mutated. The current hypothesis would predict that expression of modest levels of proteins that lack the ability to be hyperphosphorylated should be sufficient to prevent the activation of genes regulated by E2F.

There are two other potential mechanisms for the activation of E2F-regulated genes by G_1-specific Cdks. Phosphorylation of E2F-1's serine residues 332 and 337 hinders its interaction with Rb but facilitates its interaction with the adenovirus E4 protein. Both residues are largely unphosphorylated during early G_1, become phosphorylated during late G_1, and can be phosphorylated by cyclin D/Cdk4 kinase in vitro (Fagan et al. 1994). Cell-cycle-specific phosphorylation of E2F-1 by cyclin D/Cdks could therefore also contribute to the accumulation of "free" E2F in late G_1. Finally, the abundance of E2F mRNAs increases dramatically in late G_1. It has therefore been proposed that activation of Cdks in late G_1 stimulates synthesis of E2F, which could contribute to its accumulation in a form unbound to the Rb protein (Johnson et al. 1994). The effectiveness of this mechanism might depend on whether phosphorylation of Rb or E2F can actually liberate E2F from Rb. If not, then de novo synthesis of E2F in late G_1 could have a vital role and not just a helping hand in the activation of E2F-regulated genes.

The role of Rb as a key cell-cycle regulator has been questioned recently by the discovery that embryonic stem cells homozygous for Rb deletions can contribute to most tissues in the mouse; that is, cell-autonomous Rb function is not essential for the proliferation and differentiation of most types of mammalian cells (Sherr 1994b). It will be important to establish to what extent the normality of Rb-negative cells in vivo is due to the assumption of Rb's regulatory functions by other Rb-related proteins, to E2F remaining adequately regulated in the absence of any such protein, to regulation of E2F-mediated gene expression being less important than hitherto envisaged, or to deregulated entry into S phase being not particularly damaging to embryonic development because cell proliferation can be equally well if not better regulated by growth control or cell death (apoptosis). There are modest increases in the rates of proliferation within certain neuronal tissues in *rb* mutants but these seem to be compensated by increased apoptosis (Morgenbesser et al. 1994).

Comparisons to Yeast: Are Cyclin D and Rb on the Same Pathway?

The similarities between yeast and animal cells are sufficient to encourage a comparison of their factors that promote S phase. Several facts suggest that cyclin D/Cdk4 kinases perform functions that are analogous to Cln3/Cdk1. First, the patterns of synthesis of cyclin D and Cln3 are similar; neither is tightly periodic and both appear to be stimulated primarily by "growth"-promoting signals rather than chromosome cycle signals. Second, both proteins are very different from other cyclins involved later in the cycle, like the E, A, and B types, and differ greatly in their substrate specificity in that neither Cln3/Cdk1 (Tyers et al. 1992) nor cyclin D/Cdk4 readily phosphorylates histone H1 (Kato et al. 1993). Third, although the sequences of Cln3 and cyclin Ds do not greatly resemble each other, there is some evidence that they can perform equivalent tasks when expressed in yeast (Hatakeyama et al. 1994).

One of the more striking similarities between cyclin D and Cln3 is that the main, if not sole, task of both cyclins may be to activate late G_1-specific transcription. The dependence of S phase on Cln3 in yeast (i.e., its occurrence at a normal cell size) is abolished in cells that express late G_1-specific mRNAs constitutively due to hyperactivation of SBF. Likewise, entry into S phase of mammalian cell lines in which E2F cannot be regulated by Rb (due to mutations in the Rb gene) is sensitive neither to the injection of cyclin-D-specific antibodies (Lukas et al. 1994) nor to the overexpression of the cyclin-D-specific inhibitor p16INK4 (Serrano et al. 1995). Thus, both Cln3 and cyclin D appear to be "upstream" effectors of late G_1-specific transcription whose function becomes redundant when the "downstream" transcription factors are activated by other means.

There are indications that the abundance of cyclin D might be regulated by Rb. Cyclin D levels are very low in cell lines lacking Rb (Bates et al. 1994), suggesting that Rb, to which it is capable of binding in vitro, might play some role in stabilizing cyclin D protein. That cells can proliferate with very low levels of cyclin D when Rb is missing is consistent with its function being bypassed in these cells. Whether this is exclusively due to the resulting hyperactivation of E2F or to the deregulation of other S-phase-promoting events in *rb* mutants remains to be established.

What Are the Animal Equivalents of Clb5 and Clb6?

Cyclin E/Cdk2 kinases could perform functions similar to those of Cln1 and Cln2/Cdk1 kinases or those of Clb5 and Clb6/Cdk1 kinases. In contrast to Cln3 and cyclin D, accumulation of all these cyclins is tightly regulated by passage through the chromosome cycle. Certain observations suggest that an analogy between cyclin E and the "early" Clb cyclins Clb5 and Clb6 might be more appropriate. The functions of cyclin E seem highly conserved between vertebrates and arthropods, which is analogous to the conserved role of B-type cyclins (but not Cln cyclins) in regulating S phase in *S. pombe* and *S. cerevisiae*. Furthermore, the sequences of cyclins E resemble B-type cyclins much more than than they do Cln cyclins. The apparent activation of cyclin E/Cdk2 kinases 1 or 2 hours before the onset of S phase (Koff et al. 1992) is, on the other hand, more analogous to the behavior of Cln1 and Cln2, and is at first glance inconsistent with the notion that activation of cyclin E/Cdk2 is the "S phase trigger." If cyclin E/Cdk2 performs functions similar to those of Clb5/Cdk1, then cyclin A/Cdk2 could perform functions similar to those of Clb3 and Clb4/Cdk1 kinases. Cyclins A and Clb3 and Clb4 all accumulate later than cyclins E and Clb5 and Clb6, possibly too late to initiate S phase in wild-type cells, but not too late to facilitate the firing of "late" origins of replication.

Prospects for Genetic Analysis of S Phase in Animals

Analysis of the cell cycle in mammalian cells has until recently been dominated by experiments involving "gains of function"; hyperactivation of cyclins accelerate the cycle, overexpression of Rb suppresses division, cotransfection of transcription factor X can activate reporter gene Y, etc. There are two reasons for this. First, the prime motivation for many who study the mammalian cell cycle has been to understand oncogenesis, which until recently was thought to be largely due to gain-of-function mutations. Second, it is still much easier to study the effects caused by gains of function in mammalian cells than it is to study effects due to loss of function. Antisense has a discredited track record in establishing phenotype, whereas antibody injection is technically demanding and difficult to interpret unambiguously. Gain-of-function approaches only tell us what "can" function, which may be important for oncogenesis, but generally distracts us from understanding how normal cells actually function. The more recent appreciation that many, if not most, mutations that cause cancer are "loss-of-function" mutations and the development of methods to delete genes in mouse ES cells have been a very important step in the right direction. The latter has provided a reliable means of generating either animals or cell lines that are homozygous for mutations in key cell-cycle regulators.

It should not be forgotten, however, that the "reverse genetics" approach pioneered in yeast and now available in the mouse has its limitations, and there will continue to be much scope for the conventional genetic approach of isolating mutants following screening. This is possible in yeast, flies, and (hopefully) in the zebra fish, but is unlikely ever to be practical for mice. In an era when genetic analysis increasingly holds sway in experimental biology, it is seldom appreciated how difficult it is to determine gene function merely by analyzing phenotype. Gene knockouts, whether they are performed in mice or yeast, appear either to have little effect or to cause lethality. In the first case, it is difficult to ascertain what might have gone wrong without previous intuitions about the gene's function, whereas in the second case, it is difficult to distinguish primary from secondary phenotypes. Conventional mutant screens, when successful, are highly selective on both counts: Mutants are sought with specific defects in an area of biology in which the investigator is an expert and pleiotropic mutants are usually avoided. Reverse genetics will be most valuable when we have some idea where to look for defects.

STOPPING DNA REPLICATION IN G_2

Could Cell-cycle State Be Determined Solely by Cdks?

How do cells ensure that no DNA sequence is replicated more than once during S phase and that mitosis precedes the next round of replication? There is ample evidence in yeast and animal cells that the timing of S and M phases is determined by the pattern of activation of G_1- and G_2-specific Cdks. Might not just the timing but also whether a cell undergoes S or M phase be determined by the state of Cdk-cyclin complexes? Let us create an imaginary cell that possesses only two types of Cdks, an S-phase- and an M-phase-promoting form (SPF and MPF), which oscillate in abundance with identical frequencies but out of phase with each other. Let us further specify the shape of each wave such that the S-phase-promoting Cdk is inactive during the first quarter of the cycle, active during the second, but inactive again during the third and fourth quarters and such that the M-phase-promoting Cdk is inactive throughout first and second quarters, partially active during the third, and fully active during the fourth. Let us further assume that whether or not the cell undergoes chromosome duplication (S phase) or segregation (M phase) is determined entirely by the state of the two Cdks: If SPF is active and MPF inactive, then cells replicate DNA, whereas if MPF is fully active and SPF is inactive, then cells undergo mitosis. Our imaginary cells will undergo alternate S and M phases with intervening gaps equivalent to G_1 and G_2. This hypothesis, that cell-cycle state is defined solely by the state of S- and M-phase-specific Cdks (Nurse 1994), predicts that activation of G_1-specific Cdks in the absence of G_2-specific forms should trigger S phase, whereas activation of G_2-specific Cdks in the absence of G_1-specific forms should trigger M phase, irrespective of a cell's position in the cycle at the time. In other words, it should be possible, by artificial manipulation of SPF and MPF activities, to cause cells to jump between states that are not normally adjacent; that is, it should be possible to override the cell's history. Two observations in *S. pombe* are consistent with this view (Hayles et al. 1994). First, *S. pombe* mutants that lack the "mitotic" cdc13/Cdk1 kinase fail to enter M phase and instead undergo multiple rounds of DNA replication. Thus, preventing activation of an MPF allows cells to reenter the S-phase state without proceeding through the M-phase state. Second, *cdc10* mutants, which normally arrest in G_1 at their restrictive temperature due to a failure to activate late-G_1-specific transcription, can be induced instead to enter mitosis merely by overproducing the two components of the mitotic Cdk: cdc13 and Cdk1. Thus, premature activation of MPF can induce cells lacking SPF and MPF to enter M phase without any prior S phase.

If, as indicated by these experiments, the "Cdk state" model is correct, then understanding how the cell cycle runs in the correct direction (i.e., how cells ensure that chromosome duplication precedes their segregation) should be simply a question of understanding the dynamics or logic of SPF and MPF oscillations. Much progress has been made in understanding how different cyclin-dependent kinases regulate each other in *S. cerevisiae*. Consider the S-phase-promoting Cln1 and Cln2/Cdk1 kinases and the M-phase-promoting Clb1-Clb4/Cdk1 kinases. Cln1 and Cln2 promote activation of Clbs by turning off their proteolysis and by triggering destruction of p40Sic1. Clbs then repress transcription of *CLN1* and *CLN2*, promote mitosis and finally their own destruction, leading back to a state in which *CLN1* and *CLN2* genes can be activated by Cln3 in the absence of any interference by Clbs. But are such regulatory networks really solely responsible for imposing the procession of S and M phases?

The State of Nuclei Determines the Response to a Given Cdk

A major problem with the Cdk state hypothesis is that it provides no explanation for how cells avoid the repetitive firing of origins during S phase. S-phase-promoting Cdks must presumably be active throughout S phase to fire both early and late origins of replication. How do cells prevent the refiring of early origins as they fire old ones? It can be debated whether this remarkable property is relevant to the problem of how S and M phases alternate (Roberts 1993), but there is no denying that it would be nice to have a theory that could explain both.

Doubts as to whether cell-cycle state could solely be determined by the state of S- and M-phase-promoting factors were raised long before the discovery of Cdks even, when it was found that G_1 but not G_2 nuclei can be triggered into S phase by fusion of G_1 or G_2 cells with S-phase cells (Johnson and Rao 1971). However, this could still be explained by the "Cdk state" hypothesis if G_2 nuclei continued to harbor G_2-specific Cdks even after fusion to S-phase cells. The Cdk state hypothesis cannot, however, explain the recent discoveries that Clb/Cdk1 and not Cln/Cdk1 kinases are essential for S phase in *S. cerevisiae* (Schwob et al. 1994), that supposedly mitotic Clb/Cdk1 kinases can trigger S phase in G_1 cells (Amon et al. 1994), or that mutants deprived of the mitotic Clb1–Clb4/Cdk1 kinases possess high levels of the S-phase-promoting Clb5/Cdk1 kinase but arrest permanently in G_2 (Amon et al. 1993). These data show that the distinction between S- and M-phase-promoting Cdks is not a clear one. Mitotic Clb/Cdk1 kinases can also promote S phase! Cell-cycle state cannot therefore be determined solely by the presence or absence of specific S- or M-phase-promoting Cdks. It must also be influenced by a cell's history; that is, whether it has recently responded to these Cdks by undergoing S phase. Put in molecular terms, whether or not a cell responds to an S-phase-promoting Cdk by undergoing DNA replication must be determined by the state of Cdk substrates at the time of this Cdk's activation. G_1 nuclei, but not G_2 nuclei, presumably contain substrates whose phosphorylation by S-phase-promoting Cdks triggers DNA replication.

A Two-step Mechanism for S-phase Onset

Licensing Factor

What the crucial difference between a G_1 and a G_2 nucleus might be has until very recently been largely investigated using extracts from activated *Xenopus* eggs (Lohka and Masui 1984) in which exogenous DNA can undergo multiple rounds of DNA replication and where the distinction between G_1 and G_2 nuclei can be measured in vitro (see Laskey and Madine, this volume). *Xenopus* eggs are arrested in metaphase of meiosis II and contain high levels of MPFs like cyclin B/Cdk1. Fertilization, or artificial "activation," triggers destruction of cyclin B, and the eggs exit from mitosis and enter interphase. Extracts made from these "activated" eggs support replication of sperm chromatin following its assembly into nuclei (Blow and Sleeman 1990); the extracts presumably contain SPFs like cyclin E/Cdk2. Shortly after DNA replication, cyclin B/Cdk1 activity accumulates, which causes nuclear envelope breakdown and chromosome condensation, but the kinase later disappears due to cyclin B destruction, which leads to nuclear reformation and, in some cases, a second round of DNA replication (Hutchinson et al. 1987; Blow and Laskey 1988). The extracts thus undergo, at least partly, the natural sequence of events that occurs during cleavage divisions, in that cyclin B levels oscillate and DNA replication occurs only once per cell cycle (Murray and Kirschner 1989). The system's resemblance to the in vivo situation is striking in that no DNA molecules replicate more than once per cell cycle (Blow and Laskey 1986).

Addition of cycloheximide to the extracts prevents the appearance of cyclin B/Cdk1 kinase activity but presumably does not block activation of S-phase-promoting Cdks, because a single round of replication still occurs within the freshly assembled nuclei. Addition of an MPF activity to cycloheximide-treated extracts causes nuclear envelope breakdown and allows at least partial rereplication. The cycloheximide-treated extracts also cause G_1 but not G_2 nuclei from HeLa cells to enter S phase. The cycloheximide extracts therefore partly mimic the cytoplasm of S-phase HeLa cells, in that they can trigger S phase in G_1 but not in G_2 nuclei (Leno et al. 1992).

To explain these observations, it has been suggested (Blow and Laskey 1988) that initiation of DNA replication might be a two-step process. The first step would be the binding to origins of a factor needed for the initiation of DNA replication whose entry into the nucleus is restricted to M phase, when nuclear envelopes are disassembled. This factor would license origins to fire upon the subsequent arrival of SPF in the next cell cycle. To explain the lack of rereplication in the absence of MPF, it was envisaged that initiation of DNA replication destroys all licensing factor within the nucleus and that the nuclear membrane must be disassembled before licensing factor residing in the cytoplasm can regain access to origins. In an experiment that is widely regarded as an important confirmation of this hypothesis, it was found that permeabilization of the nuclear envelope permits rereplication of G_2 nuclei in the absence of MPF activation (Blow and Laskey 1988; Coverley et al. 1993). Purification of the putative licensing factor has been thwarted until now due to the lack of a good assay (but see Laskey and Madine, this volume).

Preparations for S Phase as Cells Exit from Mitosis

The *CDC46*, *MCM2*, and *MCM3* gene products in yeast have certain properties reminiscent of the putative licensing factor (Hennessy and Botstein 1991; Hennessy et al. 1991; Tye 1994). They are necessary for the initiation of DNA replication in yeast, and their access to chromatin is under cell cycle control. Cdc46 and Mcm3 proteins accumulate in the cytoplasm during G_2 and only enter nuclei as cells undergo anaphase. Vertebrate homologs of these proteins are not like their yeast counterparts excluded from nuclei during G_2, but their association with chromatin may nevertheless be under cell cycle control (Kimura et al.

1994). The P1 protein, a homolog of Mcm3 that binds to DNA polymerase-α, is distributed in the cytoplasm during metaphase and associates with chromatin only as chromosomes decondense during telophase. The behavior of Cdc46/Mcm proteins and that of the putative licensing factor differ drastically in one important respect. The Cdc46/Mcm proteins enter nuclei or associate with chromatin as cells exit from M phase (Hennessy et al. 1990), as cyclin B/Cdk1 kinase is destroyed (Surana et al. 1993), and not during M phase, when cyclin B/Cdk1 kinase is high, as originally proposed for licensing factor. This difference could be vital because, as we shall see, the cyclin B/Cdk1 kinase might be an important regulator of DNA rereplication (Hayles et al. 1994; Dahmann et al. 1995). Another example of preparations for S phase occurring at the end of M phase concerns the 70-kD subunit of RP-A, a three-subunit single-stranded DNA-binding protein needed for prepriming, whose association with subnuclear foci proposed to be prereplication centers (pre-RCs) as chromosomes decondense following exit from M phase (Adachi and Laemmli 1994) can be inhibited by cyclin B/Cdk1 kinase.

Chromatin Structure of Yeast Origins during the Cell Cycle

Yet more persuasive evidence of preparations for replication occurring at the end of mitosis have come from analyses of chromatin structure at DNA replication origins during the cell cycle (Diffley et al. 1994). So far, discrete replication origins have been well defined only in *S. cerevisiae*, where a 200-bp DNA sequence (ARS) containing a core consensus sequence called A and flanking sequences B1, B2, and B3 is sufficient for origin function on episomes and within chromosomes (Marahrens and Stillman 1992). A 250-kD protein complex called the origin recognition complex (ORC) binds to sequences A and B1 (Bell and Stillman 1992; Diffley and Cocker 1992), whereas a separate protein, Abf1, binds to the B3 element (Diffley and Stillman 1989). The state of these origins during the cell cycle has been analyzed by genomic footprinting. Both ORC and Abf1 appear to be bound throughout the cell cycle, but there are important changes in the neighboring chromatin (Diffley et al. 1994). From S phase until late anaphase, the pattern of cleavage by DNase I in vivo resembles that induced by ORC binding in vitro, which causes an adjacent DNA sequence to become hypersensitive. As cells exit from mitosis, this DNase I hypersensitive site disappears and does not reappear until cells enter S phase in the next cycle. The data indicate that a second factor binds next to ORC during G_1 but not at other stages of the cell cycle.

CDC6 *Is Necessary for Pre-RCs*

A key question is whether the cell-cycle-regulated factor that binds to yeast origins is required for DNA replication or for preventing its premature onset. If the former were true, then disappearance of DNase I hypersensitivity at the end of mitosis should depend on proteins needed for origin firing. A good candidate for such a protein is that encoded by *CDC6*, which is required for the initiation of DNA replication in *S. cerevisiae* (Bueno and Russell 1992), as is a homologous gene, *cdc18*, in *S. pombe* (Kelly et al. 1993). Minichromosomes whose replication is dependent on a single ARS sequence are much more frequently lost in *cdc6* mutants than in wild-type cells, but this defect is suppressed when the minichromosome carries multiple ARS sequences, implying that *cdc6* mutants are not defective in DNA replication per se but rather in the function of origins (Hogan and Koshland 1992). The Cdc6 protein is unstable and is synthesized in a burst at the end of mitosis (as well as in late G_1) simultaneous with the change in chromatin structure that occurs at origins (Piatti et al. 1995). More important, the disappearance of ORC-induced DNase I hypersensitivity at origins does not take place when cells exit from mitosis in the absence of Cdc6 synthesis (Cocker et al. 1996). Furthermore, the "protected" state of origins characteristic of cells arrested in G_1 (by pheromone) is temperature sensitive in *ts cdc6* mutants. Cdc6 is therefore necessary for the formation and maintenance of a chromatin configuration at origins that precedes DNA replication and is presumably essential for its initiation. The observation that Cdc6 and ORC can form complexes in vitro (Liang et al. 1995) is consistent with this view.

S-phase onset in yeast might therefore involve two temporarily distinct events at replication origins. The first would be the formation of pre-RCs, whose appearance during telophase depends on *CDC6* and coincides with the destruction of Clb/Cdk1 kinases (MPFs). The second event would be the activation of S-phase-promoting Clb/Cdk1 kinases (Schwob et al. 1994) in late G_1 (SPFs), which would trigger pre-RCs to initiate DNA replication. There is a pleasing congruence between this scheme and the arrival on chromatin at the end of mitosis of RP-A subnuclear foci and of Cdc46/Mcm proteins, which might also be components of yeast prereplicative complexes.

Control of pre-RC Formation

If the formation of pre-RCs at future origins is a prerequisite for initiating DNA replication, then control of pre-RC formation could be as vital an aspect of S-phase control as the production of S-phase-promoting factors (Cdks). In *S. cerevisiae*, activation of cyclin B/Cdk1 kinases during G_1 causes the onset of S phase, but the very same kinases are active during G_2 and yet cells do not reenter S phase. This can now be explained by the absence in G_2 cells of pre-RCs that are, as it were, the key substrates for S-phase-promoting Cdks. What then excludes the formation of pre-RCs during S, G_2, and M phases?

In yeast, the adoption by origins of a prereplicative state indicative of pre-RCs occurs with very similar kinetics at the end of mitosis as the destruction of cyclin B/Cdk1 kinases (Surana et al. 1993; Diffley et al. 1994). This suggests that cyclin B/Cdk1 kinases could have a role in preventing the de novo assembly of pre-RCs. Direct evidence that destruction of Cdks in G_2 or M phase is sufficient to trigger pre-RCs and thereby creates the potential to rereplicate upon activation of S-phase-promoting Cdks has been obtained using an inducible *SIC1* gene in *S. cerevisiae* (Dahmann et al. 1995). The microtubule-disrupting drug nocodazole prevents nuclear division in *S. cerevisiae* and causes cells to arrest with duplicated DNA, origins in a postreplicative state, and high levels of all six Clb kinases. Induction of high levels of the Clb/Cdk1 inhibitor p40Sic1 from a *GAL-SIC1* gene fusion inhibits completely both Clb2 and Clb5/Cdk1 kinases (and presumably all other Clb kinases as well), causes cells to rebud without nuclear division, causes origins to adopt a prereplicative chromatin structure, but does not induce rereplication. However, reactivation of Clb/Cdk1 kinases by subsequently shutting off p40Sic1 synthesis causes all cells to rereplicate their DNA.

Mutants That Rereplicate

Quite independent evidence implicating cyclin B/Cdk1 kinases in blocking rereplication in G_2 cells stems from the characterization of yeast mutants that undergo successive S phases in the absence of an intervening mitosis. Such mutants have been isolated on the basis of their ability to diploidize and thereby sporulate. Budding yeast mutant cells with a temperature-sensitive allele of the *ESP1* gene replicate normally at the restrictive temperature, but they subsequently fail to undergo anaphase due to an unspecified mitotic spindle defect. Despite their mitotic failure, the mutant cells proceed as usual with destruction of Clb/Cdk1 kinase, and they subsequently embark on a new cell cycle, during which they rereplicate their chromosomes (McGrew et al. 1992; Surana et al. 1993). The behavior of *esp1* mutants shows that Clb destruction is accompanied by rereplication even when mitosis has not occurred. Thus, mitosis per se is not necessary.

Many more diploidizing mutants have subsequently been analyzed in *S. pombe*. The vast majority fall into the *esp1* class, in that they attempt to enter mitosis, fail, and proceed with the next cycle. In rare mutants, however, cells rereplicate DNA without attempting mitosis (although nitrogen starvation is also required). Upon return to the permissive temperature, some of these "G_2" cells rereplicate their DNA before entering mitosis. Mutations in two genes give rise to this phenotype. Remarkably, they encode *cdc2* (*cdk1*) and *cdc13*, the two components of the cyclin B/Cdk needed for mitosis in fission yeast (Broek et al. 1991; Hayles et al. 1994). The behavior of cells completely lacking the *cdc13* gene product is even more striking. Upon inoculation into fresh medium, wild-type spores germinate, replicate their DNA, spend the next 90 minutes in G_2, and then undergo mitosis. The second round of DNA replication follows almost immediately after completion of the first mitosis. Spores lacking *cdc13* resemble wild-type cells up to the point at which the Cdc13/Cdc2 kinase should have become active. Due to its absence, the mutant cells never enter mitosis; G_2 is prolonged, but not indefinitely, because 5 hours after the first S phase, cells rereplicate their DNA and this process recurs thereafter, roughly every generation time. As a consequence, some cells eventually accumulate up to 32 copies of their genome. Therefore, it seems that Cdc13 is necessary to prevent *S. pombe* cells from entering a G_1-like state after they have spent some time in G_2. Why might this be so? One possibility is that loss of Cdc13/Cdk1 kinase encourages the formation of pre-RCs.

Cyclin B/Cdk1 kinases have also been implicated in preventing DNA replication between meiosis I and II. During maturation of *Xenopus* oocytes, cyclin B/Cdk1 is inactivated after meiosis I but, due to the action of c-*mos*, is promptly reactivated prior to meiosis II. Ablation of c-*mos* product causes maturing oocytes instead to reform nuclei and replicate DNA after the first meiotic division (Furuno et al. 1994). The same effect is observed with dominant-negative alleles of Cdk1. Thus, cyclin B/Cdk1 kinases inhibit DNA replication between meioses I and II as well as between S and M phases during vegetative cell cycles. Whether they do so by inhibiting pre-RC formation is yet to be determined.

Preventing Rereplication during S and G_2 Phases: Do S-phase-promoting Cdks Have a Dual Function?

We now have a coherent explanation for the behavior of origins during mitosis and G_1. Assembly of pre-RCs requires destruction of mitotic Cdks and the synthesis of pre-RC components, either at the end of mitosis or in late G_1. Pre-RCs are then transformed into replication forks by activation of S-phase-promoting Cdks. Thus, the lack of DNA replication during G_1 and M phases has very different causes. G_1 cells that have formed pre-RCs have the potential to enter S phase but do not do so for lack of S-phase-promoting Cdks, whereas M-phase cells could not enter S phase even if they were to express S-phase-promoting Cdks (and some types of cell indeed do) because assembly of pre-RCs is inhibited by mitotic cyclin B/Cdk1 kinases. This model cannot yet account for the behavior of S and G_2 cells. During DNA replication, S-phase-promoting Cdks are active whereas mitotic cyclin B/Cdk1 kinases are inactive, and cells should therefore be able both to form pre-RCs and to turn them into replication forks. What then prevents origins from firing more than once? A similar problem could afflict G_2 cells that have not yet activated cyclin B/Cdk1 kinases but contain Cdks like cyclin A/Cdk2 in animals or Clb1–Clb4/Cdk1 in yeast, which are capable of promoting S phase.

The phenotype of *S. cerevisiae* mutants lacking G_2-specific B-type cyclins Clb1–Clb4 may provide an important clue to this mystery. Strains lacking Clb1, Clb3, and Clb4 and kept alive with a temperature-sensitive allele of *CLB2* (conditional *clb1–clb4* mutants) can replicate normally but fail to build a mitotic spindle at the restrictive temperature. In this regard, their phenotype resembles that of *S. pombe cdc13* deletion mutants. Their phenotype differs greatly in another regard, for they arrest in G_2 indefinitely without rereplicating. Clb5/Cdk1 kinase remains active during this arrest. Is it possible that this kinase also inhibits assembly of pre-RCs and thereby prevents rereplication? Mutations in at least two genes (*SIM*1 and *SIM*2) allow *clb1-4* mutants to rereplicate (C. Dahmann et al., in prep.) Remarkably, the mutants are all defective in maintaining high levels of Clb5/Cdk1 kinase and rebud at the same time as they rereplicate, which is a sign of Clb kinase decline in vivo. Moreover, they fail to rereplicate if *CLB5* gene dosage is increased. The behavior of these mutants shows that Clb5/Cdk1 kinase acts to suppress replication in *clb1-4* mutants, which is all the more remarkable because Clb5/Cdk1 normally promotes S phase.

The apparent contradiction that Clb5/Cdk1 both promotes and inhibits S phase could be explained if the kinase had different replication functions in G_1 and G_2: promoting it during G_1 but blocking it during G_2. Might the very same kinase that promotes transformation of pre-RCs into replication forks also inhibit de novo assembly of pre-RCs (see Fig. 3)? Indeed, it seems likely that all of the six Clb/Cdk1 kinases that are capable of promoting S phase in *S. cerevisiae* also inhibit formation of pre-RCs. Clb5 and Clb6 might be primarily responsible for inhibiting

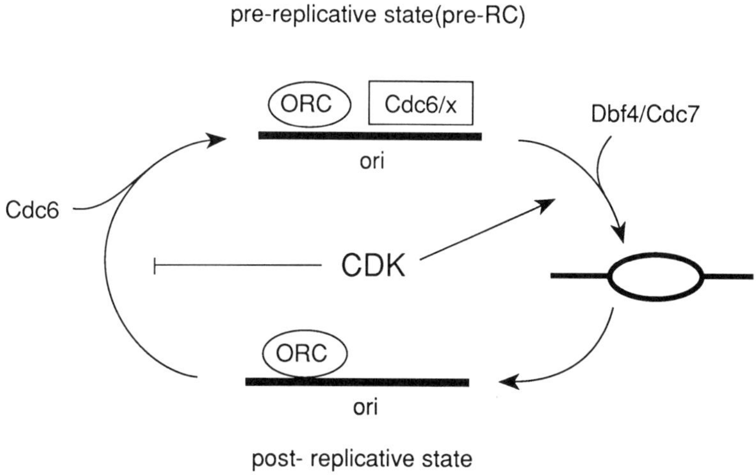

Figure 3 A dual function for Cdks: promoting initiation of DNA replication and inhibiting pre-RC assembly. Initiation is a two-step process, the first step of which is the formation of pre-RCs at future origins. This process depends on the synthesis of the unstable replication protein, Cdc6. It usually occurs at the end of mitosis when there is a burst of Cdc6 synthesis but can also occur in late G_1, when there is a second burst of Cdc6 synthesis, at least in daughter cells. The second step is activation in late G_1 of cyclin B/Cdk1 kinases along with the Cdc7 kinase. It seems that activity of the very same set of cyclin B/Cdk1 kinases prevents de novo assembly of pre-RCs. Each round of initiation therefore depends on a cycle of cyclin activation/destruction (see Fig. 4).

pre-RC assembly during S phase, Clb3 and Clb4 during early G_2, and Clb1 and Clb2 during late G_2 and during M phase, when the respective kinases are most active. If this model were correct, the different behavior of *S. pombe cdc13* mutants, which undergo multiple rounds of replication, and *S. cerevisiae clb1-4* quadruple mutants, which arrest indefinitely in G_2, might be simply one of degree. *cdc13* mutants also do not re-replicate for many hours after their first S phase, during which time other B-type cyclins, e.g., Cig1 and Cig2, might prevent the assembly of pre-RCs. The Cig1 and Cig2 B-type cyclins might either be somewhat less effective than Clb5 and Clb6 and eventually be overcome by the forces driving the assembly of pre-RCs or, for reasons not yet understood, they may simply not remain active during G_2 for as long as Clb5 and Clb6.

Pre-RCs Can Form Either at the End of M Phase or in Late G_1

At first sight, a mechanism whereby S-phase initiation depends on events that occurred at the end of the previous mitosis seems somewhat precarious. It is hard to imagine how cells ensure that prereplication complexes formed at the end of mitosis could be relied on to survive an extended arrest in G_1, as occurs during stationary phase in yeast. The same problem would afflict quiescent mammalian cells, which can spend months, if not years, arrested in G_1 before they are stimulated to reenter S phase. In fact, the chromatin structure thought to represent pre-RCs at yeast origins can form not only at the end of mitosis, as occurs in cycling cells, but also during G_1 phase when inoculation into fresh medium stimulates stationary-phase yeast cells to reenter the cell cycle (Diffley et al. 1994). This observation is not at odds with the notion that destruction of cyclin B/Cdk1 kinases is necessary for the formation of pre-RCs, according to which pre-RC assembly should be possible throughout G_1, for as long as the inhibitory cyclin B/Cdk1 kinases remain dormant. Thus, DNA replication could be seeded predominantly from pre-RCs formed at the end of mitosis in cells with short G_1 periods, as in yeast mother cells and embryonic cells undergoing cleavage divisions, whereas it could be seeded mainly by pre-RCs formed during G_1 in cells with long G_1 periods.

Democracy of the Genome

The notion that assembly of new pre-RCs might be prevented by the very same set of Cdks that promote their transition into replication forks is an interesting idea, because it explains how cells ensure that there is no stage during the cell cycle in which pre-RCs can both be assembled and be transformed into replication forks (see Fig. 4). S-phase-promoting Cdks with dual functions would preclude any possibility of origins firing more than once during the cell cycle. Furthermore, if we assume that all Cdk1 (or Cdk2) forms can hinder pre-RC formation, we could explain why G_2 nuclei cannot enter S phase even when they are placed in a cytoplasm containing S-phase-promoting Cdks (Johnson and Rao 1971). We presume, by analogy with *S. cerevisiae*, that from S phase until the end of mitosis, animal cells also continuously maintain an environment that is hostile to the formation of pre-RCs. S-phase-promoting Cdks like cyclin E/Cdk2 or cyclin A/Cdk2 might prevent their formation during S phase, whereas cyclin A/Cdk1 and cyclin B/Cdk1 assume this function during G_2 and M phase.

Primordial Cell Cycles

Although multiple Cdks may participate in the exclusion of pre-RC assembly during the S, G_2, and M phases of most present-day eukaryotic

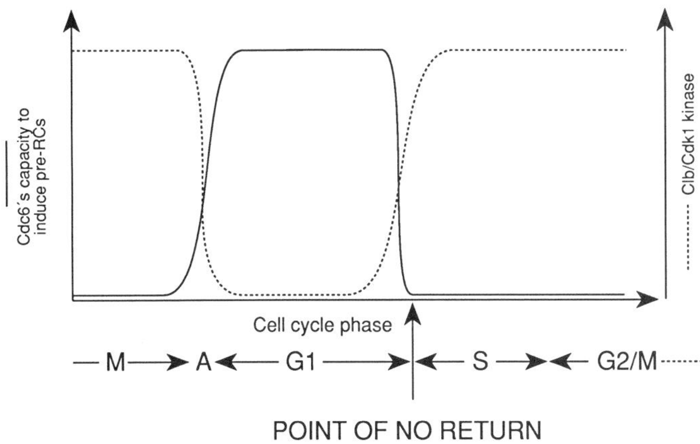

Figure 4 The pre-RC/Cdk cycle in yeast. Cyclin B/Cdk1 kinases prevent the formation of pre-RCs during S, G_2, and M phases. Pre-RCs therefore only form upon proteolysis of cyclin B molecules during anaphase. Reactivation of cyclin B/Cdk1 kinases during the subsequent G_1 period triggers initiation at origins that had formed pre-RCs during G_1, when cyclin B/Cdk1 kinases were inactive.

cells, and specialized Cdks may be employed to transform pre-RCs into replication forks, it is worth considering that discrete rounds of replication could theoretically be driven by the alternate activation and inactivation of a single "dual function" Cdk. In *S. cerevisiae*, where mitosis does not entail extensive chromosome condensation, entry into mitosis and DNA replication can co-exist. It is therefore conceivable that in the primordial eukaryotic cell, activation of a single Cdk might have triggered simultaneously transformation of pre-RCs into replication forks (i.e., S phase) and attachment of chromosomes to mitotic spindles (i.e., M phase). Subsequent inactivation of this same Cdk could have signaled both the assembly of pre-RCs and exit from mitosis. It is therefore possible to envisage how the oscillation of a single Cdk could have formed the heart of our distant ancestors' cell cycle. To function with high fidelity, such a system might have required a surveillance mechanism that ensured that destruction of Cdks was never initiated before DNA replication was complete. This might then have been the cell cycle's oldest checkpoint control.

Dependence of S on Anaphase

The hypothesis that mitotic Cdks prevent pre-RC assembly also explains why sister chromatid separation always precedes a second round of DNA replication. S-phase-promoting Cdks decline during G_2, G_2-specific Cdks like cyclin A/Cdk1 decline during metaphase, but the destruction of the last remaining Cdks capable of inhibiting the assembly of pre-RCs does not occur until mitosis-specific Cdks like cyclin B/Cdk1 are inactivated due to cyclin B proteolysis at anaphase. The latter depends on activation of a specialized multisubunit ubiquitin ligase, which is necessary also for the separation of sister chromatids and is therefore called the anaphase-promoting complex (APC) (see Fig. 5). The APC is thought to catalyze proteolysis not only of B-type cyclins but also of proteins that inhibit sister chromatid separation (Holloway et al. 1993; Irniger et al. 1995; King et al. 1995). Thus, cells cannot begin to assemble pre-RCs (a pre-condition for chromosome duplication) without having also triggered anaphase.

A Three-parameter Model for the Cell Cycle

We started this section with the notion that cell-cycle state might be determined by the activity of two key parameters: the level of S- and M-phase-promoting Cdks. This idea turns out to be untenable. At least three parameters are required to define the major cell-cycle states of most existing eukaryotic cells: the levels of S- and M-phase-promoting Cdks (SPF and MPF) and the presence or absence of pre-RCs. Early G_1 cells lack SPF and MPF but usually (although not always)`contain pre-RCs; cells embarking on S phase contain SPF and pre-RCs; cells in S phase contain SPF and are in the process of converting their pre-RCs into replication forks; G_2 cells have finished this process and may or may not contain SPF but have not fully activated MPF; M-phase cells have fully activated MPF, and anaphase cells are in the process of destroying their MPF and other proteins whose destruction is needed for chromosome segregation and are about to form pre-RCs. Only when we take pre-RCs into account can we explain why G_1 and G_2 nuclei respond differently to activation of S-phase-promoting Cdks. The three-parameter model of the cell cycle also allows us to explain the very different effects of activating mitotic Cdks in G_1-arrested *S. pombe* and *S. cerevisiae* cells; it induced replication in *S. cerevisiae* cells deprived of Clns but failed to do so in *S. pombe cdc10* mutants (Amon et al. 1994; Hayles et al. 1994). *cdc10* is needed for the synthesis of Cdc18, which is homologous to Cdc6 from *S. cerevisiae* and might therefore be necessary for the formation of pre-RCs. The presence of pre-RCs in *cln*-arrested *S. cerevisiae* cells and their absence in *S. pombe cdc10* mutants would explain their different responses.

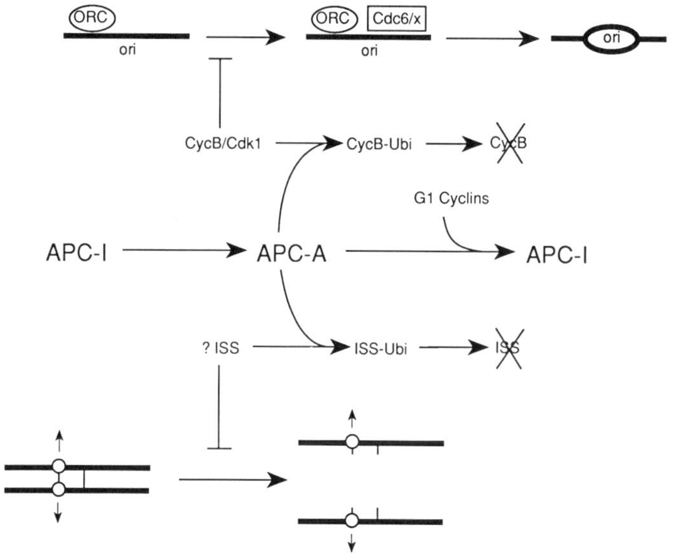

Figure 5 The APC could link S phase to anaphase. The APC is a 20S particle that contains Cdc16, Cdc23, and Cdc27 proteins (among others) and catalyzes the ligation of multiple ubiquitin molecules to cyclin B proteins containing a destruction box, targeting them for destruction by the 26S proteosome. The ubiquitin ligase reaction also requires ATP, ubiquitin, an activating enzyme (E1), and a conjugating enzyme (E2). APC components are needed not only for cyclin B ubiquitination and proteolysis, but also for the onset of anaphase. It is therefore thought that the transition from metaphase to anaphase depends on the destruction of inhibitors of sister chromatid separation (ISS). The APC is thought to be inactive from late G_1 until the onset of anaphase. Its activation then triggers anaphase and the destruction of cyclin B proteins. The latter leads to the formation of pre-RCs and thereby to a new round of DNA replication when cyclin B/Cdk1 kinases are reactivated during the subsequent G_1 period. Activation of Cln1 or Cln2/Cdk1 kinases turns the APC off and allows the accumulation of mitotic cyclin B proteins.

The inclusion of pre-RCs as a major cell-cycle parameter helps to impart a rigid directionality to our cell-cycle model. The presence of these structures is partly determined by the cell's history, but this directionality comes at a price. The timing of cell-cycle events becomes more crucial. For example, activation of S-phase-promoting Cdks before pre-RCs have been assembled, as would be possible in the outgrowth of cells that had long lost pre-RCs inherited from their previous telophase, would completely finesse the replication machinery and lead to an unreplicated state from which a cell could no longer extract itself. Much of the regulation of pre-RC components and S-phase-promoting Cdks, which is more conventionally thought to be contributing to the control of cellular proliferation, may have the more important role of helping to avoid the disastrous consequences of activating S-phase-promoting Cdks before pre-RCs have had time to form.

Why Do Most Quiescent Cells Arrest in G_1?

A final insight that stems from these ideas concerns why so many metazoan cells choose to enter a quiescent state during G_1 and not during S or G_2 phases. Cells at these latter stages must maintain active Cdks to exclude the assembly of pre-RCs. A temporary inactivation of these Cdks due to unspecified physiological disasters could lead to the assembly of pre-RCs and thereby to an unscheduled round of DNA replication if S-phase-promoting Cdks were activated before mitosis could take place. It is interesting in this regard that rereplication of Cdk1 alleles in *S. pombe* was first observed only upon extreme heat shock and nitrogen starvation (Broek et al. 1991); that is, extreme environmental conditions can contribute to the destruction of Cdks. In contrast, destruction of preassembled pre-RCs in G_1 cells should not be disastrous, as long as Cdks capable of inhibiting pre-RC assembly were not yet activated and as long as cells contained pre-RC components or were capable of entering a program that caused their synthesis. The logic of replication control may therefore dictate that the early G_1 state is one from which a cell can more readily recover its true cell-cycle state following an experience that erases its memory of cell-cycle position. It is the one stage of the cell cycle where history does not matter!

Closing the Cycle

The first section of this review dealt with the events needed to drive a G_1 cell into S phase. We treated the resting G_1 cell as a blank slate and considered how cells produce the factors that trigger it to replicate. In the second half, we considered how cells prevent replication during G_2 and M phases. We can now appreciate how events at the end of one cell cycle lead to the start of the next one; that is, we can begin to appreciate its cyclical nature. There is sufficient knowledge about this process in *S. cerevisiae* to piece together the following picture. During G_2 and M phases, Clb/Cdk1 kinases prevent transcription of S-phase-promoting kinases Cln1 and Cln2/Cdk1 by preventing the binding of SBF to SCBs in their promoters. These kinases also prevent the assembly of pre-RCs (how is not understood). Thus, destruction of Clb/Cdk1 kinases during anaphase through the activation of Clb proteolysis and the accumulation of p40Sic1 removes the key barrier to transcriptional activation of *CLN1* and *CLN2* by Cln3/Cdk1 and the formation of pre-RCs at origins (which can form at this stage or later in G_1). In the absence of Clb/Cdk1 kinases, growth of daughter cells to a threshold size facilitates transcription of *CLN1*, *CLN2*, and *CDC6*, which produces active Cln/Cdk1 kinases and yet more pre-RCs. Cln1 and Cln2/Cdk1 kinases then trigger the reactivation of Clb/Cdk1 kinases by promoting proteolysis of p40Sic1 and by turning off proteolysis of Clb cyclins, whose activity now promotes the transition of pre-RCs into replication forks.

ACKNOWLEDGMENTS

I thank Christian Koch, Nick Jones, and Gustav Ammerer for helpful comments on the manuscript.

REFERENCES

Adachi, Y. and U.K. Laemmli. 1994. Study of the cell cycle-dependent assembly of the DNA pre-replication centres in *Xenopus* egg extracts. *EMBO J.* **13:** 4153–4164.

Amon, A., S. Irniger, and K. Nasmyth. 1994. Closing the cell cycle circle in yeast: G2 cyclin proteolysis initiated at mitosis persists until the activation of G1 cyclins in the next cycle. *Cell* **77:** 1037–1050.

Amon, A., M. Tyers, B. Futcher, and K. Nasmyth. 1993. Mechanisms that help the yeast cell cycle clock tick: G2 cyclins transcriptionally activate G2 cyclins and repress G1 cyclins. *Cell* **74:** 993–1007.

Andrews, B.J. and L.A. Moore. 1992. Interaction of the yeast Swi4 and Swi6 cell cycle regulatory proteins in vitro. *Proc. Natl. Acad. Sci.* **89:** 11852–11856.

Baldin, V., J. Lukas, M.J. Marcote, M. Pagano, and G. Draetta. 1993. Cyclin D1 is a nuclear protein required for cell cycle progression in G1. *Genes Dev.* **7:** 812–821.

Baroni, M.D., P. Monti, and L. Alberghina. 1994. Repression of growth-regulated G1 cyclin expression by cyclic AMP in budding yeast (see comments). *Nature* **371:** 339–342.

Bates, S., D. Parry, L. Bonetta, K. Vousden, C. Dickson, and G. Peters. 1994. Absence of cyclin D/cdk complexes in cells lacking functional retinoblastoma protein. *Oncogene* **9:** 1633–1640.

Beach, D., B. Durkacz, and P. Nurse. 1982. Functionally homologous cell cycle control genes in budding and fission yeast. *Nature* **300:** 706–709.

Bell, S.P. and B. Stillman. 1992. ATP-dependent recognition of eukaryotic origins of DNA replication by a multiprotein complex (see comments). *Nature* **357:** 128–134.

Bjorklund, S., S. Skog, B. Tribukait, and L. Thelander. 1990. S-phase-specific expression of mammalian ribonucleotide reductase R1 and R2 subunit mRNAs. *Biochemistry* **29:** 5452–5428.

Blow, J.J. and R.A. Laskey. 1986. Initiation of DNA replication in nuclei and purified DNA by a cell-free extract of *Xenopus* eggs. *Cell* **47:** 577–587.

———. 1988. A role for the nuclear envelope in controlling DNA replication within the cell cycle. *Nature* **332:** 546–547.

Blow, J.J. and A.M. Sleeman. 1990. Replication of purified DNA in *Xenopus* egg extract is dependent on nuclear assembly. *J. Cell Sci.* **95:** 383–391.

Bodrug, S.E., B.J. Warner, M.L. Bath, G.J. Lindeman, A.W. Harris, and J.M. Adams. 1994. Cyclin D1 transgene impedes lymphocyte maturation and collaborates in lymphomagenesis with the *myc* gene. *EMBO J.* **13:** 2124–2130.

Booher, R. and D. Beach. 1988. Involvement of *cdc13*$^+$ in mitotic control in *Schizosaccharomyces pombe*: Possible interaction of the gene product with microtubules. *EMBO J.* **7:** 2321–2328.

Breeden, L. and K. Nasmyth. 1987. Similarity between cell-cycle genes of budding yeast and fission yeast and the *notch* gene of *Drosophila*. *Nature* **329:** 651–654.

Brenner, C., N. Nakayama, M. Goebl, K. Tanaka, A. Toh-e, and K. Matsumoto. 1988. CDC33 encodes mRNA cap-binding protein eIF-4E of *Saccharomyces cerevisiae*. *Mol. Cell. Biol.* **8:** 3556–3559.

Broek, D., R. Bartlett, K. Crawford, and P. Nurse. 1991. Involvement of p34^{cdc2} in establishing the dependency of S phase on mitosis (see comments). *Nature* **349:** 388–393.

Bueno, A. and P. Russell. 1992. Dual functions of CDC6: A yeast protein required for DNA replication also inhibits nuclear division. *EMBO J.* **11:** 2167–2176.

———. 1993. Two fission yeast B-type cyclins, Cig2 and Cdc13, have different functions in mitosis. *Mol. Cell. Biol.* **13:** 2286–2297.

Bueno, A., H. Richardson, S.I. Reed, and P. Russell. 1991. A fission yeast B-type cyclin functioning early in the cell cycle (published erratum in *Cell* 1993 **73:** following 1049). *Cell* **66:** 149–159.

Caligiuri, M. and D. Beach. 1993. Sct1 functions in partnership with Cdc10 in a transcription complex that activates cell cycle START and inhibits differentiation. *Cell* **72:** 607–619.

Chapman, J.W. and L.H. Johnston. 1989. The yeast gene, *DBF4*, essential for entry into S phase is cell cycle regulated. *Exp. Cell Res.* **180:** 419–428.

Cocker, J.H., S. Piatti, C. Santocanale, K. Nasmyth, and J.F.X. Diffley. 1996. An essential role for the Cdc6 protein in forming the pre-replicative complexes of budding yeast. *Nature* **379:** 180–182.

Connolly, T. and D. Beach. 1994. Interaction between the Cig1 and Cig2 B-type cyclins in the fission yeast cell cycle. *Mol. Cell. Biol.* **14:** 768–776.

Correa-Bordes, J. and P. Nurse. 1995. p25^{rum1} orders S phase and mitosis by acting as an inhibitor of the p34^{cdc2} mitotic kinase. *Cell* **83:** 1001–1009.

Coverley, D., C.S. Downes, P. Romanowski, and R.A. Laskey. 1993. Reversible effects of nuclear membrane permeabilization on DNA replication: Evidence for a positive licensing factor. *J. Cell Biol.* **122:** 985–992.

Coxon, A., K. Maundrell, and S.E. Kearsey. 1992. Fission yeast *cdc21*$^+$ belongs to a family of proteins involved in an early step of chromosome replication. *Nucleic Acids Res.* **20:** 5571–5577.

Creanor, J. and J.M. Mitchison. 1984. Protein synthesis and its relation to the DNA division cycle in the fission yeast *Schizosaccharomyces pombe*. *J. Cell Sci.* **69:** 199–210.

Cross, F.R. 1988. *DAF1*, a mutant gene affecting size control, pheromone arrest, and cell cycle kinetics of *Saccharomyces cerevisiae. Mol. Cell. Biol.* **18:** 4675–4684.

Cross, F.R. and C.M. Blake. 1993. The yeast Cln3 protein is an unstable activator of Cdc28. *Mol. Cell. Biol.* **13:** 3266–3271.

Cross, F.R. and A.H. Tinkelenberg. 1991. A potential positive feedback loop controlling *CLN1* and *CLN2* gene expression at the start of the yeast cell cycle. *Cell* **65:** 875–883.

Cross, F., J. Roberts, and H. Weintraub. 1989. Simple and complex cell cycles. *Annu. Rev. Cell Biol.* **5:** 341–395.

Dahmann, C., J.F.X. Diffley, and K. Nasmyth. 1995. S-phase-promoting cyclin-dependent kinases prevent re-replication by inhibiting the transition of replication origins to a pre-replicative state. *Curr. Biol.* **5:** 1257–1269.

Dalton, S. 1992. Cell cycle regulation of the human *cdc2* gene. *EMBO J.* **11:** 1797–1804.

Deng, C., P. Zhang, J.W. Harper, S.J. Elledge, and P. Leder. 1995. Mice lacking $p21^{CIP1/WAF1}$ undergo normal development, but are defective in G1 checkpoint control. *Cell* **82:** 675–684.

Deshaies, R., V. Chau, and M. Kirschner. 1995. Ubiquitination of the G1 cyclin Cln2p by a Cdc34p-dependent pathway. *EMBO J.* **14:** 303–312.

Diffley, J.F.X. and J.H. Cocker. 1992. Protein-DNA interactions at a yeast replication origin. *Nature* **357:** 169–172.

Diffley, J.F.X. and B. Stillman. 1989. Similarity between the transcriptional silencer binding proteins ABF1 and RAP1. *Science* **246:** 1034–1038.

Diffley, J.F.X., J.H. Cocker, S.J. Dowell, and A. Rowley. 1994. Two steps in the assembly of complexes at yeast replication origins in vivo. *Cell* **78:** 303–316.

Dirick, L. and K. Nasmyth. 1991. Positive feedback in the activation of G1 cyclins in yeast. *Nature* **351:** 754–757.

Dirick, L., T. Böhm, and K. Nasmyth. 1995. Roles and regulation of Cln-Cdc28 kinases at the start of the cell cycle in *Saccharomyces cerevisiae. EMBO J.* **14:** 4803–4813.

Dolznig, H., P. Bartunek, K. Nasmyth, E. Müllner, and H. Beug. 1995. Terminal differentiation of normal chicken, erythroid progenitors: Shortening of G_1 correlates with loss of D-cyclin/cdk4 expression and altered cell size control. *Cell Growth Differ.* **6:** 1341–1352.

Donovan, J.D., J.H. Toyn, A.L. Johnson, and L.H. Johnston. 1994. P40SDB25, a putative CDK inhibitor, has a role in the M/G1 transition in *Saccharomyces cerevisiae. Genes Dev.* **8:** 1640–1653.

Dou, Q.P., P.J. Markell, and A.B. Pardee. 1992. Thymidine kinase transcription is regulated at G1/S phase by a complex that contains retinoblastoma-like protein and a *cdc2* kinase. *Proc. Natl. Acad. Sci.* **89:** 3256–60.

Dowell, S.J., P. Romanowski, and J.F. Diffley. 1994. Interaction of Dbf4, the Cdc7 protein kinase regulatory subunit, with yeast replication origins in vivo. *Science* **265:** 1243–1246.

Dulic, V., E. Lees, and S.I. Reed. 1992. Association of human cyclin E with a periodic G1-S phase protein kinase. *Science* **257:** 1958–1961.

Duronio, R.J. and P.H. O'Farrell. 1994. Developmental control of a G1-S transcriptional program in *Drosophila. Development* **120:** 1503–1515.

Duronio, R.J., P.H. O'Farrell, J.E. Xie, A. Brook, and N. Dyson. 1995. The transcription factor E2F is required for S phase during *Drosophila* embryogenesis. *Genes Dev.* **9:** 1445–1455.

Dynlacht, B.D., O. Flores, J.A. Lees, and E. Harlow. 1994. Differential regulation of E2F transactivation by cyclin/cdk2 complexes. *Genes Dev.* **8:** 1772–1786.

El-Deiry, W.D., T. Tokino, V.E. Velcuiescu, D.B. Levy, R. Parsons, J.M. Trent, D. Lin, W.E. Mercer, K.W. Kinzler, and B. Vogelstein. 1993. *WAF1*, a potential mediator of p53 tumor suppression. *Cell* **75:** 817–825.

Epstein, C.B. and F.E. Cross. 1992. CLB5: A novel B cyclin from budding yeast with a role in S phase. *Genes Dev.* **6:** 1695–1706.

Evans, T., E.T. Rosenthal, J. Youngbloom, D. Distel, and T. Hunt. 1983. Cyclin: A protein specified by maternal mRNA in sea urchin eggs that is destroyed at each cleavage division. *Cell* **33:** 389–396.

Fagan, R., K.J. Flint, and N. Jones. 1994. Phosphorylation of E2F-1 modulates its interaction with the retinoblastoma gene product and the adenoviral E4 19 kDa protein. *Cell* **78:** 799–811.

Falconi, M.M., A. Piseri, M. Ferrari, G. Lucchini, P. Plevani, and M. Foiani. 1993. De novo synthesis of budding yeast DNA polymerase α and *POL1* transcription at the G1/S boundary are not required for entrance into S phase. *Proc. Natl. Acad. Sci.* **90:** 10519–10523.

Fang, F. and J.W. Newport. 1991. Evidence that the G1-S and G2-M transitions are controlled by different cdc2 proteins in higher eukaryotes. *Cell* **66:** 731–742.

Fesquet, D., J.C. Labbe, J. Derancourt, J.P. Capony, S. Galas, F. Girard, T. Lorca, J. Shuttleworth, M. Doree, and J.C. Cavadore. 1993. The *MO15* gene encodes the catalytic subunit of a protein kinase that activates cdc2 and other cyclin-dependent kinases (CDKs) through phosphorylation of Thr161 and its homologues. *EMBO J.* **12:** 3111–3121.

Firpo, E.J., A. Koff, M.J. Solomon, and J.M. Roberts. 1994. Inactivation of a Cdk2 inhibitor during interleukin 2-induced proliferation of human T lymphocytes. *Mol. Cell. Biol.* **14:** 4889–4901.

Fisher, R.P. and D.O. Morgan. 1994. A novel cyclin associates with MO15/CDK7 to form the CDK-activating kinase. *Cell* **78:** 713–724.

Fitch, I., C. Dahmann, U. Surana, A. Amon, K. Nasmyth, L. Goetsch, B. Byers, and B. Futcher. 1992. Characterization of four B-type cyclin genes of the budding yeast *Saccharomyces cerevisiae. Mol. Biol. Cell.* **3:** 805–818.

Forsburg, S.L. and P. Nurse. 1991a. Cell cycle regulation in the yeasts *Saccharomyces cerevisiae* and *Schizosaccharomyces pombe. Annu. Rev. Cell Biol.* **7:** 227–256.

———. 1991b. Identification of a G1-type cyclin puc1$^+$ in the fission yeast *Schizosaccharomyces pombe* (published erratum in *Nature* [1991] **352:** 648). *Nature* **351:** 245–248.

———. 1994. Analysis of the *Schizosaccharomyces pombe* cyclin puc1: Evidence for a role in cell cycle exit. *J. Cell Sci.* **107:** 601–613.

Furuno, N., M. Nishizawa, K. Okazaki, H. Tanaka, J. Iwashita, N. Nakajo, Y. Ogawa, and N. Sagata. 1994. Suppression of DNA replication via Mos function during meiotic divisions in *Xenopus* oocytes. *EMBO J.* **13:** 2399–2410.

Ghiara, J.B., H.E. Richardson, K. Sugimoto, M. Henze, D.J. Lew, C. Wittenberg, and S.I. Reed. 1991. A cyclin B homolog in *S. cerevisiae*: Chronic activation of the Cdc28 protein kinase by cyclin prevents exit from mitosis. *Cell* **65:** 163–174.

Gibson, S.I., R.T. Surosky, and B.K. Tye. 1990. The phenotype of the minichromosome maintenance mutant *mcm3* is characteristic of mutants defective in DNA replication. *Mol. Cell. Biol.* **10:** 5707–5720.

Girard, F., U. Strausfeld, A. Fernandez, and N.J. Lamb. 1991. Cyclin A is required for the onset of DNA replication in mammalian fibroblasts. *Cell* **67:** 1169–1179.

Goebl, M.G., J. Yochem, S. Jentsch, J.P. McGrath, A. Varshavsky, and B. Byers. 1988. The yeast cell cycle gene *CDC34* encodes a ubiquitin-conjugating enzyme. *Science* **241:** 1331–1335.

Grandin, N. and S.I. Reed. 1993. Differential function and expression of *Saccharomyces cerevisiae* B-type cyclins in mitosis and meiosis. *Mol. Cell. Biol.* **13:** 2113–2125.

Guan, K.-L., C.W. Jenkins, Y. Li, M.A. Nichols, X. Wu, C.L. O'Keefe, A.G. Matera, and Y. Xiong. 1994. Growth suppression by p18, a p16INK4/MTS1-and p14INK4B/MTS2-related CDK6 inhibitor, correlates with wild-type pRb function. *Genes Dev.* **8:** 2939–2952.

Hadwiger, J.A., C. Wittenberg, H.E. Richardson, M. De Barros Lopes, and S.I. Reed. 1989. A family of cyclin homologs that control the G1 phase in yeast. *Proc. Natl. Acad. Sci.* **86:** 6255–6259.

Hagan, I., J. Hayles, and P. Nurse. 1988. Cloning and sequencing the cyclin-related *cdc13*$^{+}$ gene and a cytological study of its role in fission yeast mitosis. *J. Cell Sci.* **91:** 587–596.

Hamaguchi, J.R., R.A. Tobey, J. Pines, H.A. Crissman, T. Hunter, and E.M. Bradbury. 1992. Requirement for p34^{cdc2} kinase is restricted to mitosis in the mammalian *cdc2* mutant FT210. *J. Cell Biol.* **117:** 1041–1053.

Hannon, G.J. and D. Beach. 1994. p15INK4B is a potential effector of TGF-β-induced cell cycle arrest. *Nature* **371:** 257–261.

Harper, J.W., G.R. Adami, N. Wei, K. Keyomarsi, and S.J. Elledge. 1993. The p21 Cdk-interacting protein Cip1 is a potent inhibitor of G1 cyclin-dependent kinases. *Cell* **75:** 805–816.

Hartwell, L.H. (1993). Getting started in the cell cycle. In *The early days of yeast genetics* (ed. M.N. Hall and P. Linder), pp. 307–314. Cold Spring Harbor Laboratory Press, Cold Spring Harbor, New York.

Hartwell, L.M. and M.W. Unger. 1977. Unequal division in *Saccharomyces cerevisiae* and its implications for the control of cell division. *J. Cell Biol.* **75:** 422–435.

Hartwell, L., J.Culotti, J.R. Pringle, and B.J. Reid. 1974. Genetic control of the cell division cycle in yeast. *Science* **183:** 46–51.

Hatakeyama, M., J.A. Brill, G.R. Fink, and R.A. Weinberg. 1994. Collaboration of G1 cyclins in the functional inactivation of the retinoblastoma protein. *Genes Dev.* **8:** 1759–1771.

Hayles, J., D. Fisher, A. Woollard, and P. Nurse. 1994. Temporal order of S phase and mitosis in fission yeast is determined by the state of the p34^{cdc2}-mitotic B cyclin complex. *Cell* **78:** 813–822.

Hennessy, K.M. and D. Botstein. 1991. Regulation of DNA replication during the yeast cell cycle. *Cold Spring Harbor Symp. Quant. Biol.* **56:** 279–284.

Hennessy, K.M., C.D. Clark, and D. Botstein. 1990. Subcellular localization of yeast CDC46 varies with the cell cycle. *Genes Dev.* **4:** 2252–2263.

Hennessy, K.M., A. Lee, E. Chen, and D. Botstein. 1991. A group of interacting yeast DNA replication genes. *Genes Dev.* **5:** 958–969.

Hinds, P.W. and R.A. Weinberg. 1994. Tumor suppressor genes. *Curr. Opin. Genet. Dev.* **4:** 135–141.

Hoffmann, I., G. Draetta, and E. Karsenti. 1994. Activation of the phosphatase activity of human cdc25A by a cdk2-cyclin E dependent phosphorylation at the G1/S transition. *EMBO J.* **13:** 4302–4310.

Hofmann, J.F. and D. Beach. 1994. *cdt1* is an essential target of the Cdc10/Sct1 transcription factor: Requirement for DNA replication and inhibition of mitosis. *EMBO J.* **13:** 425–434.

Hogan, E. and D. Koshland. 1992. Addition of extra origins of replication to a minichromosome suppresses its mitotic loss in *cdc6* and *cdc14* mutants of *S. cerevisiae*. *Proc. Natl. Acad. Sci.* **89:** 3098–3102.

Hollingsworth, R.E., Jr. and R.A. Sclafani. 1990. DNA metabolism gene *CDC7* from yeast encodes a serine (threonine) protein kinase. *Proc. Natl. Acad. Sci.* **87:** 6272–6276.

Holloway, S.L., M. Glotzer, R.W. King, and A.W. Murray. 1993. Anaphase is initiated by proteolysis rather than by the inactivation of maturation-promoting factor. *Cell* **73:** 1393–1402.

Howard, A. and S.R. Pelc. 1951. Synthesis of nucleoprotein in bean root cells. *Nature* **167:** 599–600.

Hubler, L., J. Bradshaw Rouse, and W. Heideman. 1993. Connections between the Ras-cyclic AMP pathway and G1 cyclin expression in the budding yeast *Saccharomyces cerevisiae*. *Mol. Cell. Biol.* **13:** 6274–6282.

Hunter, T. and J. Pines. 1994. Cyclins and cancer. II. Cyclin D and CDK inhibitors come of age. *Cell* **79:** 573–582.

Hutchinson, C.J., R. Cox, S. Drepaul, M. Gomperts, and C. Ford. 1987. Periodic DNA synthesis in cell-free extracts of *Xenopus* eggs. *EMBO J.* **6:** 2003–2010.

Iida, H. and I. Yahara. 1984. Durable synthesis of high molecular weight heat shock proteins in G0 cells of the yeast and other eukaryotes. *J. Cell Biol.* **99:** 199–207.

Irniger, S., S. Piatti, C. Michaelis, and K. Nasmyth. 1995. Genes involved in sister chromatid separation are needed for B-type cyclin proteolysis in budding yeast. *Cell* **81:** 269–277.

Jackson, A.L., P.M. Pahl, K. Harrison, J. Rosamond, and R.A. Sclafani. 1993. Cell cycle regulation of the yeast Cdc7 protein kinase by association with the Dbf4 protein. *Mol. Cell. Biol.* **13:** 2899–2908.

Johnson, D.G., K. Ohtani, and J.R. Nevins. 1994. Autoregulatory control of E2F1 expression in response to positive and negative regulators of cell cycle progression. *Genes Dev.* **8:** 1514–1525.

Johnson, P.T. and P.N. Rao. 1970. Mammalian cell fusion: Induction of premature chromosome condensation in interphase nuclei. *Nature* **226:** 717–722.

———. 1971. Nucleo-cytoplasmic interactions in the achievement of nuclear synchrony in DNA synthesis and mitosis in multinucleate cells. *Biol. Rev.* **46:** 97–155.

Kaffman, A., I. Herskowitz, R. Tjian, and E.K. O'Shea. 1994. Phosphorylation of the transcription factor PHO4 by a cyclin-CDK complex, PHO80-PHO85 (see comments). *Science* **263:** 1153–1156.

Kato, J.Y., M. Matsuoka, K. Polyak, J. Massagué, and C.J. Sherr. 1994. Cyclic AMP-induced G1 phase arrest mediated by an inhibitor (p27Kip1) of cyclin-dependent kinase 4 activation. *Cell* **79:** 487–496.

Kato, J., H. Matsushime, S.W. Hiebert, M.E. Ewen, and C.J. Sherr. 1993. Direct binding of cyclin D to the retinoblastoma gene product (pRb) and pRb phosphorylation by the cyclin D-dependent kinase CDK4. *Genes Dev.* **7:** 331–342.

Kelly, T.J., G.S. Martin, S.L. Forsburg, R.J. Stephen, A. Russo, and P. Nurse. 1993. The fission yeast *cdc18*$^{+}$ gene product couples S phase to START and mitosis (see comments). *Cell* **74:** 371–382.

Killander, D. and A. Zetterberg. 1965. A quantitative cytochemical investigation of the relationship between cell mass and the initiation of DNA synthesis in mouse fibroblasts in vitro. *Exp. Cell Res.* **40:** 12–20.

Kim, Y.-J., S. Björklund, Y. Li, M.H. Sayre, and R.D. Kornberg. 1994. A multiprotein mediator of transcriptional activation and its interaction with the C-terminal repeat domain of RNA polymerase II. *Cell* **77:** 599–608.

Kimura, H., N. Nozaki, and K. Sugimoto. 1994. DNA polymerase alpha associated protein P1, a murine homolog of yeast MCM3, changes its intranuclear distribution during the DNA synthetic period. *EMBO J.* **13:** 4311–4320.

King, R.W., J. Peters, S. Tugendreich, M. Rolfe, P. Hieter, and M.W. Kirschner. 1995. A 20S complex containing CDC27 and CDC16 catalyzes the mitosis-specific conjugation of ubiquitin to cyclin B. *Cell* **81:** 279–288.

Knoblich, J.A. and C.F. Lehner. 1993. Synergistic action of *Drosophila* cyclins A and B during the G2-M transition. *EMBO J.* **12:** 65–74.

Knoblich, J.A., K. Sauer, L. Jones, H. Richardson, R. Saint, and C.F. Lehner. 1994. Cyclin E controls S phase progression and its down-regulation during *Drosophila* embryogenesis is required for the arrest of cell proliferation. *Cell* **77:** 107–120.

Koch, C. and K. Nasmyth. 1994. Cell cycle regulated transcription in yeast. *Curr. Opin. Cell Biol.* **6:** 451–459.

Koch, C., A. Schleiffer, G. Ammerer, and K. Nasmyth. 1996. Switching transcription on and off during the yeast cell cycle: Cln/Cdc28 kinases activate bound transcription factor SBF (Swi4/Swi6) at Start, whereas Clb/Cdc28 kinases displace it from the promoter in G_2. *Genes Dev.* **10:** 129–142.

Koch, C., T. Moll, M. Neuberg, H. Ahorn, and K. Nasmyth. 1993. A role for the transcription factors Mbp1 and Swi4 in progression from G1 to S phase. *Science* **261:** 1551–1557.

Koff, A., A. Giordano, D. Desai, K. Yamashita, J.W. Harper, S. Elledge, T. Nishimoto, D.O. Morgan, B.R. Franza, and J.M. Roberts. 1992. Formation and activation of a cyclin E-cdk2 complex during the G1 phase of the human cell cycle. *Science* **257:** 1689–1994.

Kung, A.L., S.W. Sherwood, and R.T. Schimke. 1993. Differences in the regulation of protein synthesis, cyclin B accumulation, and cellular growth in response to the inhibition of DNA synthesis in Chinese hamster ovary and HeLa S3 cells. *J. Biol. Chem.* **268:** 23072–23080.

La Thangue, N.B. 1994. DP and E2F proteins: Components of a heterodimeric transcription factor implicated in cell cycle control. *Curr. Opin. Cell Biol.* **6:** 443–450.

Lehner, C.F. and P.H. O'Farrell. 1990. The roles of *Drosophila* cyclins A and B in mitotic control. *Cell* **61:** 535–547.

Leno, G.H., C.S. Downes, and R.A. Laskey. 1992. The nuclear membrane prevents replication of human G2 nuclei but not G1 nuclei in *Xenopus* egg extract. *Cell* **69:** 151–158.

Liang, C., M. Weinreich, and B. Stillman. 1995. ORC and Cdc6p interact and determine the frequency of initiation of DNA replication in the genome. *Cell* **81:** 667–676.

Lohka, M.J. and Y. Masui. 1984. Roles of cytosol and cytoplasmic particles in nuclear membrane assembly and sperm pronuclear formation in cell-free preparations from amphibian eggs. *J. Cell Biol.* **98:** 1222–1230.

Lohka, M.J., M.K. Hayes, and J.L. Maller. 1988. Purification of maturation-promoting factor, an intracellular regulator of early mitotic events. *Proc. Natl. Acad. Sci.* **85:**

3009–3013.

Lorincz, A. and B.L.A. Carter. 1979. Control of cell size at bud initiation in *Saccharomyces cerevisiae. J. Gen. Microbiol.* **113:** 287–295.

Lorincz, A.T. and S.I. Reed. 1984. Primary structure homology between the product of yeast division control gene *CDC28* and vertebrate oncogenes. *Nature* **307:** 183–185.

Lovec, H., A. Grzeschiczek, M.B. Kowalski, and T. Moroy. 1994. Cyclin D1/bcl-1 cooperates with *myc* genes in the generation of B-cell lymphoma in transgenic mice. *EMBO J.* **13:** 3487–3495.

Lucibello, F.C., A. Sewing, S. Brusselbach, C. Burger, and R. Muller. 1993. Deregulation of cyclins D1 and E and suppression of cdk2 and cdk4 in senescent human fibroblasts. *J. Cell Sci.* **105:** 123–133.

Lukas, J., H. Muller, J. Bartkova, D. Spitkovsky, A.A. Kjerulff, P. Jansen Durr, M. Strauss, and J. Bartek. 1994. DNA tumor virus oncoproteins and retinoblastoma gene mutations share the ability to relieve the cell's requirement for cyclin D1 function in G1. *J. Cell Biol.* **125:** 625–638.

Marahrens, Y. and B. Stillman. 1992. A yeast chromosomal origin of DNA replication defined by multiple functional elements. *Science* **255:** 817–823.

Masui, Y. and C. Markert. 1971. Cytoplasmic control of nuclear behavior during meiotic maturation of frog oocytes. *J. Exp. Zool.* **177:** 129–146.

Matsumoto, K., I. Uno, Y. Oshima, and T. Ishikawa. 1982. Isolation and characterization of yeast mutants deficient in adenyl cyclase and cAMP dependent protein kinase. *Proc. Natl. Acad. Sci.* **79:** 2355–2359.

Matsushime, H., M.F. Roussel, R.A. Ashmun, and C.J. Sherr. 1991. Colony-stimulating factor 1 regulates novel cyclins during the G1 phase of the cell cycle. *Cell* **65:** 701–713.

Matsushime, H., M.W. Ewen, D.K. Strom, J.Y. Kato, S.K. Hanks, M.F. Roussel, and C.J. Sherr. 1992. Identification and properties of an atypical catalytic subunit (p34PSK-J3/cdk4) for mammalian D type G1 cyclins. *Cell* **71:** 323–334.

McGrew, J.T., L. Goetsch, B. Byers, and P. Baum. 1992. Requirement for *ESP1* in the nuclear division of *S. cerevisiae. Mol. Biol. Cell* **3:** 1443–1454.

Mendenhall, M.D. 1993. An inhibitor of p34CDC28 protein kinase activity from *S. cerevisiae. Science* **259:** 216–219.

Mitchison, J.M. 1970. *The biology of the cell cycle.* Cambridge University Press, Cambridge, England.

Moreno, S. and P. Nurse. 1994. Regulation of progression through the G1 phase of the cell cycle by the *rum1*$^+$ gene. *Nature* **367:** 236–242.

Morgenbesser, S.D., B.O. Williams, T. Jacks, and R.A. DePinho. 1994. p53-dependent apoptosis produced by Rb-deficiency in the developing mouse lens. *Nature* **371:** 72–74.

Murray, A.W. and M.W. Kirschner. 1989. Cyclin synthesis drives the early embryonic cell cycle. *Nature* **339:** 275–286.

Nakashima, N., K. Tanaka, S. Sturm, and M. Okayama. 1995. Fission yeast Rep2 is a putative transcriptional activator subunit for the cell cycle "start" function of Res2-Cdc10. *EMBO J.* **14:** 4794–4802.

Nash, R., G. Tokiwa, S. Anand, K. Erickson, and A.B. Futcher. 1988. The WH11$^+$ gene of *Saccharomyces cerevisiae* tethers cell division to cell size and is a cyclin homolog. *EMBO J.* **7:** 4335–4346.

Nasmyth, K. 1993. Control of the yeast cell cycle by the Cdc28 protein kinase. *Curr.*

Opin. Cell Biol. **5:** 166–179.

Nugroho, T.T. and M.D. Mendenhall. 1994. An inhibitor of yeast cyclin-dependent protein kinase plays an important role in ensuring the genomic integrity of daughter cells. *Mol. Cell. Biol.* **14:** 3320–3328.

Nurse, P. 1975. Genetic control of cell size at cell division in yeast. *Nature* **256:** 547–551.

———. 1990. Universal control mechanism regulating onset of M-phase. *Nature* **344:** 503–508.

———. 1994. Ordering S phase and M phase in the cell cycle. *Cell* **79:** 547–550.

Nurse, P. and Y. Bissett. 1981. Gene required in G1 for commitment to cell cycle and in G2 for control of mitosis in fission yeast. *Nature* **292:** 558–560.

Nurse, P. and P. Thuriaux. 1977. Controls over the timing of DNA replication during the cell cycle of fission yeast. *Exp. Cell Res.* **107:** 376–385.

———. 1980. Regulatory genes controlling mitosis in the fission yeast *Schizosaccharomyces pombe. Genetics* **96:** 627–637.

Nurse, P., P. Thuriaux, and K. Nasmyth. 1976. Genetic control of the cell division cycle in the fission yeast *Schizosaccharomyces pombe. Mol. Gen. Genet.* **146:** 167–178.

Obara Ishihara, T. and H. Okayama. 1994. A B-type cyclin negatively regulates conjugation via interacting with cell cycle 'start' genes in fission yeast. *EMBO J.* **13:** 1863–1872.

Ohtsubo, M. and J.M. Roberts. 1993. Cyclin-dependent regulation of G1 in mammalian fibroblasts. *Science* **259:** 1908–1912.

Osmani, A.H., N. van Peij, M. Mischke, M.J. O'Connell, and S.A. Osmani. 1994. A single p34cdc2 protein kinase (encoded by nimXcdc2) is required at G1 and G2 in *Aspergillus nidulans. J. Cell Sci.* **107:** 1519–28.

Pagano, M., R. Pepperkok, F. Verde, W. Ansorge, and G. Draetta. 1992. Cyclin A is required at two points in the human cell cycle. *EMBO J.* **11:** 961–971.

Pardee, A.B. 1989. G1 events and regulation of cell proliferation. *Science* **246:** 603–608.

Paris, J., R. Le Guellec, A. Couturier, K. Le Guellec, F. Omilli, J. Camonis, S. MacNeill, and M. Philippe. 1991. Cloning by differential screening of a *Xenopus* cDNA coding for a protein highly homologous to cdc2. *Proc. Natl. Acad. Sci.* **88:** 1039–1043.

Parker, S.B., G. Eichele, P. Zhang, A. Rawls, A.T. Sands, A. Bradley, E.N. Olson, J.W. Harper, and S.J. Elledge. 1995. p53-independent expression of p21Cip1 in muscle and other terminally differentiating cells. *Science* **267:** 1024–1027.

Pearson, B.E., H.P. Nasheuer, and T.S. Wang. 1991. Human DNA polymerase alpha gene: Sequences controlling expression in cycling and serum-stimulated cells. *Mol. Cell. Biol.* **11:** 2081–2095.

Piatti, S., C. Lengauer, and K. Nasmyth. 1995. Cdc6 is an unstable protein whose de novo synthesis in G1 is important for the onset of S phase and for preventing a "reductional" anaphase in the budding yeast *Saccharomyces cerevisiae. EMBO J.* **14:** 3788–3799.

Pines, J. and T. Hunter. 1990. Human cyclin A is adenovirus E1A-associated protein p60 and behaves differently from cyclin B. *Nature* **346:** 760–763.

Polyak, K., M.H. Lee, H. Erdjument Bromage, A. Koff, J.M. Roberts, P. Tempst, and J. Massagué. 1994. Cloning of p27Kip1, a cyclin-dependent kinase inhibitor and a potential mediator of extracellular antimitogenic signals. *Cell* **78:** 59–66.

Primig, M., S. Sockanathan, H. Auer, and K. Nasmyth. 1992. Anatomy of a transcription factor important for the start of the cell cycle in *Saccharomyces cerevisiae. Nature* **358:**

593–597.

Pringle, J.R. and L.H. Hartwell. 1981. The *Saccharomyces cerevisiae* cell cycle. In *The molecular biology of the yeast* Saccharomyces: *Life cycle and inheritance* (ed. J. Strathern et al.), pp. 97–142. Cold Spring Harbor Laboratory Press, Cold Spring Harbor, New York.

Quelle, D.E., R.A. Ashmun, S.A. Shurtleff, J.Y. Kato, D. Bar Sagi, M.F. Roussel, and C.J. Sherr. 1993. Overexpression of mouse D-type cyclins accelerates G1 phase in rodent fibroblasts. *Genes Dev.* **7:** 1559–1571.

Rao, P.N. and R.T. Johnson. 1970. Mammalian cell fusion: studies on the regulation of DNA synthesis and mitosis. *Nature* **225:** 159–164.

Reed, S.I., V. Dulic, D.J. Lew, H.E. Richardson, and C. Wittenberg. 1992. G1 control in yeast and animal cells. *Ciba Found. Symp.* **170:** 7–15 (discussion 15–19).

Reid, B. and L.H. Hartwell. 1977. Regulation of mating in the cell cycle of *S. cerevisiae. J. Cell Biol.* **75:** 355–365.

Resnitzky, D., M. Gossen, H. Bujard, and S.I. Reed. 1994. Acceleration of the G1/S phase transition by expression of cyclins D1 and E with an inducible system. *Mol. Cell. Biol.* **14:** 1669–1679.

Reymond, A., J. Marks, and V. Simanis. 1993. The activity of *S. pombe* DSC-1-like factor is cell cycle regulated and dependent on the activity of $p34^{cdc2}$. *EMBO J.* **12:** 4325–4334.

Riabowol, K., G. Draetta, L. Brizuela, D. Vandre, and D. Beach. 1989. The cdc2 kinase is a nuclear protein that is essential for mitosis in mammalian cells. *Cell* **57:** 393–401.

Richardson, H.E., C. Wittenberg, R. Cross, and S.I. Reed. 1989. An essential G1 function for cyclin-like proteins in yeast. *Cell* **59:** 1127–1133.

Richardson, H., D.J. Lew, M. Henze, K. Sugimoto, and S.I. Reed. 1992. Cyclin-B homologs in *Saccharomyces cerevisiae* function in S phase and in G2. *Genes Dev.* **6:** 2021–2034.

Roberts, J.M. 1993. Turning DNA replication on and off. *Curr. Opin. Cell Biol.* **5:** 201–206.

Roy, R., J.P. Adamczewski, T. Seroz, W. Vermeulen, J.-P. Tassan, L. Schaeffer, E.A. Nigg, J.H.J. Hoeijmakers, and J.-M. Egly. 1994. The MO15 cell cycle kinase is associated with the TFIIH transcription-DNA repair factor. *Cell* **79:** 1093–1101.

Rusch, H.P., W. Sachsenmaier, K. Behrens, and V. Gruter. 1966. Synchronization of mitosis by the fusion of the plasmodia of *Physarum polycephalum. J. Cell Biol.* **31:** 204–209.

Schwob, E. and K. Nasmyth. 1993. CLB5 and CLB6, a new pair of B cyclins involved in DNA replication in *Saccharomyces cerevisiae. Genes Dev.* **7:** 1160–1175.

Schwob, E., T. Bohm, M.D. Mendenhall, and K. Nasmyth. 1994. The B-type cyclin kinase inhibitor p40SIC1 controls the G1 to S transition in *S. cerevisiae. Cell* **79:** 233–244.

Serrano, M., G.J. Hannon, and D. Beach. 1993. A new regulatory motif in cell cycle control causing specific inhibition of cyclin D/CDK4. *Nature* **366:** 704–707.

Serrano, M., E. Gomez-Lahoz, R.A. DePinho, D. Beach, and D. Bar-Sagi. 1995. Inhibition of Ras-induced proliferation and cellular transformation by $p16^{INK4}$. *Science* **267:** 249–251.

Seufert, W., B. Futcher, and S. Jentsch. 1994. Role of a ubiquitin-conjugatiing enzyme in degradation of S- and M-phase cyclins. *Nature* **373:** 78–81.

Sewing, A., C. Burger, S. Brusselbach, C. Schalk, F.C. Lucibello, and R. Muller. 1993.

Human cyclin D1 encodes a labile nuclear protein whose synthesis is directly induced by growth factors and suppressed by cyclic AMP. *J. Cell Sci.* **104:** 545–555.
Sherr, C.J. 1994a. G1 phase progression: Cycling on cue. *Cell* **79:** 551–555.
———. 1994b. The ins and outs of RB: Coupling gene expresion to the cell cycle clock. *Trends Cell Biol.* **4:** 15–18.
Sicinski, P., J.L. Donaher, S.B. Parker, T. Li, A. Fazeli, H. Gardner, S.Z. Haslam, R.T. Bronson, S.J. Elledge, and R.A. Weinberg. 1995. Cyclin D1 provides a link between development and oncogenesis in the retina and breast. *Cell* **82:** 621–630.
Solomon, M.J., J.W. Harper, and J. Shuttleworth. 1993. CAK, the $p34^{cdc2}$ activating kinase, contains a protein identical or closely related to p40MO15. *EMBO J.* **12:** 3133–3142.
Solomon, M.J., T. Lee, and M.W. Kirschner. 1992. Role of phosphorylation in $p34^{cdc2}$ activation: Identification of an activating kinase. *Mol. Biol. Cell.* **3:** 13–27.
Sudbery, P.E., A.R. Goodey, and B.L.A. Carter. 1980. Genes which control cell proliferation in the yeast *Saccharomyces cerevisiae. Nature* **288:** 401–404.
Surana, U., A. Amon, C. Dowzer, J. McGrew, B. Byers, and K. Nasmyth. 1993. Destruction of the CDC28/CLB mitotic kinase is not required for the metaphase to anaphase transition in budding yeast. *EMBO J.* **12:** 1969–1978.
Surana, U., H. Robitsch, C. Price, T. Schuster, I. Fitch, A.B. Futcher, and K. Nasmyth. 1991. The role of CDC28 and cyclins during mitosis in the budding yeast *S. cerevisiae. Cell* **65:** 145–161.
Tam, S.W., J.W. Shay, and M. Pagano. 1994. Differential expression and cell cycle regulation of the cyclin-dependent kinase 4 inhibitor p16Ink4. *Cancer Res* **54:** 5816–5820.
Tanaka, K., K. Okazaki, N. Okazaki, T. Ueda, A. Sugiyama, H. Nojima, and H. Okayama. 1992. A new cdc gene required for S phase entry of *Schizosaccharomyces pombe* encodes a protein similar to the $cdc10^+$ and *SWI4* gene products. *EMBO J.* **11:** 4923–4932.
Tassan, J.P., S.J. Schultz, J. Bartek, and E.A. Nigg. 1994. Cell cycle analysis of the activity, subcellular localization, and subunit composition of human CAK (CDK-activating kinase). *J. Cell Biol.* **127:** 467–478.
Thommes, P., R. Fett, B. Schray, R. Burkhart, M. Barnes, C. Kennedy, N.C. Brown, and R. Knippers. 1992. Properties of the nuclear P1 protein, a mammalian homologue of the yeast Mcm3 replication protein. *Nucleic Acids Res.* **20:** 1069–1074.
Tokiwa, G., M. Tyers, T. Volpe, and B. Futcher. 1994. Inhibition of G1 cyclin activity by the Ras/cAMP pathway in yeast. *Nature* **371:** 342–345.
Tsai, L.H., E. Harlow, and M. Meyerson. 1991. Isolation of the human cdk2 gene that encodes the cyclin A- and adenovirus E1A-associated p33 kinase. *Nature* **353:** 174–177.
Tsai, L.H., E. Lees, B. Faha, E. Harlow, and K. Riabowol. 1993. The cdk2 kinase is required for the G1-to-S transition in mammalian cells. *Oncogene* **8:** 1593–1602.
Tye, B.-K. 1994. The MCM2-3-5 proteins: Are they replication licensing factors? *Trends Cell Biol.* **4:** 160–165.
Tyers, M., G. Tokiwa, and B. Futcher. 1993. Comparison of the *Saccharomyces cerevisiae* G1 cyclins: Cln3 may be an upstream activator of Cln1, Cln2 and other cyclins. *EMBO J.* **12:** 1955–1968.
Tyers, M., G. Tokiwa, R. Nash, and B. Futcher. 1992. The Cln3-Cdc28 kinase complex of *S. cerevisiae* is regulated by proteolysis and phosphorylation. *EMBO J.* **11:** 1773–1784.

Valay, J.G., M. Simon, and G. Faye. 1993. The Kin28 protein kinase is associated with a cyclin in *S. cerevisiae*. *J. Mol. Biol.* **234:** 307–310.

Wittenberg, C., K. Sugimoto, and S.I. Reed. 1990. G1-specific cyclins of *S. cerevisiae*: Cell cycle periodicity, regulation by mating pheromone, and association with the p34CDC28 protein kinase. *Cell* **62:** 225–237.

Yaglom, J., M.H. Linskens, S. Sadis, D.M. Rubin, B. Futcher, and D. Finley. 1995. p34Cdc28-mediated control of Cln3 cyclin degradation. *Mol. Cell. Biol.* **15:** 731–741.

Yoon, H.J., S. Loo, and J.L. Campbell. 1993. Regulation of *Saccharomyces cerevisiae* CDC7 function during the cell cycle. *Mol. Biol. Cell* **4:** 195–208.

Zhu, Y., T. Takeda, K. Nasmyth, and N. Jones. 1994. pct1+, which encodes a new DNA-binding partner of p85cdc10, is required for meiosis in the fission yeast *Schizosaccharomyces pombe*. *Genes Dev.* **8:** 885–898.

Zindy, F., E. Lamas, X. Chenivesse, J. Sobczak, J. Wang, D. Fesquet, B. Henglein, and C. Brechot. 1992. Cyclin A is required in S phase in normal epithelial cells. *Biochem. Biophys. Res. Commun.* **182:** 1144–1154.

12
Temporal Order of DNA Replication

Itamar Simon and Howard Cedar
Hebrew University Medical School
Jerusalem, Israel 91120

THE REPLICATION CLOCK

In every organism, DNA replication takes place in an ordered physical and temporal fashion. Although this process is primarily necessary to carry out the task of copying the genome, it may also indirectly play a role in the regulation of gene expression. In *Escherichia coli*, for example, the entire 4-Mb circular chromosome undergoes bidirectional DNA synthesis that is initiated from a single origin. It takes at least 40 minutes to copy the complete bacterial genome, and as a result, genes located near the origin are replicated early in the cell cycle whereas those positioned at the opposite end of the chromosome are synthesized late in the cycle. One direct consequence of this temporal organization is that in rapidly dividing cells, the early-replicating genes themselves accumulate differentially in high copy number (von Meyenburg and Hansen 1987), and this most certainly has profound effects on the overall expression pattern of the organism.

Compared to prokaryotes, eukaryotic organisms contain a much larger genome. Although the total DNA is divided into separate chromosome molecules, each of these entities in itself carries a considerable amount of genetic information, and DNA synthesis is accomplished through the action of multiple bidirectional origins. If all of these replication units were activated simultaneously, the entire process of replication could be completed very quickly. However, replication starts are actually distributed in a programmed manner throughout the length of S phase, and, as a consequence, each origin is turned on at a specific time. In the yeast *Saccharomyces cerevisiae*, DNA located near telomeres, for example, undergoes replication late in S phase, whereas sequences associated with centromeres are early replicating (Reynolds et al. 1989). Other regions of the genome are replicated at various fixed times within the 30-minute S phase.

DNA Replication in Eukaryotic Cells

The process of DNA replication in animal cells takes place over a longer period of time, an average of 8–10 hours. Despite the complexity of these genomes (3000 Mb), a number of related methods have been developed to pinpoint the replication time of individual genes. Basically, replicating DNA is first labeled in vivo with bromodeoxyuridine (BrdU) and then purified by gradient centrifugation from cells at various stages of S phase. This fractionation is accomplished either by prior synchronization (Goldman et al. 1984; Schmidt and Migeon 1990) or by cell-sorting techniques (Braunstein et al. 1982; Gilbert 1986), and the resultant newly synthesized DNA samples are then subjected to blot hybridization with specific probes. Early-replicating genes hybridize to fractions from the beginning of S phase, whereas late-replicating sequences only hybridize with BrdU DNA from cells in the later fractions of S. By increasing the number of interval fractions in S phase, one can obtain better cell-cycle resolution of replication timing. Using this methodology, a large number of specific genes from various organisms have been analyzed, and in general, the results confirm the idea that replication proceeds progressively in an ordered programmed manner throughout S phase (for review, see Holmquist 1987).

REPLICATION TIMING AND GENE EXPRESSION

Although the biological function of replication timing control in the cell cycle is not yet clear, studies in yeast and in animal cells have revealed a striking association between replication timing and gene expression. In *S. cerevisiae*, for example, the unexpressed *HML* and *HMR* loci are located in subtelomeric regions that replicate late in S phase. When, as part of the normal process of cell-type-specific switching, they are copied to the *MAT* locus in an early-replicating region of the same chromosome, the promoters are activated (Reynolds et al. 1989). That replication timing is involved in this control mechanism is suggested by the observation that positional silencing at *HMR* requires an autonomously replicating sequence (ARS) element, and mutations in the ARS that abolish origin function also relieve transcriptional silencing (for review, see Laurenson and Rine 1992). In higher organisms, the general relationship between transcription and replication is even more straightforward. Constitutively transcribed housekeeping genes undergo relatively early DNA replication (some time during the first half of S) in a variety of cell types. In contrast, many tissue-specific genes are late-replicating (the last 50% of S) in their repressed state, but are early-replicating in cells in which they are expressed (for review, see Holmquist 1987). It thus appears that in

Table 1 Replication times for genes in vertebrate cells

Gene	Replication time
Housekeeping	
HGRPT	E
APRT	E
CAD	E
DHER	E
arginine succinate synthetase	E
G6PD	E
β-actin	E
β-tubulins	E
metallothionein-1	E
PDG X linked	E
Tyr aminotransferase	E
c-ras	E
c-myc	E
all histones tested	E
Xenopus somatic 5s	E
Tissue-specific	
apolipoprotein	E
placental lactogen	E
α_1-antitrypsin	L
β-casein	L
Phe hydroxylase	L
α-globin	E
fibronectin	L
β-globin	L*
immunoglobulin V_H	L*
Xenopus oocyte 5s	L*
cystic fibrosis	L*

Data on replication timing were either derived from standard cell-culture procedures (adapted from Holmquist 1987) or obtained by FISH analysis (e.g., cystic fibrosis genes). It has already been demonstrated that these two different techniques yield similar results (Selig et al. 1992). As shown, all housekeeping genes replicate in the first half of S phase (E), whereas many tissue-specific genes replicate in the last half (L) of S in cell types that do not express the gene. In some cases, these same gene sequences were found to replicate early in expressing cells (*). This table only includes representative examples and is not exhaustive.

these organisms, replication timing is a developmentally regulated process that is closely associated with gene expression (Table 1).

Unlike some organisms, such as *Xenopus* or *Drosophila*, that are characterized by extremely short replication cycles during early-cleavage stages of development, in mammals, programmed, cell-cycle-controlled replication is present even at the earliest cell divisions (Takagi 1974). To obtain a better picture of how the replication timing pattern of the genome is established during development, replication was analyzed in F9 teratocarcinoma cells (S. Selig, unpubl.). These studies suggest that tissue-specific genes, such as β-globin or albumin, are initially late-replicating in cells of the early embryo prior to organogenesis. It is likely that these same genes then remain late-replicating in most cell types, but shift to early replication as a part of the differentiation process that brings about gene activation. In parallel, housekeeping genes remain early-replicating and active throughout development. Almost all studies on replication timing have been carried out in rapidly growing tissue-culture cells, and much work must still be done to evaluate this process in vivo.

Tissue-specific genes, which are initially late-replicating in the early embryo, undergo conversion to early replication in specific cell types. However, other DNA sequences in the genome have an opposite pattern and become late-replicating in a developmentally regulated manner. Genes on the X chromosome, for example, shift dramatically from early to late replication in association with the inactivation process that occurs at about the time of implantation in female eutherans (Takagi 1974). At the same stage in development, centromeric mouse satellite sequences also undergo a shift to a later time of replication (Selig et al. 1988). Although the role of replication timing in this case is not known, this change probably reflects the overall stage-specific structural alterations that take place on these chromosomal regions.

REPLICATION TIME ZONES IN MAMMALIAN CELLS

Since in each cell type, individual sectors of DNA are programmed to undergo replication at specific times in S phase, the genome must be organized into replication time units. By following BrdU labeling, these can be visualized cytogenetically as an alternating pattern of early- and late-replicating regions on each chromosome (Fig. 1). In this procedure, nonsynchronized cell cultures are incubated with BrdU for fixed times, and the resulting incorporated label is then detected in metaphase chromosomes. After 4 hours of labeling, for example, all of the BrdU visualized in metaphase represents sequences that replicate late in S phase just prior to condensation in G_2 (~2 hours), and this produces a striking pattern of highly ordered late-replication bands on each chromosome. Labeling with BrdU for longer intervals reveals a continuous series of chromosomal bands that replicate in a programmed manner throughout S (Latt 1973; Wolf and Perry 1974; Vogel et al. 1986). The very existence of this visible banding pattern immediately

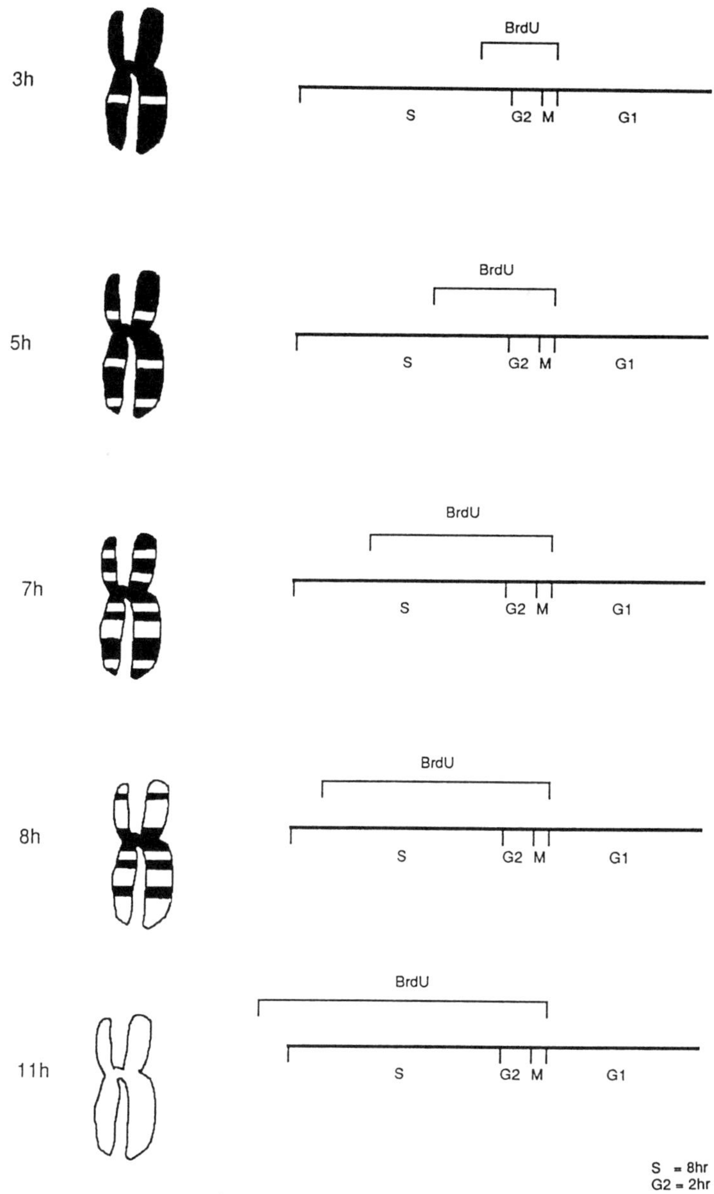

Figure 1 Replication banding. Nonsynchronous cells are cultured in the presence of BrdU for various fixed times, and nuclei are then prepared and stained (see Drouin et al. 1990). The locations of BrdU incorporation are visualized as light (quenched) bands. Since chromosomes are only observed in metaphase cells (M), the 3-hr incubation labels only those regions that replicate in the last hour of S, and the 5-hr incubation labels those that replicate in the last 3 hr of S. After 11 hr, the entire chromosome is labeled with BrdU.

suggests that DNA replication timing units are not only quite large, but also spatially well delineated. Replication bands are an integral part of the basic chromosome unit structure, and these same early-replicating DNA regions appear to comap quite precisely with Giemsa-stained light G bands (Hand 1978). Using DNase I sensitivity as a molecular probe for genome accessibility, it has been possible to map "active" regions (D bands) on metaphase chromosomes, and these also correlate with early-replicating time zones (Kerem et al. 1984). The fact that these banding patterns do not vary appreciably from cell type to cell type suggests that the early-replicating active regions may represent genomic loci rich in housekeeping genes. Taken together, these morphological observations indicate that the entire genome may be subdivided into a series of basic modules which represent units of common structure, function, and replication timing.

Original estimates for the size of each replication band suggest that these domains may cover 5–10 Mb and contain multiple replicons. High-resolution mapping of both replication bands (Drouin et al. 1990) and G bands (Yunis 1981) on extended pro-metaphase chromosomes, however, has delineated at least 2000 individual stripes, suggesting that each chromosomal unit is actually much smaller, containing, on average, 1.5 Mb of DNA (Kitsberg et al. 1991). Due to their relatively large size, most of these replication domains have not yet been mapped at the molecular level, but the availability of extensive cloned contigs, as well as recent advances in technology, should now make it possible to examine these units in detail.

The replication time of a genomic DNA fragment can be measured in almost any cell type by using fluorescence in situ hybridization (FISH) to interphase nuclei. In this method, unreplicated DNA is visualized as a single hybridization dot, and replicated DNA appears as a double dot. In a nonsynchronized population of dividing cells, a high percentage of nuclei with double hybridization signals indicates that this particular gene replicates relatively early in S phase, whereas a low count is obtained for gene sequences that replicate late in the cycle (Selig et al. 1992). Furthermore, by labeling cells with BrdU prior to harvesting, it becomes possible to specifically identify nuclei in S phase and thus to disregard single-dot nuclei from G_1 and double-dot nuclei from G_2 (Kitsberg et al. 1993b). This method is particularly accurate for measuring the replication time of one zone relative to another. In this case, each probe is labeled with a different fluorescent dye, making it possible to detect nuclei where one gene has already replicated while the other is still uncopied in the same cell (Selig et al. 1992). Using this approach, replication time zones can probably be resolved within minutes on the cell-cycle clock.

A 1.5-Mb region of DNA containing the cystic fibrosis (CF) gene has been studied using this method. By assaying a contiguous set of phage and cosmid clones in fibroblasts or lymphocytes, cells that do not express this gene, one observes two distinct, but internally uniform, replication time bands: an early/middle zone located 5′ to the gene and an extensive late zone that includes the 250-kb domain that carries the CF coding sequence. In the intestinal epithelial-derived cell line, Caco-2, in which the CF gene is transcriptionally active, the basic replication structure is quite similar, except for a distinct region of 500–700 kb around the CF gene that replicates very early (Selig et al. 1992). Thus, in this region, replication timing is organized into defined spatial domains and is developmentally regulated (Fig. 2). The human β-globin gene region has also been mapped for replication timing at the molecular level. In this case, as well, a relatively large domain containing the β-like genes is late-replicating in a number of different cell types, but early-replicating in erythroleukemia cells (Epner et al. 1988; Hatton et al. 1988). Because only 250 kb of sequence surrounding these genes has been analyzed, it has not been possible to map the full extent and outer boundaries of this time zone.

REPLICON STRUCTURE

Since animal cell replicons are, on average, 50–300 kb in length (Hand 1978), it is likely that a typical replication band is made up of multiple adjacent replicons that are coordinately regulated to undergo replication at the same time in S. This was originally derived from autoradiograph studies of DNA synthesis, which clearly reveal replicating regions containing strings of multiple pulse-labeled origins (Edenberg and Huberman 1975), and this observation has now been confirmed at the molecular level. The most clear-cut example of this phenomenon is in the mouse major histocompatibility complex (MHC) locus, where a large region of more than 2 Mb contains at least 5–6 distinct replicons that all replicate fairly synchronously as a single time zone (Spack et al. 1992). Several well-mapped replication origins near the DHFR gene also appear to fire in a coordinated manner, all at the beginning of S phase (Ma et al. 1990).

Several lines of evidence indicate that the activation time of any particular origin is not an intrinsic property of the origin sequence itself. The chromosome V origin, ARS501 in yeast, is activated in the second half of S phase and is responsible for the late replication of a 66-kb domain

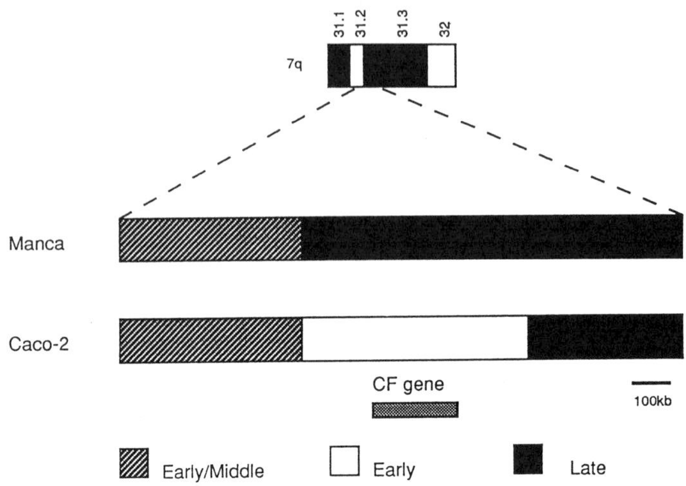

Figure 2 Replication time zones in the cystic fibrosis gene domain. A large number of plasmid, phage, and cosmid clones covering a 1.5-Mb region on human chromosome 7 were analyzed for replication timing in two cell types: Manca lymphocytes, which do not express the CF gene, and Caco-2 cells, which do (see Selig et al. 1992). This map corresponds to chromosome region 7q31.

that includes the right telomere (Ferguson et al. 1991). The time of replication of this origin is regulated by position, since late activation requires the presence of a telomere nearby, and when this origin is transplanted to another site, it replicates early. These observations show that the time at which an origin is activated can depend on its chromosomal context (Ferguson and Fangman 1992). Origins in animal cells apparently behave in a similar manner. In the human β-globin gene domain, mapping has demonstrated that a large region of at least 100 kb is part of a single replicon, with an origin located slightly 5′ to the β-globin gene sequence. Interestingly, this same site serves as the origin both in lymphocytes, where replication occurs late in S, and in erythroleukemia cells, where replication is early (Kitsberg et al. 1993a). This represents clear-cut evidence that the origin sequence serves as a topographical marker for the site of initiating DNA synthesis, whereas temporal activation is controlled by chromosomal context in *cis* and most likely by protein factors in *trans*.

REGULATION OF REPLICATION TIMING

Several experiments indicate that long-range *cis*-acting elements must play a role in the control of replication timing. In animal cells this effect is most obvious in regions that have undergone chromosomal rearrangement. The Xic region on the inactive X, for example, probably plays an important role in controlling replication timing, since its translocation to another chromosome causes adjacent sequences to become late-replicating and, in many cases, inactive (see, e.g., Disteche et al. 1979). Significant changes in replication timing also occur during the normal process of human immunoglobulin heavy chain (IgH) rearrangement in the B-cell lineage. In most somatic cells, the heavy-chain constant region is embedded within a large early-replicating domain (Brown et al. 1987), whereas distal variable sequences replicate late in S. When rearrangement takes place, however, the juxtaposed variable gene falls under control of the constant domain and assumes an early replication time (Calza et al. 1984). This shift could be part of the variable gene promoter activation process, which is ultimately responsible for antibody production in B cells. Similar changes in replication timing also occur at the site of the IgC_H-*myc* translocation in MPC11 plasmacytoma cells (Calza et al. 1984) or at other translocated loci in human leukemias (Karube and Watanabe 1988).

Genomic mapping studies in yeast have succeeded in identifying and characterizing defined *cis*-acting sequences that control replication timing. A telomere itself, for example, can induce late replication on origins located within approximately 30 kb. This effect is undoubtedly mediated through the telomeric DNA sequence itself, since small $C_{1\text{-}3}A$ tracts also shift replication timing when placed close to an origin (Ferguson and Fangman 1992). It is likely that *cis*-acting sequences in other parts of the genome can also direct the initiation time of DNA synthesis, and an element of this nature has recently been mapped near the late-replicating KEX2 ARS on yeast chromosome XIV (B.J. Brewer et al., unpubl.).

Replication timing in animal cells is also under the control of *cis*-acting sequences, and there is now good evidence for a defined replication timing regulatory element associated with the locus control region (LCR) in the human β-globin gene domain. This enhancer-like sequence, located about 20 kb upstream of the ε gene, probably plays a major role in the regulation of genes in this cluster, since patients carrying a deletion of this control region (Hispanic thalassemia) do not make any β-globin transcripts (Driscoll et al. 1989). In normal human lymphocytes, this locus is in a repressed configuration, but fusion to mouse erythroleukemia cells induces a striking activation of the domain and a concomitant shift to early replication. In contrast, the β-globin domain from Hispanic thalassemia lymphocytes fails to undergo transcriptional activation and, at the same time, remains late-replicating. It thus appears that seqences associated with, or near, the globin LCR are probably involved in replication timing control (Forrester et al. 1990).

Indirectly, these fusion studies also suggest that the donor erythroleukemia cells contain a set of protein factors which control the replication timing of the globin domain in *trans*, perhaps through specific interactions with LCR elements. Experiments in culture have shown that treatment of cells with the demethylating agent, 5-azacytidine (5azaC), dramatically shifts the replication time of both the inactive X chromosome (Shafer and Priest 1984; Jablonka et al. 1985) and mouse satellite sequences (Selig et al. 1988) from late to early in the cycle. The highly coordinated and quantum nature of this response suggests that this drug does not act independently on each of the multiple chromosomal regions themselves. Rather, 5azaC mediates the changes in replication timing either through an effect on a master *cis*-acting control element or by inducing *trans*-acting protein factors. This latter possibility is supported by cell fusion experiments. Thus, for example, when embryonic carcinoma cells are fused with female-derived thymocytes, this brings about a dramatic early shift in X-chromosome replication timing (Takagi et al. 1983). The abnormal early replication time of trisomic chromosomal region 15E seen in cases of T-cell leukemia represents another case where protein factors in the nucleus must play some role in the control of the replication clock (Somssich et al. 1984). These studies, taken together, thus support a model whereby replication timing is regulated through *cis* elements that are recognized by and interact with specific *trans*-acting factors.

ALLELE-SPECIFIC GENE EXPRESSION

Transcriptional regulation is thought to be carried out in a large part through the interaction of regulatory sequences with combinations of multiple *trans*-acting factors in the nucleus. In parallel, *cis*-acting effects, such as chromatin structure or DNA methylation, also modulate gene expression by affecting the accessibility of gene regions to these same factors. The importance of *cis*-acting mechanisms is nowhere more obvious than in cases where only one allele is expressed in a given nucleus. Thus, for example, genes on both X chromosomes in female cells are potentially exposed to the same *trans*-acting factors in the nucleus, yet structural differences render one X inaccessible and therefore excluded from the normal transcriptional machinery. DNA methylation represents one important mode for mediating allelic selection. Methyl moieties are known to inhibit transcription both by altering specific protein-DNA interactions and by modulating general chromatin structure (Eden and Cedar 1994). When only one allele is modified, this leads to selective inhibition.

A major role of DNA methylation on the X chromosome is to maintain the differential expression pattern from generation to generation (Mohandas et al. 1981). Once established, methyl moieties are conservatively preserved in daughter cells due to the action of a maintenance methylase that copies the modification pattern of the previous generation (Cedar 1988). Another parameter that differentiates between the active and inactive X is their time of replication in S phase; this may represent yet another *cis*-acting structural marker for distinguishing between two otherwise equivalent alleles in the same cell. Assuming that replication timing is set by cell-cycle-controlled factors, it is easy to see that this parameter, much like methylation, can be propagated from generation to generation and can thus be used to maintain allelic identity.

Replication timing may also be involved in the allele-specific regulation of imprinted genes. A number of these genes have now been identified and characterized in the mouse. Insulin-like growth factor 2 (*Igf2*), U2af-binding protein-related sequence, and small nuclear ribonucleoprotein polypeptide n (*Snrpn*), for example, are expressed exclusively from the paternal allele, whereas insulin-like growth factor 2 receptor (*Igf2r*) and *H19* are transcribed only from their maternal copy in somatic cells (for review, see Razin and Cedar 1994; see also Villar and Pedersen 1994). Imprinting has also been observed for some of these same genes in humans. The Prader-Willi and Angelman syndromes (PWS/AS), for example, come about as a result of either uniparental disomy or monoparental deletions on chromosome 15, and it is likely that imprinting of *SNRPN* or other gene sequences located in this region may explain the variety of symptoms seen in these diseases (Nicholls 1993).

IMPRINTED REPLICATION TIME ZONES

Much like the genes on the X chromosome, these sequences have a genomic structure that distinguishes between the two parental alleles. Thus, as might be expected, almost all of the loci that have been analyzed show differential methylation (Razin and Cedar 1994). In situ hybridization has been employed to evaluate the replication timing properties of these genes. This method is especially appropriate for detecting asynchronous replication. In any cell population, FISH analysis generally reveals mainly nuclei with either two single or two double hybridization dots, indicating that both homologs of most genes replicate

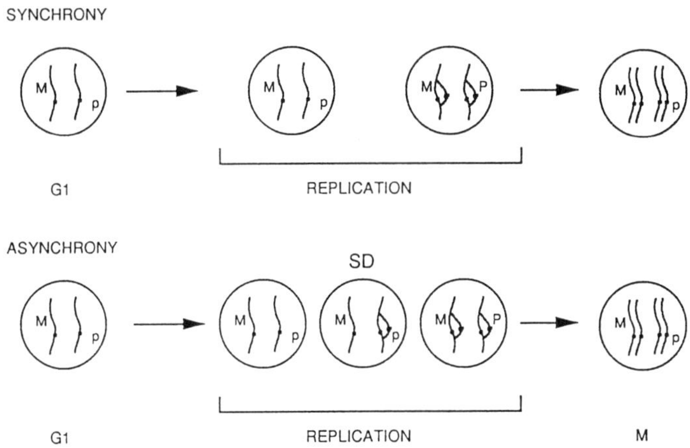

Figure 3 Allele-specific replication timing. By using FISH analysis, all gene regions are unreplicated on both alleles in G_1 and therefore show two single hybridization dots. For most sequences, both homologs replicate relatively synchronously, so nuclei in S phase either have two single dots or two double dots. For imprinted genes, each allele replicates at a different time in S, so a high percentage of nuclei show both an unreplicated (single dot) and a replicated (double dot) allele. At metaphase, all of the DNA has already replicated.

synchronously during the cell cycle. In contrast, for imprinted genes, a large number (~30%) of nuclei contain one replicated and one unreplicated gene (Fig. 3), suggesting that these alleles undergo DNA synthesis asynchronously in the cell cycle (Kitsberg et al. 1993b; Knoll et al. 1994). It appears from these studies that many imprinted genes are imbedded within large differentially replicating domains which, in some cases, can even be visualized by standard replication banding techniques (Izumikawa et al. 1991). By using polymorphic markers to identify each parental allele individually, it has been shown that most of these regions replicate coordinately in an allele-specific manner, with the paternal allele generally replicating earlier in the cell cycle (Kitsberg et al. 1993b). As a result of this regional pattern, the active allele of some imprinted genes actually replicates late in the S phase. This clearly represents an exception to the general correlation between early replication timing and gene expression and suggests that this standard structure/function relationship may be altered for imprinted genes.

REGIONAL REGULATION

The finding of domain-wide replication-timing control strongly suggests that imprinting itself may be regulated at the regional level. This idea is supported by the observation that imprinted genes are clustered. Experiments with mono-parental disomic mice initially indicated the existence of a limited number of relatively well defined chromosomal regions that may be involved in the generation of parent-of-origin defects during development (for review, see Cattanach 1986), and more recently, genomic mapping has shown that imprinted genes are indeed clustered. *Igf2* and *H19*, for example, are positioned within 90 kb of each other and are imprinted in both mouse and man (Razin and Cedar 1994). Furthermore, the nearby insulin gene has been found to be imprinted in extraembryonic tissues (Giddings et al. 1994), and *Mash-2*, a newly identified paternally transcribed gene, is located in the same chromosomal region about 500 kb upstream of the *Igf2* locus (Guillemot et al. 1995).

Even more striking is the PWS region on human chromosome 15, where at least four individual genes are transcribed exclusively from the paternal allele (Sutcliffe et al. 1994; Wevrick et al. 1994). Further evidence suggesting regional control at this locus comes from cases of PWS carrying small deletions around the *Snrpn* gene on the paternal chromosome (Buiting et al. 1995). Although these mutations are relatively small, several nearby intact imprinted genes on this chromosome become transcriptionally inactive (Sutcliffe et al. 1994), selective adjacent sites adopt a maternal pattern of methylation (Reis et al. 1994), and for at least one case, an entire 600-kb domain undergoes a dramatic change in replication timing on the paternal allele (Gunaratne et al. 1995). It is thus likely that these individuals have a defect in *cis*-acting DNA sequences directly involved in the regionwide regulation of imprinting. In general, these imprinted domains include gene sequences that are not necessarily expressed in an allele-specific manner. It thus seems likely that the regulation of imprinting is a complicated process that utilizes multiple control mechanisms and most certainly involves both local and regional effectors of expression.

The organization and control of asynchronous replication, both on the X chromosome and in imprinted domains, are not well understood. Although the two X chromosomes in female cells generally replicate asynchronously (Takagi 1974), the inactive X still retains a detailed multireplication-band structure, with each individual domain replicating at a fixed time in S, usually later than the corresponding region on the active X (Wahrman et al. 1983). Because of this structure, most genes on the active X replicate in the first half of S, but for some genes, despite their asynchrony, both alleles actually replicate in late S phase (Hansen et al. 1993; Boggs and Chinault 1994). Whereas replication timing on the X chromosome appears to be coordinated, individual replication bands may also be regulated autonomously. In the fragile X syndrome, for example, a single large domain of over 150 kb is shifted to a very late replication time while the rest of the chromosome replicates with normal timing properties (Hansen et al. 1993). Overall replication timing asynchrony on the X is most likely centrally regulated by *cis*-acting sequences at *Xic*, a locus that may itself be essential for the inactivation process (Brown et al. 1991b). Further studies have highlighted the possible role of the *Xist* gene which maps to this region and codes for an untranslated nuclear RNA species transcribed exclusively from the inactive chromosome (Brown et al. 1991a, 1992). Interestingly, this gene is embedded in a unique domain that replicates asynchronously, but unlike the rest of the chomosome, is copied earlier on the inactive X (Boggs and Chinault 1994; Torchia et al. 1994). Recent studies have shown that this gene is not required for the maintenance of X inactivation, making it unlikely that it itself plays a role in the control of replication timing (Brown and Willard 1994).

Complex patterns of replication time zones may also be characteristic of imprinted regions. Both in the Igf2r and Igf2 domains, asynchronous replication appears to be uniform and coordinated in such a way that large regions on each parental chromosome replicate together (Kitsberg et al. 1993b). In contrast, it is likely that the asynchronously replicating 3- to 4-Mb PWS/AS domain is actually made up of smaller individual subunits, each subject to its own replication timing control. Most DNA from this region clearly replicates with the paternal allele early, but it has also been demonstrated that sequences near the GABA receptor genes replicate maternally early (LaSalle and Lalande 1995), and at least one probe near the P locus replicates asynchronously, but without any allele specificity (Knoll et al. 1994).

ALLELIC EXCLUSION

The most striking aspect of asynchronous replication timing is its consistent association with the phenomenon of domain-wide allelic inactivation. Olfactory receptor gene loci may represent yet another example of this unique type of regulatory structure. The sense of smell in a number of different organisms is mediated through olfactory neurons in the nasal organ (Buck and Axel 1991). Each of these neurons evidently expresses one, or a small number, out of a repertoire of about 1000 odorant receptor genes (Vassar et al. 1993). The problem of distinguishing which receptors have been activated therefore reduces to a problem of identifying which neurons have been activated. In such a model, perception of an odor encountered by the organism would result from the identification of the subset of cells responsive to that odorant.

Recent evidence suggests that in an individual neuron the receptor gene is expressed from only one of the two parental alleles (Chess et al. 1994). In some cells it is the paternal gene which is active, whereas in others the maternal gene undergoes transcription. Since olfactory receptor genes are clustered at a number of different chromosomal sites, it has been suggested that a single gene in an array is turned on stochastically by means of a *cis*-acting locus control element. If this is the case, there must be a mechanism of allelic exclusion to prevent the second allele from activating another gene. Although there is, as yet, no direct experimental proof for this model, it has been shown that these loci indeed replicate asynchronously (Table 2), and this suggests indirectly that only one allele in each cell may be in an active conformation. Unlike imprinted gene domains, the replication pattern of the olfactory receptor loci is not allele-specific, and this is in keeping with the observation that either allele for any given receptor gene can be active. The asynchronous pattern is probably generated in the early embryo, since it has been observed in a variety of different cell types and in F9 teratocarcinoma cells (Chess et al. 1994). In a sense, this structure would thus represent a form of autosomal allelic exclusion analogous to X-chromosome inactivation. There may very well be additional regions of the genome that behave in this manner, and the replication assay provides a convenient method for detecting these.

REGULATION OF GENE EXPRESSION

Although there is a correlation between replication timing and gene expression, the cause and effect relationship between these phenomena has not yet been explained. One possibility is that shifts in replication timing are secondary to changes which occur in gene expression or chromatin structure. Since replication time zones are so large, however, this model would predict that a local modulation in the transcription of one particular gene could bring about a domain-wide shift in replication timing. It is much more likely that replication time is controlled independently by long-range *cis*-acting elements (e.g., X-inactivation center or LCRs) and the resulting domain-wide structure then influences specific genes at the local level.

Studies on *Xenopus* 5S RNA genes provide a good model for understanding how replication timing may modulate transcription. Two different sequence types for this gene are carried in the genome; one pro-

Table 2 Synchronous and asynchronous replication timing

Locus	Single-double (%)
Control loci	
c-*myc*	9
H-*ras*-1	12
IgH	10
albumin	9
p53	12
C48	9
Pfk-L	11
Imprinted gene	
Igf2	23
Igf2R	35
H19	30
Snrpn	34
Olfactory receptors	
102	34
107	35
I7	38
I28	38
Y22(MEL)	32
Y22(F9)	39
Y22(PEF)	31

A number of gene regions were tested for replication timing using FISH analysis and the number of nuclei showing asynchronous replication (see Fig. 3) was scored (% single-double). Most loci replicate relatively synchronously with approximately 10% single-double nuclei. Imprinted genes and olfactory receptor genes have a high percentage of nuclei showing asynchronous replication. In the case of imprinted genes, replication is allele-specific (Kitsberg et al. 1993b), whereas for olfactory receptor genes either allele can replicate early (Chess et al. 1994). Olfactory receptor gene replication was analyzed in mouse erythroleukemia cells (MEL), embryonal carcinoma (F9), or primary embryonic fibroblasts (PEF).

vides the major transcript in the oocyte and another is expressed mainly in somatic tissues. In the oocyte, high concentrations of transcription factors activate both gene types equally, but since the oocyte 5S RNA genes are more numerous, this species provides the overwhelming majority of RNA molecules. In somatic cells, on the other hand, the oocyte genes are in a repressed chromatin state which prevents their transcription (Wolffe and Brown 1988).

In contrast to the somatic 5S RNA genes, the oocyte genes replicate late in the cell cycle (Gilbert 1986). It has been suggested that this may put them at a disadvantage for binding the critical transcription factors in somatic cells, and through this simple mechanism, they would be maintained in an inactive state even following replication. The feasibility of this mechanism is, of course, based on the assumption that certain protein factors can only interact with the gene while it is undergoing replication and the basic chromatin structure is thus disrupted. In keeping with this prediction, transcription factors indeed stimulate oocyte 5S RNA synthesis, but only after histone H1 is removed from the chromatin (Wolffe and Brown 1988). Replication timing appears to be an important part of this model, since even the oocyte 5S RNA gene is activated in a cell line carrying a translocation that forces this sequence to undergo early replication (Guinta et al. 1986). There is no question that the control of somatic and ooctye 5S RNA gene expression is both complex and multifaceted, but it certainly appears that replication timing is an important element in this process.

Direct evidence for an effect of replication timing on gene expression has been obtained from an analysis of the *URA3* gene in the yeast *S. cerevisiae*. At its normal locus in the genome, this gene is constitutively expressed at a basal level, but it can be induced to higher levels of expression by the binding of the *trans*-activator PPR1 to the gene regulatory domain (Roy et al. 1990). When *URA3* is located in a late-replicating telomeric region, however, its basal level of expression is markedly repressed and, as a result, the cells become phenotypically ura3$^-$, unless they are rescued by the presence of *PPR1*. In an elegant series of experiments, the *PPR1* gene was put under the regulation of a galactose-inducible promoter. Cells were then arrested at the G_1/S interphase in the absence of *PPR1* expression, a condition that prevents *URA3* transcription. Surprisingly, the induction of *PPR1* did not bring about *URA3* activation, and only when the cells were allowed to continue through the later part of the cell cycle did *URA3* become transcriptionally competent (Aparicio and Gottschling 1994). These studies clearly show that the induction of this gene is dependent on cell-cycle events and strongly suggest that replication is required for the activation process.

Two models have evolved for the effect of chromatin structure on gene regulation. In one case, chromatin prevents *trans*-activating proteins from gaining access to the DNA, thus keeping the gene irreversibly repressed. However, during DNA replication the chromatin structure of the gene is perturbed, and activation factors have the opportunity to gain access and establish transcription before reassembly of the chromatin is completed. In an alternate model, the *trans*-acting proteins can induce gene transcription at any time in a replication-independent manner, effectively disrupting the repressive nature of the chromatin. Examples of both forms of regulation have been observed (Felsenfeld 1992).

Assuming that at least some genes require a replication event in order to set up an active transcription complex, the time at which replication takes place could play a big role in determining the fate of each particular sequence. Since dynamic changes take place during the cell cycle, the levels of pertinent factors may fluctuate and thus be present in the nucleus only during specific intervals of S. In this way, the activation of a target gene may only be possible if it is replicated in this specified "window of opportunity." Replication at an alternate time in S could lead to an inactive configuration. E2F proteins may represent an example of this type of transcription factor, since their availability is carefully monitored in a cell-cycle-dependent manner (Shirodkar et al. 1992).

CONCLUSIONS

DNA replication takes place in an ordered and programmed manner during S phase. In animal cells, the genome is divided into distinct Mb-size time zones whose replication timing is probably controlled by *trans*-acting factors and central *cis*-acting elements. Housekeeping genes generally replicate during the first half of S phase, whereas many tissue-specific genes are late-replicating, yet developmentally regulated to undergo early replication in their tissue of expression. Replication time is apparently associated with chromosome structure, but it is not yet clear whether it is a consequence of chromatin interactions or whether it actually plays a role in the regulation of gene expression. Alternate replication time structures can also serve to distinguish between two allelic regions in the same nucleus, as is observed for imprinted genes, for the X chromosome in female cells, and in cases of allelic exclusion.

ACKNOWLEDGMENTS

Some of the research described in this review was supported by the National Institutes of Health, the Israel Academy of Sciences, and the Israel Cancer Research Fund.

REFERENCES

Aparicio, O.M. and D.E. Gottschling. 1994. Overcoming telomeric silencing: A *trans*-activator competes to establish gene expression in a cell cycle-dependent way. *Genes Dev.* **8:** 1133–1146.

Boggs, B.A. and A.C. Chinault. 1994. Analysis of replication timing properties of human X-chromosomal loci by fluorescence in situ hybridization. *Proc. Natl. Acad. Sci.* **91:** 6083–6087.

Braunstein, D., D. Schulze, T. DelGiudice, A. Furst, and C.L. Schildkraut. 1982. The temporal order of replication of murine immunoglobulin heavy chain constant region sequences corresponds to their linear order in the genome. *Nucleic Acids Res.* **10:** 6887–6902.

Brown, C.J. and H.F. Willard. 1994. The human X-inactivation centre is not required for maintenance of X-chromosome inactivation. *Nature* **368:** 154–156.

Brown, C.J., A. Ballabio, J.L. Rupert, R.G. Lafreniere, M. Groumpe, R. Tonlorenzi, and H.F. Willard. 1991a. A gene from the region of the human X inactivation centre is expressed exclusively from the inactive X chromosome. *Nature* **349:** 38–44.

Brown, C.J., B.D. Hendrich, J.L. Rupert, R.G. Lafreniere, Y. Xing, J. Lawrence, and H.F. Willard. 1992. The human Xist gene: Analysis of 17 kb inactive X-specific RNA that contains conserved repeats and is highly localized within the nucleus. *Cell* **71:** 527–542.

Brown, C.J., R.G. Lafreniere, V.E. Powers, G. Sebastio, A. Ballabio, A.L. Pettigrew, D.H. Ledbetter, E. Levy, I.W. Craig, and H.F. Willard. 1991b. Localization of the X inactivation centre on the human X chromosome in Xq13. *Nature* **349:** 82–84.

Brown, E.H., M.A. Iqbal, S. Stuart, K.S. Hatton, J. Valinsky, and C.L. Schildkraut. 1987. Rate of replication of the murine immunoglobulin heavy-chain locus: Evidence that the region is part of a single replicon. *Mol. Cell. Biol.* **7:** 450–457.

Buck, L. and R. Axel. 1991. A novel multigene family may encode odorant receptors: A molecular basis for odor recognition. *Cell* **65:** 175–187.

Buiting, K., S. Saitoh, S. Gross, B. Dittrich, S. Schwartz, R.D. Nicholls, and B. Horsthemke. 1995. Inherited microdeletions in the Angelman and Prader-Willi syndromes define an imprinting centre on human chromosome 15. *Nat. Genet.* **9:** 395–400.

Calza, R.E., L.A. Eckardt, T. DelGiudice, and C.L. Schildkraut. 1984. Changes in gene position are accompanied by a change in time of replication. *Cell* **36:** 689–696.

Cattanach, B.M. 1986. Parental origin effects in mice. *J. Embryol. Exp. Morphol.* (suppl.) **97:** 137–150.

Cedar, H. 1988. DNA methylation and gene activity. *Cell* **53:** 3–4.

Chess, A., I. Simon, H. Cedar, and R. Axel. 1994. Allelic inactivation regulates olfactory receptor gene expression. *Cell* **78:** 823–834.

Disteche, C.M., E.M. Eicher, and S.A. Latt. 1979. Late replication in an X-autosome translocation in the mouse: Correlation with genetic inactivation and evidence for selective effects during embryogenesis. *Proc. Natl. Acad. Sci.* **76:** 5234–5238.

Driscoll, M.C., C.S. Dobkin, and B.P. Alter. 1989. γδβ-Thalassemia due to a de novo mutation deleting the 5′ β-globin gene activation-region hypersensitive sites. *Proc. Natl. Acad. Sci.* **86:** 7470–7474.

Drouin, R., N. Lemieux, and C.-L. Richer. 1990. Analysis of DNA replication during S phase by means of dynamic chromosome banding at high resolution. *Chromosoma* **99:** 273–280.

Eden, S. and H. Cedar. 1994. Role of DNA methylation in the regulation of transcription. *Curr. Opin. Genet. Dev.* **4:** 255–259.

Edenberg, H.J. and J.A. Huberman. 1975. Eukaryotic chromosome replication. *Annu. Rev. Genet.* **9:** 245–285.

Epner, E., W.C. Forrester, and M. Groudine. 1988. Asynchronous DNA replication within the human β-globin gene locus. *Proc. Natl. Acad. Sci.* **85:** 8081–8085.

Felsenfeld, G. 1992. Chromatin as an essential part of the transcriptional mechanism. *Nature* **355:** 219–224.

Ferguson, B.M. and W.L. Fangman. 1992. A position effect on the time of replication origin activation in yeast. *Cell* **68:** 333–339.

Ferguson, B.M., B.J. Brewer, A.E. Reynolds, and W.L. Fangman. 1991. A yeast origin of replication is activated late in S phase. *Cell* **65:** 507–515.

Forrester, W.C., E. Epner, M.C. Driscoll, T. Enver, M. Brice, T. Papayannopoulou, and M. Groudine. 1990. A deletion of the human β-globin locus activation region causes a major alteration in chromatin structure and replication across the entire β-globin locus. *Genes Dev.* **4:** 1637–1649.

Giddings, S.J., C.D. King, K.W. Harman, J.F. Flood, and L.R. Carnaghi. 1994. Allele specific inactivation of insulin 1 and 2, in the mouse yolk sac, indicates imprinting. *Nat. Genet.* **6:** 310–313.

Gilbert, D.M. 1986. Temporal order of replication of *Xenopus laevis* 5S ribosomal RNA genes in somatic cells. *Proc. Natl. Acad. Sci.* **83:** 2924–2928.

Goldman, M.A., G.P. Holmquist, L.A. Caston, and A. Nag. 1984. Replication timing of genes and middle repetitive sequences. *Science* **224:** 686–692.

Guillemot, F., T. Caspary, S.M. Tilghman, N.G. Copeland, D.J. Gilbert, N.A. Jenkins, D.J. Anderson, A.L. Joyner, J. Rossand, and A. Nagy. 1995. Genomic imprinting of *Mas2*, a mouse gene required for trophoblast development. *Nat. Genet.* **9:** 235–242.

Guinta, D.R., J. Yun Tso, S. Narayanswami, B.A. Hamkalo, and L.J. Korn. 1986. Early replication and expression of oocyte-type 5S RNA genes in a *Xenopus* somatic cell line carrying a translocation. *Proc. Natl. Acad. Sci.* **83:** 5150–5154.

Gunaratne, P.H., M. Nakao, D.H. Ledbetter, J.S. Sutcliffe, and A.C. Chinault. 1995. Tissue-specific and allele-specific replication timing control in the imprinted human Prader-Willi syndrome region. *Genes Dev.* **9:** 808–920.

Hand, R. 1978. Eucaryotic DNA: Organization of the genome for replication. *Cell* **15:** 317–325.

Hansen, R.S., T.K. Canfield, M.M. Lamb, S.M. Gartier, and C.D. Laird. 1993. Association of fragile X. syndrome with delayed replication of the *FMR1* gene. *Cell* **73:** 1403–1409.

Hatton, K.A., V. Dhar, E.H. Brown, M.A. Iqbal, S. Stuart, V.T. Didamo, and C.L. Schildkraut. 1988. Replication program of active and inactive multigene families in mammalian cells. *Mol. Cell. Biol.* **8:** 2149–2158.

Holmquist, G.P. 1987. Role of replication time in the control of tissue-specific gene expression. *J. Hum. Genet.* **40:** 151–173.

Izumikawa, Y., K. Naritomi, and K. Hirayama. 1991. Replication asynchrony between homologs 15q11.2: Cytogenetic evidence for genomic imprinting. *Hum. Genet.* **87:** 1–5.

Jablonka, E., R. Goitein, M. Marcus, and H. Cedar. 1985. DNA hypomethylation causes an increase in DNase-I sensitivity and an advance in the time of replication of the entire inactive X chromosome. *Chromosoma* **93:** 152–156.

Karube, T. and S. Watanabe. 1988. Analysis of the chromosomal DNA replication pattern using the bromodeoxyuridine labeling method. *Cancer Res.* **48:** 219–222.

Kerem, B.-S., R. Goitein, G. Diamond, H. Cedar, and M. Marcus. 1984. Mapping of DNAase I sensitive regions of mitotic chromosomes. *Cell* **38:** 493–499.

Kitsberg, D., S. Selig, and H. Cedar. 1991. Chromosome structure and eukaryotic gene organization. *Curr. Opin. Genet. Dev.* **1:** 534–537.

Kitsberg, D., S. Selig, I. Keshet, and H. Cedar. 1993a. Replication structure of the human β-globin gene domain. *Nature* **366:** 588–590.

Kitsberg, D., S. Selig, M. Brandeis, I. Simon, I. Keshet, D.J. Driscoll, R.D. Nicholls, and H. Cedar. 1993b. Allele-specific replication timing of imprinted gene regions. *Nature*

364: 459–463.

Knoll, J.H.M., S.-D. Cheng, and M. Lalande. 1994. Allele specificity of DNA replication timing in the Angelman/Prader-Willi syndrome imprinted chromosomal region. *Nat. Genet.* **6:** 41–46.

LaSalle, J.M. and M. Lalande. 1995. Domain organization of allele-specific replication within the GABRB3 gene cluster requires a biparental 15q11-13 contribution. *Nat. Genet.* **9:** 386–394.

Latt, S.A. 1973. Microfluorometric detection of deoxyribonucleic acid replication in human metaphase chromosomes. *Proc. Natl. Acad. Sci.* **70:** 3395–3399.

Laurenson, P. and J. Rine. 1992. Silencers, silencing, and heritable transcriptional states. *Microbiol. Rev.* **56:** 543–560.

Ma, C., T.H. Leu, and J.L. Hamlin. 1990. Multiple origins of replication in the dihydrofolate reductase amplicons of a methotrexate-resistant Chinese hamster cell line. *Mol. Cell. Biol.* **10:** 1338–1346.

Mohandas, T., R.S. Sparker, and L.J. Shapiro. 1981. Reactivation of an inactive human X chromosome: Evidence for X inactivation by DNA methylation. *Science* **211:** 393–396.

Nicholls, R.D. 1993. Genomic imprinting and candidate genes in the Prader-Willi and Angelman syndromes. *Curr. Opin. Genet. Dev.* **3:** 445–456.

Razin, A. and H. Cedar. 1994. DNA methylation and genomic imprinting. *Cell* **77:** 473–476.

Reis, A., B. Dittrich, V. Greger, K. Buiting, M. Lalande, G. Gillessen-Kaesbach, M. Anvret, and B. Horsthemke. 1994. Imprinting mutations suggested by abnormal DNA methylation patterns in familial Angelman and Prader-Willi syndromes. *Am. J. Hum. Genet.* **54:** 741–747.

Reynolds, A.E., R.M. McCarroll, C.S. Newlon, and W.L. Fangman. 1989. Time of replication of ARS elements along yeast chromosome III. *Mol. Cell. Biol.* **9:** 4488–4494.

Roy, A., F. Exinger, and R. Losson. 1990. *cis-* and *trans*-acting regulatory elements of the yeast URA 3 promoter. *Mol. Cell. Biol.* **10:** 5257–5270.

Schmidt, M. and B.R. Migeon. 1990. Asynchronous replication of homologous loci on human active and inactive X chromosome. *Proc. Natl. Acad. Sci.* **87:** 3685–3689.

Selig, S., K. Okumura, D.C. Ward, and H. Cedar. 1992. Delineation of DNA replication time zones by fluorescence in situ hybridization. *EMBO J.* **11:** 1217–1225.

Selig, S., M. Ariel, R. Goitein, M. Marcus, and H. Cedar. 1988. Regulation of mouse satellite DNA replication time. *EMBO J.* **7:** 419–426.

Shafer, D.A. and J.H. Priest. 1984. Reversal of DNA methylation with 5-aza-C alters chromosome replication patterns in human lymphocyte and fibroblast cultures. *Am. J. Hum. Genet.* **36:** 534–545.

Shirodkar, S., M. Ewen, J.A. DeCaprio, J. Morgan, D.M. Livingston, and T. Chittenden. 1992. The transcription factor E2F interacts with the retinoblastoma product and a p107-cyclin A complex in a cell cycle-regulated manner. *Cell* **68:** 157–166.

Somssich, I.E., J. Spira, H. Hameister, and G. Klein. 1984. Correlation between tumorigenicity and banding pattern of chromosome 15 in murine T-cell leukemia cells. *Chromosoma* **91:** 39–45.

Spack, E.G., E.D. Lewis, B. Paradowski, R.T. Schimke, and P.P. Jones 1992. Temporal order of DNA replication in the H-2 major histocompatibility complex of the mouse. *Mol. Cell. Biol.* **12:** 5174–5188.

Sutcliffe, J.S., M. Nakao, S. Christian, K.H. Orstavik, N. Tommerup, D.H. Ledbetter, and

A.L. Beaudet. 1994. Deletions of a differentially methylated CpG island at the SNRPN gene define a putative imprinting control region. *Nat. Genet.* **8:** 52–58.

Takagi, N. 1974. Differentiation of X chromosomes in early female mouse embryos. *Exp. Cell Res.* **86:** 127–135.

Takagi, N., M.A. Yoshida, O. Sugawara, and M. Sasaki. 1983. Reversal of X-inactivation in female mouse somatic cells hybridized with murine teratocarcinoma stem cells in vitro. *Cell* **34:** 1053–1062.

Torchia, B.S., L.M. Call, and B.R. Migeon. 1994. DNA replication analysis of FMR1, XIST, and factor 8C loci by FISH shows nontranscribed X-linked genes replicate late. *Am. J. Hum. Genet.* **55:** 96–104.

Vassar, R., J. Ngai, and R. Axel. 1993. Spatial segregation of odorant receptor expression in the mammalian olfactory epithelium. *Cell* **74:** 309–318.

Villar, A.J. and R.A. Pedersen. 1994. Parental imprinting of the Mas protooncogene in mouse. *Nat. Genet.* **8:** 373–379.

Vogel, W., M. Autenrieth, and G. Speit. 1986. Detection of bromodeoxyuridine-incorporation in mammalian chromosomes by a bromodeoxyuridine-antibody. I. Demonstration of replication patterns. *Hum. Genet.* **72:** 129–132.

von Meyenburg, K. and F.G. Hansen. 1987. Regulation of Chromosome Replication. In Escherichia coli *and* Salmonella typhimurium: *Cellular and molecular biology* (ed. F.C. Neidhardt), pp. 1555–1577. American Society for Microbiology, Washington, D.C.

Wahrman, J., C. Richler, E. Neufeld, and A. Friedman. 1983. The origin of multiple sex chromosomes in the gerbil *Gerbillus gerbillus* (Rodentia: Gerbillinae). *Cytogenet. Cell Genet.* **35:** 161–180.

Wevrick, R., J.A. Kerns, and U. Francke. 1994. Identification of a novel paternally expressed gene in the Prader-Willi syndrome region. *Hum. Mol. Genet.* **3:** 1877–1882.

Wolf, S. and P. Perry. 1974. Differential staining of sister chromatids and the study of sister chromatid exchanges without autoradiography. *Chromosoma* **48:** 341–353.

Wolffe, A.P. and D.D. Brown. 1988. Developmental regulation of two 5S ribosomal RNA genes. *Science* **241:** 1626–1632.

Yunis, J.J. 1981. Mid prophase human chromosomes. The attainment of 2000 bands. *Hum. Genet.* **56:** 293–298.

13

Changes in DNA Replication during Animal Development

Janet L. Carminati[1] and Terry L. Orr-Weaver
Department of Biology, Massachusetts Institute of Technology
and Whitehead Institute
Cambridge, Massachusetts 02142

Throughout animal growth and development, the cell cycle is modified in response to developmental signals. These alterations in the cell cycle influence the control of S phase and the properties of DNA replication (outlined in Table 1). The production of haploid gametes requires a modified cell cycle, meiosis, in which two rounds of chromosome segregation follow a single S phase. Thus, S phase must be prevented between the two meiotic divisions. In the early embryos of many animals, a rapid cell cycle occurs in which S phase oscillates with mitosis without gap phases. This S-M cycle necessitates unique controls for the entry into S phase. Later in embryogenesis in *Drosophila* and *Xenopus*, a G_1 phase is added to the cell cycle, resulting in another developmental alteration of the onset of S phase. There are numerous examples of tissues that become polytene as a consequence of a modified cell cycle with only an S phase and a gap phase.

These developmental changes in the cell cycle require special controls for entry into S phase. The regulation of the onset of S phase has been extensively investigated in the normal cell cycle with G_1-S-G_2-M phases and is reviewed by Nasymth (this volume) and Weisshart and Fanning (this volume). Cyclin-dependent kinases (cyclin-cdk complexes), including cyclins E, D, and A complexed with cdk2, 4, and 6, are all known to play a role in S-phase regulation of higher eukaryotes (Sherr 1993, 1994). G_1 cyclin kinases are thought to phosphorylate the retinoblastoma gene product, pRb, releasing it from the transcription factor E2F, thus allowing the S-phase transcriptional program (Sherr 1994). Other S-phase kinases, perhaps cyclin A kinase, might then inactivate E2F once its transcriptional program is complete (Dynlacht et al. 1994; Krek et al. 1994). In the yeast *Saccharomyces cerevisiae*, three functionally redundant

[1]Present address: Department of Biological Sciences, Stanford University, Stanford, California 94305.

DNA Replication in Eukaryotic Cells

Table 1 Alterations in regulation of DNA replication during development

Developmental change	S-phase control	Replication modifications
Meiosis	distinct entry into S phase	slower S phase *S. cerevisiae*: same origins used, asynchronous activation mouse: slower fork rate newt: fewer origins activated
Oocyte activation	*Xenopus*: Mos inhibition of S phase during meiosis	sea urchin: inactive initiation factors in the unfertilized egg
Restart of S phase at fertilization	*Drosophila*: inhibition of S phase prior to fertilization by *plu*, *png*, and *gnu*	
Early embryonic cycles	S/M cycle posttranscriptional control of S-phase genes *Drosophila*: genes that couple S and M phases	*Xenopus* and *Drosophila*: periodic spacing of origins; sequence independent and synchronous activation
Onset of transcription and addition of G_1 phase (MBT)	G_1 phase added; transcriptional control of S-phase genes	longer S phase; asynchronous origin activation and late-replicating heterochromatin
Polyploidy/polyteny	*Drosophila*: endo cell cycle (S/G); spatial and temporal pattern	*Drosophila*: removal of block to rereplication; late replication and underrepresentation of heterochromatin
Amplification		tissue and temporal control; removal of block to rereplication

cyclins, CLN 1, 2, and 3, control the G_1/S transition, and two cyclin-B homologs, CLB5 and 6, also act to control S phase (Richardson et al. 1989; Epstein and Cross 1992; Schwob and Nasmyth 1993). These cyclins all interact with the CDC28 kinase. In addition to these positive regulators, in the normal cell cycle, inhibitors of cyclin kinases exist that are able to inhibit progression of the cell cycle in response to various environmental cues (Sherr 1994; Sherr and Roberts 1995).

In the developmentally modified cell cycles discussed here, entry into S phase is regulated by different mechanisms from those of the normal cell cycle (Table 1). In yeast meiosis, the CDC28 kinase appears to act later than S phase. During the embryonic S-M cycles, entry into S phase must be controlled posttranscriptionally and must be regulated in the presence of constitutive cyclin E kinase activity. The addition of a G_1 phase during embryogenesis requires that extrinsic developmental cues influence the onset of S phase. In *Drosophila*, this is mediated at least in part through cyclin E. During the S-G cycle that produces polytene cells, entry into S phase can no longer have the completion of mitosis as a prerequisite.

In addition to differential control of the onset of S phase, the actual parameters of S phase are altered during development. By parameters of S phase, we mean the intrinsic properties of DNA replication. The parameters of DNA replication changed in modified cell cycles include replication origin usage and activation, the rate of replication fork movement, and the block to rereplication (Table 1). We discuss how these parameters are changed in each of the variant cell cycles. For a background on the normal intrinsic regulation of DNA replication, we refer the reader to DePamphilis; Heintz; Newlon; Simon and Cedar; and Blow and Chong (all this volume).

S PHASE OF THE MEIOTIC CELL CYCLE

Meiosis is a modified cell cycle in which two rounds of chromosome segregation follow a single S phase. This results in the chromosome number being reduced by half. This is essential so that the chromosome number is restored when the sperm and egg fuse at fertilization. The S phase that precedes chromosome segregation in meiosis has been termed "premeiotic S." The use of the term premeiotic S implies that the term meiosis refers only to the actual segregation of chromosomes. We refer to the S phase of the meiotic cell cycle as premeiotic S, but the reader should realize that premeiotic S phase occurs as part of the same cell cycle in which the two meiotic divisions occur.

Regulation of Entry into S Phase

Studies addressing the regulation of premeiotic S phase have been investigated mainly in the yeast, *S. cerevisiae*. Many genes that govern S phase during the mitotic cell cycle have been characterized and are covered by Nasymth (this volume). However, several genes that are required for entry into S phase during the mitotic cell cycle are not required for the control of premeiotic S phase. Interestingly, some of these genes that are involved in S phase during the mitotic cell cycle do have a role in meiosis, yet are required after replication and before the meiosis I division. One example is the mutation *cdc7*, in which mitotic cells are blocked prior to replication, yet meiotic cells are arrested following premeiotic DNA synthesis (Schild and Byers 1978; Buck et al. 1991). Recently, Cdc7 has been linked to replication origins, in that a protein that binds to and activates the Cdc7 kinase, Dbf4, also binds to *ARS* elements (Jackson et al. 1993; Dowell et al. 1994). If Cdc7 does prove to have a role at the replication origin, this might then suggest that premeiotic replication is differentially regulated. Other similar examples include the *S. cerevisiae* mutations, *cdc28* and *cdc4*, in which DNA replication is blocked in the mitotic cell cycle, yet premeiotic synthesis occurs (Simchen and Hirschberg 1977; Piggott et al. 1982; Shuster and Byers 1989; Reed and Wittenberg 1990). Cdc28 is a component of maturation promoting factor (MPF) and is required both at the G_1/S and G_2/M transitions during the mitotic cell cycle yet does not appear to be required for the G_1/S transition of premeiotic S phase. It should be noted, however, that the *Schizosaccharomyces pombe* homolog of *cdc28*, *cdc2*, is required for premeiotic DNA replication (Iino et al. 1995).

Spatial Control of Meiotic Replication Origins

Comparisons of S phase during the meiotic and mitotic cell cycles using fiber autoradiography indicated that a similar spacing of origins is present and that replication forks travel at a similar rate (Johnston et al. 1982; Newlon 1988). However, the resolution of fiber autoradiography was not precise enough to determine whether the specific origins used were the same origins used during the mitotic cycle. Given the fact that not all *ARS* elements identified by the plasmid assay are active chromosomal origins during the mitotic cell cycle, it is possible that the other *ARS* elements were specifically used during premeiotic S phase. In one study designed to examine *ARS1* function in premeiotic S phase, plasmid loss occurred in both mitosis and meiosis upon induction of transcription through the *ARS1* element (Hollingsworth and Sclafani 1993). Although this study used an indirect assay for *ARS* function, it suggests that the same origin can function in S phase in both the meiotic and mitotic cycles. In another study to identify replication origins active during premeiotic S phase, two-dimensional gel analysis was used to examine origins on chromosome III (Collins and Newlon 1994). The five premeiotic S-phase origins map to the same *ARS* elements as the origins from the mitotic cell cycle. One possible exception was an origin located at *CEN3* that had weak activity in the mitotic cycle and did not appear to have any activity in the meiotic cycle. No additional premeiotic S-phase origins were found that did not correspond to origins used in the mitotic cycle. Further analysis will be needed to determine if this conclusion is true for all chromosomes, but these studies strongly suggest that meiotic origins are indeed the same as mitotic origins.

In higher eukaryotes, perhaps this is not the case. Using fiber autoradiography in the newt *Triturus*, the spacing between origins used in meiosis is larger than that seen during the mitotic cycle, suggesting that either different origins are used or a subset of mitotic origins are activated (Callan 1974).

Temporal Control of Meiotic Origins

S phase in eukaryotes can vary widely in length, from extremely short S phases during embryonic development (several minutes) to longer S phases in somatic cells (hours) and often very prolonged S phases during premeiotic S phase (at least 24 hours in the *Drosophila* ovary) (Chandley 1966; Grell 1973). S phase seems to lengthen as development places increasingly complex controls on the cell cycle. In *S. cerevisiae*, premeiotic S phase is at least twice as long as S phase in the mitotic cycle (65 min versus 30 min) (Williamson et al. 1983). Despite this lengthening of S phase, origin usage and the rate of fork movement are the same as during mitotic S phase (Johnston et al. 1982).

Two possibilities for the longer premeiotic S phase are either that origins within a chromosome are fired asynchronously or that different chromosomes are replicated at different times. In the study of origins in the meiotic cell cycle on chromosome III mentioned above, the efficiency of usage of specific *ARS* elements and characteristic termination patterns were similar between meiosis and mitosis (Collins and Newlon 1994). This implies that within a single chromosome the kinetics of initiation and termination are roughly the same in meiosis and mitosis. Further studies are needed to determine if this is true for all chromosomes. Collins and Newlon therefore suggest that the longer premeiotic S phase is due to different initiation times for different chromosomes.

In the premeiotic S phase of mouse spermatogonia, the lengthening of S phase was studied by fiber autoradiography (Jagiello et al. 1983). As in yeast, origin spacing during S phase in the meiotic and mitotic cycle is similar. However, the rate of fork movement is slower during meiosis and might account for the longer S phase.

S PHASE DURING EARLY EMBRYOGENESIS

Inhibition of DNA Replication in the Developing Oocyte and Unfertilized Egg

Many animals undergo rapid cell divisions following fertilization, including *Xenopus*, *Drosophila*, sea urchin, starfish, and clam. DNA replication factors are stockpiled in the egg to prepare for the high demand during the rapid embryonic divisions. However, DNA replication must be repressed in the oocyte so as not to occur during meiosis or prior to fertilization. This repression could be at the level of S-phase control such that entry into S phase is blocked during meiosis, or alternatively, at the level of replication parameters, such as initiation factors that might be kept in an inactive state. Two well-characterized mechanisms of replication control in the developing oocyte and unfertilized egg have been found in *Xenopus* and sea urchin.

The inability of *Xenopus* oocytes to replicate DNA is mediated by Mos activity blocking entry into S phase during meiosis (Furuno et al. 1994). Mos is a serine/threonine kinase that is expressed at high levels in germ cells of vertebrates and is maternally loaded into the *Xenopus* oocyte (Sagata et al. 1988). Mos acts at three steps in *Xenopus* oocyte maturation: (1) It activates meiosis; (2) it represses DNA replication between the meiotic divisions; and (3) it helps maintain the metaphase II arrest. The first activity of Mos is that of a meiotic initiator. *Xenopus* immature oocytes (stage VI) are arrested at G_2/M of meiosis I. Following hormone stimulation, germinal vesicle breakdown (GVBD) occurs, and oocytes progress to metaphase II of meiosis. This maturation is dependent on the translation of Mos mRNA following hormone stimulation, after which mature oocytes remain arrested at metaphase II until fertilization.

The second Mos activity, the ability of Mos to block DNA replication during meiosis, is relevant for this review. By examining a precise time course of MPF activity throughout meiosis, it was found that MPF is inactivated early in meiosis I (early metaphase I) and is then reactivated during late metaphase I, well before meiosis II (Furuno et al. 1994; Ohsumi et al. 1994). Mos mediates this reactivation of MPF, which then suppresses DNA replication between the two meiotic divisions. Ablation studies using c-*mos* antisense RNA or Mos antibodies showed that upon inactivation during GVBD, mature oocytes enter S phase inappropriately following meiosis I (Furuno et al. 1994). The role of MPF in suppressing an intervening S phase was also confirmed by studies that block MPF activation at meiosis II, resulting in DNA replication (Furuno et al. 1994; Ohsumi et al. 1994). Once MPF is reactivated, it is stabilized by the action of cytostatic factor (CSF) at the metaphase II arrest (Minshull 1993). This is the third meiotic activity of Mos, because CSF is composed of both Mos and an unknown factor. Thus, Mos is needed to maintain the metaphase II arrest in the mature oocyte.

Mos homologs have not been identified in invertebrates, suggesting that different mechanisms might act to control replication during oocyte maturation. In contrast to the control of S-phase entry seen in *Xenopus*, the mechanism that inhibits DNA replication in the unfertilized sea urchin egg is the absence of active initiation factors responsible for DNA synthesis. Following oocyte maturation, sea urchin eggs complete meiosis and arrest at G_1 of the first mitotic cycle. Fertilization then releases this arrest, and replication factors are posttranslationally activated within a short 3 minutes following fertilization (Zhang and Ruderman 1993). The first S phase occurs 20 minutes following fertilization, even in the absence of protein synthesis.

The unfertilized egg of sea urchin is not permissive for DNA replication, as assayed by the inability of egg extracts to replicate added sperm nuclei or double-stranded DNA templates (Zhang and Ruderman 1993). Conversely, embryonic interphase extracts are capable of supporting replication of added templates. Mixing experiments of egg and embryonic extracts showed no evidence of negative factors in the egg capable of repressing replication; the egg extract did not inhibit the ability of the embryonic extract to replicate DNA. Thus, the unfertilized egg does not contain inhibitors of replication, and instead, the inhibition of replication seen is due to inactive initiation factors present at this stage of development.

Restart of S Phase at Fertilization

In many organisms, the unfertilized egg is arrested during or following meiosis, and fertilization then releases the egg from this arrest. This coupling of fertilization with the resumption of the cell cycle ensures that the female and male pronuclei can then enter the first S phase and subsequent cell cycles with the proper timing. In *Drosophila*, starfish, and sea urchin, the completion of meiosis is not coupled to fertilization. In *Drosophila*, the mature oocyte is arrested at metaphase I in the ovary. Upon passage of the egg through the uterus, the egg becomes activated, and meiosis is completed regardless of whether fertilization occurs. Fertilization is necessary to restart the cell cycle in the embryo following the completion of meiosis, and unique regulators may be needed at this developmental step.

In *Drosophila*, three genes that act at this point to couple fertilization with DNA replication are the maternal-effect genes *giant nuclei* (*gnu*), *pan gu* (*png*), and *plutonium* (*plu*). Unfertilized mutant eggs complete meiosis resulting in four meiotic products, yet then undergo improper DNA replication, resulting in large polyploid nuclei (Freeman and Glover 1987; Shamanski and Orr-Weaver 1991). These genes normally act as negative regulators of DNA replication to make the restart of S phase dependent on fertilization. Fertilization must overcome the action of these genes in order to resume the cell cycle.

In many organisms, including starfish, *Xenopus*, sea urchin, and mouse, a transient increase in intracellular calcium (Ca^{++}) occurs at fertilization and is associated with the onset of development including meiotic maturation, pronuclear migration, DNA replication, and nuclear envelope breakdown. Time-lapse confocal imaging was used to study calcium dynamics throughout starfish early development (Stricker 1995). No transient increases in intracellular calcium were observed during oocyte maturation, whereas a single prolonged transient coincided with fertilization, and repetitive calcium oscillations occurred during the early cleavage divisions. The levels of inositol triphosphate also fluctuate during fertilization and might act to trigger the calcium transients seen during sea urchin development (Ciapa et al. 1994). However, the cause and effect relationship between calcium and the second messengers in other systems remains less clear (Whitaker and Swann 1993).

Following the calcium burst in *Xenopus* mature oocytes, the calmodulin-dependent kinase, CaMKII, is responsible for the inactivation of both MPF and CSF, thus releasing the oocyte from a metaphase II arrest (Lorca et al. 1993). How this then regulates the resumption of DNA replication following meiosis is less clear. The role of calcium bursts in controlling DNA replication can be seen in starfish, where the calcium ionophore, A 23187, acts as a parthenogenic agent capable of inducing several rounds of replication in mature oocytes (Picard et al. 1987). The calcium ionophore also triggers the onset of the cell cycle in unfertilized sea urchin eggs (Whitaker and Patel 1990). Whether transient fluxes in calcium levels play a role in *Drosophila* fertilization and the restart of S phase is unclear.

Rapid Early Cycles in *Xenopus* and Flies

Many organisms undergo rapid embryonic cycles following fertilization. The early embryonic cycles of *Xenopus* and *Drosophila* have been well characterized and consist of rapid cycles of alternating S phase and M phase that are controlled by maternally supplied products present in the egg. DNA replication is differentially regulated during this time in development. Unique regulation occurs at both the entry into S phase and in the replication parameters used. S-phase entry is regulated posttranscriptionally due to the absence of zygotic transcription. DNA replication parameters are also developmentally controlled such that many synchronous origins are activated to ensure complete replication within the very rapid 3- to 10-minute S phases.

Posttranscriptional Control of S Phase

The rapid early cycles of *Xenopus* and *Drosophila* are controlled by posttranscriptional modifications of regulators during S phase and M phase. In *Drosophila*, the first 13 divisions in the fertilized embryo are rapid, synchronous nuclear divisions within a common shared cytoplasm, where S phase occurs in an extremely short 3- to 4-minute period. The early cycles in *Xenopus* are similar in that the first 12 divisions consist of rapid, synchronous cycles with S phase occurring in a brief 10 minutes. During these early embryonic cycles of both *Xenopus* and *Drosophila*, transcription of the zygotic nucleus does not occur. In *Drosophila*, maximal zygotic transcription occurs following cellularization at cycle 14, whereas in *Xenopus*, zygotic transcription occurs at the midblastula transition (Newport and Kirschner 1982b; Edgar and Schubiger 1986).

Known regulators of S phase are present at high levels in *Drosophila* and *Xenopus* embryos. In *Drosophila*, the S-phase cyclin E-cdk2 kinase is present throughout early embryonic development (Richardson et al. 1993; Knoblich et al. 1994). Similarly, a large pool of maternal cyclin E is present during the early cleavage divisions of *Xenopus* (Jackson et al. 1995). The constitutive high levels of these proteins indicate that control of entry into S phase during the early division cycles must be regulated differently than during the later cycles when S phase is preceded by a G_1 phase.

In addition to the posttranscriptional control of known S-phase regulators, unique regulators might also be used early in development for the rapid cell cycle. In *Drosophila*, three maternal-effect genes, *giant nuclei* (*gnu*), *pan gu* (*png*), and *plutonium* (*plu*), appear to be novel cell cycle regulators. As discussed above, these genes are needed to couple S phase with fertilization such that mutant unfertilized eggs undergo inappropriate DNA replication. When fertilized, these embryos also give rise to giant polyploid nuclei, suggesting that these genes control S phase during the early cycles. Although fertilized, these embryos fail to properly couple S phase and M phase, and some aspects of mitosis such as centrosome duplication continue to cycle independently from nuclear division (Freeman et al. 1986; Freeman and Glover 1987; Shamanski and Orr-Weaver 1991). *png* is unique in that several presumably leaky alleles transiently couple S phase and M phase, resulting in embryos containing many more giant nuclei. However, defects in mitotic figures and DNA condensation can be seen as early as the first division, and the uncoupling of replication and mitosis progresses as development proceeds (J. Carminati, unpubl.). The unfertilized and fertilized phenotypes can be explained by the proposal that these genes inhibit DNA replication. In unfertilized eggs they function to make S phase dependent on fertilization, whereas in fertilized embryos these gene products make S phase dependent on the proper completion of mitosis. However, at what level these genes act is unclear. They might control either the entry into S phase, the block to rereplication, or other aspects of the cell cycle, such as chromosome condensation, that link mitosis to replication.

Molecular data confirm that *plu* is a unique regulator that acts solely during the early *Drosophila* divisions and not in later canonical cell cycles. RNA null alleles are maternal-effect alleles, and expression of the *plu* transcript is not present during later stages of development (Axton et al. 1994). Plu encodes a 19-kD protein consisting of three ankyrin repeats. Interestingly, another small ankyrin repeat protein is the cdk4 inhibitor, $p16^{INK4}$, which acts to inhibit cdk4, thus disrupting its association with the S-phase cyclin D protein (Serrano et al. 1993). Plu might act analogously in its role in repressing DNA replication, possibly by inhibiting S-phase cyclin kinases until their proper time of action.

Replication Origins

Studies in both *Xenopus* and *Drosophila* suggest that in the extremely rapid cycles of the early embryo, origins are controlled by a unique mechanism perhaps involving chromosome folding or attachment to the nuclear envelope. This could ensure the complete replication of the genome during the brief S phase. This mechanism acts temporally and spatially, such that periodically spaced origins are activated synchronously at the beginning of each S phase.

Experiments using *Xenopus* eggs showed that plasmid replication upon injection into the egg is under cell cycle control. Replication is not dependent on specific sequences, but instead depends on the size of the plasmid injected (Harland and Laskey 1980; Méchali and Kearsey 1984). Two-dimensional gel analysis of replication intermediates revealed that in both *Xenopus* eggs and extracts, plasmids containing either rDNA repeats or single-copy sequences initiated and terminated replication at random sites throughout the plasmid (Hyrien and Méchali 1992; Mahbubani et al. 1992). It was also determined that although initiation could occur at random sites on the plasmid, a single initiation event gave rise to complete replication of each plasmid molecule.

Similar conclusions were reached upon examination of the replication of chromosomal rDNA repeats in early embryos prior to the midblastula transition (Hyrien and Méchali 1993). Initiation occurred at random positions, and the estimated replicon size was 9–12 kb. To complete replication in the rapid S phase, all origins were presumed to be activated synchronously at the beginning of S phase. Whereas replication initiation is random with respect to sequence, the periodic spacing of replicons suggests that initiation is not random with respect to higher-order chromatin folding. If initiation were entirely random, there could be instances where some replicons were too far apart to finish replication in the short S phase. Thus, the authors suggest that chromosomal folding might specify a periodic spacing of origins, guaranteeing complete replication within the 10-minute S phase.

A similar periodic spacing of replicons was seen in the *Drosophila* early cleavage nuclei in which replication occurs in a 3- to 4-minute S phase. By electron microscopy (EM) studies, the average replicon size was 7.9 kb with a preferred periodicity of 3.4 kb and a maximum size of 19 kb (Blumenthal et al. 1974). An estimated 20,000 bidirectional origins must be activated nearly synchronously to finish S phase in the extremely short time. The maximum replicon size correlates with the amount of DNA that can be replicated in 3–4 minutes, given the rate of fork movement observed. However, these studies did not resolve whether initiation occurs at defined or random sequences. Using two-dimensional gel analysis of replication intermediates from early embryos, it was found that random initiation occurs both within the histone repeats and within a 40-kb single-copy sequence (Shinomiya and Ina 1991). Therefore, the periodicity of replicon spacing and the sequence-independent nature of replication are similar to that found in early *Xenopus* embryos. Thus, there appears to be a specific control of origin usage in the early rapid cycles.

S-phase Slowdown

Following the rapid embryonic cell cycles, the cell cycle lengthens to allow certain developmental processes to occur, such as the onset of zygotic gene transcription and gastrulation. In both *Xenopus* and *Drosoph-*

ila, the increase of the nuclear-to-cytoplasmic ratio results in the slowing of the cell cycle (Newport and Kirschner 1982a,b; Edgar and Schubiger 1986). During this time in development, both the control of replication origins and the entry into S phase are altered. In *Xenopus*, this is a one-step process that occurs at the midblastula transition (MBT), whereas in *Drosophila* these events occur in two distinct steps. First, following cellularization at cycle 13, alterations in the temporal control of origins occur. At this time, a G_2 phase is added to the cell cycle, and zygotic transcription commences. Following three more divisions, a G_1 phase is added, permitting transcriptional control of S phase. This transcriptional control in *Drosophila* and the MBT in *Xenopus* are discussed below.

In *Drosophila*, replication proceeds from the fast synchronous S phase of the early cycles to a prolonged S phase in which origin activation becomes asynchronous. A change in condensation of both euchromatin and heterochromatin occurs during this lengthening of S phase (Foe et al. 1993). During the first 13 cycles of the *Drosophila* embryo, the euchromatin remains decondensed throughout interphase. The heterochromatic regions decondense for a short time in S phase, decondensing progressively later in interphase as the cycle slows during cycles 11 to 13. By cycle 13, S phase has lengthened from 3–4 minutes to 13 minutes. During the three post-blastoderm cycles 14 to 16, S phase is eight times as long as the early S phases, taking 35–45 minutes.

In contrast to the early cycles, at least 200 particles of highly condensed euchromatin can be seen during interphase of cycles 14, 15, and 16 (Foe et al. 1993). EM studies show that new forks appear throughout the first 20 minutes of S phase during cycle 14, indicating that origin activation becomes asynchronous during interphase of these later cycles (McKnight and Miller 1977). Heterochromatin remains condensed throughout most of interphase and is not replicated until euchromatic replication has finished (Edgar and O'Farrell 1990; Foe et al. 1993). This altered regulation of origins and concurrent lengthening of S phase correlates with the onset of zygotic transcription and tissue-specific expression of certain genes. Replication must be coordinated with other processes now occurring during the cell cycle. In the post-blastoderm cycles, replication is followed by a G_2 phase of varying length, whereas mitosis occurs in distinct temporal and spatial domains.

Mouse

Similar to *Xenopus*, mouse embryos also are able to support the replication of injected double-stranded DNA templates, whereas injected plasmids are not replicated in oocytes (Wirak et al. 1985). Replication in embryos is dependent on specific *cis*-acting origin sequences of polyomavirus (PyV) or simian virus 40 (SV40) present on the plasmids. This is in contrast to results in *Xenopus* eggs, in which replication is independent of *cis*-acting sites and initiates at random sequences.

Using mouse embryos, an added layer of control is seen when comparing arrested one-cell embryos containing the unfused male and female pronuclei with two-cell embryos containing zygotic nuclei (Martinez-Salas et al. 1988, 1989). One-cell embryos are able to replicate DNA containing a minimal origin core sequence from PyV, whereas the two-cell embryo requires enhancer sequences present in *cis* to the PyV origin. Enhancers are also required for gene expression in the zygote as opposed to the one-cell embryo (Martinez-Salas et al. 1989; Majumder et al. 1993; Melin et al. 1993). The cytoplasmic factors that act on promoters and enhancers are not present in the one-cell embryo, but rather appear during the formation of the two-cell embryo (Henery et al. 1995). Enhancers are postulated to prevent the repression of origins and promoters by altered chromatin structure that is thought to occur upon zygote formation.

Replication in the mouse embryo is distinct from that occurring in the rapid early embryonic cycles of *Xenopus* and *Drosophila*, possibly due to the difference seen in zygotic transcription in these organisms. *Xenopus* and *Drosophila* have an early period lacking zygotic transcription, so are able to support fast cycles of alternating S and M phase. In contrast, transcription occurs in the early mouse embryo following zygotic formation, and thus replication must be coordinated with transcription. Enhancers might be associated with origins to overcome chromatin effects due to a transcriptionally active genome.

DEVELOPMENTAL SHIFT TO TRANSCRIPTIONAL REGULATION OF S PHASE

Following the rapid cycles of *Xenopus* and *Drosophila* that are controlled by posttranscriptional modifications of cell cycle regulators, a developmental shift occurs and transcription of the zygotic genome becomes active. At this time in development, S-phase regulators can be controlled at a transcriptional level, a mode of regulation characteristic of the somatic cell cycle.

MBT in *Xenopus*

The cell cycle of *Xenopus* is modified during the MBT, which occurs after the first 6 hours of development. The early rapid cycles of S and M phase slow, giving rise to a longer cell cycle in which both G_1 and G_2 gap phases are added to the cell cycle. This process then allows for the resumption of zygotic transcription (Kimelman et al. 1987). This slowing of the cell cycle is thought to occur due to a mitotic initiation factor that becomes rate-limiting at the MBT (Kirschner et al. 1985; Newport et al. 1985). Blastula cleavage becomes less synchronous, cells become motile, and zygotic transcription turns on. Recent results determined that the excess of histones present in the early embryo is responsible for the repression of transcription prior to the MBT (Prioleau et al. 1994).

As mentioned previously, the developmental regulation of S phase is altered at two different levels during the MBT in *Xenopus*. First, similar to *Drosophila*, origin activation presumably becomes asynchronous, giving rise to a lengthened S phase. DNA replication becomes asynchronous between cells, as evidenced by a variation of proliferating cell nuclear antigen (PCNA) staining, with some nuclei showing a peripheral staining and others showing a homogeneous staining (Leibovici et al. 1992). This is in contrast to the early cycles in which PCNA staining is homogeneous throughout S phase. Second, the entry into S phase is altered such that a G_1 phase is added to the cell cycle. Zygotic transcription resumes, permitting S-phase regulators to become transcriptionally controlled.

Addition of G_1 in *Drosophila*

The developmental shift to asynchronous origin activation has already occurred by cycle 14 of the *Drosophila* embryo. The second developmental control placed on S phase occurs following cycle 16 when a G_1 phase is added to the cell cycle. Following cycle 16, mitotic embryonic cells either arrest in G_1 and divide later in development (imaginal cells) or continue to divide (neuronal cells). The cells giving rise to most of the larval tissues become polytene, and as described below, these cells go through additional cycles that have only S and gap phases, but no mitosis. Thus, cycle 17 is the first time during embryogenesis there is a G_1 phase and, with the exception of a small group of cells on the ventral epidermis, it occurs in cells that are becoming polytene. At this point in embryogenesis, the cyclin E gene changes from being constitutively transcribed to a dynamic pattern of transcription in which the appearance of cyclin E transcripts precedes S phase. The down-regulation of cyclin E is needed for the arrest of cells in G_1, and cyclin E is then necessary for the G_1/S transition (Richardson et al. 1993; Knoblich et al. 1994). Both the cyclin E transcript and protein are down-regulated following the last mitotic division of epidermal cells during cycle 16 and remain off in these G_1-arrested cells (Knoblich et al. 1994).

During *Drosophila* development, the transcriptional control of cyclin E during the added G_1 phase is a new mode of S-phase regulation that occurs specifically following embryonic cell cycles 1–16. Mutations in the E2F transcription factor block DNA replication beginning at cycle 17, correlating with the first appearance of G_1 (Duronio et al. 1995). In embryonic cells that have a G_1 cell cycle phase, cyclin E is the downstream target of E2F, as evidenced by the observation that ectopic expression of cyclin E can overcome the requirement for E2F for S phase (Duronio and O'Farrell 1995). Thus, E2F appears to control the developmental pattern of cyclin E transcription that is responsible for the G_1 to S transition. Cyclin E in turn activates the transcription of a set of genes encoding essential replication factors, including polymerase-α, *PCNA*, and ribonuclease reductase 1 and 2 (*RNR1, 2*) (Duronio and O'Farrell 1994). E2F and cyclin E exert a different role in the cells becoming polytene than in the neuronal cells that lack a G_1 phase (Duronio and O'Farrell 1995). In the nervous system, cyclin E continues to be constitutively expressed, and this expression is independent of E2F. Rather, in the nervous system, cyclin E appears to regulate E2F.

POLYPLOIDY/POLYTENY

Changes in S-phase Regulation

Polyploid cells exist in a number of organisms including plants, ciliates, and dipteran insects, as well as in some mammalian cell types such as the trophoblasts that give rise to the mammalian placenta. Polyploidy is often associated with cells or tissues in which a requirement for increased protein production is needed; multiple chromosome copies is one way in which evolution has met that demand. An area of current research concerns the identification of regulators that govern the developmental transition leading to polyploidy. Similar to the developmental alterations of DNA replication already discussed, polyploid replication is controlled at two basic levels. These include changes in regulation of the cell cycle as well as alterations at the level of replication origins and other parameters used during DNA synthesis. In *Drosophila*, the transition to polyteny results in an altered cell cycle, termed the endo cell cycle, which consists of an alternating S phase and gap phase (Smith and Orr-Weaver 1991). Parameters of replication also become altered, and the block to rereplication is overcome during the endo cell cycle.

Polyploid and polytene cells are defined as those in which DNA replication has become uncoupled from mitosis, giving rise to cells with greater than diploid content of DNA. The degree to which replication is uncoupled from mitotic aspects of the cell cycle can vary. Polytene cells uncouple replication from all aspects of mitosis; 1000 or more chromosome copies remain synapsed, forming the large polytene chromosomes characteristic of the *Drosophila* salivary gland. Some polyploid cells do not uncouple replication from all aspects of mitosis, and chromosome segregation or cycles of chromosome condensation still occur. This is referred to as endopolyploidy or endomitosis. Finally, multinucleate cells have been referred to as polyploid cells. The *Drosophila* larval polytene cells are among the best characterized in terms of alterations in S phase that occur during this developmental transition.

Regulators of the Endo Cell Cycle

One of the best-understood examples of polyteny is in the *Drosophila* larval tissues (Smith and Orr-Weaver 1991). Following embryogenesis, *Drosophila* larval growth is due to an increase in cell size upon polytenization, because only cells in the nervous system and imaginal tissues undergo mitosis during larval development. In *Drosophila*, most tissues enter the endo cell cycle during late embryogenesis, and this transition is temporally and spatially regulated (Smith and Orr-Weaver 1991). The first transitions to the endo cell cycle occur in tissue-specific domains that replicate at characteristic times after cycle 16, with the salivary gland being the first tissue to enter the endo cell cycle, followed by the midgut, hindgut, and Malpighian tubules. It was determined for the hindgut that cells enter the endo cell cycle from the G_1 phase of the cell cycle, whereas salivary gland cells may enter the endo cell cycle following G_2. Gap phases of the endo cell cycle can vary widely in length, from 3 hours for the midgut to 18 hours for the salivary gland (Smith and Orr-Weaver 1991).

The spatially and temporally regulated pattern seen in *Drosophila* polytene tissues argues that a novel factor controls these cycles. However, regulators of the endo cell cycle could in theory be known mitotic cell cycle regulators that also act to control the endo cell cycle. In this case, the dependency between S phase and M phase must be disrupted in this altered cell cycle. In *Drosophila*, the mitotic cell cycle regulators, cyclin A, cdc2, and the cdc25 phosphatase homolog, *string*, are not needed for the endo cell cycle. Endoreplication proceeds in embryos lacking any of these regulators (Smith and Orr-Weaver 1991; Smith et al. 1993; Stern et al. 1993).

The cyclin E gene is essential for polytenization. The endo cell cycle does not occur in embryos lacking cyclin E or E2F, and cyclin E expression parallels S phase in endoreplicating tissues (Knoblich et al. 1994; Duronio et al. 1995). Thus, developmental regulation of the transcription of the cyclin E gene can account for the pattern of polytene S phases. Cyclin E transcription in endoreplicating cells is subject to a negative feedback, because ectopic expression of cyclin E down-regulates endogenous cyclin E transcription in these cells (Sauer et al. 1995). This negative feedback loop may be responsible for the transient appearance of cyclin E transcripts in the endo cell cycle and could be necessary for cyclic, rather than continuous, DNA replication in these cells. It may simply be that periodic transcription of cyclin E in the absence of cyclins A and B results in a cycle of S phase in the absence of mitosis (Sauer et al. 1995). However, it is possible that novel regulators couple developmental signals to the regulation of the endo cell cycle.

An interesting gene, *escargot*, is a transcription factor that appears to maintain the diploid state of arrested imaginal cells in the *Drosophila* larva (Hayashi et al. 1993; Fuse et al. 1994). In certain allelic combinations, a group of imaginal cells known as the histoblast nests overreplicate. Inappropriate expression of *escargot* in the polytene salivary gland represses endoreplication. A model has been proposed by which *escargot* maintains diploidy via transcriptional repression of regulators of the endo cell cycle. However, an alternate model might be that *escargot* plays a more direct role in controlling the cell fate of imaginal cells.

In the fission yeast, *S. pombe*, the absence of cyclin B due to a *cdc13* deletion causes cells to undergo multiple rounds of S phase, resulting in high levels of polyploidy (Hayles et al. 1994). These results have led to a model whereby high levels of cyclin B-cdc2 kinase promote entry into M phase and, conversely, low levels cause the entry into S phase. Thus, by simply disrupting the cyclin B kinase activity, cells are able to reset to a G_1 phase and enter S phase. In this example, *S. pombe* has created a cell cycle leading to polyploidy solely by altering the control of mitotic cell cycle regulators. The cell cycle of *S. pombe*, however, represents a very simplified cell cycle. In higher eukaryotes a more complex set of regulators controls the cell cycle and ensures the proper coupling of replication and mitosis. Therefore, in most organisms, the transition to the endo cell cycle most likely requires more than simply inactivating G_2/M regulators, although this might be a necessary step for the transition to polyteny. Other regulatory changes must also occur, such as the alteration of checkpoints that act to couple S phase and mitosis, as well as the block to rereplication which must be removed.

Removal of the Block to Rereplication

A major alteration in the parameters of replication that must occur during the endo cell cycle is the removal of the block to rereplication. One postulated mechanism that normally limits DNA replication to once per cell cycle is the existence of licensing factor. This factor is proposed to permit one round of DNA replication and to then become inactivated following replication (Blow and Laskey 1988). Active licensing factor is then excluded from the nucleus, and reentry into the nucleus is only permitted following nuclear envelope breakdown in mitosis (see Blow and Chong, this volume). Perhaps the actions of licensing factor or other factors involved in replication initiation are altered in the endo cell cycle, but the nature of this alteration is unclear. A mutation has been identified in the *Drosophila MCM2* gene, a proposed component of licensing factor (Treisman et al. 1995). It is interesting that this mutation inhibits proliferation of the diploid imaginal and neuronal cells but has almost no effect on the polytene cells. The polytene cells grow and undergo DNA replication, but the chromosomes may be more fragile in the mutant. Thus, licensing factor may become dispensable in the endo cell cycle. In the case of the multiple rounds of replication that occur in *S. pombe* as described above, the block to rereplication was removed solely by the disruption of cyclin B without a direct effect on licensing factor.

Late-replicating Heterochromatin/Underrepresentation

Another aspect of altered replication during the endo cell cycle is the temporal control of polytene replication. Fiber autoradiography of polytene chromosomes shows a pattern of late-replicating heterochromatic regions (Spradling and Orr-Weaver 1987). In *Drosophila virilus* the replicon size in polytene chromosomes is similar to diploid brain cells, yet the rate of fork movement is three times slower (Steinemann 1981a,b). Similar studies of *Drosophila nasuta* show that shorter and slower replicons are seen in late-replicating regions (Lakhotia and Sinha 1983). Perhaps the chromatin organization of polytene heterochromatin inhibits replication fork movement. This slower replication might account for the late replication of the heterochromatin.

A second characteristic alteration of DNA replication in the endo cell cycle is that 20–30% of the genome is underrepresented, including the centric heterochromatin and rDNA and histone repeats. Less underrepresentation of rDNA is seen, however, in the polyploid nurse cells, most likely due to the function of the nurse cells in producing the rRNA for the developing oocyte. Regions of euchromatin are generally replicated to the same extent during polyploidization. In *Drosophila* polytene chromosomes, both bands and interbands also replicate to the same extent, as confirmed by quantitative Southern blots (Spierer and Spierer 1984).

Heterochromatic underrepresentation might be caused by incomplete replication of specific sequences during polytenization, or alternatively, by elimination of sequences from the chromosome (Karpen and Spradling 1990; Glaser et al. 1992). One hypothesis is that elimination of specific regions is caused by the excision of transposable elements. Interestingly, many transposable elements are found solely in heterochromatic regions. A testable prediction of the elimination model is that novel DNA junctions should be formed upon elimination. The role of underrepresentation during the endo cell cycle is unclear; perhaps these sequences are not needed for the high level of protein expression characteristic of most polytene tissues.

Amplification

Amplification of specific genomic sequences is a developmentally regulated mechanism that allows the production of large amounts of protein products in a short developmental time frame. Amplification control occurs at the level of the block to rereplication of specific sequences, whereby reinitiation of replication leads to multiple copies of genomic sequences. Amplification provides a model system to study the developmental regulation of a eukaryotic replicon. Two well-characterized examples of developmental amplification are the *Drosophila* chorion genes and the *Sciara* DNA puffs. This is reviewed more extensively by Gerbi and Urnov (this volume).

Drosophila Chorion Genes

During oogenesis, the somatic follicle cells surrounding the egg chamber are responsible for the secretion of the chorionic eggshell layers encompassing the developing oocyte. Following polyploidization of these cells, a further tissue-specific mechanism ensures an increased copy number of the chorion genes, so that proteins can be made in a rapid developmental time window. The major chorion genes are organized into two chromosomal clusters present on the X and third chromosomes, which amplify to levels of 15-fold and 60-fold, respectively (Orr-Weaver 1991). Amplification within these clusters occurs by repeated reinitiation of an origin as shown by multiple eye forms (bubbles within bubbles) in EM spreads (Osheim et al. 1988).

Studies of both clusters have identified *cis*-acting regions responsible for amplification, termed the amplification control element (ACE). ACE3 of the third chromosome cluster has been delineated to a 320-bp region that acts in a distance- and orientation-independent manner (Orr-Weaver et al. 1989). Replication intermediates in this region have been analyzed, and a predominant replication origin has been mapped 1.5 kb downstream from ACE3. This lies in a region important for high levels of amplification known as amplification enhancing region-d, (AER-d) (Delidakis and Kafatos 1989; Heck and Spradling 1990). However, initiation events also occur throughout a 12-kb region surrounding both ACE3 and AER-d.

ACE3 is able to direct the autonomous amplification of sequences when inserted throughout the genome, albeit at lower amplification levels (Carminati et al. 1992). These studies suggest a model by which ACE3 controls the reinitiation of nearby origins, perhaps by capturing limiting replication factors and overcoming a block to rereplication.

Sciara DNA Puffs

Amplification also occurs in the fungus fly *Sciara coprophila* within puff regions of the larval salivary gland polytene chromosomes. Similar to the *Drosophila* chorion genes, amplification is temporally and tissue-regulated. Amplification as well as transcription of the *Sciara* DNA puffs is developmentally regulated by the steroid hormone, ecdysone (Bienz-Tadmor et al. 1991; Gerbi et al. 1993). Amplification presumably allows the rapid production of proteins needed during late larval development such as those for the formation of the pupal case. Similarly, in *Rhynchosciara*, puffs encode polypeptides necessary for the production of the pupal cocoon and undergo amplification (Glover et al. 1982).

Amplification occurs in several major puff regions in *Sciara*, and puff expansion is due to a burst of transcription following amplification of these sequences to approximately 20-fold levels (DiBartolomeis and Gerbi 1989; Wu et al. 1993). DNA amplification within one of the major puffs, II/9A, has been well characterized and is consistent with an onion-skin mechanism, similar to the *Drosophila* chorion genes. Puff II/9A encodes two genes that share 85% sequence similarity, and a major amplification origin has been mapped to a 1-kb region lying upstream of the two genes (Liang et al. 1993; Liang and Gerbi 1994). Replication from this origin occurs bidirectionally.

SUMMARY

DNA replication is regulated by a wide variety of mechanisms that act throughout development to coordinate S phase and replication with developmental transitions. Replication can be controlled both at the level of key S-phase regulators and at the level of parameters of DNA synthesis, such as origin usage. By studying the unique ways that different developmental events control and alter S phase, we will broaden our understanding of the regulators and mechanisms involved in DNA replication.

ACKNOWLEDGMENTS

We thank Tim Hunt, Peter Jackson, Marc Kirschner, Helena Richardson, and Rob Saint for providing unpublished information. This work was supported by National Institutes of Health grant GM-39341. J.C. was supported by National Institutes of Health predoctoral training grant CA-09541.

REFERENCES

Axton, J.M., F.L. Shamanski, L.M. Young, D.S. Henderson, J.B. Boyd, and T.L. Orr-Weaver. 1994. The inhibitor of DNA replication encoded by the *Drosophila* gene *plutonium* is a small, ankyrin repeat protein. *EMBO J.* **13:** 462–470.

Bienz-Tadmor, B., H.S. Smith, and S.A. Gerbi. 1991. The promoter of DNA puff gene II/9-1 of *Sciara coprophila* is inducible by ecdysone in late prepupal salivary glands of *Drosophila melanogaster*. *Cell Regul.* **2:** 875–878.

Blow, J.J. and R.A. Laskey. 1988. A role for the nuclear envelope in controlling DNA replication within the cell cycle. *Nature* **332:** 546–548.

Blumenthal, A.B., H.J. Kriegstein, and D.S. Hogness. 1974. The units of DNA replication in *Drosophila melanogaster* chromosomes. *Cold Spring Harbor Symp. Quant. Biol.* **38:** 205–223.

Buck, V., A. White, and J. Rosamond. 1991. CDC7 protein kinase activity is required for mitosis and meiosis in *Saccharomyces cerevisiae*. *Mol. Gen. Genet.* **227:** 452–457.

Callan, H.G. 1974. DNA replication in the chromosomes of eukaryotes. *Cold Spring Harbor Symp. Quant. Biol.* **38:** 195–203.

Carminati, J.L., C.G. Johnston, and T.L. Orr-Weaver. 1992. The *Drosophila* ACE3 chorion element autonomously induces amplification. *Mol. Cell. Biol.* **12:** 2444–2453.

Chandley, A.C. 1966. Studies on oogenesis in *Drosophila melanogaster* with 3H-thymidine label. *Exp. Cell Res.* **44:** 201–215.

Ciapa, B., D. Pesando, M. Wilding, and M. Whitaker. 1994. Cell-cycle calcium transients driven by cyclic changes in inositol trisphosphate levels. *Nature* **368:** 875–878.

Collins, I. and C.S. Newlon. 1994. Chromosomal DNA replication initiates at the same origins in meiosis and mitosis. *Mol. Cell. Biol.* **14:** 3524–3534.

Delidakis, C. and F.C. Kafatos. 1989. Amplification enhancers and replication origins in the autosomal chorion gene cluster of *Drosophila*. *EMBO J.* **8:** 891–901.

DiBartolomeis, S.M. and S.A. Gerbi. 1989. Molecular characterization of DNA puff

II/9A genes in *Sciara coprophila*. *J. Mol. Biol.* **210:** 531–543.

Dowell, S.J., P. Romanowski, and J.F. Diffley. 1994. Interaction of Dbf4, the Cdc7 protein kinase regulatory subunit, with yeast replication origins in vivo. *Science* **265:** 1243–1246.

Duronio, R.J. and P.H. O'Farrell. 1994. Developmental control of a G_1-S transcriptional program in *Drosophila*. *Development* **120:** 1503–1515.

———. 1995. Developmental control of the G1 to S transition in *Drosophila*: Cyclin E is a limiting downstream target of E2F. *Genes Dev.* **9:** 1456–1468.

Duronio, R.J., P.H. O'Farrell, J.-E. Xie, A. Brook, and N. Dyson. 1995. The transcription factor E2F is required for S phase during *Drosophila* embryogenesis. *Genes Dev.* **9:** 1445–1455.

Dynlacht, B.D., O. Flores, J.A. Lees, and E. Harlow. 1994. Differential regulation of E2F *trans*-activation by cyclin/cdk2 complexes. *Genes Dev.* **8:** 1772–1786.

Edgar, B.A. and P.H. O'Farrell. 1990. The three postblastoderm cell cycles of *Drosophila* embryogenesis are regulated in G2 by *string*. *Cell* **62:** 469–480.

Edgar, B.A. and G. Schubiger. 1986. Parameters controlling transcriptional activation during early *Drosophila* development. *Cell* **44:** 871–877.

Epstein, C.B. and F.R. Cross. 1992. CLB5: A novel B cyclin from budding yeast with a role in S phase. *Genes Dev.* **6:** 1695–1706.

Foe, V.E., G.M. Odell, and B.A. Edgar. 1993. Mitosis and morphogenesis in the *Drosophila* embryo: Point and counterpoint. In *The development of* Drosophila melanogaster (ed. M. Bate and A. Martinez Arias), pp. 149–300. Cold Spring Harbor Laboratory Press. Cold Spring Harbor, New York.

Freeman, M. and D.M. Glover. 1987. The *gnu* mutation of *Drosophila* causes inappropriate DNA synthesis in unfertilized and fertilized eggs. *Genes Dev.* **1:** 924–930.

Freeman, M., C. Nüsslein-Volhard, and D.M. Glover. 1986. The dissociation of nuclear and centrosomal division in *gnu*, a mutation causing giant nuclei in *Drosophila*. *Cell* **46:** 457–468.

Furuno, N., M. Nishizawa, K. Okazaki, H. Tanaka, J. Iwashita, N. Nakajo, Y. Ogawa, and N. Sagata. 1994. Suppression of DNA replication via Mos function during meiotic divisions in *Xenopus* oocytes. *EMBO J.* **13:** 2399–2410.

Fuse, N., S. Hirose, and S. Hayashi. 1994. Diploidy of *Drosophila* imaginal cells is maintained by a transcriptional repressor encoded by *escargot*. *Genes Dev.* **8:** 2270–2281.

Gerbi, S.A., C. Liang, N. Wu, S.M. DiBartolomeis, B. Bienz-Tadmor, H.S. Smith, and F.D. Urnov. 1993. DNA amplification in DNA puff II/9A of *Sciara coprophila*. *Cold Spring Harbor Symp. Quant. Biol.* **58:** 487–493.

Glaser, R.L., G.H. Karpen, and A.C. Spradling. 1992. Replication forks are not found in a *Drosophila* minichromosome demonstrating a gradient of polytenization. *Chromosoma* **102:** 15–29.

Glover, D.M., A. Zaha, A.J. Stocker, R.V. Santelli, M.T. Pueyo, S.M. De Toledo, and F.J. Lara. 1982. Gene amplification in *Rhynchosciara* salivary gland chromosomes. *Proc. Natl. Acad. Sci.* **79:** 2947–2951.

Grell, R.F. 1973. Recombination and DNA replication in the *Drosophila melanogaster* oocyte. *Genetics* **73:** 87–108.

Harland, R.M. and R.A. Laskey. 1980. Regulated replication of DNA microinjected into eggs of *Xenopus laevis*. *Cell* **21:** 761–771.

Hayashi, S., S. Hirose, T. Metcalfe, and A.D. Shirras. 1993. Control of imaginal cell de-

velopment by the *escargot* gene of *Drosophila*. *Development* **118:** 105–115.
Hayles, J., D. Fisher, A. Woollard, and P. Nurse. 1994. Temporal order of S phase and mitosis in fission yeast is determined by the state of the p34cdc2-mitotic B cyclin complex. *Cell* **78:** 813–822.
Heck, M.M. and A.C. Spradling. 1990. Multiple replication origins are used during *Drosophila* chorion gene amplification. *J. Cell Biol.* **110:** 903–914.
Henery, C.C., M. Miranda, M. Wiekowski, I. Wilmut, and M.L. DePamphilis. 1995. Repression of gene expression at the beginning of mouse development. *Dev. Biol.* **169:** 448–460.
Hollingsworth, R., Jr. and R.A. Sclafani. 1993. Yeast pre-meiotic DNA replication utilizes mitotic origin ARS1 independently of *CDC7* function. *Chromosoma* **102:** 415–420.
Hyrien, O. and M. Méchali. 1992. Plasmid replication in *Xenopus* eggs and egg extracts: A 2D gel electrophoretic analysis. *Nucleic Acids Res.* **20:** 1463–1469.
———. 1993. Chromosomal replication initiates and terminates at random sequences but at regular intervals in the ribosomal DNA of *Xenopus* early embryos. *EMBO J.* **12:** 4511–4520.
Iino, Y., Y. Hiramine, and M. Yamamoto. 1995. The role of *cdc2* and other genes in meiosis in *Schizosaccharomyces pombe*. *Genetics* **140:** 1235–1245.
Jackson, A.L., P.M. Pahl, K. Harrison, J. Rosamond, and R.A. Sclafani. 1993. Cell cycle regulation of the yeast Cdc7 protein kinase by association with the Dbf4 protein. *Mol. Cell. Biol.* **13:** 2899–2908.
Jackson, P.K., S. Chevalier, M. Philippe, and M.W. Kirschner. 1995. Early events in DNA replication require cyclin E and are blocked by $p21^{CIP1}$. *J. Cell Biol.* **130:** 755–769.
Jagiello, G., W.K. Sung, and J. Van't Hof. 1983. Fiber DNA studies of premeiotic mouse spermatogenesis. *Exp. Cell Res.* **146:** 281–287.
Johnston, L.H., D.H. Williamson, A.L. Johnson, and D.J. Fennell. 1982. On the mechanism of premeiotic DNA synthesis in the yeast *Saccharomyces cerevisiae*. *Exp. Cell Res.* **141:** 53–62.
Karpen, G.H. and A.C. Spradling. 1990. Reduced DNA polytenization of a minichromosome region undergoing position-effect variegation in *Drosophila*. *Cell* **63:** 97–107.
Kimelman, D., M. Kirschner, and T. Scherson. 1987. The events of the midblastula transition in *Xenopus* are regulated by changes in the cell cycle. *Cell* **48:** 399–407.
Kirschner, M., J. Newport, and J. Gerhart. 1985. The timing of early developmental events in *Xenopus*. *Trends Genet.* **1:** 41–47.
Knoblich, J.A., K. Sauer, L. Jones, H. Richardson, R. Saint, and C.F. Lehner. 1994. Cyclin E controls S phase progression and its down-regulation during *Drosophila* embryogenesis is required for the arrest of cell proliferation. *Cell* **77:** 107–120.
Krek, W., M.E. Ewen, S. Shirodkar, Z. Arany, W. Kaelin, Jr., and D.M. Livingston. 1994. Negative regulation of the growth-promoting transcription factor E2F-1 by a stably bound cyclin A-dependent protein kinase. *Cell* **78:** 161–172.
Lakhotia, S.C. and P. Sinha. 1983. Replication in *Drosophila* chromosomes. X. Two kinds of active replicons in salivary gland polytene nuclei and their relation to chromosomal replication patterns. *Chromosoma* **88:** 265–276.
Leibovici, M., G. Monod, J. Geraudie, R. Bravo, and M. Méchali. 1992. Nuclear distribution of PCNA during embryonic development in *Xenopus laevis*: A reinvestigation of

early cell cycles. *J. Cell Sci.* **102:** 63–69.

Liang, C. and S.A. Gerbi. 1994. Analysis of an origin of DNA amplification in *Sciara coprophila* by a novel three-dimensional gel method. *Mol. Cell. Biol.* **14:** 1520–1529.

Liang, C., J.D. Spitzer, H.S. Smith, and S.A. Gerbi. 1993. Replication initiates at a confined region during DNA amplification in *Sciara* DNA puff II/9A. *Genes Dev.* **7:** 1072–1084.

Lorca, T., F.H. Cruzalegui, D. Fesquet, J.C. Cavadore, J. Mery, A. Means, and M. Doree. 1993. Calmodulin-dependent protein kinase II mediates inactivation of MPF and CSF upon fertilization of *Xenopus* eggs. *Nature* **366:** 270–273.

Mahbubani, H.M., T. Paull, J.K. Elder, and J.J. Blow. 1992. DNA replication initiates at multiple sites on plasmid DNA in *Xenopus* egg extracts. *Nucleic Acids Res.* **20:** 1457–1462.

Majumder, S., M. Miranda, and M.L. De Pamphilis. 1993. Analysis of gene expression in mouse preimplantation embryos demonstrates that the primary role of enhancers is to relieve repression of promoters. *EMBO J.* **12:** 1131–1140.

Martinez-Salas, E., D.Y. Cupo, and M.L. DePamphilis. 1988. The need for enhancers is acquired upon formation of a diploid nucleus during early mouse development. *Genes Dev.* **2:** 1115–1126.

Martinez-Salas, E., E. Linney, J. Hassell, and M.L. DePamphilis. 1989. The need for enhancers in gene expression first appears during mouse development with formation of the zygotic nucleus. *Genes Dev.* **3:** 1493–1506.

McKnight, S.L. and O.L.J. Miller. 1977. Electron microscopic analysis of chromatin replication in the cellular blastoderm *Drosophila melanogaster* embryo. *Cell* **12:** 795–804.

Méchali, M. and S. Kearsey. 1984. Lack of specific sequence requirement for DNA replication in *Xenopus* eggs compared with high sequence specificity in yeast. *Cell* **38:** 55–64.

Melin, F., M. Miranda, N. Montreau, M.L. De Pamphilis, and D. Blangy. 1993. Transcription enhancer factor-1 (TEF-1) DNA binding sites can specifically enhance gene expression at the beginning of mouse development. *EMBO J.* **12:** 4657–4666.

Minshull, J. 1993. Cyclin synthesis: Who needs it? *BioEssays* **15:** 149–155.

Newlon, C.S. 1988. Yeast chromosome replication and segregation. *Microbiol. Rev.* **52:** 568–601.

Newport, J. and M. Kirschner. 1982a. A major developmental transition in early *Xenopus* embryos. I. Characterization and timing of cellular changes at the midblastula stage. *Cell* **30:** 675–686.

———. 1982b. A major developmental transition in early *Xenopus* embryos. II. Control of the onset of transcription. *Cell* **30:** 687–696.

Newport, J., T. Spann, J. Kanki, and D. Forbes. 1985. The role of mitotic factors in regulating the timing of the midblastula transition in *Xenopus*. *Cold Spring Harbor Symp. Quant. Biol.* **50:** 651–656.

Ohsumi, K., W. Sawada, and T. Kishimoto. 1994. Meiosis-specific cell cycle regulation in maturing *Xenopus* oocytes. *J. Cell Sci.* **107:** 3005–3013.

Orr-Weaver, T.L. 1991. *Drosophila* chorion genes: Cracking the eggshell's secrets. *BioEssays* **13:** 97–105.

Orr-Weaver, T.L., C.G. Johnston, and A.C. Spradling. 1989. The role of ACE3 in *Drosophila* chorion gene amplification. *EMBO J.* **8:** 4153–4162.

Osheim, Y.N., O. Miller, Jr., and A.L. Beyer. 1988. Visualization of *Drosophila melanogaster* chorion genes undergoing amplification. *Mol. Cell. Biol.* **8:** 2811–2821.

Picard, A., E. Karsenti, M.C. Dabauvalle, and M. Doree. 1987. Release of mature starfish oocytes from interphase arrest by microinjection of human centrosomes. *Nature* **327:** 170–172.

Piggott, J.R., R. Rai, and B.L.A. Carter. 1982. A bifunctional gene product involved in two phases of the yeast cell cycle. *Nature* **298:** 391–393.

Prioleau, M.N., J. Huet, A. Sentenac, and M. Méchali. 1994. Competition between chromatin and transcription complex assembly regulates gene expression during early development. *Cell* **77:** 439–449.

Reed, S.I. and C. Wittenberg. 1990. Mitotic role for the *Cdc28* protein kinase of *Saccharomyces cerevisiae*. *Proc. Natl. Acad. Sci.* **87:** 5697–5701.

Richardson, H.E., L.V. O'Keefe, S.I. Reed, and R. Saint. 1993. A *Drosophila* G1-specific cyclin E homolog exhibits different modes of expression during embryogenesis. *Development* **119:** 673–690.

Richardson, H.E., C. Wittenberg, F. Cross, and S.I. Reed. 1989. An essential G_1 function for cyclin-like proteins in yeast. *Cell* **59:** 1127–1133.

Sagata, N., M. Oskarsson, T. Copeland, J. Brumbaugh, and G.F. Vande Woude. 1988. Function of *c-mos* proto-oncogene product in meiotic maturation in *Xenopus* oocytes. *Nature* **335:** 519–525.

Sauer, K., J.A. Knoblich, H. Richardson, and C.F. Lehner. 1995. Distinct modes of cyclin E/cdc2c kinase regulation and S-phase control in mitotic and endoreduplication cycles of *Drosophila* embryogenesis. *Genes Dev.* **9:** 1327–1339.

Schild, D. and B. Byers. 1978. Meiotic effects of DNA-defective cell division cycle mutations of *Saccharomyces cerevisiae*. *Chromosoma* **70:** 109–130.

Schwob, E. and K. Nasmyth. 1993. *CLB5* and *CLB6*, a new pair of B cyclins involved in DNA replication in *Saccharomyces cerevisiae*. *Genes Dev.* **7:** 1160–1175.

Serrano, M., G.J. Hannon, and D. Beach. 1993. A new regulatory motif in cell-cycle control causing specific inhibition of cyclin D/CDK4. *Nature* **366:** 704–707.

Shamanski, F.L. and T.L. Orr-Weaver. 1991. The *Drosophila plutonium* and *pan gu* genes regulate entry into S phase at fertilization. *Cell* **66:** 1289–1300.

Sherr, C.J. 1993. Mammalian G1 cyclins. *Cell* **73:** 1059–1065.

———. 1994. G1 phase progression: Cycling on cue. *Cell* **79:** 551–555.

Sherr, C.J. and J.M. Roberts. 1995. Inhibitors of mammalian G1 cyclin-dependent kinases. *Genes Dev.* **9:** 1149–1163.

Shinomiya, T. and S. Ina. 1991. Analysis of chromosomal replicons in early embryos of *Drosophila melanogaster* by two-dimensional gel electrophoresis. *Nucleic Acids Res.* **19:** 3935–3941.

Shuster, E.O. and B. Byers. 1989. Pachytene arrest and other meiotic effects of the start mutations in *Saccharomyces cerevisiae*. *Genetics* **123:** 29–43.

Simchen, G. and J. Hirschberg. 1977. Effects of the mitotic cell-cycle mutation *cdc4* on yeast meiosis. *Genetics* **86:** 57–72.

Smith, A.V. and T.L. Orr-Weaver. 1991. The regulation of the cell cycle during *Drosophila* embryogenesis: The transition to polyteny. *Development* **112:** 997–1008.

Smith, A.V., J.A. King, and T.L. Orr-Weaver. 1993. Identification of genomic regions required for DNA replication during *Drosophila* embryogenesis. *Genetics* **135:** 817–829.

Spierer, A. and P. Spierer. 1984. Similar level of polyteny in bands and interbands of *Drosophila* giant chromosomes. *Nature* **307:** 176–178.

Spradling, A. and T. Orr-Weaver. 1987. Regulation of DNA replication during

Drosophila development. *Annu. Rev. Genet.* **21:** 373–403.

Steinemann, M. 1981a. Chromosomal replication in *Drosophila virilis*. II. Organization of active origins in diploid brain cells. *Chromosoma* **82:** 267–288.

———. 1981b. Chromosomal replication in *Drosophila virilis*. III. Organization of active origins in the highly polytene salivary gland cells. *Chromosoma* **82:** 289–307.

Stern, B., G. Ried, N.J. Clegg, T.A. Grigliatti, and C.F. Lehner. 1993. Genetic analysis of the *Drosophila cdc2* homolog. *Development* **117:** 219–232.

Stricker, S.A. 1995. Time-lapse confocal imaging of calcium dynamics in starfish embryos. *Dev. Biol.* **170:** 496–518.

Treisman, J.E., P.J. Follette, P.H. O'Farrell, and G.M. Rubin. 1995. Cell proliferation and DNA replication defects in a *Drosophila MCM2* mutant. *Genes Dev.* **9:** 1709–1715.

Whitaker, M. and R. Patel. 1990. Calcium and cell cycle control. *Development* **108:** 525–542.

Whitaker, M. and K. Swann. 1993. Lighting the fuse at fertilization. *Development* **117:** 1–12.

Williamson, D.H., L.H. Johnston, D.J. Fennell, and G. Simchen. 1983. The timing of the S phase and other nuclear events in yeast meiosis. *Exp. Cell Res.* **145:** 209–217.

Wirak, D.O., L.E. Chalifour, P.M. Wassarman, W.J. Muller, J.A. Hassell, and M.L. DePamphilis. 1985. Sequence-dependent DNA replication in preimplantation mouse embryos. *Mol. Cell. Biol.* **5:** 2924–2935.

Wu, N., C. Liang, S.M. DiBartolomeis, H.S. Smith, and S.A. Gerbi. 1993. Developmental progression of DNA puffs in *Sciara coprophila*: Amplification and transcription. *Dev. Biol.* **160:** 73–84.

Zhang, H. and J.V. Ruderman. 1993. Differential replication capacities of G1 and S-phase extracts from sea urchin eggs. *J. Cell Sci.* **104:** 565–572.

14

Comparison of DNA Replication in Cells from Prokarya and Eukarya

Bruce Stillman
Cold Spring Harbor Laboratory
Cold Spring Harbor, New York 11724

It is now supposed that utilization of DNA as genetic material emerged after an RNA world existed (Gesteland and Atkins 1993) and that all contemporary organisms evolved from a common ancestor related to primitive photosynthetic bacteria (Woese and Pace 1993). It is therefore not surprising that cells in the three major kingdoms, Bacteria, Archaea, and Eukarya, use roughly similar strategies and mechanisms for genome duplication. Nevertheless, the diversity of DNA replication is evident when the varied strategies used for replication of bacteriophage, plasmid, and virus genomes in both prokaryotes and eukaryotes are considered. This diversity becomes obvious when scanning the chapters that summarize the replication of virus genomes later in this book (Hassell and Brinton; Stenlund; Hay; Challberg; Yates; Traktman; Cotmore and Tattersall; Seeger and Mason; Bisaro; Ahrens et al.). I do not attempt to review these varied mechanisms here but rather focus on comparing the replication of cellular DNA in bacteria and eukaryotes. Unfortunately, except for some studies on DNA polymerases, little is known about the process of DNA replication in Archaea, and therefore inclusion of this kingdom adds little to the discussion.

In organisms from bacteria to the multicellular eukaryotes, DNA replication is intimately coupled to the physiology of the individual cell and to global growth controls imposed on a population of cells, whether in a colony of bacteria or in a subset of cells within a developing organism. These regulatory pathways are different in bacteria and eukaryotes and also vary between different cell types within a single organism. Nevertheless, the actual process of DNA replication has striking parallels between the two groups of organisms. A universal challenge for the replication machinery is that the process must be precise and tightly controlled to ensure faithful transmission of the genetic heritage to the next generation of cells. If this goal is not achieved, disastrous consequences await the organism or its descendants.

In general, the replication of chromosomes in both groups can be separated into multiple stages. First is the recognition of specific DNA sequences within the genome by initiator proteins. This interaction is the primary determinant of the location of the origin of DNA replication. At this location, a pre-replication complex of proteins is established that renders the genome competent for replication. The two important enzyme activities that must be attracted to the origin are a helicase that will eventually unwind the template DNA and a primase that will provide the necessary primer for initiation of DNA synthesis by DNA polymerases. Activation of the pre-replicative complex is a key step that eventually leads to the unwinding of the DNA and DNA synthesis. The establishment of a multiprotein complex called a replisome at the replication fork involves the assembly of the DNA polymerases and accessory proteins, followed by fork movement. Finally, termination of DNA replication occurs prior to separation of the daughter chromosomes. It is in each of these areas that the processes in bacteria and eukaryotes will be compared. This discussion relies heavily on the reviews presented in this volume on the replication mechanisms in eukaryotes for specific examples, and much of the primary literature is cited in these other chapters.

CHROMOSOME STRUCTURE AND REPLICATION

Perhaps the most obvious difference between the two classes of cells is how the genetic material is organized into chromosomes. Bacteria have a circular chromosome that contains one principal origin of DNA replication located at the *OriC* locus in *Escherichia coli*. The rate at which the DNA polymerases move along the *E. coli* chromosome is approximately 100 kb per minute, and thus the entire chromosome can be replicated from a single origin in about 40 minutes (Kornberg and Baker 1992). The genomes of eukaryotes are much larger than the bacterial chromosomes, and the rate of replication fork movement averages about 2 kb per minute, a rate that is prohibitive for even the smallest genome to replicate from a single origin (Fangman and Brewer 1992). Moreover, DNA in eukaryotes is invariably sequestered into multiple chromosomes, which demands more than one origin per genome. In certain artificial chromosomes in *Saccharomyces cerevisiae*, one origin is sufficient to replicate a chromosome of about 100 kb in size, but if the chromosome is larger than this limit, it becomes very unstable with only one origin (Wellinger and Zakian 1989; Dershowitz and Newlon 1993; Newlon, this volume).

It has been known for more than three decades that different regions of individual eukaryotic chromosomes replicate at distinct times during S phase, demonstrating that multiple origins of DNA replication must exist (Hand 1978; Fangman and Brewer 1992; DePamphilis, this volume). Unlike bacteria, eukaryotes initiate chromosome replication in a temporally regulated manner that can change during development (Simon and Cedar, this volume). The existence of multiple origins affords considerable flexibility in the timing of replication.

In addition to the relative timing of replication of different regions of the genome, the total time taken to replicate the entire genome can vary considerably during development of an organism (Blumenthal et al. 1974; Callan 1974). This appears to be due to a change in the frequency of active origins of DNA replication distributed along the chromosome, rather than a change in the rate of replication fork movement. For example, in the rapid replication cycles of nuclei in the early *Drosophila* embryo, origins of DNA replication are located on average 3–7 kb apart, but later during embryogenesis, the frequency of replication origins changes to about 40–100 kb as cells begin to differentiate (Blumenthal et al. 1974). Interestingly, if mammalian cells are arrested for a long period of time at the G_1 to S phase, the frequency along the genome of active origins of DNA replication increases following removal of the block, and the total time taken to pass through S phase is much shorter than the time taken in the absence of the inhibitor (Taylor 1977). This suggests that a factor may accumulate in the nucleus of cells that determines the frequency of active origins of DNA replication. A good candidate protein is the mammalian homolog of the $p65^{cdc18}$ protein from *Schizosaccharomyces pombe* and the related protein Cdc6p from *S. cerevisiae* (Kelly et al. 1993; Liang et al. 1995).

A major difference between the two groups of cells emerges from these observations. The existence of a single origin in the bacterial chromosome fixes the relative order of gene duplication, except perhaps for occasional selective gene amplification. In contrast, eukaryotes have enormous flexibility in the relative time during S phase when specific genes replicate. This permits the replication of specific regions of the genome to be linked to gene expression, a topic discussed in an accompanying chapter (Simon and Cedar, this volume). This flexibility may be an important component of developmental patterning in cells of metazoan origin. Even though the number of active origins in a single bacterial chromosome does not vary, as described below in the section on control of DNA replication, the frequency of active origins within a cell can vary considerably in bacteria due to reinitiation in a single cell-division cycle.

In bacteria that have distinct cell types, developmentally imposed control of the initiation of DNA replication can occur. In *Caulobacter crescentus,* cell division can produce two morphologically different cell types that have different replicative fates (Marczynski et al. 1995). One cell type, called the stalked cell, replicates the entire chromosome after cell division, whereas the other cell type, the swarmer cell, only replicates its DNA after a delay. In the stalked cell, a single chromosomal origin of DNA replication is utilized. The replicator element, *Cori,* has recently been characterized, and in addition to the initiator protein-binding sites that are analogous to the DnaA initiator protein-binding sites, within the *E. coli OriC* replicator, *Cori* contains essential transcriptional promoter elements that are only active in the stalked cell (Marczynski et al. 1995). Thus, cell-type-specific transcription can influence replication.

It has been known for some years that transcription can facilitate initiation of DNA replication from the bacteriophage λ origin as well as the *E. coli OriC* (Kornberg and Baker 1992). The *C. crescentus* case is an example in bacteria where transcription may effect cell-type-specific initiation of DNA replication. Transcription factors in eukaryotic cells can also be intimately involved in the initiation of DNA replication and influence origin firing in a cell-type- and temporal-specific manner (van der Vliet, this volume; for review, see DePamphilis 1993). A recent example in mammalian cells is the locus controlling region (LCR) that regulates the developmentally controlled expression of the human β-globin gene. The LCR is required for the initiation of DNA replication from an origin located near the β-globin gene (Aladjem et al. 1995). These selected examples demonstrate that despite the quite different organization of the chromosomes, and the number of replication origins, bacteria and eukaryotes can utilize similar strategies to control whether or not initiation of DNA replication will occur.

Consideration of the influence of chromosome structure on DNA replication in bacteria and eukaryotes must also take into account the different organization of DNA in the cell. The bacterial chromosome is associated with the cell membrane but otherwise is exposed to the entire intracellular environment. In contrast, eukaryotic DNA replication occurs in a distinct compartment in the cell, affording the separation of proteins that may influence the initiation of DNA replication. Within the nucleus, initiation of eukaryotic DNA replication occurs at pre-replicative complexes that are established prior to the beginning of S phase, and these presumably contain the initiator and other replication proteins (Adachi and Laemmli 1994; Diffley et al. 1994). A similar initiation complex may exist at the membrane-bound bacterial replicator, since DnaA protein from *E. coli* is a lipid-binding protein and is associated with the membrane (Sekimizu and Kornberg 1988; Sekimizu et al. 1988a,b). Thus, in both cell types, initiation may actually occur on a solid-state support, albeit that the supports and environment may be quite different.

REPLICATORS

The mechanism of initiation was first proposed for replication of the bacterial chromosome in the early 1960s (Jacob et al. 1964). In this classic paper, a *cis*-acting DNA element called the replicator was proposed to be a primary determinant of the location in the genome where initiation of DNA replication was to occur. Furthermore, it was posited that initiation of replication of the bacterial genome also required a *trans*-acting factor called the initiator that interacted with the replicator. The replicator can be defined by genetic means, and, indeed, the *E. coli* replicator *OriC* was characterized by the isolation of fragments from the *E. coli* genome that could support autonomous replication of recombinant plasmids (Hirota et al. 1979; Messer et al. 1979; von Meyenburg et al. 1979). These plasmids behaved in a manner similar to the complete chromosome.

Fine-structure analysis by mutagenesis and evolutionary comparison of replicators from five enteric bacteria revealed that the bacterial replicator is highly conserved between species and is composed of multiple essential DNA sequence elements (Fig. 1A). The *E. coli* replicator has four recognition sites for the initiator protein DnaA that are flanked on one side by an A•T-rich sequence and on the other by a series of repeats that can be easily unwound in negatively supercoiled DNA (Kornberg and Baker 1992). These three repeats, or 13-mers, are locally unwound by DnaA once it is bound to *OriC* (Bramhill and Kornberg 1988). This general structure for bacterial replicators is also found in the origins of DNA replication from the lambdoid phages. Interspersed within the *E. coli* replicator are ten recognition sites for the DNA adenine methyltransferase (dam methylase), and methylation of these sites plays an important role in the control of initiation (see below).

The replicators present in eukaryotic viruses have been characterized by mutational analysis. One of the best understood of these is the simian virus 40 (SV40) replicator called the SV40 *ori* (Stillman 1989; Hassell and Brinton, this volume). Like the bacterial replicators, the SV40 *ori* contains multiple binding sites for the SV40 initiator protein, T antigen, and these binding sites are flanked by an A•T-rich sequence on one side and a partial palindromic repeated sequence (Fig. 1B) (Parsons et al.

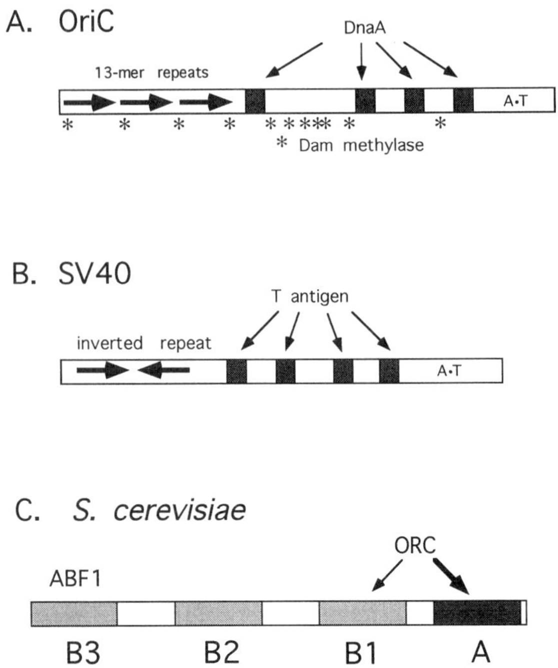

Figure 1 Comparison of the replicators from *E. coli*, SV40, and *S. cerevisiae.* (*A*) *E. coli OriC*. (*B*) The SV40 *ori*. (*C*) *S. cerevisiae ARS1*.

1990). The latter repeat also interacts with T antigen, resulting in the unwinding of 8 bp of DNA. Thus, the SV40 *ori* is remarkably similar in organization to the *E. coli* replicator.

Understanding the nature of replicators in eukaryotic cell chromosomes has been more problematic, particularly in multicellular eukaryotes. Only in the yeasts have replicators been isolated by selection of fragments that permit autonomous replication of plasmids (Campbell and Newlon 1991; DePamphilis; Newlon; both this volume). Attempts to isolate such autonomous replicating sequences (ARSs) in other eukaryotes have not been successful, and thus the replicators from metazoan cells remain poorly defined. In *S. cerevisiae,* a number of replicators have been characterized extensively, and the general picture that has emerged is that the replicators have little similarity to the bacterial and viral replicators (Fig. 1C). The best-characterized replicator is ARS1 from chromosome IV, and it consists of four DNA sequence elements, only one of which (A) is essential (Marahrens and Stillman 1992). The initiator protein for eukaryotes is the origin recognition complex (ORC), and this six-subunit protein binds to the A and B1 elements in an

ATP-dependent manner (Bell and Stillman 1992; Rao and Stillman 1995; Rowley et al. 1995). Another element, which is not essential and is only present in some replicators in the *S. cerevisiae* chromosomes, is a binding site for the transcription factor ABF1 (Marahrens and Stillman 1992). Interestingly, other transcription factors can substitute for ABF1 and activate initiation of DNA replication, similar to the stimulation of initiation of replication at virus origins by transcription factors (DePamphilis 1993; Hassell and Brinton; van der Vliet; both this volume).

In general, the yeast replicators contain DNA sequences that are easily unwound when a plasmid in which they are cloned is in a highly negatively supercoiled state (Natale et al. 1993). A similar phenomenon has been observed with the *E. coli OriC* and the SV40 *ori* (Kowalski and Eddy 1989; Lin and Kowalski 1994), and in these cases, the region that unwinds first corresponds to the region that is unwound by the binding of DnaA or SV40 T antigen, respectively. This suggests that all origins of DNA replication have a tendency to unwind, and the energy for separation of the double helix either comes from torsional strain in the DNA, binding of the initiator protein, or both. For replication of plasmid DNAs containing the phage λ or *OriC* replicators in vitro, torsional strain provided by a balance of gyrase, topoisomerase I, and the HU protein greatly facilitates origin unwinding and initiation of DNA replication (Sekimizu et al. 1988a,b; Mensa-Wilmot et al. 1989; Hwang and Kornberg 1992). Furthermore, transcription of the template DNA in the vicinity of the origin is necessary for the earliest stages of duplex strand unwinding during the initiation of DNA replication (Baker and Kornberg 1988; Learn et al. 1993). Whether transcription per se or the action of transcriptional activation domains present on sequence-specific DNA-binding proteins activates initiation at eukaryotic virus and cell origins remains to be determined.

INITIATORS

As indicated above, initiator proteins bind in a sequence-specific manner to the replicator, and together they function as the primary determinants of the location of the origin of DNA replication. The primary function of the initiator-replicator complex is as a landing pad for other replication proteins such as the helicase and the primase. Initiators can come in many different versions. All function as DNA-binding proteins, and all interact with other proteins that are required for the initiation of DNA replication. Some, such as the phage λ O protein, function only in this capacity (Dodson et al. 1985), whereas others have additional activities.

For example, the DnaA protein from *E. coli* is an ATP/ADP-binding protein with the ability to slowly hydrolyze the ATP to ADP, whereas the SV40 T antigen, the papillomavirus E1 protein, and the herpes simplex virus (HSV) UL9 initiators bind ATP, have a potent intrinsic ATPase activity, and in addition, contain DNA helicase function (Stahl et al. 1986; Sekimizu et al. 1988a,b; Bruckner et al. 1991; see Boroweic; Challberg; Hassell and Brinton; Stenlund; all this volume). One intriguing initiator from bacteriophage P4 is a sequence DNA-binding protein, an ATPase, a DNA helicase, and a DNA primase, and thus fulfills many of the functions required for initiation of DNA replication (Ziegelin et al. 1993).

A common characteristic of initiator proteins is that they often oligomerize upon binding to the DNA. The DnaA and λ O initiator proteins form complexes of 8 or about 20 proteins bound to *OriC* or *oriλ*, respectively, even though there are only four DNA recognition sequences in the replicator (Echols 1986; Kelman and O'Donnell 1994). Similarly, the SV40 T antigen has four recognition sites in the replicator, yet two hexamers of T antigen bind to the SV40 *ori* in an ATP-dependent manner (Dean et al. 1987). The oligomerization probably acts to stabilize the initiator-replicator interactions and to allow efficient interaction with other proteins. In the case of SV40 T antigen, the functional helicase unit is a hexamer. In many respects, the T-antigen hexamer resembles the RuvB ATPase that is involved in branch migration of DNA at a recombination Holliday junction (Parsons et al. 1995).

Oligomerization of an initiator protein can prevent DNA replication. The bacteriophage P1 genome can exist in a lysogenic state in its host bacteria as a plasmid, and initiation of DNA replication requires the *OriP1* replicator and the plasmid-encoded RepA initiator (Baker and Wickner 1992). The RepA protein exists as a stable dimer when not bound to the DNA. The *E. coli* chaperone proteins DnaJ, DnaK, and GrpE combine to break apart the RepA dimer into RepA monomers, thereby allowing RepA to bind to *OriP1* in cooperation with the DnaA protein to form a productive replicator-initiator complex (Baker and Wickner 1992).

In many cases, but not all, the binding of the initiator protein to the replicator causes local unwinding of a region of DNA at the origin of DNA replication. The DnaA protein induces unwinding of the 13-mer elements at *OriC*, and SV40 T antigen causes unwinding of 8 bp in the partial "early" palindrome within the SV40 *ori* (Bramhill and Kornberg 1988; Borowiec et al. 1990; Hassell and Brinton; Brush and Kelly; both this volume). Unwinding of the DNA at the origin of DNA replication is a key step in the pathway leading to initiation of DNA synthesis at the origin. This is facilitated by regions within the origin of DNA replication that have a low helical stability and thus have a tendency to unwind when placed under torsional strain (Kowalski and Eddy 1989; Natale et al. 1993; Lin and Kowalski 1994).

The assembly and disassembly of the multiprotein complex on the replicator and the ability of the complex to function as a landing pad are highly regulated processes. The function of the *E. coli* DnaA complex is regulated by membrane-bound acidic phospholipids (Crooke et al. 1992). The ATP form, but not the ADP form, of DnaA is able to unwind the 13-mers in *OriC*, and membrane-bound acidic phospholipids greatly stimulate the exchange of ADP to ATP bound to DnaA. Exchange to the active ATP•DnaA complex requires the DnaA to be bound to *OriC* because the acidic phospholipids inactivate DNA-free DnaA protein so that it cannot bind *OriC*. These results suggest a link between DnaA activity and the membrane localization of *OriC*.

Phosphorylation of proteins is not known to be a direct mechanism for control of DNA replication in bacteria, but it is a major mechanism for regulation of some eukaryotic virus initiator proteins. For example, the ability of SV40 T antigen to unwind the origin of DNA replication is positively regulated by phosphorylation of a single threonine by a cyclin-dependent protein kinase (Ismail et al. 1993; McVey et al. 1993). On the other hand, the DNA-binding activity is negatively regulated by phosphorylation of two serine residues by casein kinase I, which can be reversed by the action of phosphatase 2A (Cegielska et al. 1994a,b). In eukaryotes, phosphorylation can also control the cellular localization of proteins. For example, phosphorylation by casein kinase II of serine residues that lie adjacent to the nuclear localization signal in SV40 T antigen greatly facilitates the transport of the protein from the cytoplasm to the nucleus. It is easy to imagine that this may play a regulatory role for the initiation of DNA replication. Thus, in eukaryotes, unlike bacteria, control of the intracellular locale of proteins may affect their function. A discussion of the role of the nuclear envelope in the initiation of cellular DNA replication can be found in Laskey and Madine (this volume).

UNWINDING AT THE ORIGIN

Following DNA sequence recognition and establishment of the initiator-protein complex at the origin of DNA replication, the next important events are unwinding of the template DNA and priming DNA replication. Initiator proteins such as DnaA protein (Bramhill and Kornberg 1988) and SV40 T antigen (Borowiec et al. 1990) can locally unwind a region of the origin of DNA replication, and this region of single-stranded DNA is presumed to be the entry site for a DNA helicase and single-stranded DNA-binding proteins such as the *E. coli* DnaB helicase and SSB, respectively. A DNA helicase has to be loaded onto the initiator-DNA complex for more extensive unwinding to occur. SV40 T antigen is an unusual case because it is a DNA helicase as well as an initiator protein; thus, the helicase is automatically located in a prime site for extensive unwinding. Indeed, T antigen unwinds almost the entire SV40 genome in cooperation with the eukaryotic cell single-stranded DNA-binding protein RP-A and a topoisomerase (Borowiec et al. 1990; see Borowiec; Hassell and Brinton; both this volume). In most cases, however, the helicase is a separate polypeptide from the initiator protein and needs to be loaded onto the DNA.

Loading of the DNA helicase is a complicated affair. The DnaB helicase from *E. coli* is used to initiate DNA replication from *OriC* and is also required for replication of the bacteriophage λ genome (Baker and Wickner 1992; Stillman 1994a). In neither case can the DnaB helicase recognize the initiator protein (DnaA or λ O protein) by itself, but needs to be chaperoned by a specialized protein. The *E. coli* DnaC and the phage λ P proteins function as helicase-loading proteins because they both bind to the DnaB helicase and also bind either to the DnaA or λ O proteins, respectively (Fig. 2). These helicase-loading proteins can inhibit the DNA helicase activity, and it is necessary to remove them before the helicase can be activated to unwind the template DNA (Stillman 1994a). For initiation at the phage λ origin (*ori*λ), the *E. coli* heat shock chaperone proteins DnaJ, DnaK, and GrpE cooperate to remove λ P protein and thus activate the DnaB helicase (Baker and Wickner 1992; Stillman 1994a). Here is another example from the prokaryotic world where chaperone proteins pull apart protein complexes that are involved in the initiation of DNA replication. To date, chaperones have not been shown to play a role in disassembling protein complexes in eukaryotes, but little is known about the eukaryotic pre-replicative complex. It is likely that similar assembly and disassembly of multiprotein complexes occur at eukaryotic replicators.

The helicase-loading proteins can also facilitate the assembly of a DNA helicase onto the template DNA in the absence of an origin of DNA replication. For example, the bacteriophage T4 gene 59 protein (gp59) binds to both single-stranded DNA and the phage single-strand-binding protein, gp32 (Barry and Alberts 1994; Morrical et al. 1994).

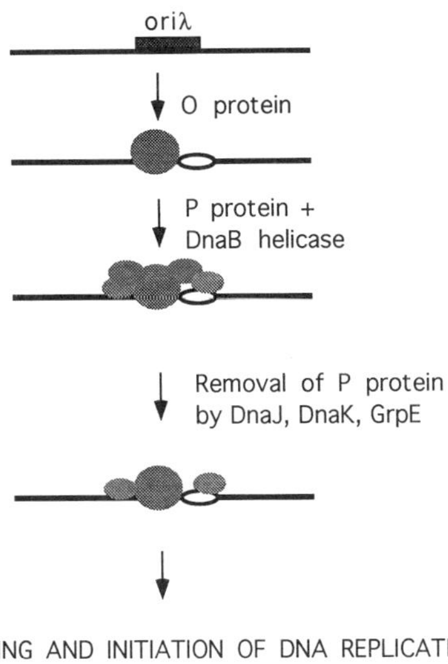

Figure 2 Loading of a DNA helicase. The bacteriophage λ P protein loads the DNA helicase onto the DNA. The λ P protein must be removed before the helicase (DnaB) can function to unwind the DNA prior to DNA replication. This is facilitated by the chaperone proteins DnaJ, DnaK, and GrpE. (Reprinted, with permission, from Stillman 1994a.)

When gp32 coats a single-stranded DNA template, the phage helicase gp41 cannot bind to the DNA. The gp59 protein loads the gp41 helicase onto the gp32-coated DNA, and the helicase can now move rapidly along the template in search of duplex DNA to unwind. This mechanism of helicase loading is almost certainly used during the elongation stages of DNA replication. In a similar manner, SV40 T antigen can bind the RP-A protein, the eukaryotic SSB, facilitating loading of a helicase onto the RP-A-coated single-stranded DNA (Collins and Kelly 1991; Melendy and Stillman 1993).

Although many DNA helicases have been identified in eukaryotes, it is not clear which are required for the initiation of DNA replication (Borowiec, this volume). A good candidate is the helicase B from a murine cell line that harbors a temperature-sensitive helicase which causes a defect in DNA replication at the nonpermissive temperature (Seki et al. 1995). Moreover, although proteins such as Cdc6p that interact with the eukaryotic initiator protein ORC are beginning to be identified (Liang et al. 1995), it is not yet clear whether the same mechanism of helicase loading occurs in eukaryotic cells as occurs in bacteria.

PRIMING DNA SYNTHESIS

Following unwinding of the template DNA surrounding the replicator, priming of DNA synthesis occurs, usually by a DNA primase. As discussed by others in this volume (Brush and Kelly; Salas et al.), priming can occur by a number of different mechanisms, including priming by a nucleotide linked to a protein such as occurs during Φ29 phage and adenovirus DNA replication. For initiation of chromosomal DNA replication in bacteria and almost certainly in eukaryotes, priming involves the assembly of a primase on the DNA. In *E. coli*, the primase exists as a free enzyme that is not always bound to the DNA polymerase (Baker and Wickner 1992; Kornberg and Baker 1992), whereas in eukaryotes, the primase enzyme is always in a complex with DNA polymerase-α (Wang, this volume).

Herein lies a significant difference between prokaryotes and eukaryotes: Bacteria contain only one principal replicative DNA polymerase (DNA polymerase III in *E. coli*) that must function on both the leading and lagging strands at a replication fork, but in eukaryotes, the DNA polymerizing activity is shared by three DNA polymerases, pol-α, pol-δ, and pol-ε (Kornberg and Baker 1992; Wang, this volume). Thus, the pol III enzyme in *E. coli* needs to be free of the primase when replicating the continuously synthesized leading strand, but must cyclically interact with the primase during lagging-strand replication (Hacker and Alberts 1994a,b; Stillman 1994b; Stukenberg et al. 1994). Eukaryotes have devised a DNA polymerase switching mechanism that involves the pol-α:primase complex in priming DNA replication at the origin of DNA replication and for every Okazaki fragment, and then a second DNA polymerase that is not associated with the primase (pol-δ or ε) completing the synthesis (Figs. 3 and 4) (Nethanel and Kaufmann 1990; Tsurimoto et al. 1990; Waga and Stillman 1994; Stillman 1994b).

A key question is how the primase finds either the origin of DNA replication or the site at which Okazaki fragments must begin. Bacterial and phage primases can exist in two different classes: those that are part of the helicase polypeptide and others that exist as proteins separate from the helicase polypeptide (Ilyina et al. 1992). If the primase activity coexists in the same polypeptide as the helicase, then the primase is associated with the replication fork by default. For bacterial chromosome replication and for the replication of some eukaryote virus genomes such as SV40 and herpesvirus, the primase-loading problem has been solved by the primase having an affinity either for the helicase that has already been loaded onto the template DNA or, alternatively, for a structure in the DNA created by the helicase (Baker and Wickner 1992; Marians

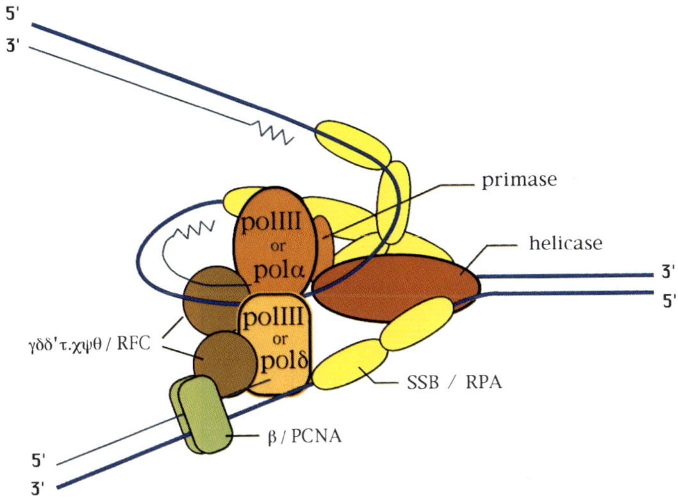

Figure 3 Replication fork proteins. Model for a dimeric DNA polymerase at the replication fork. (Reprinted, with permission, from Stillman 1994b [copyright Cell Press].)

1992; Kelman and O'Donnell 1994; Stillman 1994a; Hassell and Brinton; Challberg; both this volume). For example, the SV40 T antigen binds directly to the pol-α:primase complex and can load the primase onto an RP-A-coated template DNA by direct interaction with RP-A (Collins and Kelly 1991; Melendy and Stillman 1993; Dornreiter et al. 1993). In *E. coli,* the DnaB helicase activates the DnaG primase (Baker and Wickner 1992; Marians 1992). Once located at the priming site, the primase in both prokaryotes and eukaryotes synthesizes a short (8–12 nucleosides) RNA primer that is utilized by the DNA polymerase for synthesis of DNA.

ASSEMBLY OF THE DNA POLYMERASE HOLOENZYME

A common feature of the DNA replication fork in bacteria and eukaryotes is that DNA is synthesized in a semidiscontinuous manner, with one strand, the leading strand, polymerized in a continuous manner and the other lagging strand formed in a discontinuous manner via the synthesis of Okazaki fragments (Brush and Kelly, this volume). The DNA polymerase is loaded onto the template DNA at the site of the primer by a mechanism that is remarkably conserved between prokaryotes and eukaryotes (Table 1) (Kelman and O'Donnell 1994; Stillman 1994b). The process is accomplished by polymerase accessory

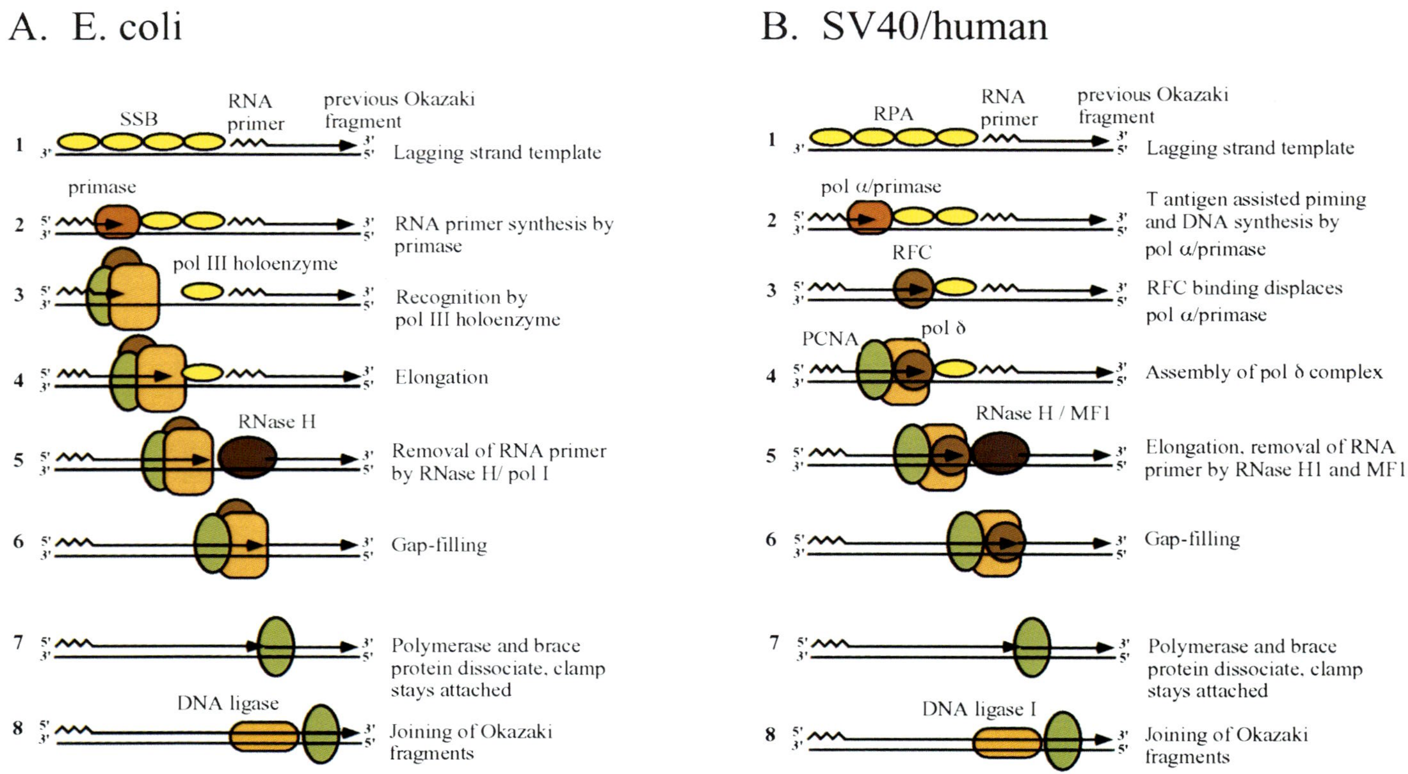

Figure 4 Replication fork proteins. Comparison between the synthesis of Okazaki fragments in bacteria and eukaryotes. (Reprinted, with permission, from Stillman 1994b [copyright Cell Press].)

Table 1 Functions of DNA replication fork proteins

E. coli	Phage T4	SV40/human	Functions
DnaB	41	T antigen	DNA helicase; stimulates priming on ssDNA
DnaC	59	T antigen	allows loading of helicase and primase onto SSB-coated DNA (primosome assembly)
SSB	32	RP-A	single-stranded DNA-binding; stimulates DNA polymerase; facilitates helicase loading
γ-Complex (γδ.δ′χψ)	44/62	RF-C	DNA-dependent ATPase; primer-template binding; stimulates DNA polymerase
τ	43	?	dimerization of holoenzyme
β	45	PCNA	stimulates DNA polymerase; stimulates DNA-dependent ATPase
pol III core (αθε)	43	pol-δ[a]	DNA polymerase 3′→5′ exonuclease
DnaG	61	primase pol-α[b]	primase
Ligase	T4 ligase	ligase I	ligation of DNA
DNA pol I	43	FEN-1/MF1	nuclease for removal of RNA primers
RNase H	RNase H	RNase H1	nuclease for removal of RNA primers

(Reprinted, with permission, from Stillman 1994b [copyright Cell Press].)

[a]DNA polymerase-δ has been shown to function in SV40 DNA replication; an essential DNA polymerase, DNA polymerase-ε, has not yet been assigned a specific function in DNA replication.

[b]The human DNA polymerase-α and primase activities function as a multiprotein complex to synthesize RNA/DNA primers. (RP-A) Replication protein A; (RF-C) replication factor C; (PCNA) proliferating cell nuclear antigen; (SSB) single-strand DNA-binding protein.

proteins. The eukaryotic proteins are discussed by Hübscher et al. (this volume). Similar mechanisms and enzymatic activities in these eukaryote proteins and the bacteriophage T4 counterparts were recognized first (Tsurimoto and Stillman 1990; Tsurimoto et al. 1990), and later, some sequence similarities were observed (O'Donnell et al. 1993).

The polymerase accessory proteins from *E. coli*, bacteriophage T4, and eukaryotes include the single-strand-binding protein (SSB, gp32, or RP-A), a polymerase clamp protein (β subunit of pol III, gp45, or PCNA), and a DNA-dependent ATPase that loads the polymerase clamp onto the DNA (γδδ′τχψ, gp44/62, and RF-C). Only in the case of the clamp-loading proteins have significant DNA sequence similarities been observed (O'Donnell et al. 1993), but there is remarkable structural similarity between the polymerase clamp proteins even though there is no primary amino acid sequence similarity (Fig. 5) (Krishna et al. 1994). The β subunit of the pol III enzyme from *E. coli* and the *S. cerevisiae* PCNA have three-dimensional structures that can be almost superimposed one on top of the other. Both form a donut or torus-like shape, and the DNA passes through the hole in the middle of the structure. The β subunit contains 366 amino acids, whereas the PCNA contains only 258 amino acids, approximately two-thirds the size of the bacterial counterpart. Two monomers of the β subunit form the torus structure, whereas three monomers of the smaller PCNA combine to form an identical structure. Each torus ring has an internal sixfold symmetry. The 228-amino acid phage T4 gp45 protein is presumed to also form a trimer.

Interestingly, the carrot plant has two highly related PCNA proteins that are similar in sequence to PCNA from human and yeast; one is 264 amino acids, similar to the "eukaryotic short" form and the other is 365 amino acids, similar to the "bacterial long" form (Kelman and O'Donnell 1995). It has been suggested that these plants have one dimer PCNA and one trimer PCNA that may function at different times during development when DNA replication rates vary. Alternatively, since PCNA also functions in DNA repair (Nichols and Sancar 1992; Shivji et al. 1992), it is possible that one PCNA is involved in DNA repair and the other in DNA replication.

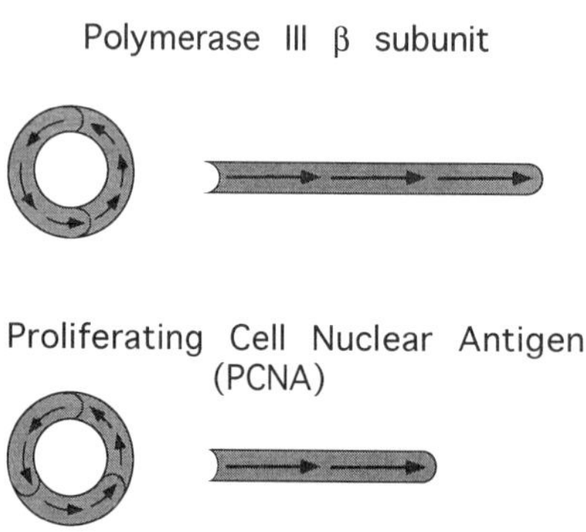

Figure 5 Comparison of the structures of the clamp proteins from bacteria and yeast. Both proteins form a torus or donut-shaped molecule with sixfold symmetry. The *E. coli* pol III β subunit forms a dimer and PCNA forms a trimer. The arrows represent a repeated structural motif in the proteins. The DNA passes through the center of the structure.

The mechanism of polymerase loading is very similar in bacteria and eukaryotes and is discussed in detail elsewhere in this volume (Hübscher et al.) and in other reviews (Kelman and O'Donnell 1994; Stillman 1994b). In both systems, an ATP-dependent clamp-loading protein recognizes a primer-template junction and assembles the polymerase clamp onto the DNA. The single-stranded DNA-binding protein that coats the single-stranded template DNA facilitates the recognition of the primer-template junction by preventing the clamp-loading protein from binding to single-stranded DNA. Once the clamp complex is assembled, a DNA polymerase is assembled onto the primer ready for DNA synthesis.

TOPOLOGY AND DNA REPLICATION

Topoisomerases in both prokaryotes and eukaryotes deal with the topological problems associated with unwinding the double helix during DNA replication and with separating the daughter chromosomes after DNA replication is completed. The different types of topoisomerase have been reviewed recently (Roca 1995), and a discussion of the eukaryotic topoisomerases appears in Hangaard Andersen et al. (this volume). The role of topoisomerases from bacteria and eukaryotes in DNA replication has been well discussed in a recent review (Ullsperger et al. 1995), and only some general points are noted here. Bacterial chromosome topology is controlled by a balance between topoisomerases that relieve torsional strain on the one hand and gyrases that induce negative supercoiling on the other. There is net negative supercoiling in the bacterial chromosome, and these negative supercoils are available to facilitate unwinding of the double helix for processes such as transcription and the initiation of DNA replication. In contrast, eukaryotic chromosomes have more complex topology. Eukaryotic chromosomal DNA has the same supercoil density as found in bacteria, but all the negative supercoils are believed to be constrained in nucleosomes (Ullsperger et al. 1995). Furthermore, the eukaryotic chromosome is folded into complex higher-order structures that the DNA replication apparatus must deal with. Thus, the DNA replication machineries in the two cell types must deal with the very different templates, although as indicated above, the basic replication proteins are quite similar.

Unwinding of the DNA helix during replication fork progression generates positive supercoils ahead of the replication fork in the unreplicated DNA. There are different mechanisms to remove these positive supercoils (Ullsperger et al. 1995). The first is to actively generate negative

Table 2 Topoisomerase activities

	Prokaryotes				Eukaryotes		
	topo I	topo II	topo III	topo IV	topo I	topo II	topo III
Name	omega	gyrase	–	–	–	–	–
Type	1	2	1	2	1	2	1
Genes	*topA*	*gyrA, gyrB*	*topB*	*parC, parE*	*top1*	*top2*	*top3*
(+) Supercoil removal in vivo	no	yes	no	yes?	yes	yes	no
Decatenation	no	yes	no	yes	no	yes	no
Relaxes only (–) supercoils	yes	no	yes	no	no	no	yes

(Adapted from Ullsperger et al. 1995 [copyright Springer-Verlag GmbH].)

supercoils in the unreplicated DNA before the positive supercoils accumulate, a role for the bacterial type II topoisomerase commonly referred to as DNA gyrase (Table 2). The bacterial gyrase can keep up with the generation of positive supercoils ahead of the replication fork by introducing negative supercoils. A second mechanism is to remove directly the positive supercoils in the unreplicated DNA; both the eukaryotic type I and type II topoisomerases can perform this function (Table 2). This mechanism is thought to occur in the early stages of replication from an origin when there is a relatively large amount of unreplicated DNA. In the later stages of DNA replication, it is more difficult to introduce positive supercoils into the unreplicated DNA, and the replicated DNA tends to become interwound or catenated (Ullsperger et al. 1995). Thus, a third mechanism for relieving the positive supercoils generated by replication fork progression is to remove catenated DNA, a job for different types of topoisomerases in bacteria and eukaryotes. The bacterial type I topoisomerase, topo III, binds to single-stranded DNA and thus can operate behind the replication fork to decatenate the replicated DNA. In contrast, the type II topoisomerases, topo II in eukaryotes and type IV in bacteria, can decatenate the replicated duplex DNA and thus segregate completely replicated daughter chromosomes (Ullsperger et al. 1995).

Finally, as pointed out by Ullsperger et al. (1995), if the replication fork contains a dimeric DNA polymerase with physically linked polymerase molecules of the type shown in Figure 3, this creates even more problems with DNA topology because the DNA polymerases are not free to rotate about the template DNA. If the DNA polymerases are connected to each other, as seems likely in bacteria and eukaryotes, then enzymatic mechanisms must exist that deal with the passage of the double helix through the polymerase complex.

FIDELITY AND DNA REPAIR

Both bacteria and eukaryotic cells have mechanisms for repairing DNA (Friedberg et al. 1995), and it is beyond the scope of this discussion to review these varied mechanisms here (for eukaryotes, see Roberts and Kunkel; Friedberg and Wood; both this volume). As discussed by Nasmyth (this volume) and others elsewhere (Li and Deshaies 1993; Murray 1994; Nurse 1994), eukaryotic cells have a monitoring system that links DNA replication with cell-cycle progression, cell growth, and DNA repair. The existence of DNA damage-checkpoint mechanisms in bacteria that temporally delay DNA replication and cell-cycle progression in response to DNA damage is not as clear as has become apparent in eukaryotes, but inhibition of DNA synthesis and recovery are known to occur (Bridges 1995). Although the mechanisms of DNA repair are similar between bacteria and eukaryotes, the mechanisms that control the cell division cycle are very different. Thus, simple parallels between prokaryotes and eukaryotes have not emerged.

CONTROL OF INITIATION

Because the pathways that control the cell division cycle are very different between bacteria and eukaryotes, the mechanisms for controlling the frequency of initiation are also different. Methylation of the bacterial origin of DNA replication by the dam methylase causes the bacterial DNA to become hemi-methylated immediately after it has replicated (Baker 1994). In the *E. coli OriC* replicator, the Dam methylation sequences (5′-GATC-3′) are clustered in the 13-mer repeats that unwind in the presence of DnaA and ATP. *OriC* is negatively regulated by a process called sequestration. For some time after the initiation of DNA replication, the hemi-methylated DNA is unable to reinitiate, thereby limiting multiple rounds of DNA replication in a cell. The product of the SeqA gene is required for sequestration, and SeqA interacts genetically with DnaA (Campbell and Kleckner 1990; Baker 1994). Recently, it has been shown that the SeqA protein binds to hemi-methylated DNA from either the *E. coli OriC* 13-mer region or from the bacteriophage P1 origin, *P1oriR* (Brendler et al. 1995). Binding to these replication origins is preferred because these sequences are somehow slow to be fully methylated and therefore remain in the hemi-methylated state for some time. These data suggest that the SeqA protein might prevent initiation of DNA replication by interacting with the newly hemi-methylated replicator, thereby preventing productive binding of the replicator by the initiator DnaA.

Methylation probably does not play a role in the regulation of initiation of DNA replication in eukaryotes. More likely, a multiprotein complex is formed on the DNA at the site of the replicator to produce a "competent" pre-replicative complex. This complex can then be acted on by other regulators of DNA replication that coordinate the initiation of DNA replication with progression through the cell division cycle (see Nasmyth, this volume). In many ways, this type of regulation may be similar to the assembly and disassembly of proteins at the phage λ replicator (Stillman 1994a). This type of regulation may also facilitate the coordination of replication from multiple origins of DNA replication that does not occur in bacteria.

TERMINATION OF DNA REPLICATION

The termination of DNA replication has been well studied in bacteria, including *E. coli* and *Bacillus subtilis* (Baker 1995). Replication of the bacterial chromosome terminates when the two replication forks that diverged from *OriC* converge in a region of genome that is 180° from the origin. There are multiple *cis*-acting sites in the DNA called Ter sites in *E. coli* and imperfect repeats (IR-I and IR-II) in *B. subtilis*. These DNA elements have directionality with respect to the advancing replication fork and interact with a protein called Tus in *E. coli* and RTP in *B. subtilis*. The Ter-Tus protein complex functions as an anti-helicase complex because it blocks the progression of DNA unwinding by a DNA helicase in the absence of DNA replication. The Tus-Ter complex has been shown to block the DnaB helicase (Khatri et al. 1989; Lee et al. 1989) and, interestingly, it can also inhibit the SV40 T-antigen helicase (Bedrosian and Bastia 1991). A more thorough description of termination is given by Bastia and Mohanty (this volume).

Replication termination in eukaryotes is not well studied, but it is clear that specific DNA sequences are not generally required. For example, the SV40 origin of DNA replication has been cloned into many plasmids, and these plasmids replicate very well when they are introduced into primate cells, yet the DNA sequences located in the region opposite the origin are very different (Lai and Nathans 1975). Similarly, replication forks that meet each other following initiation at adjacent origins in eukaryotic chromosomes are believed to terminate when the two forks meet each other, but in a sequence-independent manner.

There are, however, DNA sequences in eukaryotic genomes that cause DNA replication to arrest and occasionally terminate. For example, repeated DNA recognition sites for the Epstein-Barr virus (EBV) nuclear antigen 1 (EBNA1) cause replication forks to pause or arrest completely (Little and Schildkraut 1995). Furthermore, DNA replication fork barriers have been found in the ribosomal DNA repeats in *S. cerevisiae* (Brewer et al. 1992). Thus, it is possible that sequence-specific DNA-binding proteins may affect the passage of the eukaryotic DNA replication fork in much the same way as the Tus-Ter system in *E. coli* and that these sites may play some regulatory role that has yet to be appreciated.

REFERENCES

Adachi, Y. and U.K. Laemmli. 1994. Study of the cell cycle-dependent assembly of the DNA prereplication centres in *Xenopus* egg extracts. *EMBO* **13:** 4153–4164.

Baker, T.A. 1994. A new controller in *Escherichia coli. Curr. Opin. Genet. Dev.* **4:** 945–946.

Aladjem, M.I., M. Groudine, L.L. Brody, E.S. Dieken, R.E.K., Fournier, G.M. Wahl, and E.M. Epner. 1995. Participation of the human β-globin locus control region in initiation of DNA replication. *Science* **270:** 815–819.

———. 1995. Replication arrest. *Cell* **80:** 521–524.

Baker, T.A. and A. Kornberg. 1988. Transcriptional activation of initiation of replication from the *E. coli* chromosomal origin: An RNA-DNA hybrid near oriC. *Cell* **55:** 113–123.

Baker, T.A. and S.H. Wickner. 1992. Genetics and enzymology of DNA replication in *Escherichia coli. Annu. Rev. Genet.* **26:** 447–477.

Barry, J. and B. Alberts. 1994. Purification and characterization of bacteriophage T4 gene 59 protein. *J. Biol. Chem.* **269:** 33049–33062.

Bedrosian, C.L. and D. Bastia. 1991. *Escherichia coli* replication terminator protein impedes simian virus 40 (SV40) DNA replication fork movement and SV40 large tumor antigen helicase activity *in vitro* at a prokaryotic terminus sequence. *Proc. Natl. Acad. Sci.* **88:** 2618–2622.

Bell, S.P. and B. Stillman. 1992. ATP-dependent recognition of eukaryotic origins of DNA replication by a multiprotein complex. *Nature* **357:** 128–134.

Blumenthal, A.B., H.J. Kriegstein, and D.S. Hogness. 1974. The units of DNA replication in *Drosophila melanogaster* chromosomes. *Cold Spring Harbor Symp. Quant. Biol.* **38:** 205–223.

Borowiec, J.A., F.B. Dean, P.A. Bullock, and J. Hurwitz. 1990. Binding and unwinding—How T-antigen engages the SV40 origin of DNA replication. *Cell* **60:** 181–184.

Bramhill, D. and A. Kornberg. 1988. A model for initation at origins of DNA replication. *Cell* **54:** 915–918.

Brendler, T., A. Abeles, and S. Austin. 1995. A protein that binds to the P1 origin core and the *oriC* 13-mer region in a methylation-specific fashion is the product of the host *seqA* gene. *EMBO J.* **14:** 4083–4089.

Brewer, B., D. Lockshon, and W. Fangman. 1992. The arrest of replication forks in the rDNA of yeast occurs independently of transcription. *Cell* **71:** 267–276.

Bridges, B.A. 1995. Are there DNA damage checkpoints in *E. coli*? *BioEssays* **17:** 63–70.

Bruckner, R.C., J.J. Crute, M.S. Dodson, and I.R. Lehman, 1991. The herpes simplex virus origin binding protein: A DNA helicase. *J. Biol. Chem.* **266:** 2669–2674.

Callan, H.G. 1974. DNA replication in the chromosomes of eukaryotes. *Cold Spring Harbor Symp. Quant. Biol.* **38:** 195–203.

Campbell, J.L. and N. Kleckner. 1990. *E coli ori*C and the *dna*A gene promoter are sequestered from *dam* methyltransferase following the passage of the chromosomal replication fork. *Cell* **62:** 967–979.

Campbell, J.L. and C.S. Newlon. 1991. Chromosomal DNA replication. In *The molecular and cellular biology of the yeast* Saccharomyces: *Genome dynamics, protein synthesis, and energetics* (ed. J.R. Broach), pp. 41–146. Cold Spring Harbor Laboratory Press. Cold Spring Harbor, New York.

Cegielska, A., I. Moarefi, E. Fanning, and D.M. Virshup. 1994a. T-antigen kinase inhibits simian virus 40 DNA replication by phosphorylation of intact T antigen on serines 120 and 123. *J. Virol.* **68:** 269–275.

Cegielska, A., S. Shaffer, R. Derua, J. Goris, and D.M. Virshup. 1994b. Different oligomeric forms of protein phosphatase 2A activate and inhibit simian virus 40 DNA replication. *Mol. Cell. Biol.* **14:** 4616–4623.

Collins, K.L. and T.J. Kelly. 1991. Effects of T antigen and replication protein A on the initiation of DNA synthesis by DNA polymerase α-primase. *Mol. Cell. Biol.* **11:** 2108–2115.

Crooke, E., C. Castuma, and A. Kornberg. 1992. The chromosome origin of *Escherichia coli* stabilizes DnaA protein during rejuvenation by phospholipids. *J. Biol. Chem.* **267:** 16779–16782.

Dean, F.B., M. Dodson, H. Echols, and J. Hurwitz. 1987. ATP-dependent formation of a specialized nucleoprotein structure by simian virus 40 (SV40) large tumor antigen at the SV40 replication origin. *Proc. Natl. Acad. Sci.* **84:** 8981–8985.

DePamphilis, M.L. 1993. How transcription factors regulate origins of DNA replication in eukaryotic cells. *Trends Cell Biol.* **3:** 1161–163.

Dershowitz, A. and C.S. Newlon. 1993. The effect on chromosome stability of deleting replication origins. *Mol. Cell. Biol.* **13:** 391–398.

Diffley, J.F.X., J.H. Cocker, S.J. Dowell, and A. Rowley. 1994. Two steps in the assembly of complexes at yeast replication origins in vivo. *Cell* **78:** 303–316.

Dodson, M., J. Roberts, R. McMacken, and H. Echols. 1985. Specialized nucleoprotein structures at the origin of replication of bacteriophage λ: Complexes with λO protein and with λO, λP, and *Escherichia coli* DnaB proteins. *Proc. Natl. Acad. Sci.* **82:** 4678–4682.

Dornreiter, I., W.C. Copeland, and T.S.-F. Wang. 1993. Initiation of simian virus 40 DNA replication requires the interaction of a specific domain of human DNA polymerase α with large T antigen. *Mol. Cell. Biol.* **13:** 809–820.

Echols, H. 1986. Bacteriophage W development: Temporal switches and the choice of lysis or lysogeny. *Trends Genet.* **2:** 26–30.

Fangman, W. and B. Brewer. 1992. A question of time: Replication origins of eukaryotic chromosomes. *Cell* **71:** 363–366.

Friedberg, E.C., G.C. Walker, and W. Siede. 1995. *DNA repair and mutagenesis.* ASM

Press, Washington, D.C.

Gesteland, R.F. and J.F. Atkins, eds. 1993. *The RNA world.* Cold Spring Harbor Laboratory Press, Cold Spring Harbor, New York.

Hacker, K.J. and B.M. Alberts. 1994a. The rapid dissociation of the T4 DNA polymerase holoenzyme when stopped by a DNA hairpin helix. *J. Biol. Chem.* **269:** 24221–24228.

———. 1994b. The slow dissociation of the T4 DNA polymerase holoenzyme when stalled by nucleotide omission. *J. Biol. Chem.* **269:** 24209–24220.

Hand, R. 1978. Eucaryotic DNA: Organization of the genome for replication. *Cell* **15:** 317–325.

Hirota, Y., S. Yasuda, M. Yamada, A. Nishimura, K. Sugimoto, H. Sugisaki, A. Oka, and M. Takanami. 1979. Structural and functional properties of the *Escherichia coli* origin of DNA replication. *Cold Spring Harbor Symp. Quant. Biol.* **43:** 129–138.

Hwang, D. and A. Kornberg. 1992. Opening of the replication origin of *Escherichia coli* by DnaA protein with protein HU or IHF. *J. Biol. Chem.* **267:** 23083–23086.

Ilyina, T.V., A.E. Gorbalenya, and E.V. Koonin 1992. Organization and evolution of eubacterial and bacteriophage primase-helicase systems. *J. Mol. Evol.* **34:** 351–357.

Ismail, F., D. Small, I. Gilbert, M. Hopfner, S. Randall, C. Schneider, A. Russo, U. Ramsperger, A. Arthur, H. Stahl, T. Kelly, and E. Fanning. 1993. Mutation of the cyclin-dependent kinase phosphorylation site in simian virus 40 (SV40) large T antigen specifically blocks SV40 origin DNA unwinding. *J. Virol.* **67:** 4992–5002.

Jacob, F., S. Brenner, and F. Cuzin. 1964. On the regulation of DNA replication in bacteria. *Cold Spring Harbor Symp. Quant. Biol.* **28:** 329–348.

Kelly, T.J., G.S. Martin, S.L. Forsburg, R.J. Stephen, A. Russo, and P. Nurse. 1993. The fission yeast *cdc18*$^{+}$ gene product couples S. phase to START and mitosis. *Cell* **74:** 371–382.

Kelman, Z. and M. O'Donnell. 1994. DNA replication: Enzymology and mechanisms. *Curr. Opin. Genet. Dev.* **4:** 185–195.

———. 1995. Embryonic PCNA: A missing link? *Curr. Biol.* **5:** 814.

Khatri, G.S., T. MacAllister, P.R. Sista, and D. Bastia. 1989. The replication terminator protein of *E. coli* is a DNA sequence-specific contra-helicase. *Cell* **59:** 667–674.

Kornberg, A. and T.A. Baker. 1992. *DNA replication*, 2nd edition. W.H. Freeman, New York.

Kowalski, D. and M.J. Eddy. 1989. The DNA unwinding element: A novel, *cis*-acting component that facilitates opening of the *Escherichia coli* replication origin. *EMBO J.* **8:** 4335–4344.

Krishna, T.S.R., X.-P. Kong, S. Gary, P.M. Burgers, and J. Kuriyan. 1994. Crystal structure of the eukaryotic DNA polymerase processivity factor PCNA. *Cell* **79:** 1233–1243.

Lai, C.J. and D. Nathans. 1975. Non-specific termination of simian virus 40 DNA replication. *J. Mol. Biol.* **97:** 113–118.

Learn, B., A.W. Karzai, and R. McMacken. 1993. Transcription stimulates the establishment of bidirectional λ DNA replication in vitro. *Cold Spring Harbor Symp. Quant. Biol.* **58:** 389–402.

Lee, E.H., A. Kornberg, M. Hidaka, T. Kobayashi, and T. Horiuchi. 1989. *Escherichia coli* replication termination protein impedes the action of helicases. *Proc. Natl. Acad. Sci.* **86:** 9104–9108.

Li, J.J. and R.J. Deshaies. 1993. Exercising self-restraint: Discouraging illicit acts of S and M in eukaryotes. *Cell* **74:** 223–226.

Liang, C., M. Weinreich, and B. Stillman. 1995. ORC and Cdc6p interact and determine the frequency of initiation of DNA replication in the genome. *Cell* **81:** 667–676.

Lin, S. and D. Kowalski. 1994. DNA helical instability facilitates initiation at the SV40 replication origin. *J. Mol. Biol.* **235:** 496–507.

Little, R.D. and C.L. Schildkraut. 1995. Initiation of latent DNA replication in the Epstein-Barr virus genome can occur at sites other than the genetically defined origin. *Mol. Cell. Biol.* **15:** 2893–2903.

Marahrens, Y. and B. Stillman. 1992. A yeast chromosomal origin of DNA replication defined by multiple functional elements. *Science* **255:** 817–823.

Marczynski, G.T., K. Lentine, and L. Shapiro. 1995. A developmentally regulated chromosomal origin of replication uses essential transcription elements. *Genes Dev,* **9:** 1543–1557.

Marians, K.J. 1992. Prokaryotic DNA replication. *Annu. Rev. Biochem.* **61:** 673–719.

McVey, D., S. Ray, Y. Gluzman, L. Berger, A.G. Wildeman, D.R. Marshak, and P. Tegtmeyer. 1993. cdc2 phosphorylation of threonine 124 activates the origin-unwinding functions of simian virus 40 T antigen. *J. Virol.* **67:** 5206–5215.

Melendy, T. and B. Stillman. 1993. An interaction between replication protein A and SV40 T antigen appears essential for primosome assembly during SV40 DNA replication. *J. Biol. Chem.* **268:** 3389–3395.

Mensa-Wilmot, K., K. Carroll, and R. McMacken. 1989. Transcriptional activation of bacteriophage λ DNA replication *in vitro*: Regulatory role of histone-like protein HU of *Escherichia coli*. *EMBO J.* **8:** 2393–2402.

Messer, W., M. Meijer, H.E.N. Bergmans, F.G. Hanse, K. von Meyenburg, E. Beck, and H. Schaller. 1979. Origin of replication, *oriC*, of the *Escherichia coli* K12 chromosome: Nucleotide sequence. *Cold Spring Harbor Symp. Quant. Biol.* **43:** 139–145.

Morrical, S.W., K. Hempstead, and M.D. Morrical. 1994. The gene 59 protein of bacteriophage T4 modulates the intrinsic and single-stranded DNA-stimulated ATPase activities of gene 41 protein, the T4 replicative DNA helicase. *J. Biol. Chem.* **269:** 33069–33081.

Murray, A. 1994. Cell cycle checkpoints. *Curr. Opin. in Cell Biol.* **6:** 872–876.

Natale, D.A., R.M. Umek, and D. Kowalski. 1993. Ease of DNA unwinding is a conserved property of yeast replication origins. *Nucleic Acids Res.* **21:** 555–560.

Nethanel, T. and G. Kaufmann. 1990. Two DNA polymerases may be required for synthesis of the lagging DNA strand of simian virus 40. *J. Virol.* **64:** 5912–5918.

Nichols, A.F. and A. Sancar. 1992. Purification of PCNA as a nucleotide excision repair protein. *Nucleic Acids Res.* **20:** 2441–2446.

Nurse, P. 1994. Ordering S phase and M phase in the cell cycle. *Cell* **79:** 547–550.

O'Donnell, M., R. Onrust, F.B. Dean, M. Chen, and J. Hurwitz. 1993. Homology in accessory proteins of replicative polymerase—*E. coli* to humans. *Nucleic Acids Res.* **21:** 1–3.

Parsons, C.A., A. Stasiak, R.J. Bennett, and S.C. West. 1995. Structure of a multisubunit complex that promotes DNA branch migration. *Nature* **374:** 375–378.

Parsons, R., M.E. Anderson, and P. Tegtmeyer. 1990. Three domains in the simian virus 40 core origin orchestrate the binding, melting, and DNA helicase activities of T antigen. *J. Virol.* **64:** 509–518.

Rao, H. and B. Stillman. 1995. The origin recognition complex interacts with a bipartitie DNA binding site within yeast replicators. *Proc. Natl. Acad. Sci.* **92:** 2224–2228.

Roca, J. 1995. The mechanism of DNA topoisomerases. *Trends Biochem. Sci.* **20:** 156–160.

Rowley, A., J.H. Cocker, J. Harwood, and J.F.X. Diffley. 1995. Initiation complex assembly at budding yeast replication origins begins with the recognition of a bipartitie sequence by limiting amounts of the initiator, ORC. *EMBO J.* **14:** 2631–2641.

Seki, M., T. Kohda, T. Yano, S. Tada, J. Yanagisawa, T. Eki, M. Ui, and T. Enomoto. 1995. Characterization of DNA synthesis and DNA-dependent ATPase activity at a restrictive temperature in temperature-sensitive tsFT848 cells with thermolabile DNA helicase B. *Mol. Cell. Biol.* **15:** 165–172.

Sekimizu, K. and A. Kornberg. 1988. Cardiolipin activation of dna A protein, the initiation protein of replication in *Escherichia coli. J. Biol. Chem.* **263:** 7131–7135.

Sekimizu, K., D. Bramhill, and A. Kornberg. 1988a. Sequential early stages in the in vitro initiation of replication at the origin of the *Escherichia coli* chromosome. *J. Biol. Chem.* **263:** 7124–7130.

Sekimizu, K., B.Y. Yung, and A. Kornberg. 1988b. The dna A protein of *Escherichia coli*: Abundance, improved purification and membrane binding. *J. Biol. Chem.* **263:** 7136–7140.

Shivji, M.K.K., M.K. Kenny, and R.D. Wood. 1992. Proliferating cell nuclear antigen is required for DNA excision repair. *Cell* **69:** 367–374.

Stahl, H., P. Droge, and R. Knippers. 1986. DNA helicase activity of SV40 large tumor antigen. *EMBO J.* **5:** 1939–1944.

Stillman, B. 1989. Initiation of eukaryotic DNA replication *in vitro. Annu. Rev. Cell. Biol.* **5:** 197–245.

———. 1994a. Initiation of chromosomal DNA replication in eukaryotes. Lessons from lambda. *J. Biol. Chem.* **269:** 7047–7050.

———. 1994b. Smart machines at the DNA replication fork. *Cell* **78:** 725–728.

Stukenberg, P.T., J. Turner, and M. O'Donell. 1994. An explanation for lagging strand replication: Polymerase hopping among DNA sliding clamps. *Cell* **78:** 877–887.

Taylor, J.H. 1977. Increase in DNA replication sites in cells held at the beginning of S phase. *Chromosoma* **62:** 291–300.

Tsurimoto, T. and B. Stillman. 1990. Functions of replication factor C and proliferating cell nuclear antigen: Functional similarity of DNA polymerase accessory proteins from human cells and bacteriophage T4. *Proc. Natl. Acad. Sci.* **87:** 1023–1027.

Tsurimoto, T., T. Melendy, and B. Stillman. 1990. Sequential initiation of lagging and leading strand synthesis by two DNA polymerase complexes at the SV40 DNA replication origin. *Nature* **346:** 534–539.

Ullsperger, C.J., A.V. Vologodskii, and N.R. Cozzarelli. 1995. Unlinking of DNA by topoisomerases during DNA replication. *Nucleic Acids Mol. Biol.* **9:** 115–142.

von Meyenburg, K., F.G. Hansen, E. Riise, H.E.N. Bergmans, M. Meijer, and W. Messer. 1979. Origin of replication, *oriC*, of the *Escherichia coli* K12 chromosome: Genetic mapping and minichromosome replication. *Cold Spring Harbor Symp. Quant. Biol.* **43:** 121–128.

Waga, S. and B. Stillman. 1994. Anatomy of a DNA replication fork revealed by reconstitution of SV40 DNA replication *in vitro. Nature* **369:** 207–212.

Wellinger, R.J. and V.A. Zakian. 1989. Lack of positional requirements for autonomously replicating sequence elements on artificial yeast chromosomes. *Proc. Natl. Acad. Sci.* **86:** 973–977.

Woese, C.R. and N.R. Pace. 1993. Probing RNA structure, function, and history by com-

parative analysis. In *The RNA world* (ed. R.F. Gesteland and J.F. Atkins), pp. 91–117. Cold Spring Harbor Laboratory Press, Cold Spring Harbor, New York.
Ziegelin, G., E. Scherzinger, R. Lurz, and E. Lanka. 1993. Phage P4 α protein is multifunctional with origin recognition, helicase and primase activities. *EMBO J.* **12:** 3703.

15

Cellular DNA Polymerases

Teresa S.-F. Wang
Department of Pathology
Stanford University School of Medicine
Stanford, California 94305-5324

ENZYMATIC PROPERTIES AND CHARACTERISTICS THAT DISTINGUISH EACH DNA POLYMERASE

During the past decade, five DNA polymerases (pol) have been characterized in eukaryotic cells. These enzymes are called α, β, γ, δ, and ε (Wang 1991). Their enzymatic properties, protein composition and structure, effectors, genetic relationships, genomic locations, expression during cell growth and proliferation, and their roles in DNA replication and repair have been described in previous reviews (Challberg and Kelly 1989; Stillman 1989, 1994; Hurwitz et al. 1990; Bambara and Jessee 1991; Wang 1991; So and Downey 1992). This chapter is an updated version of these reviews.

DNA pols α, β, δ, and ε are located in the nucleus, and DNA pol-γ is localized in the mitochondria. Therefore, pol-γ is thought to be the enzyme that replicates the mitochondrial DNA. The general enzymatic properties and characteristics that distinguish each DNA pol are summarized in Table 1, and the optimal assay conditions for each enzyme, such as pH, preferred primer template, and metal activators, are summarized in Table 2. The K_m values of their substrates and K_i values of their inhibitors vary widely depending on the purity of the enzyme, the integrity of the catalytic polypeptide, and the associated subunit components of the enzyme fraction used for the assay. Nonetheless, inhibitors can be used to distinguish one polymerase from another. Pols α, δ, and ε are sensitive to aphidicolin, but pols β and γ are not. Pols α, δ, and γ are sensitive to *N*-ethylmaleimide. Pol-α is particularly sensitive to butylphenyl-dGTP and butylphenyl-dATP. Pol-β and pol-γ are sensitive to dideoxynucleoside triphosphate (ddNTPs) (Edenberg et al. 1978; Krokan et al. 1979; Kaguni and Lehman 1988). Although carbonyldiphosphonate is a potent inhibitor of pol-δ, this compound does not inhibit pol-α and only moderately inhibits pol-ε. In addition, pol-δ is stimulated by 10% dimethylsulfoxide, whereas pol-ε is inhibited by it (Syvaoja et al. 1990).

DNA Replication in Eukaryotic Cells

Table 1 Cellular DNA polymerases

	DNA polymerases				
	α	β	γ	δ	ε
Cellular location	nuclear	nuclear	mito	nuclear	nuclear
Protein structure					
mammalian cells catalytic					
polypeptides[a] (kD)	165[a]	40	140	125	255
associated subunits (kD)	70	none	35–55	48	55
	58				
	49				
Budding yeast					
catalytic polypeptides (kD)	165	68	140	124	250
associated subunits (kD)	86			48	80
					34/31
	58				29
	49				
Mammalian cDNA					
catalytic subunit					
human	yes	yes	yes	yes	yes
rodent	yes	yes	no	yes	no
associated subunits					
human	p70			p48	no
mouse	p58				
	p49				
Yeast genes					
budding yeast					
catalytic subunit	*POL1*	*POL4*	*MIP1*	*POL3*	*POL2*
associated subunits	*POL12*				*DPB2*
	Pri2				*DPB3*
	Pri1				*DPB4*
fission yeast					
catalytic subunit	polα^+	—	polγ^+	polδ^+	—
associated subunits	—		—	—	—
chromosome localization					
mammal	Xq21.3-q22.1[b]	8p12-p11[c]	15q24[d]	19q13.3-q13.4[e]	12q24.3[f]
budding yeast	XIV[g]	III[h]	XV[i]	IV[j]	—
fission yeast	I[k]	—	III[l]	II[l]	—

[a]Molecular mass derived from deduced primary sequence. [b]Wang et al. (1985) and Adler et al. (1991). [c]Matsukage et al. (1986) and McBride et al. (1987). [d]W.C. Copeland (pers. comm.). [e]Chung et al. (1991) and Kemper et al. (1992). [f]Szpirer et al. (1994). [g]Lucchini et al. (1987) and Foiani et al. (1989). [h]Leem et al. (1994). [i]Foury (1989). [j]Boulet et al. (1989). [k]Singh and Klar (1993). [l]W.C. Copeland (pers. comm.).

Table 2 Properties of cellular DNA polymerases

	DNA polymerases				
	α	β	γ	δ	ε
Optimal asssay conditions					
pH	8.0	8.9	8.5	7.5	7.5
primer template	gapped DNA (60–150 nt)	gapped DNA (14–20 nt)	gapped DNA or poly(rA) oligo(dT)	poly(dA)-oligo(dT) or poly(dA-T)	poly(dA)-oligo(dT) or gapped DNA
preferred metal activator	Mg^{++}	Mg^{++}	Mg^{++} or Mn^{++}	Mg^{++}	Mg^{++}
Inhibitors					
aphidicolin	sensitive	not sensitive	sensitive	sensitive	sensitive
N-ethylmaleimide	sensitive	not sensitive	sensitive	sensitive	—
butylphenyl-dGTP or butylphenyl-dATP	highly sensitive	not sensitive	not sensitive	moderately sensitive	moderately sensitive
carbonyldiphosphonate	not sensitive	unknown	unknown	highly sensitive	not sensitive
dideoxynucleotide triphosphate (ddNTP)	not sensitive	sensitive	sensitive	not sensitive	not sensitive
Associated activities					
3′→5′ exonuclease	none	none	yes	yes	yes
primase	yes	none	none	none	none
Auxiliary proteins	none	none	none	PCNA	none
Processivity					
inherent	moderate	low	high	low	high
with PCNA	—	—	—	high	high
Fidelity	med	low	high	high	high

Pol-α is the only enzyme with an associated DNA primase (Wang 1991). Pol-β has no associated enzymatic activities, but pols γ, δ, and ε all have an intrinsic 3′→5′ exonuclease activity associated within their catalytic subunit. Pol-δ has an auxiliary protein, proliferating cell nuclear antigen (PCNA), that serves as a DNA sliding clamp and enhances the processivity of pol-δ, but pol-γ and pol-ε have intrinsic higher processivity. However, the processivity of these enzymes and the effect of PCNA on their processivity are both influenced by the Mg^{++} concentration in the reaction. The fidelity of DNA polymerases is reviewed by Roberts and Kunkel (this volume).

PROTEIN STRUCTURES AND SUBUNIT COMPONENTS

DNA Pol-α

DNA pol-α consists of four subunits: one catalytic subunit of 165–180 kD, one 70-kD subunit (also referred to as the B subunit or p70 subunit) with no detectable enzymatic activity (Collins et al. 1993; Foiani et al. 1994), and two subunits of 49 kD and 58 kD that contain the DNA primase activity (Wang 1991; Copeland and Wang 1993b). cDNAs for the catalytic subunit from human, mouse, *Drosophila melanogaster* embryo, fission yeast, and the gene from budding yeast have been isolated and characterized (Johnson et al. 1985; Wong et al. 1988; Damagnez et al. 1991; Hirose et al. 1991; Miyazawa et al. 1993; Park et al. 1993). Based on this sequence information, the catalytic subunit for pol-α is 165 kD. The catalytic subunit of human pol-α is glycosylated (Hsi et al. 1990); it is also phosphorylated in a cell-cycle-dependent manner (Nasheuer et al. 1991). Due to these posttranslational modifications, the apparent molecular mass is 180 kD (Hsi et al. 1990; Copeland and Wang 1991; Nasheuer et al. 1991). The fission yeast pol-α catalytic subunit is also phosphorylated in a cell-cycle-dependent manner (Park et al. 1995).

cDNAs for the pol-α p70 subunit have also been isolated from human, *D. melanogaster* embryo, and mouse, as well as from the gene from budding yeast (Cotterill et al. 1992; Collins et al. 1993; Miyazawa et al. 1993; Stadlbauer et al. 1994). Based on sequence data, the p70 subunit from human cells is 69.5 kD (Collins et al. 1993). The human p70 subunit, like the catalytic subunit, is phosphorylated in a cell-cycle-dependent manner (Nasheuer et al. 1991). The B subunit of budding yeast is 86 kD, whereas its apparent mass is present in two forms of 86 and 91 kD due to its posttranslational modification (Foiani et al. 1994).

The two DNA primase subunits from human, mouse, *D. melanogaster* embryo, and budding yeast all have apparent molecular masses of

58 kD and 49 kD in agreement with the molecular masses predicted from their respective cDNAs (Foiani et al. 1989; Prussak et al. 1989; Bakkenist and Cotterill 1994; Stadlbauer et al. 1994).

DNA Pol-β

Pol-β from human, mouse, rat, and calf is a single polypeptide of 39 kD, based on purified enzyme fractions or cDNA sequences (Wang 1991). In budding yeast, an open reading frame designated YCR14C corresponds to a protein of 67 kD and has been reported to be the gene encoding a DNA pol-β homolog named POL4 (Prasad et al. 1993b; Leem et al. 1994).

DNA Pol-δ

Pol-δ from both budding and fission yeast (also named POL3), as well as from mammalian cells, has two subunits, a catalytic subunit with an apparent molecular mass of 124 kD and a 48-kD subunit (Wang 1991). Both cDNAs and genes encoding the 124-kD catalytic subunit from budding and fission yeast, as well as from mammalian cells, have been isolated (Boulet et al. 1989; Pignede et al. 1991; Zhang et al. 1991; Yang et al. 1992; Cullmann et al. 1993; Park et al. 1993). The observed molecular mass of DNA pol-δ purified from various organisms is in agreement with the molecular mass deduced from the cDNA sequence, suggesting that pol-δ is not extensively modified. The cDNA encoding the small subunit of bovine and human pol-δ has been isolated. The deduced proteins, 50,885 D and 51,289 D, respectively, are 94% identical to each other. Although the primary sequence of the catalytic subunit of pol-δ shares substantial similarity with the herpesvirus family pol catalytic subunit, the small subunit of mammalian pol-δ shows no similarity to the small subunit of either herpes simplex type 1 (HSV1) UL42 protein or the Epstein-Barr virus BMRF1 protein. The small subunit of human pol-δ is located on chromosome 7 (Zhang et al. 1995).

DNA Pol-ε

The gene of pol-ε from budding yeast (also named *POL2*) and cDNA from human cells have been cloned and contain an open reading frame that encodes the 255-kD catalytic subunit (Morrison et al. 1990; Syvaoja 1990; Kesti et al. 1993). The exact subunit composition is not yet resolved. The most purified pol-ε enzyme fraction from cultured human cells reproducibly contains a 255-kD protein and a 55-kD protein. The human 55-kD subunit is antigenically unrelated to similarly sized subunits in yeast. Partially purified human pol-ε contains a larger complex of proteins that include 85-kD, 70-kD, and 49-kD proteins (S. Linn, pers. comm.). In budding yeast, pol-ε was reported to contain a 256-kD catalytic subunit encoded by the *POL2* gene and four subunits of 80 kD, 34 kD, 31 kD, and 29 kD encoded by genes *DPB2*, *DPB3*, *DPB3*, and *DPB4*, respectively. Although *POL2* is an essential gene, *DPB2* and *DPB3* are not essential genes (Sugino 1995). Both pol-δ and pol-ε contain an intrinsic proofreading 3′→5′ exonuclease. A notable distinction between these two pols is not only their catalytic subunit size, but also their response to PCNA.

DNA Pol-γ

The gene of mitochondrial DNA pol, pol-γ, from budding yeast has been isolated and named MIPI. It encodes a protein of 143.5 kD (Foury 1989). The cDNAs of pol-γ from fission yeast, *D. melanogaster,* and human cells were recently isolated. The human DNA pol-γ cDNA encodes a protein of 1266 amino acids with a deduced molecular mass of 140 kD. The fission yeast pol-γ cDNA encodes a protein of 1018 amino acids with a deduced molecular mass of 116 kD (W.C. Copeland, pers. comm.). Studies of pol-γ purified from *D. melanogaster* embryos have revealed that the enzyme is a heterodimer comprising subunits of 125 kD and 35 kD (Kaguni and Lehman 1988).

CATALYTIC MECHANISMS

DNA pol-α interacts with its substrates in an ordered sequential manner, first interacting with template (single-stranded DNA), then with primer, and finally with dNTP. Specification of dNTP interaction is dictated by the template sequence. The minimum length of primer that can interact with pol-α effectively is eight nucleotides. The terminal three–five nucleotides of the primer must be complementary to the template. Pol-α interacts with primer terminated in either 3′-H or 3′-OH, but not with 3′-PO_4. (Fisher et al. 1981; Fisher and Korn 1981a,b). Pol-β, unlike pol-α, has a weak affinity for single-stranded DNA and no affinity for duplex DNA, but it has a high affinity for nicked duplex DNA with nicked termini bearing either 3′-OH or 3′-PO_4. Pol-β can perform limited strand-displacement synthesis at a nick (Wang and Korn 1980). Thus, this enzyme was shown to perform limited strand-displacement synthesis at the 3′-OH site of a nick generated by AP endonuclease at the 5′side of an apurinic site (Mosbaugh and Linn 1983). Pol-β from either human cells or mouse cells interacts with its substrates in a rigidly ordered bi-bi kinetics (Tanabe et al. 1979; Wang and Korn 1982). Pol-β interacts with primer template concertedly followed by dNTP binding. For catalytically productive recognition of the primer template by pol-β, the primer template must be a primer base-paired to a short length of potentially single-stranded template. The metal activators, Mg^{++} or Mn^{++}, influence the minimum length of template requirement as well as the enzyme's processivity and error rate (Wang and Korn 1982).

Pol-α is a moderately processive enzyme in Mg^{++}-catalyzed reactions (Copeland and Wang 1991). Pol-β is a distributive enzyme with long gapped DNA as a substrate (Wang and Korn 1980, 1982), but with short gaps of up to six nucleotides, pol-β synthesizes DNA processively. Gap-filling activity requires 5′-PO_4-terminated polynucleotide downstream (Singhal and Wilson 1993). Pol-β prefers to bind downstream from the 3′-end of the gap (Prasad et al. 1994).

The catalytic mechanisms of pols δ and ε have not been investigated as extensively as those of pols α and β, but their response to PCNA has been investigated. PCNA can stimulate processive DNA synthesis for pol-δ on primed, single-stranded DNA primer template (Tan et al. 1986; Prelich et al. 1987; So and Downey 1988, 1992; Downey et al. 1990; Sabatino et al. 1990). Studies of mouse pol-δ showed that the 125-kD catalytic subunit of pol-δ alone is not stimulated by PCNA. The presence of the 48-kD subunit of pol-δ appears to be necessary for the stimulation of processivity by PCNA (Goulian et al. 1990). This finding was later confirmed by the study of pol-δ from *D. melanogaster* embryo (Chiang et al. 1993). Pol-ε was originally identified as a large exonuclease containing pol that is insensitive to PCNA stimulation (Syvaoja and Linn 1989; Morrison et al. 1990; Syvaoja et al. 1990; Syvaoja 1990; Weiser et al. 1991). However, pol-ε can form a stable complex with replication factor C (RF-C), ATP, and PCNA at the primer terminus (Burgers 1991; Lee et al. 1991a,b; Podust et al. 1992).

The polymerization and exonucleolytic mechanism of mtDNA pol-γ from embryos of *D. melanogaster* have been investigated (Olson and Kaguni 1992; Williams et al. 1993; Lewis et al. 1994; Williams and Kaguni 1995). The enzyme replicates with the highest substrate specificity on single-stranded DNA of natural DNA sequence in a quasi-processive manner with a high degree of nucleotide insertion fidelity. The intrinsic mispair-specific 3′→5′exonuclease hydrolyzes 3′-terminal mispair at approximately 15-fold higher efficiency than 3′-

terminal base pairs under nonpolymerization conditions. Under DNA polymerizing conditions, pol-γ does not extend a 3′-mispair even in the presence of a large excess of the next correct nucleotide. The mechanism of transferring template primer from the exonuclease active site to the pol-γ active site is intermolecular (Olson and Kaguni 1992). Maximal activity of pol-γ is obtained when assayed under moderate salt, although the highest processivity is observed under low-salt conditions (Williams et al. 1993). In the presence of single-stranded DNA-binding protein, pol-γ exhibits concurrently maximal DNA polymerizing activity and processivity (Williams et al. 1993; Williams and Kaguni 1995).

PRIMARY SEQUENCE AND STRUCTURAL CONSERVATION OF CELLULAR DNA POLS

The availability of new protein sequence data in the last few years allows the compilation and alignment of all the DNA pol protein sequences. According to their similarities to the three *Escherichia coli* DNA pols I, II, and III, Ito and Braithwaite classified the DNA-dependent DNA pols into four families: A, B, C, and X (Ito and Braithwaite 1991). The α-like DNA pols including the cellular pols α, δ, and ε (yeast *POL1*, *POL3*, and *POL2*, respectively) were classified in family B, whereas cellular pols β and γ were classified in families X and A, respectively (Ito and Braithwaite 1991). The deduced human DNA pol-α amino acid sequence from the cDNA has identified six highly conserved domains in the eukaryotic and prokaryotic pols (Wong et al. 1988; Wang et al. 1989; Wang 1991). Pols containing these six conserved regions are designated as "α-like" DNA pols (Delarue et al. 1990). The α-like pols include the yeast pol1, *E. coli* pol2, pols from the herpesvirus family, vaccinia virus, adenovirus, the bacterial phage pols such as the *E. coli* phage PRD1 and T4, and the *Bacillus* phage ϕ29. The similarity between these conserved regions is most striking in their linear spatial arrangement on each pol polypeptide. The regions are designated I–VI by their extent of similarity, with region I being the most conserved (Wong et al. 1988). Delarue et al. designated the three highly conserved regions II, III, and I named by Wang et al. (Wang 1991) as motifs A, B, and C, respectively (Delarue et al. 1990).

Conservation of these regions in cellular pols from prokaryotic to eukaryotic cells suggests that these enzymes may have diverged from a common ancestral gene and are conserved for maintenance of basic DNA polymerizing function. Sequence alignment of motif C and motif A of α-like DNA pols with the active site of *E. coli* pol1 and HIV-1 reverse transcriptase (RT) is almost nonexistent. Predicted secondary structures of these highly conserved regions in α-like pols, however, show striking structural similarities with the secondary structures of the active sites of *E. coli* pol I and HIV-1 RT (Fig. 1). The predicted secondary structures of regions I and II (or motif C and A) of the α-like pols are similar to the secondary structures of the β strands 8, 12, 13, 9, and 14 of Klenow fragment, and to the five β strands 5, 9, 10, 6, and 11 of the "palm" subdomain of HIV RT (Kohlstaedt et al. 1992; Jacobo-Molina et al. 1993). These observations strongly support the notion that similar mechanisms in basic DNA polymerization by different pols from diverse phylogenetic sources dictate the structural conservation of the catalytic site. Thus, the three most conserved regions, I, II, and III (or motifs, C, A, and B), are components of the catalytic site of α-like DNA pols.

STRUCTURE-FUNCTION RELATIONSHIPS OF DNA POL-α AND DNA POL-β

DNA Pol-α

The overproduction of catalytically active recombinant human pol-α and the lack of an intrinsic 3′→5′ exonuclease of pol-α make this particular pol an ideal model for structure-function relationship study, particularly for identifying residues responsible for DNA synthetic fidelity. Results of the study can serve as a prototypic model for cellular pols δ and ε.

The most conserved region in the α-like pols, region I (or motif C), is centered around six highly invariant residues, Tyr-Gly-Asp-Thr-Asp-Ser (-YGDTDS-). The predicted secondary structure of this region is a pair of antiparallel β sheets with a sharp turn. With the overexpression of the recombinant human DNA pol-α catalytic subunit (Copeland and Wang 1991) and the development of a one-step immunoaffinity purification protocol (Wang et al. 1995), five site-directed mutations were introduced at the most conserved amino acids in region I (-YGDTDS-) (Copeland and Wang 1993a) based on the rationale described previously (Copeland and Wang 1993a; Copeland et al. 1995). Steady-state kinetic analysis of the mutants indicates that the effect of these conserved mutations is manifest in altered rates of catalysis, not in affinity to substrates. Moreover, two mutations, Asp-1002 to asparagine and Thr-1003 to serine, yielded mutant enzymes utilizing Mn^{++} more effectively in catalysis than Mg^{++}. This finding suggests that these amino acids (Asp-1004, Asp-1002, and Thr-1003) are involved in binding the activator metal that is critical for catalysis. The two aspartate residues, Asp-1002 and Asp-1004, may directly participate in chelating the metal ion like the aspartate residues in the active site of Klenow, HIV RT, and T7 RNA pol to form

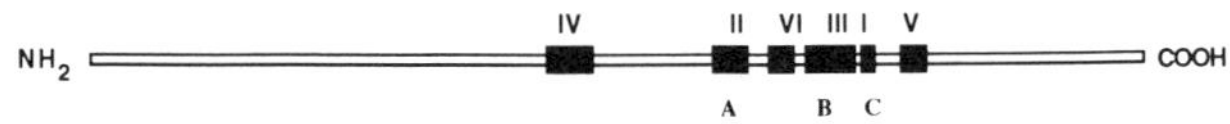


```
                       --- REGION II ----        --------- REGION III ---------        ---- REGION I ---
                                 *   * * *                    *         **                 *  *** *
Human pol α     855-FILLL D FNSLYPSIIQ    944-IRQKAL K LTA.NSM YG CLGFSYSRFY    995-NLEVIYG DTD SIMIN

E.coli pol I    700-VIVSA D Y.SQIELRIM    752-EQRRSA K AINFGLI YG MSAFGLARQL    875-RMINQVH D.E LVFEV

T7 RNA pol      532-LPLAF D GSCSGIQHFS    625-VTRSVT K RSVMTLA YG SKEFGFRQQV    805-ESFALIH D.S FGTIP

HIV-1 RT        105-SVTVL D VGDAYFSVPL                                          178-IVIYQYM D.D LYVGS

Secondary structure
Hum pol α           bbbbb b btttttaaa         aaaaaa a att.ttt tt tttttttttt          ttbbbbt ttt bbbbb
E. coli pol I       bbbbb b btttaaaaaa        aaaaaa a aaaaaaa tt tttttttttt          tbbbbbt t.t bbbbb
T7 RNA pol          bbbbb b btttaaaaaa        aaaaaa a aaaaaat tt tttttttttt          bbbbbbb t.t tbbbb
HIV-1 RT            bbbbb b bbttaaaaaa                                                bbbbbbt t.t bbbbb

                a=α-helix; b=β-strand; t=turn.
```

Figure 1 Sequence alignment and secondary structure comparison of the three most conserved regions of human DNA pol-α with the active sites of *E. coli* pol I, T7 RNA pol, and HIV-1 RT. A schematic diagram depicting the spatial relationship of the six conserved regions of human DNA pol-α is shown at the top. Amino acid residues in the three most conserved regions, I, II, and III (or A, B, and C motifs in α-like DNA polymerases) of human DNA pol-α are aligned with the amino acid residues in the active sites of *E. coli* pol I, T7 RNA pol, and HIV-1 RT. Amino acid residues of human DNA pol-α marked with asterisks are the residues targeted for mutagenesis. Predicted secondary structure of human DNA pol-α regions I, II, and III are compared with the secondary structures of the active sites of *E. coli* pol I, T7 RNA pol, and HIV-1 RT. (Reprinted, with permission, from Copeland et al. 1995.)

the "catalytic triad" (Kohlstaedt et al. 1992; Jacobo-Molina et al. 1993; Sousa et al. 1993). Results from the Thr-1003 mutation suggest that the β-methyl group of the threonine side chain might be locked in a hydrophobic pocket, preventing free rotation around the C_α-C_β bond, thus positioning the Thr-1003 hydroxyl group to form a crucial bond with the metal ion (Copeland and Wang 1993a).

Further investigation was performed with these mutants in the most conserved region on the fidelity of DNA synthesis with either Mg^{++} or Mn^{++} as the metal activator. Mn^{++} is known to cause polymerases to misinsert nucleotides and to utilize dideoxynucleotide. The two region I mutants, D1002N and T1003S, which utilize Mn^{++} in catalysis more effectively than Mg^{++}, actually have an improved misinsertion fidelity in Mn^{++}-catalyzed reactions as compared to that of the wild-type enzyme. The improved misinsertion fidelity of these two mutants in Mn^{++}-catalyzed reactions is due to K_m discrimination of the incorrect nucleotide.

Study of mispair extension showed that in reactions with Mn^{++} as metal activator, the wild-type enzyme readily extended the mispaired primer. In contrast, the two mutants, D1002N and T1003S, were unable to extend the mispaired termini in either Mg^{++}- or Mn^{++}-catalyzed reactions. These findings suggest that the side chains of these residues in region I play an essential role in the Mn^{++}-induced infidelity during DNA synthesis by human DNA pol-α (Copeland et al. 1993).

The second most conserved region of α-like DNA pols, region II (or motif A), spans a block of 40 amino acid residues centered at the core sequence -DFNSLYPSII-. The functions of Gly-841, Asp-860, Ser-863, Tyr-865, and Ser-867 in the region II (motif A) of the human DNA pol-α catalytic subunit were tested by conserved site-directed mutagenesis. Steady-state kinetic analysis of mutants G841A, D860A, D860S, D860N, Y865S, and Y865F demonstrated no significant difference in their K_m values for primer template compared to that of the wild-type enzyme. In contrast, mutants D860A, Y865S, and Y865F showed a 5- to 10-fold increase in the K_m for dNTP compared to the wild-type enzyme. DNA synthetic fidelity studies of these mutants showed that mutant Y865S but not Y865F had a much higher misinsertion efficiency than the wild-type enzyme in Mg^{++}-catalyzed as well as in Mn^{++}-catalyzed reactions. These results indicate that Asp-860 and Tyr-865 in region II (motif A) of human DNA pol-α are involved in incoming dNTP substrate binding. By using three deoxynucleotide structural analogs as probes, the structural requirement for dNTP binding with Tyr-865 is shown to be the nucleotide base. Results of these studies strongly suggest that the phenyl ring of Tyr-865 directly interacts with the nucleotide base moiety of the deoxynucleotide triphosphate and plays a critical role in the misinsertion fidelity of DNA synthesis. Mutation of Gly-841, which is highly conserved among the α-like DNA pols, appears to affect both catalysis and substrate deoxynucleotide binding. This suggests that Gly-841 is essential for the maintenance of the overall structure of the pol-α catalytic site (Dong et al. 1993a). The function of two serine residues, Ser-863 and Ser-867, in this core sequence was also investigated. Mutation of the two serine residues in the core sequence of region II to either alanine or threonine yielded mutant enzymes with similar K_m for dNTPs, k_{cat}, processivity, and misinsertion fidelity of DNA synthesis as the wild-type enzyme. Mutation of Ser-867 to alanine demonstrated a 30-fold increase in K_m for primer template and a 5-fold higher K_D for binding primer template, suggesting that Ser-867 is involved in interacting with primer template. Single-stranded DNA inhibition data suggest that removal of the hydroxyl side chain of Ser-867 affects the pol's interaction with primer and not with template. Moreover, according to footprinting experiments, the structural feature of the primer recognized by Ser-867 is the 3′-OH terminus. Mutation of Ser-867 to alanine also decreases the mutant enzyme's K_m for dNTP to extend a mispaired primer and, thus, enhances its capacity to extend a mispaired primer terminus. These findings suggest that the hydroxyl side chain of Ser-867 of human DNA pol-α is involved in primer interaction during DNA synthesis and plays an essential role in mispair-extension fidelity of DNA synthesis (Dong et al. 1993b).

The function of a highly conserved lysine residue, Lys-950, of human DNA pol-α located in the third most conserved region in a predicted α helix was also analyzed by site-directed mutation. Lys-950 was mutagenized to arginine, alanine, and asparagine. All of the three Lys-950 mutants utilized Mn^{++} as metal activator more effectively than the wild-type enzyme. The mutant enzymes, K950R, K950A, and K950N, showed an increase in the K_m values for deoxynucleotide triphosphate in either Mg^{++}- or Mn^{++}-catalyzed reactions. Although mutations of Lys-950 residue caused an increase in the K_m value for dNTP, mutations of Lys-950 to arginine, alanine, or asparagine did not alter the mutant enzymes' misinsertion efficiency as compared to that of the wild-type enzyme in Mg^{++}-catalyzed reactions, suggesting that the nucleotide base of the incoming dNTP is not the structural feature interacting with the side chain of Lys-950. However, in Mn^{++}-catalyzed reactions, all of the three Lys-950 mutants have improved misinsertion DNA synthetic fidelity. Studies with inhibitor, structural analogs of dNTP, pyrophosphate, and comparison of the affinity of the Lys-950 mutant derivatives to dCTPαS versus dCTP suggest that this highly conserved Lys-950 is involved in interacting with the oxygen group of the α-phosphate moiety of the incoming dNTP (Dong and Wang 1995).

There are no structural data for any member of the α-like DNA pols (family B). Given the structural similarity between the α-like DNA pol regions I, II, and III with the active sites of *E. coli* Klenow fragment and HIV-1 RT, it is reasonable to assume that the three most conserved regions are the components of the α-like DNA pol active site. Based on these mutational studies, a model for the way in which these highly conserved residues in the active site of α-like pols collaborate for catalysis was proposed (Fig. 2). In the active site of human DNA pol-α, two aspartic acid residues in region I, Asp-1002 and Asp-1004, chelate with the metal activator cation, Mg^{++}, which in turn chelates with the oxygen moiety of the β-phosphate and γ-phosphate of the incoming dNTP (Copeland and Wang 1993a). The phenyl ring side chain of Tyr-865

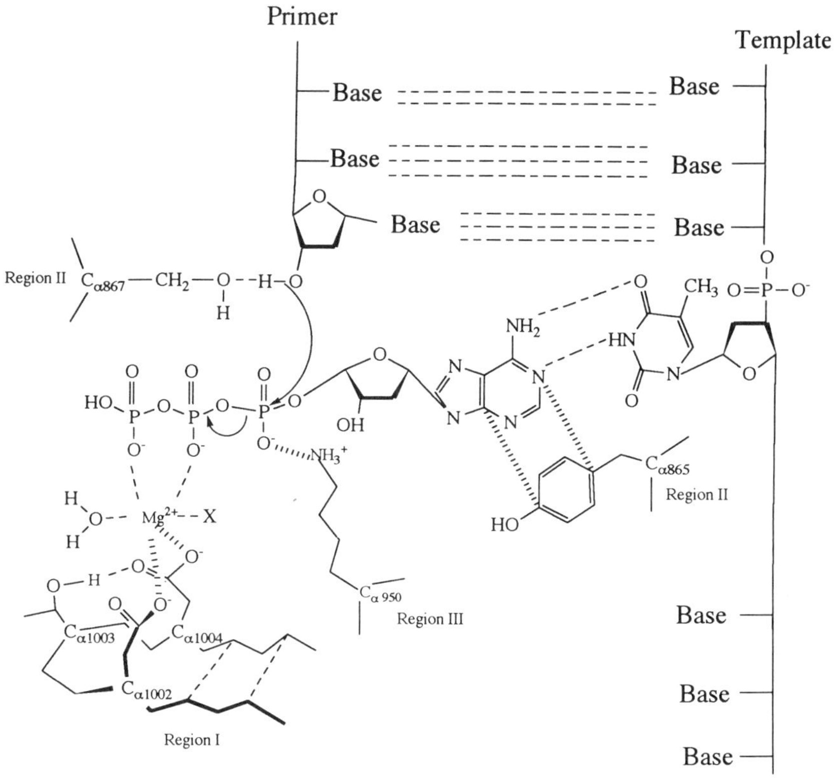

Figure 2 Model of the α-like DNA pol active-site residues that collaborate for nucleotidyl transfer. Shown are the proposed functions of residues of human DNA pol-α studied by site-directed mutagenesis. The three region I residues, Asp-1002, Thr-1003, and Asp-1004, bind to the metal-nucleotide complex. The phenyl ring of Tyr-856 in region II interacts with the nucleotide base moiety of the incoming dNTP. The hydroxyl side chain of the Ser-867 residue forms hydrogen bonds to the 3′-OH terminus of the primer. The positively charged side chain of Lys-950 of region III interacts with the oxygen group of the α-phosphate of the incoming dNTP. An X represents the unknown side chain(s) of residue(s) in the active site that might participate in chelating the Mg^{++} metal activator. (Reprinted, with permission, from Dong and Wang 1995.)

forms hydrophobic bonds with the nucleotide base of the incoming dNTP to properly position the incoming dNTP for Watson-Crick base-pairing (Dong et al. 1993a). The oxygen moiety of the Ser-867 hydroxyl side chain forms a hydrogen bond either directly or indirectly with the 3′-OH terminus of the primer. The hydrogen-bond formation might enhance the oxygen moiety at the 3′-OH-primer terminus for nucleophilic attack at the α-phosphate of the incoming dNTP (Dong et al. 1993b). The positively charged side chain of Lys-950 interacts either directly or indirectly with the oxygen group of the α-phosphate moiety of the incoming dNTP to position the α-phosphate for the nucleophilic attack by the oxygen moiety at the 3′-OH-primer terminus.

DNA Pol-β

Pol-β consists of two domains connected by a protease-sensitive hinge region. The amino-terminal region 8-kD domain (residues 1–87) is required for template binding, and the carboxy-terminal 31-kD domain (residues 88–335) possesses the catalytic activity (Kumar et al. 1990a,b). These two independent domains can be separated by proteolysis. The 8-kD domain alone has no enzymatic activity, whereas the 31-kD domain has approximately 5% of the pol activity of the 39-kD pol-β (Kumar et al. 1990a). The amino-terminal 8-kD domain has been shown by affinity labeling and photocrosslinking studies to be the single-stranded DNA-binding domain (template-binding domain) (Casas et al. 1991, 1992; Prasad et al. 1993a).

The structure of rat pol-β in complex with a DNA primer template and ddCTP was determined at 2.9 Å and 3.6 Å resolution (Pelletier et al. 1994). The crystal structure has revealed that the 3′-OH terminus of the primer, the ddCTP phosphates, and two Mg^{++} ions are all clustered around Asp-190, Asp-192, and Asp-256 in the active site of pol-β. These three aspartic acid residues occupy positions spatially similar to the proposed "catalytic triad" aspartic acid residues in Klenow fragment, HIV-1 RT, and T7 RNA pol (Kohlstaedt et al. 1992; Jacobo-Molina et al. 1993; Sousa et al. 1993). These aspartic acid residues may have similar biochemical function as the two aspartic acid residues of α-like DNA pols in the most conserved region I described above (Copeland et al. 1993; Copeland and Wang 1993a). Mutations of Asp-190 and Asp-192 yielded pol-β mutants with 0.1% of the activity of the wild type (Date et al. 1990, 1991). The conserved geometric arrangement of these aspartic acid residues in the pol-β active site with the other structurally known pols provides the physical evidence that pols from vast phylogenetic origins share a common mechanism of dNMP transfer during catalysis. Thus, the active-site structures of these pols are conserved in order to maintain the basic mechanism of nucleotidyl transfer. In this ternary complex structure of pol-β, however, the DNA template is bound to pol-β in a different manner from that proposed for the other pols.

The crystal structures of the 31-kD domain of rat pol-β and the 39-kD domain of rat pol-β were also determined at 2.3 Å resolution and 3.6 Å resolution, respectively. The structural data of the rat pol-β show that the subdomain structures can be compared to the Klenow fragment as the fingers, palm, and thumb subdomains of a right hand (Kohlstaedt et al. 1992). Although the palm region of pol-β can be superimposed on Klenow fragment and HIV-1 RT, there are notable differences in their folding patterns (Davies et al. 1994). The topology of the β sheet of the pol-β palm subdomain is different from that of Klenow fragment and HIV-1 RT. There is an additional strand in pol-β. The amino-terminal 8-kD domain is composed of two pairs of antiparallel helices and is attached to the 31-kD domain at the tip of the finger subdomain by a flexible hinge. The 31-kD domain has the structural dimensions compatible to bind B-form DNA (Sawaya et al. 1994). Since biochemical evidence has shown that the hinge is less sensitive to proteolysis when pol-β is bound to the DNA primer template (Kumar et al. 1990a), this suggests that the 8-kD domain may fold over the bound template. The structure of the ternary complex indeed has shown that the 8-kD domain moves closer to the palm subdomain than the 39-kD structure alone (Pelletier et al. 1994; Sawaya et al. 1994).

Crystal structure of rat pol-β soaked in 5 mM $MnCl_2$ was obtained at 2.8 Å resolution. An electron density map showed the presence of two large positive peaks near the side chains of Asp-190, Asp-192, and Asp-256 that were interpreted as two metal-binding sites (Davies et al. 1994). The two Mn^{++} metal ions appear to be coordinated by one aspartic acid residue, Asp-190. The side chains of Asp-192 and Asp-256 do not directly contact the metal ion, but form salt bridges with a nearby arginine residue, Arg-254. The finding of two Mn^{++} at the catalytic site of pol-β is consistent with the proposed two-metal reaction model proposed for Klenow fragment and HIV-1 RT (Joyce and Steitz 1994).

Results from extensive biochemical studies of the pol-α active site, high-resolution structure analysis of pol-β either alone or as a ternary complex, and comparison with the structure of Klenow fragment and HIV-1 RT clearly demonstrate the generality of a catalytic mechanism for nucleotidyl transfer among the template-dependent cellular pols. The presence of a few highly conserved aspartic acid residues as critical active-site structural elements to chelate metal activator for catalysis reflects a common active architecture of DNA-template-dependent DNA pols.

EXPRESSION OF DNA POL GENES

DNA Pol-α

Expression of human DNA pol-α is up-regulated during activation of quiescent cells (G_0) to enter the mitotic cell cycle at the levels of transcript, protein, and enzymatic activity (Wahl et al. 1988; Pearson et al. 1991). The expression of human pol-α at the levels of transcript, protein, and enzymatic activity is greatly reduced in cells rendered temporarily quiescent either by contact inhibition or by serum starvation (Wahl et al. 1988). The promoter sequence required for the expression of a heterologous gene in actively cycling cells and in serum-deprived cells following serum addition was investigated (Pearson et al. 1991). The upstream sequence is G+C-rich and lacks a TATA sequence but has a CCAAT sequence on the opposite strand. Studies have shown that efficient expression of the reporters in cycling cells requires 248 bp of sequence upstream from the cap site containing Sp1-, Ap1-, Ap2-, and E2F-binding sites. The promoter confers on the reporter an appropriate, late serum response, as observed in serum induction of pol-α transcript (Wahl et al. 1988). Multiple sequence elements appear to mediate the full serum response (Pearson et al. 1991).

When cells permanently exit from the cell cycle during terminal differentiation, the expression of human DNA pol-α at the level of transcript, protein, and enzymatic activity are down-regulated, and the two primase subunits are also both down-regulated at the transcript level. Down-regulation of DNA pol-α at the messenger RNA level is caused in part by a decrease of transcriptional initiation rate with no transcription elongation block (Moore and Wang 1994).

In actively cycling cells, however, the transcript, protein, and enzymatic activity of human DNA pol-α are constitutively expressed throughout the cell cycle. The protein and enzymatic activity show a nominal increase in the S phase (Wahl et al. 1988). Similar findings have been observed in several mammalian replication proteins (Sherley and Kelley 1988; Tseng et al. 1989; Morris and Mathews 1989). Although pol-α is expressed throughout the cell cycle, the human DNA pol-α catalytic subunit and the p70 subunit proteins are both phosphorylated in a cell-cycle-regulated manner (Nasheuer et al. 1991).

DNA Pol-β

Expression of DNA pol-β in cultured cells and tissues is considered to be constitutive. Pol-β enzymatic activity and steady-state transcripts are expressed at a relatively low level and are independent of cell growth and cell cycle stage (Chang and Bollum 1972; Zmudzka et al. 1988; Nowak et al. 1989). However, in rodent cells, pol-β enzymatic activity and steady-state transcript appear to be regulated in a tissue-specific manner (Hirose et al. 1989; Nowak et al. 1989) and are induced by treatment of cells with some DNA-damaging agents such as alkylating agents MMS or MNNG (Fornace et al. 1989). The promoters of pol-β genes from human, bovine, and rodent sources are G+C-rich and contain binding sites of two well-known transcriptional activators, Sp1 and ATF/CREB (Widen et al. 1988; Widen and Wilson 1991). In studies of the human pol-β core promoter, the ATF/CREB site is shown to be functionally responsive to elevated levels of intracellular cAMP (Englander et al. 1991; Englander and Wilson 1992b), and the ATF/CREB site also mediates a positive response of the cloned promoter to treatment of transfected Chinese hamster ovary (CHO) cells with MNNG (Kedar et al. 1991). This up-regulation was not shown in genetically mutated CHO cells that were deficient in protein kinase A activity. Thus, this suggests that this particular signal transduction pathway is required for the response to MNNG (Englander and Wilson 1992a,b). The pol-β core promoter is also up-regulated by expression of the activated $p21^{ras}$ protein. This suggests that activation of the protein kinase C signal transduction pathway may also have a role in pol-β gene expression (Kedar et al. 1990).

DNA Pol-δ

The mRNA and protein of pol-δ appear to be present throughout the mammalian cell cycle with a slight increase at the G_1/S junction (Zeng et al. 1994a).

Yeast DNA Pols

In budding yeast, the transcripts of genes encoding most DNA replication proteins, including pol-α (*POL1* or *CDC17*), pol-δ (*POL3* or *CDC2*), pol-ε (*POL2*), the 86-kD subunit of pol-α (B-subunit; *POL12*), and the two subunits of DNA primase (*PRI*1 and *PRI*2) (Johnson et al. 1985; Johnston et al. 1987, 1990; Bauer and Burgers 1990; Araki et al. 1992; Foiani et al. 1989, 1994), are coordinately induced at the G_1/S-phase junction. A DNA element, 5′-A/TPuACGCGTNA/T-3′, containing an *Mlu*I restriction site (termed *Mlu*I cell cycle box, MCB) was found in the upstream regions in all of these genes. These findings suggest that the cell-cycle-dependent transcription of the DNA synthesis genes in *S.*

cerevisiae may be coordinated by a common MCB-binding factor (Andrews and Herskowitz 1990; Gordon and Campbell 1991; McIntosh et al. 1991). Although the *POL1* gene is transiently transcribed during the cell cycle at the G_1/S-phase boundary, budding yeast pol-α protein is present at all stages of the cell cycle with only a slight increase following the peak of *POL1* transcript. This suggests that the de novo synthesis of pol-α protein is not rate-limiting or required for entrance of S phase of the cell cycle (Falconi et al. 1993). The 86-kD subunit of budding yeast pol-α, like its homolog in human cells, is phosphorylated in a cell-cycle-dependent manner (Foiani et al. 1994). In budding yeast, a weakly expressed 3.2-kb transcript of pol-β (*POL4*) was detected at constant level throughout the cell cycle. During meiosis, a 2.2-kb *POL4* transcript was highly induced, whereas the 3.2-kb transcript remained at a constant level (Leem et al. 1994).

The expression of two fission yeast (*Schizosaccharomyces pombe*) pols α and δ during the cell cycle was investigated (Bouvier et al. 1992; Pignede et al. 1992; Park et al. 1993). The steady-state transcripts of *S. pombe* $pol\alpha^+$ and $pol\delta^+$ genes, like those of their mammalian cell homologs, are present throughout the cell cycle. Sequence analysis of the $pol\alpha^+$ and $pol\delta^+$ genes revealed the absence of the *Mlu*I motifs in their upstream sequences (Park et al. 1993). The proteins of *S. pombe* pols α and δ are expressed constantly throughout the cell cycle. The enzymatic activity of pol-α measured by in vitro assay was found in all stages of the cell cycle (Park et al. 1993). The *S. pombe* pol-α, like its homolog in human cells, is phosphorylated in a cell-cycle-regulated manner (Park et al. 1995).

GENETIC STUDIES OF YEAST DNA POLS

DNA Pol-α

Genetic studies of budding and fission yeast DNA pols have revealed some novel functions of cellular DNA pols. During initiation of mating-type switching at the *mat1* locus of fission yeast, there is a double-strand chromosomal break (DSB). Three mating-type switch genes of *S. pombe*, *swi1*, *swi3*, and *swi7*, have been identified, and mutations of these genes reduce the level of DSB and switch efficiency. A genomic sequence that complemented the switching-defective *swi7* strain is the DNA pol-α gene. Disruption of the *swi7/pol*α gene is lethal both in switching and non-switching strains, as expected for an essential replication gene. Thus, *S. pombe* pol-α has a role in initiation of mating-type switching at *mat1*, suggesting generation of DSB coupled with DNA replication. This finding also links DNA replication in dictating the program of cellular differentiation (Singh and Klar 1993).

In a normal cell cycle, entry into mitosis is dependent on the completion of S phase, and the dependency is maintained by a checkpoint control mechanism (Enoch et al. 1991, 1992, 1993; Enoch and Nurse 1991). Results of fission yeast genetic studies (Kelly et al. 1993; Saka and Yanagida 1993; Hofmann and Beach 1994) have suggested that proteins or enzymes involved in the formation of an initiation or replication complex have a role in mitotic entry checkpoint operation. Germinating spores derived from *S. pombe* with a disrupted DNA pol-α gene have a heterogeneous phenotype. This is different from germinating spores derived from a disrupted pol-δ gene, which has a cell cycle division (cdc) phenotype (elongated phenotype) (Francesconi et al. 1993b). Further analysis has shown that germinating spores derived from spores lacking the DNA pol-α gene showed abnormal nuclear morphology with 1N DNA content (D. Bhuamik and T.S. Wang, in prep.). A multicopy suppressor of a temperature-sensitive DNA pol-α mutant named cds1$^+$ (*c*heck *D*NA *s*ynthesis) encoding a protein with a typical protein kinase motif was reported recently. Genetic data suggest that the primary role of cds1$^+$ is to monitor DNA synthesis by interacting with DNA pol-α and sending a signal to restrain the onset of mitosis during the progression of S phase (Murakami and Okayama 1995). The observations of germinating spores with pol-α gene disruption or deletion, together with the genetic results of cds1$^+$, suggest that pol-α may be involved in cell-cycle-checkpoint control.

DNA Pol-ε

A recent study of budding yeast DNA pol-ε shows that *pol2* mutants are defective for the S-phase checkpoint. By interallelic complementation and mutation analysis, the budding yeast pol-ε protein can be divided into two separate domains, an amino-terminal pol domain and a carboxy-terminal checkpoint domain. The carboxy-terminal domain of pol-ε was shown to function as a sensor for DNA replication coordinating the transcriptional and cell cycle responses to replication block (Navas et al. 1995).

ROLES OF DNA POLS IN DNA REPLICATION

The proposed functional roles of the principal replicative cellular pols α, δ, and ε are deduced from either yeast genetic data or cell-free reconstituted replication reactions of the simian virus 40 (SV40) origin containing DNA. Genetic studies of budding yeast and fission yeast have demonstrated that cellular pols α, δ, and ε are all essential for cell viability and nuclear DNA replication (Campbell 1993; Francesconi et al. 1993b). In contrast, strains harboring a deletion or mutation of POL4 (pol-β gene) are viable. Strains with POL4 gene deletion or mutation also exhibit no meiosis defect (Prasad et al. 1993b; Leem et al. 1994). Chromosomal disruption of budding yeast *MIP1* gene (pol-γ gene) is not lethal to the cell, but renders cells with total loss of mitochondrial DNA and pol-γ activity (Foury 1989). Thus, the functional role of the MIP1 gene product, pol-γ, is primarily for mitochondrial DNA synthesis.

DNA pol-α is thought to be the principal cellular pol responsible for initiation (Wang 1991). This is based on the fact that pol-α has an associated primase activity that resides in the 49-kD and 58-kD subunits. Evidence from cell-free and reconstituted in vitro SV40 DNA replication has demonstrated that pol-α with its primase subunits functions primarily for synthesis of initiator DNA (iDNA) (Waga and Stillman 1994; Waga et al. 1994). In the cell-free SV40 replication reaction, it has been shown that the catalytic subunit of pol-α (p180) physically interacts with SV40 large T antigen (Dornreiter et al. 1990, 1993). A domain of human pol-α catalytic subunit in the amino-terminal region from residue 195 to residue 313 physically interacts with T antigen. Transition of SV40 DNA replication from preinitiation stage to initiation stage requires the physical interaction of this amino-terminal domain of human pol-α catalytic subunit with T antigen in the preinitiation complex (Dornreiter et al. 1993). Furthermore, the catalytic subunit of pol-α has also been shown to physically interact with bovine papillomavirus (BPV) initiation protein E1 and is an essential component for the cell-free BPV replication reaction (Park et al. 1994). Besides the catalytic subunit of pol-α, the 70-kD subunit (B subunit) was shown to interact with the T antigen (Collins et al. 1993).

In reconstituted replication reactions of SV40 origin containing DNA, pol-α:primase functions primarily to synthesize RNA-DNA primers for initiation of DNA replication at the origin. Pol-α:primase also functions for priming each Okazaki fragment on the replication protein A (RP-A)-coated lagging-strand template (Waga and Stillman 1994; Waga et al. 1994). The synthesis of the nascent primer and the Okazaki fragment on the RP-A-coated lagging-strand template is thought to be mediated by the interaction between T antigen and pol-α:primase complex (Dornreiter et al. 1993). T antigen functions as a DNA helicase on the leading-strand template in a 3′ to 5′ direction (Borowiec et al. 1990). This allows the primase activity of the pol-α:primase complex to synthesize a short RNA primer and then to be extended by pol-α by synthesizing an iDNA (Hu et al. 1984; Nethanel and Kaufmann 1990). In vitro evidence has shown that in reconstituted SV40 DNA replication reaction, the function of pol-α:primase cannot be substituted by any other DNA pol and primase (Tsurimoto et al. 1990). Likewise, despite the sequence conservation among the pol-α of budding yeast, fission yeast, and human cells, there is a stringent species-specificity in vivo. These in vivo data strongly argue that specific protein-protein interactions between pol-α and other proteins are important not only for initiation during SV40 DNA replication, but also during chromosomal DNA replication (Tsurimoto et al. 1990; Francesconi et al. 1993a).

On the basis of the results of reconstituted SV40 replication reaction, it is proposed that, once the iDNA is synthesized, an ATP-dependent structure-specific DNA-binding protein, RF-C, binds the iDNA and then loads PCNA and pol-δ into the replication fork to displace pol-α:primase from the DNA template. DNA pol-δ thus assumes the leading-strand synthesis at the SV40 origin (Tsurimoto et al. 1990). On the lagging strand, pol-α:primase synthesizes the RNA-DNA primers for each Okazaki fragment. A polymerase switching mechanism occurs by binding the RF-C to the 3′-end of the iDNA and then helps loading PCNA and pol-δ to complete the synthesis of each Okazaki fragment. Therefore, it has been proposed that a switching between DNA pol-α and DNA pol-δ occurs during each Okazaki fragment synthesis on the lagging-strand template like that on the leading strand (Waga and Stillman 1994). In this model, pol-α:primase is responsible for initiation and synthesis of iDNA on both leading and lagging strands of DNA. Pol-δ is responsible for elongating the leading-strand DNA as well as for the completion of each Okazaki fragment synthesis (Waga and Stillman 1994; Waga et al. 1994).

The role of pol-ε in replication is not yet clear. Genetic evidence in budding yeast has shown that pol-ε is essential for completion of S phase (Araki et al. 1992). On the basis of its processive biochemical property, pol-ε was suggested as the pol for leading-strand DNA synthesis (Morrison et al. 1990; Morrison and Sugino 1992). In contrast, in an in vitro SV40 replication reaction, substitution of pol-ε for pol-δ has yielded products of insufficient length as leading-strand products. In this reaction, pol-ε can synthesize moderate length products in the presence of pol-α:primase, implying pol-ε might have a role in Okazaki fragment synthesis (Lee et al. 1991b).

It should be noted that the current concept regarding the functional role of pol-δ as the principal pol in nuclear DNA synthesis should be interpreted with caution, since the proposed model is only based on biochemical data derived from reconstituted in vitro SV40 DNA replication reactions. Numerous biochemical data have also implied that more than one kind of polymerase is involved in the completion of Okazaki fragment synthesis (Nethanel and Kaufmann 1990; Bullock et al. 1991; Burgers 1991; Nethanel et al. 1992; Podust and Hübscher 1993). Although production of form I SV40 DNA product in reconstituted reaction can be achieved with only two cellular DNA pols, α and δ, the efficiency of the form I DNA synthesis in the reconstituted replication reaction is not equal to the efficiency of DNA synthesis in either the cell-free replication system or SV40 viral DNA replication in vivo.

ROLES OF DNA POLS IN DNA REPAIR

Cellular DNA pols are also involved in nucleotide excision repair, postreplication repair, and recombination repair (Friedberg 1991, 1994; Friedberg et al. 1994). Biochemical and genetic studies of budding yeast base excision repair have suggested that pols α, δ, and ε effect excision repair, either directly or indirectly, under certain conditions. One report described that nucleotide base excision repair pathway is mainly catalyzed by pol-ε and further modulated by the presence of pol-α and pol-δ (Wang et al. 1993). In contrast, another study has shown that mutants of budding yeast pol-δ are defective in survival and repair of DNA methylation damage (Blank et al. 1994). In addition to the ongoing uncertainty regarding the role of pol-δ or pol-ε in base excision repair, another study of the cellular pol in nucleotide excision repair was performed by monitoring postirradiation molecular weight changes of cellular DNA in mutants defective in three cellular pols, α, δ, and ε, as well as Rev3 (Budd and Campbell 1995). This study has shown that a mutant defective in cellular pol-α, pol-δ, or pol-ε, as well as Rev3, is defective in repair. However, single mutations in any of these four genes do not show a repair defect. Pairwise combination of pol mutations demonstrated a repair defect only in a pol δ and ε double mutant. Since the extent of repair in this double mutant was comparable to the mutant defective in all four genes, pol-α and Rev3 appear to have only a minor, if any, role in the nucleotide base excision repair pathway. The data suggest that pols δ and ε both are able to perform repair synthesis, and that one can substitute for the other (Budd and Campbell 1995).

In mammalian cells, pol-ε was originally identified as a 10.2S and an approximately 220-kD polypeptide that complemented the repair synthesis of UV-irradiated and permeabilized cultured diploid human fibroblasts (Nishida et al. 1988). Pol-ε transcript was found to be up-regulated during induction of quiescent cells to proliferate; however, after UV-irradiation, no significant difference in the level of pol-ε transcript in HeLa cells or fibroblasts was observed. Nonetheless, several lines of evidence have recently shown that pol-ε has a role in DNA repair. Using monoclonal antibodies against human pol-ε and immunoblot, Linn and his colleagues have identified >200-, 85-, and 70-kD polypeptides in a mammalian protein complex that was shown to be involved in repairing double-strand breaks and deletions by recombination in vitro (Jessberger et al. 1993). Pol-ε is also found in a large form of a mammalian damage DNA-binding protein (Keeney et al. 1993; Reardon et al. 1993). Moreover, the 85-kD and 70-kD proteins that were immunoreactive to the monoclonal antibody in the pol-ε fractions have been demonstrated to be the subunits of Ku protein, which is a DNA-binding protein with double-strand end specificity and is a helicase implicated in recombination, repair, and transcription functions. These findings suggest that pol-ε not only has a role in DNA repair or recombinational repair, but that pol-ε is also able to form a functional complex with Ku protein (S. Linn, pers. comm.). Complete repair synthesis has recently been reconstituted by 30 polypeptides of human cells in which two subunits of pol-ε, RF-C, PCNA, and DNA ligase I were found to be the essential components (Aboussekhra et al. 1995). Together, these results strongly implicate DNA pol-ε in nucleotide base excision repair.

Although the elegant demonstration of the reconstituted nucleotide excision repair system does not require pol-δ, the capacity of a human cell extract to repair UV-damaged plasmid DNA can be inhibited by antibodies against either pol-δ or PCNA. The mRNA levels of both pol-δ and PCNA were also stimulated after UV-irradiation of cultured cells, implicating pol-δ and its accessory protein PCNA in UV-irradiation DNA damage repair synthesis (Zeng et al. 1994b).

The biochemical properties of mammalian pol-β suggest that it is involved in DNA repair. Pol-β can fill gaps of six nucleotides in a processive manner if the gap has a 5′-phosphate (Singhal and Wilson 1993). The recent isolation of POL4 from budding yeast demonstrated that strains harboring a *pol4* deletion or mutation exhibit defects in neither mitotic growth nor meiosis. This suggests that pol-β does not have a role in replication and is involved in nonessential functions in DNA metabolism (Prasad et al. 1993b; Leem et al. 1994). Strains with deletion of *pol4* exhibit mild sensitivity to MMS treatment but do not exhibit UV sensitivity. Strains with *pol4* deletion also demonstrate a hyperrecombination phenotype during intragenic recombination and a high frequency of illegitimate mating during a mating test (Leem et al. 1994). Together, these genetic results implicate that pol-β has a role in the double-strand-break repair pathway. Pol-β has been implicated in catalyzing DNA synthesis during base excision DNA repair (Matsumoto et al. 1994; Singhal et al. 1995). Pol-β also catalyzes release of 5′-terminal deoxyribose phosphate (dRP) residue from incised apurinic-apyrimidinic sites. The amino-terminal 8-kD domain of pol-β is the catalytic domain for this excision activity. The excision reaction appears to occur by β-elimination, not by hydrolysis (Matsumoto and Kim 1995).

ACKNOWLEDGMENT

Works cited from the author's laboratory are supported by grants CA-14835 and CA-54415 from the National Institutes of Health of the National Cancer Institute.

REFERENCES

Aboussekhra, A., M. Biggerstaff, M.K.K. Shivji, J.A. Vilpo, V. Monocollin, V. Podust, M. Protic, U. Hübscher, J.-M. Egly, and R.D. Wood. 1995. Mammalian DNA nucleotide excision repair reconstituted with purified protein components. *Cell* **80:** 859–868.

Adler, D.A., B.Y. Tseng, T.S.-F. Wang, and C.M. Disteche. 1991. Physical mapping of the genes for three components of the mouse DNA replication complex: Polymerase alpha to the X chromosome, primase p49 subunit to chromosome 10, and primase p58 subunit to chromosome 1. *Genomics* **9:** 642–646.

Andrews, B.J. and I. Herskowitz. 1990. Regulation of cell cycle-dependent gene expression in yeast. *J. Biol. Chem.* **265:** 14057–14060.

Araki, H., P.A. Ropp, A.L. Johnson, L.H. Johnston, A. Morrison, and A. Sugino. 1992. DNA polymerase II, the probable homolog of mammalian DNA polymerase ε, replicates chromosomal DNA in the yeast *Saccharomyces cerevisiae*. *EMBO J.* **11:** 733–740.

Bakkenist, C.J. and S. Cotterill. 1994. The 50-kDa primase subunit of *Drosophila melanogaster* DNA polymerase α. Molecular characterization of the gene and functional analysis of the overexpressed protein. *J. Biol. Chem.* **269:** 26759–26766.

Bambara, R.A. and C.B. Jessee. 1991. Properties of DNA polymerases δ and ε and their roles in eukaryotic DNA replication. *Biochim. Biophys. Acta* **1088:** 11–24.

Bauer, G.A. and P.M.J. Burgers. 1990. Molecular cloning, structure, and expression of the yeast proliferating cell nuclear antigen gene. *Nucleic Acids Res.* **18:** 261–265.

Blank, A., B. Kim, and L.A. Loeb. 1994. DNA polymerase δ is required for base excision repair of DNA methylation damage in *Saccharomyces cerevisiae*. *Proc. Natl. Acad. Sci.* **91:** 9047–9051.

Borowiec, J.A., F.B. Dean, P.A. Bullock, and J. Hurwitz. 1990. Binding and unwinding—How T antigen engages the SV40 origin of DNA replication. *Cell* **60:** 181–184.

Boulet, A., M. Simon, G. Faye, G.A. Bauer, and P.M. Burgers. 1989. Structure and function of the *Saccharomyces cerevisiae CDC2* gene encoding the large subunit of DNA polymerase III. *EMBO J.* **8:** 1849–1854.

Bouvier, D., G. Pignede, V. Damagnez, J. Tillit, A.M. de Recondo, and G. Baldacci. 1992. DNA polymerase alpha in the fission yeast *Schizosaccharomyces pombe:* Identification and tracing of the catalytic subunit during the cell cycle. *Exp. Cell Res.* **198:** 183–190.

Budd, M.E. and J.L. Campbell. 1995. DNA polymerases required for repair of UV-induced damage in *Saccharomyces cerevisiae. Mol. Cell. Biol.* **15:** 2173–2179.

Bullock, P.A., Y.S. Seo, and J. Hurwitz. 1991. Initiation of simian virus 40 DNA synthesis in vitro. *Mol. Cell. Biol.* **11:** 2350–2361.

Burgers, P.M. 1991. *Saccharomyces cerevisiae* replication factor C. II. Formation and activity of complexes with the proliferating cell nuclear antigen and with DNA polymerases δ and ε. *J. Biol. Chem.* **266:** 22698–22706.

Campbell, J.L. 1993. Yeast DNA replication. *J. Biol. Chem.* **268:** 25261–25264.

Casas, F.J., A. Kumar, R.L. Karpel, and S.H. Wilson. 1992. Mammalian DNA polymerase β: Characterization of a 16-kDa transdomain fragment containing the nucleic acid-binding activities of the native enzyme. *Biochemistry* **31:** 10272–10280.

Casas, F.J., A. Kumar, G. Morris, S.H. Wilson, and R.L. Karpel. 1991. Spectroscopic studies of the structural domains of mammalian DNA β-polymerase. *J. Biol. Chem.* **266:** 19618–19625.

Challberg, M.D. and T.J. Kelly. 1989. Animal virus DNA replication. *Annu. Rev. Biochem.* **58:** 671–717.

Chang, L.M. and F.J. Bollum. 1972. Variation of deoxyribonucleic and polymerase activities during rat liver regeneration. *J. Biol. Chem.* **247:** 7948–7950.

Chiang, C.S., P.G. Mitsis, and I.R. Lehman. 1993. DNA polymerase δ from embryos of *Drosophila melanogaster. Proc. Natl. Acad. Sci.* **90:** 9105–9109.

Chung, D.W., J.A. Zhang, C.K. Tan, E.W. Davie, A.G. So, and K.M. Downey. 1991. Primary structure of the catalytic subunit of human DNA polymerase δ and chromosomal location of the gene. *Proc. Natl. Acad. Sci.* **88:** 11197–11201.

Collins, K.L., A.A. Russo, B.Y. Tseng, and T.J. Kelly. 1993. The role of the 70 kDa subunit of human DNA polymerase α in DNA replication. *EMBO J.* **12:** 4555–4566.

Copeland, W.C. and T.S.-F. Wang. 1991. Catalytic subunit of human DNA polymerase α overproduced from baculovirus-infected insect cells. Structural and enzymological characterization. *J. Biol. Chem.* **266:** 22739–22748.

———. 1993a. Mutational analysis of the human DNA polymerase α. The most conserved region in α-like DNA polymerases is involved in metal-specific catalysis. *J. Biol. Chem.* **268:** 11028–11040.

———. 1993b. Enzymatic characterization of the individual mammalian primase subunits reveals a biphasic mechanism for initiation of DNA replication. *J. Biol. Chem.* **268:** 26179–26189.

Copeland, W.C., Q. Dong, and T.S.-F. Wang. 1995. DNA replication: Rationale for mutagenesis of DNA polymerase active sites—DNA polymerase α. *Methods Enzymol.* **262:** 294–303.

Copeland, W.C., N.K. Lam, and T.S.-F. Wang. 1993. Fidelity studies of the human DNA polymerase α. The most conserved region among α-like DNA polymerases is responsible for metal-induced infidelity in DNA synthesis. *J. Biol. Chem.* **268:** 11041–11049.

Cotterill, S., I.R. Lehman, and P. McLachlan. 1992. Cloning of the gene for the 73 kD subunit of the DNA polymerase α primase of *Drosophila melanogaster*. *Nucleic Acids Res.* **20:** 4325–4330.

Cullmann, G., R. Hindges, M.W. Berchtold, and U. Hübscher. 1993. Cloning of a mouse cDNA encoding DNA polymerase δ: Refinement of the homology boxes. *Gene* **134:** 191–200.

Damagnez, V., J. Tillit, A.M. de Recondo, and G. Baldacci. 1991. The *POL1* gene from the fission yeast, *Schizosaccharomyces pombe*, shows conserved amino acid blocks specific for eukaryotic DNA polymerases α. *Mol. Gen. Genet.* **226:** 182–189.

Date, T., S. Yamamoto, K. Tanihara, Y. Nishimoto, and A. Matsukage. 1991. Aspartic acid residues at positions 190 and 192 of rat DNA polymerase β are involved in primer binding. *Biochemistry* **30:** 5286–5292.

Date, T., S. Yamamoto, K. Tanihara, Y. Nishimoto, N. Liu, and A. Matsukage. 1990. Site-directed mutagenesis of recombinant rat DNA polymerase β: Involvement of arginine-183 in primer recognition. *Biochemistry* **29:** 5027–5034.

Davies, J., II, R.J. Almassy, Z. Hostomska, R.A. Ferre, and Z. Hostomsky. 1994. 2.3 Å crystal structure of the catalytic domain of DNA polymerase β. *Cell* **76:** 1123–1133.

Delarue, M., O. Poch, N. Tordo, D. Morase, and P. Argos. 1990. An attempt to unify the structure of polymerases. *Protein Eng.* **3:** 461–467.

Dong, Q. and T.S.-F. Wang. 1995. Mutational studies of human DNA polymerase α: Lysine^{950} in the third most conserved region of α-like DNA polymerase is involved in binding deoxynucleotide triphosphate. *J. Biol. Chem.* **270:** 21563–21570.

Dong, Q., W.C. Copeland, and T.S. Wang. 1993a. Mutational studies of human DNA polymerase α. Identification of residues critical for deoxynucleotide binding and misinsertion fidelity of DNA synthesis. *J. Biol. Chem.* **268:** 24163–24174.

———. 1993b. Mutational studies of human DNA polymerase α. Serine^{867} in the second most conserved region among α-like DNA polymerases is involved in primer binding and mispair primer extension. *J. Biol. Chem.* **268:** 24175–24182.

Dornreiter, I., W.C. Copeland, and T.S.-F. Wang. 1993. Initiation of simian virus 40 DNA replication requires the interaction of a specific domain of human DNA polymerase α with large T antigen. *Mol. Cell. Biol.* **13:** 809–820.

Dornreiter, I., A. Hoss, A.K. Arthur, and E. Fanning. 1990. SV40 T antigen binds directly to the large subunit of purified DNA polymerase α. *EMBO J.* **9:** 3329–3336.

Downey, K.M., C.K. Tan, and A.G. So. 1990. DNA polymerase δ: A second eukaryotic DNA replicase. *BioEssays* **12:** 231–236.

Edenberg, H.J., S. Anderson, and P.M. DePamphilis. 1978. Involvement of DNA polymerase α in simian virus 40 DNA replication. *J. Biol. Chem.* **253:** 3273–3280.

Englander, E.W. and S.H. Wilson. 1992a. DNA damage response of cloned DNA β-polymerase promoter is blocked in mutant cell lines deficient in protein kinase A. *Nucleic Acids Res.* **20:** 5527–5531.

———. 1992b. The cloned promoter of the human DNA β-polymerase gene contains a cAMP response element functional in HeLa cells. *DNA Cell Biol.* **11:** 61–69.

Englander, E.W., S.G. Widen, and S.H. Wilson. 1991. Mammalian β-polymerase promoter: Phosphorylation of ATF/CRE-binding protein and regulation of DNA binding. *Nucleic Acids Res.* **19:** 3369–3375.

Enoch, T. and P. Nurse. 1991. Coupling M phase and S phase: Controls maintaining the dependence of mitosis on chromosome replication. *Cell* **65:** 921–923.

Enoch, T., A.M. Carr, and P. Nurse. 1992. Fission yeast genes involved in coupling mitosis to completion of DNA replication. *Genes Dev.* **6:** 2035-2046.

———. 1993. Checkpoint check (letter). *Nature* **361:** 26.

Enoch, T., K.L. Gould, and P. Nurse. 1991. Mitotic checkpoint control in fission yeast. *Cold Spring Harbor Symp. Quant Biol.* **56:** 409–416.

Falconi, M.M., A. Piseri, M. Ferrari, G. Lucchini, P. Plevani, and M. Foiani. 1993. De novo synthesis of budding yeast DNA polymerase α and POL1 transcription at the G1/S boundary are not required for entrance into S phase. *Proc. Natl. Acad. Sci.* **90:** 10519–10523.

Fisher, P.A. and D. Korn. 1981a. Ordered sequential mechanism of substrate recognition and binding by KB cell DNA polymerase α. *Biochemistry* **20:** 4560–4569.

———. 1981b. Properties of the primer-binding site and the role of magnesium ion in primer-template recognition by KB cell DNA polymerase α. *Biochemistry* **20:** 4570–4578.

Fisher, P.A., J.T. Chen, and D. Korn. 1981. Enzymological characterization of KB cell DNA polymerase-α. Regulation of template binding by nucleic acid base composition. *J. Biol. Chem.* **256:** 133–141.

Foiani, M., C. Santocanale, P. Plevani, and G. Lucchini. 1989. A single essential gene, PRI2, encodes the large subunit of DNA primase in *Saccharomyces cerevisiae. Mol. Cell. Biol.* **9:** 3081–3087.

Foiani, M., F. Marini, D. Gamba, G. Lucchini, and P. Plevani. 1994. The B subunit of the DNA polymerase α-primase complex in *Saccharomyces cerevisiae* executes an essential function at the initial stage of DNA replication. *Mol. Cell. Biol.* **14:** 923–933.

Fornace, A.J., B. Zmudzka, M.C. Hollander, and S.H. Wilson. 1989. Induction of beta-polymerase mRNA by DNA-damaging agents in Chinese hamster ovary cells. *Mol. Cell. Biol.* **9:** 851–853.

Foury, F. 1989. Cloning and sequencing of the nuclear gene *MIP1* encoding the catalytic subunit of the yeast mitochondrial DNA polymerase. *J. Biol. Chem.* **264:** 20552–20560.

Francesconi, S., W.C. Copeland, and T.S.-F. Wang. 1993a. In vivo species specificity of DNA polymerase α. *Mol. Gen. Genet.* **241:** 457–466.

Francesconi, S., H. Park, and T.S.-F. Wang. 1993b. Fission yeast with DNA polymerase δ temperature-sensitive alleles exhibits cell division cycle phenotype. *Nucleic Acid Res.* **21:** 3821–3828.

Friedberg, E.C. 1991. Eukaryotic DNA repair: Glimpses through the yeast *Saccharomyces cerevisiae. BioEssays* **13:** 295–302.

———. 1994. DNA repair: Looking back and peering forward. *BioEssays* **16:** 645–649.

Friedberg, E.C., A.J. Bardwell, L. Bardwell, Z. Wang, and G. Dianov. 1994. Transcription and nucleotide excision repair—Reflections, considerations and recent biochemical insights. *Mutat. Res.* **307:** 5–14.

Gordon, C.B. and J.L. Campbell. 1991. A cell cycle-responsive transcriptional control element and a negative control element in the gene encoding DNA polymerase α in *Saccharomyces cerevisiae. Proc. Natl. Acad. Sci.* **88:** 6058–6062.

Goulian, M., S.M. Herrmann, J.W. Sackett, and S.L. Grimm. 1990. Two forms of DNA polymerase δ from mouse cells. Purification and properties. *J. Biol. Chem.* **265:** 16402–16411.

Hirose, F., Y. Hotta, M. Yamaguchi, and A. Matsukage. 1989. Difference in the expression level of DNA polymerase β among mouse tissues: High expression in the pachytene spermatocyte. *Exp. Cell Res.* **181:** 169–180.

Hirose, F., M. Yamaguchi, Y. Nishida, M. Masutani, H. Miyazawa, F. Hanaoka, and A. Matsukage. 1991. Structure and expression during development of *Drosophila melanogaster* gene for DNA polymerase α. *Nucleic Acids Res.* **19:** 4991–4998.

Hofmann, J.X. and D. Beach. 1994. cdt1 is an essential target of the cdc10/Sct1 transcription factor: Requirement for DNA replication and inhibition of mitosis. *EMBO J.* **13:** 425–434.

Hsi, K.L., W.C. Copeland, and T.S. Wang. 1990. Human DNA polymerase α catalytic polypeptide binds ConA and RCA and contains a specific labile site in the N-terminus. *Nucleic Acids Res.* **18:** 6231–6237.

Hu, S.Z., T.S. Wang, and D. Korn. 1984. DNA primase from KB cells. Evidence for a novel model of primase catalysis by a highly purified primase/polymerase-alpha complex. *J. Biol. Chem.* **259:** 2602–2609.

Hurwitz, J., F.B. Dean, A.D. Kwong, and S.-H. Lee. 1990. The *in vitro* replication of DNA containing the SV40 origin. *J. Biol. Chem.* **265:** 18043–18046.

Ito, J. and D.K. Braithwaite. 1991. Compilation and alignment of DNA polymerase sequences. *Nucleic Acids Res.* **19:** 4045–4057.

Jacobo-Molina, A., J. Ding, R.G. Nanni, A.D. Clark, X. Lu, C. Tantillo, R.L. Williams, G. Kamer, A.L. Ferris, P. Clark, A. Hizi, S.H. Hughes, and E. Arnold. 1993. Crystal structure of human immunodeficiency virus type I reverse transcriptase complexed with double-stranded DNA at 3.3 Å resolution shows bent DNA. *Proc. Natl. Acad. Sci.* **90:** 6320–6324.

Jessberger, R., V. Podust, U. Hübscher, and P. Berg. 1993. A mammalian protein complex that repairs double-strand breaks and deletions by recombination. *J. Biol. Chem.* **268:** 15070–15079.

Johnson, L.M., M. Snyder, L.M.-S. Chang, R.W. Davis, and J.L. Campbell. 1985. Isolation of the gene encoding yeast DNA polymerase I. *Cell* **43:** 369–377.

Johnston, L.H., J.H. White, A.L. Johnson, G. Lucchini, and P. Plevani. 1987. The yeast DNA polymerase I transcript is regulated in both the mitotic cell cycle and in meiosis and is also induced after DNA damage. *Nucleic Acids Res.* **15:** 5017–5030.

———. 1990. Expression of the yeast DNA primase gene, *PRI1* is regulated within the mitotic cell cycle and meiosis. *Mol. Gen. Genet.* **221:** 44–48.

Joyce, C.M. and T.A. Steitz. 1994. Function and structure relationship in DNA polymerases. *Annu. Rev. Biochem.* **63:** 777–822.

Kaguni, L.S. and I.R. Lehman. 1988. Eukaryotic DNA polymerase-primase: Structure, mechanism and function. *Biochim. Biophys. Acta* **950:** 87–101.

Kedar, P.S., D.R. Lowy, S.G. Widen, and S.H. Wilson. 1990. Transfected human β-polymerase promoter contains a *ras*-responsive element. *Mol. Cell. Biol.* **10:** 3852–3856.

Kedar, P.S., S.G. Widen, E.W. Englander, A.J. Fornace, and S.H. Wilson. 1991. The ATF/CREB transcription factor-binding site in the polymerase β promoter mediates the positive effect of N-methyl-N′-nitro-N-nitrosoguanidine on transcription. *Proc. Natl. Acad. Sci.* **88:** 3729–3733.

Keeney, S., G.J. Chang, and S. Linn. 1993. Characterization of a human DNA damage binding protein implicated in xeroderma pigmentosum E. *J. Biol. Chem.* **268:** 21293–300.

Kelly, T.J., G.S. Martin, S.L. Forsburg, R.J. Stephen, A. Russo, and P. Nurse. 1993. The fission yeast *cdc18*$^{+}$ gene product couples S phase to START and mitosis (see comments). *Cell* **74:** 371–382.

Kemper, R.R., E.R. Ahn, P. Zhang, M.Y. Lee, and M. Rabin. 1992. Human DNA polymerase δ gene maps to region 19q13.3-q13.4 by in situ hybridization. *Genomics* **14:** 205–206.

Kesti, T., H. Frantti, and J.E. Syvaoja. 1993. Molecular cloning of the cDNA for the catalytic subunit of human DNA polymerase ε. *J. Biol. Chem.* **268:** 10238–10245.

Kohlstaedt, L.A., J. Wang, J.M. Friedman, P.A. Rice, and T.A. Steitz. 1992. Crystal structure at 3.5 Å resolution of HIV-1 reverse transcriptase complexed with an inhibitor. *Science* **256:** 1783–1790.

Krokan, H., P. Schaffer, and M.L. DePamphilis. 1979. Involvement of eucaryotic deoxyribonucleic acid polymerases alpha and gamma in the replication of cellular and viral deoxyribonucleic acid. *Biochemistry* **18:** 4431–4443.

Kumar, A., J. Abbotts, E.M. Karawya, and S.H. Wilson. 1990a. Identification and properties of the catalytic domain of mammalian DNA polymerase β. *Biochemistry* **29:** 7156–7159.

Kumar, A., S.G. Widen, K.R. Williams, P. Kedar, R.L. Karpel, and S.H. Wilson. 1990b. Studies of the domain structure of mammalian DNA polymerase β. Identification of a discrete template binding domain. *J. Biol. Chem.* **265:** 2124–2131.

Lee, S.H., A.D. Kwong, Z.Q. Pan, and J. Hurwitz. 1991a. Studies on the activator 1 protein complex, an accessory factor for proliferating cell nuclear antigen-dependent DNA polymerase delta. *J. Biol. Chem.* **266:** 594–602.

Lee, S.H., Z.Q. Pan, A.D. Kwong, P.M. Burgers, and J. Hurwitz. 1991b. Synthesis of DNA by DNA polymerase ε in vitro. *J. Biol. Chem.* **266:** 22707–22717.

Leem, S.H., P.A. Ropp, and A. Sugino. 1994. The yeast *Saccharomyces cerevisiae* DNA polymerase IV: Possible involvement in double strand break DNA repair. *Nucleic Acids Res.* **22:** 3011–3017.

Lewis, D.L., C.L. Farr, A.L. Farquhar, and L.S. Kaguni. 1994. Sequence, organization, and evolution of the A+T region of *Drosophila melanogaster* mitochondrial DNA. *Mol. Biol. Evol.* **11:** 523–538.

Lucchini, G., S. Francesconi, M. Foiani, G. Badaracco, and P. Plevani. 1987. Yeast DNA polymerase—DNA primase complex; cloning of *PRI1*, a single essential gene related to DNA primase activity. *EMBO J.* **6:** 737–742.

Matsukage, A., M. Yamaguchi, K.R. Utsumi, Y. Hayashi, R. Ueda, and M.C. Yoshida. 1986. Assignment of the gene for human DNA polymerase β (POLB) to chromosome 8. *Jpn. J. Cancer Res.* **77:** 330–333.

Matsumoto, Y. and K. Kim. 1995. Excision of deoxyribose phosphate residues by DNA polymerase β during DNA repair. *Science* **269:** 699–701.

Matsumoto, Y., K. Kim, and D.F. Bogenhagen. 1994. Proliferating cell nuclear antigen-dependent abasic site repair in *Xenopus laevis* oocytes: An alternative pathway of base excision DNA repair. *Mol. Cell. Biol.* **14:** 6187–6197.

McBride, O.W., B.Z. Zmudzka, and S.H. Wilson. 1987. Chromosomal location of the human gene for DNA polymerase β. *Proc. Natl. Acad. Sci.* **84:** 503–507.

McIntosh, E.M., T. Atkinson, R.K. Storms, and M. Smith. 1991. Characterization of a short, *cis*-acting DNA sequence which conveys cell cycle stage-dependent transcription in *Saccharomyces cerevisiae*. *Mol. Cell. Biol.* **11:** 329–337.

Miyazawa, H., M. Izumi, S. Tada, R. Takada, M. Masutani, M. Ui, and F. Hanaoka. 1993. Molecular cloning of the cDNAs for the four subunits of mouse DNA polymerase α-primase complex and their gene expression during cell proliferation and the cell cycle. *J. Biol. Chem.* **268:** 8111–8122.

Moore, A.L. and T.S.-F. Wang. 1994. Down-regulation of genes encoding DNA replication proteins during the cell cycle exit. *Cell Growth Differ.* **5:** 485–494.

Morris, G.F. and M.B. Mathews. 1989. Regulation of proliferating cell nuclear antigen during the cell cycle. *J. Biol. Chem.* **264:** 13856–13864.

Morrison, A. and A. Sugino. 1992. Roles of *POL3*, *POL2* and *PMS1* genes in maintaining accurate DNA replication. *Chromosoma* (suppl.) **102:** S147–S149.

Morrison, A., H. Araki, A.B. Clark, R.K. Hamatake, and A. Sugino. 1990. A third essential DNA polymerase in *S. cerevisiae*. *Cell* **62:** 1143–1151.

Mosbaugh, D.W. and S. Linn. 1983. Excision repair and DNA synthesis with a combination of HeLa DNA polymerase β and DNase V. *J. Biol. Chem.* **258:** 108–118.

Murakami, H. and H. Okayama. 1995. A kinase from fission yeast responsible for blocking mitosis in S phase. *Nature* **374:** 817–819.

Nasheuer, H.-P., A. Moore, A.F. Wahl, and T.S.-F. Wang. 1991. Cell cycle-dependent phosphorylation of human DNA polymerase α. *J. Biol. Chem.* **266:** 7893–7903.

Navas, T.A., Z. Zhou, and S.J. Elledge. 1995. DNA polymerase ε links DNA replication machinery to the S phase checkpoint. *Cell* **80:** 29–39.

Nethanel, T. and G. Kaufmann. 1990. Two DNA polymerases may be required for synthesis of the lagging DNA strand of simian virus 40. *J. Virol.* **64:** 5912–5918.

Nethanel, T., T. Zlotkin, and G. Kaufmann. 1992. Assembly of simian virus 40 Okazaki pieces from DNA primers is reversibly arrested by ATP depletion. *J. Virol.* **66:** 6634–6640.

Nishida, C., P. Reinhard, and S. Linn. 1988. DNA repair synthesis in human fibroblast requires DNA polymerase δ. *J. Biol. Chem.* **263:** 501–510.

Nowak, R., J.A. Siedlecki, L. Kaczmarek, B.Z. Zmudzka, and S.H. Wilson. 1989. Levels and size complexity of DNA polymerase β mRNA in rat regenerating liver and other organs. *Biochim. Biophys. Acta* **1008:** 203–207.

Olson, M.W. and L.S. Kaguni. 1992. 3′-5′exonuclease in *Drosophila* mitochondrial DNA polymerase. Substrate specificity and functional coordination of nucleotide polymerization and mispair hydrolysis. *J. Biol. Chem.* **267:** 23136–23142.

Park, H., R.E. Davis, and T.S.-F. Wang. 1995. Studies of *Schizosaccharomyces pombe* DNA polymerase α at different stages of the cell cycle. *Nucleic Acids Res.* **23:** 4337–4344.

Park, H., S. Francesconi, and T.S.-F. Wang. 1993. Cell cycle expression of two replicative DNA polymerases α and δ from *Schizosaccharomyces pombe*. *Mol. Biol. Cell.* **4:** 145–157.

Park, P., W. Copeland, L. Yang, T. Wang, M.R. Botchan, and I.J. Mohr. 1994. The cellular DNA polymerase α-primase is required for papillomavirus DNA replication and associates with the viral E1 helicase. *Proc. Natl. Acad. Sci.* **91:** 8700–8704.

Pearson, B.E., H.-P. Nasheuer, and T.S.-F. Wang. 1991. Human DNA polymerase α gene: Sequence controlling expression in cycling and serum stimulated cells. *Mol. Cell. Biol.* **11:** 2081–2095.

Pelletier, H., M.R. Sawaya, A. Kumar, S.H. Wilson, and J. Kraut. 1994. Structures of ternary complexes of rat DNA polymerase β, a DNA template-primer, and ddCTP. *Science* **264:** 1891–1903.

Pignede, G., D. Bouvier, A.-M. Recondo, and G. Baldacci. 1991. Characterization of the *POL3* gene product from *S. pombe* indicates inter-species conservation of the catalytic subunit of DNA polymerase δ. *J. Mol. Biol.* **222:** 209–218.

Pignede, G., G. Moussy, D. Bouvier, J. Tillit, A.M de Recondo, and G. Baldacci. 1992.

Expression of the catalytic subunits of pol α and pol δ from fission yeast *Schizosaccharomyces pombe. Chromosoma* (suppl.) **102:** S128–S132.

Podust, V.N. and U. Hübscher. 1993. Lagging strand DNA synthesis by calf thymus DNA polymerases α, β, δ and ε in the presence of auxiliary proteins. *Nucleic Acids Res.* **21:** 841–846.

Podust, V.N., A. Georgaki, B. Strack, and U. Hübscher. 1992. Calf thymus RF-C as an essential component for DNA polymerase δ and ε holoenzymes function. *Nucleic Acids Res.* **20:** 4159–4165.

Prasad, R., W.A. Beard, and S.H. Wilson. 1994. Studies of gapped DNA substrate binding by mammalian DNA polymerase β. Dependence on 5′-phosphate group. *J. Biol. Chem.* **269:** 18096–18101.

Prasad, R., A. Kumar, S.G. Widen, F.J. Casas, and S.H. Wilson. 1993a. Identification of residues in the single-stranded DNA-binding site of the 8-kDa domain of rat DNA polymerase β by UV cross-linking. *J. Biol. Chem.* **268:** 22746–22755.

Prasad, R., S.G. Widen, R.K. Singhal, J. Watkins, L. Prakash, and S.H. Wilson. 1993b. Yeast open reading frame YCR14C encodes a DNA β-polymerase-like enzyme. *Nucleic Acids Res.* **21:** 5301–5307.

Prelich, G., C. Tan, M. Kostura, M.B. Mathews, A.G. So, K.M. Downey, and B. Stillman. 1987. Functional identity of proliferating cell nuclear antigen and a DNA polymerase α auxiliary protein. *Nature* **326:** 517–520.

Prussak, C.E., M.T. Almazan, and B.Y. Tseng. 1989. Mouse primase p49 subunit molecular cloning indicates conserved and divergent regions. *J. Biol. Chem.* **264:** 4957–4963.

Reardon, J.T., A.F. Nichols, S. Keeney, C.A. Smith, J.S. Taylor, S. Linn, and A. Sancar. 1993. Comparative analysis of binding of human damaged DNA-binding protein (XPE) and *Escherichia coli* damage recognition protein (UvrA) to the major ultraviolet photoproducts: T[c,s]T, T[t,s]T, T[6-4]T, and T[Dewar]T. *J. Biol. Chem.* **268:** 21301–21308.

Sabatino, R.D., T.W. Myers, and R.A. Bambara. 1990. Substrate specificity of the exonuclease associated with calf DNA polymerase. *Cancer Res.* **50:** 5340–5344.

Saka, Y. and M. Yanagida. 1993. Fission yeast *cut5*$^+$, required for S phase onset and M phase restraint, is identical to the radiation-damage repair gene *rad4*$^+$. *Cell* **74:** 383–393.

Sawaya, M.R., H. Pelletier, A. Kumar, S.H. Wilson, and J. Kraut. 1994. Crystal structure of rat DNA polymerase β: Evidence for a common polymerase mechanism. *Science* **264:** 1930–1935.

Sherley, J.L. and T.J. Kelly. 1988. Regulation of human thymidine kinase during the cell cycle. *J. Biol. Chem.* **263:** 8350–8358.

Singh, J. and A.J. Klar. 1993. DNA polymerase-α is essential for mating-type switching in fission yeast. *Nature* **361:** 271–273.

Singhal, R.K. and S.H. Wilson. 1993. Short gap-filling synthesis by DNA polymerase β is processive. *J. Biol. Chem.* **268:** 15906–15911.

Singhal, R.K., R. Prasad, and S.H. Wilson. 1995. DNA polymerase β conducts the gap-filling step in uracil-initiated base excision repair in a bovine testis nuclear extract. *J. Biol. Chem.* **270:** 949–957.

So, A.G. and K.M. Downey. 1988. Mammalian DNA polymerases α and δ: Current status in DNA replication. *Biochemistry* **27:** 4591–4595.

———. 1992. Eukaryotic DNA replication. *Crit. Rev. Biochem. Mol. Biol.* **27:** 129–155.

Sousa, R., Y.J. Chung, J.P. Rose, and B.C. Wang. 1993. Crystal structure of bacteriophage T7 RNA polymerase at 3.3 Å resolution. *Nature* **364:** 593–599.

Stadlbauer, F., A. Brueckner, C. Rehfuess, C. Eckerskorn, F. Lottspeich, V. Forster, B.Y. Tseng, and H.P. Nasheuer. 1994. DNA replication in vitro by recombinant DNA-polymerase-α-primase. *Eur. J. Biochem.* **222:** 781–793.

Stillman, B. 1989. Initiation of eukaryotic DNA replication *in vitro*. *Annu. Rev. Cell Biol.* **5:** 197–245.

———. 1994. Initiation of chromosomal DNA replication in eukaryotes. Lessons from lambda. *J. Biol. Chem.* **269:** 7047–7050.

Sugino, A. 1995. Yeast DNA polymerases and their role at the replication fork. *Trends Pharmacol. Sci.* **20**: 319–323.

Syvaoja, J.E. 1990. DNA polymerase ε: The latest member in the family of mammalian DNA polymerases. *BioEssays.* **12:** 533–536.

Syvaoja, J. and S. Linn. 1989. Characterization of a large form of DNA polymerase δ from HeLa cells that is insensitive to proliferating cell nuclear antigen. *J. Biol. Chem.* **264:** 2489–2497.

Syvaoja, J., S. Suomensaari, C. Nishida, J.S. Goldsmith, G.S. Chui, S. Jain, and S. Linn. 1990. DNA polymerases α, δ, and ε: Three distinct enzymes from HeLa cells. *Proc. Natl. Acad. Sci.* **87:** 6664–6668.

Szpirer, J., F. Pedeutour, T. Kesti, M. Riviere, J.E. Syvaoja, C.C. Turc, and C. Szpirer. 1994. Localization of the gene for DNA polymerase ε (POLε) to human chromosome 12q24.3 and rat chromosome 12 by somatic cell hybrid panels and fluorescence in situ hybridization. *Genomics* **20:** 223–226.

Tan, C.K., C. Castillo, A.G. So, and K.M. Downey. 1986. An auxiliary protein for DNA polymerase-delta from fetal calf thymus. *J. Biol. Chem.* **261:** 12310–12316.

Tanabe, K., E.W. Bohn, and S.H. Wilson. 1979. Steady-state kinetics of mouse DNA polymerase β. *Biochemistry* **18:** 3401–3406.

Tseng, B.Y., C.E. Prussak, and M.T. Almazan. 1989. Primase p49 mRNA expression is serum stimulated but does not vary with the cell cycle. *Mol. Cell. Biol.* **9:** 1940–1945.

Tsurimoto, T., T. Melendy, and B. Stillman. 1990. Sequential initiation of lagging and leading strand synthesis by two different polymerase complexes at the SV40 DNA replication origin. *Nature* **346:** 534–539.

Waga, S. and B. Stillman. 1994. Anatomy of a DNA replication fork revealed by reconstitution of SV40 DNA replication in vitro. *Nature* **369:** 207–212.

Waga, S., G. Bauer, and B. Stillman. 1994. Reconstitution of complete SV40 DNA replication with purified replication factors. *J. Biol. Chem.* **269:** 10923–10934.

Wahl, A.F., A.M. Geis, B.H. Spain, S.W. Wong, D. Korn, and T.S.-F. Wang. 1988. Gene expression of human DNA polymerase a during cell proliferation and the cell cycle. *Mol. Cell. Biol.* **8:** 5016–5025.

Wang, T.S.-F. 1991. Eukaryotic DNA polymerases. *Annu. Rev. Biochem.* **60:** 513–552.

Wang, T.S.-F. and D. Korn. 1980. Reactivity of KB cell deoxyribonucleic acid polymerases α and β with nicked and gapped deoxyribonucleic acid. *Biochemistry* **19:** 1782–1790.

———. 1982. Specificity of the catalytic interaction of human DNA polymerase β with nucleic acid substrates. *Biochemistry* **21:** 1597–1608.

Wang, T.S.-F., S.W. Wong, and D. Korn. 1989. Human DNA polymerase α: Predicted functional domains and relationships with viral DNA polymerases. *FASEB J.* **3:** 14–21.

Wang, T.S.-F., W.C. Copeland, L. Rogge, and Q. Dong. 1995. DNA replication: Purification of mammalian DNA polymerases—DNA polymerase α. *Methods Enzymol.* **262:** 77–84.

Wang, T.S.-F., B.E. Pearson, H.A. Suomalainen, T. Mohandas, L.J. Shapiro, J. Schroder, and D. Korn. 1985. Assignment of the gene for human DNA polymerase α to the X chromosome. *Proc. Natl. Acad. Sci.* **82:** 5270–5274.

Wang, Z., X. Wu, and E.C. Friedberg. 1993. DNA repair synthesis during base excision repair in vitro is catalyzed by DNA polymerase ε and is influenced by DNA polymerase α and δ in *Saccharomyces cerevisiae*. *Mol. Cell. Biol.* **13:** 1051–1058.

Weiser, T., M. Gassmann, P. Thommes, E. Ferrari, P. Hafkemeyer, and U. Hübscher. 1991. Biochemical and functional comparison of DNA polymerases α, δ, and ε from calf thymus. *J. Biol. Chem.* **266:** 10420–10428.

Widen, S.G. and S.H. Wilson. 1991. Mammalian beta-polymerase promoter: Large-scale purification and properties of ATF/CREB palindrome binding protein from bovine testes. *Biochemistry* **30:** 6296–6305.

Widen, S.G., P. Kedar, and S.H. Wilson. 1988. Human β-polymerase gene. Structure of the 5′-flanking region and active promoter. *J. Biol. Chem.* **263:** 16992–16998.

Williams, A.J. and L.S. Kaguni. 1995. Stimulation of *Drosophila* mitochondrial DNA polymerase by single-stranded DNA-binding protein. *J. Biol. Chem.* **270:** 860–865.

Williams, A.J., C.M. Wernette, and L.S. Kaguni. 1993. Processivity of mitochondrial DNA polymerase from *Drosophila* embryos. Effects of reaction conditions and enzyme purity. *J. Biol. Chem.* **268:** 24855–24862.

Wong, S.W., A.F. Wahl, P.M. Yuan, N. Arai, B.E. Pearson, K.-I. Arai, D. Korn, M.W. Hunkapiller, and T.S.-F. Wang. 1988. Human DNA polymerase α gene expression is cell proliferation dependent and its primary structure is similar to both prokaryotic and eukaryotic replicative DNA polymerases. *EMBO J.* **7:** 37–47.

Yang, C.L., L.S. Chang, P. Zhang, H. Hao, L. Zhu, N.L. Toomey, and M.Y. Lee. 1992. Molecular cloning of the cDNA for the catalytic subunit of human DNA polymerase δ. *Nucleic Acids Res.* **20:** 735–745.

Zeng, X.R., H. Hao, Y. Jiang, and M.Y. Lee. 1994a. Regulation of human DNA polymerase δ during the cell cycle. *J. Biol. Chem.* **269:** 24027–24033.

Zeng, X.R., Y. Jiang, S.J. Zhang, H. Hao, and M.Y. Lee. 1994b. DNA polymerase δ is involved in the cellular response to UV damage in human cells. *J. Biol. Chem.* **269:** 13748–13751.

Zhang, J., C.-K. Tan, B. McMullen, K.M. Downey, and A.G. So. 1995. Cloning of the cDNAs for the small subunits of bovine and human DNA polymerase δ and chromosomal location of the human gene. *Genomics* **29:** 179–186.

Zhang, J., D.W. Chung, C.-K. Tan, K.M. Downey, E.W. Davie, and A.G. So. 1991. Primary structure of the catalytic subunit of calf thymus DNA polymerase δ: Sequence similarities with other DNA polymerases. *Biochemistry* **30:** 11742–11750.

Zmudzka, B.Z., A. Fornace, J. Collins, and S.H. Wilson. 1988. Characterization of DNA polymerase β mRNA: Cell-cycle and growth response in cultured human cells. *Nucleic Acids Res.* **16:** 9587–9596.

16

Viral DNA Polymerases

Donald M. Coen
Department of Biological Chemistry and Molecular Pharmacology
Harvard Medical School
Boston, Massachusetts 02115

Because many viruses are readily amenable to molecular, genetic, and biochemical analyses, and because the replication of certain viruses has been the target of antiviral drug development, a wealth of information has accumulated regarding the structure, functions, properties, and regulation of eukaryotic viral DNA polymerases. Viral DNA polymerases have interesting similarities with eukaryotic cellular DNA polymerases (Wang, this volume) and fascinating differences, both of which shed light on mechanisms of polymerase function.

With the exception of the papovaviruses and parvoviruses, which utilize cellular DNA polymerases for their replication, all known DNA viruses that infect animal cells encode their own DNA polymerases. In each case, the DNA polymerase is essential for the replication of the virus. Why have these viruses evolved to encode and require their own DNA polymerases? (In this chapter, only DNA polymerases that strictly utilize DNA templates are considered. Information about the reverse transcriptases encoded by retroviruses or the polymerases encoded by hepadnaviruses can be found in Skalka and Goff [1993] and in Seeger and Mason [this volume].) For poxviruses, which replicate in the cytoplasm (Traktman, this volume), the answer may be that the viral genome does not gain access to the cellular polymerases. For other viruses, the answer may be that their normal life cycles entail the productive infection of nondividing cells that do not express sufficient cellular polymerase.

Despite the fact that these viruses have each evolved a unique DNA polymerase that cannot be replaced by cellular polymerases, the viral enzymes share considerable sequence homology with eukaryotic DNA polymerases α, δ, and ε as well as *Escherichia coli* DNA polymerase II and the DNA polymerases of certain bacteriophage such as T4 and ϕ29 (Wong et al. 1988; Ito and Brathwaite 1991). Moreover, based on sequence alignments among a wide variety of DNA and RNA polymerases and the crystal structures of the Klenow fragment of *E. coli* DNA

DNA Replication in Eukaryotic Cells

polymerase I, rat DNA polymerase β, human immunodeficiency virus reverse transcriptase, and T7 RNA polymerase, the case has been made that all polymerases share common structures and sequence motifs with specific functions, as discussed by Wang (this volume). Studies of viral polymerases have been helpful in understanding the functions of these motifs, as discussed in this chapter.

This chapter summarizes information on the functions, enzymology, pharmacology, genetics, protein-protein interactions, and regulation (often posttranscriptional) of selected viral DNA polymerases. The subunit composition and functions of several viral polymerases are presented in Table 1. For each polymerase, the holoenzyme is the form found in the virus-infected cell. However, the enzymes are often readily prepared via the use of heterologous expression systems such as recombinant baculovirus-infected cells, permitting the overexpression, purification, and analysis of the holoenzyme or each subunit.

For those viral DNA polymerases that have been most intensively studied, primer-template preferences, relevant inhibitors, and phenotypes of interesting mutants are presented in additional tables. For different polymerases, different properties are emphasized. The first group of polymerases to be discussed are those of herpesviruses. The focus is on the prototype herpes simplex virus (HSV) enzyme, which has been the object of more study than other viral polymerases. This is partly due to its being an excellent target for (profitable) antiviral drugs and to its being amenable to biochemical and genetic analysis, often with the aid of mutants isolated for resistance to antiviral drugs. The poxvirus DNA polymerases are represented by the vaccinia virus enzyme, whose features are outlined by Traktman (this volume). Further details, particularly regarding its genetics, are presented here. The third group of enzymes to be covered are those of adenoviruses (mainly adenovirus types 2 and 5). Much of the relevant information about adenovirus DNA polymerases can be found in Hay (this volume); thus, the presentation here is limited to aspects of its enzymology, genetics, and functions. Finally, aspects of the regulation of viral polymerases, in which there are interesting similarities among diverse systems, are summarized.

HSV DNA POLYMERASE

Herpesviruses

There are a multitude of herpesviruses that infect fish, birds, and mammals. Among the seven or more human herpesviruses are HSV (types 1 and 2), varicella zoster virus (VZV), human cytomegalovirus (CMV),

Table 1 Subunits of viral DNA polymerases

Virus	Catalytic subunit (names[a]; mass[b]; functions[c])	Other subunits (names; mass; functions)	Composition
HSV	UL30, pol; 137 kD; polymerase 3′→5′Exo, RNase H	UL42, $65K_{DBP}$; 52 kD; DNA binding, processivity	pol:UL42, 1:1
CMV	UL54, pol; 137 kD; polymerase 3′→5′Exo[d]	UL44, ICP36, p52; 46 kD; DNA binding, processivity[e]	unknown
EBV	BALF5, pol; 113 kD; polymerase, 3′→5′Exo	BMRF1, EA-D; 43 kD; DNA binding, processivity[e]	pol:BMRF1, 1:1[f]
Adeno	E2B, pol; 135 kD; polymerase 3′→5′Exo	pTP; 80 kD; protein primer, nuclear transport, origin-recognition	pol:pTP, 1:1
Vaccinia	E9, pol; 116 kD; polymerase 3′→5′Exo	none known	pol monomer
Baculo	pol; 114 kD; polymerase 3′→5′Exo[d]	none known	unknown
ASFV	G1207R, pol; 142 kD; polymerase	none known	unknown

[a]For each polypeptide, the first name listed is the name of the gene, usually the name of the open reading frame in the DNA sequence (HSV, CMV, EBV, vaccinia, ASFV). Most investigators refer to the catalytic subunits also as Pol or pol.

[b]Masses are those predicted from the DNA sequences.

[c]For catalytic subunits, only intrinsic enzymatic activities are listed. For other subunits, activities imparted on the catalytic subunit are included as well as intrinsic binding activities. Each subunit, of course, also interacts with the others.

[d]The exonuclease activity has not yet been shown rigorously to be intrinsic.

[e]Processivity has not yet been shown rigorously by template challenge.

[f]T. Tsurumi (pers. comm.).

and Epstein-Barr virus (EBV). Infections with these viruses are common, and because of their propensity to establish lifelong latent infections, these viruses often recur to cause disease. In most immunocompetent adults, herpesvirus diseases are not life-threatening but can be at least temporarily debilitating, painful, and/or emotionally troublesome. In immunocompromised adults and neonates, herpesviruses can cause severe disease and death. These clinical properties have encouraged pharmaceutical houses to develop drugs active against herpesviruses, especially HSV.

The features of herpesvirus DNA replication are reviewed by Challberg (this volume) and Yates (this volume). Germane to this chapter is that all herpesviruses encode a DNA polymerase. In each case examined thus far, the catalytic activity resides in a large subunit (110–140 kD), and there is a smaller subunit that binds DNA and stimulates polymerase activity (Table 1).

Enzymology and Pharmacology

HSV DNA polymerase consists of a catalytic subunit (pol) and a smaller subunit, UL42 (Table 1). Distinctive enzymological features of the holoenzyme and of the catalytic subunit, which has been overexpressed by itself using baculovirus expression vectors (Hernandez and Lehman 1990; Marcy et al. 1990b), are presented in Table 2. HSV DNA polymerase is able to utilize a variety of primer templates, notably poly(dC)oligo(dG), and is able to extend RNA primers. It is distinguished from most cellular polymerases by its stimulation by relatively high concentrations of salt, its relatively low K_m values for dNTPs, and its sensitivity to a variety of inhibitors. This sensitivity contributes to the success of several antiviral drugs, especially acyclovir. Acyclovir, which is a guanine base attached to an acyclic sugar moiety, is converted to the monophosphate mainly via an HSV-encoded thymidine kinase. The triphosphate, formed from the monophosphate via the action of cellular enzymes, is a much more potent inhibitor of the viral DNA polymerase than it is of cellular polymerases (Martin et al. 1994). The mechanism of inhibition entails three steps (Reardon and Spector 1989): (1) binding, which can be competed by dGTP; (2) incorporation into the growing DNA chain, leading to chain termination (acyclovir lacks a 3′ hydroxyl); (3) very potent inhibition by triphosphates complementary to the next position in the template, which prevents dissociation of the enzyme from the primer template.

Table 2 Properties of herpesvirus DNA polymerases

Form of enzyme	Salt and other optima for polymerase activity[a]	K_m for dNTPs[a]	Preferred primer templates[b]	Selected inhibitors
HSV holoenzyme (pol-UL42)	NaCl, KCl: 200–250 mM $(NH_4)_2SO_4$: 100–150 mM	0.05–0.5 μM	poly(dA)oligo(dT) poly(dT)oligo(dA)	acyclovir triphosphate phosphonoformic acid aphidicolin peptide corresponding to carboxy-terminal 36 residues of pol
HSV catalytic subunit (pol) only	less established; one report indicates lower salt concentrations are optimal (Hart and Boehme 1992)	similar	activated DNA poly(dT)oligo(rA) poly(dC)oligo(dG)	sensitive to all of the above except peptide
CMV holoenzyme (pol-UL44)	NaCl, KCl: 100–150 mM $(NH_4)_2SO_4$: 60–100 mM	0.5 μM	activated DNA poly(dA)oligo(dT) primed ss bacteriophage DNA	acyclovir triphosphate phosphonoformic acid aphidicolin
CMV catalytic subunit (pol) only	similar	less established	poly(dC)oligo(dG)	similar
EBV holoenzyme (pol-BMRF1)	$(NH_4)_2SO_4$: 100 mM	0.5 μM	poly(dA)oligo(dT) poly(dC)oligo(dG) primed ss bacteriophage DNA	acyclovir triphosphate phosphonoacetic acid aphidicolin
EBV catalytic subunit (pol) only	$(NH_4)_2SO_4$: 0 mM		activated DNA poly(dT)oligo(rA)	similar

Selected references: HSV, Weissbach et al. (1973); Derse and Cheng (1981); Derse et al. (1982); Frank et al. (1984); Reardon and Spector (1989); Gottleib et al. (1990); Hart and Boehme (1992); Hamatake et al. (1993); Digard et al. (1995). CMV, Nishiyama et al. (1983); Mar et al. (1985); Ertl and Powell (1992); Weiland et al. (1994). EBV, Chiou and Cheng (1985); Tsurumi et al. (1993a,b).

[a]Using activated DNA as primer template. Ranges reflect differences in published values and among different dNTPs. Different values are obtained with other primer templates.

[b]Those listed for catalytic subunit only are utilized with similar efficiency by the holoenzyme. Those only listed for holoenzyme are generally utilized much more efficiently by the holoenzyme than the catalytic subunit.

HSV DNA polymerase contains an intrinsic 3′→5′ exonuclease activity and an intrinsic RNase H activity (Crute and Lehman 1989; Marcy et al. 1990b; Weisshart et al. 1994). There is controversy regarding whether the RNase H functions as a 5′→3′ exonuclease (Crute and Lehman 1989) or whether it is instead a manifestation of the 3′→5′ exonuclease activity (Knopf and Weisshart 1990; Knopf and Strick 1993).

The catalytic subunit alone is much less active than the holoenzyme for polymerization on certain primer templates (Table 2), including poly(dA)oligo(dT) and singly primed M13 DNA (Gottleib et al. 1990; Hamatake et al. 1993). This is due to the ability of the small subunit, UL42, to stimulate processivity (Gottleib et al. 1990; Hernandez and Lehman 1990), which correlates with increased affinity for DNA in a primer-template configuration (Gottleib and Challberg 1994; B.D. Song; D. Herschlag; I.R. Lehman; all pers. comm.; K. Weisshart and D.M. Coen, unpubl.). UL42 is an unusual processivity factor in that it has intrinsic affinity for double-stranded DNA, which is required for processivity (Chow and Coen 1995).

UL42 also binds specifically to the catalytic subunit, and this binding too is required for processivity (Digard et al. 1993a,b). It is likely that these properties are shared by other small herpesvirus polymerase subunits like EBV BMRF1 and CMV UL44, which also have intrinsic DNA-binding activity and stimulate their cognate catalytic subunits (Tables 1 and 2). The current working model then is that UL42 simultaneously binds pol and DNA and thereby increases affinity for primer template. How this can occur without braking elongation (a problem solved by processivity factors such as proliferating cell nuclear antigen by wrapping around DNA without binding, thereby creating a "sliding clamp") is an area of active investigation.

The interaction of HSV pol and UL42 is not only required for processivity in vitro; it is also required for viral replication in vivo (Digard et al. 1993a,b). Thus, this interaction could be the target for antiviral drugs. Residues crucial for the interaction lie at the extreme carboxyl terminus of pol (Digard et al. 1993a; Stow 1993; Tenney et al. 1993b). Accordingly, certain peptides corresponding to the carboxyl terminus of pol block UL42-dependent DNA synthesis catalyzed by pol at concentrations that had no effect on synthesis by pol alone or by heterologous polymerases (Table 2) (Marsden et al. 1994; Digard et al. 1995). Thus, these peptides represent a new class of polymerase inhibitors that act by blocking accessory-subunit-dependent synthesis. They or their structures may form the basis for the synthesis of antiviral drugs.

Genetics: Drug Resistance and Conserved Regions

HSV is an excellent genetic system, and this has been advantageous in studies of HSV DNA polymerase. Many genetic studies have focused on mutations in the *pol* gene that alter the sensitivity of the virus to drugs. Such mutations alter the catalytic subunit so that it is functional, yet exhibits altered specificity. As the drugs mimic and/or compete for binding with natural polymerase substrates, the mutations can be expected to alter amino acids that are within or impinge on the binding site for the substrate. Clusters of such mutations should identify regions that are directly or indirectly involved in binding the substrates.

In the HSV system, this approach has mainly identified amino acids in six regions that are conserved among many viral and cellular polymerases of the α-like family (also known as family B; see Wang, this volume). These regions are designated as I, II, III, V, and VII, which are found in almost all α-like polymerases, and δ-region C (Zhang et al. 1991), which is found in cellular DNA polymerases δ and ε, but not α (Fig. 1, Table 3). Thus, these regions are presumably conserved for their role in polymerase catalysis. Because regions II and III contain the greatest clustering of such mutations (Table 3), it was proposed that these segments would be likely to directly interact with substrates and drugs (Gibbs et al. 1988). As reviewed by Wang (this volume), there is consid-

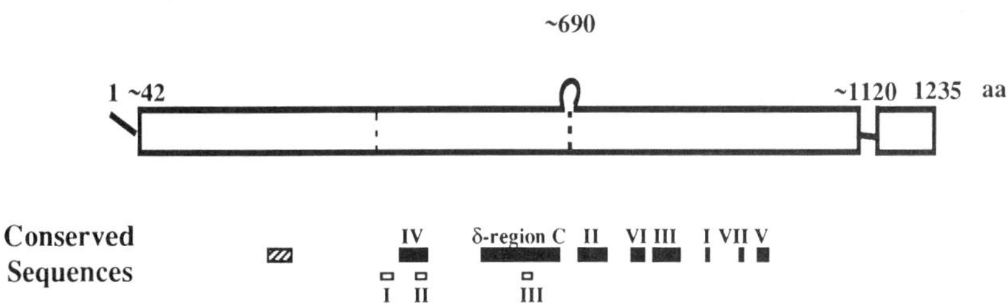

Figure 1 Locations of conserved regions and structural domains on viral DNA polymerases using the HSV DNA polymerase catalytic subunit as an example. The 1235-amino acid (aa) HSV pol polypeptide is represented as lines, indicating less-ordered regions susceptible to proteolytic digestion, with the residue numbers indicating the sites of cleavage (Weisshart et al. 1994) and boxes representing structural domains. Dashed lines indicate sites of proteolytic digestion that do not separate structural domains. Below are the locations of conserved regions I–VII shared among α-like polymerases (Wong et al. 1988; Hwang et al. 1992) and δ-region C (Zhang et al. 1991) as filled boxes; ExoI, II, and III (Bernad et al. 1989) as open boxes; and a segment with some homology to RNase H enzymes and certain α-like polymerases (Reha-Krantz 1990) as a hatched box.

Table 3 Altered drug sensitivity mutations affecting conserved regions of viral DNA polymerases

	δ-Region C (includes ExoIII and region A)	Region II
Consensus	XXXXLSSYKLNXVA..XXXEXARXXGIPhXRXhXXGQQ	YNGGXXhXPXXG..VhhDhNSLYPSIMX
	A	
HSV wild type	531	696
	DKIKLSSYKLNAVA..ELSAVARLAGINITRTIYDGQQ	YQGARVLDPTSG. .VVFDFASLYPSIIQ
HSV mutations	N K V	H G V N*
	D	
CMV mutations	I I	
VZV mutations	K	Y T
Vaccinia mutations	V	
	T	
	(wild type A)	

	Region III	Region I
Consensus	XQhhhKLXXNShYGXXGXXX..XXhhhXXhXXXGRX	hXhIYGDTDSIFV
HSV wild type	806	880
	QQAAIKVVCNSVYGFTGVQH..CLNVAATVTTIGRE	MRIIYGDTDSIFV
HSV mutations	M S* C M SS	A*CM
	C	Y
VZV mutations	S	L
Vaccinia mutations	M	

```
                    Region VII             Region V
Consensus           XKKXYhhX               KXXXKGXXXXRXXXX
HSV wild type       937                    949
                    AKKKYIGV               KMLIKGVDLVRKNNC
HSV mutations           H                              K
CMV mutations                                           G

                    ExoI                   ExoII
Consensus           hhXFDIEXXX             hXGXNXXXFDXXXIhXXL
CMV wild type       297                    404
                    CLSFDIECMS             VTGYNINSFDLKYILTRL
CMV mutations           N                          V
Vaccinia mutations           S                         G
                    (wt      F)
```

For each conserved region, the first line provides a consensus sequence showing residues shared among many family members in single-letter code with X indicating no clear consensus and h indicating a hydrophobic residue. In one case, where two amino acids predominate, both are indicated. The second line provides, in most cases, the wild-type HSV sequence and for ExoI and ExoII, the CMV sequence with the number of the first residue shown. Portions of the sequence omitted are indicated as "..". Below these are indicated residues found in *pol* mutations from viruses exhibiting altered drug sensitivity. In a few cases, it has not yet been demonstrated that the mutation observed actually causes the altered drug sensitivity phenotypes. References: HSV, Larder et al. (1987); Gibbs et al. (1988 and references therein); Hall et al. (1989); Hwang et al. (1992); Wang et al. (1992); Chiou et al. (1995). VZV, K. Biron (pers. comm.). CMV, Lurain et al. (1992); Sullivan et al. (1993); Lurain et al. (1994). Vaccinia, Earl et al. (1986); Traktman et al. (1989); Defilippes (1989); Taddie and Traktman (1991); Taddie and Traktman (1993). The asterisk (*) refers to mutations that affect drug sensitivity of HSV pol in vitro but which have not been tested in the authentic context of the virus genome. Thus, although the N in region II and the S in region III that are asterisked have been tested in the context of the virus genome, substitutions to A, D, E, H, K, Q, and T at the former position and A, D, E, G, L, Q, V, and Y (Matthews et al. 1993) at the latter position have not, but do affect drug sensitivity of HSV pol in vitro. The S to A substitution that is asterisked and italicized in region I was only tested in vitro (Dorksy and Plourde 1993).

erable support for this proposal based on structural studies of other polymerases that appear to contain homologs of these regions and from enzymological studies of mutant DNA polymerase-α and other α-like polymerases. Region I appears to be involved in catalysis, rather than affinity for substrates, so its involvement in substrate and drug recognition may be more indirect. Aside from drug-resistance mutations, certain mutations in this region can inactivate polymerase activity in vitro (Dorsky and Crumpacker 1990) and viral replication in vivo (Marcy et al. 1990a). The precise roles of regions V and VII are poorly understood at present.

The first mutations in δ-region C were suggested to identify a 60-residue segment (then known as region A) that was specific to viruses which are sensitive to antiviral drugs (Gibbs et al. 1988). Subsequently, this region was found to comprise a much longer segment (>100 residues in HSV) that is also conserved among cellular DNA polymerases δ and ε (Zhang et al. 1991). Interestingly, a drug-resistance mutation isolated by a suppressor approach maps to the amino-terminal portion of this segment (Wang et al. 1992). The mapping of the mutations suggests that δ-region C could also form part of the dNTP-binding site, even though it also contains the ExoIII segment, one of three regions (ExoI, II, and III, see Fig. 1) that are conserved among the active sites of 3′→5′ exonucleases (Bernad et al. 1989). These and other data (see below) argue for interdependence between portions of the enzyme involved in polymerization and exonuclease activities.

Of all the *pol* mutations that confer altered drug sensitivity to HSV, only two do not map to regions conserved among many α-like polymerases. One of these lies adjacent to region III and alters a proline and, thus, could conceivably act by shifting the orientation of region III (Gibbs et al. 1988). The other, isolated for suppression of phosphonoacetic acid (PAA)-hypersensitivity (Wang et al. 1992), lies in a region of sequence similarity shared among a few α-like polymerases and certain RNase Hs (Fig. 1) (Reha-Krantz 1990). The significance of this has not been established.

Several of the drug-resistant mutants exhibit an antimutator phenotype (Hall et al. 1984, 1985; Wang et al. 1992; Hwang and Chen 1995). For at least one mutant, PAA^{r}5, this appears to result from decreased misincorporation of incorrect bases during polymerization associated with higher K_m values for dNTPs rather than increased editing by the 3′→5′ exonuclease (Hall et al. 1985). At the sequence level, the wild-type polymerase mediates transition and frameshift mutations at about a 1:1 ratio in an HSV gene. Interestingly, the PAA^{r}5 enzyme exhibits an altered spectrum of mutations, being especially impaired in mediating transition mutations (Hwang and Chen 1995). The mechanisms underlying the antimutator phenotype for this and other mutants deserve further study.

Mapping of Activities

Numerous mutations have been constructed in the *pol* and *UL42* genes and tested for their effects on enzyme function in vitro and virus replication in vivo. Deletion analysis indicates that the catalytic subunit can suffer deletions as large as 227 residues at the amino terminus and 58 residues at the carboxyl terminus without loss of polymerase activity in vitro (Dorsky and Crumpacker 1988; Haffey et al. 1990), although the latter deletion and other mutations at the carboxyl terminus impair association with UL42 and its stimulation of long-chain DNA synthesis as well as viral replication (Digard et al. 1993a; Stow 1993; Tenney et al. 1993b; Marsden et al. 1994). Current data support the concept of an approximately 36-residue subdomain at the extreme carboxyl terminus of HSV pol that interacts with UL42 (Digard et al. 1995). This subdomain appears to be part of an approximately 12-kD carboxy-terminal structural domain identified by limited proteolysis (Fig. 1). This domain also has DNA-binding activity and is required for polymerase activity (Digard et al. 1993a; Weisshart et al. 1994; W.R. Bebrin et al., unpubl.). Limited proteolysis has also shown that the amino-terminal half of the pol molecule is sufficient for its 3′→5′ exonuclease and RNase H activities (Weisshart et al. 1994). However, deletions and single and clustered point mutations in many locations in the *pol* gene, including the amino-terminal half, ablate polymerase activity (Dorsky and Crumpacker 1988, 1990; Haffey et al. 1990; Gibbs et al. 1991; W.R. Bebrin et al., unpubl.). Interestingly, mutations within conserved region IV in the ExoII segment inactivate polymerization activity, yet the protein retains certain other activities (Gibbs et al. 1991). These data, taken together with the drug-resistance data cited above, tend *not* to support a model in which polymerase activity resides on a structurally and functionally independent polymerase domain à la *E. coli* DNA polymerase I (Derbyshire et al. 1993). Rather, it appears that HSV polymerase activity requires contributions from other domains and/or active sites. Related results in other viral systems are discussed below.

Limited proteolysis has also mapped sites of conformational changes that occur in pol upon DNA binding (Weisshart et al. 1993). One site is in or near the "hinge" between the 12-kD carboxy-terminal domain and the rest of the molecule (Fig. 1). The other is in the vicinity of δ-region C and region II (Fig. 1).

The UL42 subunit has been studied less extensively than pol. Its most surprising feature is an approximately 175-residue carboxy-terminal tail with no apparent function in vitro or in vivo (Digard et al. 1993b; Gao et al. 1993; Hamatake et al. 1993; Monahan et al. 1993; Tenney et al. 1993a). However, most mutations in the remaining approximately 315 amino acids either inactivate all three UL42 activities identified (DNA binding, pol binding, processivity) or have little effect on any. The important exceptions are a linker insertion mutation at codon 160, which inactivates pol binding without apparent effect on DNA binding, and linker insertions at codons 203 and 206, which diminish DNA binding without apparent effect on pol binding. Each of these mutations severely impairs the ability of UL42 to stimulate pol processivity and to support viral replication (Digard et al. 1993b; Chow and Coen 1995). Whether these mutations identify short contiguous pol- or DNA-binding sites, respectively, is not clear.

Interactions with Other Proteins

It would be surprising if there were no interactions of either pol or UL42 with other HSV replication proteins other than with each other (see Challberg, this volume). However, as yet, little direct evidence has been published for such interactions.

OTHER HERPESVIRUS DNA POLYMERASES

Enzymology and Pharmacology

The other herpesvirus DNA polymerases have not been studied as extensively as HSV DNA polymerase, but several have provided interesting insights. There is reason to suspect complications in the stoichiometries of the subunits of these enzymes. Although the EBV enzyme is evidently a 1:1 heterodimer (T. Tsurumi, pers. comm.), the EBV catalytic subunit is maximally stimulated by its accessory protein, BMRF1, at a ratio of two BMRF1 molecules to each EBV pol (Tsurumi et al. 1993a). The murine cytomegalovirus homolog of the CMV accessory protein forms disulfide-linked homodimers and higher-order forms (Loh et al. 1994). The CMV and EBV holoenzymes exhibit many of the properties of the HSV enzyme (Table 2). However, the catalytic subunits alone differ substantially in their salt optima and preferred primer templates, and these parameters strongly influence the response of the catalytic subunits to their cognate accessory proteins (Table 2 and references therein). Thus, it is not yet clear to what extent these other herpesvirus accessory proteins function similarly to HSV UL42.

Like its HSV counterpart, the EBV pol contains an intrinsic 3′→5′ exonuclease activity (Tsurumi et al. 1993b). Ribonucleoside monophosphates inhibit this activity, but not polymerase (Tsurumi 1992; Tsurumi et al. 1994), unlike the HSV holoenzyme (Frank and Cheng 1986). Interestingly, the BMRF1 accessory subunit enhances the dsDNA nuclease activity of EBV pol, but not its activity on ssDNA (Tsurumi et al. 1994).

Genetics: Drug Resistance and Conserved Regions

A number of drug-resistant VZV and CMV mutants with lesions in the *pol* gene have been isolated. The five VZV mutations are very similar in location to those of HSV (Table 3). One mutant contains an alteration in region II not previously observed, and the others contain residues altered in regions I, II, and III and in δ-region C corresponding to residues altered in HSV mutants. The five CMV mutants, however, have a rather unusual spectrum of alterations (Table 3). This may be due to their selection for resistance to the nucleoside analog, ganciclovir. None of the mutations is in regions I–III. One is in region V and two are in δ-region C, bolstering the cases that these segments are involved in substrate recognition. Most surprising, two of the mutations are in highly conserved residues within the ExoI and ExoII segments. The aspartate altered in the ExoI segment is believed to be involved in metal binding and is required for exonuclease activity in other α-like polymerases (Bernad et al. 1989; Morrison et al. 1991; Simon et al. 1991). The phenylalanine altered in ExoII is adjacent to an aspartate that has similar properties to that in ExoI (Bernad et al. 1989; Simon et al. 1991). One interpretation of these results is that the mutations may affect the dNTP-binding site of CMV pol (Lurain et al. 1992). A second interpretation is that ganciclovir resistance is due to increased exonuclease activity to remove incorporated drug; however, the mutations would be expected to decrease exonuclease activity. A third possibility is based on the ability of ganciclovir triphosphate to promote chain termination despite its having a 3′ hydroxyl residue (Reid et al. 1988). Perhaps, the polymerase is less likely to stall and come off the primer template following drug incorporation when the exonuclease is less active, leading to internal incorporation of drug and subsequent repair or proper copying. These alternatives can be addressed by biochemical analysis of mutant polymerases.

Aside from these drug-resistant mutants, Ye and Huang (1993) have engineered point and small deletion mutations into CMV pol and assessed effects following in vitro expression. Mutations in regions I, II, III, and V severely diminished enzyme activity, as expected. A mutation in region III abolished DNA binding in a co-sedimentation assay, but whether this was a specific effect is not clear. Interestingly, mutations in regions I and II retained DNA-binding activity but did not detectably bind substrate. Further quantitative studies will be required to relate these results to the functional roles of the conserved regions.

Mapping of Activities

There have been some mutational analyses of catalytic and accessory subunits in other herpesvirus systems. Fairly small deletions of the amino terminus of EBV pol severely diminish activity in vitro (Kiehl and Dorsky 1991; Lin et al. 1991). In pseudorabies virus (PRV, a pig virus similar to HSV), deletion of the last 30 residues of the catalytic subunit results in an active polymerase that, however, is not stimulated by the PRV homolog of HSV UL42 (Berthomme et al. 1995). Remarkably, despite this functional analogy of the carboxyl terminus of PRV pol to that of HSV pol, there is little or no sequence homology, and the HSV and PRV catalytic subunits are not stimulated by the noncognate accessory proteins. This provides an interesting example of evolution of protein-protein interactions.

The accessory subunits, CMV UL44 and EBV BMRF1, have been analyzed for effects on DNA binding and pol stimulation. As was the case for HSV UL42, fairly large carboxy-terminal deletions had little or no effect (Weiland et al. 1994; Kiehl and Dorsky 1995). Most deletions within the remaining approximately 310 residues destroyed pol stimulation and DNA binding, with the exception of small deletions in BMRF1 between residues 194 and 233, which permitted DNA binding, but not stimulation. This suggests the existence of a bipartite DNA-binding region for this protein (Kiehl and Dorsky 1995); however, further work will be required to determine if given deletion mutations specifically affected DNA binding or more globally affected protein folding. As yet, mutations that specifically impair binding to the cognate pol have not been published.

VACCINIA VIRUS DNA POLYMERASE

Vaccinia virus, the descendant of the virus used by Jenner to vaccinate against smallpox, is the prototype poxvirus. The poxviruses have a number of fascinating biological features and are amenable to biochemical and genetic analysis. The vaccinia virus DNA polymerase was among the first animal virus DNA polymerases to be purified to near homogeneity (Challberg and Englund 1979a) and, unlike the other viral enzymes discussed here, is a monomer (Table 1). The protein has recently been overexpressed using an engineered vaccinia virus vector (McDonald and Traktman 1994a). Molecular genetic approaches, which can be especially rapid in the vaccinia system, have yielded many insights that have reinforced, complemented, and extended related work in herpesvirus systems.

Enzymology, Pharmacology, and Protein-protein Interactions

Selected enzymological features of the single subunit vaccinia DNA polymerase are presented in Tables 1 and 4. It contains an intrinsic 3′→5′ exonuclease (Challberg and Englund 1979a) but lacks detectable 5′→3′ exonuclease activity (McDonald and Traktman 1994a). Its polymerase activity is perhaps most distinguished by a limited repertoire of preferred primer templates, with best activity on activated DNA and poly(dC)oligo(dG); less activity on other homopolymeric DNA primer templates; poor activity on primed single-stranded bacteriophage DNA where it stalls at areas of secondary structure (Challberg and Englund 1979b); and little or no activity on RNA-primed homopolymers, consistent with the proposed mechanism of viral DNA replication (see Traktman, this volume). It is sensitive to aphidicolin, but less so than HSV DNA polymerase or DNA polymerase-α (Pedrali-Noy and Spadari

Table 4 Properties of vaccinia virus DNA polymerases

Form of enzyme	Optima	K_m for dNTPs	Primer templates	Selected inhibitors
116-kD monomer	50 mM KPO_4 pH 8–9 5 mM $MgCl_2$, but more processive at 1 mM	1–4 μM	good: activated DNA, poly(dC)oligo(dG) poor: RNA primed DNA, primed ss bacteriophage DNA	aphidicolin (modest), arabinosyl CTP

Selected references: Citarella et al. (1972); Challberg and Englund (1979a,b); Pedrali-Noy and Spadari (1980); McDonald and Traktman (1994a,b).

1980). It is also sensitive to a number of other inhibitors that are active against herpesvirus DNA polymerases, such as arabinosylcytosine (araC) triphosphate.

The enzyme is not processive as a monomer except at low ionic strength and $MgCl_2$ concentrations and lacks the strand-displacement activity required by current models for viral DNA replication (Challberg and Englund 1979b; McDonald and Traktman 1994b; see Traktman, this volume). The enzyme's activity on primed M13 DNA can be greatly stimulated by *E. coli* single-strand DNA-binding protein, which presumably is due to destabilization of secondary structure rather than specific protein-protein interactions (McDonald and Traktman 1994b). The physiological significance of these observations is not yet certain, but they have provided a starting point for purification of accessory factors from virus-infected cells (see Traktman, this volume).

Genetics

A number of studies have examined drug-resistant vaccinia virus mutants with lesions in the *pol* gene and correlated phenotype (including antimutator and mutator behaviors) with amino acid changes in conserved motifs, much as has been done with herpesviruses (see above). The results are summarized in Table 3. The mutations analyzed confer resistance or hypersensitivity to aphidicolin, PAA, and/or araC. As with CMV, the mutations have a different spectrum of locations from those observed with HSV and VZV. None of the mutations is in regions I, II, or V. One mutation is in region III. Two alter a residue in δ-region C, and three others are in a nonconserved region in vaccinia virus DNA polymerase that splits δ-region C into two halves. This further bolsters the case that δ-region C is involved in substrate recognition. Finally, as was the case with a CMV mutant, one mutant, isolated for araC-resistance, is in the ExoI segment. The three possible interpretations suggested earlier for the CMV ganciclovir-resistance mutations in ExoI and ExoII apply here as well. Taddie and Traktman (1993) tended to favor the idea that the mutation may improve the ability of the exonuclease to remove araC residues. Regardless, the mapping of the mutations favors a model of interdependence between polymerase and exonuclease sites (Taddie and Traktman 1993).

As yet, there has been little analysis of the effects of mutations on different polymerase activities in vitro. With the overexpression and ready purification of enzyme (McDonald and Traktman 1994a) and the hunt for interacting proteins, these studies may develop in the near future.

ADENOVIRUS DNA POLYMERASE

Adenoviruses infect birds and mammals. In humans, they are a cause of respiratory disease and molecular biological research. The in vitro reconstitution of adenovirus DNA replication by Challberg and Kelly (1979) led to the identification of all the factors required, including a previously unrecognized DNA polymerase (see Hay, this volume). The catalytic subunit, which also has 3′→5′ exonuclease activity, is normally found in a 1:1 heterodimer with the precursor of the terminal protein (pTP). Mature terminal protein (TP) is bound covalently to both ends of virion DNA via linkage to the 5′ phosphates. This covalent linkage is accomplished by the polymerase, which utilizes pTP as a protein primer and incorporates dCTP into a pTP-dCMP initiation complex. Whereas with herpesvirus and vaccinia virus DNA polymerases much of the emphasis has been on issues of substrate and drug recognition, with adenovirus DNA polymerase the emphasis has been on the details of its role in the replication of adenovirus DNA. These are covered thoroughly by Hay (this volume), so the coverage here is brief and focuses on other aspects.

Enzymology and Pharmacology

Aside from the origin-dependent elongation reactions using various viral and cellular proteins (covered by Hay, this volume), adenovirus DNA polymerase by itself is typically assayed in two ways (Table 5): (1) incorporation of dNTPs on conventional primer templates and (2) incorporation of dCTP into the pTP-dCMP initiation complex, typically using viral origin sequences as template. A number of features distinguish adenovirus DNA polymerase from other viral and cellular enzymes (Table 5). The enzyme is rather salt-sensitive and is stimulated by ATP (Lichy et al. 1981; Field et al. 1984), evidently without hydrolysis to ADP (Lindenbaum et al. 1986). On its own, the enzyme is resistant to aphidicolin, unlike most other members of the α-polymerase family (Enomoto et al. 1981; Lichy et al. 1981).

A remarkable aspect of adenovirus DNA polymerase is how its behavior changes depending on the presence of particular proteins and/or primer templates. Addition of adenovirus DNA-binding protein (DBP) allows the enzyme to utilize poly(dT)oligo(dA) efficiently (Field et al. 1984). This is reminiscent of the stimulation of herpesvirus catalytic subunits on certain primer templates by their accessory subunits; however, in this case, DBP is not a strongly associating subunit of the polymerase. (The presence or absence of pTP seems to have little effect on the

Table 5 Features of adenovirus DNA polymerase

Assay	Optima	K_m for dNTPs	Primer templates	Inhibitor responses[a]
Elongation	pH 7.5–8, no added salt, 3–5 mM ATP	1 μM[b]	good: activated DNA, poly(dC)oligo(dG) poor[c]: poly(dT)oligo(dA), poly(dT)oligo(rA)	sensitive to ddCTP (100 μM) resistant to aphidicolin (100 μM)
pTP-dCMP formation	same as above	4 μM[d]	Ad virion DNA (with TP) ss adenoviral DNA	resistant to both ddCTP and aphidicolin

Selected references: Lichy et al. (1981); Lichy et al. (1982); Nagata et al. (1982); Field et al. (1984); Mul and van der Vliet (1993).

[a]Elongation with all the components required for replication of full-length viral DNA in vitro is relatively sensitive to aphidicolin (IC_{50} <10 μM).

[b]Value is for elongation following formation of an initiation complex on the adenovirus origin and is in the presence of DBP (Mul and van der Vliet 1993).

[c]In the presence of DBP, these primer templates are used fairly efficiently (Field et al. 1984).

[d]Value shown is in the absence of the adenovirus DBP. With DBP, the K_m for formation of pTP-dCMP drops to 0.5 μM (Mul and van der Vliet 1993).

polymerase's catalytic properties.) The enzyme's sensitivity to inhibitors is highly plastic. The enzyme is sensitive to ddCTP on activated DNA templates or during elongation on adenoviral DNA, but it is resistant to the inhibitor during incorporation of dCTP into pTP. Perhaps most surprising is the response to aphidicolin. Despite the enzyme's inherent resistance to aphidicolin, DNA synthesis by the polymerase using complete reconstituted systems or crude extracts is relatively sensitive to the drug (Krokan et al. 1979; Longiaru et al. 1979; Nagata et al. 1983). An appealing possibility is that the other proteins alter the conformation of the polymerase to render it sensitive to aphidicolin. This scenario was invoked to explain how certain HSV mutations in the HSV-DBP gene conferred hypersensitivity to aphidicolin (Chiou et al. 1985). Alternatively, the replication activities of one of the other components in the reaction may be aphidicolin-sensitive.

A fundamental difference between adenovirus DNA polymerase and the other polymerases discussed in this book is its direct role in origin recognition. This is covered in some detail by Hay (this volume).

Genetics

Adenovirus DNA polymerase shares conserved sequence motifs I–V, VII, and the ExoI–III segments with other members of the α-polymerase family. It does not appear to contain segments such as δ-region C found in cellular polymerases δ and ε and herpes and poxvirus enzymes (Zhang et al. 1991). Joung et al. (1991) tested the importance of conserved region I by site-directed mutagenesis and assays of enzyme activity from extracts of transfected cells. A variety of substitution mutations had similar effects—mainly highly deleterious—on initiation and elongation assays, with the conserved glycine being the most tolerant of alterations.

Extensive linker insertion mutagenesis of both subunits of adenovirus polymerase has been performed (Freimuth and Ginsberg 1986; Chen and Horwitz 1989; Fredman et al. 1991; Roovers et al. 1991, 1993). The major finding from these studies is that mutations at locations throughout either subunit destroy polymerase activity in both initiation and elongation assays. Although it is difficult to rule out effects of the mutations on overall protein folding, the results are in accord with the studies on herpesvirus and poxvirus enzymes which suggest that polymerase active sites require contributions from distant regions of the molecule. These mutations have also been used to investigate protein-protein interactions, as summarized by Hay (this volume).

A novel activity—attachment to nuclear matrix—has been ascribed to the terminal protein, and this is thought to be important as well for transcriptional efficiency of viral DNA, presumably by proper intranuclear localization (Schaack et al. 1990; Fredman and Engler 1993). pTP also appears to facilitate the nuclear localization of the catalytic subunit, in a study that also mapped both subunits' nuclear localization signals (Zhao and Padmanabhan 1988). A third novel activity, which remains unexplained, is the role of the polymerase in oncogenic transformation; a temperature-sensitive *pol* mutation confers a temperature-sensitive transformation phenotype on the virus toward rat cells (Miller and Williams 1987).

REGULATION OF VIRAL POLYMERASES

The expression of viral DNA polymerases is regulated during virus infection. Much of this regulation is transcriptional so that the genes are transcribed at the correct time; most polymerases are "early" gene products, expressed prior to and independent of DNA synthesis. A striking feature of many viral polymerases is that they are subject to posttranscriptional controls that down-regulate their abundance. The consequences of this kind of control are perhaps better understood in RNA virus and retrovirus (Meier et al. 1987; Felsenstein and Goff 1988; Dinman and Wickner 1992). There it appears that overexpression of polymerase can be deleterious due to inappropriate ratios of polymerase to the proteins with which it interacts. Mechanisms in these and bacteriophage systems to set low polymerase abundance relative to other proteins include partial transcriptional termination, relatively inefficient translational frameshifting, nonsense suppression, and autogenous translational regulation (Iverson and Rose 1981; Andrake et al. 1988; Atkins et al. 1990; Tuerk et al. 1990). The viruses whose polymerases are described in this chapter also utilize posttranscriptional mechanisms to down-regulate polymerase abundance.

Translational Shutoff of HSV Pol

HSV pol is encoded by a single major mRNA on which the *pol* gene is the only known function (a small open reading frame [ORF] upstream of the *pol* ORF is not important for regulation) (Yager and Coen 1988; Yager et al. 1990). At times when the synthesis of most early proteins peaks, the mRNA is poorly translated, resulting in a shutoff of polymerase expression (Yager et al. 1990; Wobbe et al. 1993). The mRNA is poorly associated with polyribosomes at these times, suggesting a block at the level of translational initiation. Neither the *cis*-acting sequences nor the *trans*-acting factors for this shutoff are known. However, translational shutoff depends on viral DNA synthesis (Wobbe et al. 1993), suggesting that a protein synthesized late in infection is required. Oddly, when DNA synthesis is inhibited by drugs such as aphidicolin or by mutation, *pol* mRNA accumulation declines, leading to no net change in polymerase accumulation (Wobbe et al. 1993). Whether this effect on mRNA accumulation is transcriptional or posttranscriptional is unclear. Similar effects of aphidicolin on the accumulation of baculovirus *pol* mRNA have been noted (Tomalski et al. 1988).

Other Herpesvirus Pols

EBV DNA polymerase expression has a number of unusual features (Furnari et al. 1992). The *pol* gene lacks a TATA box and contains multiple transcriptional start sites. There is a functional poly(A) signal near the 5′ end of the gene; this could provide a mechanism for limiting polymerase expression. In the B95-8 strain, the 3′ end of *pol* mRNA is generated in the absence of a poly(A) signal. The mRNAs nevertheless are polyadenylated but contain a series of inserted nucleotides between the *pol* sequences and the poly(A) tail (Furnari et al. 1993). CMV *pol* mRNA contains a long 5′-untranslated region (Kouzarides et al. 1987), which impairs translation in transfected cells and in vitro (Geballe and Mocarski 1988; Ye and Huang 1993).

Vaccinia Virus Pol

The regulation of vaccinia virus polymerase expression is in some ways the mirror image of HSV regulation. McDonald et al. (1992) found that under normal circumstances, vaccinia *pol* mRNA is translated with equal efficiencies at all stages of infection. However, in the presence of aphidicolin, the mRNA, rather than declining late in infection, remains at high levels, but polymerase translation is shut off. The effect of aphidicolin does not seem to be on DNA synthesis, however, as aphidicolin-resistant *pol* mutants manifest the same effects. Whether polymerase expression is regulated differently from other early genes has not yet been fully explored.

Splicing Versus Polyadenylation Regulates Adenovirus Polymerase Synthesis

The mechanisms of posttranscriptional regulation of adenovirus DNA polymerase are perhaps the best understood. Pol, pTP, and DBP share the same promoter. The mRNAs for these proteins, however, differ greatly in abundance, such that the mRNA for DBP is roughly 100 times more abundant than those for pol and pTP (Stillman et al. 1981). The different mRNAs are generated by alternative splicing of a common primary transcript (Stillman et al. 1981; Shu et al. 1988). The *dbp* mRNA uses a poly(A) signal upstream of the ORFs for the two polymerase subunits, leading to the proposal that preferential use of this signal leads to low abundance of the *pol* and *pTP* mRNAs. Given our current understanding, this implies a race between splicing to generate the downstream mRNAs on the one hand, and cleavage, polyadenylation, and perhaps transcription termination on the other, with splicing losing the race most of the time. Thus, the ratio of the mRNAs and, by extension, the two polymerase subunits to DBP is set, perhaps exclusively, by a balance between two competing processes. Given the myriad protein-protein interactions of Pol, pTP, and DBP (Hay, this volume), it is presumably important to set the correct ratios between pol, pTP, and DBP to ensure proper stoichiometries. It is very tidy to utilize alternative splicing to set these ratios rather than setting transcription rates at three different promoters.

SUMMARY

Despite the many differences in replication strategies of the different virus groups and the numerous differences in enzymatic properties, there are major similarities that unite the different virus DNA polymerases.

The different polymerases all share sequence homologies with each other and with other members of the α-like DNA polymerase family. For each of the polymerases there is evidence against the idea that the polymerase activity resides on a single modular structural and functional domain. Each of the polymerases appears to require another viral protein to stimulate elongation, but by rather different mechanisms ranging from tightly associated subunits to utilization of proteins that are not subunits. In each case, there is evidence for posttranscriptional regulation of expression. Given the evolutionary relationships, it would be surprising if some of these unifying features among viral polymerases did not also apply to cellular enzymes.

ACKNOWLEDGMENTS

Thanks to the many colleagues, especially K. Biron, M. DePamphilis, P. Digard, J. Engler, A.K. Field, R. Hay, C. Knopf, I.R. Lehman, D. Marvel, G. Rohrmann, P. van der Vliet, K. Thompson, P. Traktman, T. Tsurumi, T. Wang, and D. Yager who generously provided advice, reprints and preprints, and personal communications. Thanks also to J. Brown, C. Chow, and M. Kramer for selfless help with figures, tables, and manuscript preparation. Grant support from the National Institutes of Health (AI-19838, AI-26077, and AI-33357) is gratefully acknowledged.

REFERENCES

Andrake, M., N. Guild, T. Hsu, L. Gold, C. Tuerk, and J. Karam. 1988. DNA polymerase of bacteriophage T4 is an autogenous translational repressor. *Proc. Natl. Acad. Sci.* **85:** 7942–7946.

Atkins, J.F., R.B. Weiss, and R.F. Gesteland. 1990. Ribosome gymnastics—Degree of difficulty 9.5, style, 10.0. *Cell* **62:** 413–423.

Bernad, A., Blanco, L., Lazaro, J.M., Martin, G., and Salas, M. 1989. A conserved 3′-5′ exonuclease active site in prokaryotic and eukaryotic DNA polymerases. *Cell* **59:** 219–228.

Berthomme, H., S.J. Monahan, D.S. Parris, B. Jacquemont, and A.L. Epstein. 1995. Cloning, sequencing, and functional characterization of the two subunits of the pseudorabies virus DNA polymerase holoenzyme: Evidence for specificity of interaction. *J. Virol.* **69:** 2811–2818.

Challberg, M.D. and P.T. Englund. 1979a. Purification and properties of the deoxyribonucleic acid polymerase induced by vaccinia virus. *J. Biol. Chem.* **254:** 7812–7819.

———. 1979b. The effect of template secondary structure on vaccinia DNA polymerase. *J. Biol. Chem.* **254:** 7820–7826.

Challberg, M.D. and T.J. Kelly. 1979. Adenovirus DNA replication in vitro. *Proc. Natl. Acad. Sci.* **76:** 655–659.

Chen, M. and M.S. Horwitz. 1989. Dissection of functional domains of adenovirus DNA polymerase by linker-insertion mutagenesis. *Proc. Natl. Acad. Sci.* **86:** 6116–6120.

Chiou, H.C., S.K. Weller, and D.M. Coen. 1985. Mutations in the herpes simplex virus major DNA-binding protein gene leading to altered sensitivity to DNA polymerase inhibitors. *Virology* **145:** 213–226.

Chiou, H.C., K. Kumura, A. Hu, K.M. Kerns, and D.M. Coen. 1995. Penciclovir-resistance mutations in the herpes simplex virus DNA polymerase gene. *Antiviral Chem. Chemother.* **6:** 281–288.

Chiou, J.-F. and Y.-C. Cheng. 1985. Interaction of Epstein-Barr virus DNA polymerase and 5′-triphosphates of several antiviral nucleoside analogs. *Antimicrob. Agents Chemother.* **27:** 416–418.

Chow, C.S. and D.M. Coen. 1995. Mutations that specifically impair the DNA binding activity of the herpes simplex virus protein UL42. *J. Virol.* **69:** 6965–6971.

Citarella, R.V., R. Muller, A. Schabach, and A. Weissbach. 1972. Studies on vaccinia virus-directed deoxyribonucleic acid polymerase. *J. Virol.* **10:** 721–729.

Crute, J.J. and I.R. Lehman. 1989. Herpes simplex-1 DNA polymerase: Identification of an intrinsic 5′–3′ exonuclease with ribonuclease H activity. *J. Biol. Chem.* **264:** 19266–19270.

Defillippes, F.M. 1989. Site of the base change in the vaccinia virus DNA polymerase gene which confers aphidicolin resistance. *J. Virol.* **63:** 4060–4063.

Derbyshire, V., M. Astatke, and C.M. Joyce. 1993. Re-engineering the polymerase domain of Klenow fragment and evaluation of overproduction and purification strategies. *Nucleic Acids Res.* **21:** 5439–5448.

Derse, D. and Y.-C. Cheng. 1981. Herpes simplex virus type 1 DNA polymerase: Kinetic properties of the associated 3′–5′ exonuclease activity and its role in araAMP incorporation. *J. Biol. Chem.* **256:** 8525–8530.

Derse, D., K.F. Bastow, and Y.-C. Cheng. 1982. Characterization of the DNA polymerases induced by a group of herpes simplex type I variants selected for growth in the presence of phosphonoformic acid. *J. Biol. Chem.* **297:** 10251–10260.

Digard, P., W.R. Bebrin, K. Weisshart, and D.M. Coen. 1993a. The extreme C-terminus of herpes simplex virus DNA polymerase is crucial for functional interaction with processivity factor UL42 and for viral replication. *J. Virol.* **67:** 398–406.

Digard, P., C.S. Chow, L. Pirrit, and D.M. Coen. 1993b. Functional analysis of the herpes simplex virus UL42 protein. *J. Virol.* **67:** 1159–1168.

Digard, P., K.P. Williams, P. Hensley, C.E. Dahl, and D.M. Coen. 1995. Specific inhibition of the herpes simplex virus DNA polymerase by helical peptides corresponding to the subunit interface. *Proc. Natl. Acad. Sci.* **92:** 1456–1460.

Dinman, J.D. and R.B. Wickner. 1992. Ribosomal frameshifting efficiency and gag/gag-pol ratio are critical for yeast M_1 double-stranded RNA virus propagation. *J. Virol.* **66:** 3669–3676.

Dorsky, D.I. and C.S. Crumpacker. 1988. Expression of herpes simplex virus type 1 DNA polymerase gene by in vitro translation and effects of gene deletions on activity. *J. Virol.* **62:** 3224–3232

———. 1990. Site-specific mutagenesis of a highly conserved region of the herpes simplex virus type 1 DNA polymerase gene. *J. Virol.* **64:** 1394–1397.

Dorsky, D.I. and C. Plourde. 1993. Resistance to antiviral inhibitors caused by the mutation S889A in the highly conserved 885-GDTDS motif of the herpes simplex virus type 1 DNA polymerase. *Virology* **195:** 831–835.

Earl, P.L., E.V. Jones, and B. Moss. 1986. Homology between DNA polymerase of poxviruses, herpesviruses, and adenoviruses: Nucleotide sequence of the vaccinia virus DNA polymerase gene. *Proc. Natl. Acad. Sci.* **83:** 3659–3663.

Enomoto, T., J.H. Lichy, J.E. Ikeda, and J. Hurwitz. 1981. Adenovirus replication in vitro: Purification of the terminal protein in a functional form. *Proc. Natl. Acad. Sci.* **78:** 6779–6783.

Ertl, P.F. and K.L. Powell. 1992. Physical and functional interaction of human cytomegalovirus DNA polymerase and its accessory protein (ICP36) expressed in insect cells. *J. Virol.* **66:** 4126–4133.

Felsenstein, K.M. and S.P. Goff. 1988. Expression of the gag-pol fusion protein of Moloney murine leukemia virus without gag protein does not induce virion formation or proteolytic processing. *J. Virol.* **62:** 2179–2182.

Field, J., R.M. Gronostajski, and J. Hurwitz. 1984. Properties of the adenovirus DNA polymerase. *J. Biol. Chem.* **259:** 9487–9495.

Frank, K.B. and Y.-C. Cheng. 1986. Inhibition of herpes simplex virus DNA polymerase by purine ribonucleoside monophosphates. *J. Biol. Chem.* **261:** 1510–1513.

Frank, K.B., D.D. Derse, K.F. Bastow, and Y.-C. Cheng. 1984. Novel interaction of aphidicolin with herpes simplex virus DNA polymerase and polymerase-associated exonuclease. *J. Biol. Chem.* **259:** 13282–13286.

Fredman, J.N. and J.A. Engler. 1993. Adenovirus precursor to terminal protein interacts with the nuclear matrix in vivo and in vitro. *J. Virol.* **67:** 3384–3395.

Fredman, J.N., S.C. Pettit, M.S. Horwitz, and J.A. Engler. 1991. Linker insertion mutations in the adenovirus preterminal protein that affect DNA replication in vivo and in vitro. *J. Virol.* **65:** 4591–4597.

Freimuth, P.I. and H.S. Ginsberg. 1986. Codon insertion mutant of the adenovirus terminal protein. *Proc. Natl. Acad. Sci.* **83:** 7816–7820.

Furnari, F.B., M.D. Adams, and J.S. Pagano. 1992. Regulation of the Epstein-Barr virus DNA polymerase gene. *J. Virol.* **66:** 2837–2845.

———. 1993. Unconventional processing of the 3′ termini of the Epstein-Barr virus DNA polymerase mRNA. *Proc. Natl. Acad. Sci.* **90:** 378–382.

Gao, M., S.F. DiTusa, and M.G. Cordingley. 1993. The C-terminal third of UL42, a HSV-1 DNA replication protein, is dispensable for viral growth. *Virology* **194:** 647–653.

Geballe, A.P. and E.S. Mocarski. 1988. Translational control of cytomegalovirus gene expression is mediated by upstream AUG codons. *J. Virol.* **62:** 3334–3340.

Gibbs, J.S., H.C. Chiou, K.F. Bastow, Y.C. Cheng, and D.M. Coen. 1988. Identification of amino acids in herpes simplex virus DNA polymerase involved in substrate and drug recognition. *Proc. Natl. Acad. Sci.* **85:** 6672–6676.

Gibbs, J.S., K. Weisshart, P. Digard, A. deBruynKops, D.M. Knipe, and D.M. Coen. 1991. Polymerization activity of an α-like DNA polymerase requires a conserved 3′–5′ exonuclease active site. *Mol. Cell. Biol.* **11:** 4786–4795.

Gottlieb, J. and M.D. Challberg. 1994. Interaction of herpes simplex virus type 1 DNA polymerase and the UL42 accessory protein with a model primer template. *J. Virol.* **68:** 4937–4945.

Gottlieb, J., A.I. Marcy, D.M. Coen, and M.D. Challberg. 1990. The HSV-1 UL42 gene product: A subunit of DNA polymerase that functions to increase processivity. *J. Virol.* **64:** 5976–5987.

Haffey, M.L., J. Novotny, R.E. Bruccoleri, R.D. Carroll, J.T. Stevens, and J.T. Matthews.

1990. Structure-function studies of the herpes simplex virus DNA polymerase. *J. Virol.* **64:** 5008–5018

Hall, J.D., P.A. Furman, M.H. St. Clair, and C.W. Knopf. 1985. Reduced in vivo mutagenesis by mutant herpes simplex DNA polymerase involves improved nucleotide selection. *Proc. Natl. Acad. Sci.* **82:** 3889–3993.

Hall, J.D., Y. Wang, J. Pierpont, M.S. Berlin, S.E. Rundlett, and S. Woodward. 1989. Aphidicolin resistance in herpes simplex virus type 1 reveals features of the DNA polymerase dNTP binding site. *Nucleic Acids Res.* **17:** 9231–9244.

Hall, J.D., D.M. Coen, B.L. Fisher, M. Weisslitz, S. Randall, R.E. Almy, P.T. Gelep, and P.A. Schaffer. 1984. Generation of genetic diversity in herpes simplex virus: An antimutator phenotype maps to the DNA polymerase locus. *Virology* **132:** 26–37.

Hamatake, R.K., M. Bifano, D.J. Tenney, W.W. Hurlburt, and M.G. Cordingley. 1993. The herpes simplex virus type 1 DNA polymerase accessory protein, UL42, contains a functional protease-resistant domain. *J. Gen. Virol.* **74:** 2181–2189.

Hart, G.J. and R.E. Boehme. 1992. The effect of the UL42 protein on the DNA polymerase activity of the catalytic subunit of the DNA polymerase encoded by herpes simplex virus type 1. *FEBS Lett.* **305:** 97–100.

Hernandez, T.R. and I.R. Lehman. 1990. Functional interaction between the herpes simplex-1 DNA polymerase and UL42 protein *J. Biol. Chem.* **265:** 11227–11232.

Hwang, C.B.C. and H.J.-H. Chen. 1995. An altered spectrum of herpes simplex virus mutations mediated by an antimutator DNA polymerase. *Gene* **152:** 191–193.

Hwang, C.B.C., K.L. Ruffner, and D.M. Coen. 1992. A point mutation within a distinct conserved region of the herpes simplex virus DNA polymerase gene confers drug resistance. *J. Virol.* **66**: 1774–1776.

Ito, J. and D.K. Brathwaite. 1991. Compilation and alignment of DNA polymerase sequences. *Nucleic Acids Res.* **19:** 4045–4057.

Iverson, L.E. and J.K. Rose. 1981. Localized attenuation and discontinuous synthesis during vesicular stomatitis virus transcription. *Cell* **23:** 477–484.

Joung, I., M.S. Horwitz, and J.A. Engler. 1991. Mutagenesis of conserved region I in the DNA polymerase from human adenovirus serotype 2. *Virology* **184:** 235–241.

Kiehl, A. and D.I. Dorsky 1991. Cooperation of EBV DNA polymerase and EA-D (BMRF1) in vitro and colocalization in nuclei of infected cells. *Virology* **184:** 330–340.

———. 1995. Bipartite DNA-binding region of the Epstein-Barr virus BMRF1 product essential for DNA polymerase accessory function. *J. Virol.* **69:** 1669–1677.

Knopf, C.W. and R. Strick. 1993. Herpes simplex virus type 1 DNA polymerase: Eukaryotic model enzyme and principal target of antiviral therapy. In *Frontiers of virology* (ed. G. Darai and Y. Becker), vol. 3, pp. 87–135. Springer Verlag, Heidelberg.

Knopf, C.W. and K. Weisshart. 1990. Comparison of exonucleolytic activities of herpes simplex virus type-1 DNA polymerase and DNase. *Eur. J. Biochem.* **191:** 263–273.

Kouzarides, T., A.T. Bankier, S.C. Satchwell, K. Weston, P. Tomlinson, and B.G. Barrell. 1987. Sequence and transcription analysis of the human cytomegalovirus DNA polymerase gene. *J. Virol.* **61:** 125–133.

Krokan, H., P. Schaffer, and M.L. DePamphilis. 1979. Involvement of eucaryotic deoxyribonucleic acid polymerase α in the replication of cellular and viral deoxyribonucleic acid. *Biochemistry* **18:** 4431–4443.

Larder, B.A., S.D. Kemp, and G. Darby. 1987. Related functional domains in virus DNA polymerases. *EMBO J.* **6:** 169–175.

Lichy, J.H., J. Field, M.S. Horwitz, and J. Hurwitz. 1981. Formation of a covalent complex between the 80kDa adenovirus terminal protein and 5′-dCMP in vitro. *Proc. Natl. Acad. Sci.* **78:** 2678–2682.

———. 1982. Separation of the adenovirus terminal protein precursor from its associated DNA polymerase: Role of both proteins in the initiation of adenovirus DNA replication. *Proc. Natl. Acad. Sci.* **79:** 5225–5229.

Lin, J.-C., N.D. Sista, F. Besençon, J. Kamine, and J.S. Pagano. 1991. Identification and functional characterization of Epstein-Barr virus DNA polymerase by in vitro transcription-translation of a cloned gene. *J. Virol.* **65:** 2728–2731.

Lindenbaum, J.O., J. Field, and J. Hurwitz. 1986. The adenovirus DNA binding protein and adenovirus DNA polymerase interact to catalyze elongation of primed DNA templates. *J. Biol. Chem.* **261:** 10218–10227.

Loh, L.C., W.J. Britt, C. Raggo, and S. Laferté. 1994. Sequence analysis and expression of the murine cytomegalovirus phosphoprotein pp50, a homolog of the human cytomegalovirus UL44 gene product. *Virology* **200:** 413–427.

Longiaru, M., J.-E. Ikeda, Z. Jarkovsky, S.B. Horwitz, and M.S. Horwitz. 1979. The effect of aphidicolin on adenovirus DNA synthesis. *Nucleic Acids Res.* **6:** 3369–3386.

Lurain, N.S., L.K. Penland, and K.D. Thompson. 1994. Ganciclovir-resistant DNA polymerase mutants isolated from an HCMV clinical specimen. Abstr. 19th International Herpesvirus Workshop, #211.

Lurain, N.S., K.D. Thompson, E.W. Holmes, and G.S. Read. 1992. Point mutations in the DNA polymerase gene of human cytomegalovirus that result in resistance to antiviral agents. *J. Virol.* **66:** 7146–7152.

Mar, E.-C., J.-F. Chiou, Y.-C. Cheng, and E.-S. Huang. 1985. Inhibition of cellular DNA polymerase α and human cytomegalovirus-induced DNA polymerase by the triphosphates of 9-(2-hydroxyethoxymethyl)guanine and 9-(1,3,-dihdroxy-2-propoxymethyl)guanine. *J. Virol.* **53:** 776–780.

Marcy, A.I., C.B.C. Hwang, K.L. Ruffner, and D.M. Coen. 1990a. Engineered herpes simplex virus DNA polymerase point mutants: The most highly conserved region shared among α-like DNA polymerases is involved in substrate recognition. *J. Virol.* **64:** 5883–5890.

Marcy, A.I., P.D. Olivo, M.D. Challberg, and D.M. Coen. 1990b. Enzymatic activities of overexpressed herpes simplex virus DNA polymerase purified from recombinant baculovirus-infected insect cells. *Nucleic Acids Res.* **18:** 1207–1215.

Marsden, H.S., M. Murphy, G.L. McVey, K.A. MacEachran, A.M. Owsianka, and N.D. Stow. 1994. Role of the carboxy-terminus of HSV-1 DNA polymerase in its interaction with UL42. *J. Gen. Virol.* **75:** 3127–3135.

Martin, J.L., C.E. Brown, N. Matthews-Davis, and J.E. Reardon. 1994. Effects of antiviral nucleoside analogs on human DNA polymerases and mitochondrial DNA synthesis. *Antimicrob. Agents Chemother.* **38:** 2743–2749.

Matthews, J.T., B.J. Terry, and A.K. Field. 1993. The structure and function of the HSV replication proteins: Defining novel antiviral targets. *Antiviral Res.* **20:** 89–114.

McDonald, W.F. and P. Traktman. 1994a. Overexpression and purification of the vaccinia virus DNA polymerase. *Protein Expr. Purif.* **5:** 409–421.

———. 1994b. Vaccinia virus DNA polymerase: In vitro analysis of parameters affecting processivity. *J. Biol. Chem.* **269:** 31190–31197.

McDonald, W.F., V. Crozel-Goudot, and P. Traktman. 1992. Transient expression of the vaccinia virus DNA polymerase is an intrinsic feature of the early phase of infection and is unlinked to DNA replication and late gene expression. *J. Virol.* **66:** 534–547.

Meier, E., G.G. Harmison, and M. Schubert. 1987. Homotypic and heterotypic exclusion of vesicular stomatitis virus replication by high levels of recombinant polymerase protein L. *J. Virol.* **61:** 3133–3142.

Miller, B.W. and J. Williams. 1987. Cellular transformation by adenovirus type 5 is influenced by the viral DNA polymerase. *J. Virol.* **61:** 3630–3634.

Monahan, S.J., T.F. Barlam, C.S. Crumpacker, and D.S. Parris. 1993. Two regions of the herpes simplex virus type 1 UL42 protein are required for its functional interaction with the viral DNA polymerase. *J. Virol.* **67:** 5922–5931.

Morrison, A., J.B. Bell, T.A. Kunkel, and A. Sugino. 1991. Eukaryotic DNA polymerase amino acid sequence required for 3′–5′ exonuclease activity. *Proc. Natl. Acad. Sci.* **88:** 9473–9477.

Mul, Y.M. and P.C. van der Vliet. 1993. The adenovirus DNA binding protein effects the kinetics of DNA replication by a mechanism distinct from NF1 or Oct-1. *Nucleic Acids Res.* **21:** 641–647.

Nagata, K., R.A. Guggenheimer, and J. Hurwitz. 1983. Adenovirus DNA replication in vitro: Synthesis of full length DNA with purified proteins. *Proc. Natl. Acad. Sci.* **80:** 4266–4270.

Nagata, K., R.A. Guggenheimer, T. Enomoto, J.H. Lichy, and J. Hurwitz. 1982. Adenovirus DNA replication in vitro: Identification of a host factor that stimulates synthesis of the preterminal protein-dCMP complex. *Proc. Natl. Acad. Sci.* **79:** 6438–6442.

Nishiyama, Y., M. Koichiro, and S. Yoshida. 1983. Characterization of human cytomegalovirus-induced DNA polymerase and the associated 3′–5′ exonuclease. *Virology* **124:** 221–231.

Pedrali-Noy, G. and S. Spadari. 1980. Mechanism of inhibition of herpes simplex virus and vaccinia virus DNA polymerases by aphidicolin, a highly specific inhibitor of DNA replication in eukaryotes. *J. Virol.* **36:** 457–464.

Reardon, J.E. and T. Spector. 1989. Herpes simplex virus type 1 DNA polymerase: Mechanism of inhibition by acyclovir triphosphate. *J. Biol. Chem.* **264:** 7405–7411.

Reha-Krantz, L.J. 1990. Genetic evidence for two protein domains and a potential new activity in bacteriophage T4 DNA polymerase. *Genetics* **124:** 213–220.

Reid, R., E.-C. Mar, E.-S. Huang, and M.D. Topal. 1988. Insertion and extension of acyclic, dideoxy, and ara nucleotides by herpesviridae, human α, and human β polymerases. *J. Biol. Chem.* **263:** 3898–3904.

Roovers, D.J., F. van der Lee, J. van der Wees, and J.S. Sussenbach. 1993. Analysis of the adenovirus type 5 terminal protein precursor and DNA polymerase by linker insertion mutagenesis. *J. Virol.* **67:** 265–276.

Roovers, D.J., C.S.H. Young, H.L. Vos, and J.S. Sussenbach. 1991. Linker mutation scanning of the genes encoding the adenovirus type 5 terminal protein precursor and DNA polymerase. *Virology* **180:** 273–284.

Schaack, J., W.Y.-W. Ho, P. Freimuth, and T. Shenk. 1990. Adenovirus terminal protein mediates both nuclear matrix association and efficient transcription of adenovirus DNA. *Genes Dev.* **4:** 1197–1298.

Shu, L., S.C. Pettit, and J.A. Engler. 1988. The precise structure and coding capacity of mRNAs from early region 2B of human adenovirus serotype 2. *Virology* **165:** 348–356.

Simon, M., L. Giot, and G. Faye. 1991. The 3′–5′ exonuclease activity located in the DNA polymerase δ subunit of *Saccharomyces cerevisiae* is required for accurate replication. *EMBO J.* **8:** 2165–2170.

Skalka, A.-M. and S.P. Goff, eds. 1993. *Reverse transcriptase*. Cold Spring Harbor Laboratory Press, Cold Spring Harbor, New York.

Stillman, B.W., J.B. Lewis, L.T. Chow, M.B. Mathews, and J.E. Smart. 1981. Identification of the gene and mRNA for the adenovirus terminal protein precursor. *Cell* **23:** 497–508.

Stow, N.D. 1993. Sequences at the C-terminus of the herpes simplex virus type 1 UL30 protein are dispensible for DNA polymerase activity but not for viral origin-dependent DNA replication. *Nucleic Acids Res.* **21:** 87–92.

Sullivan, V., K.K. Biron, C. Talarico, S.C. Stanat, M. Davis, L.M. Pozzi, and D.M. Coen. 1993. A point mutation in the human cytomegalovirus DNA polymerase gene confers resistance to ganciclovir and phosphonylmethoxyalkyl derivatives. *Antimicrob. Agents Chemother.* **37:** 19–25.

Taddie, J.A. and P. Traktman. 1991. Genetic characterization of the vaccinia virus DNA polymerase: Identification of point mutations conferring altered drug sensitivities and reduced fidelity. *J. Virol.* **65:** 869–879.

———. 1993. Genetic characterization of the vaccinia virus DNA polymerase: Cytosine arabinoside resistance requires a variable lesion conferring phosphonoacetate resistance in conjunction with an invariant mutation localized to the 3′–5′ exonuclease domain. *J. Virol.* **67:** 4323–4336.

Tenney, D.J., W.W. Hurlburth, M. Bifano, J.T. Stevens, P.A. Micheletti, R.K. Hamatake, and M.G. Cordingley. 1993a. Deletions of the carboxy terminus of herpes simplex virus type 1 UL42 define a conserved amino-terminus functional domain. *J. Virol.* **67:** 1959–1966.

Tenney, D.J., P.A. Micheletti, J.T. Stevens, R.K. Hamatake, J.T. Matthews, A.R. Sanchez, W.W. Hurlburt, M. Bifano, and M.G. Cordingley. 1993b. Mutations in the C terminus of herpes simplex virus type 1 DNA polymerase can affect binding and stimulation by its accessory protein UL42 without affecting basal polymerase activity. *J. Virol.* **67:** 543–547.

Tomalski, M.D., J. Wu, and L.K. Miller. 1988. The location, sequence, transcription, and regulation of a baculovirus DNA polymerase gene. *Virology* **167:** 591–600.

Traktman, P., M. Kelvin, and S. Pacheco. 1989. Molecular genetic analysis of vaccinia virus DNA polymerase mutants. *J. Virol.* **63:** 841–846.

Tsurumi, T. 1992. Selective inhibition of the 3′–5′ exonuclease activity associated with Epstein-Barr virus DNA polymerase by ribonucleoside 5′-monophosphates. *Virology* **189:** 803–807.

Tsurumi, T., T. Daikoku, and Y. Nishiyama. 1994. Further characterization of the interaction between the Epstein-Barr virus DNA polymerase catalytic subunit and its accessory subunit with regard to the 3′–5′ exonucleolytic activity and stability of initiation complex at primer-terminus. *J. Virol.* **68:** 3354–3363.

Tsurumi, T., T. Daikoku, R. Kurachi, and Y. Nishiyama. 1993a. Functional interaction between Epstein-Barr virus DNA polymerase catalytic subunit and its accessory subunit in vitro. *J. Virol.* **67:** 7648–7653.

Tsurumi, T., A. Kobayashi, K. Tamai, T. Daikoku, R. Kurachi, and Y. Nishiyama. 1993b. Functional expression and characterization of the Epstein-Barr virus DNA polymerase catalytic subunit. *J. Virol.* **67:** 4651–4658.

Tuerk, C., S. Eddy, D. Parma, and L. Gold. 1990. Autogenous translational operator recognized by bacteriophage T4 DNA polymerase. *J. Mol. Biol.* **213:** 749–761.

Wang, Y., S. Woodward, and J.D. Hall. 1992. Use of suppressor analysis to identify

DNA polymerase mutations in herpes simplex virus which affect deoxynucleoside triphosphate substrate specificity. *J. Virol.* **66:** 1814–1816.

Weiland, K.L, N.L. Oien, F. Homa, and M.W. Wathen. 1994. Functional analysis of human cytomegalovirus polymerase accessory protein. *Virus Res.* **34:** 191–206.

Weissbach, A., S.-C.L. Hong, J. Aucker, and R. Muller. 1973. Characterization of herpes simplex virus-induced deoxyribonucleic acid polymerase. *J. Biol. Chem.* **248:** 6270–6277.

Weisshart, K., A.A. Kuo, C.B.C. Hwang, K. Kumura, and D.M. Coen. 1994. Structural and functional organization of herpes simplex virus DNA polymerase investigated by limited proteolysis. *J. Biol. Chem.* **269:** 22788–22796.

Weisshart, K., A.A. Kuo, G.R. Painter, L.L. Wright, P.A. Furman, and D.M. Coen. 1993. Conformational changes induced in herpes simplex virus DNA polymerase upon DNA binding. *Proc. Natl. Acad. Sci.* **90:** 1028–1032.

Wobbe, K.K., P. Digard, D. Staknis, and D.M. Coen. 1993. Unusual regulation of expression of the herpes simplex virus DNA polymerase gene. *J. Virol.* **67:** 5419–5425.

Wong, S.W., A.F. Wahl, P.-M. Yuan, N. Arai, B.E. Pearson, K.-I. Arai, D. Korn, M.W. Hunkapiller, and T.S.-F. Wang. 1988. Human DNA polymerase α gene expression is cell proliferation dependent and its primary structure is similar to both prokaryotic and eukaryotic replicative DNA polymerases. *EMBO J.* **7:** 37–47.

Yager, D.R. and D.M. Coen. 1988. Analysis of the transcript of the herpes simplex virus DNA polymerase gene provides evidence that polymerase expression is inefficient at the level of translation. *J. Virol.* **62:** 2007–2015.

Yager, D.R., A.I. Marcy, and D.M. Coen. 1990. Translational regulation of herpes simplex virus DNA polymerase. *J. Virol.* **64:** 2217–2225.

Ye, L.-B. and E.-S. Huang. 1993. In vitro expression of the human cytomegalovirus DNA polymerase gene: Effects of sequence alterations on enzyme activity. *J. Virol.* **67:** 6339–6347.

Zhang, J., D.W. Chung, C.-K. Tan, K.M. Downey, E.W. Davie, and A.G. So. 1991. Primary structure of the catalytic subunit of calf thymus DNA polymerase δ: Sequence similarities with other DNA polymerases. *Biochemistry* **30:** 11742–11750.

Zhao, L.-J. and R. Padmanabhan. 1988. Nuclear transport of adenovirus DNA polymerase is facilitated by interaction with preterminal protein. *Cell* **55:** 1005–1015.

17

DNA Replication Accessory Proteins

Ulrich Hübscher, Giovanni Maga, and Vladimir N. Podust
Department of Veterinary Biochemistry
University of Zürich-Irchel
CH-8057 Zürich, Switzerland

INTRODUCTION

DNA replication requires the concerted action of many enzymes, as well as other protein and non-protein cofactors. The DNA, in preparation for DNA synthesis, has to become single-strand to serve as a template for the replicative DNA polymerases (pols). It is this form of the DNA that is especially prone to damage of any kind. Nature has provided a set of proteins that support the replicative pols in performing processive, accurate, and rapid DNA synthesis. Furthermore, such proteins also prevent damage to the transient single-strand (ss) DNA. These proteins are called DNA replication accessory proteins. The three best known are the proliferating cell nuclear antigen (PCNA), replication factor C (RF-C), and replication protein A (RP-A). In this chapter, we focus on these three protein classes and compare them to their selected counterparts in eukaryotic viruses. Additional replication proteins that also assist the proper function of pols, such as the 3′→5′exonuclease, DNA primase, RNase H, 5′→3′exonuclease, DNA helicases, DNA ligases, and DNA topoisomerases, are covered in various other chapters.

Early Discovery of Replication Accessory Proteins in Prokaryotes by Genetics and Defined In Vitro Replication Systems

Fifteen years ago it was realized that bacteriophages of *Escherichia coli* provide a window to understand the cellular events of DNA replication (Kornberg and Baker 1992). By using ssDNA from ϕX174, G4, and M13 as model replicons, the requirements for a ssDNA-binding protein (SSB) and a DNA synthesis complex were identified. The latter includes the pol III holoenzyme, which consists of many polypeptides besides the pol III core. For example, it includes the homodimeric β subunit and the hetero-pentameric γ complex (containing the γ, δ, δ′, χ, and ψ subunits) (for review, see Kuriyan and O'Donnell 1993). Well-established genetics

 0-87969-459-9/96 $5 + .00

facilitated the cloning of all genes and expression of the polypeptides in bacteria. This allowed the study of the prokaryotic replication machinery in detail (see, e.g., Stukenberg et al. 1994). Similarly in bacteriophage T4, the gp45 and gp44/62 complex share analogous functions to the β subunit and γ complex, respectively (see, e.g., Young et al. 1992; Kuriyan and O'Donnell 1993).

Discovery of Replication Accessory Proteins in Eukaryotes Thanks to Defined In Vitro Replication Systems

In eukaryotes, a protein was identified that increased the processivity of pol-δ (Tan et al. 1986). It was later identified in the SV40 in vitro replication system as PCNA (Prelich et al. 1987). The same in vitro replication system brought insight into the requirement of an SSB called HSSB (Wobbe et al. 1987), replication factor A (Fairman and Stillman 1988) or RP-A (Wold and Kelly 1988), the latter name now adopted for this protein. Soon thereafter, a factor that could coordinate the synthesis of leading and lagging strands during SV40 DNA replication was identified and called RF-C (Tsurimoto and Stillman 1989). All three proteins have been found to be functional analogs of prokaryotic accessory proteins (see Table 1).

Why Are Accessory Proteins Required in DNA Replication?

DNA replication accessory proteins provide particular functions that are mandatory for replicative pols. Such functions include the recruitment of particular pols when needed, the facilitation of pol binding to the primer terminus, the increase in pol processivity, the prevention of nonproductive binding of the pol to ssDNA, the release of the pol after DNA synthesis, and the bridging of pol interactions to other replication proteins. Thus, it is not surprising that these proteins are universally found in nature (Table 1).

REPLICATION ACCESSORY PROTEINS

PCNA

PCNA is the most extensively studied cellular DNA replication accessory protein (Table 2). It was originally discovered as a cell-cycle-regulated nuclear protein whose rate of synthesis correlates with the proliferative state of normal cells and tissues, and later as a processivity

Table 1 Eukaryotic and prokaryotic DNA replication accessory proteins

	Organism				
Functional component	*E. coli*	bacteriophage T4	eukaryote (yeast to man)	HSV	adenovirus
Processivity factor or sliding clamp	β subunit	gp45	PCNA	UL 42	AdDBP[a]
Clamp loader, brace protein, or matchmaker	γ complex	gp44–gp62 complex	RF-C	?	?
SSB	SSB	SSB (*E. coli*)	RP-A	ICP8[b]	AdDBP

Other DNA replication accessory proteins include 3′→5′ exonuclease, DNA primase, RNase H, 5′→3′ exonuclease, DNA helicases, and DNA ligases. They are reviewed in other chapters in this book.

[a]The AdDBP increases the strand-displacement activity of the adenovirus-encoded DNA polymerase.

[b]ICP8 can also enhance processivity.

Table 2 Proliferating cell nuclear antigen

Number of amino acids	260,[a–i] 365,[i] 275[k]
Molecular size (kD)	29,[a–i] 40,[i] 31[k]
Stoke's radius (A)	36.5,[d] 40[e]
pI	4.8[d]
Sedimentation coefficient (S)	5[d,e]
Structure in solution	ring-shaped homotrimer
Interaction with	pol-δ and pol-ε
	RF-C
	cyclins A, B, D; CDC2, CDKs 2, 4, 5; p21 CDK-inhibitor; Gadd45
Functions	sliding clamp, increases primer binding and processivity of pol-δ and pol-ε
	stimulates ATPase activity of RF-C, and forms together with RF-C and ATP the primer recognition complex
	regulation of DNA replication (?)

[a]*D. melanogaster* (Yamaguchi et al. 1990); [b]rat (Matsumoto et al. 1987); [c]calf (Tan et al. 1986); [d]human (Almendral et al. 1987); [e]*S. cerevisiae* (Bauer and Burgers 1990); [f]*S. pombe* (Waseem et al. 1992); [g]baculovirus (O'Reilly et al. 1989); [h]*Oxytricha sativa* (Suzuka et al. 1989, 1991); [i]*D. carota* (Hata et al. 1992); [k]*Plasmodium falciparum* (Kilbey et al. 1993).

factor for pol-δ (for review, see Hübscher and Spadari 1994). The gene encoding this protein has been cloned from a variety of sources, including yeast, human, and plants, and shows a highly conserved structure (Table 2). Compared to human PCNA, the amino acid sequence similarity is 32% for lower eukaryotes (yeast and protozoa), 64% for plant, and 70% for *Drosophila melanogaster*. PCNA is a highly acidic protein encoded by a single gene in all the organisms tested, with the one exception of *Dracus carota*, in which two distinct genes encoding two different forms of PCNA were isolated. The human gene has been mapped to chromosome 20, and two pseudogenes have been localized on chromosomes X and 6. The structure of the genomic clone of human PCNA is known: It comprises six exons, five introns, and a short 3′-untranslated region. Introns 1 and 4 have been implicated in the negative regulation of transcription of the PCNA gene (Chang et al. 1990; Alder et al. 1992). The promoter sequence contains binding sites for several transcription factors such as Sp1 and E2F, suggesting that the expression of the PCNA gene may be tightly regulated (Jones et al. 1988; Mudryj et al. 1990).

The extreme structural conservation of PCNA is also seen at a functional level. With synthetic oligonucleotides or poly(dA)/oligo(dT) as

DNA templates, PCNA increases the productive binding to a primer and the processivity of pol-δ. Yeast PCNA interacts with mammalian pol-δ, and mammalian PCNA interacts with yeast pol-δ (Bauer and Burgers 1988). PCNA was also shown to stimulate the primer binding and primer elongation reactions catalyzed by pol-ε on similar templates (Maga and Hübscher 1995). If, however, a singly primed M13 DNA was used as a model for a natural template, at least two other auxiliary proteins, RF-C and RP-A, are required. It has been shown that PCNA forms a complex together with RF-C at the primer in an ATP-dependent manner (Lee and Hurwitz 1990; Burgers 1991; Tsurimoto and Stillman 1991; Podust et al. 1992). This primer recognition complex can efficiently recruit pol-δ or pol-ε (but not pol-α) to the primer terminus. These functional experiments, together with the observation that PCNA is essential for coordinated leading- and lagging-strand synthesis during in vitro SV40 DNA replication (Prelich and Stillman 1988), clearly indicate a role for PCNA in DNA replication. PCNA has also been shown to participate in DNA excision repair and to be required for the gap-filling step (Shivji et al. 1992). Recent results showed that PCNA physically interacts with Gadd45, a protein that stimulates DNA excision repair and inhibits entry into S phase in mammalian cells, thus providing a new link between PCNA and the DNA repair pathway (Smith et al. 1994).

PCNA itself does not bind to DNA. It has been shown that PCNA can be loaded onto DNA in two ways: (1) enzymatically, by RF-C and ATP, and (2) topologically, in an ATP-independent manner (Burgers and Yoder 1993). Despite less than 5% of amino-acid-sequence similarity, PCNA is functionally equivalent to two prokaryotic proteins known to function as auxiliary factors in DNA replication: the β subunit of pol III holoenzyme of *E. coli* and the product of the gene 45 of the bacteriophage T4 (Kuriyan and O'Donnell 1993). Yeast PCNA has recently been crystallized, and its structure has been determined (Krishna et al. 1994). PCNA has been shown to be a homotrimer with a closed circular structure that can encircle duplex DNA with a minimum of specific interactions, similar to the dimeric β subunit of *E. coli* pol III. The current model for the function of PCNA suggests that this protein acts as a sliding clamp that allows pols to move rapidly along the DNA while remaining topologically bound to it.

Recent results have indicated that PCNA can interact with the cell-cycle-regulatory machinery. Isolation of kinase complexes containing cyclins A, B, and D, associated with the cyclin-dependent kinases CDC2, CDK2, CDK4, or CDK5, showed that a complex of PCNA and the cyclin-dependent kinase inhibitor p21 associates with each cyclin/CDK dimer to form quaternary complexes (Xiong et al. 1992; Zhang et al. 1993). Furthermore, p21 has been shown to inhibit PCNA-dependent DNA replication in the absence of cyclins/CDKs, through a direct protein-protein interaction with PCNA (Flores-Rozas et al. 1994; Li et al. 1994; Waga et al. 1994). It is known that the major tumor suppressor protein p53 regulates the expression of p21. In addition, it has recently been shown that p53 is also implicated in the regulation of the expression of the PCNA gene (Jackson et al. 1994; Yamaguchi et al. 1994), suggesting that, during p53-mediated suppression of cell proliferation, p21 and PCNA may be involved in coordination of the cell-cycle progression, DNA replication, and DNA repair.

RF-C

RF-C as an essential replication factor had first been isolated from human 293 cells in the in vitro SV40 DNA replication system (Tsurimoto and Stillman 1989). This protein was later purified from HeLa cells (Lee et al. 1991a), yeast (Yoder and Burgers 1991; Fien and Stillman 1992), and calf thymus (Podust et al. 1992) by using a complementation assay that enabled pol-δ to replicate a primed ssDNA in the presence of PCNA and RP-A. RF-C binds preferentially to template-primer junctions on the DNA with ATP or ATPγS stimulating binding of RF-C to DNA (Lee et al. 1991a; Tsurimoto and Stillman 1991). Footprinting experiments revealed that RF-C bound at a template-primer junction covers 12 bases of ssDNA, whereas on the double-strand (ds) DNA it binds 8 bases of the template strand and 15 bases of the primer strand (Tsurimoto and Stillman 1991). RF-C possesses a DNA-stimulated ATPase activity (Tsurimoto and Stillman 1990; Yoder and Burgers 1991; Lee et al. 1991a; Podust et al. 1992). In the absence of PCNA, ssDNA stimulates the ATPase, whereas primer-template junctions have little or no effect on the activity. PCNA increases ATP hydrolysis by RF-C on multiple primed ssDNA, whereas on unprimed ssDNA, only a marginal effect was detected. These data suggested that PCNA causes the RF-C ATPase activity to become dependent on a primer terminus. All eight nucleoside triphosphates could be hydrolyzed by RF-C with preference for dATP and ATP. K_m values for ATP in the ATPase assay were 42 μM for human and 15 μM for yeast RF-C, respectively. ATPγS and AMP-PNP are competitive inhibitors of the ATPase of yeast RF-C with K_i values of 1.8 μM and 130 μM, respectively.

RF-C itself had little or no effect on DNA synthesis by pols α, δ, and ε. In contrast, the conjunction of RF-C with PCNA, RP-A, and ATP strongly inhibited pol-α and stimulated pols δ and ε. Isolation of stable pol δ and ε complexes, formed with DNA in the presence of ATP, RF-C, and PCNA (called holoenzymes), indicated that pols δ and ε directly interact with these auxiliary proteins (Lee and Hurwitz 1990; Lee et al. 1991b; Burgers 1991; Podust et al. 1992).

RF-C is a multiprotein complex composed of five subunits, including one large subunit and four small subunits (Table 3). All subunits are encoded by different genes, and all show an extensive amino-acid-sequence homology in the middle part of the protein (O'Donnell et al. 1993; Bunz et al. 1993; Li and Burgers 1994b). All the genes encoding the subunits of yeast RF-C have been shown to be essential (Howell et al. 1994; Li and Burgers 1994a,b; Noskov et al. 1994; Cullmann et al. 1995). The large subunit of human RF-C could be cross-linked to DNA (Tsurimoto and Stillman 1991), and the DNA-binding properties of this subunit might also be expected from primary sequence data (Bunz et al. 1993; Howell et al. 1994). Interestingly, this subunit was able to be cloned from human and murine cDNA libraries by virtue of its dsDNA-binding properties (Burbelo et al. 1993; Lu and Riegel 1994; Luckow et al. 1994). In yeast, the interaction with PCNA was attributed to the large RF-C subunit (McAlear et al. 1994) and in human, to one of the small subunits, hRFC40 (Chen et al. 1992b; Pan et al. 1993). hRFC40 and hRFC37 interacted directly with pol-δ and pol-ε, respectively, as well as with each other (Pan et al. 1993). Yeast Rfc3p and Rfc4p yielded a stable complex upon coexpression in *E. coli* (Li and Burgers 1994b). Finally, hRFC37 and its yeast homolog Rfc2p showed specific primer end binding (Pan et al. 1993; Noskov et al. 1994), whereas hRFC36 and Rfc3p displayed a DNA-dependent ATPase activity (Li and Burgers 1994a; H. Flores-Rozas et al., pers. comm.).

RF-C is a protein that enzymatically loads PCNA onto DNA. Biochemical studies of this reaction might be generalized by the following model (Fig. 1) (Lee and Hurwitz 1990; Burgers 1991; Stillman 1994; Podust et al. 1995): RF-C first binds to DNA in the presence of ATP. PCNA then binds to the RF-C/DNA complex, forming an unstable intermediate complex. Upon hydrolysis of ATP by RF-C, the protein/DNA intermediate undergoes a conformational change resulting in a stable complex (the PCNA clamp). Current data suggest that RF-C remains as a part of the complex after PCNA loading (Podust et al. 1995). The clamp can interact with pol-δ and pol-ε, resulting in the formation of the corresponding holoenzymes that can efficiently replicate single-strand DNA. Preferential binding of RF-C to the template primer, together with the footprinting data, suggested that assembly of the RF-C/PCNA com-

Table 3 Replication factor C

Subunit[a]	Human sequence	Size (kD)[b]	Proposed function for human subunits	Yeast subunits[c]	Yeast sequence	Size (kD)[b]	Proposed function for yeast subunits	Identity (similarity) between human and yeast
hRFC140	Bunz et al. (1993)	128.3	Nucleotide binding, DNA binding	*RFC1* (Rfc1p)	Howell et al. (1994)[d]	94.9	Nucleotide binding, DNA binding, PCNA binding	35.8% (55.3%)
hRFC40	Chen et al. (1992b)	39	ATP binding, interaction with PCNA, RFC37, and pol-δ	*RFC4* (Rfc4p)	Li and Burgers (1994b)	36.2	Interaction with Rfc3p	60.1% (77.6%)
hRFC38	Cullmann et al. (1995)	40.5	unknown	*RFC5* (Rfc5p)	Cullmann et al. (1995)	39.9	unknown	44.6% (63.4%)
hRFC37	Chen et al. (1992a)	39.6	binds to primer ends and ssDNA, interacts with pol-ε	*RFC2* (Rfc2p)	Noskov et al. (1994)	39.7	binds to primer ends	50.6% (66.6%)
hRFC36	Cullmann et al. (1995)	38.5	ATPase	*RFC3* (Rfc3p)	Li and Burgers (1994a)	38.1	DNA-dependent ATPase	50.5% (72.1%)

For further references, see text.
[a]The subunits of RF-C were ordered and named according to the apparent sizes in SDS-PAGE for the purified human RF-C complex.
[b]Calculated from the sequence.
[c]Gene: italic; protein: in parentheses.
[d]*CDC44* (Howell et al. 1994; McAlear et al. 1994) is identical to *RFC1* (Cullmann et al. 1995).

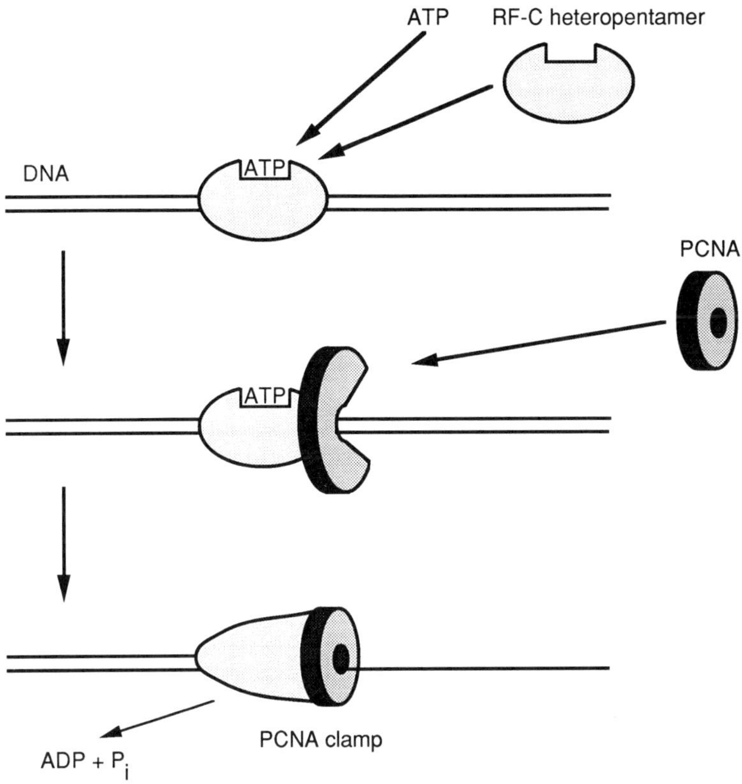

Figure 1 Model for the assembly of the RF-C/PCNA clamp. The heteropentameric RF-C first loads to dsDNA in the presence of ATP. PCNA then binds to the RF-C/DNA complex forming an unstable intermediate. This complex can slide along dsDNA until it encounters a template-primer junction. A 3′-OH group appears to be required for hydrolysis of ATP by the RF-C ATPase activity and for the conversion of the RF-C/PCNA intermediate into a catalytically competent PCNA clamp. The so-formed clamp is able to track along dsDNA. For details see text and citations therein.

plex occurred at the 3′-OH end of the primer. Recent studies showed that the assembly of the primary RF-C/PCNA complex does not require 3′-OH ends and occurs directly on dsDNA, whose topological structure and sequence do not give restriction for loading. The primary salt-sensitive RF-C/PCNA complex was suggested to slide along DNA to the 3′-OH end of a primer, where the final conversion of the intermediate complex to the catalytically competent PCNA clamp takes place (Podust et al. 1995). Using different approaches, it has been shown that the RF-

C/PCNA complex is able to track along the dsDNA (Podust et al. 1994, 1995; Tinker et al. 1994).

RP-A

RP-A is an essential protein that participates in DNA replication, DNA repair, and homologous DNA recombination (Table 4) (for review, see Hübscher and Spadari 1994). Various functional aspects of this protein are also covered in other chapters (see, e.g., Weisshart and Fanning; Brush and Kelly; Friedberg and Wood; Stillman; Borowiec; Newlon; Blow and Chong; all this volume). The protein was first discovered in human cells in the in vitro SV40 replication system (Wobbe et al. 1987) and has a heterotrimeric structure with polypeptides of molecular weights of 70K (called RP-A1), 32–34K (called RP-A2), and 11–14K (called RP-A3). All three subunits of RP-A are essential for viability of the cell (Brill and Stillman 1991) and have functions in DNA replication (Wobbe et al. 1987; Fairman and Stillman 1988; Wold and Kelly 1988), DNA repair (nucleotide excision repair) (Coverley et al. 1991), and homologous DNA recombination (Longhese et al. 1994).

The trimeric RP-A protein has many functional tasks. It is an SSB, partially sensitive to the base content with a preference for thymine-rich stretches in the SV40 origin of replication and has a low affinity to dsDNA (Kim et al. 1992). The binding to ssDNA appears to occur in two modes, since complexes covering 8–10 nucleotides and 30 nucleotides were identified (Blackwell and Borowiec 1994). RP-A assists the SV40 large T antigen in unwinding of the viral origin of replication (Wold and Kelly 1988; Kenny et al. 1989), binds to pol-α:primase (Dornreiter et al. 1992), and appears to suppress nonspecific priming events by the DNA primase activity of pol-α:primase (Collins and Kelly 1991). Furthermore, an important role was postulated in primosome assembly (Melendy and Stillman 1993). RP-A directly interacts with the DNA-binding protein XPA, which is known to be involved in the damage-recognition step of the nucleotide excision repair pathway (Matsuda et al. 1995). Binding to the tumor suppressor protein p53 and to the transcriptional activators VP16 and GAL4 suggested a role in transcription as well (He et al. 1993; Li and Botchan 1993). Finally, RP-A stimulated the four different DNA helicases A, B, C, and D from calf thymus in a species-specific way (Thömmes et al. 1992) and copurified with DNA helicase F (Georgaki et al. 1994).

What is known about the individual subunits? Human RP-A1 is located on chromosome 17q13.3 (Umbricht et al. 1994), is responsible for

Table 4 Replication protein A

Other names	replication factor A (RF-A) human ssDNA-binding protein (HSSB)
Structure	heterotrimer of 70 kD (RP-A1), 32–34 kD (RP-A2), and 11–14 kD (RP-A3)
Involvement in DNA transactions	DNA replication homologous DNA recombination nucleotide excision repair transcription?
Properties of the heterotrimeric RP-A	partially sequence-dependent SSB, binds in two different modes low affinity to dsDNA assists SV40 T antigen in unwinding of the SV40 origin of replication interacts with pol-α:primase suppresses nonspecific priming events functions in primosome assembly participates in DNA elongation binds to p53, V16, and GAL4
Properties of RP-A1	essential gene located on human chromosome 17q13.3 responsible for binding to ssDNA unwinding of DNA at the origin of SV40 replication
Properties of RP-A2	essential gene located on human chromosome 1p35 phosphorylated by a DNA-activated protein kinase hyperphosphorylated after treatment with ionizing radiation and UV light phosphorylation stimulates nonspecific DNA unwinding phosphorylation not required for in vitro replication binds to XP-A protein
Properties of RP-A3	essential gene located on chromosome 7p22 possible role in assembly of RP-A

For references, see text.

binding to ssDNA, and assists unwinding of the origin of SV40 replication by viral T antigen (Kenny et al. 1990). Furthermore, RP-A1 unwinds dsDNA nonspecifically (Georgaki and Hübscher 1993). Human RP-A2 is located on chromosome 1p35 (Umbricht et al. 1994). This subunit is phosphorylated in a cell-cycle-dependent manner on Ser-23 and Ser-29

(Din et al. 1990). Hyperphosphorylation of this subunit occurs after treatment of the cell with UV light (Carty et al. 1994) and ionizing radiation (Liu and Weaver 1993). Phosphorylation in the two serines mentioned is not required for in vitro replication (Henricksen and Wold 1994). Phosphorylation of RP-A2 is catalyzed in cyclin-A-activated extracts by a cdk/cyclin A complex and by a DNA-dependent protein kinase (Brush et al. 1994; Pan et al. 1994). Finally, the sequence-independent unwinding of DNA by RP-A1 is stimulated after phosphorylation of RP-A2 in vitro (Georgaki and Hübscher 1993). It is not clear at which step of the heterotrimer assembly the phosphorylation of RP-A2 occurs in vivo (see also Fig. 2). Human RP-A3 is located on chromosome 7p22 (Umbricht et al. 1994). Little is known about its function. A role in the assembly of the heterotrimeric RP-A has been suggested (Fig. 2) (Henricksen et al. 1994; Stigger et al. 1994).

Viral Counterparts of Eukaryotic Accessory Proteins

The two model systems of adenovirus and herpes simplex virus type 1 (HSV-1) rely on their own SSBs for their replication. The adenovirus DNA-binding protein, Ad-DBP, is described in more detail by Hay (this volume). It is a 59-kD zinc metalloprotein and has a triple function in DNA replication (Table 1): (1) binding to ssDNA, (2) helping the viral pol to perform strand-displacement DNA synthesis, and (3) helping in the initiation of DNA replication of the adenovirus. The recently reported crystal structure of the Ad-DBP nucleic-acid-binding domain allows insight into functional details of this protein (Tucker et al. 1994).

ICP8 from HSV-1 is the SSB and is a zinc metalloprotein with a molecular mass of 128 kD (for more details, see Challberg, this volume). The protein is essential for DNA replication, probably by acting as a typical SSB, and may play a key role in initiation of DNA replication by virtue of its helix-destabilizing properties and by its interaction with the initiating DNA helicase UL9 (Boehmer et al. 1994). Finally, ICP8 stimulates the cognate pol, the UL30 gene product (Ruyechan and Weir 1984).

The UL42 protein from HSV-1 is a processivity enhancing factor for the herpetic pol (UL30) and forms a complex with it (Hernandez and Lehman 1990; for more details, see Challberg, this volume). The complex of the pol, the UL42 processivity factor, the ICP8 protein, and the herpes simplex-encoded helicase:primase could be identified. It is shown to replicate dsDNA by a rolling-circle mechanism (Skaliter and Lehman 1994).

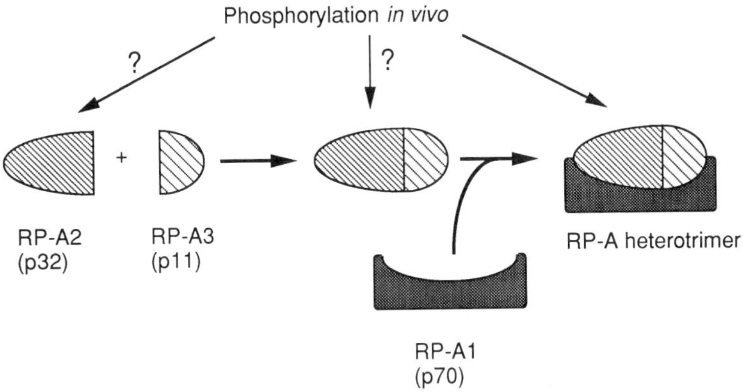

Figure 2 Model for the assembly of the RP-A trimer. RP-A2 and RP-A3 first form a complex, which provides the proper conformation for binding of the large RP-A1 subunit, resulting in a functionally active heterotrimeric RP-A form. (Adapted from Henricksen et al. 1994.)

CONCLUSIONS

The three DNA replication accessory proteins, PCNA, RF-C, and RP-A, possess a variety of essential functions in DNA replication. Their roles provide the replication machinery with physiological properties that allow the achievement of a coordinated action and of an accurate and fast polymerization mode. The functional conservation from bacteria to man is not surprising, due to their universal tasks in DNA replication. Since all subunits of PCNA, RF-C, and RP-A have been cloned and overexpressed, more insights into their biological functions in the DNA replication process will emerge soon.

ACKNOWLEDGMENTS

The work carried out in the authors' laboratory has continously been supported by the Swiss National Science Foundation. G.M. is supported by the Bundesamt für Bildung und Wissenschaft, which sponsors the Swiss part of a EU Human Capital and Mobility Program. We thank B. Stillman and J. Hurwitz for providing us with unpublished data.

REFERENCES

Alder, H., M. Yoshinouchi, M.B. Prystowsky, P. Appasamy, and R. Baserga. 1992. A conserved region in intron 1 negatively regulates the expression of the PCNA gene. *Nucleic Acids Res.* **20:** 1769–1775.

Almendral, J.M., D. Huebsch, P.A. Blundell, H. Macdonald-Bravo, and R. Bravo. 1987. Cloning and sequence of the human nuclear protein cyclin: Homology with DNA-binding proteins. *Proc. Natl. Acad. Sci.* **84:** 1575–1579.

Bauer, G.A. and P.M.J. Burgers. 1988. The yeast analog of mammalian cyclin/proliferating cell nuclear antigen interacts with mammalian DNA polymerase δ. *Proc. Natl. Acad. Sci.* **85:** 7506–7510.

———. 1990. Molecular cloning, structure and expression of the yeast proliferating cell nuclear antigen. *Nucleic Acids Res.* **18:** 261–265.

Blackwell, L.J. and J.A. Borowiec. 1994. Human replication protein A binds single-stranded DNA in two distinct complexes. *Mol. Cell. Biol.* **14:** 3993–4001.

Boehmer, P.E., M.C. Craigie, N.D. Stow, and I.R. Lehman. 1994. Association of origin binding protein and single strand DNA-binding protein, ICP8, during herpes simplex virus type 1 DNA replication *in vivo*. *J. Biol. Chem.* **269:** 29329–29334.

Brill, S.J. and B. Stillman. 1991. Replication factor-A from *Saccharomyces cerevisiae* is encoded by three essential genes coordinately expressed at S phase. *Genes Dev.* **5:** 1589–1600.

Brush, G.S., C.W. Anderson, and T.J. Kelly. 1994. The DNA-activated protein kinase is required for the phosphorylation of replication protein A during simian virus 40 DNA replication. *Proc. Natl. Acad. Sci.* **91:** 12520–12524.

Bunz, F., R. Kobayashi, and B. Stillman. 1993. cDNAs encoding the large subunit of human replication factor C. *Proc. Natl. Acad. Sci.* **90:** 11014–11018.

Burbelo, P.D., A. Utani, Z.-Q. Pan, and Y. Yamada. 1993. Cloning of the large subunit of activator 1 (replication factor C) reveals homology with bacterial DNA ligases. *Proc. Natl. Acad. Sci.* **90:** 11543–11547.

Burgers, P.M.J. 1991. *Saccharomyces cerevisiae* replication factor C. II. Formation and activity of complexes with the proliferating cell nuclear antigen and with DNA polymerases δ and ε. *J. Biol. Chem.* **266:** 22698–22706.

Burgers, P.M.J. and B.L. Yoder. 1993. ATP-independent loading of the proliferating cell nuclear antigen requires DNA ends. *J. Biol. Chem.* **268:** 19923–19926.

Carty, M.P., M. Zernik-Kobak, S. McGrath, and K. Dixon. 1994. UV light-induced DNA synthesis arrest in HeLa cells is associated with changes in phosphorylation of human single-stranded DNA-binding protein. *EMBO J.* **13:** 2114–2123.

Chang, C.-D., L. Ottavio, S. Travali, K.E. Lipson, and R. Baserga. 1990. Transcriptional and posttranscriptional regulation of the proliferating cell nuclear antigen gene. *Mol. Cell. Biol.* **10:** 3289–3296.

Chen, M., Z.-Q. Pan, and J. Hurwitz. 1992a. Studies of the cloned 37-kDa subunit of activator 1 (replication factor C) of HeLa cells. *Proc. Natl. Acad. Sci.* **89:** 5211–5215.

———. 1992b. Sequence and expression in *Escherichia coli* of the 40-kDa subunit of activator 1 (replication factor C) of HeLa cells. *Proc. Natl. Acad. Sci.* **89:** 2516–2520.

Collins, K.L. and T.J. Kelly. 1991. Effects of T antigen and replication protein A on the initiation of DNA synthesis by DNA polymerase α-primase. *Mol. Cell. Biol.* **11:** 2108–2115.

Coverley, D., M.K. Kenny, M. Munn, W.D. Rupp, D.P. Lane, and R.D. Wood. 1991. Requirement for the replication protein SSB in human DNA excision repair. *Nature* **349:** 538–541.

Cullmann, G., K. Fien, and B. Stillman. 1995. Characterization of the five replication factor C genes of *Saccharomyces cerevisiae*. *Mol. Cell. Biol.* **15:** 4661–4671.

Din, S.-U., S.J. Brill, M.P. Fairman, and B. Stillman. 1990. Cell-cycle regulated phosphorylation of DNA replication factor A from human and yeast cells. *Genes Dev.* **4:** 968–977.

Dornreiter, I., L.F. Erdile, I.U. Gilbert, D. von Winkler, T.J. Kelly, and E. Fanning. 1992. Interaction of DNA polymerase α-primase with cellular replication protein A and SV40 T antigen. *EMBO J.* **11:** 769–776.

Fairman, M.P. and B. Stillman. 1988. Cellular factors required for multiple stages of SV40 DNA replication *in vitro. EMBO J.* **7:** 1211–1218.

Fien, K. and B. Stillman. 1992. Identification of replication factor C from *Saccharomyces cerevisiae*: A component of the leading-strand DNA replication complex. *Mol. Cell. Biol.* **12:** 155–163.

Flores-Rozas, H., Z. Kelman, F.B. Dean, Z.-Q. Pan, J.W. Harper, S.J. Elledge, M. O'Donnell, and J. Hurwitz. 1994. Cdk-interacting protein 1 directly binds with proliferating cell nuclear antigen and inhibits DNA replication catalyzed by the DNA polymerase δ holoenzyme. *Proc. Natl. Acad. Sci.* **91:** 8655–8659.

Georgaki, A. and U. Hübscher. 1993. DNA unwinding by replication protein A is a property of the 70 kDa subunit and is facilitated by phosphorylation of the 32 kDa subunit. *Nucleic Acids Res.* **21:** 3659–3565.

Georgaki, A., N. Tuteja, B. Sturzenegger, and U. Hübscher. 1994. Calf thymus DNA helicase F, a replication protein A copurifying enzyme. *Nucleic Acids Res.* **22:** 1128–1134.

Hata, S., H. Kouchi, Y. Tanaka, E. Minami, T. Matsumoto, I. Suzuka, and J. Hashimoto. 1992. Identification of carrot cDNA clones encoding a second putative proliferating cell-nuclear antigen, DNA polymerase δ auxiliary protein. *Eur. J. Biochem.* **203:** 367–371.

He, Z., B.T. Brinton, J. Greenblatt, J.A. Hassel, and C.J. Ingles. 1993. The transactivator proteins VP16 and GAL4 bind replication factor A. *Cell* **73:** 1223–1232.

Henricksen, L.A. and M.S. Wold. 1994. Replication protein A mutants lacking phosphorylation sites for $p34^{cdc2}$ kinase support DNA replication. *J. Biol. Chem.* **269:** 24203–24208.

Henricksen, L.A., C.B. Umbricht, and M.S. Wold. 1994. Recombinant replication protein A: Expression, complex formation, and functional characterization. *J. Biol. Chem.* **269:** 11121–11132.

Hernandez, T.R. and I.R. Lehman. 1990. Functional interaction between herpes simplex-1 DNA polymerase and UL42 protein. *J. Biol. Chem.* **265:** 11227–11232.

Howell, E.A., M.A. McAlear, D. Rose, and C. Holm. 1994. CDC44: A putative nucleotide-binding protein required for cell cycle progession that has homology to subunits of replication factor C. *Mol. Cell. Biol.* **14:** 255–267.

Hübscher, U. and S. Spadari. 1994. DNA replication and chemotherapy. *Physiol. Rev.* **74:** 259–304.

Jackson, D.A., A.B. Hassan, R.J. Errington, and P.R. Cook. 1994. Sites in human nuclei where damage induced by ultraviolet light is repaired: Localization relative to transcription sites and concentrations of proliferating cell nuclear antigen and the tumor suppressor protein, p53. *J. Cell Sci.* **107:** 1753–1760.

Jones, N.C., P.W. Rigby, and E.B. Ziff.1988. *Trans*-acting protein factors and the regulation of eukaryotic transcription: Lessons from studies on DNA tumor viruses. *Genes Dev.* **2:** 267–281.

Kenny, M.K., S.-H. Lee, and J. Hurwitz. 1989. Multiple functions of human single-strand-DNA binding protein in simian virus 40 DNA replication: Single-strand stabilization and stimulation of DNA polymerases α and δ. *Proc. Natl. Acad. Sci.* **86:** 9757–9761.

Kenny, M.K., U. Schlegel, H. Furneau, and J. Hurwitz. 1990. The role of human single-stranded DNA binding protein and its individual subunits in simian virus 40 DNA replication. *J. Biol. Chem.* **265:** 7693–7700.

Kilbey, B.J., I. Fraser, S. McAleese, M. Goman, and R.G. Ridley. 1993. Molecular characterisation and stage-specific expression of proliferating cell nuclear antigen (PCNA) from the malarial parasite, *Plasmodium falciparum*. *Nucleic Acids Res.* **21:** 239–243.

Kim, C., R.O. Snyder, and M.S. Wold. 1992. Binding properties of replication protein A from human and yeast cells. *Mol. Cell. Biol.* **12:** 3050–3059.

Kornberg, A. and T.A. Baker. 1992. *DNA replication*. W.H. Freeman, New York.

Krishna, T.S.R., X.-P. Kong, S. Gary, P.M.J. Burgers, and J. Kuriyan. 1994. Crystal structure of the eukaryotic DNA polymerase processivity factor PCNA. *Cell* **79:** 1233–1243.

Kuriyan, J. and M. O'Donnell. 1993. Sliding clamps of DNA polymerases. *J. Mol. Biol.* **234:** 915–925.

Lee, S.-H. and J. Hurwitz. 1990. Mechanism of elongation of primed DNA by DNA polymerase δ, proliferating cell nuclear antigen, and activator 1. *Proc. Natl. Acad. Sci.* **87:** 5672–5676.

Lee, S.-H., A.D. Kwong, Z.-Q. Pan, and J. Hurwitz. 1991a. Studies on the activator 1 protein complex, an accessory factor for proliferating cell nuclear antigen-dependent DNA polymerase δ. *J. Biol. Chem.* **266:** 594–602.

Lee, S.-H., Z.-Q. Pan, A.D. Kwong, P.M.J. Burgers, and J. Hurwitz. 1991b. Synthesis of DNA by DNA polymerase ε *in vitro*. *J. Biol. Chem.* **266:** 22707–22717.

Li, R. and M.R. Botchan. 1993. The acidic transcriptional activation domains of VP16 and p53 bind the cellular replication protein A and stimulate *in vitro* BPV-1 DNA replication. *Cell* **73:** 1207–1221.

Li, R., S. Waga, G.J. Hannon, D. Beach, and B. Stillman. 1994. Differential effects by the p21 CDK inhibitor on PCNA-dependent DNA replication and repair. *Nature* **371:** 354–357.

Li, X. and P.M.J. Burgers. 1994a. Molecular cloning and expression of the *Saccharomyces cerevisiae RFC3* gene, an essential component of replication factor C. *Proc. Natl. Acad. Sci.* **91:** 868–872.

———. 1994b. Cloning and characterization of the essential *Saccharomyces cerevisiae RFC4* gene encoding the 37-kDa subunit of replication factor C. *J. Biol. Chem.* **269:** 21880–21884.

Liu, V.F. and D.T. Weaver. 1993. The ionizing radiation-induced replication protein A phosphorylation response differs between ataxia telangiectasia and normal human cells. *Mol. Cell. Biol.* **13:** 7222–7231.

Longhese, M.P., P. Plevani, and G. Lucchini. 1994. Replication factor A is required in vivo for DNA replication, repair, and recombination. *Mol. Cell. Biol.* **14:** 7884–7890.

Lu, Y. and A.T. Riegel. 1994. The human DNA-binding protein, PO-GA, is homologous to the large subunit of mouse replication factor C: Regulation by alternate 3′ processing of mRNA. *Gene* **145:** 261–265.

Luckow, B., F. Bunz, B. Stillman, P. Lichter, and G. Schütz. 1994. Cloning, expression, and chromosomal localization of the 140-kilodalton subunit of replication factor C from mice and humans. *Mol. Cell. Biol.* **14:** 1626–1634.

Maga, G. and U. Hübscher. 1995. DNA polymerase ε interacts with proliferating cell

nuclear antigen in primer recognition and elongation. *Biochemistry* **34:** 891–901.
Matsuda, T., M. Saijo, I. Kuraoka, T. Kobayashi, Y. Nakatsu, A. Nagai, T. Enjoji, C. Masutani, K. Sugasawa, F. Hanaoka, A. Yasui, and K. Tanaka. 1995. DNA repair protein XPA binds replication protein A (RPA). *J. Biol. Chem.* **270:** 4152–4157.
Matsumoto, K., T. Moriuchi, T. Koji, and P.K. Nakane. 1987. Molecular cloning of cDNA coding for rat proliferating cell nuclear antigen (PCNA)/cyclin. *EMBO J.* **6:** 637–642.
McAlear, M.A., E.A. Howell, K.K. Espenshade, and C. Holm. 1994. Proliferating cell nuclear antigen (*pol30*) mutations suppress *cdc44* mutations and identify potential regions of interaction between the two encoded proteins. *Mol. Cell. Biol.* **14:** 4390–4397.
Melendy, T. and B. Stillman. 1993. An interaction between replication protein A and SV40 T antigen appears essential for primosome assembly during SV40 DNA replication. *J. Biol. Chem.* **268:** 3389–3395.
Mudryj, M., S.W. Hiebert, and J.R. Nevins. 1990. A role for the adenovirus inducible E2F transcription factor in a proliferation dependent signal transduction pathway. *EMBO J.* **9:** 2179–2184.
Noskov, V., S. Maki, Y. Kawasaki, S.-H. Leem, B.-I. Ono, H. Araki, Y. Pavlov, and A. Sugino. 1994. The *RFC2* gene encoding a subunit of replication factor C of *Saccharomyces cerevisiae*. *Nucleic Acids Res.* **22:** 1527–1535.
O'Donnell, M., R. Onrust, F.B. Dean, M. Chen, and J. Hurwitz. 1993. Homology in accessory proteins of replicative polymerases—*E. coli* to humans. *Nucleic Acids Res.* **21:** 1–3.
O'Reilly, D.R., A.M. Crawford, and L.K. Miller. 1989. Viral proliferating cell nuclear antigen. *Nature* **337:** 606.
Pan, Z.-Q., M. Chen, and J. Hurwitz. 1993. The subunits of activator 1 (replication factor C) carry out multiple functions essential for proliferating cell nuclear antigen-dependent DNA synthesis. *Proc. Natl. Acad. Sci.* **90:** 6–10.
Pan, Z.-Q., A.A. Amin, E. Gibbs, H. Niu, and J. Hurwitz. 1994. Phosphorylation of the p34 subunit of human single-stranded-DNA-binding protein in cyclin A-activated G_1 extracts is catalyzed by cdk-cyclin A complex and DNA-dependent protein kinase. *Proc. Natl. Acad. Sci.* **91:** 8343–8347.
Podust, L.M., V.N. Podust, C. Floth, and U. Hübscher. 1994. Assembly of DNA polymerase δ and ε holoenzymes depends on the geometry of the DNA template. *Nucleic Acids Res.* **22:** 2970–2975.
Podust, L.M., V.N. Podust, J.M. Sogo, and U. Hübscher. 1995. Mammalian DNA polymerase auxiliary proteins: Analysis of replication factor C-catalyzed proliferating cell nuclear antigen loading onto circular double-stranded DNA. *Mol. Cell. Biol.* **34:** 3072–3081.
Podust, V.N., A. Georgaki, B. Strack, and U. Hübscher. 1992. Calf thymus RF-C as an essential component for DNA polymerase δ and ε holoenzymes function. *Nucleic Acids Res.* **20:** 4159–4165.
Prelich, G. and B. Stillman. 1988. Coordinated leading and lagging strand synthesis during SV40 DNA replication in vitro requires PCNA. *Cell* **53:** 117–126.
Prelich, G., M. Kostura, D.R. Marshak, M.B. Mathews, and B. Stillman. 1987. The cell-cycle regulated proliferating cell nuclear antigen is required for SV40 DNA replication *in vitro*. *Nature* **326:** 471–475.
Ruyechan, W.T. and A.C. Weir. 1984. Interaction with nucleic acids and stimulation of

the viral DNA polymerase by the herpes simplex type 1 major DNA-binding protein. *J. Virol.* **52:** 727–733.

Shivji, M.K.K., M.K. Kenny, and R.D. Wood. 1992. Proliferating cell nuclear antigen is required for DNA excision repair. *Cell* **69:** 367–374

Skaliter, R. and I.R. Lehman. 1994. Rolling circle DNA replication *in vitro* by a complex of herpes simplex virus type 1-encoded enzymes. *Proc. Natl. Acad. Sci.* **91:** 10665–10669.

Smith, M.L., I.-T. Chen, Q. Zhan, I. Bae, C.-Y. Chen, T.M. Gilmer, M.B. Kastan, P.M. O'Connor, and A.J. Fornace, Jr. 1994. Interaction of the p53-regulated protein Gadd45 with proliferating cell nuclear antigen. *Science* **266:** 1376–1380.

Stigger, E., F.B. Dean, J. Hurwitz, and S.-K. Lee. 1994. Reconstitution of functional human single-stranded DNA-binding protein from individual subunits expressed by recombinant baculoviruses. *Proc. Natl. Acad. Sci.* **91:** 579–583.

Stillman, B. 1994. Smart machines at the DNA replication fork. *Cell* **78:** 725–728.

Stukenberg, P.T., J. Turner, and M. O'Donnell. 1994. An explanation for lagging strand DNA replication: Polymerase hopping among DNA sliding clamps. *Cell* **78:** 877–887.

Suzuka, I., S. Hata, M. Matsuoka, S. Kosugi, and T. Hashimoto. 1991. Highly conserved structure of proliferating cell nuclear antigen (DNA polymerase of auxiliary protein) gene in plants. *Eur. J. Biochem.* **195:** 571–575.

Suzuka, I., H. Daidoji, M. Matsuoka, K.-I. Kadowaki, Y. Takasaki, P.K. Nakane, and T. Moriuchi. 1989. Gene for proliferating-cell nuclear antigen (DNA polymerase δ auxiliary protein) is present in both mammalian and higher plant genomes. *Proc. Natl. Acad. Sci.* **86:** 3189–3193.

Tan, C.-K., C. Castillo, A.G. So, and K.L. Downey. 1986. An auxiliary protein for DNA polymerase δ from fetal calf thymus. *J. Biol. Chem.* **261:** 12310–12316.

Thömmes, P., E. Ferrari, R. Jessberger, and U. Hübscher. 1992. Four different DNA helicases from calf thymus. *J. Biol. Chem.* **267:** 6063–6073.

Tinker, R.L., G.A. Kassavetis, and E.P. Geiduschek. 1994. Detecting the ability of viral, bacterial and eukaryotic replication proteins to track along DNA. *EMBO J.* **13:** 5330–5337.

Tsurimoto, T. and B. Stillman. 1989. Purification of a cellular replication factor, RF-C, that is required for coordinated synthesis of leading and lagging strands during simian virus 40 DNA replication *in vitro*. *Mol. Cell. Biol.* **9:** 609–619.

———. 1990. Functions of replication factor C and proliferating cell nuclear antigen: Functional similarity of DNA polymerase accessory proteins from human cells and bacteriophage T4. *Proc. Natl. Acad. Sci.* **87:** 1023–1027.

———. 1991. Replication factors required for SV40 DNA replication *in vitro*. I. DNA structure-specific recognition of a primer-template junction by eukaryotic DNA polymerases and their accessory proteins. *J. Biol. Chem.* **266:** 1950–1960.

Tucker, P.A., D. Tsernoglou, A.D. Tucker, F.E.J. Coenjaerts, H. Lennders, and P.C. van der Vliet. 1994. Crystal structure of the adenovirus DNA binding protein reveals a hook-on model for cooperative DNA binding. *EMBO J.* **13:** 2994–3002.

Umbricht, C.B., C.A. Griffin, A.L. Hawkins, K.H. Grzeschik, P. O'Connell, R. Leach, E.D. Green, and T.J. Kelly. 1994. High-resolution genomic mapping of the three human replication protein A genes (*RPA1, RPA2,* and *RPA3*). *Genomics* **20:** 249–257.

Waga, S., G.J. Hannon, D. Beach, and B. Stillman. 1994. The p21 inhibitor of cyclin-dependent kinases controls DNA replication by interaction with PCNA. *Nature* **369:** 574–578.

Waseem, N.H., K. Labib, P. Nurse, and D.P. Lane. 1992. Isolation and analysis of the fission yeast gene encoding DNA polymerase δ accessory protein PCNA. *EMBO J.* **11:** 5111–5120.

Wobbe, C.R., L. Weissbach, J.A. Boroviec, F.B. Dean, Y. Murkami, P. Bullock, and J. Hurwitz. 1987. Replication of simian virus 40 origin-containing DNA *in vitro* with purified proteins. *Proc. Natl. Acad. Sci.* **84:** 1834–1838.

Wold, M.S. and T. Kelly. 1988. Purification and characterization of replication protein A, a cellular protein required for *in vitro* replication of simian virus 40 DNA. *Proc. Natl. Acad. Sci.* **85:** 2523–2527.

Xiong, Y., H. Zhang, and D. Beach. 1992. D type cyclins associate with multiple protein kinases and the DNA replication and repair factor PCNA. *Cell* **71:** 505–514.

Yamaguchi, M., Y. Hayashi, S. Matsuoka, T. Takahashi, and A. Matsukage. 1994. Differential effect of p53 on the promoters of mouse DNA polymerase β gene and proliferating cell nuclear antigen gene. *Eur. J. Biochem.* **221:** 227–237.

Yamaguchi, M., N. Yasuyoshi, T. Moriuchi, F. Hirose, C.-C. Hui, Y. Suzuky, and A. Matsukage. 1990. *Drosophila* proliferating cell nuclear antigen (cyclin) gene: Structure, expression during development and specific binding of homeodomain proteins to its 5′-flanking region. *Mol. Cell. Biol.* **10:** 872–879.

Yoder, B.L. and P.M. Burgers. 1991. *Saccharomyces cerevisiae* replication factor C. I. Purification and characterization of its ATPase activity. *J. Biol. Chem.* **266:** 22689–22697.

Young, M.C, M.K. Reddy, and P.H. von Hippel. 1992. Structure and function of bacteriophage T4 DNA polymerase holoenzyme. *Biochemistry* **31:** 8675–8690.

Zhang, H., Y. Xiong, and D. Beach. 1993. Proliferating cell nuclear antigen and p21 are components of multiple cell cycle kinase complexes. *Mol. Biol. Cell* **4:** 897–906.

18
DNA Helicases

James A. Borowiec
Department of Biochemistry
and Kaplan Comprehensive Cancer Center
New York University Medical Center
New York, New York 10016

DNA helicases are proteins that use the energy of NTP hydrolysis to processively denature duplex DNA.[1] As such, helicases can be considered similar to other mechanochemical motors that can catalyze directional movement along a polymeric lattice (e.g., microtubule-based motors as kinesin). Acting during DNA replication, helicases convert the duplex DNA to single strands and thereby activate the DNA for nascent strand synthesis by the replicative DNA polymerases. Because of the fundamental nature of this process, helicases also play essential roles during DNA repair and recombination as well as RNA transcription and are found in virtually all cells (Thömmes and Hübscher 1992; Lohman 1993; Matson et al. 1994).

Our knowledge of the DNA helicases acting during eukaryotic DNA replication is somewhat meager. The replicative helicases functioning during chromosomal replication for any eukaryote are unknown. However, information is available from the study of animal viruses that encode replicative helicases, as well as from prokaryotic replication systems. Here I outline the current understanding of DNA helicases that function in eukaryotic DNA replication.

DNA HELICASE ASSAYS

DNA helicases are often first detected by a DNA-dependent or DNA-stimulated NTPase activity. Verification of DNA helicase activity is then accomplished using an assay that reveals the conversion of double-stranded DNA (dsDNA) into two single strands. The most common as-

[1]Not included in this category are proteins that can denature DNA by binding stoichiometrically to single-stranded DNA (ssDNA) using an NTP-independent mechanism (e.g., ssDNA binding proteins such as bacteriophage T4 gp32 protein; Alberts and Frey 1970), proteins that catalyze movement of a transient bubble through DNA (e.g., RNA polymerases), or topoisomerases that alter the linking number of covalently closed circular DNA molecules.

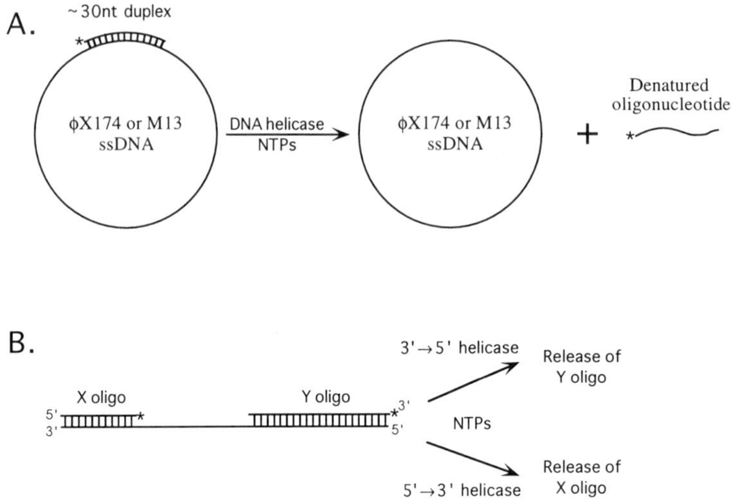

Figure 1 Schematic of assays used to detect DNA helicase activity. (*A*) The basic assay. A ^{32}P-labeled oligonucleotide is annealed to a larger ssDNA molecule such as φX174. In an NTP-hydrolysis-dependent reaction, DNA helicases denature the oligonucleotide from the ssDNA circle. Activity can be revealed by subjecting the reaction products to native gel electrophoresis and autoradiography. (*B*) The polarity assay. To determine the polarity of DNA helicase action, a substrate is employed that contains two duplex regions of distinct length flanking a central ssDNA region. Both short DNA strands (designated X oligo and Y oligo) are radioactively labeled. DNA helicase unwinding in the 5′→3′ direction would release the X oligonucleotide, and unwinding in the 3′→5′ direction would release the Y oligonucleotide.

say measures the release of a ^{32}P-labeled oligonucleotide annealed to a larger ssDNA molecule, such as φX174 or M13 viral DNA (Fig. 1A) (Venkatesan et al. 1982; Matson et al. 1983). The radioactive substrate and product molecules differ greatly in size so that native gel electrophoresis of the reactive mixture quickly reveals helicase activity.

Upon identification, DNA helicases are most simply characterized with respect to a few basic features, most notably the nucleotide requirements, specificity for DNA and RNA, polarity, and the requirement for ssDNA extensions. Nonhydrolyzable nucleotide analogs such as adenosine 5′-(γ-thio)triphosphate (ATPγS) or adenosine 5′-(β,γ-imido)triphosphate (AppNp) are usually tested to verify the requirement for nucleotide hydrolysis. Polarity, sometimes referred to as directionality, is defined as the direction of helicase movement on ssDNA, i.e., 5′→3′ or 3′→5′. Determination of helicase polarity utilizes a partially duplex substrate in which a central ssDNA region is flanked by two duplex regions of different lengths (Fig. 1B) (Matson 1986). Polarity is ascertained by identifying the duplex region that is denatured by the helicase. For helicases involved in DNA replication, the polarity of the reaction is strongly indicative of helicase placement on the leading (a 3′→5′polarity) or lagging (a 5′→3′polarity) strand. Although the assay can determine the polarity of virtually all helicases, certain helicases require or are greatly stimulated by a ssDNA extension on the strand opposite to that used for primary contacts (e.g., *Escherichia coli* DnaB; LeBowitz and McMacken 1986). Optimal activity on helicase substrates containing a DNA fork (i.e., noncomplementary ssDNA extensions on both strands) suggests that the protein interacts with both the leading- and lagging-strand templates during DNA unwinding.

The basic helicase assay has proven extremely useful because of its simplicity. However, it relies on the production of a completely denatured product and thus provides limited information on the mechanism by which the helicase unwinds DNA. The assay has therefore been modified in various ways to reveal information concerning intermediates in the process. For example, synthesis of oligonucleotides containing 2-aminopurine (2-AP) in place of adenine allows the generation of helicase substrates with distinctive fluorescent properties (Raney et al. 1994). Quenching of the 2-AP fluorescence in duplex DNA is alleviated during DNA denaturation. This assay can therefore be used to formulate kinetic models of the unwinding reaction. Other fluorescence-based assays have also been developed (Roman and Kowalczykowski 1989; Houston and Kodadek 1994).

MOLECULAR BIOLOGY OF HELICASES

Comparison of the translated sequences of genes encoding either biochemically characterized or putative DNA helicases has revealed various categories of sequence homology. A significant fraction of prokaryotic and eukaryotic helicases, able to unwind DNA in either the 5′→3′or 3′→5′direction, are members of a DNA/RNA helicase superfamily (Hodgman 1988; Gorbalenya et al. 1989). Members of the superfamily are identified by the presence of six conserved protein motifs (I–VI). A seventh motif, termed Ia, is found only in a subset of helicases within the superfamily and has a poorly defined consensus sequence. Variations of motifs I and II are found in numerous proteins able to hydrolyze purine nucleotide triphosphates. Although crystal structures of helicases are currently unavailable, there is abundant high-resolution structural information from diverse nucleotide-utilizing enzymes (Walker et al. 1982; Fry et al. 1986; Saraste et al. 1990). Such studies reveal that motif I and the closely associated motif Ia form the canonical "A-site" or "P-loop" involved in nucleotide triphosphate recognition. Mutation of motif I, particularly of the lysine in the conserved GXGK$^{T}/_{S}$ sequence, deleteriously affects hydrolysis of the nucleotide with less significant effects on nucleotide binding. Motif II (the "B-site"; ++++DEX$^{H}/_{D}$ where hydrophobic residues [I, L, V, M, F, Y, and W] are indicated by a plus, and X indicates any amino acid) also plays a significant role in nucleotide binding and is thought to interact with the divalent magnesium associated with the nucleotide phosphates. The roles of motifs III–VI are unknown. However, the effects of systematic mutation of each motif (e.g., in the herpes simplex virus type I [HSV] UL5 [Zhu and Weller 1992] and UL9 DNA helicases [Martinez et al. 1992], or the human DNA repair helicase ERCC3 [Ma et al. 1994b]) strongly support the conclusion that the motifs represent structural components that are critical for DNA helicase function.

Although virtually all helicases contain elements similar to the motifs I and II, many lack any recognizable homology with the other motifs found in the helicase superfamily. Comparison of these helicase sequences also does not reveal a second broad grouping of related sequences. However, more limited subsets of homologous sequences have been noted. For example, sequences encoding the large tumor antigen of the papovavirus family, typified by the SV40 T antigen, show significant homology over a broad region to predicted E1 proteins encoded by the papillomavirus family (Clertant and Seif 1984). Whether these helicases are structurally similar to superfamily members is a matter of debate.

REPLICATIVE VIRAL DNA HELICASES

The lack of clearly defined cell-free systems that catalyze chromosomal replication has precluded an unambiguous definition of any cellular replicative helicase (discussed below). However, various animal viruses such as SV40 or HSV encode DNA helicases with known roles in viral DNA replication (Table 1). Study of these viral enzymes has provided a useful avenue to explore the role of helicases in eukaryotic DNA replication. Furthermore, as the replication of certain viruses can be reproduced in vitro, it is possible to biochemically explore the interplay between the DNA helicase and the replication machinery, the latter often of cellular origin.

Table 1 Virally encoded DNA helicases essential for viral DNA replication

Helicase	Deduced m.w.[a]	Oligomeric state (m.w.)[b]	Nucleotide cofactors	K_m (mM)	Polarity	References
SV40 large T antigen	82,000	hexamer (492,000)	ATP, dATP, dTTP>dCTP, UTP	ATP 0.6	3′→5′	Goetz et al. 1988; Wiekowski et al. 1988; Scheffner et al. 1989
Py large T antigen	88,000	n.d.	ATP, dATP>CTP, UTP	ATP 0.64	3′→5′	Seki et al. 1990; Wang and Prives 1991
BPV E1	69,000[c]	n.d.	ATP, dATP, CTP, dCTP, UTP, dTTP, GTP, dGTP	ATP 0.20	3′→5′	Seo et al. 1993; Yang et al. 1993
HSV UL5	114,000	hetero-oligomeric complex (1:1:1) of UL5, UL52, and UL8 (293,000)	site I GTP, ATP site II ATP dATP	site I ATP 1.3 GTP 1.0 site II ATP 0.17	5′→3′	Crute et al. 1991; Earnshaw and Jarvest 1994
HSV UL9	94,000	dimer	ATP, CTP, dATP, dCTP, UTP	ATP 0.54	3′→5′	Fierer and Challberg 1992; Boehmer et al. 1993; Earnshaw and Jarvest 1994
AAV Rep68, Rep78	61,000, 71,000	dimer (Rep78)	ATP>CTP>dATP>GTP> UTP	n.d.	3′→5′	Im and Muzyczka 1990, 1992; X. Zhou and N. Muzyczka, pers. comm.
MVM NS-1	83,000	n.d.	ATP>dATP	n.d.	n.d.	Wilson et al. 1991

(n.d.) Not determined.

[a]Molecular weight deduced from the DNA sequence encoding the helicase.

[b]The most prominent species in solution is indicated. For the SV40 T antigen, the helicase-active species has been found to be a hexamer (Dean et al. 1992; SenGupta and Borowiec 1992; Wessel et al. 1992).

[c]The molecular weight of the E1 protein is assumed from translation of the E1 open reading frame.

Papovavirus

SV40 T Antigen

The viral DNA helicase perhaps best understood is the SV40 large tumor antigen (T antigen; for review, see Fanning and Knippers 1992; see also Weisshart and Fanning; Hassell and Brinton; both this volume). T antigen is an essential initiation and elongation factor for SV40 DNA replication. The protein can also regulate viral transcription and transform various cells by binding to tumor suppressor proteins (e.g., p53), although these activities are not directly involved in the replicative functions of T antigen. During the initiation of replication, T antigen binds as a double hexamer to the viral origin in an ATP-dependent reaction. In the presence of human replication protein A (hRP-A) or heterologous SSB proteins, the origin becomes denatured, and T antigen unwinds the viral DNA bidirectionally outward from the origin. The exposed single strands serve as templates for nascent-strand synthesis by the host DNA polymerase machinery (see Hassell and Brinton, this volume).

T antigen (708 amino acids; predicted M_r 82,000) contains both DNA and RNA helicase activities, each proceeding in the 3′→5′direction (Stahl et al. 1986; Goetz et al. 1988; Wiekowski et al. 1988; Scheffner et al. 1989). The two activities are distinguishable by their nucleotide requirements. The DNA helicase is most active in the presence of ATP or dATP, with dTTP, UTP, and dCTP functioning to a lesser extent (K_m [ATP] 0.6 mM; Goetz et al. 1988; Wiekowski et al. 1988; Scheffner et al. 1989). In contrast, the RNA helicase is unable to efficiently utilize ATP or dATP, but has significant RNA-unwinding activity in the presence of the other common nucleotides (Scheffner et al. 1989). T antigen most efficiently unwinds DNA substrates containing a 3′overhang 20 nucleotides in length, although a lesser stimulatory effect can be seen using single-strand extensions as small as 5 nucleotides (Wiekowski et al. 1988). The rate of DNA unwinding at 37°C was determined to be in the range of 1.5–3 bp/second (Wiekowski et al. 1988; Murakami and Hurwitz 1993).

SenGupta and Borowiec used nuclease and chemical probing techniques to examine the interaction of T antigen with model replication forks (SenGupta and Borowiec 1992; SenGupta et al. 1992). In the presence of low levels of ATP (≤100 μM), T antigen quantitatively bound the fork without appreciable denaturation of the duplex structure. T antigen significantly protected only one of the two strands from nuclease attack, that containing the 3′single-stranded end (the top strand). Approximately 15 nucleotides of DNA were protected, including the dsDNA/ssDNA junction, a result consistent with the stimulatory effects of 3′ overhangs on T antigen unwinding (Wiekowski et al. 1988). These results differ in certain aspects from a more limited study of T antigen binding to a synthetic DNA fork (Wessel et al. 1992). However, the latter study used adenosine 5′-(β,γ-methylene)triphosphate (AppCp) rather than ATP and employed conditions in which only a small fraction of the fork substrate was bound by T antigen. The particulars of this discrepancy are not yet understood.

Functional studies were also employed to determine the contacts used by T antigen during fork denaturation (SenGupta and Borowiec 1992). T antigen utilized critical contacts with six phosphates at the dsDNA/ssDNA junction of the top strand during fork denaturation. This result suggests that the T antigen achieves polarity of movement using directional cues within the sugar-phosphate backbone. These and other results also indicate that T antigen has the ability to recognize the fork structure directly, an observation useful in understanding the mechanism of helicase loading onto DNA during the initiation of replication (see, e.g., SenGupta and Borowiec 1994).

T antigen bound to the fork primarily as a hexamer, although double hexameric complexes were also noted (SenGupta and Borowiec 1992). Although T antigen can exist in a variety of oligomeric states, the presence of ATP stimulates the formation of T-antigen hexamers, entities known to be helicase-active (Dean et al. 1992; Wessel et al. 1992), and suggests that the two origin-bound hexamers separate during replication initiation. This question has been addressed by Stahl and colleagues, who examined the products of the origin-dependent DNA-unwinding reaction by electron microscopy (Wessel et al. 1992). In approximately one-quarter of the unwound molecules, T antigen bound at each fork remained joined such that two ssDNA loops ("rabbit ears") were extruded from the complex. This intriguing result implies that SV40 replication on both forks occurs simultaneously within a single mini-factory, a smaller example of the replication centers thought to duplicate the eukaryotic chromosome. Further exploration of this SV40 replication feature is obviously required.

The SV40 T antigen, as well as the polyomavirus (Py) T antigen and the papillomavirus E1 (below), are required for denaturation of the cognate origins during replication initiation. Although the denaturation reactions require ATP hydrolysis, the role of the DNA helicase activity in these processes is unclear. A threonine to alanine mutation at position 124 in the SV40 T antigen gives rise to a protein that has wild-type DNA helicase activity and can bind the origin, yet is unable to initiate the origin-dependent DNA-unwinding reaction (McVey et al. 1993; Moarefi et al. 1993). Thus, the SV40 T antigen, and by inference E1 and the Py T antigen, likely require functions in addition to the DNA helicase and origin-binding activities for origin denaturation.

T antigen has been subject to extensive analysis to determine critical regions required for the T-antigen DNA helicase and other activities (for review, see Fanning and Knippers 1992). Partial proteolysis demonstrates that the essential helicase domain is contained within amino acids 131 to about 616, although the presence of amino acids to 708 is required for wild-type activity (Wun-Kim and Simmons 1990). This domain contains the core origin-binding domain (amino acids 131–257) with various data indicating that the DNA helicase and origin substrates bind to partially, but not completely, overlapping sites (see, e.g., Mohr et al. 1989; Lin et al. 1992).

Polyomavirus T Antigen

The Py and SV40 large T antigens are approximately 60% similar at the amino acid level and are alike with respect to virtually all replication activities (Seki et al. 1990; Wang and Prives 1991). The Py T antigen (predicted M_r 88,000) is a 3′→5′ DNA helicase with Py origin-binding and origin-unwinding activities. The Py and SV40 helicase activities are equivalent, with only the nucleotide requirements found to differ slightly (Wang and Prives 1991). The Py T antigen is most active in the presence of ATP and dATP, with CTP and UTP substituting to approximately 50% efficiency (K_m [ATP] 0.64 mM).

Papillomavirus: E1

Replication of bovine papillomavirus type 1 (BPV) and most human papillomavirus (HPV) types requires the products of two viral open reading frames (ORFs) designated E1 and E2 (Ustav and Stenlund 1991; Chiang et al. 1992; Del Vecchio et al. 1992; Stenlund, this volume). Analysis of E1 ORF sequences from BPV and HPV type 6 revealed that they bore significant homology with the SV40 and Py T antigen genes over various regions (Clertant and Seif 1984). Thus, it was anticipated that the putative papillomavirus E1 proteins would have similar activities as T antigen, a hypothesis later borne out.

The BPV E1 can bind specifically to the viral origin in a reaction stimulated both by E2, a transcriptional activator, and by ATP (Mohr et al. 1990; Lusky et al. 1993). E1 is a DNA helicase and can also catalyze the large-scale unwinding of DNA molecules containing the viral origin

(Seo et al. 1993; Yang et al. 1993; Hughes and Romanos 1993). In the presence of crude cellular extracts, origin-containing DNA molecules are replicated in a reaction dependent on E1 (Yang et al. 1991). Each of these activities was demonstrated with E1 produced from recombinant baculovirus because of the very low level of the native E1 (predicted M_r 69,000) in infected cells (Santucci et al. 1990; Sun et al. 1990). However, it would be surprising if the native E1 differed significantly from the recombinant form.

E1 is a 3′→5′DNA helicase that can utilize all common ribo- and deoxyribo- NTPs (K_m[ATP] 200 μM; Seo et al. 1993; Yang et al. 1993). E1 denatures 5′-tailed oligonucleotides annealed to single-stranded ϕX174 fivefold more efficiently than non-tailed oligonucleotides (Seo et al. 1993). The demonstrated 3′→5′polarity suggests that E1 utilizes primary interactions with the leading strand during DNA denaturation, but possibly also uses significant lagging-strand contacts during viral replication. Although E2 can bind E1 in solution and can stimulate the origin-dependent DNA-unwinding reaction, the presence of E2 does not appear to have detectable effects on the E1 helicase activity in vitro (Seo et al. 1993; Yang et al. 1993).

HSV

HSV encodes two helicases essential for viral replication designated UL5 and UL9 (see Challberg, this volume). Although a bona fide cell-free replication system for HSV is not yet available, much has been learned about these helicases from a combination of biochemical and genetic studies.

UL5

The product of the UL5 gene (predicted M_r 114,000) is found in a heterotrimeric complex with products of the HSV UL8 and UL52 genes in a stoichiometry of 1:1:1 (Crute et al. 1989). In addition to being a 5′→3′DNA helicase (Crute et al. 1988), the complex (predicted M_r ~293,000) also contains DNA primase activity (Crute et al. 1989). The complex has therefore been suggested to act on the lagging strand during viral replication, using the primase activity to initiate Okasaki strand synthesis.

The UL5 gene is a member of the DNA/RNA helicase superfamily, although neither recombinant baculovirus-expressed UL5 protein nor UL5 overexpressed in mammalian cells contains detectable helicase activity (Calder and Stow 1990; Dodson and Lehman 1991). However, UL5/UL52 complexes isolated from baculovirus-infected insect cells have DNA helicase and primase activities similar to the complete heterotrimeric complex (Calder and Stow 1990; Dodson and Lehman 1991). Indeed, the native complex appears indistinguishable from the complex isolated from insect cells triply infected with baculoviruses expressing the UL5, UL8, and UL52 genes (Dodson et al. 1989; Calder and Stow 1990). In the presence of the ICP8 protein, an HSV-encoded SSB, the rate of duplex DNA unwinding by the heterotrimeric complex has been measured at approximately 2 bp/second (Crute and Lehman 1991). Although this rate is similar to that found for movement of a viral replication fork in vitro initiating from a 5′-tailed duplex substrate (~4 bp/sec; Rabkin and Hanlon 1990), it is considerably slower than the in vivo rate of viral fork movement of the related pseudorabies virus (~50 bp/sec; Ben-Porat et al. 1977). Assuming that the primary helicase during viral replication is the UL5/UL8/UL52 complex, it appears that other viral or host factors have the potential to stimulate the DNA-unwinding rate of this complex by a factor of ten.

Examination of the dimeric UL5/UL52 or heterotrimeric UL5/UL52/UL8 complexes has suggested that two distinct nucleotide triphosphatase (NTPase) sites, distinguishable by their affinity for ATP and GTP, can participate in the helicase reaction (Crute et al. 1991). Moreover, the complex was also seen to contain two effector sites for DNA, each of which independently activates the two NTPase sites. The UL5 complex thus appears to represent an uncommon example of a DNA helicase containing two structurally distinct NTPase sites independently able to drive DNA unwinding, although this conclusion has recently been challenged (Earnshaw and Jarvest 1994). Regardless, information concerning the mechanism of unwinding by the UL5-containing complexes should prove particularly interesting.

UL9

The HSV UL9 protein contains both DNA helicase (Bruckner et al. 1991; Fierer and Challberg 1992) and origin-binding activities (Elias and Lehman 1988; Olivo et al. 1988). UL9 has therefore been proposed to act during the initiation of HSV replication, perhaps facilitating origin denaturation, although such an activity has not yet been demonstrated.

This 3′→5′ helicase is most active in the presence of ATP or dATP, although CTP, dCTP, or UTP can be utilized to a nearly equivalent extent (Boehmer et al. 1993). UL9 contains a single nucleotide-binding site with K_m(ATP) found to be 0.54 mM (Earnshaw and Jarvest 1994). The protein (predicted M_r 94,000) exists primarily as a dimer in solution and appears to be multimeric as a DNA helicase, although the actual oligomeric species is not yet defined (Boehmer et al. 1993). The rate of unwinding has been measured at approximately 1.25 bp/second, with the processivity of the enzyme limited to roughly 200 nucleotides. ICP8 decreases the amount of UL9 protein required for DNA unwinding and increases the processivity of the reaction to well over 2 kb (Fierer and Challberg 1992; Boehmer et al. 1993). This stimulation is likely due, at least in part, to a physical interaction between UL9 and ICP8 (Boehmer and Lehman 1993).

Parvovirus

Adeno-associated Virus Rep68 and Rep78 Proteins

Transcription from two promoters coupled with alternative splicing leads to the production of four polypeptides from the *rep* region of adeno-associated virus (AAV) (Srivastava et al. 1983; Cotmore and Tattersall, this volume). With predicted molecular weights of 71,000, 61,000, 45,000, and 35,000 (Srivastava et al. 1983), these proteins are referred to as Rep78, Rep68, Rep52, and Rep40 from their mobility on acrylamide gels (Mendelson et al. 1986). The Rep68 and Rep78 proteins are ATP-dependent DNA helicases and can site-specifically cleave the viral DNA upon binding the viral terminal hairpin structures of AAV (*trs*; Im and Muzyczka 1989, 1990, 1992). During the cleavage reaction, the Rep protein becomes covalently attached to the 5′end of the viral DNA (Im and Muzyczka 1990). These activities are not found associated with either the Rep40 or Rep52 proteins (Im and Muzyczka 1992).

The 3′→5′helicase activity of Rep68 is most active in the presence of ATP, although other nucleotides can substitute (CTP>dATP> GTP>UTP; X. Zhou and N. Muzyczka, pers. comm.). Mutations within the *rep* gene targeting the putative ATP-binding motif abolish viral replication, *trs* endonucleolytic cleavage, and the viral transcriptional *trans*-activation activity found for *rep* gene products (Labow et al. 1989; Im and Muzyczka 1992). Because Rep68 and Rep78 form covalent linkages with the AAV DNA during *trs* cleavage, it is possible that the Rep helicase remains attached to the 5′end during DNA unwinding. This mechanism would be similar to the combined action of the *E. coli* rep DNA helicase and bacteriophage ϕX174 A protein during the synthesis of single-stranded circular ϕX174 DNA from the double-stranded replicative form I DNA (Kornberg and Baker 1992). Alternatively, the nicking and DNA-unwinding activities may be partitioned into distinct Rep complexes, so that additional AAV Rep protein binds to the preformed Rep/*trs* complex to unwind the viral DNA.

Minute Virus of Mice NS-1

The organization of the MVM and AAV genomes are similar, so that the MVM NS-1 gene appears to be the AAV *rep* homolog. Subsequent to identification of the DNA helicase activity of AAV Rep68, the NS-1 protein (predicted M_r 83,000) expressed in baculovirus was also found to contain an ATP-dependent DNA helicase (Wilson et al. 1991). Although as yet incompletely characterized, the NS-1 protein can form a covalent linkage with the viral DNA and thus appears to have site-specific endonuclease activity (Chow et al. 1986; Cotmore and Tattersall 1988).

Adenovirus: The Exception to the Rule

Although many viruses encode their own replicative helicases, it is worth noting that the requirement for a DNA helicase is not absolute. Adenovirus appears perfectly capable of replicating its DNA in the absence of a virally encoded helicase (Challberg and Kelly 1989; see Hay, this volume). Adenovirus DNA polymerase is able to elongate through lengthy (5.4 kb) stretches of dsDNA in a reaction requiring the adenovirus DNA-binding protein (the viral SSB; Lindenbaum et al. 1986). The unwinding of duplex DNA during adenovirus replication thus appears to be powered by the hydrolysis of dNTPs as they are incorporated into the growing DNA strand. The lack of need for a DNA helicase may reflect the fact that only leading-strand synthesis occurs during adenovirus DNA replication (Challberg and Kelly 1989). A continuous, unidirectional mechanism is utilized during mitochondrial DNA replication (Clayton, this volume), and thus it is also possible that a DNA helicase activity is not required in the mitochondrial system.

CELLULAR DNA HELICASES

Nuclear DNA Helicases

A handful of DNA helicases have been isolated from eukaryotic cells with defined roles in nucleic acid metabolism (see, e.g., Sung et al. 1987; Drapkin et al. 1994). Because these helicases are not known to function in DNA replication but rather in repair, recombination, or transcription, their identification has not provided clear information regarding chromosomal replication.

In addition, numerous "orphan helicases" have been isolated whose functions are unknown. In an attempt to ascertain the role of these orphan helicases, some have undergone significant biochemical scrutiny. The primary characteristics of these helicases, and of those with more clearly defined roles in nucleic acid metabolism, are listed in Table 2.

Study of viral and prokaryotic DNA replication systems has indicated that a prominent characteristic of replicative helicases is their ability to specifically interact with other elements of the replication machinery (see also below). By analogy, it can be safely predicted that cellular replicative helicases will be found to physically interact with the cognate DNA polymerases (DNA polymerase-α:DNA primase, DNA polymerase-δ), polymerase accessory factors (e.g., RF-C), or replication protein A (RP-A). Although such standards can serve only as a rough biochemical compass, their use points to various helicases whose properties are consistent with a replicative function. A few examples of such helicases are described below to indicate the state of the field, but any such list must necessarily be interpreted with caution.

Replication Factor-C-associated DNA Helicase

Replication factor C (RF-C) is a multisubunit complex initially isolated from human cells and found essential for SV40 DNA replication (Stillman 1994; see Hübscher et al., this volume). The general function of RF-C, an accessory factor for DNA polymerase-δ and -ε, appears highly conserved in eukaryotes with a RF-C homolog isolated from *Saccharomyces cerevisiae* (Yoder and Burgers 1991; Fien and Stillman 1992). Burger and colleagues have isolated a 5′→3′ DNA helicase from *S. cerevisiae* that copurifies with RF-C through numerous purification steps (Li et al. 1992a). The 60-kD (apparent) helicase, likely a dimer in solution, prefers ATP for DNA unwinding (K_m[ATP] 60 μM). The helicase retains significant activity in the presence of dATP and is only slightly stimulated by *S. cerevisiae* RP-A (scRP-A).

DNA Helicases Associated with DNA Polymerases

DNA polymerase-α-associated DNA helicases. Thömmes and Hübscher identified a 3′→5′ DNA helicase, termed DNA helicase A (apparent M_r 47,000), that followed the activity of the calf thymus DNA polymerase-α:DNA primase complex during purification (Thömmes and Hübscher 1990; Thömmes et al. 1992). The helicase efficiently utilizes ATP (K_m

Table 2 Eukaryotic nuclear DNA helicases

Helicase	Source (cell line)	Apparent (m.w.)[a]	Nucleotide cofactors[e]	Polarity	Remarks[g]	References
Human DNA helicase I (HDH I)	human (HeLa)	65,000	ATP, dATP	3′→5′	can unwind DNA:RNA hybrids	Tuteja et al. 1990
Human DNA helicase II (HDH II) Ku autoantigen	human (HeLa)	87,000, 72,000[b]	ATP, dATP	3′→5′	functions in double-strand break repair and V(D)J recombination; regulator of DNA-dependent protein kinase	Tuteja et al. 1994; Taccioli et al. 1994; Smider et al. 1994
Human DNA helicase III (HDH III)	human (HeLa)	46,000	ATP, dATP	3′→5′	—	Tuteja et al. 1992
Human DNA helicase IV (HDH IV)	human (HeLa)	100,000	ATP, dATP	5′→3′	can unwind DNA:RNA hybrids	Tuteja et al. 1991
Human DNA helicase V (HDH V)	human (HeLa)	92,000	ATP, dATP	3′→5′	—	Tuteja et al. 1993
XPD/ERCC2	human	87,000[c]	ATP, dATP	5′→3′	functions in nucleotide excision repair; component of BTF2-TFIIH transcription factor	Sung et al. 1993; Drapkin et al. 1994
XPB/ERCC3	human	89,000[c]	ATP	3′→5′	functions in nucleotide excision repair; component of BTF2-TFIIH transcription factor	Schaeffer et al. 1993; Drapkin et al. 1994; Ma et al. 1994a
DNA helicase ε	human (HeLa)	72,000[d]	ATP, dATP > CTP	3′→5′	helicase activity is dependent on hRP-A	Seo et al. 1991

DNA helicase α	human (HeLa)	110,000, 90,000	ATP, dATP > CTP, dCTP	3′→5′	—	Seo and Hurwitz 1993
RIP100	human (HeLa)	100,000	ATP, dATP	3′→5′	associated with RIP60; RIP60 binds to replication origin region of DHFR[h]	Dailey et al. 1990
DNA helicase Q1	human (HeLa)	73,000	ATP, dATP	3′→5′	gene homologous to *E. coli* RecQ gene; identical to human DNA helicase I?	Seki et al. 1994
DNA helicase Q2	human (HeLa)	100,000	ATP	5′→3′	identical to human DNA helicase IV?	Seki et al. 1994
DNA helicase A	calf thymus	47,000[d]	ATP, dATP > CTP	3′→5′	copurifies with ct DNA pol-α:primase; stimulated ~20-fold by ctRPA	Thömmes and Hübscher 1990; Thömmes et al. 1992
DNA helicase B	calf thymus	100,000	dATP, ATP > GTP, CTP, dGTP, UTP, dCTP, dTTP	5′→3′	—	Thömmes et al. 1992
DNA helicase C	calf thymus	40,000[d]	dATP, ATP	5′→3′	—	Thömmes et al. 1992
DNA helicase D	calf thymus	100,000, 45,000	dATP, ATP	5′→3′	stimulated ~10-fold by ctRPA	Thömmes et al. 1992
DNA helicase E	calf thymus	104,000	dATP, ATP	3′→5′	copurifies with DNA pol-ε	Siegal et al. 1992
DNA helicase F	calf thymus	72,000	ATP, dATP, dCTP, UTP, CTP, GTP, dGTP, dTTP	5′→3′	copurifies with RPA	Georgaki et al. 1994

Table 2 (continued)

Helicase	Source (cell line)	Apparent (m.w.)[a]	Nucleotide cofactors[e]	Polarity	Remarks[g]	References
DNA helicase I	calf thymus	200,000	ATP, dATP	3′→5′	—	Zhang and Grosse 1991
DNA helicase II	calf thymus	130,000	ATP, dATP > CTP, dCTP > GTP, dGTP > UTP, dTTP	3′→5′	can unwind dsRNA	Zhang and Grosse 1991, 1994
δ Helicase	fetal calf thymus	57,000	ATP, dATP	5′→3′	copurifies with DNA pol-δ	Li et al. 1992b
DNA helicase B	mouse (FM3A)	58,000	ATP > dATP, dGTP, GTP, CTP, UTP	5′→3′	—	Seki et al. 1988
DNA helicase C1	mouse (FM3A)	? (5.2 S)	ATP	5′→3′	—	Yanagisawa et al. 1992
DNA helicase I	*Xenopus laevis*	62,000 and 75,000?, 107,000?	ATP, dATP	3′→5′	—	Poll et al. 1994
Rad3	*S. cerevisiae*	90,000[c]	ATP, dATP	5′→3′	critical in DNA excision repair; homologous to the XPD gene; constituent of pol II transcription factor b	Sung et al. 1987; Feaver et al. 1993
Rad25	*S. cerevisiae*	95,000[c]	ATP, dATP	n.d. (3′→5′?)[f]	functions in nucleotide excision repair; homologous to XPB; required for pol II transcription	Guzder et al. 1994

Srs2	*S. cerevisiae*	130,000	ATP, dATP	3′→5′	mutants have a hyperrecombination phenotype; involved in error-prone repair?	Rong and Klein 1993
PIF1	*S. cerevisiae*	98,000[c]	ATP, dATP	5′→3′	functions in mtDNA repair and recombination; affects chromosomal telomere length	Lahaye et al. 1991, 1993; Schulz and Zakian 1994
RF-C-associated DNA helicase	*S. cerevisiae*	60,000[d]	ATP > dATP	5′→3′	copurifies with RF-C	Li et al. 1992a
DNA helicase A	*S. cerevisiae*	90,000?	ATP, dATP	5′→3′	copurifies with DNA pol-α:primase	Biswas et al. 1993a,b
DNA helicase B	*S. cerevisiae*	127,000[c]	ATP, dATP	5′→3′	stimulated by scRPA; encoded by the yORF61 gene	Biswas et al. 1995
scHelI	*S. cerevisiae*	135,000	ATP, dATP	5′→3′	—	Bean et al. 1993
DNA helicase III	*S. cerevisiae*	120,000	ATP, dATP	5′→3′	—	Shimizu and Sugino 1993

(n.d.) Not determined.
[a]Determined by SDS-PAGE, unless otherwise indicated.
[b]Exists as a heterodimer in solution.
[c]Molecular weight predicted from the sequence of the cloned gene.
[d]Exists as a homodimer in solution.
[e]Nucleotides that give rise to ≥50% maximal helicase activity are shown.
[f]The polarity of the XPB/ERCC3 DNA helicase, a human homolog of Rad25, has been reported to be 3′→5′ (Drapkin et al. 1994).
[g]"Copurifies with . . ." indicates that the helicase is found to initially copurify with the indicated factor through several chromatographic steps.
[h]RIP60 can be photocrosslinked to a region containing an origin of DNA replication near the dihydrofolate reductase gene from Chinese hamsters.

0.2 mM), dATP (K_m 0.25 mM), and CTP (K_m 0.45 mM). The isolated helicase A is poorly processive, but this can be greatly stimulated, along with the DNA helicase activity, by calf thymus RP-A (ctRP-A; Thömmes et al. 1992). As human RP-A inhibited the activity of helicase A, a physical interaction between ctRP-A and helicase A is suggested.

Similarly, purification of *S. cerevisiae* DNA polymerase-α:DNA primase, using immunoaffinity chromatography, indicated that a 5′→3′ DNA helicase could associate with the bound polymerase up to approximately 0.35 M NaCl (Biswas et al. 1993a). Photoaffinity cross-linking with radiolabeled ATP suggests that the helicase is approximately 90,000 in molecular weight. The helicase activity was stimulated up to threefold by scRP-A.

DNA polymerase-δ-associated DNA helicase. A 5′→3′ DNA helicase, termed the δ helicase, cofractionated with fetal calf thymus DNA polymerase-δ through five chromatographic steps (Li et al. 1992b). Upon separation from DNA polymerase-δ, the DNA helicase activity was found associated with an approximately 57-kD polypeptide that could utilize ATP and dATP for DNA unwinding. The use of a preformed DNA fork significantly stimulated δ helicase activity.

DNA polymerase-ε-associated DNA helicase. DNA helicase E was found by Bambara and colleagues to coelute with calf thymus DNA polymerase-ε through four chromatographic steps (Siegal et al. 1992). The 104-kD helicase unwinds DNA in the 3′→5′ direction and is most active in the presence of dATP and ATP.

RP-A-associated DNA Helicase

Calf thymus DNA helicase F has been observed to copurify with the cognate RP-A through various chromatographic steps (Georgaki et al. 1994). The 5′→3′ activity appears to reside in a 72-kD polypeptide, although gel filtration studies suggest that DNA helicase F can also exist in a dimeric form. The helicase is somewhat unusual in that it can efficiently utilize all common NTPs and dNTPs. Although the enzyme was not active on substrates containing lengthy duplex regions (e.g., 400 bp in length), unwinding was observed in the presence of ctRP-A, whereas the *E. coli* or bacteriophage T4 SSBs had no significant effect.

A variety of DNA helicases are known that either require the cognate RP-A for activity or are greatly stimulated by its addition. These include human DNA helicase ε (Seo et al. 1991), calf thymus DNA helicases A and D (Thömmes et al. 1992), and yeast DNA helicase B (Biswas et al. 1995).

Mitochondrial DNA Helicases

Two mitochondrial DNA helicases have been identified, although neither is unambiguously associated with mitochondrial DNA replication (see Clayton, this volume). The nuclear *PIF1* gene of *S. cerevisiae* encodes a 5′→3′ DNA helicase that is more likely involved in mitochondrial DNA repair and recombination than replication (Lahaye et al. 1991, 1993). More recently, the *PIF1* gene product has also been implicated as an inhibitor of both chromosomal telomere elongation and de novo telomere formation, indicating that the PIF1 helicase has an additional nuclear function (Schulz and Zakian 1994). Bovine mitochondria contain a 3′→5′ DNA helicase, indicating that it is not the mammalian counterpart of PIF1 protein (Hehman and Hauswirth 1992). The role of the bovine enzyme in mitochondrial DNA metabolism is currently unknown.

MECHANISM OF DNA UNWINDING

The mechanism of DNA unwinding by helicases is not currently understood, even though the area is one of vigorous investigation. Studies with various helicases, however, have suggested two distinct models by which DNA unwinding could occur. The two models are distinguished by the manner of helicase movement along the DNA, namely, "rolling" or "creeping" (SenGupta and Borowiec 1992; Wong and Lohman 1992). As demonstrated in Figure 2 for the SV40 large T antigen, each model postulates a DNA helicase entity that utilizes at least two distinct binding sites for the DNA substrate, and incorporates a cycle of NTP binding and hydrolysis that induces changes in the relative affinity of the binding sites for DNA.

The rolling model is most consistent with an oligomeric DNA helicase (Fig. 2A). All DNA helicases are apparently oligomeric (generally dimers or hexamers; Lohman 1993), allowing at least two DNA- and two NTP-binding sites per helicase complex. A basic postulate of this model is that monomers in the helicase complex can bind separately to the ssDNA and dsDNA behind and ahead of the fork, respectively, or to dsDNA and the ssDNA-dsDNA junction itself. The helicase denatures the DNA duplex between these two sites using an active mechanism dependent on NTP hydrolysis. In the process, the general protein/DNA complex formed by the downstream monomer is shifted forward to the upstream monomer. The cycle is completed by rolling of the helicase complex to bind a new helicase monomer to the dsDNA ahead of the fork. The model is supported by studies with the dimeric rep helicase of *E. coli*, indicating that the state of the bound nucleotide (e.g.,

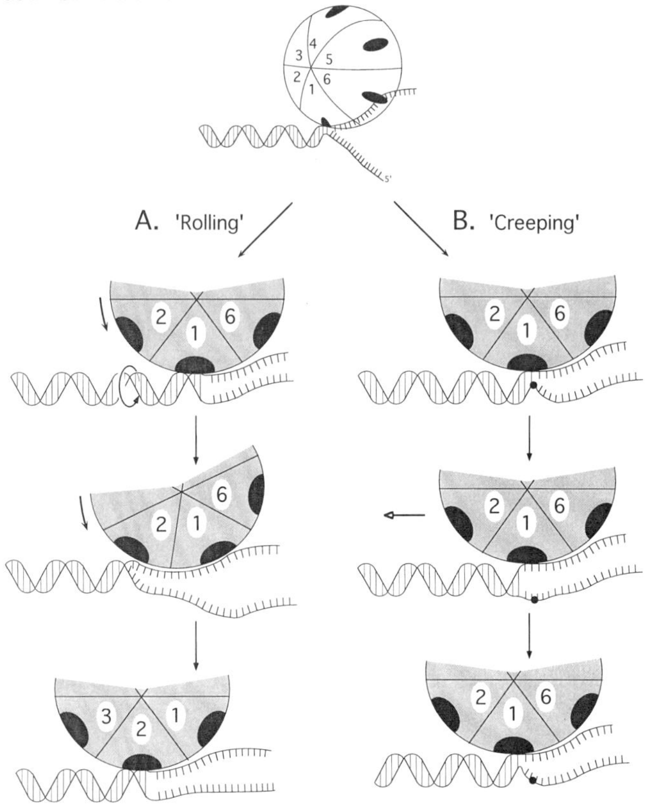

Figure 2 Models of DNA helicase action shown with the SV40 large T antigen. T antigen binds to synthetic replication forks as a hexamer, using contacts with the sugar-phosphate backbone of the strand containing the 3′ssDNA end (SenGupta and Borowiec 1992). The primary DNA-binding site for each monomer is indicated by a dark patch. Other orientations of T antigen with the fork are possible (see, e.g., SenGupta et al. 1992), but are not shown for the sake of brevity. (*A*) The rolling model. The T-antigen hexamer is initially bound to the fork using the DNA-binding site of monomer 1. In an ATP-dependent process, monomer 2 interacts with the dsDNA roughly 10–15 bp ahead of the fork, denaturing the duplex DNA between the DNA-binding sites of monomers 1 and 2. During this step, DNA rotation is required to allow complete untwisting of the duplex DNA. The hexameric helicase then rolls to regenerate the initial state, except that monomer 2 is now bound to the fork. (*B*) The creeping model. The T-antigen hexamer binds to the fork using the DNA-binding site of monomer 1. Either by an active (ATP-dependent) or passive (ATP-independent) process, a few base pairs (e.g., one or two) denature at the fork. The T-antigen hexamer moves forward along the DNA in an ATP-dependent process and prevents DNA reannealing. T-antigen movement with respect to a black marker on the DNA is shown. In contrast to the rolling model, monomer 1 remains associated with the fork through the unwinding cycle.

ATP or ADP) can alter the relative affinity of the rep subunits for ssDNA or dsDNA (Lohman 1993). The rolling model makes the prediction that the step size of helicase movement is approximately equivalent to the distance between the two DNA-binding sites used sequentially within the multimeric DNA helicase. One potential complication of the rolling model is a requirement for a significant number of DNA base pairs to be unwound per helicase step. For the *E. coli* rep helicase, the step size is predicted to be about 16 nucleotides (Wong and Lohman 1992). Determination of the dimensions of the T-antigen hexamer (see, e.g., Wessel et al. 1992) also leads to a predicted step size of about 10–15 nucleotides, assuming equivalent spacing of DNA-binding sites around the hexameric surface. A single unwinding step for each of the helicases would therefore require the hydrolysis of multiple ATP molecules and torsional rotation of the DNA helix of over 360°.

The creeping (or inchworm) model of DNA unwinding posits that the helicase has contacts with both the ssDNA and dsDNA on each side of the fork. NTP utilization changes the affinity of the dsDNA-binding site to favor the binding of ssDNA. Either by a passive or active process, one or a small number of base pairs become denatured and the helicase (and fork) advance (Fig. 2B). Alternatively, the helicase need only bind to the ssDNA behind the fork. Statistical analysis indicates that the average lifetime of a GC base pair at a replication fork is <0.1 msec (millisecond) at room temperature (Chen et al. 1992). The rate of DNA unwinding for the most rapid helicases is about 1 msec/bp (~1000 bp/sec) and, thus, DNA helicases are not required to forcibly denature the DNA duplex, but rather move passively while thermal forces cause the DNA to "breathe" at the fork. The creeping model does not require the DNA helicase to be oligomeric and instead suggests that the non-DNA-bound subunits serve to interact with other proteins at the fork. Variations of both models are possible and it would be foolhardy to predict that all helicases unwind by identical mechanisms.

A large fraction of the eukaryotic DNA is in the form of chromatin and is bound by factors regulating nucleic acid enzymology (e.g., transcription factors) and structural proteins such as histones. Such proteins have the potential to alter the dynamic properties of the DNA, and it is conceivable that protein/DNA complexes can significantly reduce the rate of duplex denaturation. Although eukaryotic DNA replication complexes have the potential to destabilize protein/DNA complexes ahead of the fork (Wolffe and Brown 1986), chromatin is a relatively inefficient template for SV40 DNA replication in vitro compared to naked DNA (see, e.g., Cheng and Kelly 1989; see Wolffe, this volume). In vitro progression of a minimal bacteriophage T4 replication fork, powered by the T4 gene 41 helicase, is halted by various protein/DNA complexes such as those formed by histones (Bonne-Andrea et al. 1990) or prokaryotic RNA polymerases or repressors (Bedinger et al. 1983). This inhibition can be relieved by addition of the T4 dda DNA helicase, which allows movement through these barriers (Bedinger et al. 1983; Bonne-Andrea et al. 1990). Thus, one area of potentially fruitful investigation in the study of eukaryotic chromosomal replication is the search for specialized helicases or helicase accessory factors that can overcome barriers caused by common protein/DNA complexes.

INTERACTION OF DNA HELICASES WITH OTHER FACTORS

DNA helicases act to open the DNA duplex for nascent-strand synthesis, and thus it is not unexpected that prokaryotic and eukaryotic replicative helicases physically interact with the cognate DNA polymerase(s), DNA primase, or SSB. In eukaryotic viral systems, both the SV40 T antigen and the BPV E1 protein bind to the human DNA polymerase-α:DNA primase complex (Collins et al. 1993; Dornreiter et al. 1993; Park et al. 1994) and T antigen also interacts with hRP-A (Dornreiter et al. 1992). As noted above, the HSV UL5 protein is contained within a heterotrimeric complex that also contains DNA primase activity and can bind the HSV ICP8. Similarly, the bacteriophage T7 gene 4 protein contains both DNA helicase and DNA primase activities in its 63-kD form and can interact with the T7 DNA polymerase (Nakai and Richardson 1986) and T7 SSB (Kim and Richardson 1994). Comparable interactions between the DNA helicases and the polymerase machinery of *E. coli* and bacteriophage T4 also occur (Marians 1992).

The observation that DNA helicases generally form complexes with the DNA polymerase machinery suggests a requirement for tight coupling of DNA unwinding and DNA synthesis. In lieu of such coupling, a block to DNA polymerase action by template damage could conceivably lead to the generation of lengthy regions of ssDNA between the moving fork and stalled polymerase. This notion suggests that the rate of replication fork progression is generally modulated primarily through the rate of nucleotide incorporation rather than the rate of DNA unwinding. Furthermore, Richardson and colleagues have demonstrated that primer synthesis by the T7 gene 4 helicase/primase slows the rate of DNA unwinding (Debyser et al. 1994). This study indicates that the modulation of the rate of DNA unwinding is also critical for the coordination of leading- and lagging-strand DNA synthesis.

Various viral DNA helicases have also been demonstrated to interact with a variety of other factors (e.g., T antigen and the tumor suppressor p53 protein), although none of these interactions yet appears to be involved in the DNA-unwinding reaction per se. Studies with prokaryotic systems indicate that termination factors exist which bind to specific DNA sites, preventing DNA unwinding by helicases through the site (Kuempel et al. 1989). Replication fork barriers are also found in higher eukaryotes, suggesting that similar factors may exist in chromatin (see Bastia and Mohanty, this volume).

CONCLUDING REMARKS

Significant information is known concerning eukaryotic DNA helicases, but much more remains to be discovered. There are numerous aspects of DNA helicase function that are worthy of study, but four of the most critical are: (1) isolation and characterization of cellular DNA helicases functioning in chromosomal DNA replication; (2) acquisition of knowledge concerning the three-dimensional structure(s) of DNA helicases; (3) characterization of the molecular mechanism(s) by which DNA helicases unwind duplex DNA; and (4) development of agents that target viral or cellular DNA helicase function. The essential nature of DNA helicase function during replication indicates that they potentially represent an Achilles' heel toward which to direct antiviral and anticancer therapy. Advancement toward each of these four points will allow a test of this possibility.

ACKNOWLEDGMENTS

I thank Leonard Blackwell, Natalia Smelkova, Eda Kapsis, and Thomas Gillette for critical comments on the manuscript, and X. Zhou and N. Muzyczka for communicating results prior to publication. Work from the author's laboratory was supported by National Institutes of Health grants AI-29963 and CA-62198, the Pew Biomedical Scholars Program (T88-00457-063), and Kaplan Cancer Center Developmental Funding and Kaplan Cancer Center Support Core Grant (NCI P30CA-16087).

REFERENCES

Alberts, B.M. and L. Frey. 1970. T4 bacteriophage gene 32: A structural protein in the replication and recombination of DNA. *Nature* **227:** 1313–1318.

Bedinger, P., M. Hochstrasser, C.V. Jongeneel, and B.M. Alberts. 1983. Properties of the

T4 bacteriophage DNA replication apparatus: The T4 dda DNA helicase is required to pass a bound RNA polymerase molecule. *Cell* **34:** 115–123.

Bean, D.W., W.E. Kallam, Jr., and S.W. Matson. 1993. Purification and characterization of a DNA helicase from *Saccharomyces cerevisiae*. *J. Biol. Chem.* **268:** 21783–21790.

Ben-Porat, T., M.L. Blankenship, J.M. DeMarchi, and A.S. Kaplan. 1977. Replication of herpesvirus DNA. III. Rate of DNA elongation. *J. Virol.* **22:** 734–741.

Biswas, E.E., P.H. Chen, and S.B. Biswas. 1993a. DNA helicase associated with DNA polymerase alpha: Isolation by a modified immunoaffinity chromatography. *Biochemistry* **32:** 13393–13398.

Biswas, E.E., C.M. Ewing, and S.B. Biswas. 1993b. Characterization of the DNA-dependent ATPase and a DNA unwinding activity associated with the yeast DNA polymerase α complex. *Biochemistry* **32:** 3020–3026.

Biswas, E.E., P.-H. Chen, J. Leszyk, and S.B. Biswas. 1995. Biochemical and genetic characterization of a replication protein A dependent DNA helicase from the yeast, *Saccharomyces cerevisiae*. *Biochem. Biophys. Res. Commun.* **206:** 850–856.

Boehmer, P.E. and I.R. Lehman. 1993. Physical interaction between the herpes simplex virus 1 origin-binding protein and single-stranded DNA-binding protein ICP8. *Proc. Natl. Acad. Sci.* **90:** 8444–8448.

Boehmer, P.E., M.S. Dodson, and I.R. Lehman. 1993. The herpes simplex virus type-1 origin binding protein. DNA helicase activity. *J. Biol. Chem.* **268:** 1220–1225.

Bonne-Andrea, C., M.L. Wong, and B.M. Alberts. 1990. In vitro replication through nucleosomes without histone displacement. *Nature* **343:** 719–726.

Bruckner, R.C., J.J. Crute, M.S. Dodson, and I.R. Lehman. 1991. The herpes simplex virus 1 origin binding protein: A DNA helicase. *J. Biol. Chem.* **266:** 2669–2674.

Calder, J.M. and N.D. Stow. 1990. Herpes simplex virus helicase-primase: The UL8 protein is not required for DNA-dependent ATPase and DNA helicase activities. *Nucleic Acids Res.* **18:** 3573–3578.

Challberg, M. and T. Kelly. 1989. Animal virus DNA replication. *Annu. Rev. Biochem.* **58:** 671–717.

Chen, Y.Z., W. Zhuang, and E.W. Prohofsky. 1992. Energy flow considerations and thermal fluctuational opening of DNA base pairs at a replicating fork: Unwinding consistent with observed replication rates. *J. Biomol. Struct. Dyn.* **10:** 415–427.

Cheng, L. and T.J. Kelly. 1989. Transcriptional activator nuclear factor I stimulates the replication of SV40 minichromosomes in vivo and in vitro. *Cell* **59:** 541–551.

Chiang, C.M., M. Ustav, A. Stenlund, T.F. Ho, T.R. Broker, and L.T. Chow. 1992. Viral E1 and E2 proteins support replication of homologous and heterologous papillomaviral origins. *Proc. Natl. Acad. Sci.* **89:** 5799–5803.

Chow, M., J.W. Bodnar, M. Polvino-Bodnar, and D.C. Ward. 1986. Identification and characterization of a protein covalently bound to DNA of minute virus of mice. *J. Virol.* **57:** 1094–1104.

Clertant, P. and I. Seif. 1984. A common function for polyoma virus large-T and papillomavirus E1 proteins? *Nature* **311:** 276–279.

Collins, K.L., A.A. Russo, B.Y. Tseng, and T.J. Kelly. 1993. The role of the 70 kDa subunit of human DNA polymerase alpha in DNA replication. *EMBO J.* **12:** 4555–4566.

Cotmore, S.F. and P. Tattersall. 1988. The NS-1 polypeptide of minute virus of mice is covalently attached to the 5′ termini of duplex replicative form DNA and progeny single strands. *J. Virol.* **62:** 851–860.

Crute, J.J. and I.R. Lehman. 1991. Herpes simplex virus-1 helicase-primase. Physical and catalytic properties. *J. Biol. Chem.* **266:** 4484–4488.

Crute, J.J., E.S. Mocarski, and I.R. Lehman. 1988. A DNA helicase induced by herpes simplex virus type 1. *Nucleic Acids Res.* **16:** 6585–6596.

Crute, J.J., R.C. Bruckner, M.S. Dodson, and I.R. Lehman. 1991. Herpes simplex-1 helicase-primase. Identification of two nucleoside triphosphatase sites that promote DNA helicase action. *J. Biol. Chem.* **266:** 21252–21256.

Crute, J.J., T. Tsurumi, L.A. Zhu, S.K. Weller, P.D. Olivo, M.D. Challberg, E.S. Mocarski, and I.R. Lehman. 1989. Herpes simplex virus 1 helicase-primase: A complex of three herpes-encoded gene products. *Proc. Natl. Acad. Sci.* **86:** 2186–2189.

Dailey, L., M.S. Caddle, N. Heintz, and N.H. Heintz. 1990. Purification of RIP60 and RIP100, mammalian proteins with origin-specific DNA-binding and ATP-dependent DNA helicase activities. *Mol. Cell. Biol.* **10:** 6225–6235.

Dean, F.B., J.A. Borowiec, T. Eki, and J. Hurwitz. 1992. The simian virus 40 T antigen double hexamer assembles around the DNA at the replication origin. *J. Biol. Chem.* **267:** 14129–14137.

Debyser, Z., S. Tabor, and C.C. Richardson. 1994. Coordination of leading and lagging strand DNA synthesis at the replication fork of bacteriophage T7. *Cell* **77:** 157–166.

Del Vecchio, A.M., H. Romanczuk, P.M. Howley, and C.C. Baker. 1992. Transient replication of human papillomavirus DNAs. *J. Virol.* **66:** 5949–5958.

Dodson, M.S. and I.R. Lehman. 1991. Association of DNA helicase and primase activities with a subassembly of the herpes simplex virus 1 helicase-primase composed of the UL5 and UL52 gene products. *Proc. Natl. Acad. Sci.* **88:** 1105–1109.

Dodson, M.S., J.J. Crute, R.C. Bruckner, and I.R. Lehman. 1989. Overexpression and assembly of the herpes simplex virus type 1 helicase-primase in insect cells. *J. Biol. Chem.* **264:** 20835–20838.

Dornreiter, I., W.C. Copeland, and T.S. Wang. 1993. Initiation of simian virus 40 DNA replication requires the interaction of a specific domain of human DNA polymerase alpha with large T antigen. *Mol. Cell. Biol.* **13:** 809–820.

Dornreiter, I., L.F. Erdile, I.U. Gilbert, D. von Winkler, T.J. Kelly, and E. Fanning. 1992. Interaction of DNA polymerase alpha-primase with cellular replication protein A and SV40 T antigen. *EMBO J.* **11:** 769–776.

Drapkin, R., J.T. Reardon, A. Ansari, J.C. Huang, L. Zawel, K. Ahn, A. Sancar, and D. Reinberg. 1994. Dual role of TFIIH in DNA excision repair and in transcription by RNA polymerase II. *Nature* **368:** 769–772.

Earnshaw, D.L. and R.L. Jarvest. 1994. Characterisation of the nucleotide and DNA coeffector binding sites of the herpes simplex virus type 1 (HSV-1) encoded helicase-primase complex and UL9 origin binding protein. *Biochem. Biophys. Res. Commun.* **199:** 1333–1340.

Elias, P. and I.R. Lehman. 1988. Interaction of origin binding protein with an origin of replication of herpes simplex virus 1. *Proc. Natl. Acad. Sci.* **85:** 2959–2963.

Fanning, E. and R. Knippers. 1992. Structure and function of simian virus 40 large tumor antigen. *Annu. Rev. Biochem.* **61:** 55–85.

Feaver, W.J., J.Q. Svejstrup, L. Bardwell, A.J. Bardwell, S. Buratowski, K.D. Gulyas, T.F. Donahue, E.C. Friedberg, and R.D. Kornberg. 1993. Dual roles of a multiprotein complex from *S. cerevisiae* in transcription and DNA repair. *Cell* **75:** 1379–1387.

Fien, K. and B. Stillman. 1992. Identification of replication factor C from *Saccharomyces cerevisiae*: A component of the leading-strand DNA replication complex. *Mol. Cell. Biol.* **12:** 155–163.

Fierer, D.S. and M.D. Challberg. 1992. Purification and characterization of UL9, the herpes simplex virus type 1 origin-binding protein. *J. Biol. Chem.* **66:** 3986–3995.

Fry, D.C., S.A. Kuby, and A.S. Mildvan. 1986. ATP-binding site of adenylate kinase: Mechanistic implications of its homology with *ras*-encoded p21, F_1-ATPase, and other nucleotide-binding proteins. *Proc. Natl. Acad. Sci.* **83:** 907–911.

Georgaki, A., N. Tuteja, B. Sturzenegger, and U. Hübscher. 1994. Calf thymus DNA helicase F, a replication protein A copurifying enzyme. *Nucleic Acids Res.* **22:** 1128–1134.

Goetz, G.S., F.B. Dean, J. Hurwitz, and S.W. Matson. 1988. The unwinding of duplex regions in DNA by the simian virus 40 large tumor antigen-associated DNA helicase activity. *J. Biol. Chem.* **263:** 383–392.

Gorbalenya, A.E., E.V. Koonin, A.P. Donchenko, and V.M. Blinov. 1989. Two related superfamilies of putative helicases involved in replication, recombination, repair, and expression of DNA and RNA genomes. *Nucleic Acids Res.* **17:** 4713–4730.

Guzder, S.N., P. Sung, V. Bailly, L. Prakash, and S. Prakash. 1994. RAD25 is a DNA helicase required for DNA repair and RNA polymerase II transcription. *Nature* **369:** 578–581.

Hehman, G.L. and W.W. Hauswirth. 1992. DNA helicase from mammalian mitochondria. *Proc. Natl. Acad. Sci.* **89:** 8562–8566.

Hodgman, T.C. 1988. A new superfamily of replicative proteins. *Nature* **333:** 22–23.

Houston, P. and T. Kodadek. 1994. Spectrophotometric assay for enzyme-mediated unwinding of double-stranded DNA. *Proc. Natl. Acad. Sci.* **91:** 5471–5474.

Hughes, F.J. and M.A. Romanos. 1993. E1 protein of human papillomavirus is a DNA helicase/ATPase. *Nucleic Acids Res.* **21:** 5817–5823.

Im, D.-S. and N. Muzyczka. 1989. Factors that bind to adeno-associated virus terminal repeats. *J. Virol.* **63:** 3095–3104.

———. 1990. The AAV origin binding protein Rep68 is an ATP-dependent site-specific endonuclease with DNA helicase activity. *Cell* **61:** 447–457.

———. 1992. Partial purification of adeno-associated virus Rep78, Rep52, and Rep40 and their biochemical characterization. *J. Virol.* **66:** 1119–1128.

Kim, Y.T. and C.C. Richardson. 1994. Acidic carboxyl-terminal domain of gene 2.5 protein of bacteriophage T7 is essential for protein-protein interactions. *J. Biol. Chem.* **269:** 5270–5278.

Kornberg, A. and T.A. Baker. 1992. *DNA Replication*. W.H. Freeman, New York.

Kuempel, P.L., A.J. Pelletier, and T.M. Hill. 1989. Tus and the terminators: The arrest of replication in prokaryotes. *Cell* **59:** 581–583.

Labow, M.A., P.L. Hermonat, and K.I. Berns. 1989. Positive and negative autoregulation of the adeno-associated virus type 2 genome. *J. Virol.* **60:** 251–258.

Lahaye, A., S. Leterme, and F. Foury. 1993. PIF1 DNA helicase from *Saccharomyces cerevisiae*. Biochemical characterization of the enzyme. *J. Biol. Chem.* **268:** 26155–26161.

Lahaye, A., H. Stahl, D. Thines-Sempoux, and F. Foury. 1991. PIF1: A DNA helicase in yeast mitochondria. *EMBO J.* **10:** 997–1007.

LeBowitz, J.H. and R. McMacken. 1986. The *Escherichia coli* dnaB replication protein is a DNA helicase. *J. Biol. Chem.* **261:** 4738–4748.

Li, X., B.L. Yoder, and P.M. Burgers. 1992a. A *Saccharomyces cerevisiae* DNA helicase associated with replication factor C. *J. Biol. Chem.* **267:** 25321–25327.

Li, X., C.-K. Tan, A.G. So and K.M. Downey. 1992b. Purification and characterization of δ helicase from fetal calf thymus. *Biochemistry* **31:** 3507–3513.

Lin, H.-J., R.H. Upson, and D.T. Simmons. 1992. Nonspecific DNA binding activity of simian virus 40 large T antigen: Evidence for the cooperation of two regions for full activity. *J. Virol.* **66:** 5443–5452.

Lindenbaum, J.O., J. Field, and J. Hurwitz. 1986. The adenovirus DNA binding protein and adenovirus DNA polymerase interact to catalyze elongation of primed DNA templates. *J. Biol. Chem.* **261:** 10218–10227.

Lohman, T.M. 1993. Helicase-catalyzed DNA unwinding. *J. Biol. Chem.* **268:** 2269–2272.

Lusky, M., J. Hurwitz, and Y.S. Seo. 1993. Cooperative assembly of the bovine papilloma virus E1 and E2 proteins on the replication origin requires an intact E2 binding site. *J. Biol. Chem.* **268:** 15795–15803.

Ma, L., E.D. Siemssen, H.M. Noteborn, and A.J. van der Eb. 1994a. The xeroderma pigmentosum group B protein ERCC3 produced in the baculovirus system exhibits DNA helicase activity. *Nucleic Acids Res.* **22:** 4095–4102.

Ma, L., A. Westbroek, A.G. Jochemsen, G. Weeda, A. Bosch, D. Bootsma, J.H. Hoeijmakers, and A.J. van der Eb. 1994b. Mutational analysis of ERCC3, which is involved in DNA repair and transcription initiation: Identification of domains essential for the DNA repair function. *Mol. Cell. Biol.* **14:** 4126–4134.

Marians, K.J. 1992. Prokaryotic DNA replication. *Annu. Rev. Biochem.* **61:** 673–719.

Martinez, R., L. Shao, and S.K. Weller. 1992. The conserved helicase motifs of the herpes simplex virus type 1 origin-binding protein UL9 are important for function. *J. Virol.* **66:** 6735–6746.

Matson, S.W. 1986. *Escherichia coli* helicase II (urvD gene product) translocates unidirectionally in a 3′ to 5′ direction. *J. Biol. Chem.* **261:** 10169–10175.

Matson, S.W., D.W. Bean, and J.W. George. 1994. DNA helicases: Enzymes with essential roles in all aspects of DNA metabolism. *BioEssays* **16:** 13–22.

Matson, S.W., S. Tabor, and C.C. Richardson. 1983. The gene 4 protein of bacteriophage T7. Characterization of helicase activity. *J. Biol. Chem.* **258:** 14017–14024.

McVey, D., S. Ray, Y. Gluzman, L. Berger, A.G. Wildeman, D.R. Marshak, and P. Tegtmeyer. 1993. cdc2 phosphorylation of threonine 124 activates the origin-unwinding functions of simian virus 40 T antigen. *J. Virol.* **67:** 5206–5215.

Mendelson, E., J.P. Trempe, and B.J. Carter. 1986. Identification of the *trans*-acting Rep proteins of adeno-associated virus by antibodies to a synthetic oligopeptide. *J. Virol.* **60:** 823–832.

Moarefi, I.F., D. Small, I. Gilbert, M. Höpfner, S.K. Randall, C. Schneider, A.A.R. Russo, U. Ramsperger, A.K. Arthur, H. Stahl, T.J. Kelly, and E. Fanning. 1993. Mutation of the cyclin-dependent kinase phosphorylation site in simian virus 40 (SV40) large T antigen specifically blocks SV40 origin DNA unwinding. *J. Virol.* **67:** 4992–5002.

Mohr, I.J., M.P. Fairman, B. Stillman, and Y. Gluzman. 1989. Large T-antigen mutants define multiple steps in the initiation of simian virus 40 DNA replication. *J. Virol.* **63:** 4181–4188.

Mohr, I.J., R. Clark, S. Sun, E.J. Androphy, P. MacPherson, and M.R. Botchan. 1990. Targeting the E1 replication protein to the papillomavirus origin of replication by complex formation with the E2 transactivator. *Science* **250:** 1694–1699.

Murakami, Y. and J. Hurwitz. 1993. Functional interactions between SV40 T antigen and other replication proteins at the replication fork. *J. Biol. Chem.* **268:** 11008–11017.

Nakai, H. and C.C. Richardson. 1986. Interactions of the DNA polymerase and gene 4 protein of bacteriophage T7. Protein-protein and protein-DNA interactions involved in RNA-primed DNA synthesis. *J. Biol. Chem.* **261:** 15208–15216.

Olivo, P.D., N.J. Nelson, and M.D. Challberg. 1988. Herpes simplex virus DNA replication: The UL9 gene encodes an origin-binding protein. *Proc. Natl. Acad. Sci.* **85:** 5414–5418.

Park, P., W. Copeland, L. Yang, T. Wang, M.R. Botchan, and I.J. Mohr. 1994. The cellular DNA polymerase alpha-primase is required for papillomavirus DNA replication and associates with the viral E1 helicase. *Proc. Natl. Acad. Sci.* **91:** 8700–8704.

Poll, E.H., J. Harrison, A. Umthun, D.L. Dobbs, and R.M. Benbow. 1994. *Xenopus laevis* ovarian DNA helicase I: A 3′ to 5′ helicase that unwinds short duplexes. *Biochemistry* **33:** 3841–3847.

Rabkin, S.D. and B. Hanlon. 1990. Herpes simplex virus DNA synthesis at a preformed replication fork in vitro. *J. Virol.* **64:** 4957–4967.

Raney, K.D., L.C. Sowers, D.P. Millar, and S.J. Benkovic. 1994. A fluorescence-based assay for monitoring helicase activity. *Proc. Natl. Acad. Sci.* **91:** 6644–6648.

Roman, L.J. and S.C. Kowalczykowski. 1989. Characterization of the helicase activity of the *Escherichia coli* RecBCD enzyme using a novel helicase assay. *Biochemistry* **28:** 2863–2873.

Rong, L. and H.L. Klein. 1993. Purification and characterization of the SRS2 DNA helicase of the yeast *Saccharomyces cerevisiae*. *J. Biol. Chem.* **268:** 1252–1259.

Santucci, S., E. Androphy, C. Bonne-Andréa, and P. Clertant. 1990. Proteins encoded by the bovine papillomavirus E1 open reading frame: Expression in heterologous systems and in virally transformed cells. *J. Virol.* **64:** 6027–6033.

Saraste, M., P.R. Sibbald, and A. Wittinghofer. 1990. The P-loop—Common motif in ATP- and GTP-binding proteins. *Trends Biochem. Sci.* **15:** 430–434.

Schaeffer, L., R. Roy, S. Humbert, V. Moncollin, W. Vermeulen, J.H.J. Hoeijmakers, P. Chambon, and J.-M. Egly. 1993. DNA repair helicase: A component of BTF2 (TFIIH) basic transcription factor. *Science* **260:** 58–63.

Scheffner, M., R. Knippers, and H. Stahl. 1989. RNA unwinding activity of SV40 large T antigen. *Cell* **57:** 955–963.

Schulz, V.P. and V.A. Zakian. 1994. The *Saccharomyces PIF1* DNA helicase inhibits telomere elongation and de novo telomere formation. *Cell* **76:** 145–155.

Seki, M., T. Enomoto, J. Yanagisawa, F. Hanaoka, and M. Ui. 1988. Further characterization of the DNA helicase activity of mouse DNA-dependent adenosinetriphosphatase B (DNA helicase B). *Biochemistry* **27:** 1766–1771.

Seki, M., J. Yanagisawa, T. Kohda, T. Sonoyama, M. Ui, and T. Enomoto. 1994. Purification of two DNA-dependent adenosinetriphosphatases having DNA helicase activity from HeLa cells and comparison of the properties of the two enzymes. *J. Biochem.* **115:** 523–531.

Seki, M., T. Enomoto, T. Eki, A. Miyajima, Y. Murakami, F. Hanaoka, and M. Ui. 1990. DNA helicase and nucleoside-5′-triphosphatase activities of polyoma virus large tumor antigen. *Biochemistry* **29:** 1003–1009.

SenGupta, D. and J.A. Borowiec. 1992. Strand-specific recognition of a synthetic DNA replication fork by the SV40 large tumor antigen. *Science* **256:** 1656–1661.

———. 1994. Strand and face: The topography of interactions between the SV40 origin of replication and T antigen during the initiation of replication. *EMBO J.* **13:** 982–992.

SenGupta, D.J., L.J. Blackwell, T. Gillette, and J.A. Borowiec. 1992. Recognition of model DNA replication forks by the SV40 large tumor antigen. *Chromosoma* **102:** S46–S51.

Seo, Y.-S. and J. Hurwitz. 1993. Isolation of helicase α, a DNA helicase from HeLa cells stimulated by a fork structure and single-stranded DNA binding proteins. *J. Biol. Chem.* **268:** 10282–10295.

Seo, Y.-S., S.-H. Lee, and J. Hurwitz. 1991. Isolation of a DNA helicase from Hela cells requiring the multisubunit human single-stranded DNA-binding protein for activity. *J. Biol. Chem.* **266:** 13161–13170.

Seo, Y.-S., F. Müller, M. Lusky, and J. Hurwitz. 1993. Bovine papilloma viral (BPV)-encoded E1 protein contains multiple activities required for BPV DNA replication. *Proc. Natl. Acad. Sci.* **90:** 702–706.

Shimizu, K. and A. Sugino. 1993. Purification and characterization of DNA helicase III from the yeast *Saccharomyces cerevisiae*. *J. Biol. Chem.* **268:** 9578–9584.

Siegal, G., J.J. Turchi, C.B. Jessee, T.W. Myers, and R.A. Bambara. 1992. A novel DNA helicase from calf thymus. *J. Biol. Chem.* **267:** 13629–13635.

Smider, V., W.K. Rathmell, M.R. Lieber, and G. Chu. 1994. Restoration of X-ray resistance and V(D)J recombination in mutant cells by Ku cDNA. *Science* **266:** 288–291.

Srivastava, A., E.W. Lusby, and K.I. Berns. 1983. Nucleotide sequence and organization of the adeno-associated virus 2 genome. *J. Virol.* **45:** 555–564.

Stahl, H., P. Dröge, and R. Knippers. 1986. DNA helicase activity of SV40 large tumor antigen. *EMBO J.* **5:** 1939–1944.

Stillman, B. 1994. Smart machines at the DNA replication fork. *Cell* **78:** 725–728.

Sun, S., L. Thorner, M. Lentz, P. MacPherson, and M. Botchan. 1990. Identification of a 68-kilodalton nuclear ATP-binding phosphoprotein encoded by bovine papillomavirus type 1. *J. Virol.* **64:** 5093–5105.

Sung, P., L. Prakash, S.W. Matson, and S. Prakash. 1987. RAD3 protein of *Saccharomyces cerevisiae* is a DNA helicase. *Proc. Natl. Acad. Sci.* **84:** 8951–8955.

Sung, P., V. Bailly, C. Weber, L.H. Thompson, L. Prakash, and S. Prakash. 1993. Human xeroderma pigmentosum group D gene encodes a DNA helicase. *Nature* **365:** 852–855.

Taccioli, G.E., T.M. Gottlieb, T. Blunt, A. Priestley, J. Demengeot, R. Mizuta, A.R. Lehmann, F.W. Alt, S.P. Jackson, and P.A. Jeggo. 1994. Ku80: Product of the *XRCC5* gene and its role in DNA repair and V(D)J recombination. *Science* **265:** 1442–1445.

Thömmes, P. and U. Hübscher. 1990. DNA helicase from calf thymus. Purification to apparent homogeneity and biochemical characterization of the enzyme. *J. Biol. Chem.* **265:** 14347–14354.

———. 1992. Eukaryotic DNA helicases: Essential enzymes for DNA transactions. *Chromosoma* **101:** 467–473.

Thömmes, P., E. Ferrari, R. Jessberger, and U. Hübscher. 1992. Four different DNA helicases from calf thymus. *J. Biol. Chem.* **267:** 6063–6073.

Tuteja, N., K. Rahman, R. Tuteja, and A. Falaschi. 1991. DNA helicase IV from HeLa cells. *Nucleic Acids Res.* **19:** 3613–3618.

———. 1993. Human DNA helicase V, a novel DNA unwinding enzyme from HeLa cells. *Nucleic Acids Res.* **21:** 2323–2329.

Tuteja, N., R. Tuteja, K. Rahman, L.Y. Kang, and A. Falaschi. 1990. A DNA helicase from human cells. *Nucleic Acids Res.* **18:** 6785–6792.

Tuteja, N., K. Rahman, R. Tuteja, A. Ochem, D. Skopac, and A. Falaschi. 1992. DNA helicase III from HeLa cells: An enzyme that acts preferentially on partially unwound

DNA duplexes. *Nucleic Acids Res.* **20:** 5329–5337.
Tuteja, N., R. Tuteja, A. Ochem, P. Taneja, N.W. Huang, A. Simoncsits, S. Susic, K. Rahman, L. Marusic, J. Chen, J. Zhang, S. Wang, S. Pongor, and A. Falaschi. 1994. Human DNA helicase II: A novel DNA unwinding enzyme identified as the Ku autoantigen. *EMBO J.* **13:** 4991–5001.
Ustav, M. and A. Stenlund. 1991. Transient replication of BPV-1 requires two viral polypeptides encoded by the E1 and E2 open reading frames. *EMBO J.* **10:** 449–457.
Venkatesan, M., L.L. Silver, and N.G. Nossal. 1982. Bacteriophage T4 gene 41 protein, required for the synthesis of RNA primers, is also a DNA helicase. *J. Biol. Chem.* **257:** 12426–12434.
Walker, J.E., M. Saraste, M.J. Runswick, and N.J. Gay. 1982. Distantly related sequences in the α- and β-subunits of ATP synthase, myosin, kinases and other ATP-requiring enzymes and a common nucleotide binding fold. *EMBO J.* **1:** 945–951.
Wang, E.H. and C. Prives. 1991. DNA helicase and duplex DNA fragment unwinding activities of polyoma and simian virus 40 large T antigen display similarities and differences. *J. Biol. Chem.* **266:** 12668–12675.
Wessel, R., J. Schweizer, and H. Stahl. 1992. Simian virus 40 T-antigen DNA helicase is a hexamer which forms a binary complex during bidirectional unwinding from the viral origin of DNA replication. *J. Virol.* **66:** 804–815.
Wiekowski, M., M.W. Schwarz, and H. Stahl. 1988. Simian virus 40 large T antigen DNA helicase. *J. Biol. Chem.* **263:** 436–442.
Wilson, G.M., H.K. Jindal, D.E. Yeung, W. Chen, and C.R. Astell. 1991. Expression of minute virus of mice major nonstructural protein in insect cells: Purification and identification of ATPase and helicase activities. *Virology* **185:** 90–98.
Wolffe, A.P. and D.D. Brown. 1986. DNA replication in vitro erases a *Xenopus* 5S RNA gene transcription complex. *Cell* **47:** 217–227.
Wong, I. and T.M. Lohman. 1992. Allosteric effects of nucleotide cofactors on *Escherichia coli* Rep helicase-DNA binding. *Science* **256:** 350–355.
Wun-Kim, K. and D.T. Simmons. 1990. Mapping of helicase and helicase substrate-binding domains on simian virus 40 large T antigen. *J. Virol.* **64:** 2014–2020.
Yanagisawa, J., M. Seki, T. Kohda, E. Enomoto, and M. Ui. 1992. DNA-dependent adenosinetriphosphatase C1 from mouse FM3A cells has DNA helicase activity. *J. Biol. Chem.* **267:** 3644–3649.
Yang, L., R. Li, I.J. Mohr, R. Clark, and M.R. Botchan. 1991. Activation of BPV-1 replication *in vitro* by the transcription factor E2. *Nature* **353:** 628–632.
Yang, L., I. Mohr, E. Fouts, D.A. Lim, M. Nohaile, and M. Botchan. 1993. The E1 protein of bovine papilloma virus 1 is an ATP-dependent DNA helicase. *Proc. Natl. Acad. Sci.* **90:** 5086–5090.
Yoder, B.L. and P.M. Burgers. 1991. *Saccharomyces cerevisiae* replication factor C. I. Purification and characterization of its ATPase activity. *J. Biol. Chem.* **266:** 22689–22697.
Zhang, S. and F. Grosse. 1991. Purification and characterization of two DNA helicases from calf thymus nuclei. *J. Biol. Chem.* **266:** 20483–20490.
———. 1994. Nuclear DNA helicase II unwinds both DNA and RNA. *Biochemistry* **33:** 3906–3912.
Zhu, L.A. and S.K. Weller. 1992. The six conserved helicase motifs of the UL5 gene product, a component of the herpes simplex virus type 1 helicase-primase, are essential for its function. *J. Virol.* **66:** 469–479.

19
DNA Ligases

Rachel Nash and Tomas Lindahl
Imperial Cancer Research Fund
Clare Hall Laboratories
South Mimms
Herts EN6 3LD, United Kingdom

DNA ligases are Mg^{++}-dependent enzymes that catalyze the formation of phosphodiester bonds at single-strand breaks in double-stranded DNA (for review and main references up to 1992, see Engler and Richardson 1982; Lindahl and Barnes 1992). The first step in the reaction is the formation of a covalent enzyme/adenylate intermediate. DNA ligases from eukaryotes, archaea, and viruses employ ATP as cofactor, whereas eubacterial DNA ligases use NAD to generate the adenylyl group. The ATP is cleaved to AMP and pyrophosphate with the adenylyl residue linked by a phosphoramidate bond to the ε-amino group of a specific lysine residue at the active site of the protein. The reaction is readily reversed in vitro by addition of pyrophosphate. Since DNA ligases contain an unusually reactive lysine residue in their active site, a Schiff base can be formed with pyridoxal phosphate. In consequence, the activities of DNA ligases, e.g., mammalian DNA ligases I and II, are inhibited in vitro by pyridoxal phosphate. ATP-dependent DNA ligases can employ certain cofactor analogs such as dATP, but the anomalous enzyme/nucleotide complexes formed by some ligases appear to function poorly in subsequent steps of the DNA-joining reaction. The activated AMP residue of the DNA ligase/adenylate intermediate is transferred to the 5′-phosphate terminus of a single-strand break in double-stranded DNA to generate a covalent DNA-AMP complex with a 5′–5′ phosphoanhydride bond. In the final step of DNA ligation, unadenylylated DNA ligase is required for the generation of a phosphodiester bond and catalyzes displacement of the AMP residue through attack by the adjacent 3′-hydroxyl group on the adenylylated site (Fig. 1).

The active site of mammalian DNA ligase I has been identified by isolation and amino acid sequencing of the tryptic peptide containing a lysine-[^{3}H]AMP moiety and localized by comparison with the complete peptide sequence deduced from the cDNA. The amino acid sequence of the peptide containing the active-site lysine residue has also been

DNA Replication in Eukaryotic Cells

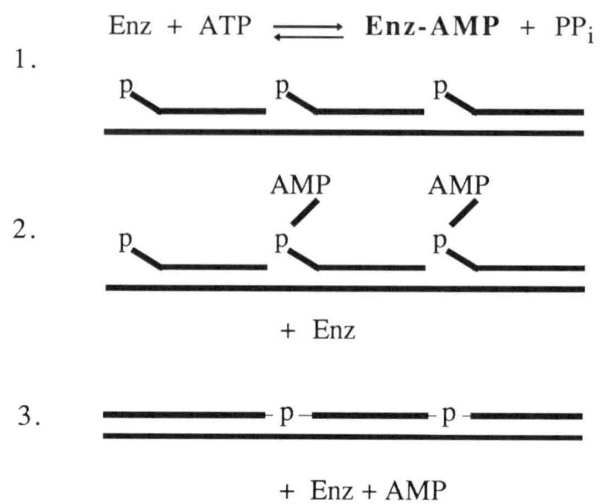

Figure 1 The reaction catalyzed by DNA ligases. The enzyme/adenylate intermediate is highlighted in bold.

determined for mammalian DNA ligase II (Wang et al. 1994). Comparisons with nucleic acid sequences encoding DNA ligases of other species have identified a ubiquitous K*X*DG active-site motif. This motif has also been found at the active site of the vaccinia virus mRNA-capping enzyme that catalyzes transfer of GMP from GTP to the 5′ terminus of RNA and forms an enzyme/guanylate reaction intermediate with a lysine/GMP phosphoramidate bond. The observed sequence conservation indicates a common mechanism of covalent catalysis in these nucleotidyl transfer reactions (Cong and Shuman 1993).

Some DNA ligases are able to catalyze blunt-end joining in addition to ligating nicks in double-stranded DNA. Blunt-end joining proceeds less efficiently than ligation of nicks or cohesive ends. Mammalian DNA ligase I effectively seals single-strand breaks in DNA and joins restriction enzyme DNA fragments with staggered ends. The enzyme is also able to catalyze blunt-end joining of DNA. As for other DNA ligases, the latter reaction is stimulated greatly by the macromolecular crowding conditions inflicted by the presence of polyethylene glycol in reaction mixtures. DNA ligase I is more effective at blunt-end joining than mammalian DNA ligases II and III but is less efficient in this regard than bacteriophage T4 DNA ligase.

ASSAYS

Formation of an Enzyme/Adenylate Intermediate

This assay depends on the enzyme's ability to covalently bind AMP. If the DNA ligase is incubated with [α-^{32}P]ATP, the enzyme/AMP intermediate is radiolabeled. These complexes are stable in the absence of pyrophosphate or DNA substrate. The enzyme/adenylate can then be separated by SDS-PAGE and visualized using autoradiography. This method is useful in determining which DNA ligases are present in partially purified protein fractions, as the respective enzyme/AMP intermediates migrate differently according to their molecular masses. Figure 2 illustrates this point by showing mammalian DNA ligases I, II, and III, (lanes a, b, and c, respectively), labeled at their active sites with [α-^{32}P]ATP. The proteins are then fractionated by SDS-gel electrophoresis to reveal apparent molecular masses of 125 kD, 69 kD, and 100 kD, respectively. Due to anomalously slow migration of DNA ligase I, the first number is a slight overestimate in comparison with data deduced from the cDNA open reading frame.

Specific Dissociation of an Enzyme/Adenylate Intermediate by Nicked DNA

The adenylate group can be specifically removed from the complex either by transfer to a 5′-phosphate acceptor of a polynucleotide substrate or by incubation with pyrophosphate, reversing the initial reaction to

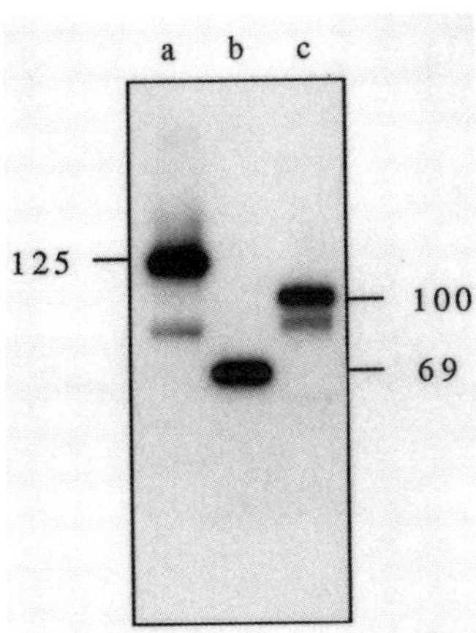

Figure 2 Adenylylation of DNA ligases I, II, and III. DNA ligase I (*a*), DNA ligase II (*b*), and DNA ligase III (*c*) were incubated for 15 min at room temperature in a standard reaction mixture (10 μl), containing 60 mM Tris-HCl (pH 8.0), 10 mM $MgCl_2$, 5 mM DTT, 50 μg/ml bovine serum albumin, and 0.5 μCi [α-^{32}P]ATP (3000 Ci/mmole, Amersham). After the addition of 5 μl of SDS sample buffer, reaction mixtures were heated at 90°C for 10 min. Proteins were separated by SDS-PAGE. Gels were fixed for 10 min in 10% acetic acid, and dried. Adenylylated proteins were detected by autoradiography.

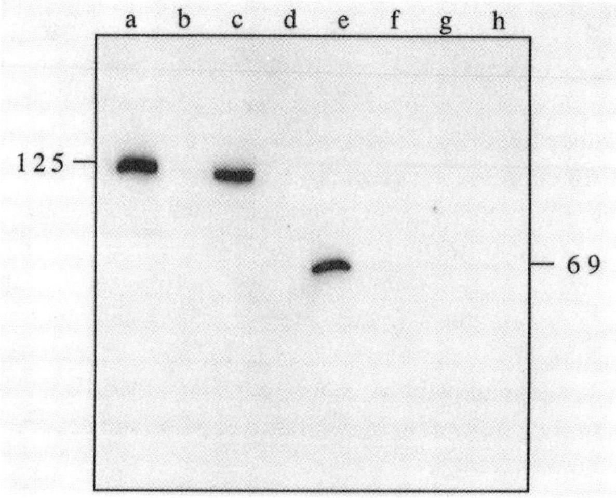

Figure 3 Reactivity of the adenylylated forms of DNA ligases I and II. DNA ligase I (*a–d*) and DNA ligase II (*e–h*) adenylate intermediates were formed, as described in Fig. 2, in a final volume of 40 μl. They were then further incubated with unlabeled polynucleotide substrate or sodium pyrophosphate. Aliquots (10 μl) were incubated with no addition (*a, e*); 0.8 μg unlabeled oligo(dT)•poly(dA) (*b, f*); 0.8 μg unlabeled oligo(dT)•poly(rA) (*c, g*); or 10 nmoles of sodium pyrophosphate (*d, h*) for 1 hr at 37°C. The reactions were stopped by the addition of SDS sample buffer, and adenylylated polypeptides were detected as described in Fig. 2.

regenerate ATP. Figure 3 illustrates the differing abilities of mammalian DNA ligases I and II to react with two polynucleotide substrates. Lanes a–d refer to DNA ligase I, and lanes e–h refer to DNA ligase II. The first lane in each set of four shows the respective enzyme/adenylate intermediate prior to addition of any substrate. Lanes b and f represent the reaction of the radiolabeled intermediate with oligo(dT)•poly(dA). Both DNA ligases I and II are able to ligate this substrate, as shown by the loss of radiolabeled enzyme/AMP intermediate in each case, resulting from transfer of the radiolabeled AMP group to the substrate. DNA ligase II (lane g), but not DNA ligase I (lane c), is also able to react with an oligo(dT)•poly(rA) substrate. Lanes d and h show reversal of the first step in the reaction by the addition of pyrophosphate.

Formation of Joined Oligonucleotides

In this case, the same substrates are used as in the previous assay, except that the oligonucleotide is radioactively labeled at its 5′ end with ^{32}P.

The enzyme is incubated, in the presence of Mg^{++} and ATP, with the substrate, e.g., 5′-^{32}P-labeled oligo(dT) hybridized to either poly(dA) or poly(rA). Multimers of the oligonucleotides are subsequently detected by autoradiography after electrophoresis through a denaturing polyacrylamide gel.

Ligation of Nicked Circular DNA to Covalently Closed Circular DNA

This system provides a striking and convincing demonstration of DNA ligase activity. However, in contrast to the assays described above, it does not provide additional information on either the size of the DNA ligase or its substrate specificity with regard to joining DNA-RNA hybrids. To generate the appropriate substrate, covalently closed circular DNA is treated with a very low concentration of pancreatic DNase I, or restriction endonuclease *Eco*RI in the presence of ethidium bromide. The nicked DNA is purified free from remaining covalently closed circular DNA and employed as substrate. The reaction is stopped by addition of EDTA and proteinase K. Rejoining of the DNA strand break by a DNA ligase may be detected by agarose gel electrophoresis in the presence of ethidium bromide.

Blunt-end Joining

The ability of a DNA ligase to join blunt ends can be assayed by the joining of DNA restriction fragments generated by restriction endonucleases such as *Hpa*I that give DNA fragments with fully base-paired ends. Inclusion of 10–20% polyethylene glycol 6000 in standard reaction mixtures may promote the reaction.

Assay Pitfalls

A number of factors can cause problems with the various assays employed to detect DNA ligase activity. Small amounts of contaminating DNase activity in crude cell extracts, or partly purified DNA ligase fractions, may interfere seriously with attempts to detect joining of DNA or oligonucleotides. The assay employing nicked circular DNA as a substrate is particularly sensitive to traces of endonuclease activity. Problems can also be encountered in methods that measure the level of DNA ligation. The presence of exonuclease activities prevents accurate quantification due to the conversion of nicks to small gaps, which are no longer substrates for DNA ligases. In particular, double-stranded DNA-RNA substrates such as oligo(dT)•poly(rA) are uniquely susceptible to RNase H degradation; RNase H activities are often abundant in crude enzyme fractions. Thus, contamination of a partly purified DNA ligase with RNase H may give the erroneous impression that the DNA ligase can join single-strand breaks in a DNA•DNA, but not a DNA•RNA, double helix.

The formation of a covalent enzyme/adenylate complex in the presence of [α-^{32}P]ATP is a fairly robust and very useful assay, but high ATPase concentrations interfere. Moreover, this is a less sensitive and quantitative method than the DNA ligation assay, since formation of a stoichiometric complex is measured rather than catalytic events. In addition, proteins other than DNA ligases in crude enzyme fractions may become adenylated. In practice, this is rarely a serious problem, but it is imperative to demonstrate that the complex is a DNA ligase/AMP reaction intermediate. This can be shown by establishing that the radioactive moiety is specifically released from the protein by addition of nicked DNA or oligo(dT)•poly(dA) to reaction mixtures.

DNA LIGASE I

An enzyme of this type has been found in all eukaryotic cells investigated. The main function of human DNA ligase I is probably the joining of Okazaki fragments during lagging-strand DNA replication (Waga et al. 1994). The enzyme is also involved in DNA excision repair (Prigent et al. 1994). Similarly to DNA polymerase-α, DNA ligase I is induced 10- to 15-fold in S phase in mammalian cells. Thus, it is present at much higher concentrations in proliferating tissues. A less impressive 3-fold induction has been observed 24–48 hours after cellular exposure to ultraviolet radiation (Montecucco et al. 1992). Immunofluorescence studies with antibodies against DNA ligase I have shown that the enzyme is present in cell nuclei, exhibiting the granular staining also seen for several other replication factors. This may reflect the occurrence of DNA ligase I in "replication factories." A 21S DNA replication complex purified from HeLa cells contained DNA ligase I together with DNA polymerase-α:primase, proliferating cell nuclear antigen, RNase H, and a small number of other proteins (Congjun et al. 1994).

The cloned cDNA for human DNA ligase I encodes a 102-kD protein of 919 amino acid residues. The amino-terminal region of the protein is not required for DNA ligation activity in vitro. This amino-terminal part of the enzyme is hydrophilic and contains several stretches of either negatively or positively charged amino acid residues. DNA ligase I is a phosphoprotein, and most or all of the phosphate residues are localized to the amino-terminal region. Furthermore, the amino-terminal part is highly susceptible to proteolysis, so a 78-kD active fragment of mammalian DNA ligase I, comprising the catalytic domain of the enzyme, is often generated as a preparation artifact due to endogenous degradation during enzyme purification.

Large amounts of active human DNA ligase I can be obtained by expression of the cDNA in a baculovirus expression system (Wang et al. 1994). The cDNA has also been expressed in *Escherichia coli*, but it is controversial whether such an unphosphorylated form of the enzyme retains all the key features of the native enzyme purified from mammalian cells. However, the subcloned catalytic domain of human DNA ligase I is sufficient to complement a conditional-lethal DNA ligase mutant of *E. coli*. The properties of the amino-terminal region of DNA ligase I indicate that it may serve in specific binding of the enzyme to other DNA replication and repair proteins. It is not required for interactions with either nicked DNA or the ATP cofactor. The active-site lysine residue is centrally located within the catalytic domain of the enzyme. cDNA deletion experiments have shown that although 250 amino acid residues can be removed from the amino-terminal region of the protein without loss of enzyme activity, only 16 residues can be deleted at the carboxyl terminus before an essential, highly conserved motif is encountered.

The human gene encoding DNA ligase I is located at chromosome 19q13.2-13.3. The gene covers 53 kb and contains 28 exons (Noguiez et al. 1992). Unusually for an enzyme involved in DNA replication, an individual has been identified with an inherited defect in DNA ligase I (Barnes et al. 1992). The clinical symptoms were severe immunodeficiency, stunted growth, and sun sensitivity. The individual died at a young age from an abdominal lymphoma with an associated acute infection (Webster et al. 1992). A fibroblast cell strain, 46BR, derived from the individual shows deleterious mutations in both alleles of the DNA ligase I gene. One has an inactivating mutation in the active-site region, whereas the other has a mutation in a conserved part of the carboxy-terminal region that allows a small amount of residual DNA ligase activity.

The leakiness of the latter point mutation probably accounts for the fact that a mutation in a DNA replication enzyme is compatible with cellular viability. An SV40-transformed subline, 46BR•1G1, is available that contains only the latter, leaky mutation expressed in hemizygous or homozygous form. 46BR cells show inefficient joining of Okazaki fragments and anomalous gap filling during excision repair (Prigent et al. 1994). However, V(D)J joining seems to proceed normally in 46BR cells (Hsieh et al. 1993). Since the 46BR patient was sun sensitive, and the late steps of lagging-strand DNA replication and nucleotide excision repair share many factors (Aboussekhra et al. 1995), it seems likely that DNA ligase I is responsible for the final DNA-joining step in nucleotide excision repair.

The *Saccharomyces cerevisiae CDC9* and *Schizosaccharomyces pombe cdc17*$^{+}$ gene products appear to be the yeast counterparts of mammalian DNA ligase I. Thus, the purified *S. cerevisiae* and *S. pombe* enzymes have catalytic properties practically identical to those of the mammalian enzyme, whereas they differ from mammalian DNA ligases II and III (Tomkinson et al. 1992; P. Schär, unpubl.). The 86–87-kD yeast enzymes also have an amino-terminal region, shorter than that of the mammalian enzyme, which is very susceptible to proteolysis and nonessential for DNA ligation activity in vitro. The overall sequence homology between human DNA ligase I and the yeast enzymes is about 40%; however, no similarity is detected between the sequences of the amino-terminal regions of these three enzymes, whereas high homology (70%) is observed in the region surrounding the active-site lysine residue.

Conditional-lethal yeast mutants with a temperature-sensitive DNA ligase accumulate non-joined DNA fragments during abortive replication at the restrictive temperature and show defects in DNA repair and recombination at the permissive temperature. Presently, it is unclear whether *S. cerevisiae* and *S. pombe* contain enzymes similar to mammalian DNA ligases II and III. They may not yet have been detected, but it is also possible that their functions are not required in yeast cells or that the *CDC9*- and *cdc17*$^{+}$-encoded enzymes also perform the corresponding roles of mammalian DNA ligases II and III.

DNA ligase I from *Drosophila melanogaster* and *Xenopus laevis* also has biochemical properties very similar to those of mammalian DNA ligase I. Both these enzymes resemble mammalian DNA ligase I in having a protease-sensitive region not required for ligation activity. The intact *Drosophila* enzyme is an 83- to 86-kD protein, similar in size to the yeast enzyme. In contrast, the *Xenopus* enzyme has a catalytic domain of about the same size as that of the mammalian enzyme but an unusually long, phosphorylated amino-terminal region. Thus, *Xenopus* DNA ligase I migrates as an apparent 180-kD protein on SDS-PAGE (Aoufouchi et al. 1992).

DNA LIGASE II

This approximately 69-kD DNA ligase can be distinguished from DNA ligase I by its ability to join an oligo(dT)•poly(rA) substrate. DNA ligase II is not induced on cell proliferation, and its cellular role is not clear. It is the major DNA ligase activity in certain nonproliferating tissues, e.g., adult liver. The enzyme is more firmly retained in cell nuclei than DNA ligase I and requires buffers of moderate or high salt concentration for efficient extraction. Neutralizing rabbit antibodies against DNA ligase I do not detectably recognize or inhibit DNA ligase II, and, similarly, neutralizing antibodies against DNA ligase II do not inactivate DNA ligase I. The two enzymes generate entirely different fragmentation patterns on proteolysis, and there is no indication of a protease-sensitive amino-terminal region (Roberts et al. 1994). Recently, the amino acid sequences of the active-site peptide and several other peptides of bovine DNA ligase II were determined (Wang et al. 1994). These show only about 30% homology between mammalian DNA ligases I and II, although main active-site features common to all ATP-dependent DNA ligases are present. Like DNA ligase I and most nuclear proteins, DNA ligase II has a blocked amino-terminal amino acid residue. Interestingly, mammalian DNA ligase II is much more similar to the vaccinia virus-encoded DNA ligase than to DNA ligase I (Wang et al. 1994). *D. melanogaster* also has a DNA ligase II, which differs from DNA ligase I in being present in similar amounts during developmental stages with high (early embryos) or low (pupae, adults) DNA replication rates.

DNA LIGASE III

DNA ligase III is a mammalian DNA ligase of 103 kD. A cDNA has been cloned, and the gene for DNA ligase III has been mapped to human chromosome 17q11.2-12 (Wei et al. 1995). The enzyme resembles DNA ligase I in having a protease-sensitive amino-terminal region not required for DNA ligation activity in vitro, but it resembles DNA ligase II in its ability to join an oligo(dT)•poly(rA) substrate. Sequences of DNA ligase III and the smaller DNA ligase II seem identical or closely similar, so both these forms of the enzyme may be derived from the same gene (Roberts et al. 1994; Husain et al. 1995; Wei et al. 1995). Although DNA ligase III is of similar size to DNA ligase I, it elutes well before the asymmetric DNA ligase I on gel filtration of crude enzyme fractions, indicating that it is a dimer or binds a separate nuclear protein in a salt-resistant interaction. A prime candidate for this partner of DNA ligase III is the 70-kD XRCC1 protein, because the two proteins bind each other in vitro (Caldecott et al. 1994). Moreover, the Chinese hamster cell line EM9, which is defective in XRCC1 protein, also exhibits an anomalously low level of DNA ligase III activity (Caldecott et al. 1994; Ljungquist et al. 1994). The phenotype of XRCC1 involves a high level of sister-chromatid exchange and hypersensitivity to alkylating agents and ionizing radiation. The latter observations suggest a role for DNA ligase III in base excision repair. A mammalian protein complex that performs recombination events in vitro has been reported to contain DNA ligase III but no detectable DNA ligase I (Jessberger et al. 1993).

DNA LIGASE IV

A fourth mammalian DNA ligase has been detected recently by a search of expressed sequence tags with short sequence motifs highly conserved in eukaryotic DNA ligases (Wei et al. 1995). The complete cDNA encodes a 96-kD protein with DNA ligase activity that exhibits partial sequence homology with ligases I–III. The human gene for DNA ligase IV is localized on chromosome 13q33-34. Northern blots indicate that the enzyme is expressed in thymus and testis and at a very low level in several other tissues, but its physiological role is not known. An unusual feature of DNA ligase IV, in comparison with the other mammalian DNA ligases, is an extended carboxy-terminal region of more than 300 amino acids that shows no homology with other proteins in databases. This region may conceivably be involved in protein-protein interactions that could functionally distinguish this enzyme from the other DNA ligases. On SDS-PAGE, DNA ligases III and IV migrate at indistinguishable rates as approximately 100-kD proteins. Thus, an enzyme/AMP complex of this size could be due to either ligase III or IV, making further identification necessary by immunological methods or by exploring the interaction with the XRCC1 protein unique to DNA ligase III. A comparison of the properties of DNA ligases I, II, III, and IV is shown in Table 1.

VIRUS-ENCODED DNA LIGASES

Herpesviruses and smaller DNA and RNA animal viruses do not encode a DNA ligase. However, a distinct virus-encoded enzyme is produced by vaccinia virus and a number of other poxviruses. The 63-kD vaccinia protein only shows weak homology (~30%) with DNA ligase I from human cells, *S. pombe*, or *S. cerevisiae*. Although poxviruses replicate in the host-cell cytoplasm, the vaccinia DNA ligase is nonessential for viral

Table 1 Properties of four mammalian DNA ligases

	I	II	III	IV
Molecular mass estimated				
by cDNA sequence	102 kD	—	103 kD	96 kD
by SDS-PAGE	125 kD	70 kD	100 kD	100 kD
Chromosomal localization	19q13.2–13.3	—	17q11.2–12	13q33–34
Ligation of				
oligo(dT)•poly(dA)	yes	yes	yes	yes
oligo(dT)•poly(rA)	no	yes	yes	yes
oligo(rA)•poly(dT)	yes	no	yes	no
K_m for ATP	1 μM	40 μM	2 μM	—
Function	lagging-strand replication, nucleotide excision repair	—	base excision repair, recombination	—

DNA replication and growth. However, a virus DNA ligase-deficient mutant shows attenuated virulence in vivo and is anomalously sensitive to DNA-damaging agents during infection, implying a role for the enzyme in viral DNA repair. A DNA ligase with a surprisingly different sequence is encoded by African swine fever virus, another large DNA virus replicating in the cytoplasm of infected cells (Hammond et al. 1992). The latter data imply that African swine fever virus is only distantly related to vaccinia virus.

REFERENCES

Aboussekhra, A., M. Biggerstaff, M.K.K. Shivji, J.A. Vilpo, V. Moncollin, V.N. Podust, M. Protic, U. Hübscher, J.-M. Egly, and R.D. Wood. 1995. Mammalian DNA nucleotide excision repair reconstituted with purified protein components. *Cell* **80:** 859–868.

Aoufouchi, S., C. Prigent, N. Theze, M. Philippe, and P. Thiebauld. 1992. Expression of DNA ligase I and II during oogenesis and early development of *Xenopus laevis. Dev. Biol.* **152:** 199–202.

Barnes, D.E., A.E. Tomkinson, A.R. Lehmann, A.D.B. Webster, and T. Lindahl. 1992. Mutations in the DNA ligase I gene of an individual with immunodeficiencies and cellular hypersensitivity to DNA-damaging agents. *Cell* **69:** 495–503.

Caldecott, K.W., C.K. McKeown, J.D. Tucker, S. Ljungquist, and L.H. Thompson. 1994. An interaction between the mammalian DNA repair protein XRCC1 and DNA ligase III. *Mol. Cell. Biol.* **14:** 68–76.

Cong, P. and S. Shuman. 1993. Covalent catalysis in nucleotidyl transfer. *J. Biol. Chem.* **268:** 7256–7260.

Congjun L., J. Goodchild, and E.F. Baril. 1994. DNA ligase I is associated with the 21 S complex of enzymes for DNA synthesis in HeLa cells. *Nucleic Acids Res.* **22:** 632–638.

Engler, M.J. and C.C. Richardson. 1982. DNA ligases. In *The enzymes*, 3rd edition (ed. P.D. Boyer), vol. 15B, pp. 1–29. Academic Press, New York.

Hammond, J.M., S.M. Kerr, G.L. Smith, and L.K. Dixon. 1992. An African swine fever virus gene with homology to DNA ligases. *Nucleic Acids Res.* **20:** 2667–2671.

Hsieh, C.-L., C.F. Arlett, and M.R. Lieber. 1993. V(D)J recombination in ataxia telangiectasia, Bloom's syndrome, and a DNA ligase I-associated immunodeficiency disorder. *J. Biol. Chem.* **268:** 20105–20109.

Husain, I., A.E. Tomkinson, W.A. Burkhart, M.B. Moyer, W. Ramos, Z.B. Mackey, J.M. Besterman, and J. Chen. 1995. Purification and characterization of DNA ligase III from bovine testes. *J. Biol. Chem.* **270:** 9683–9690.

Jessberger, R., V. Podust, U. Hübscher, and P. Berg. 1993. A mammalian protein complex that repairs double-strand breaks and deletions by recombination. *J. Biol. Chem.* **268:** 15070–15079.

Lindahl, T. and D.E. Barnes. 1992. Mammalian DNA ligases. *Annu. Rev. Biochem.* **61:** 251–281.

Ljungquist, S., K. Kenne, L. Olsson, and M. Sandstrom. 1994. Altered DNA ligase III activity in the CHO EM9 mutant. *Mutat. Res.* **314:** 177–186.

Montecucco, A., G. Biamonti, E. Savini, F. Focher, S. Spadari, and G. Ciarrocchi. 1992. DNA ligase I gene expression during differentiation and cell proliferation. *Nucleic Acids Res.* **20:** 6209–6214.

Noguiez, P., D.E. Barnes, H.W. Mohrenweiser, and T. Lindahl. 1992. Structure of the human DNA ligase I gene. *Nucleic Acids Res.* **20:** 3845–3850.

Prigent, C., M.S. Satoh, G. Daly, D.E. Barnes, and T. Lindahl. 1994. Aberrant DNA repair and DNA replication due to an inherited enzymatic defect in human DNA ligase I. *Mol. Cell. Biol.* **14:** 310–317.

Roberts, E., R.A. Nash, P. Robins, and T. Lindahl. 1994. Different active sites of mammalian DNA ligases I and II. *J. Biol. Chem.* **269:** 3789–3792.

Tomkinson, A.E., N.J. Tappe, and E.C. Friedberg. 1992. DNA ligase I from *Saccharomyces cerevisiae*: Physical and biochemical characterisation of the *CDC9* gene product. *Biochemistry* **31:** 11762–11771.

Waga, S., G. Bauer, and B. Stillman. 1994. Reconstitution of complete SV40 DNA replication with purified replication factors. *J. Biol. Chem.* **269:** 10923–10934.

Wang, Y.-C.J., W.A. Burkhart, Z.B. Mackey, M.B. Moyer, W. Ramos, I. Husain, J. Chen, J.M. Besterman, and A.E. Tomkinson. 1994. Mammalian DNA ligase II is highly homologous with vaccinia DNA ligase. *J. Biol. Chem.* **269:** 31923–31928.

Webster, A.D.B., D.E. Barnes, C.F. Arlett, A.R. Lehmann, and T. Lindahl. 1992. Growth retardation and immunodeficiency in a patient with mutations in the DNA ligase I gene. *Lancet* **339:** 1508–1509.

Wei, Y.-F., P. Robins, K. Carter, K. Caldecott, D.J.C. Pappin, G.-L. Yu, R.-P. Wang, B.K. Shell, R.A. Nash, P. Schär, D.E. Barnes, W.A. Haseltine, and T. Lindahl. 1995. Molecular cloning and expression of human cDNAs encoding a novel DNA ligase IV and DNA ligase III, an enzyme active in DNA repair and recombination. *Mol. Cell. Biol.* **15:** 3206–3216.

20
DNA Topoisomerases

Anni Hangaard Andersen, Christian Bendixen, and Ole Westergaard
Department of Molecular and Structural Biology
University of Aarhus
8000 Århus C, Denmark

The double-helical nature of DNA and the anchoring of DNA to nuclear structures result in a number of topological problems during replication and transcription, mainly due to DNA-tracking polymerases and helicases. These activities cause the accumulation of positive supercoils ahead of the moving polymerase and negative supercoils behind it (Liu and Wang 1987; Brill and Sternglanz 1988; Giaever and Wang 1988; Wu et al. 1988). (By definition, DNA becomes positively supercoiled when there is a decrease in the number of base pairs per helical turn below 10.3. Likewise, an increase in the number of base pairs per turn above 10.3 results in negatively supercoiled DNA.) The topological imbalance will, if not leveled, ultimately present an impenetrable energy barrier to the tracking protein complexes (Gartenberg and Wang 1992). Enzymes that influence the topological state of DNA thus play a crucial role in controlling the physiological functions of DNA.

In the eukaryotic cell, the topological structure of DNA is modulated by two groups of ubiquitous enzymes known as type I and type II topoisomerases. The enzymes alter the DNA linking number, which is the number of times the two strands are interwound. Type I enzymes (topoisomerase I and the evolutionarily distinct topoisomerase III) interconvert different topological forms of DNA by breaking and rejoining a single strand of the DNA double helix, changing the linking number in steps of one. Type II enzymes (topoisomerase II), however, catalyze topology changes by reversibly breaking both strands of the DNA double helix, resulting in a linking number change of two (Vosberg 1985; Wang 1987).

DNA Replication in Eukaryotic Cells

TYPE I TOPOISOMERASES

Physical Characteristics

The first eukaryotic topoisomerase I activity was purified from rat liver cells in 1972 as an untwisting enzyme capable of relaxing negatively supercoiled DNA (Champoux and Dulbecco 1972). Since then, topoisomerase I activities have been purified from a broad range of eukaryotic organisms and tissues (for review, see Vosberg 1985). The enzyme is present as a monomer in solution with a molecular mass of approximately 90–110 kD.

Topoisomerase I coding sequences have been reported for a number of eukaryotic species, including yeasts (Thrash et al. 1985; Uemura et al. 1987a), *Drosophila* (Hsieh et al. 1992), mouse (Koiwai et al. 1993), and human (D'Arpa et al. 1988). Sequence comparisons show that a central region of 400–450 amino acids and a carboxy-terminal region of about 70 amino acids are highly conserved in eukaryotic topoisomerase I. The active site is located in the carboxy-terminal region, and the tyrosine involved in the transient covalent linkage to DNA has been mapped in *Saccharomyces cerevisiae* and *Schizosaccharomyces pombe* to residues 727 (Lynn et al. 1989) and 771 (Eng et al. 1989), respectively. Alsner et al. (1992) have found that a small amino-terminal region, not required for topoisomerase I activity in vitro, is responsible for nuclear localization, but apart from this, little is known about the functional organization of the enzyme.

Kunze et al. (1990) have characterized the promoter region of the human *TOP1* gene. It holds the characteristics of other housekeeping genes and, accordingly, the enzyme level does not vary significantly during the cell cycle (Heck et al. 1988). Generally, the enzyme is constantly expressed at a level of 10^5 to 10^6 copies per cell (Champoux and McConaughy 1976; Liu and Miller 1981).

Topoisomerase I is posttranslationally modified and presumably positively regulated by phosphorylation. The kinases involved have been proposed to be protein kinase C (Pommier et al. 1990; Cardellini and Durban 1993) and nuclear casein kinase II (Durban et al. 1985).

Enzymology of Topoisomerase I

Topoisomerase I relaxes both negatively and positively supercoiled DNA to completion with no requirement for divalent cations or ATP (Table 1 and Fig. 1), and the catalytic cycle can be separated into DNA binding, cleavage, strand passage, religation, and turnover (Fig. 2).

Table 1 Characteristics of eukaryotic DNA topoisomerases

Enzyme	Subunit structure, size	Prevalence	Covalent linkage	Catalytic activity	ATP requirement	Mg^{++} requirement
Topoisomerase I	monomer, 90–110 kD	all eukaryotes examined; some viruses (vaccinia, Shope fibroma)	3′-P	relaxation of negative and positive supercoils; decatenation of nicked substrates	no	no
Topoisomerase II	homodimer, 160–180 kD/ monomer	all eukaryotes examined	5′-P	relaxation of negative and positive supercoils; catenation/decatenation, knotting/unknotting of double-stranded DNA	yes	yes
Topoisomerase III	monomer, ~75 kD	yeast (*S. cerevisiae*)	5′-P	weak relaxation of negative supercoils; preference for single-stranded DNA	no	yes

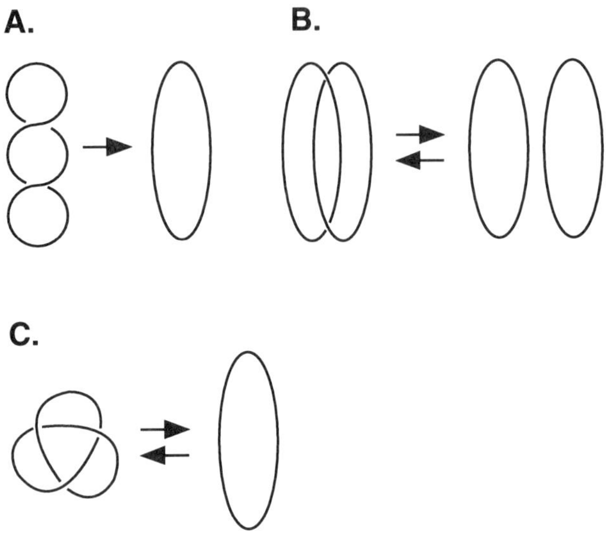

Figure 1 Reactions catalyzed by DNA topoisomerases. DNA topoisomerases catalyze the relaxation of positively and negatively supercoiled DNA (*A*), catenation/decatenation of DNA (*B*), and knotting/unknotting of DNA (*C*). For enzyme specificities and other details, see Table 1.

The substrate specificity of topoisomerase I has been widely studied, and it appears that the enzyme recognizes double-stranded DNA, relatively independent of sequence. A highly degenerate consensus sequence has been reported for the four base pairs located immediately 5′ to the site of topoisomerase-I-mediated cleavage (Edwards et al. 1982; Been et al. 1984; Jaxel et al. 1991). Despite the apparent promiscuity in binding-site selection, sites have been described to which topoisomerase I binds with high affinity. Thus, a site initially described in the rDNA of *Tetrahymena thermophila* is a site of preferential binding and cleavage in vivo and in vitro (Bonven et al. 1985; Andersen et al. 1985; Thomsen et al. 1987), as well as a site mediating preferential catalytic activity (Busk et al. 1987). Several studies suggest that topoisomerase I recognizes gross structural features of the DNA double helix rather than base sequence per se. Particularly, it seems that bent DNA is the substrate of choice. This is reflected in the preferential action on highly supercoiled or otherwise curved DNA (Camilloni et al. 1988, 1989; Caserta et al. 1989,1990) and by the finding that the high-affinity sequence described above exhibits static intrinsic curvature (Krogh et al. 1991).

Footprinting and interference studies of topoisomerase I show protection and interference in the "consensus" region 5′ to the nick, but interac-

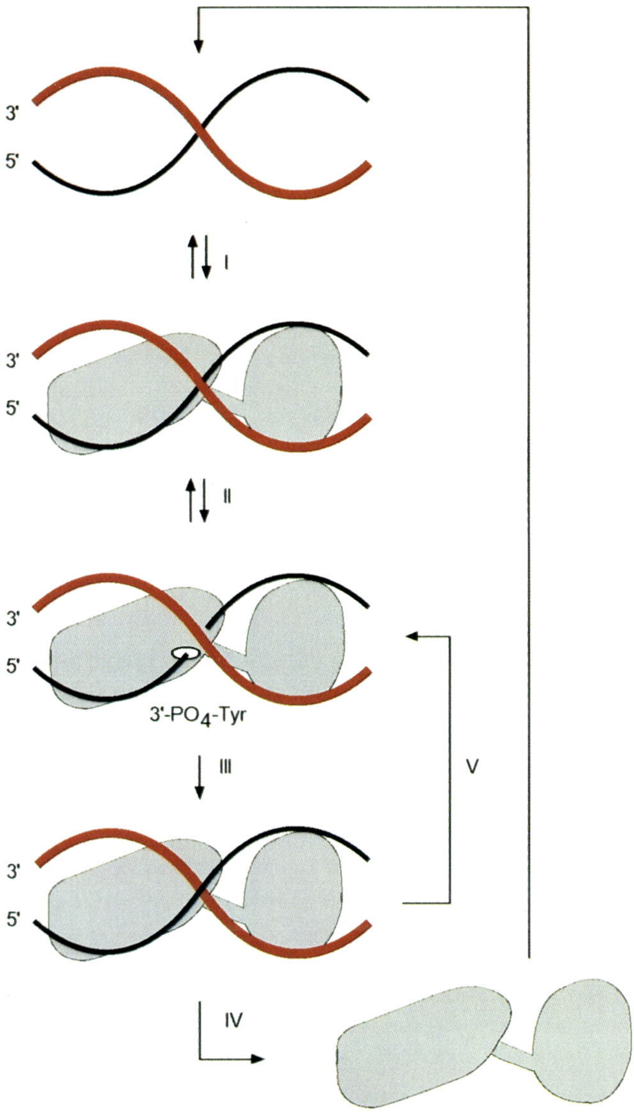

Figure 2 Steps in the catalytic cycle of eukaryotic DNA topoisomerase I. DNA topoisomerase I binds noncovalently to DNA (*I*) and cleaves one of the DNA strands, forming a covalent linkage between the newly generated 3′-phosphate and the active-site tyrosine (*II*). DNA strand passage then takes place, followed by DNA ligation (*III*). The enzyme can interact with DNA in either a distributive or a processive mode (McConaughy et al. 1981). In the distributive mode (*IV*), the enzyme dissociates from the DNA after ligation, whereas in the processive mode (*V*), the enzyme reenters the catalytic cycle without dissociation from the DNA.

tion is also observed 3′ to the nick (Stevnsner et al. 1989; Bendixen et al. 1990; Krogh et al. 1991). This additional 3′ contact was later confirmed by the use of oligonucleotide substrates (Christiansen et al. 1993; Christiansen and Westergaard 1994), and it has been proposed to operate as a topology sensor (Krogh et al. 1991). The topoisomerase-I-mediated DNA cleavage reaction has traditionally been studied by detergent trapping of cleavage complexes (Vosberg 1985) but has recently been investigated using suicidal DNA substrates; i.e., partially single-stranded DNA or oligonucleotide duplexes (Been and Champoux 1981; Svejstrup et al. 1991; Christiansen et al. 1993). Mechanistically, topoisomerase I acts by directing a nucleophilic attack from the hydroxyl group of an internal tyrosine into a phosphodiester bond in the DNA backbone, resulting in nicking of one strand of the DNA double helix and formation of a covalent linkage between the 3′-phosphate and the tyrosine hydroxyl group.

The mechanism of strand passage has not been described in detail, but two different models have been proposed: (1) the "strand passage model," where the intact strand passes through the nicked strand, requiring local DNA denaturation, and (2) the "swiveling model," where the free 5′-OH end is able to rotate around the intact strand, either freely or in a stepwise fashion. Following topoisomerization, the nick is religated by a nucleophilic attack of the 5′-OH end on the O^4-phosphotyrosine bond, releasing the enzyme (for reviews, see Maxwell and Gellert 1986; Osheroff 1989b). When oligonucleotide DNA duplexes are used as substrates for topoisomerase I, the generated cleavage complex contains an active enzyme, which is able to ligate the cleaved DNA to a variety of DNA fragments containing a free 5′-OH end. Using this system, it has been found that the enzyme preferentially ligates the cleaved DNA to single-stranded DNA having the ability to base-pair, although religation to substrates with incomplete base-pairing or to DNA with blunt ends has also been described (Christiansen et al. 1993; Christiansen and Westergaard 1993, 1994).

TYPE II TOPOISOMERASES

Physical Characteristics

All type II topoisomerases studied so far are structurally related and show similar physical properties. The gene for topoisomerase II has been cloned from a variety of eukaryotic organisms, including yeasts (Giaever et al. 1986; Uemura et al. 1986), protozoans (Strauss and Wang 1990; Fragoso and Goldenberg 1992; Pasion et al. 1992), *Drosophila* (Wyckoff et al. 1989), mouse (Adachi et al. 1992), rat (Park et al. 1993), and human (Tsai-Pflugfelder et al. 1988; Jenkins et al. 1992; Austin et al. 1993). The enzyme is encoded by a single-copy gene, but it exists in solution as a homodimer (Vosberg 1985) with molecular masses ranging from about 300 kD to 360 kD. Only one topoisomerase II form has been found in lower eukaryotes, whereas two forms, the α form and the β form, exist in humans and probably in all higher eukaryotes (Drake et al. 1989a). The two isoforms have different DNA-binding/cleavage site preferences and different DNA dissociation rates and also show differences in their susceptibility to drugs (Drake et al. 1989b).

Sequence homology studies have shown that type II topoisomerases from various eukaryotic sources are highly conserved, with the amino acid sequence of the enzymes sharing an overall homology of about 50%. Homology is extensive in the amino-terminal approximately 1200 amino acids, but the carboxy-terminal 200–300 amino acids show no conservation (Wyckoff et al. 1989). The conserved state of the enzymes is further seen from the ability of several eukaryotic topoisomerase II genes to complement a yeast *top2*^{ts} mutant or a *top2* null mutant (Uemura et al. 1987b; Wyckoff and Hsieh 1988; Adachi et al. 1992).

Eukaryotic topoisomerase II shows significant homology with the bacterial type II topoisomerase enzyme, DNA gyrase. The latter is tetrameric (A_2B_2), consisting of two distinct subunits, GyrA and GyrB, where GyrA contains the active site for DNA-cleavage/religation and GyrB holds the ATPase activity (for review, see Reece and Maxwell 1991; Orphanides and Maxwell 1994). Based on the homology of the amino-terminal approximately 650 amino acids in the eukaryotic enzyme to GyrB, this region has been inferred to be responsible for ATP hydrolysis, in agreement with the observation that point mutations in this region abolish ATP binding (Uemura et al. 1987b; Lindsley and Wang 1991). Likewise, the amino-terminal part of GyrA holding the active site is homologous to the region from amino acids approximately 650 to 900 in the eukaryotic enzyme, correlating well with the identification of the active-site tyrosine to Y783 in *S. cerevisiae* (Worland and Wang 1989). The final carboxy-terminal region from amino acid 1200 has no counterpart in DNA gyrase. The region is rich in highly charged amino acids, suggesting an involvement in DNA binding (Uemura et al. 1986). Nuclear localization signals have been found in the nonconserved region, although the outermost 200–250 amino acids are dispensable for cell viability (Shiozaki and Yanagida 1991, 1992; Crenshaw and Hsieh 1993; Caron et al. 1994).

The promoter region of the gene for human topoisomerase IIα shares several features of traditional housekeeping gene promoters, but it contains in addition a regulatory region with a structure similar to that found in proliferation-specific genes (Hochhauser et al. 1992). This is in agreement with observations showing that the amount of topoisomerase IIα varies with respect to the proliferative state of the cell and the position in the cell cycle (Heck and Earnshaw 1986; Fairman and Brutlag 1988; Swedlow et al. 1993). Actively proliferating cells contain about 10^5 copies per cell, whereas the enzyme is hardly detectable in quiescent or terminally differentiated cells (Heck and Earnshaw 1986; Fairman and Brutlag 1988; Hsiang et al. 1988). The nuclear level of enzyme raises steadily from the end of mitosis until prophase and drops rapidly during metaphase and anaphase (Swedlow et al. 1993), suggestive of a mitotic involvement of the enzyme. In contrast to the α form, immunohistochemical studies of topoisomerase IIβ expression have shown that this enzyme is mainly expressed in G_0 cells, and the amount of the enzyme is more or less constant once the cells have entered the cell cycle (Drake et al. 1989a; Woessner et al. 1991; Prosperi et al. 1992).

Topoisomerase II has been shown to be posttranslationally modified by phosphorylation (Cardenas and Gasser 1993), with the major phosphorylation sites located in the extreme carboxy-terminal domain. Phosphorylation modulates the overall catalytic activity of topoisomerase II by stimulating the rate-limiting ATP hydrolysis step and, thus, the catalytic turnover rate of the enzyme (Corbett et al. 1992, 1993). Several lines of evidence have suggested casein kinase II as the kinase responsible for phosphorylation in vivo (Ackerman et al. 1988; Heck et al. 1989; Cardenas et al. 1992; Saijo et al. 1992), but protein kinase C and $p34^{cdc2}$ kinase may play a role as well (Rottmann et al. 1987; Cardenas et al. 1992; DeVore et al. 1992). Cell cycle variations exist in the level of topoisomerase II phosphorylation (Heck et al. 1988, 1989; Cardenas et al. 1992; Saijo et al. 1992), with the enzyme being hyperphosphorylated during mitosis. Furthermore, the phosphorylation patterns vary throughout the cell cycle for both topoisomerase II isoforms (Burden and Sullivan 1994; Kimura et al. 1994).

Enzymology of Topoisomerase II

Eukaryotic topoisomerase II introduces transient double-stranded breaks into the DNA backbone and changes the linking number of DNA in steps of two, via a double-stranded energy-requiring DNA passage mechanism (Vosberg 1985; Maxwell and Gellert 1986; Osheroff 1989b). The enzyme catalyzes relaxation of negatively as well as positively supercoiled DNA (Osheroff et al. 1983; Schomburg and Grosse 1986) and is able to interconvert different topological forms of double-stranded DNA by catenation/decatenation (Hsieh and Brutlag 1980; Goto and Wang 1982) and knotting/unknotting (see Table 1 and Fig. 1) (Liu et al. 1980; Hsieh 1983).

Topoisomerase II requires the presence of divalent cations (Goto and Wang 1982; Goto et al. 1984; Osheroff 1987; Andersen et al. 1989; Lee et al. 1989) and ATP as an energy source (Shelton et al. 1983; Goto et al. 1984; Osheroff 1986) for all catalytic reactions. The enzyme relaxes supercoiled DNA in a highly processive manner, as the rate of relaxation is considerably faster than the rate of dissociation of the enzyme/DNA complex (Osheroff et al. 1983; Schomburg and Grosse 1986).

A number of mechanistic studies on topoisomerase II have broken the reaction into at least six discrete steps, as shown in Figure 3 (for review, see Osheroff et al. 1991; Orphanides and Maxwell 1994). A current model (Roca and Wang 1992, 1994) suggests that the enzyme acts as a molecular clamp, which is open in the absence of ATP and closed in the presence of ATP or an ATP analog. In the absence of ATP, the enzyme can bind noncovalently to a linear or circular DNA, the gate segment or G-segment (step 1). Another DNA segment, the transported segment or T-segment, can now pass in and out of the clamp as long as the clamp stays in the open form (step 2). Still in the absence of ATP, topoisomerase II can mediate a double-stranded cleavage of the G-segment, which can be religated without any change in the overall DNA topology (step 3). The cleavage/religation process is an equilibrium, which normally is shifted toward religation. However, when ATP is present and binds to the DNA-bound enzyme, a conformational change occurs, whereby the protein clamp closes. If a T-segment has entered the clamp before closure, it will be trapped in the clamp and transported through the transiently cleaved G-segment (step 4). How the final exit of the T-segment from the complex takes place has been a subject of much debate. A two-gate operation of the enzyme has been proposed to explain how the T-segment leaves the enzyme/DNA complex through a second opening in the enzyme (Roca and Wang 1994). According to this, the second protein gate recloses upon the exit of the T-segment (step 5), and after hydrolysis of ATP the clamp returns to its initial conformation (step 6). The enzyme is then able to enter a new catalytic cycle or dissociate from the DNA.

The steps 1 (noncovalent DNA binding) and 3 (cleavage/religation) have been studied in detail. Noncovalent DNA binding can be analyzed in the absence of divalent cations (Osheroff 1987; Sander et al. 1987). The enzyme binds preferentially to negatively supercoiled DNA (Osher-

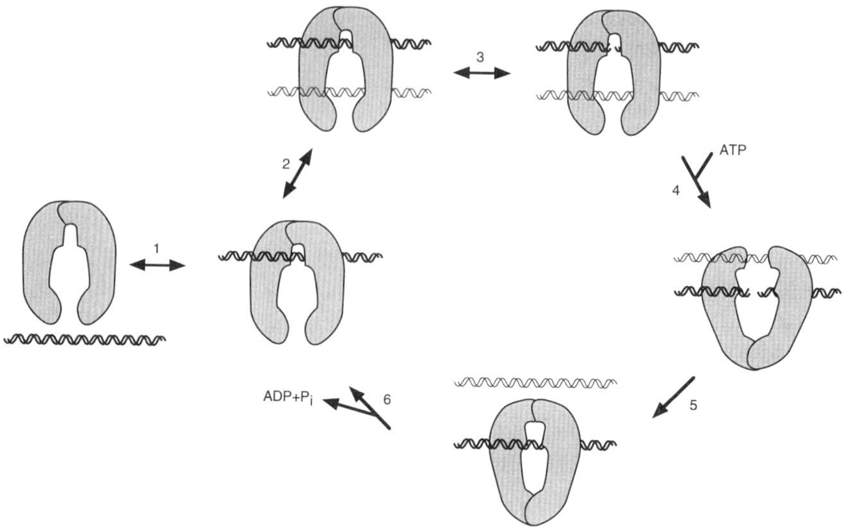

Figure 3 Steps in the catalytic cycle of the homodimeric, eukaryotic topoisomerase II enzyme. DNA topoisomerase II binds noncovalently to the G-segment (step 1) and another DNA segment, the T-segment, can pass in and out of the molecular clamp formed by the enzyme (step 2) (Roca and Wang 1992). The enzyme introduces a transient double-stranded cleavage in the G-segment (step 3). In the presence of ATP, the enzyme undergoes a conformational change allowing the T-segment to pass through the break in the G-segment and leave the complex via a second opening in the enzyme (step 4). After reclosure of this opening (step 5), ATP is hydrolyzed, and the enzyme returns to its initial conformation (step 6). The dark and light helices represent the G- and T-segments, respectively.

off and Brutlag 1983; Osheroff 1986), and footprinting has shown that the enzyme protects about 25 bp (Lee et al. 1989; Thomsen et al. 1990).

Subsequent to DNA binding, topoisomerase II will cleave the DNA and enter a cleavage/religation equilibrium if divalent cations are present. The homodimeric enzyme makes a 4-bp staggered cleavage of the DNA duplex (for reviews, see Vosberg 1985; Maxwell and Gellert 1986) and becomes linked to the protruding 5′ ends through O^4-phosphotyrosine bonds (Liu et al. 1983; Sander and Hsieh 1983). The cleavage complex has been studied in detail by treatment of the equilibrium with a strong denaturing agent, which traps the covalently linked enzyme/DNA complex (for reviews, see Vosberg 1985; Maxwell and Gellert 1986; Osheroff 1989b). Identical cleavage complexes can be generated in an active form, when suicidal DNA substrates are used for topoisomerase II, i.e., single-stranded DNA or duplex DNA 5′recessed on one strand (Gale and Osheroff 1990; Andersen et al. 1991). Investigation of cleavage complex formation has shown that eukaryotic topoisomerase II acts at preferred sites and shows some nucleotide sequence specificity in its cleavage reaction (Spitzner and Muller 1988; Andersen et al. 1989; Spitzner et al. 1990), although only rather weak consensus sequences have been obtained (Sander and Hsieh 1985; Spitzner and Muller 1988). The minimum duplex region required for topoisomerase-II-mediated cleavage has been established to be 16 bp located symmetrically around the 4-bp stagger (Lund et al. 1990). DNA cleavage absolutely requires the presence of divalent cations (Sander and Hsieh 1983; Osheroff 1987; Andersen et al. 1991), usually magnesium, but calcium can be substituted (Osheroff and Zechiedrich 1987; Andersen et al. 1989; Zechiedrich et al. 1989). Furthermore, the reaction has no requirement for ATP, although the energy cofactor stimulates cleavage two- to threefold (Sander and Hsieh 1983; Osheroff 1986).

When suicidal DNA substrates are used, the cleavage complex contains a kinetically competent topoisomerase II enzyme able to perform ligation of the cleaved DNA to a free 3′-hydroxyl end on another DNA strand (Gale and Osheroff 1990, 1992; Andersen et al. 1991). Studies of the religation half-reaction per se have shown that the efficiency of the ligation reaction is enhanced if the ligating DNA is able to hybridize to the cleaved DNA (Andersen et al. 1991; Schmidt et al. 1994). Dinucleotides capable of base-pairing are ligated, but the ligation reaction displays a strong preference for double-stranded DNA with a four-base, single-stranded region (Andersen et al. 1991; Schmidt et al. 1994). Like the cleavage reaction, the religation half-reaction absolutely requires divalent cations, but it takes place in the absence of ATP (Gale and Osheroff 1990; Andersen et al. 1991).

INHIBITORS OF DNA TOPOISOMERASES

DNA topoisomerases have been identified as the cellular target for a number of clinically relevant antineoplastic drugs (Table 2) (for reviews, see Liu 1989; Osheroff et al. 1991; Pommier 1993; Wang 1994). The drugs interfere with the catalytic cycle of topoisomerase I and II and can be divided into two groups on the basis of their interaction mode. Drugs belonging to the first group (group I) influence the cleavage/religation equilibrium (step 3 in Fig. 3), stabilizing the covalently linked topoisomerase/DNA intermediate by stimulation of the forward cleavage

Table 2 Inhibitors of DNA topoisomerases

Drug classes	Examples	Drug effect	References
		Group I	
Topoisomerase I inhibitors			
camptothecins	camptothecin topotecan CPT-11	complex formation; inhibits catalytic activity	Liu (1989); Hsiang et al. (1985)
Topoisomerase II inhibitors			
epipodophyllotoxins	VM-26 VP-16	complex formation; inhibits religation and catalytic activity	Chen et al. (1984); Osheroff (1989a)
isoflavonoids	genistein	complex formation; inhibits strand passage and ATP hydrolysis	Marcovits et al. (1989)
acridines	m-AMSA	complex formation; inhibits religation, strand passage, and ATP hydrolysis	Robinson and Osheroff (1990); Sørensen et al. (1992); Robinson et al. (1991, 1993)
anthracenediones	mitoxantrone	complex formation	Fox and Smith (1990)
anthracyclines	adriamycin (doxorubicin)	complex formation	Capranico et al. (1990)

ellipticines	9-hydroxy-iso-ellipticine	complex formation	Tewey et al. (1984)
2-nitroimidazoles	Ro 15-0216	complex formation; no inhibition of catalytic activity	Sørensen et al. (1990, 1992)
quinolones	CP-115,953	complex formation; stimulates cleavage; inhibits strand passage and ATP hydrolysis	Robinson et al. (1991, 1993)
		Group II	
anthracenyl peptides	merbarone	inhibits catalytic activity	Drake et al. (1989b)
anthracyclines	aclarubicin	antagonizes complex formation by other drugs; inhibits catalytic activity	Jensen et al. (1990)
bisdioxopiperazines	ICRF-193	antagonizes complex formation by other drugs; inhibits catalytic activity	Tanabe et al. (1991); Roca et al. (1994)
hexasulfated naphthylurea	suramin	antagonizes complex formation by other drugs; inhibits catalytic activity	Bojanowski et al. (1992)

reaction and/or inhibition of the religation half-reaction (Osheroff 1989a; Robinson and Osheroff 1990; Sørensen et al. 1992). Camptothecins, which for the time being are the most potent topoisomerase I inhibitors, belong to this group (Kunimoto et al. 1987; Johnson et al. 1989). Group I also includes the most potent topoisomerase II drugs, which can broadly be categorized into DNA intercalators including amsacrine (mAMSA), ellipticine, and mitoxantrone; and non-intercalators like etoposide (VP16), teniposide (VM26), genistein, and the quinolone, CP115,953 (Liu 1989; Osheroff et al. 1991).

The second group (group II) includes agents that interfere with the catalytic cycle of topoisomerases without trapping the covalent topoisomerase/DNA complex. Members of this group are merbarone (Drake et al. 1989b), aclarubicin (Jensen et al. 1990), ICRF-193 (Tanabe et al. 1991; Roca et al. 1994), and suramin (Bojanowski et al. 1992), all of which target topoisomerase II.

Drugs belonging to both groups have been widely used as tools to probe the reaction mechanisms of DNA topoisomerases. Group I drugs have been used in the mapping of topoisomerase/DNA interaction sites in vitro and in vivo. Drug-trapping of the enzyme covalently linked to cleaved DNA has allowed studies of the DNA sequence specificities of the enzymes, although sites cleaved by topoisomerases are stimulated differently by drug treatment (Kjeldsen et al. 1988).

Studies have shown that topoisomerase inhibitors exert different base preferences, indicating drug interaction with the preferred base (Pommier et al. 1993). This has been further demonstrated by Kreuzer and coworkers, who have shown cross-linking of a photoreactive mAMSA derivative to one of the residues in the base pair 5′to the cleaved phosphodiester bond (Freudenreich and Kreuzer 1994).

Information on the mechanistic activity of group II drugs is scarce. However, ICRF-193 has been found to inhibit topoisomerase II via a novel mechanism (Roca et al. 1994). In the presence of ATP, the drug locks the enzyme in a salt-stable conformation, analogous to the ATP-bound closed-clamp form of topoisomerase II. After formation of the closed form, ATP can be removed without opening of the clamp, suggesting that ICRF-193 acts by inhibiting the interconversion between the open and closed form of the enzyme (steps 5 and 6 in Fig. 3).

The final result of cell treatment with most topoisomerase-targeting drugs is cell killing, but little is known about the underlying causes of this event (for review, see Zhang et al. 1990). A major determinant of group I drug cytotoxicity is suggested to be an interconversion of the drug-stimulated, enzyme-mediated cleavages into irreversible double-stranded DNA breaks. This model is supported by the fact that mutations in *RAD52*, which is involved in the repair of double-stranded breaks, increase the cytotoxicity of these drugs considerably (Nitiss and Wang 1988). Some of the likely candidates responsible for the conversion are the replicational and transcriptional machineries through their collision with the drug-induced cleavage complexes (Snapka 1986; Snapka et al. 1988; Hsiang et al. 1989; Bendixen et al. 1990; Thomsen et al. 1990). Cell killing by group II drugs targeting topoisomerase II has been suggested to result from chromosomal breakage and aneuploidy due to progression through mitosis in the absence of a functional topoisomerase II enzyme (Holm et al. 1989; Downes et al. 1991; Wang 1994) (see below).

BIOLOGICAL FUNCTIONS OF DNA TOPOISOMERASES

Topoisomerases as Modulators of Transcription

In 1987, Liu and Wang proposed that the tracking of an RNA polymerase along a DNA helix would generate a positively supercoiled domain in front of it and a negatively supercoiled domain behind it. Experiments supporting this model have been reported in both prokaryotes (Wu et al. 1988) and eukaryotes (Giaever and Wang 1988). The so-called "twin supercoil domain" model clearly demonstrates a need for topoisomerases in transcription.

Topoisomerase I has been shown to be preferentially associated with genes actively transcribed by RNA polymerase II (Gilmour and Elgin 1987; Stewart and Schutz 1987; Kroeger and Rowe 1989, 1992; Stewart et al. 1990), and RNA elongation has been found to be inhibited by antibodies directed against topoisomerase I (Egyhazi and Durban 1987). Furthermore, a kinetic study of RNA polymerase II and topoisomerase I activities on the c-*fos* proto-oncogene after activation has revealed a close linkage of the two activities (Stewart et al. 1990).

Besides a role of topoisomerase I in transcriptional elongation, a number of recent reports argue that topoisomerase I plays a direct role in the regulation of transcriptional initiation by acting as a repressor of the basal-level transcription (Choder 1991; Merino et al. 1993). A hint toward a specific mechanism for transcriptional repression was given by the finding that topoisomerase I might interact directly with the TATA-binding protein (TBP), thereby competing with the binding of TFIIA and other TBP-binding factors. The repression obtained by this interaction seems independent of topoisomerase I enzymatic activity, because active-site mutants behave similarly (Merino et al. 1993).

An involvement of topoisomerase II in RNA polymerase II transcription has been inferred by the preferential presence of topoisomerase II cleavage sites in DNase-hypersensitive regions 5′ or 3′ to actively transcribed genes (Reitman and Felsenfeld 1990; Kroeger and Rowe 1992) and by the localization of topoisomerase-II-mediated cleavage sites in transcriptional enhancer regions (Cockerill and Garrard 1986; Gasser and Laemmli 1986).

The transcription of rDNA by RNA polymerase I constitutes around 80% of the total transcription in *S. cerevisiae,* and accumulation of 18S and 25S rRNA is critically dependent on the presence of either topoisomerase I or topoisomerase II (Brill et al. 1987). An abundance of topoisomerase I in nucleoli has been documented by immunofluorescence and by SDS cleavage studies (Fleischmann et al. 1984; Muller et al. 1985; Zhang et al. 1988). Furthermore, high-affinity binding sites for topoisomerase I have been reported in the upstream spacer region in the rDNA from *Tetrahymena* and *Dictyostelium* (Gocke et al. 1983; Andersen et al. 1985; Bonven et al. 1985; Ness et al. 1986; Bettler et al. 1988).

In contrast to RNA polymerases I and II, RNA polymerase III is apparently unaffected by topoisomerase activity in vivo, as shown by Brill et al. (1987), who reported that the amount of tRNA produced in an *S. cerevisiae top1-top2*^{ts} double mutant was indistinguishable from the amount in wild-type cells.

DNA Topoisomerases in Replication, Mitosis, and Meiosis

Ongoing replication requires the removal of positive supercoils generated ahead of the replication fork (for review, see Nitiss 1994), and on the basis of studies on the replication of SV40 DNA (Sundin and Varshavsky 1981; Yang et al. 1987; Snapka et al. 1988; Porter and Champoux 1989), it has been proposed that topoisomerase I provides the major swivel activity during replicational elongation. The enzyme is, however, lost from the DNA before replication has finished, due to steric hindrance from the two progressing replication forks. In the absence of swivel activity, continued replication fork movement results in daughter molecule intertwinings, which subsequently are removed by topoisomerase-II-mediated decatenation (Nelson et al. 1986). The model has been supported by data obtained with different yeast mutants. When the replicational elongation process is studied in yeast, *top1* mutants, but not *top2* mutants, show a distinct delay in replicational elongation (Kim and Wang 1989a). Furthermore, topoisomerase II appears essential for cell survival (Goto and Wang 1984), where the onset of inviability coincides with the time of mitosis (Holm et al. 1985). Inviability can, however, be prevented if formation of the mitotic spindle is inhibited by nocodazole, suggesting an involvement of topoisomerase II during chromosome segregation. *Top1* mutants show a dispensability of topoisomerase I activity for yeast survival (Goto and Wang 1985; Thrash et al. 1985), whereas *top1-top2*^{ts} double mutants show phenotypes strikingly different from those of the single mutants in that they rapidly stop growth at the nonpermissive temperature, irrespective of the cell cycle stage (Uemura and Yanagida 1984; Goto and Wang 1985; Holm et al. 1985; Brill et al. 1987).

Studies with the yeast mutants have shown that topoisomerase-II-mediated decatenation of interwound sister chromatids takes place at mitosis, rather than at the end of replication. This has been proposed to result from a requirement of a directional force for topoisomerase II. During mitosis, a force is given by the pulling of the mitotic spindle, which can direct the untangling activity of the enzyme (Holm et al. 1989; Holm 1994). There seems to be no regulatory checkpoint for the untangling process in yeast. Thus, in the absence of topoisomerase II activity, chromosome segregation still proceeds, resulting in chromosome breakage, nondisjunction, and finally, cell death (Uemura and Yanagida 1986; Uemura et al. 1987b; Holm et al. 1989; Spell and Holm 1994). In mammalian cells, however, strong evidence for a topoisomerase-II-dependent checkpoint in G_2 has recently been described (Downes et al. 1994).

During meiosis I, topoisomerase II is required at the time of chromosome segregation for the resolution of recombined chromosomes (Rose et al. 1990). *Top2*^{ts} yeast mutants induced to sporulate at the nonpermissive temperature complete early meiotic events, including premeiotic DNA synthesis and commitment to meiotic levels of recombination, but arrest in meiosis I prior to spindle formation, before suffering significant DNA damage. Meiotic cells thus have a checkpoint mechanism, in contrast to mitotic cells, preventing an otherwise lethal chromosome segregation (Rose and Holm 1993). In meiosis II, topoisomerase II plays a role similar to that in mitosis (Rose et al. 1990).

Topoisomerases in Recombination

Mutant strains of *S. cerevisiae* lacking topoisomerase I, II, or III display a dramatic increase in homologous recombination. Christman et al. (1988) reported that strongly increased levels of mitotic recombination occur in *top1* and *top2*^{ts} single mutants. The effect was only observed when measuring the rate of loss of a marker embedded in the rDNA array. No measurable increase or decrease in recombination was detected when direct repeat recombination was studied elsewhere in the genome (Christman et al. 1988). A strong effect upon recombination has also been observed in yeast cells deficient for topoisomerase III, where a more than 70-fold increase in the frequency of rDNA marker loss is seen (Gangloff et al. 1994). In contrast to the observation with *top1* and *top2* mutants, *top3* mutants have also been found to be hyperrecombinogenic for mitotic recombination events occurring outside the rDNA locus (Wallis et al. 1989; Bailis et al. 1992). It appears that it is the combined action of topoisomerases I, II, and III that stabilizes the genomic rDNA array, performing different, but overlapping, functions. The exact cause for the elevated recombination rates has not been defined, but it has been suggested that highly supercoiled DNA structures in the topoisomerase mutants are processed into recombinogenic lesions by the excision repair machinery (Kim and Wang 1989b; Bailis et al. 1992). A recent finding by Fink and coworkers supports the notion that aberrant DNA structures do arise due to the lack of DNA topoisomerase activity, as they have demonstrated an rDNA-linked chromosome abnormality in *S. cerevisiae top1* mutants, but not in *top2* mutants (Christman et al. 1993). Whether these structures are a result of recombination or aberrant replication is still unclear. Although topoisomerases act as suppressors of recombination, none of the topoisomerase mutants has so far been shown to have deficiencies in the mechanism of homologous recombination.

Both topoisomerases I and II are able to rearrange DNA fragments in vitro when suicidal DNA substrates are used for cleavage (Been and Champoux 1981; Champoux et al. 1984; Bae et al. 1988; Gale and Osheroff 1990, 1992; Andersen et al. 1991; Christiansen et al. 1993). The covalent complexes generated with these substrates are energetically active and religate intra- and intermolecularly to free hydroxyl ends. Topoisomerases are thus possible candidates as mediators of illegitimate recombination under the assumption that similar aberrant DNA structures might be present in vivo and physically available to the action of topoisomerases. Studies of the junction sequences resulting from illegitimate recombination events have indicated a high occurrence of sites matching the consensus sequence of topoisomerase I, at or near the junctions, both in mammalian cells (Bullock et al. 1985; Konopka 1988; Wang and Rogler 1991) and in yeast (Schiestl et al. 1993). Using a similar approach, it has also been suggested that topoisomerase II promotes illegitimate recombination in vivo (Sperry et al. 1989; Han et al. 1993) and in vitro (Schmidt et al. 1994).

CONCLUDING REMARKS

DNA topoisomerases are ubiquitous enzymes that change the topological state of DNA without altering the overall outline of the DNA double helix (Vosberg 1985; Maxwell and Gellert 1986; Watt and Hickson 1994). In the past decade, the biological importance of these enzymes has been manifested through genetic and biochemical studies that have described roles of the enzymes in transcription, replication, recombination, and chromosome dynamics. Recently, new forms of both type I and type II topoisomerases have been discovered, and it is hoped that the future will elucidate the distinct roles of the different eukaryotic topoisomerase enzymes.

An interesting area in the topoisomerase field is the interaction of topoisomerases with other proteins. The interaction between topoisomerase II and the scaffold protein ScII (Ma et al. 1993), and the newly discovered interaction between topoisomerase III and a DNA helicase (Gangloff et al. 1994), are examples of such interactions. Interacting proteins might be instrumental in unraveling the biological functions of eukaryotic DNA topoisomerases.

Note Added in Proof

The crystal structures of fragments of topoisomerase I and II have now been determined (Lue et al. 1995; Berger et al. 1996), and together with the biochemical and genetic data, they will contribute significantly to an understanding of the detailed mechanism of action of these enzymes.

ACKNOWLEDGMENTS

We thank Dr. Kent Christiansen for his critical reading of the manuscript and Lumir Krejcí for assistance with the drawings. The work is supported by the Danish Cancer Society (grants 93-004 and 78-5000), the Danish Centre for Human Genome Research, and the Danish Centre for Respiratory Physiological Adaptation.

REFERENCES

Ackerman, P., C.V. Glover, and N. Osheroff. 1988. Phosphorylation of DNA topoisomerase II *in vivo* and in total homogenates of *Drosophila* Kc cells. The role of casein kinase II. *J. Biol. Chem.* **263:** 12653–12660.

Adachi, N., M. Miyaike, H. Ikeda, and A. Kikuchi. 1992. Characterization of cDNA encoding the mouse DNA topoisomerase II that can complement the budding yeast *top2* mutation. *Nucleic Acids Res.* **20:** 5297–5303.

Alsner, J., J.Q. Svejstrup, E. Kjeldsen, B.S. Sørensen, and O. Westergaard. 1992. Identification of an N-terminal domain of eukaryotic DNA topoisomerase I dispensable for catalytic activity but essential for *in vivo* function. *J. Biol. Chem.* **267:** 12408–12411.

Andersen, A.H., E. Gocke, B.J. Bonven, O.F. Nielsen, and O. Westergaard. 1985. Topoisomerase I has a strong binding preference for a conserved hexadecameric sequence in the promoter region of the rRNA gene from *Tetrahymena pyriformis. Nucleic Acids Res.* **13:** 1543–1557.

Andersen, A.H., K. Christiansen, E.L. Zechiedrich, P.S. Jensen, N. Osheroff, and O. Westergaard. 1989. Strand specificity of the topoisomerase II mediated double-stranded DNA cleavage reaction. *Biochemistry* **28:** 6237–6244.

Andersen, A.H., B.S. Sørensen, K. Christiansen, J.Q. Svejstrup, K. Lund, and O. Westergaard. 1991. Studies of the topoisomerase II-mediated cleavage and religation reactions by use of a suicidal double-stranded DNA substrate. *J. Biol. Chem.* **266:** 9203–9210.

Austin, C.A., J.H. Sng, S. Patel, and L.M. Fisher. 1993. Novel HeLa topoisomerase II is the II beta isoform: Complete coding sequence and homology with other type II topoisomerases. *Biochim. Biophys. Acta* **1172:** 283–291.

Bae, Y.S., I. Kawasaki, H. Ikeda, and L.F. Liu. 1988. Illegitimate recombination mediated by calf thymus DNA topoisomerase II *in vitro. Proc. Natl. Acad. Sci.* **85:** 2076–2080.

Bailis, A.M., L. Arthur, and R. Rothstein. 1992. Genome rearrangement in *top3* mutants of *Saccharomyces cerevisiae* requires a functional *RAD1* excision repair gene. *Mol. Cell. Biol.* **12:** 4988–4993.

Been, M.D. and J.J. Champoux. 1981. DNA breakage and closure by rat liver type 1 topoisomerase: Separation of the half-reactions by using a single-stranded DNA substrate. *Proc. Natl. Acad. Sci.* **78:** 2883–2887.

Been, M.D., R.R. Burgess, and J.J. Champoux. 1984. Nucleotide sequence preference at rat liver and wheat germ type I DNA topoisomerase breakage sites in duplex SV40 DNA. *Nucleic Acids Res.* **12:** 3097–3114.

Bendixen, C., B. Thomsen, J. Alsner, and O. Westergaard. 1990. Camptothecin-stabilized topoisomerase I-DNA adducts cause premature termination of transcription. *Biochemistry* **29:** 5613–5619.

Berger, J.M., S.J. Gamblin, S.C. Harrison, and J.C. Wang. 1996. Structure and mechanism of DNA topoisomerase II. *Nature* **379:** 225–232.

Bettler, B., P.J. Ness, S. Schmidlin, and R.W. Parish. 1988. The upstream limit of nuclease-sensitive chromatin in *Dictyostelium* rRNA genes neighbors a topoisomerase I-like cluster. *J. Mol. Biol.* **204:** 549–559.

Bojanowski, K., S. Lelievre, J. Markovits, J. Couprie, A. Jacquemin-Sablon, and A.K. Larsen. 1992. Suramin is an inhibitor of DNA topoisomerase II *in vitro* and in Chinese hamster fibrosarcoma cells. *Proc. Natl. Acad. Sci.* **89:** 3025–3029.

Bonven, B.J., E. Gocke, and O. Westergaard. 1985. A high affinity topoisomerase I binding sequence is clustered at DNase I hypersensitive sites in *Tetrahymena* R-chromatin. *Cell* **41:** 541–551.

Brill, S.J. and R. Sternglanz. 1988. Transcription-dependent DNA supercoiling in yeast DNA topoisomerase mutants. *Cell* **54:** 403–411.

Brill, S.J., S. DiNardo, K. Voelkel-Meiman, and R. Sternglanz. 1987. Need for DNA topoisomerase activity as a swivel for DNA replication and for transcription of ribosomal RNA. *Nature* **326:** 414–416.

Bullock, P., J.J. Champoux, and M. Botchan. 1985. Association of crossover points with topoisomerase I cleavage sites: A model for nonhomologous recombination. *Science* **230:** 954–958.

Burden, D.A. and D.M. Sullivan. 1994. Phosphorylation of the α- and β-isoforms of the DNA topoisomerase II is quantitatively different in interphase and mitosis in Chinese hamster ovary cells. *Biochemistry* **33:** 14651–14655.

Busk, H., B. Thomsen, B.J. Bonven, E. Kjeldsen, O.F. Nielsen, and O. Westergaard. 1987. Preferential relaxation of supercoiled DNA containing a hexadecameric recognition sequence for topoisomerase I. *Nature* **327:** 638–640.

Camilloni, G., E. Di-Martino, M. Caserta, and E. Di Mauro. 1988. Eukaryotic DNA topoisomerase I reaction is topology dependent. *Nucleic Acids Res.* **16:** 7071–7085.

Camilloni, G., E. Di-Martino, E. Di Mauro, and M. Caserta. 1989. Regulation of the function of eukaryotic DNA topoisomerase I: Topological conditions for inactivity. *Proc. Natl. Acad. Sci.* **86:** 3080–3084.

Capranico, G., F. Zunino, K. Kohn, and Y. Pommier. 1990. Sequence-selective topoisomerase II inhibition by anthracycline derivatives in SV40 DNA: Relationship with DNA binding affinity and cytotoxicity. *Biochemistry* **29:** 562–569.

Cardellini, E. and E. Durban. 1993. Phosphorylation of human topoisomerase I by protein kinase C *in vitro* and in phorbol 12-myristate 13-acetate-activated HL-60 promyelocytic leukaemia cells. *Biochem. J.* **291:** 303–307.

Cardenas, M.E. and S.M. Gasser. 1993. Regulation of topoisomerase II by phosphorylation: A role for casein kinase II. *J. Cell Sci.* **104:** 219–225.

Cardenas, M.E., Q. Dang, C.V. Glover, and S.M. Gasser. 1992. Casein kinase II phosphorylates the eukaryote-specific C-terminal domain of topoisomerase II in vivo. *EMBO J.* **11:** 1785–1796.

Caron, P.R., P. Watt, and J.C. Wang. 1994. The C-terminal domain of *Saccharomyces cerevisiae* DNA topoisomerase II. *Mol. Cell. Biol.* **14:** 3197–3207.

Caserta, M., A. Amadei, G. Camilloni, and E. Di Mauro. 1990. Regulation of the function of eukaryotic DNA topoisomerase I: Analysis of the binding step and of the catalytic constants of topoisomerization as a function of DNA topology. *Biochemistry* **29:** 8152–8157.

Caserta, M., A. Amadei, E. Di Mauro, and G. Camilloni. 1989. *In vitro* preferential topoisomerization of bent DNA. *Nucleic Acids Res.* **17:** 8463–8474.

Champoux, J.J. and R. Dulbecco. 1972. An activity from mammalian cells that untwists superhelical DNA—A possible swivel for DNA replication. *Proc. Natl. Acad. Sci.* **69:** 143–146.

Champoux, J.J. and B.L. McConaughy. 1976. Purification and characterization of the DNA untwisting enzyme from rat liver. *Biochemistry* **15:** 4638–4642.

Champoux, J.J., W.K. McCoubrey, and M.D. Been. 1984. DNA structural features that lead to strand breakage by eukaryotic type I topoisomerase. *Cold Spring Harbor Symp. Quant. Biol.* **49:** 435–442.

Chen, G.L., L. Yang, T.C. Rowe, B.D. Halligan, K.M. Tewey, and L.F. Liu. 1984. Nonintercalative antitumor drugs interfere with the breakage-reunion reaction of mammalian DNA topoisomerase II. *J. Biol. Chem.* **259:** 13560–13566.

Choder, M. 1991. A general topoisomerase I-dependent transcriptional repression in the stationary phase in yeast. *Genes Dev.* **5:** 2315–2326.

Christiansen, K. and O. Westergaard. 1993. Involvement of eukaryotic topoisomerase I in illegitimate recombination: Generation of deletions and insertions in DNA repair mechanisms. *Proc. Alfred Benzon Symp.* **35:** 361–371.

———. 1994. Characterization of intra- and inter molecular DNA ligation mediated by eukaryotic topoisomerase I. *J. Biol. Chem.* **269:** 721–729.

Christiansen, K., A.B. Svejstrup, A.H. Andersen, and O. Westergaard. 1993. Eukaryotic topoisomerase I-mediated cleavage requires bipartite DNA interaction. Cleavage of DNA substrates containing strand interruptions implicates a role for topoisomerase I in illegitimate recombination. *J. Biol. Chem.* **268:** 9690–9701.

Christman, M.F., F.S. Dietrich, and G.R. Fink. 1988. Mitotic recombination in the rDNA of *S. cerevisiae* is suppressed by the combined action of DNA topoisomerases I and II. *Cell* **55:** 413–425.

Christman, M.F., F.S. Dietrich, N.A. Levin, B.U. Sadoff, and G.R. Fink. 1993. The rDNA-encoding DNA array has an altered structure in topoisomerase mutants of *Saccharomyces cerevisiae. Proc. Natl. Acad. Sci.* **90:** 7637–7641.

Cockerill, P.N. and W.T. Garrard. 1986. Chromosomal loop anchorage of the kappa immunoglobulin gene occurs next to the enhancer in a region containing topoisomerase II sites. *Cell* **44:** 273–282.

Corbett, A.H., R.F. DeVore, and N. Osheroff. 1992. Effect of casein kinase II-mediated phosphorylation on the catalytic cycle of topoisomerase II. Regulation of enzyme activity by enhancement of ATP hydrolysis. *J. Biol. Chem.* **267:** 20513–20518.

Corbett, A.H., A.W. Fernald, and N. Osheroff. 1993. Protein kinase C modulates the catalytic activity of topoisomerase II by enhancing the rate of ATP hydrolysis: Evidence for a common mechanism of regulation by phosphorylation. *Biochemistry* **32:** 2090–2097.

Crenshaw, D.G. and T.-S. Hsieh. 1993. Function of the hydrophilic carboxyl terminus of type II DNA topoisomerase from *Drosophila melanogaster. J. Biol. Chem.* **268:** 21335–21345.

D'Arpa, P., P.S. Machlin, H. Ratrie, N.F. Rothfield, D.W. Cleveland, and W.C. Earnshaw. 1988. cDNA cloning of human DNA topoisomerase I: Catalytic activity of a 67.7-kDa carboxyl-terminal fragment. *Proc. Natl. Acad. Sci.* **85:** 2543–2547.

DeVore, R.F., A.H. Corbett, and N. Osheroff. 1992. Phosphorylation of topoisomerase II by casein kinase II and protein kinase C: Effects on enzyme-mediated DNA cleavage/religation and sensitivity to the antineoplastic drugs etoposide and 4′-(9-acridinylamino)methane- sulfon-m-anisidide. *Cancer Res.* **52:** 2156–2161.

Downes, C.S., A.M. Mullinger, and R.T. Johnson. 1991. Inhibitors of DNA topoisomerase II prevent chromatid separation in mammalian cells but do not prevent exit from mitosis. *Proc. Natl. Acad. Sci.* **88:** 8895–8899.

Downes, C.S., D.J. Clarke, A.M. Mullinger, J.F. Giménez-Abián, A.M. Creighton, and R.T. Johnson. 1994. A topoisomerase II-dependent G2 cycle checkpoint in mammalian cells. *Nature* **372:** 467–470.

Drake, F.H., G.A. Hofmann, H.F. Bartus, M.R. Mattern, S.T. Crooke, and C.K. Mirabelli. 1989a. Biochemical and pharmacological properties of p170 and p180 forms of topoisomerase II. *Biochemistry* **28:** 8154–8160.

Drake, F.H., G.A. Hofmann, S.-M. Mong, J. O. Bartus, R.P. Hertzberg, R.K. Johnson, M.R. Mattern, and C.K. Mirabelli. 1989b. *In vitro* and intracellular inhibition of topoisomerase II by the antitumor agent merbarone. *Cancer Res.* **49:** 2578–2583.

Durban, E., M. Goodenough, J. Mills, and H. Busch. 1985. Topoisomerase I phosphorylation in vitro and in rapidly growing Novikoff hepatoma cells. *EMBO J.* **4:**

2921–2926.

Edwards, K.A., B.D. Halligan, J.L. Davis, N.L. Nivera, and L.F. Liu. 1982. Recognition sites of eukaryotic DNA topoisomerase I: DNA nucleotide sequencing analysis of topo I cleavage sites on SV40 DNA. *Nucleic Acids Res.* **10:** 2565–2576.

Egyhazi, E. and E. Durban. 1987. Microinjection of anti-topoisomerase I immunoglobulin G into nuclei of *Chironomus tentans* salivary gland cells leads to blockage of transcription elongation. *Mol. Cell. Biol.* **7:** 4308–4316.

Eng, W.K., S.D. Pandit, and R. Sternglanz. 1989. Mapping of the active site tyrosine of eukaryotic DNA topoisomerase I. *J. Biol. Chem.* **264:** 13373–13376.

Fairman, R. and D.L. Brutlag. 1988. Expression of the *Drosophila* type II topoisomerase is developmentally regulated. *Biochemistry* **27:** 560–565.

Fleischmann, G., G. Pflugfelder, E.K. Steiner, K. Javaherian, G.C. Howard, J.C. Wang, and S.C. Elgin. 1984. *Drosophila* DNA topoisomerase I is associated with transcriptionally active regions of the genome. *Proc. Natl. Acad. Sci.* **81:** 6958–6962.

Fox, M.E. and P.J. Smith. 1990. Long-term inhibition of DNA synthesis and the persistence of trapped topoisomerase II complexes in determining the toxicity of the antitumor DNA intercalators mAMSA and mitoxantrone. *Cancer Res.* **50:** 5813–5818.

Fragoso, S.P. and S. Goldenberg. 1992. Cloning and characterization of the gene encoding *Trypanosoma cruzi* DNA topoisomerase II. *Mol. Biochem. Parasitol.* **55:** 127–134.

Freudenreich, C. and K.N. Kreuzer. 1994. Localization of an aminoacridine antitumor agent in a type II topoisomerase-DNA complex. *Proc. Natl. Acad. Sci.* **91:** 11007–11011.

Gale, K.C. and N. Osheroff. 1990. Uncoupling the DNA cleavage and religation activities of topoisomerase II with a single-stranded nucleic acid substrate: Evidence for an active enzyme-cleaved DNA intermediate. *Biochemistry* **29:** 9538–9545.

———. 1992. Intrinsic intermolecular DNA ligation activity of eukaryotic topoisomerase II. Potential roles in recombination. *J. Biol. Chem.* **267:** 12090–12097.

Gangloff, S., J.P. McDonald, C. Bendixen, L. Arthur, and R. Rothstein. 1994. The yeast type I topoisomerase *Top3* interacts with *Sgs1*, a DNA helicase homolog: A potential eukaryotic reverse gyrase. *Mol. Cell. Biol.* **14:** 8391–8398.

Gartenberg, M.R. and J.C. Wang. 1992. Positive supercoiling of DNA greatly diminishes mRNA synthesis in yeast. *Proc. Natl. Acad. Sci.* **89:** 11461–11465.

Gasser, S.M. and U.K. Laemmli. 1986. Cohabitation of scaffold binding regions with upstream enhancer elements of three developmentally regulated genes of *D. melanogaster*. *Cell* **46:** 521–530.

Giaever, G.N. and J.C. Wang. 1988. Supercoiling of intracellular DNA can occur in eukaryotic cells. *Cell* **55:** 849–856.

Giaever, G.N., R. Lynn, T. Goto, and J.C. Wang. 1986. The complete nucleotide sequence of the structural gene *TOP2* of yeast DNA topoisomerase II. *J. Biol. Chem.* **261:** 12448–12454.

Gilmour, D.S. and S.C. Elgin. 1987. Localization of specific topoisomerase I interactions within the transcribed region of active heat shock genes by using the inhibitor camptothecin. *Mol. Cell. Biol.* **7:** 141–148.

Gocke, E., B.J. Bonven, and O. Westergaard. 1983. A site and strand specific nuclease activity with analogies to topoisomerase I frames the rRNA gene of *Tetrahymena*. *Nucleic Acids Res.* **11:** 7661–7678.

Goto, T. and J.C. Wang. 1982. Yeast DNA topoisomerase II. An ATP-dependent type II topoisomerase that catalyzes the catenation, decatenation, unknotting, and relaxation of double-stranded DNA rings. *J. Biol. Chem.* **257:** 5866–5872.

———. 1984. Yeast DNA topoisomerase II is encoded by a single-copy, essential gene. *Cell* **36:** 1073–1080.

———. 1985. Cloning of yeast *TOP1*, the gene encoding DNA topoisomerase I, and construction of mutants defective in both DNA topoisomerase I and DNA topoisomerase II. *Proc. Natl. Acad. Sci.* **82:** 7178–7182.

Goto, T., P. Laipis, and J.C. Wang. 1984. The purification and characterization of DNA topoisomerases I and II of the yeast *Saccharomyces cerevisiae. J. Biol. Chem.* **259:** 10422–10429.

Han, Y.-H., M.J.F. Austin, Y. Pommier, and L.F. Povirk. 1993. Small deletion and insertion mutations induced by the topoisomerase II inhibitor teniposide in CHO cells and comparison with sites of drug-stimulated DNA cleavage *in vitro. J. Mol. Biol.* **229:** 52–66.

Heck, M.M. and W.C. Earnshaw. 1986. Topoisomerase II: A specific marker for cell proliferation. *J. Cell Biol.* **103:** 2569–2581.

Heck, M.M., W.N. Hittelman, and W.C. Earnshaw. 1988. Differential expression of DNA topoisomerases I and II during the eukaryotic cell cycle. *Proc. Natl. Acad. Sci.* **85:** 1086–1090.

———. 1989. *In vivo* phosphorylation of the 170-kDa form of eukaryotic DNA topoisomerase II. Cell cycle analysis. *J. Biol. Chem.* **264:** 15161–15164.

Hochhauser, D., C.A. Stanway, A.L. Harris, and I.D. Hickson. 1992. Cloning and characterization of the 5′-flanking region of the human topoisomerase II alpha gene. *J. Biol. Chem.* **267:** 18961–18965.

Holm, C. 1994. Coming Undone: How to untangle a chromosome. *Cell* **77:** 955–957.

Holm, C., T. Stearns, and D. Botstein. 1989. DNA topoisomerase II must act at mitosis to prevent nondisjunction and chromosome breakage. *Mol. Cell. Biol.* **9:** 159–168.

Holm, C., T. Goto, J.C. Wang, and D. Botstein. 1985. DNA topoisomerase II is required at the time of mitosis in yeast. *Cell* **41:** 553–563.

Hsiang, Y.H., M.G. Lihou, and L.F. Liu. 1989. Arrest of replication forks by drug-stabilized topoisomerase I-DNA cleavable complexes as a mechanism of cell killing by camptothecin. *Cancer Res.* **49:** 5077–5082.

Hsiang, Y.H., H.Y. Wu, and L.F. Liu. 1988. Proliferation-dependent regulation of DNA topoisomerase II in cultured human cells. *Cancer Res.* **48:** 3230–3235.

Hsiang, Y.H., R. Hertzberg, S. Hecht, and L.F. Liu. 1985. Camptothecin induces protein-linked DNA breaks via mammalian DNA topoisomerase I. *J. Biol. Chem.* **260:** 14873–14878.

Hsieh, T.S. 1983. Knotting of the circular duplex DNA by type II DNA topoisomerase from *Drosophila melanogaster*. *J. Biol. Chem.* **258:** 8413–8420.

Hsieh, T.S. and D. Brutlag. 1980. ATP-dependent DNA topoisomerase from *D. melanogaster* reversibly catenates duplex DNA rings. *Cell* **21:** 115–125.

Hsieh, T.S., S.D. Brown, P. Huang, and J. Fostel. 1992. Isolation and characterization of a gene encoding DNA topoisomerase I in *Drosophila melanogaster. Nucleic Acids Res.* **20:** 6177–6182.

Jaxel, C., G. Capranico, D. Kerrigan, K.W. Kohn, and Y. Pommier. 1991. Effect of local DNA sequence on topoisomerase I cleavage in the presence or absence of camptothecin. *J. Biol. Chem.* **266:** 20418–20423.

Jenkins, J.R., P. Ayton, T. Jones, S.L. Davies, D.L. Simmons, A.L. Harris, D. Sheer, and I.D. Hickson. 1992. Isolation of cDNA clones encoding the beta isozyme of human

DNA topoisomerase II and localisation of the gene to chromosome 3p24. *Nucleic Acids Res.* **20:** 5587–5592.

Jensen, P.B., B.S. Sørensen, E.J.F. Demant, M. Sehested, P.S. Jensen, L. Vindeløv, and H.H. Hansen. 1990. Antagonistic effect of aclarubicin on the cytotoxicity of etoposide and 4′-(9-acridinyl-amino)methanesulfon-m-aniside in human small cell lung cancer cells and on topoisomerase II-mediated DNA cleavage. *Cancer Res.* **50:** 3311–3316.

Johnson, R.K., F.L. McCabe, L.F. Faucette, R.P. Hertzberg, W.D. Kinsbury, J.C. Boehm, M.J. Caranfa, and K.G. Holden. 1989. SK&F 10864, a water-soluble analog of camptothecin with broad-spectrum activity in preclinical tumor models. *Proc. Am. Assoc. Cancer Res.* **30:** 623–628.

Kim, R.A. and J.C. Wang. 1989a. Function of DNA topoisomerases as replication swivels in *Saccharomyces cerevisiae*. *J. Mol. Biol.* **208:** 257–267.

———. 1989b. A subthreshold level of DNA topoisomerases leads to the excision of yeast rDNA as extrachromosomal rings. *Cell* **57:** 975–985.

Kimura, K., N. Nozaki, M. Saijo, A. Kikuchi, M. Ui, and T. Enomoto. 1994. Identification of the nature of modification that causes the shift of DNA topoisomerase IIβ to apparent higher molecular forms in the M phase. *J. Biol. Chem.* **269:** 24523–24526.

Kjeldsen, E., S. Mollerup, B. Thomsen, B.J. Bonven, L. Bolund, and O. Westergaard. 1988. Sequence-dependent effect of camptothecin on human topoisomerase I DNA cleavage. *J. Mol. Biol.* **202:** 333–342.

Koiwai, O., Y. Yasui, Y. Sakai, T. Watanabe, K. Ishii, S. Yanagihara, and T. Andoh. 1993. Cloning of the mouse cDNA encoding DNA topoisomerase I and chromosomal location of the gene. *Gene* **125:** 211–216.

Konopka, A.K. 1988. Compilation of DNA strand exchange sites for non-homologous recombination in somatic cells. *Nucleic Acids Res.* **16:** 1739–1758.

Kroeger, P.E. and T.C. Rowe. 1989. Interaction of topoisomerase I with the transcribed region of the *Drosophila* HSP 70 heat shock gene. *Nucleic Acids Res.* **17:** 8495–8509.

———. 1992. Analysis of topoisomerase I and II cleavage sites on the *Drosophila* actin and Hsp70 heat shock genes. *Biochemistry* **31:** 2492–2501.

Krogh, S., U.H. Mortensen, O. Westergaard, and B.J. Bonven. 1991. Eukaryotic topoisomerase I-DNA interaction is stabilized by helix curvature. *Nucleic Acids Res.* **19:** 1235–1241.

Kunimoto, T., K. Nitta, T. Tanaka, N. Uebuara, H. Baba, M. Takeuchi, T. Yokokura, S. Sawada, T. Miyasaka, and M. Mutai. 1987. Antitumor activity of 7-ethyl-10-[4-(1-piperidino)-1-piperidino]carbonyloxy-camptothecin, a novel water-soluble derivative of camptothecin, against murine tumors. *Cancer Res.* **47:** 5944–5949.

Kunze, N., M. Klein, A. Richter, and R. Knippers. 1990. Structural characterization of the human DNA topoisomerase I gene promoter. *Eur. J. Biochem.* **194:** 323–330.

Lee, M.P., M. Sander, and T.S. Hsieh. 1989. Nuclease protection by *Drosophila* DNA topoisomerase II. Enzyme/DNA contacts at the strong topoisomerase II cleavage sites. *J. Biol. Chem.* **264:** 21779–21787.

Lindsley, J.E. and J.C. Wang. 1991. Proteolysis patterns of epitopically labeled yeast DNA topoisomerase II suggests an allosteric transition in the enzyme induced by ATP binding. *Proc. Natl. Acad. Sci.* **88:** 10485–10489.

Liu, L.F. 1989. DNA topoisomerase poisons as antitumor drugs. *Annu. Rev. Biochem.* **58:** 351–375.

Liu, L.F. and K.G. Miller. 1981. Eukaryotic DNA topoisomerases: Two forms of type I DNA topoisomerases from HeLa cell nuclei. *Proc. Natl. Acad. Sci.* **78:** 3487–3491.

Liu, L.F. and J.C. Wang. 1987. Supercoiling of the DNA template during transcription. *Proc. Natl. Acad. Sci.* **84:** 7024–7027.

Liu, L.F., C.C. Liu, and B.M. Alberts. 1980. Type II DNA topoisomerases: Enzymes that can unknot a topologically knotted DNA molecule via a reversible double-strand break. *Cell* **19:** 697–707.

Liu, L.F., T.C. Rowe, L. Yang, K.M. Tewey, and G.L. Chen. 1983. Cleavage of DNA by mammalian topoisomerase II. *J. Biol. Chem.* **258:** 15365–15370.

Lue, N., A. Sharma, A. Mondragón, and J.C. Wang. 1995. A 26 kDa yeast topoisomerase I fragment: Crystallographic structure and mechanism implications. *Structure* **3:** 1315–1322.

Lund, K., A.H. Andersen, K. Christiansen, J.Q. Svejstrup, and O. Westergaard. 1990. Minimal DNA requirement for topoisomerase II-mediated cleavage in vitro. *J. Biol. Chem.* **265:** 13856–13863.

Lynn, R.M., M.A. Bjornsti, P.R. Caron, and J.C. Wang. 1989. Peptide sequencing and site-directed mutagenesis identify tyrosine-727 as the active site tyrosine of *Saccharomyces cerevisiae* DNA topoisomerase I. *Proc. Natl. Acad. Sci.* **86:** 3559–3563.

Ma, X., N. Saitoh, and P.J. Curtis. 1993. Purification and characterization of a nuclear DNA-binding factor complex containing topoisomerase II and chromosome scaffold protein 2. *J. Biol. Chem.* **268:** 6182–6188.

Markovits, J., C. Linassier, P. Fossé, J. Couprie, J. Pierre, A. Jacquemin-Sablon, J.-M. Saucier, J.-B. Le Pecq, and A.K. Larsen. 1989. Inhibitory effects of the tyrosine kinase inhibitor genistein on mammalian DNA topoisomerase II. *Cancer Res.* **49:** 5111–5117.

Maxwell, A. and M. Gellert. 1986. Mechanistic aspects of DNA topoisomerases. *Adv. Protein Chem.* **38:** 69–107.

McConaughy, B.L., L.S. Young, and J.J. Champoux. 1981. The effect of salt on the binding of the eukaryotic DNA nicking-closing enzyme to DNA and chromatin. *Biochim. Biophys. Acta* **655:** 1–8.

Merino, A., K.R. Madden, W.S. Lane, J.J. Champoux, and D. Reinberg. 1993. DNA topoisomerase I is involved in both repression and activation of transcription. *Nature* **365:** 227–232.

Muller, M.T., W.P. Pfund, V.B. Mehta, and D.K. Trask. 1985. Eukaryotic type I topoisomerase is enriched in the nucleolus and catalytically active on ribosomal DNA. *EMBO J.* **4:** 1237–1243.

Nelson, W.G., L.F. Liu, and D.S. Coffey. 1986. Newly replicated DNA is associated with DNA topoisomerase II in cultured rat prostatic adenocarcinoma cells. *Nature* **322:** 187–189.

Ness, P.J., R.W. Parish, and T. Koller. 1986. Mapping of endogenous nuclease-sensitive regions and of putative topoisomerase sites of action along the chromatin of *Dictyostelium* ribosomal RNA genes. *J. Mol. Biol.* **188:** 287–300.

Nitiss, J.L. 1994. Roles of DNA topoisomerases in chromosomal replication and segregation. *Adv. Pharmacol.* **29A:** 103–134.

Nitiss, J.L. and J.C. Wang. 1988. DNA topoisomerase-targeting antitumor drugs can be studied in yeast. *Proc. Natl. Acad. Sci.* **85:** 7501–7505.

Orphanides, G. and A. Maxwell. 1994. In one gate, out the other. *Curr. Biol.* **4:** 1006–1009.

Osheroff, N. 1986. Eukaryotic topoisomerase II. Characterization of enzyme turnover. *J. Biol. Chem.* **261:** 9944–9950.

———. 1987. Role of the divalent cation in topoisomerase II mediated reactions.

Biochemistry **26:** 6402–6406.

———. 1989a. Effect of antineoplastic agents on the DNA cleavage/religation reaction of eukaryotic topoisomerase II: Inhibition of DNA religation by etoposide. *Biochemistry* **28:** 6157–6160.

———. 1989b. Biochemical basis for the interactions of type I and type II topoisomerases with DNA. *Pharmacol. Ther.* **41:** 223–241.

Osheroff, N. and D.L. Brutlag. 1983. Recognition of supercoiled DNA by *Drosophila* topoisomerase II. In *Mechanism of DNA replication and recombination* (ed. N.R. Cozzarelli), pp. 55–64. A.R. Liss, New York.

Osheroff, N. and E.L. Zechiedrich. 1987. Calcium-promoted DNA cleavage by eukaryotic topoisomerase II: Trapping the covalent enzyme-DNA complex in an active form. *Biochemistry* **26:** 4303–4309.

Osheroff, N., E.R. Shelton, and D.L. Brutlag. 1983. DNA topoisomerase II from *Drosophila melanogaster*. Relaxation of supercoiled DNA. *J. Biol. Chem.* **258:** 9536–9543.

Osheroff, N., E.L. Zechiedrich, and K.C. Gale. 1991. Catalytic function of DNA topoisomerase II. *BioEssays* **13:** 269–273.

Park, S.H., J.H. Yoon, Y.D. Kwon, and S.D. Park. 1993. Nucleotide sequence analysis of the cDNA for rat DNA topoisomerase II. *Biochem. Biophys. Res. Commun.* **193:** 787–793.

Pasion, S.G., J.C. Hines, R. Aebersold, and D.S. Ray. 1992. Molecular cloning and expression of the gene encoding the kinetoplast-associated type II DNA topoisomerase of *Crithidia fasciculata*. *Mol. Biochem. Parasitol.* **50:** 57–67.

Pommier, Y. 1993. DNA topoisomerase I and II in cancer chemotherapy: Update and perspectives. *Cancer Chemother. Pharmacol.* **32:** 103–108.

Pommier, Y., D. Kerrigan, K.D. Hartman, and R.I. Glazer. 1990. Phosphorylation of mammalian DNA topoisomerase I and activation by protein kinase C. *J. Biol. Chem.* **265:** 9418–9422.

Pommier, Y., K.W. Kohn, G. Capranico, and C. Jaxel. 1993. Base sequence selectivity of topoisomerase inhibitors suggest a common model for drug action. In *Molecular biology of DNA topoisomerases and its application to chemotherapy* (ed. T. Andoh et al.), pp. 215–227. CRC Press, Boca Raton, Florida.

Porter, S.E. and J.J. Champoux. 1989. Mapping in vivo topoisomerase I sites on simian virus 40 DNA: Asymmetric distribution of sites on replicating molecules. *Mol. Cell. Biol.* **9:** 541–550.

Prosperi, E., E. Sala, C. Negri, C. Oliani, R. Supino, G.B. Astraldi Ricotti, and G. Bottiroli. 1992. Topoisomerase II alpha and beta in human tumor cells grown *in vitro* and *in vivo*. *Anticancer Res.* **12:** 2093–2099.

Reece, R.J. and A. Maxwell. 1991. DNA gyrase: Structure and function. *Crit. Rev. Biochem. Mol. Biol.* **26:** 335–375.

Reitman, M. and G. Felsenfeld. 1990. Developmental regulation of topoisomerase II sites and DNase I-hypersensitive sites in the chicken β-globin locus. *Mol. Cell. Biol.* **10:** 2774–2786.

Robinson, M.J. and N. Osheroff. 1990. Stabilization of the topoisomerase II-DNA cleavage complex by antineoplastic drugs: Inhibition of enzyme-mediated DNA religation by 4′-(9-acridinylamino)methanesulfon-m-anisidide. *Biochemistry* **29:** 2511–2515.

Robinson, M.J., A.H. Corbett, and N. Osheroff. 1993. Effects of topoisomerase II-targeted drugs on enzyme-mediated DNA cleavage and ATP hydrolysis: Evidence for distinct drug interaction domains on topoisomerase II. *Biochemistry*. **32:** 3638–3643.

Robinson, M.J., B.A. Martin, T.D. Gootz, P.R. McGuirk, M. Moynihan, J.A. Sutcliff, and N. Osheroff. 1991. Effects of quinolone derivatives on eukaryotic topoisomerase II: A novel mechanism for enhancement of enzyme-mediated DNA cleavage. *J. Biol. Chem.* **266:** 14585–14592.

Roca, J. and J.C. Wang. 1992. The capture of a DNA double helix by an ATP-dependent protein clamp: A key step in DNA transport by type II DNA topoisomerases. *Cell* **71:** 833–840.

———. 1994. DNA transport by a type II topoisomerase: Evidence in favor of a two-gate mechanism. *Cell* **77:** 609–616.

Roca, J., R. Ishida, J.M. Berger, T. Andoh, and J.C. Wang. 1994. Antitumor bis-dioxopiperazines inhibit yeast DNA topoisomerase II by trapping the enzyme in the form of a closed protein clamp. *Proc. Natl. Acad. Sci.* **91:** 1781–1785.

Rose, D. and C. Holm. 1993. Meiosis-specific arrest revealed in DNA topoisomerase II mutants. *Mol. Cell. Biol.* **13:** 3445–3455.

Rose, D., W. Thomas, and C. Holm. 1990. Segregation of recombined chromosomes in meiosis I requires DNA topoisomerase II. *Cell* **60:** 1009–1017.

Rottmann, M., H.C. Schröder, M. Gramzow, K. Renneisen, B. Kurelec, A. Dorn, U. Friese, and W.E.G. Muller. 1987. Specific phosphorylation of proteins in pore complex-laminae from the sponge *Geodia cydonium* by the homologous aggregation factor and phorbol ester. Role of protein kinase C in the phosphorylation of DNA topoisomerase II. *EMBO J.* **6:** 3939–3944.

Saijo, M., M. Ui, and T. Enomoto. 1992. Growth state and cell cycle dependent phosphorylation of DNA topoisomerase II in Swiss 3T3 cells. *Biochemistry* **31:** 359–363.

Sander, M. and T.S. Hsieh. 1983. Double strand DNA cleavage by type II DNA topoisomerase from *Drosophila melanogaster*. *J. Biol. Chem.* **258:** 8421–8428.

———. 1985. *Drosophila* topoisomerase II double-strand DNA cleavage: Analysis of DNA sequence homology at the cleavage site. *Nucleic Acids Res.* **13:** 1057–1072.

Sander, M., T.S. Hsieh, A. Udvardy, and P. Schedl. 1987. Sequence dependence of *Drosophila* topoisomerase II in plasmid relaxation and DNA binding. *J. Mol. Biol.* **194:** 219–229.

Schiestl, R.H., M. Dominska, and T.D. Petes. 1993. Transformation of *Saccharomyces cerevisiae* with nonhomologous DNA: Illegitimate integration of transforming DNA into yeast chromosomes and *in vivo* ligation of transforming DNA to mitochondrial DNA sequences. *Mol. Cell. Biol.* **13:** 2697–2705.

Schmidt, V.K., B.S. SØrensen, H.V. SØrensen, J. Alsner, and O. Westergaard. 1994. Intramolecular and intermolecular DNA ligation mediated by topoisomerase II. *J. Mol. Biol.* **241:** 18–25.

Schomburg, U. and F. Grosse. 1986. Purification and characterization of DNA topoisomerase II from calf thymus associated with polypeptides of 175 and 150 kDa. *Eur. J. Biochem.* **160:** 451–457.

Shelton, E.R., N. Osheroff, and D.L. Brutlag. 1983. DNA topoisomerase II from *Drosophila melanogaster*. Purification and physical characterization. *J. Biol. Chem.* **258:** 9530–9535.

Shiozaki, K. and M. Yanagida. 1991. A functional 125-kDA core polypeptide of fission yeast DNA topoisomerase II. *Mol. Cell. Biol.* **11:** 6093–6102.

———. 1992. Functional dissection of the phosphorylated termini of fission yeast DNA topoisomerase II. *J. Cell. Biol.* **119:** 1023–1036.

Snapka, R.M. 1986. Topoisomerase inhibitors can selectively interfere with different stages of simian virus 40 DNA replication. *Mol. Cell. Biol.* **6:** 4221–4227.

Snapka, R.M., M.A. Powelson, and J.M. Strayer. 1988. Swiveling and decatenation of replicating simian virus 40 genomes *in vivo*. *Mol. Cell. Biol.* **8:** 515–521.

Sørensen, B.S., J. Sinding, A.H. Andersen, J. Alsner, P.B. Jensen, and O. Westergaard. 1992. Mode of action of topoisomerase II targeting agents at a specific DNA sequence: Uncoupling the DNA binding, cleavage and religation events. *J. Mol. Biol.* **228:** 778–786.

Sørensen, B.S., P.S. Jensen, A.H. Andersen, K. Christiansen, J. Alsner, B. Thomsen, and O. Westergaard. 1990. Stimulation of topoisomerase II mediated DNA cleavage at specific sequence elements by the 2-nitroimidazole Ro 15-0216. *Biochemistry* **29:** 9507–9515.

Spell, R.M. and C. Holm. 1994. Nature and distribution of chromosomal intertwinings in *Saccharomyces cerevisiae*. *Mol. Cell. Biol.* **14:** 1465–1476.

Sperry, A.O., V.C. Blasquez, and W.T. Garrard. 1989. Dysfunction of chromosomal loop attachment sites: Illegitimate recombination linked to matrix association regions and topoisomerase II. *Proc. Natl. Acad. Sci.* **86:** 5497–5501.

Spitzner, J.R. and M.T. Muller. 1988. A consensus sequence for cleavage by vertebrate DNA topoisomerase II. *Nucleic Acids Res.* **16:** 5533–5556.

Spitzner, J.R., I.K. Chung, and M.T. Muller. 1990. Eukaryotic topoisomerase II preferentially cleaves alternating purine-pyrimidine repeats. *Nucleic Acids Res.* **18:** 1–11.

Stevnsner, T., U.H. Mortensen, O. Westergaard, and B.J. Bonven. 1989. Interactions between eukaryotic DNA topoisomerase I and a specific binding sequence. *J. Biol. Chem.* **264:** 10110–10113.

Stewart, A.F. and G. Schutz. 1987. Camptothecin-induced in vivo topoisomerase I cleavages in the transcriptionally active tyrosine aminotransferase gene. *Cell* **50:** 1109–1117.

Stewart, A.F., R.E. Herrera, and A. Nordheim. 1990. Rapid induction of *c-fos* transcription reveals quantitative linkage of RNA polymerase II and DNA topoisomerase I enzyme activities. *Cell* **60:** 141–149.

Strauss, P.R. and J.C. Wang. 1990. The *TOP2* gene of *Trypanosoma brucei*: A single-copy gene that shares extensive homology with other *TOP2* genes encoding eukaryotic DNA topoisomerase II. *Mol. Biochem. Parasitol.* **38:** 141–150.

Sundin, O. and A. Varshavsky. 1981. Arrest of segregation leads to accumulation of highly intertwined catenated dimers: Dissection of the final stages of SV40 DNA replication. *Cell* **25:** 659–669.

Svejstrup, J.Q., K. Christiansen, I.I. Gromova, A.H. Andersen, and O. Westergaard. 1991. New technique for uncoupling the cleavage and religation reactions of eukaryotic topoisomerase I. The mode of action of camptothecin at a specific recognition site. *J. Mol. Biol.* **222:** 669–678.

Swedlow, J.R., J.W. Sedat, and D.A. Agard. 1993. Multiple chromosomal populations of topoisomerase II detected in vivo by time-lapse, three-dimensional wide-field microscopy. *Cell* **73:** 97–108.

Tanabe, K., Y. Ikegami, R. Ishida, and T. Andoh. 1991. Inhibition of topoisomerase II by antitumor agents bis(2,6-dioxopiperazine) derivatives. *Cancer Res.* **51:** 4903–4908.

Tewey, K.M., G.L. Chen, E.M. Nelson, and L.F. Liu. 1984. Intercalative antitumor drugs interfere with the breakage-reunion reaction of mammalian DNA topoisomerase II. *J. Biol. Chem.* **259:** 9182–9187.

Thomsen, B., C. Bendixen, K. Lund, A.H. Andersen, B.S. Sørensen, and O. Westergaard. 1990. Characterization of the interaction between topoisomerase II and DNA by transcriptional footprinting. *J. Mol. Biol.* **215:** 237–244.

Thomsen, B., S. Mollerup, B.J. Bonven, R. Frank, H. Blocker, O.F. Nielsen, and O. Westergaard. 1987. Sequence specificity of DNA topoisomerase I in the presence and absence of camptothecin. *EMBO J.* **6:** 1817–1823.

Thrash, C., A.T. Bankier, B.G. Barrell, and R. Sternglanz. 1985. Cloning, characterization, and sequence of the yeast DNA topoisomerase I gene. *Proc. Natl. Acad. Sci.* **82:** 4374–4378.

Tsai-Pflugfelder, M., L.F. Liu, A.A. Liu, K.M. Tewey, J. Whang-Peng, T. Knutsen, K. Huebner, C.M. Croce, and J.C. Wang. 1988. Cloning and sequencing of cDNA encoding human DNA topoisomerase II and localization of the gene to chromosome region 17q21-22. *Proc. Natl. Acad. Sci.* **85:** 7177–7181.

Uemura, T. and M. Yanagida. 1984. Isolation of type I and II DNA topoisomerase mutants from fission yeast: Single and double mutants show different phenotypes in cell growth and chromatin organization. *EMBO J.* **3:** 1737–1744.

———. 1986. Mitotic spindle pulls but fails to separate chromosomes in type II DNA topoisomerase mutants: Uncoordinated mitosis. *EMBO J.* **5:** 1003–1010.

Uemura, T., K. Morikawa, and M. Yanagida. 1986. The nucleotide sequence of the fission yeast DNA topoisomerase II gene: Structural and functional relationships to other DNA topoisomerases. *EMBO J.* **5:** 2355–2361.

Uemura, T., K. Morino, S. Uzawa, K. Shiozaki, and M. Yanagida. 1987a. Cloning and sequencing of *Schizosaccharomyces pombe* DNA topoisomerase I gene, and effect of gene disruption. *Nucleic Acids Res.* **15:** 9727–9739.

Uemura, T., H. Ohkura, Y. Adachi, K. Morino, K. Shiozaki, and M. Yanagida. 1987b. DNA topoisomerase II is required for condensation and separation of mitotic chromosomes in *S. pombe*. *Cell* **50:** 917–925.

Vosberg, H.P. 1985. DNA topoisomerases: Enzymes that control DNA conformation. *Curr. Top. Microbiol. Immunol.* **114:** 19–102.

Wallis, J.W., G. Chrebet, G. Brodsky, M. Rolfe, and R. Rothstein. 1989. A hyper-recombination mutation in *S. cerevisiae* identifies a novel eukaryotic topoisomerase. *Cell* **58:** 409–419.

Wang, J.C. 1987. Recent studies of DNA topoisomerases. *Biochim. Biophys. Acta* **909:** 1–9.

———. 1994. DNA topoisomerases as targets of therapeutics: An overview. *Adv. Pharmacol.* **29A:** 1–19.

Wang, H.P. and C.E. Rogler. 1991. Topoisomerase I-mediated integration of hepadnavirus DNA in vitro. *J. Virol.* **65:** 2381–2392.

Watt, P.M. and I.D. Hickson. 1994. Structure and function of type II DNA topoisomerases. *Biochem. J.* **303:** 681–695.

Woessner, R.D., M.R. Mattern, C.K. Mirabelli, R.K. Johnson, and F.H. Drake. 1991. Proliferation- and cell cycle-dependent differences in expression of the 170 kilodalton and 180 kilodalton forms of topoisomerase II in NIH-3T3 cells. *Cell Growth Differ.* **2:** 209–214.

Worland, S.T. and J.C. Wang. 1989. Inducible overexpression, purification, and active site mapping of DNA topoisomerase II from the yeast *Saccharomyces cerevisiae*. *J.*

Biol. Chem. **264:** 4412–4416.

Wu, H.Y., S.H. Shyy, J.C. Wang, and L.F. Liu. 1988. Transcription generates positively and negatively supercoiled domains in the template. *Cell* **53:** 433–440.

Wyckoff, E. and T.-S. Hsieh. 1988. Functional expression of a *Drosophila* gene in yeast: Genetic complementation of DNA topoisomerase II. *Proc. Natl. Acad. Sci.* **85:** 6272–6276.

Wyckoff, E., D. Natalie, J.M. Nolan, M. Lee, and T.S. Hsieh. 1989. Structure of the *Drosophila* DNA topoisomerase II gene. Nucleotide sequence and homology among topoisomerases II. *J. Mol. Biol.* **205:** 1–13.

Yang, L., M.S. Wold, J.J. Li, T.J. Kelly, and L.F. Liu. 1987. Roles of DNA topoisomerases in simian virus 40 DNA replication in vitro. *Proc. Natl. Acad. Sci.* **84:** 950–954.

Zechiedrich, E.L., K. Christiansen, A.H. Andersen, O. Westergaard, and N. Osheroff. 1989. Double-stranded DNA cleavage/religation reaction of eukaryotic topoisomerase II: Evidence for a nicked DNA intermediate. *Biochemistry* **28:** 6229–6236.

Zhang, H., P. D'Arpa, and L.F. Liu. 1990. A model for tumor cell killing by topoisomerase poisons. *Cancer Cells* **2:** 23–27.

Zhang, H., J.C. Wang, and L.F. Liu. 1988. Involvement of DNA topoisomerase I in transcription of human ribosomal RNA genes. *Proc. Natl. Acad. Sci.* **85:** 1060–1064.

21
DNA Telomerases

Carol W. Greider, Kathleen Collins, and Chantal Autexier
Cold Spring Harbor Laboratory
Cold Spring Harbor, New York 11724

Telomerase is a specialized DNA polymerase that synthesizes telomeric DNA sequences onto chromosome ends. Linear DNA genomes pose a special problem for the DNA replication machinery. Because DNA polymerases function in the 5′ to 3′ direction and require a primer, part of the lagging strand is predicted to be lost at each round of division. For most eukaryotic chromosomes, sequence loss is balanced by the de novo addition of telomeric sequence by telomerase (for review, see Blackburn 1991).

Telomere DNA sequences in ciliates consist of simple GT-rich tandem repeats, and this motif is generally conserved in eukaryotes, although the exact repeat sequence is species-specific. For example, *Tetrahymena* contains tandem TTGGGG repeats, whereas mammals have TTAGGG repeats (for review, see Blackburn 1991; Henderson 1995). Telomerase synthesizes these short repeats by using an integral RNA component of the enzyme as a template (Fig. 1) (Greider and Blackburn 1989). Telomerase activity has been identified from *Tetrahymena*, *Euplotes*, *Oxytricha*, yeast, *Xenopus*, and mammals. In each case, the sequence synthesized by telomerase in vitro corresponds to the telomere sequence of the organism from which it was isolated (Table 1), and the RNA template contains more than one repeat of the sequence (Table 2) (Greider and Blackburn 1985; Zahler and Prescott 1988; Morin 1989; Shippen-Lentz and Blackburn 1989; Prowse et al. 1993; Lingner et al. 1994; Mantell and Greider 1994; Cohn and Blackburn 1995; Lin and Zakian 1995; Lue and Wang 1995).

TETRAHYMENA TELOMERASE

The telomerase mechanism has been best studied in *Tetrahymena*, where enzyme activity was first identified. A typical assay utilizes a synthetic DNA d(TTGGGG)$_3$ primer to mimic the end of a chromosome,

 0-87969-459-9/96 $5 + .00

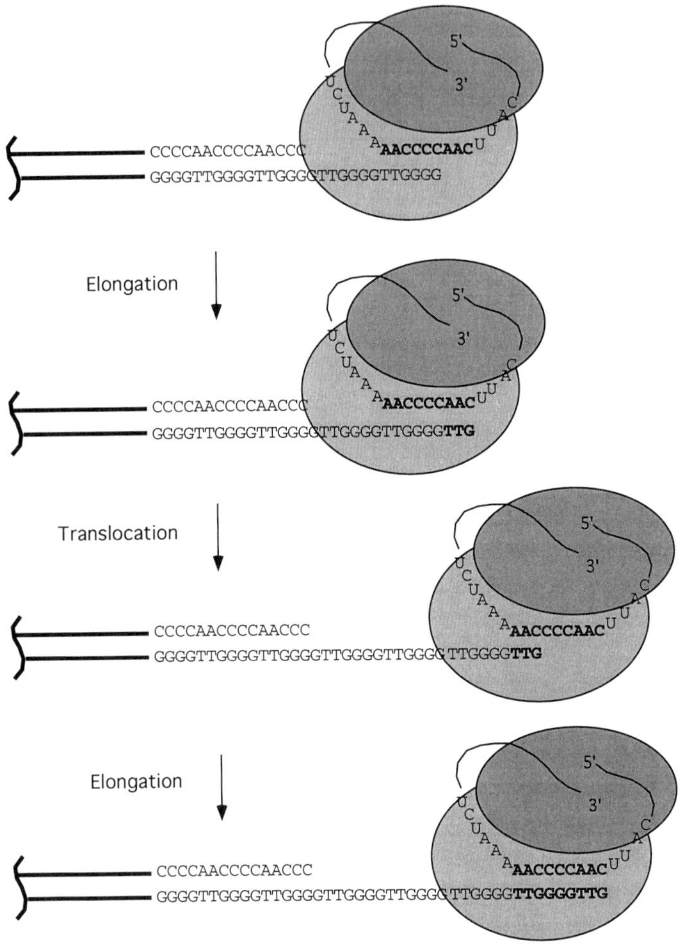

Figure 1 Telomerase elongation model. (1) Telomerase recognizes the telomere substrate, and the terminal d(TTGGGG) repeat is base-paired with the CAACCCCAA RNA template. (2) The RNA is copied to the end of the template region. (3) Translocation repositions the terminal TTGGGGTTG sequence and exposes additional template sequences. (4) Another round of template copying produces additional TTGGGG repeats. (Adapted from Greider and Blackburn 1989.)

[^{32}P]dGTP, and TTP. After incubation with telomerase extracts, the radiolabeled products are resolved on a denaturing gel or precipitated and counted. When resolved on a gel, the extension of the input 18-mer oligonucleotide by the addition of hundreds of d(TTGGGG) repeats can be directly visualized (Greider and Blackburn 1985). The nucleotides are

Table 1 Telomerase activity in various eukaryotes

Organism	Telomere sequence	RNA identified?	Reference
Tetrahymena	TTGGGG	yes	Greider and Blackburn (1985)
Oxytricha	TTTTGGGG	yes	Zahler and Prescott (1988)
Euplotes	TTTTGGGG	yes	Shippen-Lentz and Blackburn (1989)
Human	TTAGGG	yes	Morin (1989)
Mouse	TTAGGG	yes	Prowse et al. (1993)
Xenopus	TTAGGG	no	Mantell and Greider (1994)
S. cerevisiae	$TG_{(1-3)}$	yes	Cohn and Blackburn (1995); Lin and Zakian (1995)
S. castellii	$TCTGGG(TG)_{1-4}$	no	Cohn and Blackburn (1995)

added one base at a time, and the products show a distinct six-base periodicity. This periodicity is due to both pausing and enzyme dissociation during the addition of each repeat predominantly at the first G within the sequence TTGGGG (Greider 1991; Lee and Blackburn 1993). Thus, the sequence added in vitro can be deduced directly from the products of the reaction. Elongation of circularly permuted telomere sequence oligonucleotides results in banding patterns that are offset: The primer $d(TTGGGG)_3$ generates bands at positions primer +3, +9, +15, etc., whereas the primer $d(GTTGGG)_3$ generates bands at positions +4, +10, +16, etc. Telomerase binds and elongates telomere sequence oligonucleotide primers and, at a lower affinity, nontelomeric sequence primers (Greider and Blackburn 1987; Blackburn et al. 1989; Harrington and Greider 1991). A recent modification to the telomerase assay reaction in-

Table 2 Telomerase RNA components

Organism	Telomere sequence	RNA template sequence	Size	References
Tetrahymena	TTGGGG	CAACCCCAA	160	1
Euplotes	TTTTGGGG	CAAAACCCCAAAACC	190	2
Oxytricha	TTTTGGGG	CAAAACCCCAAAACC	190	3,4
Human	TTAGGG	CUAACCCUAAC	450	5
Mouse	TTAGGG	CCUAACCCU	450	6
S. cerevisiae	$TG_{(1-3)}$	CACCACACCCACACAC	1300	7
K. lactis	TTTGATTAGGTATG TGGTGTACGGA	UCAAAUCCGUACACCAC ~AUACCUAAUCAAA	1300	8

References: (1) Greider and Blackburn (1989); (2) Shippen-Lentz and Blackburn (1990); (3) Lingner et al. (1994); (4) Melek et al. (1994); (5) Feng et al. (1995); (6) Blasco et al. (1995); (7) Singer and Gottschling (1994); (8) McEachern and Blackburn (1995).

volves the detection of telomerase products by polymerase chain reaction (PCR) amplification (Kim et al. 1994). This assay is sensitive and can detect low levels of telomerase activity.

Identification and characterization of the *Tetrahymena* telomerase RNA component showed that the RNA sequence serves as a template for the synthesis of telomeric repeats in vitro and in vivo (Greider and Blackburn 1987, 1989; Yu et al. 1990; Autexier and Greider 1994). The *Tetrahymena* telomerase RNA was identified by its copurification with telomerase activity in a series of six chromatographic steps. The RNA contains within it the sequence 5′-CAACCCCAA-3′, which is complementary to one and a half repeats of the *Tetrahymena* telomere sequence d(TTGGGG). Using RNase H and oligonucleotides complementary to different regions of the RNA, this 159-nucleotide RNA was shown to be essential for telomerase function (Greider and Blackburn 1989).

SECONDARY STRUCTURE OF THE TETRAHYMENINE TELOMERASE RNAs

To identify potential functional domains in the *Tetrahymena* telomerase RNA, a secondary structure model was derived from a phylogenetic sequence comparison of telomerase RNAs from different tetrahymenine ciliates (Romero and Blackburn 1991; McCormick-Graham and Romero 1995). A potential secondary structure for the *Tetrahymena* telomerase RNA was proposed that includes four conserved helices, numbered from 5′ to 3′ as helices I–IV (Fig. 2). Telomerase RNAs were subsequently cloned from hypotrichous ciliates and shown to have a secondary structure similar to the proposed *Tetrahymena* structure, although, strikingly, the primary sequence is not conserved (Shippen-Lentz and Blackburn 1990; Lingner et al. 1994; Melek et al. 1994). In the proposed structure, helix I involves long-range base-pairing and is the most conserved of the helices. Abolishing this potential stem only partially decreased telomerase activity reconstituted in vitro, suggesting that these interactions are not essential (C. Autexier and C. W. Greider, unpubl.). Helix II is not conserved in all ciliates (Lingner et al. 1994; McCormick-Graham and Romero 1995). An unstructured region is predicted between helices II and III and includes the template region. Nucleotides in the loop of helix III can base-pair with a single-stranded region just upstream of the helix III stem to form a pseudoknot (ten Dam et al. 1991). Part of helix II contains a conserved sequence, 5′-(C)UGUCA-3′, which appears to play a role in defining the 5′ boundary of the template domain (Lingner et al. 1994; Autexier and Greider 1995). Chemical modification studies

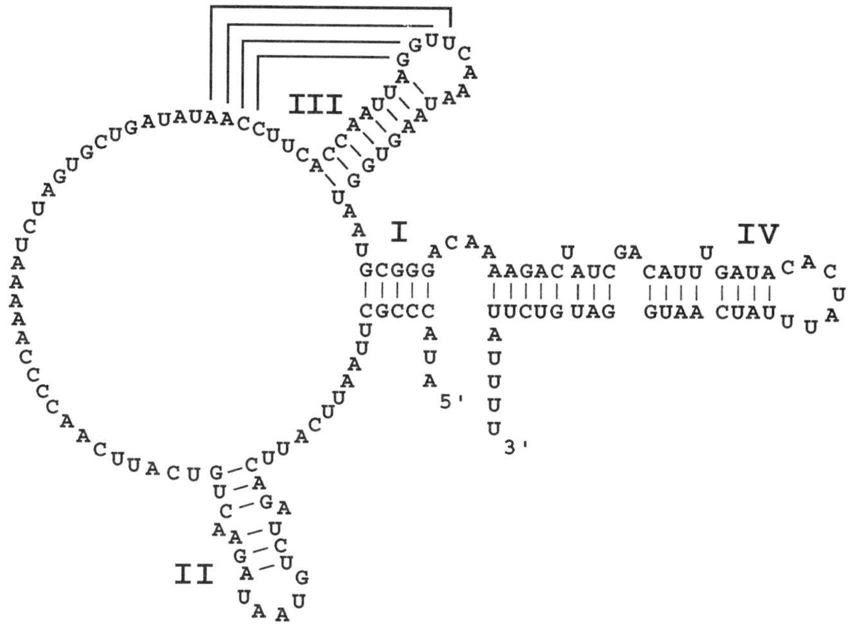

Figure 2 A secondary structure model was proposed for the *Tetrahymena* telomerase RNA based on conserved RNA sequences in different *Tetrahymena* species (Romero and Blackburn 1991). This figure shows the sequence of the *Tetrahymena thermophila* RNA. Other ciliate RNAs also fold into a similar secondary structure, although the primary sequence varies (Lingner et al. 1994). It is not yet known if other telomerase RNAs adopt a similar secondary structure.

of telomerase RNA in vivo and in vitro support the phylogenetically derived model, although some differences are seen between naked RNA and the RNP complex (Bhattacharyya and Blackburn 1994; Zaug and Cech 1995). In addition to conserved sequences within the *Tetrahymena* telomerase RNAs, there are conserved promoter regions found upstream of the telomerase RNA genes (Shippen-Lentz and Blackburn 1990; Romero and Blackburn 1991; Lingner et al. 1994).

TELOMERASE TEMPLATE FUNCTION

To dissect the functional regions of the *Tetrahymena* telomerase RNA, an in vitro reconstitution assay was developed (Autexier and Greider 1994). Telomerase extract is treated with micrococcal nuclease to digest endogenous telomerase RNA. After inactivation of the nuclease, the extract is incubated with in-vitro-transcribed *Tetrahymena* telomerase RNA to restore telomerase activity. The elongation products of telomerase reconstituted with telomerase RNAs mutated in the template domain were analyzed to determine whether all nine nucleotides of the template region serve to direct the synthesis of telomeric repeats. The analysis of several mutants suggested that the first six nucleotides of the sequence 5′-CAACCCCAA-3′ serve as template residues and that the 3′-most three residues serve to align substrate and product sequences. Under certain conditions, residue 49 can be used as a template nucleotide (Gilley et al. 1995; Autexier and Greider 1995). For instance, if residue 43 is deleted, telomerase can still synthesize wild-type GGGGTT repeats by using residue 49 as a template nucleotide (Autexier and Greider 1995).

The template region boundaries of the telomerase RNAs from hypotrichous ciliates have been analyzed using template-complementary primers and analysis of the elongation product patterns. Initially, in *Euplotes crassus*, oligonucleotides complementary to the telomerase RNA were tested for their ability to act as primers for telomerase or to inhibit the telomerase elongation reaction (Shippen-Lentz and Blackburn 1990). Preincubation with oligonucleotides that could base-pair to the putative template and adjacent sequences inhibited telomerase reactions, presumably by blocking access of the d(TTTTGGGG)$_4$ primer to the template sequence. These results were similar to the results obtained with analogous primers complementary to the *Tetrahymena* telomerase RNA (Greider and Blackburn 1989).

Later studies with *Oxytricha nova* reached different conclusions about which regions of the template provide sequence information. Using template-complementary oligonucleotides, Lingner et al. (1994) defined the template regions of the *O. nova* and *Stylonychia* species at positions 37–44 and 38–45, respectively, 5′-C/GAAAACCCCAAAACC-3′ and that of *Euplotes aediculayus* at positions 36–44, 5′-CAAAACCCC AAAACC-3′. Melek et al. concluded that the template domain of the *O. nova* telomerase RNA ranged from position 42 to 50, 5′-CAAAAC CCCAAAACC-3′ (Melek et al. 1994). The resolution of which positions specify nucleotide addition awaits mutational analysis of the telomerase RNA in these species.

YEAST TELOMERASE RNAs

A gene encoding the telomerase RNA component from *Saccharomyces cerevisiae* was recently cloned in a screen for genes affecting telomeric silencing (Singer and Gottschling 1994). Telomeric silencing is a phenomenon whereby genes placed near telomeres in yeast are transcriptionally repressed (Gottschling et al. 1990). A screen was developed to identify high-copy suppressors of telomere silencing. One of the genes identified, *TLC1*, contained an expressed RNA but no open reading frame. However, *TLC1* did contain a potential yeast telomerase RNA template motif predicted from de novo telomere addition during chromosome healing (Kramer and Haber 1993). To confirm that *TLC1* encodes the telomerase RNA, a *Hae*III site was engineered into the putative template domain, and this mutant allele was used to replace one of the *TLC1* alleles in a diploid strain. To determine if the altered sequence was incorporated into newly formed telomeres, genomic DNA was digested with *Hae*III. Specific digestion of telomere repeats was seen in the mutant but not the wild-type strains, showing that *TLC1* encodes the telomerase RNA. Northern blot analysis indicated that the wild-type strain of yeast contains a relatively abundant 1.3-kb RNA that hybridizes to a *TLC1* probe.

The telomerase RNA was also cloned from a distinct species of budding yeast (McEachern and Blackburn 1995). Unlike the short 6- to 8-bp telomeric repeats in many eukaryotes, the telomeric repeats of certain yeasts are relatively long. For instance, telomere repeats in *Candida albicans* and *Kluveromyces lactis* are 23 bp and 25 bp in length, respectively (McEachern and Hicks 1993; McEachern and Blackburn 1994). It was hypothesized that the template region of the telomerase RNAs from these species would be at least the length of the telomeric repeats and that it should be possible to detect these RNAs using repeats of the telomere sequences as probes in hybridization experiments. However, to identify such genes, the abundant telomere repeats had to first be removed. This was done by BAL 31 nuclease digestion. BAL 31 is an exonuclease that removes the terminal sequence from genomic DNA. With this method, telomerase RNA gene candidates in *K. lactis*, *C. albicans*, and *Candida glabrata* were identified (McEachern and Blackburn 1994). To clone the gene from *K. lactis*, libraries were made from BAL-31-digested genomic DNA, and a potential RNA gene clone was identified. To verify that the cloned gene encoded the telomerase RNA, the sequence of the putative template domain was changed by the introduction of a *Bgl*II site. Yeast cells in which the wild-type locus was replaced by the *Bgl*II-containing mutation had the altered sequence at their telomeres, indicating that the cloned gene encodes the telomerase RNA component. Northern blot analysis showed that the *K. lactis* telomerase RNA is 1.3 kb, like the *S. cerevisiae* telomerase RNA. Due to the large size of these RNAs, it is not yet clear if the conserved secondary structures found in the ciliate telomerase RNAs are conserved in either of the yeast telomerase RNAs (Singer and Gottschling 1994; McEachern and Blackburn 1995). Recently, a telomerase activity assay has been reported from *S. cerevisiae* cell extracts (Cohn and Blackburn 1995; Lin and Zakian 1995; Lue and Wang 1995). The availability of this in vitro assay will allow a detailed analysis of yeast telomerase activity and components.

MAMMALIAN TELOMERASE RNAs

In mammals, both the number of telomeres and the level of telomerase activity are lower than in ciliates, suggesting telomerase components might not be abundant. To clone the gene encoding the human telomerase RNA, a strategy was employed to amplify the telomerase RNA (Feng et al. 1995). A randomly primed cDNA library was prepared from either partially purified telomerase fractions or size-selected total RNA. The cDNA library was subjected to several rounds of selection for sequences encoding a putative telomerase RNA template domain 5′-CUAACCCUA-3′, using a biotinylated telomeric oligonucleotide and PCR. The telomerase RNA candidates were evaluated by determining whether they copurified with telomerase activity and whether antisense oligonucleotides inhibited telomerase activity. The genomic copy of the best candidate was cloned, and the putative template domain was mutated to encode d(TTGGGG) or d(TTTGGG) instead of d(TTAGGG) repeats. In vitro, telomerase assays using telomerase extracted from cells stably transfected with the altered gene indicated that the altered telomeric repeats were synthesized, which showed that this gene encodes the human telomerase RNA (hTR) component.

The hTR is approximately 450 nucleotides long, in contrast to both the shorter ciliate telomerase RNAs and the longer yeast RNAs. Also in contrast to the ciliate RNAs, evidence suggests that the hTR is an RNA polymerase II, not a polymerase III, transcript (Greider and Blackburn 1989; Yu et al. 1990; Avilion 1995; Feng et al. 1995). Using the human hTR gene as a probe in hybridization experiments, the mouse telomerase RNA (mTR) was identified, and its gene was cloned (Blasco et al. 1995). To confirm that this RNA is an essential component of mouse telomerase, oligonucleotide inhibition experiments and template alteration experiments were done. Interestingly, the mouse telomerase RNA has only 61% sequence identity to the human telomerase RNA, in contrast to other small RNAs that are more highly conserved between human and mouse. For example, the U6 snRNA has 100% and U7 has 85% identity between the two organisms (Shumyatsky and Reddy 1992). The template regions of the two mammalian telomerase RNAs also differ; the mouse has a shorter template domain than the human. This shorter template domain in the mouse RNA could in part explain the decreased processivity of the mouse enzyme compared to human (Prowse et al. 1993), although there are also other biochemical differences between telomerases in the two species, such as the affinity for dGTP (see below).

IN VIVO ANALYSIS OF MUTANT TELOMERASE RNAs

In vivo expression of mutant telomerase RNA genes in *Tetrahymena*, and the corresponding alteration of telomere sequences, demonstrated that telomerase synthesizes telomeric repeats in vivo (Yu et al. 1990). The telomerase RNA gene template region was mutated and introduced into *Tetrahymena* cells. The RNA template was changed from 5′-CAACCCCAA-3′ to 5′-C<u>G</u>ACCCCAA-3′, resulting in the synthesis of d(GGGGT<u>C</u>) repeats; whereas changing the template domain to 5′-CAA<u>CCCCC</u>AA-3′ resulted in the synthesis of d(<u>GGGGG</u>TT) repeats at telomeres. A more efficient method of transformation by electroporation has recently been developed. Thus, it is now possible to make extracts from *Tetrahymena* cells expressing mutant telomerase RNAs and to analyze their elongation properties in vitro (Gilley et al. 1995).

In addition to causing changes in telomeric sequences, overexpression of mutant telomerase RNAs in *Tetrahymena* also causes changes in telomere length and in cell morphology and viability (Yu et al. 1990). In the transformants expressing telomerase RNA specifying d(GGGGGTT) or d(GGGGTC) telomere repeats, telomere lengths were longer than in wild-type cells. In one transformant in which mutant telomere sequences were not detected, telomere length was shorter. Expression of each of the three mutant RNAs eventually caused striking morphological changes and cell death. Before the cells died, most were large and arrested in cell division, had rounded or irregular shapes, and had enlarged and irregularly shaped macronuclei. Only loss of the mutant telomerase gene caused a reversion of these phenotypes. This was the first evidence establishing an essential role for telomerase in vivo. A more extensive analysis of cells expressing mutant telomerase RNAs indicates the mutants fall into three classes: some that have a wild-type phenotype, some that are early lethals, and some that are delayed lethals (Gilley et al. 1995).

A gene disruption of *TLC1* in yeast causes telomere shortening, as predicted for cells that can no longer synthesize telomeric DNA de novo (Singer and Gottschling 1994). In addition, these cells display a gradual increase in division time by generation 65, accompanied by a 50% drop in viability after 75 generations. These phenotypes are similar to those reported for *est1* strains (Lundblad and Szostak 1989). *tlc1* cultures survive only when the population is overtaken by faster-growing cells. Presumably, these cells are similar to the *est1* survivors, where telomere recombination provides some level of telomere maintenance (Lundblad and Blackburn 1993).

Mutants in the template region of *K. lactis* telomerase RNA are divided into three classes: Some have wild-type telomere length, some have telomeres that are short and stable, and some manifest dramatic telomere lengthening (McEachern and Blackburn 1995). Mutations that fall into each of these categories are interspersed throughout the 23-bp template domain. Complete deletion of the telomerase RNA gene also has *est1*-like phenotypes in *K. lactis*. A strain overexpressing the *Bgl*II-containing telomerase RNA showed a decreasing number of wild-type repeats with increasing rounds of cell division. At about 400 generations, when all wild-type sequence may have been replaced with mutant repeats, there was a drastic increase in telomere length and the cells died, suggesting that sequence alteration of all telomere repeats has a drastic effect on both telomere function and cell viability (McEachern and Blackburn 1995).

BIOCHEMISTRY OF TELOMERASES

Telomerase activity has been assayed in vitro using extracts from a variety of organisms and cell lines including ciliates (*Tetrahymena*, *Euplotes*, and *Oxytricha*), yeasts (*Saccharomyces*), vertebrates (*Xenopus*), and mammals (human and mouse) (Table 1). Although all telomerase activities are sensitive to pretreatment of the extract with RNase, other requirements for DNA synthesis vary. Currently the interpretation of differences in these telomerase activities is complicated by the lack of purified enzyme preparations. However, despite this limitation, clear differences are found in processivity, and primer and nucleotide specificity, between telomerases from *Tetrahymena*, hypotrichs, and mammalian cells.

Primer Specificity

The sequence requirements for primer elongation by *Tetrahymena* telomerase are minimal. Although telomere sequences are preferred, at very high oligonucleotide concentrations, telomerase elongates almost any oligonucleotide primer (Greider and Blackburn 1987; Blackburn et al. 1989; Harrington and Greider 1991; Lee and Blackburn 1993). The addition of telomeric repeats onto nontelomeric primers greatly increases the affinity of telomerase for a synthetic primer. In fact, the presence of only four guanosine residues at the primer 3′ end is sufficient to promote efficient elongation of a nonspecific oligonucleotide (Harrington and Greider 1991). Similar results are observed for human (Morin 1989), *Euplotes* (Shippen-Lentz and Blackburn 1990; Melek et al. 1994), and *Oxytricha* (Lingner et al. 1994; Melek et al. 1994) telomerases. These results suggest that in vitro, primer recognition is relatively promiscuous for telomerase enzymes.

The K_m values measured for *Tetrahymena* telomerase primer binding vary with the purity of the extract, the length and sequence of the primers, and the assay used. When processive products were measured, the apparent K_m was around 100 nM (Greider 1991). However, in this processive assay, multiple catalytic events are occurring. If elongation assays are used to determine K_m, measuring the first nucleotide added is most accurate, since only one catalytic step is required after binding for product formation. Using this assay, the $K_{m\ (app)}$ values for the 18-nucleotide primers $(GGGGTT)_3$ and $(GGGTTG)_3$, were measured at 1.5 μM. However, the $K_{m\ (app)}$ for dideoxy-terminated higher products in these reactions was between 10 nM and 50 nM (Collins and Greider 1993). This suggests that terminal dG addition onto these primers may represent several different steps. Lee and Blackburn were careful to show telomerase turnover under their reaction conditions, indicating that elongation to the first nucleotide did not represent stalled telomerase product complex. In these experiments measuring the addition of the first dT residue onto oligonucleotide primers, the $K_{m\ (app)}$ for $(TTGGGG)_3$ was 20 nM (Lee and Blackburn 1993).

To more directly assay primer binding to *Tetrahymena* telomerase, a direct primer-binding assay was developed (Harrington et al. 1995). This assay has the advantage that catalysis of dNTP addition is not required to measure primer binding. In this gel shift assay, the sequence specificity of primer binding is similar to that previously determined for primer elongation, suggesting that elongation specificity is determined in part at the level of binding.

Despite the different results with different assays, in all assays, the affinity for primers increases as primer length decreases: 18 nucleotides > 12 nucleotides > 10 nucleotides >> 6 nucleotides, such that a 6-nucleotide primer requires 1 μM to 5 μM concentration for elongation (Collins and Greider 1993; Lee and Blackburn 1993). If the inefficient 6-nucleotide telomeric primer is extended by additional specific or nonspecific sequence at the primer 5′ end, elongation is improved. In contrast, if nonspecific sequence is added to the 3′ end of 6 nucleotides of telomeric sequence, elongation is reduced. Thus, for optimal binding, the primer 3′ end should be complementary to the template RNA; however, the sequences in the primer 5′ region also affect elongation (Harrington and Greider 1991; Collins and Greider 1993; Lee and Blackburn 1993).

These results and others suggest that *Tetrahymena* telomerase contains two (or more) distinct primer-binding sites (Collins and Greider 1993; Lee and Blackburn 1993). One primer-binding site, termed the template site, aligns the primer 3′ end with the template RNA; primer binding at this site increases with complementarity to the template RNA. The second site, termed the anchor site, binds a more 5′ region of the primer (Fig. 3). Binding at the anchor site is enhanced with telomeric or G-rich sequences, but other sequences can serve as an anchor for processive elongation (Collins and Greider 1993; Lee and Blackburn 1993). It is unlikely that a folded primer structure such as a hairpin or a G-quartet is recognized by the anchor site (Blackburn et al. 1989). Such structures inhibit priming by *Oxytricha* telomerase (Zahler et al. 1991), and a 24-nucleotide-long primer containing only six 3′ terminal d(TTGGGG) bases is elongated efficiently, although it cannot form intramolecular hairpins (Harrington and Greider 1991). In addition, G-rich sequence primers with dispersed G residues are also elongated (Blackburn et al. 1989; Harrington 1993).

Analysis of primer K_m values for telomerases other than *Tetrahymena* is complicated by the presence of competing primer-binding activities in crude extracts. However, it is clear from primer titrations (Morin 1989) and the standard conditions used in most laboratories that the affinity of mammalian telomerases for telomeric primers is likely to be at most 10-fold higher than that of the *Tetrahymena* enzyme. For human telomerase, longer primers are more efficiently elongated than primers of 1–1.5 repeats. At saturating primer concentrations, nonspecific sequence added to the 5′ end of inefficient primers increases elongation, but nonspecific sequence added to a telomeric primer 3′ end is inhibitory (Morin 1989). These similarities of human and *Tetrahymena* telomerase primer specificity suggest that human telomerase also utilizes a second primer-binding site in addition to the template for optimal elongation (Morin 1989).

Nucleotide Specificity

The specificity of nucleotide incorporation by *Tetrahymena* telomerase can be altered in vivo or in vitro by mutation of the RNA template sequence. Thus, a large component of the nucleotide specificity of

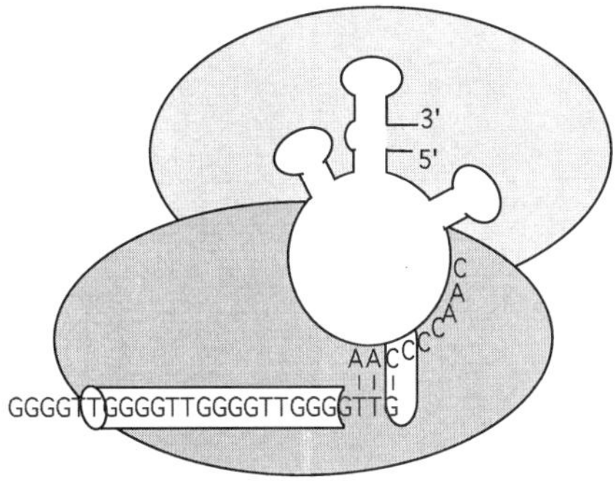

Figure 3 Two-site model of telomerase primer binding and elongation. Telomeric primer oligonucleotides interact with telomerase at two distinct sites, the 3′ end aligns and base-pairs with the template region allowing additions in the catalytic site (white oval), and the body of the primer is bound by a second site, the anchor site (white tube). This two-site model is similar to that proposed for RNA polymerases, to allow for processive elongation. To utilize the 9-nucleotide template regions, there must be some conformational change in the RNP complex. Either the catalytic site moves along the template or the template RNA moves relative to the catalytic site.

telomerase is determined by base-pairing with the template (Yu et al. 1990; Autexier and Greider 1994). However, quantitation of telomerase activity with titrations of individual nucleotides, assayed in partially purified extract (Greider and Blackburn 1987; Greider 1991; Lee and Blackburn 1993) or with highly purified telomerase (Collins and Greider 1995), showed *Tetrahymena* telomerase prefers guanosine nucleotides. Only 1 μM to 10 μM dGTP is required for optimal elongation, but TTP is required at 100 μM to 500 μM. This represents a 50- to 100-fold difference in the affinity of telomerase for its two natural dNTP substrates. Template-complementary ddNTPs are efficiently incorporated at concentrations of 100 μM to 500 μM (Greider 1991; Autexier and Greider 1994).

Euplotes and *Oxytricha* telomerases processively elongate telomeric primers at nucleotide concentrations similar to those used for *Tetrahymena* (Zahler and Prescott 1988; Shippen-Lentz and Blackburn 1989; Lingner et al. 1994). Like the *Tetrahymena* enzyme, *Oxytricha* telomerase requires significantly less dGTP than TTP for elongation (Zahler and Prescott 1988). Although human telomerase assays use 10-fold higher dNTP levels than the ciliates, human telomerase also appears to have a different affinity for dGTP versus TTP and dATP (Morin 1989). The dideoxynucleotides ddATP, ddTTP, and ddGTP are also incorporated by mouse telomerase (Prowse et al. 1993).

Processivity

The processivity of a polymerase represents the probability of dissociation after each successive nucleotide addition. Using telomeric $(TTGGGG)_n$ primers with three or more repeats, the *Tetrahymena* telomerase is highly processive (Greider 1991). Primers with shorter length and those that diverge from the telomeric $d(TTGGGG)_n$ sequence are elongated less processively (Collins and Greider 1993; Lee and Blackburn 1993). Nucleolytic cleavage has been observed for *Tetrahymena* telomerase, and a role for cleavage in normal processive elongation is suggested by analogy with DNA-dependent RNA polymerases (for review, see Kassavetis and Geiduschek 1993). The primer length requirements for processive elongation by *Tetrahymena* telomerase support the two-site model for primer binding. Products not bound to the anchor site are more likely to dissociate during repositioning of the product 3′ end. This model is similar to that proposed for processive elongation by RNA polymerase (Chamberlin 1992).

In contrast to the processive telomerases from ciliates and human cells, mouse and *Xenopus* telomerases are relatively nonprocessive. Using the reaction conditions originally described, hundreds of repeats are added onto a primer by human telomerase, whereas predominantly 1–3 repeats are added by mouse and *Xenopus* telomerases (Morin 1989; Prowse et al. 1993; Mantell and Greider 1994). Some of this difference arises from the requirement of the nonprocessive enzymes for higher concentrations of dGTP: When 10- to 100-fold higher concentrations of dGTP are used, synthesis of at least 10 repeats is observed for *Xenopus*, rat, and mouse telomerases (L.L. Mantell et al., unpubl.). In addition, the shorter template region in mouse telomerase RNA may affect processivity (Blasco et al. 1995).

TETRAHYMENA RIBONUCLEOPROTEIN STRUCTURE

The protein components of telomerase have recently been identified and the genes have been cloned from *Tetrahymena* (Collins et al. 1995). Two polypeptides of 80 kD and 95 kD were identified that reproducibly copurified with telomerase activity. Using peptide sequence from these proteins, the genes were both cloned from *Tetrahymena*. To determine if p80 and p95 were essential components of telomerase, antipeptide antibodies were generated against these two proteins. Telomerase activity was immunoprecipitated using an antibody directed against p80, but an unrelated antibody that did not recognize p80 or p95 did not immunoprecipitate telomerase activity. These experiments indicated that p80 is a functional component of telomerase. Western analysis of the pellet fraction showed that p95 coprecipitated with p80 in fractions that had activity. In addition, the p80 and p95 proteins copurified with each other over all columns tested, comigrated in a native gel, and cofractionated in a glycerol gradient even after RNase treatment. These data suggest that p80 and p95 bind to each other directly and are required for telomerase activity. Based on the predicted mass of telomerase, the active enzyme is likely to consist of one subunit each of the telomerase RNA, p80, and p95 (Collins et al. 1995).

The sequences of p80 and p95 do not show a high degree of similarity with other proteins in available databases. The p80 protein contains a zinc finger that may be involved in RNA or DNA binding. A region in the p95 protein can be aligned with the active-site motifs from two different classes of polymerases: RNA-dependent RNA polymerases and DNA polymerases α and β (Poch et al. 1989; Delarue et al. 1990). The lack of any other sequence homology with known polymerases, together with these motif regions, suggests that telomerase represents a new class of polymerase (Collins et al. 1995). RNA-dependent RNA polymerases are thought to be some of the earliest polymerases (Gesteland and Atkins 1993). If telomerase is indeed most closely related to this class of polymerase, it suggests that telomerase may have evolved from an ancestral polymerase.

Protein components have not yet been identified from organisms other than *Tetrahymena*. Nevertheless, it is possible to compare RNP composition through information obtained during enzyme purification. *Tetrahymena* telomerase fractionates during glycerol gradient sedimentation and gel filtration with a molecular weight of 270,000 or 200,000 (Greider and Blackburn 1987; Collins and Greider 1993; Harrington 1993; Collins et al. 1995). In contrast, telomerase from human cells fractionates by velocity sedimentation or gel filtration with a molecular weight around 750,000, and the mouse enzyme appears to fractionate even slightly larger than the human (K. Collins et al., unpubl.). Although differences in the sizes of telomerase RNAs may contribute somewhat to the increased molecular mass of the mammalian enzymes, there may be additional (or larger) protein components as well. Human and mouse telomerases appear to be less stable to column chromatography than *Tetrahymena*, perhaps due to the increased mass of the mammalian enzymes. Detailed comparison of telomerase RNP structures awaits the purification and cloning of additional telomerase protein components.

TELOMERASE ACTIVITY IN HUMAN CANCER CELLS

Unlike telomeres in most single-cell eukaryotes, telomeres in human somatic tissues are not maintained at a constant average length. In fibroblasts, blood cells, and other tissues, telomeres shorten with each round of cell division both in vitro and in vivo, whereas germ-line cells are immune to this progressive shortening (Cooke and Smith 1986; Harley et al. 1990; Hastie et al. 1990; Lindsey et al. 1991; Allsopp et al. 1992; Vaziri et al. 1993). Telomerase activity was not detected in many primary fibroblast cultures, suggesting that telomere shortening is due to the absence of telomerase (Counter et al. 1992, 1994; Kim et al. 1994). Telomere shortening may be due to the inability of conventional polymerase to replicate chromosome ends. Using a recently developed sensitive PCR assay, telomerase activity was also not detected in a number of primary human tissue samples (Kim et al. 1994). However, recent evidence indicates that some primary cells such as white blood cells do contain telomerase activity (Broccoli et al. 1995; Counter et al. 1995; Hiyama et al. 1995).

In contrast to some normal human tissues, cancer cells from tissue culture and those taken directly from tumors contain detectable telomerase activity (for review, see Bacchetti and Counter 1995). Evidence suggests that telomerase may be required for the growth of tumor cells. Thus, telomerase inhibition is potentially a prime target for cancer chemotherapy (Harley et al. 1990; Hastie et al. 1990; Counter et al. 1992; Kim et al. 1994; Klingelhutz et al. 1994; Mehle et al. 1994).

Unlike many human tissues that appear to lack telomerase activity, many primary mouse tissues express telomerase activity (Chadeneau et al. 1995; Prowse and Greider 1995). One explanation for this difference may be that mouse cells lack a repressor or inhibitor of telomerase that is present in human cells. Primary human cells rarely, if ever, spontaneously immortalize, whereas mouse cells frequently spontaneously immortalize (Macieira-Coelho and Azzarone 1988). If human cells have additional negative controls on telomerase that mouse cells lack, it may take fewer "hits" to activate expression in mouse cells than in human cells (Prowse and Greider 1995).

What role telomerase activation plays in cellular immortalization and cancer is an important question for future studies. To determine the role of telomerase in cancer, and to help design potential inhibitors, it is essential to have a thorough understanding of the telomerase mechanism. With the detailed groundwork of telomerase biochemistry laid and with the recent cloning of the yeast and mammalian RNAs and the *Tetrahymena* proteins, the stage is set to explore the telomerase polymerase mechanism and the role of telomerase in cell biology, development, and tumorigenesis.

REFERENCES

Allsopp, R.C., H. Vaziri, C. Patterson, S. Goldstein, E.V. Younglai, A.B. Futcher, C.W. Greider, and C.B. Harley. 1992. Telomere length predicts the replicative capacity of human fibroblasts. *Proc. Natl. Acad. Sci.* **89:** 10114–10118.

Autexier, C. and C.W. Greider. 1994. Functional reconstitution of wild type and mutant *Tetrahymena* telomerase. *Genes Dev.* **8:** 563–575.

———. 1995. Boundary elements of the *Tetrahymena* telomerase RNA template and alignment domains. *Genes Dev.* **15:** 2227–2239.

Avilion, A.A. 1995. *"Identification and characterization of human telomerase RNA."* Ph.D. thesis, State University of New York, Stony Brook.

Bacchetti, S. and C.M. Counter. 1995. Telomeres and telomerase in human cancer: Review. *Int. J. Oncol.* **7:** 423–432.

Bhattacharyya, A. and E.H. Blackburn. 1994. Architecture of telomerase RNA. *EMBO J.* **13:** 5521–5531.

Blackburn, E.H. 1991. Structure and function of telomeres. *Nature* **350:** 569–573.

Blackburn, E.H., C.W. Greider, E. Henderson, M. Lee, J. Shampay, and D. Shippen-Lentz. 1989. Recognition and elongation of telomeres by telomerase. *Genome* **31:** 553–560.

Blasco, M., W. Funk, B. Villepounteau, and C.W. Greider. 1995. Functional characterization and developmental regulation of mouse telomerase RNA. *Science* **269:** 1267–1270.

Broccoli, D., J.W. Young, and T. de Lange. 1995. Telomerase activity in normal and malignant haematopoietic cells. *Proc. Natl. Acad. Sci.* **92:** 9082–9086.

Chadeneau, C., P. Siegel, C.B. Harley, W. Muller, and S. Bacchetti. 1995. Telomerase activity in normal and malignant murine tissues. *Oncogene* **11:** 893–898.

Chamberlin, M.J. 1992. New models for the mechanism of transcription elongation and its regulation. *Harvey Lect.* **88:** 1–21.

Cohn, M. and E.H. Blackburn. 1995. Telomerase in yeast. *Science* **269:** 396–400.

Collins, K. and C.W. Greider. 1993. *Tetrahymena* telomerase catalyzes nucleolytic cleavage and non-processive elongation. *Genes Dev.* **7:** 1364–1376.

———. 1995. Elongation of RNA primers and incorporation of ribonucleotides by *Tetrahymena* telomerase. *EMBO J.* **14:** 5422–5432.

Collins, K., R. Koybayashi, and C.W. Greider. 1995. Purification of *Tetrahymena* telomerase and cloning of the genes for the two protein components of the enzyme. *Cell* **81:** 677–686.

Cooke, H.J. and B.A. Smith. 1986. Variability at the telomeres of human X/Y pseudo-autosomal region. *Cold Spring Harbor Symp. Quant. Biol.* **51:** 213–219.

Counter, G.M., H.W. Hirte, S. Bacchetti, and C.B. Harley. 1994. Telomerase activity in human ovarian carcinoma. *Proc. Natl. Acad. Sci.* **91:** 2900–2904.

Counter, C.M., J. Gupta, C.B. Harley, B. Leber, and S. Bacchetti. 1995. Telomerase activity in normal leukocytes and in hematologic malignancies. *Blood* **85:** 2315–2320.

Counter, C.M., A.A. Avilion, C.E. LeFeuvre, N.G. Stewart, C.W. Greider, C.B. Harley, and S. Bacchetti. 1992. Telomere shortening associated with chromosome instability is arrested in immortal cells which express telomerase activity. *EMBO J.* **11:** 1921–1929.

Delarue, M., O. Poch, N. Tordo, and P. Argos. 1990. An attempt to unify the structure of polymerases. *Protein Eng.* **3:** 461–467.

Feng, J., W. Funk, S. Wang, S. Weinrich, A. Avilion, C.-P. Chiu, R. Adams, E. Chang, J. Yu, S. Le, M. West, C.B. Harley, W. Andrews, C.W. Greider, and B. Villeponteau. 1995. The human telomerase RNA component. *Science* **269:** 1236–1241.

Gesteland, R.F. and J.F. Atkins, eds. 1993. *The RNA world.* Cold Spring Harbor Laboratory Press, Cold Spring Harbor, New York.

Gilley, D., M. Lee, and E.H. Blackburn. 1995. Altering specific telomerase RNA template residues affects active site function. *Genes Dev.* **9:** 2214–2226.

Gottschling, D.E., O.M. Aparicio, B.L. Billington, and V.A. Zakian. 1990. Position effect at *S. cerevisiae* telomeres: Reversible repression of Pol II transcription. *Cell* **63:** 751–762.

Greider, C.W. 1991. Telomerase is processive. *Mol. Cell. Biol.* **11:** 4572–4580.

Greider, C.W. and E.H. Blackburn. 1985. Identification of a specific telomere terminal transferase activity in *Tetrahymena* extracts. *Cell* **43:** 405–413.

———. 1987. The telomere terminal transferase of *Tetrahymena* is a ribonucleoprotein enzyme with two kinds of primer specificity. *Cell* **51:** 887–898.

———. 1989. A telomeric sequence in the RNA of *Tetrahymena* telomerase required for telomere repeat synthesis. *Nature* **337:** 331–337.

Harley, C.B., A.B. Futcher, and C.W. Greider. 1990. Telomeres shorten during ageing of human fibroblasts. *Nature* **345:** 458–460.

Harrington, L.A. 1993. *Characterization and purification of Tetrahymena telomerase.* Ph. D. thesis, State University of New York, Stony Brook.

Harrington, L.A. and C.W. Greider. 1991. Telomerase primer specificity and chromosome healing. *Nature* **353:** 451–454.

Harrington, L.H., C. Hull, J. Crittenden, and C.W. Greider. 1995. Gel shift and UV crosslinking analysis of *Tetrahymena* telomerase. *J. Biol. Chem.* **270:** 8893–8901.

Hastie, N.D., M. Dempster, M.G. Dunlop, A.M. Thompson, D.K. Green, and R.C. Allshire. 1990. Telomere reduction in human colorectal carcinoma and with ageing. *Nature* **346:** 866–868.

Henderson, E.R. 1995. Telomeric DNA sequences. In *Telomeres* (ed. E.H. Blackburn and C.W. Greider) pp. 11–34. Cold Spring Harbor Laboratory Press, Cold Spring Harbor, New York.

Hiyama, K., Y. Hirai, S. Kyoizumi, M. Akiyama, E. Hiyama, M.A. Piatyszek, J. Shay, S. Ishioka, and M. Yamakido. 1995. Activation of telomerase in human lymphocytes and hematopoietic progenitor cells. *J. Immunol.* **155:** 3711–3715.

Kassavetis, G.A. and E.P. Geiduschek. 1993. RNA polymerase marching forward. *Science* **259:** 944–945.

Kim, N.W., M.A. Piatyszek, K.R. Prowse, C.B. Harley, M.D. West, P.L. Ho, G.M. Coviello, W.E. Wright, S.L. Weinrich, and J.W. Shay. 1994. Specific association of human telomerase activity with immortal cells and cancer. *Science* **266:** 2011–2014.

Klingelhutz, A.J., S. Barber, P.P. Smith, K. Dyer, and J.K. McDougall. 1994. Restoration of telomeres in human papillomavirus-immortalized human anogenital epithelial cells. *Mol. Cell Biol.* **14:** 961–969.

Kramer, K.M. and J.E. Haber. 1993. New telomeres in yeast are initiated with a highly selected subset of TG_{1-3} repeats. *Genes Dev.* **7:** 2345–2356.

Lee, M.S. and E.H. Blackburn. 1993. Sequence-specific DNA primer effects on telomerase polymerization activity. *Mol. Cell Biol.* **13:** 6586–6599.

Lin, J.-J., and V.A. Zakian. 1995. An in vitro assay for *Saccharomyces* telomerase requires *EST1*. *Cell* **81:** 1127–1135.

Lindsey, J., N.I. McGill, L.A. Lindsey, D.K. Green, and H.J. Cooke. 1991. In vivo loss of telomeric repeats with age in humans. *Mutat. Res.* **256:** 45–48.

Lingner, J., L.L. Hendrick, and T.R. Cech. 1994. Telomerase RNAs of different ciliates have a common secondary structure and a permuted template. *Genes Dev.* **8:** 1984–1998.

Lue, N.F. and J.C. Wang. 1995. ATP-dependent processivity of a telomerase activity from *Saccharomyces cerevisiae*. *J. Biol. Chem.* **270:** 21453–21456.

Lundblad, V. and E.H. Blackburn. 1993. An alternative pathway for yeast telomere maintenance rescues *est1*$^{-}$ senescence. *Cell* **73:** 347–360.

Lundblad, V. and J.W. Szostak. 1989. A mutant with a defect in telomere elongation leads to senescence in yeast. *Cell* **57:** 633–643.

Macieira-Coelho, A. and B. Azzarone. 1988. The transition from primary culture to sponteaneous immortalization in mouse fibroblast populations. *Anticancer Res.* **8:** 669–676.

Mantell, L.L. and C.W. Greider. 1994. Telomerase activity in germline and embryonic cells of *Xenopus*. *EMBO J.* **13:** 3211–3217.

McCormick-Graham, M. and D.P. Romero. 1995. Ciliate telomerase RNA structural features. *Nucleic Acids Res.* **23:** 1091–1097.

McEachern, M.J. and E.H. Blackburn. 1994. A conserved sequence motif within the exceptionally diverse telomeric sequences of budding yeasts. *Proc. Natl. Acad. Sci.* **91:** 3453–3457.

———. 1995. Runaway telomere elongation caused by telomerase RNA gene mutations in the yeast *K. lactis*. *Nature* **376:** 403–409.

McEachern, M.J. and J.B. Hicks. 1993. Unusually large telomeric repeats in the yeast *Candida albicans*. *Mol. Cell. Biol.* **13:** 551–560.

Mehle, C., B. Ljungberg, and G. Roos. 1994. Telomere shortening in renal cell carcinoma. *Cancer Res.* **54:** 236–241.

Melek, M., B. Davis, and D.E. Shippen. 1994. Oligonucleotides complementary to the *Oxytricha nova* telomerase RNA delineate the template domain and uncover a novel mode of primer utilization. *Mol. Cell. Biol.* **14:** 7827–7838.

Morin, G.B. 1989. The human telomere terminal transferase enzyme is a ribonucleoprotein that synthesizes TTAGGG repeats. *Cell* **59:** 521–529.

Poch, O., I. Sauvaget, M. Delarue, and N. Tordo. 1989. Identification of four conserved motifs among the RNA-dependent polymerase encoding elements. *EMBO J.* **8:** 3867–3874.

Prowse, K.R. and C.W. Greider. 1995. Developmental and tissue specific regulation of mouse telomerase and telomere length. *Proc. Natl. Acad. Sci.* **92:** 4818–4822.

Prowse, K.R., A.A. Avilion, and C.W. Greider. 1993. Identification of a nonprocessive telomerase activity from mouse cells. *Proc. Natl. Acad. Sci.* **90:** 1493–1497.

Romero, D.P. and E.H. Blackburn. 1991. A conserved secondary structure for telomerase RNA. *Cell* **67:** 343–353.

Shippen-Lentz, D. and E.H. Blackburn. 1989. Telomere terminal transferase activity from *Euplotes crassus* adds large numbers of TTTTGGGG repeats onto telomeric primers. *Mol. Cell. Biol.* **9:** 2761–2764.

———. 1990. Functional evidence for an RNA template in telomerase. *Science* **247:** 546–552.

Shumyatsky, G. and R. Reddy. 1992. Compilation of small RNA sequence. *Nucleic Acids Res.* **20:** 2159–2165.

Singer, M.S. and D.E. Gottschling. 1994. *TLC1:* Template RNA component of *Saccharomyces cerevisiae* telomerase. *Science* **266:** 404–409.

ten Dam, E., A. Van Belkum, and C.W.A. Pleij. 1991. A conserved pseudoknot in telomerase RNA. *Nucleic Acids Res.* **19:** 6951.

Vaziri, H., F. Schaechter, I. Uchida, L. Wei, Z. Xiaoming, R. Effros, D. Choen, and C.B. Harley. 1993. Loss of telomeric DNA during aging of normal and trisomy 21 human lymphocytes. *Am. J. Hum. Genet.* **52:** 661–667.

Yu, G.-L., J.D. Bradley, L.D. Attardi, and E.H. Blackburn. 1990. *In vivo* alteration of telomere sequences and senescence caused by mutated *Tetrahymena* telomerase RNAs. *Nature* **344:** 126–132.

Zahler, A.M. and D.M. Prescott. 1988. Telomere terminal transferase activity in the hypotrichous ciliate *Oxytricha nova* and a model for replication of the ends of linear DNA molecules. *Nucleic Acids Res.* **16:** 6953–6972.

Zahler, A.M., J.R. Williamson, T.R. Cech, and D.M. Prescott. 1991. Inhibition of telomerase by G-quartet DNA structures. *Nature* **350:** 718–720.

Zaug, A.J. and T.R. Cech. 1995. Analysis of the structure of *Tetrahymena* nuclear RNAs in vivo: Telomerase RNA, self-splicing rRNA intron and U2 snRNA. *RNA* **1:** 363–374.

22

SV40 and Polyomavirus DNA Replication

John A. Hassell and Bradford T. Brinton
Cancer Research Group
Institute for Molecular Biology and Biotechnology
McMaster University
Hamilton, Ontario
Canada L8S 4K1

The papovavirus family comprises two virus subgroups: the polyomaviruses and the papillomaviruses. The polyomavirus subgroup consists of a dozen members and includes simian virus 40 (SV40), the murine polyomavirus (PyV), and two human isolates, JCV and BKV (for review, see Tooze 1980). The polyomaviruses display strict species specificity for their replication, but otherwise share a common architecture, genome organization, and replication cycle. SV40 replicates in simian cells and to a lesser extent in human cells, whereas PyV replicates in cells of murine origin. The circular double-stranded DNA genomes of SV40 and PyV comprise about 5300 bp. The viral DNA is assembled into chromatin both in virions and in infected cells, and apart from the absence of histone H1, it is structurally indistinguishable from that of the cell.

This review focuses principally on the replication of SV40 DNA and draws on the PyV literature to highlight differences between these two papovaviruses where appropriate. Several excellent reviews on this subject have appeared recently (DePamphilis and Bradley 1986; Challberg and Kelly 1989; Stillman 1989; Hurwitz et al. 1990; Melendy and Stillman 1992). We refer the reader to these and to other chapters in this volume for more in-depth discussion of topics we are unable to describe here and for more comprehensive citations to the primary literature.

Since their discovery in the late 1950s, SV40 and PyV have served as paradigms to understand cellular processes as diverse as DNA replication, transcription, and oncogenic transformation. The rationale for employing these viruses as models for DNA replication is mainly that their simple genomes are replicated by mechanisms similar to those of the cell, principally through the action of cellular enzymes. Only a single viral protein, large tumor (T) antigen, is required for viral DNA replica-

DNA Replication in Eukaryotic Cells

tion. Furthermore, like cellular DNA, the viral genome is packaged into chromatin, and its replication occurs only during the S phase of the cell cycle.

Large T antigen, the viral initiator protein for DNA replication, is a sequence-specific DNA-binding protein that also possesses intrinsic DNA helicase and ATPase activity. These activities of large T antigen are required directly for its ability to effect viral DNA replication. Large T antigen also functions indirectly to effect viral DNA replication by stimulating the synthesis of cellular enzymes whose activity is needed for viral DNA replication. For example, large T antigen binds to $p105^{Rb}$/E2F complexes, resulting in the release and activation of the transcription activator E2F, whose target genes encode enzymes such as thymidine kinase, dihydrofolate reductase, and DNA polymerase-α:primase.

Initiation of viral DNA replication takes place within the origin (*ori*), proceeds bidirectionally, and terminates when the two replication forks meet approximately 180° from where they started. Replication of the leading strand occurs by a continuous mechanism, whereas that of the lagging strand proceeds discontinuously (see Brush and Kelly, this volume). Hence, both SV40 and PyV DNA replication are semi-discontinuous. Large T antigen binds to repeats of its recognition element within *ori* and unwinds this region to create a replication bubble. Large T antigen also interacts with cellular replication proteins to assemble the primosome at *ori* and functions as a helicase at replication forks.

Replication of the viral genome requires the interplay among *cis*-acting sequences defining the functional *ori*, large T antigen, and a number of cellular proteins, including transcription activators and replication proteins. In the sections that follow, we summarize what is known about these three major components of the replication apparatus and suggest a model by which their interaction results in the replication of the viral genome.

VIRAL SEQUENCES AND PROTEINS REQUIRED FOR DNA REPLICATION

Origin for DNA Replication

The genetically defined *ori* for SV40 or PyV DNA replication is located in the noncoding region of the viral genome intermingled with early and late promoter elements (for review, see DePamphilis 1988). It comprises two functional components: a core region, referred to as *ori*-core (the minimal origin), and auxiliary regions (Fig. 1). *Ori*-core is required for replication under all conditions and encompasses the sites where DNA replication initiates (origin for bidirectional DNA replication [OBR]).

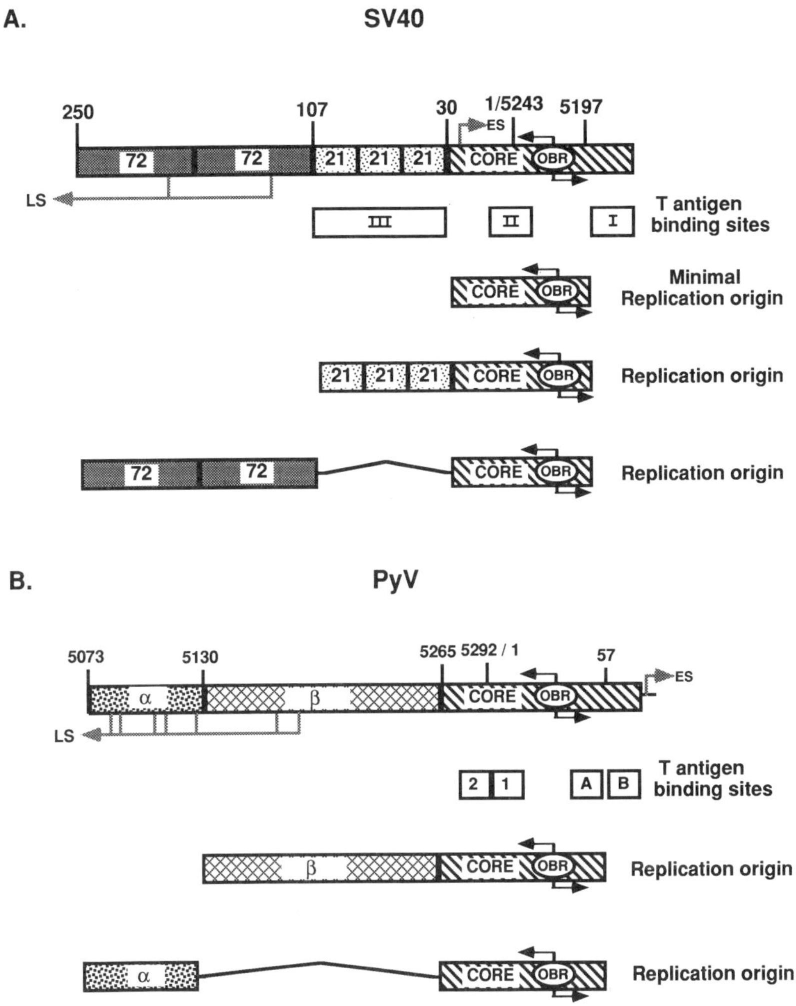

Figure 1 (*A*) Structure of the SV40 *ori* for DNA replication. The locations of the 72-bp repeat enhancer and the 21-bp repeat upstream promoter elements are shown. The *ori*-core contains the *ori* of bidirectional DNA replication (OBR). Stippled arrowheads represent the principal late (LS) and early (ES) start sites for transcription. The locations of the three large-T-antigen-binding sites are displayed below the origin region. The SV40 *ori*-core is sufficient to constitute a minimal *ori* for DNA replication in vivo; efficient replication *ori*s are composed of the *ori*-core and either the 21-bp repeats or the 72-bp repeats. (*B*) Structure of the PyV *ori* for DNA replication. The positions of the α and β elements of the PyV enhancer are shown. A functional PyV replication *ori* requires the *ori*-core and either the α or the β element located at the late border of *ori*-core.

The auxiliary regions increase the efficiency and cell-type specificity of viral DNA replication. The extent to which auxiliary elements enhance DNA replication differs between SV40 and PyV: SV40 DNA replication is stimulated 5- to 25-fold by these elements, whereas PyV DNA replication is enhanced between 100- and 1000-fold (Hassell et al. 1986).

The *ori*-core of each virus is about 65 bp and comprises three sets of sequences: an A/T-rich stretch at the late border; a central G/C-rich palindrome, and a region referred to as the early palindrome at the early border (Fig. 2) (Parsons et al. 1990 and references therein). All polyomaviruses, including distantly related members like SV40 and PyV, display significant sequence similarity in their *ori*-core. The A/T-rich stretch flanks the late border of the core and comprises 20 bp including a string of 8 consecutive A and corresponding T residues on each strand. The number of these residues is critical for *ori* function; deletion of only a few base pairs from this region compromises replication in vivo and in vitro (Gerard and Gluzman 1986). In SV40, the early TATA transcriptional element is located within the A/T-rich region. However, the PyV early TATA box is not found within its *ori*-core, suggesting that this transcriptional motif per se is not required for *ori* function. The A/T-rich region is a natural site of DNA bending (Deb et al. 1986); the binding of large T antigen to the central palindrome in the *ori*-core further distorts the structure of this region (Parsons et al. 1990).

The central 30-bp palindrome contains the four pentanucleotide (5′-GAGGC-3′) large-T-antigen-binding sites (DeLucia et al. 1983; Pomerantz and Hassell 1984). The sequence, arrangement, and spacing of these pentanucleotides are critical for *ori* function (Dean et al. 1987; Deb et al. 1987). This is consistent with the preservation of these features in the *ori*-cores of all polyomavirus subgroup members. The large-T-antigen-binding sites within the central palindrome are arranged differently than are those outside this region: The pentanucleotide-binding sites outside the *ori*-core are oriented in the same direction, whereas those within *ori*-core are arranged as two inverted pairs (Fig. 2). The arrangement of these pentanucleotide motifs may determine whether interaction of large T antigen with these sequences results in initiation of DNA replication or repression of early transcription.

The approximately 20-bp early palindrome comprises the remainder of *ori*-core. This sequence too is conserved among those polyomaviruses that replicate in cells from the same or similar species (i.e., monkey viruses SV40 and SA17, and the human papovaviruses BK and JC). Primer mapping experiments suggest that the early palindrome lies near the center of the replication bubble (Hay and DePamphilis 1982;

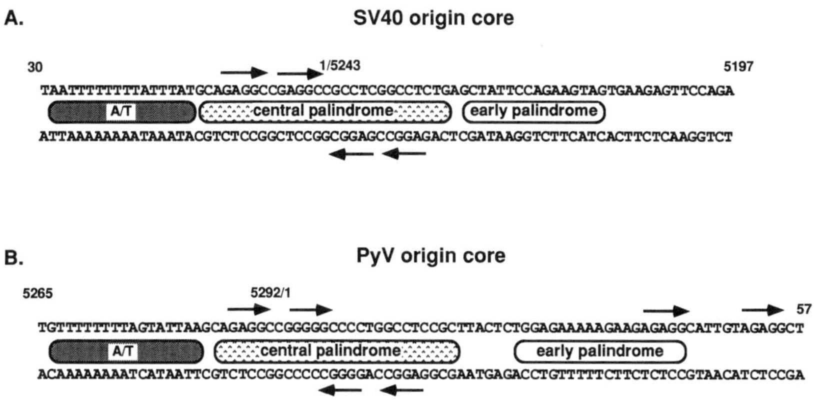

Figure 2 Sequence of the SV40 (*A*) and PyV (*B*) *ori*-cores for DNA replication. The three principal sequence elements of the *ori*-core are shown: the A/T-rich region at the late border, the central G/C-rich palindrome, and the early palindrome at the early border. Arrowheads depict the position and orientation of the 5′-GAGGC-3′ large-T-antigen-binding motifs.

Hendrickson et al. 1987). This region may be functionally analogous to the DNA unwinding element (DUE) first recognized in yeast *ori*s (Lin and Kowalski 1994). Surprisingly, at high concentrations, large T antigen is capable of binding with specificity to the early palindrome despite the absence of a recognizable pentanucleotide element (Parsons et al. 1990).

Two classes of auxiliary elements constitute the replication *ori*s of SV40 and PyV. One class comprises large-T-antigen-binding sites. In SV40 these sites flank both borders of *ori*-core (sites I and III); in PyV this class of auxiliary elements maps the early border of the PyV *ori*-core (sites A and B) (Fig. 1). The affinity of large T antigen for these sites in vitro exceeds or parallels that for those binding sites within *ori*-core. Hence, this class of auxiliary elements may stimulate DNA replication by recruiting large T antigen to the vicinity of *ori*-core and thereby facilitate large-T-antigen-mediated *ori* unwinding (Gutierrez et al. 1990). Deletion of site I in SV40 reduces *ori* activity but does not affect bidirectionality of DNA replication (Guo et al. 1991).

The other category of auxiliary elements flanks the late border of each *ori*-core; these elements are coincident with the transcriptional motifs that make up the viral enhancer and promoters (Hassell et al. 1986; Muller et al. 1988). Auxiliary elements of this type augment the replicative potential of the *ori*-core to a much greater extent than do those composed of large-T-antigen-binding sites (for review, see DePamphilis and Bradley 1986; DePamphilis 1988, 1993). There are several functionally redundant sets of sequences in both the SV40 and PyV regulatory regions that can act as auxiliary elements (Fig. 1). For example, either one of two regions, termed alpha or beta, functions as an auxiliary element for PyV DNA replication (Muller et al. 1988). Similarly, the 21-bp or the 72-bp repeat of SV40 can functionally substitute for each other in the SV40 *ori*. Furthermore, the SV40 auxiliary elements can substitute for those of PyV. Indeed, repeats of many different types of transcriptional elements can replace the PyV auxiliary regions. Notably, binding sites for GAL4, a transcriptional activator from budding yeasts, can act as auxiliary elements for PyV DNA replication as long as GAL4 is provided in *trans* (Bennett-Cook and Hassell 1991; Guo and DePamphilis 1992). Both the DNA-binding domain of GAL4 and its transcription activation domains are required to enhance viral DNA replication (Bennett-Cook and Hassell 1991). GAL4 chimeras bearing a diversity of activation domains from other transcription factors are also capable of activating PyV DNA replication, albeit to differing extents, suggesting that this property is not unique to particular classes of activation domains (Bennett-Cook and Hassell 1991; Guo and DePamphilis 1992; B.T. Brinton et al., in prep.). In this regard, it is of interest that the activation domains of three different transcription activators bind to replication protein A (RP-A), a critical component of the preinitiation complex (He et al. 1993; Li and Botchan 1993).

Several mechanisms have been suggested to account for the ability of transcriptional activators to stimulate SV40 and PyV DNA replication (for review, see DePamphilis; van der Vliet; both this volume). These include relieving the repressive effects of nucleosomes (Cheng and Kelly 1989; Cheng et al. 1992), stimulating the binding and/or the activity of large T antigen at *ori* (Guo et al. 1989; Gutierrez et al. 1990), and recruiting RP-A and/or enhancing its function at *ori* (He et al. 1993; Li and Botchan 1993). These mechanisms are not mutually exclusive; the recruitment of RP-A to *ori* by transcription activators could enhance large T antigen's ability to unwind the *ori*-core.

Large T Antigen

The large T antigens of SV40 and PyV are multifunctional, nuclear phosphoproteins (for review, see Prives 1990; Fanning 1992; Fanning and Knippers 1992; Pipas 1992; Manfredi and Prives 1994). SV40 and PyV large T antigen are composed of 708 and 785 amino acids, respectively. These two proteins are related in primary structure (~36% sequence identity) and share the same biochemical activities (Fig. 3). These

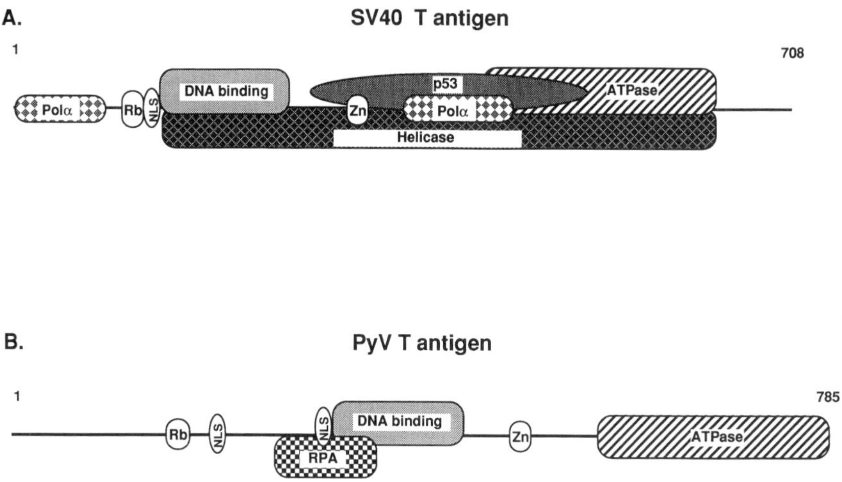

Figure 3 Schematic representation of the location of protein-binding sites, sequence motifs, and functional domains on a linear representation of SV40 (*A*) and PyV (*B*) large T antigen. Two regions in SV40 T antigen are required for interaction with DNA polymerase-α:primase (polα). Regions required for large T antigen binding to p105^{Rb} (Rb), RP-A, and p53 are also shown. The nuclear localization signals (NLS) and putative Zn finger (Zn) are illustrated. The minimal functional domains of large T antigen required for DNA binding, and ATPase and helicase activity are shown. The regions in PyV large T antigen required for binding to the various subunits of DNA polymerase-α:primase have not been identified, nor has its helicase domain been localized. Similarly, the domain in SV40 large T antigen required for interaction with RP-A has not been reported. The first and last amino acids in each large T antigen are numbered.

include sequence-specific double-stranded and single-stranded DNA binding, and ATPase and helicase activity. Discrete functional domains located in similar regions of the two proteins are required to mediate these various functions. In SV40 large T antigen, the sequence-specific DNA-binding domain occupies 115 amino acids between residues 135 and 249 (for review, see Pipas 1992), whereas the corresponding domain in PyV large T antigen maps to a 116-amino-acid stretch between residues 282 and 398 (Sundstrom et al. 1991). Interestingly, both large T antigens recognize and bind to the same pentanucleotide sequence motif, 5′-G(A/G)GGC-3′(Pomerantz and Hassell 1984). This is consistent with the fact that their DNA-binding domains display considerable sequence similarity (Sundstrom et al. 1991). A solitary pentanucleotide is not sufficient for large T antigen binding; rather, repeats of this motif, as occur in the various large-T-antigen-binding sites in the viral genome, are required for efficient sequence-specific DNA binding (Pomerantz and Hassell 1984). The binding of large T antigen to its recognition sites is also influenced by the nature and composition of the spacer region between pentanucleotide repeats (DeLucia et al. 1983; Dean et al. 1987; Deb et al. 1987). Large T antigen also binds to single-stranded DNA; this activity may be integral to its ability to unwind DNA at the replication fork (for review, see Fanning and Knippers 1992). Large T antigen hydrolyzes ATP, an activity that is required to act as a DNA helicase. The ATP-binding/ATPase domain of large T antigen occupies a region within the carboxyl terminus of the protein. Unlike many other helicases, large T antigen unwinds double-stranded *ori* DNA by translocating in a 3′to 5′direction along the leading-strand template to advance the replication fork (Goetz et al. 1988; Wiekowski et al. 1988).

A number of cellular proteins interact with SV40 and PyV large T antigens (Fig. 3). These include the DNA replication proteins, DNA polymerase-α:primase and single-stranded DNA-binding protein RP-A, and transcription regulatory proteins $p105^{Rb}$ and p53. Large T antigen also contains one (SV40) or more (PyV) nuclear localization signals (NLS) and several conserved sequence motifs associated with its replication functions (Fig. 3). One highly conserved sequence element, HPDKGG, in the amino-terminal region, is present in 11 of the 12 polyomavirus subgroup large T antigens. This sequence is known to be required for SV40 DNA replication (Peden and Pipas 1992; Pipas 1992); it may bridge the association of large T antigen with heat shock proteins (T. Roberts, pers. comm.). Another sequence element conserved among the large T antigens is a putative Zn-finger region (Zn) located between the DNA-binding and ATPase domains. This region is thought to be required for the formation of the double hexamer of large T antigen that assembles at the *ori*-core prior to the initiation of viral DNA replication (Deb and Tegtmeyer 1987; Mastrangelo et al. 1989; Loeber et al. 1991; Parsons et al. 1991).

The activity of SV40 large T antigen is regulated both positively and negatively by phosphorylation (for review, see Fanning 1994; see Weisshart and Fanning, this volume). Whereas large T antigen is phosphorylated in vivo at a total of nine serine and threonine residues clustered at its amino and carboxyl termini (Scheidtmann et al. 1991b), genetic (Schneider and Fanning 1988) and biochemical analyses suggest that the modification of only three of these sites, S120, S123, and T124, plays a critical role in regulating its activity (McVey et al. 1989, 1993; Cegielska et al. 1994). Phosphorylation of T124 positively influences the activity of large T antigen by increasing its ability to assemble into double hexamers at *ori* and to bidirectionally unwind DNA (McVey et al. 1993; Moarefi et al. 1993). The cyclin-dependent kinase, cdc2, in association with cyclin B, phosphorylates T124 in vitro (McVey et al. 1989, 1993). Interestingly, the S-phase cyclin-dependent kinase, cyclin A/cdk2, also phosphorylates large T antigen at T124. Cyclin A/cdk2 is physically associated with large T antigen in SV40-infected cells (Adamczewski et al. 1993). This and the fact that SV40 and PyV DNA replication takes place only during S phase when cyclin A/cdk2 is active suggest that this cyclin-dependent kinase may be one of the physiologically relevant kinases that regulates large T antigen activity (Adamczewski et al. 1993).

Phosphorylation of S120 and S123 negatively regulates the replicative capacity of large T antigen in vitro (Scheidtmann et al. 1991b; Virshup et al. 1992; Cegielska et al. 1994). Paradoxically, genetic analyses imply that phosphorylation of these residues is required for large T antigen function because their mutation to alanine results in an inactive large T antigen (Schneider and Fanning 1988). These findings can be reconciled if phosphorylation-dephosphorylation of these sites plays a regulatory role during the virus life cycle, and if this is not manifest in the in vitro DNA replication system. An isoform of casein kinase I (Cegielska and Virshup 1993) phosphorylates these sites in vitro, and the catalytic subunit of protein phosphatase 2A dephosphorylates them (Sheidtmann et al. 1991a). It is still unclear whether these enzymes regulate the activity of large T antigen in vivo; neither is known to be regulated during the cell cycle, and one of them (phosphatase 2A) is located principally in the cytoplasm.

The capacity of large T antigen to assemble into double hexamers and to unwind *ori* DNA is regulated by phosphorylation, but the helicase activity of large T antigen is unaffected by this modification (Virshup et al. 1992; Moarefi et al. 1993; McVey et al. 1993). Helicase activity measures the ability of a single hexamer of large T antigen to unwind a partially single-stranded template, whereas *ori* unwinding requires that a double hexamer of large T antigen unwinds double-stranded *ori* DNA bidirectionally. Electron microscopic studies suggest that DNA unwinding is accomplished by reeling the DNA through the double-hexamer complex rather than by independent progress of each hexamer bidirectionally from the center of *ori*-core (Wessel et al. 1992; Fanning 1994).

CELLULAR PROTEINS REQUIRED FOR VIRAL DNA REPLICATION

The development by Li and Kelly (1984) of an in vitro system capable of faithfully replicating SV40 DNA represented a significant breakthrough.

This system has been used to identify numerous cellular replication proteins, thus providing insight into the mechanism and regulation of viral and cellular DNA replication. To date, a minimum of ten cellular proteins have been identified that are required together with large T antigen and *ori*-bearing plasmids to replicate SV40 DNA in vitro (Table 1). These proteins form complexes that act cooperatively to accomplish the replication of the viral genome. Examples of such macromolecular assemblies include the preinitiation complex or primosome comprising large T antigen, RP-A and DNA polymerase-α:primase, an elongation complex composed of replication factor C (RF-C), proliferating cell nuclear antigen (PCNA), and DNA polymerase-δ and a maturation complex constituting maturation factor I, ribonuclease H1, and DNA ligase I. Elucidation of the structure and function of these proteins has led to the realization that distantly related organisms employ structurally related proteins to achieve replication of their genomes by similar mechanisms (see Stillman, this volume).

Replication Protein A

RP-A, which is also known as replication factor A (RFA) and human single-stranded DNA-binding protein (HSSB), was originally identified as an essential single-stranded DNA-binding protein required for SV40 DNA replication in vitro (Wobbe et al. 1987; Wold and Kelly 1988; Fairman and Stillman 1988). RP-A functions in concert with large T antigen and DNA polymerase-α:primase to initiate DNA replication at *ori*. RP-A is also needed during the elongation phase, acting to stimulate the activity of both DNA polymerase-α:primase and DNA polymerase-δ at replication forks (for review, see Melendy and Stillman 1992). The subunit composition of RP-A and the sequence of each of its subunits are highly conserved among diverse species. Gene disruption experiments in the budding yeast *Saccharomyces cerevisiae* demonstrate that each subunit is required for viability (Heyer et al. 1990; Brill and Stillman 1991). In this organism, and likely all eukaryotes, RP-A participates not only in DNA replication, but also in recombination and repair (Longhese et al. 1994; Firmenich et al. 1995; Smith and Rothstein 1995).

Human RP-A is a heterotrimer of 70-kD, 32-kD, and 14-kD subunits (Fairman and Stillman 1988; Wold and Kelly 1988). The 70-kD subunit is independently capable of binding to single-stranded DNA (Brill and Stillman 1989; Wold et al. 1989; Kenny et al. 1990), to DNA polymerase-α:primase (Dornreiter et al. 1992), and to the activation domains of several transcription activators (He et al. 1993; Li and

Table 1 Proteins required for SV40 DNA replication in vitro

Factor	Molecular mass (kD)	Function
Large T antigen	90	origin recognition protein; unwinds DNA; helicase; primosome loading protein
RP-A	70 32 14	binds single-stranded DNA; promotes origin unwinding; stimulates DNA polymerase-α:primase; cooperates with RF-C and PCNA to stimulate DNA polymerase-δ
DNA polymerase-α:primase	180 70 58 48	initiates leading- and lagging-strand synthesis
DNA polymerase-δ	125 48	completes leading- and lagging-strand synthesis
DNA polymerase-ε	255 55	potential DNA polymerase for completing lagging-strand synthesis
RF-C	140 40 38 37 36.5	auxiliary factor for DNA polymerase-δ and ε; loads PCNA onto DNA; DNA-dependent ATPase
PCNA	36	auxiliary factor for DNA polymerase-δ and ε, increases processivity
Topoisomerase I	110	relieves torsional strain in front of replication forks
Topoisomerase IIa	170	segregates progeny strands by removing
Topoisomerase IIb	180	catenated intertwines behind forks
MF-I	45	5′→3′ exonuclease removes terminal monoribonucleotide from Okazaki fragments
RNase H1	89	endonuclease; cleaves RNA primer
DNA ligase I	125	ligates Okazaki fragments

Botchan 1993). No activity or function has yet been ascribed to the other two subunits, but the 70-kD subunit is incapable of substituting for the trimeric protein in SV40 DNA replication, implying that one or both of them are important in this process (Erdile et al. 1991). Antibodies specific for each of the subunits inhibit SV40 DNA replication in vitro, suggesting that all the subunits are required (Erdile et al. 1990, 1991; Kenny et al. 1990; Umbricht et al. 1993). The 32-kD and 14-kD subunits form a complex, which may serve to recruit the larger subunit (Hendrickson et al. 1994).

The 32-kD species from budding yeast and humans is phosphorylated in a cell-cycle-dependent manner (Din et al. 1990). Phosphorylation occurs at the G_1 to S-phase transition, and dephosphorylation occurs during mitosis. The kinase(s) responsible for these phosphorylation events in vivo is not known, but both cyclin B/cdc2 (Dutta and Stillman 1992) and cyclin A/cdk2 phosphorylate RP-A in vitro (Elledge et al. 1992; Pan et al. 1993). Exposure of human cells to ionizing radiation (Liu and Weaver 1993) or UV light also results in phosphorylation of the 32-kD subunit (Carty et al.1994). During SV40 replication in vitro, RP-A is phosphorylated by a DNA-activated kinase (Fotedar and Roberts 1992; Brush et al. 1994; Hendrickson et al. 1994). However, none of these phosphorylations demonstrably affects the activity of RP-A as assessed in vitro. Indeed, mutation of the sites in the 32-kD subunit that are targets for the cyclin-dependent kinases does not alter the capacity of RP-A to participate in SV40 DNA replication in vitro (Hendrickson and Wold 1994). Hence, phosphorylation of RP-A may affect a function that has yet to be determined or cannot be observed in the cell-free replication system.

RP-A binds to large T antigen in a partially species-specific manner. For example, SV40 large T antigen binds to vertebrate sources of RP-A, but not to yeast RP-A (Dornreiter et al. 1992; Schneider et al. 1992; Melendy and Stillman 1993). This likely accounts for the fact that all mammalian sources of RP-A tested functionally substitute for human RP-A in SV40 DNA replication, albeit to differing extents (Schneider et al. 1992). In general, the greater the relatedness between the species of origin of RP-A and that in which the virus replicates, the greater the affinity of large T antigen for that particular source of RP-A (Schneider et al. 1992). Hence, yeast and trypanosome RP-A cannot functionally substitute for human, simian, or bovine sources of RP-A to replicate SV40 DNA in vitro (Brill and Stillman 1989; Brown et al. 1992; Schneider et al. 1992; Melendy and Stillman 1993). The RP-A subunit that contacts large T antigen has not been identified. However, the large subunit alone is insufficient to mediate this interaction (Dornreiter et al. 1992). The region in SV40 large T antigen required for binding RP-A also has not been determined; a region in PyV large T antigen that overlaps with its DNA-binding domain interacts with human and mouse RP-A in vitro (Fig. 3) (P. Desjardins and J.A. Hassell, unpubl.).

RP-A binds in vitro to the activation domains of at least three transcription activators, including that of GAL4, VP16, and p53 (He et al. 1993; Li and Botchan 1993). The 70-kD subunit is sufficient for direct binding to the activation domains of these transcription factors. The physiological relevance of these findings is suggested by the observation that transcription activators are required for efficient replication of SV40 and PyV DNA in vivo (see DePamphilis; van der Vliet; both this volume). Thus, the binding of transcription activators to the *ori* auxiliary elements may recruit and/or stabilize the interaction of RP-A with the *ori* through protein:protein interactions, thereby facilitating assembly of the primosome. Alternatively, the function of RP-A at this site may be stimulated by direct contact with transcription activators. RP-A also interacts physically with DNA polymerase-α:primase, which is required to initiate SV40 and PyV DNA replication. These various protein:protein interactions are consistent with a role for RP-A in the initiation of DNA replication.

DNA Polymerase-α:Primase

DNA polymerase-α:primase is the only DNA polymerase in mammalian cells capable of initiating DNA synthesis de novo through the synthesis of RNA primers (for review, see Wang 1991 and this volume). Studies with inhibitors and inactivating antibodies to the protein reveal that it is required to replicate SV40 and PyV DNA (Melendy and Stillman 1992 and references therein). Currently, it is thought that the role of DNA polymerase-α:primase is to initiate both leading-strand synthesis at the *ori* and Okazaki fragment synthesis on the lagging-strand template at replication forks (Waga and Stillman 1994a; Waga et al. 1994 and references therein).

Human DNA polymerase-α:primase is composed of four subunits of molecular mass 180, 70, 58, and 48 kD; the subunits of the mouse protein are similar in size (180, 68, 54, and 46 kD) (Wang 1991; Miyazawa et al. 1993). DNA polymerase activity is intrinsic to the 180-kD subunit, whereas primase activity resides in the 48-kD subunit (Copeland and Wang 1993). The 58-kD subunit binds to the 48-kD subunit and appears to stabilize the primase activity of the smaller subunit. The role of the 70-kD subunit is not clear. It associates with the 180-kD subunit (Collins et al. 1993) and is phosphorylated in a cell-cycle-dependent manner, suggesting that it may play a regulatory role (Nasheuer et al. 1991). These attributes of the human DNA polymerase-α:primase subunits are conserved in their murine counterparts.

The replication of SV40 and PyV DNA is highly species-specific (Bennett et al. 1989 and references therein). SV40 DNA replication takes place in monkey or human cells, but not in mouse cells. In contrast, mouse cells, but not primate cells, support the replication of PyV DNA. Species-specific DNA replication in vivo requires large T antigen, the cognate virus *ori*, and factors from permissive cells (Bennett et al. 1989). Cell-free replication systems recapitulate this species specificity and have proven invaluable in identifying permissive factors. The addition of purified mouse DNA polymerase-α:primase to human cell extracts renders them permissive for PyV DNA replication, and correspondingly, the addition of the human enzyme to mouse cell extracts allows the mouse extracts to support SV40 DNA replication (Murakami et al. 1986). These experiments suggest that DNA polymerase-α:primase is the primary host determinant for species-specific viral DNA replication. Further fractionation of the DNA polymerase revealed that the primase subunits (a heterodimer of the 54-kD and 46-kD subunits) from mouse DNA polymerase-α:primase enabled human extracts to replicate PyV DNA (Eki et al. 1991). Extension of these experiments with DNA polymerase-α:primase, reconstituted in insect cells after infection with recombinant baculoviruses encoding individual subunits of human or mouse origin, showed that the small mouse 46-kD primase subunit is sufficient to mediate species-specific replication of PyV DNA in vitro in human cell extracts (Bruckner et al. 1995). Surprisingly, similar experiments with these reconstituted DNA polymerase-α:primase enzymes of mixed human-mouse origin showed that SV40 DNA replication in vitro required a different subunit, namely, the p180 catalytic subunit of the DNA polymerase from human cells (Stadlbauer et al. 1996). Presumably, this reflects a difference in the functional interaction between the DNA polymerase and the large T antigen of each virus.

The large T antigens of both SV40 and PyV physically interact with several of the subunits of DNA polymerase-α:primase in vitro (Smale and Tjian 1986; Gannon and Lane 1987; Dornreiter et al. 1990; Moses and Prives 1994: Schneider et al. 1994; Bruckner et al. 1995). SV40 large T antigen binds independently to both the 180-kD and the 70-kD subunits of human DNA polymerase-α:primase (Dornreiter et al. 1992; Collins et al. 1993), although the principal interaction appears to be with the 70-kD subunit (Collins et al. 1993). Amino acids 195–313 in the 180-kD subunit (Dornreiter et al. 1993) and residues 1–239 of the 70-kD subunit are sufficient for interaction with SV40 large T antigen (Collins et al. 1993). Interestingly, two discontinuous regions in SV40 large T antigen are required to bind to the human holoenzyme (Fig. 3), but it is not known whether each region binds to a different subunit (Smale and Tjian 1986; Gannon and Lane 1987; Dornreiter et al. 1990). PyV large T antigen binds to the 180-kD subunit of mouse DNA polymerase-α:primase, to a complex containing the murine 180-kD and 68-kD subunits, and to the 46-kD primase subunit (Bruckner et al. 1995). It is not known whether PyV large T antigen binds independently to the 68-kD subunit (Bruckner et al. 1995). Hence, the large T antigens of SV40 and PyV make contact with multiple subunits of DNA polymerase-α:primase, and at least one of these represents a functional species-specific interaction.

The binding of SV40 large T antigen to human DNA polymerase-α:primase increases both the primase and DNA polymerase activity of the holoenzyme (Collins and Kelly 1991; Erdile et al. 1991). This stimulatory effect requires the 70-kD subunit (Collins and Kelly 1991). Large T antigen is thought to bridge the interaction between DNA polymerase-α:primase and the DNA template, thereby facilitating DNA replication (Collins et al. 1993).

DNA polymerase-α:primase also binds to RP-A (Dornreiter et al. 1992; Melendy and Stillman 1993). This interaction occurs between the dimeric primase component of DNA polymerase-α:primase (the 48- to 58-kD complex) and the largest (70-kD) subunit of RP-A (Dornreiter et al. 1992). The association appears to be species-specific; single-stranded DNA-binding proteins encoded by prokaryotes and their viruses do not associate with mammalian sources of DNA polymerase-α:primase (Dornreiter et al. 1992). Furthermore, DNA polymerase-α:primase from human sources does not functionally interact with *Drosophila melanogaster* RP-A to effect SV40 DNA replication in a cell-free system (Kamakaka et al. 1994). These findings are consistent with reports demonstrating that only mammalian sources of RP-A function in the cell-free systems to replicate SV40 DNA. One consequence of the interaction of these proteins is the stimulation of the DNA polymerase and primase activity of DNA polymerase-α:primase on artificial DNA templates in vitro (Erdile et al. 1991; Melendy and Stillman 1993). Hence, multiple and species-specific protein:protein interactions occur among the three major protein components (large T antigen, RP-A, and DNA polymerase-α:primase) of the primosome, resulting in the stimulation of those enzymatic activities required for the initiation of DNA replication at *ori*.

Proliferating Cell Nuclear Antigen

The simultaneous demonstration that a chromatographic fraction (named replication factor B) required for SV40 DNA replication in vitro corresponded by many criteria, including sequence analysis, to a 36-kD protein known as proliferating cell nuclear antigen (PCNA) (Prelich et al. 1987), and that this protein is identical to a polymerase-δ auxiliary factor (Bravo et al. 1987; Prelich et al. 1987), led to the realization that DNA polymerase-α:primase is not the only DNA polymerase required to replicate SV40 DNA. PCNA colocalizes with sites of active DNA replication, suggesting that it participates in cellular DNA replication. Although PCNA is synthesized shortly before the onset of cellular DNA replication and its abundance in the nucleus is highest during S phase, it is a stable protein with a half-life of about 20 hours that is present throughout the cell cycle. PCNA has been found in all eukaryotes that have been examined; its primary structure has been highly conserved among diverse species including yeasts, plants, and mammals. In *S. cerevisiae*, PCNA is essential for viability (Bauer and Burgers 1988).

Analyses of SV40 replication intermediates synthesized in vitro in the presence and absence of PCNA showed that it is not required for the synthesis of nascent DNA fragments at *ori*, but acts at the elongation phase to stimulate leading-strand synthesis (Prelich and Stillman 1988). It is thought that DNA polymerase-α:primase initiates the synthesis of nascent DNA in a reaction that does not require PCNA. As discussed below, RF-C recognizes the primer-template junction and loads PCNA onto the DNA, which in conjunction with DNA polymerase-δ effects the complete synthesis of the leading strand (Tsurimoto and Stillman 1990; Lee et al. 1991). Lagging-strand synthesis requires PCNA as well; here too, DNA polymerase-α:primase initiates synthesis that is completed by either DNA polymerase-δ or perhaps DNA polymerase-ε (Waga and Stillman 1994a). Hence, PCNA together with RF-C plays a critical role in coordinating leading- and lagging-strand synthesis (for review, see Stillman 1994).

PCNA is the first cellular protein component of the eukaryotic replicative machinery whose crystal structure has been determined (Kong et al. 1992; Krishna et al. 1994). It is a trimer, in keeping with the prediction of its molecular mass in solution (Bauer and Burgers 1988), and forms a donut-shaped closed circular ring that is thought to surround duplex DNA (Krishna et al. 1994). This structure fits nicely with its role as a sliding clamp that tracks along duplex DNA in conjunction with DNA polymerase-δ (for review, see Kelman and O'Donnell 1994; Stillman 1994). In this regard, it is noteworthy that there are striking structural and functional similarities between PCNA and the *Escherichia coli* β protein and bacteriophage T4 45 protein, which also function as sliding clamps during DNA replication in prokaryotes (see Stillman, this volume).

PCNA is required not only for DNA replication, but also for recombination (Jessberger et al. 1993) and repair (Shivji et al. 1989). DNA repair is mediated in part by DNA polymerase-ε (Nishida et al. 1988). A functional interaction occurs between PCNA and DNA polymerase-ε (Burgers 1991; Lee et al. 1991; Podust et al. 1992). PCNA stimulates the binding of DNA polymerase-ε to DNA primer/template junctions and augments the rate of DNA synthesis on these templates (Maga and Hübscher 1995). Hence, PCNA appears to be an auxiliary protein for both DNA polymerase-δ and ε.

In mammalian cells, PCNA is found in a quaternary complex with components of the cell-cycle machinery. These complexes include a cyclin (A, B, D1, or D3), its cognate cyclin-dependent kinase (cdc2, cdk2, cdk4, or cdk5), and the 21-kD (p21) inhibitor of these kinases, also known as Cip1 or Waf 1 (Xiong et al. 1992, 1993; Zhang et al. 1993). PCNA interacts directly and independently with the carboxy-terminal region of p21 (Waga et al. 1994; Chen et al. 1995). In the SV40 in vitro system this interaction between PCNA and p21 is sufficient to inhibit DNA replication (Flores-Rozas al. 1994; Waga et al. 1994). p21 is a p53 target gene and, hence, under conditions of p53-mediated growth arrest, p21 may block cellular DNA replication by targeting PCNA for inactivation. In contrast to its ability to inhibit SV40 DNA replication in vitro, p21 does not inhibit DNA repair (Li et al. 1994). The repair synthesis of short gaps in duplex DNA by DNA polymerase-ε is insensitive to inhibition by p21, perhaps because under these circumstances, PCNA is needed principally to load the DNA polymerase onto the primer/template junction and not to stimulate DNA synthesis (Li et al. 1994). In this regard, it is noteworthy that PCNA also interacts directly with another p53 target gene product, GADD45, whose synthesis, like that of p21, increases in response to DNA damage (Smith et al. 1994). GADD45 may play a positive role in DNA repair. It has been suggested that GADD45 releases PCNA from p21/cdk complexes and facilitates its interaction with components of the DNA repair machinery of the cell (Smith et al. 1994). PCNA is not known to be modified posttranslationally, and hence its activity in DNA replication and repair appears to be regulated principally by interaction with other proteins such as p21 and GADD45. The physical interaction between PCNA and the p21-cyclin-dependent kinase inhibitor illustrates a novel means of regulating the activity of elements of the DNA replication machinery and of coordinating DNA replication and repair with cell-cycle progression.

Replication Factor C

Human RF-C (also called activator 1 or A1) is a multisubunit DNA polymerase accessory protein composed of five independent subunits of molecular mass 140, 40, 38, 37, and 36.5 kD. RF-C functions in conjunction with PCNA to increase the DNA binding and processivity of DNA polymerases δ and ε (Tsurimoto and Stillman 1989; Lee et al. 1991). It binds to the 3′terminus of DNA primers annealed to single-stranded DNA templates and, in a reaction requiring ATP, loads PCNA onto the DNA. This complex is recognized by DNA polymerase-δ or ε, which then elongates the primer (Lee and Hurwitz 1990; Tsurimoto and Stillman 1990, 1991b; Fien and Stillman 1992). Whereas PCNA can independently bind to the ends of linear DNA and clamp onto the template, its loading onto circular DNA requires RF-C (Podust et al. 1995). Interestingly, RF-C is similar in subunit composition, primary structure, and function to the 5-subunit "clamp loader" proteins of *E. coli* (γ complex) and phage T4 (gene 44/62 complex), suggesting that eukaryotic RF-C functions as a clamp loader or brace protein and that PCNA acts as a sliding clamp during DNA synthesis (O'Donnell et al. 1993; Kelman and O'Donnell 1994; Stillman 1994 and this volume).

RF-C possesses DNA-dependent ATPase activity, which is stimulated by PCNA and RP-A, and is required for its function in SV40 DNA replication (Tsurimoto and Stillman 1990, 1991a,b; Lee et al. 1991). ATPase activity is not required for translocation of the DNA polymerase but is needed to load PCNA onto DNA at replication forks (O'Donnell et al. 1993). The 40-kD subunit in the holoenzyme can be cross-linked to ATP, suggesting that this subunit possesses ATPase activity (Tsurimoto and Stillman 1991a). RF-C also recognizes and binds to primer/template junctions; this activity is stimulated by ATP and further stabilized by PCNA (Tsurimoto and Stillman 1990; Lee et al. 1991). The DNA-binding activity of the RF-C holoenzyme has been localized by UV cross-linking experiments to the 140-kD subunit (Tsurimoto and Stillman 1991a).

cDNAs encoding all the subunits of mammalian RF-C have been isolated and sequenced (Chen et al. 1992a,b; Bunz et al. 1993; Burbelo et al. 1993; O'Donnell, et al. 1993), and most recently, genes encoding all five subunits of *S. cerevisiae* RF-C have also been identified and analyzed (Howell et al. 1994; Li and Burgers 1994a,b; Noskov et al. 1994; Cullman et al. 1995). Gene disruption experiments in yeast reveal that each of the five subunits is essential for viability (Cullman et al. 1995 and references therein).

Biochemical analyses of the activities of the mammalian and yeast RF-C subunits expressed in *E. coli* have begun to shed light on their individual roles in DNA replication. Complete cDNAs have been isolated and the encoded proteins characterized for three of the mammalian subunits (140, 40, and 37 kD); only partial cDNAs are available for the 38-kD and 36.5-kD species. The 140-kD subunit binds to double-stranded oligonucleotides likely by recognizing a structure that mimics a primer/template junction (Tsurimoto and Stillman 1991a; Burbelo et al. 1993; Lu et al. 1993). The 37-kD subunit also is capable of binding to primer/template junctions, but its specific DNA-binding activity is only 0.1% that of the RF-C holoenzyme, and hence its role in primer binding is unclear (Pan et al. 1993; Noskov et al. 1994). Whereas all of the RF-C subunits bear the putative ATP-binding motif, only the 40-kD subunit is known to bind to ATP (Pan et al. 1993).

To accomplish its role in DNA replication, the RF-C subunits must interact with each other, with PCNA, and with DNA polymerase-δ and ε. To date, two interactions among the various RF-C subunits have been described: The 40-kD subunit associates with the 37-kD subunit in vitro (Pan et al. 1993), and the yeast Rfc4p subunit (equivalent to 40-kD mammalian subunit) binds to the Rfc3p subunit (the homolog of the mammalian 36.5-kD subunit) during coexpression in *E. coli*. Interaction of RF-C with PCNA is accomplished by direct binding of the latter to the 40-kD subunit (Pan et al. 1993). Genetic evidence in yeasts suggests that the Rfc1p subunit (homologous to the mammalian 140-kD subunit) interacts with PCNA (Howell et al. 1994). There is also indirect evidence supporting interactions between RF-C and DNA polymerase-δ and ε. For example, the isolated 37-kD subunit stimulates DNA polymerase-ε activity in vitro under high-salt conditions, and the 40-kD subunit inhibits DNA polymerase-δ activity (Pan et al. 1993). The molecular cloning of complete cDNAs encoding all the mammalian subunits, coupled with more complete analyses of their interaction with each other, with PCNA, and with the various subunits of DNA polymerase-δ and ε, should lend insight into the mechanism of action of this very important multisubunit enzyme.

DNA Polymerases δ and ε

The demonstration that PCNA stimulates SV40 DNA replication in vitro and that it acts as a processivity factor for DNA polymerase-δ strongly suggested that this DNA polymerase is required to replicate SV40 DNA. This prediction has been borne out by biochemical experiments demonstrating that DNA polymerase-δ effects leading-strand synthesis in vitro (Waga and Stillman 1994a and references therein). DNA polymerase-δ is a heterodimer comprising a 125-kD catalytic subunit and a 55-kD subunit of unknown function (for review, see Linn 1991). Unlike DNA polymerase-α:primase, DNA polymerase-δ has 3′→5′ proofreading exonuclease activity. The gene encoding the catalytic subunits of *S. cerevisiae* and *Schizosaccharomyces pombe* DNA polymerase-δ has been cloned and shown to be required for viability (for review, see Campbell 1993; see also Wang, this volume). Human cDNAs encoding the catalytic subunit also have been isolated (Yang et al. 1992).

DNA polymerase-δ binds poorly to DNA, and hence PCNA is required to mediate its interaction with the DNA template. Deletion of the first 220 amino acids of the DNA polymerase-δ catalytic subunit from yeast renders it insensitive to PCNA stimulation, suggesting that this region is required for association with PCNA (Brown and Campbell 1993). Zhang et al. (1995) have provided direct evidence for an association between human DNA polymerase-δ and PCNA that is mediated by the amino-terminal 182 amino acids of the 125-kD DNA polymerase catalytic subunit.

The potential that DNA polymerase-ε might be required for SV40 and PyV DNA replication has begun to receive increased attention for a number of reasons. First, the gene encoding the catalytic subunit of DNA polymerase-ε is essential, and its protein product is required for chromosomal DNA replication in yeast (Morrison et al. 1990). Second, in vivo (Nethanel et al. 1988) and in vitro in unfractionated mammalian cell lysates with SV40 DNA as template, DNA polymerase-α:primase synthesizes DNA fragments less than 40 nucleotides in length, which is considerably shorter than Okazaki fragments (~200 nucleotides), implying that another DNA polymerase synthesizes these fragments (Bullock et al. 1991). These findings suggest that DNA polymerase-ε may be required for lagging-strand replication.

DNA polymerase-ε was first isolated from budding yeast (known as DNA polymerase II) and later from mammalian sources. Highly purified mammalian DNA polymerase-ε is composed of two subunits of molecular mass 215 and 55 kD, whereas the budding yeast enzyme comprises subunits of 255, 80, 34, 30, and 29 kD (for review, see Linn 1991; Hübscher and Spadari 1994). The mammalian enzyme may possess additional subunits that dissociate from the core DNA polymerase during purification. cDNAs encoding the large catalytic subunit of human DNA polymerase-ε have been cloned (Kesti et al. 1993), as have the genes encoding the 255-, 80-, 34-, and 30-kD subunits of the *S. cerevisiae* enzyme (for review, see Campbell 1993). The largest subunit possesses DNA polymerase and 3′→5′exonuclease activity; the function of the smaller subunits is not known. Given the conservation of subunit composition among the various eukaryotic DNA polymerases, it seems likely that additional mammalian subunits of DNA polymerase-ε exist. It will be interesting to learn whether this is so and to determine what their role might be in DNA replication. One possibility is that these subunits are involved in coordinating leading- and lagging-strand synthesis at replication forks.

DNA polymerase-ε is similar to DNA polymerase-δ in possessing proofreading 3′→5′ exonuclease activity, but unlike the latter, it is processive in the absence of PCNA. Despite its processivity, DNA polymerase-ε interacts with PCNA and RF-C to form a stable complex at primer/template junctions (Burgers 1991); PCNA increases the rate of DNA synthesis by DNA polymerase-ε on synthetic primer templates in vitro (Maga and Hübscher 1995). Furthermore, DNA polymerase-ε activity is inhibited by physiological salt concentrations, but its activity is restored upon addition of RP-A, PCNA, and RF-C (Yoder and Burgers 1991). Hence, like DNA polymerase-δ, DNA polymerase-ε likely requires these auxiliary proteins for its proper functioning in vivo.

Topoisomerases I and II

DNA topoisomerases are enzymes that alter the topology of DNA (for review, see Watt and Hickson 1994; see Hangaard Andersen et al., this volume). They have been classified as types I and II according to their catalytic mechanism of action. Type I topoisomerases are nicking-closing enzymes that transiently cleave one strand of the DNA duplex, allowing the other strand to go through the nick before ligating the nicked strand. These enzymes do not require ATP for their action, but their activity is stimulated by Mg^{++} ions in vitro. Type II topoisomerases also transiently cleave DNA, but they make double-stranded breaks so that a double-stranded segment of DNA can go through the break before it is sealed. Type II enzymes require ATP for their activity. During SV40 DNA replication, topoisomerase I or II is required to unwind the positive supercoils that accumulate ahead of the replication forks (Yang et al. 1987), whereas topoisomerase II is needed for the decatenation (segregation) of the progeny DNA molecules after replication is completed (Yang et al. 1987; Ishimi et al. 1991).

Type I topoisomerases are single-chain enzymes that function as monomers. *S. cerevisiae* possesses two genes that encode type I topoisomerase; human cells have only a single gene. Neither of the yeast genes is essential for viability. The human enzyme has a measured molecular mass of 100 kD and comprises 765 amino acids. Topoisomerase I functions not only in DNA replication and recombination, but also in transcription, where it may relax positive supercoils ahead of the transcribing RNA polymerases.

Topoisomerase II is encoded by a single essential gene in *S. cerevisiae* and two genes in mammals (topoisomerase IIa and IIb). Human topoisomerase IIa comprises 1530 amino acids (molecular mass 170 kD), whereas IIb is made up of 1621 amino acids (180 kD). Topoisomerase IIa is likely involved in chromosome condensation and separation of daughter chromosomes during mitosis; the function of topoisomerase IIb is not known, but this protein is localized primarily to the nucleolus.

Maturation Factor I, Ribonuclease H1, and DNA Ligase I

The complete synthesis of covalently closed circular SV40 DNA requires, in addition to the various cellular proteins described earlier, RNase H1, MF-I (also known as FEN-1), and DNA ligase I. These enzymes function coordinately to remove ribonucleotide primers from Okazaki fragments and to ligate these to form completed lagging strands (Turchi et al. 1994; Waga and Stillman 1994b; Waga et al. 1994).

RNase H specifically degrades the RNA strand of an RNA:DNA heteroduplex (for review, see Crouch 1990). In both prokaryotes and eukaryotes, RNase H is thought to remove RNA primers from Okazaki fragments prior to their ligation. This activity is also encoded by retroviruses, where it constitutes a domain in a dual domain protein that possesses reverse transcriptase activity. Two mammalian RNase H activities have been characterized; RNase H1 (~60–100 kD) and RNase H2 (~30–40 kD), but cDNAs encoding these enzymes have not yet been isolated (Busen 1980). A role for RNase H1 in DNA replication is suggested by the observation that its activity increases in accordance with mammalian cell DNA synthesis, and by the finding that RNase H1 purified from *D. melanogaster* stimulates DNA synthesis by DNA polymerase-α:primase and forms a complex with the DNA polymerase (DiFrancesco and Lehman 1985). Most recently, Huang et al. (1994) have provided compelling evidence that RNase H1 removes RNA primers from Okazaki fragments during DNA replication through an endonucleolytic cleavage of the RNA, resulting in its release intact from the DNA fragment. This leaves a monoribonucleotide at the RNA/DNA junction, which is removed by the exonuclease activity of MF-I.

MF-I is a single-subunit, double-strand specific, 5′→3′exonuclease that degrades either double-stranded DNA or the RNA moiety of an DNA/RNA heteroduplex from its 5′terminus (Ishimi et al. 1988; Goulian et al. 1990; Siegal et al. 1992; Turchi and Bambara 1993; Turchi et al. 1994; Waga et al. 1994). MF-I was described as a 44-kD activity from human HeLa cells required to achieve synthesis of form I SV40 DNA in vitro using solely DNA polymerase-α:primase as the replicative DNA polymerase (Ishimi et al. 1988). This enzyme is the same as three other exonucleases that have been described, including DNase IV (Lindahl 1970), AF-II (Goulian et al. 1987, 1990), pL (Guggenheimer et al. 1984; Kenny et al. 1988), and a 56-kD 5′→3′exonuclease found in association with DNA polymerase-ε from calf thymus (Siegal et al. 1992).). MF-I is required to synthesize covalently closed circular DNA using single-stranded circular DNA as template in a system with purified DNA polymerase-α:primase that emulates lagging-strand synthesis. Using synthetic templates, Turchi et al. (1994) have provided compelling evidence that MF-I functions in conjunction with RNase H1 to remove RNA primers from Okazaki fragments. Furthermore, Murante et al. (1994) have shown that MF-I activity is stimulated by DNA polymerase-α:primase, DNA polymerase-δ, and DNA polymerase-ε. The model that has emerged from these experiments is that RNase H1 makes an endonucleolytic cut within the RNA primer one nucleotide from the DNA/RNA junction of Okazaki fragments, the RNA primer dissociates as an intact species from the DNA template, MF-I removes the remaining monoribonucleotide, DNA polymerase-δ or ε fills in the gap using an adjacent Okazaki fragment as primer, and DNA ligase I seals the nick (Turchi et al. 1994). Hence, the combined action of these enzymes leads to the completion of lagging-strand synthesis and ultimately to the circularization of newly replicated SV40 DNA.

Four DNA ligases (DNA ligase I, II, III, and IV) have been identified in mammalian cells that can be discriminated principally on the basis of their biochemical and physical properties (Wei et al. 1995; see Nash and Lindahl, this volume). During SV40 DNA replication, DNA ligase I covalently joins Okazaki fragments arising from discontinuous lagging-strand synthesis (Ishimi et al. 1988; Waga et al. 1994). Neither mammalian DNA ligase III nor bacteriophage T4 DNA ligase can substitute for human DNA ligase I during SV40 DNA replication in vitro, suggesting that the mammalian enzyme makes specific protein:protein contacts with other replication factors (Waga et al. 1994). RF-C inhibits the activity of DNA ligase I in vitro, and hence these proteins may functionally interact during lagging-strand synthesis (Waga et al. 1994).

MODEL FOR SV40 AND PYV DNA REPLICATION

Knowledge of the structure of SV40 and PyV DNA replication intermediates occurring in vivo, and of the mechanism of action of the principal viral and cellular proteins that participate in DNA replication in vitro, suggests a model for replication of these viral genomes. The temporal sequence of events can be broadly defined as large T antigen binding to and unwinding of *ori*, assembly of the primosome, initiation of leading- and lagging-strand synthesis, DNA polymerase switching and elongation of leading and lagging strands, maturation of lagging DNA strands, and replication termination and segregation of the daughter DNA molecules.

The first presynthesis event leading to the initiation of DNA replication is the binding of large T antigen to the *ori*-core (for review, see Borowiec et al. 1990; see also SenGupta and Borowiec 1994 and references therein). Large T antigen binds as a monomer to each of the four pentanucleotide repeats within *ori*-core. Additional large T antigen binds to this complex and assembles into two hexamers that straddle each side of *ori*-core. Each hexameric complex contacts the sugar-phosphate DNA backbone on different strands. ATP stimulates the binding of large T antigen to *ori*-core and is required for the assembly of hexameric large T antigen, but hydrolysis of ATP is not needed for either process. Large T antigen then effects structural alterations in the *ori*-core, including untwisting of the A/T-rich region and melting of the early palindrome. RP-A joins the large-T-antigen/*ori* complex facilitating more extensive bidirectional unwinding of the *ori* through the intrinsic $3' \rightarrow 5'$ helicase activity of large T antigen. This process requires ATP hydrolysis and leads to the establishment of a replication bubble within the *ori*. It is noteworthy that whereas many different single-stranded DNA-binding proteins can facilitate these early reactions, subsequent events at *ori* require mammalian RP-A. RP-A may be recruited to *ori* through interaction with transcription factors bound to the auxiliary regions, by direct interaction with large T antigen bound to the *ori*-core, or by both mechanisms. The RP-A/large-T-antigen complex in turn recruits DNA polymerase-α:primase to form a preinitiation complex or primosome. As recounted previously, protein:protein interactions among all the constituents of this complex serve to stabilize the primosome. Hence, during the initiation of DNA replication, large T antigen plays a number of distinct roles; it alters the structure of *ori* DNA, unwinds DNA ahead of the replication fork, and nucleates the assembly of the primosome at *ori*.

The steps leading to the initiation of SV40 and PyV DNA replication are thought to be very similar, with one notable difference. PyV large T antigen binds poorly to its *ori*-core in comparison to the binding of SV40 large T antigen to its cognate *ori*. Therefore, it is possible that assembly of the primosome at the PyV *ori* is more dependent on transcription activators than is assembly of the SV40 primosome. This may account for the greater requirement for transcription activators for PyV DNA replication than for SV40 DNA replication.

Initiation of leading-strand synthesis at *ori* is carried out by DNA polymerase-α:primase. RNA-primed DNA synthesis occurs at one of several possible sites within an initiation zone and proceeds in the direction of the early region, using the early mRNA strand as a template (Hay and DePamphilis 1982; Hendrickson et al. 1987). RNA primers are extended by DNA polymerase-α:primase, yielding RNA/DNA primers for the leading-strand DNA polymerase, DNA polymerase-δ. The RNA/DNA primers synthesized by DNA polymerase-α:primase are about 35 nucleotides long. Elongation of these to about 200 nucleotides by DNA polymerase-δ exposes a single-stranded region on the retrograde template (the second initiation zone). It is thought that the average distance between nucleosomes arranged on the viral genome dictates the size of the initiation zones. This allows the priming of leading-strand DNA synthesis by DNA polymerase-α:primase in the opposite direction toward the late region, thus establishing another replication fork moving in the opposite direction. Following the switch from DNA polymerase-α:primase to DNA polymerase-δ, DNA synthesis proceeds continuously in the direction of fork movement.

Initiation of lagging-strand synthesis at each replication fork is also effected by DNA polymerase-α:primase and appears to be mechanistically indistinguishable from the initiation of leading-strand synthesis. Lagging-strand synthesis proceeds discontinuously away from the overall direction of fork movement, resulting in the synthesis of Okazaki fragments that are about 200 nucleotides long. Like the size of the initiation zones at *ori*, the length of Okazaki fragments likely is determined by the distance between nucleosomes. Analyses of Okazaki fragments synthesized in vivo have revealed that the RNA moiety attached to these fragments averages 9 bases, although this value can vary from as few as 3 to as many as 12 nucleotides (Hay and DePamphilis 1982; Hay et al. 1984). The 5′ end of the RNA moiety of these fragments begins with a purine (generally A), followed by an A and, less frequently, a G at the second position, and an A or T at the third position. Examination of the DNA templates complementary to these RNA primers reveals a preference in the DNA template for a purine at position –1 preceding the RNA start site (defined as +1). These authors suggest that 3′-PuTT-5′ in the

DNA template constitutes an initiation signal for DNA polymerase-α:primase.

In part, these findings contrast with those of Bullock et al. (1994), who analyzed RNA/DNA primers synthesized in a 5-second pulse in an in vitro system containing SV40 DNA, large T antigen, and a HeLa cell nuclear extract. The majority of RNA/DNA primers synthesized under these conditions are between 24 and 34 nucleotides long, are derived largely from the lagging-strand template, and are composed of an RNA moiety of 8 or 9 nucleotides. These particular observations are in good agreement with those made by analyzing Okazaki fragments synthesized in vivo. However, examination of the DNA template giving rise to these RNA/DNA primers in vitro led to the inference that initiation occurs with any one of the four ribonucleotide triphosphates. There was a strong preference for A in the second and third positions, in agreement with the results from the in vivo analyses. Although the general features of the RNA moiety of the primers synthesized in vivo and in vitro are similar, there is a clear discrepancy in assignment of the initiating nucleotide. It is not immediately obvious what accounts for this difference. It is of interest to note that if one assumes that the in vitro data suffer from a consistent error of one nucleotide, then the in vitro assignment of the initiating nucleotide is compatible with all the in vivo findings. Whatever the explanation, Bullock et al. (1994) suggest a model for the initiation of DNA synthesis whereby DNA polymerase-α:primase translocates along the DNA template until it recognizes an initiation signal (3′-NTT-5′) and binds an ATP molecule opposite the T residue at position +2 in the DNA template. Then a second ribonucleotide triphosphate complementary to the +1 position is bound by the primase, a dinucleotide is formed, and the RNA chain is extended. Therefore, it would seem that the sole role of DNA polymerase-α:primase is to synthesize RNA/DNA primers for leading- and lagging-strand synthesis (Waga and Stillman 1994a).

Extension of leading- and lagging-strand primers can be accomplished by DNA polymerase-δ in vitro (for review, see Stillman 1994; see also Waga and Stillman 1994a). DNA polymerase switching is achieved when the RF-C/PCNA complex, which is capable of tracking around the circular viral genome, recognizes the 3′end of a primer/template junction. DNA polymerase-α:primase is displaced from the template, and DNA polymerase-δ is recruited in its place. Leading-strand synthesis proceeds continuously in a processive manner after the switch of DNA polymerases, whereas lagging-strand synthesis occurs in a discontinuous fashion. RF-C and PCNA are required for both leading-

and lagging-strand synthesis. As discussed earlier, the synthesis of Okazaki fragments from retrograde templates may require DNA polymerase-ε, but its precise role, if any, in this process remains to be established. Hence, at least two, and perhaps as many as three, DNA polymerases are required for SV40 and PyV DNA replication.

These events lead to the establishment of two replication forks that move in opposite directions at about the same rate around the circular viral genome. Propagation of both replication forks requires the helicase function of large T antigen, which translocates on the forward arm of the parental DNA template. This is consistent with the observation that SV40 large T antigen is located at replication forks in vivo, and with the finding that no other helicase is known to be required for SV40 DNA replication in vitro. Topoisomerase I or topoisomerase II is required to unwind positive supercoils in the DNA template that accumulate ahead of the replication forks.

The maturation of Okazaki fragments requires removal of the RNA primer by RNase H1 and the 5′→3′exonuclease activity of MF-I, gap-filling by one of the DNA polymerases, and ligation of Okazaki fragments by DNA ligase I (Turchi et al. 1994; Waga et al. 1994). By extrapolation from in vitro studies with model lagging-strand templates, it would seem that the role of RNase H1 in SV40 and PyV DNA replication is to make an endonucleolytic cut one nucleotide 5′of the RNA/DNA junction in the RNA primer (Turchi et al. 1994). MF-I then eliminates the remaining monoribonucleotide from the initiator RNA. Hence, both RNase HI and MF-I are required and act cooperatively to remove RNA primers. In vitro gap-filling can be accomplished by any one of the replicative DNA polymerases (α, δ, or ε). However, it seems unlikely that DNA polymerase-α:primase fulfills this role because RP-A, PCNA, and RF-C block maturation of Okazaki fragments by DNA polymerase-α:primase in vitro (Waga and Stillman 1994b). There is a specific requirement for DNA ligase I for the completion of lagging-strand synthesis; neither mammalian DNA ligase III nor bacteriophage T4 DNA ligase is capable of substituting for DNA ligase I (Waga et al. 1994). Therefore, the completion of lagging-strand synthesis and joining of Okazaki fragments appears to require specific contacts among the proteins of the lagging-strand machinery.

The mechanism accounting for the coordinated synthesis of both leading and lagging strands at replication forks is unknown. Recently, Waga and Stillman (1994a) proposed a model whose central thesis posits a dimer between the leading- and lagging-strand DNA polymerases analogous to the dimeric DNA polymerase III at the *E. coli* replication fork. These authors provide evidence that DNA polymerase-δ is the principal replicative enzyme required for both leading- and lagging-strand synthesis in vitro, and suggest that this DNA polymerase dimerizes at replication forks. However appealing this hypothesis may be, there is currently no evidence to support the occurrence of DNA polymerase-δ homodimers in solution. However, the possibility remains that a heretofore undiscovered cellular protein mediates dimerization of the DNA polymerases at the replication fork. It will be interesting to learn whether any of these or some other protein mediates dimerization of the DNA polymerases at replication forks.

Termination of SV40 and PyV DNA replication occurs on average 180° from the *ori* when the two replication forks converge; the small genomes of these viruses do not contain replication termination signals (for review, see DePamphilis and Bradley 1986; see Bastia and Mohanty, this volume). When replication is about 90% completed, fork movement is impeded and replication intermediates accumulate, indicating that the completion of replication and the segregation of progeny molecules comprise a slow step in the DNA replication cycle. If DNA replication and unwinding mediated by topoisomerase II continue concurrently, then the progeny molecules separate, yielding double-stranded circular DNAs with a small gap of about 50 nucleotides. However, should DNA replication be completed without concomitant unwinding, then catenated dimers result whose resolution also requires topoisomerase II. Examination of the structure of SV40 and PyV replication intermediates suggests that both pathways can be used to complete replication of these viral genomes.

FUTURE PROSPECTS

The last decade has witnessed a dramatic increase in our understanding of SV40 and PyV DNA replication, which can be attributed primarily to the development of a cell-free system for the replication of these viral genomes. Biochemical fractionation and complementation assays have led to the isolation and characterization of many cellular replication proteins from several mammalian sources, and to the molecular cloning of cDNAs and genes that encode these proteins from diverse organisms. Genetic studies in yeasts have validated the role of these proteins in cellular DNA replication. This progress notwithstanding, much remains to be learned. For example, we are only now beginning to appreciate the link between cell-cycle control and the regulation of viral DNA replication. Both large T antigen and several of the cellular replication proteins are regulated by cyclins, cyclin-dependent kinases, and their protein inhibitors. Regulation occurs at the level of phosphorylation of replication proteins and by direct interactions between cell-cycle components and those of the replication machinery. Similarly, the mechanism whereby transcription activators function to stimulate DNA replication is incompletely understood. Our understanding of the mechanism of replication of SV40 and PyV DNA is based largely on the use of in vitro systems with naked DNA. Little is known concerning the replication of chromatin, the natural state of the viral genome in cells. All the replicative DNA polymerases have not been unambiguously defined, and the mechanism that accounts for coordinated synthesis of the leading and lagging strands at replication forks remains to be elucidated. Finally, there are very likely additional cellular replication proteins that remain to be identified. These are but a few of the more obvious areas for future research.

The primary rationale for studying papovavirus DNA replication is to gain greater insight into cellular DNA replication. These viruses have served us well in this regard. Now we appear to stand on the threshold of discovering what is for some the "holy grail" in this field; namely, the identity of functional mammalian replication origins and the initiator proteins that regulate their utilization. The knowledge we have gained from studies of papovavirus DNA replication will serve as a solid foundation for these endeavors.

REFERENCES

Adamczewski, J.P., J.V. Gannon, and T. Hunt. 1993. Simian virus large T antigen associates with cyclin A and p33^{cdk2}. *J. Virol.* **67:** 6551–6557.

Bauer, G.A. and P.M.J. Burgers. 1988. The yeast analog of mammalian cyclin/proliferating-cell nuclear antigen interacts with mammalian DNA polymerase δ. *Proc. Natl. Acad. Sci.* **85:** 7506–7510.

Bennett, E.R., M. Naujokas, and J.A. Hassell. 1989. Requirements for species-specific papovavirus DNA replication. *J. Virol.* **63:** 5371–5385.

Bennett-Cook, E.R. and J.A. Hassell. 1991. Activation of polyomavirus DNA replication by yeast GAL4 is dependent on its transcriptional activation domains. *EMBO J.* **10:** 959–969.

Borowiec, J.A., F.B. Dean, P.A. Bullock, and J. Hurwitz. 1990. Binding and unwinding-how T antigen engages the SV40 origin of DNA replication. *Cell* **60:** 959–969.

Bravo, R., R. Frank, P.A. Blundell, and H. MacDonald-Bravo. 1987. Cyclin/PCNA is the auxiliary protein of polymerase-δ. *Nature* **326:** 515–517.

Brill, S. and B. Stillman. 1989. Yeast replication factor-A functions in the unwinding of the SV40 origin of DNA replication. *Nature* **342:** 92–95.

———. 1991. Replication factor-A from *Saccharomyces cerevisiae* is encoded by three essential genes coordinately expressed at S phase. *Genes Dev.* **5:** 1589–1600.

Brown, G.W., T. Melendy, and D.S. Ray. 1992. Conservation of structure and function of DNA replication protein A in trypanasomitid *Crithidia fasciculata*. *Proc. Natl. Acad. Sci.* **89:** 10227–10231.

Brown, W.C. and J.L. Campbell. 1993. Interaction of proliferating cell nuclear antigen with yeast DNA polymerase δ. *J. Biol. Chem.* **268:** 21706–21710.

Bruckner, A., F. Stadlbauer, L.A. Guarino, A. Brunahl, C. Schneider, C. Rehfuess, C. Prives, E. Fanning, and H.P. Nasheuer. 1995. The mouse DNA polymerase α-primase subunit p48 mediates species-specific replication of polyomavirus DNA in vitro. *Mol. Cell. Biol.* **15:** 1716–1724.

Brush, G.S., C.W. Anderson, and T.J. Kelly. 1994. The DNA-activated protein kinase is required for the phosphorylation of replication protein A during simian virus 40 DNA replication. *Proc. Natl. Acad. Sci.* **91:** 12520–12524.

Bunz, F., R. Kobayashi, and B. Stillman. 1993. cDNAs encoding the large subunit of human replication factor C. *Proc. Natl. Acad. Sci.* **90:** 11014–11018.

Bullock, P.A., Y.S. Seo, and J. Hurwitz. 1991. Initiation of simian virus 40 DNA synthesis *in vitro*. *Mol. Cell. Biol.* **11:** 2350–2361.

Bullock, P.A., S. Tevosian, C. Jones, and D. Denis. 1994. Mapping initiation sites for simian virus 40 DNA synthesis events *in vitro*. *Mol. Cell. Biol.* **14:** 5043–5055.

Burbelo, P.D., A. Utani, Z.-Q. Pan, and Y. Yamada. 1993. Cloning of the large subunit of activator 1 (replication factor C) reveals homology with bacterial DNA ligases. *Proc. Natl. Acad. Sci.* **90:** 11543–11547.

Burgers, P.M.J. 1991. *Saccharomyces cerevisiae* replication factor C. *J. Biol. Chem.* **266:** 22698–22706.

Busen, W. 1980. Purification, subunit structure, and serological analysis of calf thymus ribonuclease H I. *J. Biol. Chem.* **255:** 9434–9443.

Campbell, J.L. 1993. Yeast DNA replication. *J. Biol. Chem.* **268:** 25261–25264.

Carty, M.P., M. Zernick-Kobak, S. McGrath, and K. Dixon. 1994. UV light-induced DNA synthesis arrest in HeLa cells is associated with changes in phosphorylation of human single-stranded DNA-binding protein. *EMBO J.* **13:** 2114–2123.

Cegielska, A. and D.M. Virshup. 1993. Control of simian virus 40 replication by the HeLa cell nuclear kinase I. *Mol. Cell. Biol.* **13:** 1202–1211.

Cegielska, A., I. Moarefi, E. Fanning, and D.M. Virshup. 1994. T-antigen inhibits simian virus 40 DNA replication by phosphorylaton of intact T antigen on serines 120 and 123. *J. Virol.* **68:** 269–275.

Challberg, M.D. and T.J. Kelly. 1989. Animal virus DNA replication. *Annu. Rev. Biochem.* **58:** 671–717.

Chen, J., P.K. Jackson, M.W. Kischner, and A. Dutta. 1995. Separate domains of p21 involved in the inhibition of Cdk kinase and PCNA. *Nature* **374:** 386–388.

Chen, M., Z.-Q. Pan, and J. Hurwitz. 1992a. Sequence and expression in *Escherichia coli* of the 40-kDa subunit of activator 1 (replication factor C) of HeLa cells. *Proc. Natl. Acad. Sci.* **89:** 2516–2520.

———. 1992b. Studies of the cloned 37 kDa subunit of activator 1 (replication factor C) of HeLa cells. *Proc. Natl. Acad. Sci.* **89:** 5211–5215.

Cheng, L. and T.J. Kelly. 1989. Transcriptional activator nuclear factor I stimulates the replication of SV40 minichromosomes *in vivo* and *in vitro*. *Cell* **89:** 541–551.

Cheng, L., J.L. Workman, R.E. Kingston, and T.J. Kelly. 1992. Regulation of DNA replication *in vitro* by the transcriptional activator GAL4-VP16. *Proc. Natl. Acad. Sci.* **89:** 589–593.

Collins, K.L. and T.J. Kelly. 1991. Effects of T antigen and replication protein A on the initiation of DNA synthesis by DNA polymerase α-primase. *Mol. Cell. Biol.* **11:** 2108–2115.

Collins, K.L., A.A.R. Russo, B.Y. Tseng, and T.J. Kelly. 1993. The role of the 70 kd subunit of human DNA polymerase α in DNA replication. *EMBO J.* **12:** 4555–4566.

Copeland, W.C. and T.S.F. Wang. 1993. Enzymatic characterization of the individual mammalian primase subunits reveals a biphasic mechanism for initiation of DNA replication. *J. Biol. Chem.* **268:** 26179–26189.

Crouch, R.J. 1990. Ribinonuclease H: From discovery to 3D structure. *New Biol.* **2:** 771–777.

Cullman, G., K. Fien, R. Kobayashi, and B. Stillman. 1995. Characterization of five replication factor C genes of *Saccharomyces cerevisiae. Mol. Cell. Biol.* **15:** 4661–4671.

Dean, F.B., J.A. Borowiec, Y. Ishim, S. Deb, P. Tegtmeyer, and J. Hurwitz. 1987. Simian virus 40 large tumor antigen requires three core replication domains for DNA unwinding and replication *in vitro. Proc. Natl. Acad. Sci.* **84:** 8267–8271.

Deb, S.P. and P. Tegtmeyer. 1987. ATP enhances the binding of simian virus 40 large T antigen to the origin of replication. *J. Virol.* **61:** 3649–3654.

Deb, S.P., A.L. DeLucia, C. Baur, A. Koff, and P. Tegtmeyer. 1986. Domain and structure of the simian virus 40 core origin of replication. *Mol. Cell. Biol.* **6:** 1663–1670.

Deb, S.P., S. Tsui, A. Koff, A.L. DeLucia, R. Parsons, and P. Tegtmeyer. 1987. The T-antigen-binding protein domain of the simian virus 40 core origin of replication. *J. Virol.* **61:** 2143–2149.

DeLucia, A.L., B.A. Lewton, R. Tjian, and P. Tegtmeyer. 1983. Topography of simian virus 40 A protein-DNA complexes: Arrangement of pentanucleotide interaction sites at the origin of replication. *J. Virol.* **46:** 143–150.

DePamphilis, M.L. 1988. Transcriptional elements as components of eukaryotic origins of DNA replication. *Cell* **52:** 635–638.

———. 1993. How transcription factors regulate origins of DNA replication in eukaryotic cells. *Trends Cell Biol.* **3:** 161–167.

DePamphilis, M.L. and M. Bradley. 1986. Replication of SV40 and polyomavirus chromosomes. In *The papovaviridae* (ed. N. Salzman), vol. 1, pp. 99–246. Plenum Press, New York.

DiFrancesco, R.A. and I.R. Lehman. 1985. Interaction of ribonuclease H from *Drosophilia melanogaster* embryos with DNA polymerase-primase. *J. Biol. Chem.* **260:** 14764–14770.

Din, S., S.J. Brill, M.P. Fairman, and B. Stillman. 1990. Cell-cycle-regulated phosphorylation of DNA replication factor A from human and yeast cells. *Genes Dev.* **4:** 968–977.

Dornreiter, I., W.C. Copeland, and T.S.F. Wang. 1993. Initiation of simian virus 40 DNA replication requires the interaction of a specific domain of human DNA polymerase α with large T antigen. *Mol. Cell. Biol.* **13:** 809–820.

Dornreiter, I., A. Hoss, A.K. Arthur, and E. Fanning. 1990. SV40 T antigen binds directly to the large subunit of purified DNA polymerase α. *EMBO J.* **9:** 3329–3336.

Dornreiter, I., L.F. Erdile, I.U. Gilbert, D. von Winkler, T.J. Kelly, and E. Fanning. 1992. Interaction of DNA polymerase α-primase with cellular replication protein A and SV40 T antigen. *EMBO J.* **11:** 769–776.

Dutta, A. and B. Stillman. 1992. Cdc2 family kinases phosphorylate a human cell DNA replication factor, RPA, and activate DNA replication. *EMBO J.* **11:** 2189–2199.

Eki, T., T. Enomoto, C. Masutani, A. Miyajima, R. Takada, Y. Murikami, T. Ohno, F. Hanaoka, and M. Ui. 1991. Mouse DNA primase plays a principal role in determination of permissiveness for polyomavirus DNA replication. *J. Virol.* **65:** 4874–4881.

Elledge, S.J., R. Richman, F.L. Hall, R.T. Williams, N. Lodgson, and J.W. Harper. 1992. CDK2 encodes a 33-kDa cyclin A-associated protein kinase and is expressed before CDC2 n the cell cycle. *Proc. Natl. Acad. Sci.* **89:** 2907–2911.

Erdile, L.F., M.S. Wood, and T.J. Kelly. 1990. The primary structure of the 32 kDa subunit of human replication protein A. *J. Biol. Chem.* **265:** 3177–3182.

Erdile, L.F., W.D. Heyer, R. Kolodner, and T.J. Kelly. 1991. Characterization of a cDNA encoding the 70 kDa single stranded DNA-binding subunit of human replication protein A and the role of the protein in DNA replication. *J. Biol. Chem.* **266:** 12090–12098.

Fairman, M.P. and B. Stillman. 1988. Cellular factors required for multiple stages of SV40 DNA replication *in vitro*. *EMBO J.* **7:** 1211–1218.

Fanning, E. 1990. Minireview—Simian virus 40 large T antigen: The puzzle, the pieces, and the emerging picture. *J. Virol.* **66:** 1289–1293.

———. 1994. Control of SV40 DNA replication by protein phosphorylation: A model for cellular DNA replication. *Trends Cell Biol.* **4:** 250–254.

Fanning, E. and R. Knippers. 1992. Structure and function of simian virus 40 large tumor antigen. *Annu. Rev. Biochem.* **61:** 55–85.

Fien, K. and B. Stillman. 1992. Identification of replication factor C from *Saccharomyces cerevisiae:* A component of the leading strand DNA replication complex. *Mol. Cell. Biol.* **12:** 155–163.

Firmenich, A., M. Elias-Arnanz, and P. Berg. 1995. A novel allele of *Saccharomyces cerevisiae* that is deficient in recombination and repair and suppressible by *RAD*52. *Mol. Cell. Biol.* **15:** 1620–1631.

Flores-Rozas, H., Z. Kelman, F.B. Dean, Z.-Q. Pan, J.W. Harper, S.J. Elledge, M. O'Donnell, and J. Hurwitz. 1994. Cdk-interacting protein 1 directly binds with proliferating cell nuclear antigen and inhibits DNA replication catalyzed by the DNA polymerase δ holoenzyme. *Proc. Natl. Acad. Sci.* **91:** 8655–8659.

Fotedar, R. and J.M. Roberts. 1992. Cell cycle regulated phosphorylation of RPA-32 occurs within the replication initiation complex. *EMBO J.* **11:** 2177–2187.

Gannon. J.V. and D.P. Lane. 1987. p53 and DNA polymerase α compete for binding to SV40 T antigen. *Nature* **329:** 456–458.

Gerard, R. and Y. Gluzman. 1986. Functional analysis of the role of the A + T rich region and upstream flanking sequences in simian virus 40 DNA replication. *Mol. Cell. Biol.* **6:** 4570–4577.

Goetz, G.S., F.B. Dean, J. Hurwitz, and S.W. Matson. 1988. The unwinding of duplex regions in DNA by the simian virus 40 large tumor antigen-associated DNA helicase activity. *J. Biol. Chem.* **263:** 383–392.

Goulian, M., C. Carton, L. deGrandpre, B. Olinger, and S. Richard. 1987. Processive discontinuous DNA synthesis by mammalian DNA polymerase α-primase. In *DNA replication and recombination* (ed. T. Kelly and R. McMacken), pp. 101–109. A.R. Liss, New York.

Goulian, M., S.H. Richards, C.J. Heard, and B.M. Bigsby. 1990. Discontinous DNA synthesis by purified mammalian proteins. *J. Biol. Chem.* **265:** 18461–18471.

Guggenheimer, R.A., K. Nagata, M. Kenny, and J. Hurwitz. 1984. Protein-primed replication of plasmids containing the terminus of the adenovirus genome. *J. Biol. Chem.* **259:** 7815–7825.

Guo, Z.-S. and M.L. DePamphilis. 1992. Specific transcription factors stimulate simian virus 40 and polyomavirus origins of DNA replication. *Mol. Cell. Biol.* **12:** 2514–2524.

Guo, Z-S., Y. Heine, and M.L. DePamphilis. 1991. T-antigen binding to site I facilitates initiation of SV40 DNA replication, but does not affect bidirectionality. *Nucleic Acids Res.* **19:** 7081–7088.

Guo, Z.-S., C. Gutierrez, U. Heine, J.M. Sogo, and M.L. DePamphilis. 1989. Origin auxiliary sequences can facilitate initiation of simian virus 40 DNA replication *in vitro* as they do *in vivo*. *Mol. Cell. Biol.* **9:** 3593–3602.

Gutierrez, C., Z.S. Guo, J.Roberts, and M.L. DePamphilis. 1990. Simian virus 40 origin auxiliary sequences weakly facilitates T-antigen binding but strongly facilitate DNA unwinding. *Mol. Cell. Biol.* **10:** 1719–1728.

Hassell J.A., W.J. Muller, and C.R. Mueller. 1986. The dual role of the polyomavirus enhancer in transcription and DNA replication. *Cancer Cells* **6:** 561–569.

Hay, R.T. and M.L. DePamphilis. 1982. Initiation of SV40 DNA replication *in vivo*: Location and structure of 5′ends of DNA synthesized in the *ori* region. *Cell* **28:** 767–779.

Hay, T.T., E.A. Hendrickson, and M.L. DePamphilis. 1984. Sequence specificity for the initiation of RNA-primed simian virus 40 DNA synthesis *in vivo*. *J. Mol. Biol.* **175:** 131–157.

He, Z., B.T. Brinton, J. Greenblatt, J.A. Hassell, and C.J. Ingles. 1993. The transcription proteins VP16 and GAL4 bind replication factor A. *Cell* **73:** 1223–1232.

Hendrickson, E.A., C.E. Fritze, W.R. Folk, and M.L. DePamphilis. 1987. The origin of bidirectional DNA replication in polyomavirus. *EMBO J.* **6:** 2011–2018.

Hendrickson, L.A. and M.S. Wold. 1994. Replication protein A mutants lacking phosphorylation sites for $p34^{cdc2}$ kinase support DNA replication. *J. Biol. Chem.* **269:** 24203–24208.

Hendrickson, L.A., C.B. Umbricht, and M.S. Wold. 1994. Recombinant replication protein A: Expression, complex formation, and functional characterization. *J. Biol. Chem.* **269:** 11121–11132.

Heyer, W.D., M.R.S. Rao, L.F. Erdile, T.J. Kelly, and R.D. Kolodner. 1990. An essential *Saccharomyces cerevisiae* single-stranded DNA binding protein is homologous to the large subunit of human RP-A. *EMBO J.* **9:** 2321–2329.

Howell, E.A., M.A. McAlear, D. Rose, and C. Holm. 1994. CDC44: A putative nucleotide-binding protein required for cell cycle progression that has homology to subunits of replication protein C. *Mol. Cell. Biol.* **14:** 255–267.

Huang, L., Y. Kim, J.J. Turchi, and R.A. Bambara. 1994. Structure-specific cleavage of the RNA primer from Okazaki fragments by calf thymus RNAse H1. *J. Biol. Chem.* **269:** 25922–25927.

Hübscher, U. and S. Spadari. 1994. DNA replication and chemotherapy. *Physiol. Rev.* **74:** 259–304.

Hurwitz, J., F.B. Dean, A.D. Kwong, and S.H. Lee. 1990. The *in vitro* replication of DNA containing the SV40 origin. *J. Biol. Chem.* **265:** 18043–18046.

Ishimi, Y., A. Claude, P. Bullock, and J. Hurwitz. 1988. Complete enzymatic synthesis containing the SV40 origin of replication. *J. Biol. Chem.* **263:** 19723–19733.

Ishimi, Y., K. Sugasawa, F. Hanaoka, and A. Kikuchi. 1991. Replication of the simian virus 40 chromosome with purified proteins. *J. Biol. Chem.* **266:** 16141–16148.

Jessberger, R., V. Podust, U. Hübscher, and P. Berg. 1993. A mammalian protein complex that repairs double-strand breaks and deletions by recombination. *J. Biol. Chem.* **268:** 15070–15079.

Kamakaka, R.T., P.D. Kaufman, B. Stillman, P.G. Mitsis, and J.T. Kadonaga. 1994. Simian virus 40 origin- and T-antigen-dependent DNA replication with *Drosophila* factors *in vitro*. *Mol. Cell. Biol.* **14:** 5114–5122.

Kelman, Z. and M. O'Donnell. 1994. DNA replication: Enzymology and mechanisms. *Curr. Opin. Genet. Dev.* **4:** 185–195.

Kenny, M.K., L. Balogh, and J. Hurwitz. 1988. Initiation of adenovirus DNA replication. *J. Biol. Chem.* **263:** 9801–9808.

Kenny, M.K., U. Schlegel, H. Furneaux, and J. Hurwitz. 1990. The role of human single-stranded-DNA binding protein and its individual subunits in simian virus 40 DNA replication. *J. Biol. Chem.* **265:** 7693–7700.

Kesti, T., H. Frantti, and J.E. Syvaoja. 1993. Molecular cloning of the cDNA for the catalytic subunit of human DNA polymerase ε. *J. Biol. Chem.* **268:** 10238–10245.

Krishna, T.S.R., X-P. Kong, S. Gary, P.M. Burgers, and J. Kuriyan. 1994. Crystal structure of the eukaryotic DNA polymerase processivity factor PCNA. *Cell* **79:** 1233–1243.

Kong, X.P., R. Ontrust, M. O'Donnell, and J. Kuriyan. 1992. Three-dimensional structure of the β subunit of *E. coli.* DNA polymerase III holoenzyme: A sliding DNA clamp. *Cell* **69:** 425–437.

Lee, S.H. and J. Hurwitz. 1990. Mechanism of elongation of primed DNA by DNA polymerase δ, proliferating cell nuclear antigen, and activator 1. *Proc. Natl. Acad. Sci.* **87:** 5672–5676.

Lee, S.H., A.D. Kwong, Z.H. Pan, and J. Hurwitz. 1991. Studies on the activator 1 protein complex, an accessory factor for proliferating cell nuclear antigen-dependent DNA polymerase δ. *J. Biol. Chem.* **266:** 594–602.

Li, J.J. and T.J. Kelly. 1984. Simian virus 40 DNA replication *in vitro*. *Proc. Natl. Acad. Sci.* **81:** 6973–6977.

Li, R. and M.R. Botchan. 1993. The acidic transcriptional activation domains of VP16 and p53 bind the cellular replication protein A and stimulate *in vitro* BPV-1 DNA replication. *Cell* **73:** 1207–1221.

Li, R., S. Waga, G.J. Hannon, D. Beach, and B. Stillman. 1994. Differential effects by the p21 CDK inhibitor on PCNA-dependent DNA replication and repair. *Nature* **371:** 534–537.

Li, X. and P. Burgers. 1994a. Molecular cloning and expression of the *Saccharomyces cerevisiae RFC3* gene, an essential component of replication factor C. *Proc. Natl. Acad. Sci.* **91:** 868–872.

———. 1994b. Cloning and characterization of the essential *Saccharomyces cerevisiae RFC4* gene encoding the 37 kDa subunit of replication factor C. *J. Biol. Chem.* **269:** 21880–21884.

Lin, S. and D. Kowalski. 1994. DNA instability facilitates initiation at the SV40 replication origin. *J. Mol. Biol.* **235:** 496–507.

Linn, S. 1991. How many pols does it take to replicate nuclear DNA? *Cell* **66:** 185–187.

Lindahl, T. 1970. An endonuclease specific for double-stranded DNA: Deoxyribonuclease IV from rabbit tissues. *Methods Enzymol.* **21:** 148–153.

Liu, V.F. and D.T. Weaver. 1993. The ionizing radiation-induced replication protein A phosphorylation response differs between ataxia telangiectasia and normal human cells.

Mol. Cell. Biol. **13:** 7222–7231.

Loeber, G., J.E. Stenger, S. Ray, R.E. Parsons, M.E. Anderson, and P. Tegtmeyer. 1991. The zinc finger region of simian virus 40 large T antigen is needed for hexamer assembly and origin melting. *J. Virol.* **65:** 3167–3174.

Longhese, M.P., P. Plevani, and G. Lucchini. 1994. Replication factor A is required *in vivo* for DNA replication, repair, and recombination. *Mol. Cell. Biol.* **14:** 7884–7890.

Lu, Y., A.S. Zeft, and A.T. Riegal. 1993. Cloning and expression of a novel human DNA binding protein, PO-GA. *Biochem. Biophys. Res. Commun.* **193:** 779–786.

Maga, G. and U. Hübscher. 1995. DNA polymerase ε interacts with proliferating cell nuclear antigen in primer recognition and elongation. *Biochemistry* **34:** 891–901.

Manfredi, J. and C. Prives. 1994. The transforming activity of simian virus 40 large tumor antigen. *Biochim. Biophys. Acta* **1198:** 65–83.

Mastrangelo, I.A., P.V.C. Hough, J.S. Wall, M. Dodson, F.B. Dean, and J. Hurwitz. 1989. ATP-dependent assembly of double hexamers of SV40 T antigen at the viral origin of DNA replication. *Nature* **338:** 658–662.

McVey, D., L. Brizuela, I. Mohr, D.R. Marshak, Y. Gluzman, and D. Beach. 1989. Phosphorylation of large tumour antigen by cdc2 stimulates SV40 DNA replication. *Nature* **341:** 503–507.

McVey, D., S. Ray, Y. Gluzman, L. Berger, A. Wildeman, D.R. Marshak, and P. Tegtmeyer. 1993. cdc2 phosphorylation of threonine 124 activates the origin-unwinding functions of simian virus 40 T antigen. *J. Virol.* **67:** 5206–5215.

Melendy, T. and B. Stillman. 1992. SV40 DNA replication. *Nucleic Acids Mol. Biol.* **6:** 129–158.

———. 1993. An interaction between replication protein A and SV40 T antigen appears essential for primosome assembly during SV40 DNA replication. *J. Biol. Chem.* **268:** 3389–3395.

Miyazawa, H., M. Izumi, S. Tada, R. Takada, M. Masutani, M. Ui, and F. Hanoaka. 1993. Molecular cloning of the cDNAs for the four subunits of mouse DNA polymerase α-primase complex and their gene expression during cell proliferation and the cell cycle. *J. Biol. Chem.* **268:** 8111–8122.

Moarefi, I.F., D. Small, I. Gilbert, M. Hopfner, S. K. Randall, C. Schneider, A.A.R. Russo, U. Ramsperger, A.K. Arthur, H. Stahl, T. Kelly, and E. Fanning. 1993. Mutation of the cyclin-dependent kinase phosphorylation site in simian virus 40 (SV40) large T antigen specifically blocks SV40 origin DNA unwinding. *J. Virol.* **67:** 4992–5002.

Morrison, A., H. Araki, A.B. Clark, R.K. Hamatake, and A. Sugino. 1990. A third essential DNA polymerase in *S. cerevisiae*. *Cell* **62:** 1143–1151.

Moses, K. and C. Prives. 1994. A unique subpopulation of murine DNA polymerase α-primase specifically interacts with polyomavirus T antigen and stimulates DNA replication. *Mol. Cell. Biol.* **14:** 2767–2776.

Muller, W.J., D. Dufort, and J.A. Hassell. 1988. Multiple subelements within the polyomavirus enhancer function synergistically to activate DNA replication. *Mol. Cell. Biol.* **8:** 5000–5015.

Murakami, Y., C.R. Wobbe, L. Weissbach, F.B. Dean, and J. Hurwitz. 1986. Role of DNA polymerase α and DNA primase in simian virus 40 DNA replication *in vitro*. *Proc. Natl. Acad. Sci.* **83:** 2869–2873.

Murante, R.S., L. Huang, J.J. Turchi, and R. A. Bambara. 1994. The calf 5′- to 3′-exonuclease is also an endonuclease with both activities dependent on primers annealed upstream of the point of cleavage. *J. Biol. Chem.* **269:** 1191–1196.

Nasheuer, H.P., A. Moore, A. Wahl, and T.S.F. Wang. 1991. Cell cycle-dependent phosphorylation of human DNA polymerase α. *J. Biol. Chem.* **266:** 7893–7903.

Nethanel, T.S., G. Reisfeld, G.D. Gottlieb, and G. Kaufmann. 1988. An Okazaki piece of simian virus 40 may be synthesized by ligation of shorter precursor chains. *J. Virol.* **62:** 2867–2873.

Noskov, V., S. Maki, Y. Kawasaki, S.-H. Leem, B.-I. Ono, H. Araki, Y. Pavlov, and A. Sugino. 1994. The *RFC2* gene encoding a subunit of replication factor C of *Saccharomyces cerevisiae*. *Nucleic Acids Res.* **22:** 1527–1535.

Nishida, C., P. Reinhard, and S. Linn. 1988. DNA repair synthesis in human fibroblasts requires DNA polymerase. *J. Biol. Chem.* **263:** 501–510.

O'Donnell, M., R. Ontrust, F. Dean, M. Chen, and J. Hurwitz. 1993. Homology in accessory proteins of replicative polymerases—*E. coli* to humans. *Nucleic Acids Res.* **21:** 1–3.

Pan, Z.Q., M. Chen, and J. Hurwitz. 1993. The subunits of activator 1 (replication factor C) carry out multiple functions essential for proliferating-cell nuclear antigen-dependent DNA synthesis. *Proc. Natl. Acad. Sci.* **90:** 6–10.

Parsons, R.E., M. Anderson, and P. Tegtmeyer. 1990. Three domains in the simian virus 40 core origin orchestrate the binding, melting, and helicase activities of T antigen. *J. Virol.* **64:** 509–518.

Parsons, R.E., J.E. Stenger, S. Ray, R. Welker, M.E. Anderson, and P. Tegtmeyer. 1991. Cooperative assembly of simian virus 40 T-antigen hexamers on functional halves of the replication origin. *J. Virol.* **65:** 2798–2806.

Peden, K.W.C. and J.M. Pipas. 1992. Simian virus 40 mutants with amino-acid substitutions near the amino terminus of large T antigen. *Virus Genes* **6:** 107–118.

Pipas, J.M. 1992. Common and unique features of T antigens encoded by the polyomavirus group. *J. Virol.* **66:** 3979–3985.

Podust, L.M., V.N. Podust, C. Floth, and U. Hübscher. 1992. Assembly of DNA polymerase δ and ε holoenzymes depends on the geometry of the DNA template. *Nucleic Acids Res.* **22:** 2970–2975.

Podust, L.M., V.N. Podust, J.M. Sogo, and U. Hübscher. 1995. Mammalian DNA polymerase auxiliary proteins: Analysis of replication factor C-catalyzed proliferating cell nuclear antigen loading onto circular double-stranded DNA. *Mol. Cell. Biol.* **15:** 3072–3081.

Pomerantz, B.J. and J.A. Hassell. 1984. Polyomavirus and simian virus 40 large T antigens bind to common DNA sequences. *J. Virol.* **49:** 925–937.

Prelich, G. and B. Stillman. 1988. Coordinated leading and lagging strand synthesis during SV40 DNA replication *in vitro* requires PCNA. *Cell.* **53:** 117–126.

Prelich, G., M. Kostra, D.R. Marshak, M.B. Mathews, and B. Stillman. 1987. The cell-cycle regulated proliferating cell nuclear antigen is required for SV40 DNA replication *in vitro*. *Nature* **326:** 471–475.

Prives, C. 1990. The replication functions of SV40 T antigen are regulated by phosphorylation. *Cell* **61:** 725–738.

Scheidtmann, K.H., D.M. Virshup, and T.J. Kelly. 1991a. Protein phosphatase 2A dephosphorylates simian virus 40 large T antigen specifically at residues involved in regulation of DNA-binding activity. *J. Virol.* **65:** 2098–2101.

Scheidtmann, K.H., M. Buck, J. Schneider, D. Kalderon, E. Fanning, and A.E. Smith. 1991b. Biochemical characterization of phosphorylation site mutants of simian virus 40 large T antigen: Evidence for interaction between amino- and carboxy-terminal domains. *J. Virol.* **65:** 1479–1490.

Schneider, C., K. Weisshart, L.A. Guarino, I. Dornreiter, and E. Fanning. 1994. Species specific functional interactions of DNA polymerase α-primase with simian virus 40 (SV40) T antigen require SV40 origin DNA. *Mol. Cell. Biol.* **14:** 3176–3185.

Schneider, C., D. von Winkler, L. Dornreiter, H.-P. Nasheuer, I. Gilbert, S. Dehda, A.K. Arthur, and E. Fanning. 1992. Bovine RP-A functions in SV40 DNA replication *in vitro*, but bovine polymerase α-primase inhibits replication competitively. In *DNA replication: The regulatory mechanism* (ed. P. Hughes et al.), pp. 385–397. Springer Verlag, Berlin.

Schneider, J. and E. Fanning. 1988. Mutations in the phosphorylation sites of simian virus 40 (SV40) T antigen alter its origin DNA-binding specificity for sites I or II and affect SV40 DNA replication activity. *J. Virol.* **65:** 1598–1605.

SenGupta, D. and J.A. Borowiec. 1994. Strand and face: The topography of interactions between SV40 origin of replication and T-antigen during the initiation of replication. *EMBO J.* **13:** 982–992.

Shivji, M.K.K., S.J. Grey, U.P. Strausfeld, R.D. Wood, and J.J. Blow. 1989. Cip1 inhibit DNA replication but not PCNA-dependent nucleotide excision-repair. *Curr. Biol.* **4:** 1062–1068.

Siegal, G., J.J. Turchi, T.W. Myers, and R.A. Bambara. 1992. A 5′ to 3′ exonuclease functionally interacts with calf DNA polymerase ε. *Proc. Natl. Acad. Sci.* **89:** 9377–9381.

Smale, S.T. and R. Tjian. 1986. T-antigen-DNA polymerase α complex implicated in simian virus 40 DNA replication. *Mol. Cell. Biol.* **6:** 4077–4087.

Smith, M.L., I.-T. Chen, Q. Zhan, I. Bae, C.-Y. Chen., T.M. Gilmer, M.B. Kastan, P.M. O'Conner, and A.J. Fornace, Jr. 1994. Interaction of the p53-regulated protein GADD45 with proliferating cell nuclear antigen. *Science* **266:** 1376–1379.

Smith, J. and R. Rothstein. 1995. A mutation in the gene encoding the *Saccharomyces cerevisiae* single-stranded DNA-binding protein Rfa1 stimulates a *RAD*52-independent pathway for direct-repeat recombination. *Mol. Cell. Biol.* **15:** 1632–1641.

Stadlbauer, F., C. Voitenleitner, A. Bruckner, E. Fanning, and H.-P. Nasheuer. 1996. Species-specific replication of simian virus 40 DNA *in vitro* requires the p180 subunit of human DNA polymerase α-primase. *Mol. Cell. Biol.* **16:** 94–104.

Stillman, B. 1989. Initiation of eukaryotic DNA replication *in vitro*. *Annu. Rev. Cell Biol.* **5:** 197–245.

———. 1994. Smart machines at the DNA replication fork. *Cell* **78:** 725–728.

Sundstrom, N.A., N.H. Acheson, and J.A. Hassell. 1991. Determination of the origin-specific DNA-binding domain of polyomavirus large T antigen. *J. Virol.* **65:** 6998–7003.

Tooze, J., ed. 1980. *Molecular biology of tumor viruses,* 2nd edition: *DNA tumor viruses*. Cold Spring Harbor Laboratory, Cold Spring Harbor, New York.

Tsurimoto, T. and B. Stillman. 1989. Purification of a cellular replication factor, RF-C, that is required for coordinated synthesis of leading and lagging strands during simian virus 40 DNA replication *in vitro*. *Mol. Cell. Biol.* **9:** 609–619.

———. 1990. Functions of replication factor C and proliferating-cell nuclear antigen: Functional similarity of DNA polymerase accessory proteins from human cells and bacteriophage T4. *Proc. Natl. Acad. Sci.* **87:** 1023–1027.

———. 1991a. Replication factors required for SV40 DNA replication *in vitro*. *J. Biol.*

Chem. **266:** 1950–1960.

———. 1991b. Replication factors required for SV40 DNA replication *in vitro*. *J. Biol. Chem.* **266:** 1961–1968.

Turchi, J.J. and R.A. Bambara. 1993. Completion of mammalian lagging strand DNA replication using purified proteins. *J. Biol. Chem.* **268:** 15136–15141.

Turchi, J.J., L. Huang, R.S. Murante, Y. Kim, and R.A. Bambara. 1994. Enzymatic completion of mammalian lagging strand DNA replication. *Proc. Natl. Acad. Sci.* **91:** 9803–9807.

Umbricht, C.B., L.F. Erdile, E.W. Jabs, and T.J. Kelly. 1993. Cloning, overexpression, and genomic mapping of the 14-kDa subunit of human replication protein A. *J. Biol. Chem.* **268:** 6131–6138.

Virshup, D.M., A. Russo, and T.J. Kelly. 1992. Mechanism of activation of simian virus 40 DNA replication by protein phosphatase 2A. *Mol. Cell. Biol.* **12:** 4883–4895.

Waga, S. and B. Stillman. 1994a. Anatomy of a DNA replication fork revealed by reconstitution of SV40 DNA replication *in vitro*. *Nature* **369:** 207–212.

———. 1994b. Reconstitution of complete SV40 DNA replication with purified replication factors. *J. Biol. Chem.* **269:** 10923–10934.

Waga, S., G.J. Hannon, D. Beach, and B. Stillman. 1994. The p21 inhibitor of cyclin-dependent kinases controls DNA replication by interaction with PCNA. *Nature* **369:** 574–578.

Wang, T.S.F. 1991. Eukaryotic DNA polymerases. *Annu. Rev. Biochem.* **60:** 513–552.

Watt P.M. and I.D. Hickson. 1994. Structure and function of type II DNA topoisomerases. *Biochem. J.* **303:** 681–695.

Wei, Y.-F., P. Robins, K. Carter, K. Caldecott, D. Pappin, G.-L. Yu, R.-P. Wang, B. Shell, R. Nash, P. Schar, D. Barnes, W. Haseltine, and T. Lindahl. 1995. Molecular cloning and expression of human cDNAs encoding a novel DNA ligase IV and DNA ligase III, an enzyme active in DNA repair and recombination. *Mol. Cell. Biol.* **15:** 3206–3216.

Wiekowski, M., M.W. Schwarz, and H. Stahl. 1988. Simian virus 40 large T antigen DNA helicase. *J. Biol. Chem.* **263:** 436–442.

Wessel, R., J. Schweizer, and H. Stahl. 1992. Simian virus 40-T antigen DNA helicase is a hexamer which forms a binary complex during bidirectional unwinding from the viral origin of DNA replication. *J. Virol.* **66:** 804–815.

Wobbe, C.R., L. Weissbach, J.A. Borowiec, F. Dean, Y. Murakami, P. Bullock, and J. Hurwitz. 1987. Replication of simian virus 40 origin-containing DNA *in vitro* with purified proteins. *Proc. Natl. Acad. Sci.* **84:** 1834–1838.

Wold, M.S. and T.J. Kelly. 1988. Purification and characterization of replication protein A, a cellular protein required for *in vitro* replication of simian virus 40 DNA. *Proc. Natl. Acad. Sci.* **85:** 2523–2527.

Wold, M.S., D.H. Weinberg, D.M. Virshup, J.J. Li, and T.J. Kelly. 1989. Identification of cellular proteins required for simian virus 40 DNA replication. *J. Biol. Chem.* **264:** 2801–2809.

Xiong, Y., H. Zhang, and D. Beach. 1992. D type cyclins associate with multiple protein kinases and the DNA replication and repair factor PCNA. *Cell* **71:** 505–514.

Xiong, Y., G.J. Hannon, H. Zhang, D. Casso, R. Kobayashi, and D. Beach. 1993. p21 is a universal inhibitor of cyclin kinases. *Nature* **366:** 701–704.

Yang, C.-L., M.S. Wold, J.J. Li, T.J. Kelly, and L.F. Liu. 1987. Roles of DNA topoisomerases in simian virus 40 DNA replication *in vitro*. *Proc. Natl. Acad. Sci.* **84:**

950–954.
Yang, C.-L. L.-S. Chang, P. Zhang, H. Hao, Z. Lingyn, N.L. Toomey, and M.Y.W.T. Lee. 1992. Molecular cloning of the cDNA for the catalytic subunit of human DNA polymerase δ. *Nucleic Acids Res.* **20:** 735–745.
Yoder, B.L. and P.M.J. Burgers. 1991. *Saccharomyces cerevisiae* replication factor C. *J. Biol. Chem.* **266:** 22689–22697.
Zhang, H., Y. Xiong, and D. Beach. 1993. Proliferating cell nuclear antigen and p21 are components of multiple cell cycle kinase complexes. *Mol. Biol. Cell* **4:** 897–906.
Zhang, S.J., X.R. Zeng, P. Zang, N.L. Toomey, R.Y. Chuang, L.S. Chang, and M.Y.W.T. Lee. 1995. A conserved region in the amino terminus of DNA polymerase δ is involved in proliferating cell nuclear antigen binding. *J. Biol. Chem.* **270:** 7988–7992.

23
Papillomavirus DNA Replication

Arne Stenlund
Cold Spring Harbor Laboratory
Cold Spring Harbor, New York 11724

The papillomaviruses constitute a large family of viruses that cause benign tumors (warts) in their hosts. In recent years, these viruses have been found to have substantial clinical importance (Syrjanen et al. 1987; zur Hausen 1991). The replication properties of papillomaviruses have been studied for a number of years; however, many results obtained in the early studies concerning both *trans*-acting factors and *cis*-acting elements required for replication have not been reproducible. Therefore, a clear picture of the viral replicon has emerged only during the last 5 years and the understanding of papillomavirus replication is relatively primitive compared to, for example, the well-studied SV40 system. For technical and historical reasons, the field has been dominated by studies of one particular member of the papillomavirus family, bovine papillomavirus (BPV), which has served as a prototype for the papillomaviruses. More recently, largely because of information obtained from the BPV system, it has become possible to develop systems to study the replication properties of several human papillomaviruses. In this chapter, I describe the BPV system in some detail and, from this point of reference, discuss the variations and differences that have been observed for replication of other papillomaviruses.

The natural life cycle of papillomaviruses is complex and not understood in detail. However, in the basal cells of the papilloma, viral DNA is maintained in a latent form characterized by episomal replication at a relatively low copy number. At this stage, expression of viral capsid proteins cannot be detected. As the cells differentiate and migrate toward the outer layers of the papilloma, substantially higher copy numbers and assembled virions can be detected (Stoler et al. 1990; Dollard et al. 1992; Meyers et al. 1992). The basis for all papillomavirus replication systems originates in the early observation that BPV can transform mouse cells in culture and that in the transformed cells, the viral DNA replicates as a plasmid with a modest copy number (Dvoretsky et al. 1980; Lancaster 1981; Law et al. 1981). Furthermore, the viral DNA is maintained over

DNA Replication in Eukaryotic Cells

many cell generations such that permanent cell lines with episomal viral DNA can be established. The replication that takes place in these transformed cells is believed to resemble the replication occurring in the basal layers of the papilloma. Early studies indicated that replication of BPV was under the control of chromosomal once-per-cell-cycle replication. These results have not been reproduced, and more recent experiments demonstrate that selection of molecules for replication takes place at random (Gilbert and Cohen 1987). However, replication of BPV is regulated by the cell cycle in that replication of the viral DNA is strictly confined to S phase (Ravnan et al. 1992).

The behavior of BPV in tissue-culture cells illustrates some clear distinctions between the replication properties of papillomaviruses and other papovaviruses. Whereas, for example, SV40 and polyomavirus rapidly replicate to very high copy numbers in permissive cells, resulting in death and lysis of the infected cell, the papillomaviruses appear to limit their DNA replication to a relatively low copy number, and viral DNA replication has no obvious deleterious effect on the cell (Law et al. 1981; Dvoretsky et al. 1980). In addition, despite a high degree of host and cell-type specificity for infection, little host-cell specificity appears to exist for papillomaviruses at the level of DNA replication, in contrast to the strict specificity observed for SV40 and polyomavirus (Tooze 1981; Murakami et al. 1986). When the required viral genes are expressed from heterologous expression vectors, replication of a variety of papillomaviruses has been achieved in a number of cell lines of different mammalian origins (Ustav et al. 1991; Chiang et al. 1992b; Del Vecchio et al. 1992; Lu et al. 1993). A further distinction is that papillomaviruses require a virus-encoded protein in addition to the initiator for viral DNA replication. This protein, the transcription factor E2, appears to be the counterpart of the auxiliary factors involved in replication of SV40 and polyomavirus, as discussed below (see also van der Vliet, this volume).

These differences appear to represent variations on a common theme rather than fundamental differences between papillomaviruses and other papovaviruses. When DNA replication is examined at the molecular level, substantial similarities can be observed. These include similarities between the respective viral factors that are required for replication and extend to the interactions with cellular factors. For example, what was originally observed as a very limited sequence homology between the BPV E1 protein and SV40 and polyomavirus T antigen is now well established as a functional homology (Clertant and Seif 1984); E1 is clearly the counterpart of T antigen, providing recognition of the origin of DNA replication (*ori*) and *ori*-unwinding activity (Ustav et al. 1991;

Wilson and Ludes-Meyers 1991; Yang et al. 1991, 1993; Seo et al. 1993a). E1 also appears to interact directly with the replication machinery through physical interactions with DNA polymerase-α, which is an interaction partner also for T antigen (Park et al. 1994; Bonne-Andrea et al. 1995; see also Borowiec, this volume). The cellular proteins required at the replication fork also appear to be similar, if not identical, to the factors required for SV40 replication; a minimal replication system for SV40 consisting of factors purified from HeLa cells can function to replicate BPV (Muller et al. 1994). However, a purified system consisting of factors purified from 293 cells that is fully functional for SV40 replication fails to replicate BPV, indicating that some difference in the requirement for cellular factors exists (Melendy et al. 1995; see also Brush and Kelly, this volume).

ASSAYS FOR REPLICATION

DNA replication of papillomaviruses has been studied in three different experimental systems. Since the types of questions that have been addressed in these systems are to some degree different and the results obtained also differ to some extent, a brief summary of the different systems and the experimental results is given here.

Stable (Long-term) Replication Assays

The earliest studies of papillomavirus replication were performed by taking advantage of the transforming ability of BPV. After transfection of the viral genome, DNA could be prepared from transformed foci, and replicating viral DNA could be measured by Southern analysis (Lancaster 1981; Law et al. 1981). Using this method, at least five genes were initially thought to be involved in stable replication (Lusky and Botchan 1984, 1985; Sarver et al. 1984; Berg et al. 1986). Two of these, the genes E1 and E2, have withstood the test of time. However, the conclusions that could be drawn about the function of these genes was relatively limited due to the nature of the assay. Although it was obvious that DNA replication was required, it was by no means clear that DNA replication was sufficient to maintain the viral DNA over the extended periods of time that were required for these assays. In addition, since the assay required transfection of the entire viral genome, it was difficult to determine whether a given gene had a direct involvement in replication or was required indirectly, for example, as a regulator of viral gene expression.

Transient (Short-term) Replication Assays

A clear picture of BPV replication was first obtained using a short-term replication system (Ustav and Stenlund 1991). This system was based on the prediction that if a sufficiently large fraction of the cells could be transfected with viral DNA, detection of replication at early times (1–3 days) after transfection should be possible without the need to select for transformed foci. This assay would have the advantage of measuring replication with only a minimal contribution of mechanisms that would affect, for example, partitioning. It would also be more convenient for genetic analysis, since it would allow simple complementation experiments. Using this system for BPV, it was possible to establish unequivocally that two viral proteins, the full-length products from the E1 and E2 open reading frames, were required for viral DNA replication (Ustav and Stenlund 1991). Provision of these two factors from heterologous expression vectors also allowed the identification of viral *cis*-acting elements that were required for replication (Ustav et al. 1991). This type of assay has now become a standard method for analysis of papillomavirus replication and has been applied to a variety of papillomaviruses.

Cell-free Replication

The identification of the two viral proteins required for replication in vivo has allowed the development of cell-free replication for BPV along the lines of the systems previously established for SV40 and polyomavirus (Yang et al. 1991; Seo et al. 1993a; Bonne-Andrea et al. 1995). This in vitro replication system consists of a nuclear extract preparation from a permissive cell line supplemented with the viral E1 and E2 proteins overexpressed and purified from baculovirus-infected insect cells or from *Escherichia coli*. With the use of this system, a number of important properties of BPV replication have been determined, including the mapping of *ori* sequences active for replication in vitro, demonstration of bidirectional replication, and the requirement for cellular factors. Recently, a similar in vitro system has also been established for the human papillomavirus HPV-11 (Kuo et al. 1994).

REQUIREMENTS FOR VIRAL DNA REPLICATION IN VIVO

The requirements for replication for a number of papillomaviruses have been established using the transient replication assay. The results from these assays demonstrate that the requirements for replication are similar for most, if not all, the different virus types that have been tested, consistent with the high degree of overall similarity within the papillomavirus group. Papillomaviruses in general require the well-conserved E1 and E2 proteins and also a small *cis*-acting sequence element that is conserved both in terms of location and DNA sequence. Some variation as to the relative importance of the two viral factors, as well as the different elements of the *ori*, exists between different viruses. These variations are discussed below.

Viral Origin of Replication

The *cis*-acting sequences required for replication have been defined for a number of papillomaviruses using transient replication assays where the respective E1 and E2 proteins have been expressed from heterologous expression vectors (Ustav et al. 1991; Chiang et al. 1992b; Del Vecchio et al. 1992; Remm et al. 1992; Lu et al. 1993; Sverdrup and Khan 1995). In all cases, these experiments have resulted in the identification of a small fragment located between the early open reading frames and a region called the long control region (LCR) or upstream regulatory region (URR), which contains regulatory signals for viral transcription. This position appears to be completely conserved between a number of different virus types. The use of a larger fragment containing additional sequences from the URR region commonly results in higher replication activity than for a smaller fragment, but whether these effects are due to specific elements in the URR is unknown. None of these *ori* sequences, including the BPV *ori*, has been examined in great detail, but on the basis of a combination of homology, deletion analysis, and linker substitution analysis (Ustav et al. 1991; Chiang et al. 1992b; Remm et al. 1992; Lu et al. 1993; Zhu et al. 1994; Russell and Botchan 1995), several different elements can be discriminated. The papillomavirus sequences show distinctive similarities to other *ori* sequences that have been described (DePamphilis 1993 and this volume) and can be classified according to two schematic variants (see Fig. 1). Some animal papillomaviruses including BPV-1 are organized as illustrated in Figure 1A, where two E2-binding sites (E2BSs) flank an A+T-rich sequence and the palindromic sequence that constitutes the E1-binding site (E1BS). The majority of the human papillomaviruses have a slightly different organization, as shown in Figure 1B. The E1BS is located between two A+T-rich sequences that in turn are flanked by three E2BSs, one on the upstream (left) side and two on the downstream (right) side. Some *ori*s contain slight variations on this theme, lacking, for example, one of the two

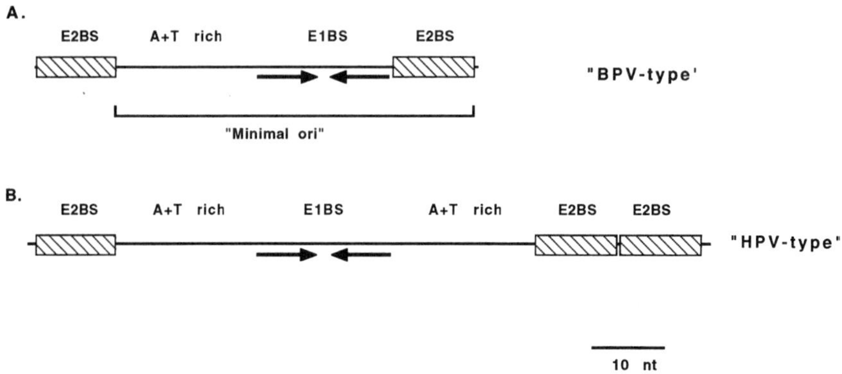

Figure 1 Schematic representation of the origin sequences of papillomaviruses classified in two groups. The "BPV-type" represents the organization found in the small number of animal papillomaviruses that have been examined. The "HPV-type" represents the organization found in virtually all sequenced human papillomaviruses. (E1BS) Binding site for the viral replication initiator E1. (E2BS) Binding site for the viral transcriptional activator E2.

downstream E2BSs. The A+T-rich sequences for both of the groups are of variable length with little sequence conservation.

The definition of minimal *ori* fragments has been complicated by apparent redundancies, especially concerning the E2BS. For example, in the BPV *ori*, a fragment comprising the A+T-rich sequence, the E1BS, and the downstream (right) E2BS was originally defined as the minimal *ori* (Ustav et al. 1991). However, another fragment containing the upstream (left) E2BS, the A+T sequence, and the E1BS is also active for replication, albeit at lower levels. Definition of a minimal HPV *ori* is even more complex in this regard; deletion analysis of the HPV-11 and HPV-18 *ori*s demonstrates that even nonoverlapping fragments have replication activity, indicating redundancy (Chiang et al. 1992a; Lu et al. 1993; Sverdrup and Khan 1995). Furthermore, a significant replication activity can be observed with an *ori* containing only the two high-affinity E2BSs (Lu et al. 1993; Sverdrup and Khan 1995). These results indicate that a search for a minimal sequence that can function as an *ori* may be less informative than a more quantitative approach where all elements that affect *ori* activity are identified. Presumably, in the context of the viral genome, conservation of a given element indicates function. In this regard, it is important to note that in the replication assays used for these studies, the viral E1 and E2 proteins are highly expressed from expression vectors. This may mask the requirement for some elements that are important for function at lower levels of E1 and E2.

The E2BS(s)

All papillomavirus origins of replication include one or more binding sites for the virus-encoded transcription factor E2, which binds to the consensus sequence ACCGNNNNCGGT (Androphy et al. 1987; Hawley-Nelson et al. 1988; Hirochika et al. 1988; Li et al. 1989). Results from studies of the BPV *ori* as well as several HPV *ori*s indicate that the requirement for an E2BS is absolute, but that a great deal of flexibility exists in terms of both the position and the affinity of the E2BS (Chiang et al. 1992a; Remm et al. 1992; Lu et al. 1993; Ustav et al. 1993; Sverdrup and Khan 1994). One single exception to this rule has been described: HPV-1 is capable of replication in the absence of E2 and E2BS; however, E2 is stimulatory for replication (Gopalakrishnan and Khan 1994). For BPV it is apparent that a single E2BS of very low affinity is sufficient for an active *ori*, at least under conditions where E1 and E2 are overexpressed (Ustav et al. 1991; Ustav et al. 1993). Moreover, a relationship appears to exist between the affinity of the E2BS and the ability to function at a distance from the binding site for E1 (Ustav et al. 1993). In multimerized form, the E2BSs are capable of exerting their function even when placed at a distance of several kilobases from the rest of the *ori*.

The human papillomaviruses generally show a stricter requirement with regard to the properties of the E2BS. In the natural context, most human papillomaviruses have three high-affinity E2BSs located in the *ori* fragment. Deletion or mutation of any one of these sites results in a small reduction of replication (Chiang et al. 1992a; Remm et al. 1992; Lu et al. 1993; Sverdrup and Khan 1994, 1995; Russell and Botchan 1995). Deletion or mutation of any two of these sites results in a severe drop in replication activity, indicating that at least two E2BSs are required for efficient replication. The requirement for specific spacing has not been tested directly in the HPV system.

The E1BS

The binding site for the E1 replication helicase has been defined in BPV as an imperfect palindromic sequence 18 nucleotides in length with the sequence ATTGTTGTTAACAATAAT (Wilson and Ludes-Meyers 1991; Ustav et al. 1991; Yang et al. 1991; Holt et al. 1993). A related sequence is present in all papillomaviruses examined, but binding of other E1s to these putative E1BSs has not been defined in detail (Bream et al. 1993; Lu et al. 1993; Liu et al. 1995). Therefore, the identity of the E1BS in HPV *ori* is largely based on homology with the BPV E1BS. The dependence on the E1BS for replication varies between different viruses.

Results from BPV demonstrate that single point mutations in the BPV E1BS can reduce binding of E1, as well as replication, more than 20-fold (Sedman and Stenlund 1995). In HPV-11, linker substitution mutations in the putative E1BS have less severe effects, indicating that sequence-specific binding of E1 may be of less importance for HPV replication (Russell and Botchan 1995). However, as stated above, the lack of a significant effect of mutations in the E1BS could be related to the levels of expression of the E1 and E2 proteins. Obviously, the conservation of the E1BS argues that it is likely to play a role in replication of the viral genome.

The A+T-rich Region

A+T-rich sequences are important elements of both prokaryotic and eukaryotic origins of replication; for SV40 and polyomavirus origins, various activities have been assigned to these elements, including DNA bending and facilitation of the unwinding activity by T antigen (Deb et al. 1987; Bramhill and Kornberg 1988; DePamphilis 1993 and this volume). The A+T-rich regions in papillomaviruses are very poorly characterized, both in terms of the function and the actual sequence requirement. The A+T-rich sequence in the context of the minimal BPV *ori* is clearly important for replication in vivo, since a deletion of this sequence results in a precipitous drop in replication activity (Ustav et al. 1991). The spacing relative to the E1BS also appears to be important, since insertions between the E1BS and the A+T-rich region result in loss of replication activity. In contrast, both deletion analysis and linker substitution analysis in HPV have failed to show significant effects of mutations in these elements (Lu et al. 1993; Russell and Botchan 1995). A possible function for the A+T-rich region in the BPV *ori* has been suggested by the induction of structural changes in this sequence upon binding of E1 to the *ori* in the presence of ATP (Gillette et al. 1994). It has also been suggested that E1 in the presence of ATP and E2 binds in a biphasic manner, first to the E1BS and at higher concentrations to the A+T-rich region. One possibility is therefore that the A+T-rich region serves as a secondary site for binding of E1 that may be involved in the formation of a larger ATP-dependent multimeric E1/*ori* complex.

Viral Proteins

The two viral proteins that are required for BPV replication are encoded from the full-length E1 and E2 open reading frames. E1 is an initiator protein based on a number of criteria and has the expected helicase, ATPase, and unwinding activities (Seo et al. 1993a; Yang et al. 1993;

Gillette et al. 1994; MacPherson et al. 1994; Park et al. 1994). The E2 protein falls in the category of accessory or auxiliary factors simply because this factor is not required for replication under all conditions; *ori*-specific replication in vitro can clearly take place with E1 alone. Both of these factors are discussed in greater detail in the appropriate chapters dealing with general properties of initiator proteins and accessory factors (see Coen; Hübscher et al.; both this volume). However, it should be noted that in contrast to other papovaviruses which show a high degree of specificity for the cognate initiator, for papillomaviruses, E1 and E2 proteins appear to be largely interchangeable for replication of different virus types, suggesting a high degree of functional relatedness and a low degree of specificity (Chiang et al. 1992b; Del Vecchio et al. 1992; Lu et al. 1993; Sverdrup and Khan 1994; Park et al. 1994).

REQUIREMENTS FOR REPLICATION IN VITRO

The requirements for replication of BPV in cell-free replication systems differ in two important respects from the requirements for replication in vivo. The E1 protein is clearly by itself sufficient to generate *ori*-specific replication in a cell-free replication extract (Yang et al. 1991, 1993; Seo et al. 1993a; Lusky et al. 1994; Muller et al. 1994). Consistent with this finding, an E2BS is not a required component for an *ori* functional for in vitro replication (Park et al. 1994). Furthermore, cell-free replication under these conditions shows only limited *ori* specificity; at high levels of E1, efficient *ori*-independent replication can be observed (Seo et al. 1993b; Yang et al. 1993). An interesting difference exists between cell-free replication of BPV and HPV-11 in a similar cell-free replication system. HPV-11 shows a strong dependence on E2 and an E2BS for replication in vitro (Kuo et al. 1994). Possible explanations for this difference are discussed below.

DNA Binding of E1 and E2

Both E1 and E2 are DNA-binding proteins, and in the BPV system it has been demonstrated that both proteins can bind to their respective binding sites at the origin of replication (Ustav et al. 1991; Wilson and Ludes-Meyers 1991; Yang et al. 1991; Seo et al. 1993b; Gillette et al. 1994; Lusky et al. 1994). E2 is very well characterized in this regard, and the X-ray structure of a protein/DNA co-crystal has been solved (Hedge et al. 1992). E2 binds to a palindromic site as a dimer with base contacts in both the major and minor grooves. In contrast, very little is known about

DNA binding by E1. Because of the palindromic nature of the E1BS and the large protection that is observed in DNase footprints, it has been assumed that E1 binds in a multimeric form, and an estimate of the stoichiometry of binding has also been made (Lusky et al. 1994). A number of observations have demonstrated that the E1 and E2 proteins can interact specifically with each other in vitro (Mohr et al. 1990; Blitz and Laiminis 1991; Lusky and Fontane 1991; Thorner et al. 1993; Frattini and Laiminis 1994; Benson and Howley 1995). Consistent with these observations, binding of E1 and E2 to the *ori* is cooperative, i.e., binding of either E1 or E2 takes place at substantially lower concentrations in the presence of the other protein, provided that binding sites for both proteins are present (Yang et al. 1991; Seo et al. 1993b; Gillette et al. 1994). It has therefore been suggested that E2 may serve to recruit E1 to the origin of replication.

More detailed analysis of binding of E1 to the *ori*, by itself and in combination with E2, has revealed that the situation is more complicated than was originally believed. The complex formed by E1 alone on the *ori* is substantially larger than the complex that also contains E2, indicating that the E1 protein is capable of binding to the *ori* in one form by itself and in a different form in the presence of E2 (Yang et al. 1991; Lusky et al. 1994; Sedman and Stenlund 1995). The most likely explanation for these results is that the E1 protein can bind to DNA in different multimeric forms; a larger form is capable of binding to *ori* by itself and forms an E1/*ori* complex, whereas binding of the smaller form of E1 takes place in cooperation with E2, forming an E1/E2/*ori* complex. Interestingly, binding of the smaller form of E1 in the absence of E2 has not been detected, indicating that cooperation with E2 is required for DNA-binding activity of this form. A biochemical function has not been assigned to either of these two complexes directly; however, the conditions under which the E1/*ori* complex forms are also the conditions under which replication in vitro can be detected (Lusky et al. 1994).

Role of E2 in Replication

A peculiarity of papillomavirus DNA replication is the absolute requirement in vivo for the viral transcription factor E2. This requirement is highly specific, and various other transcription factors that have been tested are inactive (Ustav et al. 1991; Li and Botchan 1993). This is in contrast to the auxiliary transcription factors required for replication of other papovaviruses. SV40 and polyomavirus are not limited to any one particular factor, but both viruses show some specificity in that not all transcription factors are capable of providing the auxiliary function (DePamphilis 1993; see also van der Vliet, this volume). A number of activities have been proposed for the function of transcriptional activators in replication in general, and these activities may be involved in the activation of BPV replication. For example, it has been demonstrated that E2 can interact with the cellular single-stranded binding protein RP-A (Li and Botchan 1993), and E2 appears to be capable of counteracting repression of replication by nucleosomes (Li and Botchan 1994). However, these functions do not explain the specific requirement for E2, since other transcription factors have these activities but are not functional for BPV replication in vivo (Li and Botchan 1994).

A possible explanation for the requirement of E2 for DNA replication in vivo could obviously be that the generation of the specific E1/E2/*ori* complex that can be observed in vitro is required for replication. Recent analysis of the BPV *ori* with mutations affecting the binding sites for E1 and E2 and the spacing between the two sites has revealed that a very good correlation exists between the ability to form the E1/E2/*ori* complex and replication activity (Sedman and Stenlund 1995). In contrast, mutations that only affect the formation of the E1 complex have minimal effects on replication. Furthermore, the mutational data also demonstrate that, consistent with a strong interaction between the two proteins on the *ori*, an increase in the affinity of the E2BS can compensate for mutations in the E1BS that reduce the affinity of E1 and vice versa (Spalholz et al. 1993; Sedman and Stenlund 1995). These results indicate that the ability to form the E1/E2/*ori* complex, and hence replication activity, is dependent on at least three different interactions: E1 and E2 with their respective binding sites and a direct interaction between E1 and E2.

One consequence of the cooperative binding between E1 and E2 is a substantial increase in the specificity by which E1 can recognize the *ori* (Sedman and Stenlund 1995). Binding of E1 in the form of the E1/*ori* complex can readily be competed by modest quantities of nonspecific DNA, indicating that the sequence specificity of this particular complex is low. The resistance to competitor is significantly increased when E1 binds to the *ori* in combination with E2, demonstrating greater sequence specificity. This effect on sequence specificity by E2 provides a possible mechanism for the function of E2 as an auxiliary factor. This possibility has recently been examined through modification of the in vitro replication conditions by addition of competitor DNA (Sedman and Stenlund 1995). As predicted, E1-dependent replication can be suppressed by addition of competitor DNA. Addition of E2 under these conditions can restore replication completely, in a manner that is dependent on a binding site for E2 at the *ori*, closely reproducing the characteristics of E2 dependence for replication in vivo. A function for E2 as a specificity factor is also consistent with the observation that E1 alone in vivo, even at very high levels, is incapable of supporting replication (Ustav and Stenlund 1991).

This model for the function of E2 in replication of papillomaviruses, where E2 serves as a specificity factor for binding of the initiator E1, shows striking similarities to the proposed function of the transcription factor NFI in replication of adenovirus (Mul and van der Vliet 1992). The DNA-binding domain of NFI interacts with and promotes binding of the adenovirus preterminal protein/DNA polymerase (pTP/pol) complex to the adenovirus *ori*. The clear distinction between these systems is that for the interaction between E1 and E2, the *trans*-activation domain of E2 is specifically required. Thus, with regard to the requirement for auxiliary transcription factors, papillomaviruses provide an interesting link between SV40 and polyomavirus on the one hand and adenovirus on the other. Both SV40 and polyomavirus require the activation domain of the transcription factor, but the mechanism by which the activation domain exerts its activity is not known. Adenovirus utilizes the DNA-binding domain of a transcription factor that interacts with and affects the binding of the pTP/pol complex. Papillomaviruses utilize the activation domain of the transcription factor E2 for what appears to be the same purpose. This suggests that the auxiliary factors in SV40 and polyomaviruses may also have a function related to facilitation of binding of the initiator to the origin of replication.

A Model for *ori* Recognition

The importance of the E1/E2/*ori* complex in replication of BPV provides a simple explanation for some of the variability concerning the requirement for E1 and E2 and also the *cis*-acting sequences that has been observed for some human papillomaviruses (Fig. 2). It has been demonstrated for HPV-11 and HPV-18 that at least when the E1 and E2 proteins are overexpressed, a duplicated binding site for E2 can serve as an *ori* (Lu et al. 1993; Sverdrup and Khan 1995). HPV-1, on the other hand, appears to be capable of replication using E1 alone, at least when E1 is expressed at high levels (Gopalakrishnan and Khan 1994). Both of these results can be explained by simply allowing for different contributions from the three different interactions "1," "2," and "3" that we have listed. If, for example, interaction 1 between E1 and the E1BS is sufficiently specific, the other two interactions may be of lesser importance,

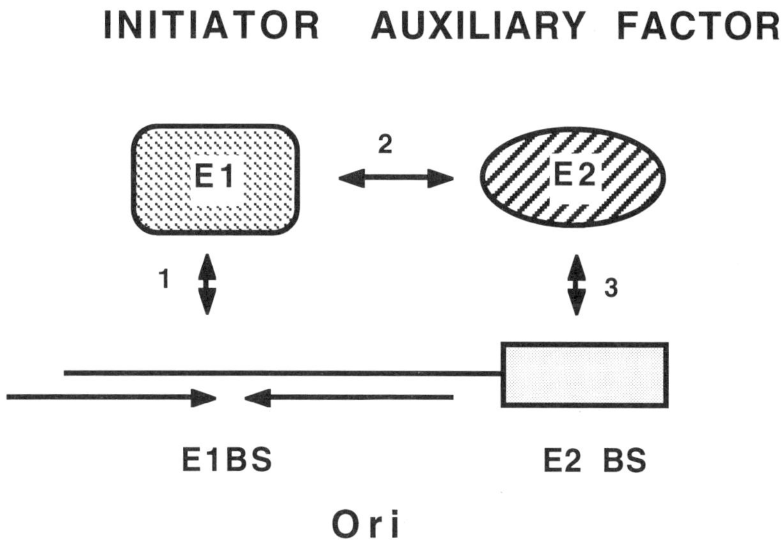

Figure 2 Schematic model illustrating the interactions between E1 and DNA, E2 and DNA, and E1 and E2 that are required for the formation of a replication-competent *ori* complex based on genetic and biochemical analyses. See text for details.

especially at high concentrations of E1(HPV-1). If, on the other hand, the interaction between E1 and DNA is of low specificity, a strong interaction between E2 and E2BSs (2) and between E1 and E2 (3), would be required to bring E1 to the *ori* (HPV-11). A low-specificity interaction between E1 and DNA in HPV-11 is consistent with the recent observation that an in vitro replication system utilizing E1 and *ori* from HPV-11 requires E2 for full activity (Kuo et al. 1994; Liu et al. 1995). This model indicates that the deposition of the viral initiator onto the origin of replication, to form an active initiation complex, can take place via different routes. Origin recognition can be performed by the initiator (E1) itself, or by E2, or by a combination of the two, indicating that E1 is likely to be capable of providing the enzymatic activities required for initiation of replication on DNA sequences lacking an apparent E1BS, consistent with the low *ori* specificity observed for replication in vitro (Yang et al. 1993).

Maintenance

An unusual and interesting feature of papillomavirus DNA replication is the ability of the virus to maintain the viral DNA as a nuclear plasmid in transformed cells over long periods of time (Lancaster 1981; Law et al.

1981). This property, which has been termed maintenance, is one of the clear distinctions between papillomaviruses and other members of the papovavirus group and resembles the latent replication of Epstein-Barr virus (see Yates, this volume). For BPV and some HPVs, this behavior can be observed in transformed cells in culture and is believed to constitute a part of the normal viral life cycle, perhaps corresponding to replication in the basal layer of the wart. This aspect of viral DNA replication has been difficult to study in the context of the viral genome for technical reasons. To circumvent these problems, a cell line that constitutively expresses E1 and E2 proteins has been constructed. *ori*-containing plasmids can be replicated in this cell line, which essentially is a BPV counterpart of the COS cells used for SV40 replication. The use of this cell line has established that long-term maintenance can be achieved in this artificial system and that the only viral gene products that are required are the E1 and E2 proteins (Piirsoo et al. 1996). However, replication of a plasmid per se is not sufficient for stable maintenance; the minimal *ori*, despite being fully replication-competent in transient replication assays, is not maintained in long-term assays. A sequence element located in the E2-dependent enhancer can provide the maintenance activity, and maintenance activity can also be supplied by multimerized E2BSs linked to the *ori*. The mechanism of action of this element is unclear, but a likely possibility is that, as has been proposed for the family of repeats of *oriP* (Yates et al. 1984; Krysan et al. 1989; Wysokenski and Yates 1989), nuclear retention or partitioning at cell division is affected by the presence of this element.

Interestingly, density-labeling experiments using these cell lines with stably replicating plasmids demonstrate that replication of BPV, unlike *oriP* (Yates and Guan 1991), is not controlled by once-per-cell-cycle replication, and consequently, in mammalian cells different strategies can be utilized to achieve stable maintenance of extrachromosomal elements (Piirsoo et al. 1996).

CONCLUSIONS

The early studies addressing the replication properties of papillomaviruses seemed to indicate that DNA replication of papillomaviruses was very different from DNA replication of the well-studied SV40 and polyomaviruses. Both the basic mechanisms and the regulatory circuit by which replication was regulated appeared different, and fairly elaborate models were constructed inspired by bacterial plasmid copy number control systems. However, as the understanding of papillomavirus replication progressively has improved, it has become clear that the similarities between papillomaviruses and other papovaviruses clearly outweigh the differences, and that at the molecular level these systems are in fact very similar. These similarities include the utilization of an identical or near identical set of cellular factors, as well as viral initiator proteins that are clearly homologous. These factors act on viral *cis*-acting sequences that show clear similarities to other papovaviruses. At present, the regulation of papillomavirus DNA replication has not been studied extensively; however, it seems likely that control of papillomavirus replication is achieved at the level of expression of the viral replication factors E1 and E2, since no indication exists that other viral gene products are required for either transient or long-term replication, at least not in the currently used tissue-culture systems.

ACKNOWLEDGMENTS

This work was supported by National Institutes of Health grant CA-13106.

REFERENCES

Androphy, E.J., D.R. Lowy, and J.T. Schiller. 1987. Bovine papillomavirus E2 *trans*-activating gene product binds to specific sites in papillomavirus DNA. *Nature* **325:** 70–73.

Benson, J.D. and P.M. Howley. 1995. Amino-terminal domains of the bovine papillomavirus type 1 E1 and E2 proteins participate in complex formation. *J. Virol.* **69:** 4364–4372.

Berg, L., K. Singh, and M. Botchan. 1986. Complementation of a BPV low copy number mutant: Evidence for a temporal requirement of the complementing gene. *Mol. Cell. Biol.* **6:** 859–869.

Blitz, I. and L. Laiminis. 1991. The 68-kilodalton E1 protein of bovine papillomavirus is a DNA binding phosphoprotein which associates with the E2 transcriptional activator in vitro. *J. Virol.* **65:** 649–656.

Bonne-Andrea, C., S. Santucci, P. Clertant, and F. Tillier. 1995. Bovine papillomavirus E1 protein binds specifically DNA polymerase a but not replication protein A. *J. Virol.* **69:** 2341–2350.

Bramhill, D. and A. Kornberg. 1988. Duplex opening by dnaA protein at novel sequences in initiation of replication at the origin of the *E. coli* chromosome. *Cell* **52:** 743–755.

Bream, G.L., C.-A. Ohmstede, and W. Phelps. 1993. Characterization of human papillomavirus type 11 E1 and E2 proteins expressed in insect cells. *J. Virol.* **67:** 2655–2663.

Chiang, C.M., G. Dong, T.R. Broker, and L.T. Chow. 1992a. Control of human papillomavirus type 11 origin of replication by the E2 family of transcription regulatory proteins. *J. Virol.* **66:** 5224–5231.

Chiang, C.M., M. Ustav, A. Stenlund, T.F. Ho, T.R. Broker, and L.T. Chow. 1992b. Viral E1 and E2 proteins support replication of homologous and heterologous papillomaviral origins. *Proc. Natl. Acad. Sci.* **89:** 5799–5803.

Clertant, P. and I. Seif. 1984. A common function for polyoma virus large-T and papillomavirus E1 proteins? *Nature* **311:** 276–279.

Deb, S., A.L. DeLucia, C.-P. Baur, A. Koff, and P. Tegtmeyer. 1987. The adenine-thymine domain of the simian virus 40 core origin directs DNA bending and coordinately regulates replication. *Mol. Cell. Biol.* **6:** 45478–4584.

DePamphilis, M.L. 1993. Eukaryotic DNA replication: Anatomy of an origin. *Annu. Rev. Biochem.* **62:** 29–63.

Del Vecchio, A., H. Romanczuk, P.M. Howley, and C.C. Baker. 1992. Transient replication of human papillomavirus DNAs. *J. Virol.* **66:** 5949–5958.

Dollard, S.C., J.L. Wilson, L.M. Demeter, W. Bonnez, R.C. Reichman, T.R. Broker, and L.T. Chow. 1992. Production of human papillomavirus and modulation of the infectious program in epithelial raft cultures. *Genes Dev.* **6:** 1131–1142.

Dvoretsky, I., R. Shober, S.K. Chattopadhyay, and D.R. Lowy. 1980. A quantiative in vitro focus assay for bovine papillomavirus. *Virology* **103:** 369–375.

Frattini, M.G. and L.A. Laiminis. 1994. Binding of the human papillomavirus E1 origin-recognition protein is regulated through complex formation with the E2 enhancer binding protein. *Proc. Natl. Acad. Sci.* **91:** 12398–12402.

Gilbert, D.M. and S.N. Cohen. 1987. Bovine papillomavirus plasmids replicate randomly in mouse fibroblasts throughout S phase of the cell cycle. *Cell* **50:** 59–68.

Gillette, T.G., M. Lusky, and J.A. Borowiec. 1994. Induction of structural changes in the bovine papillomavirus type 1 origin of replication by the viral E1 and E2 proteins. *Proc. Natl. Acad. Sci.* **91:** 8846–8850.

Gopalakrishnan, V. and S.A. Khan. 1994. E1 protein of human papillomavirus type 1a is sufficient for initiation of viral DNA replication. *Proc. Natl. Acad. Sci.* **91:** 9597–9601.

Hawley-Nelson, P., E.J. Androphy, D.R. Lowy, and J.T. Schiller. 1988. The specific DNA recognition sequence of the bovine papillomavirus E2 protein is an E2-dependent enhancer. *EMBO J.* **7:** 525–531.

Hedge, R.S., S.R. Grossman, R.B. Pepinsky, L.A. Laiminis, and P.B. Sigler. 1992. Crystal structure at 1.7 Å of the bovine papillomavirus-1 E2 DNA-binding domain bound to its DNA target. *Nature* **359:** 505–512.

Hirochika, H., R. Hirochika, T.R. Broker, and L.T. Chow. 1988. Functional mapping of the human papillomavirus type 11 transcriptional enhancer and its interaction with the *trans*-acting E2 proteins. *Genes Dev.* **2:** 54–67.

Holt, S.E., G. Schuller, and V.G. Wilson. 1993. DNA binding specificity of the bovine papillomavirus E1 protein is determined by the sequences contained within an 18-base-pair inverted repeat element at the origin of replication. *J. Virol.* **68:** 1094–1102.

Krysan, P.J., S.B. Haase, and M.P. Calos. 1989. Isolation of human sequences that replicate autonomously in human cells. *Mol. Cell. Biol.* **9:** 1026–1033.

Kuo, S.-R., J.-S. Liu, T.R. Broker, and L.T. Chow. 1994. Cell-free replication of the human papillomavirus DNA with homologous viral E1 and E2 proteins and human cell extracts. *J. Biol. Chem.* **269:** 24058–24065.

Lancaster, W.D. 1981. Apparent lack of integration of bovine papillomavirus DNA in virus-induced equine and bovine tumor cells and virus transformed mouse cells. *Virology* **108:** 251–255.

Law, M.-F., D.R. Lowy, I. Dvoretsky, and P.M. Howley. 1981. Mouse cells transformed by bovine papillomavirus contain only extrachromosomal viral DNA sequences. *Proc. Natl. Acad. Sci.* **78:** 2727–2731.

Li, R. and M. Botchan. 1993. The acidic transcriptional activation domains of VP16 and p53 bind the cellular replication protein A and stimulate in vitro BPV-1 DNA replication. *Cell* **73:** 1207–1221.

———. 1994. Acidic transcription factors alleviate nucleosome-mediated repression of DNA replication of bovine papillomavirus type 1. *Proc. Natl. Acad. Sci.* **91:** 7051–7055.

Li, R., J. Knight, G. Bream, A. Stenlund, and M. Botchan. 1989. Specific recognition nucleotides and their DNA context determine the affinity of E2 protein for 17 binding sites in the BPV genome. *Genes Dev.* **3:** 510–526.

Liu, J.-S., S.-R. Kuo, T.R. Broker, and L.T. Chow. 1995. The functions of human papillomavirus type 11 E1, E2, and E2C proteins in cell-free replication. *J. Biol. Chem.* **270:** 27283–27291.

Lu, J.Z.-J., Y.-N. Sun, R.C. Rose, W. Bonnez, and D.J. McCance. 1993. Two E2 binding sites (E2BS) alone or one E2BS plus an A/T rich region are minimal requirements for the replication of the human papillomavirus type 11 origin. *J. Virol.* **67:** 7131–7139.

Lusky, M. and M.R. Botchan. 1984. Characterization of the bovine papillomavirus plasmid maintenance sequences. *Cell* **36:** 391–401.

———. 1985. Genetic analysis of bovine papillomavirus plasmid type 1 *trans*-acting replication factors. *J. Virol.* **53:** 955–965.

Lusky, M. and E. Fontane. 1991. Formation of the complex of bovine papillomavirus E1 and E2 proteins is modulated by E2 phosphorylation and depends on sequences within the carboxyl terminus of E1. *Proc. Natl. Acad. Sci.* **88:** 6363–6367.

Lusky, M., J. Hurwitz, and Y.-S. Seo. 1994. The bovine papillomavirus E2 protein modulates the assembly but is not stably maintained in a replication-competent multimeric E1-replication origin complex. *Proc. Natl. Acad. Sci.* **91:** 8895–8899.

MacPherson, P., L. Thorner, and M. Botchan. 1994. The bovine papillomavirus E1 protein has ATPase activity essential to viral DNA replication and efficient transformation in cells. *Virology* **204:** 403–408.

Melendy, T., J. Sedman, and A. Stenlund. 1995. Cellular factors required for papillomavirus DNA replication. *J. Virol.* **69:** 7857–7867.

Meyers, C., M.G. Frattini, J.B. Hudson, and L.A. Laiminis. 1992. Biosynthesis of human papillomavirus from a continous cell line upon epithelial differentiation. *Science* **257:** 971–973.

Mohr, I.J., R. Clark, S. Sun, E.J. Androphy, P. MacPherson, and M. Botchan. 1990. Targeting the E1 replication protein to the papillomavirus origin of replication by complex formation with the E2 transactivator. *Science* **250:** 1694–1699.

Mul, Y.M. and P.C. van der Vliet. 1992. Nuclear factor I enhances adenovirus DNA replication by increasing the stability of a preinitiation complex. *EMBO J.* **11:** 751–760.

Muller, F., Y.-S. Seo, and J. Hurwitz. 1994. Replication of bovine papillomavirus type 1 origin-containing DNA in crude extracts and with purified factors. *J. Biol. Chem.* **269:** 17086–17094.

Murakami, Y., T. Eki, M.-A. Yamada, C. Prives, and J. Hurwitz. 1986. Species-specific in vitro synthesis of DNA containing the polyoma virus origin of replication. *Proc. Natl. Acad. Sci.* **83:** 6347–6351.

Park, P., W. Copeland, L. Yang, T. Wang, M.R. Botchan, and I. Mohr. 1994. The cellular

DNA polymerase α-primase is required for papillomavirus DNA replication and associates with the viral E1 helicase. *Proc. Natl. Acad. Sci.* **91:** 8700–8704.

Piirsoo, M., E. Ustav, T. Mandel, A. Stenlund, and M. Ustav. 1996. *Cis*- and *trans*-requirements for stable episomal maintenance of the BPV-1 replicator. *EMBO J.* **14:** 1–11.

Ravnan, J.-B., D.M. Gilbert, K.G. Ten Hagen, and S.N. Cohen. 1992. Random-choice replication of extrachromosomal bovine papillomavirus (BPV) molecules in heterogenous, clonally derived BPV-infected cell lines. *J. Virol.* **66:** 6946–6952.

Remm, M., R. Brain, and J.R. Jenkins. 1992. The E2 binding sites determine the efficiency of replication for the origin of human papillomavirus type 18. *Nucleic Acids Res.* **20:** 6015–6021.

Russell, J.A. and M.R. Botchan. 1995. *cis*-acting components of human papillomavirus (HPV) DNA replication: Linker substitution analysis of the HPV type 11 origin. *J. Virol.* **69:** 651–660.

Sarver, N., M.S. Rabson, Y.-C. Yang, J.C. Byrne, and P.M. Howley. 1984. Localization and analysis of bovine papillomavirus type 1 transforming functions. *J. Virol.* **52:** 377–388.

Sedman, J. and A. Stenlund. 1995. Co-operative interaction between the initiator E1 and the transcriptional activator E2 is required for replicator specific DNA replication of bovine papillomavirus in vivo and in vitro. *EMBO J.* **14:** 6218–6228.

Seo, Y.-S., F. Muller, M. Lusky, and J. Hurwitz. 1993a. Bovine papillomavirus (BPV)-encoded E1 protein contains multiple activities required for BPV DNA replication. *Proc. Natl. Acad. Sci.* **90:** 702–706.

Seo, Y.-S., F. Muller, M. Lusky, E. Gibbs, H.-Y. Kim, B. Phillips, and J. Hurwitz. 1993b. Bovine papilloma virus (BPV)-encoded E2 protein enhances binding of E1 protein to the BPV replication origin. *Proc. Natl. Acad. Sci.* **90:** 2865–2869.

Spalholz, B.A., A.A. McBride, T. Sarafi, and J. Quintero. 1993. Binding of bovine papillomavirus E1 to the origin is not sufficient for DNA replication. *Virology* **193:** 201–212.

Stoler, M.H., A. Whitbeck, S.M. Wolinsky, T.R. Broker, L.T. Chow, M.K. Howett, and J.W. Kreider. 1990. Infectious cycle of human papillomavirus type 11 in human foreskin xenografts in nude mice. *J. Virol.* **64:** 3310–3318.

Sverdrup, F. and S.A. Khan. 1994. Replication of human papillomavirus (HPV) DNAs supported by the HPV type 18 E1 and E2 proteins. *J. Virol.* **68:** 505–509.

———. 1995. Two E2 binding sites alone are sufficient to function as the minimal origin of replication of human papillomavirus type 18 DNA. *J. Virol.* **69:** 1319–1323.

Syrjanen, K., L. Gissman, and L.G. Koss, eds. 1987. *Papillomaviruses and human disease.* Springer-Verlag, New York.

Thorner, L., D. Lim, and M. Botchan. 1993. DNA binding domain of bovine papillomavirus type 1 E1 helicase: Structural and functional aspects. *J. Virol.* **67:** 6000–6014.

Tooze, J., ed. 1981. *Molecular biology of tumor viruses.* Cold Spring Harbor Laboratory, Cold Spring Harbor, New York.

Ustav, E., M. Ustav, P. Szymanski, and A. Stenlund. 1993. The bovine papillomavirus origin of replication requires a binding site for the E2 transcriptional activator. *Proc. Natl. Acad. Sci.* **90:** 898–902.

Ustav, M. and A. Stenlund. 1991. Transient replication of BPV-1 requires two viral polypeptides encoded by the E1 and E2 open reading frames. *EMBO J.* **10:** 449–457.

Ustav, M., E. Ustav, P. Szymanski, and A. Stenlund. 1991. Identification of the origin of replication of bovine papillomavirus and characterization of the viral origin recognition factor E1. *EMBO J.* **10:** 4321–4329.

Wilson, V.G. and J. Ludes-Meyers. 1991. A bovine papillomavirus E1-related protein binds specifically to bovine papillomavirus DNA. *J. Virol.* **65:** 5314–5322.

Wysokenski, D.A. and J.L. Yates. 1989. Multiple EBNA1-binding sites are required to form an EBNA1-dependent enhancer and to activate a minimal replicative origin within ori P of Epstein-Barr virus. *J. Virol.* **63:** 2657–2666.

Yang, L., R. Li, I. Mohr, R. Clark, and M.R. Botchan. 1991. Activation of BPV-1 replication in vitro by the transcription factor E2. *Nature* **353:** 628–633.

Yang, L., I. Mohr, E. Fouts, D.A. Lim, M. Nohaile, and M. Botchan. 1993. The E1 protein of the papillomavirus BPV-1 is an ATP dependent DNA helicase. *Proc. Natl. Acad. Sci.* **90:** 5086–5090.

Yates, J.L. and N. Guan. 1991. Epstein-Barr virus-derived plasmids replicate only once per cell cycle and are not amplified after entry into cells. *J. Virol.* **65:** 483–488.

Yates, J., N. Warren, D. Resiman, and B. Sugden. 1984. A *cis*-acting element from the Epstein-Barr viral genome that permits stable replication of recombinant plasmids in latently infected cells. *Proc. Natl. Acad. Sci.* **81:** 3806–3810.

Zhu, Q.-L., T.F. Smith, E.J. Lefkowitz, L.T. Chow, and T.R. Broker. 1994. *Nucleic acid and protein sequence alignments of human papillomaviruses constrained by functional sites.* University of Alabama, Birmingham.

zur Hausen, H. 1991. Viruses in human cancers. *Science* **254:** 1167–1173.

24

Adenovirus DNA Replication

Ronald T. Hay
School of Biological and Medical Sciences
University of St. Andrews
St. Andrews, Scotland KY16 9AL, United Kingdom

The adenovirus genome can be replicated in vitro by the combined action of three viral proteins, two cellular transcription factors, and topoisomerase I (Table 1). Early viral transcription results in expression of three viral genes, preterminal protein (pTP), DNA polymerase (Ad pol), and DNA-binding protein (DBP), that genetic evidence indicates are required for viral DNA replication. With the development of in vitro systems (Challberg and Kelly 1979), it was soon shown that cellular factors were required in addition to the three viral proteins. One of these proteins could be replaced by calf thymus topoisomerase and was required for complete elongation of the genome (Nagata et al. 1982). In contrast, the other factors were both sequence-specific DNA-binding proteins that recognized DNA within the inverted terminal repeats (ITRs) of the adenovirus genome (Nagata et al. 1983; Pruijn et al. 1986). Because all of these proteins have been cloned, overexpressed, and purified, the adenovirus genome represents one of the few eukaryotic systems in which all the components for DNA replication have been precisely defined.

Within virions, the adenovirus genome is maintained in a highly condensed form by small basic proteins, but after infection of susceptible cells, the virion is uncoated and viral DNA is released into the cell nucleus. Adenovirus genomes are linear double-stranded DNA molecules of 35–36 kb with ITRs of about 100 bp, the exact size depending on serotype. DNA replication initiates at either end of the linear 36-kbp viral genome by a protein-priming mechanism in which a covalent linkage is formed between the α-phosphoryl group of the terminal residue, dCMP, and the β-OH group of a serine residue in pTP. This reaction is catalyzed by Ad pol, which then utilizes the 3′-OH group of the pTP/dCMP complex as a primer for synthesis of the nascent strand. DNA synthesis is accompanied by displacement of the non-template strand, which can form a partial duplex by base-pairing of the ITRs on which a second round of DNA synthesis may be initiated (Hay et al. 1984; Stow 1982). Alterna-

Table 1 Properties of proteins involved in Ad DNA replication

Protein	Size (kD)	Domain required for DNA replication	Interacting protein(s)	Activities
Ad pol	140	all	Ad pTP, NFI, NFIII	DNA polymerase 3′→5′ proofreading exonuclease; as pTP/pol binds to the core of the origin; transfers dCMP onto pTP during initiation of DNA replication; catalyzes strand-displacement synthesis during elongation phase of DNA replication
Ad pTP	80	all	Ad pol	protein primer for initiation of DNA replication; as pTP/pol binds to the core of the origin; substrate for the adenovirus protease
DBP	59	DNA-binding domain aa 174–529	—	enhances initiation and elongation during DNA replication; binds single- and double-stranded DNA; stimulates binding of NFI to origin; reduces K_m for dCTP; increases processivity of Ad pol; acts as a helix-destabilizing protein
NFI/CTFI	55	DNA-binding domain aa 1–220	Ad pol	binding to auxiliary origin enhanced by DBP; enhances initiation of DNA replication
NFII	30	all	—	topoisomerase I; required for synthesis of genome-length molecules
NFIII/Oct-1	90	DNA-binding domain aa 270–440	Ad pol	binds to auxiliary origin; enhances initiation of DNA replication

tively, displaced complementary strands may simply reanneal to form double-stranded products. This proposal is based on the in vitro properties of DBP, which promotes intermolecular reannealing but inhibits intramolecular reannealing (Zijderveld et al. 1993). The accepted model, based on analysis of replication intermediates, for adenovirus DNA synthesis is discussed by Brush and Kelly (this volume).

The subject of adenovirus DNA replication has been reviewed extensively (Hay and Russell 1989; Stillman 1989; van der Vliet 1990; Salas 1991), but new developments in the field warrant further evaluation. In this chapter, I describe the components of the adenovirus replicon and analyze how these components assemble into a nucleoprotein complex that precedes initiation of viral DNA replication.

ADENOVIRUS ORIGINS OF DNA REPLICATION

The *cis*-acting DNA sequences that define *ori*, the origin of DNA replication, are located at the ends of the genome within the ITRs. Covalently bound to each 5′end of the DNA is a terminal protein (TP), which may well be an additional *cis*-acting component of *ori*. Comparisons of *ori*s from a wide range of adenoviruses and extensive mutational analysis of *ori* DNA have defined four regions within the terminal 50 bp of the Ad2 genome that contribute to *ori* activity in vitro and in vivo. As with many other *ori*s, that of Ad2 consists of an essential core region and auxiliary regions that enhance the efficiency of DNA replication (Fig. 1). The terminal 18 bp of the viral genome represents the minimal replication origin and contains a 10-bp region (bp 9–18, Fig. 1) conserved in all human adenoviruses (Stillman et al. 1982). Although the integrity of this region is required for *ori* function, in isolation it supports only a limited, basal level of initiation of DNA replication (Tamanoi and Stillman 1983; Van Bergen et al. 1983; Challberg and Rawlins 1984; Guggenheimer et al. 1984; Lally et al. 1984; Hay 1985a; Wides et al. 1987). Separated from the minimal *ori* by a spacer region of defined length (Adyha et al. 1986; Wides et al. 1987; Bosher et al. 1990) is the DNA-binding site of the cellular protein nuclear factor I (NFI) or CAAT transcription factor (Nagata et al. 1982; Jones et al. 1987). Distal to the NFI-binding site is the DNA recognition site for another cellular protein, octamer-binding protein (Oct-1) or nuclear factor III (NFIII) (Pruijn et al. 1986; Rosenfeld et al. 1987; Wides et al. 1987; O'Neill and Kelly 1988; Sturm et al. 1988). In contrast to Ad2, Ad4 requires only the terminal 18 bp of the viral genome to efficiently initiate DNA replication in vivo (Hay 1985b) and in vitro (Harris and Hay 1988; Temperley and Hay

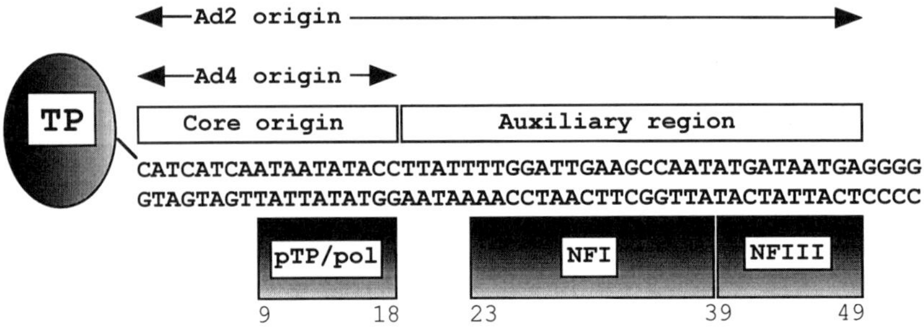

Figure 1 Structure and organization of the Ad2 and Ad4 origins of DNA replication. The binding sites for pTP/Ad pol, NFI, and NFIII/Oct-1 are indicated.

1991; Temperley et al. 1991). Ad4 therefore appears to have circumvented the need for the host factors NFI and NFIII/Oct-1. The ITR does not contain an NFI recognition site, and although it does have a binding site for NFIII/Oct-1, neither factor is required for DNA replication in vivo or is capable of stimulating DNA replication in vitro (Hay et al. 1988). The reasons for the different requirements of Ad2 and Ad4 DNA replication are not obvious, and DNA-binding assays have indicated that the affinity of baculovirus-expressed Ad4 pTP/Ad pol for the core origin is not substantially different from that of the Ad2 pTP/Ad pol (I.R. Leith and R.T. Hay, unpubl.)

ROLE OF pTP, TP, AND PROTEASE IN THE REPLICATION OF VIRAL DNA

Although mature adenovirus DNA contains a 55-kD protein covalently attached to both 5′termini (TP), early in vitro studies indicated the presence of an 80-kD precursor terminal protein (pTP) covalently attached to 5′ends of nascent DNA (Challberg et al. 1980). Ad2 pTP is cleaved to TP via an intermediate (iTP), and because the cleavage sites are conserved in all sequenced pTPs, it is likely that cleavage of pTP to TP, via iTP, plays an important role in the infectious cycle (Webster et al. 1989). Analysis of an Ad2 mutant (H2ts1) with a temperature-sensitive protease (Ad protease) that replicates efficiently at the restrictive temperature, but whose progeny virions are not infectious, revealed that in contrast to a wild-type virus infection, pTP was not processed to TP (Stillman et al. 1981). Thus, pTP is a substrate for Ad protease, which is essential for viral infectivity and accumulates to high levels late in the infectious cycle. Site-directed mutagenesis and sequence alignments of viral proteases indicate that the enzyme may represent a new subclass of cysteine proteases (Rancourt et al. 1994). Ad protease is synthesized in an inactive form but is activated by a disulfide-linked peptide (Mangel et al. 1993; Webster et al. 1993) derived from the carboxyl terminus of the pVI virion protein (Webster et al. 1993). Although it was proposed that DNA was a cofactor in the protease reaction (Mangel et al. 1993), subsequent studies demonstrated that, although DNA can stimulate protease activity under certain defined conditions, it is not required for protease activity (Webster et al. 1993, 1994).

In adenovirus-infected cells, pTP exists in the form of a heterodimer in which it is tightly bound to Ad pol (Lichy et al. 1982; Stillman et al. 1982). Work on pTP has been facilitated by cloning of cDNA copies of the gene (Pettit et al. 1988) into vaccinia virus (Stunnenberg et al. 1988) or baculovirus (Bosher et al. 1990) for high-level expression in human or insect cells. Although a variety of approaches have been employed to define functional domains in pTP that are required for viral DNA replication, it has not proved possible to ascribe specific activities to domains within the protein. Genetic and biochemical experiments have indicated that the amino-terminal region, removed during processing by Ad protease, is required for specific DNA binding, interaction with Ad pol, and initiation of DNA replication (Freimuth and Ginsberg 1986; Pettit et al. 1989; Fredman et al. 1991; Roovers et al. 1993; Webster et al. 1994). The sequence-specific DNA-binding activity of pTP that allows it to recognize the origin of DNA replication (Temperley and Hay 1992) may be a target for cellular regulatory mechanisms, since it has been demonstrated that origin binding by pTP is dependent on the phosphorylation state of the protein (Kusukawa et al. 1994).

In considering the role of the pTP processing by Ad protease in DNA replication, free pTP and pTP/TP that is covalently attached to the 5′ end of DNA should be regarded as distinct functional entities. Whereas the role of free pTP is to act as the protein primer for DNA replication, the significance of the covalently attached pTP/TP remains to be fully elucidated. However, it has been known for some time that in transfection experiments, the infectivity of TP-DNA is orders of magnitude greater than that of naked DNA. Genome-bound pTP/TP may protect viral DNA from exonucleases, allow attachment of virus DNA to the nuclear matrix (Schaack et al. 1990), and participate in unwinding of the DNA duplex at the origin of replication. In addition, genome-bound TP serves to stabilize binding of the incoming pTP-pol heterodimer at the origin of DNA replication (Pronk and van der Vliet 1993) and induces a conformational change in origin DNA. However, the relationship between these two phenomena has yet to be established. It should be emphasized that in vivo the template for early transcription and the first round of DNA replication is TP-DNA, whereas the template for subsequent rounds of replication is pTP-DNA irrespective of whether infections are carried out with wild-type Ad2 or Ad2ts1 at the permissive or nonpermissive temperatures. It has been suggested that cleavage of pTP-DNA to TP-DNA is required to free TP-DNA from the nuclear matrix prior to assembly of virus particles (Fredman and Engler 1993). However, because pTP-DNA is packaged within virus particles during Ad2ts1 infections at nonpermissive temperatures, this would appear less likely (Weber 1990). It cannot be ruled out, however, that the nuclear matrix-binding properties of TP-DNA and pTP-DNA are distinct and that this has some bearing on the transcription/replication sites of viral DNA. Another possibility is that processing of pTP-DNA to TP-DNA is used to create different templates for either early transcription or the first round of DNA replication. Since pTP binds to DNA and Ad pol, and TP does not (Webster et al. 1994), it would not be surprising to find differences between the properties of TP-DNA and pTP-DNA as templates for DNA replication.

ADENOVIRUS DNA POLYMERASE

Purification of an activity required for viral replication from an Ad2-infected cytosolic extract yielded two polypeptide species: the previously described pTP and a 140-kD protein with a unique DNA polymerase activity (Enomoto et al. 1981). Complementation experiments with temperature-sensitive mutants of Ad5 (group N mutants) defective in both the initiation and elongation of DNA replication in vivo demonstrated the viral origin of the DNA polymerase (Stillman et al. 1982; Van Bergen et al. 1983). Extracts prepared from nuclei of cells infected with these mutant viruses were unable to support initiation of DNA replication, but activity was restored by addition of Ad pol (Stillman et al. 1982). More recent experiments (Chen et al. 1994) have defined the biochemical defects associated with the temperature-sensitive phenotype. Insertion of pol cDNAs (Shu et al. 1987) into recombinant vaccinia virus (Stunnenberg et al. 1988) and baculovirus (Watson and Hay 1990) has allowed high-level expression of the protein in human and insect cells.

Recognizable in Ad pol are the conserved domains that are characteristic of DNA polymerases from across the evolutionary spectrum (Coen, this volume) and like many DNA polymerases, Ad pol also possesses an intrinsic 3′→5′ proofreading exonuclease activity. However, Ad pol also appears to contain additional domains that may contribute to the unique properties of the protein. pTP, NFI, and NFIII all independently contact Ad pol in the preinitiation complex. Although it is likely that these proteins bind to spatially distinct domains on the surface of Ad pol, their precise location has yet to be determined. In addition, two potential zinc-binding motifs have been identified which mutational analyses have indicated are important for DNA binding and initiation of DNA replication (Joung and Engler 1992). DBP was shown to have a profound effect on both DNA polymerase and exonuclease activities of Ad pol (Field et al. 1984). DNA synthesis on poly(dT):oligo(rA) was stimulated 10- to 100-fold by the presence of DBP due to an increase in processivity of Ad pol. Other single-stranded binding proteins (SSBs), such as *Escherichia coli* SSB, were unable to substitute for Ad DBP. Likewise, Ad DBP could not substitute for human SSB in stimulating the activity of human DNA polymerase-α. It is thought that DBP may act by stabilizing the interaction between Ad pol and the DNA template. In contrast to DNA polymerase activity, the proofreading exonuclease activity of Ad pol is inhibited by DBP. Inhibition by DBP is a direct effect on the exonuclease activity of Ad pol rather than binding to and protecting DNA ends (Lindenbaum et al. 1986). Although DBP does increase the thermostability of Ad pol (Lindenbaum et al. 1986), it has not yet been possible to demonstrate a direct physical association between the two proteins.

Intriguingly, a role for phosphorylation in the activity of Ad pol has been suggested (Ramachandra et al. 1993) by the observation that Ad pol is phosphorylated on S67 by a stably associated "cdc2-like" histone H1 kinase (Ramachandra and Padmanabhan 1993). Dephosphorylation of the protein altered its ability to participate in the initiation reaction.

ADENOVIRUS DNA-BINDING PROTEIN

The large amount of DBP that accumulates during the adenovirus infectious cycle (up to 5×10^6 molecules per cell) was responsible for its being the first of the viral nonstructural proteins to be identified and subsequently purified (van der Vliet and Levine 1973). An essential role of DBP in viral infectivity was demonstrated genetically by isolation of temperature-sensitive mutations H5ts125 and H5ts107 in the DBP gene. Nuclear extracts from both of these mutants are defective for full-length Ad DNA replication on both exogenous and endogenous templates (Friefeld et al. 1983). In both cases, activity is restored by the addition of purified wild-type DBP. DBP displays multiple biochemical activities that are required for efficient viral DNA synthesis (Brough et al. 1993) and can be found in a large protein complex associated with cellular proteins (Ricigliano et al. 1994). However, the functional significance of this interaction has yet to be determined.

Although the role of DBP in the elongation phase of DNA replication, both in vivo and in vitro, has been well documented, its specific role in the initiation of Ad DNA replication has been more contentious. However, studies using purified components reveal a stimulatory effect on the level of initiation when DBP is present (Kenny and Hurwitz 1988; Cleat and Hay 1989; Mul and van der Vliet 1993). This may be a consequence of a functional interaction between DBP and NFI. The ability of NFI to stimulate initiation in vitro is dependent on the concentration of DBP (De Vries et al. 1985). DBP increases the affinity of NFI for its recognition site in the adenovirus origin of DNA replication (Cleat and Hay 1989; Stuiver and van der Vliet 1990). Because a direct interaction between the two proteins has not been detected, the observed effect is likely a consequence of DBP's ability to alter the structure of bound DNA. In fact, DBP can remove the tertiary structure of double-stranded DNA fragments upon binding (Stuiver et al. 1992). It therefore seems likely that in the DBP/DNA complex, the relative positions of hydrogen-bond donor and acceptor groups in the major grooves are altered, causing a fine-tuning of the contacts between NFI and DNA that leads to a higher-affinity interaction. DBP can also stimulate initiation of DNA replication by decreasing the K_m of the polymerase for the initiator nucleotide dCTP (Mul and van der Vliet 1993).

Partial chymotryptic digestion of purified DBP yields a carboxy-terminal fragment of around 40 kD and a highly phosphorylated 27-kD amino-terminal fragment (Klein et al. 1979; Cleghon et al. 1993). The amino-terminal fragment appears to be dispensable for DNA replication, because the carboxy-terminal domain of the molecule can fully substitute for the intact molecule during Ad DNA replication in vitro. Within the carboxy-terminal region are three highly conserved motifs (Kitchingman et al. 1985; Vos et al. 1988; Eagle and Klessig 1992) that mutational studies have identified as important for binding to single-stranded DNA (Neale and Kitchingman 1990). Photocrosslinking has revealed that the two residues, Met-299 and Phe-418 in the carboxy-terminal domain, play an important part in DBP's ability to bind single-stranded DNA (Cleghon and Klessig 1992). Determination of the three-dimensional structure of the carboxy-terminal DNA-binding domain of DBP has revealed that the conserved motifs described above are important elements in the architecture of the DBP molecule. Two zinc atoms stabilize the overall folding of the molecule, and although structural data are not yet available for the

DNA/protein complex, a positively charged region incorporating the cross-linked F418 has been suggested as the surface that mediates interactions with DNA (Tucker et al. 1994). Biochemical experiments have implicated the carboxy-terminal region of DBP as important for cooperative binding to DNA, and this is confirmed in the crystal structure. A carboxy-terminal extension of 17 amino acids from one molecule "hooks onto" a second molecule, thus forming a protein chain around which a DNA chain is proposed to wind (Tucker et al. 1994).

Ad DBP has been shown to possess the properties of a helix-destabilizing protein (Monaghan et al. 1994; Zijderveld and van der Vliet 1994). On DNA templates containing a large proportion of single-stranded DNA, the double-stranded portion is efficiently unwound in a highly cooperative reaction. Entirely double-stranded templates are also unwound, but this reaction is restricted by the length and G+C content of the DNA fragment (Monaghan et al. 1994). Like other helix-destabilizing proteins, DBP-catalyzed DNA-unwinding requires neither ATP nor $MgCl_2$. In fact, the latter is inhibitory to the process, as are other agents, such as NaCl, which increase the stability of duplex DNA molecules. These properties are very similar to those of the calf thymus and herpes simplex virus coded single-strand-specific DNA-binding proteins that appear to be involved in cellular and viral DNA replication (Georgaki et al. 1992; Boehmer and Lehman 1993; Georgaki and Hübscher 1993). Two alternatives could explain the ability of DBP to unwind completely double-stranded DNA. DBP could first bind to double-stranded DNA (Cleat and Hay 1989; Stuiver and van der Vliet 1990) then invade the DNA duplex and bind to the exposed single strands in a stable fashion. Alternatively, DBP could bind transiently to single-stranded regions of DNA exposed during "breathing" of short double-stranded DNA molecules. However, when double-stranded DNA fragments are tightly bound by either NFI or NFIII/Oct-1 at their cognate recognition sites, DBP is unable to unwind the DNA. This could be because the bound proteins stabilize the DNA duplex, or the bound proteins may interfere with the ability of DBP to form a continuous protein chain on the DNA. Unlike DNA helicases that translocate unidirectionally on DNA, there does not appear to be a strict directionality to the unwinding reaction catalyzed by DBP. The ability of DBP to impose a rigid structure on DNA is responsible for the ability of DBP to inhibit intramolecular renaturation but promote intermolecular renaturation (Zijderveld et al. 1993). The denaturation reaction described above is inhibited by high concentrations of monovalent and divalent cations, but the renaturation reaction is highly resistant to their presence (Zijderveld et al. 1993).

The ability of DBP to destabilize double-stranded DNA duplexes could be utilized at a number of points in viral DNA replication. As discussed above, DBP stimulates the initiation of Ad2 DNA replication by decreasing the K_m for transfer of dCMP onto pTP and by increasing the binding of NFI to the replication origin. However, it is likely that neither of these reactions involves melting duplex DNA, and DBP may stimulate initiation of DNA replication by additional activities. Ad4 DBP dramatically stimulates initiation of DNA replication, but in this case, the extent of stimulation is independent of the concentrations of NFI and dCTP (Temperley and Hay 1991). One possibility is that DBP may participate with other replication proteins in unwinding the DNA double helix at the termini of the genome prior to initiation. TP (Pronk et al. 1992; Pronk and van der Vliet 1993), NFI (De Vries et al. 1987; Mul and van der Vliet 1992), and NFIII/Oct-1 (Verrijzer et al. 1991) can all modify the structure of viral DNA. In combination with the ability of DBP to distort and unwind short, fully duplex DNA, this property could allow adenovirus replication proteins to destabilize and unwind *ori* DNA. The role of DBP in elongation is likely to be a consequence of DBP's stabilizing displaced single strands and altering the previously described properties of the Ad pol.

ROLE OF CELLULAR PROTEINS INVOLVED IN AD DNA REPLICATION

Limited initiation of DNA replication in vitro occurs in the presence of virally encoded pTP/Ad pol and *ori.* However, addition of a crude nuclear extract of uninfected HeLa stimulated initiation activity. This stimulatory effect was due to the presence of two cellular transcription factors NFI and NFIII/Oct-1. Since these factors are also discussed by van der Vliet (this volume), only their relevance to the formation of an Ad preinitiation complex is discussed here.

NFI

Uninfected HeLa cell nuclear extract can stimulate the efficiency of initiation and elongation obtained in vitro with infected cytosol or purified pTP/pol heterodimer and DBP (Ikeda et al. 1982; Lichy et al. 1982). This stimulatory activity, purified from the nuclear extract, contained a single major protein species with a size of 47 kD and was termed NFI (Nagata et al. 1982). Initial characterization of NFI concentrated on its ability to specifically bind the sequence 5′-TGGC (N_6) GCCAA-3′ present in a number of adenovirus origins of DNA replication (Leegwater et al. 1985). Interaction of NFI with this DNA sequence of the genome increases the frequency of initiation of viral DNA replication both in vivo and in vitro (Nagata et al. 1983; Guggenheimer et al. 1984; Rawlins et al. 1984; De Vries et al. 1985; Hay 1985a,b). By a number of criteria, NFI proteins are indistinguishable from CTF transcription factors (Jones et al. 1987), a family of proteins involved in cellular gene transcription that are distinct from the CCAAT-binding proteins (Zorbas et al. 1992). Analysis of cDNAs from various human NFI/CTF mRNAs have shown that they all originate from a single gene, giving rise to multiple mRNA transcripts by differential splicing of a precursor molecule (Paonessa et al. 1988; Santoro et al. 1988; Meisterernst et al. 1989). Mutagenesis studies on the cDNA of the largest NFI protein (CTF-1) identified two functional domains: a highly conserved amino-terminal domain, which contains the functions for DNA binding, dimerization, and DNA replication (Santoro et al. 1988; Mermod et al. 1989; Gounari et al. 1990; Bosher et al. 1991; Bandyopadhyay and Gronostajski 1994), and a less highly conserved, carboxy-terminal domain, which contains a transcriptional activation function (Mermod et al. 1989; Altmann et al. 1994).

The mechanism of stimulation of Ad DNA replication by NFI in vitro is complex (Mul et al. 1990). Stimulation is strongly dependent on the concentration of pTP/Ad pol. At low pTP/Ad pol concentrations, NFI or the DNA-binding domain of NFI stimulated replication up to 50-fold, whereas at high concentrations of pTP/Ad pol, stimulation was less than 2-fold. Thus, in vitro, the need for NFI can be overcome by increased pTP/Ad pol concentrations. This result provides an explanation for the different levels of stimulation by NFI reported previously (De Vries et al. 1985; Adhya et al. 1986), as well as a strong indication that NFI and pTP/Ad pol interact directly. The orientation and spacing between the NFI site and the core origin sequence (Fig. 1) are critical in Ad DNA replication. Insertion of additional sequence between these two sequence regions abolished NFI-mediated stimulation of DNA replication in vitro (Adhya et al. 1986; Wides et al. 1987; Coenjaerts et al. 1991) and in vivo (Bosher et al. 1990). This suggested that a strict constraint on the spatial arrangement between the core origin sequence and the NFI site is necessary to allow specific protein-protein interactions between NFI and other replication proteins. A direct interaction between NFI and Ad pol was subsequently demonstrated (Bosher et al. 1990; Chen et al. 1990; Mul et al. 1990), the function of which is to stabilize the relatively weak interaction between the pTP/Ad pol heterodimer and its recognition site in the adenovirus origin of DNA replication. Further dissection of this interaction by mutational analysis has demonstrated that the DNA-binding and Ad pol interaction activities of NFI can be separated, although both are required for stimulation of viral DNA replication (Armentero et al. 1994).

NFIII

The host-encoded cellular factor NFIII was initially identified in HeLa nuclear extracts through its ability to stimulate the initiation of Ad DNA replication in vitro in the presence of NFI (Pruijn et al. 1986). Purified NFIII has a size of 92 kD and, when bound to its recognition site in the adenovirus origin, stimulates the level of initiation in vitro 3- to 7-fold (O'Neill and Kelly 1988). A combination of DNase I footprinting and methylation protection studies on Ad2 identified the core binding site as 5′-TATGATAAT-3′, which is situated between nucleotides 39 and 48 in the Ad2 ITR. NFIII is indistinguishable from Oct-1 (O'Neill et al. 1988; Pruijn et al. 1989). As with other octamer-binding proteins, NFIII/Oct-1 contains a conserved DNA-binding region known as the POU domain, which, like the amino-terminal domain of NFI, is sufficient for the in vitro stimulation of Ad2 DNA replication (Verrijzer et al. 1990). In vivo, the situation is less clear, as mutant Ad2 viral genomes with a deleted NFIII/Oct-1 site are as infectious as wild-type virus with an intact NFIII/Oct-1 site (Hay and McDougall 1986), although the site may be functionally redundant with sequences adjacent to the ITR (Hatfield and Hearing 1993). However, transfection assays on plasmids containing only the core origin of replication linked to an NFIII/Oct-1 site (no NFI site) indicated that these templates could replicate more efficiently than plasmids containing only the core origin of replication (Hay et al. 1988). This confirmed in vitro replication studies by Mul et al. (1990), who found that both the NFIII/Oct-1 and NFI-binding sites were needed for optimal stimulation of replication. Since the binding sites of the two proteins overlap, they could interact in a cooperative manner to stimulate DNA replication. However, this does not appear to be the case. It seems that both proteins can, by binding to their respective recognition sites, stimulate initiation independently (Mul et al. 1990). NFIII/Oct-1 can induce bending of the DNA at the origin of DNA replication, which may stimulate DNA replication by promoting interactions between the various components in the preinitiation complex (Verrijzer et al. 1991). In fact, direct interactions between the POU homeodomain and the pTP/Ad pol heterodimer are responsible for stabilization of the preinitiation complex formed at *ori* (Coenjaerts et al. 1994), and it is likely that both mechanisms contribute to NFIII/Oct-1-induced stimulation of Ad DNA replication.

A MODEL FOR THE INITIATION OF AD DNA REPLICATION

On the basis of the observations described above, a model outlined in Figure 2 has been formulated to describe events that lead to the initiation of Ad DNA replication. Ad2 DBP, originally classified as a single-stranded DNA-binding protein, but which also binds to double-stranded DNA, is produced in such large amounts that it is likely that in vivo all template molecules are coated with the protein. This has the effect of increasing the affinity of NFI for its recognition site in the Ad2 *ori* and thus ensuring that the NFI-binding site in *ori* is fully occupied (Cleat and Hay 1989; Stuiver and van der Vliet 1990). A direct protein-protein interaction between NFI and Ad pol then recruits the pTP/Ad pol heterodimer into the preinitiation complex (Bosher et al. 1990; Chen et al. 1990; Mul et al. 1990; Armentero et al. 1994). This interaction is further stabilized by direct interactions between pTP/Ad pol and the POU domain of NFIII (Coenjaerts et al. 1994). Correct positioning of the pTP/Ad pol heterodimer at the origin is accomplished by an interaction between pTP/Ad pol and the DNA sequence from bp 9–18 that is perfectly conserved in all human adenoviruses sequenced to date and is regarded as the core of the Ad2 *ori* (Mul and van der Vliet 1992; Temperley and Hay 1992). Further stabilization of this complex may be accomplished by interactions between the incoming pTP/Ad pol and genome-bound TP (Pronk and van der Vliet 1993). At this stage, the origin must unwind to expose the single-stranded DNA that is the template for DNA synthesis. The ATP-independent helix-destabilizing properties of DBP may be particularly important in catalyzing this reaction, and it is worth noting that under certain conditions, Ad DNA replication in vitro can proceed in the absence of ATP (Pronk et al. 1994). Unlike most other systems, this unwinding reaction does not appear to require the activity of a bona fide helicase. Chemical modification experiments have recently demonstrated that this unwinding takes place in the highly conserved region within the core of the viral origin of DNA replication (I.R. Leith and R.T. Hay, unpubl.). The pTP/Ad pol heterodimer must then be repositioned for initiation such that the acceptor serine in pTP is positioned opposite the GTA at positions 4–6 in the genome rather than the same sequence present at positions 1–3. A pTP-CAT intermediate is then synthesized which remarkably "jumps back" to occupy positions 1–3 before elongation proceeds (King and van der Vliet

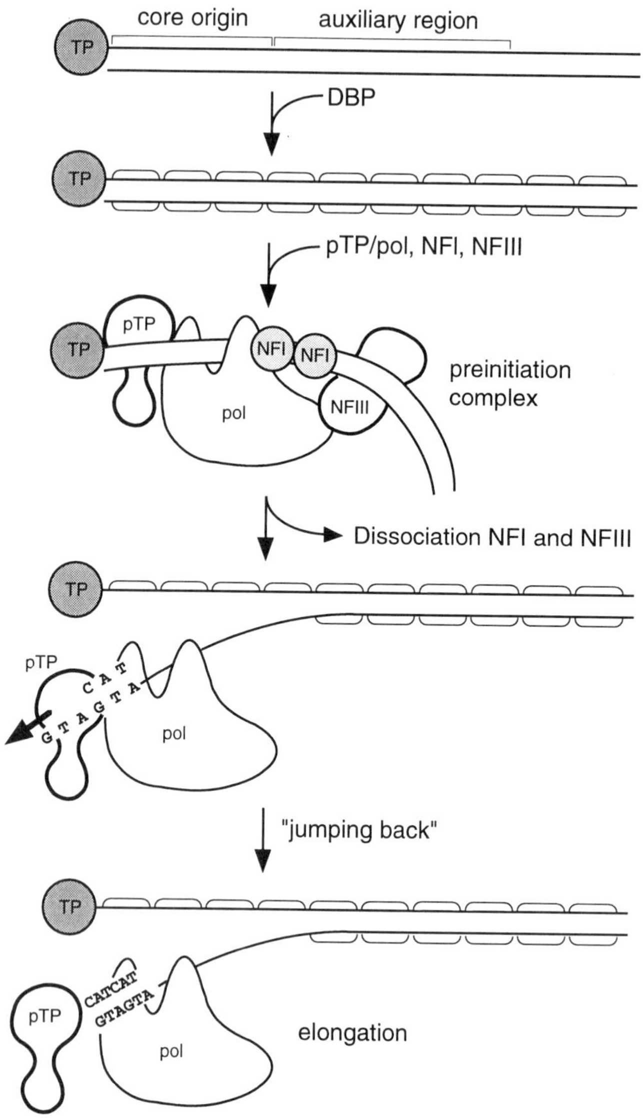

Figure 2 Model for the initiation of Ad2 DNA synthesis showing formation of the preinitiation complex, helix opening at the origin of DNA replication, and the "jumping back" mechanism of initiation.

1994). This type of mechanism appears to have been adopted as an evolutionary solution to the problem of initiation on linear protein-linked genomes and is discussed in detail elsewhere in this volume (Salas et al.). The events that follow the transition from initiation to elongation are not well defined, although NFI dissociates from the origin at an early stage in the process (Coenjaerts and van der Vliet 1994).

ACKNOWLEDGMENTS

I acknowledge the valuable contributions made by Ailsa Webster, Jane Parker, Ian Leith, Alasdair Freeman, and Alan Monaghan. Work in the author's laboratory was supported by the Medical Research Council and the Wellcome Trust.

REFERENCES

Adhya, S., P.S. Schneidman, and J. Hurwitz. 1986. Reconstruction of adenovirus replication origins with a human nuclear factor 1 binding site. *J. Biol. Chem.* **261:** 3339–3346.

Altmann, H., W. Wendler, and E.L. Winnacker. 1994. Transcriptional activation by CTF proteins is mediated by bipartite low-proline domain. *Proc. Natl. Acad. Sci.* **91:** 3901–3905.

Armentero, M.T., M. Horwitz, and N. Mermod. 1994. Targetting of DNA-polymerase to the adenovirus origin of DNA-replication by interaction with nuclear factor-I. *Proc. Natl. Acad. Sci.* **91:** 11537–11541.

Bandyopadhyay, S. and R.M. Gronostajski. 1994. Identification of a conserved oxidation-sensitive cysteine residue in the NFI family of DNA-binding proteins. *J. Biol. Chem.* **269:** 29949–29955.

Boehmer, P.E. and I.R. Lehman. 1993. Herpes simplex virus type I ICP8: Helix-destabilising properties. *J. Virol.* **67:** 711–715.

Bosher, J., E.C. Robinson, and R.T. Hay. 1990. Interactions between the adenovirus type 2 polymerase and the DNA binding domain of nuclear factor I. *New Biol.* **2:** 1083–1090.

Bosher, J., I.R. Leith, S.M. Temperley, M. Wells, and R.T. Hay. 1991. The DNA-binding domain of nuclear factor I is sufficient to cooperate with the adenovirus type 2 DNA-binding protein in viral replication. *J. Gen. Virol.* **72:** 2975–2980.

Brough, D.E., G. Droguett, M.S. Horwitz, and D.F. Klessig. 1993. Multiple functions of the adenovirus DNA-binding protein are required for efficient viral DNA synthesis. *Virology* **196:** 269–281.

Challberg, M.D. and T.J. Kelly. 1979. Adenovirus DNA replication in vitro. *Proc. Natl. Acad. Sci.* **76:** 655–659.

Challberg, M.D. and D.R. Rawlins. 1984. Template requirements for the initiation of adenovirus DNA replication. *Proc. Natl. Acad. Sci.* **81:** 100–104.

Challberg, M.D., S.V. Desiderio, and T.J. Kelly. 1980. Adenovirus DNA replication in vitro: Characterisation of a protein covalently linked to nascent DNA strands. *Proc. Natl. Acad. Sci.* **77:** 5105–5109.

Chen, M., N. Mermod, and M.S. Horwitz. 1990. Protein-protein interactions between adenovirus DNA polymerase and nuclear factor I mediate formation of the DNA replication pre-initiation complex. *J. Biol. Chem.* **265:** 18634–18642.

Chen, H.F., M. Ramachandra, and R. Padmanabhan. 1994. Biochemical-characterization of a temperature-sensitive adenovirus DNA-polymerase. *Virology* **205:** 364–370.

Cleat, P.H. and R.T. Hay. 1989. Co-operative interactions between NF-I and the adenovirus DNA binding protein at the origin of DNA replication. *EMBO J.* **8:** 1841–1848.

Cleghon, V. and D.F. Klessig. 1992. Characterization of the nucleic acid binding region of adenovirus DNA binding protein by partial proteolysis and photochemical cross-linking. *J. Biol. Chem.* **267:** 17872–17881.

Cleghon, V., A. Piderit, D.E. Brough, and D.F. Klessig. 1993. Phosphorylation of the adenovirus DNA-binding protein and epitope mapping of monoclonal antibodies against it. *Virology* **197:** 564–575.

Coenjaerts, F.E.J. and P.C. van der Vliet. 1994. Early dissociation of nuclear factor-1 from the origin during initiation of adenovirus DNA-replication studied by origin immobilization. *Nucleic Acids Res.* **22:** 5235–5240.

Coenjaerts, F.E.J., J.A.W.M. van Oosterhout, and P.C. van der Vliet. 1994. The Oct-1 Pou domain stimulates adenovirus DNA-replication by a direct interaction beween the viral precursor terminal protein-DNA polymerase complex and the Pou homeodomain. *EMBO J.* **13:** 5401–5409.

Coenjaerts, F.E.J., E. De Vries, G.J.M. Pruijn, W. van Driel, S.M. Bloemers, N.M.T. van der Lugt, and P.C. van der Vliet. 1991. Enhancement of DNA replication by transcription factors NFI and NFIII/Oct-1 depends critically on the positions of their binding sites in the adenovirus origins of DNA replication. *Biochim. Biophys. Acta* **1090:** 61–69.

De Vries, E., W. van Driel, M. Tromp, J. van Boom, and P.C. van der Vliet. 1985. Adenovirus DNA replication in vitro: Site directed mutagenesis of the nuclear factor I binding site of the Ad. origin. *Nucleic Acids Res.* **13:** 4943–4952.

De Vries, E., W. van Driel, S.J.L. van den Heuvel, and P.C. van der Vliet. 1987. Contact point analysis of the HeLa nuclear factor I recognition site reveals symmetrical binding at one side of the helix. *EMBO J.* **6:** 161–168.

Eagle, P.A. and D.F. Klessig. 1992. A zinc-binding motif located between amino acids 273 and 286 in the adenovirus DNA-binding protein is necessary for ssDNA binding. *Virology* **187:** 777–787.

Enomoto, T., J.H. Lichy, J.E. Ikeda, and J. Hurwitz. 1981. Adenovirus replication in vitro: Purification of the terminal protein in a functional form. *Proc. Natl. Acad. Sci.* **78:** 6779–6783.

Field, J., R.M. Gronostajski, and J. Hurwitz. 1984. Properties of the adenovirus DNA polymerase. *J. Biol. Chem.* **259:** 9487–9495.

Fredman, J.N. and J.A. Engler. 1993. Adenovirus precursor to terminal protein interacts with the nuclear matrix in vivo and in vitro. *J. Virol.* **67:** 3384–3395.

Fredman, J.N., S.C. Pettit, M.S. Horwitz, and J.A. Engler. 1991. Linker insertion mutations in the adenovirus preterminal protein that affect DNA replication in vivo and in vitro. *J. Virol.* **65:** 4591–4597.

Freimuth, P.I. and H.S. Ginsberg. 1986. Codon insertion mutants of the adenovirus terminal protein. *Proc. Natl. Acad. Sci.* **83:** 7816–7820.

Friefeld, B.R., M.D. Krevolin, and M.S. Horwitz. 1983. Effect of the adenovirus H5ts125 and Hts107 DNA binding proteins on DNA replication in vitro. *Virology* **124:** 380–389.

Georgaki, A. and U. Hübscher. 1993. DNA unwinding by replication factor A is a property of the 70KDa subunit and is facilitated by phosphorylation of the 32KDa subunit. *Nucleic Acids Res.* **21:** 3659–3665.

Georgaki, A., B. Strack, V. Podust, and U. Hubscher. 1992. DNA unwinding activity of replication protein A. *FEBS Lett.* **308:** 240–244.

Gounari, F., R. De Francesco, J. Schmitt, P.C. van der Vliet, R. Cortese, and H. Stunnenberg. 1990. Amino terminal domain of NF-I binds to DNA as a dimer and activates adenovirus DNA replication. *EMBO J.* **9:** 559–566.

Guggenheimer, R.A., K. Nagata, M. Kenny, and J. Hurwitz. 1984. Protein primed replication of plasmids containing the terminus of the adenovirus genome 2: Purification and characterisation of a host protein required for the replication of DNA templates devoid of the terminal protein. *J. Biol. Chem.* **259:** 7815–7825.

Harris, M.P.G. and R.T. Hay. 1988. DNA sequences required for the initiation of adenovirus type 4 DNA replication. *J. Mol. Biol.* **201:** 57–67.

Hatfield, L. and P. Hearing. 1993. The NFIII/Oct-1 binding site stimulates adenovirus DNA replication in vivo and is functionally redundant with adjacent sequences. *J. Virol.* **67:** 3931–3939.

Hay, R.T. 1985a. The origin of adenovirus DNA replication: Minimal DNA sequence in vivo. *EMBO J.* **4:** 421–426.

———. 1985b. The origin of adenovirus DNA replication: Role of the nuclear factor I site in vivo. *J. Mol. Biol.* **186:** 129–136.

Hay, R.T. and I.M. McDougall. 1986. Viable viruses with deletions in the left inverted terminal repeat define the origin of Ad.2 DNA replication. *J. Gen. Virol.* **6:** 321–332.

Hay, R.T. and W.C. Russell. 1989. Recognition mechanisms in the synthesis of animal virus DNAs. *Biochem J.* **258:** 3–16.

Hay, R.T., N.D. Stow, and I.M. McDougall. 1984. Replication of adenovirus mini chromosomes. *J. Mol. Biol.* **175:** 493–510.

Hay, R.T., L. Clark, P.H. Cleat, M.P.G. Harris, E.C. Robertson, and C.J. Watson. 1988. Requirements for the initiation of adenovirus types 2 and 4 DNA replication. *Cancer Cells* **6:** 71–75.

Ikeda, J-E., T. Enomoto, and J. Hurwitz. 1982. Adenovirus protein primed initiation of DNA chains in vitro. *Proc. Natl. Acad. Sci.* **79:** 2442–2446.

Jones, K.A., J.T. Kadonaga, P.J. Rosenfeld, T.J. Kelly, and R. Tjian. 1987. A cellular DNA binding protein that activates eukaryotic transcription and DNA replication. *Cell* **48:** 79–89.

Joung, I. and J.A. Engler. 1992. Mutations in two cysteine-histidine-rich clusters in adenovirus type 2 DNA polymerase affect DNA binding. *J. Virol.* **66:** 5788–5796.

Kenny, M.K. and J. Hurwitz. 1988. Initiation of adenovirus DNA replication. II. Structural requirements using synthetic oligonucleotide adenovirus templates. *J. Biol. Chem.* **263:** 9801–9808.

King, A.J. and P.C. van der Vliet. 1994. A precursor terminal protein trinucleotide intermediate during initiation of adenovirus DNA-replication—Regeneration of molecular ends in vitro by a jumping back mechanism. *EMBO J.* **13:** 5786–5792.

Kitchingman, G.R. 1985. Sequence of the DNA binding protein of a human subgroup E adenovirus (type 4): Comparisons with subgroup A (type 12), subgroup B (type 7) and subgroup C (type 5). *Virology* **146:** 90–101.

Klein, H., W. Maltzman, and A.J. Levine. 1979. Structure function relationships of the adenovirus DNA binding protein. *J. Biol. Chem.* **254:** 11051–11060.

Kusukawa, J., M. Ramachandra, R. Nakano, and R. Padmanabhan. 1994. Phosphorylation-dependent interaction of adenovirus preterminal protein with the viral origin of DNA replication. *J. Biol. Chem.* **269:** 2189–2196.

Lally, C., T. Dorper, W. Groger, G. Antoine, and E.-L. Winnacker. 1984. A size analysis of the adenovirus replicon. *EMBO J.* **3:** 333–337.

Leegwater, P.A.J., W. van Driel, and P.C. van der Vliet. 1985. Recognition site of nuclear factor I, a sequence specific DNA binding protein from HeLa cells that stimulates adenovirus DNA replication. *EMBO J.* **4:** 1515–1521.

Lichy, J. H., J. Field, M.S. Horwitz, and J. Hurwitz. 1982. Separation of the adenovirus terminal protein precursor from its associated DNA polymerase: Role of both proteins in the initiation of adenovirus DNA replication. *Proc. Natl. Acad. Sci.* **79:** 5225–5229.

Lindenbaum, J.O., J. Field, and J. Hurwitz. 1986. The adenovirus DNA binding protein and adenovirus DNA polymerase interact to catalyse elongation of primed DNA templates. *J. Biol. Chem.* **261:** 10218–10227.

Mangel, W.F., W.J. McGrath, D. Toledo, and C.W. Anderson. 1993. Viral DNA and a viral peptide are cofactors of adenovirus virion proteinase activity. *Nature* **361:** 274–275.

Meisterernst, M., L. Rogge, R. Foeckler, M. Karaghiosoff, and E.-L. Winnacker. 1989. Structural and functional organisation of a porcine gene coding for nuclear factor I. *Biochemistry* **28:** 8191–8200.

Mermod, N., E.A. O'Neill, T.J. Kelly, and R. Tjian. 1989. The proline rich transcriptional activator of CTF/NF-I is distinct from the replication and DNA binding domain. *Cell* **58:** 741–753.

Monaghan, A., A.G. Webster, and R.T. Hay. 1994. Adenovirus DNA binding protein: Helix destabilising property. *Nucleic Acids Res.* **5:** 742–748.

Mul, Y.M. and P.C. van der Vliet. 1992. Nuclear factor I enhances Adenovirus DNA replication increasing the stability of a preinitiation complex. *EMBO J.* **11:** 751–760.

———. 1993. The adenovirus DNA binding protein effects the kinetics of DNA replication by a mechanism distinct from NF-I or Oct-I. *Nucleic Acids Res.* **21:** 641–647.

Mul, Y.M., C.P. Verrijzer, and P.C. van der Vliet. 1990. Transcription factors NF-I and NF-III/Oct-1 function independently, employing different mechanisms to enhance adenovirus DNA replication. *J. Virol.* **64:** 5510–5518.

Nagata, K., R.A. Guggenheimer, and J. Hurwitz. 1983. Specific binding of acellular DNA replication protein to the origin of replication of adenovirus DNA. *Proc. Natl. Acad. Sci.* **80:** 6177–6181.

Nagata, K., R.A. Guggenheimer, T. Enomoto, J.H. Lichy, and J. Hurwitz. 1982. Adenovirus DNA replication in vitro: Identification of a host factor that stimulates synthesis of the preterminal protein-dCMP complex. *Proc. Natl. Acad. Sci.* **79:** 6438–6442.

Neale, G.A.M. and G.R. Kitchingman. 1990. Conserved region 3 of the adenovirus type 5 DNA binding protein is important for interaction with single-stranded DNA. *J. Virol.* **64:** 630–638.

O'Neill, E.A. and T.J. Kelly. 1988. Purification and characterisation of nuclear factor III (origin recognition protein C), a sequence specific DNA binding protein required for efficient initiation of adenovirus DNA replication. *J. Biol. Chem.* **263:** 931–937.

O'Neill, E.A., C. Fletcher, C.R. Burrow, N, Heintz, R.G. Roeder, and T.J. Kelly. 1988. The transcription factor OTF-1 is functionally identical to the adenovirus DNA replication factor NF-III. *Science* **241:** 1210–1213.

Paonessa, G., F. Gounari, R. Frank, and R. Cortese. 1988. Purification of an NFI-like binding protein from rat liver and cloning of the corresponding cDNA. *EMBO J.* **7:** 3115–3123.

Pettit, S.C., M.S. Horwitz, and J.A. Engler. 1988. Adenovirus preterminal protein synthesised in COS cells from cloned DNA is active in DNA replication in vitro. *J. Virol.* **62:** 496–500.

———. 1989. Mutations in the precursor to the terminal protein of adenovirus serotypes 2 and 5. *J. Virol.* **63:** 5344–5250.

Pronk, R. and P.C. van der Vliet. 1993. The adenovirus terminal protein influences binding of replication proteins and changes the origin structure. *Nucleic Acids Res.* **21:** 2293–2300.

Pronk, R., M.H. Stuiver, and P.C. van der Vliet. 1992. Adenovirus DNA replication: The function of the covalently bound terminal protein. *Chromosoma* **102:** 39–45.

Pronk, R., W. van Driel, and P.C. van der Vliet. 1994. Replication of adenovirus DNA in vitro is ATP-independent. *FEBS Lett.* **337:** 33–38.

Pruijn, G.J.M., W. Van Driel, and P.C. van der Vliet. 1986. A novel sequence specific DNA binding protein from HeLa cells stimulating adenovirus DNA replication. *Nature* **322:** 656–659.

Pruijn, G.J.M., P.C. van der Vliet, N.A. Dathan, and I.W. Mattaj. 1989. Anti-OTF-1 antibodies inhibit NF-III stimulation in vitro adenovirus DNA replication. *Nucleic Acids Res.* **17:** 1845–1863.

Ramachandra, M. and R. Padmanabhan. 1993. Adenovirus DNA polymerase is phosphorylated by a stably associated histone H1 kinase. *J. Biol. Chem.* **268:** 17448–17456.

Ramachandra, M., R. Nakano, P.M. Mohan, A.B. Rawitch, and R. Padmanabhan. 1993. Adenovirus DNA polymerase is a phosphoprotein. *J. Biol. Chem.* **268:** 442–448.

Rancourt, C., K. Tihanyi, M. Bourbonniere, and J.M. Weber. 1994. Identification of active-site residues of the adenovirus endopeptidase. *Proc. Natl. Acad. Sci.* **91:** 844–847.

Rawlins, D.R., P.J. Rosenfeld, R.J. Wides, M.D. Challberg, and T.J. Kelly. 1984. Structure and function of the adenovirus origin of DNA replication. *Cell* **37:** 309–319.

Ricigliano, J.W., D.E. Brough, and D.F. Klessig. 1994. Identification of a high-molecular-weight cellular protein complex containing the adenovirus DNA-binding protein. *Virology* **202:** 715–723.

Roovers, D.J., F.M. van der Lee, J. van der Wees, and J.S. Sussenbach. 1993. Analysis of the adenovirus type 5 terminal protein precursor and DNA polymerase by linker insertion mutagenesis. *J. Virol.* **67:** 265–276.

Rosenfeld, P.J., E.A. O'Neill, R.J. Wides, and T.J. Kelly. 1987. Sequence specific interactions between cellular DNA-binding proteins and the adenovirus origin of replication. *Mol. Cell. Biol.* **7:** 875–886.

Salas, M. 1991. Protein-priming of DNA replication. *Annu. Rev. Biochem.* **60:** 39–71.

Santoro, C., N. Mermod, P.C. Andrews, and R. Tjian. 1988. A family of human CCAAT box binding proteins active in transcription and DNA replication: Cloning and expression of multiple cDNA's. *Nature* **334:** 218–224.

Schaack, J., W.Y.-H. Ho, P. Freimuth, and T. Shenk. 1990. Adenovirus terminal protein mediates both nuclear matrix association and efficient transcription of adenovirus DNA. *Genes Dev.* **4:** 1197–1208.

Shu, L., M.S. Horwitz, and J.A. Engler. 1987. Expression of enzymatically active adenovirus DNA polymerase from cloned cDNA requires sequences upstream of the main open reading frame. *Virology* **161:** 520–526.

Stillman, B. 1989. Initiation of eukaryotic DNA replication in vitro. *Annu. Rev. Cell.*

Biol. **5:** 197–245.
Stillman, B., F. Tamanoi, and M.B. Matthews. 1982. Purification of an adenovirus-coded DNA polymerase that is required for initiation of DNA replication. *Cell* **31:** 613–623.
Stillman, B., J.B. Lewis, L.T. Chow, M.B. Mathews, and J.E. Smart. 1981. Identification of the gene and mRNA for the adenovirus terminal protein precursor. *Cell* **23:** 497–508.
Stow, N.D. 1982a. The infectivity of adenovirus genomes lacking DNA sequences from their left hand termini. *Nucleic Acids Res.* **10:** 5105–5119.
Stuiver, M.H. and P.C. van der Vliet. 1990. Adenovirus DNA binding protein forms a multimeric protein complex with double-stranded DNA and enhances binding of nuclear factor I. *J. Virol.* **65:** 379–389.
Stuiver, M.H., W.G. Bergsma, A.C. Arnberg, H. van Amerongen, R. van Grondelle, and P.C. van der Vliet. 1992. Structural alterations of double-stranded DNA in complex with the adenovirus DNA-binding protein: Implications for its function in DNA replication. *J. Mol. Biol.* **225:** 999–1011.
Stunnenberg, H.G., H. Lange, L. Philipson, R.T. van Miltenburg, and P.C. van der Vliet. 1988. High expression of functional adenovirus DNA polymerase and precursor terminal protein using recombinant vaccinia virus. *Nucleic Acids Res.* **16:** 2431–2444
Sturm, R.A., G. Das, and W. Herr. 1988. The ubiquitous octamer binding protein oct–1 contains a POU domain and a homeobox subdomain. *Genes Dev.* **2:** 1582–1599.
Tamanoi, F. and B. Stillman. 1983. Initiation of adenovirus DNA replication in vitro requires a specific DNA sequence. *Proc. Natl. Acad. Sci.* **80:** 6446–6450.
Temperley, S.M. and R.T. Hay. 1991. Replication of adenovirus type 4 DNA by a purified fraction from infected cells. *Nucleic Acids Res.* **19:** 3243–3249.
———. 1992. Recognition of the adenovirus type 2 origin of replication by the virally encoded DNA polymerase and preterminal proteins. *EMBO. J.* **11:** 761–768.
Temperley, S.M., C.R. Burrow, T.J. Kelly, and R.T. Hay. 1991. Identification of two distinct regions within the adenovirus minimal origin of replication that are required for adenovirus type 4 DNA replication in vitro. *J. Virol.* **65:** 5037–5044.
Tucker, P.A., D. Tsernoglou, A.D. Tucker, F.E.J. Coenjaerts, H. Leenders, and P.C. van der Vliet. 1994. X-ray crystal structure of the 39KDa C-terminal adenovirus DNA binding protein. *EMBO J.* **13:** 2994–3002
Van Bergen, B.G.M., P.A. van der Ley, W. van Driel, A.D.M. van Mansfeld, and P.C. van der Vliet. 1983. Replication of origin containing adenovirus DNA fragments that do not carry the terminal protein. *Nucleic Acids Res.* **11:** 1975–1989.
van der Vliet, P.C. 1990. Adenovirus DNA replication in vitro. In *The eukaryotic nucleus* (ed. P. Straus and S.H. Wilson), pp. 1–32. Telford Press, Caldwell, New Jersey.
Van der Vliet, P.C. and A.J. Levine. 1973. DNA binding proteins specific for cells infected by adenovirus. *Nat. New Biol.* **246:** 170–174.
Verrijzer, C.P., A.J. Kal, and P.C. van der Vliet. 1990. The DNA binding domain (POU domain) of transcription factor Oct-1 suffices for stimulation of DNA replication. *EMBO J.* **9:** 1883–1888.
Verrijzer, C.P., J.A.W.M. van Oosterhout, W.W. van Weperen, and P.C. van der Vliet. 1991. POU proteins bend DNA via the POU-specific domain. *EMBO. J.* **10:** 3007–3014.
Vos, H.L., F.M. van der Lee, and J.S. Sussenbach. 1988. The binding of in vitro synthesized adenovirus DNA binding protein to single-stranded DNA is stimulated by zinc ions. *FEBS Lett.* **239:** 251–254.

Watson, C.J. and R.T. Hay. 1990. Expression of adenovirus type 2 DNA polymerase in insect cells infected with a recombinant baculovirus. *Nucleic Acids Res.* **18:** 1167–1173.

Weber, J. 1990. The adenovirus proteinases. *Semin. Virol.* **1:** 379–384.

Webster, A., R.T. Hay, and G.D. Kemp. 1993. The adenovirus protease is activated by a virus-coded disulphide-linked peptide. *Cell* **72:** 97–104.

Webster, A., I.R. Leith, and R.T. Hay. 1994. Activation of adenovirus-coded protease and processing of preterminal protein. *J. Virol.* **68:** 7292–7300.

Webster, A., S. Russell, P. Talbot, W.C. Russell, and G.D. Kemp. 1989. Characterisation of the adenovirus protease: Substrate specificity. *J. Gen. Virol.* **70:** 3225–3234.

Wides, R.J., M.D. Challberg, D.R. Rawlins, and T.J. Kelly. 1987. Adenovirus origin of replication: Sequence requirements for replication in vitro. *Mol. Cell. Biol.* **7:** 864–874.

Zijderveld, D.C. and P.C. van der Vliet. 1994. Helix-destabilizing properties of the adenovirus DNA-binding protein. *J. Virol.* **68:** 1158–1164.

Zijderveld, D.C., M.H. Stuiver, and P.C. van der Vliet. 1993. The adenovirus DNA binding protein enhances intermolecular renaturation but inhibits intramolecular DNA renaturation. *Nucleic Acids Res.* **21:** 641–647.

Zorbas, H., T. Rein, A. Krause, K. Hoffman, and E.-L. Winnacker. 1992. Nuclear factor I (NFI) binds to an NFI site but not to the CCAAT site in the human α-globin gene promoter. *J. Biol. Chem.* **267:** 8478–8484.

25

Herpesvirus DNA Replication

Mark Challberg
Laboratory of Viral Diseases
National Institute of Allergy and Infectious Diseases
National Institutes of Health
Bethesda, Maryland 20892

The genomes of herpesviruses are linear double-stranded DNA molecules ranging in size from 120 kb to more than 200 kb. Although the many different herpesviruses display a wide variety of tissue tropisms and vary enormously in the way in which they interact with their natural hosts, one common feature of the biology of all herpesviruses is the mechanism by which they replicate their genomes during the lytic phase of the replication cycle. Lytic DNA replication in every herpesvirus studied occurs by a mechanism that generates long head-to-tail concatemers of viral genomes that are cleaved to unit-length genomes during the process of encapsidation. This common mode of lytic DNA replication reflects a conserved set of viral genes encoding the basic components of the replication machinery. Another common feature of herpesvirus biology is the capacity to remain latent in the infected host, but, unlike the case with lytic DNA replication, the mechanism by which the viral genomes are maintained during latency apparently differs considerably among the herpesviruses. The cells that harbor latent genomes are different for the different viruses, and perhaps the more intimate relationship between viral and host chromosomal replication during latency accounts for the greater diversity of mechanism. For example, Epstein-Barr virus (EBV), which is latent in dividing B cells, is replicated during latency from a latency-specific origin called *oriP* that is distinct from the origin of lytic DNA replication. Replication in this system requires a single virus-encoded protein (EBNA1) and is apparently carried out by the chromosomal replication machinery (see Yates, this volume). On the other hand, herpes simplex virus (HSV) is latent in postmitotic neurons, and there is no evidence for any viral DNA replication during latency. In this review, I focus on lytic DNA replication, with an emphasis on studies of HSV DNA replication, the herpesvirus system about which most is known.

 0-87969-459-9/96 $5 + .00

MODE OF REPLICATION—IN VIVO STUDIES

The general picture of herpesvirus lytic replication derives from analyses of replication intermediates, which have two characteristic properties. First, DNA pulse-labeled in vivo with [^{3}H]thymidine sediments more rapidly than unit length viral DNA (Jacob and Roizman 1977). Second, pulse-labeled DNA is "endless"; i.e., the molecular termini of mature viral DNA are fused together (Jacob et al. 1979; Jongeneel and Bachenheimer 1981). On the basis of these two observations, it has been proposed that parental linear viral DNA is circularized shortly after entry into the host cell and that replication takes place predominantly by a rolling-circle mechanism, generating linear concatemers of tandemly repeated viral genomes (Jacob et al. 1979). In support of this model is the finding that circularization of input HSV genomes does take place, by a mechanism apparently involving direct ligation of the termini (Mocarski and Roizman 1982; Poffenberger and Roizman 1985; Garber et al. 1993). Since circularization occurs in the absence of viral gene expression, one of the host-cell ligases is likely to be responsible. Circularization of parental viral genomes prior to the onset of DNA synthesis is the simplest explanation for the complete lack of genomic termini in replicating DNA, and it seems likely that replication involves a rolling-circle intermediate. Nevertheless, recent experiments utilizing pulsed-field gel electrophoresis suggest that both the structure of replication intermediates and the replication process itself may be more complex than a simple rolling-circle model would predict (Bataille and Epstein 1994; McVoy and Adler 1994; Severini et al. 1994; Zhang et al. 1994). A number of investigators have observed that newly replicated DNA has a very slow electrophoretic mobility when analyzed by pulsed-field gel electrophoresis—the majority of nascent DNA will not enter a gel unless it is highly fragmented. This low electrophoretic mobility is partially retained even after digestion of the DNA with a restriction enzyme that cuts once per unit-length genome (Severini et al. 1994). These results, in conjunction with earlier electron microscopic studies (Jacob and Roizman 1977), suggest that newly replicated viral DNA is composed of highly branched, complex networks. The origin of these branched structures is not known, although the most plausible hypothesis is that they are derived at least in part by homologous recombination. Several studies have shown that HSV DNA in infected cells undergoes high levels of homologous recombination (Schaffer et al. 1974; Honess et al. 1980; Smiley et al. 1980) and that this high rate of recombination is closely linked to DNA synthesis (Schaffer et al. 1974; Honess et al. 1980; Smiley et al. 1980; Weber et al. 1988, 1990; Dutch et al. 1992, 1995). By analogy with bacteriophage T4, which produces replication intermediates with very similar properties, it is possible that the networks are formed by a combination of DNA synthesis and recombination and that resolution of recombination intermediates is a slow step. There are at least two key questions that need to be answered: (1) Is recombination an important mechanism for generating replication forks once DNA replication is under way? and (2) What is the mechanism by which DNA replication and recombination are linked? It is possible that replication intermediates are inherently recombinogenic due to unligated strands or partially single-stranded regions. On the other hand, it is also possible that one (or more) virus-encoded protein acts as a recombinase. Although there is no biochemical evidence that suggests the existence of a herpesvirus-encoded recombinase, the virus-encoded single-stranded DNA-binding protein promotes the annealing of single strands (Bortner et al. 1993; Dutch and Lehman 1993), and the existence of a recombinase clearly cannot yet be ruled out.

ORIGINS OF DNA REPLICATION

The existence of *cis*-acting replication origins was first inferred from the structure of defective genomes (Frenkel et al. 1975, 1976; Schroder et al. 1975) that arise during serial passage of HSV at high multiplicities of infection. Plasmids containing origin sequences derived either from these defective genomes or from wild-type genomes are amplified when introduced into HSV-infected cells by transfection, and this transient plasmid amplification assay has been utilized both to identify the lytic origins of a number of other herpesviruses and as the basis for functional analyses of cloned origin sequences. Although, as noted above, the overall mode of lytic DNA replication is conserved among the herpesviruses, the sequence and structure of origins vary considerably among the various herpesvirus groups; by implication, therefore, it is likely that the mechanism of replication initiation varies as well. Schematic models of the lytic origins of HSV and human cytomegalovirus (HCMV) are presented in Figure 1, and their structure and function are described in more detail below. (The structure and function of the EBV lytic origin are reviewed by Yates, this volume.) These three viruses represent examples of all three major herpesvirus groups. It is readily apparent that there is no common arrangement of structural or functional elements, and the lack of similarity is underscored further by the differences in the virus-encoded proteins that interact with their cognate origins. As discussed below, HSV encodes a protein, UL9, that appears to be a classic initiator protein: It binds specifically to the HSV origins of replication and is a

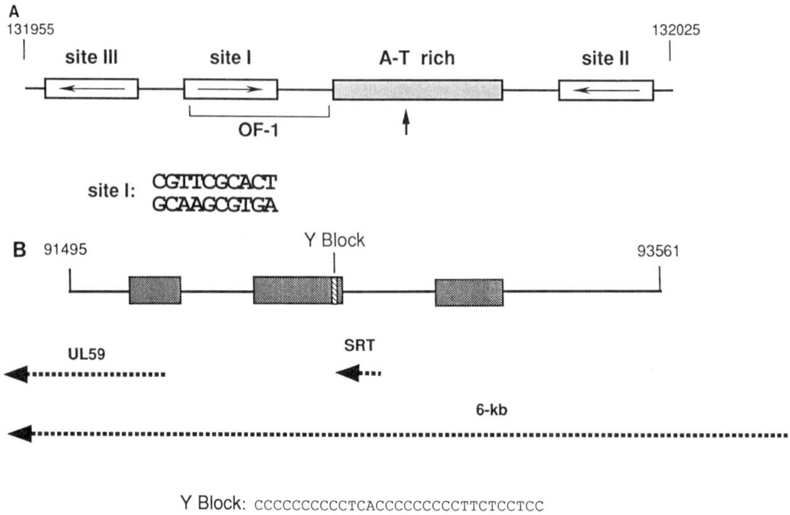

Figure 1 (*A*) Functional domains of the core *ori*$_S$ of HSV. See text for details. (*B*) Functional domains of the core *oriLyt* of HCMV (D. Anders, pers. comm.). The shaded areas represent regions into which insertion of a kanamycin-resistance cassette causes a >50-fold reduction in origin function in transfection assays. The dashed lines represent RNAs that either begin in or extend into the core origin region; SRT is a family of short transcripts that terminate within the Y block. See text for details.

helicase. On the other hand, the EBV-encoded protein that binds to the EBV *oriLyt* is the transcriptional activator protein BZLF1, a key regulatory protein that controls the switch from the latent to lytic transcription program and a protein with no known helicase activity. The CMV *oriLyt* has a number of possible sites for binding of cellular transcription factors, but no virus-encoded proteins that interact with the origin have been identified. The one feature that all of the herpesvirus origins do have in common is the requirement for the binding of transcriptional activator proteins for optimal origin activity. As discussed elsewhere in this volume and below, this is a feature that herpesvirus origins have in common with many eukaryotic origins. The mechanism(s) by which transcription factors enhance replication efficiency is not yet known.

HSV *ori*$_S$ and *ori*$_L$

The HSV genome contains two classes of origin sequences: *ori*$_S$, located in the inverted repeat segment flanking the short component of the genome (and therefore present in the genome in two copies) and ori_L, the sequence of which comprises a 144-bp perfect palindrome, located between the *ul29* (single-stranded DNA-binding protein) and the *ul30* (*pol*) genes near the middle of the long unique component of the genome (Spaete and Frenkel 1982; Stow 1982; Vlasny and Frenkel 1982; Stow and McMonagle 1983; Weller et al. 1985). The sequences of ori_S and ori_L are closely related (Murchie and McGeoch 1982; Weller et al. 1985; Knopf et al. 1986). Both contain an extensive inverted repeat sequence, the central 18 bp of which are exclusively AT base pairs. As described below, the minimum sequences required for the function of ori_S correspond well to the region of highest similarity with ori_L (Lockshon and Galloway 1988). The binding sites for viral and cellular proteins have been shown to be equivalent, and, in transfection assays at least, ori_S and ori_L behave similarly (Hardwicke and Schaffer 1995). The significance of having three origins of replication in the HSV genome is not clear. Mutant viruses lacking ori_L or one or both copies of ori_S have been isolated and have no obvious growth defect either in cultured cells or in animal models (Poffenberger and Roizman 1985; Polvino-Bodnar et al. 1987; Igarashi et al. 1993). A virus lacking ori_L and containing two partially defective copies of ori_S has been constructed and has been shown to have a significant growth defect, whereas an analogous virus containing a wild-type ori_L has no growth defect (S. Wong and P. Schaffer, pers. comm.). These data constitute the strongest genetic evidence to date that ori_S and ori_L serve as essential *cis*-acting functions during the replication of viral DNA in vivo and provide additional support for the idea that ori_S and ori_L are at least partially redundant functionally. Recent studies have shown that the few nucleotide differences between the core regions of ori_S and ori_L result in the presence of a glucocorticoid response element that is not present in ori_S. Preliminary experiments suggest that in cells of neural origin, the presence of this binding site is responsible for a glucocorticoid-induced increase in efficiency of replication from ori_L, and its absence results in a glucocorticoid-induced decrease in efficiency from ori_S (M. Hardwicke and P. Schaffer, pers. comm.). Thus, it is possible that ori_L may have an important role in the initial rounds of DNA replication that occur upon the reactivation of latent genomes. It is clear, however, that ori_L is not absolutely required for reactivation (Polvino-Bodnar et al. 1987). Additional experiments will be required to determine the precise role of multiple origin sequences and to answer the question of whether initiation occurs predominantly at one class of origin sequence during particular phases of replication.

Several laboratories have carried out mutational analyses of plasmids containing *ori*$_S$, with similar results (Lockshon and Galloway 1988; Deb and Deb 1989; Weir and Stow 1990; Hernandez et al. 1991). *ori*$_S$ can be divided into two components: the core origin containing UL9-binding sites, and flanking sequences that increase the efficiency of replication by 50-fold or more (Wong and Schaffer 1991). As mentioned, the minimal core origin sequence corresponds well to the region of highest similarity with *ori*$_L$. The flanking sequences contain a number of consensus binding sites for transcriptional activator proteins such as SP1 and NFI, and the available evidence suggests that the binding of one or more of these transcriptional regulators to sites in close juxtaposition to the core sequence may be critical for enhancing DNA replication.

The core origin sequence is composed of at least five functional domains: two high-affinity UL9-binding sites (called site I and site II or box I and box II) (Elias and Lehman 1988; Olivo et al. 1988; Weir et al. 1989; Elias et al. 1990); an AT-rich region; a sequence homologous to site I, but with much lower affinity for UL9, called site III (or box III) (Elias et al. 1990; Dabrowski and Schaffer 1991); and a binding site for an as-yet-uncharacterized cellular protein(s) called OF-1 (Dabrowski and Schaffer 1991; Dabrowski et al. 1994). Sites I and II are located on the arms of an imperfect 46-bp palindrome, separated by the AT-rich region. Site III and site I form the arms of another, shorter palindrome, with site III present just 5′ of site I. Nuclease protection, chemical modification, and saturation mutagenesis studies (Koff and Tegtmeyer 1988; Elias et al. 1990; Hazuda et al. 1991) have shown that the high-affinity recognition sequence for UL9 is contained in the sequence 5′-CGTTCGCACT. In *ori*$_S$ (but not in *ori*$_L$), site II differs from site I at two positions, resulting in a reduced binding affinity for UL9 to about one-fifth that of site I (Elias and Lehman 1988; Elias et al. 1990; Hazuda et al. 1991). Site III differs in sequence from site I at only one position, but its affinity for UL9 is reduced to less that one-thousandth that of site I (Hazuda et al. 1991). The sequence of *ori*$_L$ comprises a perfect palindrome (Weller et al. 1985). As a consequence of the greater symmetry in *ori*$_L$ relative to *ori*$_S$, the two high-affinity binding sites (equivalent to site I in *ori*$_S$) for UL9 are identical, and *ori*$_L$ contains two, rather than one, low-affinity binding sites (equivalent to site III in *ori*$_S$). The functional consequences of these differences, if any, are not known.

Genetic experiments indicate that these five domains of the core origin sequence are essential for optimal efficiency in promoting DNA replication in transient assays (Lockshon and Galloway 1988; Weir and Stow 1990; Hernandez et al. 1991). Mutations in site I or site II that abolish UL9 binding eliminate or greatly reduce the replication efficiency of plasmids containing these sequences, and mutations in site III reduce replication efficiency by a factor of 5. Deletions or substitution of GC pairs for AT pairs in the central AT-rich region significantly decreases replication efficiency (Lockshon and Galloway 1988; Werstuck et al. 1990). Finally, mutations that eliminate OF-1 binding also diminish the replication efficiency of the origin-containing plasmids (Dabrowski et al. 1994).

Both ori_S and ori_L are located between divergently transcribed genes, and, as mentioned above, there is evidence that the binding of one or more transcriptional activator proteins to enhancer sequences flanking the core origin sequence does have a pronounced stimulatory effect on the extent of DNA replication (Wong and Schaffer 1991). In this regard, HSV DNA replication is similar to other better-characterized viral replication systems such as the adenoviruses and the papovaviruses. In the case of HSV, there is almost no information on the mechanism by which these stimulatory factors act. It seems unlikely that the stimulatory effect depends on transcription per se, since the stimulatory sequences do not include the known start sites for transcription of the surrounding genes, although this point will have to be more closely investigated in the future. In this context, it would be of some interest to know whether there is a difference in the time following infection at which initiation takes place at ori_S and ori_L. ori_S is flanked by the immediate-early class of genes, that is, genes that are transcribed immediately from the infecting viral genome, whereas ori_L is flanked by two delayed-early genes, the transcription of which depends on the presence of one or more immediate-early gene products. There is a transcript in HSV-infected cells that extends through ori_S (Hubenthal-Vass et al. 1987). Neither the 5′ nor 3′ end of this transcript is located within sequences that have an effect on origin function. It has been suggested that this transcript may play a role in regulating DNA replication, but there is no evidence in support of this speculation. Finally, since HSV DNA is thought not to be complexed with cellular histones following uncoating in the nucleus (Leinbach and Summers 1980), it seems unlikely that the role of transcription factors in DNA replication is to exclude the formation of nucleosomes in the vicinity of the origin, as has been suggested in the case of SV40 (Cheng and Kelly 1989). At this point, the available data seem to be most consistent with a model in which the binding of transcription factors near the origin stimulates the binding of critical replication factors to the origin, much as has been proposed for the role of NFI in adenovirus DNA replication. This model will be testable once more information is gained concerning the events that occur at the core origin sequence.

HCMV *oriLyt*

The origin of HCMV lytic DNA replication was identified by a novel method utilizing an inhibitor of the HCMV DNA polymerase, ganciclovir, to terminate DNA synthesis shortly after initiation (Hamzeh et al. 1990), as well as by the more conventional approach of transient transfection assay (Anders and Punturieri 1991; Masse et al. 1992). Deletion analysis of the cloned *oriLyt* has shown that the core functional unit spans more than 2.0 kb of DNA near the middle of the U_L region of the viral genome (Anders et al. 1992; Y. Zhu and D. Anders, pers. comm.). As with HSV ori_S, the sequences flanking the core origin also contribute to the overall efficiency of replication. The structure of the 2.0-kb core origin is highly complex. The sequence contains a number of repeated elements, some of which include transcription factor binding motifs, as well as regions of dyad symmetry. Extensive mutational analysis of this sequence has revealed several structural elements that are critical for function, but the only non-repeated element that is absolutely required is an oligopyrimidine stretch (on the top strand) named the Y block (see Fig. 1B). This sequence motif is similar to the "downstream region" in the *oriLyt* of EBV (Schepers et al. 1993b) and is conserved in simian and murine CMV. As noted earlier, no HCMV-encoded protein has yet been shown to bind specifically to the *oriLyt* sequence. Recently, however, a family of short, non-polyadenylated transcripts (named the "short replicator transcripts"; SRT) expressed within *oriLyt* have been identified and characterized (L. Huang et al., pers. comm.). The 5′ ends of these RNAs are located about 200 bp to the right of the Y block (at position 92685), and the 3′ ends all map within or just to the left of the Y block. A very similar arrangement of transcript and oligopyrimidine sequence occurs in the mitochondrial heavy-strand replicator (Xu and Clayton 1995; see Clayton, this volume). In mitochondrial DNA, the oligopyrimidine sequence inhibits the release of the nascent transcript and promotes the formation of a displacement loop; the RNA is then processed and used as a primer for DNA synthesis. On the basis of this similarity, it has been suggested that a critical specific step in the initiation of CMV replication may be the synthesis of the SRT rather than, or in addition to, the site-specific binding of virus-encoded protein (D. Anders, pers. comm.). According to this model, the SRT either would act directly as a primer for DNA synthesis or would mediate a local unwinding of the DNA, which would then serve as an entry point for the virus-encoded replication machinery. This model awaits experimental verification.

VIRAL REPLICATION PROTEINS

The complete set of viral genes that are required for HSV DNA replication were identified by means of a transient complementation assay in which cloned segments of HSV DNA were tested for the ability to support the replication of a cotransfected plasmid containing ori_S or ori_L (Challberg 1986; Wu et al. 1988). Seven genes were found to be both necessary and sufficient for origin-dependent DNA replication. The results of the transfection assay are supported by detailed mapping of available *ts* mutants with clear DNA-negative phenotypes and by the construction of viruses containing targeted null mutations (insertions or deletions) in these seven genes (Purifoy et al. 1977; Chartrand et al. 1980; Conley et al. 1981; Purifoy and Powell 1981; Coen et al. 1984; Orberg and Schaffer 1987; Weller et al. 1987; Carmichael et al. 1988; Goldstein and Weller 1988; Marchetti et al. 1988; Zhu and Weller 1988, 1992a; Carmichael and Weller 1989; Marcy et al. 1990a; Yamada et al. 1990; Johnson et al. 1991). The success of the systematic transfection assay in locating the essential replication genes in HSV has now been repeated for both HCMV (Pari and Anders 1993; Pari`et al. 1993) and EBV (Fixman et al. 1992, 1995). The results of these studies, which are summarized in Table 1, as well as nucleotide sequence analyses of a number of other herpesviruses, lead to the following general conclusions. The genes required for lytic DNA replication include a set of six genes that are conserved in all herpesviruses. As discussed in detail below, the products of these six genes in HSV (UL5, UL8, UL29, UL30, UL42, and UL52) comprise the core biochemical activities expected at a replication fork: a processive DNA polymerase, helicase, primase, and a single-stranded DNA-binding protein. Although very little biochemistry has been carried out on the homologs of these genes in other herpesviruses, the available evidence (Ertl and Powell 1992; Tsurumi 1993; Tsurumi et al. 1993a,b,1994) suggests that their function is essentially the same as the HSV proteins. It therefore seems likely, as mentioned earlier, that the characteristic features of herpesvirus lytic DNA replication are due to the conserved enzymological features of this set of six conserved genes. Each herpesvirus also encodes other proteins required for DNA replication. These additional proteins fall into two groups: (1) Transcriptional regulatory proteins required for the expression of the other replication

Table 1 Herpesvirus genes required for lytic DNA replication

HSV[a,b]	EBV[c,d]	HCMV[e,f]	Function
UL30	BALF5	UL54	DNA polymerase
UL42	BMRF1	UL44	polymerase processivity
UL29	BALF2	UL57	single-stranded DNA binding
UL5	BBLF4	UL105	helicase/primase
UL52	BSLF1	UL70	helicase/primase
UL8	BBLF2/3	UL101–102	helicase/primase
UL9			origin binding/helicase
	BZLF1		origin binding/transcription activation
		UL112–113	transcription regulation
		IE1/IE2	transcription regulation
		IRS1	transcription regulation
		UL84	?

The data are summarized from the following sources: [a]Challberg 1986; [b]Wu et al. 1988; [c]Fixman et al. 1992; [d]Fixman et al. 1995; [e]Pari and Anders 1993; [f]Pari et al. 1993.

genes. In the case of HSV, expression of the nonregulatory genes from a constitutive promoter eliminated the requirement for the regulatory genes (Heilbronn and zur Hausen 1989). In the case of EBV, constitutive expression of all of the genes did not eliminate the requirement for the transcriptional activator, BZLF1 (Fixman et al. 1995). This experiment has not yet been completed in the case of HCMV. (2) Origin recognition proteins such as the HSV UL9 protein. In the case of EBV, the requirement for the transcriptional activator protein is due to the fact that BZLF1 also serves as the origin recognition protein (Schepers et al. 1993a). As mentioned earlier, the lack of conservation of origin recognition proteins mirrors the diversity of structure of the origins of different herpesvirus groups.

It should also be noted that the various herpesviruses encode a number of other gene products that play a critical role in DNA replication in the natural host but which are not essential for DNA replication in the rather artificial system of actively dividing immortalized cells in culture. These functions include a number of enzymes involved in nucleotide biosynthesis such as ribonucleotide reductase, thymidine kinase, thymidylate kinase, and thymidylate synthetase. Additionally, it is clear that host proteins play a role in DNA replication as well. As noted, all of the herpesvirus origins contain binding sites for host transcription factors. Moreover, there is evidence in the case of HSV that the virus-encoded origin-binding protein, UL9, interacts specifically with the host DNA polymerase-α (Lee et al. 1995). Even if there are no cellular proteins that participate directly in the synthetic events at the replication fork, there are other activities that would be predicted to be required for DNA replication that may well be supplied by the host cell. For example, neither a ligase activity nor a topoisomerase activity has been associated with any of the purified viral replication proteins, and there is no report that such activities are induced following infection. One of the major challenges in the future will be to understand how the virus- and host-encoded proteins are integrated into a functional replication system.

ORIGIN BINDING PROTEIN (UL9)

The *ul9* gene encodes a polypeptide of 851 amino acids with a predicted molecular mass of 94 kD. The protein contains at least two functional domains: The amino-terminal two-thirds of the protein mediates dimerization (Elias et al. 1992; Hazuda et al. 1992) as well as a DNA-dependent helicase activity (Bruckner et al. 1991; Fierer and Challberg 1992; Martinez et al. 1992; Boehmer et al. 1993), and the carboxy-terminal domain of 317 amino acids mediates sequence-specific DNA binding (Weir et al. 1989; Hazuda et al. 1991; Arbuckle and Stow 1993; Martinez and Edwards 1993; Perry et al. 1993; Martin et al. 1994; Fierer and Challberg 1995). The amino-terminal two-thirds of UL9 contains six motifs that make it a member of a superfamily of helicases (Gorbalenya et al. 1988, 1989). Genetic evidence has demonstrated that these motifs are essential for viral DNA replication: Mutations in five of the six motifs render the gene unable to support DNA replication in a transient replication system (Martinez et al. 1992). UL9 has been purified to homogeneity both from HSV-infected cells and from insect cells infected with a recombinant baculovirus expressing UL9. The purified protein exhibits DNA-dependent ATPase activity and helicase activity (Bruckner et al. 1991; Fierer and Challberg 1992; Boehmer et al. 1993; Dodson and Lehman 1993). The helicase has a 3′ to 5′ polarity, is able to unwind partially duplex DNA of nonspecific sequence, and does not require a single-stranded tail (Fierer and Challberg 1992; Boehmer et al. 1993). As discussed below, UL9 has not been shown to unwind fully duplex DNA, however, nor to preferentially unwind origin-containing DNA (Fierer and Challberg 1992; Boehmer et al. 1993).

Purified UL9 is a homodimer in solution (Bruckner et al. 1991; Fierer and Challberg 1992). The dimerization domain appears to reside predominantly in the amino-terminal two-thirds of the molecule. Thus, the full-length UL9 is a tightly associated homodimer in solution with an upper limit for the dissociation constant of approximately 1×10^{-10} M. The DNA-binding domain, however, which comprises the carboxy-terminal

317 amino acids (t-UL9), is predominantly a monomer in solution with a dissociation constant of approximately 1 x 10^{-5} M (D.S. Fierer et al., unpubl.). Several different approaches have been used to determine the number of monomeric UL9 DNA-binding domains that interact with a single binding site, with conflicting results. In one approach, Fab fragment directed at the carboxyl terminus of UL9 was added to a mixture of purified DNA-binding domain and a single UL9-binding site and analyzed by gel mobility shift assay (Stabell and Olivo 1993). Increasing concentrations of the Fab resulted in the appearance of two bands of lower electrophoretic mobility, suggesting that two t-UL9 molecules bind to the site. In another approach (Martin et al. 1994; Gustafsson et al. 1995), two slightly different-sized versions of the UL9 DNA-binding domain were produced. In gel shift experiments in which both of these forms of UL9 were present, only two protein/DNA complexes were observed, the mobilities of which corresponded to the mobilities of the complexes observed with each form of the DNA-binding domain alone. This result suggests that no heterodimers between the larger and smaller polypeptides were formed and, therefore, that the DNA-binding domain binds to a single binding site as a monomer. Finally, the stoichiometry of binding was measured directly by using a double-label gel shift assay (Fierer and Challberg 1995), and the results of this approach showed that UL9 binds to a single site as a dimer. Additional work will be required to reconcile these conflicting results.

There is strong evidence that binding of UL9 to the core origin results in the formation of a complex, higher-order nucleoprotein structure (Elias et al. 1990; Koff et al. 1991; Fierer and Challberg 1992; Gustafsson et al. 1994). The binding of UL9 to sites I and II is cooperative (Elias et al. 1990, 1992; Hazuda et al. 1992), suggesting some form of interaction between the proteins bound at the two sites. DNase I footprinting experiments using derivatives of the origin in which the distance between sites I and II was lengthened have shown that the binding of UL9 results in the formation of a periodic pattern of DNase hypersensitivity in the DNA between the two binding sites, with an interval between hypersensitive sites of 10 bp (Koff et al. 1991; Fierer and Challberg 1992; Stabell and Olivo 1993). This pattern of hypersensitivity is a clear indication that the DNA between the two UL9-binding sites must be held in some sort of fixed conformation, and it has been suggested that protein-protein interaction between UL9 bound at the two sites holds the DNA between the two sites in a loop. In contrast, neither cooperative binding nor the periodic hypersensitivity pattern was observed when the carboxy-terminal UL9-binding domain was used instead of full-length UL9 (Elias et al. 1990; Stabell and Olivo 1993). Taken together, these results indicate that the amino-terminal two-thirds of UL9 is required for the interaction between UL9 molecules bound to the origin and support the view that protein-protein interactions between UL9 molecules bound at the two sites hold the DNA in a fixed conformation. The current data are consistent with such protein-protein interactions occurring between dimers of UL9 bound at sites I and II, although more complicated models are also possible.

The events following the binding of UL9 to the core origin sequence that lead to the actual initiation of DNA synthesis are less well-understood than the structure of the UL9/DNA complex. Clearly, the two parental strands must be unwound as a prelude to the initiation of daughter-strand synthesis. A reasonable model is that UL9 binds specifically to the origin and unwinds the AT-rich region to allow or direct assembly of the replication machinery. As noted earlier, however, UL9 has not been demonstrated to exhibit origin-unwinding activity, and attempts by several investigators to demonstrate origin-specific unwinding by UL9 have proven unsuccessful (Fierer and Challberg 1992; Boehmer and Lehman 1993). UL9-induced distortions of the DNA in the vicinity of the origin were detected by permanganate footprinting experiments (Koff et al. 1991), but these distortions did not require ATP hydrolysis, as might be expected if they were the result of unwinding catalyzed by the helicase activity of UL9. It seems likely that these distortions reflect the formation of a specific nucleoprotein structure, rather than local unwinding at a site within the origin. Nevertheless, despite the lack of any biochemical confirmation to date, the idea that the helicase activity of UL9 plays an essential role in initiation by unwinding DNA at the origin is still an attractive hypothesis (Martinez et al. 1992). It is possible that the experimental conditions necessary to detect such an unwinding event in vitro have not yet been discovered, or it is possible that there is another component required for the reaction, a cellular protein perhaps, that has not yet been identified. As mentioned, it has been reported that mutations in ori_S that eliminate the binding of the cellular protein, OF-1, diminish the replication efficiency of ori_S in a transient DNA replication assay (Dabrowski and Schaffer 1991; Dabrowski et al. 1994). It has also been reported recently that UL9 interacts with the cellular replication protein DNA polymerase-α (Lee et al. 1995). Future work on UL9 will have to be directed toward understanding the exact role of its helicase activity, the role of the complex nucleoprotein structure that is formed upon binding of the protein to the origins, and the role of cellular proteins that also interact with the origin.

In the model of UL9-initiated HSV DNA replication suggested above, it would be expected that UL9 would interact with the HSV replication proteins present at the newly formed replication fork. Such interactions have been demonstrated between UL9 and at least three other HSV replication proteins. The ssDNA-binding protein, ICP8, has been shown to stimulate the UL9 helicase activity (Fierer and Challberg 1992; Boehmer et al. 1993; Boehmer and Lehman 1993). This stimulation depends on an interaction between ICP8 and the carboxy-terminal DNA-binding domain of UL9. Moreover, deletion of the carboxy-terminal 27 amino acids of UL9 eliminated the interaction between UL9 and ICP8 without affecting the other activities of the protein, and this mutant UL9 did not support DNA replication in an in vivo assay, strongly suggesting that the interaction is critically important for DNA replication (Boehmer et al. 1994). Recently, it has been demonstrated by immunoprecipitation experiments that the helicase/primase complex interacts with UL9 via specific contacts with the UL8 subunit (McLean et al. 1994). The DNA-binding domain of UL9 was found to be dispensable for this interaction; therefore it is possible, at least in principle, for UL9 to interact with both ICP8 and the helicase/primase simultaneously. It has also been reported that UL42, the processivity factor subunit of the DNA polymerase (see below), can be co-immunoprecipitated with UL9 (D. Parris, pers. comm.), and the catalytic subunit of the polymerase has been shown to interact specifically with UL5, the helicase component of the helicase/primase enzyme (J. Crute, pers. comm.). Thus, there is evidence for the interaction of each of the seven essential viral replication proteins with at least one other member of the group. Indeed, it has been reported that the HSV replication proteins can be purified as a large, multimolecular complex when co-expressed from recombinant baculoviruses in insect cells (Skaliter and Lehman 1994). Moreover, the multiple interactions of UL9 with the other replication proteins, in conjunction with the specific DNA-binding activity of UL9, appear to be suited to recruit these seven polypeptides into a large multiprotein "replisome" at an origin of replication.

THE REPLICATION FORK

DNA Polymerase

The HSV polymerase was the first virus-encoded replication protein to be detected and purified, and a great deal of work has gone into the characterization of this enzyme. The polymerase, as isolated from HSV-infected cells, exists exclusively as a heterodimeric complex of UL30 (140 kD), the catalytic subunit (pol), and UL42 (52 kD), a phosphoprotein that binds tightly to double-stranded DNA (Vaughan et al. 1985; Marsden et al. 1987; Gallo et al. 1988; Crute and Lehman 1989; Gottlieb et al. 1990; Hernandez and Lehman 1990). The catalytic subunit contains an intrinsic 3′→5′ exonuclease activity that probably serves a proofreading function to increase the fidelity of DNA synthesis (Powell and Purifoy 1977; Knopf 1979; O'Donnell et al. 1987; Marcy et al. 1990b). It has been reported that the polymerase also contains an intrinsic 5′→3′ exonuclease/RNase H activity, which is presumed to play a role in the removal of primers from the Okazaki fragments made on the lagging strand during semi-discontinuous synthesis (Crute and Lehman 1989). More recently, however, it has been reported that the polymerase can be separated from 5′→3′ exonuclease activity (J. Hall, pers. comm.) and that this activity may be carried out by another virus-encoded polypeptide, the alkaline exonuclease (UL12) (Knopf and Weisshart 1990).

The interaction between the two subunits of the polymerase has been studied in detail. Mutational analysis of UL42 has demonstrated that the amino-terminal two-thirds of the polypeptide is sufficient for interaction with the polymerase as well as for overall function (Digard et al. 1993b; Gao et al. 1993; Hamatake et al. 1993; Tenney et al. 1993a). It has not been possible, however, to identify a single small peptide sequence or subdomain within the amino-terminal functional domain that is both necessary and sufficient for interaction with pol (Monahan et al. 1993). In contrast, mutational analysis of the pol polypeptide has revealed a short peptide sequence located near the carboxyl terminus of the protein that is sufficient for the interaction of pol with UL42 (Digard et al. 1993a; Tenney et al. 1993b). Deletion of this sequence eliminated the ability of pol to support DNA replication in vivo, despite the fact that the deletion was shown to have no effect on the activity of the enzyme in standard in vitro assays (Digard et al. 1993a; Stow 1993; Tenney et al. 1993b). It seems likely, therefore, that the interaction of pol and UL42 is critical for DNA replication.

The viral pol/UL42 polymerase holoenzyme is a highly processive enzyme, able to synthesize long strands without dissociating from the template strand, and it is the UL42 subunit of the polymerase that is responsible for the high processivity of the enzyme (Gottlieb et al. 1990; Hernandez and Lehman 1990). The mechanism of this increased processivity has been examined using a model primer template composed of a hairpin duplex region with a 5′ single-stranded sequence (Gottlieb and Challberg 1994). Nuclease protection experiments showed that pol bound this substrate at the single-stranded/double-stranded junction (mimicking the primer-template junction), whereas pol/UL42 had a footprint that extended further into the downstream duplex region of the substrate. In contrast to pol alone, UL42 alone did not interact specifically with a primer-template junction but rather interacted nonspecifically with the double-stranded portion of the substrate and did not interact to a detectable degree with the single-stranded region. These findings are consistent with previous studies on the DNA-binding properties of UL42, which showed that this protein is a double-stranded-specific DNA-binding protein (Vaughan et al. 1985; Gallo et al. 1988) In addition, the pol/UL42 complex had an overall 10-fold higher affinity for the substrate than pol alone. Taken together, these data suggest that the double-stranded DNA-binding property of the UL42 subunit acts to increase the processivity of the polymerase by forming a "sliding clamp," anchoring the polymerase to the newly synthesized duplex DNA and decreasing the probability that the polymerase dissociates from the template strand after each round of catalysis. Mutational analysis of the two polypeptides also supports this model. First, there is a tight correlation between processivity and the ability of the two proteins to interact (Digard and Coen 1990; Digard et al. 1993a,b; Gao et al. 1993; Hamatake et al. 1993; Monahan et al. 1993; Tenney et al. 1993a,b). Second, mutants of UL42 that lack double-stranded DNA-binding activity but still retain the ability to interact with pol also do not support processive synthesis (Chow and Coen 1995).

The available data thus suggest that UL42 and its homologs in other herpesviruses represent a novel class of polymerase processivity factors whose mechanism is quite different from some of the better-characterized proteins such as the β subunit of *Escherichia coli* Pol III, the eukaryotic replication factor PCNA, and the gene 45 protein for T4 phage DNA polymerase (for review, see Kuriyan and O'Donnell 1993). The mechanism by which UL42 acts differs from these other proteins in a number of ways. First, UL42 has high intrinsic affinity for DNA. Second, it forms a tight heterodimeric complex with the DNA polymerase, and there is no evidence that UL42 is assembled into a multimeric torus around DNA. Finally, the function of UL42 does not require the participation of a primer recognition protein nor does it require ATP. UL42 appears to be most similar to the thioredoxin subunit of the T7 DNA polymerase, which is thought to function by causing a decrease in the dissociation rate of the polymerase from the primer template (Tabor et al. 1987; Marians 1992). A key question that now remains to be answered concerns the mechanistic basis for sliding along DNA by proteins like UL42 that have a high intrinsic affinity for DNA. It seems likely that the answer to this question will require detailed structural information.

Single-stranded DNA-binding Protein ICP8

Infected-cell protein 8 (ICP8; product of UL29 gene) was recognized many years ago as an abundant HSV-induced protein of about 130 kD that binds tightly to single-stranded DNA cellulose columns (Honess and Roizman 1973; Bayliss et al. 1975; Powell and Courtney 1975; Powell and Purifoy 1976; Powell et al. 1981). ICP8 has many of the properties that are characteristic of helix-destabilizing proteins. It binds more tightly to single-stranded DNA than to double-stranded DNA (Ruyechan and Weir 1984), and its binding to single-stranded DNA is cooperative and independent of sequence (Ruyechan 1983; Ruyechan and Weir 1984). Purified ICP8 destabilizes short duplex DNA segments and promotes the renaturation of long complementary strands (Bortner et al. 1993; Dutch and Lehman 1993). It seems reasonable to assume that the function of ICP8 is analogous to that of other single-stranded DNA-binding proteins: to bind to the single-stranded DNA formed at a replication fork by the unwinding of the parental duplex DNA and to facilitate the use of these strands as templates for DNA polymerase. It has been reported that purified ICP8 has a small (no greater than 2-fold) stimulatory effect on the activity of purified HSV DNA polymerase using activated DNA as template (Ruyechan and Weir 1984), and using a single-stranded M13 DNA template, ICP8 was shown to be required for the processive synthesis of long DNA strands (Hernandez and Lehman 1990).

There is also evidence suggesting that ICP8 may interact specifically with other replication proteins. As noted earlier, ICP8, but not heterologous SSBs (Fierer and Challberg 1992; Boehmer and Lehman 1993), stimulates the helicase activity of UL9 on substrates having long duplex regions. Several *ts* mutants with defects in ICP8 have been shown to display altered sensitivities (at the permissive temperature) to drugs that inhibit the viral DNA polymerase (Chiou et al. 1985). ICP8 stimulates both the helicase and primase activities of the HSV helicase/primase (R. Hamatake and D. Tenney, pers. comm., D. Klinedinst and M. Challberg, unpubl.). Finally, protein affinity chromatography experiments have suggested that ICP8 interacts with UL42 and with the virus-encoded nuclease, UL12 (Vaughan et al. 1984; Thomas et al. 1992). The functional significance of the interaction between ICP8 and the UL12 nuclease is not clear, although it is worth noting that the nuclease appears to be necessary for the processing of branched replication intermediates (Shao et al. 1993) and may be responsible for removing the RNA primers at the 5′ end of Okazaki fragments (Knopf and Weisshart 1990). It seems likely, therefore, that ICP8 plays a role in organizing a multiprotein complex at the replication fork by providing specific contacts with the other replication proteins. It is also possible that at least some of these observations reflect a role for ICP8 in organizing or maintaining the large subnuclear structures, known as replication compartments, in which viral DNA replication takes place (Quinlan et al. 1984; de Bruyn Kops and Knipe 1988). More detailed molecular genetic analyses of ICP8, as well as the continued development and analysis of in vitro reactions involving combinations of the viral replication proteins, should help to clarify some of these issues.

Helicase/Primase UL5/UL8/UL52

Infection of cells with HSV induces novel helicase and primase activities (Crute et al. 1988). These two activities have been purified to homogeneity; both helicase and primase activities are components of a three-subunit enzyme comprising the products of the *ul5, ul8,* and *ul52* genes (Crute et al. 1989; Dodson et al. 1989; Crute and Lehman 1991).

The helicase activity of this enzyme has been extensively characterized. The helicase can utilize either ATP or GTP as a co-factor for unwinding (Crute et al. 1988; Crute and Lehman 1991). The activity of the helicase on model substrates suggests that it moves in the 5′ to 3′ direction on the strand to which it is bound and requires a single-stranded tail on the fragment which is being displaced (Crute et al. 1988). The predicted UL5 amino acid sequence contains the six sequence motifs that are characteristic of helicases (Gorbalenya et al. 1988, 1989; Hodgman 1988; Lane 1988; McGeoch et al. 1988), and there is genetic evidence that mutations in highly conserved residues within each of these six motifs render the protein inactive in DNA synthesis (Zhu and Weller 1992b). Even though UL5 can be purified in soluble form, the isolated UL5 polypeptide has only weak, if any, activity, as a helicase in the absence of UL52. Co-expression of UL5 and UL52, however, results in the formation of a heterodimeric complex that has full helicase activity on standard substrates (Calder and Stow 1990; Dodson and Lehman 1991; Sherman et al. 1992). The UL8 subunit, therefore, is not required for helicase activity per se, although as indicated below, UL8 may affect helicase activity in more complex assays.

The primase activity of this enzyme has also been extensively characterized. It was recently shown that the UL52 polypeptide is responsible for the primase activity of the helicase/primase complex (Klinedinst and Challberg 1994; Dracheva et al. 1995). A sequence motif in UL52 that is highly conserved in the UL52 homologs of other herpesviruses contains a DXD element that is similar to the divalent metal-binding site of DNA polymerases and several DNA primases. Mutation of either of the aspartic acid residues in this motif abolished the ability of the protein to support replication in vivo, and completely eliminated primase activity in vitro, with no effect on the ATPase and helicase activities of the enzyme. It has not been possible to obtain sufficient quantities of soluble UL52 to determine whether the isolated protein has enzymatic activity in the absence of UL5, although the heterodimeric UL5/UL52 complex does have detectable primase activity (Dodson and Lehman 1991; Sherman et al. 1992; Tenney et al. 1994). Elimination of helicase activity by specific mutation of the nucleoside triphosphate binding site in UL5 had no effect on primase activity (J. Crute and J. Gottlieb, pers. comm.). The predominant products of the primase are oligoribonucleotides 8–10 residues in length (Crute and Lehman 1991; Sherman et al. 1992; Tenney et al. 1994, 1995). Unlike the primase activity of the eukaryotic pol-α:primase enzyme, the HSV primase has a strong preference for certain template sequences (Tenney et al. 1995; J. Gottlieb and M. Challberg, unpubl.). The current understanding of the rules for template recognition and primer synthesis are as follows: (1) The minimal required sequence consists of a deoxyguanosine followed (3′–5′) by a stretch of at least five pyrimidines. The five pyrimidines can consist entirely of deoxycytidines, but not entirely of thymidines. (2) The deoxyguanosine residue at the 3′ end of this sequence is absolutely required for primase recognition, but it is not a template for the first nucleotide of the primer; primer synthesis begins at the second nucleotide in the template sequence. (3) Deoxyadenosine or deoxyguanosine substitutions are tolerated after the second pyrimidine (GYY...), although they do reduce overall activity. (4) Primer synthesis terminates after the incorporation of 7–15 nucleotides, regardless of sequence; purines in the template increase the probability that primer synthesis will terminate prematurely.

The function of UL8 is currently not well understood. As mentioned earlier, although UL8 is not absolutely required for primase activity, the absence of UL8 does reduce primer synthesis about 10-fold on long single-stranded DNA templates (Tenney et al. 1994); UL8 does not, however, appear to have an effect on the template requirements of the enzyme or on the nature of the product oligoribonucleotides. Recent findings suggest that UL8 has an additional activity that is distinguishable biochemically from its effect on the primase (J. Gottlieb and M. Challberg, unpubl.). DNA synthesis was reconstituted with purified HSV fork proteins and a preformed replication fork substrate consisting of a single-stranded DNA circle primed with an oligonucleotide containing an unpaired 5′ single-stranded tail. In this system, efficient leading-strand synthesis, in which the product DNA strands ranged in size from 1 to more than 10 times the unit length of the substrate, required the HSV DNA polymerase, the single-stranded DNA-binding protein ICP8, and all three subunits of the UL5/UL8/UL52 helicase/primase; reactions lacking UL8 failed to carry out displacement synthesis. The mechanism of these two effects of UL8 is not known. It is possible that the two activities reflect two distinct functions of the UL8 polypeptide, or that both activities reflect the same underlying function. One possibility, for example, is that UL8 increases the processivity of the helicase/primase. If so, then the mechanism of increased processivity is more complex than simply to act as a clamp as UL42 does for the DNA polymerase, since although UL8 obviously interacts with the catalytic subunits of the enzyme, UL8 alone does not bind appreciably to single-stranded DNA (Parry et al. 1993).

CONCLUDING REMARKS

The biochemical functions of the herpesvirus-encoded replication proteins are now reasonably well understood, although clearly there are many outstanding questions. It seems likely that continued enzymological and molecular genetic studies will provide many answers to these questions in the future. Two key questions concern the mechanism of initiation and the potential involvement of cellular proteins in the lytic replication process. As indicated, some progress has been made in these areas, but a major stumbling block to additional progress continues to be the lack of good in vitro systems that clearly mimic the events that occur in vivo. There are now several reports of in vitro systems that carry out origin-independent rolling-circle-type replication that is dependent on the virus-encoded fork proteins (Rabkin and Hanlon 1990; Skaliter and Lehman 1994). Continued analysis of these systems should shed additional light on the ways in which the viral proteins interact at a replication fork to increase the overall efficiency of the individual components and may provide insight into the role of cellular proteins in the events that follow initiation. What is still missing is an in vitro system in which origin-dependent initiation takes place, and a complete understanding of herpesvirus replication will depend on the development of such a system.

REFERENCES

Anders, D.G. and S.M. Punturieri. 1991. Multicomponent origin of cytomegalovirus lytic-phase DNA replication. *J. Virol.* **65:** 931–937.

Anders, D.G., M.A. Kacica, G. Pari, and S.M. Punturieri. 1992. Boundaries and structure of human cytomegalovirus oriLyt, a complex origin for lytic-phase DNA replication. *J. Virol.* **66:** 3373–3384.

Arbuckle, M.I. and N.D. Stow. 1993. A mutational analysis of the DNA-binding domain of the herpes simplex virus type 1 UL9 protein. *J. Gen. Virol.* **74:** 1349–1355.

Bataille, D. and A. Epstein. 1994. Herpes simplex virus replicative concatemers contain L components in inverted orientation. *Virology* **203:** 384–388.

Bayliss, G.J., H.S. Marsden, and J. Hay. 1975. Herpes simplex virus proteins: DNA binding proteins in infected cells and in the virus structure. *Virology* **68:** 124–134.

Boehmer, P.E. and I.R. Lehman. 1993. Physical interaction between the herpes simplex virus 1 origin-binding protein and single-stranded DNA-binding protein ICP8. *Proc. Natl. Acad. Sci.* **90:** 8444–8448.

Boehmer, P.E., M.S. Dodson, and I.R. Lehman. 1993. The herpes simplex virus type I origin binding protein: DNA helicase activity. *J. Biol. Chem.* **268:** 1220–1225.

Boehmer, P., M. Craigie, N. Stow, and I.R. Lehman. 1994. Association of origin binding protein and single strand DNA-binding protein, ICP8, during herpes simplex virus type 1 DNA replication in vivo. *J. Biol. Chem.* **269:** 29329–29334.

Bortner, C., T.R. Hernandez, I.R. Lehman, and J. Griffith. 1993. Herpes simplex virus 1 single-stranded DNA-binding protein (ICP8) will promote homologous pairing and strand transfer. *J. Mol. Biol.* **231:** 241–250.

Bruckner, R.C., J.J. Crute, M.S. Dodson, and I.R. Lehman. 1991. The herpes simplex virus 1 origin binding protein: A DNA helicase. *J. Biol. Chem.* **266:** 2669–2674.

Calder, J.M. and N.D. Stow. 1990. Herpes simplex virus helicase-primase: The UL8 protein is not required for DNA-dependent ATPase and DNA helicase activities. *Nucleic Acids Res.* **25:** 3573–3578.

Carmichael, E.P. and S.K. Weller. 1989. Herpes simplex virus type 1 DNA synthesis requires the product of the UL8 gene: Isolation and characterization of an ICP6:lacZ insertion mutation. *J. Virol.* **63:** 591–599.

Carmichael, E.P., M. Kosovosky, and S.K. Weller. 1988. Isolation and characterization of herpes simplex virus type 1 host range mutants defective in viral DNA synthesis. *J. Virol.* **62:** 91–99.

Challberg, M.D. 1986. A method for identifying the viral genes required for herpesvirus DNA replication. *Proc. Natl. Acad. Sci.* **83:** 9094–9098.

Chartrand, P., C.S. Crumpacker, P.A. Schaffer, and N.M. Wilkie. 1980. Physical and genetic analysis of the herpes simplex virus DNA polymerase locus. *Virology* **103:** 311–326.

Cheng, L. and T.J. Kelly. 1989. Transcriptional activator nuclear factor I stimulates the replication of SV40 minichromosomes in vivo and in vitro. *Cell* **59:** 541–551.

Chiou, H.C., S.K. Weller, and D.M. Coen. 1985. Mutations in the herpes simplex virus major DNA-binding protein gene leading to altered sensitivity to DNA polymerase inhibitors. *Virology* **145:** 213–226.

Chow, C.S. and D.M. Coen. 1995. Mutations that specifically impair DNA binding activity of the herpes simplex virus protein UL42. *J. Viol.* **69:** 6965–6971.

Coen, D.M., D.P. Aschman, P.T. Gelep, M.J. Retondo, S.K. Weller, and P.A. Schaffer. 1984. Fine mapping and molecular cloning of mutations in the herpes simplex virus DNA polymerase locus. *J. Virol.* **49:** 236–247.

Conley, A.J., D.M. Knipe, P. Jones, and B. Roizman. 1981. Molecular genetics of herpes simplex virus. VII. Characterization of a temperature-sensitive mutant produced by in vitro mutagenesis and defective in DNA synthesis and accumulation of gamma polypeptides. *J. Virol.* **37:** 191–206.

Crute, J.J. and I.R. Lehman. 1989. Herpes simplex-1 DNA polymerase. Identification of an intrinsic 5′ to 3′ exonuclease with ribonuclease H activity. *J. Biol. Chem.* **264:** 19266–19270.

———. 1991. Herpes simplex virus-1 helicase-primase. Physical and catalytic properties. *J. Biol. Chem.* **266:** 4484–4488.

Crute, J.J., E.S. Mocarski, and I.R. Lehman. 1988. A DNA helicase induced by herpes simplex virus type 1. *Nucleic Acids Res.* **16:** 6585–6596.

Crute, J.J., T. Tsurumi, L. Zhu, S.K. Weller, P.D. Olivo, M.D. Challberg, E.S. Mocarski, and I.R. Lehman. 1989. Herpes simplex virus 1 helicase-primase: A complex of three herpes-encoded gene products. *Proc. Natl. Acad. Sci.* **86:** 2186–2189.

Dabrowski, C.E. and P.A. Schaffer. 1991. Herpes simplex virus type 1 origin-specific binding protein: oriS-binding properties and effects of cellular proteins. *J. Virol.* **65:** 3140–3150.

Dabrowski, C.E., P.J. Carmillo, and P.A. Schaffer. 1994. Cellular protein interactions with herpes simplex virus type 1 oriS. *Mol. Cell. Biol.* **14:** 2545–2555.

de Bruyn Kops, A. and D.M. Knipe. 1988. Formation of DNA replication structures in herpes virus-infected cells requires a viral DNA binding protein. *Cell* **55:** 857–868.

Deb, S. and S.P. Deb. 1989. Analysis of Ori-S sequence of HSV-1: Identification of one functional DNA binding domain. *Nucleic Acids Res.* **17:** 2733–2752.

Digard, P. and D.M. Coen. 1990. The binding site on the herpes simplex virus polymerase for the viral UL42 protein. *J. Biol. Chem.* **265:** 17393–17396.

Digard, P., W.R. Bebrin, K. Weisshart, and D.M. Coen. 1993a. The extreme C terminus of herpes simplex virus DNA polymerase is crucial for functional interaction with processivity factor UL42 and for viral replication. *J. Virol.* **67:** 398–406.

Digard, P., C.S. Chow, L. Pirrit, and D.M. Coen. 1993b. Functional analysis of the herpes simplex virus UL42 protein. *J. Virol.* **67:** 1159–1168.

Dodson, M.S. and I.R. Lehman. 1991. Association of DNA helicase and primase activities with a subassembly of the herpes simplex virus 1 helicase-primase composed of the UL5 and UL52 gene products. *Proc. Natl. Acad. Sci.* **88:** 1105–1109.

———. 1993. The herpes simplex virus type I origin binding protein. *J. Biol. Chem.*. **268:** 1213–1219.

Dodson, M.S., J.J. Crute, R.C. Bruckner, and I.R. Lehman. 1989. Overexpression and assembly of the herpes simplex virus type-1 helicase-primase in insect cells. *J. Biol. Chem.* **264:** 20835–20838.

Dracheva, S., E. Koonin, and J.J. Crute. 1995. Identification of the primase active site of the herpes simplex virus type 1 helicase–primase. *J. Biol. Chem.* **270:** 14148–14153.

Dutch, R.E. and I.R. Lehman. 1993. Renaturation of complementary DNA strands by herpes simplex virus type 1 ICP8. *J. Virol.* **67:** 6945–6949.

Dutch, R.E., V. Bianchi, and I.R. Lehman. 1995. Herpes simplex virus type 1 DNA replication is specifically required for high-frequency homologous recombination between repeated sequences. *J. Virol.* **69:** 3084–3089.

Dutch, R.E., R.C. Bruckner, E.S. Mocarski, and I.R. Lehman. 1992. Herpes simplex virus type 1 recombination: Role of DNA replication and viral a sequences. *J. Virol.* **66:** 277–285.

Elias, P. and I.R. Lehman. 1988. Interaction of origin binding protein with an origin of replication of herpes simplex virus 1. *Proc. Natl. Acad. Sci.* **85:** 2959–2963.

Elias, P., C.M. Gustafsson, and O. Hammarsten. 1990. The origin binding protein of herpes simplex virus 1 binds cooperatively to the viral origin of replication. *J. Biol. Chem.* **265:** 17167–17173.

Elias, P., C.M. Gustafsson, O. Hammarsten, and N.D. Stow. 1992. Structural elements required for the cooperative binding of the herpes simplex virus origin binding protein to oriS reside in the N-terminal part of the protein. *J. Biol. Chem.* **267:** 17424–17429.

Ertl, P.F. and K.L. Powell. 1992. Physical and functional interaction of human cytomegalovirus DNA polymerase and its accessory protein (ICP36) expressed in insect cells. *J. Virol.* **66:** 4126–4133.

Fierer, D.S. and M.D. Challberg. 1992. Purification and characterization of UL9, the herpes simplex virus type 1 origin-binding protein. *J. Virol.* **66:** 3986–3995.

———. 1995. The stoichiometry of binding of the herpes simplex virus type 1 origin binding protein, UL9, to oriS. *J. Biol. Chem.* **270:** 7330–7334.

Fixman, E.D., G.S. Hayward, and S.D. Hayward. 1992. *trans*-Acting requirements for replication of Epstein-Barr virus ori-Lyt. *J. Virol.* **66:** 5030–5039.

———. 1995. Replication of Epstein-Barr virus oriLyt: Lack of a dedicated virally encoded origin-binding protein and dependence on Zta in cotransfection assays. *J. Virol.* **69:** 2998–3006.

Frenkel, N., H. Locker, W. Batterson, G.S. Hayward, and B. Roizman. 1976. Anatomy of herpes simplex virus DNA. VI. Defective DNA originates from the S component. *J. Virol.* **20:** 527–531.

Frenkel, N., R.J. Jacob, R.W. Honess, G.S. Hayward, H. Locker, and B. Roizman. 1975. Anatomy of herpes simplex virus DNA. III. Characterization of defective DNA molecules and biological properties of virus populations containing them. *J. Virol.* **16:** 153–167.

Gallo, M.L., D.H. Jackwood, M. Murphy, H.S. Marsden, and D.S. Parris. 1988. Purification of the herpes simplex virus type 1 65-kilodalton DNA-binding protein: Properties of the protein and evidence of its association with the virus-encoded DNA polymerase. *J. Virol.* **62:** 2874–2883.

Gao, M., S.F. DiTusa, and M.G. Cordingley. 1993. The C-terminal third of UL42, a HSV-1 DNA replication protein, is dispensable for viral growth. *Virology* **194:** 647–653.

Garber, D.A., S.M. Beverley, and D.M. Coen. 1993. Demonstration of circularization of herpes simplex virus DNA following infection using pulsed field gel electrophoresis. *Virology* **197:** 459–462.

Goldstein, D.J. and S.K. Weller. 1988. An ICP6::lacZ insertion mutagen is used to demonstrate that the *UL52* gene of herpes simplex virus type 1 is required for virus growth and DNA synthesis. *J. Virol.* **62:** 2970–2977.

Gorbalenya, A.E., E.V. Koonin, A.P. Donchenko, and V.M. Blinov. 1988. A conserved NTP-motif in putative helicases. *Nature* **333:** 22.

———. 1989. Two related superfamilies of putative helicases involved in replication, repair and expression of DNA and RNA genomes. *Nucleic Acids Res.* **17:** 4713–4730.

Gottlieb, J. and M.D. Challberg. 1994. Interaction of herpes simplex virus type 1 DNA polymerase and the UL42 accessory protein with a model primer template. *J. Virol.* **68:** 4937–4945.

Gottlieb, J., A.I. Marcy, D.M. Coen, and M.D. Challberg. 1990. The herpes simplex virus type 1 *UL42* gene product: A subunit of DNA polymerase that functions to increase processivity. *J. Virol.* **64:** 5976–5987.

Gustafsson, C.M., O. Hammarsten, M. Falkenberg, and P. Elias. 1994. Herpes simplex virus DNA replication: A spacer sequence directs the ATP-dependent formation of a nucleoprotein complex at oriS. *Proc. Natl. Acad. Sci.* **91:** 4629–4633.

Gustafsson, C.M., M. Falkenberg, S. Simonsson, H. Valadi, and P. Elias. 1995. The DNA ligands influence the interactions between the herpes simplex virus 1 origin binding protein and the single strand DNA-binding protein, ICP-8. *J. Biol. Chem.* **270:** 19028–19034.

Hamatake, R.K., M. Bifano, D.J. Tenney, W.W. Hurlburt, and M.G. Cordingley. 1993. The herpes simpex virus type 1 DNA polymerase accessory protein, UL42, contains a functional protease-resistant domain. *J. Gen. Virol.* **74:** 2181–2189.

Hamzeh, F.M., P.S. Lietman, W. Gibson, and G.S. Hayward. 1990. Identification of the lytic origin of DNA replication in human cytomegalovirus by a novel approach utilizing ganciclovir-induced chain termination. *J. Virol.* **64:** 6184–6195.

Hardwicke, M.A. and P.A. Schaffer. 1995. Cloning and characterization of herpes simplex virus type 1 oriL: Comparison of replication and protein-DNA complex formation by oriL and oriS. *J. Virol.* **69:** 1377–1388.

Hazuda, D.J., H.C. Perry, and W.L. McClements. 1992. Cooperative interactions between replication origin-bound molecules of herpes simplex virus origin-binding protein are mediated via the aminoterminus of the protein. *J. Biol. Chem.* **267:** 14309–14315.

Hazuda, D.J., H.C. Perry, A.M. Naylor, and W.L. McClements. 1991. Characterization of the herpes simplex virus origin binding protein interaction with oriS. *J. Biol. Chem.* **266:** 24621–24626.

Heilbronn, R. and H. zur Hausen. 1989. A subset of herpes simplex virus replication genes induces DNA amplification within the host cell genome. *J. Virol.* **63:** 3683–3692.

Hernandez, T.R. and I.R. Lehman. 1990. Functional interaction between the herpes simplex-1 DNA polymerase and UL42 protein. *J. Biol. Chem.* **265:** 11227–11232.

Hernandez, T.R., R.E. Dutch, I.R. Lehman, C. Gustafsson, and P. Elias. 1991. Mutations in a herpes simplex virus type 1 origin that inhibit replication. *J. Virol.* **65:** 1649–1652.

Hodgman, T.C. 1988. A new superfamily of replicative proteins. *Nature* **333:** 22–23.

Honess, R.W. and B. Roizman. 1973. Proteins specified by herpes simplex virus. XI. Identification and relative molar rates of synthesis of structural and nonstructural herpes virus polypeptides in the infected cell. *J. Virol.* **12:** 1346–1365.

Honess, R.W., A. Buchan, I.W. Halliburton, and D.H. Watson. 1980. Recombination and linkage between structural and regulatory genes of herpes simplex virus type 1: Study of the functional organization of the genome. *J. Virol.* **34:** 716–742.

Hubenthal-Vass, J., L. Starr, and B. Roizman. 1987. The herpes simplex virus origins of DNA synthesis in the S component are each contained in a transcribed open reading frame. *J. Virol.* **61:** 3349–3355.

Igarashi, K., R. Fawl, R.J. Roller, and B. Roizman. 1993. Construction and properties of a recombinant herpes simplex virus 1 lacking both S-component origins of DNA synthesis. *J. Virol.* **67:** 2123–2132.

Jacob, R. and B. Roizman. 1977. Anatomy of herpes simplex virus DNA. VIII. Properties of the replicating DNA. *J. Virol.* **23:** 394–411.

Jacob, R.J., L.S. Morse, and B. Roizman. 1979. Anatomy of herpes simplex virus DNA. XII. Accumulation of head-to-tail concatemers in nuclei of infected cells and their role in the generation of the four isomeric arrangements of viral DNA. *J. Virol.* **29:** 448–457.

Johnson, P.A., M.G. Best, T. Friedman, and D.S. Parris. 1991. Isolation of a herpes

simplex virus type 1 mutant deleted for the essential *UL42* gene and characterization of its null phenotype. *J. Virol.* **65:** 700–710.
Jongeneel, C.V. and S.L. Bachenheimer. 1981. Structure of replicating herpes simplex virus DNA. *J. Virol.* **39:** 656–660.
Klinedinst, D.K. and M.D. Challberg. 1994. Helicase-primase complex of herpes simplex virus type 1: A mutation in the UL52 subunit abolishes primase activity. *J. Virol.* **68:** 3693–3701.
Knopf, C.W. and K. Weisshart. 1990. Comparison of exonucleolytic activities of herpes simplex virus type-1 DNA polymerase and DNase. *Eur. J. Biochem.* **191:** 263–273.
Knopf, C.W., B. Spies, and H.C. Kaerner. 1986. The DNA replication origins of herpes simplex virus type 1 strain Angelotti. *Nucleic Acids Res.* **14:** 8655–8667.
Knopf, K.W. 1979. Properties of herpes simplex virus DNA polymerase and characterization of its associated exonuclease activity. *Eur. J. Biochem.* **98:** 231–244.
Koff, A. and P. Tegtmeyer. 1988. Characterization of major recognition sequences for a herpes simplex virus type 1 origin-binding protein. *J. Virol.* **62:** 4096–4103.
Koff, A., J.F. Schwedes, and P. Tegtmeyer. 1991. Herpes simplex virus origin-binding protein (UL9) loops and distorts the viral replication origin. *J. Virol.* **65:** 3284–3292.
Kuriyan, J. and M. O'Donnell. 1993. Sliding clamps of DNA polymerases. *J. Mol. Biol.* **234:** 915–925.
Lane, D. 1988. Enlarged family of putative helicases. *Nature* **334:** 478.
Lee, S.S.-K., Q. Dong, T.S.-F. Wang, and I.R. Lehman. 1995. Interaction of herpes simplex virus 1 origin-binding protein with DNA polymerase α. *Proc. Natl. Acad. Sci.* **92:** 7882–7886.
Leinbach, S.S. and W.C. Summers. 1980. The structure of herpes simplex virus type 1 DNA as probed by micrococcal nuclease digestion. *J. Gen. Virol.* **51:** 45–59.
Lockshon, D. and D.A. Galloway. 1988. Sequence and structural requirements of a herpes simplex viral DNA replication origin. *Mol. Cell. Biol.* **8:** 4018–4027.
Marchetti, M.E., C.A. Smith, and P.A. Schaffer. 1988. A temperature-sensitive mutation in an HSV-1 gene required for viral DNA synthesis maps to coordinates 0.609-0.614 in UL. *J. Virol.* **62:** 715–721.
Marcy, A.I., C.B.C. Hwang, K.L. Ruffner, and D.M. Coen. 1990a. Engineered herpes simplex virus DNA polymerase point mutants: The most highly conserved region shared among alpha-like DNA polmerases is involved in substrate recognition. *J. Virol.* **64:** 5883–5890.
Marcy, A.I., P.D. Olivo, M.D. Challberg, and D.M. Coen. 1990b. Enzymatic activities of overexpressed herpes simplex virus DNA polymerase purified from recombinant baculovirus-infected insect cells. *Nucleic Acids Res.* **18:** 1207–1215.
Marians, K.J. 1992. Prokaryotic DNA replication. *Annu. Rev. Biochem.* **61:** 673–719.
Marsden, H.S., E.M. Cambell, L. Haarr, M.C. Frame, D.S. Parris, M. Murphy, R.G. Hope, M.T. Muller, and C.M. Preston. 1987. The 65,000 M_r DNA-binding and virion trans-inducing proteins of herpes simplex virus type 1. *J. Virol.* **61:** 2428–2437.
Martin, D.W., R.M. Muñoz, D. Oliver, M.A. Subler, and S. Deb. 1994. Analysis of the DNA-binding domain of the HSV-1 origin-binding protein. *Virology* **198:** 71–80.
Martinez, R. and C.A. Edwards. 1993. Expression, purification, and functional characterization of the DNA-binding domain of the herpes simplex virus type 1 UL9 protein. *Protein Expr. Purif.* **4:** 32–37.
Martinez, R., L. Shao, and S.K. Weller. 1992. The conserved helicase motifs of the herpes simplex virus type 1 origin-binding protein UL9 are important for function. *J.*

Virol. **66:** 6735–6746.

Masse, M.J., S. Karlin, G.A. Schachtel, and E.S. Mocarski. 1992. Human cytomegalovirus origin of DNA replication (oriLyt) resides within a highly complex repetitive region. *Proc. Natl. Acad. Sci.* **89:** 5246–5250.

McGeoch, D.J., M.A. Dalrymple, A. Dolan, D. McNab, L.J. Perry, P. Taylor, and M.D. Challberg. 1988. Structures of herpes simplex virus type 1 genes required for replication of virus DNA. *J. Virol.* **62:** 444–453.

McLean, G.W., A.P. Abbotts, M.E. Parry, H.S. Marsden, and N.D. Stow. 1994. The herpes simplex virus type 1 origin-binding protein interacts specifically with the viral UL8 protein. *J. Gen. Virol.* **75:** 2699–2706.

McVoy, M.A. and S.P. Adler. 1994. Human cytomegalovirus DNA replicates after early circularization by concatemer formation, and inversion occurs within the concatemer. *J. Virol.* **68:** 1040–1051.

Mocarski, E.S. and B. Roizman. 1982. Structure and role of the herpes simplex virus DNA termini in inversion, circularization and generation of virion DNA. *Cell* **31:** 89–97.

Monahan, S.J., T.F. Barlam, C.S. Crumpacker, and D.S. Parris. 1993. Two regions of the herpes simplex virus type 1 UL42 protein are required for its functional interaction with the viral DNA polymerase. *J. Virol.* **67:** 5922–5931.

Murchie, M.J. and D.J. McGeoch. 1982. DNA sequence analysis of an immediate-early gene region of the herpes simplex virus type 1 genome (map coordinates 0.950 to 0.978). *J. Gen. Virol.* **62:** 1–15.

O'Donnell, M.E., P. Elias, and I.R. Lehman. 1987. Processive replication of single-stranded DNA templates by the herpes simplex virus-induced DNA polymerase. *J. Biol. Chem.* **262:** 4252–4259.

Olivo, P.D., N.J. Nelson, and M.D. Challberg. 1988. Herpes simplex virus DNA replication: The UL9 gene encodes an origin binding protein. *Proc. Natl. Acad. Sci.* **85:** 5414–5418.

Orberg, P.K. and P.A. Schaffer. 1987. Expression of herpes simplex virus type 1 major DNA-binding protein, ICP8, in transformed cell lines: Complementation of deletion mutants and inhibition of wild-type virus. *J. Virol.* **61:** 1136–1146.

Pari, G.S. and D.G. Anders. 1993. Eleven loci encoding *trans*-acting factors are required for transient complementation of human cytomegalovirus oriLyt-dependent DNA replication. *J. Virol.* **67:** 6979–6988.

Pari, G.S., M.A. Kacica, and S.D.G. Ander. 1993. Open reading frames UL44, IRS1/TRS1, and UL36-38 are required for transient complementation of human cytomegalovirus oriLyt-dependent DNA synthesis. *J. Virol.* **67:** 2575–2582.

Parry, M.E., N.D. Stow, and H.S. Marsden. 1993. Purification and properties of the herpes simplex virus type 1 UL8 protein. *J. Gen. Virol.* **74:** 607–612.

Perry, H.C., D.J. Hazuda, and W.L. McClements. 1993. The DNA binding domain of herpes simplex virus type 1 origin binding protein is a transdominant inhibitor of virus replication. *Virology* **193:** 73–79.

Poffenberger, K.L. and B. Roizman. 1985. A noninverting genome of a viable herpes simplex virus 1: Presence of head-to-tail linkages in packaged genomes and requirements for circularization after infection. *J. Virol.* **53:** 587–595.

Polvino-Bodnar, M., P.D. Orberg, and P.A. Schaffer. 1987. Herpes simplex virus type 1 oriL is not required for virus replication or for the establishment and reactivation of latent infection in mice. *J. Virol.* **61:** 3528–3535.

Powell, K.L. and R.J. Courtney. 1975. Polypeptides synthesized in herpes simplex virus type 2-infected Hep-2 cells. *Virology* **66:** 217–228.

Powell, K.L. and D.J.M. Purifoy. 1976. DNA-binding proteins induced by herpes simplex virus type 2 in Hep-2 cells. *J. Virol.* **19:** 717–731.

———. 1977a. Nonstructural proteins of herpes simplex virus. I. Purification of the induced DNA polymerase. *J. Virol.* **24:** 618–626.

———. 1977b. Identification of the herpes simplex virus DNA polymerase gene. *Nature* **269:** 621–623.

Powell, K.L., E. Littler, and D.J.M. Purifoy. 1981. Nonstructural proteins of herpes simplex virus. II. Major virus-specific DNA-binding protein. *J. Virol.* **39:** 894–902.

Purifoy, D.J.M. and K.L. Powell. 1981. Temperature-sensitive mutants in two distinct complementation groups of herpes simplex virus type 1 specify thermolabile DNA polymerase. *J. Gen. Virol.* **54:** 219–222.

Purifoy, D.J.M., R.B. Lewis, and K.L. Powell. 1977. Identification of the herpes simplex virus DNA polymerase gene. *Nature* **269:** 621–623.

Quinlan, M.P., L.B. Chen, and D.M. Knipe. 1984. The intranuclear location of a herpes simplex virus DNA-binding protein is determined by the status of viral DNA replication. *Cell* **36:** 857–868.

Rabkin, S.D. and B. Hanlon. 1990. Herpes simplex virus DNA synthesis at a preformed replication fork in vitro. *J. Virol.* **64:** 4957–4967.

Ruyechan, W.T. 1983. The major herpes simplex virus DNA-binding protein holds single-stranded DNA in an extended configuration. *J. Virol.* **46:** 661–666.

Ruyechan, W.T. and A.C. Weir. 1984. Interaction with nucleic acids and stimulation of the viral DNA polymerase by the herpes simplex virus type 1 major DNA-binding protein. *J. Virol.* **52:** 727–733.

Schaffer, P.A., M.J. Tevethia, and M. Benyesh-Melnick. 1974. Recombination between temperature-sensitive mutants of herpes simplex virus type 1. *Virology* **58:** 219–228.

Schepers, A., D. Pich, and W. Hammerschmidt. 1993a. A transcription factor with homology to the AP-1 family links RNA transcription and DNA replication in the lytic cycle of Epstein-Barr virus. *EMBO J.* **12:** 3921–3929.

Schepers, A., D. Pich, J. Mankertz, and W. Hammerschmidt. 1993b. *cis*-Acting elements in the lytic origin of DNA replication of Epstein-Barr virus. *J. Virol.* **67:** 4237–4245.

Schroder, C.H., B. Stegmann, H.F. Lauppe, and H.C. Kaerner. 1975. An unusual defective genotype derived from herpes simplex virus strain ANG. *Intervirology* **46:** 270–284.

Severini, A., A.R. Morgan, D.R. Tovell, and D.L. Tyrrell. 1994. Study of the structure of replicative intermediates of HSV-1 DNA by pulsed-field gel electrophoresis. *Virology* **200:** 428–435.

Shao, L., L.M. Rapp, and S.K. Weller. 1993. Herpes simplex virus 1 alkaline nuclease is required for efficient egress of capsids from the nucleus. *Virology* **196:** 146–162.

Sherman, G., J. Gottlieb, and M.D. Challberg. 1992. The UL8 subunit of the herpes simplex virus helicase-primase complex is required for efficient primer utilization. *J. Virol.* **66:** 4884–4892.

Skaliter, R. and I.R. Lehman. 1994. Rolling circle DNA replication in vitro by a complex of herpes simplex virus type 1-encoded enzymes. *Proc. Natl. Acad. Sci.* **91:** 10665–10669.

Smiley, J.R., M.J. Wagner, W.P. Summers, and W.C. Summers. 1980. Genetic and physical evidence for the polarity of transcription of the thymidine kinase gene of herpes simplex virus. *Virology* **102:** 83–93.

Spaete, R.R. and N. Frenkel. 1982. The herpes simplex virus amplicon: A new eucaryotic defective-virus cloning-amplifying vector. *Cell* **30:** 295–304.

Stabell, E.C. and P.D. Olivo. 1993. A truncated herpes simplex virus origin binding protein which contains the carboxyl teminal origin binding domain binds to the origin of replication but does not alter its conformation. *Nucleic Acids Res.* **21:** 5203–5211.

Stow, N.D. 1982. Localization of an origin of DNA replication within the TRs/IRs repeated region of the herpes simplex virus type 1 genome. *EMBO J.* **1:** 863–867.

———. 1993. Sequences at the C-terminus of the herpes simplex virus type1 UL30 protein are dispensable for DNA polymerase activity but not for viral origin-dependent DNA replication. *Nucleic Acids Res.* **21:** 87–92.

Stow, N.D. and E.C. McMonagle. 1983. Characterization of the TRs/IRs origin of DNA replication of herpes simplex virus type 1. *Virology* **130:** 427–438.

Tabor, S., H.E. Huber, and C.C. Richardson. 1987. *Escherichia coli* thioredoxin confers processivity on the DNA polymerase activity of the gene 5 protein of bacteriophage T7. *J. Biol. Chem.* **262:** 16212–16223.

Tenney, D.J., W.W. Hurlburt, P. Micheletti, M. Bifano, and R.K. Hamatake. 1994. The UL8 component of the herpes simplex virus helicase-primase complex stimulates primer synthesis by a subassembly of the UL5 and UL52 components. *J. Biol. Chem.* **269:** 5030–5035.

Tenney, D.J., K. Sheaffer, W.W. Hurlburt, M. Bifano, and R. Hamatake. 1995. Sequence-dependent primer synthesis by the herpes simplex virus helicase-primase complex. *J. Biol. Chem.* **270:** 9129–9136.

Tenney, D.J., W.W. Hurlburt, M. Bifano, J.T. Stevens, P.A. Micheletti, R.K. Hamatake, and M.G. Cordingley. 1993a. Deletions of the carboxy terminus of herpes simplex virus type 1 UL42 define a conserved amino-terminal functional domain. *J. Virol.* **67:** 1959–1966.

Tenney, D.J., P.A. Micheletti, J.T. Stevens, R.K. Hamatake, J.T. Matthews, A.R. Sanchez, W.W. Hurlburt, M. Bifano, and M.G. Cordingley. 1993b. Mutations in the C-terminus of herpes simplex virus type 1 DNA polymerase can affect binding and stimulation by its accessory protein UL42 without affecting basal polymerase activity. *J. Virol.* **67:** 543–547.

Thomas, M.S., M. Gao, D.M. Knipe, and K.L. Powell. 1992. Association between the herpes simplex virus major DNA-binding protein and alkaline nuclease. *J. Virol.* **66:** 1152–1161.

Tsurumi, T. 1993. Purification and characterization of the DNA-binding activity of the Epstein-Barr virus DNA polymerase accessory protein BMRF1 gene products, as expressed in insect cells by using the baculovirus system. *J. Virol.* **67:** 1681–1687.

Tsurumi, T., T. Daikoku, and Y. Nishiyama. 1994. Further characterization of the interaction between the Epstein-Barr virus DNA polymerase catalytic subunit and its accessory subunit with regard to the 3′-to-5′ exonucleolytic activity and stability of initiation complex at primer terminus. *J. Virol.* **68:** 3354–3363.

Tsurumi, T., T. Daikoku, K. Ryutaro, and Y. Nishiyama. 1993a. Functional interaction between Epstein-Barr virus DNA polymerase catalytic subunit and its accessory subunit in vitro. *J. Virol.* **67:** 7648–7653.

Tsurumi, T., A. Kobayashi, K. Tamai, T. Daikoku, R. Kurachi, and Y. Nishiyama. 1993b. Functional expression and characterization of the Epstein-Barr virus DNA polymerase catalytic subunit. *J. Virol.* **67:** 4651–4658.

Vaughan, P.J., D.J.M. Purifoy, and K.L. Powell. 1985. DNA-binding protein associated with herpes simplex virus DNA polymerase. *J. Virol.* **53:** 501–508.

Vaughan, P.J., L.M. Banks, D.J.M. Purifoy, and K.L. Powell. 1984. Interactions between herpes simplex virus DNA-binding proteins. *J. Gen. Virol.* **65:** 2033–2041.

Vlasny, D.A. and N. Frenkel. 1982. Replication of herpes simplex virus DNA: Localization of replication recognition signals within defective virus genomes. *Proc. Natl. Acad. Sci.* **78:** 742–746.

Weber, P.C., M. Levine, and J.C. Glorioso. 1990. Recombinogenic properties of herpes simplex virus type-1 DNA sequences resident in simian virus-40 minichromosomes. *J. Virol.* **64:** 300–306.

Weber, P.C., M.D. Challberg, N.J. Nelson, M. Levine, and J.C. Glorioso. 1988. Inversion events in the HSV-1 genome are directly mediated by the viral DNA replication machinery and lack sequence specificity. *Cell* **54:** 369–381.

Weir, H.M. and N.D. Stow. 1990. Two binding sites for the herpes simplex virus type 1 UL9 protein are required for efficient activity of the oriS replication origin. *J. Gen. Virol.* **71:** 1379–1385.

Weir, H.M., J.M. Calder, and N.D. Stow. 1989. Binding of the herpes simplex virus type 1 UL9 gene product to an origin of viral DNA replication. *Nucleic Acids Res.* **17:** 1409–1425.

Weller, S.K., E.P. Carmichael, D.P. Aschman, D.J. Goldstein, and P.A. Schaffer. 1987. Genetic and phenotypic characterization of mutants in four essential genes that map to the left half of HSV-1 UL DNA. *Virology* **161:** 198–210.

Weller, S.K., A. Spadaro, J.E. Schaffer, A.W. Murray, A.M. Maxam, and P.A. Schaffer. 1985. Cloning, sequencing, and functional analysis of oriL, a herpes simplex virus type 1 origin of DNA synthesis. *Mol. Cell. Biol.* **5:** 930–942.

Werstuck, G., P. Bilan, and J.P. Capone. 1990. Enhanced infectivity of herpes simplex virus type-1 viral DNA in a cell line expressing the *trans*-inducing factor Vmw65. *J. Virol.* **64:** 984–991.

Wong, S.W. and P.A. Schaffer. 1991. Elements in the transcriptional regulatory region flanking herpes simplex virus type 1 oriS stimulate origin function. *J. Virol.* **65:** 2601–2611.

Wu, C.A., N.J. Nelson, D.J. McGeoch, and M.D. Challberg. 1988. Identification of herpes simplex virus type 1 genes required for origin-dependent DNA synthesis. *J. Virol.* **62:** 435–443.

Xu, B. and D.A. Clayton. 1995. A persistent RNA-DNA hybrid is formed during transcription at a phylogenetically conserved mitochondrial DNA sequence. *Mol. Cell. Biol.* **15:** 580–589.

Yamada, Y., N. Yamamoto, K. Maeno, and Y. Nishiyama. 1990. Role of DNA topoisomerase-I in the replication of herpes simplex virus type-2. *Arch. Virol.* **110:** 121–127.

Zhang, X., S. Efstathiou, and A. Simmons. 1994. Identification of novel herpes simplex virus replicative intermediates by field inversion gel electrophoresis: Implications for viral DNA amplification strategies. *Virology* **202:** 530–539.

Zhu, L. and S.K. Weller. 1988. UL5, a protein required for HSV DNA synthesis: Genetic analysis, overexpression in *Escherichia coli*, and generation of polyclonal antibodies. *Virology* **166:** 366–378.

———. 1992a. The UL5 gene of herpes simplex virus type 1: Isolation of a lacZ insertion mutant and association of UL5 gene product with other members of the helicase-

primase complex. *J. Virol.* **66:** 458–468.
———. 1992b. The six conserved helicase motifs of the UL5 gene product, a component of the herpes simplex virus type 1 helicase-primase, are essential for its function. *J. Virol.* **66:** 469–479.

26
Epstein-Barr Virus DNA Replication

John L. Yates
Department of Human Genetics
Roswell Park Cancer Institute
Buffalo, New York 14263

Epstein-Barr virus (EBV) and its close relatives are the only known viruses that have two fully independent systems for replicating their genomes, one that supports virus production during active or "lytic" infection, and a second one that operates during latent stages of infection to allow the circular viral genomes to be duplicated during each cell division cycle. Although EBV's system of lytic DNA replication is for the most part a heritage of the herpesvirus family, its system for replication during latency is entirely its own. During latent infection, while the genes required for lytic replication and virus production are silent, the circular EBV chromosome is replicated by the cell's DNA replication machinery, apparently only once during each S phase. This arrangement of having the host cell replicate its genome during latency is probably related to the fact that EBV establishes latency in cells that are prone to divide. EBV appears to maintain its life-long infection of people by residing within B cells (Klein 1994; Miyashita et al. 1995), and, during the initial phase of latent infection of B cells, EBV expresses a small set of genes that together cause the cells to proliferate (Kieff and Leibowitz 1990; Miller 1990). EBV DNA replication during latency might also be important for the eventual productive infection of epithelial cells, because EBV appears to be most able to infect undifferentiated epithelial cells but to replicate lytically in differentiated cells (Sixbey 1989).

EBV replication during latency relies almost entirely on the cell. A single EBV-encoded protein, EBNA1, apparently guides replication to initiate at *oriP* on the EBV genome—more specifically, at one of two regions of *oriP* to which EBNA1 binds (Fig. 1). All enzymatic steps of replication, including the initial unwinding of duplex DNA, must be performed by cellular factors under cellular control. Initiation of replication on the EBV chromosome has been found to occur not only at *oriP*, but also at sites well away from *oriP*, over a broad region not dissimilar from the zones of initiation that have been observed at replication origins of mammalian chromosomes (Little and Schildkraut 1995). This and other

DNA Replication in Eukaryotic Cells

evidence suggests that the specification of an initiation site by *oriP* may not be essential for replication of EBV genomes in latently infected cells. Instead, a second function of *oriP,* which provides for prolonged survival of DNAs in cells independently of replication, could be the central mechanism that stably maintains EBV chromosomes in proliferating cells.

Lytic replication for EBV initiates at specific lytic-phase origins and yields thousands of copies of viral genomes per cell for assembly into virus particles. EBV encodes an ensemble of lytic replication enzymes and accessory factors common to all herpes family viruses. For this reason, it used to be assumed that, at this final stage of infection, EBV could amplify its genome without depending on cellular factors. Yet recent studies have led to the conclusion that EBV lacks the means to initiate its lytic replication and, instead, relies heavily on the cell to conduct this critical step to end latency.

EBNA1-DEPENDENT PLASMID REPLICATION AND MAINTENANCE

Multiple Functions of *oriP* and EBNA1

oriP was identified by systematically testing cloned segments of the EBV genome for the ability to support stable autonomous maintenance

Figure 1 EBV replication during latent infection. (*A*) The positions of *oriP* and the two copies of *oriLyt* on the EBV circularized chromosome are shown relative to the joined terminal repeats (TR). During latent infection, replication initiates at *oriP*, but initiation has also been observed at sites dispersed over a large region, as indicated by the hyphenated arcs at the lower left. In non-B cells and in a restricted state of latency in B cells, the EBNA1 gene is expressed from an autoregulated promoter (*bent arrow*) located over 40 kbp upstream of its coding region. In EBV-immortalized B cells, promoters much farther upstream (*pair of bent arrows*) give rise to regulated levels of EBNA1 synthesis. (*B*) The bipartite structure of *oriP* and a comparison between EBV and a close relative, HVP. (DS) dyad symmetry region; (FR) family of repeats. Each black circle, or oval in the enlargements, represents a binding site for an EBNA1 dimer. At the OBR of both viruses, the spacing of 21 bp exists between neighboring pairs of binding sites. The two viruses have different arrangements of EBNA1-binding sites at the left end of *oriP*. The EBNA1-binding sites were predicted for the HVP sequence using the binding requirements that were determined experimentally for EBNA1 of EBV (Ambinder et al. 1990). The first site, at position 2089 (Loeb et al. 1990), might have low affinity and thus is indicated with an open circle. A partial tenth binding site was originally indicated by Loeb et al. at the left end of this cluster (position 2332), but it is not shown here because it is not expected to have significant affinity for EBNA1.

of plasmids under selection in cells that were latently infected by EBV (Yates et al. 1984). A single EBV-encoded protein, EBNA1, was found to be both necessary and sufficient for *oriP* function (Yates et al. 1985; Lupton and Levine 1985). Stable maintenance of *oriP*-containing plasmids in proliferating cells seems to involve two EBNA1-dependent functions: initiation of DNA replication and a distinct, plasmid-maintenance function that can be demonstrated to operate in the absence of replication. Correspondingly, *oriP* has two essential regions, each containing multiple EBNA1-binding sites and separated by almost 1000 bp of nonessential DNA (Reisman et al. 1985; Lupton and Levine 1985). *oriP* of herpesvirus papio (HVP), an EBV-like virus that infects baboons, has two analogous clusters of EBNA1-binding sites, implying a conserved, bipartite functional arrangement (Fig. 1B) (Loeb et al. 1990).

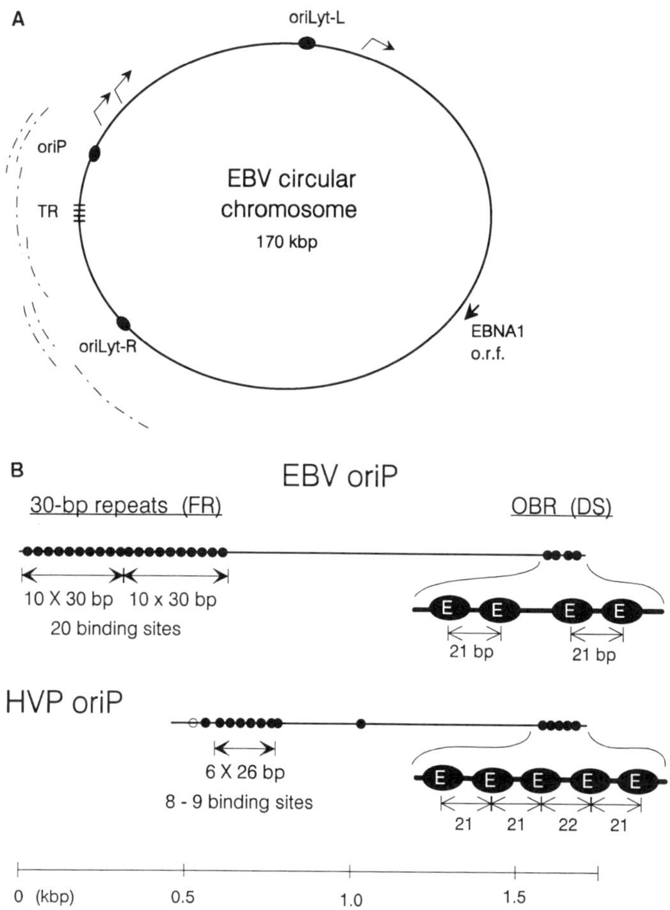

Figure 1 (See facing page for legend.)

One of *oriP*'s essential regions contains four EBNA1-binding sites, suitably arranged to serve as an origin of bidirectional replication, or OBR (Gahn and Schildkraut 1989; Wysokenski and Yates 1989; Harrison et al. 1994). The other essential region is an array of high-affinity EBNA1-binding sites within a family of 30-bp repeats, which form a large, stable complex with EBNA1 (Jones et al. 1989; Ambinder et al. 1990). The 30-bp repeats activate replication initiation at the OBR (Reisman et al. 1985; Wysokenski and Yates 1989), and they are also responsible for the nonreplicational activities associated with *oriP*. The 30-bp repeats, in the presence of EBNA1, prevent plasmids from being lost rapidly from dividing cells, and indications are that this replication-independent maintenance function is as important as replication for episomal maintenance in mammalian cells (Reisman et al. 1985; Krysan et al. 1989). The 30-bp repeats also act as an EBNA1-dependent transcriptional enhancer, a property of unknown significance to replication or maintenance of the viral genome (Reisman and Sugden 1986; Sugden and Warren 1989; Gahn and Sugden 1995).

Replication from *oriP*

The position within *oriP* from which replication forks originate and diverge bidirectionally, the OBR, was found by Gahn and Schildkraut, using two-dimensional gel electrophoresis, to coincide with the functional element of *oriP* containing four EBNA1-binding sites, within a resolution of several hundred bp (Gahn and Schildkraut 1989). This functional element was originally named the "dyad symmetry region" for a large inverted repeat that spans two of the EBNA1-binding sites (Reisman et al. 1985). In the homologous region of *oriP* of HVP, no significant amount of dyad symmetry is evident except that which exists among the EBNA1-binding sites due to the inherent symmetry of each site, as shown in Figure 1. I prefer OBR as a more descriptive term for this functional element of *oriP*, keeping in mind the possibility that the physical origin and this functional cluster of EBNA1-binding sites might not coincide precisely.

Why does replication initiate at the OBR and not at the 30-bp repeats where EBNA1-binding sites are more numerous? Evidence favors a requirement for a proper arrangement of EBNA1-binding sites, as well as a preference for an optimal number of sites (Wysokenski and Yates 1989; Platt et al. 1993; Harrison et al. 1994). To date, no evidence has been found that any protein other than EBNA1 must interact with the OBR in a sequence-specific manner to support replication.

oriP of EBV and of HVP are likely to function through similar mechanisms, since *oriP* of either virus is active when supported by EBNA1 from either virus (Loeb et al. 1990; J.L. Yates et al., unpubl.). Interestingly, the only obvious features that are common to the OBRs of the two viruses are (1) the presence of EBNA1-binding sites and (2) a center-to-center spacing of 21 nucleotides for adjacent pairs of sites. For EBV, the four EBNA1-binding sites have spacings of 21 bp, 30 bp, and 21 bp, and thus can be viewed as two pairs of sites with each pair having a 21-bp spacing. For HVP, the five EBNA1-binding sites have spacings of 21 bp, 21 bp, 22 bp, and 21 bp for neighboring sites (Fig. 1). The B-form DNA double helix completes two full turns in 21 bp, so EBNA1 dimers bind to these adjacent sites on the same side of the double helix. Mutational studies have shown for EBV that either the left or the right pair of sites spaced 21 bp apart can support a significant level of OBR activity in the absence of the other pair of sites, whereas two improperly spaced sites (16 bp, 26 bp, or 30 bp apart) are inactive (Harrison et al. 1994). However, unpublished studies in my own laboratory have shown that the proximity of a third site to either active pair (regardless of the precise distance) improves the efficiency of replication and that all four sites are required for full activity. Binding of EBNA1 to the OBR causes a distortion of the DNA double helix, presumably a bend or a twist, at each pair of 21-bp-spaced EBNA1 sites, as revealed by increased reactivity of specific thymine bases to oxidation by permanganate, both in vitro and in vivo (Frappier and O'Donnell 1992; Hearing et al. 1992; Hsieh et al. 1993).

The above observations suggest that a specific geometric arrangement of EBNA1 molecules and bound DNA forms a substrate which allows cellular proteins to initiate DNA unwinding and replication. DNA of the OBR is relatively easily unwound, a feature common to replication origins of several organisms (Williams and Kowalski 1993).

Most studies have found that *oriP*-specific replication requires the 30-bp repeats, which activate the OBR with little regard to distance or orientation between the two functional elements (Reisman et al. 1985; Wysokenski and Yates 1989). One study found that when certain cell lines were used for transient transfection assays, the OBR was fully active without being linked to the 30-bp repeats (Harrison et al. 1994). Several attempts to reproduce this result in my own laboratory have failed, but the result suggests that there might be conditions under which EBNA1 can activate the OBR without the aid of the 30-bp repeats. EBNA1 binds to the OBR with lower affinity than to the 30-bp repeats

(Jones et al. 1989; Ambinder et al. 1990; Frappier and O'Donnell 1991a). EBNA1 dimers bound to the 30-bp repeats appear to coalesce into a stable spherical mass as viewed by electron microscopy, and this complex interacts with EBNA1 bound at the OBR to form a DNA "loop." This interaction between the two EBNA1/DNA complexes stabilizes the association of EBNA1 with the OBR (Frappier and O'Donnell 1991b; Su et al. 1991; Middleton and Sugden 1992). The 30-bp repeats act in a highly cooperative manner to activate replication and to form a transcriptional enhancer, with at least seven repeats (EBNA1-binding sites) needed to function effectively. In contrast to the spatial requirements of the OBR, a sufficient number of EBNA1-binding sites can assume the activities of the 30-bp repeats with substantial indifference to their spatial arrangement (Chittenden et al. 1989; Wysokenski and Yates 1989). The homologous cluster of EBNA1-binding sites at the left end of *oriP* of HVP is arranged quite differently than within the family repeats of *oriP* of EBV (Fig. 1).

Pausing and Termination

The large, stable complex formed by EBNA1 at the 30-bp repeats causes replication forks to slow down when they reach it, and the accumulation of forked molecules is easily recognized by two-dimensional gel analysis (Gahn and Schildkraut 1989; Dhar and Schildkraut 1991). Fork progression through the 30-bp repeats is slow enough that, for recombinant plasmids, the fork that has reached the 30-bp repeats by traversing the short distance from the OBR is often met by the opposite fork that has traveled several kilobase pairs around the plasmid, leading replication to terminate there. Although the 30-bp repeats delay fork progression, they do not seem to act as a specific terminator. Placement of an additional set of 30-bp repeats on an *oriP*-replicated plasmid caused only a slight decrease in plasmid stability (Kirchmaier and Sugden 1995). Pausing and termination structures containing bi-forked molecules have also been detected at the 30-bp repeats of EBV genomes in latently infected cells (Little and Schildkraut 1995). Interestingly, in this study, pausing and termination were also detected near or at the OBR of *oriP* on chromosomes for which *oriP* was being replicated passively because initiation had occurred elsewhere (see below). Pausing was also detected at or near each of the genes for the EBV-encoded small RNAs, EBER1 and EBER2, which are located just to the left of *oriP* on the EBV chromosome (Little and Schildkraut 1995).

Regulation of Initiation of Replication at *oriP*

EBV genomes in latently infected cells and plasmids that are replicated from *oriP* are duplicated in apparent synchrony with cell DNA and are only replicated once per S phase (Hampar et al. 1974; Adams 1987; Yates and Guan 1991). Plasmid copy levels are insensitive to increased levels of EBNA1 in cells and are insensitive to placement of an additional copy of *oriP* on the plasmid, indicating that it is neither the availability of EBNA1 nor the inherent activity of *oriP* that limits copy number (Sugden and Warren 1988; Yates and Gaun 1991). In addition, all EBNA1-binding sites at *oriP* appear to be fully occupied by EBNA1 in asynchronously dividing cells, as revealed by in vivo footprinting (Hsieh et al. 1993). Thus, EBNA1 appears to remain bound to both elements of *oriP* throughout most of the cell cycle, as does the origin recognition complex, ORC, at replication origins of budding yeast (Diffley and Cocker 1992). Initiation of replication from *oriP* presumably awaits the regulatory events that initiate replication from cellular origins and seems to be fully subservient to the cellular controls that limit chromosome replication to once during S phase.

Selectable plasmids carrying *oriP* can attain high copy levels in cells during the transfection process. However, it is not clear how EBV chromosomes increase in number from a single infecting genome to several copies or to hundreds of copies in cells that are latently infected. Unequal segregation of replicated genomes during mitosis clearly occurs and could provide a slow increase. Transient and reversible activation of lytic replication might account for high copy levels. A more explicit consideration of these issues has been presented elsewhere (Yates 1993).

Plasmid Maintenance Function of the 30-bp Repeats

When the chromosomes of latently infected cells are viewed at metaphase using fluorescence in situ hybridization to detect EBV DNA, EBV chromosomes are seen decorating the condensed human chromosomes at seemingly random positions (Hurley et al. 1991; see also Harris et al. 1985). By associating with human chromosomes during mitosis, EBV chromosomes are ensured of being included within daughter nuclei when nuclear membranes re-form around the segregated human chromosomes. The association of EBV chromosomes with human chromosomes during mitosis must have an EBV determinant, since extrachromosomal DNA elements of cellular origin (e.g., double minutes) do not share this property. EBNA1 protein appears to localize exclusively to the condensed human chromosomes during mitosis (Grogan et al.

1983). It often has been assumed that it is EBNA1, bound to the 30-bp repeats of *oriP*, that causes EBV chromosomes to associate with human chromosomes during mitosis. This notion is due to an interesting activity of the 30-bp repeats and EBNA1, a plasmid maintenance activity which is often called a "nuclear retention" or "segregation" function.

Plasmids carrying the 30-bp repeats of *oriP* but lacking the OBR replicate very poorly but show prolonged survival when introduced into EBNA1-containing cells (Reisman et al. 1985; Krysan et al. 1989). This form of specific plasmid maintenance has been inferred to occur without any requirement for DNA replication (Reisman et al. 1985; Krysan et al. 1989; Middleton and Sugden 1994), and unpublished experiments in my own laboratory have demonstrated this directly. If EBNA1 in a complex with the 30-bp repeats were to keep plasmids from being lost to the cytoplasm during each mitosis, as described above, this would explain the maintenance effect if, and only if, it were also assumed that plasmids are less likely to find the nucleus by other means and that plasmids left in the cytoplasm are more prone to degradation. However, transfected plasmids appear to reach the nuclei of cells and to be retained in the nuclear fraction for a few cell divisions equally well whether supported by the 30-bp repeats and EBNA1 or not (our unpublished data), so EBNA1 bound to the 30-bp repeats might protect plasmids from destruction in some other way. Perhaps related to either possibility is the observation by Jankelevich et al. (1992) that the region of the EBV genome including the 30-bp repeats, and only this region, attaches to the nuclear matrix (isolated from interphase cells). Regardless of the mechanisms involved, observations described below imply that the nonreplicative maintenance function of the 30-bp repeats and EBNA1 could be the central feature of EBV's system for maintaining its chromosomes in latently infected cells.

EBNA1

EBNA1 is presumed to carry out all of its functions at *oriP* through interactions with cellular factors. Purified EBNA1 lacks ATPase or DNA helicase activities (Frappier and O'Donnell 1991a; Middleton and Sugden 1992). To initiate replication at *oriP*, EBNA1 requires a species-specific factor that is not required for enhancer activation or for nonreplicative plasmid maintenance, indicating that EBNA1 makes at least two distinct functional interactions with cellular factors (Wysokenski and Yates 1989; Krysan and Calos 1993). It has not been possible to separate EBNA1's three activities by introducing deletions into the protein (Yates and Camiolo 1988 and unpubl.), but consideration of the structural properties of EBNA1 might explain this failure.

EBNA1 dimerizes in solution and binds as a dimer to its symmetric binding sites on DNA, perhaps changing its conformation in the process (Ambinder et al. 1991; Frappier and O'Donnell 1991a; Shah et al. 1992). Approximately 160 amino acids of the carboxy-terminal domain of EBNA1 comprise a structurally complex DNA-binding/dimerization domain (Chen et al. 1993). Crystallographic studies of this part of EBNA1 have recently revealed a structure strikingly similar to that of the E2 DNA-binding protein of bovine papillomavirus, with four antiparallel β strands of each molecule combining to form an eight-stranded β barrel with α helices available for contacting the DNA (Bochkarev et al. 1995). EBNA1 dimers that are bound to DNA, particularly at clusters of sites, are prone to adhere to other EBNA1/DNA complexes. Thus, EBNA1 brings the 30-bp repeats and the OBR of *oriP* together, forming a DNA loop (Frappier and O'Donnell 1991b; Su et al. 1991; Middleton and Sugden 1992). This "looping" or "linking" activity is mediated by two separate regions of the amino-terminal two-thirds of EBNA1 (Fig. 2) (Goldsmith et al. 1993; Mackey et al. 1995) and appears to be required for all three activities associated with *oriP:* DNA replication, plasmid maintenance, and activation of transcription (J.L. Yates and S.M. Camiolo, unpubl.).

One-third of the EBNA1 polypeptide is a peculiar, repetitive arrangement of glycines and alanines over 200 amino acids long that is not important for any known function of EBNA1 (Yates et al. 1985; Yates and Camiolo 1988). A recent study found that attaching the glycine, alanine repeats to a heterologous protein would prevent cells expressing the protein from being recognized and lysed by specific cytotoxic T lymphocytes, accounting for the apparent lack of a cellular immune response against EBNA1 (Klein 1994; Levitskaya et al. 1995). In EBV's most restricted state of latent infection, EBNA1 may be the only EBV-encoded protein that is synthesized (Tierney et al. 1994; Miyashita et al. 1995), so in such cells the infection is truly *latent* (from Latin, meaning hidden).

INITIATION OF LATENT REPLICATION AWAY FROM *oriP*

Is the Initiation Site Function of *oriP* Essential?

Because *oriP* was shown to be the only element in the viral genome (strain B95-8) capable of supporting efficient, stable replication of recombinant plasmids introduced into cells under selection (Yates et al. 1984), for a long time it was assumed that *oriP* is essential for replication of EBV chromosomes during latent infection. *oriP* may yet prove to be

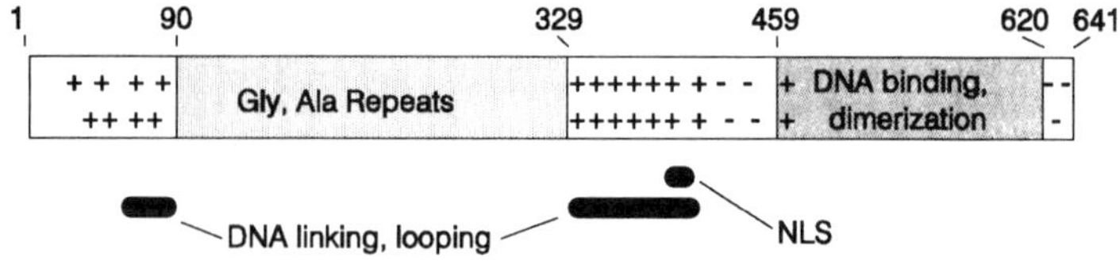

Figure 2 Structure and function of the EBNA1 polypeptide. The regions involved in DNA linking and looping are indicated beneath a rectangular depiction of the polypeptide chain. Most of the polypeptide chain outside of the glycine, alanine repetitious region is involved in binding to DNA and linking it into higher-order structures. Regions of significant net positive or negative charge are indicated with multiple plus and minus signs. (NLS) Nuclear localization sequence (Ambinder et al. 1991).

essential for maintaining circular EBV genomes in proliferating, latently infected cells, but it now appears that its function as an initiation site for replication is not always needed.

Transcriptional studies of an EBV-transformed B-cell line known as X50-7 a few years ago revealed that the EBV strain (of unknown origin) carried by these cells had a genomic rearrangement that deleted all of *oriP* except for the 30-bp repeats (Yandava and Speck 1992). The variant EBV genomes of X50-7 cells are maintained as circular molecules (Cho and Tran 1993) and are capable of transforming human B lymphocytes and, thus, of establishing new latent infections (Gradoville et al. 1990). We cannot be sure that a short sequence derived from *oriP* is not present elsewhere in the viral genome and functioning as an EBNA1-dependent OBR, but otherwise, these results suggest that the replication initiation function of *oriP* is not required for latent replication of EBV genomes in B cells.

One possible explanation for the replication of X50-7 EBV genomes is that replication of large DNA molecules in mammalian cells may not require highly specific initiation sites. Support for this idea came from an attempt to isolate sequences from the human genome that could function as replication origins when inserted into a plasmid that carried only the 30-bp repeats of *oriP*. In the presence of EBNA1, the 30-bp repeats would allow the plasmid to be retained by cells, but without the OBR of *oriP*, replication would be much too inefficient for sustained maintenance unless a segment of human DNA, from a library of randomly inserted fragments, could provide that function. The surprising outcome of such studies was that most randomly tested fragments of human DNA of about 10 kb or longer would allow the plasmids to replicate rather stably in cells (Heinzel et al. 1991). Replication was found to initiate at multiple sites over most regions of the plasmids (Krysan and Calos 1991). If EBNA1 and the 30-bp repeats contribute to the replication of these plasmids, the contribution is probably indirect. Although EBNA1 and *oriP* cannot support replication in cell lines derived from rodents, the maintenance function of the 30-bp repeats and EBNA1 does function in rodent cells (Wysokenski and Yates 1989; J.L. Yates and N. Guan, unpubl.). Plasmids whose OBR has been substituted by large fragments of human DNA replicate stably in rodent cells (Krysan and Calos 1993), implying that the mechanism for initiating replication on such plasmids is different from that of *oriP*-replicated plasmids and does not require a specific replication function of EBNA1. It should be mentioned that the OBR-substituted plasmids are not as stable as *oriP*-replicated plasmids even under selection, and they generally work well only in certain laboratory cell lines such as 293 (transformed by adenovirus) and HeLa (transformed by human papillomavirus) (J.L. Yates and N. Guan, unpubl.). Nevertheless, the results clearly show that, in the proper context, replication can initiate rather nonspecifically with adequate efficiency in mammalian cells.

Initiation of Replication at *oriP* and Elsewhere on the EBV Chromosome

Clearly, it is important to determine just where replication initiates on EBV chromosomes during latent infection, and Little and Schildkraut (1995) have begun to do so using two-dimensional gels to analyze replication intermediates. With four different latently infected B-cell lines, replication intermediates representing initiation events at *oriP* were seen in all cases. But intermediates due to passive replication by forks that had initiated outside of the region indicated that replication was initiating at sites away from *oriP* a significant fraction of the time in all four cell lines. In one of the cell lines, Raji, most initiations appeared to occur elsewhere. As a control, cells carrying a 19-kb, *oriP*-replicated plasmid were analyzed in parallel, and in this case the expected pattern consistent with predominant or exclusive initiation at *oriP* was observed. Using Raji cells, EBV sequences to the left of *oriP* were examined, and weak signals indicating infrequent initiation events were detected in every region examined over an expanse of at least 50 kb. One region, centered near position 153,000 (B95-8 coordinate), seemed to contain a preferred site of initiation, but initiation was not detected in this region when another cell line was examined. Not all regions of the EBV genome were examined, but at least one region, about 30 kb to the right of *oriP*, lacked detectable initiation events, indicating that initiation does not occur over the entire EBV chromosome in Raji cells. Consistent with these findings, an earlier electron microscopic study of replicating EBV chromosomes of Raji cells revealed some molecules with more than two forks, implying that initiation had occurred at more than one place on those DNA molecules (Gussander and Adams 1984).

The delocalized initiation events detected with Raji cells are similar to the broad initiation zones that have been observed at replication origins on mammalian chromosomes (Vaughn et al. 1990; Dijkwel and Hamlin 1992; Little et al. 1993; Heintz, this volume). Might EBV have a replication origin of the mammalian chromosomal kind, with a *cis*-acting replicator (distinct from *oriP*), a preferred site of initiation, and a broad initiation zone? Genetic studies are needed to determine whether EBV has a distinct function of this kind, to what extent initiation of replication at *oriP* is important, whether specific initiation sites are required at all, and whether a nuclear anchoring function of the 30-bp repeats and EBNA1 is indeed the most essential feature of autonomous chromosome maintenance for EBV.

REPLICATION DURING LYTIC INFECTION

Components of EBV Lytic Replication: Different Strategies of Initiation for EBV and HSV

Whereas EBV DNA replication during latency serves to duplicate the number of circular viral genomes prior to cell division, during lytic infection viral genomes are amplified to thousands of copies per cell in long concatemers, perhaps through a rolling-circle mechanism, for assembly into virus particles. In contrast to its complete dependence on the cell for its replication during latency, EBV, like other herpesviruses, is much more self-reliant at replicating its genome during lytic infection. All herpesviruses specify six common, conserved replication proteins: DNA polymerase and an accessory factor, a single-stranded DNA-binding protein, and a helicase/primase complex with an associated factor (Table 1). These replication proteins are expected to function similarly for all herpesviruses, and they have been extensively investigated in the case of HSV (Challberg, this volume). However, cellular factors must perform a central role during initiation of EBV lytic replication, and this may be true for other herpesviruses as well, although the different herpesviruses have diverged greatly in their lytic initiation mechanisms.

At a glance, the viral components required for initiation of lytic replication for EBV and HSV bear little resemblance to each other. For

Table 1 Lytic replication proteins of EBV and HSV

EBV ORF	HSV gene	Protein, function
BALF5	UL30	DNA polymerase
BMRF1	UL42	polymerase processivity factor
BALF2	UL29	single-stranded DNA-binding protein
BBLF4	UL5	helicase
BSLF1	UL52	primase
BBLF2/3	UL8	primase-associated factor
—	UL9	HSV origin-binding protein; helicase
BZLF1	—	Z transcriptional activator (EBV); activates *oriLyt* promoter, replication

HSV, the protein product of the UL9 gene binds to each HSV lytic origin at two sites and is essential for replication; the UL9 protein has ATPase and DNA helicase activities, consistent with a role in unwinding DNA at the origin (Challberg, this volume). The EBV genes that are required for lytic replication were identified using the "Challberg approach," that is, by determining which EBV genomic segments must be transfected into noninfected cells in order to activate specific replication of a test plasmid carrying an EBV lytic replication origin, *oriLyt* (Fixman et al. 1992). Because EBV-encoded regulatory proteins are necessary for expression of all "early" genes, including lytic replication genes, the list was then refined by expressing the EBV replication genes using heterologous promoters, so that the EBV early gene *trans*-activators could be omitted. Under these conditions, one EBV transcriptional activator, Z (also called ZEBRA, EB1, BZLF1, and Zta), was still required for replication (Fixman et al. 1995). Z binds to and activates a strong promoter that serves as an integral component of *oriLyt* (Fig. 3). Z and the set of EBV homologs of the six conserved herpesvirus replication proteins complete the list of EBV gene products that are necessary and sufficient for replication from *oriLyt* (Table 1). EBV has no homolog of UL9, and HSV has no homolog of Z; the two proteins are unrelated functionally and phylogenetically.

The lytic replication origins of EBV and HSV do not share common recognizable sequence elements and appear, on initial inspection, to be organized quite differently. The EBV genome carries two lytic replication origins, *oriLyt*, which are determined by two essentially identical copies of a 1055-bp "duplicated segment," located approximately opposite each other on the circularized viral chromosome (Fig. 1A). EBV *oriLyt* is composed of multiple contributing elements (Fig. 3) (Hammerschmidt and Sugden 1988; Schepers et al. 1993b). These include two essential, or *core*, components, separated by about 400 bp, and several

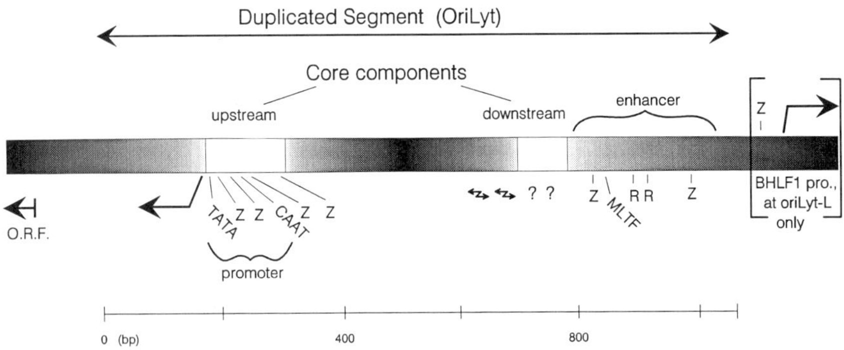

Figure 3 Functional components of *oriLyt*. *oriLyt* is contained within the 1055-bp duplicated segment, which is present at two locations on the EBV genome. The "upstream" and "downstream" core components of *oriLyt* are represented by the indicated white rectangles; the auxiliary regions are shaded. Bent arrows indicate the position and direction of transcription initiation. Binding sites for Z, R, MLTF, and for presumptive cellular factors (?) are indicated; a TATA-like element (GATAAAA) and a CAAT transcriptional element are also indicated. Zig-zag arrows indicate inverted repeats. The arrow at the lower left indicates the initial coding regions for BHLF1 and LF3 (also called *Not*I-repeat protein and *Pst*-repeat protein, respectively), which are synthesized from mRNA initiated from either the left or the right copy of *oriLyt*, respectively, on the EBV genome. Rightward transcription away from *oriLyt* occurs only for the left copy of *oriLyt* (Parker et al. 1990). The upstream core component was originally represented so as to exclude the two Z-binding sites most distal from the transcription start site (Schepers et al. 1993b). However, recent data indicate that the third and fourth ZREs support replication as effectively as the first two (Schepers et al. 1996), so the entire promoter region is presented here as a functional unit.

auxiliary components that contribute to activity but are not essential under all circumstances (see DePamphilis, this volume). The core components of EBV *oriLyt* lack significant activity in the absence of the auxiliary components, and the auxiliary effect of flanking DNA extends at least for several hundred base pairs to each side, making the functional boundaries of *oriLyt* rather ambiguous. One core component, at the left or "upstream" part of *oriLyt*, is a strong early promoter that is activated by Z, which binds to the promoter at four sites (Rooney et al. 1989; Leiberman et al. 1990); replication from *oriLyt* requires that Z interact with these same four sites (Schepers et al. 1993a,b). The transcriptional enhancer at the right end of *oriLyt* performs an auxiliary role that can be replaced by a heterologous enhancer (Hammerschmidt and Sugden

1988). The other core component of *oriLyt,* the "downstream" component, is a region less than 90 bp in length that is critical for replication but seems to play no role in transcription, consistent with its functioning directly in the initiation reaction. The factors that functionally interact with this critical region have not been identified, but the genetic studies mentioned above imply that they must be provided by the cell.

The HSV lytic origin, ori_S, has a single core component, to which the UL9 protein binds, flanked by divergent transcriptional control regions which serve auxiliary functions for replication (Challberg, this volume). EBV *oriLyt* contains two core components, but since one core component is a promoter, this might be viewed as a variation of the design used by HSV, with the "real" (downstream) core of EBV simply having greater dependence on its auxiliary components. The core of HSV ori_S in fact shows more autonomy from its auxiliary components than do the two EBV core components combined, since the HSV core can function as a cloned fragment as small as 150 bp, and the combined EBV core components cannot function alone. Studies to be described below, however, have revealed that the *oriLyt* promoter, or upstream core component, contributes to replication through interactions with Z in a more specific way than is typical of auxiliary components of replication origins.

A Specific Replication Function for Z; The Possible Involvement, or Entanglement, of Other Transcription Factors with *oriLyt*

The *oriLyt* promoter is quite efficient once activated by Z, yielding the two most abundant EBV early mRNAs, via transcription outward from each copy of *oriLyt* (Hummel and Kieff 1982; Freese et al. 1983). The fact that this strong promoter colocalizes with an essential component of *oriLyt* raises two questions. First, what promoter elements, if any, support replication activity? Second, do transcription factors aid replication directly through interactions among proteins, or indirectly by excluding nucleosomes from the region, or might transcription itself be part of the mechanism? (For general discussions of the involvement of transcriptional control elements with replication, see DePamphilis; van der Vliet; both this volume.)

Z is the only transcription factor that is known to be essential for the replication function of the *oriLyt* promoter (Schepers et al. 1993a). Replication activity was seen to be diminished, but not abolished, by deleting small regions of this core component (Schepers et al. 1993b). The results can be explained partly by the fact that all four Z-binding sites are not required for replication, although all apparently contribute to it; abolishing either the two Z-binding sites closest to the transcriptional start site or the two farthest away by substitution mutations reduced replication activity to 20–30% of the wild-type level (Schepers et al. 1996). However, deleting from the TATA element through the CAAT element, or simultaneously deleting these two elements while leaving three of the four Z-binding sites, had the effect of completely abolishing replication activity (Schepers et al. 1993b). The study of these and other mutants indicates that the Z-binding sites alone might not be sufficient for the replication function of this region of *oriLyt* and that other promoter elements may contribute. Z has been shown to promote the stable association of transcription factors TFIID and TFIIA with the *oriLyt* promoter and to bind directly to TATA-binding protein (TBP), a component of TFIID (Lieberman and Berk 1991, 1994), so the *oriLyt* promoter might contribute to replication through a Z-containing transcriptional preinitiation complex.

The results of "domain-swapping" experiments involving Z suggest that Z contributes to replication in a way that other transcription factors cannot (Schepers et al. 1993a). The Z protein has two separable domains: an activation domain, which binds to TFIID (Lieberman and Berk 1991; Giot et al. 1991; Flemington et al. 1992), and a DNA-binding/dimerization domain, which is structurally related to the equivalent region of the AP1 family of cellular transcription factors (Z binds well to AP1 sites, but AP1 does not recognize ZREs; Farrell et al. 1989; Urier et al. 1989; Leiberman et al. 1990). Using altered versions of *oriLyt*, in which all Z-binding sites had been replaced with binding sites for either GAL4 (of yeast) or E2 (of BPV), Schepers et al. asked whether the activation domain from Z or from unrelated transcriptional activators would stimulate replication from *oriLyt*. (It was not possible simply to replace the activation domain of Z with other activation domains because wild-type Z is needed for induction of EBV replicative gene expression.) The result was that the *trans*-activation domain of Z could activate replication if bound to *oriLyt* through the binding domains of either GAL4 or E2, although the efficiency was much lower than with wild-type Z bound to *oriLyt* in the natural way. However, the transcriptional activation domains of several unrelated transcription factors, including c-fos, were completely inactive for support of replication, although they activated the *oriLyt* promoter effectively. Because the *trans*-activating domain of Z was unique in being able to activate replication, it is reasonable to suspect that replication and *trans*-activation might be distinct functions specified by the same region of the Z polypeptide.

It is clear from the domain-swapping experiments that efficient transcription from the *oriLyt* promoter is not a sufficient contribution from this component of *oriLyt* to support replication. Unfortunately, we still cannot conclude that transcription from this promoter is not a prerequisite for replication. For example, the GAL4-binding domain linked to the Z activation domain activated transcription from the *oriLyt* promoter only weakly, but replication was correspondingly weak (although it was specific) when compared to the wild-type situation. It seems simpler to assume that transcription and replication are independent processes with shared elements, but genetic manipulations that might allow efficient replication without transcription from the *oriLyt* promoter have not yet been achieved.

The enhancer at the right end of *oriLyt* is activated by Z and by another EBV early gene *trans*-activator, R (Chevallier-Greco et al. 1989; Cox et al. 1990). This region of *oriLyt* performs an auxiliary function for replication that can be replaced by the enhancer for the major immediate-early gene of human cytomegalovirus (Hammerschmidt and Sugden 1988; Schepers et al. 1993b). However, the results of the transfection studies mentioned above imply that R itself does not contribute significantly to replication from *oriLyt* (Fixman et al. 1995), and the Z-binding sites in this region do not contribute measurably to replication, either (Schepers et al. 1993b). A binding site for the ubiquitous transcription factor MLTF (USF) is largely responsible for constitutive activity of the *oriLyt* enhancer (Ryon et al. 1993), but the contribution of this site to replication has not been determined.

Little is known about the auxiliary functions of other sequences between the two core components of *oriLyt* or flanking them, except that sequences throughout *oriLyt* are about 90% identical between EBV and HVP, and thus among the most highly conserved regions of these viral genomes (Ryon et al. 1993). Deletion of a pair of inverted repeat sequences that exist to the left of the downstream core component (Fig. 3) reduced replication by 40% (Schepers et al. 1993b) or by 80% (Ryon et al. 1993) in two different assays.

Initiation of EBV Lytic Replication by Cellular Factors: A Common Mechanism for Herpesviruses?

Discovering what factors, presumably of cellular origin, functionally interact with the "downstream" core component will very probably be the key to learning how *oriLyt* is activated. The downstream component is less than 100 bp long and contains a central 40-bp region that is highly sensitive to substitution mutations (Schepers et al. 1993b; Gruffat et al. 1995). This extreme sensitivity to substitution mutations contrasts with the additive nature of contributing elements at the *oriLyt* promoter and suggests an exactness of essential protein/DNA interactions. Deletions in this region either had no effect on transcription from the *oriLyt* promoter or caused a moderate increase in transcription, perhaps by disrupting association between proteins ordinarily bound to the downstream component and Z and transcription factors bound to the promoter (Schepers et al. 1993b).

Initiation of replication at HSV origins undoubtedly involves the UL9-encoded origin-binding protein, which has DNA helicase activity (Challberg, this volume). Since EBV encodes no equivalent function, it is natural to expect that a cellular helicase might be recruited by EBV to act at *oriLyt*. Despite these differences between the way the two viruses appear to initiate lytic replication, experiments performed by Fixman, Hayward, and Hayward indicate that the intermediates that are formed for initiation of replication by the two viruses are fundamentally similar. In a transfection assay for replication of an *oriLyt*-containing plasmid, providing the set of six replication proteins from HSV rather than from EBV gave full replication activity so long as the Z protein of EBV was provided (Fixman et al. 1995). This remarkable result would appear to indicate that HSV replication proteins do not need to recognize UL9 at the HSV origins and, because the HSV proteins probably cannot recognize Z, that the EBV replication proteins do not need to recognize Z at the EBV origins. The result suggests that the replication intermediates generated during initiation by Z and cellular proteins at the EBV lytic origins are similar or identical to those produced during initiation by UL9 (and cellular proteins?) at the HSV origins and that it is these replication intermediates that are recognized by the herpesvirus family replication proteins.

ACKNOWLEDGMENTS

It is appropriate to acknowledge that significant contributions of many investigators, although important in bringing us to our current level of understanding, have not been cited for the sake of brevity. I thank Wolfgang Hammerschmidt and Bill Sugden for providing manuscripts in advance of publication, and I thank Joel Huberman and Bill Sugden for their thoughtful comments on this chapter.

REFERENCES

Adams, A. 1987. Replication of latent Epstein-Barr virus genomes in Raji cells. *J. Virol.* **61:** 1743–1746.

Ambinder, R.F., M.A. Mullen, Y.-N. Chang, G.S. Hayward, and S.D. Hayward. 1991. Functional domains of Epstein-Barr virus nuclear antigen EBNA-1. *J. Virol.* **65:** 1466–1478.

Ambinder, R.F., W.A. Shah, D.R. Rawlins, G.S. Hayward, and S.D. Hayward. 1990. Definition of the sequence requirements for binding of the EBNA-1 protein to its palindromic target sites in Epstein-Barr virus DNA. *J. Virol.* **64:** 2369–2379.

Bochkarev, A., J.A. Barwell, R.A. Pfuetzner, W. Furey, Jr., A.M. Edwards, and L. Frappier. 1995. Crystal structure of the DNA-binding domain of the Epstein-Barr virus origin-binding protein EBNA1. *Cell* **83:** 39–46.

Chen, M.-R., J.M. Middeldorp, and S.D. Hayward. 1993. Separation of the complex DNA binding domain of EDNA-1 into DNA recognition and dimerization subdomains of novel structure. *J. Virol.* **67:** 4875–4885.

Chevallier-Greco, A., H. Gruffat, E. Manet, A. Calender, and A. Sergeant. 1989. The Epstein-Barr virus (EBV) enhancer contains two functionally different domains: Domain A is constitutive and cell specific, domain B is transactivated by the EBV early protein R. *J. Virol.* **63:** 615–623.

Chittenden, T., S. Lupton, and A.J. Levine. 1989. Functional limits of *oriP*, the Epstein-Barr virus plasmid origin of replication. *J. Virol.* **63:** 3016–3025.

Cho, M.-S. and V.-M. Tran. 1993. A concatenated form of Epstein-Barr viral DNA in lymphoblastoid cell lines induced by transfection with BZLF1. *Virology* **194:** 838–842.

Cox, M.A., J. Leahy, and J.M. Hardwick. 1990. An enhancer within the divergent promoter of Epstein-Barr virus responds synergistically to the R and Z transactivators. *J. Virol.* **64:** 313–321.

Dhar, V. and C.L. Schildkraut. 1991. Role of EBNA-1 in arresting replication forks at the Epstein-Barr virus *oriP* family of repeats. *Mol. Cell. Biol.* **11:** 6268–6278.

Diffley, J.F.X. and J.H. Cocker. 1992. Protein-DNA interactions at a yeast replication origin. *Nature* **357:** 169–172.

Dijkwel, P.A. and J.L. Hamlin. 1992. Initiation of DNA replication in the dihydrofolate reductase locus is confined to the early S period in CHO cells synchronized with the plant amino acid mimosine. *Mol. Cell. Biol.* **12:** 3715–3722.

Farrell, P.J., D.T. Rowe, C.M. Rooney, and T. Kouzarides. 1989. Epstein-Barr virus BZLF1 *trans*-activator specifically binds to a consensus AP-1 site and is related to c-*fos*. *EMBO J.* **8:** 127–132.

Fixman, E., G.S. Hayward, and S.D. Hayward. 1992. *trans*-Acting requirements for replication of Epstein-Barr virus ori-Lyt. *J. Virol.* **66:** 5030–5039.

———. 1995. Replication of Epstein-Barr virus oriLyt: Lack of a dedicated virally encoded origin-binding protein and dependence on Zta in cotransfection assays. *J. Virol.* **67:** 2998–3006.

Flemington, E., A. Borras, J. Lytle, and S. Speck. 1992. Characterization of the Epstein-Barr virus BZLF1 protein transactivation domain. *J. Virol.* **66:** 922–929.

Frappier, L. and M. O'Donnell. 1991a. Overproduction, purification, and characterization of EBNA-1, the origin binding protein of Epstein-Barr virus. *J. Biol. Chem.* **266:** 7819–7826.

———. 1991b. Epstein-Barr nuclear antigen-1 mediates a DNA loop within the latent replication origin of Epstein-Barr virus. *Proc. Natl. Acad. Sci.* **88:** 10875–10879.

———. 1992. EBNA1 distorts *oriP*, the Epstein-Barr virus latent replication origin. *J. Virol.* **66:** 1786–1790.

Freese, U.K., G. Laux, J. Hudewentz, E. Schwarz, and G.W. Bornkamm. 1983. Two distant clusters of partially homologous small repeats of Epstein-Barr virus are transcribed upon induction of an abortive or lytic cycle of the virus. *J. Virol.* **48:** 731–743.

Gahn, T.A. and C.L. Schildkraut. 1989. The Epstein-Barr virus origin of plasmid replication, oriP, contains both the initiation and termination sites of DNA replication. *Cell* **58:** 527–535.

Gahn, T.A. and B. Sugden. 1995. An EBNA-1-dependent enhancer acts from a distance of 10 kilobase pairs to increase expression of the Epstein-Barr virus LMP gene. *J. Virol.* **69:** 2633–2636.

Giot, J.-F., I. Mikaelian, M. Buisson, E. Manet, I. Joab, J.-C. Nicolas, and A. Sergeant. 1991. Transcriptional interference between the EBV transcription factors EB1 and R: Both DNA-binding and activation domains of EB1 are required. *Nucleic Acids Res.* **19:** 1251–1258.

Goldsmith, K., L. Bendell, and L. Frappier. 1993. Identification of EBNA1 amino acid sequences required for the interaction of the functional elements of the Epstein-Barr virus latent origin of DNA replication. *J. Virol.* **67:** 3418–3426.

Gradoville, L., E. Grogan, N. Taylor, and G. Miller. 1990. Differences in the extent of activation of Epstein-Barr virus replicative gene expression among four nonproducer cell lines stably transformed by oriP/BZLF1 plasmids. *Virology* **178:** 345–354.

Grogan, E.A., W.P. Summers, S. Dowling, D. Shedd, L. Gradoville, and G. Miller. 1983. Two Epstein-Barr viral nuclear neoantigens distinguished by gene transfer, serology, chromosome binding. *Proc. Natl. Acad. Sci.* **80:** 7650–7653.

Gruffat, H., O. Renner, D. Pich, and W. Hammerschmidt. 1995. Cellular proteins bind to the downstream component of the lytic origin of DNA replication of Epstein-Barr. *J. Virol.* **69:** 1878–1886.

Gussander, E. and A. Adams. 1984. Electron microscopic evidence for replication of circular Epstein-Barr virus genomes in latently infected Raji cells. *J. Virol.* **53:** 549–556.

Hammerschmidt, W. and B. Sugden. 1988. Identification and characterization of *oriLyt*, a lytic origin of DNA replication of Epstein-Barr virus. *Cell* **55:** 427–433.

Hampar, B., A. Tanaka, M. Nonayama, and J.G. Derge. 1974. Replication of the resident repressed Epstein-Barr virus genome during the early S phase (S-1 period) of nonproducer Raji cells. *Proc. Natl. Acad. Sci.* **71:** 631–635.

Harris, A., B.D. Young, and B.E. Griffin. 1985. Random association of Epstein-Barr virus genome with host cell metaphase chromosomes in Burkitt's lymphoma-derived cell lines. *J. Virol.* **56:** 328–332.

Harrison, S., K. Fisenne, and J. Hearing. 1994. Sequence requirements of the Epstein-Barr virus latent origin of DNA replication. *J. Virol.* **68:** 1913–1925.

Hearing, J., Y. Mülhaupt, and S. Harper. 1992. Interaction of Epstein-Barr virus nuclear antigen 1 with the latent origin of replication. *J. Virol.* **66:** 694–705.

Heinzel, S.S., P.J. Krysan, C.T. Tran, and M.P. Calos. 1991. Autonomous DNA replication in human cells is affected by the size and the source of the DNA. *Mol. Cell. Biol.* **11:** 2263–2272.

Hsieh, D.-J., S.M. Camiolo, and J.L. Yates. 1993. Constitutive binding of EBNA1 protein to the Epstein-Barr virus replication origin, oriP, with distortion of DNA structure during latent infection. *EMBO J.* **12:** 4933–4944.

Hummel, M. and E. Kieff. 1982. Epstein-Barr virus RNA. VIII. Viral RNA in permissively infected B95-8 cells. *J. Virol.* **43:** 262–272.

Hurley, E.A., S. Agger, J.A. McNeil, J.B. Lawrence, A. Calendar, G. Lenoir, and D.A. Thorley-Lawson. 1991. When Epstein-Barr virus persistently infects B-cell lines, it frequently integrates. *J. Virol.* **65:** 1245–1254.

Jankelevich, S., J.L. Kolman, J.W. Bodnar, and G. Miller. 1992. A nuclear matrix attachment region organizes the Epstein-Barr viral plasmid in Raji cells into a single DNA domain. *EMBO J.* **11:** 1165–1176.

Jones, C.H., S.D. Hayward, and D.R. Rawlins. 1989. Interaction of lymphocyte-derived Epstein-Barr virus nuclear antigen (EBNA-1) with its DNA-binding sites. *J. Virol.* **63:** 101–110.

Kieff, E. and D. Liebowitz. 1990. Epstein-Barr virus and its replication. In *Virology* (ed. B.N. Fields et al.), pp. 1889–1920. Raven Press, New York.

Kirchmaier, A.L. and B. Sugden. 1995. Plasmid maintenance of derivatives of oriP of Epstein-Barr virus. *J. Virol.* **69:** 1280–1283.

Klein, G. 1994. Epstein-Barr virus strategy in normal and neoplastic B cells. *Cell* **77:** 791–793.

Krysan, P.J. and M.P. Calos. 1991. Replication initiates at multiple locations on an autonomously replicating plasmid in human cells. *Mol. Cell. Biol.* **11:** 1464–1472.

———. 1993. Epstein-Barr virus-based vectors that replicate in rodent cells. *Gene* **136:** 137–143.

Krysan, P.J., S.B. Haase, and M.P. Calos. 1989. Isolation of human sequences that replicate autonomously in human cells. *Mol. Cell. Biol.* **9:** 1026–1033.

Levitskaya, J., M. Coram, V. Levitsky, S. Imreh, P.M. Steigerwald-Mullen, G. Klein, M.G. Kurilla, and M.G. Masucci. 1995. Inhibition of antigen processing by the internal repeat region of the Epstein-Barr virus nuclear antigen-1. *Nature* **375:** 685–688.

Lieberman, P.M. and A.J. Berk. 1991. The Zta *trans*-activator protein stabilizes TFIID association with promoter DNA by direct protein-protein interaction. *Genes Dev.* **5:** 2441–2454.

———. 1994. A mechanism for TAFs in transcriptional activation: Activation domain enhancement of TFIID-TFIIA–promoter DNA complex formation. *Genes Dev.* **8:** 995–1006.

Lieberman, P.M., J.M. Hardwick, J. Sample, G.S. Hayward, and S.D. Hayward. 1990. The Zta transactivator involved in induction of lytic cycle gene expression in Epstein-Barr virus-infected lymphocytes binds to both AP-1 and ZRE sites in target promoter and enhancer regions. *J. Virol.* **64:** 1143–1155.

Little, R. and C. Schildkraut. 1995. Initiation of latent DNA replication in the Epstein-Barr virus genome can occur at sites other than the genetically defined origin. *Mol. Cell. Biol.* **15:** 2893–2903.

Little, R.D., T.H.K. Platt, and C.L. Schildkraut. 1993. Initiation and termination of DNA replication in human rRNA genes. *Mol. Cell. Biol.* **13:** 6600–6613.

Loeb, D.D., N. Schwab Sung, R.L. Pesano, C.J. Sexton, C. Hutchison III, and J.S. Pagano. 1990. Plasmid origin of replication of herpes virus Papio: DNA sequence and enhancer function. *J. Virol.* **64:** 2876–2883.

Lupton, S. and A.J. Levine. 1985. Mapping genetic elements of Epstein-Barr virus that facilitate extrachromosomal persistence of Epstein-Barr virus derived plasmids in human cells. *Mol. Cell. Biol.* **5:** 2533–2542.

Mackey, D., T. Middleton, and B. Sugden. 1995. Multiple regions of EBNA1 can link DNAs. *J. Virol.* **69:** 6199–6208.

Middleton, T. and B. Sugden. 1992. EBNA1 can link the enhancer element to the initiator element of the Epstein-Barr virus plasmid origin of DNA replication. *J. Virol.* **66:** 489–495.

———. 1994. Retention of plasmid DNA in mammalian cells is enhanced by binding of the Epstein-Barr virus replication protein EBNA1. *J. Virol.* **68:** 4067–4071.

Miller, G. 1990. Epstein-Barr virus, biology, pathogenesis, and medical aspects. In *Virology* (ed. B.N. Fields et al.), pp. 1921–1957. Raven Press, New York.

Miyashita, E.M., B. Yang, K.M. Lam, D.H. Crawford, and D.A. Thorley-Lawson. 1995. A novel form of Epstein-Barr virus latency in normal B cells in vivo. *Cell* **80:** 593–601.

Parker, B.D., A. Bankier, S. Satchwell, B. Barrell, and P.J. Farrell. 1990. Sequence and transcription of Raji Epstein-Barr virus DNA spanning the B95-8 deletion region. *Virology* **179:** 339–346.

Platt, T.H.K., I. Tcherepanova, and C. Schildkraut. 1993. Effect of number and position of EBNA-1 binding sites in Epstein-Barr virus *oriP* on sites of initiation, barrier formation, and termination of replication. *J. Virol.* **67:** 1739–1745.

Reisman, D. and B. Sugden. 1986. *trans*-Activation of an Epstein-Barr viral transcriptional enhancer by the Epstein-Barr viral nuclear antigen 1. *Mol. Cell. Biol.* **6:** 3838–3846.

Reisman, D., J.L. Yates, and B. Sugden. 1985. A putative origin of replication of plasmids derived from Epstein-Barr virus is composed of two *cis*-acting components. *Mol. Cell. Biol.* **5:** 1822–1832.

Rooney, C.M., D.T. Rowe, T. Ragot, and P.J. Farrell. 1989. The spliced BZLF1 gene of Epstein-Barr virus (EBV) transactivates an early EBV promoter and induces the virus productive cycle. *J. Virol.* **63:** 3109–3116.

Ryon, J.J., E.D. Fixman, C. Houchens, J. Zong, P.M. Lieberman, Y.-N. Chang, G.S. Hayward, and S.D. Hayward. 1993. The lytic origin of herpesvirus papio is highly homologous to Epstein-Barr virus ori-Lyt: Evolutionary conservation of transcriptional activation and replication signals. *J. Virol.* **67:** 4006–4016.

Schepers, A., D. Pich, and W. Hammerschmidt. 1993a. A transcription factor with homology to the AP-1 family links RNA transcription and DNA replication in the lytic cycle of Epstein-Barr virus. *EMBO J.* **12:** 3921–3929.

———. 1996. Activation of *oriLyt*, the lytic origin of DNA replication of Epstein-Barr virus, by BSLF1. *Virology* (in press).

Schepers, A., D. Pich, J. Mankertz, and W. Hammerschmidt. 1993b. *cis*-Acting elements in the lytic origin of DNA replication of Epstein-Barr virus. *J. Virol.* **67:** 4237–4245.

Shah, W.A., R.F. Ambinder, G.S. Hayward, and S.D. Hayward. 1992. Binding of EBNA-1 to DNA creates a protease-resistant domain that encompasses the DNA recognition and dimerization functions. *J. Virol.* **66:** 3355–3362.

Sixbey, J. 1989. Epstein-Barr virus and epithelial cells. *Adv. Viral Oncol.* **8:** 187–202.

Su, W., T. Middleton, B. Sugden, and H. Echols. 1991. DNA looping between the origin of replication of Epstein-Barr virus and its enhancer site: Stabilization of an origin complex with Epstein-Barr nuclear antigen 1. *Proc. Natl. Acad. Sci.* **88:** 10870–10874.

Sugden, B. and N. Warren. 1988. Plasmid origin of replication of Epstein-Barr virus, *oriP*, does not limit replication in *cis*. *Mol. Biol. Med.* **5:** 85–94.

———. 1989. A promoter of Epstein-Barr virus which can function during latent infection can be *trans*-activated by EBNA1, a viral protein required for DNA replication during latent infection. *J. Virol.* **63:** 2644–2649.

Tierney, R.J., N. Steven, L.S. Young, and A.B. Rickinson. 1994. Epstein-Barr virus latency in blood mononuclear cells: Analysis of viral gene transcription during primary infection and in the carrier state. *J. Virol.* **68:** 7374–7385.

Urier, G., M. Buisson, P. Chambard, and A. Sergeant. 1989. The Epstein-Barr virus early protein EB1 activates transcription from different responsive elements including AP-1 binding sites. *EMBO J.* **8:** 1447–1453.

Vaughn, J. P., P.A. Dijkwel, and J.L. Hamlin. 1990. Replication initiates in a broad zone in the amplified CHO dihydrofolate reductase domain. *Cell* **61:** 1075–1087.

Williams, D.L. and D. Kowalski. 1993. Easily unwound DNA sequences and hairpin structures in the Epstein-Barr virus origin of plasmid replication. *J. Virol.* **67:** 2707–2715.

Wysokenski, D.A. and J.L. Yates. 1989. Multiple EBNA1-binding sites are required to form an EBNA1-dependent enhancer and to activate a minimal replicative origin within *oriP* of Epstein-Barr virus. *J. Virol.* **63:** 2657–2666.

Yandava, C.N. and S.H. Speck. 1992. Characterization of the deletion and rearrangement in the BamHI C region of the X50-7 Epstein-Barr virus genome, a mutant viral strain which exhibits constitutive BamHI W promoter activity. *J. Virol.* **66:** 5646–5650.

Yates, J.L. 1993. EBV DNA replication: How do viral genomes become amplified during latent infection? *Colloq. INSERM* **225:** 267–272.

Yates, J.L. and S.M. Camiolo. 1988. Dissection of DNA replication and enhancer activation function of Epstein-Barr virus nuclear antigen 1. *Cancer Cells* **6:** 197–205.

Yates, J.L. and N. Guan. 1991. Epstein-Barr virus-derived plasmids replicate only once per cell cycle and are not amplified after entry into cells. *J. Virol.* **65:** 483–488.

Yates, J.L., N. Warren, and B. Sugden. 1985. Stable replication of plasmids derived from Epstein-Barr virus in various mammalian cells. *Nature* **313:** 812–815.

Yates, J.L., N. Warren, D. Reisman, and B. Sugden. 1984. A *cis*-acting element from the Epstein-Barr viral genome that permits stable replication of recombinant plasmids in latently infected cells. *Proc. Natl. Acad. Sci.* **81:** 3806–3810.

27
Poxvirus DNA Replication

Paula Traktman
Departments of Cell Biology and Microbiology
Cornell University Medical College
New York, New York 10021

THE POXVIRAL LIFE CYCLE

Poxviruses, of which vaccinia is the prototype, are complex DNA viruses that display an unusual degree of physical and genetic autonomy from the host cell. The vaccinia genome encodes most, if not all, of the functions required for three temporally regulated phases of transcription and for DNA replication (Moss 1990a,b, 1994; Traktman 1990a,b, 1991). In addition, the 192-kb genome encodes numerous proteins that interact with and modulate the inflammatory and immune responses of the host (Smith 1993). The virus also encodes two protein kinases and a protein phosphatase (Traktman et al. 1989a; Guan et al. 1991; Banham and Smith 1992; Lin et al. 1992; Rempel and Traktman 1992; Lin and Broyles 1994), suggesting that its life cycle, like that of eukaryotic cells, is regulated by dynamic networks of protein phosphorylation.

Infection is initiated by the attachment of virions to an unknown receptor on the cell surface. After direct fusion of cellular and viral membranes, the inner viral core is delivered into the cytoplasmic compartment, where it remains intact for several hours. Early gene expression initiates immediately and is directed by the encapsidated transcriptional machinery. This virally encoded apparatus includes a heterodimeric transcription factor (vETF), a seven-subunit RNA polymerase, the RAP94 polymerase accessory protein, the heterodimeric capping enzyme/termination protein, and the heterodimeric poly(A) polymerase, one of whose subunits also functions as a methyltransferase during mRNA cap maturation (Moss 1990a). An encapsidated RNA helicase is also essential for transcription (Shuman 1992), and two other encapsidated NTPases are reported to affect transcription in vivo and/or in vitro (Kunzi and Traktman 1989; Kahn and Esteban 1990; Diaz-Guerra and Esteban 1993; Simpson and Condit 1994; Bayliss and Condit 1995). Early mRNAs are extruded from the core and translated on host polysomes. A secondary uncoating event then ensues, which releases the viral genome into the cytoplasm where it is accessible to the newly synthe-

 0-87969-459-9/96 $5 + .00

sized replication proteins. DNA replication, in addition to providing progeny genomes, is also an obligate step for the switch to intermediate gene expression. Some feature of the replicated template is required in *cis* for the activation of intermediate promoters (Vos and Stunnenberg 1988; Keck et al. 1990) by intermediate transcription factors. Among the products of intermediate gene expression are the transcription factors required for the expression of late genes. Late gene products include encapsidated enzymes and structural proteins.

Virion morphogenesis is a complex process that leads to the production of two major classes of mature virions. Intracellular mature virions (IMV) remain within the cytoplasm of the infected cell and, in the wild, are presumed to play a role in horizontal spread of poxviruses among individuals. They are delimited by a single membrane derived from the intermediate compartment between the endoplasmic reticulum and the Golgi apparatus (Sodeik et al. 1993). A subset of virions acquire an additional wrapping from the *trans*-Golgi network (Schmelz et al. 1994), fuse with the plasma membrane, and are released by exocytosis. Some of these enveloped virions remain tethered to the cell surface (cell-associated virions, CEV) and mediate spread to proximal cells (Blasco and Moss 1992). CEV are responsible for the ability of vaccinia virus to form discrete and localized plaques under liquid overlay in tissue culture. The remainder of the enveloped virions are fully released from the infected cell (extracellular enveloped virions, EEV) and are responsible for distal spread within the host.

DISCUSSION OF POXVIRAL DNA REPLICATION

Spatial and Temporal Localization

The localization of poxvirus replication to the cytoplasm of the infected cell is unique among DNA viruses. DNA replication occurs in viral factories or virosomes, localized regions within the cytoplasm which are devoid of cellular organelles and serve as foci of DNA synthesis and virion morphogenesis. Release of the viral genome into the cytoplasm (Sarov and Joklik 1972) occurs during secondary uncoating and brings the DNA template and the newly synthesized replication proteins into contact. At high multiplicities of infection, the onset of DNA replication is detectable at approximately 2 hpi, as measured by the incorporation of [^{3}H]thymidine into macromolecular, cytoplasmic species. Although thymidine incorporation declines by 4–5 hpi, other assays of viral DNA accumulation reveal that DNA synthesis continues until 10–12 hpi (Rempel et al. 1990; Traktman 1990a, 1991). Changes in nucleotide pools and feedback inhibition of thymidine kinase (TK) have been invoked to explain this discrepancy. It is now generally accepted that viral DNA accumulation can be most accurately quantitated by dot-blot filter hybridization. Genetic analyses reveal that late gene expression is not required for DNA replication. However, when an inhibitor of protein synthesis is added to infected cultures at 3–5 hpi, the rate of subsequent DNA accumulation is compromised. It is possible that optimal replication requires the continuous production of viral and/or cellular proteins which are also synthesized at intermediate and/or late times after infection. Replication can be categorized as having two functional components, one involving the synthesis of nascent genomes which are present as concatemeric intermediates (see Figs. 1, 2), and the other involving the resolution of these intermediates to mature, monomeric genomes (see Fig. 3). These two processes are separable both temporally and genetically. Replication does not appear to be coupled to genome encapsidation within assembling virions, since it proceeds normally when the latter process is inhibited. Moreover, significantly more DNA is synthesized than is encapsidated into progeny virions.

The Genomic Template: Sequence and Nucleoprotein Complex

Telomeric Sequence Organization

The vaccinia genome is a linear duplex of 192 kbp that has covalently closed hairpin termini (Baroudy et al. 1982; Goebel et al. 1990). The DNA is extremely AT-rich (67%) and is not methylated. The hairpin termini (104 nucleotides) contain almost exclusively A and T residues, and each contains 12 extrahelical residues (10 on one strand and 2 on the other) in addition to 44 fully paired bases and a 4-nucleotide terminal loop. This general organization is retained in all poxviruses, although the number and position of the extrahelical bases vary. Two isoforms of the hairpin are found in any population of genomes; they represent inverted complements of one another and are termed *flip* and *flop*. The maintenance of the extrahelical and mismatched bases within these hairpin termini, and their presence in both isoforms, must be addressed in any replication model proposed.

Internal to the hairpins is an additional 87 bp of sequence which is found at both genomic telomeres. This sequence contains elements required both for replication of exogenous templates and for resolution of concatemeric intermediates (discussed in detail below). Beyond this region are found clusters of tandemly repeated sequences. The vaccinia genome (strain WR) contains 13 copies of a 70-bp element, a 325-bp

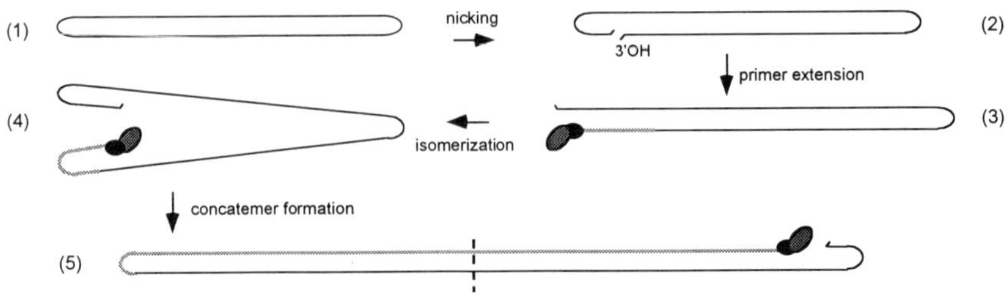

Figure 1 Schematic representation of a working model for poxviral DNA replication. The parental genome, a linear duplex with covalently closed termini, is shown at the top left (*1*). Replication begins with the introduction of a nick near one of the termini; nicking leaves a free 3′-OH group, as shown (*2*). The polymerase/processivity factor complex (shaded and filled ovals, respectively) then performs strand-displacement primer extension (*3*). The nascent strand is shown as a stippled line. This nascent DNA and its template are both self-complementary, and therefore the intermediate can isomerize to the form shown in *4*. The polymerase complex is then poised to continue synthesis along the length of the genome, around the hairpin, and then along the complementary strand. The intermediate generated by this process is a tail/tail dimer, as shown in *5*. The axis of symmetry in this dimeric intermediate is shown by a vertical dashed line.

spacer, an additional 18 copies of the 70-bp element, 2 copies of a 125-bp element, and 8 copies of a 54-bp element. Again, the general scheme is conserved among poxviruses, although the precise number and organization of the repeats do differ. These repeats are thought to participate in frequent inter- and intragenomic recombinational events.

The Nucleoprotein Complex

The encapsidated template is thought to be wrapped by at least two DNA-binding proteins of 25 kD and 11 kD. The 25-kD protein is encoded by the L4 gene and is proteolytically processed during virion morphogenesis by removal of 32 amino-terminal amino acids from the 28-kD primary translation product (Yang and Bauer 1988a,b). This abundant protein (7000 molecules per virion) possesses DNA-binding activity and, under physiological salt conditions, exhibits a preference for single-stranded DNA. In the absence of the L4 protein, normal numbers of virions are assembled, but they are severely compromised in infectivity (Wilcock and Smith 1994). It has been hypothesized that the L4 protein stabilizes a subset of the genomic sequences in a single-stranded conformation. The 11-kD protein is encoded by the F18 gene; it is not subject to proteolytic maturation but is phosphorylated on two serine

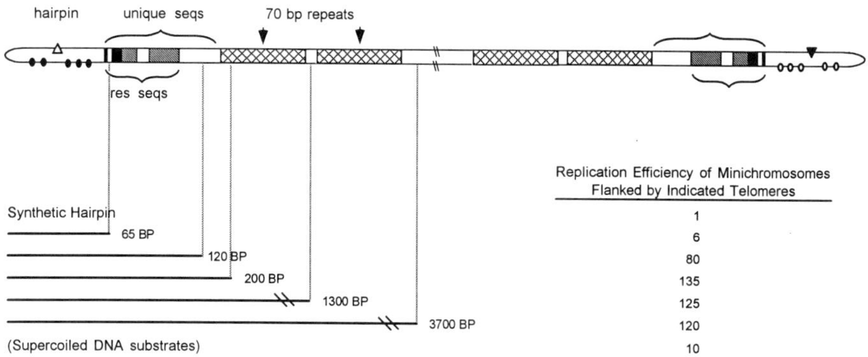

Figure 2 Organization of vaccinia virus genomic termini and analysis of the sequences involved in template activity. The genome is shown at the top, with features of the terminal organization shown in detail. The ovals and triangle shown within the 104-nt hairpin represent the extrahelical bases. The elements comprising the resolution sequence (res seqs) are shown as filled (motifs IA and I) and shaded (motifs II and III) boxes. The extent of the unique sequence block (unique seqs) beyond these resolution motifs is also indicated. The two blocks of 70-bp repeats (13 and 18 copies, respectively) are shown as cross-hatched boxes separated by an intervening spacer sequence. The lower part of the figure represents an analysis of the replication efficiency of exogenous substrates introduced into infected cells. Minichromosome substrates contained a central core of plasmid sequence and telomeres derived from the vaccinia genome. A minichromosome containing synthetic, nonviral hairpin termini was used as a control and assigned a replication level of 1. The substrates tested contained approximately 65, 120, 200, 1300, or 3700 bp of viral sequences at each terminus, as shown; the relative efficiency to which these substrates replicated is shown in the column to the right. The relative length of the telomeric viral sequences contained in each substrate is shown with a horizontal black line; the stippled gray lines extending upward toward the drawing of the genome delimit the terminal elements contained in each substrate. The final group of substrates tested were supercoiled plasmid preparations. In each case, the values shown represent the average of three independent experiments; the differences observed in replication efficiency were statistically significant with the exception of 135, 125, and 120, which were not statistically different from one another. (Data and figure adapted from S. Du and P. Traktman, in prep.)

residues (Kao et al. 1981; Nowakowski et al. 1978; Kao and Bauer 1987; Zhang and Moss 1991). It shows a preference for binding to superhelical DNA substrates in vitro and is thought be present at 27,000 copies per virion. In the absence of the F18 protein, virion maturation is blocked at a stage that appears to coincide with genome encapsidation. The DNA-binding properties of F18, as well as its abundance and its tendency to

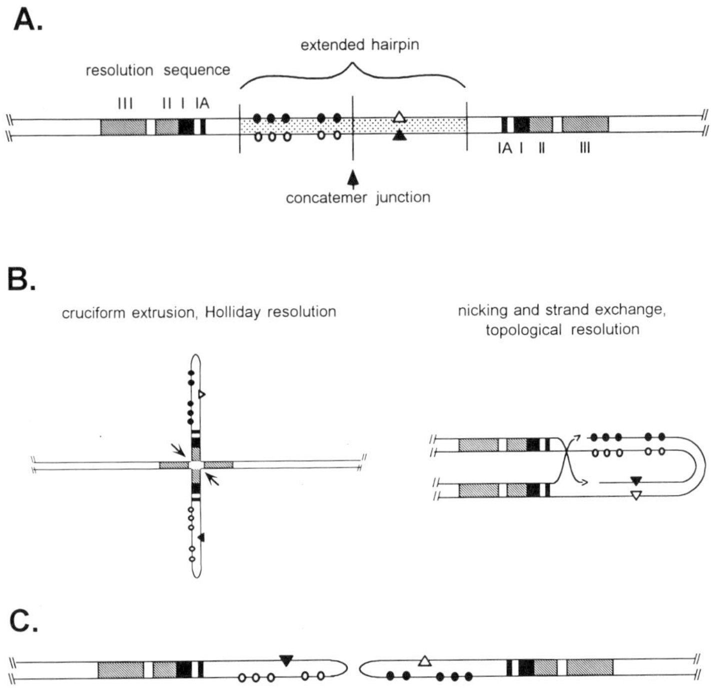

Figure 3 Resolution of concatemeric intermediates to progeny genomes. The central region of the dimeric replication intermediate is shown in panel *A*. The organization of the resolution sequence motifs (IA, I, II, III) on either side of the concatemer junction is illustrated. The extrahelical bases present in the hairpin of the parental genome are shown as ovals or triangles. The hairpin sequences are present in a fully duplexed conformation in the dimeric intermediate; note, however, that there is asymmetry between the hairpin sequences found on the left and right sides of the concatemer junction. Panel *B* shows two possible mechanisms whereby the two copies of the resolution sequences are brought together and resolved. The drawing on the left illustrates extrusion of the hairpins into a cruciform conformation. Cruciform extrusion is shown as extending through box II of the resolution sequence; only extrusion through box I is mandated by the available data. The base of the cruciform would be recognized as a Holliday junction structure, and resolution would be accomplished by asymmetric cleavage (*arrows*) and religation. The drawing on the right illustrates an alternative model involving nicking, strand exchange, and topological resolution. For simplicity, strand exchange is shown as initiating between the hairpin and box IA; however, we know that exchange extends into the genome at least as far as box I of the resolution motif. Panel *C* illustrates the telomeric structure of the mature progeny genomes formed by resolution of the intermediate shown in *A* via either of the mechanisms shown in *B*. Restoration of the hairpin termini with extrahelical bases is indicated; the two forms of the hairpin shown, which are inverted complements of one another, are referred to as "flip" and "flop."

multimerize, have led to the hypothesis that it serves as the major genomic condensing protein. There is some evidence in the literature that these two components of the nucleoprotein complex may be linked by inter-protein disulfide bridges (Ichihashi et al. 1984). Although both of these proteins are bound to the encapsidated genome, it has not been proven rigorously that they remain on the genome after secondary uncoating, and therefore it can only be hypothesized that either or both are present on the DNA replication template.

A Working Model for Poxvirus Replication

DNA replication is thought to begin with the introduction of a nick in sequences proximal to a genomic terminus (see Fig. 1). Formation of the nick has been inferred from the increased topological freedom and altered sedimentation properties that are associated with intracellular viral genomes after infection is under way (Pogo 1977). Strand cleavage leaves a free 3′-OH group to serve as a primer terminus for the viral DNA polymerase. Strand displacement synthesis proceeds toward the hairpin telomere, yielding extended template and product strands that are both self-complementary. This duplex terminus can isomerize to form two intrastrand duplexes. The 3′ terminus of the nascent strand is then poised to permit strand-displacement synthesis through the entire genome and around the distal hairpin terminus. The primary product would be a dimeric tail-to-tail concatemer; if the newly synthesized strand again folded back on itself, a tetramer could be formed, and so on. These concatemeric intermediates have been detected by electrophoretic analysis of replicating DNA (Moyer and Graves 1981; Merchlinsky et al. 1988; DeLange 1989; DeLange and McFadden 1990).

The simplest model for genome replication invokes only leading-strand synthesis. Reports of significant levels of single-stranded genomic DNA during virus replication are consistent with this model of strand-displacement synthesis (Esteban and Holowczak 1977; Pogo et al. 1981). However, both leading- and lagging-strand synthesis might occur, with the two meeting at the hairpin loop. Although no recent studies have provided evidence for such a model, early reports intimated that short nascent strands could be chased into larger forms, which was construed as evidence for Okazaki fragment synthesis (Esteban and Holowczak 1977). It also remains unknown whether reinitiation occurs on incompletely synthesized and/or resolved progeny genomes, leading to the formation of rather complex and branched genomic structures. Some evidence for such structures, which fail to enter gels under most conditions of electrophoresis, has been obtained (DeLange 1989; Merchlinsky and Moss 1989b).

Sequences Required for Template Replication

A definition of which *cis*-acting sequences are responsible for vaccinia DNA replication, and/or where DNA synthesis initiates, has remained elusive for many years. When cells infected with *ts* DNA^- mutants were held at the nonpermissive temperature to synchronize infection and then shifted to the permissive temperature, [^{3}H]thymidine was first incorporated within 150 nucleotides of the termini (Pogo et al. 1984). Because both telomeres were labeled, it was supposed that initiation could occur near the left and/or right telomere; the observation that concatemers arise as replication intermediates suggests that, within a given template, one site of initiation is likely.

Attempts to define an origin that could serve as an autonomous replication sequence were not successful and changed the general view of initiation within the field. Any supercoiled, but not linear, plasmid introduced into vaccinia- or Shope fibroma virus-infected cells was seen to replicate to some extent (as judged by the recovery of *Dpn*I-resistant material) (DeLange and McFadden 1986; Merchlinsky and Moss 1988). Replication was not stimulated by the inclusion of any viral sequences within these plasmids. Together, these data suggested that replication might initiate from random nicks present in transfected plasmids and that there might not be any viral sequences that recruited the replication machinery to the viral genome. This unusual feature was in part attributed to the cytoplasmic localization of viral infection, with the rationale that specificity was less important when a virus was not competing with an excess of nuclear DNA.

Clarity regarding the *cis*-acting sequences required for replication has emerged very recently with the analysis of another class of exogenous templates (S. Du and P. Traktman, in prep.). Minichromosomes containing a central region of plasmid DNA and telomeres derived from the vaccinia genome were constructed and introduced into infected cells (see Fig. 2). At 24 hpi, cytoplasmic DNA was harvested and the replication of the minichromosomes was analyzed. The level of replication was normalized to that seen with a minichromosome containing synthetic, nonviral, hairpin termini (assigned a replication level of 1). A sixfold increase in replicated product was seen when the telomeric 65 nucleotides of the viral genome were used; the level of replication increased to 80 when 120 bp of viral sequences were used. A maximum replication level of 120–135 was seen when 200 bp, 1300 bp, or 3700 bp of telomeric sequences were included. In contrast, several supercoiled substrates replicated to a value of 10. These data suggest that a linear template with hairpin termini containing 200 bp of viral telomeric sequences replicates to a maximum level within vaccinia-infected cells. Within these 200 bp are the hairpin loop and an additional 87 bp of unique sequence which contains boxes IA, I, II, and III of the resolution sequence (58 bp) as well as an additional 29 bp. Comparative analysis of the replication efficiency of the various substrates tested implies that sequence and/or structural elements extending throughout the 87 bp between the hairpin and the 70-bp repeats are necessary and sufficient for efficient replication of an exogenous minichromosome by the vaccinia replication machinery.

Resolution of Concatemeric Intermediates

It is now accepted that concatemeric intermediates arise during poxvirus replication. Because it appears that the viral protein(s) that directs the resolution of these intermediates is a late gene product (see below), the intermediates are more easily detected during the early stages of DNA replication. When late gene expression is impaired genetically or pharmacologically, these concatemers accumulate at the expense of mature genomes (Merchlinsky et al. 1988; DeLange 1989; Merchlinsky and Moss 1989b). The sequence obtained for the genomic junction within these concatemers is consistent with a tail-tail or head-head structure, as predicted previously (Moyer and Graves 1981; Merchlinsky et al. 1988). When resolution is blocked and the half-life of intermediates is therefore extended, head-tail isoforms are generated by homologous intergenomic recombination.

Analysis of the sequences required for resolution (see Fig. 3) was significantly aided by the demonstration that plasmids containing a cloned concatemeric junction fragment were resolved into linear minichromosomes with authentic hairpin termini following transfection into infected cells (DeLange et al. 1986; Merchlinsky and Moss 1986, 1989a; DeLange and McFadden 1987; Merchlinsky et al. 1988; Merchlinsky 1990a,b). Targeted mutagenesis of the terminal sequences permitted a thorough understanding of the sequences required for resolution. Several key regions have been defined: domains IA and I that are absolutely required for resolution, and domains II and III that augment the efficiency of resolution. This entire sequence is reiterated in an inverted repeat conformation on the other side of the resolution axis (see Fig. 3A). The sequence between the two resolution motifs (i.e., the extended duplex copy of the hairpin loop) must be a palindrome of less than 200 bp. Most of the natural palindrome can be deleted, however, and replaced with a synthetic palindrome. Therefore, neither the precise hairpin sequence nor the characteristic extrahelical bases are important for resolution. Experimental evidence suggests that the two copies of the resolution sequence are brought together during the recognition and resolution event. Sequence parity and spacing between the two elements are crucial. Within each of the two copies of the sequence, the two strands are fully complementary, one having served as the template for the synthesis of the other. On the other hand, the two copies are not identical, since they were independently derived from the two parental strands, which were not fully complementary due to the presence of extrahelical base pairs. Analysis of the resolution of a synthetic concatemeric junction in which one element contains an *Xba*I site in box I and the other element does not (due to a single base substitution) has revealed that the resolved telomeres are each heteroduplexes. Each contains one strand with the *Xba*I recognition sequence and one with the mutated sequence. These studies suggest that either the resolving concatemeric junction fragment extrudes in a cruciform configuration such that resolution at its base via a Holliday resolvase yields the two incompletely base-paired telomeres (Fig. 3B, left), or that recombination and branch migration accomplishes the same event (Fig. 3B, right). Either of these models would explain the need for a central palindromic sequence between the two copies of the resolution sequence. Resolution restores both the flip and the flop forms of the hairpin. Moreover, the recovery of the *Xba*I site in a heteroduplex form marks this site as the most telomeric position at which the resolving nick can occur (within box I, 16 nucleotides interior to the border of the central palindrome).

Site-directed mutagenesis has been employed to arrive at the consensus resolution sequence of boxes I and IA. The sequence $5'T_{6\text{-}7}N_{7\text{-}9}{}^{T}/_{C}AAA{}^{T}/_{A}\ 3'$, when present in inverted repeat conformation on either side of a double-stranded copy of the hairpin loop, is sufficient to direct telomere resolution. Examination of this sequence reveals that it contains the transcriptional start site associated with late viral genes (TAAAT). Indeed, the authentic Shope fibroma virus resolution sequence has been tested and found to function as a strong late promoter (Stuart et al. 1991), and transcription through the telomeric region has been detected in vivo (Parsons and Pickup 1990). Current models propose that late transcription through the concatemer junction could play a role in opening up the duplex region to facilitate entry of resolution proteins or topological rearrangement. Previously, the observation that resolution did not proceed when late viral genes were not expressed was interpreted as implicating a late protein in the resolution reaction. The overlap between the resolution sequence and late transcriptional start sites suggests that late transcription per se may be important for resolution. This finding may explain the unexpected observation that a *ts* mutant with a lesion in the capping enzyme appears to be deficient in telomere resolution at the nonpermissive temperature (Carpenter and DeLange 1991).

To date, the proteins responsible for recognizing the resolution sequence and/or executing resolution have not been identified, nor has the resolution reaction been reconstituted in vitro. It is presumed that one or more late proteins may be involved in this facet of DNA metabolism. An enzyme capable of concerted DNA nicking and cross-linking has been proposed as a possible participant in resolution (Traktman 1990a, 1991), but no clear data have emerged to confirm such an involvement.

Proteins Involved in Poxviral DNA Replication

A variety of genetic, biochemical, and molecular biological techniques have been applied to the identification of proteins involved in poxviral DNA replication. Confidence that most if not all of these proteins would be virally encoded was strengthened by the cytoplasmic localization of poxvirus replication and the observation that replication proceeded normally upon infection of enucleated cells. Three complementation groups of *ts* mutants have been shown to have primary and severe defects in DNA synthesis at the nonpermissive temperature; the E9, D5, and B1 gene products were identified by analysis of these mutants (Traktman et al. 1984; Evans and Traktman 1987; Roseman and Hruby 1987; Rempel et al. 1990). One additional mutant (D4 gene) appears to exhibit diminished DNA synthesis when infections are initiated and maintained at the nonpermissive temperature (Millns et al. 1994). Other genes have been identified by reverse genetics or by recognition of conserved motifs in the predicted amino acid sequence of genomic regions. A description of these proteins is provided below; a summary of the proteins is shown in Table 1.

DNA Polymerase

The 116-kD catalytic DNA polymerase is encoded by the E9 gene. Conserved domains I–VII common to the α family of polymerases are found within the enzyme, as are sequences diagnostic of the δ polymerases

Table 1 Vaccinia virus proteins involved in DNA replication

Protein	Gene	Molecular mass (kD)	Expression	Genetic analysis
DNA polymerase	E9	116	early	*ts* mutants exhibit fast-stop DNA^- phenotype
Processivity factor	A20	49	early	
D5 NTPase	D5	90	early	*ts* mutants exhibit fast-stop DNA^- phenotype, recombination defect
B1 protein kinase	B1	34	early	*ts* mutants exhibit DNA^- phenotype; severity varies with host cell
DNA ligase	A50	63	early	deletion confers resistance to etoposide and sensitivity to UV and bleomycin in tissue culture and attenuation in vivo
Thymidine kinase	J2	19	early	nonessential in tissue culture; deletion confers BrdU resistance in tissue culture and attentuation in vivo
Thymidylate kinase	A48	23	early	nonessential in tissue culture
Ribonucleotide reductase	F4, 14	34, 87	early	nonessential in tissue culture; HU-resistance conferred by amplification of small subunit; deletion causes attenuation in vivo
Topoisomerase I	H6	36	late, encapsid	essential
Uracil DNA glycosylase	D4	25	early	*ts* mutant exhibits reduced DNA synthesis only if replication is initiated at the nonpermissive temperature
dUTPase	F2	16	early	virus with large deletion encompassing F2 is viable, but precise deletion of F2 has not been possible

(Wang et al. 1989; Traktman 1990a; Taddie and Traktman 1993). The presence of domains Exo I, Exo II, and Exo III is consistent with the observation that the enzyme has an intrinsic 3′→5′ exonuclease activity (Challberg and Englund 1979; McDonald and Traktman 1994a). The enzyme has been overexpressed within the context of a vaccinia virus infection, and biochemical analysis indicates that it is a highly distributive enzyme under most assay conditions (McDonald and Traktman 1994a,b). It is not capable of strand displacement in vitro. Two *ts* mutants have been characterized, and aphidicolin^r, phosphonoacetic^r, and cytosine arabinoside^r mutants have also been generated (Defilippes 1984, 1989; Traktman et al. 1989b; Taddie and Traktman 1991, 1993). Among the drug^r mutants are those that exhibit mutator and antimutator phenotypes in vivo.

Processivity Factor

Unlike the purified E9 polymerase, the activity present within cytoplasmic extracts of vaccinia-infected cells is highly processive (W.F. McDonald and Traktman, unpubl.). This processive polymerase activity can be reconstituted by mixing recombinant E9 protein and extracts from cells infected with a *ts* mutant carrying a defective E9 protein. Activity could not be reconstituted using uninfected cell extracts, strongly suggesting that the processivity factor was virally encoded. The processivity activity has recently been purified from infected cell extracts and appears to have a native molecular mass of 45 kD. Amino acid analysis of the purified material has led to the assignment of the A20 open reading frame (ORF) as the processivity factor (N. Klemperer and P. Traktman, unpubl.). The structure of the gene is consistent with its being expressed early during infection, and the predicted translation product of the ORF has a molecular mass of 49 kD.

D5 NTPase

Four *ts* mutants with lesions in the D5 gene (*ts*17, 24, 69, and 6389) have a DNA$^-$ phenotype at the nonpermissive temperature (Evans and Traktman 1987, 1992; Roseman and Hruby 1987). When cultures infected with *ts*17 are shifted to the nonpermissive temperature in the midst of DNA replication, further synthesis of DNA ceases within 5 minutes. Together, these data suggest that D5 plays an essential and intimate role in the process of DNA elongation. Three of the *ts* mutants with lesions in the amino-terminal portion of the 90-kD protein also show defects in homologous recombination. In addition, the feedback inhibition of the viral TK that normally develops during vaccinia infection was not seen during nonpermissive infections with *ts*17 or *ts*24. This phenotype was not common to all infections in which DNA synthesis was blocked, suggesting that D5 might play a role in nucleotide biosynthesis. Recently, the D5 protein has been overexpressed within the context of vaccinia virus infection and purified to homogeneity. The D5 protein possesses intrinsic nucleic-acid-independent NTPase activity (Evans et al. 1995). All of the common ribonucleotides and deoxyribonucleotides are hydrolyzed, and no stimulation with any RNA or DNA cofactor has been observed. In light of the possible correlation of D5 inactivation with altered dNTP metabolism, purified D5 was tested for nucleoside kinase activity, but none was found. Further insight into the functional contribution of D5 to DNA synthesis has proven elusive, suggesting perhaps that interactions with other, as yet unidentified, proteins may be important.

B1 Protein Kinase

Members of the third complementation group of *ts* mutants with a DNA^- phenotype (*ts*2, *ts*25) contain lesions within the B1 gene (Rempel et al. 1990), which encodes a 34-kD serine/threonine protein kinase expressed only at early times after infection (Traktman et al. 1989a; Banham and Smith 1992; Lin et al. 1992; Rempel and Traktman 1992). Recombinant B1 protein produced in bacteria is enzymatically active in the absence of any other viral proteins. The only viral protein identified as a B1 substrate to date is the H5 protein (Beaud et al. 1995). This amphipathic protein shows affinity for both membranes and DNA, but its functional role in vivo is unknown. The phenotype of the B1 mutants is complex and confusing (Rempel and Traktman 1992). In vitro, recombinant *ts*2 protein has no detectable kinase activity, whereas *ts*25 protein has 3% of the specific kinase activity of the *wt* protein. Neither protein is temperature-sensitive. In mouse L cells, both *ts*2 and *ts*25 are strictly DNA^- at the nonpermissive temperature; this restriction is much leakier in primate BSC40 cells. However, whereas the *wt* protein is stable throughout infection, the mutant proteins are extremely labile in both cell types at both permissive and nonpermissive temperatures. Because B1 synthesis ceases by 3 hpi, the B1 protein is effectively absent by the midstages of infection. Nevertheless, when cultures were shifted to the nonpermissive temperature at 7 hpi in the midst of DNA replication, further DNA synthesis ceased after a lag of approximately 30 minutes. These data have elicited a model which proposes that one or more components of the replication complex behave in a temperature-sensitive fashion in the absence of appropriate phosphorylation by B1. There is ample precedent to support the hypothesis that phosphorylation might be required to stabilize protein-protein interactions at 39.5°C to a far greater extent than at 32°C.

DNA Ligase

Vaccinia virus encodes a DNA ligase that is not essential for virus replication in tissue culture (Kerr and Smith 1989, 1991; Colinas et al. 1990). The 63-kD enzyme is expressed at early times after infection and utilizes an ATP-dependent mechanism of catalysis. The ability of ligase-deficient viruses to replicate normally in culture suggests either that the mode of DNA synthesis is independent of ligase action, arguing against the existence of lagging-strand synthesis, or that one of the cellular ligases can substitute for the viral enzyme. Support for the latter possibility comes from the observation that ligase-deficient isolates are attenuated in vivo (Kerr et al. 1991). Viruses lacking the DNA ligase appear to be sensitive to DNA-damaging agents in tissue culture, suggesting that the ligase may play an important role in DNA repair. This observation is interesting in light of the recent observation that the human ligase II, a repair enzyme, shows significant homology with the viral gene product (Wang et al. 1994). Another mystery involving the viral DNA ligase is the recent observation that viral mutants resistant to etoposide, a drug most frequently identified as an inhibitor of type II topoisomerases and also shown to inhibit the viral topoisomerase in vitro (J. Countryman and P. Traktman, unpubl.), contain lesions within the DNA ligase gene (DeLange et al. 1995). Presence of a *wt* ligase gene is correlated with etoposide^{s}, and null alleles of the ligase confer resistance. Particular amino-terminal point mutations confer an even higher level of resistance. No mechanistic explanation for this unusual set of observations has been put forward, but it is clear that the role of the viral ligase remains to be elucidated.

Thymidine Kinase and Thymidylate Kinase

Vaccinia encodes two enzymes involved in thymidine biosynthesis, both of which are dispensable in tissue culture. The TK (Hruby and Ball 1982; Weir et al. 1982; Hruby et al. 1983; Weir and Moss 1983) is synthesized early after infection and functions as a homotetramer. The enzymatic activity of the enzyme declines as infection proceeds because of a sensitivity to feedback inhibition by accumulating dTDP and dTTP (Hruby 1985). This sensitivity to feedback inhibition is diagnostic of type II TK enzymes (as distinct from the type I enzymes encoded by herpesviruses); mutational studies indicate that a single amino acid replacement in conserved domain IV can eliminate feedback inhibition without impairing enzyme activity (Black and Hruby 1992). The substrate specificity of the viral TK is limited to thymidine and BrdU. The ability to phosphorylate BrdU has been exploited for the construction of vaccinia virus recombinants; insertion of foreign DNA into the TK locus inactivates TK and leads to a BrdU-resistant phenotype. Viruses deficient in TK activity are attenuated in vivo (Buller et al. 1985). The second step of dTTP biosynthesis is accomplished by the viral thymidylate kinase enzyme (Hughes et al. 1991). The vaccinia genome also contains sequences bearing homology with known guanylate kinases, but no intact reading frame remains (Smith et al. 1991).

Ribonucleotide Reductase

Vaccinia encodes a heterodimeric ribonucleotide reductase (RR) that directs the conversion of rNDPs to dNDPs (Slabaugh et al. 1988; Tengelsen et al. 1988). The structure of the viral enzyme is highly homologous to that of its mammalian and bacterial counterparts. The large regulatory subunit is responsible for the allosteric modulation of the enzyme by nucleotides. The small, catalytic subunit contains the active-site *tyrosine* and is sensitive to the action of hydroxyurea (HU). Amplification of the gene encoding this subunit confers HU^r. Phenotypic resistance to HU can also be accomplished by inclusion of deoxyadenosine in the culture medium, consistent with the observation that dATP pools are decreased most dramatically by HU treatment (Slabaugh et al. 1991). Viral mutants lacking an intact RR are severely attenuated in vivo (Child et al. 1990); in tissue culture, the cellular enzyme is apparently sufficient to support the normal replication of these mutants.

Topoisomerase I

Vaccinia virus encodes and encapsidates a type I topoisomerase (Bauer et al. 1977; Shaffer and Traktman 1987; Shuman and Moss 1987). The 36-kD product of the H6 gene directs the relaxation of supercoiled DNA in a manner analogous to the yeast and mammalian enzymes. Transient cleavage of a single DNA strand occurs at a highly conserved consensus site, $^{C}/_{T}$CCTT↓ (Shuman and Prescott 1990; Shuman 1991). The gene is apparently essential for virus viability in tissue culture (Shuman et al. 1989a), but the absence of conditionally lethal mutants with H6 lesions has hampered definitive analyses of the biological roles played by the enzyme. By analogy with other systems, a role in relieving topological strain during transcription and DNA replication is posited. Participation in recombination has also been proposed. In contrast to the paucity of data regarding function in vivo, the small size of the vaccinia enzyme and the ability to produce active recombinant protein have facilitated the accumulation of a significant body of data regarding structure/function considerations. The active-site *tyrosine* has been identified (Shuman et al. 1989b), and numerous mutations that impair DNA cleavage and relaxation have been characterized (Morham and Shuman 1990, 1992; Klemperer and Traktman 1993; Wittschieben and Shuman 1994).

Presumed Functions

At a minimum, it is also to be expected that ongoing viral DNA synthesis will involve a single-strand DNA-binding protein (SSB) and a DNA helicase. A helicase is invoked to permit the processive DNA polymerase complex to travel through a duplex substrate. However, no data regarding the nature of the enzyme involved have so far been obtained. Regarding an SSB, the best candidate to date is the 34-kD phosphoprotein encoded by the I3 gene. This protein shows a strong preference for single- versus double-stranded DNA, and gel-shift assays have been used to derive a binding site size of approximately 10 nt/I3 molecule (S.C. Rochester and P. Traktman, unpubl.). The protein is essential (S.C. Rochester and P. Traktman, unpubl.) and is synthesized at high levels both before and during DNA synthesis. A potential interaction between the I3 protein and the ribonucleotide reductase has also been observed (Davis and Mathews 1993). Additional proteins are expected to be involved in the initiation of replication and the resolution of concatemeric intermediates; no compelling candidates have been identified for either function.

A Potential Link between DNA Replication and DNA Repair

Comparative analysis of distinct vaccinia strains, as well as isolates maintained in different laboratories, has indicated that the genomes of poxviruses have a high degree of sequence stability. The DNA polymerase has an active proofreading exonuclease and exhibits a high level of fidelity. The ability of some polymerase lesions to confer mutator or antimutator phenotypes suggests that the fidelity of the polymerase plays a significant role in determining the net mutation rate of the virus (Taddie and Traktman 1991, 1993). Efforts to detect virus-specific mismatch repair enzymes have not been successful. However, the virus is known to encode two enzymes that are classically associated with DNA repair. A DNA replication defect is seen in a *ts* mutant with a lesion in one of these repair enzymes, suggesting that replication and repair might be physically and/or functionally linked.

dUTPase

The vaccinia virus F2 gene is expressed early after infection and encodes a functional enzyme that has homology with diverse dUTPases and shows a strict preference for dUTP as substrate (Broyles 1993). By analogy with other systems, the dUTPase is presumed to lower the intracellular concentration of dUTP and so prevent its incorporation into nascent viral genomes. The observation that a viral isolate containing an extensive terminal deletion encompassing the F2 gene (as well as many others) is viable (Perkus et al. 1991) led to the conclusion that the F2 gene is not essential for virus replication in tissue culture. However, efforts to isolate a virus whose sole alteration is a null mutation in the F2 gene have not been successful (N. Roseman, pers. comm.), suggesting that this conclusion might have been erroneous. It is possible that the elimination of other genes in the original, extensive deletion somehow obviated the need for the viral dUTPase.

Uracil DNA Glycosylase

The vaccinia virus D4 gene is also expressed at early times after infection and encodes a functional uracil DNA glycosylase (Stuart et al. 1993; Upton et al. 1993). This enzyme hydrolyzes incorporated UMP residues at the N-glycosylic bond; removal of the uracil moiety targets the site for cleavage by an apurinic/apyrimidinic (AP) nuclease and repair by a DNA polymerase. The viral enzyme is apparently essential, as shown both by the inability to isolate a viral recombinant containing an insertionally inactivated D4 allele and by the existence of a *ts* mutant containing a D4 lesion (Stuart et al. 1993; Millns et al. 1994). Interestingly, this mutant (*ts*4149) shows a partial DNA^- phenotype. When infections are initiated and maintained at the nonpermissive temperature, 1–2% of permissive levels of DNA are synthesized, and these DNA products do not mature into monomer genomes. When infections are initiated at the permissive temperature, a shift to nonpermissive temperature in the midst of DNA replication does not lead to a cessation of DNA synthesis. These data suggest that either (1) the enzyme plays a role only prior to, or during the initiation of, DNA replication or, more likely, (2) the enzyme is assembled into a multiprotein complex, and the thermolability of the mutant protein is masked once complex formation is complete. The latter model suggests that a bifunctional replication/repair complex might track along the DNA during viral DNA synthesis. Further elucidation of the potential role of this enzyme in replication and/or repair awaits the analysis of additional *ts* alleles. No data are available regarding AP endonucleases that might act subsequent to the action of the uracil DNA glycosylase during DNA repair.

ACKNOWLEDGMENTS

I thank the members of the poxvirus field who generously communicated their unpublished results to me; I also apologize for the inevitable lapses in providing sufficient acknowledgment of published work. The scientific contributions and support provided by the members of my own laboratory have been invaluable. Some of the work cited in this chapter was supported by a grant to P.T. from the National Institutes of Health (AI-21758); the generous support of Dr. David Cofrin and a special group of donors from the Dorothy Rodbell Cohen foundation is also acknowledged.

REFERENCES

Banham, A.H. and G.L. Smith. 1992. Vaccinia virus gene B1R encodes a 34-kDa serine/threonine protein kinase that localizes in cytoplasmic factories and is packaged into virions. *Virology* **191:** 803–812.

Baroudy, B.M., S. Vankatesan, and B. Moss. 1982. Incompletely base-paired flip-flop terminal loops link the two DNA strands of the vaccinia virus genome into one uninterrupted polynucleotide chain. *Cell* **28:** 315–324.

Bauer, W.R., E.C. Ressner, J. Kates, and J.V. Patzke. 1977. A DNA nicking-closing enzyme encapsidated in vaccinia virus: Partial purification and properties. *Proc. Natl. Acad. Sci.* **74:** 1841–1845.

Bayliss, C.D. and R.C. Condit. 1995. The vaccinia virus A18R gene product is a DNA-dependent ATPase. *J. Biol. Chem.* **270:** 1550–1556.

Beaud, G., R. Beaud, and D.P. Leader. 1995. Vaccinia virus gene H5R encodes a protein that is phosphorylated by the multisubstrate vaccinia virus B1R protein kinase. *J. Virol.* **69:** 1819–1826.

Black, M.E. and D.E. Hruby. 1992. A single amino acid substitution abolishes feedback inhibition of vaccinia virus thymidine kianse. *J. Biol. Chem.* **267:** 9743–9748.

Blasco, R. and B. Moss. 1992. Role of cell-associated enveloped vaccinia virus in cell-to-cell spread. *J. Virol.* **66:** 4170–4179.

Broyles, S.S. 1993. Vaccinia virus encodes a functional dUTPase. *Virology* **195:** 863–865.

Buller, R.M., G.L. Smith, K. Cremer, A.L. Notkins, and B. Moss. 1985. Decreased virulence of recombinant vaccinia virus expression vectors is associated with a thymidine kinase-negative phenotype. *Nature* **317:** 813–815.

Carpenter, M.S. and A.M. DeLange. 1991. A temperature-sensitive lesion in the small subunit of the vaccinia virus-encoded mRNA capping enzyme causes a defect in viral telomere resolution. *J. Virol.* **65:** 4042–4050.

Challberg, M.D. and P.T. Englund. 1979. Purification and properties of the deoxyribonucleic acid polymerase induced by vaccinia virus. *J. Biol. Chem.* **254:**

7812–7819.

Child, S.J., G.J. Palumbo, R.M. Buller, and D.E. Hruby. 1990. Insertional activation of the large subunit of ribonucleotide reducase is associated with reduced virulence in vivo. *Virology* **174:** 625–629.

Colinas, R.J., S.J. Goebel, S.W. Davis, G.P. Jonson, E.K. Norton, and E. Paoletti. 1990. A DNA ligase gene in the Copenhagen strain of vaccinia virus is nonessential for viral replication and recombination. *Virology* **179:** 267–275.

Davis, R.E. and C.K. Mathews. 1993. Acidic C terminus of vaccinia virus DNA binding protein interacts with ribonucleotide reductase. *Proc. Natl. Acad. Sci.* **90:** 745–749.

Defilippes, F.M. 1984. Effect of aphidicolin vaccinia virus: Isolation of an aphidicolin-resistant mutant. *J. Virol.* **52:** 474–482.

———. 1989. Site of the base change in the vaccinia virus DNA polymerase gene which confers aphidicolin resistance. *J. Virol.* **63:** 4060–4063.

DeLange, A.M. 1989. Identification of temperature-sensitive mutants of vaccinia virus that are defective in conversion of concatemeric replicative intermediates to the mature linear DNA genome. *J. Virol.* **63:** 2437–2444.

DeLange, A.M. and G. McFadden. 1986. Sequence-nonspecific replication of transfected plasmid DNA in poxvirus-infected cells. *Proc. Natl. Acad. Sci.* **83:** 614–618.

———. 1987. Efficient resolution of replicated poxvirus telomeres to native hairpin structures requires two inverted symmetrical copies of a core target DNA sequence. *J. Virol.* **61:** 1957–1963.

———. 1990. The role of telomeres in poxvirus DNA replication. *Curr. Top. Microbiol. Immunol.* **163:** 71–92.

DeLange, A.M., M.S. Carpenter, J. Choy, and V.E. Newsway. 1995. An etoposide-induced block in vaccinia virus telomere resolution is dependent on the virus-encoded DNA ligase. *J. Virol.* **69:** 2082–2091.

DeLange, A.M., M. Reddy, D. Scraba, C. Upton, and G. McFadden. 1986. Replication and resolution of cloned poxvirus telomeres in vivo generates linear minichromosomes with intact viral hairpin termini. *J. Virol.* **59:** 249–259.

Diaz-Guerra, M. and M. Esteban. 1993. Vaccinia virus nucleoside triphosphate phosphohydrolase I controls early and late gene expression by regulating the rate of transcription. *J. Virol.* **67:** 7561–7562.

Esteban, M. and J.A. Holowczak. 1977. Replication of vaccinia DNA in mouse L cells. III. Intracellular forms of viral DNA. *Virology* **82:** 308–322.

Evans, E. and P. Traktman. 1987. Molecular genetic analysis of a vaccinia virus with an essential role in DNA replication. *J. Virol.* **61:** 3152–3162.

———. 1992. Characterization of vaccinia virus DNA replication mutants with lesions in the D5 gene. *Chromosoma* **102:** S72–S82.

Evans, E., N. Klemperer, R. Ghosh, and P. Traktman. 1995. The essential replication protein encoded by the vaccinia virus D5 gene is a nucleic acid independent NTPase. *J. Virol.* **69:** 5353–5361.

Goebel, S.J., G.P. Johnson, M.E. Perkus, S.W. David, J.P. Winslow, and E. Paoletti. 1990. The complete DNA sequence of vaccinia virus. *Virology* **179:** 247–266.

Guan, K., S.S. Broyles, and J.E. Dixon. 1991. A tyr/ser phosphatase encoded by vaccinia virus. *Nature* **350:** 359–362.

Hruby, D.E. 1985. Inhibition of vaccinia virus thymidine kinase by the distal products of its own metabolic pathway. *Virus Res.* **2:** 151–156.

Hruby, D.E. and L.A. Ball. 1982. Mapping and identification of the vaccinia virus thymidine kinase gene. *J. Virol.* **43:** 403–409.

Hruby, D.E., R.A. Maki, D.B. Miller, and L.A. Ball. 1983. Fine structure analysis and nucleotide sequence of the vaccinia virus thymidine kinase gene. *Proc. Natl. Acad. Sci.* **80:** 3411–3415.

Hughes, S.J., L.H. Johnston, A. deCarlos, and G.L. Smith. 1991. Vaccinia virus encodes an active thymidylate kinase that complements a cdc8 mutant of *Saccharomyces cerevisiae. J. Biol. Chem.* **266:** 20103–20109.

Ichihashi, Y., M. Oie, and T. Tsuruhara. 1984. Location of DNA-binding proteins and disulphide-linked proteins in vaccinia virus structural elements. *J. Virol.* **50:** 929–938.

Kahn, J.S. and M. Esteban. 1990. Identification of the point mutations in two vaccinia virus nucleoside triphosphate phosphohydrolase I temperature-sensitive mutants and role of this DNA-dependent ATPase enzyme in virus gene expression. *Virology* **174:** 459–471.

Kao, S.Y. and W.R. Bauer. 1987. Biosynthesis and phosphorylation of vaccinia virus structural protein VP11. *Virology* **159:** 399–407.

Kao, S., E. Ressner, J. Kates, and W.R. Bauer. 1981. Purification and characterization of a superhelix binding protein from vaccinia virus. *Virology* **111:** 500–508.

Keck, J.G., C.J. Baldick, Jr., and B. Moss. 1990. Role of DNA replication in vaccinia virus gene expression: A naked template is required for transcription of three late trans-activator genes. *Cell* **61:** 801–809.

Kerr, S.M. and G.S. Smith. 1989. Vaccinia virus encodes a polypeptide with DNA ligase activity. *Nucleic Acids Res.* **17:** 9039–9050.

———. 1991. Vaccinia virus DNA ligase is nonessential for virus replication: Recovery of plasmids from virus-infected cells. *Virology* **180:** 625–632.

Kerr, S.M., L.H. Johnston, M. O'Dell, S.A. Duncan, K.M. Law, and G.L. Smith. 1991. Vaccinia DNA ligase complements *Saccharomyces cerevisiae* cdc9, localizes in cytoplasmic factories and affects virulence and virus sensitivity to DNA damaging agents. *EMBO J.* **10:** 4343–4350.

Klemperer, N. and P. Traktman. 1993. Biochemical analysis of mutant alleles of the vaccinia virus topoisomerase I carrying targeted substitutions in a highly conserved domain. *J. Biol. Chem.* **268:** 15887–15899.

Kunzi, M.S. and P. Traktman. 1989. Genetic evidence for involvement of vaccinia virus DNA-dependent ATPase I in intermediate and late gene expression. *J. Virol.* **63:** 3999–4010.

Lin, S. and S.S. Broyles. 1994. Vaccinia protein kinase 2: A second essential serine-threonine protein kinase encoded by vaccinia virus. *Proc. Natl. Acad. Sci.* **91:** 7653–7657.

Lin, S., W. Chen, and S.S. Broyles. 1992. The vaccinia virus B1R gene product is a serine/threonine protein kinase. *J. Virol.* **66:** 2717–2723.

McDonald, W.F. and P. Traktman. 1994a. Overexpression and purification of the vaccinia virus DNA polymerase. *Protein Express. Purif.* **5:** 409–421.

———. 1994b. Vaccinia virus DNA polymerase: In vitro analysis of parameters affecting processivity. *J. Biol. Chem.* **269:** 31190–31197.

Merchlinsky, M. 1990a. Mutational analysis of the resolution sequence of vaccinia virus DNA: Essential sequence consists of two separate AT-rich regions highly conserved. *J. Virol.* **64:** 5029–5035.

———. 1990b. Resolution of poxvirus telomeres: Processing of vaccinia virus concatemer junctions by conservative strand exchange. *J. Virol.* **64:** 3437–3446.

Merchlinsky, M. and B. Moss. 1986. Resolution of linear minichromosomes with hairpin ends from circular plasmids containing vaccinia virus concatemer junctions. *Cell* **45:** 879–884.

———. 1988. Sequence-independent replication and sequence-specific resolution of plasmids containing the vaccinia virus concatemer junction: Requirements for early and late *trans*-acting factors. *Cancer Cells* **6:** 87–93.

———. 1989a. Nucleotide sequence required for resolution of the concatemer junction of vaccinia virus DNA. *J. Virol.* **63:** 4354–4361.

———. 1989b. Resolution of vaccinia virus DNA concatemer junctions requires late gene expression. *J. Virol.* **63:** 1595–1603.

Merchlinsky, M., C.F. Garon, and B. Moss. 1988. Molecular cloning and sequence of the concatemer junction from vaccinia virus replicative DNA. *J. Mol. Biol.* **199:** 399–413.

Millns, A.K., M.S. Carpenter, and A.M. DeLange. 1994. The vaccinia virus-encoded uracil DNA glycosylase has an essential role in viral DNA replication. *Virology* **198:** 504–513.

Morham, S.G. and S. Shuman. 1990. Phenotypic selection and characterization of mutant alleles of a eukaryotic DNA topoisomerase I. *Genes Dev.* **4:** 515–524.

———. 1992. Covalent and noncovalent DNA binding by mutants of vaccinia DNA topoisomerase I. *J. Biol. Chem.* **267:** 15984–15992.

Moss, B. 1990a. Regulation of vaccinia virus transcription. *Annu. Rev. Biochem.* **59:** 661–688.

———. 1990b. Poxviridae and their replication. In *Virology*, 2nd edition. (ed. B.N. Fields and D.M. Knipe), pp. 2079–2111. Raven Press, New York.

———. 1994. Vaccinia virus transcription. In *Transcription: Mechanisms and regulation* (ed. R.C. Conaway and J.W. Conaway), pp. 185–206. Raven Press, New York.

Moyer, R.W. and R.L. Graves. 1981. The mechanism of cytoplasmic orthopoxvirus DNA replication. *Cell* **27:** 391–401.

Nowakowski, M., W. Bauer, and J. Kates. 1978. Characterization of a DNA-binding phosphoprotein from vaccinia virus replication complex. *Virology* **86:** 217–225.

Parsons, B.L. and D.J. Pickup. 1990. Transcription of orthopoxvirus telomeres at late times during infection. *Virology* **175:** 69–80.

Perkus, M.E., S.J. Goebel, S.W. Davis, G.P. Johnson, I.K. Norton, and E. Paoletti. 1991. Deletion of 55 open reading frames from the termini of vaccinia virus. *Virology* **180:** 406–410.

Pogo, B.G. 1977. Elimination of naturally occurring crosslinks in vaccinia virus DNA after viral penetration into cells. *Proc. Natl. Acad. Sci.* **74:** 1739–1742.

Pogo, B.G.T., E,M. Berkowitz, and S. Dales. 1984. Investigation of vaccinia virus DNA replication employing a conditional lethal mutant defective in DNA. *Virology* **132:** 436–444.

Pogo, B.G.T, M. O'Shea, and P. Freimuth. 1981. Initiation and termination of vaccinia virus DNA replication. *Virology* **108:** 241–248.

Rempel, R. and P. Traktman. 1992. Vaccinia virus B1 kinase: Phenotypic analysis of temperature-sensitive mutants and enzymatic characterization of recombinant proteins. *J. Virol.* **66:** 4413–4426.

Rempel, R.E., M.K. Anderson, E. Evans, and P. Traktman. 1990. Temperature-sensitive vaccinia virus mutants identify a gene with an essential role in viral replication. *J. Virol.* **64:** 574–583.

Roseman, N.A. and D.E. Hruby. 1987. Nucleotide sequence and transcript organization of a region of the vaccinia virus genome which encodes a constitutively expressed gene required for DNA replication. *J. Virol.* **61:** 1398–1406.

Sarov, I. and W.K. Joklik. 1972. Characterization of intermediates in the uncoating of vaccinia virus DNA. *Virology* **50:** 593–602.

Schmelz, M., B. Sodeik, M. Ericsson, E.J. Wolffe, H. Shida, G. Hiller, and G. Griffiths. 1994. Assembly of vaccinia virus: The second wrapping is derived from the trans Golgi network. *J. Virol.* **68:** 130–147.

Shaffer, R.S. and P. Traktman. 1987. Vaccinia encapsidates a novel topoisomerase with the properties of a eukaryotic type I enzyme. *J. Biol. Chem.* **262:** 9309–9315.

Shuman, S. 1991. Site-specific DNA cleavage by vaccinia virus DNA topoisomerase I. *J. Biol. Chem.* **266:** 1796–1803.

———. 1992. Vaccinia virus RNA helicase: An essential enzyme related to the DE-H family of RNA-dependent NTPases. *Proc. Natl. Acad. Sci.* **89:** 10935–10939.

Shuman, S. and B. Moss. 1987. Identification of a vaccinia virus gene encoding a type I DNA topoisomerase. *Proc. Natl. Acad. Sci.* **84:** 7478–7482.

Shuman, S. and J. Prescott. 1990. Specific DNA cleavage and binding by vaccinia virus DNA topoisomerase I. *J. Biol. Chem.* **265:** 17826–17836.

Shuman, S., M. Golder, and B. Moss. 1989a. Insertional mutagenesis of the vaccinia virus gene encoding a type I topoisomerase: Evidence that the gene is essential for virus growth. *Virology* **170:** 302–306.

Shuman, S., E.M. Kane, and S.G. Morham. 1989b. Mapping the active-site tyrosine of vaccinia virus DNA topoisomerase I. *Proc. Natl. Acad. Sci.* **86:** 9793–9797.

Simpson, D.A. and R.C. Condit. 1994. The vaccinia virus A18R protein plays a role in viral transcription during both the early and late phases of infection. *J. Virol.* **68:** 3642–3649.

Slabaugh, M.B., N. Roseman, and C. Mathews. 1988. Vaccinia virus encoded ribonucleotide reductase: Sequence conservation of the gene for the small subunit and its amplification in hydroxyurea resistant mutants. *J. Virol.* **62:** 519–527.

Slabaugh, M.B., M.L. Howell, Y. Wang, and C.K. Mathews. 1991. Deoxyadenosine reverses hydroxyurea inhibition of vaccinia virus growth. *J. Virol.* **65:** 2290–2298.

Smith, G.L. 1993. Vaccinia virus glycoproteins and immune evasion. *J. Gen. Virol.* **74:** 1725–1740.

Smith, G.L., S.Y. Chan, and S.T. Howard. 1991. Nucleotide sequence of 42 kbp of vaccinia virus strain WR from near the left inverted terminal repeat. *J. Gen. Virol.* **72:** 1349–1376.

Sodeik, B., R.W. Doms, M. Ericsson, G. Hiller, C.E. Machamer, W. van't Hof, G. van Meer, B. Moss, and G. Griffiths. 1993. Assembly of vaccinia virus: Role of the intermediate compartment between the endoplasmic reticulum and the Golgi stacks. *J. Cell Biol.* **121:** 521–541.

Stuart, D., K. Graham, M. Schreiber, C. Macaulay, and G. McFadden. 1991. The target DNA sequence for resolution of poxvirus replicative intermediates is an active late promoter. *J. Virol.* **65:** 61–70.

Stuart, D.T., C. Upton, M.A. Higman, E.G. Niles, and G. McFadden. 1993. A poxvirus-encoded uracil DNA glycosylase is essential for virus viability. *J. Virol.* **67:** 2503–2512.

Taddie, J.A. and P. Traktman. 1991. Genetic characterization of the vaccinia virus DNA polymerase: Identification of point mutations conferring altered drug sensitivities and reduced fidelity. *J. Virol.* **65:** 869–879.

———. 1993. Genetic characterization of the vaccinia virus DNA polymerase: Cytosine arabinoside resistance requires a variable lesion conferring phosphonoacetate resistance in conjunction with an invariant mutation localized to the 3′-5′ exonuclease domain. *J. Virol.* **67:** 4323–4336.

Tengelsen, L.A., M.B. Slabaugh, J.K. Bibler, and D.E. Hruby. 1988. Nucleotide sequence and molecular genetic analysis of the large subunit of ribonucleotide reductase encoded by vaccinia virus. *Virology* **164:** 121–131.

Traktman, P. 1990a. The enzymology of poxvirus DNA replication. *Curr. Top. Microbiol. Immunol.* **163:** 931–123.

———. 1990b. Poxviruses: An emerging portrait of biological strategy. *Cell* **62:** 621–626.

———. 1991. Molecular genetic and biochemical analysis of poxvirus DNA replication. *Semin. Virol.* **2:** 291–304.

Traktman, P., M.K. Anderson, and R.E. Rempel. 1989a. Vaccinia virus encodes an essential gene with strong homology to protein kinases. *J. Biol. Chem.* **264:** 21458–21461.

Traktman, P., M. Kelvin, and S. Pacheco. 1989b. Molecular genetic analysis of vaccinia virus DNA polymerase mutants. *J. Virol.* **63:** 841–846.

Traktman, P., P. Sridhar, R.C. Condit, and B.E. Roberts. 1984. Transcriptional mapping of the DNA polymerase gene of vaccinia virus. *J. Virol.* **49:** 125–131.

Upton, C., D.T. Stuart, and G. McFadden. 1993. Identification of a poxvirus gene encoding a uracil DNA glycosylase. *Proc. Natl. Acad. Sci.* **90:** 4518–4522.

Vos, J.C. and H.G. Stunnenberg. 1988. Derepression of a novel class of vaccinia virus genes upon DNA replication. *EMBO J.* **7:** 3487–3492.

Wang, T.S.-F., S.W. Wong, and D. Korn. 1989. Human DNA polymerase α: Predicted functional domains and relationships with viral DNA polymerases. *FASEB J.* **3:** 14–21.

Wang, Y.C., W.A. Burkhart, Z.B. Mackey, M.B. Moyer, W. Ramos, I. Husain, J. Chen, J.M. Besterman, and A.E. Tomkinson. 1994. Mammalian DNA ligase II is highly homologous with vaccinia DNA ligase. Identification of the DNA ligase II active site for enzyme-adenylate formation. *J. Biol. Chem.* **169:** 31923–31928.

Weir, J. and B. Moss. 1983. Nucleotide sequence of the vaccinia virus thymidine kinase gene and the nature of spontaneous frameshift mutations. *J. Virol.* **46:** 530–537.

Weir, J.P., G. Bajszar, and B. Moss. 1982. Mapping of the vaccinia virus thymidine kinase gene by marker rescue and by cell free translation of selected mRNA. *Proc. Natl. Acad. Sci.* **79:** 1210–1214.

Wilcock, D. and G.L. Smith. 1994. Vaccinia virus core protein VP8 is required for virus infectivity, but not for core protein processing or for INV and EEV formation. *Virology* **202:** 294–304.

Wittschieben, J. and S. Shuman. 1994. Mutational analysis of vaccinia DNA topoisomerase defines amino acid residues essential for covalent catalysis. *J. Biol. Chem.* **269:** 29978–29983.

Yang, W.-P. and W.R. Bauer. 1988a. Biosynthesis and post-translational cleavage of vaccinia virus structural protein VP8. *Virology* **167:** 585–590.

———. 1988b. Purification and characterization of vaccinia virus structural protein VP8. *Virology* **167:** 578–584.

Zhang, Y. and B. Moss. 1991. Vaccinia virus morphogenesis is interrupted when expression of a gene encoding an 11-kilodalton phosphorylated protein is prevented by the *Escherichia coli* lac repressor. *J. Virol.* **65:** 6101–6110.

28

Parvovirus DNA Replication

Susan F. Cotmore[1] and Peter Tattersall[1,2]
Departments of [1]Laboratory Medicine and [2]Genetics
Yale University School of Medicine
New Haven, Connecticut 06510

Parvoviruses are unique among all known viruses in having single-stranded DNA genomes which are linear. Virions are non-enveloped, containing a single copy of the small (4–6 kb) viral chromosome encapsidated in a rugged icosahedral protein capsid 18–26 nm in diameter. Although lacking associated enzymes or nucleosomal proteins, the particles have been shown, in some cases, to contain polyamines such as spermidine, spermine, and putrescine (Berns et al. 1995). The family Parvoviridae contains a broad spectrum of physically similar viruses that replicate in the nuclei of both invertebrate and vertebrate hosts. Viruses infecting mammalian cells form the subfamily Parvovirinae and include a number of helper-independent viruses, represented in this review by minute virus of mice (MVM), and the adeno-associated (AAV) viruses, represented here by AAV2, which, in general, only replicate in cells coinfected with a helper adenovirus or herpes virus. Further background information on the structure and biology of the parvoviruses can be obtained from reviews published elsewhere (Cotmore and Tattersall 1987; Berns 1990, 1995; Tjissen 1990; Muzyczka 1992). Space constraints allow only a representative selection of the most directly pertinent references to be cited herein.

Whereas AAV encapsidates, in separate virions, DNA strands of either sense, MVM selectively encapsidates (to 99%) strands that are minus sense with respect to transcription. Both viruses encode all of their known proteins from a single-sense DNA strand, and each encodes two separate gene complexes. Transcripts from one half of the genome, designated by convention the right-hand side, program synthesis of an overlapping set of capsid polypeptides, whereas the left half gives rise to a series of nonstructural proteins essential for viral DNA replication. These latter are designated the Rep proteins in AAV and the NS (nonstructural) proteins in MVM. Genome usage in all parvoviruses is remarkably efficient, so that some protein sequences are encoded in overlapping reading frames, and others contain regulatory elements for transcription or mRNA splicing. Nevertheless, the viruses' limited genetic capacity ultimately dictates that they adopt the replication machinery of the host cell, augmented and diverted by one or two specialized virally coded proteins.

ROLLING HAIRPIN REPLICATION

Parvoviral DNA replication in many ways resembles the single-strand-specific "rolling-circle" mechanisms previously characterized in prokaryotic replicons, such as the single-stranded coliphages and certain bacterial plasmids (Kornberg and Baker 1991). Incoming viral particles contain a single copy of the linear, non-permuted DNA genome which is first converted to a duplex replication intermediate. The relatively long single-stranded coding region in viral DNA is bracketed by short (121–421 base) palindromic terminal sequences capable of folding into hairpin duplexes. These telomeres play a central role in viral replication, containing most of the *cis*-acting information required for both replication and encapsidation, and their complexity and remarkable diversity in different viral genera suggest that they must serve multiple functions in the life cycle (Tjissen 1990; Berns et al. 1995). The hairpins pair the terminal 3′ nucleotide of incoming virion DNA with an internal base and, in so doing, create a DNA primer that allows a host polymerase to synthesize the first complementary DNA strand. Although unusual in eukaryotic replication, DNA primers created by the introduction of a single-strand nick into a duplex intermediate are the hallmark of prokaryotic rolling-circle mechanisms. The palindromic terminal sequences of parvoviruses allow this mechanism to be adapted for the replication of linear single-stranded progeny DNA, since a free 3′ hydroxyl group, generated in the previous host cell by the introduction of a site-specific nick into a duplex intermediate, is folded back and paired with an internal base in the progeny viral genome. Thus primed, complementary strand synthesis generates a monomer-length, duplex intermediate in which the two strands are covalently cross-linked at one end via a single copy of the viral 3′ telomere (step i in Fig. 1).

In some viruses (such as AAV2), this cross-linked 3′ structure creates a replication origin that can be activated by the Rep initiator endonuclease in a process called "terminal resolution," discussed below. In others, such as MVM, this structure does not appear to function as a replication origin. In either case, major viral DNA amplification proceeds through a series of concatemeric duplex intermediates via a unidirectional, single-strand-specific mechanism dubbed "rolling-

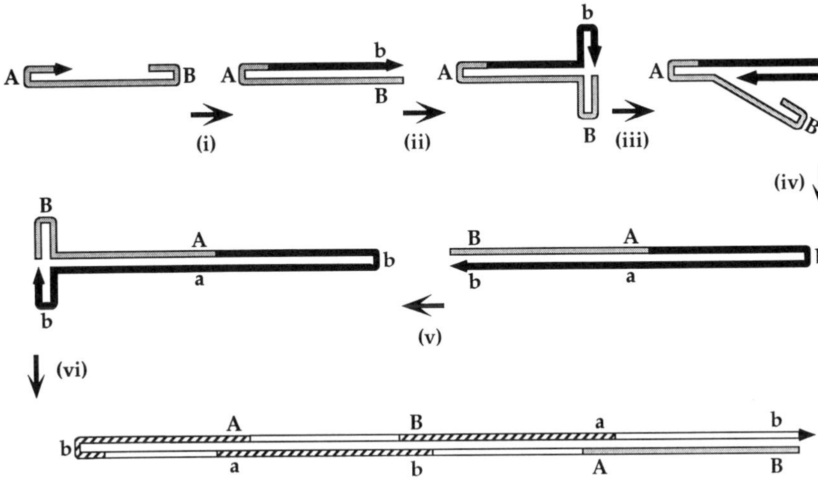

Figure 1 Rolling-hairpin replication. The sequence of the parental parvoviral genome is represented by a shaded bar. In steps *i* through *v*, newly synthesized DNA is shown as a black bar with an arrow at its 3′ end. *A* and *B* depict the palindromic sequences at each terminus, with their complements represented by *a* and *b*, respectively. Step *vi* produces a tetramer in which, for MVM, there are three progeny genomes, shown cross-hatched, in addition to the parental sequence. These overlap and are distributed throughout the molecule on alternate strands.

hairpin" synthesis. Instead of progressing continuously around circular templates as in prokaryotic rolling-circle systems, in rolling-hairpin synthesis the unidirectional replication fork appears to shuttle back and forth along the linear genome, changing direction as a result of the sequential synthesis and rearrangement of the palindromic viral termini, as diagrammed in Figure 1. Current models suggest that during this process, terminal hairpins are first unwound and copied by strand-displacement synthesis to create "extended-form" termini containing a single new DNA strand (step i) and these are then melted-out and reformed into hairpinned "rabbit-ear" structures (step ii) to provide the base-paired 3′ nucleotide needed to prime synthesis of additional linear sequences (step iii). As a result, the coding sequences of the virus are copied twice as often as the termini, and palindromic dimeric (step iv) and tetrameric (step vi) concatemers accumulate in which the unit-length duplex genomes within them are fused in left end:left end and right end:right end orientations.

Individual genomes are excised from these concatemers, and their telomeres are regenerated, by the introduction of single-strand nicks into replication origins generated at either end of each genome during the rolling-hairpin process. Excision is accomplished by the pleiotropic virally coded initiator proteins Rep68 and Rep78 (AAV) or NS1 (MVM) and leads to the establishment of new unidirectional replication forks at the nick sites, which then duplicate the required terminal sequences. Extended-form termini created by this process are equivalent to those seen in Figure 1 (steps i and iv) and can theoretically be melted out and reformed into rabbit-ear structures capable of priming additional rounds of DNA synthesis. Subsequent displacement of progeny single strands requires ongoing viral DNA synthesis and only occurs in the presence of competent capsid proteins. Since all progeny DNA is found encapsidated, its synthesis likely involves direct sequestration of displaced strands into preformed, or partially formed, capsids.

THE DEPENDOVIRUS REPLICATION STRATEGY

The AAV Genome

The AAV2 genome is 4675 nucleotides in length, with a long single-stranded coding region bracketed between identical, but inverted, 145-nucleotide terminal repeat sequences. The distal 125 bases of each repeat form a complex palindrome capable of folding into the T-shaped hairpin structure shown in Figure 2. In the absence of a coinfecting herpesvirus or adenovirus, the infecting AAV genome integrates into the host genome with high efficiency and remains latent (for review, see Berns 1995). Viruses with an intact REP gene frequently integrate into a small region of chromosome 19q13-qter (Kotin et al. 1991), which contains a strong Rep:DNA-binding site (Weitzman et al. 1994), whereas viruses deleted for the REP gene integrate at many sites in the host genome (Walsh et al. 1993). Rescue from the latent state requires coinfection with a helper virus, not primarily because the helper supplies unique gene products required by AAV, but rather because it modifies the internal cellular environment in such a way as to permit productive AAV replication. Indeed, limited replication in the absence of a helper virus can be achieved if host cells are exposed to various forms of genotoxic stress (Yakobson et al. 1987; Yakinoglu et al. 1988). Model systems suggest that excision from the host DNA is mediated by Rep proteins (Samulski et al. 1983; Hong et al. 1992), which are expressed to high level only in the presence of a helper virus (Beaton et al. 1989).

The virus has transcriptional promoters at map units 5 (P5) and 19 (P19), accessing a single REP open reading frame. An intron within the 3′ end of the gene allows expression of either a spliced or unspliced

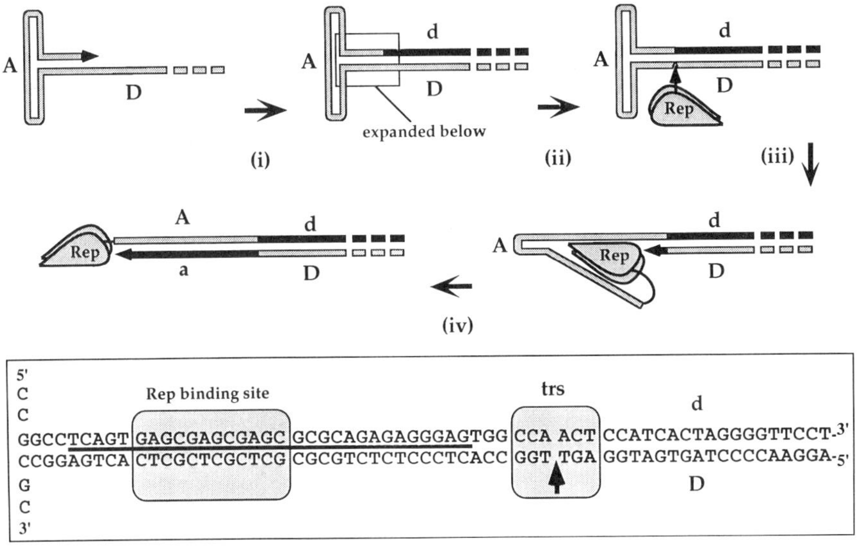

Figure 2 Terminal resolution reaction for AAV. The diagram is structured as in Fig. 1. In addition, *D* and *d* denote sequences that are contained within the terminal repeat but not within the hairpin. The double image of the Rep protein indicates that the active form is believed to be a dimer. In the expanded box, the sequence protected from DNase I digestion by Rep in an isolated linear origin (Chiorini et al. 1994) is indicated by the underline. The terminal resolution site is denoted *trs*, and the vertical arrow indicates the position of the Rep-induced nicking and 5′-attachment site.

version of each transcript, resulting in a total of four REP gene products. The two larger proteins, Rep78 and Rep68, are pleiotropic replication initiator proteins (Im and Muzyczka 1990, 1992), which appear to function almost identically in vitro (Ni et al. 1994). They are site-specific DNA-binding proteins, recognizing the sequence $(GAGC)_3$ in the stem region of the viral hairpin (Chiorini et al. 1994; McCarty et al. 1994) and variations on this sequence both within and outside the viral genome (Weitzman et al. 1994). Binding involves homodimers, if not higher-order multimers, of Rep, and such complexes can bind simultaneously to more than one DNA-binding site (McCarty et al. 1994). These proteins initiate replication by serving as site-specific endonucleases and are thought to be retained in the replication fork where they provide the necessary helicase activity (Im and Muzyczka 1990, 1992). The smaller Rep proteins, Rep52 and Rep40, are not required for duplex DNA replication but are implicated in progeny single-strand synthesis (Chejanovsky and Carter 1989).

Terminal Resolution and the Structure of AAV Origins

Duplex forms of the AAV genome that are covalently continuous at one end by a single copy of the terminal palindrome can be resolved to an extended-form configuration containing two copies of the palindrome by a process dubbed "terminal resolution." First suggested by Cavalier-Smith (1974) as a theoretical solution to the problem of maintaining the sequence of linear chromosome ends during DNA replication, this type of resolution has been recapitulated in vitro using AAV substrates (Im and Muzyczka 1990; Snyder et al. 1990a,b), and a similar process is thought to occur at the MVM 5′, but not the 3′, terminus (Cotmore and Tattersall 1992).

The AAV terminus, shown in Figure 2, is composed of three palindromic sequences which fold into a T-shaped structure in virion DNA in a way that allows the 3′ nucleotide to prime complementary strand synthesis. Rep68 and Rep78 bind to sequences centered on the $(GAGC)_3$ repeat in the hairpin stem and introduce a nick 20 bases away, at a specific site termed the *trs* (Snyder et al. 1993). The sequence boxed in Figure 2 can support limited Rep-mediated, site-specific nuclease activity in vitro, but a third recognition element, involving the arms of the hairpin and generally referred to as the "secondary structure" element, is also required for efficient initiation (Chiorini et al. 1994; McCarty et al. 1994). These T-fork sequences, which contain degenerate forms of the $(GAGC)_n$ motif, appear to alter the affinity of the substrate for Rep and to change the DNase I protection pattern, but not the number of molecules bound (McCarty et al. 1994), presumably by providing second-site interactions within multimeric Rep:DNA complexes. Cleavage results in the formation of a phosphodiester bond between the phosphoryl group of the 5′ terminal thymidine residue and an aromatic hydroxyl group from a Rep tyrosine, in a process that requires ATP (Snyder et al. 1990a, 1993). The nick effectively inverts the original complex palindrome onto the progeny strand (Fig. 2, step iii) while providing a new base-paired 3′ hydroxyl to prime synthesis of its complement. This process thus replaces the original sequence of the hairpin (dubbed the "flip" sequence) with its inverted complement, "flop" (Fig. 2, step iv). Since this inversion is repeated with every round of resolution, progeny genomes contain equal numbers of termini in both sequence orientations. The terminal palindromes of all parvoviruses are imperfect, so that flip and flop sequence orientations can be readily identified. The absence of such heterogeneity at the 3′ terminus of MVM first alerted Astell and her colleagues (1983a) to the existence of an alternate mechanism for telomere regeneration, dubbed "junction resolution," discussed below.

THE MVM REPLICATION STRATEGY

The MVM Genome

The negative-sense MVM genome is 5172 nucleotides long, with 4805 nucleotides of single-stranded DNA bracketed between unique terminal palindromes of 121 (3′) and 246 (5′) nucleotides. These are capable of folding into Y-shaped (Fig. 3) and cruciform hairpin structures, respectively. MVM does not require coinfection with a helper virus for its own productive replication, but it is unable to induce resting cells to enter S phase. However, such viruses can remain in infected G_0 cells for prolonged periods without causing apparent toxicity, and these cells can still enter S phase upon induction (for review, see Tjissen 1990). Viral transcription is not activated until after entry into S phase (Clemens and Pintel 1988) and, as the early viral gene products accumulate, host cell DNA replication is terminated. Progression through the cell cycle is suspended, and cells continue actively synthesizing viral DNA until subsequent lysis or apoptosis results in the release of progeny virus, generally within 24 hours (Cotmore and Tattersall 1987).

Exons accessed from the MVM P4 promoter give rise to a series of alternatively spliced transcripts encoding two types of nonstructural proteins, designated NS1 and NS2. NS1, the replication initiator, is an abundant and long-lived nuclear phosphoprotein of 83 kD, with helicase and ATPase activities (Wilson et al. 1991), and is the only NS protein which is essential for productive replication in all cell types (Naeger et al. 1990; Li and Rhode 1991). It is a site-specific DNA-binding protein, recognizing the sequence $(ACCA)_{1\text{-}3}$, present in the viral origins, and reiterated at multiple sites throughout the viral genome (Cotmore et al. 1995). Since most MVM sequences of 100 bp or more contain at least one copy of this recognition sequence, and some regions contain multiple tandem and inverted reiterations, NS1 can bind throughout MVM replicative-form DNA (S.F. Cotmore and P. Tattersall, unpubl.). This not only suggests a potential role for NS1 in viral nucleosome structure and progeny strand packaging, but also indicates that previously identified *cis*-acting sequences thought to interact with cellular proteins to instigate transcriptional *trans*-activation (Gu and Rhode 1992; cf. Christensen et al. 1995) or to enhance replication (Tam and Astell 1993) must be reassessed for their ability to bind and sequester NS1.

Unlike Rep:DNA binding, association and dissociation of the NS1:DNA complex is markedly dynamic under physiological conditions, modulated by the binding of ATP. Experimentally, the interaction can only be demonstrated in the presence of ATP or by cross-linking the NS1 molecules with antibodies directed against their amino- or carboxy-

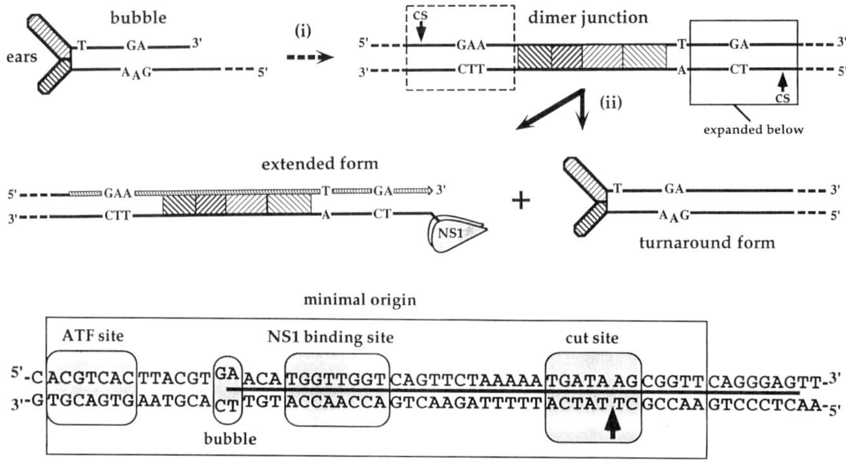

Figure 3 Junction resolution reaction for the left end of MVM. Step *i* condenses the in vivo steps *i–iv* of Fig. 1, showing greater detail of the organization of the left-end hairpin sequences within the dimer junction. Cross-hatched boxes represent the palindromic sequences that fold to give the internal "ears" in the hairpin form of the 3′ end of the genome. Step *ii* represents the in vitro resolution reaction described in the text. The potential cut sites are denoted *cs*, and newly synthesized DNA is depicted as a hatched bar with an arrowhead at its 3′ end. The domains within the minimal origin described in the text are boxed, and the underline represents the sequences protected by NS1 from DNase I digestion (Christensen et al. 1995).

terminal peptides (Christensen et al. 1995). Since NS1 appears to form multimers in vivo (Nüesch and Tattersall 1993), this suggests that ATP may induce NS1 to oligomerize and that only multimeric forms of the protein bind efficiently to its cognate site. NS1 is the initiator endonuclease for MVM DNA replication (Cotmore et al. 1992, 1993; Nüesch et al. 1995), containing critical sequence motifs encoding a putative metal coordination site and an active-site tyrosine that have been recognized in prokaryotic rolling-circle initiators (Koonin and Ilyina 1993). These two elements are thought to comprise the catalytic site of the nickase. NS1 becomes linked to the 5′ end of the displaced strand via a tyrosine residue at amino acid position 210 (Nüesch et al. 1995) and, like Rep, is presumed to remain in the replication fork providing the necessary helicase activity. NS2 polypeptides are only required for productive replication in cells of murine origin. In these cells, NS2 clearly influences multiple steps in the viral life cycle, including DNA replication, by currently undefined mechanisms (Naeger et al. 1993).

Structure of the MVM 3′ Origin

MVM termini are unique, differing from each other in size, primary sequence, and predicted structure (Astell et al. 1983b). At the 5′ end of the genome, a hairpin resolution process, thought to resemble the one recapitulated in vitro for AAV, results in the generation of equal numbers of termini in flip and flop configurations, whereas at the 3′ telomere a more restricted mechanism, dubbed junction resolution, generates termini in a single orientation (Astell et al. 1985). Although all parvoviruses probably use junction resolution to resolve concatemeric intermediates, for MVM a critical asymmetry in the stem of the 3′ hairpin restricts this mechanism to the synthesis of a single sense DNA strand (Cotmore and Tattersall 1994).

In MVM, the 121-nucleotide 3′ telomere (Fig. 3) exists as a single flip sense sequence containing internal palindromes designated the "ears" and an asymmetric, mismatch "bubble" in the stem where the triplet GAA on the inboard arm is opposed to the doublet GA on the outboard strand (Astell et al. 1979, 1983b). During rolling-hairpin replication (Fig. 1) this sequence is extended and copied to form the palindromic duplex junction that links unit-length genomes in the dimer intermediate, and it is this duplex structure, rather than the terminal hairpin, that contains the replication origin. Figure 3 shows how the asymmetries in the hairpin are organized in the junction sequence. Since it contains two copies of the terminal palindrome, one in each sequence orientation, it also contains two candidate cut sites (cs) located on opposite DNA strands and on each arm of the palindrome. However, only the cut site on the arm carrying the GA/TC doublet functions as a replication origin (Cotmore and Tattersall 1994). The use of such a replication strategy in MVM appears to be an adaptation which, by restricting NS1-mediated nicking to the outboard arm of the telomere, allows similar configurations of these same sequences on the inboard arm, and in the mismatched, hairpinned configuration of the terminus, to be dedicated to transcription (Faisst et al. 1994; Gu et al. 1995). Since the replication fork initiated on the TC arm is unidirectional and supports the synthesis of a single, continuous DNA strand (S.F. Cotmore and P. Tattersall, unpubl.), use of a single cut site inevitably results in the synthesis of DNA in a single sequence orientation. However, as discussed below, a resolution process in which replication initiated at a nick site on the TC/TTC strand leads to the selective synthesis of a GAA/GA strand must be complex and likely involves rearrangement into a cruciform structure.

The minimal active MVM 3′ origin (boxed in Fig. 3) is a multidomain structure of approximately 50 bp derived from the stem region of the hairpin. It extends from an activated transcription factor (ATF) consensus binding site to a position some 7 bp beyond the cut site. An $(ACCA)_2$ NS1-binding sequence is positioned 17 nucleotides from the cut site, separated from the ATF consensus by the TC bubble dinucleotide. The actual sequence of the bubble dinucleotide appears unimportant, suggesting that it is not an essential part of the binding site for a cellular replication factor, but insertion of any third residue here, as in the GAA arm of the dimer junction, totally inactivates the origin (Cotmore and Tattersall 1994). This suggests that the bubble represents a critical spacer element, presumed to control the interaction of a cellular protein, binding at the ATF consensus, with $(ACCA)_2$-bound NS1. This precisely controlled interaction appears to activate the endonuclease function of NS1, since purified NS1 alone does not nick the viral origin efficiently (Nüesch et al. 1995). Mutations in the ATF consensus impair origin function, but purified homodimers of CREB protein or ATF 1, 2, or 3, despite binding avidly to this site, actually impair, rather than stimulate, origin-specific replication in vitro. In contrast, a highly enriched cellular protein fraction from 293 cells contains a factor(s) that can cooperate with NS1 to mediate 3′ origin-specific nicking and replication in vitro (J. Christensen et al., in prep.). This fraction, parvovirus initiation factor (PIF), contains a 120-kD protein that shows sequence-specific cross-linking to a double-stranded oligonucleotide containing the MVM ATF-binding site region. PIF is the subject of much interest, but its identity and its role in the biology of the host cell have yet to be determined.

Relatively little is known about which other cellular replication proteins are required by MVM. Since the replication fork initiated on the 3′ origin is aphidicolin-sensitive, unidirectional, and supports the synthesis of a single, continuous DNA strand, it is likely that polymerase-δ and its accessory proteins are required for its synthesis. This is supported by the observation that neutralizing monoclonal antibodies directed against polymerase-α:primase fail to inhibit MVM replication in vitro, whereas biochemical studies clearly indicate an absolute requirement for PCNA and RP-A (J. Christensen et al., in prep.). Since MVM gene expression coincides with a rapid reduction in the rate of cellular DNA replication (Cotmore and Tattersall 1987), it seems likely that the virus inactivates a critical cellular replication protein which it does not itself require. This hypothesis is currently under active investigation.

Junction Resolution at the MVM 3′ Origin

As shown in Figure 3, MVM left-end junction sequences are resolved and replicated asymmetrically in vitro into two viral telomeres, one in the extended configuration containing a duplex copy of the entire palindrome and one containing a single copy of the palindrome in the covalently continuous, turnaround configuration (Cotmore et al. 1993; Liu et al. 1994). One of the strands of the extended-form terminus is newly synthesized, whereas its unreplicated complement is now covalently attached through its 5′ end to an NS1 molecule. How such resolution occurs is uncertain. It was originally suggested that NS1 might carry out a sequence of nicking and joining reactions analogous to those performed by the gene A protein of bacteriophage ϕX174 (Astell et al. 1983a). According to this model, the initiating nick would need to be introduced at the GAA-proximal cut site (shown as cs in Fig. 3 in the dotted box), thus liberating a 3′ hydroxyl capable of priming synthesis of the required flip-sense DNA strand across the axis of the palindrome. The NS1 molecule attached to the 5′ end of the displaced strand, remaining in the replication fork as it traversed the junction, would then encounter the "latent" nick site on the opposite, TC, strand and terminate the reaction at this second site by exchanging itself across analogous phosphodiester bonds (Astell et al. 1983a, 1985). This was an attractive model, since it simply recapitulated processes employed in prokaryotic rolling-circle replicons. Unfortunately, it is incompatible with the finding that replication initiates in the TC arm of the palindrome, since use of this cut site in the above scheme would lead to the synthesis of DNA strands with the flop (TC/TCC) sequence, rather than the flip (GAA/GA) sequence observed in vivo.

In order to synthesize a DNA strand in the flip configuration from a fork initiating on the TC arm, we suggest that the junction is reconfigured into a cruciform structure at a point soon after establishment of the replication fork, as shown in Figure 4, step ii. This would allow the fork to copy sequences in the flop orientation, producing a new flip-sense strand via a series of discrete linear and cruciform intermediates (step i and steps ii–v, respectively). The limited experimental data available tend to support such a mechanism. Replication forks can be established on linear sequences derived from a single arm of the junction (Cotmore and Tattersall 1994), but resolution of dimer junction substrates in murine cell extracts leads to the accumulation of nicked and partially replicated junction intermediates having structures compatible with those produced by steps i and iv in Figure 4 (Cotmore et al. 1993). Moreover, resolution of structures containing one copy of the GAA arm of the junction only occurs if this sequence is palindromic with, and hence able to form a cruciform with, the opposed, origin-containing arm (S.F. Cotmore and P. Tattersall, unpubl.). Since NS1 binds equally well to its cognate

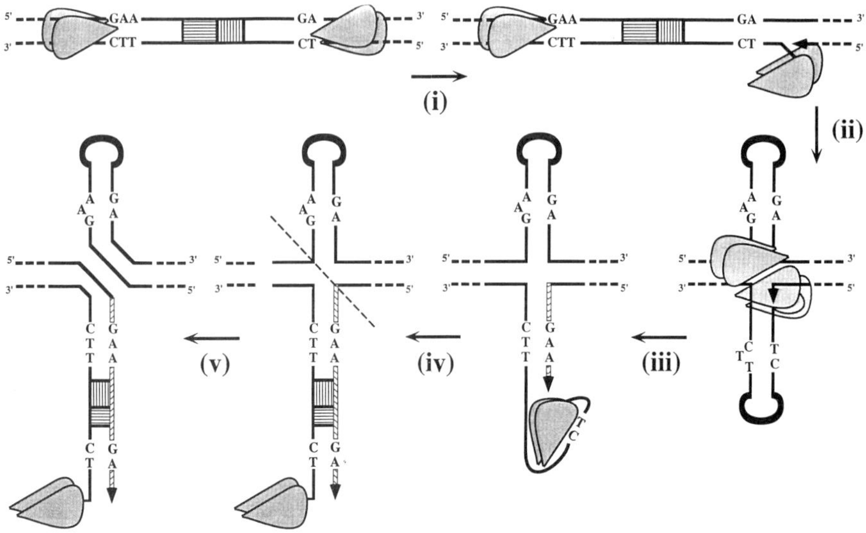

Figure 4 Hetero-cruciform model for MVM dimer junction resolution. Duplex forms of the internal "ear" palindromes are represented by hatched boxes, and hairpinned forms of these sequences are shown as thick black lines. NS1 proteins are shown bound to each arm of the junction as dimers, and from step *iii* on, the complex on the "inactive" arm is omitted for clarity. The dashed diagonal line indicates the preferred resolution axis of the cellular recombinase hypothesized to recognize the hetero-cruciform formed in step *iv* as a "Holliday structure" substrate and to resolve it in step *v*.

site on both arms of the junction palindrome (Cotmore et al. 1995), multimerization of this protein could well drive cruciform formation. Although it thus seems probable that resolution proceeds through a cruciform intermediate, how this structure is subsequently resolved is a matter of speculation. We suggest the recombinase-mediated strategy shown in Figure 4, step v, to provide what Crick has called a "don't worry" demonstration (Crick 1989). This implies that although we don't know how the strategy works, cellular mechanisms exist that could easily explain the observed products.

CONCLUDING REMARKS

In several important ways, parvoviral DNA replication resembles the single-strand-specific rolling-circle mechanisms previously characterized in prokaryotic systems. The single-stranded viral genome is replicated through multimeric duplex replication intermediates by a quasi-circular, continuous-strand synthesis mechanism. As in prokaryotic rolling-circle synthesis, the next phase is initiated by a replicon-encoded endonuclease, rather than by RNA priming. The nuclease introduces a single-strand nick into specific origin sequences, itself becoming covalently attached to the 5′ end of the DNA in the process via a phosphodiester bond. This process liberates a base-paired 3′ deoxynucleotide that primes synthesis of the daughter DNA strand via a leading-strand-specific replication fork. Similarity with prokaryotic replicons extends to details such as protein sequence motifs involved in the catalytic site of the nickase and the nucleotide requirements for nicking, but there is as yet no evidence to suggest that parvoviral initiators retain the ability to carry out the type of nicking/joining reaction exhibited by their prokaryotic counterparts.

ACKNOWLEDGMENT

The authors were supported by U.S. Public Health Service grants AI-26109 and CA-29303 from the National Institutes of Health.

REFERENCES

Astell, C.R., M.B. Chow, and D.C. Ward. 1985. Sequence analysis of the termini of virion and replicative forms of minute virus of mice DNA suggests a modified rolling hairpin model for autonomous parvovirus DNA replication. *J. Virol.* **54:** 171–177.

Astell, C.R., M. Smith, M.B. Chow, and D.C. Ward. 1979. Structure of the 3′ hairpin termini of four rodent parvovirus genomes: Nucleotide sequence homology at origins of DNA replication. *Cell* **17:** 691–703.

Astell, C.R., M. Thomson, M.B. Chow, and D.C. Ward. 1983a. Structure and replication of minute virus of mice DNA. *Cold Spring Harbor Symp. Quant. Biol.* **47:** 751–762.

Astell, C.R., M. Thomson, M. Merchlinsky, and D.C. Ward. 1983b. The complete DNA sequence of minute virus of mice, an autonomous parvovirus. *Nucleic Acids Res.* **11:** 999–1018.

Beaton, A., P. Palumbo, and K.I. Berns. 1989. Expression from the adeno-associated virus p5 and p19 promoters is negatively regulated in trans by the rep protein. *J. Virol.* **63:** 4450–4454.

Berns, K.I. 1990. Parvovirus replication. *Microbiol. Rev.* **54:** 316–329.

———. 1995. Parvoviridae and their replication. In *Fields virology*, 3rd edition. (ed. B.N. Fields et al.), 1743–1763. Raven Press, New York.

Berns, K.I., M. Bergoin, M. Bloom, M. Lederman, N. Muzyczka, G. Siegl, J. Tal, and P. Tattersall. 1995. Parvoviridae. *Arch. Virol. Suppl.* **10:** 169–178.

Cavalier-Smith, T. 1974. Palindromic base sequences and replication of eukaryotic chromosome ends. *Nature* **250:** 467–470.

Chejanovsky, N. and B.J. Carter. 1989. Mutagenesis of an AUG codon in the adeno-associated virus rep gene: Effects on viral DNA replication. *Virology* **173:** 120–128.

Chiorini, J.A., S.M. Wiener, R.A. Owens, S.R.M. Kyostio, R.M. Kotin, and B. Safer. 1994. Sequence requirements for stable binding and function of Rep68 on the adeno-associated virus type 2 inverted terminal repeat. *J. Virol.* **68:** 7448–7457.

Christensen, J., P. Tattersall, and S.F. Cotmore. 1995. The minute virus of mice transcriptional activator protein, NS1, binds directly to the *trans*-activating region (*tar*) of the viral P38 promoter in a strictly ATP-dependent manner. *J. Virol.* **69:** 5422–5430.

Clemens, K.E. and D. Pintel. 1988. The two transcription units of the autonomous parvovirus minute virus of mice are transcribed in a temporal order. *J.Virol.* **62:** 1448–1451.

Cotmore, S.F. and P. Tattersall. 1987. The autonomously replicating parvoviruses of vertebrates. *Adv. Virus Res.* **33:** 91–174.

———. 1992. In vivo resolution of circular plasmids containing concatemer junction fragments from minute virus of mice DNA and their subsequent replication as linear molecules. *J. Virol.* **66:** 420–431.

———. 1994. An asymmetric nucleotide in the parvoviral 3′ hairpin directs segregation of a single active origin of DNA replication. *EMBO J.* **13:** 4145–4152.

Cotmore, S.F., J.P.F. Nüesch, and P. Tattersall. 1992. In vitro excision and replication of 5′ telomeres of minute virus of mice DNA from cloned palindromic concatemer junctions. *Virology* **190:** 365–377.

———. 1993. Asymmetric resolution of a parvovirus palindrome in vitro. *J. Virol.* **67:** 1579–1589.

Cotmore, S.F., J. Christensen, J.P.F. Nüesch, and P. Tattersall. 1995. The NS1 polypeptide of the murine parvovirus minute virus of mice binds to DNA sequences containing the motif [ACCA]2-3. *J. Virol.* **69:** 1652–1660.

Crick, F. 1989. *What mad pursuit.* Weidenfeld and Nicolson, London.

Faisst, S., M. Perros, L. Deleu, N. Spruyt, and J. Rommelaere. 1994. Mapping of upstream regulatory elements in the P4 promoter of parvovirus minute virus of mice. *Virology* **202:** 466–470.

Gu, M.L. and S.L. Rhode. 1992. *trans*-Activation of H-1 parvovirus P38 promoter is correlated with increased binding of cellular protein(s) to the *trans*-activation responsive element (tar). *Virology* **190:** 116–123.

Gu, Z., S. Plaza, M. Perros, C. Cziepluch, J. Rommelaere, and J.J. Cornelis. 1995. NF-Y controls transcription of the minute virus of mice P4 promoter through interaction with an unusual binding site. *J. Virol.* **69:** 239–246.

Hong, G., P. Ward, and K.I. Berns. 1992. In vitro replication of adeno-associated virus DNA. *Proc. Natl. Acad. Sci.* **89:** 4673–4677.

Im, D.S. and N. Muzyczka. 1990. The AAV origin binding protein Rep68 is an ATP-dependent site-specific endonuclease with DNA helicase activity. *Cell* **61:** 447–457.

———. 1992. Partial purification of adeno-associated virus Rep78, Rep52, and Rep40 and their biochemical characterization. *J. Virol.* **66:** 1119–1128.

Koonin, E.V. and T.V. Ilyina. 1993. Computer-assisted dissection of rolling circle DNA replication. *Biosystems* **30:** 241–268.

Kornberg, A. and T.A. Baker. 1991. *DNA replication.* W.H. Freeman, New York.

Kotin, R.M., J.C. Menninger, D.C. Ward, and K.I. Berns. 1991. Mapping and direct visualization of a region-specific viral DNA integration site on chromosome 19q13-qter. *Genomics* **10:** 831–834.

Li, X. and S. Rhode III. 1991. Nonstructural protein NS2 of parvovirus H-1 is required for efficient viral protein synthesis and virus production in rat cells in vivo and in vitro. *Virology* **184:** 117–130.

Liu, Q., C.B. Yong, and C.R. Astell. 1994. In vitro resolution of the dimer bridge of the minute virus of mice (MVM) genome supports the modified rolling hairpin model for MVM replication. *Virology* **201:** 251–262.

McCarty, D.M., J.H. Ryan, S. Zolotukhin, X. Zhou, and N. Muzyczka. 1994. Interaction of the adeno-associated virus Rep protein with a sequence within the A palindrome of the viral terminal repeat. *J. Virol.* **68:** 4998–5006.

Muzyczka, N. 1992. Use of adeno-associated virus as a general transduction vector for mammalian cells. *Curr. Top. Microbiol. Immunol.* **158:** 97–129.

Naeger, L.K., J. Cater, and D.J. Pintel. 1990. The small nonstructural protein (NS2) of the parvovirus minute virus of mice is required for efficient DNA replication and infectious virus production in a cell-type-specific manner. *J. Virol.* **64:** 6166–6175.

Naeger, L.K., N. Salome, and D.J. Pintel. 1993. NS2 is required for efficient translation of viral mRNA in minute virus of mice-infected murine cells. *J. Virol.* **67:** 1034–1043.

Ni, T.H., X. Zhou, D.M. McCarty, I. Zolotukhin, and N. Muzyczka. 1994. In vitro replication of adeno-associated virus DNA. *J. Virol.* **68:** 1128–1138.

Nüesch, J.P. and P. Tattersall, 1993. Nuclear targeting of the parvoviral replicator molecule NS1: Evidence for self-association prior to nuclear transport. *Virology* **196:** 637–651.

Nüesch, J.P.F., S.F. Cotmore, and P. Tattersall. 1995. Sequence motifs in the replicator protein of parvovirus MVM essential for nicking and covalent attachment to the viral origin: Identification of the linking tyrosine. *Virology* **209:** 122–136.

Samulski, R.J., A. Srivastava, K.I. Berns, and N. Muzyczka. 1983. Rescue of adeno-associated virus from recombinant plasmids: Gene correction within the terminal repeats of AAV. *Cell* **33:** 135–143.

Snyder, R.O., D.S. Im, and N. Muzyczka. 1990a. Evidence for covalent attachment of the adeno-associated virus (AAV) rep protein to the ends of the AAV genome. *J. Virol.* **64:** 6204–6213.

Snyder, R.O., R.J. Samulski, and N. Muzyczka. 1990b. In vitro resolution of covalently joined AAV chromosome ends. *Cell* **60:** 105–113.

Snyder, R.O., D.S. Im, T. Ni, X. Xiao, R.J. Samulski, and N. Muzyczka. 1993. Features of the adeno-associated virus origin involved in substrate recognition by the viral Rep protein. *J. Virol.* **67:** 6096–6104.

Tam, P. and C.R. Astell. 1993. Replication of minute virus of mice minigenomes: Novel elements required for MVM DNA replication. *Virology* **193:** 812–824.

Tjissen, P. 1990. *Handbook of parvoviruses*, vol. 1 and vol. 2. CRC Press, Boca Raton, Florida.

Walsh, C.E., J.M. Liu, J.L. Miller, A.W. Nienhuis, and R.J. Samulski. 1993. Gene therapy for human hemoglobinopathies. *Proc. Soc. Exp. Biol. Med.* **204:** 289–300.

Weitzman, M.D., S.R. Kyostio, R.M. Kotin, and R.A. Owens. 1994. Adeno-associated virus (AAV) Rep proteins mediate complex formation between AAV DNA and its integration site in human DNA. *Proc. Natl. Acad. Sci.* **91:** 5808–5812.

Wilson, G.M., H.K. Jindal, D.E. Yeung, W. Chen, and C.R. Astell. 1991. Expression of minute virus of mice major nonstructural protein in insect cells: Purification and identification of ATPase and helicase activities. *Virology* **185:** 90–98.

Yakinoglu, A.O., R. Heilbronn, A. Burkle, J.R. Schlehofer, and H. zur Hausen. 1988. DNA amplification of adeno-associated virus as a response to cellular genotoxic stress. *Cancer Res.* **48:** 3123–3129.

Yakobson, B., T. Koch, and E. Winocour. 1987. Replication of adeno-associated virus in synchronized cells without the addition of a helper virus. *J. Virol.* **61:** 972–981.

29

Replication of the Hepatitis Virus Genome

Christoph Seeger and William S. Mason
Fox Chase Cancer Center
Philadelphia, Pennsylvania 19111

Five strains of hepadnavirus have been identified. The prototype is human hepatitis B virus (HBV), first described as a primary cause of post-transfusion hepatitis (Blumberg et al. 1967). Very similar viruses have been isolated from woodchucks (Summers et al. 1978) and Beechey ground squirrels (Marion et al. 1980), and more distantly related viruses have been described in domestic ducks (Mason et al. 1980) and gray herons (Sprengel et al. 1988). With a genome size of 3 kbp, the hepadnaviruses are among the smallest of animal viruses. Despite this, these viruses are able to reproduce to high levels and to maintain a chronic, productive infection, often in the face of a vigorous immune response. Infection and replication take place primarily in the hepatocyte, the major parenchymal cell of the liver; are noncytopathic; and may persist for the lifetime of the host. All of these viruses are blood-borne and are poorly transmittable except by contact with the blood or blood-contaminated products from an infected individual. Nonetheless, infection can efficiently spread through a population, as revealed by the fact that more than 200 million people are currently infected by human hepatitis B virus.

Hepadnaviral DNA replication is of considerable interest because it occurs via reverse transcription of a viral RNA (Summers and Mason 1982), the pregenome, but in virtually all other details, differs from the reverse transcription pathway evolved by retroviruses. In fact, the hepadnavirus provirus is an episomal, covalently closed circular (CCC) DNA and not the integrated DNA of the retrovirus, and there is no evidence for a role of integration in hepadnavirus replication. Since most retroviruses are only able to complete provirus integration in dividing cells, utilization of an episomal template for transcription of hepadnaviral RNAs probably reflects the fact that the host cell is normally in G_0, dividing every 3–6 months to facilitate replacement of hepatocytes dying through random, cytopathic events (MacDonald 1960; Grisham 1962; Fourel et

 0-87969-459-9/96 $5 + .00

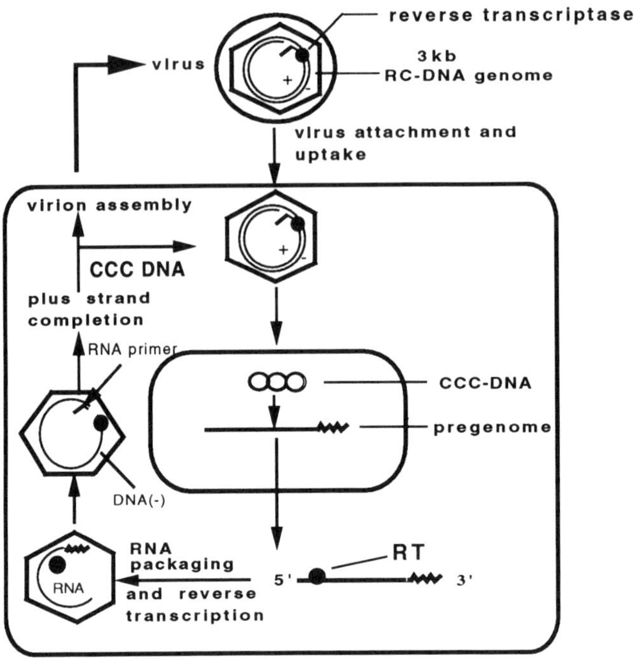

Figure 1 Reverse transcription in the replication of hepadnaviruses. The details of this replication cycle are described in the text. More detailed models of DNA replication intermediates are presented in Figs. 2–4.

al. 1994; Kajino et al. 1994). Schematically, hepadnavirus replication takes place through the series of steps shown in Figure 1. Virus, with an open circular, partially double-stranded DNA genome, infects the hepatocyte. The DNA is transported to the nucleus and converted to the covalently closed form, which serves as a template for the transcription of three (avian hepadnavirus) or four (mammalian hepadnavirus) classes of viral RNA, including the greater-than-genome-length molecule referred to as the pregenome. These RNAs are then transported to the cytoplasm, where viral assembly and DNA synthesis take place. The pregenome is packaged into subviral core particles along with a virus-encoded reverse transcriptase (RT), and viral DNA is then synthesized. Core particles with mature DNA then interact with viral envelope proteins, and virions are formed by budding into the endoplasmic reticulum, from where they are transported to the cell surface and released. Alternatively, the core particles may migrate to the nucleus to amplify the copy number of CCC DNA (Tuttleman et al. 1986), a process that is highly regulated, apparently to prevent virus replication from reaching levels that are cytocidal (Fig. 1) (Summers et al. 1990, 1991).

In this chapter, the process of DNA synthesis within virus nucleocapsids is discussed in detail. This is then related to factors that control CCC DNA amplification. Finally, the structure of the viral RT is reviewed, with respect both to known activities of this polypeptide and to functions that may potentially be carried out by this enzyme.

REVERSE TRANSCRIPTION OF THE VIRAL GENOME

Packaging and Protein Priming of Minus-strand DNA Synthesis

Replication of the viral genome occurs in subviral core particles that are present in the cytoplasm of infected cells. It depends on the expression of core proteins that assemble into icosahedral capsids with T=3 and T=4 icosahedral symmetry (Onodera et al. 1982; Nassal and Schaller 1993; Crowther et al. 1994), RT polypeptide, and the presence of *cis*-acting sequences on pregenomic RNA. Assembly of subviral core particles is triggered by the formation of a complex between pregenomic RNA, the template for reverse transcription, and the viral polymerase (Bartenschlager et al. 1990; Hirsch et al. 1990). To initiate the packaging reaction, the polymerase first binds to the packaging signal, termed ε, which has been proposed to fold into an RNA hairpin with a loop and a bulge (Fig. 2) (Junker-Niepmann et al. 1990). Although pregenomic RNA bears copies of ε at either end, a consequence of a terminal redundancy, it is the copy at the 5′ end of the RNA template that acts as a signal for RNA packaging. Fusion of pregenomic RNA sequences of HBV that comprise ε with an unrelated RNA segment leads to the encapsidation of the hybrid RNA into core particles, provided that the polymerase is expressed in the same cell (Junker-Niepmann et al. 1990); in addition, specific mutations of the 5′, but not the 3′, copy of ε have been shown to prevent packaging (Knaus and Nassal 1993; Pollack and Ganem 1993). In the avian viruses, packaging depends on additional sequences on pregenomic RNA that are located approximately 1 kb downstream from ε (Hirsch et al. 1991; Calvert and Summers 1994). The role of these sequences in virus assembly remains to be determined.

The requirement for a DNA polymerase in RNA packaging is unique among retroviruses. How the polymerase facilitates the packaging reaction is not yet understood, although genetic studies have shown that the polymerase activity per se is not involved (Bartenschlager et al. 1990; Chang et al. 1990; Hirsch et al. 1990; Roychoudhury et al. 1991; Chen et al. 1992). Genetic evidence suggests that expression of the polymerase polypeptide and RNA packaging occur in *cis* (Bartenschlager et al. 1990;

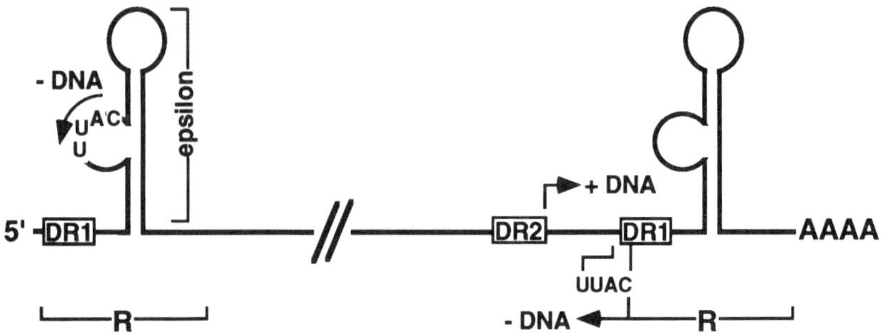

Figure 2 cis-Acting signals on pregenomic RNA. The RNA bears terminal repetitions (R) that contain copies of the packaging signal (ε), but only the 5′ copy has functional activity in vivo. The predicted stem-loop structure of ε is depicted as described by Junker-Niepmann et al. (1990). The positions of the 5′ ends of minus- and plus-strand DNAs and the UUAC motifs important for minus-strand DNA synthesis within ε and at DR1 are indicated.

Hirsch et al. 1990). This is probably a consequence of the dual role of pregenomic RNA as the mRNA for the translation of the RT and the template for reverse transcription (Huang and Summers 1991). Therefore, pregenomic RNAs encoding a structurally intact polymerase are preferred substrates for RNA packaging over RNAs with defective *pol* genes. Thus, the binding of polymerase with ε RNA may occur during or shortly after the translation of the *pol* gene product and induce the assembly of core particles. The core subunit, which is also translated from pregenomic RNA, and has RNA- and DNA-binding domains (Hatton et al. 1992), appears to function in packaging equally well in *cis* and in *trans*.

Following assembly of subviral core particles, minus-strand DNA is reverse-transcribed (Summers and Mason 1982). This step leads to the formation of a covalent linkage between the 5′ end of minus-strand DNA and protein (Gerlich and Robinson 1980). By analogy with the protein-priming mechanism first described for adenovirus DNA replication, it has long been speculated that this terminal protein was the primer for reverse transcription of hepadnavirus DNA (Rekosh et al. 1977; Gerlich and Robinson 1980; Molnar-Kimber et al. 1983; Bartenschlager and Schaller 1988). However, an understanding of the mechanism that controls the priming of viral DNA synthesis has only recently been obtained, owing to the discovery that enzymatically active RT of duck hepatitis B virus (DHBV) can be produced in a cell-free system (Wang and Seeger 1992). The major finding has been that ε RNA is not solely a signal for

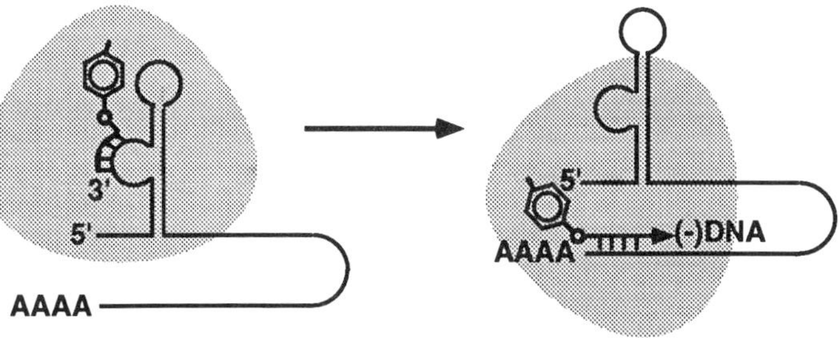

Figure 3 Model for the protein-priming reaction. Priming of DNA synthesis occurs at the RNA packaging signal ε near the 5′ end of pregenomic RNA with the help of a tyrosine residue of the polymerase polypeptide. After the polymerization of four nucleotides, the nascent DNA strand is transferred to the 3′ end of the RNA template, where minus-strand DNA synthesis continues.

RNA packaging, but actually serves as a template for the synthesis of a short 3- to 4-nucleotide-long DNA oligomer. The template for this oligomer is provided by nucleotides in the bulge region of ε (Figs. 2 and 3) (Wang and Seeger 1993). The primer for DNA synthesis is provided by a tyrosine residue located near the amino terminus of the polymerase polypeptide (Weber et al. 1994; Zoulim and Seeger 1994). As a consequence of the protein-priming mechanism, the RT remains covalently linked to the 5′ end of minus-strand DNA during all subsequent steps of viral DNA synthesis (Fig. 1). This conclusion is in good agreement with the observation that nascent DNA strands in subviral core particles, as short as 30 nucleotides in length, are covalently linked to protein (Molnar-Kimber et al. 1983).

Subsequent to the protein-priming reaction, the short DNA oligomer is transferred to the 3′ end of pregenomic RNA, where it base-pairs with a complementary sequence motif located at a 10- to 12-nucleotide-long region known as DR1 (Figs. 2 and 3). However, the 3- to 4-base acceptor site by itself is too short to specify the origin of minus-strand DNA synthesis. Genetic experiments with woodchuck hepatitis virus (WHV) and DHBV demonstrated that mutations or deletions of the natural acceptor site that prevent base-pairing with the DNA oligomer lead to the synthesis of minus-strand DNA with 5′ ends mapping to positions on pregenomic RNA that can serve as alternate acceptor sites (Condreay et al. 1992; Seeger and Maragos 1990, 1991). The selection of the natural site is most likely facilitated by the structural arrangement of pregenomic RNA in the nucleocapsid. We envision a scenario whereby the acceptor site and ε RNA are held in close physical proximity to facilitate the DNA transfer reaction. The *cis*-acting signals on pregenomic RNA that may be required for such a structural arrangment have not yet been identified. However, deletion analysis with WHV suggested that a region extending approximately 1 kb upstream of DR1 bears the signals that specify the acceptor site (Seeger and Maragos 1991). Since polymerase requires pregenomic RNA as a vehicle for incorporation into core particles, it is likely that a single polymerase polypeptide carries out both the protein-priming and the DNA synthesis reactions (see below) (Bartenschlager and Schaller 1992).

An important prediction of the proposed model for DNA priming is that base changes in the bulge of ε should appear at the 5′ end of minus-strand DNA as a consequence of DNA replication. Transfection of tissue-culture cells with DHBV DNA that carried mutations in the ε region, which did not prevent elongation of DNA synthesis from the natural acceptor site at DR1, directly demonstrated the expected transfer of genetic information across the viral genome (Wang and Seeger 1993). Similar results were obtained with enzymatically active DHBV RT expressed with a yeast Ty-1 vector (Tavis and Ganem 1993). Insertion of a nucleotide into the bulge region of ε resulted in the synthesis of minus strands with an extra nucleotide at their 5′ ends (Tavis et al. 1994).

Following the translocation reaction, minus-strand DNA synthesis continues all the way to the 5′ end of the RNA template. RNA is concomitantly degraded by an RNase H activity on the polymerase polypeptide (Summers and Mason 1982; Radziwill et al. 1990). Because of the relative location of DR1 within the terminal redundancy on the pregenome, the completed minus-strand DNA bears a short, 9-nucleotide-long terminal redundancy, which plays a role, as described below, in the circularization of the viral genome (Seeger et al. 1986; Lien et al. 1987; Will et al. 1987).

Priming of Plus-strand DNA Synthesis

Unlike the situation in retroviruses, plus-strand DNA synthesis does not initiate until minus-strand DNA synthesis has been completed. This is a direct consequence of the mechanism for the priming of plus-strand DNA synthesis. It relies on the formation of an RNA primer by the viral RNase H activity. The primer is derived from the 5′ end of pregenomic RNA, from the cap through DR1, and hence cannot be created prior to the completion of minus-strand DNA synthesis (Fig. 4) (Lien et al. 1986; Loeb et al. 1991). To prime plus-strand DNA synthesis, the RNA primer is first translocated and hybridized with sequences near the 5′ end of

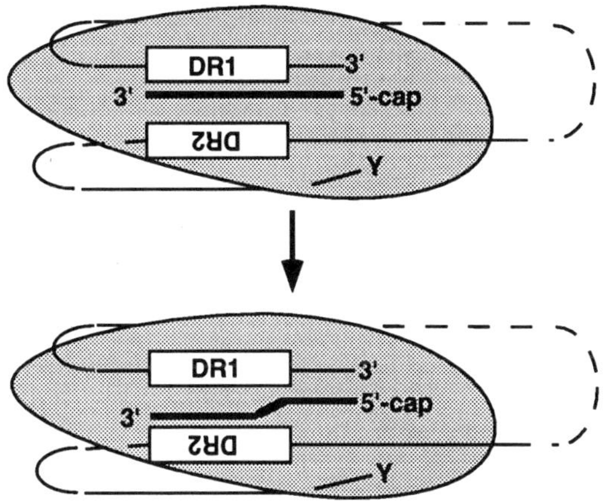

Figure 4 Model for plus-strand priming. Minus-strand DNA is arranged to facilitate the transfer of the capped RNA primer from DR1 to DR2. The polymerase is covalently linked through a tyrosine residue (Y) to the 5′ end of minus-strand DNA.

minus-strand DNA. This step is facilitated by a short 11- to 12-nucleotide-long sequence homology between DR1 and DR2. DR2 is located, depending on the species of hepadnavirus, about 50–200 bases downstream from the 5′ end of the minus-strand DNA (Figs. 2 and 4). The exact details of this reaction, which is reminiscent of the translocation reaction that occurs during the priming of minus-strand DNA synthesis, are not known, but there is ample biochemical and genetic evidence to support it. Biochemical analysis of the 5′ end of plus-strand DNA revealed the presence of an 18-nucleotide-long RNA primer of the expected sequence with a cap structure, and genetic experiments revealed that mutations introduced into the 5′ copy of DR1 appear on the RNA primer at the 5′ end of plus-strand DNA as a consequence of viral DNA replication (Lien et al. 1986; Seeger et al. 1986; Strapans et al. 1991). It is likely that the RNA transfer reaction is facilitated through the spatial arrangement of the minus-strand DNA; i.e., through the juxtaposition of the regions encompassing DR1 and DR2 on minus-strand DNA (Fig. 4). Such a structure may invoke forces that help to stabilize the 3′-OH group of the RNA primer in a position that allows the polymerase to initiate DNA synthesis at DR2 rather than at DR1. In fact, disruption of the homology between DR1 and DR2 favors an in situ DNA-priming reaction at DR1 that leads to the synthesis of double-stranded linear genomes as compared to the relaxed circular DNA species, which is the predominant species produced by wild-type virus (Strapans et al. 1991; Condreay et al. 1992). However, preparations of wild-type virus do contain a small fraction of particles with linear genomes. Although the latter are not known to play an important role in the hepadnavirus life cycle, they are infectious, as evidenced by the formation of CCC DNA and subsequent viral DNA synthesis (Yang and Summers 1995).

Once plus-strand DNA synthesis has progressed from DR2 to the 5′ end of minus-strand DNA, a template switch (i.e., circularization) is required for the continuation of DNA synthesis. The terminally redundant sequences of minus-strand DNA are believed to promote the strand-transfer reaction from the 5′ to the 3′ end. The structural requirements for this reaction must be complicated, since the polymerase has to accommodate both ends of minus-strand DNA in close proximity while it is still covalently attached to the 5′ end. The template transfer then leads to the circularization of the viral genome. In mammalian hepadnaviruses, the RT then extends the plus-strand DNA to approximately half the genome length, but in avian viruses, the polymerase elongates plus strands to nearly full length (Robinson et al. 1974; Summers et al. 1975; Lien et al. 1987). It is notable, however, that the polymerase does not displace the RNA primer and the 5′ end of plus-strand DNA (i.e., the RNA primer from DR2, Fig. 1), an event that would lead to the formation of a terminally redundant, linear genome. Linear genomes are observed in pools of viral DNA, but these arise as a consequence of the in situ plus-strand-priming event described above. The cause and significance of the premature termination of plus strands synthesized in the mammalian viruses remain obscure. It is conceivable that the steric factors imposed by the capsid and by the RT prevent the completion of this reaction during virion morphogenesis. Alternatively, capsids with incomplete plus strands may be more readily packaged into viral envelopes by the mammalian than by the avian viruses. This latter possibility may be supported by the observation that the polymerase activity present in intact virus cores can repair the single-stranded region in an in vitro reaction (Kaplan et al. 1973; Summers et al. 1975). However, core particles assembled from capsid proteins with truncated carboxyl termini lead to the accumulation of virion DNA with incomplete plus strands, indicating that capsid structure can influence elongation of plus-strand DNA (Yu and Summers 1991; Nassal 1992).

Formation and Amplification of CCC DNA from Viral DNA

Following maturation of viral DNA within cytoplasmic core particles, these particles may enter one of two pathways (Fig. 1). In the first, they interact with viral envelope proteins to form virions. In the second, they or their content of viral DNA is transported to the nucleus, where this DNA is processed to amplify the copy number of CCC DNA (Tuttleman et al. 1986; Wu et al. 1990). Entry into the second pathway only occurs if there is an inadequate cytoplasmic concentration of viral envelope proteins. Thus, CCC DNA amplification normally occurs only during the first few days of an infection, prior to the initiation of virus assembly and release. The final copy number per nucleus, in vivo, is usually between 5 and 30 (Jilbert et al. 1992; Kajino et al. 1994).

Transport to the nucleus is not mediated by viral envelope proteins, since amplification occurs when hepatocytes are infected with viral mutants that are unable to synthesize these proteins (Summers et. al. 1990, 1991). The route and mechanism of transport are unknown. It appears, however, that this route is regulated by the signals generated on core particles during their maturation, perhaps the same signals that facilitate differential assembly of nucleocapsids containing mature DNA into virions (Yu and Summers 1991, 1994a,b; Guidotti et al. 1994). The carboxyl terminus of the core protein of cytoplasmic nucleocapsids can be phosphorylated at up to four sites (Schlicht et al. 1989a; Machida et al. 1991; Yeh and Ou 1991; Yu and Summers 1994b), whereas virion nucleocapsids are probably only phosphorylated at one of these sites (Pugh et al. 1989), suggesting that differential phosphorylation may signal intracellular trafficking of nucleocapsids.

What remains completely unclear is how and when viral DNA is processed to form CCC DNA. This step requires removal of the RT from the 5′ end of minus-strand DNA, trimming of the 3′ repeat from the terminus of minus-strand DNA, and ligation of the two ends. In addition, plus-strand DNA synthesis must be completed, RNA removed from the 5′ end, and the ends ligated. It seems probable that the viral DNA polymerase could be involved in plus-strand elongation up to the primer-binding site, DR2. However, the polymerase does not appear capable of displacing this primer in intact nucleocapsids (Lien et al. 1987), suggesting that elongation through DR2 is mediated subsequent to the release of viral DNA from core particles. An important issue is the role of the viral polymerase in the cleavage and joining reactions that are involved in formation of CCC DNA versus the role of cellular enzymes, such as that proposed for topoisomerase I (Rogler 1991). Current knowledge of the functional capacity of the polymerase polypeptide is discussed below.

STRUCTURE AND FUNCTION OF RT

Hepadnaviral RTs are encoded by the viral polymerase (*pol*) gene and have an approximate molecular mass of 90 kD. Synthesis of polymerase polypeptides occurs through internal initiation of translation from an AUG codon, which is located near the 3′ end of the core gene, allowing for a short overlap between the two open reading frames. Because of amino acid homologies with retroviral RTs, it has been possible to identify the DNA polymerase and RNase H domains encoded by hepadnaviral *pol* genes (Fig. 5) (Toh et al. 1983; Radziwill et al. 1990). Alignment of conserved residues between the polymerase polypeptides of the two virus families shows that the hepadnaviral polymerase polypeptides bear an amino-terminal domain, also referred to as the terminal protein (TP) region (Fig. 5) (Bartenschlager and Schaller 1988; Radziwill et al. 1990). This domain bears the tyrosine residue utilized for the protein-priming reaction. It is separated from the RT domain by a spacer region that is not essential for any of the known activities of the RT, as indicated by observations that RT functions are refractory to mutagenesis of this region (Bartenschlager and Schaller 1988; Chang et al. 1990).

Genetic analyses of the RT of DHBV expressed either in reticulocyte lysates, in frog oocytes, or with the help of the transposon Ty-1 in yeast showed that the polymerase can exhibit protein-priming and DNA-polymerization activities in the absence of other viral polypeptides, most notably the viral capsid proteins (Wang and Seeger 1992; Seifer and Standring 1993; Tavis and Ganem 1993). Furthermore, mutations altering residues in the catalytic site of the polymerase abolish the protein-priming activity, demonstrating that an enzymatically active RT domain is required for this reaction (Wang and Seeger 1992). Interestingly, unlike the situation with other systems where a protein is known to act as a primer for DNA synthesis (i.e., adenovirus and bacteriophage ϕ29 [Salas 1991]), in hepadnaviruses the protein primer and the DNA polymerase reside on the same polypeptide. As suggested from genetic analyses of the DHBV polymerase expressed in vitro, it appears that both the TP and RT domains are required for the interaction of the polymerase with ε RNA (Pollak and Ganem 1994; Wang et al. 1994). Although it is likely that the polymerase binds directly to ε, it is conceivable that host factors are required to support this interaction, perhaps similar to the scenario proposed for the binding of HIV Tat to the TAR response element (Marciniak et al. 1990; Madore and Cullen 1993). Mutational analysis of ε suggested that binding of the RNA hairpin to the polymerase alone may not be sufficient for RNA packaging and that one or more cellular proteins may indeed be required for this reaction (Pollak and Ganem 1994). The observation that the DHBV RT expressed in wheat germ extracts fails to bind to ε RNA, unless supplemented with a factor(s) present in rabbit reticulocyte lystates, also supports the view that host fac-

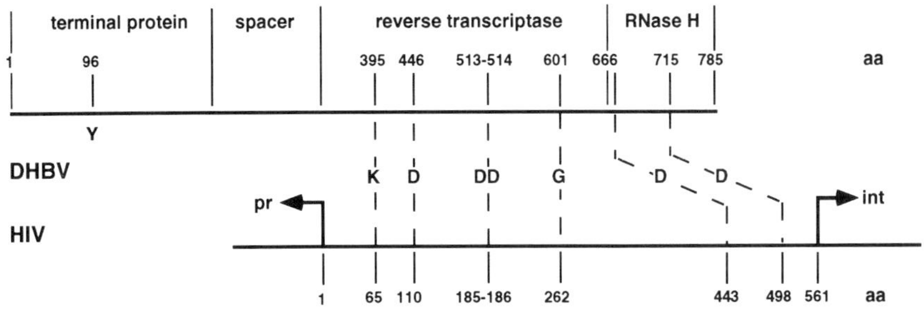

Figure 5 Structural comparison of hepadnaviral and retroviral reverse transcriptases. Shown are the linear maps of the DHBV and HIV *pol* gene products. The maps were aligned with the help of amino acids that are conserved among RTs. Position 1 on the DHBV polymerase corresponds to the AUG codon at position 170 on the DHBV genome (Mandart et al. 1984; Chang et al. 1989; Schlicht et al. 1989b). Amino acid positions on the HIV RT map were adapted from Jacobo-Molina et al. (1993). (pr) Protease, (int) integrase.

tor(s) may be involved in the expression of enzymatically active enzyme (Hu and Seeger 1996).

It is not known whether the polymerase functions within cores as a monomer or as a dimer and whether more than one functional polymerase molecule is required for viral DNA synthesis. However, it is likely that a single polymerase polypeptide may catalyze one complete round of the DNA replication cycle. This view is consistent with the observation that assembly of the RT into core particles depends on its interaction with ε sequences on pregenomic RNA, which would indicate that polymerase and RNA templates may be present at equimolar amounts in subviral particles (Bartenschlager and Schaller 1992). Quantitation of radioactively labeled RT of HBV present in core particles revealed the molar ratio of 0.7 polymerase molecule per virion DNA, which is in good agreement with the proposed mechanism for particle assembly and DNA synthesis (Bartenschlager and Schaller 1992; Bartenschlager et al. 1992).

Given this stoichiometry, it is still possible to envision, at least in a schematic sense, how the RT, RNase H, and protein-primer domains might function in minus-strand synthesis. What is more difficult to understand is how the polymerase facilitates elongation of the plus strand, prior to the strand switch, virtually all the way to the point of attachment to the protein primer (Fig. 4). A priori, this would seem to require either multiple copies of the polymerase polypeptide within virions or an unusual amount of flexibility and accessibility between different domains of the protein. It is also difficult to understand how the primer is removed by cleavage of the phosphotyrosine bond during formation of CCC DNA from virion DNA. Both these considerations, and particularly the latter, suggest that there may be additional functional domains or activities of the polymerase polypeptide that remain to be identified.

SUMMARY AND PERSPECTIVES

It has been little more than a decade since the discovery that hepadnaviruses replicate via reverse transcription. In that time, there has evolved a highly detailed model for the major steps in viral DNA synthesis. It has also become clear that reverse transcription is involved not only in the production of progeny virus DNA, but also, through a highly regulated process, in the synthesis and maintenance of CCC DNA, a species which, despite its nuclear location, lacks the regulatory sequences that would facilitate its reproduction via semi-conservative DNA synthesis. As should be clear from the discussion above, the major emphasis of most investigations has been on those steps of reverse transcription that lead to the creation of the mature virion DNA. The process of CCC DNA synthesis has been characterized; however, the molecular details underlying this process are not at all understood. Moreover, the connection between the various viral DNA forms that are associated with productive infections and the integrated DNA that accumulates in the liver and, at least in some instances, plays a major role in hepatocellular carcinogenesis via oncogene activation through region-specific integration, is completely unexplored. Finally, the role of host proteins in viral DNA synthesis, CCC DNA amplification, and virion assembly are only beginning to be addressed. These issues will define important avenues of future research, and it is hoped they will lead not only to a better understanding of how to control and eliminate chronic infections, but also to an understanding of how virus replication is spontaneously shut down both in chronically and in transiently infected individuals.

ACKNOWLEDGMENTS

The authors acknowledge support from the National Institutes of Health and the Commonwealth of Pennsylvania.

REFERENCES

Bartenschlager, R. and H. Schaller. 1988. The amino-terminal domain of the hepadnaviral P-gene encodes the terminal protein (genome-linked protein) believed to prime reverse transcription. *EMBO J.* **7:** 4185–4192.

———. 1992. Hepadnaviral assembly is initiated by polymerase binding to the encapsidation signal in the viral RNA genome. *EMBO J.* **11:** 3413–3420.

Bartenschlager, R., M. Junker-Niepmann, and H. Schaller. 1990. The P gene product of hepatitis B virus is required as a structural component for genomic RNA encapsidation. *J. Virol.* **64:** 5324–5332.

Bartenschlager, R., C. Kuhn, and H. Schaller. 1992. Expression of the P-protein of the human hepatitis B virus in a vaccinia virus system and detection of the nucleocapsid-associated P-gene product by radiolabelling at newly introduced phosphorylation sites. *Nucleic Acids Res.* **20:** 195–202.

Blumberg, B.S., B.J.S. Gerstley, D.A. Hungerford, W.T. London, and A.I. Sutnik. 1967. A serum antigen (Australai antigen) in Down's syndrome, leukemia and hepatitis. *Ann. Intern. Med.* **66:** 924–931.

Calvert, J. and J. Summers. 1994. Two regions of an avian hepadnavirus RNA pregenome are required in *cis* for encapsidation. *J. Virol.* **68:** 2084–2090.

Chang, L.J., R.C. Hirsch, D. Ganem, and H.E. Varmus. 1990. Effects of insertional and point mutations on the functions of the duck hepatitis B virus polymerase. *J. Virol.* **64:** 5553–5558.

Chang, L.J., P. Pryciak, D. Ganem, and H.E. Varmus. 1989. Biosynthesis of the reverse transcriptase of hepatitis B viruses involves de novo translational initiation not ribosomal frameshifting. *Nature* **337:** 364–368.

Chen, Y., W.S. Robinson, and P.L. Marion. 1992. Naturally occurring point mutation in the C terminus of the polymerase gene prevents duck hepatitis B virus RNA packaging. *J. Virol.* **66:** 1282–1287.

Condreay, L.D., T.-T. Wu, C.E. Aldrich, M.A. Delaney, J. Summers, C. Seeger, and W.S. Mason. 1992. Replication of DHBV genomes with mutations at the sites of initiation of minus- and plus-strand DNA synthesis. *Virology* **188:** 208–216.

Crowther, R.A., N.A. Kiselev, B. Bottcher, J.A. Berriman, G.P. Borisova, V. Ose, and P. Pumpens. 1994. Three-dimensional strucuture of hepatitis B virus core particles determined by electron cryomicroscopy. *Cell* **77:** 943–950.

Fourel, I., J.M. Cullen, J. Saputelli, C.E. Aldrich, P. Schaffer, D. Averett, J. Pugh, and W.S. Mason. 1994. Evidence that hepatocyte turnover is required for rapid clearance of DHBV during antiviral therapy of chronically infected ducks. *J. Virol.* **68:** 8321–8330.

Gerlich, W. and W. S. Robinson. 1980. Hepatitis B virus contains protein covalently attached to the 5′ terminus of its complete DNA strand. *Cell* **21:** 801–809.

Grisham, J.W. 1962. A morphologic study of deoxyribonucleic acid synthesis and cell proliferation in regenerating rat liver; Autoradiography with thymidine-H3. *Cancer Res.* **22:** 842–849.

Guidotti, L.G., V. Martinez, Y.-T. Loh, C.E. Rogler, and F.V. Chisari. 1994. Hepatitis B virus nucleocapsid particles do not cross the hepatocyte nuclear membrane in transgenic mice. *J. Virol.* **68:** 5469–5475.

Hatton, T., S. Zhou, and D.N. Standring. 1992. RNA- and DNA-binding activities in hepatitis B virus capsid protein: A model for their roles in viral replication. *J. Virol.* **66:** 5232–5241.

Hirsch, R.C., D.D. Loeb, J.R. Pollack, and D. Ganem. 1991. *cis*-Acting sequences required for encapsidation of duck hepatitis B virus pregenomic RNA. *J. Virol.* **65:** 3309–3316.

Hirsch, R.C., J.E. Lavine, L.J. Chang, H.E. Varmus, and D. Ganem. 1990. Polymerase gene products of hepatitis B viruses are required for genomic RNA packaging as well as for reverse transcription. *Nature* **344:** 552–555.

Hu, J.-M. and C. Seeger. 1996. Hsp90 is required for the activity of a hepatitis B virus reverse transcriptase. *Proc. Natl. Acad. Sci.* (in press).

Huang, M. and J. Summers. 1991. Infection initiated by the RNA pregenome of a DNA virus. *J. Virol.* **65:** 5435–5439.

Jacobo-Molina, A., J. Ding, R.G. Nanni, A.D. Clark, Jr., X. Lu, C. Tantillo, R.L. Williams, G. Kramer, A.L. Ferris, P. Clark, A. Hizi, S.H. Hughes, and E. Arnold. 1993. Crystal structure of human immunodeficiency virus type 1 reverse transcriptase complexed with double-stranded DNA at 3.0 Å resolution shows bent DNA. *Proc. Natl. Acad. Sci.* **90**: 6320–6324.

Jilbert, A.R., T.-T. Wu, J.M. England, P.M. Hall, N.Z. Carp, A.P. O'Connell, and W.S. Mason. 1992. Rapid resolution of duck hepatitis B virus infections occurs after massive hepatocellular involvement. *J. Virol.* **66:** 1377–1388.

Junker-Niepmann, M., R. Bartenschlager, and H. Schaller. 1990. A short *cis*-acting sequence is required for hepatitis B virus pregenome encapsidation and sufficient for packaging of foreign RNA. *EMBO J.* **9:** 3389–3396.

Kajino, K., A.R. Jilbert, J. Saputelli, C.E. Aldrich, J. Cullen, and W.S. Mason. 1994. A study of woodchuck hepatitis virus infections revealed very rapid recovery after a prolonged viremia and infection of virtually every hepatocyte. *J. Virol.* **68:** 5792–5803.

Kaplan, P.M., R.L. Greenman, J.L. Gerin, R.H. Purcell, and W.S. Robinson. 1973. DNA polymerase associated with human hepatitis B antigen. *J. Virol.* **12:** 995–1005.

Knaus, T. and M. Nassal. 1993. The encapsidation signal on the hepatitis B virus RNA pregenome forms a stem-loop structure that is critical for its function. *Nucleic Acids Res.* **21:** 3967–3975.

Lien, J.M., C.E. Aldrich, and W.S. Mason. 1986. Evidence that a capped oligoribonucleotide is the primer for duck hepatitis B virus plus-strand DNA synthesis. *J. Virol.* **57:** 229–236.

Lien, J.-M., D.J. Petcu, C.E. Aldrich, and W.S. Mason. 1987. Initiation and termination of duck hepatitis B virus DNA synthesis during virus maturation. *J. Virol.* **61:** 3832–3840.

Loeb, D.L., R.C. Hirsch, and D. Ganem. 1991. Sequence-independent RNA cleavages generate the primers for plus strand DNA synthesis in hepatitis B viruses: Implications for other reverse transcribing elements. *EMBO J.* **10:** 3533–3540.

MacDonald, R.A. 1960. "Lifespan" of liver cells. *Arch. Intern. Med.* **107:** 335–343.

Machida, A., H. Ohnuma, F. Tsuda, A. Yoshikawa, Y. Hoshi, T. Tanaka, S. Kishimoto, Y. Akahane, Y. Miyakawa, and M. Mayumi. 1991. Phosphorylation in the carboxyl-terminal domain of the capsid protein of hepatitis B virus: Evaluation with a monoclonal antibody. *J. Virol.* **65:** 6024–6030.

Madore, S.J. and B.R. Cullen. 1993. Genetic analysis of the cofactor requirement for human immunodeficiency virus type 1 tat function. *J. Virol.* **67:** 3703–3711.

Mandart, E., A. Kay, and F. Galibert. 1984. Nucleotide sequence of a cloned duck hepatitis B virus genome: Comparison with woodchuck and human hepatitis B virus sequences. *J. Virol.* **49:** 782–792.

Marciniak, R.A., B.J. Calnan, A.D. Frankel, and P.A. Sharp. 1990. HIV-1 tat protein *trans*-activates transcription in vitro. *Cell* **63:** 791–802.

Marion, P.L., L.S. Oshiro, D.C. Regnery, G.H. Scullard, and W.S. Robinson. 1980. A virus in Beechey ground squirrels which is related to hepatitis B virus of man. *Proc. Natl. Acad. Sci.* **77:** 2941–2945.

Mason, W.S., G. Seal, and J. Summers. 1980. Virus of Pekin ducks with structural and biological relatedness to human hepatitis B virus. *J. Virol.* **36:** 829–836.

Molnar-Kimber, K., J. Summers, J.M. Taylor, and W.S. Mason. 1983. Protein covalently bound to minus-strand DNA intermediates of duck hepatitis B virus. *J. Virol.* **45:** 165–172.

Nassal, M. 1992. The arginine-rich domain of the hepatitis B virus core protein is required for pregenome encapsidation and productive viral positive-strand DNA synthesis but not for virus assembly. *J. Virol.* **66:** 4107–4116.

Nassal, M. and H. Schaller. 1993. Hepatitis B virus nucleocapsid assembly. In *Virus strategies* (ed. W. Doerfler and P. Bohm), pp. 42–74. VCH, Weinheim, Germany.

Onodera, S., H. Ohori, M. Yamaki, and N. Ishida. 1982. Electron microscopy of human hepatitis B virus cores by negative staining-carbon film technique. *J. Med. Virol.* **10:** 147–155.

Pollack, J.R., and D. Ganem. 1993. An RNA stem-loop structure directs hepatitis B virus genomic RNA encapsidation. *J. Virol.* **67:** 3254–3263.

———. 1994. Site-specific RNA binding by a hepatitis B virus reverse transcriptase initiates two distinct reactions: RNA packaging and DNA synthesis. *J. Virol.* **68:** 5579–5587.

Pugh, J., A. Zweidler, and J. Summers. 1989. Characterization of the major duck hepatitis B virus core particle protein. *J. Virol.* **63:** 1371–1376.

Radziwill, G., W. Tucker, and H. Schaller. 1990. Mutational analysis of the hepatitis B virus P gene product: Domain structure and RNAse H activity. *J. Virol.* **64:** 613–620.

Rekosh, D.M.K., W.C. Russell, A.J.D. Bellet, and A.J. Robinson. 1977. Identification of a protein linked to the ends of adenovirus DNA. *Cell* **11:** 283–295.

Robinson, W.S., D.A. Clayton, and R.L. Greenman. 1974. DNA of a human hepatitis B virus candidate. *J. Virol.* **14:** 384–391.

Rogler, C.E. 1991. Cellular and molecular mechanisms of hepatocarcinogenesis associated with hepadnavirus infection. *Curr. Top. Microbiol. Immunol.* **168:** 104–140.

Roychoudhury, S., A.F. Faruqi, and C. Shih. 1991. Pregenomic RNA encapsidation analysis of eleven missense and nonsense polymerase mutants of human hepatitis B virus. *J. Virol.* **65:** 3617–3624.

Salas, M. 1991. Protein-priming of DNA replication. *Annu. Rev. Biochem.* **60:** 39–71.

Schlicht, H.J., R. Bartenschlager, and H. Schaller. 1989a. The duck hepatitis B virus core protein contains a highly phosphorylated C terminus that is essential for replication but not for RNA packaging. *J. Virol.* **63:** 2995–3000.

Schlicht, H.J., G. Radziwill, and H. Schaller. 1989b. Synthesis and encapsidation of duck hepatitis B virus reverse transcriptase do not require formation of core-polymerase fusion proteins. *Cell* **56:** 85–92.

Seeger, C. and J. Maragos. 1990. Identification and characterization of the woodchuck hepatitis virus origin of DNA replication. *J. Virol.* **64:** 16–23.

———. 1991. Identification of a signal necessary for initiation of reverse transcription of the hepadnavirus genome. *J. Virol.* **65:** 5190–5195.

Seeger, C., D. Ganem, and H.E. Varmus. 1986. Biochemical and genetic evidence for the hepatitis B virus replication strategy. *Science* **232:** 477–484.

Seifer, M. and D.N. Standring. 1993. Recombinant human hepatitis B virus reverse transcriptase is active in the absence of the nucleocapsid or the viral replication origin, DR1. *J. Virol.* **67:** 4513–4520.

Sprengel, R., E.F. Kaleta, and H. Will. 1988. Isolation and characterization of a hepatitis

B virus endemic in herons. *J. Virol.* **62:** 932–937.

Strapans, S., D.D. Loeb, and D. Ganem. 1991. Mutations affecting hepadnavirus plus-strand DNA synthesis dissociate primer cleavage from translocation and reveal the origin of linear viral DNA. *J. Virol.* **65:** 1255–1262.

Summers, J. and W.S. Mason. 1982. Replication of the genome of a hepatitis B-like virus by reverse transcription of an RNA intermediate. *Cell* **29:** 403–415.

Summers, J., A. O'Connell, and I. Millman. 1975. Genome of hepatitis B virus: Restriction enzyme cleavage and structure of DNA extracted from Dane particles. *Proc. Natl. Acad. Sci.* **72:** 4597–4601.

Summers, J., P.M. Smith, and A.L. Horwich. 1990. Hepadnaviral envelope proteins regulate covalently closed circular DNA amplification. *J. Virol.* **64:** 2819–2824.

Summers, J., J.M. Smolec, and R. Snyder. 1978. A virus similar to human hepatitis B virus associated with hepatitis and hepatoma in woodchucks. *Proc. Natl. Acad. Sci.* **75:** 4533–4537.

Summers, J., P.M. Smith, M. Huang, and M. Yu. 1991. Morphogenetic and regulatory effects of mutations in the envelope proteins of an avian hepadnavirus. *J. Virol.* **65:** 1310–1317.

Tavis, J.E. and D. Ganem. 1993. Expression of functional hepatitis B virus polymerase in yeast reveals it to be the sole viral protein required for correct initiation of reverse transcription. *Proc. Natl. Acad. Sci.* **90:** 4107–4111.

Tavis, J.E., S. Perri, and D. Ganem. 1994. Hepadnavirus reverse transcription initiates within the stem-loop of the RNA packaging signal and employs a novel strand transfer. *J. Virol.* **68:** 3536–3543.

Toh, H., H. Hayashida, and T. Miyata. 1983. Sequence homology between retroviral reverse transcriptase and putative polymerases of hepatitis B virus and cauliflower mosaic virus. *Nature* **305:** 827–829.

Tuttleman, J.S., C. Pourcel, and J. Summers. 1986. Formation of the pool of covalently closed circular viral DNA in hepadnavirus-infected cells. *Cell* **47:** 451–460.

Wang, G.H., and C. Seeger. 1992. The reverse transcriptase of hepatitis B virus acts as a protein primer for viral DNA synthesis. *Cell* **71:** 663–670.

———. 1993. Novel mechanism for reverse transcription in hepatitis B viruses. *J. Virol.* **67:** 6507–6512.

Wang, G.H., F. Zoulim, E.H. Leber, J. Kitson, and C. Seeger. 1994. Role of RNA in enzymatic activity of the reverse transcriptase of hepatitis B viruses. *J. Virol.* **68:** 8437–8442.

Weber, M., V. Bronsema, H. Bartos, A. Bosserhoff, R. Bartenschlager, and H. Schaller. 1994. Hepadnavirus P protein utilizes a tyrosine residue in the TP domain to prime reverse transcription. *J. Virol.* **68:** 2994–2999.

Will, H., W. Reiser, T. Weimer, E. Pfaff, M. Buscher, R. Sprengel, R. Cattaneo, and H. Schaller. 1987. Replication strategy of human hepatitis B virus. *J. Virol.* **61:** 904–911.

Wu, T.-T., L. Coates, C.E. Aldrich, J. Summers, and W.S. Mason. 1990. In hepatocytes infected with duck hepatitis B virus, the template for viral RNA synthesis is amplified by an intracellular pathway. *Virology* **175:** 255–261.

Yang, W. and J. Summers. 1995. Illegitimate replication of linear hepadnaviral DNA through non-homologous recombination. *J. Virol.* **69:** 4029–4036.

Yeh, C.T. and J.H. Ou. 1991. Phosphorylation of hepatitis B virus precore and core proteins. *J. Virol.* **65:** 2327–2331.

Yu, M.-S. and J. Summers. 1991. A domain of the hepadnavirus capsid protein is specifi-

cally required for DNA maturation and virus assembly. *J. Virol.* **65:** 2511–2517.
———. 1994a. Multiple functions of capsid protein phosphorylation in duck hepatitis B virus replication. *J. Virol.* **68:** 4341–4348.
———. 1994b. Phosphorylation of the duck hepatitis B virus capsid protein associated with conformational changes in the C terminus. *J. Virol.* **68:** 2965–2969.
Zoulim, F. and C. Seeger. 1994. Reverse transcription in hepatitis B viruses is primed by a tyrosine residue of the polymerase. *J. Virol.* **68:** 6–13.

30

Geminivirus DNA Replication

David M. Bisaro
Plant Biotechnology Center and
Department of Molecular Genetics
The Ohio State University
Columbus, Ohio 43210

Diseases caused by geminiviruses have long been recognized as a limitation to the cultivation of several important crops, including maize, cassava, bean, squash, cucurbits, and tomato, in tropical and subtropical regions of the world. More recently, geminivirus diseases, particularly those transmitted by whiteflies, have become an even greater threat to agriculture due to the appearance of a new and more aggressive whitefly biotype (Brown et al. 1995). This has renewed interest in the study of geminivirus pathogenesis and epidemiology and has stimulated work on the development of virus-resistant crop plants. For quite different reasons, geminiviruses have also attracted the attention of researchers concerned with fundamental aspects of plant molecular biology. In particular, their small DNA genomes and extensive reliance on host biosynthetic machinery make these viruses ideal model systems for the study of plant DNA replication and gene expression. Geminivirus genomes also have considerable potential as vectors for the expression of foreign genes in plants, although this potential has yet to be fully exploited. Because of their significance to plant pathology, plant molecular biology, and plant biotechnology, it is not surprising that geminiviruses have become the subjects of intensive research. As a result of efforts concentrated on a few model viruses, remarkable progress has been made toward understanding the mechanisms of viral replication and pathogenesis in the dozen years or so since the first molecular studies were initiated. Several recent reviews of this progress are available (Stanley 1991; Lazarowitz 1992; Mullineaux et al. 1992; Bisaro 1994). In this review, I focus on what is presently known about geminivirus DNA replication, with emphasis on the sequence requirements for replication and on the roles of viral proteins in this process.

DNA Replication in Eukaryotic Cells
 0-87969-459-9/96 $5 + .00

THE GEMINIVIRIDAE

General Features

The geminivirus family derives its name from the unusual twin icosahedral (geminate) capsid structure of its members. Each paired particle encapsidates a single molecule of covalently closed, circular, single-stranded DNA (ssDNA) varying in size from 2.5 kb to 3.0 kb, depending on the virus. The genomic ssDNA is replicated in the nucleus of the host cell by a rolling-circle mechanism utilizing double-stranded DNA (dsDNA) intermediates in a manner similar to the ssDNA-containing bacteriophages. Viral dsDNAs, which include open and covalently closed (supercoiled) circular forms, are organized as minichromosomes apparently complexed with histone proteins (Pilartz and Jeske 1992). The nucleus is also the site of virus assembly, and nuclear inclusions consisting of large numbers of virus particles have been observed in infected cells by transmission electron microscopy (see, e.g., Esau 1977; Kim et al. 1978; Rushing et al. 1987).

Viral Genome Organization

The geminivirus family is diverse and can be divided into at least three subgroups based on genome organization, host range, and type of insect vector (Matthews 1991). Subgroup I comprises viruses that have a single component genome, are transmitted by leafhoppers, and usually infect monocotyledonous plants. Well-characterized subgroup I pathogens include maize streak virus (MSV) and wheat dwarf virus (WDV) (Fig. 1).

The second subgroup until recently contained only a single member, beet curly top virus (BCTV). BCTV also has a monopartite genome and is transmitted by a leafhopper vector, but it infects only dicotyledonous plants. The genome organization of BCTV also differs in significant ways from MSV and its relatives (Fig. 1).

Subgroup III contains viruses with monopartite genomes (~2.8 kb) as well as bipartite viruses that have two genome components, each of similar size (~2.6 kb). In the latter, the two genomic DNAs, designated A and B, differ in sequence except for a common region of 200–250 bp that is nearly identical in the genome components of any given virus, but differs between viruses. Both genome components are required for infectivity (Hamilton et al. 1983; Stanley 1983), although DNA A contains all viral information necessary for replication and encapsidation (Rogers et al. 1986; Townsend et al. 1986; Sunter et al. 1987). The B component provides genes whose functions are required for cell-to-cell and systemic spread of the virus from the inoculation site. In contrast, the movement

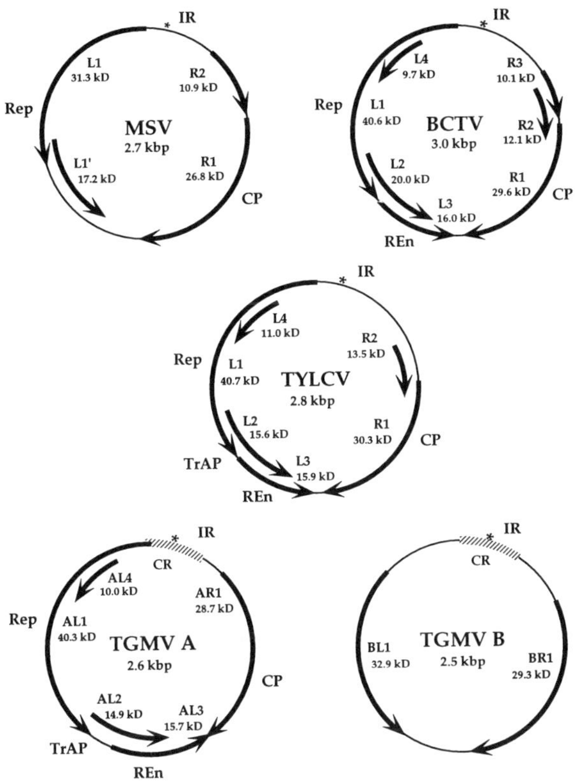

Figure 1 Geminivirus genome organization. The diagrams depict the double-stranded replicative forms of maize streak virus (MSV, subgroup I), beet curly top virus (BCTV, subgroup II), tomato yellow leaf curl virus (TYLCV, monopartite, subgroup III), and tomato golden mosaic virus (TGMV, A and B, bipartite, subgroup III). The solid arrows indicate the positions of viral genes with the approximate molecular mass of each encoded protein given in kilodaltons. Viral genes are designated by number and the direction of transcription: leftward (L, complementary sense) or rightward (R, viral sense). Certain viral genes are also indicated by name, including Rep (replication initiator protein), TrAP (transcriptional activator protein), REn (replication enhancer), and CP (coat or capsid protein). The position of the conserved hairpin is indicated by an asterisk within the intergenic region (IR). The common region (CR), a sequence of ~ 230 bp that is nearly identical in TGMV DNAs A and B, is indicated by a hatched box. (Sequence references cited in Lazarowitz 1992.)

protein genes of viruses with monopartite genomes belonging to subgroups I, II, and III are, of necessity, encoded in the single genome component. All subgroup III viruses are transmitted by a single species of whitefly (*Bemisia tabaci* Genn.) and individually have relatively narrow host ranges within the dicots. The most intensively investigated bipartite viruses include tomato golden mosaic virus (TGMV), African cassava mosaic virus (ACMV, formerly cassava latent virus), and squash leaf curl virus (SqLCV) (Fig. 1). Tomato yellow leaf curl virus (TYLCV) is the best studied of the monopartite subgroup III viruses.

All geminivirus genome components possess an intergenic region (IR) from which viral genes diverge in both the viral and complementary sense (Fig. 1). The IR contains divergent RNA polymerase II-type promoters responsible for the expression of viral genes and also contains sequence elements necessary for the replication of viral DNA. A striking feature within the IR is a conserved inverted repeat that is capable of forming a hairpin. In the dicot-infecting viruses, this sequence element has the consensus GGCCAT/ACCGNT/AA/TTAATATTACCGGA/TT GGCC (Lazarowitz 1992). The invariant sequence TAATATTAC (underlined), located in the loop of the hairpin, has been found in all geminivirus genomes sequenced to date.

Two systems of gene nomenclature are currently in use. Both designate genes and gene products by number, but one denotes genes as viral or complementary sense (V or C), whereas the other indicates genes as oriented in the rightward (R; viral sense; clockwise) or leftward (L; complementary sense; counterclockwise) direction on the genome map (Fig. 1). Efforts to standardize the nomenclature are in progress, but in the interim, the use of gene and gene product names based on function will serve to reduce confusion. Throughout this review, the R and L system will be employed and functional names will be used where appropriate.

Rolling-circle Replication

A considerable body of evidence supports the idea that geminivirus replication occurs by a rolling-circle mechanism (Saunders et al. 1991; Stenger et al. 1991 and references therein). Rolling-circle replication (RCR) is also employed by the ssDNA-containing coliphages (e.g., ϕX174) (Kornberg and Baker 1992) and certain bacterial plasmids (Koepsel et al. 1985; Gros et al. 1987). The parvoviruses employ an analogous rolling-hairpin mechanism (Im and Muzyczka 1990). A characteristic feature of RCR is the involvement of a replication initiator protein (Rep) with a nicking-closing activity similar to that found in topoisomerases. RCR occurs in three stages. In the first stage (SS→RF synthesis), viral ssDNA (plus strand) enters the cell and is converted into a covalently closed dsDNA replicative form (RF) in a process involving host-directed, RNA-primed synthesis of a complementary (minus) strand. The RF serves as template for viral transcription as well as a template for further replication. The purpose of the second stage of RCR (RF→RF synthesis) is to generate additional RF DNA. This step is initiated by viral Rep protein, gene A protein (gpA) in the case of ϕX174, whose function is to nick the plus strand at a specific sequence. Following phosphodiester bond cleavage, Rep protein covalently binds to the 5′ terminus via a phosphotyrosine linkage. The 3′-OH terminus is used as a primer for the synthesis of nascent plus strand, which displaces the parental plus strand from the intact minus-strand template. Synthesis again is carried out by host replication proteins. Completion of the nascent plus strand regenerates the origin of replication, which again is nicked by Rep, this time acting as a terminase to release the displaced unit-length plus strand, which is simultaneously ligated to circular form by the closing activity. In the process, Rep is transferred to the newly created 5′ terminus. Early in the replication cycle, the circularized ssDNA is used as template for synthesis of minus-strand DNA, resulting in the amplification of RF. The third stage of RCR (RF→SS synthesis), which occurs late in the replication cycle, is responsible for the accumulation of viral genomes for encapsidation. This stage is similar to RF→RF synthesis, except that priming is prevented and ssDNA is the predominant product.

REPLICATION-ASSOCIATED PROTEINS

Geminiviruses do not encode a gene product with polymerase activity, but instead rely on the machinery of the host to replicate their chromosomes. The host proteins required for geminivirus DNA synthesis have yet to be identified, but presumably DNA polymerase-α and -δ-like activities are involved, as well as replication accessory factors commonly found in other systems (SSB, helicase, etc.). Viral proteins that function in replication serve as specificity and initiation factors, replication enhancers, or apparently as regulators of ssDNA versus dsDNA synthesis and/or accumulation.

AL1 (Rep) Protein

All geminiviruses encode Rep (replication initiator protein), which carries out several distinct functions in viral replication. This highly conserved approximately 40-kD polypeptide is the product of the AL1 gene (also AC1, C1, or L1) in subgroup II and subgroup III viruses. Among the monocot-infecting geminiviruses of subgroup I, Rep is expressed from a spliced mRNA that fuses the L1 and L1′ ORFs (also C1 and C2), which correspond to its amino- and carboxy-terminal halves, respectively (Fig. 1) (Schalk et al. 1989; Mullineaux et al. 1990). Rep is the only viral protein that is absolutely required for viral DNA replication (Elmer et al. 1988) and, in association with host proteins, Rep is sufficient to support the replication in *trans* of DNA molecules that contain a compatible origin of replication (Hayes and Buck 1989; Hanley-Bowdoin et al. 1990).

On the basis of homologies noted between Rep and the replication initiator proteins of eubacterial plasmids and bacteriophages, it was proposed that Rep initiates RCR by functioning as a site- and strand-specific endonuclease (Ilyina and Koonin 1992; Koonin and Ilyina 1992). Three conserved sequence motifs were identified in the amino-terminal half of Rep (Fig. 2). The functions of motifs I and II are not known, but motif II contains two invariant histidine residues that may bind Mg^{++} or Mn^{++}. Motif III contains a tyrosine residue predicted to participate in phosphodiester bond cleavage and to covalently bind the 5′ terminus exposed by nicking (van Mansfeld et al. 1986; Yasukawa et al. 1991; Noirot-Gros et al. 1994).

Recent studies have confirmed that the AL1 gene product is a replication initiator protein. Rep from TYLCV and WDV, expressed in *Escherichia coli* and partially purified, were shown in vitro to introduce a

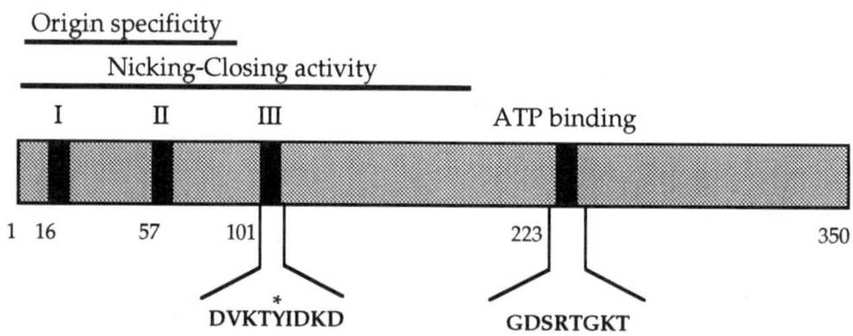

Figure 2 Functional-domains and motifs of geminivirus Rep. A diagram of Rep, combining information obtained from TGMV, TYLCV, WDV, and BCTV, is shown. The approximate amino acid coordinates and amino acid sequences given are from TGMV Rep. The positions of motifs I, II, and III, which are conserved between Rep and the rep proteins of plasmids and phage that replicate by RCR, are indicated (Koonin and Ilyina 1992). The sequence of motif III, which contains the putative catalytic tyrosine (*asterisk*), and the sequence of the ATP-binding site (Desbiez et al. 1995) are shown. A region responsible for origin specificity, which likely contains the dsDNA-binding domain, resides near the amino-terminal end of the protein (Choi and Stenger 1995). A region that retains nicking-closing activity is also shown (Heyraud-Nitschke et al. 1995).

specific nick in the plus strand within the invariant sequence TAATATT↓AC (Laufs et al. 1995; Heyraud-Nitschke et al. 1995). This sequence was previously identified by genetic analysis as the replication initiation site (Stenger et al. 1991; Heyraud et al. 1993b). The reaction is dependent on Mg^{++} or Mn^{++} and requires that the substrate be single-stranded. After cleavage, Rep remains covalently attached to the adenine residue at the newly created 5′ terminus. A closing activity capable of ligating upstream and downstream cleavage products was also demonstrated (Laufs et al. 1995). Both the nicking and closing activities reside in the amino-terminal region of the protein (Heyraud-Nitschke et al. 1995).

Because Rep contains a consensus NTP-binding motif, it has been proposed that it might possess an ATPase and associated DNA helicase activity (Gorbalenya and Koonin 1989; Gorbalenya et al. 1990). A putative geminivirus helicase could unwind and displace viral-strand DNA from the minus-strand template in advance of the replication fork, or unwind and expose the origin of replication to proteins of the replication apparatus. Studies with TYLCV Rep revealed an intrinsic, DNA-independent ATPase activity (Desbiez et al. 1995). Mutation of the conserved NTP-binding motif, which resembles a phosphate-binding fold or P-loop, led to either loss or reduction of ATPase activity. The same mutations, when introduced into the TYLCV genome, either abolished or reduced replication in protoplasts (Desbiez et al. 1995). Although Rep ATPase activity is essential for replication, its precise role remains unclear. A helicase activity has yet to be demonstrated (J. Laufs and B. Gronenborn, pers. comm.), and the DNA-independent nature of the Rep ATPase suggests it is unlikely to support one. The ATPase activity also is not required for the nicking-closing reaction (Heyraud-Nitschke et al. 1995). Thus, for the present, Rep ATPase remains an activity in search of a function.

In studies using extracts from transgenic plants or insect cells expressing the TGMV AL1 gene, Rep was shown to be a dsDNA-binding protein that interacts specifically with a 13-bp element located in the IR, 34 bp upstream of the conserved hairpin (Fig. 3) (Fontes et al. 1992, 1994a). The element, referred to as a high-affinity Rep-binding site, contains two 5-bp direct repeats separated by a central core of 3 bp (5′-<u>GGTAG</u>TAA<u>GGTAG</u>). Competitive DNA-binding studies suggested that Rep has a somewhat greater affinity for the 3′ repeat. Mutation of the Rep-binding site confirmed that it is an essential *cis*-acting element required for viral DNA replication (Fontes et al. 1994a). However, the Rep-binding site is not required for specific nicking at the invariant se-

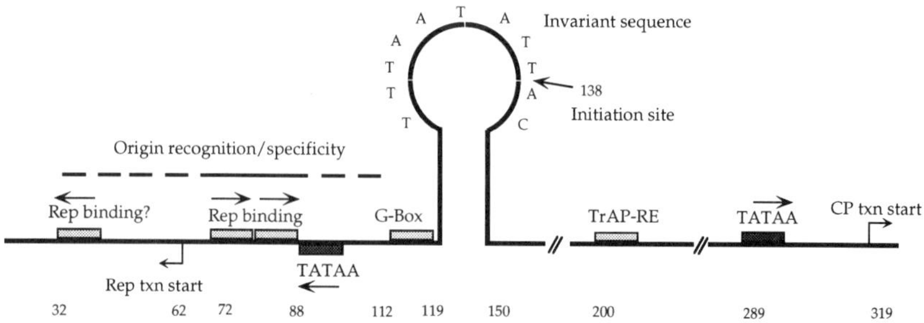

Figure 3 Organization of a geminivirus replication origin. A diagram of the TGMV replication origin is presented. Shown are the relative positions of Rep-binding sites, the invariant sequence (TAATATTAC), and the site where plus-strand synthesis initiates. Sequences involved in origin recognition/specificity are also depicted (Fontes et al. 1994b). The locations of sequence elements that interact with the transcription machinery, including TATAA sequences, Rep and CP transcription start sites (references cited in Lazarowitz 1992), a putative binding site for G-box family transcription factors, and a putative TrAP response element (the conserved late element; Argüello-Astorga et al. 1994) are also indicated. A sequence that appears to be an additional Rep-binding site in inverted orientation has been identified by sequence analysis (Argüello-Astorga et al. 1994), but its function has yet to be determined. Nucleotide coordinates are from TGMV DNA A.

quence (TAATATTAC), as single-stranded oligonucleotides containing the initiation site but lacking the GGTAG repeats are effective substrates for Rep endonuclease (Laufs et al. 1995; Heyraud-Nitschke et al. 1995). In this context, it is interesting to note that in vitro binding studies using purified, *E. coli*-expressed TGMV Rep with double- and single-stranded IR DNAs revealed that Rep has a significantly greater affinity (more than fourfold) for ssDNA than for dsDNA (Thömmes et al. 1993). The ssDNA-binding activity is apparently specific, but it is not known whether specificity is imparted by sequence elements in the IR, or by structure (e.g., the hairpin). The separation of nicking (which may depend on an ssDNA-binding activity) from binding-site recognition in the dsRF points to at least two distinct functions for Rep in replication initiation.

Attempts to construct pseudorecombinants composed of the A and B genome components from different bipartite viruses have only been successful when the heterologous genome components were derived from different strains of the same virus or from very closely related viruses. The inviability of other combinations is a consequence of the inability of Rep to sponsor replication of the heterologous genome component. The roles of specific *cis*-acting elements (including the Rep-binding site) and *trans*-acting proteins (Rep) in specific origin recognition have been investigated. Rep from TGMV and bean golden mosaic virus (BGMV) interact only with their own binding sites in vitro. However, TGMV and BGMV DNA B mutants in which Rep-binding sites were exchanged by site-directed mutagenesis were not replicated by either TGMV or BGMV DNA A, suggesting that Rep binding at these sites is necessary but not sufficient for specific origin recognition (Fontes et al. 1994b). The additional required sequences were not identified. Studies using chimeric viral genomes composed of sequences from the Logan and CFH strains of BCTV, which exhibit distinct replication specificity, demonstrated that Rep amino acids 3–89 contain the domain responsible for specific origin recognition (Choi and Stenger 1995).

The Rep-binding site lies between the TATA box and the AL1 transcription start site (Fig. 3), suggesting that binding to initiate replication may have the additional consequence of interfering with expression of the Rep gene itself. This is, in fact, the case. In protoplast cotransfection experiments, the expression of a reporter gene replacing the TGMV AL1 ORF in DNA A was reduced or abolished by the addition of a construct constitutively expressing the AL1 ORF (Sunter et al. 1993). Therefore, Rep regulates its own expression, a property it shares with other viral proteins that also are involved in replication initiation (Reed et al. 1976; Hansen et al. 1981). Autoregulation of TGMV Rep expression occurs at the level of transcription, and Rep can repress a heterologous promoter when its binding site is inserted between the TATA box and transcription start site (Eagle et al. 1994). Interestingly, recent work suggests that the TGMV AL4 protein, which is encoded by a small gene that lies entirely within AL1 but in a different reading frame (Fig. 1), also contributes to suppression of AL1 expression (Gröning et al. 1994). However, the in vivo role of AL4 protein is unclear, as some AL4 mutants have no apparent effect on viral replication and pathogenesis (Elmer et al. 1988; Etessami et al. 1991).

Transcriptional repression and replication initiation appear to be distinct activities, because certain Rep mutants that are replication defective can still fully repress transcription. Transcriptional repression also does not require a functional origin of replication (Eagle et al. 1994). The mechanism by which Rep represses its own expression is not known but is likely to involve interference with the assembly of a transcription initiation complex (Buratowski and Sharp 1992). It should perhaps be mentioned here that Rep does not repress expression of the BL1 gene, despite identical AL1 (Rep) and BL1 promoters and binding sites in the common regions of the A and B genome components (Sunter et al. 1993). Differential regulation may be due to the use of additional downstream BL1 transcription start sites mapped in TGMV DNA B (Sunter and Bisaro 1989). Thus, the bipartite viruses have devised a way to autoregulate an early gene required for replication without repressing a movement function required later in the replication cycle.

Rep from the subgroup I virus WDV was shown to stimulate transcription of the viral capsid protein gene (V1). The studies of Hofer et al. (1992) found that a functional Rep gene is required both for replication and for measurable coat protein promoter activity. Certain IR sequences, including the putative Rep-binding site and the hairpin region, were required for replication and for coat protein expression. Other IR sequences were necessary only for replication or for rightward transcription, suggesting that these are independent, Rep-mediated processes (Hofer et al. 1992). At this time, it is not known whether WDV Rep also regulates its own expression. However, that a single protein might initiate DNA replication as well as regulate early and late gene expression in subgroup I viruses suggests a simple and elegant means of coordinating the viral multiplication cycle. In contrast, the regulatory circuitry is much more complex in the dicot-infecting geminiviruses, where replication and transcription are also influenced by other viral proteins (see below). Relevant to this discussion, it has been shown in at least one dicot geminivirus (TGMV) that Rep has little or no effect on expression of the coat protein gene (Gröning et al. 1994).

To date, several distinct activities have been identified for Rep. Some of these, including the recognition of a specific binding site in the RF, are important for origin recognition, whereas others, such as ssDNA binding, a nicking-closing activity, and an ATPase activity, may be essential for replication initiation or termination. Additional activities, which may prove to be subgroup-specific, include autoregulation (repression) of early gene expression and stimulation of late (coat protein) gene expression. The picture that is beginning to emerge from biochemical and genetic studies is that different domains of Rep are responsible for its various biochemical activities (Fig. 2). How these activities might be temporally regulated during the replication cycle is unknown. If previous experience with other DNA viruses is any guide, a likely mechanism is posttranslational modification, and in particular, phosphorylation (Fanning and Knippers 1992). TYLCV and TGMV Rep are phosphoproteins (cited in Laufs et al. 1995; M.D. Hartitz and D.M. Bisaro, in prep.). Whether phosphorylation is a mechanism by which the functions of Rep are regulated remains to be seen.

AL3 (REn) Protein

The AL3 gene (also AC3, C3, or L3) is found in all subgroup II and III geminiviruses, but is absent from the genomes of viruses belonging to subgroup I. Comparatively little is known about the function of the approximately 15-kD AL3 gene product. Mutational analysis showed that an intact AL3 gene is not required for infectivity, but AL3 mutants elicit delayed and attenuated symptoms in inoculated plants (Elmer et al. 1988). AL3 mutants also accumulate reduced (as much as 50-fold) amounts of viral DNA in plants and protoplasts (Sunter et al. 1990; Etessami et al. 1991; Stanley et al. 1992). Because of this interesting phenotype, the AL3 gene product is often referred to as a replication enhancer (REn). Exactly how the AL3 protein enhances the replication (or the accumulation) of viral DNA is not known, but AL3 activity is not virus-specific (Sunter et al. 1994; Hormuzdi and Bisaro 1995).

A clue to the mechanism by which AL3 protein acts comes from the studies of Fontes et al. (1994a), who observed that certain DNA molecules harboring mutations in the 5′-GGTAG repeat of the Rep-binding site could be induced to replicate only in the presence of both Rep and AL3 protein. One interpretation of this finding is that AL3 functions via an interaction with Rep, perhaps by stabilizing Rep/DNA interactions (Fontes et al. 1994a).

Little is known about the AL3 protein itself, except that its sequence predicts a high degree of hydrophobic character. In infected plants, AL3 protein is found in both the soluble and organelle fractions in amounts comparable to Rep (Pedersen and Hanley-Bowdoin 1994). This suggests that AL3 functions catalytically rather than by acting to stabilize viral DNA, a mechanism that would likely require a large quantity of protein.

AL2 (TrAP) Protein

A key step in the control of viral DNA synthesis follows the production of circular plus-strand DNA by RCR. The nascent ssDNA may either reenter the replication pool with subsequent priming and minus-strand synthesis (RF→RF synthesis), or it may be removed from the replication pool by encapsidation or other means (RF→SS synthesis). In the subgroup III geminiviruses, this step appears to be regulated by transcriptional activator protein, or TrAP (also AL2, AC2, or C2).

All subgroup III geminiviruses encode an allele of the AL2 gene, which is not found in subgroup I virus genomes. An ORF resembling AL2 is found in BCTV (subgroup II), but recent work suggests it has a different function (Hormuzdi and Bisaro 1995). TGMV AL2 mutants are unable to infect plants, yet they retain the ability to synthesize DNA in transient leaf disc and protoplast replication assays (Elmer et al. 1988). Analysis of transgenic plants expressing the leftward TGMV proteins showed that the AL2 ORF is required for the accumulation of ssDNA (Hayes and Buck 1989). Protoplast studies further showed that TGMV AL2 mutants accumulate reduced amounts of ssDNA and do not accumulate capsid protein (Sunter et al. 1990). These observations suggested that capsid protein is required to stabilize ssDNA, and that one function of TrAP is to activate expression of the capsid protein gene. Activation was subsequently demonstrated (Sunter and Bisaro 1991; Gröning et al. 1994). It was later shown that TrAP is also required to activate the BR1 gene, and that activation occurs at the level of transcription (Sunter and Bisaro 1992). The noninfectious nature of TGMV AL2 mutants is therefore due to insufficient accumulation of BR1 protein, which is necessary for systemic spread. The BR1 protein binds ssDNA and localizes to the nucleus of infected cells (Pascal et al. 1994). Because both capsid protein and BR1 protein are capable of interacting with ssDNA and removing it from the replication pool, placing the expression of these genes under the control of TrAP ensures they are not made prematurely. However, the possibility that TrAP also plays a more direct role in regulating a switch from RF→RF synthesis to RF→SS synthesis cannot be excluded at this time. Among the members of subgroup I, this switch may be under the control of Rep, which is required in these viruses for coat protein expression (see above). In the subgroup II geminiviruses, the available evidence suggests the involvement of yet another protein (R2, see below).

The A genome components of different bipartite geminiviruses can complement a TGMV AL2^{-} mutant, proving that TrAP function is not virus specific (Sunter et al. 1994). Activation of the capsid protein and BR1 promoters appears to be mediated by a sequence element within the IR that is conserved among most subgroup III geminiviruses (Argüello-Astorga et al. 1994). However, attempts to demonstrate specific TrAP binding to DNA fragments containing this sequence (or any other) have so far been unsuccessful, although TrAP binds both ssDNA and dsDNA in a non-sequence-specific manner (M.D. Hartitz and D.M. Bisaro, in prep.).

R2 Protein

The R2 gene (also V2) of BCTV encodes a protein with a unique function, despite the fact that it shares a common designation with genes found in some other types of geminivirus genomes. The results of mutational studies suggest that the small product of the BCTV R2 gene (~12 kD) plays a direct role in regulating a switch from RF→RF synthesis to RF→SS synthesis. In infected plants and leaf discs, R2 mutants accumulate reduced amounts of ssDNA and increased amounts of dsDNA relative to wild-type virus (Stanley et al. 1992; Hormuzdi and Bisaro 1993). In protoplasts, the reduction is large (~10-fold) and is accompanied by a similar increase in dsDNA levels (Hormuzdi and Bisaro 1993). Clearly, the absence of active R2 protein results in the overamplification of dsDNA at the expense of ssDNA. The molecular basis of this phenotype is unknown, but an interaction between R2 protein and ssDNA or the replication machinery may prevent minus-strand synthesis. Alternatively, R2 protein may facilitate virion assembly, in which case encapsidation is the principal mechanism for withdrawing ssDNA from the replication pool. However, the distinctly different phenotypes of R2 mutants and capsid protein mutants make this scenario less attractive. Although capsid mutants accumulate reduced amounts of ssDNA, they do not overproduce dsDNA. Additional work is needed to elucidate the function of BCTV R2 protein.

THE ORIGIN OF REPLICATION

The IR is a complex mosaic of controlling elements containing divergent promoters for viral and complementary sense transcription, as well as the plus-strand origin of replication. The replication origin appears to be modular and resembles in some respects those found in mammalian DNA viruses and other simple eukaryotic genomes (DePamphilis 1993), but there are also elements unique to RCR that lend a distinctly prokaryotic flavor to the geminivirus origin core.

Eukaryotic origins frequently contain promoter elements, and the transcription factors that bind to them often function as auxiliary replication factors that enhance replication or impart tissue specificity (DePamphilis 1988). To what extent certain of the geminivirus IR elements are involved in both replication and transcription remains to be determined. It is worth noting that most geminivirus genomes contain a GC-rich element immediately 5′ of the conserved hairpin that, in some cases, resembles a binding site for the G-box family of plant transcription factors (Argüello-Astorga et al. 1994). The GC-rich sequence in MSV stimulates rightward transcription and binds maize nuclear factors (Fenoll et al. 1988, 1990).

The Plus-strand Origin

Because most studies have used subgroup III virus origins, these are discussed in detail. The recent sequence analysis of Argüello-Astorga and colleagues (1994), which identified iterative elements within geminivirus IRs that appear to correspond to Rep-binding sites, indicates that the organization of these elements is similar in most dicot-infecting geminiviruses. Viruses belonging to subgroup I, however, are quite different with respect to the relative locations of their putative Rep-binding sites and the conserved hairpin. Whether this reflects fundamental differences in mechanisms of origin recognition and initiation of DNA synthesis is not known.

Using chimeric DNA molecules containing IR sequences of TGMV and SqLCV, Lazarowitz and coworkers delimited the replication origin to a sequence of about 90 nucleotides encompassing the conserved hairpin and about 50 nucleotides of 5′ upstream sequence, including the Rep-binding site (Fig. 3) (Lazarowitz et al. 1992). This sequence was sufficient to drive the Rep-dependent replication of nonviral sequences to which it was joined, and both ssDNA and dsDNA products were synthesized. Recent experiments with chimeric BCTV Logan and BCTV CFH genomes yielded similar results (Choi and Stenger 1995). The relatively small origin fragments defined by these studies must contain the origin core, the minimal sequence required for replication.

The origin recognition element (ORE) is a critical component of the origin core (DePamphilis 1993 and this volume). This element is responsible for binding specific origin recognition proteins, which are required for replication initiation. In the geminiviruses, Rep appears to function as an origin recognition protein, and the high-affinity Rep-binding site as an ORE. Additional as-yet-undefined sequences are probably also involved in origin recognition (Fontes et al. 1994b).

The consequences of Rep-Rep-binding-site interactions are not yet known, but one can envision several outcomes which are not mutually exclusive. First, Rep may participate in origin unwinding, either by acting as a helicase or by assisting a cellular helicase (Stahl et al. 1986; Dodson et al. 1987; Im and Muzyczka 1990). Second, Rep binding could lead to distortion of the double helix, which in turn might promote localized duplex melting or result in extrusion of the conserved hairpin (Noirot et al. 1990). A third possibility is that Rep recruits cellular replication proteins to the origin. This could occur indirectly as a consequence of duplex melting or unwinding, or, more directly, through specific interactions with cellular replication proteins (Dornreiter et al. 1990). Whatever the precise mechanism, the primary function of sequence-specific Rep binding may be to prepare the origin for interaction with host replication proteins. It may also prepare the origin for interaction with an additional Rep molecule or complex that recognizes the invariant sequence, either as ssDNA or as part of the single-stranded hairpin loop, as a substrate for cleavage.

The conserved hairpin defines the 3′ boundary of the origin core. Nuclease S1 mapping provided evidence that the hairpin (cruciform in dsDNA) structure exists in duplex DNA (Sunter et al. 1985), although direct evidence that it is present in viral chromosomes in vivo has not yet been obtained. The importance of this sequence for replication, however, is well established (Revington et al. 1989; Lazarowitz et al. 1992). The Rep endonuclease cleavage site was initially mapped in vivo to a 20-bp region containing the conserved hairpin (Stenger et al. 1991) and later to the invariant TAATATTAC sequence in the loop of the hairpin (Heyraud et al. 1993a,b; Stanley 1995). Biochemical studies confirmed that Rep cleavage is between the last T and A of the invariant sequence, thereby defining the site of plus-strand initiation (Heyraud-Nitschke et al. 1995; Laufs et al. 1995).

The sequence requirements for nicking have been investigated only to a limited extent. TYLCV Rep is able to nick, with reduced efficiency, the origin sequence of WDV (and vice versa) even though the hairpin stems of TYLCV and WDV differ considerably in length and sequence (Heyraud-Nitschke et al. 1995; Laufs et al. 1995). TYLCV Rep can also nick, again with reduced efficiency, oligonucleotides containing substitutions in positions immediately flanking the cleavage site in the invariant sequence. A hairpin structure also is not required since oligonucleotides lacking either the 5′ or 3′ portions of the stem are efficiently cleaved (Heyraud-Nitschke et al. 1995; Laufs et al. 1995). Because it appears that neither the hairpin structure nor the exact invariant sequence is required to direct cleavage in vitro, precisely how the cleavage site is defined is not yet clear.

The situation in vivo is somewhat different. Although some point mutations in the stem that disrupt base-pairing appear to be tolerated (Roberts and Stanley 1994), the entire hairpin is necessary for replication (Lazarowitz et al. 1992). These observations suggest that the hairpin structure and not sequence per se is required for replication. There is some natural variation in length and sequence between the hairpin stems of different geminiviruses, but the extent to which these differences might contribute to replication efficiency or specificity has not been examined. In addition, the significance of each individual position in the invariant sequence has not yet been assessed. However, in ACMV, the third T could be mutated to a C without compromising the ability of the virus to infect plants, although disease development was delayed and attenuated. In contrast, alteration of the second A residue was not tolerated (Roberts and Stanley 1994). Limited natural sequence variation exists in the few nucleotides of the loop that are not part of the invariant sequence. The identity of these nucleotides does not appear to contribute in a significant way to replication efficiency or specificity (Fontes et al. 1994b).

A WDV mutant which is replication competent, but unable to process unit-length genomes from replicative intermediates, was constructed by deletion of the conserved hairpin. This mutant accumulated only high-molecular-weight, concatemeric forms of the viral genome in transfected protoplasts (Kammann et al. 1991). Replication initiation in this case apparently occurred at an alternate site in the IR that is highly homologous to the right half of the conserved hairpin (Heyraud et al. 1993b). The behavior of this unusual deletion mutant indicates that the requirements for initiation differ from those of termination, with the latter being somewhat more stringent. This is supported by in vitro experiments which suggest that Rep-catalyzed joining, but not nicking, is stimulated when substrate molecules contain sequences capable of base-pairing to form a hairpin (Heyraud-Nitschke et al. 1995).

The Minus-strand Origin

That viral ssDNA (plus strand) purified from virions is infectious is taken as evidence that RNA priming and minus-strand synthesis can be accomplished entirely by host enzymes (Goodman 1977; Hamilton et al. 1981). Interestingly, the encapsidated genomic ssDNA of subgroup I viruses is associated with a small, complementary (minus strand) ssDNA about 80 nucleotides long that contains several ribonucleotides covalently linked to the 5′ terminus (Donson et al. 1984, 1987; Hayes et al. 1988). These complementary DNAs are capable of priming minus-strand DNA synthesis in vitro, and the ribonucleotides are believed to represent remnants of the authentic in vivo primer. In MSV, WDV, and Digitaria streak virus, the complementary DNAs are annealed to sequences located in a small intergenic region opposite the larger IR on the circular genome map (Fig. 1). Similar complementary DNAs are not associated with the genomic ssDNA of subgroup II and subgroup III geminiviruses. However, analysis of replicative intermediates in ACMV-infected cells identified a heterogeneous population of complementary DNAs that also appear to contain ribonucleotides. Hybridization analysis suggested that the putative RNA primer initiates within the IR (Saunders et al. 1992). This is in agreement with the observation that the minimal origin described for subgroup III viruses by Lazarowitz et al. (1992) supports the synthesis of both single- and double-stranded DNA.

FUTURE DIRECTIONS

This is an exciting time for geminivirus research. Past work has taught us much about the mechanism of viral DNA replication and about the general roles of viral proteins in this process and has paved the way for new and important questions. What are the precise functions and activities of viral proteins involved in replication? How are these activities regulated in multifunctional viral proteins? How do viral proteins interact with host proteins during replication, and how do these interactions subvert the host replication machinery for the purpose of viral DNA replication? Geminivirus replication, as might be expected, appears to occur preferentially in cells that are actively synthesizing cellular DNA (Accotto et al. 1993). Recent evidence suggests that Rep can induce the accumulation of proliferating cell nuclear antigen (PCNA, a polymerase-δ processivity factor) in transgenic tobacco plants (Nagar et al. 1995). Do geminiviruses, like the mammalian DNA tumor viruses, possess the ability to prepare the host for viral DNA replication by stimulating normally quiescent cells to enter S phase? If so, what other viral and host proteins are involved? These and other questions are currently under investigation in a number of laboratories. What is learned during the next few years promises to teach us a great deal more about the mechanisms of viral DNA replication and pathogenesis and will provide new insights into host-pathogen interactions and the fundamental mechanisms of plant DNA replication and its control.

ACKNOWLEDGMENTS

I thank Bruno Gronenborn, Linda Hanley-Bowdoin, Jürgen Laufs, John Stanley, and Drake Stenger for preprints and information prior to publication. I also thank Judith Brown, Linda Hanley-Bowdoin, and Garry Sunter for critical reading of the manuscript and suggestions for its improvement. Financial support from the U.S. Department of Agriculture/National Research Initiative Competitive Grants Program (grant 93-37301-9076) is gratefully acknowledged.

REFERENCES

Accotto, G.-P., P.M. Mullineaux, S.C. Brown, and D. Marie. 1993. Digitaria streak geminivirus replicative forms are abundant in S-phase nuclei of infected cells. *Virology* **195:** 257–259.

Argüello-Astorga, G.R., R.G. Guevara-Gonzalez, L.R. Herrera-Estrella, and R.F. Rivera-Bustamante. 1994. Geminivirus replication origins have a group-specific organization of iterative elements: A model for replication. *Virology* **203:** 90–100.

Bisaro, D.M. 1994. Recombination in geminiviruses: Mechanisms for maintaining genome size and generating genomic diversity. In *Homologous recombination and gene silencing in plants* (ed. J. Paszkowski), pp. 39–60. Kluwer, Dordrecht.

Brown, J.K., D.R. Frohlich, and R. Rosell. 1995. The sweetpotato/silverleaf whiteflies: Biotypes of *Bemisia tabaci* Genn., or a species complex? *Annu. Rev. Entomol.* **40:** 511–534.

Buratowski, S., and P.A. Sharp. 1992. Initiation of transcription by RNA polymerase II. In *Transcriptional regulation* (ed. S.L. McKnight), pp. 227–246. Cold Spring Harbor Laboratory Press, Cold Spring Harbor, New York.

Choi, I.-R. and D.C. Stenger. 1995. Strain-specific determinants of beet curly top geminivirus DNA replication. *Virology* **206:** 904–912.

DePamphilis, M.L. 1988. Transcriptional elements as components of eukaryotic origins of replication. *Cell* **52:** 635–638.

———. 1993. Eukaryotic DNA replication: Anatomy of an origin. *Annu. Rev. Biochem.* **62:** 29–63.

Desbiez, C., C. David, A. Mettouchi, J. Laufs, and B. Gronenborn. 1995. Rep protein of tomato yellow leaf curl geminivirus (TYLCV) has an ATPase activity required for viral DNA replication. *Proc. Natl. Acad. Sci.* **92:** 5640–5644.

Dodson, M., F.B. Dean, P. Bullock, H. Echols, and J. Hurwitz. 1987. Unwinding of duplex DNA from the SV40 origin of replication by T antigen. *Science* **238:** 964–967.

Donson, J., G.P. Accotto, M.I. Boulton, P.M. Mullineaux, and J.W. Davies. 1987. The nucleotide sequence of a geminivirus from *Digitaria sanguinalis*. *Virology* **161:** 160–169.

Donson, J., B.A.M. Morris-Krsinich, P.M. Mullineaux, M. Boulton, and J.W. Davies. 1984. A putative primer for second-strand synthesis of maize streak virus is virion-associated. *EMBO J.* **3:** 3069–3073.

Dornreiter, I., A. Hoss, A.K. Arthur, and E. Fanning. 1990. SV40 T antigen binds directly to the large subunit of purified DNA polymerase alpha. *EMBO J.* **9:** 3329–3336.

Eagle, P.A., B.M. Orozco, and L. Hanley-Bowdoin. 1994. A DNA sequence required for geminivirus replication also mediates transcriptional regulation. *Plant Cell* **6:** 1157–1170.

Elmer, J.S., L. Brand, G. Sunter, W.E. Gardiner, D.M. Bisaro, and S.G. Rogers. 1988. Genetic analysis of tomato golden mosaic virus. II. The product of the AL1 coding sequence is required for replication. *Nucleic Acids Res.* **16:** 7043–7060.

Esau, K. 1977. Virus-like particles in nuclei of phloem cells in spinach leaves infected with the curly top virus. *J. Ultrastruct Res.* **61:** 78–88.

Etessami, P., K. Saunders, J. Watts, and J. Stanley. 1991. Mutational analysis of complementary-sense genes of African cassava mosaic virus DNA A. *J. Gen. Virol.* **72:** 1005–1012.

Fanning, E. and R. Knippers. 1992. Structure and function of simian virus 40 large tumor antigen. *Annu. Rev. Biochem.* **61:** 55–85.

Fenoll, C., D.M. Black, and S.H, Howell. 1988. The intergenic region of maize streak virus contains promoter elements involved in rightward transcription of the viral genome. *EMBO J.* **7:** 1589–1596.

Fenoll, C., J.J. Schwartz, D.M. Black, M. Schneider, and S.H. Howell. 1990. The intergenic region of maize streak virus contains a GC-rich element that activates rightward transcription and binds maize nuclear factors. *Plant Mol. Biol.* **15:** 865–877.

Fontes, E.P.B., V.A. Luckow, and L. Hanley-Bowdoin. 1992. A geminivirus replication protein is a sequence-specific DNA binding protein. *Plant Cell* **4:** 597–608.

Fontes, E.P.B., P.A. Eagle, P.S. Sipe, V.A. Luckow, and L. Hanley-Bowdoin. 1994a. Interaction between a geminivirus replication protein and origin DNA is essential for viral replication. *J. Biol. Chem.* **269:** 8459–8465.

Fontes, E.P.B., H.J. Gladfelter, R.L. Schaffer, I.T.D. Petty, and L. Hanley-Bowdoin. 1994b. Geminivirus replication origins have a modular organization. *Plant Cell* **6:** 405–416.

Goodman, R.M. 1977. Infectious DNA from a whitefly-transmitted virus of *Phaseolus vulgaris*. *Nature* **266:** 54–55.

Gorbalenya, A.E. and E.V. Koonin. 1989. Viral proteins containing the purine NTP-binding sequence pattern. *Nucleic Acids Res.* **17:** 8413–8440.

Gorbalenya, A.E., E.V. Koonin, and Y.I. Wolf. 1990. A new superfamily of putative NTP-binding domains encoded by genomes of small DNA and RNA viruses. *FEBS Lett.* **262:** 145–148.

Gröning, B.R., R.J. Hayes, and K.W. Buck. 1994. Simultaneous regulation of tomato golden mosaic virus coat protein and AL1 gene expression: Expression of the AL4 gene may contribute to suppression of the AL1 gene. *J. Gen. Virol.* **75:** 721–726.

Gros, M.F., H. te Riele, and S.D. Ehrlich. 1987. Rolling circle replication of single-stranded DNA plasmid pC194. *EMBO J.* **6:** 3863–3869.

Hamilton, W.D.O., D.M. Bisaro, R.H.A. Coutts, and K.W. Buck. 1983. Demonstration of the bipartite nature of the genome of a single-stranded DNA plant virus by infection with the cloned DNA components. *Nucleic Acids Res.* **11:** 7387–7396.

Hamilton, W.D.O., R.C. Sanders, R.H.A. Coutts, and K.W. Buck. 1981. Characterisation of tomato golden mosaic virus as a geminivirus. *FEMS Microbiol. Lett.* **11:** 263–267.

Hanley-Bowdoin, L., J.S. Elmer, and S.G. Rogers. 1990. Expression of functional replication protein from tomato golden mosaic virus in transgenic tobacco. *Proc. Natl. Acad. Sci.* **87:** 1446–1450.

Hansen, U., D. Tenen, D.M. Livingston, and P.A. Sharp. 1981. T antigen repression of SV40 early transcription from two promoters. *Cell* **27:** 603–612.

Hayes, R.J. and K.W. Buck. 1989. Replication of tomato golden mosaic virus DNA B in transgenic plants expressing open reading frames (ORFs) of DNA A: Requirement of ORF AL2 for production of single-stranded DNA. *Nucleic Acids Res.* **17:** 10213–10222.

Hayes, R.J., H. Macdonald, R.H.A. Coutts, and K.W. Buck. 1988. Priming of complementary DNA synthesis *in vitro* by small DNA molecules tightly bound to virion DNA of wheat dwarf virus. *J. Gen. Virol.* **69:** 1345–1350.

Heyraud, F., V. Matzeit, S. Schaefer, J. Schell, and B. Gronenborn. 1993a. The conserved nonanucleotide motif of the geminivirus stem-loop sequence promotes replicational release of virus molecules from redundant copies. *Biochimie* **75:** 605–615.

Heyraud, F., V. Matzeit, M. Kammann, S. Schaefer, J. Schell, and B. Gronenborn. 1993b. Identification of the initiation sequence for viral-strand DNA synthesis of wheat dwarf virus. *EMBO J.* **12:** 4445–4452.

Heyraud-Nitschke, F., S. Schumacher, J. Laufs, S. Schaefer, J. Schell, and B. Gronenborn. 1995. Determination of the origin cleavage and joining domain of geminivirus rep proteins. *Nucleic Acids Res.* **23:** 910–916.

Hofer, J.M.I., E.L. Dekker, H.V. Reynolds, C.J. Woolston, B.S. Cox, and P.M. Mullineaux. 1992. Coordinate regulation of replication and virion sense gene expression in wheat dwarf virus. *Plant Cell* **4:** 213–223.

Hormuzdi, S.G. and D.M. Bisaro. 1993. Genetic analysis of beet curly top virus: Evidence for three virion sense genes involved in movement and regulation of single-and double-stranded DNA levels. *Virology* **193:** 900–909.

———. 1995. Genetic analysis of beet curly top virus: Examination of the roles of L2 and L3 genes in viral pathogenesis. *Virology* **206:** 1044–1054.

Ilyina, T.V. and E.V. Koonin. 1992. Conserved sequence motifs in the initiator proteins for rolling circle DNA replication encoded by diverse replicons from eubacteria, eucaryotes, and archaebacteria. *Nucleic Acids Res.* **20:** 3279–3285.

Im, D.-S. and N. Muzyczka. 1990. The AAV origin binding protein Rep68 is an ATP-dependent site-specific endonuclease with DNA helicase activity. *Cell* **61:** 447–457.

Kammann, M., H.-J. Schalk, V. Matzeit, D. Schaefer, J. Schell, and B. Gronenborn. 1991. DNA replication of wheat dwarf virus, a geminivirus, requires two *cis*-acting signals. *Virology* **184:** 786–790.

Kim, K.S., T.L. Shock, and R.M. Goodman. 1978. Infection of *Phaseolus vulgaris* by bean golden mosaic virus: Ultrastructural aspects. *Virology* **89:** 22–33.

Koepsel, R.R., R.W. Murray, W.D. Rosenblum, and S.A. Khan. 1985. The replication initiator protein of plasmid pT181 has sequence-specific endonuclease and topoisomerase activity. *Proc. Natl. Acad. Sci.* **82:** 6845–6849.

Koonin, E.V. and T.V. Ilyina. 1992. Geminivirus replication proteins are related to prokaryotic plasmid rolling circle DNA replication initiator proteins. *J. Gen. Virol.* **73:** 2763–2766.

Kornberg, A. and T.A. Baker. 1992. *DNA replication.* W.H. Freeman, San Francisco.

Laufs, J., W. Traut, F. Heyraud, V. Matzeit, S.G. Rogers, J. Schell, and B. Gronenborn. 1995. *In vitro* cleavage and ligation at the viral origin of replication by the replication protein of tomato yellow leaf curl virus. *Proc. Natl. Acad. Sci.* **92:** 3879–3883.

Lazarowitz, S.G. 1992. Geminiviruses: Genome structure and function. *Crit. Rev. Plant Sci.* **11:** 327–349.

Lazarowitz, S.G., L.C. Wu, S.G. Rogers, and J.S. Elmer. 1992. Sequence-specific interaction with the viral AL1 protein identifies a geminivirus DNA replication origin. *Plant Cell* **4:** 799–809.

Matthews, R.E.F. 1991. *Plant virology.* Academic Press, San Diego.

Mullineaux, P.M., J.W. Davies, and C.J. Woolston. 1992. Geminiviruses as gene vectors. In *Genetic engineering with plant viruses* (ed. T.M.A. Wilson and J.W. Davies), pp. 187–215. CRC Press, Boca Raton, Florida.

Mullineaux, P.M., F. Guerineau, and G.P. Accotto. 1990. Processing of complementary sense RNAs of Digitaria streak virus in its host and in transgenic tobacco. *Nucleic Acids Res.* **18:** 7259–7265.

Nagar, S., T.J. Pedersen, K. Carrick, L. Hanley-Bowdoin, and D. Robertson. 1995. A geminivirus induces expression of a host DNA synthesis protein in terminally differentiated plant cells. *Plant Cell* **7:** 705–719.

Noirot, P., J. Bagonetti, and R. Novick. 1990. Initiation of rolling-circle replication in pT181 plasmid: Initiator protein enhances cruciform extrusion at the origin. *Proc. Natl. Acad. Sci.* **87:** 8560–8564.

Noirot-Gros, M.F., V. Bidnenko, and S.D. Ehrlich. 1994. Active site of the replication protein of the rolling circle plasmid pC194. *EMBO J.* **13:** 4412–4420.

Pascal, E., A.A. Sanderfoot, B. Ward, R. Medville, R. Turgeon, and S.G. Lazarowitz. 1994. The geminivirus BR1 movement protein binds single-stranded DNA and localizes to the cell nucleus. *Plant Cell* **6:** 995–1006.

Pedersen, T.J. and L. Hanley-Bowdoin. 1994. Molecular characterization of the AL3 protein encoded by a bipartite geminivirus. *Virology* **202:** 1070–1075.

Pilartz, M. and H. Jeske. 1992. Abutilon mosaic virus double-stranded DNA is packed into minichromosomes. *Virology* **189:** 800–802.

Reed, S.I., G.R. Stark, and J.C. Alwine. 1976. Autoregulation of SV40 gene A by T antigen. *Proc. Natl. Acad. Sci.* **73:** 3083–3087.

Revington, G.N., G. Sunter, and D.M. Bisaro. 1989. DNA sequences essential for replication of the B genome component of tomato golden mosaic virus. *Plant Cell* **1:** 985–992.

Roberts, S. and J. Stanley. 1994. Lethal mutations within the conserved stem-loop of African cassava mosaic virus DNA are rapidly corrected by genomic recombination. *J. Gen. Virol.* **75:** 3203–3209.

Rogers, S.G., D.M. Bisaro, R.B. Horsch, R.T. Fraley, N.L. Hoffmann, L. Brand, J.S. Elmer, and A.M. Lloyd. 1986. Tomato golden mosaic virus A component DNA replicates autonomously in transgenic plants. *Cell* **45:** 593–600.

Rushing, A.E., G. Sunter, W.E. Gardiner, R.R. Dute, and D.M. Bisaro. 1987. Ultrastructural aspects of tomato golden mosaic virus infection in tobacco. *Phytopathology* **77:** 1231–1236.

Saunders, K., A. Lucy, and J. Stanley. 1991. DNA forms of the geminivirus African cassava mosaic virus consistent with a rolling circle mechanism of replication. *Nucleic Acids Res.* **19:** 2325–2330.

———. 1992. RNA-primed complementary-sense DNA synthesis of the geminivirus African cassava mosaic virus. *Nucleic Acids Res.* **20:** 6311–6315.

Schalk, H.-J., V. Matzeit, B. Schiller, J. Schell, and B. Gronenborn. 1989. Wheat dwarf virus, a geminivirus of graminaceous plants, needs splicing for replication. *EMBO J.* **8:** 359–364.

Stahl, H., P. Droge, and R. Knippers. 1986. DNA helicase activity of SV40 large tumor antigen. *EMBO J.* **5:** 1939–1944.

Stanley, J. 1983. Infectivity of the cloned geminivirus genome requires sequences from both DNAs. *Nature* **305:** 643–645.

———. 1991. The molecular determinants of geminivirus pathogenesis. *Semin. Virol.* **2:** 139–149.

———. 1995. Analysis of African cassava mosaic virus recombinants suggests strand nicking occurs within the conserved nonanucleotide motif during the initiation of rolling circle DNA replication. *Virology* **206:** 707–712.

Stanley, J., J. Latham, M.S. Pinner, I. Bedford, and P.G. Markham. 1992. Mutational analysis of the monopartite geminivirus beet curly top virus. *Virology* **191:** 396–405.

Stenger, D.C., G.N. Revington, M.C. Stevenson, and D.M. Bisaro. 1991. Replicational release of geminivirus genomes from tandemly repeated copies: Evidence for rolling circle replication of a plant viral DNA. *Proc. Natl. Acad. Sci.* **88:** 8029–8033.

Sunter, G. and D.M. Bisaro. 1989. Transcription map of the B genome component of tomato golden mosaic virus and comparison with A component transcripts. *Virology* **173:** 647–655.

———. 1991. Transactivation in a geminivirus: AL2 gene product is needed for coat protein expression. *Virology* **180:** 416–419.

———. 1992. Transactivation of geminivirus AR1 and BR1 gene expression by the viral AL2 gene product occurs at the level of transcription. *Plant Cell* **4:** 1321–1331.

Sunter, G., K.W. Buck, and R.H.A. Coutts. 1985. S1-sensitive sites in the supercoiled double-stranded form of tomato golden mosaic virus DNA component B: Identification of regions of potential alternative secondary structure and regulatory function. *Nucleic Acids Res.* **13:** 4645–4659.

Sunter, G., M.D. Hartitz, and D.M. Bisaro. 1993. Tomato golden mosaic virus leftward gene expression: Autoregulation of geminivirus replication protein. *Virology* **195:** 275–280.

Sunter, G., D.C. Stenger, and D.M. Bisaro. 1994. Heterologous complementation by geminivirus AL2 and AL3 genes. *Virology* **203:** 203–210.

Sunter, G., W.E. Gardiner, A.E. Rushing, S.G. Rogers, and D.M. Bisaro. 1987. Independent encapsidation of tomato golden mosaic virus A component DNA in transgenic plants. *Plant Mol. Biol.* **8:** 477–484.

Sunter, G., M.D. Hartitz, S.G. Hormuzdi, C.L. Brough, and D.M. Bisaro. 1990. Genetic analysis of tomato golden mosaic virus. ORF AL2 is required for coat protein accumulation while ORF AL3 is necessary for efficient DNA replication. *Virology* **179:** 69–77.

Thömmes, P., T.A.M. Osman, R.J. Hayes, and K.W. Buck. 1993. TGMV replication protein AL1 preferentially binds to single-stranded DNA from the common region. *FEBS Letters* **319:** 95–99.

Townsend, R., J. Watts, and J. Stanley. 1986. Synthesis of viral DNA forms in *Nicotiana plumbaginafolia* protoplasts inoculated with cassava latent virus (CLV); evidence for independent replication of one component of the CLV genome. *Nucleic Acids Res.* **14:** 1253–1265.

van Mansfeld, A.D.M., H.A.A.M. Teeffelen, P.D. Baas, and H.S. Jansz. 1986. Two juxtaposed tyrosyl-OH groups participate in ϕX174 gene A protein catalysed cleavage and ligation of DNA. *Nucleic Acids Res.* **14:** 4229–4238.

Yasukawa, H., T. Hase, A. Sakai, and Y. Masamune. 1991. Rolling-circle replication of the plasmid pKYM isolated from a gram-negative bacterium. *Proc. Natl. Acad. Sci.* **88:** 10282–10286.

31

Baculovirus DNA Replication

Christian H. Ahrens, Douglas J. Leisy, and George F. Rohrmann
Department of Agricultural Chemistry
Agricultural and Life Sciences 1007
Oregon State University
Corvallis, Oregon 97331-7301

The *Baculoviridae* are a diverse family of viruses pathogenic for arthropods, particularly insects of the order Lepidoptera (Blissard and Rohrmann 1990). They are characterized by large, rod-shaped virions that contain double-stranded, supercoiled DNA genomes ranging in size from 88 kb to more than 160 kb, depending on the viral strain. The virions are enclosed in large proteinaceous structures known as occlusion bodies, which function to protect virions and allow them to remain viable for extended periods. Baculoviruses are divided into two genera on the basis of occlusion body morphology; the nuclear polyhedrosis viruses (NPVs) are characterized by many virions present in each polyhedron-shaped occlusion body, whereas the granulosis viruses (GVs) normally have a single virion in much smaller occlusion bodies. Baculoviruses have drawn widespread interest due to their remarkable ability to over-express heterologous genes under the control of the strong polyhedrin gene promoter (Smith et al. 1983; Pennock et al. 1984). They are also being investigated for incorporation into insect pest management programs as alternatives to chemical insecticides (Leisy and van Beek 1992).

The best-characterized baculovirus, the *Autographa californica* multinucleocapsid NPV (AcMNPV), has a genome of 134 kb and is estimated to contain 154 genes (Fig. 1) (Ayres et al. 1994). A distinctive feature of the AcMNPV genome is the presence of eight homologous regions (*hr*s) composed of multiple repeated sequences that are dispersed throughout the genome (Cochran and Faulkner 1983; Guarino et al. 1986; Guarino and Summers 1986b; Ayres et al. 1994). *hr*s, which are prominently featured in this review, have been shown to act as enhancers of baculovirus early gene expression and are implicated as origins of baculovirus DNA replication.

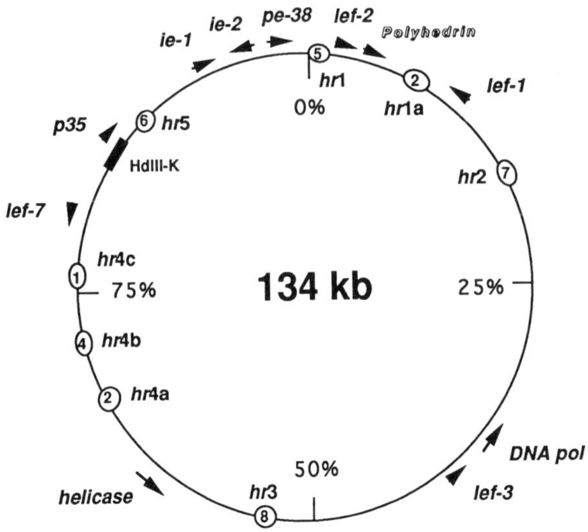

Figure 1 Location of homologous regions and genes involved in DNA replication on the AcMNPV genome. The polyhedrin gene is shown as a point of reference. The numbers within the circles indicate the number of repeated sequences within each *hr*. The arrows indicate the location and transcriptional direction of the genes involved in DNA replication. The location of the non-*hr* putative replication origin (HdIII-K) is indicated. Data for this figure are from Ayres et al. (1994).

Baculovirus genes are expressed in a transcriptional cascade in which each phase is dependent on the expression of genes during the previous phase. They can be divided into the general categories of early genes, which are transcribed by the host RNA polymerase II (Fuchs et al. 1983; Huh and Weaver 1990; Hoopes and Rohrmann 1991), and late genes, which are transcribed by a virus-specific RNA polymerase with a unique subunit composition (Yang et al. 1991). This polymerase is resistant to α-amanitin (an RNA pol II toxin) (Grula et al. 1981) and tagetitoxin (an RNA pol III toxin) (Glocker et al. 1993) and initiates transcription from within a 5-bp late promoter element with the sequence A/G/T TAAG (Blissard and Rohrmann 1990). Late gene expression is dependent on viral DNA replication and is not observed when DNA replication is inhibited (e.g., by aphidicolin) (Friesen and Miller 1986; Rice and Miller 1986–1987). Most late genes are transcribed after the onset of DNA replication with the levels of expression declining at later times after infection. However, expression of two genes termed "very late genes" reaches high levels very late in the infection. These genes are the polyhedrin gene, which encodes the major occlusion body protein, and p10, which encodes a small poorly conserved protein that may be involved in occlusion body formation or cell lysis (van Oers et al. 1993; Gross et al. 1994).

Recently, a number of advances have been made in the characterization of the replication of baculovirus DNA, including the identification of putative replication origins and the development of assays that allowed the identification of the genes involved in DNA replication. In this review we summarize these data; a more extensive review is also available (Kool et al. 1995).

BACULOVIRUS REPLICATION ORIGINS

Two different strategies have been used to identify potential baculovirus replication origins. One employed the generation of defective interfering (DI) virus particles, which contain major genomic deletions but retain essential *cis*-acting sequences for DNA replication, and the other approach relied on testing the ability of cloned baculovirus DNA sequences to undergo DNA replication when transfected into infected insect cells.

Defective Genomes

After undiluted serial passaging of AcMNPV, DI particles containing viral genomes with major deletions have been observed. Restriction enzyme analysis of viral DNA isolated after 40 serial passages showed the presence of seven supermolar *Eco*RI fragments (Kool et al. 1993a) that hybridized to DNA sequences flanking *hr*s. This suggested that the *hr*s were selectively retained in the defective genomes. In similar studies (Lee and Krell 1992, 1994), defective genomes analyzed after 81 passages were found to be heterogeneous in size with the majority migrating at about 50 kbp as determined by pulsed-field gel electrophoresis. In contrast to the results from Kool et al. (1993a), these defective genomes appeared to retain multiple repeats of a non-*hr*-containing sequence of less than 2.8 kb derived from the *Hin*dIII-K region of the parent AcMNPV genome (Fig. 1). The identification of different putative replication origins in these two studies suggests that both *hr* and non-*hr* sequences may act as replication origins in vivo.

hr Sequences as Baculovirus Replication Origins

AcMNPV contains eight homologous regions (*hr*1, *hr*1a, *hr*2, *hr*3, *hr*4a, *hr*4b, *hr*4c, and *hr*5) (Fig. 1) that vary in size from about 0.2 kb to 1.0 kb. Each *hr* is composed of one to eight repeats of a highly conserved 72-bp sequence element, which contains a 30-bp imperfect palindrome near its center (Fig. 2a). If converted to a cruciform structure, these palindromes would each have two mismatched regions outside a highly conserved 12-

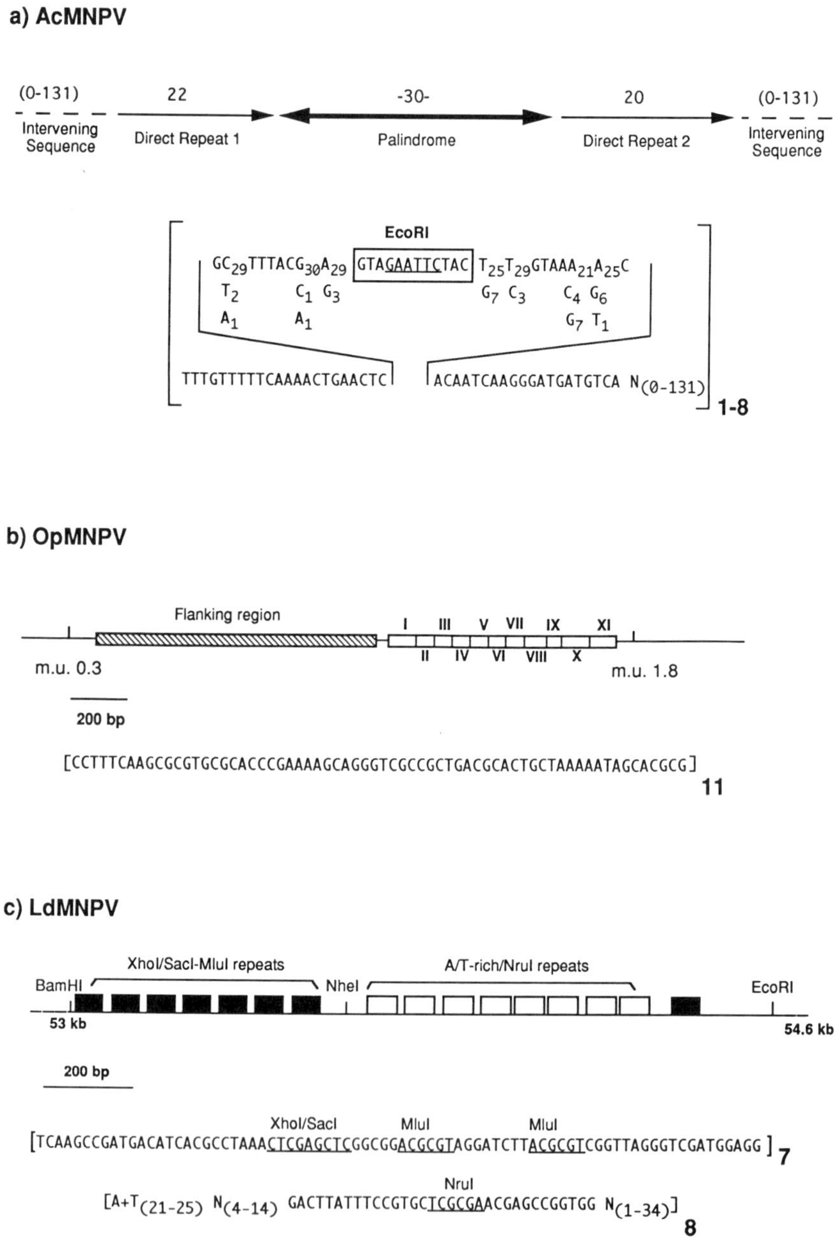

Figure 2 (*See facing page for legend.*)

bp core sequence containing an *Eco*RI site at its center. Individual repeats are separated by intervening sequences 0–131 nucleotides in length (Fig. 2a).

A transient assay employing *Dpn*I (Peden et al. 1980) has been used to examine infection-dependent replication of plasmids containing baculovirus DNA after their transfection into virus-infected insect cells (Pearson et al. 1992; Kool et al. 1993a). When transfected into AcMNPV-infected *Spodoptera frugiperda* cells, individual plasmids containing different *hr*s demonstrated the ability to replicate (Kool et al. 1993b; Leisy and Rohrmann 1993). (An *hr* with only one palindrome and flanking sequences, *hr*4c [Fig. 1], has also been reported [Ayres et al. 1994], but it lacks an *Eco*RI site and has not been functionally characterized.) Although the relative levels of replication for different *hr*s appear to be independent of the number of palindromes (e.g., *hr*4b, which contains four palindromes, replicated to a higher level than *hr*3, which contains eight palindromes) (Leisy and Rohrmann 1993), deletion mutagene-

Figure 2 Sequence organization of putative replication origins from three baculoviruses. (*a*) AcMNPV origin. A schematic representation of the general architecture of an *hr* is shown at the top. Below is shown the highly conserved 72-bp sequence element, which is present in 1–8 copies in the different homologous regions. The 30-bp palindrome is depicted above a consensus sequence of the conserved direct repeats found in *hr*s. The highly conserved 12-bp core sequence is boxed and the *Eco*RI site is underlined. The frequencies of the nucleotides in the variable positions of the palindrome (representative of 32 of the 35 palindromes) are shown as subscripts below each nucleotide. The letter N followed by subscripts indicates the length of the intervening sequences that are found between the repeats in different homologous regions. The consensus sequence for the repeated region flanking the palindrome is derived from 20 of the most conserved sequences. (*b*) OpMNPV *hr*-like origin. A schematic diagram showing the location of an origin containing a repeated element and essential flanking sequences. The efficiency of replication is dependent on the length of the flanking sequences (Ahrens et al. 1995b). The consensus sequence of the repeated region is shown below the diagram (Theilmann and Stewart 1992). (*c*) LdMNPV origin. A schematic diagram of an origin containing two repeated domains. This origin is composed of a domain containing seven repeats with overlapping *Xho*I/*Sac*I sites (*black boxes*), and *Mlu*I sites linked to a second domain containing eight repeats of an AT-rich sequence linked to a sequence containing a *Nru*I site (*open boxes*) (Pearson and Rohrmann 1995). A consensus sequence of these two domains is shown below the diagram. The letter N, followed by subscripts, indicates the nucleotides separating internal regions or separating the repeated consensus sequences.

sis of *hr*5, which contains six palindromes, indicated that the efficiency of replication was correlated with the number of palindromes present (Pearson et al. 1992).

Sequences similar to AcMNPV *hr*s have been found in a number of other baculoviruses, including *Bombyx mori* NPV, a close relative of AcMNPV (Maeda and Majima 1990; Majima et al. 1993), and *Choristoneura fumiferana* MNPV (CfMNPV) (Arif and Doerfler 1984; Kuzio and Faulkner 1984), which is less related to AcMNPV but has at least one set of repeated palindromes, some of which are more than 75% identical to AcMNPV palindromes (Xie et al. 1995). The genome of the *Orgyia pseudotsugata* MNPV (OpMNPV) was shown to contain five homologous regions by DNA-DNA hybridization. One OpMNPV *hr* has been characterized and found to act as an enhancer of early gene transcription (Theilmann and Stewart 1992) and to function as a replication origin in transient assays when linked to flanking sequences (Ahrens et al. 1995b). However, it lacks well-defined palindromes (Fig. 2b) and has only 50% sequence identity to AcMNPV *hr* sequences. The *Lymantria dispar* MNPV (LdMNPV), which is more distantly related to AcMNPV than BmNPV, CfMNPV, and OpMNPV, contains *hr*s located in eight regions of its genome (Pearson and Rohrmann 1995). Two of the *hr*s have been characterized and are composed of repeats of about 80 bp that include a series of palindromes unrelated to AcMNPV palindromes. Plasmids containing these sequences alone replicate poorly. However, LdMNPV *hr*4 is adjacent to a second domain composed of eight partially repeated sequences of 60–100 bp, each containing a 15- to 25-bp AT-rich sequence and a 6- to 10-bp palindrome containing an *Nru*I site (Fig. 2c). In combination, the two domains cause plasmids to replicate to high levels in LdMNPV-infected cells (Pearson and Rohrmann 1995).

Non-*hr*-containing Origins

In addition to *hr* sequences that may function as origins of DNA replication, two other non-*hr* putative origins have been characterized. The AcMNPV *Hin*dIII-K fragment, which lacks sequences closely related to AcMNPV *hr*s (Kool et al. 1994a), was initially implicated as a replication origin by the studies involving the generation of DI virus particles (Fig. 1) (see above). Subsequently, it was shown to replicate in an infection-dependent replication assay (Kool et al. 1993b; Lee and Krell 1994), although at less than 20% the level of *hr*-containing plasmids (Leisy and Rohrmann 1993). Similarily, the OpMNPV *Hin*dIII-N fragment (Pearson et al. 1993), which lacks *hr*-related sequences, was shown to replicate at significant levels. However, *Hin*dIII-N replicates at only about 25% the efficiency of the OpMNPV *hr*-containing sequence (Ahrens et al. 1995b). Investigations of non-*hr*-containing replication origins from AcMNPV and other baculoviruses are not complete, and it is likely that a number of other such regions occur.

Form of Transiently Replicated Plasmid DNA

A circular topology is a prerequisite for transient replication of origin-containing plasmids (Kool et al. 1993b). Linearized origin-containing plasmids were not replicated. The *hr*-containing plasmid DNA replicated in AcMNPV-infected cells was shown to be of high molecular weight, suggesting that replication does not lead to the production of an exact replica of the input circular plasmid DNA. Partial digestion of the replicated DNA with a restriction enzyme that cut the input plasmid at a unique site led to the production of a "stepladder" pattern of fragments, consistent with the DNA being organized as a linear concatemer containing multiple copies of the plasmid (Leisy and Rohrmann 1993). Although the form of replicating genomic DNA in infected cells has not been described, a linear concatemeric structure of origin-containing plasmids may indicate that viral DNA replicates via a rolling circle. Defective genomes consisting of concatemers of sequences from the *Hin*dIII-K fragment (Lee and Krell 1992, 1994) also support a rolling-circle model for baculovirus DNA replication. With such a model, a single replication initiation event would lead to the production of multiple copies of the genome. The mechanism by which such a structure might be resolved into unit-length, circular genome segments is not known, but it could involve cleavage and religation to form monomeric circles before or during packaging.

Possible Role of Multiple Replication Origins

Eight putative origins have been identified in the AcMNPV genome. The role these origins play in viral DNA replication, and whether they are active simultaneously, is not clear. Deletion of *hr*5 from the AcMNPV or BmNPV genome had no apparent effect on the replication of these viruses (Rodems and Friesen 1993; Majima et al. 1993). Subsequently, Wu and Carstens (1995) showed that deletion of other individual *hr*s had no effect on replication of AcMNPV in cell culture. It is possible that the presence of multiple *hr* sequences in the AcMNPV genome may reflect their role as transcriptional enhancers of early genes rather than their role in replication. The accelerated expression of early genes may be essential for the successful establishment of an infection. In contrast, only a single origin may be required per replication initiation event, and rolling-circle replication could lead to the production of multiple genomes. If factors required for the initiation of replication are limiting, the formation of an initiation complex may be rare and may occur only once on each DNA molecule. The selection of which origin initiates replication may occur at random. In such circumstances, functionally redundant origins may increase the probability of the formation of a functional preinitiation complex and thus increase the speed of the infection cycle. The limitation of replication factors would reduce the likelihood of multiple initiations on a single genome, which could lead to abortive replication via the production of complex branched structures that may not yield viable genomes.

GENES INVOLVED IN BACULOVIRUS DNA REPLICATION

A replication assay similar to that employed for the identification of the herpes simplex virus I replication genes (Challberg 1986) was developed for the identification of baculovirus replication genes. A set of overlapping cosmid clones encompassing the complete genome was transfected into uninfected insect cells along with a reporter plasmid containing a putative replication origin. A minimal set of cosmids was identified that replicated the plasmid DNA, and the specific genes were determined using subclones of the essential cosmids. This protocol was used to identify both essential and stimulatory genes for transient DNA replication of two baculoviruses, AcMNPV (Kool et al. 1994c; Lu and Miller 1995) and OpMNPV (Ahrens and Rohrmann 1995a,b; Ahrens et al. 1995a). The genes that have been implicated in DNA replication are described below and are summarized for AcMNPV in Table 1. Their location on the AcMNPV genome is shown in Figure 1.

DNA polymerase

In early studies (Kelly and Lescott 1981; Miller et al. 1981; Kelly 1982), it was noted that a new DNA polymerase activity was induced in baculovirus-infected cells. A 3′→5′ exonuclease activity specific for single-stranded DNA was found to be tightly associated with *B. mori* NPV DNA polymerase (Mikhailov et al. 1986). Subsequently, the location and sequence of genes from six different baculoviruses that encode predicted proteins of about 115 kD and that contain motifs conserved among a number of DNA polymerases were determined (Tomalski et al.

Table 1 AcMNPV replication genes

Gene	MW[a]	Function(s)	Essential(E) Stimulatory (S)
DNA pol	114.3		E/S
helicase	143.2		E
ie-1	66.9	transcriptional activator binds *hr* sequences	E
lef-1	30.8		E
lef-2	23.9		E
lef-3	44.6	SSB[b]	E
p35[c]	34.8	inhibits apoptosis transcriptional activator	E/S
ie-2	47.0	transcriptional activator	S
pe-38	37.4	transcriptional activator	S
lef-7	26.6		S

For details see text.
[a]Molecular weights of predicted proteins are from Ayres et al. (1994).
[b]Hang et al. (1995).
[c]*p35* is not present in OpMNPV.

1988; Bjornson et al. 1992; Cowan et al. 1994; Chaeychomsri et al. 1995; Liu and Carstens 1995; Ahrens and Rohrmann 1996). Although baculovirus DNA polymerase genes were shown to be essential for DNA replication in two studies (Pearson et al. 1993; Kool et al. 1994c), in another study (Lu and Miller 1995), low levels of replication activity were observed in the absence of the baculovirus DNA polymerase gene. This suggests that under differing assay conditions a host DNA polymerase may act in combination with the other essential baculovirus replication genes to replicate plasmid DNA.

helicase

A baculovirus gene with limited homology to helicases was originally identified by sequencing an ORF containing a temperature-sensitive mutation that resulted in virus defective for DNA synthesis (Lu and Carstens 1991). This gene encodes a predicted protein of 143 kD that contains a variety of amino acid motifs common to a number of helicases, including NTP-binding and DNA/RNA-unwinding motifs. In addition, sequences in the helicase protein that are involved in specifying host range have been identified (Maeda et al. 1993; Croizier et al. 1994), which could indicate that this region might require association with a host protein in order to function.

ie-1

Immediate early gene 1 (*ie-1*) is the only baculovirus gene to date for which both spliced and unspliced transcripts have been identified (Chisholm and Henner 1988). The spliced forms produce a protein product with an additional 54 amino acids at the amino terminus. Plasmids expressing unspliced *ie-1* have been shown to be essential for baculovirus DNA replication (Kool et al. 1994c; Ahrens and Rohrmann 1995b; Lu and Miller 1995). Deletion analysis of AcMNPV *ie-1* indicates the presence of distinct transcriptional-activation and DNA-binding domains (Kovacs et al. 1992).

ie-1 has been shown to activate a variety of baculovirus early gene promoter-reporter constructs when they are co-transfected into uninfected insect cells (Guarino and Summers 1986a; Blissard and Rohrmann 1991; Lu and Carstens 1993; Rodems and Friesen 1993), and this activation is greatly enhanced when these constructs are linked to *hr* sequences. Extracts from *S. frugiperda* cells transfected with *ie-1* and IE-1 produced in in vitro transcription-translation reactions cause retardation of *hr*-containing DNA in gel shift assays (Choi and Guarino 1995).

In the baculovirus system, *ie-1* may be essential because it activates the expression of the other replication genes to levels sufficient to produce detectable amounts of replicated DNA. Alternatively, it may be directly involved in DNA replication by binding to an origin and catalyzing early steps that lead to the assembly of a replication complex.

lef-1, lef-2, lef-3

Whereas information on possible functions of *ie-1* and the DNA polymerase and helicase genes is available because of functional studies or their homologies with well-characterized genes from other organisms, little is known about three of the essential replication genes called late expression factors (*lef-1, 2*, and *3*). These genes were originally implicated as being essential for late gene expression (Todd et al. 1995). However, this may have been due to the dependence of the late gene expression assay on DNA replication. Several investigations indicate that these genes are directly involved in DNA replication of both AcMNPV and OpMNPV (Kool et al. 1994c; Ahrens and Rohrmann 1995a,b; Ahrens et al. 1995a; Lu and Miller 1995). Yeast two-hybrid analysis indicates that LEF-1 and LEF-2 form hetero-oligomers (Leisy and Rohrmann 1995). A motif found in single-stranded DNA-binding proteins is present in OpMNPV and AcMNPV LEF-3 (Ahrens et al. 1995a), and AcMNPV LEF-3 has been shown to bind to single-stranded DNA agarose columns with high affinity and elute only at salt concentrations greater than 0.9 M (Hang et al. 1995).

p35

The AcMNPV *p35* gene was found to greatly stimulate replication in one study (Kool et al. 1994c), but using different assay conditions, other investigators suggest that it is essential for replication (Lu and Miller 1995). *p35* is an inhibitor of AcMNPV-induced apoptosis in *S. frugiperda* cells (Clem et al. 1991) and may also act as a transcriptional activator of early genes (Gong and Guarino 1994). It is possible that the direct effect of *p35* in transient replication assays through transcriptional activation is minor in comparison to the elevation of the replication signal due to inhibition of apoptosis, which would prevent the cells from dying. This theory is supported by the fact that OpMNPV, which shows a high degree of genome similarity to AcMNPV, lacks a homolog of the *p35* gene (Gombart et al. 1989). However, OpMNPV encodes a functional analog of *p35*, called Op-*iap* (*i*nhibitor of *ap*optosis), which bears no sequence identity with p35 but is able to inhibit AcMNPV-induced apoptosis in Sf-9 cells (Birnbaum et al. 1994). A related gene (Cp-*iap*) with similar properties is found in the genome of the *Cydia pomonella* GV (Crook et al. 1993). Both these genes can substitute for the *p35* gene in replication assays employing the other AcMNPV replication genes (Lu and Miller 1995). If apoptosis is induced by the replication of the transfected DNA, it would suggest that the cell has a mechanism that discriminates between normal chromosomal DNA replication and replication of foreign DNA or DNA replication not linked to the cell cycle. Alternatively, expression of one or more of the genes involved in DNA replication may directly induce apoptosis.

ie-2 and *pe-38*

Efficient DNA replication in the transient assay system requires additional genes that stimulate replication. Two of the stimulatory genes, *ie-2* (previously called *ie-n*) (Carson et al. 1988) and *pe-38* (Krappa and Knebel-Mörsdorf 1991), encode *trans*-activators of early gene transcription. In particular, *pe-38* has been shown to activate expression of the baculovirus *helicase* homolog. Furthermore, *ie-2* has been shown to stimulate *pe-38* expression (Lu and Carstens 1993) and *ie-1* expression (Yoo and Guarino 1994). *ie-2* and *pe-38* may selectively stimulate certain of the replication genes to higher levels than are necessary for minimal detection.

lef-7

lef-7 was first shown to stimulate the expression of a CAT construct under the control of either the p39-capsid or polyhedrin promoters (Morris et al. 1994) and was subsequently determined to stimulate DNA replication (Lu and Miller 1995).

pcna

The AcMNPV genome contains a gene with 42% amino acid sequence identity to rat proliferating cell nuclear antigen (PCNA) (Crawford and Miller 1988; O'Reilly et al. 1989), which is an essential component of several eukaryotic replication systems (Kornberg and Baker 1992). However, this gene is lacking in the closely related BmNPV genome (Gomi et al. 1994), and it has no effect on transient DNA replication (Kool et al. 1994c). Therefore, the role this gene plays in AcMNPV biology is not clear.

TRANSIENT REPLICATION OF DNA IN TRANSFECTED CELLS IS ORIGIN-INDEPENDENT

Although plasmid replication in virus-infected cells is sequence-dependent, replication becomes viral origin-independent when transfected naked DNA, instead of intact virions, is used to supply essential *trans*-acting factors. Plasmids containing the most efficient origins, as determined by the replication assay employing virus-infected cells, still show the strongest replication signals in transfected cells; however, other plasmids (including those lacking inserts) replicate to detectable levels (Kool et al. 1994b,c). In other eukaryotic systems, chromatin structure represses initiation of both transcription and replication (Wolffe 1991). Although the organization of DNA-binding proteins associated with baculovirus genomes is unclear, there is evidence for the association of a small, very basic protein with baculovirus DNA in virions (Wilson et al. 1987). If baculovirus DNA is packaged in a chromatin-like structure, transfection of DNA lacking these proteins may result in the deregulation and high-level expression of the genes involved in replication. High gene copy number may also play a role in origin-independent replication. Cells infected with virus at low moi on average contain only one to a few copies of each gene. In contrast, cells transfected with cloned viral replication genes contain much higher gene doses, potentially resulting in an overabundance of replication proteins, which may cause saturation of specific origin sequences by these proteins. The excess protein molecules may bind to DNA sequences nonspecifically and, hence, cause initiation of replication from any plasmid. High levels of recombination in transfected cells could also influence these observations.

SPECIFICITY OF BACULOVIRUS DNA REPLICATION

As indicated above, some baculoviruses contain several putative replication origins that may differ in structure within the same virus and between viruses. The ability of two different baculoviruses to replicate heterologous origin-containing plasmids has been examined. It was found that transfection of OpMNPV-infected *L. dispar* cells with an AcMNPV *hr*5-containing plasmid resulted in only trace amounts of plasmid replication (Pearson et al. 1993). Conversely, only minimal levels of replication of plasmids containing the putative OpMNPV replication origins *Hin*dIII-N or an OpMNPV *hr*-containing plasmid were detected in AcMNPV-infected *S. frugiperda* cells (Ahrens et al. 1995b). These data indicate that although baculoviruses are often similar in genome structure and organization, the DNA sequences that constitute an origin of replication and the viral/host proteins that recognize and interact with these sequences may differ significantly. Recent investigations suggest that some replication genes are fully interchangeable between the two systems, whereas others show high degrees of specificity (C.H. Ahrens, unpubl.).

FUTURE INVESTIGATIONS

The identification of baculovirus replication origins along with the development of assays for monitoring transient DNA replication, and the identification of genes that are essential or that stimulate DNA replication, have opened a number of avenues of research. These include characterization of the role of specific DNA sequences in origin structure and of the biochemical function(s) of the replication proteins in the initiation of replication and during DNA synthesis. The functions of some of the essential replication gene products can be predicted by comparison with those from other organisms. For example, a DNA polymerase normally requires a processivity factor that keeps the polymerase core tightly associated with the template, thereby preventing it from dissociating. Another essential component may include an origin-binding protein that recognizes an origin and catalyzes the assembly of the replication complex, and a primase that would be involved in the synthesis of the RNA primers during replication. In addition, if, as the data with *hr*-containing plasmids suggest, replication results in large genome concatemers, understanding the mechanism by which such structures are resolved into unit-length genomes is of particular interest. Contribution of host factors to the replication process also cannot be ruled out. Of high priority is the development of an in vitro replication system that can be used to define the role of each component of the replication complex in replicating DNA.

ACKNOWLEDGMENTS

We thank A. Lu, L. Miller, E. Carstens, and L. Guarino for preprints and information prior to publication, and J. Russell for critically reading this manuscript. This project was supported in part by a grant from the Study Foundation of the German People to C.H.A. and grants 91-37302-6310 (U.S. Department of Agriculture) and MCB-9219227 (National Science Foundation) to G.F.R. This is Technical Report No. 10749 from the Oregon State University Agricultural Experiment Station.

REFERENCES

Ahrens, C.H. and G.F. Rohrmann. 1995a. Identification of essential *trans*-acting regions required for DNA replication of the *Orgyia pseudotsugata* multinucleocapsid nuclear polyhedrosis virus: *lef-1* is an essential replication gene. *Virology* **207:** 417–428.

———. 1995b. Replication of *Orgyia pseudotsugata* baculovirus DNA: *ie-1* and *lef-2* are essential and *ie-2 p34* and Op-*iap* are stimulatory genes. *Virology* **212:** 650–662.

———. 1996. The DNA polymerase and helicase genes of a baculovirus of *Orgyia pseudotsugata*. *J. Gen. Virol.* (in press).

Ahrens, C.H., C. Carlson, and G.F. Rohrmann. 1995a. Identification, sequence and transcriptional analysis of *lef-3*, a gene essential for *Orgyia pseudotsugata* baculovirus DNA replication. *Virology* **210:** 372–382.

Ahrens, C.H., M.N. Pearson, and G.F. Rohrmann. 1995b. Identification and characterization of a second putative origin of DNA replication in a baculovirus of *Orgyia pseudotsugata*. *Virology* **207:** 572–576.

Arif, B. and W. Doerfler. 1984. Identification and localization of reiterated sequences in the *Choristoneura fumiferana* MNPV genome. *EMBO J.* **3:** 525–529.

Ayres, M.D., S.C. Howard, J. Kuzio, M. Lopez-Ferber, and R.D. Possee. 1994. The complete DNA sequence of *Autographa californica* nuclear polyhedrosis virus. *Virology* **202:** 586–605.

Birnbaum, M.J., R.J. Clem, and L.K. Miller. 1994. An apoptosis-inhibiting gene from a nuclear polyhedrosis virus encoding a polypeptide with Cys/His sequence motifs. *J. Virol.* **68:** 2521–2528.

Bjornson, R.M., B. Glocker, and G.F. Rohrmann. 1992. Characterization of the nucleotide sequence of the *Lymantria dispar* nuclear polyhedrosis virus DNA polymerase gene region. *J. Gen. Virol.* **73:** 3177–3183.

Blissard, G.W. and G.F. Rohrmann. 1990. Baculovirus diversity and molecular biology.

Annu. Rev. Entomol. **35:** 127–155.
———. 1991. Baculovirus gp64 gene expression: Analysis of sequences modulating immediate early transcription and transactivation by IE1. *J. Virol.* **65:** 5820–5827.
Carson, D.D., L.A. Guarino, and M.D. Summers. 1988. Functional mapping of an AcNPV immediate early gene which augments expression of the IE-1 *trans*-activated 39 k gene. *Virology* **162:** 444–451.
Chaeychomsri, S., M. Ikeda, and M. Kobayashi. 1995. Nucleotide sequence and transcriptional analysis of the DNA polymerase gene of *Bombyx mori* nuclear polyhedrosis virus. *Virology* **206:** 435–447.
Challberg, M.D. 1986. A method for identifying the viral genes required for herpesvirus DNA replication. *Proc. Natl. Acad. Sci.* **83:** 9094–9098.
Chisholm, G.E. and D.J. Henner. 1988. Multiple early transcripts and splicing of the *Autographa californica* nuclear polyhedrosis virus IE-1 gene. *J. Virol.* **62:** 3193–3200.
Choi, J. and L.A. Guarino. 1995. The baculovirus transactivator IE-1 binds to viral enhancer elements in the absence of insect cell factors. *J. Virol.* **69:** 4548–4551.
Clem, R.J., M. Fechheimer, and L.K. Miller. 1991. Prevention of apoptosis by a baculovirus gene during infection of insect cells. *Science* **254:** 1388–1390.
Cochran, M.A. and P. Faulkner. 1983. Location of homologous DNA sequences interspersed at five regions in the baculovirus AcMNPV genome. *J. Virol.* **45:** 961–970.
Cowan, P., D. Bulach, K. Goodge, A. Robertson, and D.E. Tribe. 1994. Nucleotide sequence of the polyhedrin gene region of *Helicoverpa zea* single nucleocapsid nuclear polyhedrosis virus: Placement of the virus in lepidopteran nuclear polyhedrosis virus group II. *J. Gen. Virol.* **75:** 3211–3218.
Crawford, A.M. and L.K. Miller. 1988. Characterization of an early gene accelerating expression of late genes of the baculovirus *Autographa californica* nuclear polyhedrosis virus. *J. Virol.* **62:** 2773–2781.
Croizier, G., L. Croizier, O. Argaud, and D. Poudevigne. 1994. Extension of *Autographa californica* nuclear polyhedrosis virus host range by interspecific replacement of a short DNA sequence in the p143 helicase gene. *Proc. Natl. Acad. Sci.* **91:** 48–52.
Crook, N.E., R.J. Clem, and L.K. Miller. 1993. An apoptosis-inhibiting baculovirus gene with a zinc finger-like motif. *J. Virol.* **67:** 2168–2174.
Friesen, P.D. and L.K. Miller. 1986. The regulation of baculovirus gene expression. *Curr. Top. Microbiol. Immunol.* **131:** 31–49.
Fuchs, Y.L., M.S. Woods, and R.F. Weaver. 1983. Viral transcription during *Autographa californica* nuclear polyhedrosis virus infection: A novel RNA polymerase induced in infected *Spodoptera frugiperda* cells. *J. Virol.* **48:** 641–646.
Glocker, B., R.R. Hoopes Jr., L. Hodges, and G.F. Rohrmann. 1993. In vitro transcription from baculovirus late gene promoters: Accurate mRNA initiation by nuclear extracts prepared from infected *Spodoptera frugiperda* cells. *J. Virol.* **67:** 3771–3776.
Gombart, A.F., G.W. Blissard, and G.F. Rohrmann. 1989. Characterization of the genetic organization of the *Hin*dIII-M region of the multicapsid nuclear polyhedrosis virus of *Orgyia pseudotsugata* reveals major differences among baculoviruses. *J. Gen. Virol.* **70:** 1815–1828.
Gomi, S.J., J.W. Chua, and S. Maeda. 1994. Nucleotide sequence and gene-deletion analyses of the *Bombyx mori* nuclear polyhedrosis virus (BmNPV) genome. In *Scientific Program and Abstracts* (13th Annual Meeting of the American Society for Virology, Madison, Wisconsin), p. 168.
Gong, M. and L.A. Guarino. 1994. Expression of the 39k promoter of *Autographa*

californica nuclear polyhedrosis virus is increased by the apoptotic suppressor p35. *Virology* **204:** 38–44.

Gross, C.H., R.L.Q. Russell, and G.F. Rohrmann. 1994. The *Orgyia pseudotsugata* baculovirus p10 and polyhedron envelope protein genes: Analysis of their relative expression levels and role in polyhedron structure. *J. Gen. Virol.* **75:** 1115–1123.

Grula, M.A., P.L. Buller, and R.F. Weaver. 1981. Alpha amanitin-resistant viral RNA synthesis in nuclei isolated from nuclear polyhedrosis virus-infected *Heliothis zea* larvae and *Spodoptera frugiperda* cells. *J. Virol.* **38:** 916–921.

Guarino, L.A. and M.D. Summers. 1986a. Functional mapping of a *trans*-activating gene required for expression of a baculovirus delayed-early gene. *J. Virol.* **57:** 563–571.

———. 1986b. Interspersed homologous DNA of *Autographa californica* nuclear polyhedrosis virus enhances delayed-early gene expression. *J. Virol.* **60:** 215–223.

Guarino, L.A., M.A. Gonzalez, and M.D. Summers. 1986. Complete sequence and enhancer function of the homologous DNA regions of *Autographa californica* nuclear polyhedrosis virus. *J. Virol.* **60:** 224–229.

Hang, X., W. Dong, and L.A. Guarino. 1995. The *lef-3* gene of *Autographa californica* nuclear polyhedrosis virus encodes a single-stranded DNA-binding protein. *J. Virol.* **69:** 3924–3928.

Hoopes, R.R., Jr. and G.F. Rohrmann. 1991. *In vitro* transcription of baculovirus immediate early genes: Accurate mRNA initiation by nuclear extracts from both insect and human cells. *Proc. Natl. Acad. Sci.* **88:** 4513–4517.

Huh, N.E. and R.F. Weaver. 1990. Identifying the RNA polymerases that synthesize specific transcripts of the *Autographa californica* nuclear polyhedrosis virus. *J. Gen. Virol.* **71:** 195–201.

Kelly, D.C. 1982. Baculovirus replication. *J. Gen. Virol.* **63:** 1–13.

Kelly, D.C. and T. Lescott. 1981. Baculovirus replication: Protein synthesis in *Spodoptera frugiperda* cells infected with *Trichoplusia ni* nuclear polyhedrosis virus. *Microbiologica* **4:** 35–47.

Kool, M., R.W. Goldbach, and J.M. Vlak. 1994a. A putative non-hr origin of DNA replication in the *Hin*dIII-K fragment of *Autographa californica* nuclear polyhedrosis virus. *J. Gen. Virol.* **75:** 3345–3352.

Kool, M., J.T.M. Voeten, and J.M. Vlak. 1994b. Functional mapping of regions of the *Autographa californica* nuclear polyhedrosis viral genome required for DNA replication. *Virology* **198:** 680–689.

Kool, M., C.H. Ahrens, J.M. Vlak, and G.F. Rohrmann. 1995. Replication of baculovirus DNA. *J. Gen. Virol.* **76:** 2103–2118.

Kool, M., C. Ahrens, R.W. Goldbach, G.F. Rohrmann, and J.M. Vlak. 1994c. Identification of genes involved in DNA replication of the *Autographa californica* baculovirus. *Proc. Natl. Acad. Sci.* **91:** 11212–11216.

Kool, M., P.M.M.M. Van Den Berg, J. Tramper, R.W. Goldbach, and J.M. Vlak. 1993a. Location of two putative origins of DNA replication of *Autographa californica* nuclear polyhedrosis virus. *Virology* **192:** 94–101.

Kool, M., J.T.M. Voeten, R.W. Goldbach, J. Tramper, and J.M. Vlak. 1993b. Identification of seven putative origins of *Autographa californica* MNPV DNA replication. *J. Gen. Virol.* **74:** 2661–2668.

Kornberg, A. and T.A. Baker. 1992. *DNA replication*, 2nd edition, W.H. Freeman, New York.

Kovacs, G.R., J. Choi, L.A. Guarino, and M.D. Summers. 1992. Functional dissection of the *Autographa californica* nuclear polyhedrosis virus immediate-early 1 transcriptional regulatory protein. *J. Virol.* **66:** 7429–7437.

Krappa, R. and D. Knebel-Mörsdorf. 1991. Identification of the very early transcribed baculovirus gene PE-38. *J. Virol.* **65:** 805–812.

Kuzio, J. and P. Faulkner. 1984. Region of repeated DNA in the genome of *Choristoneura fumiferana* nuclear polyhedrosis virus. *Virology* **139:** 185–188.

Lee, H.Y. and P.J. Krell. 1992. Generation and analysis of defective genomes of *Autographa californica* nuclear polyhedrosis virus. *J. Virol.* **66:** 4339–4347.

———. 1994. Reiterated DNA fragments in defective genomes of *Autographa californica* nuclear polyhedrosis virus are competent for AcMNPV-dependent DNA replication. *Virology* **202:** 418–429.

Leisy, D.J. and G.F. Rohrmann. 1993. Characterization of the replication of plasmids containing *hr* sequences in baculovirus-infected *Spodoptera frugiperda* cells. *Virology* **196:** 722–730.

———. 1995. Interactions of baculovirus replication proteins: Application of a yeast two-hybrid system. In *Scientific Program and Abstracts* (14th Annual Meeting of the American Society for Virology, Austin, Texas), p. 105.

Leisy, D.J. and N. van Beek. 1992. Baculovirus: Possible alternatives to chemical insecticides. *Chem. Ind.* **42:** 250–254.

Liu, J.J. and E.B. Carstens. 1995. Identification, localization, transcription, and sequence analysis of the *Choristoneura fumiferana* nuclear polyhedrosis virus DNA polymerase gene. *Virology* **209:** 538–549.

Lu, A. and E.B. Carstens. 1991. Nucleotide sequence of a gene essential for viral DNA replication in the baculovirus *Autographa californica* nuclear polyhedrosis virus. *Virology* **181:** 336–347.

———. 1993. Immediate-early baculovirus genes transactivate the p143 gene promoter of *Autographa californica* nuclear polyhedrosis virus. *Virology* **195:** 710–718.

Lu, A. and L.K. Miller. 1995. The roles of eighteen baculovirus late expression factor genes in transcription and DNA replication. *J. Virol.* **69:** 975–982.

Maeda, S. and K. Majima. 1990. Molecular cloning and physical mapping of the genome of *Bombyx mori* nuclear polyhedrosis virus. *J. Gen. Virol.* **71:** 1851–1855.

Maeda, S., S.G. Kamita, and A. Kondo. 1993. Host range expansion of *Autographa californica* nuclear polyhedrosis virus (NPV) following recombination of a 0.6-kilobase-pair DNA fragment originating from *Bombyx mori* NPV. *J. Virol.* **67:** 6234–6238.

Majima, K., R. Kobara, and S. Maeda. 1993. Divergence and evolution of homologous regions of *Bombyx mori* nuclear polyhedrosis virus. *J. Virol.* **67:** 7513–7521.

Mikhailov, V.S., K.A. Marlyev, J.O. Ataeva, P.K. Kullyev, and A.M. Atrazhev. 1986. Characterization of 3′ to 5′ exonuclease associated with DNA polymerase of silkworm nuclear polyhedrosis virus. *Nucleic Acids Res.* **14:** 3841–3856.

Miller, L.K., J.E. Jewell, and D. Browne. 1981. Baculovirus induction of a DNA polymerase. *J. Virol.* **40:** 305–308.

Morris, T.D., J.W. Todd, B. Fisher, and L.K. Miller. 1994. Identification of *lef-7*: A baculovirus gene affecting late gene expression. *Virology* **200:** 360–369.

O'Reilly, D.R., A.M. Crawford, and L.K. Miller. 1989. Viral proliferating cell nuclear antigen. *Nature* **337:** 606.

Pearson, M.N. and G.F. Rohrmann. 1995. *Lymantria dispar* nuclear polyhedrosis virus homologous regions: Characterization of their ability to function as replication origins.

J. Virol. **69:** 213–221.

Pearson, M.N., R.M. Bjornson, C. Ahrens, and G.F. Rohrmann. 1993. Identification and characterization of a putative origin of DNA replication in the genome of a baculovirus pathogenic for *Orgyia pseudotsugata*. *Virology* **197:** 715–725.

Pearson, M.N., R.M. Bjornson, G.D. Pearson, and G.F. Rohrmann. 1992. The *Autographa californica* baculovirus genome: Evidence for multiple replication origins. *Science* **257:** 1382–1384.

Peden, K.W.C., J.M. Pipas, S. Pearson-White, and D. Nathans. 1980. Isolation of mutants of an animal virus in bacteria. *Science* **209:** 1392–1396.

Pennock, G.D., C. Shoemaker, and L.K. Miller. 1984. Strong and regulated expression of *Escherichia coli* β-galactosidase in insect cells with a baculovirus vector. *Mol. Cell. Biol.* **4:** 399–406.

Rice, W.C. and L.K. Miller. 1986–1987. Baculovirus transcription in the presence of inhibitors and in nonpermissive *Drosophila* cells. *Virus Res.* **6:** 155–172.

Rodems, S.M. and P.D. Friesen. 1993. The hr5 transcriptional enhancer stimulates early expression from the *Autographa californica* nuclear polyhedrosis virus genome but is not required for virus replication. *J. Virol.* **67:** 5776–5785.

Smith, G.E., M.D. Summers, and M.J. Fraser. 1983. Production of human beta interferon in insect cells infected with a baculovirus expression vector. *Mol. Cell. Biol.* **3:** 2156–2165.

Theilmann, D.A. and S. Stewart. 1992. Tandemly repeated sequence at the 3′ end of the IE-2 gene of the baculovirus *Orgyia pseudotsugata* multicapsid nuclear polyhedrosis virus is an enhancer element. *Virology* **187:** 97–106.

Todd, J.W., A.L. Passarelli, and L.K. Miller. 1995. Eighteen baculovirus genes, including *lef-11, p35, 39K,* and *p47,* support late gene expression. *J. Virol.* **69:** 968–974.

Tomalski, M.D., J. Wu, and L.K. Miller. 1988. The location, sequence, transcription, and regulation of a baculovirus DNA polymerase gene. *Virology* **167:** 591–600.

van Oers, M.M., J.T.M. Flipsen, C.B.E.M. Reusken, E.L. Sliwinsky, R.W. Goldbach, and J.M. Vlak. 1993. Functional domains of the p10 protein of *Autographa californica* nuclear polyhedrosis virus. *J. Gen. Virol.* **74:** 563–574.

Wilson, M.E., T.H. Mainprize, P.D. Friesen, and L.K. Miller. 1987. Location, transcription, and sequence of a baculovirus gene encoding a small arginine-rich polypeptide. *J. Virol.* **61:** 661–666.

Wolffe, A. 1991. Implications of DNA replication for eukaryotic gene expression. *J. Cell Sci.* **99:** 201–206.

Wu, Y.-T. and E.B. Carstens. 1995. Analysis of the *in vivo* effects of deleting individual homologous repeat sequences (*hr*s) from the AcMNPV genome. In *Scientific Program and Abstracts* (14th Annual Meeting of the American Society for Virology, Austin, Texas), p. 158.

Xie, W.-D., B. Arif, P. Dobos, and P.J. Krell. 1995. Identification and analysis of a putative origin of DNA replication in the *Choristoneura fumiferana* multinucleocapsid nuclear polyhedrosis virus genome. *Virology* **209:** 409–419.

Yang, C.L., D.A. Stetler, and R.F. Weaver. 1991. Structural comparison of the *Autographa californica* nuclear polyhedrosis virus-induced RNA polymerase and the three nuclear RNA polymerases from the host, *Spodoptera frugiperda*. *Virus Res.* **20:** 251–264.

Yoo, S. and L.A. Guarino. 1994. The *Autographa californica* nuclear polyhedrosis virus *ie-2* gene encodes a transcriptional regulator. *Virology* **202:** 746–753.

32

DNA Replication in Yeast

Carol S. Newlon
Department of Microbiology and Molecular Genetics
UMDNJ-New Jersey Medical School
Newark, New Jersey 07103

The details of chromosome replication are better understood in the budding yeast, *Saccharomyces cerevisiae*, than in any other eukaryotic organism. *cis*-Acting replicator sequences required for chromosomal replication origin function were identified on the basis of their ability to promote the extrachromosomal maintenance of plasmids. These autonomously replicating sequence (ARS) elements have been dissected using the plasmid assay. The ease with which chromosomal sequences can be replaced by homologous recombination in this yeast has facilitated analysis of the effects of ARS mutations on chromosomal origin activity, which can be assayed by two-dimensional (2D) gel analysis. Both the isolation and analysis of mutants defective in DNA replication and the use of reverse genetics to identify and mutate genes encoding proteins thought to participate in DNA replication have yielded insights into proteins required for the initiation and elongation steps of DNA replication.

The fission yeast, *Schizosaccharomyces pombe*, offers many of the same advantages for the study of chromosomal DNA replication as *S. cerevisiae*. A similar plasmid assay has been used to identify ARS elements that appear to be associated with chromosomal replication origins. The sequences required for ARS activity in the plasmid assay have not been dissected thoroughly, but preliminary indications suggest that *S. pombe* ARS elements, like *S. pombe* centromeres, are larger than those of *S. cerevisiae*. In addition, the genes encoding a number of *S. pombe* replication proteins have been isolated and characterized.

Despite the enormous progress in characterizing the replicators and proteins required for yeast chromosomal DNA replication, there is still not an in vitro DNA replication system that depends on bona fide replicators for the initiation of replication. The SV40 in vitro replication system has partially filled the gap caused by the lack of a yeast in vitro

DNA Replication in Eukaryotic Cells

system. As required proteins have been identified in the SV40 system, yeast structural and functional homologs have been identified and the genes encoding them have been cloned. The construction and analysis of mutations in the yeast genes has allowed assessment of the in vivo roles of these proteins.

PROTEINS AT THE REPLICATION FORK

The biochemical properties and structural features of yeast proteins that function at the replication fork and are highly homologous to proteins in other eukaryotes are covered in the section on replication proteins. New insights about the roles of these proteins in DNA metabolism, gained largely through the analysis of conditional mutations in the genes encoding these enzymes in *S. cerevisiae*, are emphasized in this section. Table 1 provides a list of *S. cerevisiae* proteins implicated to function at the replication fork.

DNA Polymerases

A major contribution of the yeast system was the unexpected finding that three DNA polymerases, pol-α, pol-δ, and pol-ε, are all essential for life. Much ongoing research is directed at understanding the essential role(s) of each of these polymerases.

pol-α, which contains an intrinsic primase activity, is required for the initiation of both leading-strand and lagging-strand synthesis in the SV40 in vitro system. This polymerase is thought to synthesize the RNA primer for both the leading and lagging strands and to extend the RNA primer by synthesizing a short initiator DNA (Waga and Stillman 1994). As has been found for other pol-α:primase enzymes, *S. cerevisiae* DNA primase activity copurifies with the tightly associated 58-kD and 48-kD subunit complex, which can be separated from the two larger subunits. Although it has not been possible to separate it in active form from the heterodimeric complex with the 58-kD subunit, free 48-kD subunit present in yeast extracts is capable of RNA primer synthesis (Santocanale et al. 1993). Consistent with this observation, the properties of pol-α immunopurified from strains carrying ts mutations in *PRI1* or *PRI2* suggest that the 48-kD subunit has a major role in primase activity. The same immunopurification experiments suggest that the 58-kD subunit mediates or stabilizes the interaction of the 48-kD subunit with the four-subunit pol-α complex (see Table 1).

DNA polymerase activity is associated with the 167-kD subunit of pol-α. Some ts mutations in the gene encoding this subunit cause an immediate cessation of DNA synthesis at the nonpermissive temperature, suggesting that other DNA polymerases cannot continue synthesis in the absence of pol-α. This "quick-stop" phenotype may reflect strong coupling between the activities of replicative polymerases. Alternatively, these mutations may, at the nonpermissive temperature, cause a conformation change in, or instability of, pol-α that causes the replication fork to fall apart. The gene encoding the catalytic subunit of *S. pombe* pol-α has been cloned and sequenced (Damagnez et al. 1991; Park et al. 1993) and shows substantial similarity to pol-α subunits from other organisms (see Wang, this volume).

No enzymatic activity has been associated with the fourth subunit of pol-α, encoded by the *POL12* gene of *S. cerevisiae*. This subunit appears to mediate the interaction of pol-α:primase with T antigen in the SV40 in vitro system, suggesting a possible role in the initiation of both leading and lagging strands (Collins et al. 1993). The characterization of a strain carrying a ts mutation in *POL12* has revealed that this subunit plays an essential role in an initial stage of DNA replication that is complete before the hydroxyurea-sensitive step, presumably chain elongation (Foiani et al. 1994). This subunit shows a cell-cycle-regulated phosphorylation, becoming dephosphorylated as cells complete mitosis and rephosphorylated at the G_1/S boundary (Foiani et al. 1995). The ts mutant phenotype and the cell-cycle-dependent phosphorylation pattern suggest a regulatory role for this subunit. A speculative model is that the nonphosphorylated form of this subunit is required for loading pol-α:primase at replication origins by direct or indirect interactions with the origin recognition complex (Foiani et al. 1994).

Additional proteins that interact with the pol-α catalytic subunit have been sought by protein affinity chromatography. Six polypeptides were identified that bound to a matrix carrying immobilized pol-α but not to a control matrix (Miles and Formosa 1992a). One of these polypeptides, called POB1 (for *p*olymerase *o*ne *b*inding), is encoded by a gene that was identified in a screen for mutants that cause an increased rate of mitotic loss of chromosome III, *CTF4/CHL15* (Table 1) (Kouprina et al. 1992; Miles and Formosa 1992b). Although *CTF4* is not essential, null mutants have phenotypes suggesting a role in DNA metabolism. These phenotypes include elevated rates of chromosome loss and genetic recombination, enhanced temperature sensitivity of strains carrying mutations in the pol-α catalytic subunit, and accumulation in the population of cells with large buds and undivided nuclei. The other five pol-α-binding polypeptides have not been characterized further.

Table 1 Proteins at the yeast replication fork

Protein	Gene	Gene product	Remarks	References
	POL1 (*CDC17*, *HPR3*)	167-kD catalytic subunit	essential; some ts alleles cause "quick stop" phenotype; some alleles isolated in screen for hyperrecombination	reviewed by Campbell and Newlon (1991)
	POL12	86-kD (apparent) subunit	essential; predicted 79-kD protein; ts mutations cause defect in replication initiation; phosphorylated in cell-cycle-dependent manner	Foiani et al. (1994, 1995)
pol-α	*PRI1*	48-kD subunit	essential; ts mutants defective in DNA synthesis; has primase activity in vitro	Lucchini et al. (1987); Francesconi et al. (1991); Santocanale et al. (1993)
	PRI2	58-kD subunit	essential; tightly associated with 48-kD subunit; ts mutants defective in DNA synthesis; mediates or stabilizes binding of 48-kD subunit to catalytic subunit	Foiani et al. (1989); Francesconi et al. (1991); Longhese et al. (1993)
POB1	*CTF4* (*POB1*, *CHL15*)	105 kD	not essential; interacts in vitro with catalytic subunit of pol-α; mutants show increased rates of chromosome loss and mitotic recombination	Kouprina et al. (1992); Miles and Formosa (1992a,b)
pol-δ	*POL3* (*CDC2*, *HPR6*)	124-kD catalytic subunit	essential; interacts with and processivity stimulated by PCNA; one ts mutant deficient in synthesis of high molecular weight daughter strands; mutations in 3′→5′ proofreading exonuclease domain cause mutator phenotype	Boulet et al. (1989); Sitney et al. (1989); Burgers (1991); Budd and Campbell (1993); Simon et al. (1991); Brown et al. (1993); Morrison et al. (1993)

	?	55-kD subunit	function unknown	Bauer et al. (1988)
PCNA	*POL30*	29-kD monomer	essential; processivity factor for pol-δ and -ε; crystal structure of homotrimer reveals closed ring capable of encircling duplex DNA with striking similarity to *E. coli* pol III β subunit; *pol30* mutations suppress ts *cdc44* (*rfc1*) mutations	Bauer and Burgers (1990); Burgers (1991); Krishna et al. (1994); McAlear et al. (1994)
pol-ε	*POL2*	255-kD catalytic subunit	essential; ts mutants deficient in DNA synthesis; stimulated by PCNA; extracts from ts mutants deficient in excision repair synthesis in vitro; mutations in C-terminus cause deficiency in S-phase checkpoint; mutations in 3′→5′ exonuclease domain cause mutator phenotype	Morrison et al. (1990, 1991); Burgers (1991); Araki et al. (1992); Budd and Campbell (1993); Wang et al. (1993); Navas et al. (1995)
	DPB2	80-kD subunit	essential; ts mutants partially defective in DNA synthesis	Araki et al. (1991a)
	DPB3	30-kD and 34-kD subunits	deletion mutant viable but exhibits increased spontaneous mutation rate	Araki et al. (1991b)
	?	29-kD subunit	function unknown	
RP-A	*RFA1*	70.4-kD subunit	essential; binds DNA independently of other two subunits; ts mutations cause defects in DNA replication and repair; also involved in recombination	Heyer et al. (1990); Brill and Stillman (1989, 1991); Longhese et al. (1994); Smith and Rothstein (1995); Firmenich et al. (1995)

(*Continued on following pages.*)

Table 1 (continued)

Protein	Gene	Gene product	Remarks	References
RP-A	*RFA2*	29.9-kD subunit	essential; cell-cycle-regulated phosphorylation at G_1/S and dephosphorylation at M	Brill and Stillman (1991); Din et al. (1990)
	RFA3	13.8-kD subunit	essential	Brill and Stillman (1991)
RF-C	*RFC1* (*CDC44*)	95-kD Rfc1p	essential; contains DNA-binding domain and shows significant homology with other four RF-C subunits; in combination with other subunits functions to load PCNA at RNA primer terminus	Burgers (1991); Howell et al. (1994); Cullmann et al. (1995)
	RFC2	37.9-kD Rfc3p	essential; significant homology with other yeast and human RF-C subunits	Noskov et al. (1994); Cullmann et al. (1995)
	RFC3	38.2-kD Rfc3p	essential; homology with other RF-C subunits; purified protein has ATPase activity stimulated by ssDNA	Li and Burgers (1994b); Cullmann et al. (1995)
	RFC4	36.2-kD Rfc4p	essential; homology with other RF-C subunits; no ATPase activity found	Li and Burgers (1994a); Cullmann et al. (1995)
	RFC5	39.9-kD Rfc5p	essential; homology with other RF-C subunits	Cullmann et al. (1995)
Chl12p	*CHL12* (*CTF18*)	84-kD	not essential; predicted protein contains NTP-binding domain and shows homology with RF-C subunits; null mutations cause chromosome loss, increased rates of mitotic recombination, slow growth, cold-sensitivity	Kouprina et al. (1994)

Topoisomerase I	*TOP1* (*MAK1*)	90-kD	not essential; deletion mutations cause reduction in growth rate, show transient accumulation of ~5-kb daughter strands during DNA replication; *top1 top2* double mutants defective in DNA elongation	Goto and Wang (1985); Thrash et al. (1984); Brill et al. (1987); Kim and Wang (1989)
Topoisomerase II	*TOP2*	150-kD homodimer	essential; required for decatenation of chromosomes at mitosis; either *TOP1* or *TOP2* required for DNA chain elongation	DiNardo et al. (1984); Goto and Wang (1984); Holm et al. (1985, 1989); Brill et al. (1987); Spell and Holm (1994)
DNA ligase	*CDC9*	87-kD protein	essential; required for completion of DNA replication and DNA repair	Johnston and Nasmyth (1978); Johnston (1979, 1983)
5′→3′-exo-nuclease (flap endonuclease)	*YKL510* (*RTH1*, *RAD27*, *ERC11*)	45-kD protein	not essential, but null mutations cause temperature-sensitive lethality; protein has structure-specific endonuclease and 5′→3′-exonuclease activity; null mutants exhibit lengthened S phase, accumulation of large-budded cells with a single nucleus, and increased plasmid loss rates, MMS sensitivity; mutant allele isolated as synthetic lethal with null mutations of *cln1*, *cln2*	Harrington and Lieber (1994); Sommers et al. (1995); Reagan et al. (1995); Vallen and Cross (1995)
DNA2 helicase	*DNA2* (*DNA154*)	171-kD protein	essential; quick-stop replication defect in permeabilized cells; no high molecular weight DNA synthesized in vivo at nonpermissive temperature; DNA-stimulated ATPase and 3′→5′-helicase activity	Kuo et al. (1983); Budd and Campbell (1995a)

The other two essential DNA polymerases differ from pol-α in several ways. Both pol-δ and pol-ε have intrinsic "proofreading" 3′→5′ exonuclease activities, the processivity of both polymerases is stimulated by PCNA, and neither has DNA primase activity. pol-δ appears to be a two-subunit enzyme (Table 1), but the gene encoding the smaller subunit has not been cloned, and little is known of its function. The *S. pombe* gene encoding the catalytic subunit of pol-δ has been cloned and sequenced (Pignede et al. 1991; Park et al. 1993). Purified pol-ε has five subunits (Table 1); the two largest subunits are encoded by essential genes. The third and fourth subunits are encoded by a single nonessential gene. These two subunits may represent different posttranslational modifications of the primary gene product, or one may be a proteolytic degradation product of the other. The gene encoding the smallest subunit has not been identified.

A critical, as yet unanswered, question is what is the essential, nonoverlapping function of each of these PCNA-dependent polymerases. One possibility is that both polymerases function as part of the replication fork, one responsible for leading-strand synthesis and the second responsible for completing Okazaki fragments on the lagging strand. Consistent with this idea, strains carrying some ts mutations in the catalytic subunit of pol-ε exhibit a "quick stop" elongation defect when shifted to the nonpermissive temperature, and under some conditions ts mutations in the catalytic subunit of pol-δ cause a similar defect (Budd and Campbell 1993). However, pol-δ is clearly able to complete both leading- and lagging-strand replication in the SV40 system (Waga and Stillman 1994). A second possibility is that pol-ε plays an essential role in DNA repair, perhaps in correcting errors made by the replicative polymerase. Consistent with a role for pol-ε in repair, extracts prepared from *pol2* mutants are inactive in base excision repair synthesis in vitro (Wang et al. 1993). However, either pol-δ or pol-ε is sufficient for the in vivo repair of UV damage (Budd and Campbell 1995b).

Whatever the essential role(s) of its catalytic activity, the catalytic subunit of pol-ε also has been implicated strongly as a component of the cell-cycle checkpoint that prevents entry into mitosis in the presence of incompletely replicated or damaged DNA, the S-phase checkpoint (Navas et al. 1995). Mutations causing defects in this checkpoint map to the carboxyl terminus of the protein, in a domain distinct from the catalytic domain. The checkpoint-defective mutants are temperature-sensitive for growth and have defects in DNA replication, suggesting that the carboxyl terminus is also required for DNA synthesis. It has been proposed that the role of pol-ε in the checkpoint is to act as a sensor of

DNA replication that coordinates the transcriptional and cell-cycle responses to replication blocks.

Proliferating Cell Nuclear Antigen

PCNA, encoded by the essential *POL30* gene of *S. cerevisiae* (Table 1) and the *pcn1*$^+$ gene of *S. pombe* (Waseem et al. 1992), is a processivity factor for both pol-δ and pol-ε (see Table 1). Although PCNA shows no amino acid sequence similarity to the β subunit of *Escherichia coli* DNA polymerase III, and is a homotrimer rather than a homodimer, the crystal structures of these two proteins are strikingly similar. The trimer is a ring-shaped complex with a central hole large enough to accommodate a DNA duplex. These proteins, along with the bacteriophage T4 gene 45 protein, thus function as sliding clamps that anchor their interacting polymerases to DNA (for review, see Stillman 1994). These sliding clamps normally require assistance from additional factors for loading onto DNA, which likely requires opening of the ring. Both biochemical and genetic analysis implicates replication factor C (RF-C) as the protein required to load PCNA (for review, see Krishna et al. 1994). The genetic evidence is that mutations in *POL30* suppress the phenotype of ts mutations in the gene encoding the large subunit of RF-C, *CDC44* (McAlear et al. 1994; Ayyagari et al. 1995). These mutations change amino acids that lie throughout the entire protein and fail to identify a small domain of interaction between PCNA and RF-C. The suppression of phenotypes resulting from defective RF-C complexes by these PCNA mutations is likely to result from weakened interactions between PCNA subunits that make the circular clamp easier to open for loading onto DNA.

PCNA also interacts with proteins that are not part of the replication apparatus. The human cyclin-dependent protein kinase inhibitor, p21, binds to PCNA and inhibits its DNA replication functions but not its DNA repair functions (Flores-Rozas et al. 1994; Li et al. 1994; Waga et al. 1994). The identification of mutant derivatives of PCNA that support normal growth rates and interact normally with RF-C and pol-δ and pol-ε, but show defects in one or more DNA repair processes, strongly suggests that PCNA interacts with repair-specific protein(s) (Ayyagari et al. 1995).

RF-C

Like the bacteriophage T4 accessory protein complex gp44/gp62 and the *E. coli* γτ complex, RF-C appears to bind to primer/template structures and load the ring-shaped sliding clamp, PCNA. The five RF-C subunits show considerable amino acid sequence similarity, including a putative purine nucleotide-binding region (see Table 1 for references). Moreover, the five RF-C subunits show significant homology with the functionally related proteins of bacteriophage T4 (gp44) and *E. coli* (the γτ subunits of the DNA polymerase III holoenzyme), and the large subunit, encoded by the *RFC1* gene, contains a region of significant homology with prokaryotic DNA ligases and poly(ADP-ribose)-polymerases of eukaryotes (Cullmann et al. 1995). It is not clear whether all five subunits of RF-C associate in a single complex, or whether several complexes with different subunit compositions act in different aspects of DNA replication or repair. Another unresolved issue is what is (are) the specific function(s) of each RF-C subunit. Finally, the observation that loss of the *CHL12* gene product, which shows significant homology with subunits of RF-C, results in a cold-sensitive phenotype consistent with a leaky defect in DNA replication raises the possibility that additional RF-C-like proteins function in some aspect of DNA replication (Kouprina et al. 1994).

Replication Protein A

RP-A is a heterotrimeric single-stranded DNA-binding protein required for DNA replication, recombination, and repair (see Table 1). It appears to be the functional analog of T4 gene 32 protein and *E. coli* SSB. Although each of the three subunits is encoded by an essential gene, the activities of the subunits are poorly understood. The large subunit is capable of binding DNA independently of the other two subunits (Brill and Stillman 1989), and specific mutant alleles of *RPA1* implicate this subunit in interactions with recombination and repair enzymes (Firmenich et al. 1995; Smith and Rothstein 1995). The 34-kD subunit is phosphorylated in a cell-cycle-dependent manner (Din et al. 1990), suggesting a regulatory activity. The inability of yeast RP-A to efficiently substitute for human RP-A in the complete SV40 replication system suggests that RP-A makes specific protein contacts with one or more subunits of the replication complex (Brill and Stillman 1989).

Topoisomerases

The unwinding of DNA in advance of replication forks introduces positive supercoils, which must be removed by a swivel. The phenotypes of topoisomerase mutants suggest that either a type I or a type II topoisomerase can provide the essential function (see Table 1). The observation that *top1* mutants of *S. cerevisiae* transiently accumulate relatively short daughter strands during DNA replication suggests that topoisomerase I normally provides the replication swivel (Kim and Wang 1989).

A second topological problem resulting from DNA replication occurs near the end of the elongation step, when replication forks converge. The sister duplexes either end up as a pair of catenated helices that must be resolved by a type II topoisomerase, or as a pair of gapped molecules that can be resolved by either a type I or a type II topoisomerase (for review, see Wang 1991). Recent analysis of a *top3* mutant of budding yeast, deficient in a type I topoisomerase, and an extragenic suppressor of the mutant, *sgs1*, which encodes a putative helicase, is consistent with the notion that this pair of gene products has a role in the resolution of daughter helices following the completion of replication (Gangloff et al. 1994).

In both fission and budding yeast, the essential function of topoisomerase II appears to be in the untangling of sister chromatids at mitosis, where it plays a role in resolving tangles during both chromosome condensation and segregation (Holm et al. 1985, 1989; Uemura and Yanagida 1986; Funabiki et al. 1993; Spell and Holm 1994). It is interesting that Sgs1p also interacts with topo II, and that *sgs1* null mutants have a chromosome segregation defect, implicating this putative helicase in chromosome segregation as well as in the completion of DNA replicons (Watt et al. 1995).

Other Proteins

Several other proteins, including an enzyme that removes RNA primers, a DNA polymerase to fill in the gaps left by removal of primers, a DNA ligase to close single-strand nicks, and one or more helicases to drive movement of the replication fork, have a role at the replication fork. In *E. coli*, the 5′→3′ exonuclease activity of DNA polymerase I appears to remove the RNA primers attached to the 5′ ends of newly replicated DNA (Konrad and Lehman 1974). In the SV40 in vitro system, both an RNase H and a 45-kD 5′→3′ exonuclease, probably the FEN-1 nuclease, are required for primer removal (Ishimi et al. 1988; Goulian et al. 1990; Turchi and Bambara 1993; Turchi et al. 1994). Deletion of the yeast homolog of FEN-1 nuclease, encoded by the *YKL510* gene, causes phenotypes consistent with defects in both DNA replication and DNA repair (Table 1). The observation that *ykl510* null mutations cause temperature-sensitive lethality suggests that another protein can substitute for the *YKL510* gene product at 30°C but not at 37°C.

The genes encoding DNA ligase, *CDC9* of *S. cerevisiae* and *cdc17*$^{+}$ of *S. pombe*, were among the first cell-cycle genes to be associated with a protein product (Nasmyth 1977; Johnston and Nasmyth 1978). The yeast DNA ligases show significant homology with each other and with human DNA ligase I (Barker et al. 1987) and also have catalytic properties very similar to the mammalian DNA ligase I enzyme (Tompkinson et al. 1992). Temperature-sensitive DNA ligase mutants are defective in DNA replication, failing to join Okazaki fragments and completed replicons (Johnston and Nasmyth 1978; Johnston 1983), and are also defective in DNA repair and mitotic recombination (for review, see Johnston 1983). No activities corresponding to mammalian DNA ligases II and III have been reported in yeast.

A strong candidate for a replicative helicase has recently been identified as the product of the *DNA2* gene (Table 1) (Budd and Campbell 1995a). This protein has both a DNA-stimulated ATPase activity and $3' \rightarrow 5'$ helicase activity that depends on the ATPase. Interestingly, it copurifies with a nuclease that shows a substrate specificity similar to the nuclease encoded by the *YKL510* gene, and the temperature sensitivity of the *dna2* mutant is suppressed by a high-copy-number plasmid carrying *YKL510*, suggesting that the copurification may reflect a physiologically significant interaction between these two proteins. A number of other helicases have been identified in yeast, but their specific roles in DNA replication have not yet been identified (Li et al. 1992; Bean et al. 1993; Shimizu and Sugino 1993).

ORIGINS OF REPLICATION

It has been possible to define chromosomal origins of replication more precisely in yeasts than in any other eukaryotic organism. The replicator sequences necessary for replication origin activity were identified by their ability to promote the extrachromosomal maintenance of plasmids in *S. cerevisiae* and called autonomously replicating sequence (ARS) elements (for review, see Campbell and Newlon 1991). *S. pombe* sequences with similar properties were identified soon thereafter (Beach and Nurse 1981). The development of 2D gel methods for the analysis of replication intermediates made it possible to demonstrate that replication origins coincide with ARS elements in plasmids (Brewer and Fangman 1987; Huberman et al. 1987) and in chromosomes (for review, see Newlon and Theis 1993).

ARS Structure

Our current view of ARS structure, gained from the detailed analysis of a small number of ARS elements, is summarized in Figure 1. An essential region of approximately 17 bp, called domain A, contains a match to an 11-bp sequence (5′-[A/T]TTTA[T/C][A/G]TTT[A/T]-3′), the ARS consensus sequence (ACS). This ACS serves as a point of reference for other essential or stimulatory sequences, with domain B on the 3′ side of the T-rich strand of the ACS and domain C on the 5′ side. Deletion analysis has revealed that the minimal sequences required for ARS activity, defined by the ability to promote high-frequency transformation and extrachromosomal maintenance of plasmids, include domain A and a variable number of nucleotides in domain B. Additional sequences in domain B, and sometimes in domain C, increase the efficiency of ARS function, measured by plasmid stability assays.

The ACS is the only highly conserved DNA sequence in ARS elements, and the sequence conservation reflects its functional significance. Point mutations in the ACS either abolish or reduce ARS activity (Van Houten and Newlon 1990; Rivier and Rine 1992; Li and Herskowitz 1993). Although ARS elements almost always contain multiple matches to the ACS, usually a single ACS is essential, with mutations in the other matches having little or no effect on ARS activity (Marahrens and Stillman 1992; Huang and Kowalski 1993; Miller and Kowalski 1993; Shirahige et al. 1993; Rao et al. 1994; Theis and Newlon 1994). However, in 3 of the 21 ARS elements studied, either of 2 overlapping or closely spaced ACSs can provide the essential function (Van Houten and Newlon 1990; Shirahige et al. 1993; J.F. Theis and C.S. Newlon, in prep.). In about half of the ARS elements studied, the essential ACS is an exact match to the consensus; most others match at 10 of 11 positions (for review, see Newlon and Theis 1993). The extreme cases are *ARS121* and *ARS309*, which have essential 9 of 11 matches to the ACS (Walker et al. 1990; J.F. Theis and C.S. Newlon, in prep.). Domain A is the core of the recognition region of a multiprotein complex, origin recognition complex (ORC) (see below).

In contrast to domain A, there is no highly conserved DNA sequence within domain B. However, linker scan analysis of several ARS elements suggests that domain B is modular. *ARS1* contains 3 domain B elements, B1, B2 and B3, in which linker substitutions reduce ARS activity (Marahrens and Stillman 1992). The presence of any 2 of these 3 elements is sufficient for ARS activity. *ARS307* contains 2 domain B elements (Rao et al. 1994; Theis and Newlon 1994), and a similar pattern of elements is apparent in the linker scan analysis of 3 ARS elements from chromosome VI, *ARS605*, *ARS607*, and *ARS609* (Rashid et al. 1994). Although the B elements of *ARS307* share little or no sequence homology with the B elements of *ARS1*, the B1 elements of the 2 ARSs are functionally interchangeable, as are the B2 elements (Rao et al. 1994). As described below, the B1 and B3 elements function as protein-binding sites, but the function of the B2 element is unclear.

A saturation mutagenesis of the B1 element of *ARS1* revealed that point mutations in either of two adjacent base pairs caused dramatic decreases in plasmid stability (Rao et al. 1994). The B1 element of *ARS307* contains a pair of identical base pairs at a similar distance from the ACS which are important for B1 function (Rao et al. 1994; Theis and Newlon 1994). This region of the B1 element is important for ORC binding, both in vitro and in vivo (Rao and Stillman 1995; Rowley et al. 1995). Interestingly, the analysis of point mutations in the B1 element of

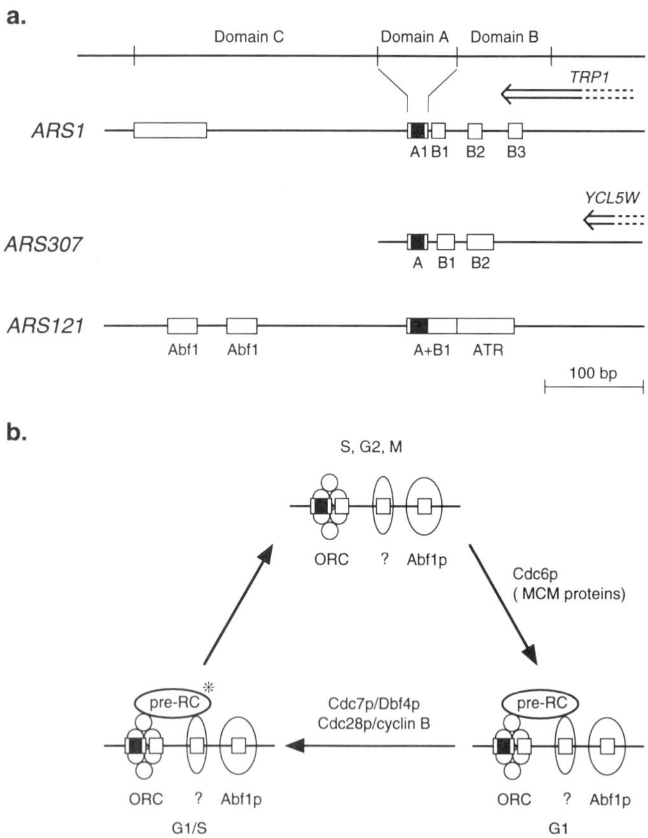

Figure 1 (*See facing page for legend.*)

ARS1 revealed that mutations in only one of the two nucleotides with significant effects on B1 function reduced ORC binding, suggesting that this element has other functions in replication initiation beyond its role in ORC-DNA interactions (Rao and Stillman 1995).

The B3 element of *ARS1* is the binding site for the multifunctional transcription and replication protein, ARS-binding factor 1 (Abf1p, also called OBF1; see ARS-binding factors below). Abf1p binding sites are found in some but not all ARS elements, where they function in a position- and orientation-independent fashion as replication enhancers (Walker et al. 1990, 1991). The B3 element of *ARS1* can be replaced by binding sites for other transcription factors, Gal4p and Rap1p, suggesting that other transcriptional activators might serve as replication enhancers (Marahrens and Stillman 1992).

The role(s) of the B2 element is not yet clearly defined. Although 8-bp linker substitutions in the B2 elements of *ARS1* and *ARS307* affected ARS activity, single base-pair changes in the B2 element of *ARS1* showed no phenotype (Marahrens and Stillman 1992; Rao et al. 1994;

Figure 1 ARS structure. (*a*) Schematic drawings of three *S. cerevisiae* ARS elements. Boxes represent functional elements identified by mutational analysis. The essential ACS of each ARS element is represented by the filled box within domain A. The domain A and domain B elements of *ARS1* were defined by Marahrens and Stillman (1992) and domain C by Strich et al. (1986). *ARS307* was dissected by Palzkill and Newlon (1988), Van Houten and Newlon (1990), Theis and Newlon (1994), and Rao et al. (1994). The *ARS121* diagram is based on data from Walker et al. (1991). The core region of *ARS121*, identified by deletion analysis, contains the essential ACS and, presumably, the B1 element. The AT-rich region (ATR) stimulates the activity of the core, and by analogy with *ARS1* and *ARS307*, is likely to contain the B2 element. (*b*) Cell-cycle-regulated protein interactions with *ARS1*. Diagram is based on data from Diffley et al. (1994) and Cocker et al. (1996). ORC, shown as a six-subunit protein, and Abf1p, shown as a monomer, are bound throughout the cell cycle. Origin activation appears to be a two-step process. In the first step, at the end of mitosis, additional factors are recruited to *ARS1* to form the prereplicative complex (pre-RC). The assembly and maintenance of the pre-RC depends on Cdc6p, which is likely to be a component of the pre-RC. In the second step, at the transition from G_1 to S phase, the pre-RC is activated (indicated by an asterisk) and replication initiates. Activation of the pre-RC requires the activity of two protein kinases (Cdc7p and Cdc28p) and their regulatory subunits (Dbf4p and a B-type cyclin). During S phase, the additional factors that form the pre-RC are dissociated and/or degraded. The factor shown interacting with the B2 element is hypothetical (see text).

Theis and Newlon 1994). These observations are consistent with a role for B2 as a structural element or as a protein-binding site with loose sequence specificity. One attractive hypothesis is that B2 functions as a DNA unwinding element (DUE). In well-studied replication origins, replication initiation is facilitated by the unwinding of a small region adjacent to the initiator protein-binding site where the remainder of the replication apparatus assembles (for review, see Kornberg and Baker 1992). The observed correlation between the progressive loss of ARS activity and increases in helical stability caused by deletions into domain B, and the demonstration that domain B can be functionally substituted by heterologous, easily unwound sequences, support the idea that a DUE is contained within domain B (Umek and Kowalski 1988; Natale et al. 1992). However, two lines of evidence suggest that B2 has a function beyond acting as a DUE. First, some linker scan mutations in B2 that reduce plasmid stability are not predicted to affect the helical stability of domain B (Theis and Newlon 1994). Second, mutations in B2 reduce the ability of *ARS1* to recruit Dbf4p (see below), suggesting that Dbf4p interacts directly or indirectly with B2 (Dowell et al. 1994).

Fewer than half of the ARS elements characterized have stimulatory sequences in domain C. The Abf1p binding sites in *ARS121* are the only well-studied domain C elements (Walker et al. 1990, 1991). The domain C sequences of *ARS1* are located approximately 200 bp from domain A and are not well characterized (Strich et al. 1986). Five of the nine chromosome VI ARS elements have been reported to have stimulatory sequences in domain C, but only in *ARS605* has the position of these sequences been defined (Shirahige et al. 1993; Rashid et al. 1994).

S. pombe ARS elements have not been studied as extensively as those of *S. cerevisiae*. An 11-bp consensus sequence, the PACS, was found by the comparison of the DNA sequences of nine ARS-containing fragments (Maundrell et al. 1988). However, unlike the *S. cerevisiae* ACS, this sequence does not appear to play an essential role in ARS function (Maundrell et al. 1988; Zhu et al. 1994). Both the sizes of *S. pombe* ARS-containing fragments and deletion analysis of two *S. pombe* ARS elements, *ARS3002* and *ARS3003*, suggest that the sequences required for full ARS activity are approximately ten times the length of *S. cerevisiae* ARS elements (Maundrell at al. 1988; Zhu et al. 1994). As in *S. cerevisiae*, *S. pombe* ARS elements clearly function as replication origins on plasmids (Caddle and Calos 1994; Wohlgemuth et al. 1994).

ARS elements have also been identified in chromosomal DNA of *Kluyveromyces lactis* (Fabiani et al. 1990). The detailed analysis of one of these ARSs revealed an essential core of approximately 40 bp, containing a sequence related to the ACS of *S. cerevisiae*, and flanked by sequences that stimulate activity (Fabiani et al. 1996).

Chromosomal Replication Origins

All *S. cerevisiae* chromosomal replication origins, identified by 2D gel analysis, coincide with known ARS elements. Moreover, no additional origins were found in systematic searches for origins of replication not associated with ARS elements on chromosome III (Greenfeder and Newlon 1992; Huberman et al. 1992; Collins and Newlon 1994). The well-documented finding that the effects of mutations that reduce or abolish ARS activity on plasmids have corresponding effects on chromosomal origin activity demonstrates directly that ARS elements are the *cis*-acting sequences required for chromosomal replication origin activity (Deshpande and Newlon 1992; Rivier and Rine 1992; Huang and Kowalski 1993; Marahrens and Stillman 1994; Theis and Newlon 1994). Similarly, the deletion of ARS elements from near the *ura4* locus of *S. pombe* was found to inactivate replication initiation events in their vicinity (Dubey et al. 1994).

The issue of the exact location relative to ARS elements of the actual replication initiation sites, i.e., the positions of leading-strand and lagging-strand primers, is not clearly resolved. If the B2 element is actually a DUE, then initiations would be expected to occur in and around B2. The accuracy of 2D gel origin mapping techniques is sufficient only to place initiation sites within a few hundred base pairs of ARS elements. Preliminary data suggest that the 5′ ends of leading strands map primarily to domain B of *ARS307* (S. Jenab and C.S. Newlon, in prep.).

ARS-binding Proteins

The activity of ARS elements as chromosomal replicators is almost certainly mediated by proteins that interact with them. Consistent with this idea, positioning a nucleosome over the ACS of *ARS1* reduces plasmid stability, presumably by blocking access of other proteins to the ARS element (Simpson 1990). The extreme sensitivity of the ACS to point mutations makes it the best candidate binding site for an initiator protein. After more than 10 years of effort in several laboratories, the multisubunit ORC that binds to the ACS was identified 3 years ago (Bell and Stillman 1992). The second DNA-binding protein that interacts with ARS elements, Abf1p, binds to the B3 element.

ORC

Initially identified by DNA footprinting assays on fractionated *S. cerevisiae* nuclear extracts, this protein with subunits of 120, 72, 62, 56, 53, and 50 kD requires ATP for DNA binding (Bell and Stillman 1992). It has been estimated that there are approximately 600 copies of Orc2p per cell (Rowley et al. 1995), which translates to approximately one ORC per replication origin. The ORC footprint includes domain A and extends into domain B of several ARS elements. The features of the footprint include protection of the domain A sequence and a strong hypersensitive site in element B1, with additional hypersensitive sites at approximately 10-bp intervals extending toward the B2 element (Bell and Stillman 1992; Bell et al. 1993; Micklem et al. 1993). The observed disruption of ORC binding by mutations in the ACS that abolish replicator activity and the reduction in the efficiency of ORC binding by mutations in B1 demonstrate the biological significance of the ORC/DNA interaction (Bell and Stillman 1992; Rao and Stillman 1995; Rowley et al. 1995). The similarity of the genomic footprint of proteins bound to *ARS1* and the 2 μm ARS in permeabilized cells to the in vitro footprint of purified ORC adds further support to the idea that ORC binds to replication origins in vivo (Diffley and Cocker 1992; Diffley et al. 1994). The pattern of DNase I hypersensitive sites and the sensitivity of domain B to copper-phenanthroline cleavage suggest that the DNA of domain B is wrapped on the ORC protein surface and is under torsional stress (Diffley and Cocker 1992).

Genetic analysis of ORC mutants has provided both additional support for the function of ORC in the initiation of DNA replication and evidence that ORC is involved in the transcriptional silencing of the silent mating loci. *ORC6* was identified in a screen for proteins that interact with the ACS and shows genetic interactions with several proteins implicated in the initiation of replication (Li and Herskowitz 1993; see below). Mutations in *ORC2* (Foss et al. 1993; Micklem et al. 1993) and *ORC5* (Loo et al. 1995) cause a plasmid maintenance defect that can be suppressed by additional replication origins, and ts *orc2* mutants are defective in the G_1/S transition at the nonpermissive temperature (Bell et al. 1993). Moreover, the efficiency of initiation at chromosomal replication origins, determined by 2D gel analysis, is reduced in *orc2* and *orc5* mutants at the permissive temperature and is further reduced at the nonpermissive temperature (Fox et al. 1995; Liang et al. 1995).

A role for ORC in transcriptional silencing was made obvious by the isolation of the *orc2* and *orc5* mutants discussed above in screens that made use of the *HMR E* silencer element that participates in the transcriptional repression (silencing) of the *HMR* locus (Foss et al. 1993; Micklem et al. 1993; Loo et al. 1995). One of the three *cis*-acting sequences required for the function of this silencer is an ACS, and the silencer is active as a chromosomal replication origin (Rivier and Rine 1992). In addition to their defects in DNA replication, these Orc$^-$ mutants are defective in transcriptional silencing. Two lines of evidence suggest that the roles of ORC in DNA replication and silencing are independent. First, the *HMR E* silencer can be functionally replaced by Gal4p DNA-binding sites when a hybrid Sir1p-Gal4p DNA-binding domain protein is expressed (Chien et al. 1993). This observation suggests that the role of ORC and other silencer-binding proteins may be to recruit Sir1p to the silencer. Second, alleles of *ORC5* have been isolated that are proficient for DNA replication but defective in silencing (Fox et al. 1995).

Aside from its DNA-binding activity, no biochemical activities have been attributed to ORC. ATP or an ATP analog with a hydrolyzable β-γ bond is required for DNA binding by ORC, suggesting that ATP hydrolysis might be required (Bell and Stillman 1992). However, no ATPase activity has been identified. SV40 T antigen and *E. coli* DnaA protein both require ATP for untwisting DNA (Bramhill and Kornberg 1988; Borowiec et al. 1990). Orc5p has a predicted guanine nucleotide-binding site, but mutation of a highly conserved lysine that alters ATP binding or hydrolysis by other proteins with a similar binding site was not lethal (Loo et al. 1995). The mutant strain did exhibit slow growth at elevated temperatures, suggesting that the putative ATP-binding motif plays a role in Orc5p function.

ARS-binding Factor 1

Abf1p was identified in several laboratories on the basis of its binding to ARS elements, transcriptional silencers, or promoters (for review, see Campbell and Newlon 1991). In *ARS1* it interacts with the B3 element to stimulate replicator activity (Diffley and Stillman 1988; Marahrens and Stillman 1992), and it has been shown to function as a replication enhancer at *ARS121* (Walker et al. 1990). Not all ARS elements have Abf1p binding sites. Binding sites for this protein have also been demonstrated to function in transcriptional activation, transcriptional repression, and plasmid segregation (for review, see Campbell and Newlon 1991). The observations that *abf1* mutants show defects in plasmid maintenance (Rhode et al. 1992) and that genomic footprints of the B3 element of *ARS1* look similar to footprints with purified Abf1p in vitro (Diffley and

Cocker 1992; Rowley et al. 1995) strongly suggest that Abf1p interacts with ARS elements in vivo.

It is not clear how Abf1p enhances replicator activity. Its binding could have a direct role, for example, by inducing a conformational change in the ARS element or by interacting with another component of the replication apparatus. The observation that the mutations in element B3 of *ARS1* have no effect on ORC binding in vitro or in vivo indicates that Abf1p is unlikely to interact with ORC (Rao and Stillman 1995; Rowley et al. 1995). The possibility that RP-A is a target is suggested by the observation that the Gal4p and VP16 transcription factors interact with RP-A in vitro (He et al. 1993; Li and Botchan 1993). Consistent with this idea, mutations in the B3 element of *ARS1* appear to affect a late step in replicator activation, acting after the assembly of the prereplicative complex discussed below (Rowley et al. 1995). Alternatively, Abf1p could play an indirect role.

Other Proteins?

Using *ARS121* as a target, Eisenberg and colleagues have sought ARS-binding proteins. In addition to Abf1p (Eisenberg et al. 1988; Francesconi and Eisenberg 1991), two additional factors have been identified that are required for the formation of a maximally retarded *ARS121* complex in gel shift assays (Estes et al. 1992). OBF2 interacts with Abf1p and DNA to form a complex of intermediate mobility. In the presence of ATP, this complex is then competent to bind the third factor, core binding factor (CBF). The requirement of ATP and an ACS for the binding of CBF to *ARS121* suggests that CBF may be the same as ORC. However, in contrast to the reported properties of CBF, the binding of ORC to DNA does not require Abf1p (Rao and Stillman 1995; Rowley et al. 1995).

The B2 element is also a potential protein-binding site. The lack of DNA sequence conservation in this region of ARS elements, and the insensitivity of the B2 element of *ARS1* to point mutations, suggest that any proteins binding to B2 should have little sequence specificity. However, interaction with ORC could provide the specificity needed to recruit such a protein to replicators.

Other Proteins Required for Replication Initiation

The clarity of the ORC genomic footprint in cells from asynchronous cultures indicated that ORC is bound to replicators throughout most of the cell cycle (Diffley and Cocker 1992), suggesting that replication initiation is unlikely to be mediated by the cell-cycle-regulated assembly of ORC at replication origins. It is possible that ORC activity is regulated by a cell-cycle-dependent modification. The presence of potential cyclin-dependent kinase phosphorylation sites in Orc2p and Orc6p (Li and Herskowitz 1993; Micklem et al. 1993) and the recent demonstration that B-type cyclins encoded by *CLB5* and *CLB6* normally trigger S phase (Epstein and Cross 1992; Schwob and Nasmyth 1993; Schwob et al. 1994; see Nasmyth, this volume) suggest one attractive model for the regulation of ORC activity.

Alternatively, one or more additional proteins may interact with ORC to regulate its activity. Evidence in support of this possibility is provided by the observation that the genomic footprints at *ARS1* and the 2 μm ARS change as a function of the cell cycle (Diffley et al. 1994). The footprints during S, G_2, and early M resemble the footprints generated by purified ORC and Abf1p in vitro. Beginning at anaphase and continuing through G_1, an additional region of protection appears that overlaps the ORC footprint and extends well into the B2 element, suggesting that additional protein(s) assembles on replicators at mitosis and remains there until S phase is triggered. The kinetics of appearance and disappearance of this prereplicative complex fit very well with observed properties of "licensing factor" in *Xenopus* and the presence of an S-phase-promoting factor deduced from mammalian cell fusion experiments (for review, see Su et al. 1995).

Excellent candidates for proteins that regulate the initiation of replication and models for how they work have recently emerged from studies of gene products identified by several genetic screens designed to target replication proteins.

CDC7 *and* DBF4

CDC7 encodes a 58-kD protein kinase having homology with cyclin-dependent protein kinases (Patterson et al. 1986). Although its target(s) has not been identified, Cdc7p is capable of phosphorylating both itself and histone H1 in vitro. It is required for several aspects of DNA metabolism, including mitotic DNA replication, meiotic DNA recombination, and replication-dependent DNA repair (for review, see Sclafani and Jackson 1994). Cdc7p appears to act just at the G_1/S boundary in mitotic cells, after completion of the protein synthesis required for DNA replication but before replication intermediates are produced (Hereford and Hartwell 1974; Petes and Newlon 1974). Although the level of Cdc7p is constant, its kinase activity is regulated during the cell cycle and peaks at the G_1/S boundary (Jackson et al. 1993; Yoon et al. 1993).

Both genetic and biochemical data suggest that Cdc7p interacts with the product of the *DBF4/DNA52* gene, mutations in which cause a cell-cycle arrest similar to *cdc7* (Johnston and Thomas 1982a,b; Soloman et al. 1992). *DBF4* acts as a multicopy suppressor of *cdc7*, and vice versa, and the *cdc7dbf4* double mutant is inviable (Kitada et al. 1992). Cdc7p kinase activity is reduced in *dbf4* mutants, and Dbf4p and Cdc7p physically interact (Kitada et al. 1992; Jackson et al. 1993; Dowell et al. 1994). In contrast to *CDC7*, *DBF4* is periodically expressed, with mRNA levels peaking at the G_1/S boundary (Chapman and Johnston 1989). These observations suggest that Cdc7p kinase activity could be activated by association with Dbf4p in a manner similar to the activation of the Cdc28p kinase by association with cyclins.

The interaction of the Cdc7p kinase with Dbf4p is of particular interest because Dbf4p has recently been shown to interact with ARS elements (Dowell et al. 1994). The latter interaction was found in a "one-hybrid" genetic screen for hybrid proteins fused to the Gal4p activation domain capable of activating transcription of a reporter gene placed downstream from an ARS element. The recruitment of Dbf4p to *ARS1* requires domain A and is reduced by mutations in B1 and B2 but not B3, suggesting that Dbf4p interacts either directly or indirectly with ORC. The domain of Dbf4p that interacts with ARS elements is separable from the domain that interacts with Cdc7p, indicating that Dbf4p is likely to recruit Cdc7p to replication origins. Taken together, these observations suggest that the target of the Cdc7p kinase is a protein in prereplicative complexes whose phosphorylation may trigger the initiation of replication.

A *CDC7* homolog has recently been identified in *S. pombe*, and named *hsk1*$^+$ (Masai et al. 1995). Disruption of the gene is lethal, and analysis of germinating spores carrying the disruption suggests that DNA replication is inhibited. Moreover, a fraction of these spores undergo an aberrant mitosis, suggesting either that Hsk1p is required for inhibiting mitosis until S phase is completed or that it has an additional role in the proper execution of mitosis. The isolation of conditional alleles of *hsk1* should help to resolve this issue.

CDC6

Cdc6p was initially implicated in the initiation of DNA replication by two observations. First, early reciprocal shift experiments suggested that its function was required for an early event in DNA replication, prior to the hydroxyurea-sensitive step (Hartwell 1976). Second, *cdc6* mutants exhibit a plasmid maintenance defect that can be suppressed by the inclusion of multiple ARS elements on the plasmid (Hogan and Koshland 1992). Recently, 2D gel analysis has been used to demonstrate directly that replication initiation at chromosomal origins is inefficient in *cdc6* mutants at the permissive temperature, and that the defect is more pronounced after a short time at the nonpermissive temperature (Liang et al. 1995). The inefficient initiation of replication observed in a ts *cdc6* strain is likely to result from partial function of the mutant protein, because the depletion of Cdc6p results in a strikingly different phenotype. In this case, both *S. pombe* strains depleted of Cdc18p and *S. cerevisiae* strains depleted of Cdc6p undergo a transient arrest at the G_1/S boundary, and then undergo an aberrant mitosis in the absence of DNA replication that results in cell death (Kelly et al. 1993; Piatti et al. 1995). Therefore, it is likely that Cdc6p and its *S. pombe* homolog Cdc18p are required both for the initiation of DNA replication and to delay mitosis until S phase is completed.

Genetic interactions between *CDC6* and *ORC6* (Li and Herskowitz 1993), the synthetic lethality of *cdc6* and *orc5* or *orc6*, and the identification of *CDC6* as a multicopy suppressor of *orc5-1* (Liang et al. 1995) suggested that Cdc6p might interact with ORC. Although the physical interaction of ORC with Cdc6p has not been directly demonstrated in yeast cells, the observation that Cdc6p expressed in insect cells could be immunoprecipitated by monoclonal antibodies against ORC following addition of purified ORC to the insect cell extracts provides biochemical evidence that the proteins interact (Liang et al. 1995).

CDC6 is expressed periodically during the cell cycle, primarily during late mitosis, but with a second burst of synthesis late during G_1 (Zhou and Jong 1990; Bueno and Russell 1992; Zwerschke et al. 1994; Piatti et al. 1995). Constitutive expression of *CDC6* at high levels causes a delay in entry into mitosis, indicating that the periodic expression pattern is important for normal function (Bueno and Russell 1992). The mitotic delay is not seen if *CDC6* is overexpressed by placing the gene under the control of its own promoter on a high copy plasmid, implying that Cdc6p either is unstable or is specifically degraded at some point in S or G_2 phase. Recent analysis using epitope-tagged Cdc6p has demonstrated that Cdc6p is unstable and disappears soon after the beginning of S phase (Piatti et al. 1995).

The timing of its expression and its interaction with ORC suggest that Cdc6p could be a component of, or required for, the assembly of the prereplicative complex assembled at replication origins. The finding that a *cdc6* mutant gives a postreplicative footprint at the nonpermissive temperature is consistent with this idea (Diffley et al. 1994), which is further supported by the observation that depletion of wild-type Cdc6p causes a defect in the formation and maintenance of the prereplicative complex (Cocker et al. 1996). The biochemical function of Cdc6p is unclear. It is predicted to be a 58-kD protein with a purine nucleotide-binding site (Zhou et al. 1989), and it has been reported to bind and hydrolyze ATP and GTP in a DNA-independent reaction (Zwerschke et al. 1994).

MCM Proteins

Genetic screens for mutations that influence plasmid stability have identified a number of candidates for proteins that function in the initiation of DNA replication (for review, see Campbell and Newlon 1991). Those most clearly implicated are a subset of the minichromosome maintenance (*mcm*) mutants that decrease the stability of plasmids carrying different ARS elements to varying extents (Maine et al. 1984; Gibson et al. 1987). These ARS-specific mutants identified a family of genes encoding related proteins and a gene encoding a transcription factor, Mcm1p.

The family of related proteins now includes the products of five *S. cerevisiae* genes, *MCM2*, *MCM3*, *CDC46/MCM5*, *CDC47*, and *CDC54* (Hennessy et al. 1990; Yan et al. 1991; Chen et al. 1992; Dalton and Whitbread 1995; Whitbread and Dalton 1995), as well as homologs from *S. pombe* (Coxon et al. 1992; Miyake et al. 1993), *Xenopus* (Coxon et al. 1992; Kubota et al. 1995), and a number of other eukaryotes (for review, see Su et al. 1995). In *S. cerevisiae* each of the five genes encoding members of this family is essential, suggesting that their functions are not completely overlapping. However, genetic interactions, both high copy suppression and synthetic lethality, suggest that the gene products interact with each other (Gibson et al. 1990; Hennessy et al. 1991; Yan et al. 1991) and with ORC (Li and Herskowitz 1993; Liang et al. 1995; Loo et al. 1995).

Several lines of evidence implicate these gene products in DNA replication. In addition to the ARS-specific plasmid maintenance defect, the mutants show hyperrecombination and chromosome loss phenotypes characteristic of DNA replication mutants (Sinha et al. 1986; Yan et al. 1991; Chen et al. 1992). More direct evidence includes the observations that *ARS1* is used inefficiently as a replication origin even at the permissive temperature in *mcm2* and *mcm3* mutants (Yan et al. 1993) and that *cdc46* mutants accumulate chromosomal DNA molecules that are unable to enter pulsed-field gels, almost certainly because they are partially replicated (Hennessy et al. 1991). The intracellular localization of these proteins changes in an interesting fashion as well. They accumulate in the cytoplasm during S and G_2 and enter the nucleus at mitosis, remaining there until the G_1/S boundary, when they are either degraded or become cytoplasmic again (Hennessy et al. 1990; Yan et al. 1993; Dalton and Whitbread 1995).

The predicted amino acid sequence of these proteins places them in a superfamily of ATPases that includes the replication initiator proteins *E. coli* DnaA and SV40 T antigen (Koonin 1993). However, none of these proteins has yet been demonstrated to have ATPase or any other biochemical activity. These proteins are also relatively abundant, making it unlikely that they interact only with replicators.

MCM1 is an essential gene that encodes a transcription factor having homology with human serum response factor (Norman et al. 1988; Passmore et al. 1988, 1989). Mcm1p interacts with at least four other regulatory proteins and possibly also independently to modulate the transcription of diverse genes including cell-type-specific genes and Ty elements (for review, see Kuo and Grayhack 1994). Mcm1p is also implicated in DNA replication by its plasmid maintenance defect and its chromosome loss and hyperrecombination phenotypes (Elble and Tye 1992). Moreover, a ts *mcm1* mutant causes the temperature-dependent accumulation of large budded cells with single nuclei and partially replicated genomes, a phenotype shared with many DNA replication mutants. How the effects of Mcm1p on DNA replication are mediated is not known. One possibility is that it binds to ARS elements, perhaps by interaction with other ARS-binding proteins. Alternatively, it may be required for the expression of one or more proteins required for DNA replication.

Determinants of Origin Use

Whereas ARS elements clearly function as chromosomal replicators, their efficiency of use is highly variable, ranging from initiating replication in every cell cycle to initiating at an undetectable frequency (Brewer and Fangman 1988; Linskens and Huberman 1988; Dubey et al. 1991; Greenfeder and Newlon 1992; Newlon et al. 1993). As described above, mutations in ARS elements and the *trans*-acting factors that interact with them reduce the efficiency with which chromosomal replicators are used. However, the observation that chromosomal replicator efficiency does not correlate with ARS efficiency, measured by plasmid stability, suggests that some aspect of the chromosomal environment beyond the *cis*-

acting sequences intrinsic to ARS elements influences replicator activity. These "chromosomal context" effects can increase or decrease chromosomal replicator efficiency relative to the efficiency of the same replicator in plasmids. For example, *ARS307* functions relatively inefficiently in plasmids (Palzkill and Newlon 1988; Van Houten and Newlon 1990; Theis and Newlon 1994) but initiates replication in virtually every cell cycle in the chromosome (Deshpande and Newlon 1992; Greenfeder and Newlon 1992). In contrast, *ARS301* is an efficient plasmid replicator, but is inactive as a chromosomal replicator (Dubey et al. 1991).

One clear example of the influence of chromosomal context on replicator activity is the phenomenon called origin interference. The basic observation is that placement of efficient replicators close to each other in a chromosome or in a plasmid creates a situation in which replication initiates at only one of the closely spaced replicators in any particular cell cycle. Thus, when *ARS1* was inserted 6.5 kb away from *ARS501* on chromosome V, replication initiated at either *ARS1* or *ARS501* in any given cell, but not at both. A similar result was found for three copies of *ARS1* at the normal position of *ARS1* on chromosome IV; again, replication initiated at only one copy of *ARS1* in any particular cell, but all three copies were used in the population (Brewer and Fangman 1993). A strikingly different result was obtained with a construct containing two copies of *ARS1* at its normal position on chromosome IV (Marahrens and Stillman 1994). In this case, only one of the copies of *ARS1* was used in the population, and the second copy remained silent. An inactivating point mutation in the ACS of the dominant copy of *ARS1* caused the second copy to become active. Origin interference has also been observed in plasmids. From the pattern of plasmid replication intermediates, it had been inferred that only a single replicator was active in multimeric plasmids from yeast (Brewer and Fangman 1987), mammalian cells (Schvartzman et al. 1990), and *E. coli* (Martin-Parras et al. 1992). However, only recently has a plasmid in which the two ARS elements could be distinguished been studied (Brewer and Fangman 1994). In this case, only one of the two equally spaced copies of *ARS1* was used on any given plasmid, with one copy used approximately four times more frequently than the other. The sequences that establish replicator preference were mapped to the *URA3* gene on the plasmid. Thus, it appears that an active replicator can somehow suppress the activity of nearby replicators and that the selection of the active replicator is influenced by local context.

The mechanism of origin interference has not been established. It is possible that an active replicator actually inhibits initiation at a nearby replicator, perhaps by steric hindrance of the formation of a second initiation complex by the first or by transmission of some sort of topological signal along the DNA molecule. Alternatively, the apparent inhibition could simply reflect slight differences in the timing of initiation. In this case, a fork from the origin that was activated first would replicate the second replicator before it was activated. Once the fork had passed, activation of the second replicator would be prevented by the mechanism that ensures that each chromosome is replicated once and only once per S phase.

The extent to which origin interference can account for the inefficient use of some chromosomal replicators and the apparent inactivation of others is not certain. However, the available information suggests that other mechanisms must exist. *ARS308*, which functions inefficiently in the chromosome, initiating replication in only 10–15% of cell cycles, is not activated by deletion of an efficient replicator (*ARS307*) located approximately 5 kb away (Greenfeder and Newlon 1992). Moreover, five inactive ARS elements on the left end of chromosome III are not activated by deletion of all active replicators from the adjacent 200-kb region of the chromosome, demonstrating that they must be inactivated by a mechanism other than origin interference (A. Dershowitz and C.S. Newlon, in prep.). Another important question is the distance over which origin interference can act. The average distance between active replicators in *S. cerevisiae* chromosomes is approximately 40 kb. The distances over which origin interference has been observed do not exceed 6.5 kb.

The active and inactive replicators on chromosome III are not interspersed but, rather, are present in blocks. The left 40 kb of the chromosome contains five inactive replicators, and two other inactive replicators are present in a 20- to 30-kb region in the middle of the right arm (Dubey et al. 1991; Newlon et al. 1993). This arrangement suggests that the chromosome might be organized into domains, some of which are permissive for replicator activity and some not. An attractive explanation for the inactivity of replicators at the left end of the chromosome was based on studies of transcriptional silencing by telomeres and by the silencer elements associated with the silent mating-type loci in *S. cerevisiae*. Mutations in a common set of approximately ten genes, including the genes encoding the core histones and *SIR2*, *SIR3*, and *SIR4*, are known to abrogate silencing by both telomeres and silencer elements (for review, see Laurenson and Rine 1992). The *SIR1* gene product functions at silencer elements but not at telomeres. The observation that mutations in the genes encoding histones abrogate silencing suggests that silencing is mediated through the assembly of a repressive chromatin structure. The presence of both a telomere and a silent mating-type locus, *HML*, at or near the left end of chromosome III suggested that the putative repressive chromatin structure inactivated the replicators. Finding that the replicators on the left end of chromosome III are not active in a *sir1* or a *sir4* mutant or in a circular derivative of the chromosome that lacks telomeres (Dubey et al. 1991) demonstrates that it is unlikely that replicator inactivation is mediated by telomeres or transcriptional silencers. The observation that an active replicator is associated with the *HMR* locus on the right arm of chromosome III (Rivier and Rine 1992) is also inconsistent with inactivation of replicators by silenced chromatin.

Another possibility is that the activity of replicators is influenced by the physiological state of the cell. For example, it is clear that *S. cerevisiae* diploids respond to nitrogen deprivation by undergoing meiosis and sporulation. Several lines of evidence suggested that mitotic S phase and premeiotic S phase were regulated differently, leading to the idea that replicator use might be different in meiotic cells. However, examination of replication origin use on chromosome III revealed that the same origins are used in meiotic S phase and mitotic S phase, demonstrating that the inactive ARS elements at the left end of the chromosome are not meiotic origins (Collins and Newlon 1994).

In the case of *S. pombe*, it is not yet clear to what extent chromosomal replicator use varies. Replication bubble-containing replication intermediates have been found in chromosomal DNA fragments containing each of several different ARS elements (Caddle and Calos 1994; Dubey et al. 1994; Wohlgemuth et al. 1994). In every case examined, the replication intermediates included both Y-shaped molecules and bubble-containing molecules of all sizes, suggesting that replication initiates inefficiently at these replicators. In the single region that has been well-studied, it has been shown that at least three separable ARS elements contribute to the initiation events and that origin interference certainly contributes to the inefficiency of use of individual replicators (Dubey et al. 1994).

In summary, it seems likely that some aspect of structure or context influences chromosomal replicator activity in *S. cerevisiae*. The level at which these effects are mediated is presently unclear. One possible model suggests that *cis*-acting elements recruit activators or repressors that interact with or prevent the formation of prereplicative complexes at replicators. Other alternatives are that chromatin structure or chromosome organization within the nucleus functions as a mediator of these effects. A clear challenge for the future is to understand the mechanism(s) of these controls.

Temporal Control of Replication

It had been known for many years that there is a reproducible temporal pattern of replication of mammalian chromosomes (see Simon and Cedar, this volume). Because intact *S. cerevisiae* chromosomal DNAs are about the size of domains of mammalian chromosomes whose replication timing varies, and S phase is short, it was not clear whether there would be a reproducible temporal pattern of replication in yeast chromosomes. The earliest indication of a specific temporal order of replication was given by experiments that made use of nitrosoguanidine, a mutagen thought to act at the replication fork, to induce mutations in synchronous cultures. The finding that different genes were susceptible to mutation at different times suggested that they were replicated at different times (Burke and Fangman 1975; Sim and Haber 1975). Direct measures of replication timing, either by the incorporation of radioactive precursors into particular DNA molecules or by the use of density-shift experiments, demonstrated that particular DNA fragments replicate at reproducibly different times (Zakian et al. 1979; Brewer et al. 1980; Fangman et al. 1983; McCarroll and Fangman 1988).

Do replicators initiate at different times, or do all replicators initiate early and does replication timing simply reflect distance from an active replicator? In a systematic examination of chromosome III, it was found that fragments containing ARS elements are replicated at different times (Reynolds et al. 1989), and direct analysis of replication intermediates isolated from synchronous cultures has confirmed that replication initiates at *ARS305* and *ARS306* 5–10 minutes earlier than at *ARS307* and *ARS309* (S. Jenab and C.S. Newlon, unpubl.). There is a 28-minute difference between the times of initiation at the earliest and latest origins identified so far, suggesting that initiation events occur throughout most of S phase (Brewer et al. 1993).

Initiation timing appears to be determined by sequences extrinsic to ARS elements. This conclusion is based on experiments that made use of *ARS1*, which initiates early in its usual location on chromosome IV, and *ARS501*, which initiates late in its usual position on chromosome V. When a fragment containing *ARS1* was placed near *ARS501* on chromosome V, *ARS1* initiated late. In contrast, when a 14-kb fragment containing *ARS501* was cloned in a circular plasmid, *ARS501* initiated early (Ferguson et al. 1991; Ferguson and Fangman 1992). These observations rule out models in which different replicators are activated at different times because they have different affinities for an initiation factor or because they are activated by different initiation factors that are synthesized sequentially.

Several observations suggest that late timing may be conferred by telomere proximity. The early-replicating circular plasmid containing *ARS501* was turned into a late-replicating linear plasmid by the addition of telomeres (Ferguson and Fangman 1992). The 2 μm plasmid ARS also initiates 10–15 minutes later on a linear plasmid that in its normal context in a circular plasmid (Wellinger et al. 1993). The presence of telomeric repeats rather than linearity per se appears to be the determinant of late replication, because a linear plasmid produced by the introduction of a specific double-stranded break in vivo still replicates early (Raghuraman et al. 1994).

Telomeres do not appear to be the only late-replication determinant. The *KEX2 ARS*, which is the latest-initiating chromosomal replicator discovered so far, is more than 200 kb from the nearest telomere. Moreover, a 16-kb fragment containing this ARS causes a circular plasmid to replicate late. Therefore, a late-timing determinant must reside within this fragment, perhaps within the ARS itself (Brewer et al. 1993).

The studies described above demonstrate that late-replicating regions of yeast chromosomes arise in two ways. The region can be far from an early replicator, as is the case for the latest-replicating region of chromosome III, the left telomere (Reynolds et al. 1989). The left telomere is replicated by a fork that initiates at *ARS305*, which is the earliest-initiating replicator identified so far and is approximately 40 kb away (Newlon et al. 1993). Alternatively, the region can contain its own late-activated replicators, e.g., *ARS501* and the *KEX2 ARS*.

Replication Termination

Replication intermediates containing converging replication forks characteristic of replication termination can be distinguished by 2D gel analysis from replication intermediates containing replication bubbles or single forks. The available evidence suggests that replication termination in *S. cerevisiae* occurs throughout broad regions rather than at specific sites. Analysis of a 61-kb circular derivative of chromosome III that contains three ARS elements revealed three replication termination zones located approximately midway between the three pairs of replicators (Greenfeder and Newlon 1992). Altering the sites of replication initiation by deletion of *ARS307* from the circular chromosome created a new termination region, demonstrating that termination is not controlled by specific *cis*-acting sequences, but rather that termination regions are determined by the positions of active replicators. Two other termination zones have been examined. Replication termination intermediates were found over a broad region midway between *ARS305* and *ARS306* on chromosome III (Zhu et al. 1992). The termination zone centromere-proximal to *ARS501* is so broad that termination intermediates associated with any particular restriction fragment were too faint to be seen in standard 2D gel analysis. Instead, the termination zone was identified by a gradual shift in the direction of replication fork movement through the region (Brewer et al. 1993). The excellent stability of chromosomes from which active replicators were deleted provides additional evidence against the existence of specific replication termini (Dershowitz and Newlon 1993; Huang and Kowalski 1993; Newlon et al. 1993). The existence of specific replication termini should prevent the replication of a region from which all origins were deleted and cause decreased chromosome stability.

ACKNOWLEDGMENTS

I thank members of the Newlon laboratory for helpful discussions; Drs. Judith Campbell, John Diffley, James Theis, and Chris Schaefer for helpful comments on the manuscript; and many colleagues for supplying preprints in advance of publication. Work from my laboratory was supported by a grant (GM-35679) from the National Institutes of Health.

REFERENCES

Araki, H., R.K. Hamatake, L.H. Johnston, and A. Sugino. 1991a. *DPB2*, the gene encoding DNA polymerase II subunit B, is required for chromosomal replication in *Saccharomyces cerevisiae*. *Proc. Natl. Acad. Sci.* **88:** 4601–4605.

Araki, H., R.K. Hamatake, A. Morrison, A.L. Johnson, L.H. Johnston, and A. Sugino. 1991b. Cloning *DPB3*, the gene encoding the third subunit of DNA polymerase II of *Saccharomyces cerevisiae*. *Nucleic Acids Res.* **19:** 4867–4872.

Araki, H., P.A. Ropp, A.L. Johnson, L.H. Johnston, A. Morrison, and A. Sugino. 1992. DNA polymerase II, the probable homolog of mammalian DNA polymerase ε, replicates chromosomal DNA in *Saccharomyces cerevisiae*. *EMBO J.* **11:** 733–740.

Ayyagari, R., K.J. Impellizzeri, B.L. Yoder, S.L. Gary, and P.M.J. Burgers. 1995. A mutational analysis of the yeast proliferating cell nuclear antigen indicates distinct roles in DNA replication and DNA repair. *Mol. Cell. Biol.* **15:** 4420–4429.

Bauer, G.A. and P.M.J. Burgers. 1990. Molecular cloning, structure, and expression of the yeast proliferating cell nuclear antigen. *Nucleic Acids Res.* **18:** 261–265.

Bauer, G.A., H.M. Heller, and P.M.J. Burgers. 1988. DNA polymerase III from *Saccharomyces cerevisiae*. I. Purification and characterization. *J. Biol. Chem.* **263:** 917–924.

Barker, D.G., J.H.M. White, and L.H. Johnston. 1987. Molecular characterisation of the DNA ligase gene, *CDC17*, from the fission yeast *Schizosaccharomyces pombe*. *Eur. J. Biochem.* **162:** 659–667.

Beach, D.H. and P. Nurse. 1981. High frequency transformation of the fission yeast *Schizosaccharomyces pombe*. *Nature* **290:** 140–142.

Bean, D.W., W.E. Kallam, Jr., and S.W. Matson. 1993. Purification and characterization of a DNA helicase from *Saccharomyces cerevisiae*. *J. Biol. Chem.* **268:** 21783–21790.

Bell, S.P. and B. Stillman. 1992. ATP-dependent recognition of eukaryotic origins of replication by a multiprotein complex. *Nature* **357:** 128–134.

Bell, S.P., R. Kobayashi, and B. Stillman. 1993. Yeast origin recognition complex functions in transcription silencing and DNA replication. *Science* **262:** 1844–1849.

Borowiec, J.A., F.B. Dean, P.A. Bullock, and J. Hurwitz. 1990. Binding and unwinding—How T antigen engages the SV40 origin of DNA replication. *Cell* **60:** 181–184.

Boulet, A., M. Simon, G. Faye, G.A. Bauer, and P.M.J. Burgers. 1989. Function of the *S. cerevisiae CDC2* gene encoding the large subunit of DNA polymerase III. *EMBO J.* **8:** 1849–1854.

Bramhill, D. and A. Kornberg. 1988. A model for initiation at origins of DNA replication. *Cell* **54:** 915–918.

Brewer, B.J. and W.L. Fangman. 1987. The localization of replication origins on ARS plasmids in *S. cerevisiae*. *Cell* **51:** 463–471.

———. 1988. A replication fork barrier at the 3′ end of yeast ribosomal RNA genes. *Cell* **55:** 637–643.

———. 1993. Initiation at closely spaced replication origins in a yeast chromosome. *Science* **262:** 1728–1731.

———. 1994. Initiation preference at a yeast origin of replication. *Proc. Natl. Acad. Sci.* **91:** 3418–3422.

Brewer, B.J., V.A. Zakian, and W.L. Fangman. 1980. Replication and meiotic transmission of yeast ribosomal RNA genes. *Proc. Natl. Acad. Sci.* **77:** 6739–6743.

Brewer, B.J., J.D. Diller, K.L. Friedman, K.M. Kolor, M.K. Raghuraman, and W.L. Fangman. 1993. The topography of chromosome replication in yeast. *Cold Spring Harbor Symp. Quant. Biol.* **58:** 425–434.

Brill, S.J. and B. Stillman. 1989. Yeast replication factor-A functions in the unwinding of the SV40 origin of DNA replication. *Nature* **342:** 92–95.

———. 1991. Replication factor-A from *Saccharomyces cerevisiae* is encoded by three essential genes coordinately expressed at S phase. *Genes Dev.* **5:** 1589–1600.

Brill, S.J., S. DiNardo, K. Voekel-Meiman, and R. Sternglanz. 1987. Need for DNA topoisomerase activity as a swivel for DNA replication and for transcription of ribosomal RNA. *Nature* **326:** 414–416.

Brown, W.C., J.A. Duncan, and J.L. Campbell. 1993. Purification and characterization of the *Saccharomyces cerevisiae* DNA polymerase δ overproduced in *E. coli*. *J. Biol. Chem.* **268:** 982–990.

Budd, M.E. and J.L. Campbell. 1993. DNA polymerases δ and ε are required for chromosomal replication in *Saccharomyces cerevisiae*. *Mol. Cell. Biol.* **13:** 496–505.

———. 1995a. A yeast gene required for DNA replication encodes a protein with homology to DNA helicases. *Proc. Natl. Acad. Sci.* **92:** 7642–7646.

———. 1995b. DNA polymerases required for repair of UV-induced damage in *Saccharomyces cerevisiae*. *Mol. Cell. Biol.* **15:** 2173–2179.

Bueno, A. and P. Russell. 1992. Dual functions of *CDC6*: A yeast protein required for DNA replication also inhibits nuclear division. *EMBO J.* **11:** 2167–2176.

Burgers, P.M.J. 1991. *Saccharomyces cerevisiae* replication factor C. II. Formation and activity of complexes with proliferating cell nuclear antigen and with DNA

polymerases δ and ε. *J. Biol. Chem.* **266:** 22698–22706.
Burke, W. and W.L. Fangman. 1975. Temporal order in yeast chromosome replication. *Cell* **5:** 263–269.
Caddle, M.S. and M.P. Calos. 1994. Specific initiation at an origin of replication from *Schizosaccharomyces pombe. Mol. Cell. Biol.* **14:** 1796–1805.
Campbell, J.L. and C.S. Newlon. 1991. Chromosomal DNA replication. In *The Molecular and Cellular Biology of the Yeast* Saccharomyces: *Genome Dynamics, Protein Synthesis, and Energetics* (ed. J.R. Broach et al.), pp. 41–146. Cold Spring Harbor Laboratory Press, Cold Spring Harbor, New York.
Chapman, J.W. and L.H. Johnston. 1989. The yeast gene, *DBF4*, essential for entry into S phase, is cell cycle regulated. *Exp. Cell Res.* **180:** 419–428.
Chen, Y., K.M. Hennessy, D. Botstein, and B.-K. Tye. 1992. CDC46/MCM5, a yeast protein whose subcellular localization is cell cycle-regulated, is involved in DNA replication at autonomously replicating sequences. *Proc. Natl. Acad. Sci.* **89:** 10459–10463.
Chien, C.-T., S. Buck, R. Sternglanz, and D. Shore. 1993. Targeting of SIR1 protein establishes transcriptional silencing. *Cell* **75:** 531–541.
Cocker, J.L., S. Piatti, C. Santocanale, K. Nasmyth, and J.F.X. Diffley. 1996. An essential role for the Cdc6 protein in forming the pre-replicative complexes of budding yeast. *Nature* **379:** 180–182.
Collins, I. and C.S. Newlon. 1994. Chromosomal DNA replication initiates at the same origins in meiosis and mitosis. *Mol. Cell. Biol.* **14:** 3524–3534.
Collins, K.L., A.A.R. Russo, B.Y. Tseng, and T.J. Kelly. 1993. The role of the 70 kDa subunit of human DNA polymerase α in DNA replication. *EMBO J.* **12:** 4555–4566.
Coxon, A., K. Maundrell, and S.E. Kearsey. 1992. Fission yeast *cdc21*$^{+}$ belongs to a family of proteins involved in an early step of chromosome replication. *Nucleic Acids Res.* **20:** 5571–5577.
Cullmann, G., K. Fein, R. Kobayashi, and B. Stillman. 1995. Characterization of the five replication factor C genes from *Saccharomyces cerevisiae. Mol. Cell. Biol.* **15:** 4661–4671.
Dalton, S. and L. Whitbread. 1995. Cell cycle-regulated nuclear import and export of Cdc47, a protein essential for initiation of DNA replication in budding yeast. *Proc. Natl. Acad. Sci.* **92:** 2514–2518.
Damagnez, V., J. Tillit, A.-M. Recondo, and G. Baldacci. 1991. The *POL1* gene from fission yeast, *Schizosaccharomyces pombe*, shows conserved amino acid blocks specific for eukaryotic DNA polymerase alpha. *Mol. Gen. Genet.* **266:** 182–189.
Dershowitz, A. and C.S. Newlon. 1993. The effect on chromosome stability of deleting replication origins. *Mol. Cell. Biol.* **13:** 391–398.
Deshpande, A.M. and C.S. Newlon. 1992. The ARS consensus sequence is required for chromosomal origin function in *Saccharomyces cerevisiae. Mol. Cell. Biol.* **12:** 4305–4313.
Diffley, J.F.X. and J.H. Cocker. 1992. Protein-DNA interactions at a yeast replication origin. *Nature* **357:** 169–172.
Diffley, J.F.X. and B. Stillman. 1988. Purification of a yeast protein that binds to origins of DNA replication and a transcriptional silencer. *Proc. Natl. Acad. Sci.* **85:** 2120–2124.
Diffley, J.F.X., J. Cocker, S.J. Dowell, and A. Rowley. 1994. Two steps in the assembly of complexes at yeast replication origins in vivo. *Cell* **78:** 303–316.

Din, S., S.J. Brill, M.P. Fairman, and B. Stillman. 1990. Cell-cycle-regulated phosphorylation of DNA replication factor A from human and yeast cells. *Genes Dev.* **4:** 968–977.

DiNardo, S., K. Voelkel, and R. Sternglanz. 1984. DNA topoisomerase II mutant of *Saccharomyces cerevisiae*: Topoisomerase II is required for segregation of daughter molecules at termination of DNA replication. *Proc. Natl. Acad. Sci.* **81:** 2616–2620.

Dowell, S.J., P. Romanowski, and J.F.X. Diffley. 1994. Interaction of DBF4, the Cdc7 protein kinase regulatory subunit, with yeast replication origins in vivo. *Science* **265:** 1243–1246.

Dubey, D.D., J. Zhu, D.L. Carlson, K. Sharma, and J.A. Huberman. 1994. Three ARS elements contribute to the *ura4* replication origin region in the fission yeast, *Schizosaccharomyces pombe*. *EMBO J.* **13:** 3638–3647.

Dubey, D.D., L.R. Davis, S.A. Greenfeder, L.Y. Ong, J. Zhu, J.R. Broach, C.S. Newlon, and J.A. Huberman. 1991. Evidence suggesting that the ARS elements associated with silencers of the yeast mating type locus, *HML*, do not function as chromosomal DNA replication origins. *Mol. Cell. Biol.* **11:** 5346–5355.

Eisenberg, S., C. Civalier, and B.-K. Tye. 1988. Specific interaction between a *Saccharomyces cerevisiae* protein and a DNA element associated with certain autonomously replicating sequences. *Proc. Natl. Acad. Sci.* **85:** 743–746.

Elble, R. and B.-K. Tye. 1992. Chromosome loss, hyperrecombination, and cell cycle arrest in a yeast *mcm1* mutant. *Mol. Biol. Cell* **3:** 971–980.

Epstein, C.B. and F. Cross. 1992. CLB5: A novel B cyclin from budding yeast with a role in S phase. *Genes Dev.* **6:** 1695–1706.

Estes, H.G., B.S. Robinson, and S. Eisenberg. 1992. At least three distinct proteins are necessary for the reconstitution of a specific multiprotein complex at a eukaryotic chromosomal origin of replication. *Proc. Natl. Acad. Sci.* **89:** 11156–11160.

Fabiani, L., M. Aragona, and L. Frontali. 1990. Isolation and sequence analysis of a *K. lactis* chromosomal element able to autonomously replicate in *S. cerevisiae* and *K. lactis*. *Yeast* **6:** 69–76.

Fabiani, L., L. Frontali, and C.S. Newlon. 1996. Identification of an essential core element and stimulatory sequences in a *Kluyveromyces lactis* ARS element, KARS101. *Mol. Microbiol.* **19:** 757–766.

Fangman, W.L., R.H. Hice, and E. Chlebowicz-Sledziewska. 1983. ARS replication during the yeast S phase. *Cell* **32:** 831–838.

Ferguson, B.M. and W.L. Fangman. 1992. A position effect on the time of replication origin activation in yeast. *Cell* **68:** 333–339.

Ferguson, B.M., B.J. Brewer, A.E. Reynolds, and W.L. Fangman. 1991. A yeast origin of replication is activated late in S phase. *Cell* **65:** 507–515.

Firmenich, A., E.-A. Montserrat, and P. Berg. 1995. A novel allele of *Saccharomyces cerevisiae RFA1* that is deficient in recombination and repair and suppressible by *RAD52*. *Mol. Cell. Biol.* **15:** 1620–1631.

Flores-Rozas, H., Z. Kelman, F.B. Dean, Z.-Q. Pan, J.W. Harper, S.J. Elledge, M. O'Donnell, and J. Hurwitz. 1994. Cdk-interacting protein 1 directly binds with proliferating cell nuclear antigen and inhibits DNA replication catalyzed by the DNA polymerase δ holoenzyme. *Proc. Natl. Acad. Sci.* **91:** 8655–8659.

Foiani, M., G. Liberi, G. Lucchini, and P. Plevani. 1995. Cell cycle-dependent phosphorylation and dephosphorylation of the yeast DNA polymerase α-primase B subunit. *Mol. Cell. Biol.* **15:** 883–891.

Foiani, M., C. Santocanale, P. Plevani, and G. Lucchini. 1989. A single essential gene, *PRI2*, encodes the large subunit of DNA primase in *Saccharomyces cerevisiae*. *Mol. Cell. Biol.* **9:** 3081–3087.

Foiani, M., F. Marini, D. Gamba, G. Lucchini, and P. Plevani. 1994. The B subunit of the DNA polymerase α-primase complex in *Saccharomyces cerevisiae* executes an essential function at the initial stage of DNA replication. *Mol. Cell. Biol.* **14:** 923–933.

Foss, M., F.J. McNally, P. Laurenson, and J. Rine. 1993. Origin recognition complex (ORC) in transcriptional silencing and DNA replication in *S. cerevisiae*. *Science* **262:** 1838–1844.

Fox, C.A., S. Loo, A. Dillin, and J. Rine. 1995. The origin recognition complex has essential functions in transcriptional silencing and chromosomal replication. *Genes Dev.* **9:** 991–924.

Francesconi, S.C. and S. Eisenberg. 1991. The multifunctional protein OBF1 is phosphorylated at serine and threonine residues in *Saccharomyces cerevisiae*. *Proc. Natl. Acad. Sci.* **88:** 4089–4093.

Francesconi, S., M.P. Longhese, A. Piseri, C. Santocanale, G. Lucchini, and P. Plevani. 1991. Mutations in conserved yeast DNA primase domains impair DNA replication in vivo. *Proc. Natl. Acad. Sci.* **88:** 3877–3881.

Funabiki, H., I. Hagan, S. Uzawa, and M. Yanagida. 1993. Cell cycle-dependent specific positioning and clustering of centromeres and telomeres in fission yeast. *J. Cell Biol.* **121:** 961–976.

Gangloff, S., J.P. McDonald, C. Bendixen, L. Arthur, and R. Rothstein. 1994. The yeast type topoisomerase Top3 interacts with Sgs1, a DNA helicase homolog: A potential eukaryotic reverse gyrase. *Mol. Cell. Biol.* **14:** 8391–8398.

Gibson, S.I., R.T. Surosky, and B.-K. Tye. 1990. The phenotype of the minichromosome maintenance mutant *mcm3* is characteristic of mutants defective in DNA replication. *Mol. Cell. Biol.* **10:** 5707–5720.

Gibson, S., F. Surosky, P. Sinha, G. Maine, and B.-K. Tye. 1987. Complexity of the enzyme system for the initiation of DNA replication in yeast. *UCLA Symp. Mol. Cell. Biol. New Ser.* **47:** 341–354.

Goto, T. and J.C. Wang. 1984. Yeast DNA topoisomerase II is encoded by a single-copy, essential gene. *Cell* **36:** 1073–1080.

———. 1985. Cloning of yeast *TOP1*, the gene encoding topoisomerase I, and construction of mutants defective in both DNA topoisomerase I and DNA topoisomerase II. *Proc. Natl. Acad. Sci.* **82:** 7178–7182.

Goulian, M., S.H. Richards, C.J. Heard, and B.M. Bigsby. 1990. Discontinuous DNA synthesis by purified mammalian proteins. *J. Biol. Chem.* **265:** 18461–18471.

Greenfeder, S.A. and C.S. Newlon. 1992. A replication map of a 61kb circular derivative of *Saccharomyces cerevisiae* chromosome III. *Mol. Biol. Cell* **3:** 999–1013.

Harrington, J.J. and M.J. Lieber. 1994. Functional domains within FEN-1 and RAD2 define a family of structure-specific endonucleases: Implications for nucleotide excision repair. *Genes Dev.* **8:** 1344–1355.

Hartwell, L.H. 1976. Sequential function of gene products relative to DNA synthesis in the cell cycle. *J. Mol. Biol.* **104:** 803–817.

He, Z., B.T. Brinton, J. Greenblatt, J.A. Hassell, and C.J. Ingles. 1993. The transactivator proteins VP16 and GAL4 bind replication factor A. *Cell* **73:** 1223–1232.

Hennessy, K.M., C.D. Clark, and D. Botstein. 1990. Subcellular localization of yeast CDC46 varies with the cell cycle. *Genes Dev.* **4:** 2252–2263.

Hennessy, K.M., A. Lee, E. Chen, and D. Botstein. 1991. A group of interacting yeast DNA replication genes. *Genes Dev.* **5:** 958–969.

Hereford, L.M. and L.H. Hartwell. 1974. Sequential gene function in the initiation of *S. cerevisiae* DNA synthesis. *J. Mol. Biol.* **84:** 445–481.

Heyer, W.-D., M.R.S. Rao, L.F. Erdile, T.J. Kelly, and R.D. Kolodner. 1990. An essential *Saccharomyces cerevisiae* single-stranded DNA binding protein is homologous to the large subunit of human RP-A. *EMBO J.* **7:** 2321–2329.

Hogan, E. and D. Koshland. 1992. Addition of extra origins of replication to a minichromosome suppresses its mitotic loss in *cdc6* and *cdc14* mutants of *Saccharomyces cerevisiae. Proc. Natl. Acad. Sci.* **89:** 3098–3102.

Holm, C., T. Stearns, and D. Botstein. 1989. DNA topoisomerase II must act at mitosis to prevent nondisjunction and chromosome breakage. *Mol. Cell. Biol.* **9:** 159–168.

Holm, C., T. Goto, J.C. Wang, and D. Botstein. 1985. DNA topoisomerase II is required at the time of mitosis in yeast. *Cell* **41:** 553–563.

Howell, E.A., M.A. McAlear, D. Rose, and C. Holm. 1994. CDC44: A putative nucleotide-binding protein required for cell cycle progression that has homology to subunits of replication factor C. *Mol. Cell. Biol.* **14:** 255–267.

Huang, R-Y. and D. Kowalski. 1993. A DNA unwinding element and an ARS consensus comprise a replication origin within a yeast chromosome. *EMBO J.* **12:** 4521–4531.

Huberman, J.A., L.D. Spotila, K.A. Nawotka, S.M. El-Assouli, and L.R. Davis. 1987. The in vivo replication origin of the yeast 2 μm plasmid. *Cell* **51:** 473–481.

Huberman, J.A., D.D. Dubey, K.A. Nawotka, G. Russev, J.A. Sanchez, Y. Yoon, and M.H.K. Linskens. 1992. Directions of DNA replication in yeast and mammalian cells. In *DNA replication: The regulatory mechanisms* (ed. P. Hughes et al.), pp. 83–95. Springer-Verlag, Paris.

Ishimi, Y., A. Claude, P. Bullock, and J. Hurwitz. 1988. Complete enzymatic synthesis of DNA containing the SV40 origin of replication. *J. Biol. Chem.* **263:** 19723–19733.

Jackson, A.L., P.M. Pahl, K. Harrison, J. Rosamond, and R.A. Sclafani. 1993. Cell cycle regulation of the yeast Cdc7 protein kinase by association with Dbf4 protein. *Mol. Cell. Biol.* **13:** 2899–2908.

Johnston, L.H. 1979. The repair capability of *cdc9*, the *Saccharomyces cerevisiae* mutant defective in DNA ligase. *Mol. Gen. Genet.* **170:** 89–92.

———. 1983. The *cdc9* ligase joins completed replicons in baker's yeast. *Mol. Gen. Genet.* **190:** 315–317.

Johnston, L.H. and K.A. Nasmyth. 1978. *Saccharomyces cerevisiae* cell cycle mutant *cdc9* is defective in DNA ligase. *Nature* **274:** 891–893.

Johnston, L.H. and A.P. Thomas. 1982a. Isolation of new DNA synthesis mutants in the yeast *Saccharomyces cerevisiae. Mol. Gen. Genet.* **186:** 439–444.

———. 1982b. A further two mutants defective in initiation of the S phase in the yeast *Saccharomyces cerevisiae. Mol. Gen. Genet.* **186:** 445–448.

Kelly, T.J., G.S. Martin, S.L. Forsburg, R.J. Stephen, A. Russo, and P. Nurse. 1993. The fission yeast *cdc18*$^{+}$ gene product couples S phase to START and mitosis. *Cell* **74:** 371–382.

Kim, R.A. and J.C. Wang. 1989. The function of DNA topoisomerases as replication swivels in *Saccharomyces cerevisiae. J. Mol. Biol.* **208:** 257–267.

Kitada, K., L.H. Johnston, T. Sugino, and A. Sugino. 1992. Temperature-sensitive *cdc7* mutations of *Saccharomyces cerevisiae* are suppressed by the *DBF4* gene, which is required for the G1/S cell cycle transition. *Genetics* **131:** 21–29.

Konrad, E.B. and I.R. Lehman. 1974. A conditional lethal mutant of *Escherichia coli* K12 defective in the 5′-3′ exonuclease associated with DNA polymerase I. *Proc. Natl. Acad. Sci.* **71:** 2048–2051.

Koonin, E.V. 1993. A common set of conserved motifs in a vast variety of putative nucleic acid-dependent ATPases including MCM proteins involved in the initiation of eukaryotic DNA replication. *Nucleic Acids Res.* **21:** 2541–2547.

Kornberg, A. and T.A. Baker. 1992. *DNA replication, 2nd edition.* W.H. Freeman, New York.

Kouprina, N., E. Kroll, A. Kirillov, V. Bannikov, V. Zakharyev, and V. Larionov. 1994. *CHL12*, a gene essential for the fidelity of chromosome transmission in the yeast *Saccharomyces cerevisiae*. *Genetics* **138:** 1067–1069.

Kouprina, N., E. Kroll, V. Bannikov, V. Bliskovsky, R. Gizatullin, A. Kirillov, B. Shestopalov, V. Zakharyev, P. Hieter, F. Spencer, and V. Larionov. 1992. *CTF4* (*CHL15*) mutants exhibit defective DNA metabolism in *Saccharomyces cerevisiae. Mol. Cell. Biol.* **12:** 5736–5747.

Krishna, T.S., X.P. Kong, S. Gary, P.M. Burgers, and J. Kuriyan. 1994. Crystal structure of the eukaryotic DNA polymerase processivity factor PCNA. *Cell* **79:** 1233–1243.

Kubota, Y., S. Mimura, S. Nishimoto, H. Takisawa, and H. Nojima. 1995. Identification of the yeast MCM3-related protein as a component of *Xenopus* licensing factor. *Cell* **81:** 601–609.

Kuo, C.-L., N.-H. Huang, and J.L. Campbell. 1983. Isolation of DNA replication mutants in permeabilized cells. *Proc. Natl. Acad. Sci.* **80:** 6465–6469.

Kuo, M.-H. and E. Grayhack. 1994. A library of yeast genomic MCM1 binding sites contains genes involved in cell cycle control, cell wall and membrane structure, and metabolism. *Mol. Cell. Biol.* **14:** 348–359.

Laurenson, P. and J. Rine. 1992. Silencers, silencing, and heritable transcriptional states. *Microbiol. Rev.* **56:** 543–560.

Li, J.J. and I. Herskowitz. 1993. Isolation of *ORC6*, a component of the yeast origin recognition complex, by a one-hybrid system. *Science* **262:** 1870–1874.

Li, R. and M.R. Botchan. 1993. The acidic transcriptional activation domains of VP16 and p53 bind the cellular replication protein A and stimulate in vitro BPV-1 DNA replication. *Cell* **73:** 1207–1221.

Li, R., S. Waga, G.J. Hannon, D. Beach, and B. Stillman. 1994. Differential effects by the p21 CDK inhibitor on PCNA-dependent DNA replication and repair. *Nature* **371:** 534–537.

Li, X. and P.M.J. Burgers. 1994a. Cloning and characterization of the essential *Saccharomyces cerevisiae RFC4* gene encoding the 37 kDa subunit of replication factor C. *J. Biol. Chem.* **269:** 21880–21884.

———. 1994b. Molecular cloning and expression of the *Saccharomyces cerevisiae RFC3* gene, an essential component of replication factor C. *Proc. Natl. Acad. Sci.* **91:** 868–872.

Li, X., B.L. Yoder, and P.M.J. Burgers. 1992. A *Saccharomyces cerevisiae* DNA helicase associated with replication factor C. *J. Biol. Chem.* **267:** 25321–25327.

Liang, C., M. Weinreich, and B. Stillman. 1995. ORC and Cdc6p interact and determine the frequency of initiation of DNA replication in the genome. *Cell* **81:** 667–676.

Linskens, M.H.K. and J.A. Huberman. 1988. Organization of replication of ribosomal DNA in *Saccharomyces cerevisiae. Mol. Cell. Biol.* **8:** 4927–4935.

Longhese, M.P., P. Plevani, and G. Lucchini. 1994. Replication factor A is required in vivo for DNA replication, repair, and recombination. *Mol. Cell. Biol.* **14:** 7884–7890.

Longhese, M.P., L. Jovine, P. Plevani, and G. Lucchini. 1993. Conditional mutations in the yeast DNA primase genes affect different aspects of DNA metabolism and interactions in the DNA polymerase α-primase complex. *Genetics* **133:** 183–191.

Loo, S., C.A. Fox, J. Rine, R. Kobayashi, B. Stillman, and S. Bell. 1995. The origin recognition complex in silencing, cell cycle progression, and DNA replication. *Mol. Biol. Cell* **6:** 741–756.

Lucchini, G., S. Francesconi, M. Foiani, G. Badaracco, and P. Plevani. 1987. The yeast DNA polymerase-primase complex: Cloning of *PRI1*, a single essential gene related to DNA primase activity. *EMBO J.* **6:** 737–742.

Maine, G.T., P. Sinha, and B.-K. Tye. 1984. Mutants of *S. cerevisiae* defective in the maintenance of minichromosomes. *Genetics* **106:** 365–385.

Marahrens, Y. and B. Stillman. 1992. A yeast chromosomal origin of DNA replication defined by multiple functional elements. *Science* **255:** 817–823.

———. 1994. Replicator dominance in a eukaryotic chromosome. *EMBO J.* **13:** 3395–3400.

Martin-Parras, L., P. Hernandez, M.L. Martinez-Robles, and J.B. Schvartzman. 1992. Initiation of DNA replication in ColE1 plasmids containing multiple potential origins of replication. *J. Biol. Chem.* **267:** 22496–22505.

Masai, H., T. Miyake, and K. Arai. 1995. *hsk1*$^+$, a *Schizosaccharomyces pombe* gene related to *Saccharomyces cerevisiae CDC7*, is required for chromosomal replication. *EMBO J.* **14:** 3094–3104.

Maundrell, K., A. Hutchison, and S. Shall. 1988. Sequence analysis of ARS elements in fission yeast. *EMBO J.* **7:** 2203–2209.

McAlear, M.A., E.A. Howell, K.K. Espenshade, and C. Holm. 1994. Proliferating cell nuclear antigen (*pol30*) mutations suppress *cdc44* mutations and identify potential regions of interaction between the two encoded proteins. *Mol. Cell. Biol.* **14:** 4390–4397.

McCarroll, R.M. and W.L. Fangman. 1988. Time of replication of yeast centromeres and telomeres. *Cell* **54:** 505–513.

Micklem, G., A. Rowley, J. Harwood, K. Nasmyth, and J.F.X. Diffley. 1993. Yeast origin recognition complex is involved in DNA replication and transcriptional silencing. *Nature* **366:** 87–89.

Miles, J. and T. Formosa. 1992a. Protein affinity chromatography with purified yeast DNA polymerase α detects proteins that bind to yeast DNA polymerase. *Proc. Natl. Acad. Sci.* **89:** 1276–1280.

———. 1992b. Evidence that POB1, a *Saccharomyces cerevisiae* protein that binds to DNA polymerase α, acts in DNA metabolism in vivo. *Mol. Cell. Biol.* **12:** 5724–5735.

Miller, C.A. and D. Kowalski. 1993. *cis*-Acting components in the replication origin from ribosomal DNA of *Saccharomyces cerevisiae*. *Mol. Cell. Biol.* **13:** 5360–5369.

Miyake, S., N. Okishio, I. Samejima, Y. Hiraoka, T. Toda, I. Saitoh, and M. Yanagida. 1993. Fission yeast genes *nda1*$^+$ and *nda4*$^+$, mutations of which lead to S-phase block, chromatin alteration, and Ca^{2+} suppression, are members of the *CDC46/MCM2* family. *Mol. Biol. Cell* **4:** 1003–1015.

Morrison, A., J.B. Bell, T.A. Kunkel, and A. Sugino. 1991. Eukaryotic DNA polymerase amino acid sequence required for 3′→5′ exonuclease activity. *Proc. Natl. Acad. Sci.* **88:** 9473–9477.

Morrison, A., A.L. Johnson, L.H. Johnston, and A. Sugino. 1993. Pathway correcting

DNA replication errors in *Saccharomyces cerevisiae*. *EMBO J.* **12:** 1467–1473.
Morrison, A., H. Araki, A.B. Clark, R.K. Hamatake, and A. Sugino. 1990. A third essential DNA polymerase in *S. cerevisiae*. *Cell* **62:** 1143–1151.
Nasmyth, K.A. 1977. Temperature-sensitive lethal mutants in the structural gene for DNA ligase in the yeast *Schizosaccharomyces pombe*. *Cell* **12:** 1109–1120.
Natale, D.A., A.E. Schubert, and D. Kowalski. 1992. DNA helical stability accounts for mutational defects in a yeast replication origin. *Proc. Natl. Acad. Sci.* **89:** 2654–2658.
Navas, T.A., Z. Zhou, and S.J. Elledge. 1995. DNA polymerase ε links the DNA replication machinery to the S phase checkpoint. *Cell* **80:** 29–39.
Newlon, C.S. and J.F. Theis. 1993. The structure and function of yeast ARS elements. *Curr. Opin. Genet. Dev.* **3:** 752–758.
Newlon, C.S., I. Collins, A.M. Dershowitz, S.A. Greenfeder, L.Y. Ong, and J.F. Theis. 1993. Analysis of replication origin function on chromosome III of *Saccharomyces cerevisiae*. *Cold Spring Harbor Symp. Quant. Biol.* **58:** 415–423.
Norman, C., M. Runswick, R. Pollock, and R. Treisman. 1988. Isolation and properties of cDNA clones encoding SRF, a transcription factor that binds to the c-*fos* serum response element. *Cell* **55:** 989–1003.
Noskov, V., S. Maki, Y. Kawasaki, S.H. Leem, B. Ono, H. Araki, Y. Pavlov, and A. Sugino. 1994. The *RFC2* gene encoding a subunit of replication factor C of *Saccharomyces cerevisiae*. *Nucleic Acids Res.* **22:** 1527–1535.
Palzkill, T.G. and C.S. Newlon. 1988. A yeast replication origin consists of multiple copies of a small conserved sequence. *Cell* **53:** 441–450.
Park, H., S. Francesconi, and T.S.-F. Wang. 1993. Cell cycle expression of two replicative polymerases α and δ from *Schizosaccharomyces pombe*. *Mol. Biol. Cell* **4:** 145–157.
Passmore, S., R. Elble, and B.-K. Tye. 1989. A protein involved in minichromosome maintenance in yeast binds a transcriptional enhancer conserved in eukaryotes. *Genes Dev.* **3:** 921–935.
Passmore, S., G.T. Maine, R. Elble, C. Christ, and B.-K. Tye. 1988. *Saccharomyces cerevisiae* protein involved in plasmid maintenance is necessary for mating of MATα cells. *J. Mol. Biol.* **204:** 593–606.
Patterson, M., R.A. Sclafani, W.L. Fangman, and J. Rosamond. 1986. Molecular characterization of cell cycle gene *CDC7* from *Saccharomyces cerevisiae*. *Mol. Cell. Biol.* **6:** 1590–1598.
Petes, T.D. and C.S. Newlon. 1974. Structure of DNA in DNA replication mutants of yeast. *Nature* **251:** 637–639.
Piatti, S., C. Lengauer, and K. Nasmyth. 1995. Cdc6 is an unstable protein whose de novo synthesis in G1 is important for the onset of S phase and for preventing a "reductional" anaphase in the budding yeast *Saccharomyces cerevisiae*. *EMBO J.* **14:** 3788–3799.
Pignede, G., D. Bouvier, A.-M. Recondo, and G. Baldacci. 1991. Characterization of the POL3 gene product from *S. pombe* indicates interspecies conservation of the catalytic subunit of DNA polymerase δ. *J. Mol. Biol.* **222:** 209–218.
Raghuraman, M.K., B.J. Brewer, and W.L. Fangman. 1994. Activation of a yeast replication origin near a double-stranded DNA break. *Genes Dev.* **8:** 554–562.
Rao, H. and B. Stillman. 1995. The origin recognition complex interacts with a bipartite DNA binding site within yeast replicators. *Proc. Natl. Acad. Sci.* **92:** 2224–2228.
Rao, H., Y. Marahrens, and B. Stillman. 1994. Functional conservation of multiple ele-

ments in yeast chromosomal replicators. *Mol. Cell. Biol.* **14:** 7643–7651.

Rashid, M.B., K. Shirahige, N. Ogasawara, and H. Yoshikawa. 1994. Anatomy of the stimulative sequences flanking the ARS consensus sequence of chromosome VI in *Saccharomyces cerevisiae*. *Gene* **150:** 213–220.

Reagan, M.S., C. Pittenger, W. Siede, and E.C. Friedberg. 1995. Characterization of a mutant strain of *Saccharomyces cerevisiae* with a deletion of the *RAD27* gene, a structural homolog of the *RAD2* nucleotide excision repair gene. *J. Bacteriol.* **177:** 364–371.

Reynolds, A.E., R.M. McCarroll, C.S. Newlon, and W.L. Fangman. 1989. Time of replication of ARS elements along yeast chromosome III. *Mol. Cell. Biol.* **10:** 4488–4494.

Rhode, P.R., S. Elsasser, and J.L. Campbell. 1992. Role of multifunctional autonomously replicating sequence binding factor 1 in the initiation of DNA replication and transcriptional control in *Saccharomyces cerevisiae*. *Mol. Cell. Biol.* **12:** 1064–1077.

Rivier, D.H. and J. Rine. 1992. An origin of DNA replication and a transcriptional silencer require a common element. *Science* **256:** 659–663.

Rowley, A., J.H. Cocker, J. Harwood, and J.F.X. Diffley. 1995. Initiation complex assembly at budding yeast replication origins begins with the recognition of a bipartite sequence by limiting amounts of the initiator, ORC. *EMBO J.* **14:** 2631–2641.

Santocanale, C., M. Foiani, G. Lucchini, and P. Plevani. 1993. The isolated 48,000-dalton subunit of yeast DNA primase is sufficient for RNA primer synthesis. *J. Biol. Chem.* **268:** 1343–1348.

Schvartzman, J.B., S. Adolph, L. Martin-Parras, and C.L. Schildkraut. 1990. Evidence that replication initiates at only some of the potential origins in each oligomeric form of bovine papillomavirus type 1 DNA. *Mol. Cell. Biol.* **10:** 3078–3086.

Schwob, E. and K. Nasmyth. 1993. *CLB5* and *CLB6*, a new pair of B cyclins involved in DNA replication in *Saccharomyces cerevisiae*. *Genes Dev.* **7:** 1160–1175.

Schwob, E., T. Boehm, M.D. Mendenhall, and K. Nasmyth. 1994. The B-type cyclin kinase inhibitor $p40^{sic1}$ controls the G1 to S transition in *S. cerevisiae*. *Cell* **79:** 233–244.

Sclafani, R.A. and A.L. Jackson. 1994. Cdc7: A protein kinase for DNA metabolism comes of age. *Mol. Microbiol.* **11:** 805–810.

Shimizu, K. and A. Sugino. 1993. Purification and characterization of DNA helicase III from the yeast *Saccharomyces cerevisiae*. *J. Biol. Chem.* **268:** 9578–9584.

Shirahige, K., T. Iwasaki, M.B. Rashid, N. Ogasawara, and H. Yoshikawa. 1993. Localization and characterization of autonomously replicating sequences (ARSs) from chromosome VI of *Saccharomyces cerevisiae*. *Mol. Cell. Biol.* **13:** 5043–5056.

Sim, G.K. and J.E. Haber. 1975. Cell cycle-dependent induction of mutations along a yeast chromosome. *Proc. Natl. Acad. Sci.* **72:** 1179–1183.

Simon, M., L. Giot, and G. Faye. 1991. The 3′ to 5′ exonuclease activity located in the DNA polymerase δ subunit of *Saccharomyces cerevisiae* is required for accurate DNA replication. *EMBO J.* **10:** 2165–2170.

Simpson, R.T. 1990. Nucleosome positioning can affect the function of a *cis*-acting DNA element in vivo. *Nature* **343:** 387–390.

Sinha, P., V. Chang, and B.-K. Tye. 1986. A mutant that affects the function of autonomously replicating sequences in yeast. *J. Mol. Biol.* **192:** 805–814.

Sitney, K.C., M.E. Budd, and J.L. Campbell. 1989. DNA polymerase III, a second essential DNA polymerase, is encoded by the *Saccharomyces cerevisiae CDC2* gene. *Cell* **56:** 599–605.

Smith, J. and R. Rothstein. 1995. A mutation in the gene encoding the *Saccharomyces cerevisiae* single-stranded DNA-binding protein Rfa1 stimulates a *RAD52*-independent pathway for direct-repeat recombination. *Mol. Cell. Biol.* **15:** 1632–1641.

Soloman, N.A., M.B. Wright, S. Chang, A.M. Buckley, L.B. Dumas, and R.F. Gaber. 1992. Genetic and molecular analysis of *DNA43* and *DNA52*: Two new cell-cycle genes in *Saccharomyces cerevisiae. Yeast* **8:** 273–289.

Sommers, C.H., E.J. Miller, B. Dujon, S. Prakash, and L. Prakash. 1995. Conditional lethality of null mutations in *RTH1* that encodes the yeast counterpart of a mammalian 5′- to 3′-exonuclease required for lagging strand DNA synthesis in reconstituted systems. *J. Biol. Chem.* **270:** 4193–4196.

Spell, R.M. and C. Holm. 1994. Nature and distribution of chromosomal intertwinings in *Saccharomyces cerevisiae. Mol. Cell. Biol.* **14:** 1465–1476.

Stillman, B. 1994. Initiation of chromosomal DNA replication in eukaryotes. *J. Biol. Chem.* **269:** 7047–7050.

Strich, R., M. Woontner, and J.F. Scott. 1986. Mutations in *ARS1* increase the rate of simple loss of plasmids in *Saccharomyces cerevisiae. Yeast* **2:** 169–178.

Su, T.T., P.J. Follette, and P.H. O'Farrell. 1995. Qualifying for the license to replicate. *Cell* **81:** 825–828.

Theis, J.F. and C.S. Newlon. 1994. Domain B of *ARS307* contains two functional elements and contributes to chromosomal replication origin function. *Mol. Cell. Biol.* **14:** 7652–7659.

Thrash, C., K. Voelkel, S. DiNardo, and R. Sternglanz. 1984. Identification of *Saccharomyces cerevisiae* mutants deficient in DNA topoisomerase I activity. *J. Biol. Chem.* **259:** 1375–1377.

Tompkinson, A.E., N.J. Tappe, and E.C. Friedberg. 1992. DNA ligase I from *Saccharomyces cerevisiae*: Physical and biochemical characterization of the *CDC9* gene product. *Biochemistry* **31:** 11762–11771.

Turchi, J.J. and R.A. Bambara. 1993. Completion of mammalian lagging strand DNA replication using purified proteins. *J. Biol. Chem.* **268:** 15136–15141.

Turchi, J.J., L. Huang, R.S. Murante, Y. Kim, and R.A. Bambera. 1994. Enzymatic completion of mammalian lagging-strand DNA replication. *Proc. Natl. Acad. Sci.* **91:** 9803–9807.

Uemura, T. and M. Yanagida. 1986. Mitotic spindle pulls but fails to separate chromosomes in type II DNA topoisomerase mutants: Uncoordinated mitosis. *EMBO J.* **5:** 1003–1010.

Umek, R.M. and D. Kowalski. 1988. The ease of DNA unwinding as a determinant of initiation of yeast replication origins. *Cell* **52:** 559–567.

Vallen, E.A. and F.R. Cross. 1995. Mutations in *RAD27* define a potential link between G1 cyclins and DNA replication. *Mol. Cell. Biol.* **15:** 4291–4302.

Van Houten, J.V. and C.S. Newlon. 1990. Mutational analysis of the consensus sequence of a replication origin from yeast chromosome III. *Mol. Cell. Biol.* **10:** 3917–3925.

Waga, S. and B. Stillman. 1994. Anatomy of a DNA replication fork revealed by reconstitution of SV40 DNA replication in vitro. *Nature* **369:** 207–212.

Waga, S., G.J. Hannon, D. Beach, and B. Stillman. 1994. The p21 inhibitor of cyclin-dependent kinases controls DNA replication by interaction with PCNA. *Nature* **369:** 574–578.

Walker, S.S., S.C. Francesconi, and S. Eisenberg. 1990. A DNA replication enhancer in *Saccharomyces cerevisiae. Proc. Natl. Acad. Sci.* **87:** 4665–4669.

Walker, S.S., A.K. Malik, and S. Eisenberg. 1991. Analysis of the interactions of functional domains of a nuclear origin of replication from *Saccharomyces cerevisiae. Nucleic Acids Res.* **19:** 6255–6262.

Wang, J.C. 1991. DNA topoisomerases: Why so many? *J. Biol. Chem.* **266:** 6659–6662.

Wang, Z., X. Wu, and E.C. Friedberg. 1993. DNA repair synthesis during base excision repair in vitro is catalyzed by DNA polymerase ε and is influenced by DNA polymerases α and δ in *Saccharomyces cerevisiae. Mol. Cell. Biol.* **13:** 1051–1058.

Waseem, N.H., K. Labib, P. Nurse, and D.P. Lane. 1992. Isolation and analysis of the fission yeast gene encoding polymerase δ accessory protein PCNA. *EMBO J.* **11:** 5111–5120.

Watt, P.M., E.J. Louis, R.H. Borts, and I.D. Hickson. 1995. Sgs1: A eukaryotic homolog of *E. coli* RecQ that interacts with topoisomerase II in vivo and is required for faithful chromosome segregation. *Cell* **81:** 253–260.

Wellinger, R.J., A.J. Wolf, and V.A. Zakian. 1993. Origin activation and formation of single-strand $TG_{1\text{-}3}$ tails occur sequentially in late S phase on a yeast linear plasmid. *Mol. Cell. Biol.* **13:** 4057–4065.

Whitbread, L.A. and S. Dalton. 1995. Cdc54 belongs to the Cdc46/Mcm3 family of proteins which are essential for initiation of eukaryotic DNA replication. *Gene* **155:** 113–117.

Wohlgemuth, J.G., G.H. Bulboaca, M. Moghadam, M.S. Caddle, and M.P. Calos. 1994. Physical mapping of origins of replication in the fission yeast *Schizosaccharomyces pombe. Mol. Biol. Cell* **5:** 839–849.

Yan, H., S. Gibson, and B.-K. Tye. 1991. Mcm2 and Mcm3, two proteins important for ARS activity, are related in structure and function. *Genes Dev.* **5:** 944–957.

Yan, H., A.M. Merchant, and B.-K. Tye. 1993. Cell cycle-regulated nuclear localization of MCM2 and MCM3 which are required for the initiation of DNA synthesis at chromosomal replication origins in yeast. *Genes Dev.* **7:** 2149–2160.

Yoon, H.J., S. Loo, and J.L. Campbell. 1993. Regulation of *Saccharomyces cerevisiae CDC7* function during the cell cycle. *Mol. Biol. Cell* **4:** 195–208.

Zakian, V.A., B.J. Brewer, and W.L. Fangman. 1979. Replication of each copy of the 2 μm plasmid occurs during the S phase. *Cell* **17:** 923–934.

Zhou, C. and A. Jong. 1990. *CDC6* mRNA fluctuates periodically in the yeast cell cycle. *J. Biol. Chem.* **265:** 19904–19909.

Zhou, C., S.-H. Huang, and A.Y. Jong. 1989. Molecular cloning of *Saccharomyces cerevisiae CDC6* gene. *J. Biol. Chem.* **264:** 9022–9029.

Zhu, J., C.S. Newlon, and J.A. Huberman. 1992. Localization of a DNA replication origin on chromosome III of *Saccharomyces cerevisiae. Mol. Cell. Biol.* **12:** 4733–4741.

Zhu, J., D.L. Carlson, D.D. Dubey, K. Sharma, and J.A. Huberman. 1994. Comparison of the two major ARS elements of the *ura4* replication origin region with other ARS elements in the fission yeast, *Schizosaccharomyces pombe. Chromosoma* **103:** 414–422.

Zwerschke, W., H.-W. Rottakob, and H. Kuntzel. 1994. The *Saccharomyces cerevisiae CDC6* gene is transcribed at late mitosis and encodes a ATP/GTPase controlling S phase initiation. *J. Biol. Chem.* **269:** 23351–23356.

33

DNA Replication in *Tetrahymena*

Geoffrey M. Kapler
Department of Medical Biochemistry and Genetics
Texas A&M Health Science Center
College Station, Texas 77845

Drena L. Dobbs
Department of Zoology and Genetics
Iowa State University
Ames, Iowa 50011

Elizabeth H. Blackburn
Department of Microbiology and Immunology
University of California, San Francisco
San Francisco, California 94143-0414

The linear ribosomal RNA gene (rDNA) minichromosome in the somatic macronucleus of *Tetrahymena thermophila* serves as a paradigm for multiple mechanisms of regulation of chromosomal DNA replication. Whereas chromosomal DNA replication in most eukaryotes is under strict cell cycle control, additional mechanisms appear to be involved in the regulation of rDNA replication in vegetatively growing *Tetrahymena* cells (Anderson 1972; Engberg et al. 1972; Truett and Gall 1977). All macronuclear chromosomes, including the rDNA minichromosome, segregate randomly and thus appear to lack centromeres. The macronucleus divides by a poorly understood amitotic mechanism, and DNA is often distributed unequally between daughter macronuclei (Doerder 1979). To explain how cells compensate for inequalities that arise from such divisions, it was proposed that a copy number control mechanism is superimposed on cell cycle control of the macronuclear DNA (Preer and Preer 1979; for review, see Larson et al. 1991). In addition, different allelic forms of the rDNA compete with each other for replication in the same macronucleus, and additional copies of the origin region confer a replication advantage to rDNA molecules. These observations are at odds with strict cell cycle control. Therefore, it was concluded that copy number control facilitates the maintenance of rDNA levels in vegetative cells (Larson et al. 1986).

Cell cycle control of DNA replication is also suppressed during developmentally programmed rDNA amplification (Yao et al. 1974). *cis*-Acting determinants controlling the formation and propagation of the rDNA minichromosome have been identified by classic genetic and DNA transformation approaches (Larson et al. 1986; Yaeger et al. 1989; Yao et al. 1990; Yasuda and Yao 1991; Kapler and Blackburn 1994; Kapler et al. 1994). The entire minichromosome has been sequenced (Engberg and Nielson 1990), and both the vegetative replication origin (Cech and Brehm 1981; Palen and Cech 1984; R.C. Gallagher and E.H. Blackburn, unpubl.) and telomere-associated regions (Budarf and Blackburn 1986; P.D. Cohen and E.H. Blackburn, in prep.) have been shown to contain highly ordered chromatin structures.

Formation of the rDNA minichromosome exemplifies the changes in chromosome structure that are part of the process leading to nuclear dimorphism of ciliated protozoa. Ciliates contain two types of nuclei within a single cell: the germ-line micronucleus and the somatic macronucleus (for review, see Orias 1986). The transcriptionally silent micronucleus divides by normal mitosis during vegetative cell divisions and undergoes meiosis, serving as the reservoir of genetic material, during cell mating (conjugation). The macronucleus is transcriptionally active, conferring the cellular phenotype. The macronucleus is destroyed during conjugation, and one of the progeny cell micronuclei differentiates to form the new macronucleus. This process, termed macronuclear development, involves a massive rearrangement of the genome (for review, see Karrer 1986). In the developing macronucleus of *T. thermophila*, the five pairs of intact precursor chromosomes are fragmented into about 200 pieces (Altschuler and Yao 1985). Telomeres are added de novo to the newly exposed chromosome ends (Spangler et al. 1988; for review, see Yao 1989). Flanking DNA sequences are lost during fragmentation. In addition, some internal sequences are eliminated via DNA deletion and religation (Godiska and Yao 1990). Macronuclear chromosomes are replicated to about 50 copies and the rDNA minichromosome, encoding the 17S, 5.8S, and 26S ribosomal RNA genes, is amplified to 10^4 copies (Yao et al. 1974). Cultures of *Tetrahymena* can be induced to undergo macronuclear development as relatively synchronous cell populations. Consequently, the reorganization of the macronuclear genome can be monitored by molecular analyses. During subsequent vegetative growth, macronuclear chromosomes are replicated on average once per cell cycle (Engberg et al. 1972); see above.

EVENTS IN THE DEVELOPING MACRONUCLEUS

Formation of the rDNA Minichromosome

The rRNA gene is present as a single copy gene in the germ-line micronucleus. To form the macronuclear rDNA minichromosome, the rDNA region is excised from the germ-line chromosome (Fig. 1 and Table 1). This process is directed by at least one known class of *cis*-acting elements: the 15-bp chromosome breakage sequence (Cbs) elements that flank each end of the rDNA, as well as each of the regions that will become the other macronuclear chromosomes (Yao et al. 1987). Cbs-mediated chromosome breakage has been studied most extensively for the rDNA minichromosome. The dependence of rDNA excision on Cbs elements was demonstrated directly by transformation of developing macronuclei by microinjection of rDNA constructs with and without Cbs elements (Yao et al. 1990). A genetic screen for mutants impaired in rDNA amplification (Kapler and Blackburn 1994) also identified a point mutation in the Cbs at one end of the micronuclear rDNA (the *rmm 11/6* mutation). This Cbs mutation caused an approximately 75% reduction in the frequency of excision of the affected rDNA allele (see below). Some aspect of the Cbs-mediated chromosome fragmentation process appears to be required for the de novo addition of telomeric repeats to newly ex-

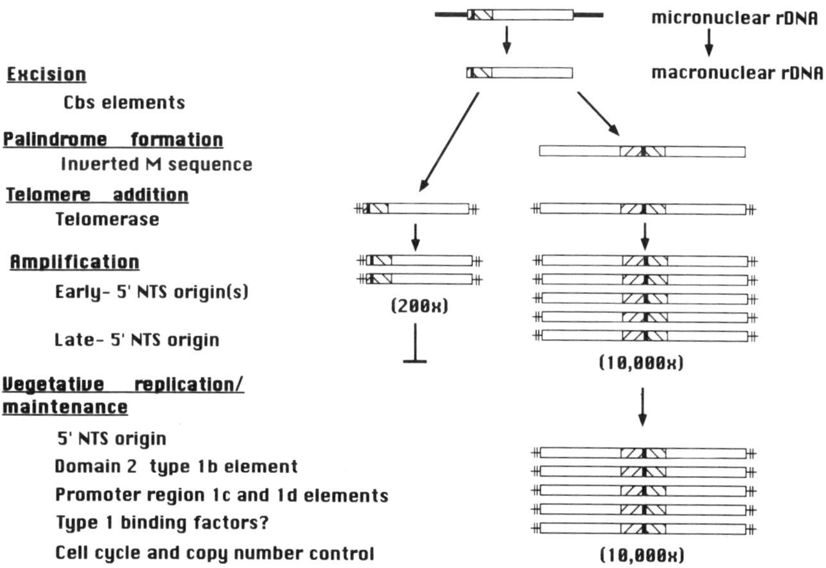

Figure 1 Model for the formation of the replication-competent palindrome. See text for explanations.

Table 1 Developmental pathway for amplification and rDNA minichromosome maintenance

Process	rDNA form	Control elements
Excision	Excised without telomeres	Cbs elements
Palindrome formation	21-kb minichromosome	inverted M repeats
Telomere addition	21-kb and 11-kb with telomeres	(telomerase)
Amplification	11 kb (10^2 copies)	
	21 kb (10^4 copies)	early-5′NTS origin(s) (complex initiation pattern) late-5′NTS origin
Vegetative replication	21 kb (10^4 copies)	5′NTS origin (single initiation event/molecule)

posed chromosome ends in vivo: Introducing a restriction endonuclease-cut DNA end by microinjection into the developing macronucleus was not sufficient to promote de novo telomere addition onto this cut end (Yao et al. 1990). This finding suggests a possible interaction between the chromosome breakage machinery and the specialized DNA polymerase, telomerase, during "chromosome healing." Telomerase has been shown to be responsible for de novo telomere addition onto rDNA and other macronuclear DNA ends during macronuclear development (Yu and Blackburn 1991), as well as for telomere maintenance during vegetative cell divisions (Yu et al. 1990; see Greider et al., this volume).

During macronuclear development, the rDNA monomer is rearranged into a 21-kb head-to-head palindrome (Fig. 1) (Karrer and Gall 1976). It is not known whether this occurs during or after excision of the rDNA from its flanking micronuclear chromosomal sequences. Free, linear monomeric rDNA minichromosomes, with telomeres at both ends, are formed in the developing macronucleus, but they do not normally persist during vegetative growth (Pan and Blackburn 1981; Challoner and Blackburn 1986). It is not known whether these linear 11-kb rDNA monomers are precursors to palindromic rDNA, or whether they are by-products of rDNA excision. Transformation of developing macronuclei with monomeric rDNA constructs has shown that palindrome formation is directed by the pair of inverted 42-bp elements (M sequences) present at the 5′ end of the rDNA monomer (Yasuda and Yao 1991). The inverted M sequences are separated by a 29-bp spacer. It has been proposed that they form a hairpin structure by intra-strand base-pairing in excised monomeric rDNA (Fig. 1). This small hairpin structure could promote the formation of a full-length monomeric rDNA hairpin by intramolecular recombination or by serving as a primer for replication. The giant hairpin molecule would then be converted into an rDNA palindrome during the next round of DNA replication. Transformation experiments with *T. thermophila* revealed that a variety of sequences could substitute for the M repeat sequences in this species, so long as they were present in the correct palindromic arrangement (Yasuda and Yao 1991). However, the M repeat sequences are evolutionarily conserved among a variety of *Tetrahymena* species (Engberg et al. 1972) and *Glaucoma chattoni* (Challoner and Blackburn 1986). Therefore, the M sequences may have other, currently unknown, roles.

Regulation of rDNA Amplification

The palindromic rDNA minichromosome is amplified to 10^4 copies in the developing macronucleus (Fig. 1). By comparison, other macronuclear chromosomes attain a copy number of about 45. In contrast to non-rDNA chromosomes, which arrest at 8–16 copies if starved throughout macronuclear development, macronuclear rDNA is replicated to high levels under these conditions (Kapler and Blackburn 1994). This implies that the control of rDNA replication in the developing macronucleus is not subject to the same checkpoint controls as non-rDNA chromosomes. The *rmm11* mutation revealed that rDNA excision is necessary, but not sufficient, for correct amplification (Kapler and Blackburn 1994). This *cis*-acting mutation resides within the 3′Cbs element and has only a weak effect on rDNA excision (a 75% reduction in the frequency of cells excising the rDNA). Surprisingly, amplification of correctly excised mutant alleles is severely impaired, so that the frequency of cells which amplify the mutant rDNA allele sufficiently for viability is about one in 10^5. Because the Cbs element containing this point mutation becomes separated from the rDNA molecule upon its excision, the downstream effect on amplification exerted by the *rmm11* mutation must be indirect. The available genetic and molecular evidence suggests that rDNA amplification is restricted to a specific developmental window, such that in this mutant even correctly excised rDNA molecules are unable to replicate to a high level in the developing macronucleus (Kapler and Blackburn 1994).

rDNA amplification in developing *Tetrahymena* macronuclei involves a minimum of 12 rounds of replication to reach the level of 10^4 molecules, if amplification is entirely geometric. Amplification occurs within a 12- to 14-hour window. The *cis*-acting sequences that directly regulate rDNA amplification are not yet well characterized. A natural genetic variant rDNA allele found in strain *B* of *T. thermophila* has a small deletion in domain 2 of the 5′ nontranscribed spacer (5′ NTS; see below and Fig. 2) relative to the strain *C3* rDNA allele. PCR experiments on individual cells undergoing macronuclear development revealed that this deletion sometimes results in decreased amplification of this rDNA allelic form (Orias and Bradshaw 1992). As described below, mutations in this element also affect vegetative replication of the rDNA (Larson et al. 1986), suggesting that amplification and vegetative replication share some common *cis*-acting determinants. Neutral-neutral two-dimensional gel revealed a complex pattern of replication intermediates in cells undergoing rDNA amplification (G.M. Kapler and E.H. Blackburn, unpubl.). Late in amplification, the predominant class of replication intermediates is identical to that of vegetative cells. This pattern is indicative of replication initiation in the central domain 1/domain 2 region of the 5′NTS (Table 1). Despite the presence of two copies of the 5′NTS, initiation occurs on only one side of the rDNA palindrome in these molecules. Since initiation from this origin is seen throughout most of the amplification period, cell cycle control of the 5′NTS origin must be suppressed in amplifying cells. A second pattern of replication intermediates is detected in amplifying cells. These novel intermediates generate

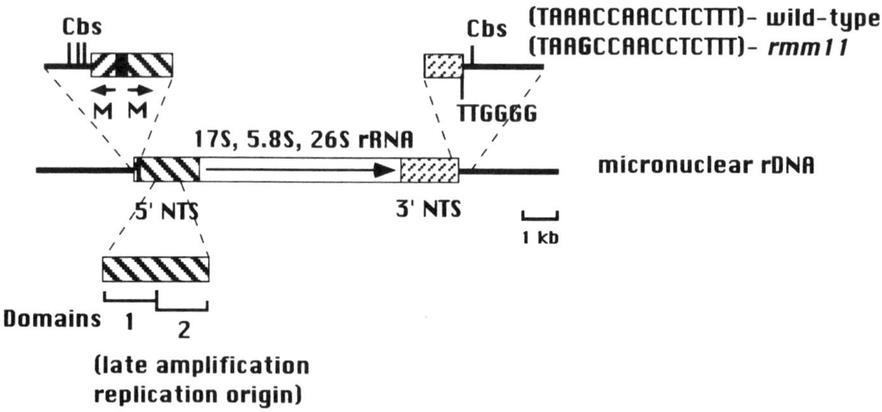

Figure 2 Micronuclear rDNA locus in *T. thermophila*. *cis*-Acting elements specifically required for rDNA maturation steps. (Cbs) Chromosome breakage sequence; (M) M repeats; (TTGGGG) telomeric repeat unit present at the site of new telomere addition. Sequences of wild-type (wt) and mutant (*rmm11*) Cbs element at 3′ end of rDNA are shown at upper right. See text for explanations.

the sole pattern seen at the earliest stage of amplification. Preiiminary studies suggest that initiation is also occurring in the vicinity of the 5′NTS; however, multiple initiation events can be seen in these rDNA palindromes. It remains to be determined whether these intermediates initiate at the central region or at new sites in the rDNA.

MAINTENANCE OF THE rDNA MINICHROMOSOME IN THE VEGETATIVE MACRONUCLEUS

Replication of Palindromic and Monomeric rDNA

In contrast to rDNA palindromes, which amplify to 10^4 copies during new macronuclear development, rDNA monomers replicate to about 200 copies in the developing macronucleus and do not persist in vegetative cells (Fig. 1) (Pan and Blackburn 1981). It is possible that proximity of the 5′NTS to telomeric DNA confers a replication disadvantage to rDNA monomers. In *Saccharomyces cerevisiae*, proximity to a telomere delays the activation of an early-firing replication origin; the efficiency of firing is not necessarily affected (Ferguson et al. 1991). In *Tetrahymena*, it is possible that rDNA palindrome formation provides a mechanism for distancing the rDNA replication origin from a repressive chromatin environment at the telomere. However, rDNA constructs defective in palindrome formation replicate stably as linear monomers in strains partially defective for maintenance of the endogenous palindrome (Yasuda and Yao 1991). Therefore, palindromic configuration is not required for origin activity. In the related holotrichous ciliate *Glaucoma chattoni*, and in hypotrichous ciliates, macronuclear rDNA exists exclusively as monomeric linear molecules (Katzen et al. 1981 and references therein). Thus, although palindrome formation may confer a replication and/or maintenance advantage in *T. thermophila*, it is not required for long-term rDNA maintenance in the ciliate macronucleus.

Sequence and Structural Features of the rDNA Origin Region

As shown in Figure 3, in the 21-kb palindromic rDNA molecule the two rRNA transcription units are separated by about 3.8 kb of noncoding sequence (two inverted identical 1.9-kb sequences) spanning the center of symmetry of the molecule. Several approaches—physical mapping, genetic analyses, and macronuclear transformation—have shown that the 1.9-kb 5′NTS contains the vegetative replication origin, *cis*-acting sequences involved in the vegetative replication and/or maintenance of the

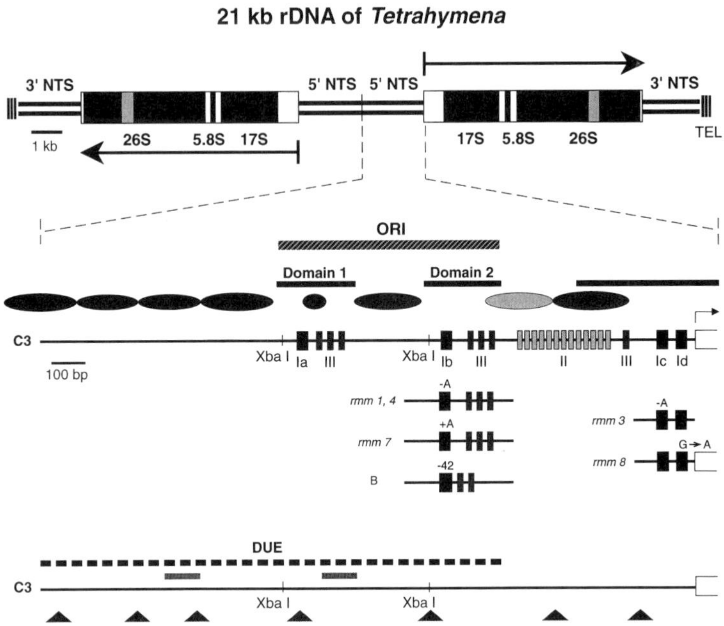

Figure 3 Structural and functional features of the macronuclear palindromic rDNA. (*Top*) Thick bar, rRNA transcription unit and coding regions (*black*); thin open bars, 5′NTS and 3′NTS; vertical lines, telomeric DNA repeat region. (*Middle*) Line, expanded version of 5′NTS of *C3* strain rDNA. Open bar at right, 5′ end of rRNA transcription unit; vertical black rectangles, type I and type III repeats; tall stippled rectangles, type II repeats; base sequence changes in various mutants (*rmm1, 3, 4, 7, 8*), or *B* strain rDNA (*B*) are shown below the line; ovals, nucleosomes; black horizontal bars, DNase I hypersensitive regions, as indicated; origin, region to which functional origin has been mapped. (*Bottom*) Structural and sequence features of bare rDNA. Hatched bar, DUE (calculated); gray bars, mung bean hypersensitive regions; filled triangles, positions of bent DNA. See text for explanations.

rDNA minichromosome, and the late-amplification replication origin described above (for review, see Larson et al. 1991; Kapler 1993). The vegetative origin has been physically mapped both by electron microscopy and by two-dimensional gel electrophoresis. Based on the assumptions that initiation occurs in the center of replication bubble structures and proceeds bidirectionally at equal rates, the electron microscopic mapping places the origin within an approximately 600-bp interval located about 650 bp from the center of the palindrome in the *B* rDNA strain (Cech and Brehm 1981). Analysis of rDNA replicative intermediates using two-dimensional gel electrophoresis shows that all detectable initiation events (i.e., > 95% of initiation events) occur within the 5′NTS (G.M. Kapler and E.H. Blackburn, unpubl.).

The rDNA replication origin region contains two approximately 270-bp nuclease hypersensitive domains, designated domain 1 and domain 2 (see Fig. 3), that flank an array of seven precisely positioned nucleosomes centered at the axis of symmetry of the palindrome. A third nuclease hypersensitive domain and a distinctive DNase I footprint encompass the rRNA promoter and transcription initiation region (Bonven and Westergaard 1982; Palen and Cech 1984; Pan et al. 1995; R. Gallagher and E. Blackburn, unpubl.). Domains 1 and 2 are each contained within an approximately 400-bp sequence that is in direct tandem repeat; there is some sequence divergence between the two repeats (see Fig. 3). In another *Tetrahymena* species, *T. pyriformis*, the corresponding sequence and its hypersensitive domain are present only once in the 5′NTS (Niles et al. 1981; Palen and Cech 1984). Although domain 1 (located between 640 and 935 nucleotides from the palindromic center) most closely coincides with the vegetative origin mapped by electron microscopy, domain 2 also lies within the broad region of the origin defined by two-dimensional gel electrophoresis of replicating intermediates. The 5′NTS also contains three families of evolutionarily conserved repeated sequences referred to as type I, type II, and type III repeats (Fig. 3) (Niles et al. 1981; Challoner et al. 1985). Each domain contains one type I and three type III repeats in a spatial arrangement along the DNA that is highly conserved between the domains and between holotrichous ciliate species (Challoner et al. 1985). The type I repeats were identified in genetic studies (Larson et al. 1986; Yaeger et al. 1989) as *cis*-acting replication control elements that are also implicated in the regulation of transcription (see below): The type I repeat immediately upstream of the transcription start site has been shown to be essential for rRNA transcription in vitro (Miyahara et al. 1993; R.R. Pearlman, pers. comm.). The type III repeats are binding and cleavage sites for DNA topoisomerase I (Bonven et al. 1985). Type II repeats are located between the promoter region and repeat 2. Two of these are required for maintenance of the rDNA as an episome (M.-C. Yao, pers. comm.).

Additional specific DNA sequence motifs and structural elements shared with other eukaryotic chromosomal origins of replication have been identified in *Tetrahymena* rDNA (Dobbs et al. 1994). The rDNA origin comprises a large region of easily unwound DNA (a putative DNA unwinding element, DUE), coincident with predicted bent DNA segments, nuclear matrix attachment consensus sequences (MARs/SARs), and yeast ARS consensus sequences. Three predicted structural features have been verified experimentally: (1) Two mung bean nuclease-hypersensitive sites in supercoiled plasmid DNA are located within the major DUE-like element predicted by thermodynamic analyses; (2) three restriction fragments of the 5′NTS region which are predicted to contain bent DNA segments do exhibit anomalous migration characteristic of bent DNA during electrophoresis on polyacrylamide gels; (3) restriction fragments of the 5′NTS region bind nuclear matrices in an in vitro binding assay, consistent with an association of the replication origin region with the nuclear matrix in vivo (Du et al. 1995). Although functional roles for DUE-like elements, bent DNA, and matrix-associated sequences in *Tetrahymena* rDNA replication have not been directly demonstrated, the association of these three structural elements with other eukaryotic replication origins is in accordance with the hypothesis that they may contribute to origin function in vivo (Benbow et al. 1992; Dobbs et al. 1994).

Studies of rDNA Replication by Genetics and Transformation

A classic genetic approach has been used to identify *cis*-acting sequences involved in rDNA origin function (Larson et al. 1986; Yaeger et al. 1989; Kapler and Blackburn 1994; Kapler et al. 1994; R.C. Gallagher et al., unpubl.). These studies exploited the observation that the rDNA allele of inbred strain *C3* has a replication and/or maintenance advantage over that of inbred strain *B* when both alleles reside in the same macronucleus (Pan et al. 1982). Thirteen *rmm* (*r*DNA *m*aturation and *m*aintenance) mutations that abolish the advantage of *C3* rDNA have been characterized (Larson et al. 1986; Yaeger et al. 1989; Kapler and Blackburn 1994; Kapler et al. 1994; R.C. Gallagher et al., unpubl.). These mutations fall into two major classes: (1) maturation mutants (*rmm6, 10, 11, 12*, and *13*) that show defects during excision, palindrome formation, and amplification in the developing macronucleus (see above) and (2) maintenance mutants (*rmm1, 2, 3, 4, 5, 7, 8,* and *9*) that appear to affect rDNA replication and/or maintenance during vegetative growth. Several mutations of the second class have been mapped to the 5′NTS (Larson et al. 1986; Yaeger et al. 1989; R.C. Gallagher et al., unpubl.) both within the physically mapped origin region and in the rRNA promoter (Fig. 3). The replication disadvantage exhibited by the *B* rDNA allele relative to *C3* rDNA is the result of a 42-bp deletion that removes sequences immediately downstream from the type Ib repeat in domain 2. Three mutations (*rmm1, 4,* and *7*) alter the same type I repeat (Ib) in the origin region, whereas a fourth (*rmm3*) alters a type I repeat (Ic) located less than 100 bp from the transcription initiation site (Fig. 2). On the basis of these studies, it was proposed that competition for limiting *trans*-acting factors that interact with the type I repeats could explain the differential replication properties of rDNA alleles (Larson et al. 1986). In addition, the *rmm8* mutation has recently been shown to alter another conserved promoter sequence (R.C. Gallagher and E.H. Blackburn, unpubl.). The *rmm8* and *rmm3* promoter changes are about 0.6 kb from the mapped origin region, suggesting the possibility of long-range interactions between these regions.

The importance of the 5′NTS region in rDNA replication and/or maintenance has been directly demonstrated in experiments using rDNA constructs to transform macronuclei by microinjection or electroporation (Tondravi and Yao 1986; Yu et al. 1988; Yu and Blackburn 1989, 1990; Gaertig and Gorovsky 1992; Pan and Blackburn 1995). Circular constructs containing only one half of the palindrome or one copy of the 5′NTS alone do transform macronuclei, but they do so by homologous recombination with the endogenous rDNA (Yu et al. 1988). In contrast, constructs containing a tandem duplication of the 5′NTS adjacent to the rest of the rRNA gene exhibit both autonomous replication properties and a replication advantage over the endogenous palindromic molecules: Intermolecular homologous recombination events between the transformed construct and the endogenous rDNA give rise to rDNA replicons containing progressively larger numbers of tandem 5′NTS repeats (Yu and Blackburn 1990; Romero and Blackburn 1995). Cells eventually accumulate linear rDNA minichromosomes carrying the rRNA coding sequences linked to more than 30 copies of the 5′NTS repeats. These rDNA replicons are maintained at high copy number and often completely replace the endogenous rDNA allele. Similarly, constructs containing the same tandem duplication of the 5′NTS unlinked to rRNA gene sequences also replicate autonomously and give rise to plasmids containing long tandem arrays of solely 5′NTS repeats, lacking the original bacterial vector sequence (Pan and Blackburn 1995). These tandem 5′NTS arrays exhibit a strong replication advantage in the absence of selection and eventually accumulate to about 10% of the total cellular DNA.

A "minimal" origin of replication for the rDNA minichromosome has not yet been clearly defined. The limitation in currently available systems lies in the fact that any introduced replicon has to compete with the rRNA genes required in high copy numbers for viability. However, the genetic analysis described above, together with recent deletion analyses, suggests that a large portion of the 5′NTS region is required for replication and/or maintenance of the rDNA in the macronucleus (M.-C. Yao, pers. comm.). Hence, the functional origin of replication in *Tetrahymena* rDNA may be considerably larger than the compact chromosomal origins of *S. cerevisiae* (Marahrens and Stillman 1994). Moreover, as with many eukaryotic viral origins, replication initiation and transcription are likely to share functional controlling elements.

Inhibition of Origin Function in rDNA Constructs by Transcription through the Origin

The replication advantage displayed by the tandem duplication of the 5′NTS plus promoter segment occurs only when the upstream copies of the 5′NTS carry an inactivated rRNA promoter. Such inactivation was first proposed for a spontaneous mutation that occurred in a highly conserved sequence in the rRNA promoter (+G at position –20, see Fig. 3) (Yu and Blackburn 1989) and allowed this tandemly repeated region to be maintained. Subsequently, the +G promoter mutation was shown directly to destroy promoter function (Pan et al. 1995). Failure to maintain tandem wild-type 5′NTS-plus-promoter segments appears to be due to transcription through the origin, because the replication advantage is restored by mutating or deleting the rRNA promoter of this 5′NTS (Yu and Blackburn 1989) or by inserting a transcriptional termination signal immediately downstream from the wild-type promoter (Pan et al. 1995).

Regulation of Vegetative rDNA Replication

Although competition for limiting amounts of a *trans*-acting replication factor that binds type I repeat sequences may be an important factor in rDNA maintenance control, it cannot be the sole determinant of rRNA gene copy number. When additional copies of the rDNA origin region were introduced into macronuclei, either linked or unlinked to rRNA genes, the overall copy number of rRNA coding sequences remained constant, even though rDNA molecules with up to 30 tandem copies of the origin region accumulated over time (Yu and Blackburn 1990). The rRNA gene copy number also remained constant in the presence of a threefold excess of tandem repeats of the origin region located on separate molecules from the rRNA genes themselves (Pan and Blackburn 1995). Thus, rRNA gene dosage is not controlled solely by monitoring the copy number of the origin region, but must involve additional sequences either in the rRNA coding region or the 3′NTS (Pan and Blackburn 1995).

DNA/Protein Interactions in the rDNA Origin Region

DNA/protein interactions in the replication origin region have been characterized by DNase I and Fe-EDTA footprinting of partially purified factors on origin DNA (Umthun et al. 1994; Hou et al. 1995) and DNase I, DMS, and $KMnO_4$ footprinting of this region in nuclei (Pan et al. 1995; R.C Gallagher and E.H. Blackburn, unpubl.). These studies reveal a complex pattern of protections and hypersensitivities throughout the origin and promoter regions, with pronounced protection of nucleotides within and adjacent to the 33-nucleotide conserved type I repeat sequence. Two classes of proteins that bind the type I repeat in vitro have been identified and characterized using synthetic oligonucleotides in electrophoretic mobility shift assays (Umthun et al. 1994). One of these, ds-TIBF, binds preferentially to duplex DNA and exhibits only moderate specificity for type I repeat sequences. In contrast, an abundant single-stranded DNA-binding protein, ssA-TIBF, specifically recognizes the A-rich strand of the type I repeat sequence. Quantitative binding competition experiments demonstrate that sequences in 3′flanking DNA, in addition to those in the 3′end of the conserved type I repeat, are important determinants for binding (Hou et al. 1995). Recently, the most highly purified fractions (~70,000-fold enriched) of ssA-TIBF have been shown to contain an ATP-dependent DNA helicase activity, suggesting the exciting possibility that the ssA-TIBF type I repeat-binding activity and the copurifying DNA helicase activity are functionally associated in vivo (C. Du et al., unpubl.).

The ssA-TIBF exhibits differential affinity for *C3* versus *B* rDNA sequences in vitro, which parallels differences in the ability of the *C3* and *B* alleles to compete for replication and/or maintenance in the macronucleus. There is at least a 25-fold higher affinity of ssA-TIBF for *C3* versus *B* rDNA sequences (Hou et al. 1995). It should be noted, however, that several induced *C3-rmm* mutations are due to a single base pair deletion or insertion in the central tract of 11 A residues in copies of the type I repeat (Larson et al. 1986; Yaeger et al. 1989; W.-L. Shaiu and D.L. Dobbs; R.C. Gallagher et al., unpubl.). DNase I footprinting experiments using purified ssA-TIBF indicate that it strongly protects nucleotides at the 3′ end of, and flanking, the type I repeat, but makes few contacts in the central A+T-rich region (Hou et al. 1995). In contrast, both strands of the A+T tract are protected in footprinting experiments performed with S100 extracts of *Tetrahymena* cells (Umthun et al. 1994). It seems likely, therefore, that additional proteins that recognize type I repeat sequences could be involved in the replication phenotypes of these *rmm* mutants. The abundance and high affinity of ssA-TIBF for type I repeats suggests that it may have an important function in the cell. Since no reliable sequence-dependent in vitro replication system has been reported in *Tetrahymena*, it is not yet possible to directly assess the role of ssA-TIBF in rDNA replication.

The conserved 16-bp type III repeats have been identified as DNA topoisomerase I binding sites in vitro and in vivo (Bonven et al. 1985). The association of topoisomerase I with rDNA sequences was first detected in experiments in which treatment of rDNA chromatin with SDS prior to high ionic strength buffer resulted in site-specific single-strand cleavages by the endogenous topoisomerase I, which remained covalently attached to the 3′ end generated by cleavage (Bonven and Westergaard 1982). Three topoisomerase I cleavage sites map within each of the three nuclease hypersensitive domains in the 5′NTS (domains 1 and 2 and the rRNA promoter; see Fig. 3); a fourth maps near the transcription termination site (Gocke et al. 1983; Bonven et al. 1985). Since these topoisomerase I binding sites are found in the replication origin region and at both ends of the transcription unit, topoisomerase I/type III repeat interactions may play a dual role in the replication and transcription of the rDNA (Bonven et al. 1985).

SUMMARY AND FUTURE DIRECTIONS

The rDNA of *Tetrahymena* has some unique advantages as a model eukaryotic system for studying replication control. The presence of the rRNA gene as a single-copy germ-line gene, making Mendelian genetics possible, is most unusual among both eukaryotes and prokaryotes. The rDNA minichromosome is generated by a set of closely regulated, developmentally programmed events, which can be induced to occur synchronously in large cultures, allowing for molecular analysis. The small size of this DNA, its high level of amplification, and its highly ordered chromatin facilitate molecular and biochemical characterizations, making it comparable to viral replicons in these respects. However, the *Tetrahymena* rDNA represents an unusual situation in that it is a nuclear cellular replicon under both cell-cycle control and copy-number control. The copy-number control of the replicon permits competition between replicons in vivo, allowing very small differences in such competition to be detected over several vegetative divisions with high sensitivity.

Although additional determinants of rDNA origin function very likely include aspects of chromatin structure that have not yet been investigated, the biochemical, molecular, and classic genetic studies of rDNA replication suggest that the functional rDNA origin in vegetatively dividing cells is contained within an approximately 1.9-kb region corresponding to the 5′NTS. *cis*-Acting sequences involved in replication control appear to be functionally redundant and dispersed throughout a region of low intrinsic helical stability. The type I repeat, for example, is present in four copies within the 5′NTS. As one of very few genetically defined replication control elements so far described in eukaryotic cellular origins, it is interesting that the type I repeat is also an rRNA gene promoter component. Hence, the involvement of transcription factor binding sites in origin function, which has been clearly demonstrated in eukaryotic viruses and yeast, also appears to be a feature of the *Tetrahymena* rDNA origin. No putative replication initiator proteins or binding sites have been identified in *Tetrahymena*.

The recent development of rDNA origin-containing vectors with selectable markers that replicate autonomously in macronuclei (Gaertig et al. 1994; Pan and Blackburn 1995), and progress toward developing an in vitro replication system (D. Dobbs, unpubl.), should facilitate further delineation of the *cis*-acting elements and other functional components in rDNA origin function.

ACKNOWLEDGMENTS

We thank Renata Gallagher for helpful discussions, Zhen Hou for assistance with Figure 3, and the National Institutes of Health for support for research in the laboratories of E.H.B. and D.L.D.

REFERENCES

Altschuler, M.I. and M.-C. Yao. 1985. Macronuclear DNA of *Tetrahymena thermophila* exists as defined subchromosomal-sized molecules. *Nucleic Acids Res.* **13:** 5817–5831.

Anderson, H.A. 1972. Requirements for DNA replication preceding cell division in *Tetrahymena pyriformis*. *Exp. Cell Res.* **75:** 89–94.

Benbow, R.M., J. Zhao, and D.D. Larson. 1992. On the nature of origins of DNA replication in eukaryotes. *BioEssays* **14:** 661–670.

Bonven, B.J. and O. Westergaard. 1982. DNAase I hypersensitive regions correlate with a site-specific endogenous nuclease activity on the r-chromatin in *Tetrahymena*. *Nucleic Acids Res.* **10:** 7593–7608.

Bonven, B.J., E. Gocke, and O. Westergaard. 1985. A high affinity topoisomerase I binding sequence is clustered at DNAase I hypersensitive sites in *Tetrahymena* r-chromatin. *Cell* **41:** 541–551.

Budarf, M.L. and E.H. Blackburn. 1986. Chromatin structure of the telomeric region and 3′ nontranscribed spacer of *Tetrahymena* ribosomal RNA genes. *J. Biol. Chem.* **261:** 363–369.

Cech, T.R. and S.L. Brehm. 1981. Replication of the extrachromosomal ribosomal RNA genes of *Tetrahymena thermophila*. *Nucleic Acids Res.* **9:** 3531–3543.

Challoner, P.B. and E.H. Blackburn. 1986. Conservation of sequences adjacent to the telomeric C_4A_2 repeats of ciliate macronuclear ribosomal RNA gene molecules. *Nucleic Acids Res.* **14:** 6299–6311.

Challoner, P.B., A.A. Amin, R.E. Pearlman, and E.H. Blackburn. 1985. Conserved arrangements of repeated DNA sequences in nontranscribed spacers of ciliate ribosomal RNA genes: Evidence for molecular coevolution. *Nucleic Acids Res.* **13:** 2661–2680.

Dobbs, D.L., W.-L. Shaiu, and R.M. Benbow. 1994. Modular sequence elements associated with origin regions in eukaryotic chromosomal DNA. *Nucleic Acids Res.* **22:** 2479–2489.

Doerder, R.P. 1979. Regulation of macronuclear DNA content in *Tetrahymena thermophila*. *J. Protozool.* **26:** 28–35.

Du, C., R. Sanzgiri, W.L. Shaiu, J.-K. Choi, Z. Hou, R.M. Benbow, and D.L. Dobbs. 1995. Modular structural elements in the replication origin region of *Tetrahymena* rDNA. *Nucleic Acids Res.* **23:** 1766–1774.

Engberg, J. and H. Nielsen. 1990. Complete sequence of the extrachromosomal rDNA molecule of the ciliate *Tetrahymena thermophila* strain B1868VII. *Nucleic Acids Res.* **18:** 6915–6919.

Engberg, J., D. Mowatt, and R.E. Pearlman. 1972. Preferential replication of the ribosomal RNA genes during a nutritional shift-up in *Tetrahymena pyriformis*. *Biochim. Biophys. Acta* **272:** 312–320.

Ferguson, B.M., B.J. Brewer, A.E. Reynolds, and W.L. Fangman. 1991. A yeast origin of replication is activated late in S phase. *Cell* **65:** 507–515.

Gaertig, J. and M.A. Gorovsky. 1992. Efficient mass transformation of *Tetrahymena thermophila* by electroporation of conjugants. *Proc. Natl. Acad. Sci.* **89:** 9196–9200.

Gaertig, J., L, Gu, B. Hai, and M. Gorovsky. 1994. Gene replacement and high frequency vector-mediated electrotransformation in *Tetrahymena*. *Nucleic Acids Res.* **22:** 5391–5398.

Gocke, E., B.J. Bonven, and O. Westergaard. 1983. A site and strand specific nuclease activity with analogies to topoisomerase I frames the rRNA gene of *Tetrahymena*. *Nucleic Acids Res.* **11:** 7661–7678.

Godiska, R. and M.-C. Yao. 1990. A programmed site-specific DNA rearrangement in *Tetrahymena thermophila* requires flanking polypurine tracts. *Cell* **61:** 1237–1246.

Hou, Z., A.R. Umthun, and D.L. Dobbs. 1995. A single-stranded DNA binding protein that specifically recognizes *cis*-acting sequences in the replication origin and transcriptional promoter region of *Tetrahymena* rDNA. *Biochemistry* **34:** 4583–4592.

Kapler, G.M. 1993. Developmentally regulated processing and replication of the *Tetrahymena* rDNA minichromosome. *Curr. Opin. Genet. Dev.* **3:** 730–735.

Kapler, G.M. and E.H. Blackburn. 1994. A weak germ-line excision mutation blocks developmentally controlled amplification of the rDNA minichromosome of *Tetrahymena thermophila*. *Genes Dev.* **8:** 84–95.

Kapler, G.M., E. Orias, and E.H. Blackburn. 1994. *Tetrahymena thermophila* mutants defective in the developmentally programmed maturation and maintenance of the rDNA minichromosome. *Genetics* **137:** 455–466.

Karrer, K.M. 1986. The nuclear DNAs of holotrichous ciliates. In *The molecular biology of ciliated protozoa* (ed. J.G. Gall), pp. 85–110. Academic Press, Orlando, Florida.

Karrer, K.M. and J.G. Gall. 1976. The macronuclear ribosomal DNA of *Tetrahymena pyriformis* is a palindrome. *J. Mol. Biol.* **104:** 421–453.

Katzen, G.M., G.M. Cann, and E.H. Blackburn. 1981. Sequence-specific fragmentation of macronuclear DNA in a holotrichous ciliate. *Cell* **24:** 313–320.

Larson, D.D., A.R. Umthun, and W.-L. Shaiu. 1991. Copy number control in the *Tetrahymena* macronuclear genome. *J. Protozool.* **38:** 258–263.

Larson, D.D., E.H. Blackburn, P.C. Yaeger, and E. Orias. 1986. Control of rDNA replication in *Tetrahymena* involves a *cis*-acting upstream repeat of a promoter element. *Cell* **47:** 229–240.

Marahrens, Y. and B. Stillman. 1994. Replicator dominance in a eukaryotic chromosome. *EMBO J.* **13:** 3395–3400.

Miyahara, K., N. Hashimoto, T. Higashinakagawa, and R.E. Pearlman. 1993. Common sequence elements are important for transcription and replication of the extrachromosomal rRNA-encoding genes of *Tetrahymena. Gene* **127:** 209–213.

Niles, E.G., J. Sutiphong, and S. Haque. 1981. Structure of the *Tetrahymena pyriformis* rRNA gene: Nucleotide sequence of the transcription initiation region. *J. Biol. Chem.* **256:** 12849–12856.

Orias, E. 1986. Ciliate conjugation. In *The molecular biology of ciliated protozoa* (ed. J.G. Gall), pp. 45–84. Academic Press, Orlando, Florida.

Orias, E. and A.D. Bradshaw. 1992. Stochastic developmental variation in the ratio of allelic rDNAs among newly differentiated, heterozygous macronuclei of *Tetrahymena thermophila. Dev. Genet.* **13:** 87–93.

Palen, T.E. and T.R. Cech. 1984. Chromatin structure at the replication origins and transcription initiation regions of the ribosomal RNA genes of *Tetrahymena. Cell* **36:** 933–942.

Pan, W.-J. and E.H. Blackburn. 1981. Single extrachromosomal ribosomal RNA gene copies are synthesized during amplification of the rDNA in *Tetrahymena. Cell* **23:** 459–466.

———. 1995. Tandem repeats of the 5′ non-transcribed spacer of *Tetrahymena* rDNA function as high copy number autonomous replicons in the macronucleus but do not prevent rDNA gene dosage regulation. *Nucleic Acids. Res.* **23:** 1561–1569.

Pan, W.-J., R.C. Gallagher, and E.H. Blackburn. 1995. Replication of an rDNA origin plasmid in the *Tetrahymena* macronucleus is prevented by transcription through the origin from an active RNA polymerase I promoter. *Mol. Cell. Biol.* **15:** 3372–3381.

Pan, W.C., E. Orias, M. Flacks, and E.H. Blackburn. 1982. Allele-specific, selective amplification of a ribosomal RNA gene in *Tetrahymena thermophila. Cell* **28:** 595–604.

Preer, J.R. and L.B. Preer. 1979. The size of macronuclear DNA and its relationship to model for maintaining genic balance. *J. Protozool.* **26:** 14–18.

Romero, D.P. and E.H. Blackburn. 1995. Circular rDNA replicons persist in *Tetrahymena thermophila* transformants synthesizing GGGGTC telomeric repeats. *J. Eukaryot. Microbiol.* **42:** 32–43.

Spangler, E.A., T. Ryan, and E.H. Blackburn. 1988. Developmentally regulated telomere addition in *Tetrahymena thermophila. Nucleic Acids Res.* **16:** 5569–5585.

Tondravi, M.M. and M.-C. Yao. 1986. Transformation of *Tetrahymena thermophila* by microinjection of ribosomal RNA genes. *Proc. Natl. Acad. Sci.* **83:** 4369–4373.

Truett, M.A. and J.G. Gall. 1977. The replication of ribosomal DNA in the macronucleus of *Tetrahymena. Chromosoma* **64:** 295–303.

Umthun, A.R., Z. Hou, Z.A. Sibenaller, W.-L. Shaiu, and D.L. Dobbs. 1994. Identification of DNA-binding proteins that recognize a conserved type I repeat sequence in the replication origin region of *Tetrahymena* rDNA. *Nucleic Acids Res.* **22:** 4432–4440.

Yaeger, P.C., E. Orias, W.-L. Shaiu, D.D. Larson, and E.H. Blackburn. 1989. The replication advantage of a free linear rRNA gene is restored by somatic recombination in *Tetrahymena thermophila. Mol. Cell. Biol.* **9:** 452–460.

Yao, M.-C. 1989. Site-specific chromosome breakage and DNA deletion in ciliates. In *Mobile DNA* (ed. D.E. Berg and M.M. Howe), pp. 714–734. American Society for Microbiology, Washington, D.C.

Yao, M.-C., A.R. Kimmel, and M.A. Gorovsky. 1974. A small number of cistrons for ribosomal RNA in the germ-line nucleus of a eukaryote, *Tetrahymena pyriformis. Proc. Natl. Acad. Sci.* **71:** 3082–3086.

Yao, M.-C., C.-H. Yao, and B. Monks. 1990. The controlling sequence for site-specific chromosome breakage in *Tetrahymena. Cell* **63:** 763–772.

Yao, M.-C., K. Zheng, and C.-H. Yao. 1987. A conserved nucleotide sequence at the site of developmentally regulated chromosomal breakage in *Tetrahymena. Cell* **48:** 779–788.

Yasuda, L.F. and M.-C. Yao. 1991. Short inverted repeats at a free end signal large palindromic DNA formation in *Tetrahymena. Cell* **67:** 505–516.

Yu, G.-L. and E.H. Blackburn. 1989. Transformation of *Tetrahymena thermophila* with a mutated circular ribosomal DNA plasmid vector. *Proc. Natl. Acad. Sci.* **86:** 8487–8491.

———. 1990. Amplification of tandemly repeated origin control sequences confers a replication advantage on rDNA replicons in *Tetrahymena thermophila. Mol. Cell. Biol.* **10:** 2070–2080.

———. 1991. Developmentally programmed healing of chromosomes by telomerase in *Tetrahymena. Cell* **67:** 823–832.

Yu, G.-L., M. Hasson, and E.H. Blackburn. 1988. Circular ribosomal DNA plasmids transform *Tetrahymena thermophila* by homologous recombination with endogenous macronuclear ribosomal DNA. *Proc. Natl. Acad. Sci.* **85:** 5151–5155.

Yu, G.-L., J.D. Bradley, L.D. Attardi, and E.H. Blackburn. 1990. *In vivo* alteration of telomeric sequences and senescence caused by mutated *Tetrahymena* telomerase RNAs. *Nature* **344:** 126–132.

34
DNA Replication in *Physarum*

Gérard Pierron and Marianne Bénard
Laboratoire Organisation Fonctionelle du Noyau, UPR 9044 - CNRS
Institut de Recherches sur le Cancer
Villejuif 94801, France

The natural synchrony of the nuclear cycle within the plasmodium of the myxomycete *Physarum polycephalum* provides an opportunity to study, without the need for any cell treatment, the complex and transient patterns of replication intermediates generated during chromosomal replication. In this review, we focus on the parameters of replication kinetics from the synthesis of Okazaki fragments to the chromosome-sized progeny molecules and summarize data suggesting that their appearance at some specific loci is precisely programmed and intimately linked to the transcriptional activity of the cell.

STRUCTURE OF S PHASE

The Plasmodium

In one phase of their life cycle, the myxomycetes develop a particular multinucleated cell type, the plasmodium, that results from nuclear divisions in the absence of cell division (Rusch 1980). Deposited on a filter paper, small plasmodia cultivated in shaken liquid cultures fuse spontaneously into a single cell (5–10 cm in diameter) that contains more than 10^8 diploid nuclei (Fig. 1A). Being in a common cytoplasm, these nuclei divide synchronously every 10 hours.

Their division is characterized by the persistence of the nuclear envelope throughout the mitotic process. This "closed" mitosis should not be seen as a loss of capability for cell division, but rather as the result of a developmental process, since in the amoebal, haploid vegetative phase of *Physarum*, the uninucleated cells divide by a conventional mitosis in which the nuclear envelope disintegrates in prophase. Thus, the *Physarum* genome encodes different cell types, with the plasmodial stage particularly suitable for DNA replication studies.

The Synchronous Nuclear Cycle

The synchrony of mitosis within the plasmodium was first described by Howard in 1932 and was extended to the synchrony of S phase by the

DNA Replication in Eukaryotic Cells
 0-87969-459-9/96 $5 + .00

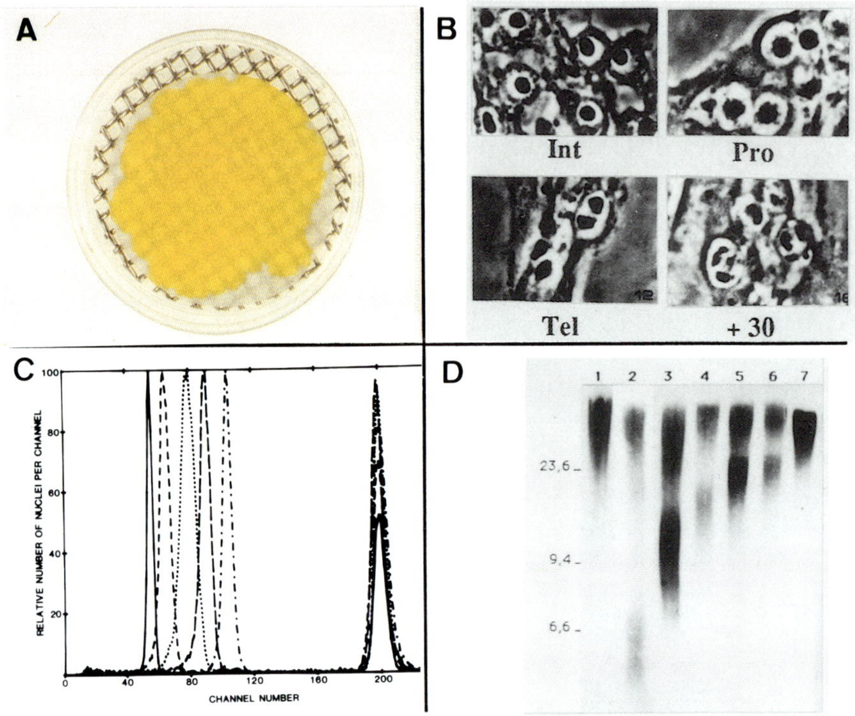

Figure 1 Synchrony of the nuclear cycle in the plasmodium of *Physarum*. (*A*) Macroscopic aspect of a plasmodium containing about 4 x 10^8 synchronous nuclei. Abruptly, in early S phase, an amount of DNA corresponding to 600 km of double-helix is synthesized per min in that cell. (*B*) Cell-cycle stage determined by phase-contrast microscopy on ethanol-fixed smears. Interphasic (Int) nuclei are characterized by a large, centrally located nucleolus, and condensed chromosomes are apparent in prometaphase (Pro), and a few minutes later in telophase (Tel). In the absence of G_1, the onset of S phase is morphologically defined by observation of telophase. Fusion of the prenucleolar bodies occurs in early S phase (+30). (*C*) Flow-cytometric analysis of the natural synchrony of S phase within a plasmodium. Internal standard chicken erythrocyte histograms overlap at channel 200. Lefthand peaks represent Hoechst-stained *Physarum* nuclei isolated at 5 min, 30 min, 1 hr, 2 hr, and 7 hr after the onset of S phase. From the positions of these peaks, the rate of DNA synthesis has been estimated at 5000 kb/min/nucleus in early S phase. (Reprinted, with permission, from Kubbies and Pierron 1983.) (*D*) Nascent-strand elongation of the LAV 1–2 replicon on alkaline gel. The growth of the replicon is perceptible in total DNA samples taken every 5 min in early S phase at 5, 10, 15, 25, 30, and 40 min post-telophase (lanes *2–7*). In contrast, only the parental DNA hybridized to the probe in the G_2-phase sample (lane *1*, above 23.6 kb). The kinetics of elongation of this particular replicon is estimated at 1 kb/min. The smearing of the nascent strands suggests a 5-min asynchrony within the 600-min-long nuclear cycle. (Reprinted, with permission, from Bénard and Pierron 1990.)

use of radioactive precursors (Nygaard et al. 1960; Braun et al. 1965). During this procedure, a remarkable feature of the nuclear cycle was discovered: the absence of a G_1 phase. Thus, the synchronous onset of S phase is morphologically defined by the observation of telophase (Fig. 1B). The 3-hour S phase is followed by a 6- to 7-hour G_2 phase, leading to the next mitosis (Fig. 1B).

The degree of synchrony has been estimated by flow cytometry of Hoechst-stained nuclei isolated at various time points after mitosis (Fig. 1C). Narrow and symmetrical histograms of DNA fluorescence were found, leaving no doubt that 99% of the nuclei enter into and exit from S phase simultaneously. This unique level of synchrony can be appreciated by measuring, at one time point, the heterogeneity in length of the 10^8 copies of a replicon. In Figure 1D, the kinetics of elongation of a specific replicon is measured at 1 kb/min up to a size of 30 kb, yet the dispersion of the nascent strands at a given time point is limited to about 5 kb, suggesting a 5-minute asynchrony within the 600-minute-long nuclear cycle, i.e., a 1% temporal window (Bénard and Pierron 1990).

Kinetics of DNA Synthesis

In *Physarum*, the C value is equal to 0.3 pg, defining a genome size about 20 times larger than yeast or one-tenth of the human genome (Mohberg et al. 1973). Nevertheless, renaturation kinetics indicate that two-thirds of the DNA is single-copy, the rest being composed of interspersed middle-repetitive sequences (Hardman et al. 1980). Incorporation studies (Braun et al. 1965) and flow-cytometric measurements (Fig. 1C) reproducibly indicate that within 5 minutes of the uncoiling of telophase chromosomes, a maximal rate of DNA synthesis of about 5000 kb/min per nucleus is reached and maintained for 90 minutes. At this point, 75% of the genome is replicated. It takes another 90 minutes to replicate the late-replicating quarter of the genome, due to a slowdown of DNA synthesis at about 1200 kb/min per nucleus. However, the mean size and the elongation rate of the replicons appear to be constant throughout S phase (Funderud et al. 1978b).

Structure of the Replicons

The replicons were found to be heterogeneous in size at 12–60 kb, with a mean size of about 35 kb, by sedimentation on alkaline sucrose gradients (Fig. 2A), electron microscopic observations, and DNA fiber autoradiography (Funderud et al. 1978a,b, 1979). The progression of the replica-

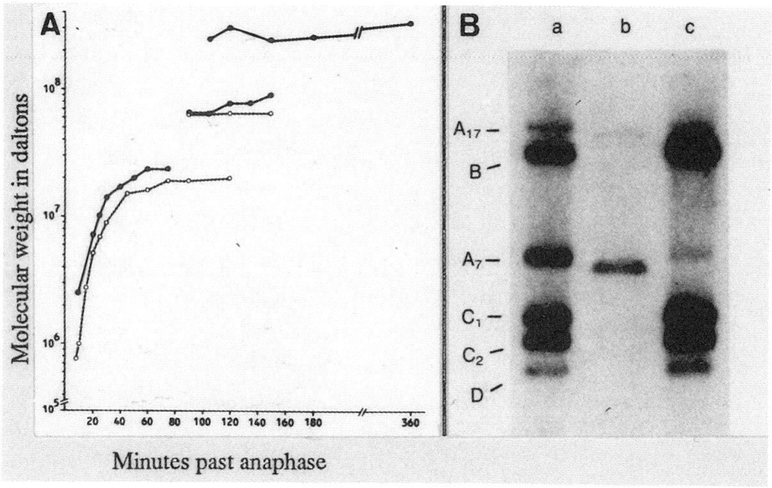

Figure 2 Structure of S phase in *Physarum*. (*A*) DNA chain growth as seen by sedimentation on alkaline sucrose gradients. Open circles correspond to the peak DNA molecules pulse-labeled at various time points in S phase, closed circles depict molecules labeled at the onset of S phase. A constant increase during the first 40–60 min of S phase, at a rate of about 1–2 kb/min/replicon is followed by a two-step increase that reveals clustering of the early activated replicons. (Reprinted, with permission, from Funderud et al. 1978b [copyright Cell Press].) (*B*) Temporal order of replication at the four unlinked actin gene loci. Following a BrdU treatment during the first 40 min of S phase, the replicated heavy–light (lane *c*) and unreplicated light–light (lane *b*) DNA fractions were isolated on cesium chloride gradients and their content in actin genes was compared to total DNA (lane *a*). Filter hybridization indicates that the *Hin*dIII restriction fragments containing the actin *ard*B gene, the two alleles of the *ard*C gene (C1 and C2), and the *ard*D gene have replicated, whereas the two *Hin*dIII fragments that define the *ard*A locus have not. This order of replication is invariant within the plasmodium. (Reprinted, with permission, from Pierron et al. 1984.)

tion forks, in which Okasaki fragments of 200 bp primed by short oligoribonucleotides (Waqar and Huberman 1975) are involved, has been evaluated at 1.2 kb/min/replicon in independent investigations (Funderud et al. 1978b; 1979; Hunt and Vogelstein 1981). Within a replicon, replication is bidirectional, with equal rates on both forks. This was shown by photolysing DNA that had been substituted with bromodeoxyuridine (BrdU) for 30 seconds at the onset of S phase. Following a chase, the nascent strands were cut into two equal pieces by UV irradiation, indicating that the BrdU-substituted origins are centrally located within the growing replicons and providing evidence for initiation limited to only one site in every replicon (Funderud et al. 1978a). This conclusion was strengthened by the observation of discrete replicating eyes in the electron microscope (EM), at a mean frequency of 1 bubble every 525 kb of early S phase DNA (Funderud et al. 1979; Haugli et al. 1982). However, a conflicting EM observation of clusters of microbubbles as the most abundant replication intermediates in early S phase of *Physarum* has also been reported (Hardman and Gillespie 1980). In this latter case, larger bubbles were also observed, but were 10 times less abundant. It was concluded that microbubbles were transient precursors of the more extended bubbles (Hardman and Gillespie 1980). From the comparison of the two sets of data, we suggest an alternative interpretation.

Macrobubbles Versus Microbubbles in *Physarum*

In observations by N. Hardman, only 1% of the DNA molecules contained large bubbles. The mean size of these DNA molecules, as inferred from measurements of distances between successive clusters of microbubbles, should be in the range of 100 kb. If so, this corresponds to 1 large bubble every 10,000 kb of DNA, much too low a number as compared to observations by F. Haugli at the same stage in S phase. Considering this lack of extended bubbles, it is tempting to speculate that microbubble clusters are generated by a partial reassociation of the parental DNA within large bubbles. This would explain that the size of the microbubble clusters and their center-to-center distances coincide with the size and the distribution of the growing replicons seen by F. Haugli and his collaborators. This interpretation would also account for comparable numbers of replicons in the two studies, rationalize the frequent and puzzling single-stranded nature of the microbubbles, and explain their absence from DNA extracted from G_2-phase or mitotic plasmodia. Finally, it is of note that microbubbles, occasionally described as the most abundant replicating structures by observation of naked DNA (Hardman and Gillespie 1980; Micheli et al. 1982), have never been observed on the corresponding chromatin spreads either in blastoderm embryos of *Drosophila* (McKnight and Miller 1977) or in early S phase of *Physarum* (Pierron et al. 1982).

Patterns of DNA Chain Growth

Using the size of the genome and the mean size of the replicons, the kinetics of DNA synthesis at 5,000 kb/min/nucleus, and the elongation rate of 1.2 kb/min/replicon, one can calculate that 4,000 of the 18,000 replicons are activated at the onset of S phase. Joining of the early activated replicons takes place by steps (Fig. 2A) that provide the first indication of their clustering (Funderud et al. 1978b). The early S-phase clusters are composed of 3–4 replicons that are completed by 40 minutes in S phase, but that fuse only at 60 minutes. A second transition in the maturation of the progeny molecules occurs at 120 minutes and further suggests a clustering of the replicon clusters. These very defined steps in the elongation of the DNA chains suggest that the same initiation sites are used in the many nuclei of a plasmodium. The invariant temporal order of replication implies that these sites are also used in consecutive S phases.

TEMPORAL ORDER OF REPLICATION IN *PHYSARUM*

Invariant Replication Timing of DNA Sequences

The first demonstration of a defined order of replication within eukaryotic chromosomes came from labeling replicons in two consecutive, synchronous S phases of the same plasmodium. In one cycle, the late-replicating DNA was labeled by [^{3}H]thymidine incorporation and, after a chase, the replicating DNA of the following S phase was density-labeled by BrdU. The tritiated late-replicating DNA was not density-shifted and therefore not replicated in the early compartment of the ensuing S phase but became denser in late S phase, demonstrating a temporal order of replication of the eukaryotic genome (Braun et al. 1965).

Studies of the timing of replication of specific genes confirmed the programmed replication of the genome. By standard density-shift experiments (Fig. 2B), it was shown that three of the four unlinked actin gene loci had replicated at 40 minutes in S phase. Measurements of the relative intensity of the hybridization bands from DNA extracted at different time points of the cell cycle on a standard Southern blot provide a control experiment, i.e., a "gene dosage" analysis, that further establishes that three actin gene loci have replicated at 8–10 minutes of S phase and that the fourth actin gene is replicated at 80–100 minutes of S phase (Pierron et al. 1984). This invariant temporal order of replication was extended to the two histone H4 genes and to a set of genes that are duplicated early or late (Jalouzot et al. 1985; Pierron et al. 1989b). On the basis of a thorough gene dosage analysis, it was recently observed that the two alleles of the tubulin *altB* gene replicate concurrently in early S phase whereas, surprisingly, the two alleles of the *altA* gene replicate in a very broad temporal window in mid S phase, with each allele having its own timing of replication. It was concluded that the control of early replication is stringent, whereas the replication schedule in mid S phase would be relaxed (Cunningham and Dove 1993). These results contrast with the apparent strict timing of the late-replicating genes (Pierron et al. 1984, 1989b). This suggests diversified late-replication patterns, whereas early patterns are constant and precisely defined temporally (Diller and Sauer 1993; Bénard et al. 1996).

Chronology of Replication and Gene Expression

The functional importance of the early-replicating DNA has been recognized for some time in *Physarum*. Inhibition of DNA replication at the onset of S phase had drastic effects on the transcriptional activity and little effect when carried out in the second half of S phase. It was inferred that the early S phase newly replicated DNA is rapidly transcribed and that the late DNA is relatively inactive (Rao and Gontcharoff 1969; Fouquet et al. 1975). To better define the relationship between genes and replicons in early S phase, a direct visualization was conducted on chromatin spreads (Pierron et al. 1982). Observation of nascent replicons undergoing active transcription provided direct evidence for a rapid activation of the newly synthesized genes (Fig. 3A). In contrast, no transcripts were found between replicating eyes, demonstrating a preferential transcription of the early replicons. Finally, the most striking feature of the majority of these early-replicating genes was their location at the center of the nascent replicons, indicating a close proximity, if not a coincidence, of these genes with their replication origin.

This enrichment of the early replicons in active genes is further confirmed by comparing the pattern of expression and the replication timing of 18 genes. The 5 genes that are inactive within the plasmodium were not confined to one compartment of S phase; 1 was found to be early, 2 replicate in mid S phase, and 2 replicate very late (Pierron et al. 1989b). Among the 13 active genes, 10 replicate very early in S phase and 1 replicates in mid S phase (the tubulin *altA* gene), whereas 2 genes encoding abundant mRNA were found both by BrdU incorporation and by gene dosage to be duplicated late in S phase: at 100–120 minutes for the LAV 3-2 gene (Pierron et al. 1989b) and later than 90 minutes for the *php* gene, a single-copy gene whose two alleles replicate late (Bénard et al. 1992).

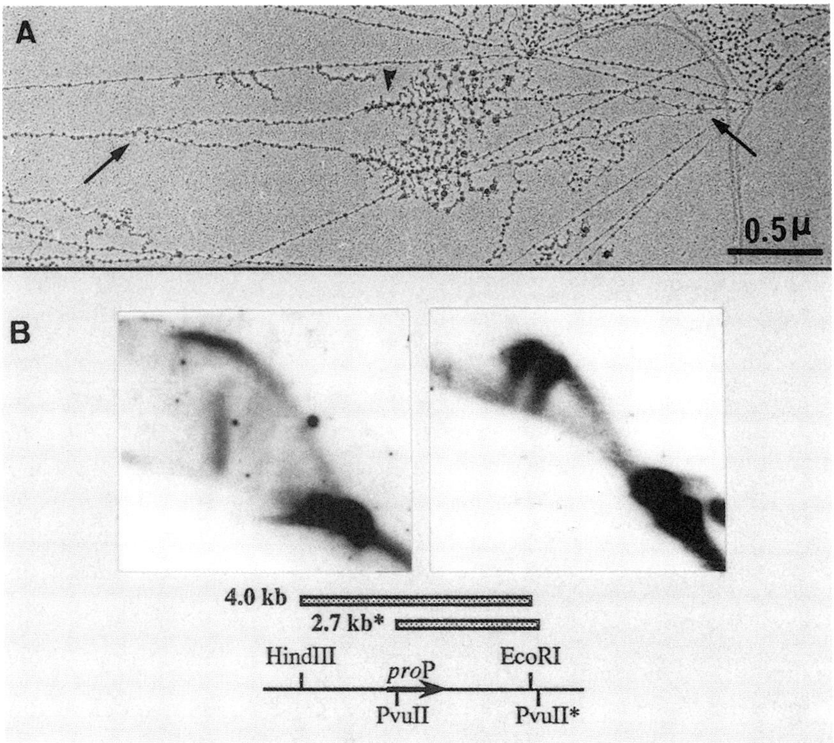

Figure 3 Tight linkage between genes and replication origins in *Physarum*. (*A*) Visualization at 15 min in S phase of a newly replicated gene at the center of a growing replicon. Assuming a bidirectional fork-rate movement, the presumptive location of the origin of replication coincides with the gene. Arrows indicate replication forks, arrowhead marks the center of the replicon. (Reprinted, with permission, from Pierron et al. 1982.) (*B*) Coincidence of a profilin gene and a site-specific replication origin. (*Left*) Two-dimensional agarose gel electrophoresis analysis of replication, on total DNA extracted at 5 min in S phase, of the 4-kb *Hin*dIII–*Eco*RI fragment that contains the profilin P gene. The distinct transition from a bubble arc to a Y arc reveals the presence of a bidirectional, site-specific origin slightly asymetrically located. (*Right*) Replication of the *Pvu*II fragment that covers the 3′ side of the gene (see map). Owing to a variable restriction site (*asterisk*), the two alleles are shown to be coordinately replicated, implying a simultaneous activation of the allelic origins. The Y-arc pattern excludes the origin from the 3′ side of the gene. (Reprinted, with permission of Oxford University Press, from Bénard and Pierron 1992.)

MAPPING OF REPLICATION ORIGINS IN *PHYSARUM*

Chromosomal Origins

In these studies, the early-replicating active genes that were examined were most likely closely associated with a replication origin. A linkage analysis, based on the assumption that the shortest nascent strands of a replicon detected by hybridization with a cDNA probe contain both the gene and the replication origin, confirmed a close association between five genes that have been studied and their respective origins (Bénard and Pierron 1990, 1992; Bénard et al. 1996). These replication origins can be further defined by two-dimensional gel analysis. In one case, the LAV 1-2 developmentally regulated gene, a composite pattern of replication intermediates was obtained, suggestive of two relatively inefficient origins, located about 10 kb apart, on both sides of the gene. These origins would be simultaneously activated, generating forks converging and stalling on the gene (Diller and Sauer 1993). This complex pattern could not have been predicted from the analysis of the nascent-strand elongation of that replicon shown in Figure 1D. In the four other cases, two-dimensional gel analysis established the presence of an efficient, site-specific origin closely associated with the promoter region of the genes (Fig. 3B). These results confirm the EM observations of replication-transcription coupling at early-replicating loci (Fig. 3A). The proximity of an origin to a gene was shown for a developmentally regulated profilin P gene (Bénard and Pierron 1992; Diller and Sauer 1993), the two unlinked and constitutively expressed actin genes, ardB and ardC (Bénard et al. 1996), and a cell-cycle-regulated histone H4 gene (G. Pierron et al., in prep.). It seems likely that these site-specific origins are genetically determined. This view is reinforced by the analysis of the replication of the rDNA of *Physarum*.

rDNA Replication

In *Physarum*, the rRNA genes are found on about 150 copies of a 60-kb linear palindromic molecule (Fig. 4B). It has long been known that these minichromosomes replicate throughout the cell cycle except during mitosis and the first hour of S phase. EM visualization of 37 replicating molecules by Vogt and Braun (1977) revealed a bidirectional elongation from either one or the other of two discrete initiation sites, slightly off-center at 33% and 45% of the molecules (Fig. 4A). Taking into account the palindromic structure of the molecules, it was concluded that only one of four potential origins was active on a given molecule. The detailed analysis of the central nontranscribed spacer structure strengthened this four-origins hypothesis, as it is composed of a sequence motif, reiterated four times in inverted orientations, at locations compatible with the origins mapped by EM (Ferris 1985). Recent data obtained by two-dimensional gel electrophoresis confirmed that there is a site-specific origin at 33% (Fig. 4C), and indirect evidence was found for a more efficient origin located at 45% (Bénard et al. 1995). A plasmid containing the origin located at 33% was shown to replicate in vitro in a cell-free extract, with the initiations being restricted to the origin region. This provides the first demonstration of the in vitro replication of a eukaryotic chromosomal origin (Daniel and Johnson 1989). All of these studies pinpoint the initiation site to a few hundred base pairs centered on the junction between a series of 31-bp repeats and a sequence of 900 bp that contains potential stem-loop structures (Fig. 4C). Although the role of these different structures in the initiation of rDNA replication is unknown, the presence of four site-specific origins in a region comprising a quadruplicate arrangement of a conserved sequence strongly supports the notion of origins defined by a specific *cis*-acting sequence.

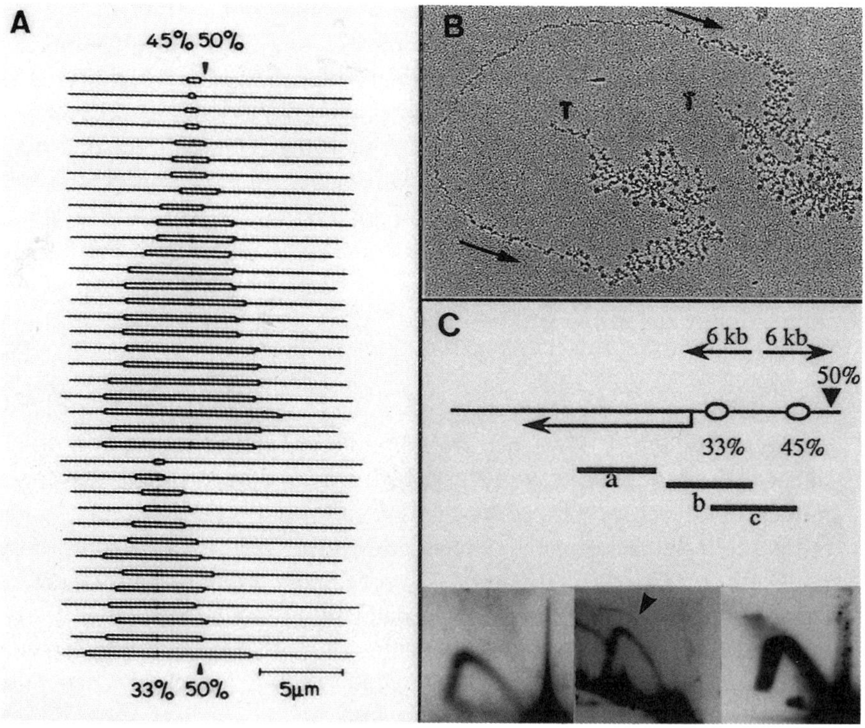

Figure 4 (*See facing page for legend.*)

CONCLUSIONS

Preliminary evidence for site-specific replication origins in *Physarum* is consistent with the wealth of data showing a precise pattern of growth of the nascent DNA strands and an invariant temporal order of replication. Considering that *Physarum* is evolutionarily distant from *Saccharomyces cerevisiae*, this tends to enlarge the base of organisms that rigidly control the initiation of DNA replication. It will be of interest to search for conserved sequences or structures within the replication origins of these two distantly related organisms. Due to a lack of resolution of the mapping of these origins and to an absence of obvious homologies, no consensus sequence has yet been established in *Physarum*. It may also be that the origin activity is dependent on the chromatin architecture of the promoter regions or the proposed association of these origins with the nuclear

Figure 4 Replication of the 60-kb linear palindromic rDNA chromosomes of *Physarum*. (*A*) Alignment of replicating rDNA molecules as seen in the EM. Bidirectional elongation from two discrete initiation sites at 33% and 45% was concluded by Vogt and Braun. Considering the palindromic structure of the molecules, they proposed the presence of four site-specific origins within the central nontranscribed spacer (CNTS). (Reprinted, with permission, from Vogt and Braun 1977.) (*B*) Visualization of an actively transcribed rDNA minichromosome on chromatin spread. Arrows depict polarity of transcription through the two divergent rRNA genes, illustrating the palindromic structure of the molecule and the extension of the 24-kb CNTS. (T) Telomeres. (Reprinted, with permission, from Pierron et al. 1989a.) (*C*) Neutral-neutral two-dimensional-gel analysis of rDNA replication. Scheme depicts half of a palindrome. Sequence analysis revealed that the CNTS is composed of a 6-kb motif of low DNA complexity, reiterated four times in inverted orientations. This motif begins with a series of 78 copies of a 31-bp repeat (TGAGACGGGGAAATTTCTTCAGCC-GGTTGAA), continues with a more complex 900-bp sequence that contains potential stem-loop structures, and ends with multiple copies of a 300-bp repeat (Ferris 1985). Probing an intragenic 5-kb fragment (a) resulted in a Y-arc pattern consistent with the expected passive replication of the genes (*left* panel). In contrast, a distinct bubble arc (*arrowhead*) is obtained when a 7-kb fragment (b) that encompasses the presumptive origin at 33% is tested (*central* panel). The site-specificity of that origin is inferred from the absence of bubble arc in the overlapping fragment (c) (*right* panel). Because of a lack of appropriate restriction sites, no direct evidence for an origin at 45% could be obtained by two-dimensional gels; however, both the presence of a weak termination signal in the right panel and the ratio of bubble arc to Y arc in the central panel are consistent with the proposed four site-specific origins of Vogt and Braun (1977) at 33%, 45%, 55%, and 67% (Bénard et al. 1995).

matrix (Wanka 1991), emphasizing the importance of the functional organization of the nucleus in regulating simultaneously various genomic activities.

Indeed, one lesson learned from *Physarum* concerns the tight association between replication origins and active genes (Pierron et al. 1982). As a consequence, active genes are clearly overrepresented in the early-replicating fraction of the genome. Nevertheless, identification of late-replicating and active genes tends to rule out models in which the chronology of DNA replication of chromosomal domains would only be defined by the transcriptional status of the genes. It remains to be seen whether or not these late-replicating genes are linked to active origins.

Replication origins linked to genes have also been found in various loci in metazoan cell types, as for example, the *Drosophila* chorion genes (Heck and Spradling 1990), the human β-globin gene (Kitsberg et al. 1993), or the DNA puff II/9a in *Sciara* (Liang et al. 1993), illustrating that the same functional constraints are at work in shaping eukaryotic genomes.

PROSPECTS

Organisms having site-specific replication origins provide powerful tools for the elucidation of regulatory mechanisms, as exemplified in yeast and eukaryotic viruses. The plasmodium form of *Physarum polycephalum* is an abundant source of such origins, whose dissection will permit a better understanding of the interactions between genes and replicons at a molecular level. The recent development of stable DNA transformation of *Physarum* offers opportunities for assaying autonomous replication of the plasmodial origins (Burland et al. 1993). This would facilitate the definition of a *cis*-acting regulatory sequence and provide a straightforward analysis of their developmental regulation in the amoebal-plasmodial transition. Finally, the availability of discrete replication origins and the possibility of preparing protein extracts from synchronous cultures should stimulate the development of an efficient in vitro DNA replication system.

ACKNOWLEDGMENTS

We thank Jacqueline Pedron for dedicated art work, Dominick Pallotta, Edmond Puvion, and Helmut Sauer for critical reading. The authors are supported by general funding from the CNRS and by grant 1301 from Association pour la Recherche sur le Cancer.

REFERENCES

Bénard, M. and G. Pierron. 1990. Physical relationship between a gene and its origin of replication in *Physarum polycephalum*. *Exp. Cell Res.* **186:** 299–305.

———. 1992. Mapping of a *Physarum* chromosomal origin of replication tightly linked to a developmentally-regulated profilin gene. *Nucleic Acids Res.* **20:** 3309–3315.

Bénard, M., C. Lagnel, and G. Pierron. 1995. Site specific initiation of DNA replication within the non transcribed spacer of *Physarum* rDNA. *Nucleic Acids Res.* **23:**1447–1453.

Bénard, M., D. Pallotta, and G. Pierron. 1992. Structure and identity of a late-replicating and transcriptionally active gene. *Exp. Cell Res.* **201:** 506–513.

Bénard, M., C. Lagnel, D. Pallotta, and G. Pierron. 1996. Mapping of a replication origin within the promoter region of two unlinked, abundantly transcribed actin genes of *Physarum polycephalum*. *Mol. Cell. Biol.* **16:** (in press).

Braun, R., C. Mittermayer, and H.P. Rusch. 1965. Sequential temporal replication of DNA in *Physarum polycephalum*. *Proc. Natl. Acad. Sci.* **53:** 924–927.

Burland, T.G., J. Bailey, D. Pallotta, and W.F. Dove. 1993. Stable, selectable, integrative DNA transformation in *Physarum*. *Gene* **132:** 207–212.

Cunningham, D. and W.F. Dove. 1993. Two alleles of a developmentally regulated α-tubulin locus in *Physarum polycephalum* replicate on different schedules. *Mol. Cell. Biol.* **23:** 449–461.

Daniel, D. and E. Johnson. 1989. Selective initiation of replication at origin sequences of rDNA molecules of *Physarum polycephalum* using synchronous plasmodial extracts. *Nucleic Acids Res.* **17:** 8343–8362.

Diller, J. and H.W. Sauer. 1993. Two early replicated, developmentally controlled genes of *Physarum* display different patterns of DNA replication by two-dimensional agarose gel electrophoresis. *Chromosoma* **102:** 563–574.

Ferris, P.J. 1985. Nucleotide sequence of the central non transcribed spacer region of *Physarum polycephalum* rDNA. *Gene* **39:** 203–211.

Fouquet, H., R. Bohme, R. Wick, H.W. Sauer, and K. Scheller. 1975. Some evidence for replication-transcription coupling in *Physarum polycephalum*. *J. Cell Sci.* **18:** 27–34.

Funderud, S., R. Andreassen, and F. Haugli. 1978a. DNA replication in *Physarum polycephalum*: Bidirectional replication of DNA within replicons. *Nucleic Acids Res.* **5:** 713–721.

———. 1978b. Size distribution and maturation of newly replicated DNA through the S and G2 phases of *Physarum polycephalum*. *Cell* **15:** 1519–1527.

———. 1979. DNA replication in *Physarum polycephalum*: Electron microscopic and autoradiographic analysis of replicating DNA from defined stages of the S period. *Nucleic Acids Res.* **6:** 1417–1427.

Hardman, N. and D.A. Gillespie. 1980. DNA replication in *Physarum polycephalum*. *Eur. J. Biochem.* **106:** 161–169.

Hardman, N., P.L. Jack, R.C. Fergie, and L.M. Gerrie. 1980. Sequence organisation in nuclear DNA from *Physarum polycephalum*: Interspersion of repetitive and single-copy sequences. *Eur. J. Biochem.* **103:** 247–254.

Haugli, F., R. Andreassen, and S. Funderud. 1982. DNA replication *in Physarum polycephalum*: Electron microscopic analysis of patterns of DNA replication in the presence of cycloheximide. *J. Cell Biol.* **95:** 323–331.

Heck, M.M.S. and A.C. Spradling. 1990. Multiple replication origins are used during *Drosophila* chorion gene amplification. *J. Cell Biol.* **4:** 903–914.

Howard, R.W. 1932. Nuclear division in plasmodia of *Physarum*. *Ann. Bot.* **46:** 461–477.
Hunt, B. and B. Vogelstein. 1981. Association of newly replicated DNA with the nuclear matrix of *Physarum polycephalum*. *Nucleic Acids Res.* **9:** 348–350.
Jalouzot, R., B. Toublan, M. Wilhelm, and F.X. Wilhelm. 1985. Replication timing of the H4 histone genes in *Physarum polycephalum*. *Proc. Natl. Acad. Sci.* **82:** 6475–6479.
Kitsberg, D., S. Selig, I. Keshet, and H. Cedar. 1993. Replication structure of the human β-globin gene domain. *Nature* **366**: 588–590.
Kubbies, M. and G. Pierron. 1983. Mitotic cell cycle control in *Physarum*. *Exp. Cell Res.* **149:** 149–158.
Liang, C., D. Spitzer, H. Smith, and S. Gerbi. 1993. Replication initiates at a confined region during DNA amplification in *Sciara* DNA puff II/9A. *Genes Dev.* **7:** 1072–1084.
McKnight, S.L. and O.L. Miller. 1977. Electron microscopic analysis of chromatin replication in the cellular blastoderm *Drosophila melanogaster* embryo. *Cell* **12:** 795–804.
Micheli, G., C. Baldari, M.T. Carri, G. Di Cello, and M. Buongiornio-Nardelli. 1982. An electron microscope study of chromosomal DNA replication in different eukaryotic systems. *Exp. Cell Res.* **137:** 127–140.
Mohberg, J., K.L. Babcock, F.B. Haugli, and H.P. Rusch. 1973. Nuclear DNA content and chromosome numbers in the myxomycete *Physarum polycephalum*. *Dev. Biol.* **34:** 228–245.
Nygaard, O.F., S. Guttes, and H.P. Rusch. 1960. Nucleic acid metabolism in a slime mold with synchronous mitosis. *Biochim. Biophys. Acta* **38:** 298–306.
Pierron, G., D. Durica, and H.W. Sauer. 1984. Invariant temporal order of replication of four actin gene loci during the naturally synchronous mitotic cycle of *Physarum polycephalum*. *Proc. Natl. Acad. Sci.* **81:** 6393–6397.
Pierron, G., J. Pedron, M. Schelling, and M. Christensen. 1989a. Immunoelectron microscopic localization of the nucleolar protein B-36 (fibrillarin) during the cell cycle of *Physarum*. *Biol. Cell* **65:** 119–126.
Pierron, G., H.W. Sauer, B. Toublan, and R. Jalouzot. 1982. Physical relationship between replicons and transcription units in *Physarum polycephalum*. *Eur. J. Cell Biol.* **29**: 104–113.
Pierron, G., M. Bénard, E. Puvion, R. Flanagan, H.W. Sauer, and D. Pallotta. 1989b. Replication timing of 10 developmentally regulated genes in *Physarum polycephalum*. *Nucleic Acids Res.* **17:** 553–566.
Rao, B. and M. Gontcharoff. 1969. Functionality of newly synthesized DNA as related to RNA synthesis during mitotic cycle in *Physarum polycephalum*. *Exp. Cell Res.* **56:** 269–275.
Rusch, H.P. 1980. The search. In *Growth and Differentiation in* Physarum polycephalum. (ed. W.F. Dove and H.P. Rusch), pp. 1–8. Princeton University Press, Princeton, New Jersey.
Vogt, V. and R. Braun. 1977. The replication of ribosomal DNA in *Physarum polycephalum*. *Eur. J. Biochem.* **80:** 557–566.
Waqar, M. and J.A. Huberman. 1975. Covalent linkage between RNA and nascent DNA in the slime mold *Physarum polycephalum*. *Biochim. Biophys. Acta* **383:** 410–417.
Wanka, F. 1991. Control of eukaryotic DNA replication at the chromosomal level. *BioEssays* **13:** 613–618.

35
Differential DNA Replication in Insects

Susan A. Gerbi and Fyodor D. Urnov
Division of Biology and Medicine
Department of Molecular Biology, Cell Biology and Biochemistry
Brown University, Providence, Rhode Island 02912

Insect chromosomes provide wonderful model systems to unravel the mysteries of DNA replication. Data from studies of conventional replication of insect chromosomes are compatible with a mode of semi-discontinuous replication by a set of enzymes similar to those in mammals. Normally, replication initiates just once per S phase in each replicon. During various stages of insect development, however, origins of DNA replication (ORI) at certain loci become the targets of complex developmental signals that lead to their selective activation or repression. As a result, insect chromosomes exhibit several molecular phenomena such as DNA amplification and underrepresentation that are in direct violation of the "rule of DNA constancy."

We believe that these phenomena may yield insights into more general principles for DNA replication control; understanding how a given ORI is driven to fire more than once (or not at all) during S phase may define the molecular mechanisms that direct an ORI to fire once and only once per cell cycle during conventional replication. In this chapter, we focus on the use of differential DNA replication in insect chromosomes as a model system to identify ORIs and explore the regulation of their activation.

Many insect tissues become polyploid or polytene, and certain cell-cycle controls are overridden during these processes. The chapter by Carminati and Orr-Weaver (this volume) describes studies on *Drosophila* mutants that have provided insights into controls for progression through the cell cycle.

CONVENTIONAL DNA REPLICATION IN INSECTS

Replicons with Bidirectional Replication

Much of our understanding of the mechanism for bidirectional replication is based on observations in *Drosophila* embryos, where the rapid

DNA Replication in Eukaryotic Cells
 0-87969-459-9/96 $5 + .00

successive rounds of DNA replication and cell division made it possible to visualize many replication bubbles, or "eye-forms" (Wolstenholme 1973; Kriegstein and Hogness 1974; Zakian 1976). The appearance of a single-stranded gap on one side of each replication fork, which can disappear due to branch migration and result in a single-stranded tail of 200 nucleotides on the other side of the fork, suggested that the gap was where the next Okazaki fragment would be placed and supported the concept of semi-discontinuous replication. Okazaki fragments, primed by an octaribonucleotide, have indeed been isolated from *Drosophila* (Blumenthal and Clark 1977; Kitani et al. 1984); it is postulated that a 61-nucleotide replication intermediate gives rise to 125-nucleotide and 240-nucleotide molecules.

The early *Drosophila* embryo has been the target of much biochemical study on the enzymology of DNA replication, since it has a molecular stockpile of enzymes and protein cofactors needed to ensure rapid rounds of chromosome duplication. Recent results of this work are summarized in Table 1, which updates a similar table published in a previous review (Spradling and Orr-Weaver 1987). Since SV40 ORI-dependent replication can be carried out by a *Drosophila* extract supplemented with T antigen (Kamakaka et al. 1994), it appears that the *Drosophila* replication machinery is highly similar to that of other well-studied vertebrate systems (see appropriate chapters in this book).

We hypothesize that even though most of the proteins have been isolated from embryos, it may be that these same factors are used in all subsequent replication events (normal replication during development, replication during polytenization, or DNA amplification). The differential DNA synthesis characteristic for some of these processes could be due to control of ORI activity by factors that do not belong to the basic subset listed in Table 1.

DNA fiber autoradiography data from *Drosophila* have shown that there are many replicons in the genome, which can be visualized to carry out DNA synthesis in a bidirectional manner from an ORI located in the center of each replicon (Steinemann 1981a,b). In *Drosophila* brain cells and tissue-culture cells, replication forks move at 0.35–1.0 μm/min (Blumenthal et al. 1974; Ananiev et al. 1977; Steinemann 1981a), which is up to ten times faster than the rate in polytene chromosomes (Cordeiro and Meneghini 1973; Steinemann 1981b; Lakhotia and Sinha 1983).

Nucleotide-level mapping of a nuclear ORI of bidirectional replication (i.e., the transition point between continuous and discontinuous DNA synthesis) has not been achieved yet in any metazoan, with the exception of some metazoan viruses (Hay and DePamphilis 1982;

Table 1 DNA replication enzymes and protein cofactors isolated from early *Drosophila* embryos

Protein	Subunit (function)	Gene cloned?	References
DNA polymerase-δ	138 (catalytic),	no	Aoyagi et al. (1994)
	or 120 (catalytic)	no	Chiang et al. (1993)
	47	no	Aoyagi et al. (1994)
PCNA	36 (pol-δ cofactor)	yes	Yamaguchi et al. (1990) Ng et al. (1990)
DNA polymerase-α	50 (primase)	yes	Bakkenist and Cotterill (1994)
	60	no	Lehman and Kaguni (1989)
	170 (catalytic)	yes	Hirose et al. (1991)
	or 180 (catalytic)	yes	Melov et al. (1992)
	73	yes	Cotterill et al. (1992)
DNA ligase 1	125 (replication)	no	Rabin et al. (1986) Takahashi and Senshu (1987)
DNA ligase 2	70 (repair?)	no	Takahashi and Tomizawa (1990)
Topoisomerase 1	135	yes	Hsieh et al. (1992)
Topoisomerase 2	170	yes	Wyckoff et al. (1989)
RNase H	2 x 49, 2 x 39	no	DiFrancesco and Lehman (1985)
RP-A	66, 31, 8	no	Marton et al. (1994)
	or 70, 30, 8	no	Mitsis et al. (1993)

Weights are given in kilodaltons, and two references are listed when there is a discrepancy in the literature on the apparent molecular weight of the protein. Proteins associated with types of DNA metabolism other than replication (e.g., repair) are not listed.

Hendrickson et al. 1987). ORIs used in DNA amplification have, however, been mapped to zones of one or several kilobases in insect chromosomes (Delidakis and Kafatos 1989; Heck and Spradling 1990; Liang et al. 1993; Liang and Gerbi 1994), as described in more detail below.

Regulation of Initiation during S Phase

DNA fiber autoradiography of samples from *Drosophila* (Blumenthal et al. 1974; Steinemann 1981b) revealed that the spacing between ORIs varies during development (the mean replicon length in *Drosophila* embryos is 7.9–10 kb; see Blumenthal et al. 1974; McKnight and Miller 1977). This suggested that many ORIs may be activated during early development resulting in a short S phase and allowing rapid cell division. Later, as tissues differentiate and the length of the cell cycle increases, many of these ORIs would become inactive, and the stretch of DNA surrounding such silent ORIs would be replicated by forks that were initiated at adjacent loci. Consequently, the replicon length would be increased.

This hypothesis presumes that there are specific sequences used as ORIs. There is little information on this point. Alternatively, if there is lack of sequence specificity for initiation, other factors such as chromatin structure or the availability of replication enzymes could determine the spacing between ORIs and hence the length of the replicon. However, even if there is lack of sequence specificity to define the position of initiation of conventional chromosomal replication, controls must still exist to prevent re-initiation in the same S phase, since nested replication forks have not been observed in electron micrographs of replicating *Drosophila* embryonic DNA (Blumenthal et al. 1974; Kriegstein and Hogness 1974).

Cytological studies of [^{3}H]thymidine uptake in giant polytene chromosomes of the larval salivary glands of dipteran insects indicated that replicons can vary in the time during S phase when they are actively synthesizing DNA (Keyl and Pelling 1963; Plaut 1963; Gabrusewycz-Garcia 1964; Plaut et al. 1966; Hägele 1970, 1973; Arcos-Terán 1972; Steinemann 1981b). Replicons in heterochromatin are active later in S than are those in euchromatin, and diploid cells also abide by this conclusion (Steinemann 1980).

Deviations from the Rule of DNA Constancy

Even though replicons can initiate DNA replication at different times, generally, by the completion of S phase all of the genomic DNA will have been replicated once and only once. This satisfies the rule of DNA constancy, which states that all cells in an organism have the same DNA content per genome equivalent (Boivin et al. 1948; Swift 1950a,b; Mirsky and Ris 1951). Insect chromosomes, however, present an exception to this rule. As described below, centromeric heterochromatin as well as the ribosomal DNA (rDNA) locus are both underrepresented in *Drosophila* polytene chromosomes; on the other hand, extra DNA is synthesized during amplification of *Drosophila* follicle cell chorion genes and *Sciara* polytene chromosome DNA puffs.

DNA sequence underrepresentation and intrachromosomal DNA amplification have both been ascribed to differential replication. It has been proposed (Laird 1973; Laird et al. 1974) that continuity of DNA through the chromatid is made possible by nested replication forks (Fig. 1); this "onion skin" model was experimentally confirmed in the case of *Drosophila* chorion genes by electron microscopy (Osheim and Miller 1983; Osheim et al. 1988). On the basis of cytological quantitation it was also suggested that nested replication forks might occur at each polytene band/interband boundary (Sorsa 1974; Laird 1980), but except for a few constrictions, this is not supported by molecular hybridization data (Lifschytz 1983; Spierer and Spierer 1984; Lamb and Laird 1987).

Failure to find replication forks at the junction of euchromatin and underrepresented heterochromatic sequences in a mini-polytene chromosome has led to an alternate model according to which the underrepresented sequences are physically excised during polytenization (Karpen and Spradling 1990; Glaser et al. 1992). In fact, although most of the literature presumes that underrepresented sequences have been underreplicated, there are few data to rule out the alternate model of excision.

UNDERREPRESENTED SEQUENCES IN DIPTERAN CHROMOSOMES

In dipteran salivary glands, and to a lesser extent in most other larval tissues (Ashburner 1970), chromosomes undergo polytenization—rounds of endoduplication without intervening cell division (Swift 1962; Rasch 1970b)—with the sister chromatids remaining synapsed together and not separating as they do in polyploid nuclei. Parts of the genome, such as the heterochromatic Y chromosome that is not represented in *Drosophila* salivary gland polytene chromosomes (Painter 1933; Lindsley and Grell 1967; Holmquist 1975; Zhang and Spradling 1995), can escape the endo-

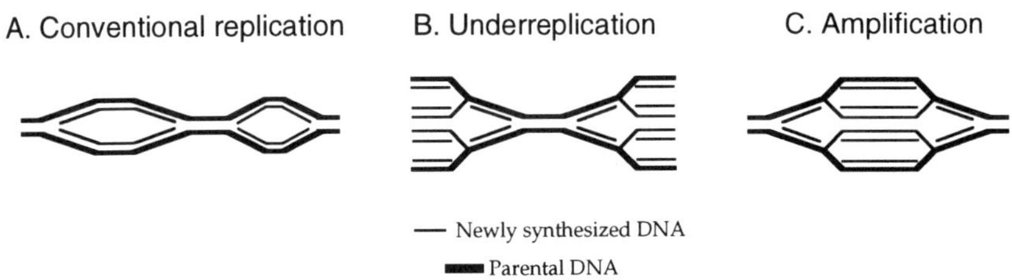

Figure 1 Models of two adjacent eye-forms for conventional replication (*A*), nested replication forks for underreplication (*B*) and DNA amplification (*C*), according to the "onion skin" model (Laird 1973; Laird et al. 1974).

duplication process. Not only whole chromosomes, but also individual loci can be subject to regulation of their representation in the genome, as exemplified by satellite DNAs and rDNA as described below.

Centromeric Satellite DNA

Heitz (1934) was the first to propose that during polytenization the amount of centromeric heterochromatin is reduced relative to euchromatin. For simple sequence satellite DNA, found in the heterochromatin of *Drosophila* polytene chromosomes (Gall and Atherton 1974; Peacock et al. 1974; Steinemann 1976), this was verified by comparing polytene and diploid cells in three ways: (1) by microspectrophotometry (Berendes and Keyl 1967; Mulder et al. 1968; Rudkin 1969), (2) by in situ hybridization (Gall et al. 1971; Peacock et al. 1974), and (3) by analytical ultracentrifugation in CsCl gradients (Dickson et al. 1971; Gall et al. 1971; Cordeiro et al. 1975). In contrast, the satellite DNA of Chironomids remains fully represented in the genome during polytenization (Walter 1973; Steinemann 1978).

Underrepresentation of satellite DNA relative to the rest of the genome is not limited to salivary gland polytene chromosomes but can occur in other tissues as well, such as nurse cells and follicle cells (Hammond and Laird 1985). When comparing DNA from *Drosophila virilis* diploid cells (embryos or adult head and brain) to adult tissues that are often polyploid, it was observed that the satellite DNAs were underrepresented to different levels in the latter (Blumenfeld and Forrest 1972; Schweber 1974; Endow and Gall 1975). Underrepresentation of satellite DNA also occurs in some other insects (Kunz and Eckhardt 1974; Redfern 1981), but in some cases satellite DNA is fully present in ovarian nurse cells (Gall et al. 1969; Zacharias 1979; Nazimiec and Beckingham 1986). Interestingly, satellite DNA can also be overrepresented relative to the rest of the genome, as is the case for pupal ovaries of *D. virilis* (Endow and Gall 1975), although with time after eclosion of the adult fly it becomes underrepresented (Endow and Gall 1975; Renkawitz-Pohl and Kunz 1975). In general, either underreplication or excision can be used equally well to account for all instances of satellite DNA underrepresentation in polytene or polyploid cells.

Ribosomal DNA

rDNA Underrepresentation during Polytenization

There is some underrepresentation of rDNA relative to the rest of the genome during polytenization of salivary gland chromosomes in *D. hydei*

(Hennig and Meer 1971) and *D. melanogaster* (Spear and Gall 1973; Spear 1974; Szabo et al. 1977), where it appears to lag behind by an average of three polytene rounds. This is not as severe as simple sequence satellite DNA, which is hardly represented at all in polytene cells. rDNA underrepresentation can also occur in polyploid tissues, such as the ovary of adult *Drosophila*, resulting in almost half its level in diploid cells (Endow and Gall 1975; Renkawitz and Kunz 1975). In contrast, there is no underrepresentation of rDNA in salivary gland polytene chromosomes of *Chironomus tentans*, which lacks heterochromatin (Hollenberg 1976); in Sciarid flies the conclusions are less clear (Gerbi 1971; Gambarini and Meneghini 1972). Moreover, in *D. hydei* nonheterochromatic moderately repetitive sequences like the 5S RNA genes appear to replicate with the rest of the genome (Renkawitz-Pohl 1978).

rDNA is less severely underrepresented in nurse cells that synthesize all the rRNA of the oocyte (Renkawitz and Kunz 1975; Beckingham and Thompson 1982; Hammond and Laird 1985), suggesting a functional selection for its presence. In addition, those rDNA repeat units that contain insertions and are not transcribed (for review, see Beckingham 1982) are underrepresented even more in polytene and polyploid cells than the intron$^-$-functional rDNA units (Endow and Glover 1979; Beckingham and Thompson 1982). Once again, the downward adjustment in copy number could be the result of either failure to replicate or sequence excision.

rDNA Amplification: Compensation and Magnification

Under certain circumstances rDNA loci can also undergo amplification. Specifically, *Drosophila* that are XO rather than XX and have one instead of two nucleolus organizer regions contain the same amount of rDNA in salivary glands as do XX individuals (Spear and Gall 1973). This somatic compensation may be due to regulation of the polyteny level of the nucleolus organizer region by overreplication (Spear and Gall 1973; Endow and Glover 1979; Endow 1980, 1982) and could also account for the relative rDNA increase in adult XO flies (Tartof 1971, 1973).

rDNA levels can also change in the germ line. The standard constitution of the nucleolus organizer region on either the X or the Y chromosome of *Drosophila* is about 200 tandemly repeated rDNA copies. The *bobbed* phenotype reflects a deficiency in the number of rDNA copies (Ritossa et al. 1966). When male flies have a deficiency in the number of rDNA copies on both the X (X^{bb}) and Y chromosomes (Y^{bb-}), they transmit X chromosomes with more rDNA than was originally carried, and after a few generations the standard level of rDNA in the nucleolus organizer region is restored. This process was named "magnification" by Ritossa (1968). Both the intron$^+$ and intron$^-$ rDNA copies are equally amplified (de Cicco and Glover 1983). Magnification can also occur in female flies deficient in rDNA provided that they carry the tip of the long arm of the Y chromosome (Komma and Endow 1986, 1987).

rDNA magnification has been hypothesized to result from replication and subsequent reintegration of extrachromosomal rDNA rings that were derived from the chromosomal nucleolus organizer region (Ritossa 1972; Locker and Prud'homme 1973; Ritossa et al. 1974; Locker 1976). However, Tartof (1974a,b) proposed an alternate hypothesis of unequal sister chromatid exchange, and recent data support this model (Endow et al. 1984; Endow and Atwood 1988; Komma and Atwood 1994). The reciprocal product of unequal exchange for magnification (bb^m) is the reduced genotype (bb^r) of a still greater deficiency in rDNA copies; the reduced genotype is rarely observed, perhaps due to selective pressure against it.

INTRACHROMOSOMAL DNA AMPLIFICATION IN DIPTERA

The examples described below of intrachromosomal DNA amplification of the *Drosophila* chorion loci and *Sciara* DNA puffs both occur as a normal, developmentally regulated event in terminally differentiated tissue. In fact, such significant deviations from the rule of DNA constancy may be a luxury enjoyed only by systems where cells are destined to degenerate shortly thereafter (the follicle cells degenerate after stage 14 and the larval salivary glands are destroyed in the subsequent pupal stage) and will not have to carry the burden of extra DNA in their genome for a long time after amplification.

It is conceivable that intrachromosomal gene amplification could also occur in the germ line, in which case it would be passed on to the resulting offspring as a stably maintained increase in that gene. One example of this was just described for rDNA magnification, although this appears to result from recombination rather than DNA overreplication. Another example is found when comparing 30 polytene bands between *Chironomus thummi thummi* and *Chironomus thummi piger* (Keyl 1965a,b). The *piger* line shows increased DNA relative to the *thummi* line in a geometric series, suggesting that extra rounds of replication occurred in the germ-line replicons of the *piger* ancestor. This phenomenon could be related perhaps to the finding that the tandemly repeated Cla elements are dispersed throughout all chromosomes of *C. thummi thummi* but are limited to the centromeric regions of *C. thummi piger* (Hankeln et al. 1994).

These exceptions are all examples of *intra*chromosomal DNA amplification of a locus relative to its usual representation in the genome. *Extra*chromosomal gene amplification can also occur (e.g., rDNA in amphibian oocytes; see Brown and Dawid 1968; Gall 1968), but a return to the standard composition of the genome is easily accomplished by an apparent failure to replicate and dilution out of the extra rDNA copies during subsequent cell divisions of the early embryo.

DNA Amplification in the Chorion Loci of *Drosophila*

During the last 16 hours of oogenesis in *D. melanogaster*, there is amplification in the ovarian follicle cells of the two clusters of chorion genes, on the X and third chromosome. The amplified genes provide extra template for the ensuing extensive production of eggshell proteins. The chorion cluster on the X chromosome amplifies 16- to 20-fold, and the cluster on the third chromosome amplifies 60- to 80-fold (Spradling and Mahowald 1980; Spradling 1981). Re-initiation is frequent (2.5-hr doubling time) and the replication forks elongate slowly (50–100 bp/min) (Spradling and Leys 1988). For both chorion clusters there is a gradient of amplification spanning almost 100 kb (Spradling 1981), suggesting that the peak of amplification in the center of the gradient may correspond to a repeatedly firing origin.

Delidakis and Kafatos (1989) and Heck and Spradling (1990) have mapped the ORI for amplification of the chorion locus on the third chromosome by two-dimensional gels (see Fig. 2 for schematic map). A major initiation zone of 7.7 kb and a minor initiation zone of at least 4.2 kb were revealed. Furthermore, it was speculated that a 1-kb region within the 7.7-kb zone may be the major ORI.

A search has begun for *trans*-acting factors that regulate chorion gene amplification. Seven female-sterile mutants with defects in chorion gene amplification have been found (Komitopoulou et al. 1986; Orr et al. 1984; Kelley and Spradling 1986; Snyder et al. 1986; Underwood et al. 1990), but clones encoding these factors have not yet been isolated, and their identity remains unknown.

Some of these unidentified *trans*-acting factors presumably bind to the *cis*-regulatory sequences important for *Drosophila* chorion gene

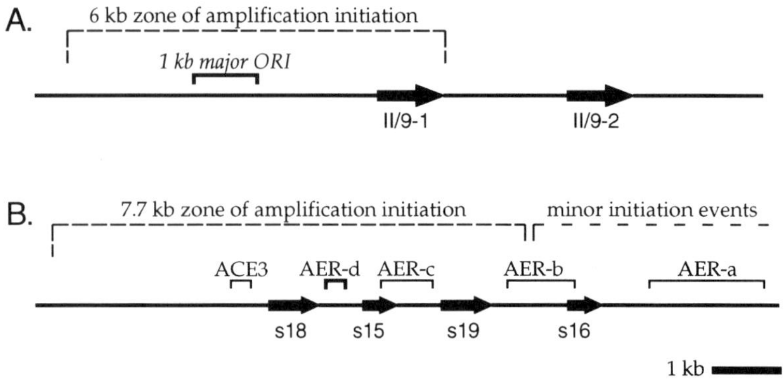

Figure 2 Schematic maps of (*A*) amplified loci for *Sciara* DNA puff II/9B (Liang et al. 1993; Liang and Gerbi 1994) and (*B*) *Drosophila* chorion genes (Delidakis and Kafatos 1989; Heck and Spradling 1990). Major origin activity is presumed to lie in AER-d (also called ORI-β), denoted by a darkened bracket. See text for explanations of abbreviations.

amplification. One such sequence, ACE3 (*a*mplification *c*ontrol *e*lement on the third chromosome) has been delimited by P-element transformation to a 320-bp region upstream of gene s18 and was separate from the transcription control region (Orr-Weaver et al. 1989). ACE3 contains an evolutionarily conserved 71-bp motif, although additional nonconserved sequences are also needed for its function (Swimmer et al. 1990). Similarly, ACE1 that regulates chorion gene amplification on the X chromosome has an essential 467-bp region with repeated sequences necessary for amplification but not for transcription (Spradling et al. 1987). Transposons bearing 1, 9, or 18 copies of 400 bp of ACE3 in a tandem array are sufficient to drive amplification, but only to low levels (Carminati et al. 1992). It remains unclear whether ACE3 contains an ORI or whether it stimulates replication from an adjacent ORI.

In addition to ACE3, there are four *a*mplification *e*nhancing *r*egions (AER a–d) whose deletion reduces but does not eliminate amplification in transformant lines (Delidakis and Kafatos 1987, 1989). Some transposons containing the AERs but lacking ACE3 can amplify to low levels (Swimmer et al. 1989). Data from such an approach need to be interpreted with caution due to the sensitivity of chorion gene amplification to position effects (these become more severe as the size of the DNA fragment to be tested decreases; de Cicco and Spradling 1984). The precise functional roles and interrelationships of ACE3 and the AERs remain to be elucidated.

DNA Amplification in Sciarid DNA Puffs

Poulson and Metz (1938) were the first to note the large DNA puffs that appear in salivary gland polytene chromosomes of sciarid flies by the end of the fourth instar. Unlike the RNA puffs of *Drosophila* or Balbiani rings of *Chironomus* that exhibit intense RNA transcription without gene amplification (Rudkin 1955; Hägele 1970; Rasch 1970b), the DNA puffs were hypothesized to contain "extra" DNA on the basis of increased Feulgen staining (Breuer and Pavan 1955) and [^{3}H]thymidine uptake (Ficq and Pavan 1957); microspectrophotometry provided definitive proof (Rudkin and Corlette 1957; Swift 1962; Crouse and Keyl 1968; Rasch 1970a).

There are nine major and nine minor DNA puffs in the giant salivary gland chromosomes of the fungus fly, *Sciara coprophila* (Gabrusewycz-Garcia 1964, 1971). We chose puff 9A on chromosome II (II/9A), one of the largest and among the first to appear, for our molecular studies (see schematic map in Fig. 2). There are two transcription units, II/9-1 and II/9-2; these are 85% similar to one another in sequence and encode a putative α-helical coiled-coil protein preceded by a signal sequence for secretion (DiBartolomeis and Gerbi 1989). It has been deduced that the DNA puffs of *Rhynchosciara* encode specific polypeptides to build the pupal cocoon (Winter et al. 1977a,b; de Toledo and Lara 1978; Ferreira and Amabis 1983; Laicine et al. 1984). Presumably, DNA puff amplification occurs to accommodate the need for massive amounts of proteins that form the pupal case.

DNA puff amplification is superimposed on the last chromosomal endoduplication (Crouse and Keyl 1968; Rasch 1970a), and there are about four extra rounds of replication to result in 16- to 22-fold more DNA (Glover et al. 1982; Paçó-Larson et al. 1992; Wu et al. 1993). The extra DNA appears to remain at the DNA puff loci even in very late larval stages that were examined by microspectrophotometry (Crouse and Keyl 1968; Rasch 1970a). DNA amplification precedes maximum puffing; morphological puffing is correlated with an increase in the amount of mRNA that was presumably transcribed off the amplified DNA (Paçó-Larson et al. 1992; Wu et al. 1993). Thus, transcription occurs after amplification has reached its final level in all cases studied except for the C8 DNA puff of *Rhynchosciara* (Santelli et al. 1991; Wu et al. 1993). As might be deduced from the different timing of amplification and puffing during development, inhibition of DNA synthesis by hydroxyurea does not inhibit puff formation (Sauaia et al. 1971).

We have mapped the origin of amplification in DNA puff II/9A by two different two-dimensional gel methods (Liang et al. 1993) and by a three-dimensional gel method (Liang and Gerbi 1994). Initiation is confined to a 6-kb zone, with the majority of initiation events occurring in a 1-kb region that is 2 kb upstream of gene II/9-1. Replication forks move outward in a bidirectional manner from the amplification ORI (Liang et al. 1993), consistent with, but not proving, an onion-skin model of DNA amplification. It is not yet clear what the boundaries of the amplification gradient are in *Sciara*; however, preliminary data from DNA puff C4 of *Bradysia hygida* suggest that the DNA amplification gradient may be considerably shorter than the 100 kb found in the *Drosophila* chorion loci (Coelho et al. 1993; Monesi et al. 1995).

It would appear from three-dimensional gels that only one and not several clustered initiations occur in the ORI zone of a single DNA molecule (Liang and Gerbi 1994). Preliminary data suggest that the same region is used for initiation of conventional DNA replication earlier in development, as for DNA puff amplification (C. Liang; F.D. Urnov; S.A. Gerbi; all unpubl.).

In a manner similar to the studies on the chorion loci, the search has begun for putative *trans*-acting factors that may be responsible for the activation of the DNA puff amplification. These data are discussed in the following section.

FUTURE DIRECTIONS AND PERSPECTIVES

Our attempts to understand how regulation of replication is achieved such that each replicon initiates once and only once per cell cycle may be illuminated by an examination of exceptions to the rule from insect systems. For instance, if underreplication is the mechanism for underrepresentation, then one insight that arises from the observations listed is that regulation can be executed at the level of the individual replicon. Thus, satellite DNAs could underreplicate to different levels from one another in the same nucleus. Similarly, rDNA units could vary in their replication level depending on whether they carry an intervening sequence or not, and in the ovary, satellite DNA is underrepresented and rDNA amplified in the same cell. It would thus appear that merely placing a given replicon in a heterochromatic domain may be important but not sufficient to inhibit its replication, and some additional regulatory mechanisms reside within each replicon (e.g., rDNA and satellite DNA).

How are the boundaries for differential replication established? In underreplication, if an ORI does not fire, why is the DNA of that replicon not replicated by forks traveling into it from adjacent replicons? Are there discrete termination sites that prevent forks from moving into this region, or is this regulated by boundaries set up perhaps by chromatin architecture? Similarly, are there boundaries that prevent the extra rounds of replication in amplified DNA puffs from sweeping outward into adjacent chromosomal regions? It is known that re-initiation at the DNA puff II/9A ORI stops by late larval life (Wu et al. 1993), but whether the replication forks continue to move outward until the salivary gland is histolyzed during pupation has not been studied.

It would appear that there is selection for the number of DNA templates needed for transcription in a given cell type. For instance, although the rDNA level may be magnified to even a higher level than normal, ultimately the rDNA copy number is corrected to the normal level (de Cicco and Glover 1983). In another example, the nonfunctional intron$^+$ rDNA copies are less prevalent than intron$^-$ rDNA copies in polytene and polyploid cells, although it is not yet proven whether this is due to underreplication. Finally, extra templates are provided by over-replication of the *Drosophila* chorion loci and sciarid DNA puffs as a result of selective pressure to produce more template for intense transcription from these loci.

Is the ORI a specific sequence, and how is its activity regulated? Specifically, how do ORI silencing mechanisms compare between instances of underreplication and developmental ORI control? Is such silencing an active process requiring an inhibitor for ORI function, or is it simply the lack of an activator for that particular ORI? Similarly, for intra-chromosomal amplification, how are the ORIs regulated such that they fire more than once relative to replication of the rest of the genome? Does this process require an activator specific for that ORI, or does it reflect repression or absence of an inhibitor that normally prevents an ORI from firing more than once?

An answer to these questions necessitates the identification of *trans*-acting factors involved, and preliminary data for *Sciara* DNA puffs may shed light on this problem. Specifically, as was deduced in earlier cytological studies (Crouse 1968; Stocker and Pavan 1974; Berendes and Lara 1975; Fresquez 1979; Ferreira and Amabis 1980; Amabis and Amabis 1984a,b; Stocker et al. 1984; Dessen and Perondini 1985; Alvarenga et al. 1991) and shown by quantitative Southern blots (M. Batra; D. Alam; C. Liang; H.S. Smith; S.A. Gerbi; all unpubl.), DNA puff amplification can be prematurely induced in cultured salivary glands from young larvae by the steroid hormone, ecdysone, which also regulates transcription of DNA puff genes (for references, see above and Bienz-Tadmor et al. 1991). Several examples are known in viral systems where the same protein regulates both replication and transcription. Is the ecdysone receptor acting in a similar way in DNA puffs? Models on how this might be accomplished have been proposed (Lara et al. 1991; Gerbi et al. 1993) and await results from future experiments for their verification.

A full understanding of the target of regulatory *trans*-acting factors requires a better characterization of the ORI itself. The endogenous pattern of DNA replication is not maintained when *Drosophila* polytene nuclei replicate in *Xenopus* egg and oocyte extracts (Sleeman et al. 1992). *D. melanogaster* DNA fragments that function as autonomously replicating sequences (ARS) in yeast fail to do so in *Drosophila* embryos (Roth 1991), and plasmids containing *D. melanogaster* DNA appear to replicate without sequence specificity for initiation in cultured *Drosophila* cells (J.G. Smith and M.P. Calos, in prep.). Therefore, the ARS assay that has been so powerful in yeast does not seem fruitful for identifying critical regions for ORI function in insect DNA. However, data from these experiments should not be taken to imply that there is no sequence specificity for the ORI, since two- and three-dimensional gel analyses on DNA from the *Drosophila* chorion loci and *Sciara* DNA puffs demonstrate localizaton of the ORI for amplification to defined regions. Perhaps P-element transformation where the DNA to be tested is in a chromosomal context rather than on an extrachromosomal plasmid may be the best way to further dissect the properties of these ORIs. Future studies are needed to determine what additional molecular elements are necessary to produce amplification or underreplication, beyond the normal controls that a cell exerts over an ORI for conventional replication.

ACKNOWLEDGMENTS

We thank Allan Spradling and Melvin DePamphilis for helpful comments and Susan Cotterill for checking Table 1. This work was supported by National Institutes of Health grant GM-35929 to S.A.G.

REFERENCES

Alvarenga, C.A., C.E. Winter, A.J. Stocker, M.T. Pueyo, and F.J. Lara. 1991. *In vivo* effects of ecdysterone on puff formation and RNA and protein synthesis in the salivary glands of *Rhynchosciara americana*. *Braz. J. Med. Biol. Res.* **24:** 985–1002.

Amabis, D.C. and J.M. Amabis. 1984a. Effects of ecdysterone in polytene chromosomes of *Trichosia pubescens*. *Dev. Biol.* **102:** 1–9.

———. 1984b. Hormonal control of gene amplification and transcription in the salivary chromosomes of *Trichosia pubescens*. *Dev. Biol.* **102:** 10–20.

Ananiev, E.V., L.C. Polukarova, and Y.B. Yurov. 1977. Replication of chromosomal DNA in diploid *Drosophila melanogaster* cells cultured *in vitro*. *Chromosoma* **59:** 259–272.

Aoyagi, N., S. Matsuoka, A. Furunobu, A. Matsukage, and K. Sakaguchi. 1994. *Drosophila* DNA polymerase delta. Purification and characterization. *J. Biol. Chem.* **269:** 6045–6050.

Arcos-Terán, L. 1972. DNS-Replikation und die Natur der spät replizierenden Orte im X-Chromosom von *Drosophila melanogaster*. *Chromosoma* **37:** 233–296.

Ashburner, M. 1970. Function and structure of polytene chromosomes during insect development. *Adv. Insect Physiol.* **7:** 1–96.

Bakkenist, C.J. and S. Cotterill. 1994. The 50 kDa primase subunit of *Drosophila melanogaster* DNA polymerase alpha. Molecular characterization of the gene and functional analysis of the overexpressed protein. *J. Biol. Chem.* **269:** 26759–26766.

Beckingham, K. 1982. Insect rDNA. In *The cell nucleus, rDNA*, part A (ed. H. Busch and L. Rothblum), vol. 10, pp. 206–269. Academic Press, New York.

Beckingham, K. and N. Thompson. 1982. Under-replication of intron$^+$ rDNA cistrons in polyploid nurse cell nuclei of *Calliphora erythrocephala*. *Chromosoma* **87:** 177–196.

Berendes, H.D. and H-G. Keyl. 1967. Distribution of DNA in heterochromatin and euchromatin of polytene nuclei of *Drosophila hydei*. *Genetics* **57:** 1–13.

Berendes, H.D. and F.J.S. Lara. 1975. RNA synthesis: A requirement for hormone induced DNA amplification in *Rhynchosciara americana*. *Chromosoma* **50:** 259–274.

Bienz-Tadmor, B., H.S. Smith, and S.A. Gerbi. 1991. The promoter of DNA puff gene II/9-1 of *Sciara coprophila* is inducible by ecdysone in late prepupal salivary glands of *Drosophila melanogaster*. *Cell Regul. (Mol. Biol. Cell)* **2:** 875–888.

Blumenthal, M. and E.J. Clark. 1977. Discrete sizes of replication intermediates in *Drosophila* cells. *Cell* **12:** 183–189.

Blumenthal, M. and H. Forrest. 1972. Differential under-replication of satellite DNAs during *Drosophila* development. *Nature* **239:** 170–172.

Blumenthal, A.B., H.J. Kriegstein, and D.S. Hogness. 1974. The units of DNA replication in *Drosophila melanogaster* chromosomes. *Cold Spring Harbor Symp. Quant. Biol.* **38:** 205–224.

Boivin, A., R. Vendreley, and C. Vendreley. 1948. L'acide désoxyribonucleique du noyau cellulaire dépositaire des caractères héréditaires; arguments d'ordre analytique. *C.R. Acad. Sci.* **226:** 1061–1063.

Breuer, M.E. and C. Pavan. 1955. Behavior of polytene chromosomes of *Rhynchosciara angelae* at different stages of larval development. *Chromosoma* **7:** 371–386.

Brown, D.D. and I.B. Dawid. 1968. Specific gene amplification in oocytes. *Science* **160:** 272–280.

Carminati, J.L., C.G. Johnston, and T.L. Orr-Weaver. 1992. The *Drosophila* ACE3 chorion element autonomously induces amplification. *Mol. Cell. Biol.* **12:** 2444–2453.

Chiang, C.S., P.G. Mitsis, and I.R. Lehman. 1993. DNA polymerase delta from embryos of *Drosophila melanogaster*. *Proc. Natl. Acad. Sci.* **90:** 9105–9109.

Coelho, P.S.R., N. Monesi, J.C. De Almeida, F. Toledo, G. Buttin, and M.L. Paçó-Larson. 1993. DNA puff C4 of *Bradysia hygida* (Diptera: Sciaridae) contains genes unequally amplified and differentially expressed during development. *Chromosome Res.* **1:** 121–126.

Cordeiro, M. and R. Meneghini. 1973. The rate of DNA replication in the polytene chromosomes of *Rhynchosciara angelae*. *J. Mol. Biol.* **78:** 261–274.

Cordeiro, M., L. Wheeler, C.S. Lee, C.D. Kastritsis, and R.H. Richardson. 1975. Heterochromatic chromosomes and satellite DNAs of *Drosophila nasutoides*. *Chromosoma* **51:** 65–73.

Cotterill, S., I.R. Lehman, and P. McLachlan. 1992. Cloning of the gene for the 73 kD subunit of the DNA polymerase alpha primase of *Drosophila melanogaster*. *Nucleic Acids Res.* **20:** 4325–4330.

Crouse, H.V. 1968. The role of ecdysterone in DNA puff formation and DNA synthesis in the polytene chromosomes of *Sciara coprophila*. *Proc. Natl. Acad. Sci.* **61:** 971–978.

Crouse, H.V. and H.-G. Keyl. 1968. Extra replications in the "DNA puffs" of *Sciara coprophila*. *Chromosoma* **25:** 357–364.

de Cicco, D.V. and D.M. Glover. 1983. Amplification of rDNA and type 1 sequences in *Drosophila* males deficient in rDNA. *Cell* **32:** 1217–1225.

de Cicco, D. and A. Spradling. 1984. Localization of a *cis*-acting element responsible for the developmentally regulated amplification of *Drosophila* chorion genes. *Cell* **38:** 45–54.

Delidakis, C. and F.C. Kafatos. 1987. Amplification of a chorion gene cluster in *Drosophila* is subject to multiple *cis*-regulatory elements and to long range position effects. *J. Mol. Biol.* **197:** 11–26.

———. 1989. Amplification enhancers and replication origins in the autosomal chorion gene cluster of *Drosophila melanogaster*. *EMBO J.* **8:** 891–901.

Dessen, E.M.B. and A.L.P. Perondini. 1985. Hormonal control of RNA and DNA puffs in polytene chromosomes of *Sciara ocellaris*. *Rev. Bras. Genet.* **4:** 669–679.

de Toledo, S.M. and F.J.S. Lara. 1978. Translation of messages transcribed from the "DNA puffs" of *Rhynchosciara*. *Biochem. Biophys. Res. Commun.* **85:** 160–166.

DiBartolomeis, S.M. and S.A. Gerbi. 1989. Molecular characterization of DNA puff II/9A genes in *Sciara coprophila*. *J. Mol. Biol.* **210:** 531–540.

Dickson, E., J.B. Boyd, and C.D. Laird. 1971. Sequence diversity of polytene chromosome DNA from *Drosophila hydei*. *J. Mol. Biol.* **61:** 615–627.

DiFrancesco, R.A and I.R. Lehman. 1985. Interaction of ribonuclease H from *Drosophila melanogaster* embryos with DNA polymerase-primase. *J. Biol. Chem.* **260:** 14674–14670.

Endow, S.A. 1980. On ribosomal gene compensation in *Drosophila*. *Cell* **22:** 149–155.

———. 1982. Polytenization of the ribosomal genes on the X and Y chromosomes of *Drosophila melanogaster*. *Genetics* **100:** 375–385.

Endow, S.A. and K.C. Atwood. 1988. Magnification: Gene amplification by an inducible system of sister chromatid exchange. *Trends Genet.* **4:** 348–351.

Endow, S.A. and J.G. Gall. 1975. Differential replication of satellite DNA in polyploid tissues of *Drosophila virilis*. *Chromosoma* **50:** 175–192.

Endow, S.A. and D.M. Glover. 1979. Differential replication of ribosomal gene repeats in polytene nuclei of *Drosophila*. *Cell* **17:** 597–605.

Endow, S.A., D.J. Komma, and K.C. Atwood. 1984. Ring chromosomes and rDNA magnification in *Drosophila*. *Genetics* **108:** 969–983.

Ferreira, J.F. and J.M. Amabis. 1980. Estudo de pufes de DNA em *Trichosia pubescens* (Diptera, Sciaridae): efeito da ecdisterona "*in vitro*" no processo de transcriçao gênica. *Congr. Bras. Biol. Cell.* 222–223.

———. 1983. Correlation between a DNA puff and a polypeptide fraction in the salivary gland of *Trichosia pubescens* larvae. *Arq. Biol. Technol.* **27:** 119.

Ficq, A. and C. Pavan. 1957. Autoradiography of polytene chromosomes of *Rhyncho-*

sciara angelae at different stages of larval development. *Nature* **180:** 983–984.
Fresquez, C.L. 1979. Nucleic acid synthesis in *Rhynchosciara hollaenderi* polytene chromosomes. I. Dose response and temporal sequence after injection of 20-hydroxyecdysone. *Insect Biochem.* **9:** 517–523.
Gabrusewycz-Garcia, N. 1964. Cytological and autoradiographic studies in *Sciara coprophila* salivary gland chromosomes. *Chromosoma* **15:** 312–344.
———. 1971. Studies in polytene chromosomes of Sciarids. I. The salivary gland chromosomes of *Sciara (Lycoriella) pauciseta* (II), Felt. *Chromosoma* **33:** 421–435.
Gall, J.G. 1968. Differential synthesis of the genes for ribosomal RNA during amphibian oogenesis. *Proc. Natl. Acad. Sci.* **60:** 553–560.
Gall, J.G. and D. Atherton. 1974. Satellite DNA sequences in *Drosophila virilis. J. Mol. Biol.* **85:** 633–664.
Gall, J.G., E.H. Cohen, and M.L. Polan. 1971. Repetitive DNA sequences in *Drosophila. Chromosoma* **33:** 319–344.
Gall, J.G., H.C. Macgregor, and M.E. Kidston. 1969. Gene amplification in the oocytes of Dytiscid water beetles. *Chromosoma* **26:** 169–187.
Gambarini, A.G. and R. Meneghini. 1972. Ribosomal RNA genes in salivary gland and ovary of *Rhynchosciara angelae. J. Cell Biol.* **54:** 421–425.
Gerbi, S.A. 1971. Localization and characterization of the ribosomal RNA cistrons in *Sciara coprophila. J. Mol. Biol.* **58:** 499–511.
Gerbi, S.A., C. Liang, N. Wu, S.M. Di Bartolomeis, B. Bienz-Tadmor, H.S. Smith, and F.D. Urnov. 1993. DNA amplification in DNA puff II/9A of *Sciara coprophila. Cold Spring Harbor Symp. Quant. Biol.* **58:** 487–494.
Glaser, R.L., G.H. Karpen, and A.C. Spradling. 1992. Replication forks are not found in a *Drosophila* minichromosome demonstrating a gradient of polytenization. *Chromosoma* **102:** 15–19.
Glover, D.M., A. Zaha, A.J. Stocker, R.V. Santelli, M.T. Pueyo, S.M. de Toledo, and F.J.S. Lara. 1982. Gene amplification in *Rhynchosciara* salivary gland chromosomes. *Proc. Natl. Acad. Sci.* **79:** 2947–2951.
Hägele, K. 1970. DNA-Replikationsmuster der Speicheldrüsen-Chromosomen von Chironomiden. *Chromosoma* **31:** 91–138.
———. 1973. Komplementäre DNA-Replikationsmuster bei *Drosophila melanogaster. Chromosoma* **41:** 231–236.
Hammond, M.P. and C.D. Laird. 1985. Chromosome structure and DNA replication in nurse and follicle cells of *Drosophila melanogaster. Chromosoma* **91:** 279–286.
Hankeln, T., A. Rohwedder, B. Weich, and E.R. Schmidt. 1994. Transposition of minisatellite-like DNA in *Chironomus* midges. *Genome* **37:** 542–549.
Hay, R.T. and M.L. DePamphilis. 1982. Initiation of SV40 DNA replication *in vivo:* Location and structure of 5′ends of DNA synthesized in the ori region. *Cell* **28:** 767–779.
Heck, M.M.S. and A.C. Spradling. 1990. Multiple replication origins are used during *Drosophila* chorion gene amplification. *J. Cell Biol.* **110:** 903–914.
Heitz, E. 1934. Über α- und β-Heterochromatin sowie Konstanz und Bau der Chromomeren bei *Drosophila. Biol. Zentralbl.* **54:** 588–609.
Hendrickson, E.A., C.E. Fritze, W.R. Folk, and M.L. DePamphilis. 1987. Polyoma virus DNA replication is semi-discontinuous. *Nucleic Acids Res.* **15:** 6369–6385.
Hennig, W. and B. Meer. 1971. Reduced polyteny of ribosomal RNA cistrons in giant chromosomes of *Drosophila hydei. Nat. New Biol.* **233:** 70–72.

Hirose, F., M. Yamaguchi, Y. Nishida, M. Masutani, H. Miyazawa, F. Hanaoka, and A. Matsukage. 1991. Structure and expression during development of *Drosophila melanogaster* gene for DNA polymerase alpha. *Nucleic Acids Res.* **19:** 4991–4998.

Hollenberg, C.P. 1976. Proportionate representation of rDNA and Balbiani ring DNA in polytene chromosomes of *Chironomus tentans. Chromosoma* **57:** 185–197.

Holmquist, G. 1975. Hoechst 3328 fluorescent staining of *Drosophila* chromosomes. *Chromosoma* **49:** 333–356.

Hsieh, T.S., S.D. Brown, P. Huang, and J. Fostel. 1992. Isolation and characterization of a gene encoding DNA topoisomerase I in *Drosophila melanogaster. Nucleic Acids Res.* **20:** 6177–6182.

Kamakaka, R.T., P.D. Kaufman, B. Stillman, P.G. Mitsis, and J.T. Kadonaga. 1994. Simian virus 40 origin- and T antigen-dependent DNA replication with *Drosophila* factors *in vitro. Mol. Cell. Biol.* **14:** 5114–5122.

Karpen, G.H. and A.C. Spradling. 1990. Reduced DNA polytenization of a minichromosome region undergoing position-effect variegation in *Drosophila. Cell* **63:** 97–107.

Kelley, R. and A. Spradling. 1986. Regulation of eggshell genes. *Carnegie Inst. Washington Year Book*, pp. 19–22.

Keyl, H.G. 1965a. A demonstrable local and geometric increase in the chromosomal DNA of *Chironomus. Experientia* **21:** 191–193.

———. 1965b. Duplikationen von Untereinheiten der chromosomalen DNS während der Evolution von *Chironomus thummi. Chromosoma* **17:** 139–180.

Keyl, H-G. and C. Pelling. 1963. Differentielle DNA-Replikation in den Speicheldrüsenchromosomen von *Chironomus thummi. Chromosoma* **14:** 347–359.

Kitani, T., K. Yoda, and T. Okazaki. 1984. Discontinuous DNA replication of *Drosophila melanogaster* is primed by octaribonucleotide primer. *Mol. Cell. Biol.* **4:** 1591–1596.

Komitopoulou, K., S. Kouyanou, and F.C. Kafatos. 1986. A temperature-sensitive mutant affecting the process of chorion gene amplification in *Drosophila melanogaster. Dev. Genet.* **7:** 75–80.

Komma, D.J. and K.C. Atwood. 1994. Magnification in *Drosophila:* Evidence for an inducible rDNA-specific recombination system. *Mol. Gen. Genet.* **242:** 321–326.

Komma, D.J. and S.A. Endow. 1986. Magnification of the ribosomal genes in female *Drosophila melanogaster. Genetics* **114:** 859–874.

———. 1987. Incomplete Y chromosomes promote magnification in male and female *Drosophila. Proc. Natl. Acad. Sci.* **84:** 2382–2386.

Kriegstein, H.J. and D.S. Hogness. 1974. Mechanism of DNA replication in *Drosophila* chromosomes: Structure of replication forks and evidence for bidirectionality. *Proc. Natl. Acad. Sci.* **71:** 135–139.

Kunz, W. and R.A. Eckhardt. 1974. Chromosomal distribution of satellite DNA in germline and somatic tissues of the gall midge, *Heteropeza pygmaea. Chromosoma* **47:** 1–19.

Laicine, E.M., M.A.R. Alves, J.C. de Almeida, E. Rizzo, W.C. Albernaz, and H. Sauaia. 1984. Development of DNA puffs and patterns of polypeptide synthesis in the salivary glands of *Bradysia hygida. Chromosoma* **89:** 280–284.

Laird, C.D. 1973. DNA of *Drosophila* chromosomes. *Annu. Rev. Genet.* **7:** 177–204.

———. 1980. Structural paradox of polytene chromosomes. *Cell* **22:** 869–874.

Laird, C.D., W.Y. Chooi, E.H. Cohen, E. Dickson, N. Hutchinson, and S.H. Turner. 1974. Organization and transcription of DNA in chromosomes and mitochondria of

Drosophila. *Cold Spring Harbor Symp. Quant. Biol.* **38:** 311–327.

Lakhotia, S.C. and P. Sinha. 1983. Replication in *Drosophila* chromosomes. X. Two kinds of active replications in salivary gland polytene nuclei and their relation to chromosomal replication patterns. *Chromosoma* **88:** 265–276.

Lamb, M.M. and C.D. Laird. 1987. Three euchromatic DNA sequences under-replicated in polytene chromosomes of *Drosophila* are localized in constrictions and ectopic fibers. *Chromosoma* **95:** 227–235.

Lara, F.J.S., A.J. Stocker, and J.M. Amabis. 1991. DNA sequence amplification in Sciarid flies: Results and perspectives. *Braz. J. Med. Biol. Res.* **24:** 233–248.

Lehman, I.R. and L.S. Kaguni. 1989. DNA polymerase alpha. *J. Biol. Chem.* **264:** 4265–4268.

Liang, C. and S.A. Gerbi. 1994. Analysis of an origin of DNA amplification in *Sciara coprophila* by a novel three-dimensional gel method. *Mol. Cell. Biol.* **14:** 1520–1529.

Liang, C., J.D. Spitzer, H.S. Smith, and S.A. Gerbi. 1993. Replication initiates at a confined region during DNA amplification in *Sciara* DNA puff II/9A. *Genes Dev.* **7:** 1072–1084.

Lifschytz, E. 1983. Sequence replication and banding organization in the polytene chromosomes of *Drosophila melanogaster*. *J. Mol. Biol.* **164:** 17–34.

Lindsley, D.L. and E.H. Grell. 1967. Genetic variations of *Drosophila melanogaster*. *Carnegie Inst. Washington Publ.* **627**.

Locker, D. 1976. Instability at the *bobbed* locus following magnification in *Drosophila melanogaster*. *Mol. Gen. Genet.* **143:** 261–268.

Locker, D. and N. Prud'homme. 1973. Étude de plusieurs facteurs faisant varier la fréquence de reversion au locus *bobbed* chez *Drosophila melanogaster*. *Mol. Gen. Genet.* **124:** 11–19.

Marton, R.F., P. Thommes, and S. Cotterill. 1994. Purification and characterisation of dRP-A: A single stranded DNA binding protein from *Drosophila melanogaster*. *FEBS Lett.* **342:** 139–144.

McKnight, S.L. and O.L. Miller. 1977. Electron microscopic analysis of chromatin replication in the cellular blastoderm *Drosophila melanogaster* embryo. *Cell* **12:** 795–804.

Melov, S., H. Vaughan, and S. Cotterill. 1992. Molecular characterization of the gene for the 180 kDa subunit of the DNA polymerase-primase of *Drosophila melanogaster*. *J. Cell Sci.* **102:** 847–856.

Mirsky, A.E. and H. Ris. 1951. The desoxyribonucleic acid content of animal cells and its evolutionary significance. *J. Gen. Physiol.* **34:** 451–462.

Mitsis, P.G., S.C. Kowalczykowski, and I.R. Lehman. 1993. A single stranded DNA binding protein from *Drosophila melanogaster:* Characterization of the heterotrimeric protein and its interaction with single stranded DNA. *Biochemistry* **32:** 5257–5266.

Monesi, N., M.A. Fernandez, A.M. Fontes, L.R. Basso, Y. Nakanishi, B. Baron, G. Buttin, and M.L. Paçó-Larson. 1995. Molecular characterization of an 18 kb segment of DNA puff C4 of *Bradysia hygida* (Diptera, sciaridiae). *Chromosoma* **103:** 715–724.

Mulder, M.P., P. Duijn, and H.J. van Gloor. 1968. The replicative reorganization of DNA in polytene chromosomes of *Drosophila hydei*. *Genetica* **39:** 385–428.

Nazimiec, M. and K. Beckingham. 1986. 3B55: A repetitious sequence family which is transcribed and proportionately replicated in germ-line polyploid nuclei of *Calliphora erythrocephala*. *Dev. Biol.* **115:** 398–406.

Ng, L., G. Prelich, C.W. Anderson, B. Stillman, and P.A. Fisher. 1990. *Drosophila* proliferating cell nuclear antigen. Structural and functional homology with its mammalian counterpart. *J. Biol. Chem.* **265:** 11948–11954.

Orr, W., K. Komitopoulou, and F.C. Kafatos. 1984. Mutants suppressing *trans* chorion gene amplification in *Drosophila. Proc. Natl. Acad. Sci.* **81:** 3773–3777.

Orr-Weaver, T.L., C.G. Johnston, and A.C. Spradling. 1989. The role of ACE3 in *Drosophila* chorion gene amplification. *EMBO J.* **8:** 4153–4162.

Osheim, Y.N. and O.L. Miller. 1983. Novel amplification and transcriptional activity of chorion genes in *Drosophila melanogaster. Cell* **33:** 543–553.

Osheim, Y.N., O.L. Miller, and A.L. Beyer. 1988. Visualization of *Drosophila melanogaster* chorion genes undergoing amplification. *Mol. Cell. Biol.* **8:** 2811–2821.

Paçó-Larson, M.L., J.C. deAlmeida, J.-E. Edström, and H. Sauaia. 1992. Cloning of a developmentally amplified gene sequence in the DNA puff C4 of *Bradysia hygida* (Diptera: Sciaridae) salivary glands. *Insect Biochem. Mol. Biol.* **22:** 439–446.

Painter, T.S. 1933. A new method for the study of chromosome rearrangements and the plotting of chromosome maps. *Science* **78:** 585–586.

Peacock, W.J., D. Brutlag, E. Goldring, R. Appels, C.W. Hinton, and D.L. Lindsley. 1974. The organization of highly repeated DNA sequences in *Drosophila melanogaster* chromosomes. *Cold Spring Harbor Symp. Quant. Biol.* **38:** 405–416.

Plaut, W. 1963. On the replicative organization of DNA in the polytene chromosome of *Drosophila melanogaster. J. Mol. Biol.* **7:** 632–635.

Plaut, W., D. Nash, and T. Fanning. 1966. Ordered replication of DNA in polytene chromosomes of *Drosophila melanogaster. J. Mol. Biol.* **16:** 85–93.

Poulson, D.F. and C.W. Metz. 1938. Studies on the structure of nucleolus-forming regions and related structures in the giant salivary gland chromosomes of Diptera. *J. Morphol.* **63:** 363–395.

Rabin, B.A., R.S. Hawley, and J.W. Chase. 1986. DNA ligase from *Drosophila melanogaster* embryos. Purification and physical characterization. *J. Biol. Chem.* **261:** 10637–10645.

Rasch, E.M. 1970a. Two-wavelength cytophotometry of *Sciara* salivary gland chromosomes. In *Introduction to quantitative cytochemistry* (ed. G.L. Wied and G.F. Bahr), vol. 2, pp. 335–355. Academic Press, New York.

———. 1970b. DNA cytophotometry of salivary gland nuclei and other tissue systems in dipteran larvae. In *Introduction to quantitative cytophotometry* (ed. G.L. Wied and G.F. Bahr), vol. 2, pp. 357–397. Academic Press, New York.

Redfern, C.P.F. 1981. Satellite DNA of *Anopheles stephensi* Liston (Diptera: Culcidae). Chromosomal location and underreplication in polytene nuclei. *Chromosoma* **82:** 561–581.

Renkawitz, R. and W. Kunz. 1975. Independent replication of the ribosomal DNA genes in the polytrophic-meroistic ovaries of *Calliphora erythrocephala, Drosophila hydei,* and *Sarcophaga barbata. Chromosoma* **53:** 131–140.

Renkawitz-Pohl, R. 1978. Number of the repetitive euchromatic 5S RNA genes in polyploid tissues of *Drosophila hydei. Chromosoma* **66:** 249–258.

Renkawitz-Pohl, R. and W. Kunz. 1975. Under-replication of satellite DNAs in polyploid ovarian tissue in *Drosophila virilis. Chromosoma* **49:** 375–382.

Ritossa, F.M. 1968. Unstable redundancy of genes for ribosomal RNA. *Proc. Natl. Acad. Sci.* **60:** 509–516.

———. 1972. Procedure for magnification of lethal deletions of genes for ribosomal RNA. *Nat. New Biol.* **240:** 109–111.

Ritossa, F.M., K.C. Atwood, and S. Spiegelman. 1966. A molecular explanation of the *bobbed* mutants of *Drosophila* as partial deficiencies of "ribosomal" DNA. *Genetics* **54:** 819–834.

Ritossa, F.M., F. Scalenghe, N. DiTuri, and A.M. Contini. 1974. On the cell stage of X-Y recombination during rDNA magnification in *Drosophila. Cold Spring Harbor Symp. Quant. Biol.* **38:** 483–490.

Roth, G.E. 1991. Replication analysis of plasmid DNAs injected into *Drosophila* embryos. *Chromosoma* **100:** 267–277.

Rudkin, G.T. 1955. The ultraviolet absorption of puffed and unpuffed homologous regions in the salivary gland chromosomes of *Drosophila melanogaster. Genetics* **40:** 593.

———. 1969. Non replicating DNA in *Drosophila. Genetics* (suppl.) **61:** 227–237.

Rudkin, G. and S.L. Corlette. 1957. Disproportionate synthesis of DNA in a polytene chromosome region. *Proc. Natl. Acad. Sci.* **43:** 964–968.

Santelli, R.V., G.M. Machado-Santelli, M.T. Pueyo, L.D. Navarro-Cattapan, and F.J.S. Lara. 1991. Replication and transcription in the course of DNA amplification of the C3 and C8 DNA puffs of *Rhynchosciara americana. Mech. Dev.* **36:** 59–65.

Sauaia, H., E.M. Laicine, and M.A.R. Alves. 1971. Hydroxyurea induced inhibition of DNA puff development in the salivary gland chromosome of *Bradysia hygida. Chromosoma* **34:** 129–151.

Schweber, M.S. 1974. The satellite bands of the DNA of *Drosophila virilis. Chromosoma* **44:** 371–382.

Sleeman, A.M., G.H. Leno, A.D. Mills, M.P. Fairman, and R.A. Laskey. 1992. Patterns of DNA replication in *Drosophila* polytene nuclei replicating in *Xenopus* egg and oocyte extracts. *J. Cell Sci.* **101:** 509–515.

Snyder, P., V. Galanopoulos, and F.C. Kafatos. 1986. *Trans*-acting amplification mutants and other eggshell mutants of the third chromosome in *Drosophila melanogaster. Proc. Natl. Acad. Sci.* **83:** 3341–3345.

Sorsa, V. 1974. Organization of replicative units in salivary gland chromosome bands. *Hereditas* **78:** 298–302.

Spear, B.B. 1974. The genes for ribosomal RNA in diploid and polytene chromosomes of *Drosophila melanogaster. Chromosoma* **48:** 159–179.

Spear, B.B. and J.G. Gall. 1973. Independent control of ribosomal gene replication in polytene chromosomes of *Drosophila melanogaster. Proc. Natl. Acad. Sci.* **70:** 1359–1363.

Spierer, A. and P. Spierer. 1984. Similar level of polyteny in bands and interbands of *Drosophila* giant chromosomes. *Nature* **307:** 176–178.

Spradling, A. 1981. The organization and amplification of two clusters of *Drosophila* chorion genes. *Cell* **27:** 193–202.

Spradling, A.C. and E. Leys. 1988. Slow replication fork movement during *Drosophila* chorion gene amplification. In *Eukaryotic DNA replication* (ed. T.J. Kelly and B. Stillman), pp. 305–309. Cold Spring Harbor Laboratory, Cold Spring Harbor, New York.

Spradling, A.C. and A.P. Mahowald. 1980. Amplification of genes for chorion proteins during oogenesis in *Drosophila melanogaster. Proc. Natl. Acad. Sci.* **77:** 1096–1100.

Spradling, A.C. and T. Orr-Weaver. 1987. Regulation of DNA replication during *Drosophila* development. *Annu. Rev. Genet.* **21:** 373–403.

Spradling, A.C., D.V. de Cicco, B.T. Wakimoto, J.F. Levine, L.J. Kalfayan, and L.

Cooley. 1987. Amplification of the X-linked chorion gene cluster requires a region upstream from the s38 chorion gene. *EMBO J.* **6:** 1045–1053.

Steinemann, M. 1976. The *in situ* formation of DNA-DNA duplexes of *Drosophila virilis* satellite DNA. *Chromosoma* **54:** 339–348.

———. 1978. Co-replication of satellite DNA of *Chironomus melanotus* with mainband DNA during polytenization. *Chromosoma* **66:** 127–139.

———. 1980. Chromosomal replication in *Drosophila virilis.* I. Diploid karyotype of brain cells. *Chromosoma* **78:** 211–223.

———. 1981a. Chromosomal replication in *Drosophila virilis.* II. Organization of active origins in diploid brain cells. *Chromosoma* **82:** 267–288.

———. 1981b. Chromosomal replication in *Drosophila virilis.* III. Organization of active origins in the highly polytene salivary gland cells. *Chromosoma* **82:** 289–307.

Stocker, A.J. and C. Pavan. 1974. The influence of ecdysterone on gene amplification, DNA synthesis and puff formation in the salivary gland chromosomes of *Rhynchosciara hollaenderi. Chromosoma* **45:** 295–319.

Stocker, A.J., M.T. Pueyo, S.D. Pereira, and F.J.S. Lara. 1984. Ecdysteroid titers and changes in the chromosomal activity in the salivary glands of *Rhynchosciara americana. Chromosoma* **90:** 26–38.

Swift, H. 1950a. The deoxyribose nucleic acid content of animal nuclei. *Physiol. Zool.* **23:** 169–198.

———. 1950b. The constancy of deoxyribose nucleic acid in plant nuclei. *Proc. Natl. Acad. Sci.* **36:** 643–654.

———. 1962. Nucleic acids and cell morphology in dipteran salivary glands. In *Molecular control of cellular activity* (ed. J.M. Allen), pp. 73–125. McGraw-Hill, New York.

Swimmer, C., C. Delidakis, and F.C. Kafatos. 1989. Amplification control element ACE-3 is important but not essential for autosomal chorion gene amplification. *Proc. Natl. Acad. Sci.* **86:** 8823–8827.

Swimmer, C., M.G. Fenerjian, J.C. Martinez-Cruzado, and F.C. Kafatos. 1990. Evolution of the autosomal chorion cluster in *Drosophila.* III. Comparison of the s18 gene in evolutionarily distant species and heterospecific control of chorion gene amplification. *J. Mol. Biol.* **215:** 225–235.

Szabo, P., R. Elder, D.M. Steffensen, and O.C. Uhlenbeck. 1977. Quantitative *in situ* hybridization of ribosomal RNA species to polytene chromosomes of *Drosophila melanogaster. J. Mol. Biol.* **115:** 539–563.

Takahashi, M. and M. Senshu. 1987. Two distinct DNA ligases from *Drosophila melanogaster* embryos. *FEBS Lett.* **213:** 345–352.

Takahashi, M. and K. Tomizawa. 1990. Purification and characterization of DNA ligase II from *Drosophila melanogaster. Eur. J. Biochem.* **192:** 735–740.

Tartof, K.D. 1971. Increasing the multiplicity of ribosomal RNA genes in *Drosophila melanogaster. Science* **171:** 294–297.

———. 1973. Regulation of ribosomal RNA gene multiplicity in *Drosophila melanogaster. Genetics* **73:** 57–71.

———. 1974a. Unequal mitotic sister chromatid exchange and disproportionate replication as mechanisms regulating ribosomal RNA gene redundancy. *Cold Spring Harbor Symp. Quant. Biol.* **38:** 491–500.

———. 1974b. Unequal mitotic sister chromatid exchange as the mechanism of ribosomal RNA gene magnification. *Proc. Natl. Acad. Sci.* **71:** 1272–1276.

Underwood, E.M., A.S. Briot, K.Z. Doll, R.L. Ludwiczak, D.C. Otteson, J. Tower, K.B. Vessey, and K. Yu. 1990. Genetics of 51D–52A, a region containing several maternal-effect genes and two maternal-specific transcripts in *Drosophila*. *Genetics* **126:** 639–650.

Walter, L. 1973. Synthesprozesse an den Riesenchromosomen von *Glyptotendipes*. *Chromosoma* **41:** 327–360.

Winter, C.E., A.G. de Bianchi, W.R. Terra, and F.J.S. Lara. 1977a. Relationships between newly synthesized proteins and DNA puff patterns in salivary glands of *Rhynchosciara americana*. *Chromosoma* **61:** 193–206.

———. 1977b. The giant DNA puffs of *Rhynchosciara americana* code for polypeptides of the salivary gland secretion. *J. Insect Physiol.* **23:** 1455–1460.

Wolstenholme, D.R. 1973. Replicating DNA molecules from eggs of *Drosophila melanogaster*. *Chromosoma* **43:** 1–18.

Wu, N., C. Liang, S.M. DiBartolomeis, H.S. Smith, and S.A. Gerbi. 1993. Developmental progression of DNA puffs in *Sciara coprophila:* Amplification and transcription. *Dev. Biol.* **160:** 73–84.

Wyckoff, E., D. Natalie, J.M. Nolan, M. Lee, and T. Hsieh. 1989. Structure of the *Drosophila* DNA topoisomerase II gene. Nucleotide sequence and homology among topoisomerases II. *J. Mol. Biol.* **205:** 1–13.

Yamaguchi, M., Y. Nishida, T. Moriuchi, F. Hirose, C.C. Hiu, Y. Suzuki, and A. Matsukage. 1990. *Drosophila* proliferating cell nuclear antigen (cyclin) gene: Structure, expression during development, and specific binding of homeodomain proteins to its 5′-flanking region. *Mol. Cell. Biol.* **10:** 872–879.

Zacharias, H. 1979. Underreplication of a polytene chromosome arm in the Chironomid *Prodiamesa olivacea*. *Chromosoma* **72:** 23–51.

Zakian, V.A. 1976. Electron microscopic analysis of DNA replication in main band and satellite DNAs of *Drosophila virilis*. *J. Mol. Biol.* **108:** 305–331.

Zhang, P. and A. C. Spradling. 1995. The *Drosophila* salivary gland chromocenter contains highly polytenized subdomains of mitotic heterochromatin. *Genetics* **139:** 659–670.

36

DNA Replication in *Xenopus*

J. Julian Blow and James P.J. Chong
ICRF Clare Hall Laboratories
South Mimms, Potters Bar
Herts EN6 3LD, United Kingdom

Cell-free extracts of eggs of the South African clawed toad *Xenopus laevis* support complete chromosome replication under normal cell cycle control (Blow and Laskey 1986). Apart from a similar system derived from embryos of the fruit fly *Drosophila melanogaster* (Crevel and Cotterill 1991), it is currently the only eukaryotic cell-free system that supports efficient chromosome replication in vitro. Since the *Xenopus* cell-free system progresses through the complete cell cycle in vitro (Hutchison et al. 1987; Murray and Kirschner 1989), it offers a unique opportunity to study the way that DNA replication is coordinated with other cell-cycle events. This subject has recently been reviewed in detail (Blow 1996); the major conclusions are summarized here.

When *Xenopus* eggs are crushed by low-speed centrifugation, the resultant low-speed supernatants, which contain abundant particulate material (including nuclear envelope precursors), can support all the major activities of the early embryonic cell cycle. During each in vitro cell cycle there is an ordered sequence of events as follows:

1. Activation of certain essential replication proteins and their assembly onto decondensing chromosomes
2. Assembly of chromosomal DNA into nuclei with an intact nuclear envelope
3. Import of proteins required for the initiation and progression of replication
4. Termination of replication forks and inactivation of certain replication proteins to prevent re-replication of DNA in the current cell cycle.

THE ROLE OF NUCLEAR ASSEMBLY IN DNA REPLICATION

The *Xenopus* system can replicate a wide range of different DNA templates, including demembranated *Xenopus* sperm nuclei (sperm chromatin: the natural substrate for DNA replication in the egg) and naked DNA. In each case the template DNA is assembled into interphase nuclei by the cell-free system, involving the assembly of nuclear pores and a double unit nuclear envelope around a chromatin mass (Lohka and Masui 1983, 1984; Vigers and Lohka 1992). Once nuclear assembly is complete, selective nuclear protein accumulation rapidly occurs, an early consequence of which is the assembly of a nuclear lamina (Newport 1987; Newport et al. 1990; Meier et al. 1991; Jenkins et al. 1993). Assembly of template DNA into a functional nucleus is crucial for the way that replication is controlled.

When low-speed supernatants are centrifuged hard to remove particulate material, the resultant high-speed supernatants neither assemble interphase nuclei nor initiate DNA replication (Lohka and Masui 1984; Newport 1987; Sheehan et al. 1988; Blow and Sleeman 1990). Both these activities can be restored by re-addition of pelleted membrane material to the supernatants. When naked DNA is incubated in low-speed supernatants, only a fraction is assembled into nuclei, and only this DNA is replicated (Blow and Sleeman 1990). These results strongly suggest that nuclear assembly is required before DNA replication can occur. Since high-speed supernatants support the elongation stage of DNA replication (Méchali and Harland 1982; Blow and Laskey 1986; Cox 1992; Shivji et al. 1994), it appears that nuclear assembly is specifically required for the initiation of DNA replication. A similar dependence of DNA replication on nuclear assembly is seen in extracts of *Drosophila* embryos (Crevel and Cotterill 1991).

One explanation for this nuclear envelope requirement is that it permits the selective nuclear accumulation of proteins involved in initiation. Consistent with this, when nuclear protein import is prevented in egg extract, the initiation of DNA replication does not occur (Cox 1992). Nuclear assembly may also provide structural components of the nucleus required for DNA replication. DNA synthesis in sperm nuclei replicating in *Xenopus* extract localizes to approximately 100–200 discrete foci in the nuclear interior, each containing about 1000 replication forks (Mills et al. 1989). Replication foci were also seen within nuclei assembled from naked DNA (Cox and Laskey 1991). The need to assemble replication forks into these foci may be part of the reason that initiation of replication is dependent on nuclear assembly. Sequential assembly of replication proteins into these foci is observed. RP-A associates with prereplication foci prior to nuclear assembly (Adachi and Laemmli 1992, 1994). PCNA and DNA polymerase-α are observed in these foci once nuclear assembly has been completed, just prior to the initiation of replication (Hutchison and Kill 1989).

Extracts immunodepleted of lamin B_3 do not assemble a lamina, nor do they support DNA replication, although functional nuclear envelopes are assembled (Newport et al. 1990; Meier et al. 1991; Jenkins et al. 1993). The involvement of the lamina in the initiation of DNA replication is unexpected, since its position underneath the nuclear envelope places it far from the replication foci in the center of the nucleus (Mills et al. 1989). In somatic cells, B-type lamins may colocalize to replication foci (Moir et al. 1994), which could give them a direct role in DNA replication (Hutchison et al. 1994).

Nuclear Structure and Replication Origins

Xenopus eggs and egg extracts replicate a wide variety of DNA templates introduced into them (Harland and Laskey 1980; Méchali and Kearsey 1984; Blow and Laskey 1986; Newport 1987). When normalized for size, the DNA sequence of the template DNA has little effect on the efficiency with which it is replicated (Méchali and Kearsey 1984). Neutral/neutral two-dimensional gel analysis (Hyrien and Méchali 1992; Mahbubani et al. 1992) and electron microscopy (McTiernan and Stambrook 1984) showed that different copies of replicating plasmid molecules contained single initiation bubbles at many different locations. Similar results were obtained by two-dimensional gel analysis of replicating sperm chromatin, showing initiation bubbles scattered throughout the rDNA of sperm chromatin (Hyrien and Méchali 1993). These results suggest that chromosomal DNA is replicated by a series of semi-discontinuous forks (Blow and Laskey 1986) initiated at sites that are not primarily dictated by DNA sequence (Fig. 1A).

Some mechanism must exist to regulate replicon size, because if initiation events occurred at random sites on the genome, there would be some excessively large replicons (Mahbubani et al. 1992). Instead of being dictated by DNA sequence, replicon size may be directly controlled by chromosome structure (Fig. 1B). Chromosomal loop size correlates well with the average replicon size as this increases during *Xenopus* development (Buongiorno Nardelli et al. 1982). In particular, each copy of rDNA appears to comprise one supercoiled loop (Marilley and Gassend Bonnet 1989), and each supports only a single initiation event (Mahbubani et al. 1992; Hyrien and Méchali 1993). Consistent with a role for nuclear structure in determining origin usage, intact hamster nuclei incubated in *Xenopus* extract continued to use the dihydrofolate reductase origin of replication, although naked DNA containing this region showed no preferential initiation (Gilbert et al. 1993, 1995).

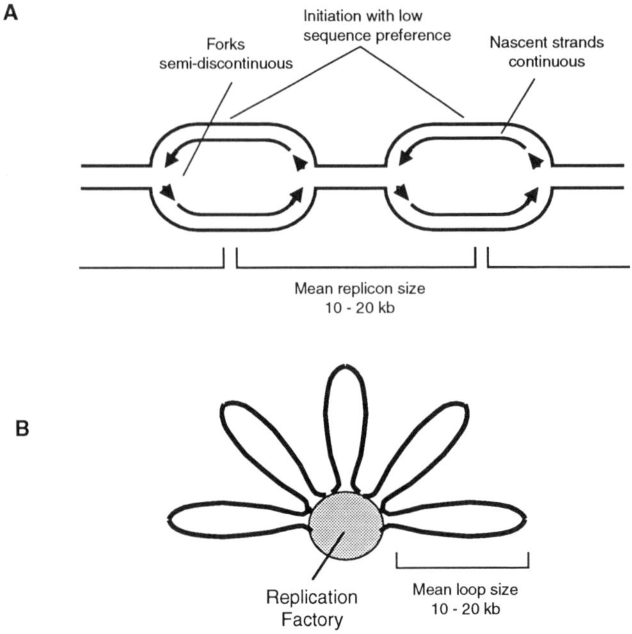

Figure 1 Replication origin usage in the *Xenopus* system. (*A*) Cartoon summarizing experiments analyzing replication origin usage and the structure of replicative intermediates in the *Xenopus* system. (*B*) Cartoon showing a possible role for chromosome looping in generating a relatively constant replicon size in the *Xenopus* system. See text for more details. (Redrawn from Blow 1996.)

The Nucleus as a Fundamental Unit of DNA Replication

Initiation occurs virtually as soon as nuclear assembly has been completed. Since nuclei are not all assembled at the same rate, this means that different nuclei may start to replicate at different times. Analysis of replication kinetics showed that nuclei act as individual units, each receiving a signal to replicate from the cytoplasm, which causes them to undergo once and only once a burst of near-synchronous initiation events (Blow and Watson 1987). The feature that defines this unit of replication is likely to be the nuclear envelope, since all the DNA surrounded by an intact nuclear envelope starts to replicate at the same time, even if this DNA was originally derived from more than one nucleus (Leno and Laskey 1991).

As outlined in Figure 2, these results provide a model for the way in which DNA replication is controlled in the *Xenopus* cell cycle (Blow and Watson 1987; Blow and Laskey 1988; Blow 1996). Replication control

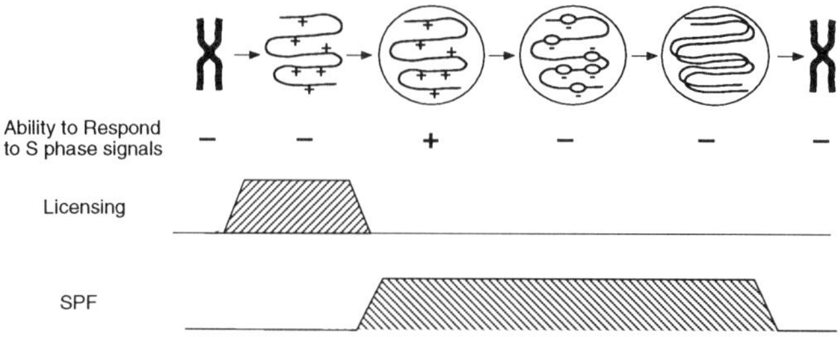

Figure 2 Model to explain replication control in the *Xenopus* system. A single nucleus is shown as it passes through a complete cell cycle. During late mitosis, prior to nuclear envelope assembly, the DNA becomes "licensed" (+) to undergo DNA replication. Once assembled into a nucleus, the licensed DNA is capable of initiating DNA replication in response to the presence of the S-phase inducer SPF. However, the license is destroyed (–) as the DNA is replicated. Only following passage through mitosis does the nucleus once again become competent to undergo further DNA replication. (Redrawn from Blow 1996.)

is divided into two distinct components: the DNA template in the nucleus, and activities present in the cytoplasm that act on this nuclear substrate. The nucleus can be in one of two states: either capable or incapable of responding to S-phase inducers by undergoing DNA replication. The ability of nuclei to respond to the S-phase inducers requires DNA to have been "licensed" for DNA replication during the previous mitosis. The cytoplasm also provides an S-phase-promoting factor that acts on intact licensed nuclei to induce them to initiate DNA replication. The license is then inactivated or destroyed in the process of DNA replication.

ROLE OF LICENSING FACTOR IN REPLICATION CONTROL

The inability of replicated nuclei to respond to S-phase inducers present in *Xenopus* extracts can be demonstrated directly. Replicated G_2 nuclei that were transferred to fresh extract did not undergo further replication (Blow and Laskey 1988). However, if these nuclei were allowed to pass into mitosis, the DNA efficiently re-replicated on transfer to interphase extract. The effect of passage through mitosis could be mimicked by agents that caused nuclear envelope permeabilization, such as lysolecithin or phospholipase (Blow and Laskey 1988). Similar results were obtained in *Drosophila* extracts (Crevel and Cotterill 1991) and on addition of nuclei from mammalian tissue-culture cells into *Xenopus* eggs (De Roeper et al. 1977) or egg extracts (Leno et al. 1992; Coverley et al. 1993).

Figure 2 provides an explanation of these results (Blow and Laskey 1988; Blow 1993; Chong et al. 1996). An essential replication factor called replication licensing factor (RLF) binds DNA during late mitosis before nuclear assembly has occurred. RLF cannot cross the nuclear envelope, so once nuclear assembly is complete, RLF is only present in the nucleus where bound to DNA. On entry into S phase, RLF bound to DNA supports a single initiation event, after which it is inactivated or destroyed. Thus, in G_2, no active RLF remains in the nucleus and the nuclear envelope must be transiently permeabilized (as normally occurs during mitosis) to allow a further round of DNA replication to be licensed.

Identification of Licensing Factor Components

RLF has recently been subjected to biochemical fractionation by exploiting the ability of protein kinase inhibitors to block the activation of RLF that normally occurs during mitosis (Blow 1993; Kubota and Takisawa 1993; Vesely et al. 1994). RLF activity resolved into two components, RLF-M and RLF-B, both of which were required for licensing (Chong et al. 1995). RLF-M was purified to apparent homogeneity and consisted of a complex of at least three polypeptides, with molecular masses of 92 kD, 106 kD, and 115 kD. The 106-kD polypeptide is the product of the *Xenopus* MCM3 gene, and the other polypeptides seem likely to be other members of the MCM family (Chong et al. 1995). Both RLF-B and RLF-M were required for each successive round of DNA replication. *Xenopus* Mcm3 associated with chromatin in G_1 but was removed during replication, consistent with its involvement in the RLF system. Furthermore, the rebinding of Mcm3 to replicated chromatin was dependent on prior nuclear envelope permeabilization (Chong et al. 1995). Immunodepletion of *Xenopus* Mcm3 also resulted in replication defects consistent with these results (Chong et al. 1995; Kubota et al. 1995; Madine et al. 1995). The role of nuclear envelope permeabilization is currently unclear. Although RLF activity does not cross the nuclear envelope in *Xenopus* (De Roeper et al. 1977; Blow and Laskey 1988; Leno et al. 1992; Coverley et al. 1993; Blow 1993; Chong et al. 1995) or *Drosophila* (Crevel and Cotterill 1991), Mcm proteins are nuclear throughout the mammalian cell cycle (Thömmes et al. 1992; Kimura et al. 1994; Todorov et al. 1994), and Mcm3 is imported into intact nuclei in *Xenopus* extracts (Madine et al. 1995). One possibility is that active RLF-B is required for RLF-M to associate with chromatin, and that RLF-B is incapable of crossing the nuclear envelope. Purification and characterization of the RLF-B fraction should elucidate this point.

SPF: THE SIGNAL TO INITIATE REPLICATION

Soon after nuclear assembly is complete, each nucleus undergoes a coordinated burst of initiation events. The intranuclear signal that generates this burst of initiation appears to be closely associated with cyclin-dependent kinases (cdks). Cdks are small protein kinase subunits, activated by complexing with a cyclin partner, that play an important role in cell-cycle regulation. In *Xenopus*, two classes of cdk (cdc2 and cdk2) and three classes of cyclin (A, B, and E) have been identified. The cdc2-cyclin B complex forms the mitotic inducer maturation promoting factor (MPF), which has a role in the activation of RLF during mitosis as described above. Other cdks generate the S-phase promoting factor (SPF) signal required for licensed nuclei to enter S phase.

Xenopus extracts affinity-depleted of cdks with the cell-cycle protein $p13^{suc1}$ were specifically unable to support the initiation of DNA replication (Blow and Nurse 1990). In extracts treated with protein synthesis inhibitors, SPF activity appears to be dependent on cdk2, as immunodepletion of cdk2 blocked DNA replication (Fang and Newport 1991), whereas the cdk inhibitor $p21^{Cip1}$ inhibited replication at concentrations comparable to that of endogenous cdk2 (Strausfeld et al. 1994; Chen et al. 1995; Jackson et al. 1995). No effect on replication fork movement or complementary strand synthesis was seen in cdk-inhibited extracts, implying that SPF function is specifically required for the initiation of replication (Blow and Nurse 1990; Fang and Newport 1991; Strausfeld et al. 1994).

Cyclins A and B are predominantly found complexed with cdc2 in the early embryo (Minshull et al. 1990; Howe et al. 1995; U.P. Strausfeld et al. in prep.), whereas cyclin E is found exclusively complexed with cdk2 (Rempel et al. 1995). Immunodepletion of cyclin E blocked DNA replication, further suggesting that a cdk2-cyclin E complex provides SPF activity (Jackson et al. 1995). However, DNA replication could be restored to cdk-defective extracts by both A- or E-type cyclins, but not B-type cyclins (Strausfeld et al. 1994 and in prep.; Jackson et al. 1995). Only low cyclin A concentrations could rescue DNA synthesis, however, as higher levels generated MPF activity and drove extracts into mitosis (U.P. Strausfeld et al., in prep.).

Figure 3 shows a model to integrate the roles of cyclins A, B, and E in controlling DNA replication in the *Xenopus* cell cycle. Both cyclin A and cyclin E can provide SPF activity (Strausfeld et al. 1994 and in prep.; Jackson et al. 1995). Since initiation occurs almost immediately once template DNA has been assembled into interphase nuclei, SPF does not appear to be rate-limiting for this process (Fig. 3, "nuclear assembly").

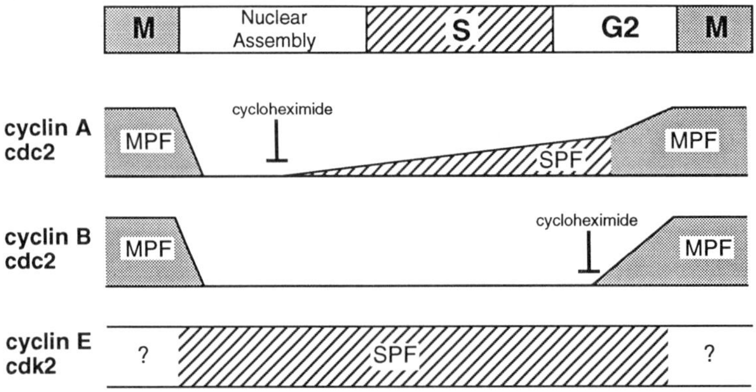

Figure 3 Cartoon showing the proposed roles of cyclins A, B, and E in *Xenopus* egg extracts. The top panel shows the cell-cycle events taking place in *Xenopus* extracts. Kinase activities of cyclin A-cdc2, cyclin B-cdc2, and cyclin E-cdk2 at the different times are shown by the boxed regions below. In the presence of cycloheximide, levels of cyclins A and B remain low, whereas cyclin E is largely unaffected. See text for more details. MPF (*stippled*) and SPF (*diagonal lines*) activities associated with these kinases are also indicated; SPF activity of cyclin E-cdk2 during mitosis is undetermined. (Redrawn from U.P. Strausfeld et al., in prep.)

On exit from mitosis in *Xenopus*, cyclin A but not cyclin E is degraded (Minshull et al. 1990; Gabrielli et al. 1992; Rempel et al. 1995), so that extracts prepared in the presence of protein synthesis inhibitors contain no cyclin A, and all SPF activity is provided by the cdk2-cyclin E complex (Fig. 3). However, cyclin A is abundantly translated and can form an active kinase with cdc2 (Minshull et al. 1990) to provide additional SPF activity (Strausfeld et al. 1994 and in prep.). Cyclin A is capable of inducing DNA synthesis even when added after nuclear assembly is complete (U.P. Strausfeld et al., in prep.), and its role may be to induce initiation at any replicons that have not already fired. Consistent with this, DNA replication in translationally active suc1-depleted extracts (which are likely to contain significantly more cyclin A than E) is more efficiently restored by addition of cdc2 mRNA than by cdk2 mRNA (Chevalier et al. 1995). As cyclin A and cyclin B kinase levels build up later in the cell cycle, this becomes sufficient to induce entry into mitosis. Correct passage through mitosis is necessary for DNA to become relicensed for DNA replication in the next cell cycle (Blow and Laskey 1988; Blow 1993). The coordination of the different cdks is therefore responsible for the regulated replication of DNA during the cell cycle.

ACKNOWLEDGMENT

J.J.B. is a Lister Institute Research Fellow.

REFERENCES

Adachi, Y. and U.K. Laemmli. 1992. Identification of nuclear pre-replication centers poised for DNA synthesis in *Xenopus* egg extracts: Immunolocalization study of replication protein A. *J. Cell Biol.* **119:** 1–15.

———. 1994. Study of the cell cycle-dependent assembly of the DNA pre-replication centres in *Xenopus* egg extracts. *EMBO J.* **13:** 4153–4164.

Blow, J.J. 1993. Preventing re-replication of DNA in a single cell cycle: Evidence for a replication licensing factor. *J. Cell Biol.* **122:** 993–1002.

———. 1996. DNA replication in *Xenopus*. In *Eukaryotic DNA replication* (ed. J.J. Blow). Oxford University Press, Oxford, United Kingdom. (In press.)

Blow, J.J. and R.A. Laskey. 1986. Initiation of DNA replication in nuclei and purified DNA by a cell-free extract of *Xenopus* eggs. *Cell* **47:** 577–587.

———. 1988. A role for the nuclear envelope in controlling DNA replication within the cell cycle. *Nature* **332:** 546–548.

Blow, J.J. and P. Nurse. 1990. A cdc2-like protein is involved in the initiation of DNA replication in *Xenopus* egg extracts. *Cell* **62:** 855–862.

Blow, J.J. and A.M. Sleeman. 1990. Replication of purified DNA in *Xenopus* egg extracts is dependent on nuclear assembly. *J. Cell Sci.* **95:** 383–391.

Blow, J.J. and J.V. Watson. 1987. Nuclei act as independent and integrated units of replication in a *Xenopus* cell-free system. *EMBO J.* **6:** 1997–2002.

Buongiorno Nardelli, M., G. Micheli, M.T. Carri, and M. Marilley. 1982. A relationship between replicon size and supercoiled loop domains in the eukaryotic genome. *Nature* **298:** 100–102.

Chen, J., P.K. Jackson, M.W. Kirschner, and A. Dutta. 1995. Separate domains of p21 involved in the inhibiton of Cdk kinase and PCNA. *Nature* **374:** 386–388.

Chevalier, S., J.-P. Tassan, R. Cox, M. Philippe, and C. Ford. 1995. Both cdc2 and cdk2 promote S phase initiation in *Xenopus* egg extracts. *J. Cell Sci.* **108:** 1831–1841.

Chong, J.P.J., P. Thömmes, and J.J. Blow. 1996. The role of MCM/P1 proteins in the licensing of DNA replication. *Trends Biochem. Sci.* (in press).

Chong, J.P.J., M.H. Mahbubani, C.-Y. Khoo, and J.J. Blow. 1995. Purification of an Mcm-containing complex as a component of the DNA replication licensing system. *Nature* **375:** 418–421.

Coverley, D., C.S. Downes, P. Romanowski, and R.A. Laskey. 1993. Reversible effects of nuclear membrane permeabilization on DNA replication: Evidence for a positive licensing factor. *J. Cell Biol.* **122:** 985–992.

Cox, L.S. 1992. DNA replication in cell-free extracts from *Xenopus* eggs is prevented by disrupting nuclear envelope function. *J. Cell Sci.* **101:** 43–53.

Cox, L.S. and R.A. Laskey. 1991. DNA replication occurs at discrete sites in pseudonuclei assembled from purified DNA in vitro. *Cell* **66:** 271–275.

Crevel, G. and S. Cotterill. 1991. DNA replication in cell-free extracts from *Drosophila melanogaster*. *EMBO J.* **10:** 4361–4369.

De Roeper, A., J.A. Smith, R.A. Watt, and J.M. Barry. 1977. Chromatin dispersal and DNA synthesis in G1 and G2 HeLa cell nuclei injected into *Xenopus* eggs. *Nature* **265:** 469–470.

Fang, F. and J.W. Newport. 1991. Evidence that the G1-S and G2-M transitions are controlled by different cdc2 proteins in higher eukaryotes. *Cell* **66:** 731–742.

Gabrielli, B.G., L.M. Roy, J. Gautier, M. Philippe, and J.L. Maller. 1992. A cdc2-related kinase oscillates in the cell cycle independently of cyclins G2/M and cdc2. *J. Biol. Chem.* **267:** 1969–1975.

Gilbert, D.M., H. Miyazawa, and M.L. DePamphilis. 1995. Site-specific initiation of DNA replication in *Xenopus* egg extract requires nuclear structure. *Mol. Cell. Biol.* **15:** 2942–2954.

Gilbert, D.M., H. Miyazawa, F.S. Nallaseth, J.M. Ortega, J.J. Blow, and M.L. DePamphilis. 1993. Site-specific initiation of DNA replication in metazoan chromosomes and the role of nuclear organization. *Cold Spring Harbor Symp. Quant. Biol.* **58:** 475–485.

Harland, R.M. and R.A. Laskey. 1980. Regulated DNA replication of DNA microinjected into eggs of *Xenopus laevis*. *Cell* **21:** 761–771.

Howe, J.A., M. Howell, T. Hunt, and J.W. Newport. 1995. Identification of a developmental timer regulating the stability of embryonic cyclin A and a new somatic A-type cyclin at gastrulation. *Genes Dev.* **9:** 1164–1176.

Hutchison, C., and I. Kill. 1989. Changes in the nuclear distribution of DNA polymerase α and PCNA/cyclin during the progress of the cell cycle, in a cell-free extract of *Xenopus* eggs. *J. Cell Sci.* **93:** 605–613.

Hutchison, C.J., J.M. Bridger, L.S. Cox, and I.R. Kill. 1994. Weaving a pattern from disparate threads: Lamin function in nuclear assembly and DNA replication. *J. Cell Sci.* **107:** 3259–3269.

Hutchison, C.J., R. Cox, R.S. Drepaul, M. Gomperts, and C.C. Ford. 1987. Periodic DNA synthesis in cell-free extracts of *Xenopus* eggs. *EMBO J.* **6:** 2003–2010.

Hyrien, O. and M. Méchali. 1992. Plasmid replication in *Xenopus* eggs and egg extracts: A 2D gel electrophoretic analysis. *Nucleic Acids Res.* **20:** 1463–1469.

———. 1993. Chromosomal replication initiates and terminates at random sequences but at regular intervals in the ribosomal DNA of *Xenopus* early embryos. *EMBO J.* **12:** 4511–4520.

Jackson, P.K., S. Chevalier, M. Phillipe, and M.W. Kirschner. 1995. Early events in DNA replication require cyclin E and are blocked by p21Cip1. *J. Cell Biol.* **130:** 755–769.

Jenkins, H., T. Holman, C. Lyon, B. Lane, R. Stick, and C. Hutchison. 1993. Nuclei that lack a lamina accumulate karyophilic proteins and assemble a nuclear matrix. *J. Cell Sci.* **106:** 275–285.

Kimura, H., N. Nozaki, and K. Sugimoto. 1994. DNA polymerase α associated protein P1, a murine homolog of yeast MCM3, changes its intranuclear distribution during the DNA synthetic period. *EMBO J.* **13:** 4311–4320.

Kubota, Y. and H. Takisawa. 1993. Determination of initiation of DNA replication before and after nuclear formation in *Xenopus* egg cell free extracts. *J. Cell Biol.* **123:** 1321–1331.

Kubota, Y., S. Mimura, S. Nishimoto, H. Takisawa, and H. Nojima. 1995. Identification of the yeast MCM3-related protein as a component of *Xenopus* DNA replication licensing factor. *Cell* **81:** 601–609.

Leno, G.H. and R.A. Laskey. 1991. The nuclear membrane determines the timing of DNA replication in *Xenopus* egg extracts. *J. Cell Biol.* **112:** 557–566.

Leno, G.H., C.S. Downes, and R.A. Laskey. 1992. The nuclear membrane prevents replication of human G2 nuclei but not G1 nuclei in *Xenopus* egg extract. *Cell* **69:** 151–158.

Lohka, M.J. and Y. Masui. 1983. Formation in vitro of sperm pronuclei and mitotic chromosomes induced by amphibian ooplasmic components. *Science* **220:** 719–721.

———. 1984. Roles of cytosol and cytoplasmic particles in nuclear envelope assembly and sperm pronuclear formation in cell-free preparations from amphibian eggs. *J. Cell Biol.* **98:** 1222–1230.

Madine, M.A., C.-Y. Khoo, A.D. Mills, and R.A. Laskey. 1995. MCM3 complex required for cell cycle regulation of DNA replication in vertebrate cells. *Nature* **375:** 421–424.

Mahbubani, H.M., T. Paull, J.K. Elder, and J.J. Blow. 1992. DNA replication initiates at multiple sites on plasmid DNA in *Xenopus* egg extracts. *Nucleic Acids Res.* **20:** 1457–1462.

Marilley, M. and G. Gassend Bonnet. 1989. Supercoiled loop organization of genomic DNA: A close relationship between loop domains, expression units, and replicon organization in rDNA from *Xenopus laevis*. *Exp. Cell Res.* **180:** 475–489.

McTiernan, C.F. and P.J. Stambrook. 1984. Initiation of SV40 DNA replication after microinjection into *Xenopus* eggs. *Biochim. Biophys. Acta* **782:** 295–303.

Méchali, M. and R.M. Harland. 1982. DNA synthesis in a cell-free system from *Xenopus* eggs: Priming and elongation on single-stranded DNA in vitro. *Cell* **30:** 93–101.

Méchali, M. and S. Kearsey. 1984. Lack of specific sequence requirement for DNA replication in *Xenopus* eggs compared with high sequence specificity in yeast. *Cell* **38:** 55–64.

Meier, J., K.H. Campbell, C.C. Ford, R. Stick, and C.J. Hutchison. 1991. The role of lamin LIII in nuclear assembly and DNA replication, in cell-free extracts of *Xenopus* eggs. *J. Cell Sci.* **98:** 271–279.

Mills, A.D., J.J. Blow, J.G. White, W.B. Amos, D. Wilcock, and R.A. Laskey. 1989. Replication occurs at discrete foci spaced throughout nuclei replicating in vitro. *J. Cell Sci.* **94:** 471–477.

Minshull, J., R. Golsteyn, C.S. Hill, and T. Hunt. 1990. The A- and B-type cyclin associated cdc2 kinases in *Xenopus* turn on and off at different times in the cell cycle. *EMBO J.* **9:** 2865–2875.

Moir, R.D., M. Montag Lowy, and R.D. Goldman. 1994. Dynamic properties of nuclear lamins: Lamin B is associated with sites of DNA replication. *J. Cell Biol.* **125:** 1201–1212.

Murray, A.W. and M.W. Kirschner. 1989. Cyclin synthesis drives the early embryonic cell cycle. *Nature* **339:** 275–280.

Newport, J. 1987. Nuclear reconstitution in vitro: Stages of assembly around protein-free DNA. *Cell* **48:** 205–217.

Newport, J.W., K.L. Wilson, and W.G. Dunphy. 1990. A lamin-independent pathway for nuclear envelope assembly. *J. Cell Biol.* **111:** 2247–2259.

Rempel, R.E., S.B. Sleight, and J.L. Maller. 1995. Maternal *Xenopus* cdk2-cyclin E complexes function during meiotic and early embryonic cell cycles that lack a G1 phase. *J. Biol. Chem.* **270:** 6843–6855.

Sheehan, M.A., A.D. Mills, A.M. Sleeman, R.A. Laskey, and J.J. Blow. 1988. Steps in the assembly of replication-competent nuclei in a cell-free system from *Xenopus* eggs. *J. Cell Biol.* **106:** 1–12.

Shivji, M.K.K., S.J. Grey, U.P. Strausfeld, R.D. Wood, and J.J. Blow. 1994. Cip1 inhibits

DNA replication but not PCNA-dependent nucleotide excision-repair. *Curr. Biol.* **4:** 1062–1068.

Strausfeld, U.P., M. Howell, R. Rempel, J.L. Maller, T. Hunt, and J.J. Blow. 1994. Cip1 blocks the initiation of DNA replication in *Xenopus* extracts by inhibition of cyclin-dependent kinases. *Curr. Biol.* **4:** 876–883.

Thömmes, P., R. Fett, B. Schray, R. Burkhart, M. Barnes, C. Kennedy, N.C. Brown, and R. Knippers. 1992. Properties of the nuclear P1 protein, a mammalian homologue of the yeast Mcm3 replication protein. *Nucleic Acids Res.* **20:** 1069–1074.

Todorov, I.T., R. Pepperkok, R.N. Philipova, S.E. Kearsey, W. Ansorge, and D. Werner. 1994. A human nuclear protein with sequence homology to a family of early S phase proteins is required for entry into S phase and for cell division. *J. Cell Sci.* **107:** 253–265.

Vesely, J., L. Havlicek, M. Strnad, J.J. Blow, A. Donnella-Deana, L. Pinna, D.S. Letham, J. Kato, L. Detivaud, S. Leclerc, and L. Meijer. 1994. Inhibition of cyclin-dependent kinases by purine analogues. *Eur. J. Biochem.* **224:** 771–786.

Vigers, G.P. and M.J. Lohka. 1992. Regulation of nuclear envelope precursor functions during cell division. *J. Cell Sci.* **102:** 273–284.

37

DNA Replication in Mammals

Nicholas H. Heintz
Department of Pathology
University of Vermont College of Medicine
Burlington, Vermont 05405

Formulated in the early 1960s, the replicon hypothesis proposed that DNA synthesis is controlled by the interaction of initiator proteins with genetic elements termed replicators (Jacob et al. 1964). Study of the interaction of purified initiator proteins with *cis*-acting replicators from bacterial chromosomes, bacteriophage, and eukaryotic viruses has led to a general model for the assembly of replication forks at physically defined sites, or origins of replication (Bramhill and Kornberg 1988). In prokaryotes and viruses, the genome commonly consists of a single replicon, with a single replicator controlling initiation of DNA synthesis under all growth conditions. In higher eukaryotes, the genome consists of hundreds to thousands of replicons, and the order of replicon synthesis may vary as a function of growth conditions or cellular differentiation. Since the decision to initiate DNA synthesis is a critical point of control in the cell cycle, there is considerable interest in the structure of eukaryotic replicators and the factors that regulate their selection and activity under various conditions. Although the replicators of mammalian chromosomes and their cognate initiators have yet to be described, the basic unit of DNA synthesis remains the replicon.

REPLICONS AND REPLICON CLUSTERS IN MAMMALIAN CELLS

DNA fiber autoradiography has shown that mammalian DNA replication is usually bidirectional, with equivalent rates of synthesis at each of the two forks in a replication bubble; that initiation begins at multiple, dispersed sites spaced 50–250 kb apart; and that initiation within clusters of contiguous replicons tends to be coordinately controlled (Huberman and Riggs 1968; see also Hand 1978). Thus, at the level of the replicon, DNA synthesis in mammalian cells initiates at a single site, proceeds bidirectionally through the agency of two replication forks, and terminates when forks of neighboring replicons converge.

DNA Replication in Eukaryotic Cells

Localization of replicating DNA in S-phase nuclei by autoradiography or immunohistochemical techniques shows that DNA synthesis occurs in discrete foci, each of which contains several hundred replication forks (Nakamura et al. 1986; Berezney 1991b; Fox et al. 1991; Hassan and Cook 1993). Models for the organization of replication foci suggest that the periodic attachment of replication origins to the nuclear scaffold (or matrix) generates topologically constrained DNA loops that are reeled through fixed sites of synthesis during DNA replication (Pardoll et al. 1980; Nakamura et al. 1986; Berezney 1991a; Cook 1991). In metaphase chromosomes, DNA is also organized in loops that are anchored to a proteinaceous scaffold (Saitoh and Laemmli 1994 and references therein). Although there is a general correlation between the sizes of chromosome loops and replicons (Buongiorno-Nardelli et al. 1982), the precise relationship between replicons, interphase DNA loops, and metaphase chromosome structure is not well understood. To date, no protein involved in the organization of chromosome loops has been linked directly to initiation of DNA synthesis, nor has a functional relationship between scaffold attachment sequences and replication origins been demonstrated.

ASSEMBLY OF REPLICATION COMPLEXES

There is considerable interest in the order of assembly of replication complexes at initiation sites and how the assembly process is regulated during the cell cycle. Immunolocalization of replication factors during the cell cycle suggests that replication complexes are assembled in multiple steps (Kill et al. 1991). Entry into the S phase is accompanied by the redistribution of some replication factors that are constitutively synthesized, such as the DNA polymerase accessory factor proliferating cell nuclear antigen (PCNA) (Bravo and Macdonald-Bravo 1987), to replication foci, whereas other replication factors, such as DNA polymerase-α (Bensch et al. 1982) and cyclin A (Cardoso et al. 1993), are preferentially expressed in G_1 and then recruited to replication sites. In nuclear reconstitution studies with *Xenopus* egg extracts, replication foci are generated during nuclear assembly prior to DNA synthesis (Adachi and Laemmli 1994). Assembly of replication foci in synthetic nuclei requires the single-stranded DNA-binding protein RP-A and inactivation of the mitotic cdc2/cyclin B kinase complex (Adachi and Laemmli 1994). These results suggest that selection of initiation sites occurs postmitotically and is dictated by factors that participate in nuclear reorganization during the transition from mitosis to the G_1 phase.

Assembly of replication foci may not require specific DNA sequences. *Xenopus* egg extracts are able to assemble DNA from any source into synthetic nuclei (Blow and Laskey 1986), which contain replication foci indistinguishable from those of native nuclei (Mills et al. 1989). Moreover, the DNA in synthetic nuclei is duplicated once and only once per nuclear assembly cycle (Blow and Laskey 1988). These results suggest metazoans may not require specific sequence elements to initiate DNA synthesis or to limit replication to once per cell cycle. Nonetheless, localization of DNA methyltransferase within mammalian replication foci requires a distinct targeting sequence (Leonhardt et al. 1992). Thus, as in other replication systems, assembly of mammalian replication complexes occurs through an ordered series of protein-protein interactions. In contrast to other systems, however, the assembly process may be directed by initiation complexes that are not associated with specific DNA sequences.

THE ORDER OF REPLICON SYNTHESIS

There is significant evidence for an ordered, but malleable, program of DNA replication in mammalian cells (see Simon and Cedar, this volume). Transcription appears to have a particularly strong influence on the time during S phase that a sequence is replicated: Genes that are transcriptionally competent tend to be replicated in the first quarter of S phase, whereas those that are not tend to replicate later (Hatton et al. 1988). How might transcription influence replication timing? In early S phase, replication sites coincide with sites of transcription in late G_1 (Hassan et al. 1994), suggesting that, as for some prokaryotic origins, activation of mammalian origins could be linked to the act of transcription (Hassan and Cook 1994). However, the colocalization of transcription and replication sites could be related to the structural organization of transcriptionally active chromatin within the nucleus rather than the process of transcription itself. Alternatively, regulatory transcription factors could participate in the assembly of initiation complexes at both gene promoters and replication origins (for review, see Pederson and Heintz 1994), and this class of DNA-binding proteins could serve to link the two processes in the cell cycle (see van der Vliet, this volume).

THE SEARCH FOR REPLICATION ORIGINS IN MAMMALIAN CELLS

The genetic elements involved in initiation of DNA synthesis are well described in only one eukaryote—the yeast *Saccharomyces cerevisiae.*

Minimal replicators from *S. cerevisiae* contain at least two functional domains: Domain A encompasses a core consensus sequence (the ACS) that is necessary, but not sufficient, for origin activity, and domain B harbors heterogeneous accessory sequences that enhance the efficiency of origin utilization (see Newlon, this volume). The fact that heterogeneous accessory elements act in concert with the yeast origin core suggests that yeast replicators resemble gene promoters (Heintz et al. 1992), a notion reinforced by the observation that transcription factors stimulate yeast origin activity (for review, see Pederson and Heintz 1994).

The time of activation during S phase and the frequency of origin utilization in yeast are subject to chromosomal context (Fangman and Brewer 1991), suggesting that long-range chromatin determinants or other factors (i.e., transcription) influence the distribution or activity of yeast initiator proteins. Indeed, a strong candidate for the *S. cerevisiae* initiator protein, the origin recognition complex (ORC), functions in both DNA replication (Bell and Stillman 1992) and the repression of transcription (Foss et al. 1993). ORC is bound to the yeast origin *ARS1* throughout most of the cell cycle (Diffley and Cocker 1992), and proteins that interact with ORC in a cell-cycle-dependent manner appear to control origin activation during entry into S phase (Diffley et al. 1994; Dowell et al. 1994). Presumably, the proteins that interact with ORC to achieve repression of transcription differ from those that interact with ORC during initiation of DNA replication.

Given the high degree of conservation of replication enzymes and the kinase cascades that control entry into the S phase in eukaryotes, it would be surprising if animal cells did not contain homologs of ORC. Nonetheless, replication origins in mammalian cells clearly differ from those in *S. cerevisiae*. In *S. cerevisiae*, replication of episomal plasmids requires specific DNA sequences, thereby allowing dissection of the *cis*-acting elements involved in replicator function. The identification of mammalian replicators by plasmid replication assays has proven far less successful. Regions of the c-*myc* promoter (McWhinney and Leffak 1990), the adenosine deaminase gene region (Virta-Pearlman et al. 1993), hsp70 gene promoter (Taira et al. 1994), and several anonymous sequences (Frappier and Zannis-Hadjopoulos 1987; Masukata et al. 1993; Nielson et al. 1994) have been reported to support the episomal replication of recombinant plasmids. In several instances, the sequences that promote plasmid replication have been shown to serve as initiation sites in the chromosome (see DePamphilis, this volume). However, significant variations in the results of plasmid transfection experiments have been observed in different laboratories (Burhans et al. 1990; Caddle and

Calos 1992; Matsukata et al. 1993), suggesting that the assay may not accurately detect *cis*-acting sequences of chromosomal replicators. Study of initiation activity with an assay based on the nuclear retention of plasmids by the Epstein-Barr virus EBNA1 protein indicates that size, and not primary sequence composition, is the most critical factor affecting the efficiency of plasmid replication in animal cells (Krysan et al. 1989). Additional studies in the EBNA1 system indicate that initiation occurs at random sites in plasmid DNA (Krysan and Calos 1991) and is precluded only from transcribed regions (Haase et al. 1994).

ARS activity in plasmids may reflect the relative propensity of different DNA sequences to unwind under superhelical stress. When supercoiled plasmids are incubated with DNA polymerase-α:primase, RP-A, T antigen, and gyrase in vitro, initiation is observed at the c-*myc* promoter, the yeast origin *ARS1*, and, to a lesser extent, the DHFR *ori*β region (Ishimi et al. 1994). A survey of human DNA for the ability to support priming of DNA synthesis by DNA polymerase-α:primase in the presence of RP-A in vitro showed that site-specific synthesis occurs preferentially at sequences containing the dinucleotide repeat dA:dT (Pack and Tsurimoto 1994). Since poly(dA:dT) repeats have a low free energy for unwinding, these sequences tend to become single-stranded in supercoiled plasmids (Caddle et al. 1990b; Pack and Tsurimoto 1994), a property that accounts for their preferential use as in vitro initiation sites. Although DNA unwinding elements (DUEs) are important components of replication origins (see DePamphilis, this volume), further work is required before a relationship between sequences that promote plasmid DNA replication and those that function as replicators in the chromosome can be established.

INITIATION SITES IN MAMMALIAN CHROMOSOMES

Without a definitive genetic assay, mapping of initiation events to specific sites in mammalian chromosomes has been largely limited to methods that are based on the conserved features of replication forks. Replication forks in animal cells have the same organization as those in other organisms: DNA synthesis is semi-discontinuous, with one template strand duplicated by a highly processive polymerase complex, and the other copied by a discontinuous mechanism through the agency of Okazaki fragments (see, e.g., Waga and Stillman 1994). Since mammalian DNA replication is semi-discontinuous and bidirectional, origin-mapping methods are based on assays that assess the template specificity of leading- and lagging-strand synthesis, polarity of replication fork movement, stability of DNA in replication bubbles, size of nascent DNA strands, or electrophoretic migration of replication forks and bubbles in two-dimensional (2D) gels (for review, see Vassilev and DePamphilis 1992). Similarly, labeling of DNA during entry into the S phase has been used to isolate initiation sites. However, as has been noted previously (Stillman 1994), mapping of initiation sites is not synonymous with mapping the *cis*-acting elements of replicators. For example, mapping of transcription start sites would not necessarily identify *cis*-acting enhancer elements that act at a distance to influence gene expression.

Using a variety of techniques, origin-mapping experiments have located initiation sites near the c-*myc* promoter region (Vassilev and Johnson 1990), two different regions of the murine adenosine deaminase gene locus (Carroll et al. 1993; Virta-Pearlman et al. 1993), the immunoglobulin heavy-chain enhancer (Ariizumi et al. 1993), the rhodopsin gene locus (Gale et al. 1992), the nontranscribed portion of the rDNA repeats (Little et al. 1993), the RPS14 gene (Tasheva and Roufa 1994a), the lamin B2 gene (Giacca et al. 1994), the hsp70 gene (Taira et al. 1994), the human β-globin locus (Kitsberg et al. 1993), and the Chinese hamster dihydrofolate reductase (DHFR) domain (see below). These initiation sites show no obvious spatial relationship to transcription units, and features of these sequences that contribute to their use as initiation sites are difficult to discern. Since regulatory transcription factors have been shown to act as accessory factors in the initiation of viral and yeast DNA replication (for review, see Pederson and Heintz 1994), the distribution of transcription factor binding sites is potentially interesting. However, the consensus sequences for transcription factor binding sites tend to be short and tolerate a number of substitutions, and therefore these sequences tend to be widely distributed in genomic DNA (Dobbs et al. 1994). Binding sites for the sequence-specific single-stranded DNA-binding protein Pur are found in gene promoters and mammalian initiation sites at a higher frequency than expected by chance (Bergemann and Johnson 1992; Dobbs et al. 1994). A Pur-binding site juxtaposed to the JC virus core replication origin has been shown to be important for both viral DNA replication (Chang et al. 1994) and early gene transcription (Chen et al. 1995), suggesting that Pur could also play a role in mammalian replication. Although comparison of six initiation sites has identified a degenerate consensus sequence (Dobbs et al. 1994), the consensus has not been shown to have a function analogous to the ACS of *S. cerevisiae* replicators.

Given the lack of a definitive assay, demonstration that specific sequences or DNA-binding proteins are involved in initiation in mammalian cells is certain to be difficult. An additional factor that confounds the identification of replicators and initiators in mammalian cells is the apparent decentralization of initiation events, a phenomenon that has been documented most thoroughly in the DHFR domain of Chinese hamster cells.

THE CHINESE HAMSTER DHFR AMPLICON

During the past decade, several laboratories have studied replication of the DHFR domain of Chinese hamster ovary (CHO) cells in an effort to identify the sequences that regulate initiation of DNA synthesis in mammalian cells. Since the CHO DHFR system has been examined with every available replication assay, it provides a good perspective of the search for mammalian replicators. In addition, recent studies link the promoter of the DHFR gene to initiation of DNA synthesis, an observation that reinforces the suggestion that DNA replication and gene expression are intimately intertwined processes.

Study of the Chinese hamster DHFR domain began with the observation that the mitotic chromosomes of methotrexate-resistant CHO cells contain large regions that display a monotonous Giemsa staining pattern. In the methotrexate-resistant cell strain CHOC 400, homogeneously staining regions (HSRs) were shown to encompass about 1000 copies of the gene that encodes DHFR, the enzyme inhibited by methotrexate (Milbrandt et al. 1981). Molecular cloning showed that each amplified DHFR gene of CHOC 400 cells is embedded in a tandemly repeated unit (or amplicon) of about 250 kb (Looney and Hamlin 1987). Each CHOC 400 amplicon contains at least three transcription units: the 25-kb DHFR gene (Milbrandt et al. 1983), the 65-kb Rep3 gene that encodes a *MutS* homolog (Mitchell et al. 1986), and the 34-kb 2BE2121 gene that encodes a protein of unknown function (Foreman and Hamlin 1989). The organization of the CHOC 400 DHFR amplicon is shown in Figure 1.

Retroactive labeling of mitotic chromosomes showed that replication of the HSRs begins at the onset of the CHOC 400 S phase (Milbrandt et al. 1981), suggesting that each DHFR amplicon contains at least one replication origin that is activated early in the S phase. Pulse-labeling studies in synchronized CHOC 400 cells identified a series of amplified restriction fragments that begin replication early in the S phase (Heintz and Hamlin 1982). Cells synchronized with cytosine arabinoside, aphidicolin, or hydroxyurea showed the same pattern of fragment labeling in early S (Heintz and Hamlin 1982, 1983), suggesting metabolic inhibitors per se do not influence the selection of initiation sites. Cloning

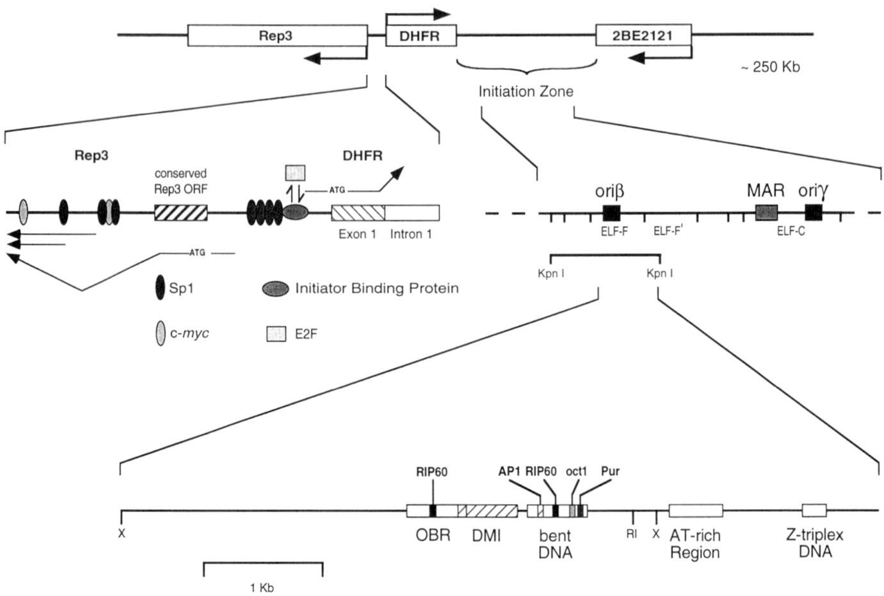

Figure 1 Organization of the Chinese hamster DHFR amplicon. The direction of transcription through the DHFR, Rep3, and 2BE2121 transcription units (*boxes*) within the 250-kb DHFR amplicon is indicated by arrows. Genomic footprinting of protein-DNA interactions within the bidirectional DHFR/rep3 promoter region shows that E2F regulates induction of DHFR transcription during traversal of the G_1 phase (*right* panel). Early-labeled *Eco*RI fragments F, F′, and C from the intergenic initiation zone are shown relative to *ori*β, *ori*γ, and a matrix attachment region (MAR). Sequence elements near the origin of bidirectional DNA replication (OBR)/*ori*β include a densely methylated island (DMI), a region of stably bent DNA that binds multiple protein factors, a 450-bp AT-rich region, and a Z-triplex DNA motif. Binding sites for RIP60, AP1, Oct-1, and Pur within these sequence motifs are shown. In the lower panel, X indicates the boundaries of the 4.3-kb *Xba*I fragment discussed in the text.

and mapping showed the early-labeled restriction fragments (ELFs) are located between the convergently transcribed DHFR and 2BE2121 transcription units (Fig. 1) (Heintz et al. 1983).

A number of approaches have been used in an effort to more precisely map initiation events in the intergenic ELF region. Hybridization of DNA that had been pulse-labeled with radiolabeled thymidine during various intervals of the CHOC 400 S phase to immobilized plasmids suggested initiation occurs with high frequency in a 4.3-kb *Xba*I fragment that maps largely within ELF-F (Fig. 1) (Burhans et al. 1986a). Preferential labeling of this fragment was also observed when cytosine arabinoside was used to stall replication forks as cells entered S phase (Burhans et al. 1986b), or UV-induced cross-linking of the template with psoralen was used to inhibit the movement of replication forks away from initiation sites (Anachkova and Hamlin 1989). Using an in-gel-renaturation technique to enrich for signals from amplified restriction fragments, preferential labeling during the onset of the S phase was shown to occur within two short regions (termed *ori*β and *ori*γ) that are separated by approximately 20 kb (Leu and Hamlin 1989). *ori*β maps within the 4.3-kb *Xba*I fragment of ELF-F, and *ori*γ maps within ELF-C (Fig. 1).

More precise mapping methods based on the template specificity of leading- and lagging-strand synthesis also indicate that initiation occurs with high frequency at *ori*β. Since the two forks within a replication bubble travel in opposite directions, and chromosomal DNA synthesis is semi-discontinuous, lagging- and leading-strand synthesis must switch template strands at the center of the replication bubble where DNA synthesis begins. Template switch points that physically demarcate leading- and lagging-strand initiation sites have been operationally defined as origins of bidirectional DNA replication (OBRs) (Vassilev and DePamphilis 1992). Using a nuclear run-on replication assay (Heintz and Stillman 1988), Okazaki fragments from CHOC 400 cells were labeled with [^{32}P]dNTPs in vitro and hybridized to strand-specific M13 clones from the DHFR ELF region. Quantitation of the strand-specific hybridization signals showed an OBR is located within a 450-bp region of the 4.3-kb early-replicating *Xba*I fragment that contains *ori*β (Burhans et al. 1990).

Mapping of an OBR to *ori*β with Okazaki fragments is supported by studies on the template specificity of leading-strand intermediates. When cells are treated with the protein synthesis inhibitor emetine, Okazaki fragment synthesis is rapidly inhibited, leading to the preferential accumulation of leading-strand replication intermediates (Burhans et al. 1991). Mapping of the strand specificity of leading-strand intermediates showed that DNA replication begins preferentially at two sites within the DHFR initiation region, one of which maps at the OBR/*ori*β, and the second near *ori*γ (Handeli et al. 1989). A significant finding from the mapping of leading-strand intermediates is that the initiation sites that are used in CHOC 400 cells also are used in CHO cells that contain only two copies of the DHFR domain, a finding confirmed with PCR-based assays that measure the amount and length of nascent DNA strands relative to specific chromosomal positions (Vassilev et al. 1990; Tasheva and Roufa 1994b; M. Giacco, pers. comm.).

Although there is convincing evidence that initiation of DNA synthesis occurs with high frequency at the *ori*β/OBR located in ELF-F, the conclusion that initiation occurs at a limited number of preferred sites in the DHFR origin region is not supported by the analysis of cellular replication intermediates with 2D gels. With neutral-neutral 2D gels, replication bubbles have been detected in every restriction fragment spanning the 55 kb of DNA that lies between the DHFR and 2BE2121 transcription units (Vaughn et al. 1990; Dijkwel and Hamlin 1992; Dijkwel et al. 1994). Neutral-alkaline 2D gel analysis indicates replication forks travel in both directions through many fragments in the initiation zone (Vaughn et al. 1990; Dijkwel et al. 1994), a finding that also indicates initiation occurs at many dispersed sites. Although it is difficult to compare the relative frequency of initiation events at different sites by 2D gels, taken together, the 2D gel results indicate initiation occurs at many sites, perhaps even randomly, throughout the entire DHFR initiation zone (Fig. 1).

There is strong evidence that the 2D gel patterns from CHO cells reflect true chromosomal replication intermediates. The migration patterns of replication intermediates from yeast and CHO cells are indistinguishable (J.A. Huberman and J.L. Hamlin, pers. comm.), replication intermediates are limited to S phase (Dijkwel and Hamlin 1992; Dijkwel et al. 1994), and inhibitors of DNA synthesis alter the migration of replication intermediates in 2D gels in a predictable fashion (Levenson and Hamlin 1993). 2D gel analysis of replication intermediates from the ribosomal DNA locus in human cells also suggests initiation occurs at multiple sites within nontranscribed spacer regions (Little et al. 1993). Models that have been proposed to account for the apparent discrepancies between 2D gel and other mapping techniques have not been supported by experimental evidence. Resolution of this apparent paradox (i.e., random versus site-specific initiation) will require the examination of other initiation regions by multiple techniques, and a stronger understanding of the relationship between 2D gel patterns and the transient replication intermediates detected by labeling studies or PCR-based techniques.

GENETIC ASSAYS FOR DHFR ORIGIN ACTIVITY

Cloning of the DHFR initiation region led to transfection experiments to identify autonomously replicating sequences analogous to yeast ARS elements. Because these experiments have been largely unsuccessful, they are not widely reported in the literature (see, e.g., Burhans et al. 1990).

Nonetheless, plasmids carrying fragments that span 40 kb of the DHFR initiation zone (including the OBR) have been tested for autonomous replication after transfection by several methods into CHO, CHOC 400, HeLa, 293, and COS-7 cells, all without success (N.H. Heintz, unpubl.; J.L. Hamlin, pers. comm.). Recently, the DHFR OBR was reported to support autonomous plasmid replication in vivo and in vitro (Zannis-Hadjopoulos et al. 1995). This latter study, however, did not directly compare ARS activity of the DHFR OBR to size-matched controls. Using the neutral-neutral 2D gel assay, Caddle and Calos (1992) tested the ability of a 14-kb *Kpn*I genomic fragment that encompasses the OBR to support site-specific initiation in plasmids that are retained in the nucleus by the viral protein EBNA1. In these experiments, 2D gel analysis showed that initiation occurred at random throughout the DHFR sequences, with no detectable preference for the OBR. Unfortunately, PCR or strand-specific mapping assays were not compared to the 2D gel results in this study.

In *Drosophila melanogaster*, developmentally regulated origins of replication known as amplification control elements (ACE) mediate the amplification of flanking sequences when transferred to new chromosomal positions (Orr-Weaver et al. 1989). Using a plasmid transfection assay, Stolzenburg et al. (1994) have reported that the DHFR OBR promotes the amplification of plasmid DNA concatemers in long-term transfection experiments. To determine if the DHFR origin region enhances the rate of chromosomal gene amplification in a manner similar to *Drosophila* ACEs, the 14-kb *Kpn*I fragment that encompasses the OBR was linked to a DHFR minigene and transfected into *dhfr*⁻ CHO-DUKX cells (Brinton 1991). After selection for DHFR enzyme activity, individual cell clones were selected for methotrexate resistance. Amplification of the integrated DHFR minigene was found to be dependent on chromosomal position; inclusion of the DHFR initiation region did not mediate position-independent gene amplification, nor did these sequences influence the rate at which the DHFR minigene was amplified (Brinton 1991). Although heterologous genetic assays have failed to detect site-specific initiation or amplification activity in the 14-kb *Kpn*I fragment, these sequences must have distinctive properties in the chromosome, for they are the only portion of the DHFR amplicon that is entirely refractory to UV-induced DNA repair (Ho et al. 1989).

FEATURES OF THE DHFR OBR

Although a *cis*-acting replicator for the DHFR gene has not been described, every origin-mapping study (including those utilizing 2D gels) indicates that initiation of replication occurs with high frequency within the 4.3-kb *Xba*I fragment that contains *ori*β/OBR. A 6-kb region surrounding *ori*β/OBR has been sequenced (Caddle et al. 1990b) and examined for DNA sequences with unusual structure, DNA unwinding elements, and protein-binding sites. Located near the 3′ end of the sequence is a 180-bp stretch of simple alternating dinucleotide repeats that forms left-handed Z and triple-stranded (or triplex) DNA when subjected to superhelical stress in vitro (Bianchi et al. 1990; Caddle et al. 1990b). Primer extension studies show that DNA polymerases traverse this element more readily in the Z-to-triplex than triplex-to-Z DNA direction in vitro (Brinton et al. 1991), suggesting that the orientation of this sequence may influence the rate of replication fork travel in vivo. Although orientation and position of the Z-triplex motif has dramatic effects on the rate of SV40-origin-dependent plasmid replication in COS-7 cells (Brinton et al. 1991), 2D gel analysis showed the Z-triplex sequence is not an absolute fork barrier, but rather causes the accumulation of late-replication intermediates in plasmids. Neutral-neutral 2D gel analysis has not detected a replication fork barrier at the Z triplex motif in situ (Vaughn et al. 1990; Dijkwel and Hamlin 1992; Dijkwel et al. 1994), and the function of this motif in the genome, if any, is unknown.

The DHFR initiation region contains several tracts of AT-rich sequences, including a prominent 450-bp segment that is 75% AT and contains multiple near-matches to the *S. cerevisiae* ACS (Fig. 1). A hierarchy of DNA unwinding elements in the DHFR initiation region was determined by digesting a panel of supercoiled plasmids with mung bean nuclease (MBN) as described by Kowalski and Eddy (1989). In the MBN assay, the DNA that was most readily unwound under superhelical stress is a simple alternating dA:dT repeat of 46 bp that maps about 1000 bp 3′ to the OBR (Caddle et al. 1990b). As discussed earlier, poly(dA:dT) sequences support DNA unwinding and priming of DNA synthesis in vitro (Caddle et al. 1990b; Pack and Tsurimoto 1994). DNA unwinding was not detected in the 450-bp AT-rich region, nor was the OBR susceptible to nuclease scission, even when isolated from flanking sequences (Caddle et al. 1990b). Located approximately 700 bp downstream from the OBR is a region of stably bent DNA that binds multiple protein factors, including the transcription factors AP1 and Oct-1 (Held et al. 1992), the DNA-binding protein RIP60 (Dailey et al. 1990), and the single-stranded DNA-binding protein Pur (Bergemann and Johnson 1992). The bent DNA AP1 site binds c-jun/c-jun and c-fos/c-jun heterodimers in vitro (Held et al. 1992), a result of interest since c-fos/c-jun heterodimers have been shown to be potent enhancers of SV40 and polyoma core origin activity in transient replication assays (Guo and DePamphilis 1992). Recently, the DHFR bent DNA motif was shown to increase the activity of the polyoma core origin 20-fold in transient replication assays, and the level of replication increased to 50-fold when the test plasmid was cotransfected with expression vectors for c-fos and c-jun (K. Murakami et al., in prep.).

Footprinting of the bent DNA region with HeLa cell nuclear extract in vitro identified a 60-kD DNA-binding protein called RIP60 that binds to an ATT-repeat motif 20 bp 3′ to the bent DNA AP1 site (Fig. 1) (Dailey et al. 1990). RIP60 bends DNA toward the major groove about 82° (Caddle et al. 1990a; M. Soultanakis and N. H. Heintz, unpubl.) and, when bound to the native bent DNA motif, counteracts the direction of DNA-directed bending. Further work showed that RIP60 binds to a second site within the 450 bp of the 4.3-kb *Xba*I fragment that includes the OBR (P. Held and N. H. Heintz, unpubl.). Dimers of RIP60 bound to both sites mediate the formation of a 710-bp twisted DNA loop in vitro (Mastrangelo et al. 1993). Thus, DNA looping by RIP60 links an AP1-dependent replication enhancer to the OBR in vitro. Recently, the sequences between the two RIP60-binding sites were shown to encompass a large densely methylated island (DMI) in which virtually all dC residues are methylated (Tasheva and Roufa 1994b). Methylation of the DMI is prominent in cycling cells, is lost in quiescent cells, and reappears within 48 hours after quiescent cells are stimulated to reenter the cell cycle. A second DMI has been mapped near an initiation site associated with the RPS14 gene (Tasheva and Roufa 1994a). These reports suggest that growth-regulated changes in DNA methylation may influence initiation of DNA replication, and they raise the possibility that the failure of transfection assays to identify mammalian origin elements may be due to lack of proper methylation of the DNA template, a post-synthetic modification that is absent in yeast.

REPLICATION OF THE DHFR AMPLICON IN NUCLEI

Cell fusion experiments show that S-phase mammalian cells contain a diffusible signal that induces DNA synthesis in G_1 nuclei. In an attempt to reproduce this finding in vitro, a nuclear run-on replication assay was developed for CHOC 400 cells (Heintz and Stillman 1988). Incubation of nuclei with cytosol from cells staged at various points in the cell cycle showed that S-phase extracts support higher rates of DNA synthesis in S-phase nuclei than do G_1 extracts, and that G_1 extracts do not contain a dominant inhibitor of DNA replication. However, under these conditions,

S-phase extracts do not induce initiation of DNA synthesis in CHOC 400 nuclei isolated from G_1 cells.

In contrast, *Xenopus* egg extracts that support nuclear assembly and DNA replication in vitro are able to induce DNA synthesis in nuclei isolated from CHOC 400 cells arrested in G_1 (Gilbert et al. 1993). Synthesis begins in the DHFR initiation zone, suggesting that the protein factors and nuclear structure that dictate selection of initiation sites reside in CHOC 400 G_1 nuclei, and that *Xenopus* extracts can activate mammalian initiation complexes (Gilbert et al. 1995). Interestingly, induction of DNA synthesis is sensitive to the protein kinase inhibitor 6-DMAP (Gilbert et al. 1995), a finding that indicates kinases in the *Xenopus* extract are able to phosphorylate factors in the extract or in the G_1 nucleus required for entry into the S phase. Fractionation of the *Xenopus* extract may lead to the identification of kinase activities required to activate mammalian replication complexes.

LINKS BETWEEN TRANSCRIPTION AND REPLICATION IN THE DHFR AMPLICON

Density shift (Caddle and Heintz 1990) and 2D gel analysis of replication intermediates (Vaughn et al. 1990; Dijkwel and Hamlin 1992) show that initiation occurs in only one of ten DHFR amplicons during the typical cell cycle. Since the DHFR amplicons share the same primary sequence, failure to initiate at every amplicon could be due to several factors: (1) The initiation region of each amplicon may not be associated with the nuclear matrix, a notion supported by preliminary experimental evidence (Dijkwel and Hamlin 1988); (2) there may be insufficient amounts of initiation proteins and replication factors for the assembly of replication complexes at all amplicons; and (3) initiation may be linked to transcription of the DHFR gene, and each copy of the DHFR gene may not be transcribed during every cell cycle.

A link between transcription of the DHFR gene and initiation is strengthened by the observation that chromosomal deletions that remove the DHFR promoter, but not the body of the DHFR gene or the 3′ origin region, delay or abolish initiation in the early-replicating region (V. Levenson et al., pers. comm.). Genomic footprinting of the bidirectional DHFR/Rep3 promoter region during the cell cycle shows that in G_1 the DHFR promoter is bound by a basal transcription complex that includes the transcription factor Sp1 and a protein bound at the transcription start site, and that elevated expression of DHFR during traversal of G_1 is correlated with occupancy of an E2F transcription factor binding site (see

Fig. 1) (J. Wells et al., in prep.). The promoter footprinting studies show that all 1000 copies of the DHFR promoter in CHOC 400 cells are occupied by a basal transcription complex, suggesting that failure to initiate replication in each amplicon during every cell cycle is not due to differences in access to general or regulatory transcription factors.

Evidence from another system suggests that gene promoters may influence replication by mechanisms other than inducing transcription per se. Deletion of an 8-kb segment of the human β-globin domain alters the template specificity of leading-strand DNA synthesis in the globin domain from a divergent, bidirectional pattern to a unidirectional pattern (Kitsberg et al. 1993), suggesting that the deletion has removed one or more elements of a replicator required for initiation of bidirectional replication at a specific site. The globin deletion alters the replication pattern whether or not the locus is transcribed, suggesting that in addition to regulating transcription, promoter sequences may have additional functions, such as promoting the association of the globin domain with the nuclear matrix. Other deletions in the globin locus, far removed from the initiation site, change the replication timing and long-range chromatin structure of this region, alterations that also influence globin gene expression (Forrester et al. 1990).

SUMMARY

Given the lack of a reliable genetic assay; the fact that specific sequence elements are not required for nuclear assembly, initiation of DNA synthesis, or limiting replication to once per cell cycle in *Xenopus* egg extracts; and the heterogeneity of sequences that appear to function as initiation sites in mammalian cells, a perplexing picture of mammalian replicators has emerged. How might initiation sites be selected in animal cells? First, the deposition of initiation complexes on DNA may be determined by a combination of primary sequence elements, chromatin organization, and nuclear structure. Study of the flanking sequences that affect the time and frequency of activation of replication origins in yeast may provide important clues about the determinants that influence the selection of mammalian initiation sites. Second, fixed *cis*-acting replicator sequences could act at a distance (perhaps by DNA looping) to inspire initiation at multiple sites. If mammalian replicators resemble gene promoters, *cis*-acting replicator elements could be dispersed over thousands of kilobase pairs and overlap with those elements required for regulation of transcription and chromatin organization within the same replicon. This possibility is attractive, for it suggests that factors which influence nuclear structure could account for the strong link between replication and transcription control. Third, as proposed by Benbow and colleagues (Benbow et al. 1992), DNA unwinding could commence at specific sites but be uncoupled from the assembly of replication forks, leading to a decentralized initiation pattern. Given the complexity of development and differentiation in mammals, it is not surprising that there are significant differences between replication origins in mammalian cells and yeast. The task at hand is to understand the significance of these differences in the context of eukaryotic evolution.

ACKNOWLEDGMENTS

I thank Kevin Murakami for assistance with Figure 1, and B. Ravenelle for help preparing the manuscript.

REFERENCES

Adachi, Y. and U.K. Laemmli. 1994. Study of the cell cycle-dependent assembly of the pre-replication centers in *Xenopus* egg extracts. *EMBO J.* **13:** 4153–4164.

Anachkova, B. and J.L. Hamlin. 1989. Replication in the amplified dihydrofolate reductase domain in CHO cells may initiate at two distinct sites, one of which is a repetitive sequence element. *Mol. Cell. Biol.* **9:** 532–540.

Ariizumi, K., Z. Wang, and P.W. Tucker. 1993. Immunoglobulin heavy chain enhancer is located near or in an initiation zone of chromosomal DNA replication. *Proc. Natl. Acad. Sci.* **90:** 3695–3699.

Bell, S.P. and B. Stillman. 1992. ATP-dependent recognition of eukaryotic origins of replication by a multi-protein complex. *Nature* **357:** 128–134.

Benbow, R.M., J. Zhao, and D.D. Larson. 1992. On the nature of origins of DNA replication in eukaryotes. *BioEssays* **14:** 661–670.

Bensch, K.G., S. Tanaka, S.-Z. Hu, T.S.-F. Wang, and D. Korn. 1982. Intracellular localization of human polymerase α with monoclonal antibodies. *J. Biol. Chem.* **257:** 8391–8396.

Berezney, R. 1991a. The nuclear matrix: A heuristic model for investigating genomic organization and function in the cell nucleus. *J. Cell. Biochem.* **47:** 109–123.

———. 1991b. Visualizing DNA replication sites in the cell nucleus. *Semin. Cell Biol.* **2:** 103–115.

Bergemann, A.D. and E.M. Johnson. 1992. The HeLa Pur factor binds single-stranded DNA at a specific element conserved in gene flanking regions and origins of DNA replication. *Mol. Cell. Biol.* **12:** 1257–1265.

Bianchi, A., R.D. Wells, N.H. Heintz, and M.S. Caddle. 1990. Sequences near the origin of replication at the *dhfr* locus of Chinese hamster ovary cells adopt left-handed Z-DNA and triplex structures. *J. Biol. Chem.* **265:** 21789–21796.

Blow, J.J. and R.A. Laskey. 1986. Initiation of DNA replication in nuclei and purified DNA by a cell-free extract of *Xenopus* eggs. *Cell* **47:** 577–587.

———. 1988. A role for the nuclear envelope in controlling DNA replication within the cell cycle. *Nature* **332:** 546–548.

Bramhill, D. and A. Kornberg. 1988. A model for initiation at origins of DNA replication. *Cell* **54:** 915–918.

Bravo, R. and H. Macdonald-Bravo. 1987. Existence of two populations of cyclin/proliferating cell nuclear antigen during the cell cycle: Association with DNA replication sites. *J. Cell Biol.* **105:** 1549–1554.

Brinton, B. 1991. "Effect of chromosomal position on the mode and frequency of gene amplification in mammalian cells." Ph.D. thesis, University of Vermont, Burlington.

Brinton, B., M.S. Caddle, and N.H. Heintz. 1991. Position and orientation-specific effects of a Z-triplex DNA motif on episomal DNA replication in COS-7 cells. *J. Biol. Chem.* **266:** 5153–5161.

Buongiorno-Nardelli, M., M.G. Micheli, M.T. Carri, and M. Marilley. 1982. A relationship between replicon size and supercoiled loop domains in the eukaryotic genome. *Nature* **298:** 100–102.

Burhans, W.C., J.E. Selegue, and N.H. Heintz. 1986a. Replication intermediates formed during initiation of DNA synthesis in methotrexate-resistant CHOC 400 cells are enriched for sequences from a specific, amplified restriction fragment. *Biochemistry* **25:** 441–449.

———. 1986b. Isolation of the origin of replication associated with the amplified Chinese hamster ovary dihydrofolate reductase domain. *Proc. Natl. Acad. Sci.* **83:** 7790–7794.

Burhans, W.C., L.T. Vassilev, M.S. Caddle, N.H. Heintz, and M.L. DePamphilis. 1990. Identification of an origin of bidirectional DNA replication in mammalian chromosomes. *Cell* **62:** 955–965.

Burhans, W.C., L.T. Vassilev, J. Wu, J.M. Sogo, F.S. Nallaseth, and M.L. DePamphilis. 1991. Emetine allows identification of origins of mammalian DNA replication by imbalanced DNA synthesis, not through conservative nucleosome segregation. *EMBO J.* **10:** 4351–4360.

Caddle, M.S. and M.P. Calos. 1992. Analysis of the autonomous replication behavior in human cells of the DHFR putative chromosomal origin of replication. *Nucleic Acids Res.* **20:** 5971–5978.

Caddle, M.S. and N.H. Heintz. 1990. The replication timing of the amplified dihydrofolate reductase genes in the Chinese hamster ovary cell line CHOC 400. *Biochem. Biophys. Res. Commun.* **170:** 134–139.

Caddle, M.S., L. Dailey, and N.H. Heintz. 1990a. RIP60, a mammalian origin-binding protein, enhances DNA bending near the dihydrofolate reductase origin of replication. *Mol. Cell. Biol.* **10:** 6236–6243.

Caddle, M.S., R.H. Lussier, and N.H. Heintz. 1990b. Intramolecular DNA triplexes, bent DNA and DNA unwinding elements in the initiation region of an amplified dihydrofolate reductase replicon. *J. Mol. Biol.* **211:** 19–33.

Cardoso, M.C., H. Leonhardt, and B. Nidal-Ginard. 1993. Reversal of terminal differentiation and control of DNA replication: Cyclin A and cdk2 specifically localize at subnuclear sites of DNA replication. *Cell* **74:** 979–992.

Carroll, S.M., M.L. DeRose, J.L. Kolman, G.H. Nonet, R.E. Kelly, and G.M. Wahl. 1993. Localization of a bidirectional DNA replication origin in the wild type and in episomally amplified murine ADA loci. *Mol. Cell. Biol.* **13:** 2971–2981.

Chang, C.-F., H. Tada, and K. Khalili. 1994. The role of a pentanucleotide repeat sequence, AGGGAAGGGA, in the regulation of JC virus DNA replication.

Gene **148:** 309–314.

Chen, N.N., C.-F. Chang, G.L. Gallia, D.A. Kerr, E.M. Johnson, C.P. Krachmarov, S.M. Barr, R.J. Frisque, B. Bollage, and K. Khalili. 1995. Cooperative action of the cellular proteins YB-1 and Purα with the tumor antigen of the human JC polyomavirus determines their interaction with the viral lytic control element. *Proc. Natl. Acad. Sci.* **92:** 1087–1091.

Cook, P.R. 1991. The nucleoskeleton and the topology of replication. *Cell* **66:** 627–635.

Dailey, L., M.S. Caddle, N. Heintz, and N.H. Heintz. 1990. Purification of RIP60 and RIP100, mammalian proteins with origin specific DNA binding and ATP-dependent DNA helicase activities. *Mol. Cell. Biol.* **10:** 6225–6235.

Diffley, J.F.X. and J.H. Cocker. 1992. Protein-DNA interactions at a yeast replication origin. *Nature* **357:** 169–172.

Diffley, J.F.X., J.H. Cocker, S.J. Dowell, and A. Rowley. 1994. Two steps in the assembly of complexes at yeast replication origins in vivo. *Cell* **78:** 303–316.

Dijkwel, P.A. and J.L. Hamlin. 1988. Matrix attachment regions are positioned near replication initiation sites, genes, and an interamplicon junction in the amplified dihydrofolate reductase domain of Chinese hamster ovary cells. *Mol. Cell. Biol.* **8:** 5398–5409.

———. 1992. Initiation of DNA replication in the DHFR locus is confined to the early S-period in CHO cells synchronized with the plant amino acid mimosine. *Mol. Cell. Biol.* **12:** 3715–3722.

Dijkwel, P.A., J.P. Vaughn, and J. L. Hamlin. 1994. Replication initiation sites are distributed widely in the amplified CHO dihydrofolate reductase domain. *Nucleic Acids Res.* **22:** 4989–4996.

Dobbs, D.L., W.-L. Shaiu, and R.M. Benbow. 1994. Modular sequence elements associated with origin regions in eukaryotic chromosomal DNA. *Nucleic Acids Res.* **22:** 2479–2489.

Dowell, S.J., P. Romanowski, and J. F. Diffley. 1994. Interaction of Dbf4, the Cdc7 protein kinase regulatory subunit, with yeast replication origins in vivo. *Science* **265:** 1243–1246.

Fangman, W.L. and B.S. Brewer. 1991. Activation of replication origins within yeast chromosomes. *Annu. Rev. Cell Biol.* **7:** 375–402.

Foreman, P.K. and J.L. Hamlin. 1989. Identification and characterization of a gene that is coamplified with dihydrofolate reductase in a methotrexate-resistant CHO cell line. *Mol. Cell. Biol.* **9:** 1137–1147.

Forrester, W.C., E. Epner, M.C. Driscoll, T. Enver, M. Brice, T. Papayannopoulou, and M. Groudine. 1990. A deletion of the human β-globin locus activation region causes a major alteration in chromatin structure and replication across the entire β-globin region. *Genes Dev.* **4:** 1637–1649.

Foss, M., F.J. McNally, P. Laurenson, and J. Rine. 1993. Origin recognition complex (ORC) in transcriptional silencing and DNA replication in *S. cerevisiae*. *Science* **262:** 1838–1844.

Fox, M.H., D.J. Arndt-Jovin, T.M. Jovin, P.H. Baumann, and M. Robert-Nicoud. 1991. Spatial and temporal distribution of DNA replication sites localized by immunofluorescence and confocal microscopy in MCF-7 cells. *J. Cell. Sci.* **99:** 247–253.

Frappier, L. and M. Zannis-Hadjopoulos. 1987. Autonomous replication of plasmids bearing monkey DNA origin-enriched sequences. *Proc. Natl. Acad. Sci.* **84:** 6668–6672.

Gale, J.M., R.A. Tobey, and J.A. D'Anna. 1992. Localization and DNA sequence of a replication origin in the rhodopsin gene locus of Chinese hamster cells. *J. Mol. Biol.* **224:** 343–358.

Giacca, M., L. Zentilin, P. Norio, S. Diviacco, D. Dimitrova, G. Contreras, G. Biamonti, G. Perini, F. Weighardt, S. Riva, and A. Falaschi. 1994. Fine mapping of a replication origin of human DNA. *Proc. Natl. Acad. Sci.* **91:** 7119–7123.

Gilbert, D.M., H. Miyazawa, and M.L. DePamphilis. 1995. Site-specific initiation of DNA replication in *Xenopus* egg extract requires nuclear structure. *Mol. Cell. Biol.* **15:** 2942–2954.

Gilbert, D.M., H. Miyazawa, F.S. Nallaseth, J.M. Ortega, J.J. Blow, and M.L. DePamphilis. 1993. Site-specific initiation of DNA replication in metazoan chromosomes and the role of nuclear organization. *Cold Spring Harbor Symp. Quant. Biol.* **58:** 475–485.

Guo, Z.-S. and M.L. DePamphilis. 1992. Specific transcription factors stimulate both simian virus 40 and polyomavirus origins of DNA replication. *Mol. Cell. Biol.* **12:** 2514–2524.

Haase, S.B., S.S. Heinzel, and M.P. Calos. 1994. Transcription inhibits the replication of autonomously replicating plasmids in human cells. *Mol. Cell. Biol.* **14:** 2516–2524.

Hand, R. 1978. Eukaryotic DNA: Organization of the genome for replication. *Cell* **15:** 317–325.

Handeli, S., A. Klar, M. Meuth, and H. Cedar. 1989. Mapping replication units in animal cells. *Cell* **57:** 909–920.

Hassan, A.B. and P.R. Cook. 1993. Visualization of replication sites in unfixed human cells. *J. Cell Sci.* **105:** 541–550.

———. 1994. Commentary: Does transcription by RNA polymerase play a direct role in the initiation of replication? *J. Cell Sci.* **107:** 1381–1387.

Hassan, A.B., R.J. Errington, N.S. White, D.A. Jackson, and P.R. Cook. 1994. Replication and transcription sites are colocalized in human cells. *J. Cell Sci.* **107:** 425–434.

Hatton, K.S., V. Dhar, E. H. Brown, M.A. Iqbal, S. Stuart, V.T. Didamo, and C.L. Schildkraut. 1988. Replication program of active and inactive gene families in mammalian cells. *Mol. Cell. Biol.* **8:** 2149–2158.

Heintz, N.H. and J.L. Hamlin. 1982. An amplified chromosomal sequence that includes the gene for dihydrofolate reductase initiates replication within specific restriction fragments. *Proc. Natl. Acad. Sci.* **79:** 4083–4087.

———. 1983. *In vivo* effects of cytosine arabinoside on DNA replication in Chinese hamster ovary cells. II. Cytosine arabinoside affects the rate of synthesis but not the pattern of labeling of an amplified chromosomal sequence at the onset of the S period. *Biochemistry* **22:** 3557–3562.

Heintz, N.H. and B.W. Stillman. 1988. Nuclear DNA synthesis in vitro is mediated via stable replication forks assembled in a temporally specific fashion in vivo. *Mol. Cell. Biol.* **8:** 1923–1931.

Heintz, N.H., L. Dailey, P. Held, and N. Heintz. 1992. Eukaryotic replication origins as promoters of bidirectional DNA synthesis. *Trends Genet.* **8:** 376–381.

Heintz, N.H., J.D. Milbrandt, K S. Greisen, and J.L. Hamlin. 1983. Cloning the initiation region of a mammalian chromosomal replicon. *Nature* **302:** 439–441.

Held, P., E. Soultanakis, L. Dailey, T. Kouzarides, N. Heintz, and N.H. Heintz. 1992. Protein-induced alterations in DNA structure at the dhfr origin of replication. *Colloq. Ges. Biol. Chem. Mosbach* **43:** 113–126.

Ho, L., V.A. Bohr, and P.C. Hanawalt. 1989. Demethylation enhances removal of pyrimidine dimers from the overall genome and from specific DNA sequences in Chinese hamster ovary cells. *Mol. Cell. Biol.* **9:** 1594–1603.
Huberman, J.A. and A.D. Riggs. 1968. On the mechanism of DNA replication in mammalian chromosomes. *J. Mol. Biol.* **32:** 327–341.
Ishimi, Y., K. Matsumoto, and R. Ohba. 1994. DNA replication from initiation zones of mammalian cells in a model system. *Mol. Cell. Biol.* **14:** 6489–6496.
Jacob, F., S. Brenner, and F. Cuzin. 1964. On the regulation of DNA replication in bacteria. *Cold Spring Harbor Symp. Quant. Biol.* **28:** 329–348.
Kill, I.R., J.M. Bridger, K.H.S. Campbell, G. Maldonado-Codina, and C.J. Hutchinson. 1991. The timing of the formation and usage of replicase clusters in S-phase nuclei of human diploid fibroblasts. *J. Cell Sci.* **100:** 869–876.
Kitsberg, D., S. Selig, I. Keshet, and H. Cedar. 1993. Replication structure of the human β-globin domain. *Nature* **366:** 588–590.
Kowalski, D. and M.J. Eddy. 1989. The DNA unwinding element: A novel *cis*-acting component that facilitates opening of the *Escherichia coli* replication origin. *EMBO J.* **8:** 4335–4344.
Krysan, P.J. and M.P. Calos. 1991. Replication initiates at multiple locations on an autonomously replicating plasmid in human cells. *Mol. Cell. Biol.* **11:** 1464–1472.
Krysan, P.J., S.B. Haase, and M.P. Calos. 1989. Isolation of human sequences that replicate autonomously in human cells. *Mol. Cell. Biol.* **9:** 1026–1033.
Leonhardt, H., A.W. Page, H.-U. Weier, and T.H. Bestor. 1992. A targeting sequence directs DNA methyltransferase to sites of DNA replication in mammalian nuclei. *Cell* **71:** 865–873.
Leu, T.-H. and J.L. Hamlin. 1989. High-resolution mapping of replication fork movement through the amplified dihydrofolate reductase domain in CHO cells by in-gel renaturation analysis. *Mol. Cell. Biol.* **9:** 523–531.
Levenson, V. and J.L. Hamlin. 1993. A general protocol for evaluating the specific effects of DNA replication inhibitors. *Nucleic Acids Res.* **12:** 3997–4004.
Little, R.D., T.H.K. Platt, and C.L. Schildkraut. 1993. Initiation and termination of DNA replication in human ribosomal RNA genes. *Mol. Cell. Biol.* **13:** 6600–6613.
Looney, J.E. and J.L. Hamlin. 1987. Isolation of the amplified dihydrofolate reductase domain from methotrexate-resistant Chinese hamster ovary cells. *Mol. Cell. Biol.* **7:** 569–577.
Mastrangelo, I.A., P.G. Held, L. Dailey, J.S. Wall, P.V.C. Hough, N. Heintz, and N.H. Heintz. 1993. RIP60 dimers and multiples of dimers assemble link structures at an origin of bidirectional replication in the dihydrofolate reductase amplicon of Chinese hamster ovary cells. *J. Mol. Biol.* **232:** 766–778.
Masukata, H., H. Satoh, C. Obuse, and T. Okazaki. 1993. Autonomous replication of human chromosomal DNA fragments in human cells. *Mol. Biol. Cell* **4:** 1121–1132.
McWhinney, C. and M. Leffak. 1990. Autonomous replication of a DNA fragment containing the chromosomal replication origin of the human c-*myc* gene. *Nucleic Acids Res.* **18:** 1233–1242.
Milbrandt, J.D., J.C. Azizkhan, K.S. Greisen, and J.L. Hamlin. 1983. Organization of a Chinese hamster ovary dihydrofolate reductase gene identified by phenotypic rescue. *Mol. Cell. Biol.* **3:** 1266–1273.
Milbrandt, J.D., N.H. Heintz, W.C. White, S.M. Rothman, and J.L. Hamlin. 1981. Methotrexate-resistant Chinese hamster ovary cells have amplified a 135-kilobase-pair

region that includes the gene for dihydrofolate reductase. *Proc. Natl. Acad. Sci.* **78:** 6043–6047.

Mills, A.D., J.J. Blow, J.G. White, W.B. Amos, D. Wilcock, and R.A. Laskey. 1989. Replication occurs at discrete foci spaced throughout nuclei replicating in vitro. *J. Cell Sci.* **94:** 471–477.

Mitchell, P.J., A.M. Carothers, J.H. Han, J.D. Harding, E. Kas, L. Venolia, and L.A. Chasin. 1986. Multiple transcription start sites, DNase I-hypersensitive sites, and an opposite-strand exon in the 5′ region of the CHO *dhfr* gene. *Mol. Cell. Biol.* **6:** 425–440.

Nakamura, H., T. Morita, and C. Sato. 1986. Structural organization of replicon domains during DNA synthesis phase in the mammalian nucleus. *Exp. Cell Res.* **165:** 291–297.

Nielson T., D. Bell, C. Lamoureux, M. Zannis-Hadjopoulos, and G. Price. 1994. A reproducible method for identification of human genomic DNA autonomously replicating sequences. *Mol. Gen. Genet.* **242:** 280–288.

Orr-Weaver, T.L., C.G. Johnston, and A.C. Spradling. 1989. The role of ACE3 in *Drosophila* chorion gene amplification. *EMBO J.* **8:** 4153–4162.

Pack, R.A. and T. Tsurimoto. 1994. Isolation of human DNA-unwinding elements as sites of DNA polymerase α/primase entry. *Gene* **148:** 277–284.

Pardoll, D.M., B. Vogelstein, and D.S. Coffey. 1980. A fixed site of DNA replication in eukaryotic cells. *Cell* **19:** 527–536.

Pederson, D.S. and N.H. Heintz. 1994. *Transcription factors and DNA replication.* R.G. Landes, Austin, Texas.

Saitoh, Y. and U.K. Laemmli. 1994. Metaphase chromosome structure: Bands drive from a differential folding path of the highly AT-rich scaffold. *Cell* **76:** 609–622.

Stillman, B. 1994. Initiation of chromosomal DNA replication in eukaryotes. Lessons from lambda. *J. Biol. Chem.* **269:** 7047–7050.

Stolzenburg, F., R. Gerwig, E. Dinkl, and F. Grummt. 1994. Structural homologies and functional similarities between mammalian origins of replication and amplification promoting sequences. *Chromosoma* **103:** 209–214.

Taira, T., M. Sanae, M. Iguchi-Ariga, and H. Ariga. 1994. A novel DNA replication origin identified in the human heat shock protein 70 gene promoter. *Mol. Cell. Biol.* **14:** 6386–6397.

Tasheva, E.S. and D.J. Roufa. 1994a. A mammalian origin of bidirectional DNA replication within the Chinese hamster *RPS14* locus. *Mol. Cell. Biol.* **14:** 5628–5635.

———. 1994b. Densely methylated DNA islands in mammalian chromosomal replication origins. *Mol. Cell. Biol.* **14:** 5636–5644.

Vassilev, L.T. and M.L. DePamphilis. 1992. Guide to identification of origins of DNA replication in eukaryotic cell chromosomes. *Crit. Rev. Biochem. Mol. Biol.* **27:** 445–472.

Vassilev, L.T. and E.M. Johnson. 1990. An initiation zone of chromosomal DNA replication located upstream of the c-*myc* gene in proliferating HeLa cells. *Mol. Cell. Biol.* **10:** 4899–4904.

Vassilev, L.T., W.C. Burhans, and M.L. DePamphilis. 1990. Mapping an origin of DNA replication at a single copy locus in exponentially proliferating mammalian cells. *Mol. Cell. Biol.* **10:** 4685–4689.

Vaughn, J.P., P.A. Dijkwel, and J.L. Hamlin. 1990. Replication initiates in a broad zone in the amplified CHO dihydrofolate reductase domain. *Cell* **61:** 1075–1087.

Virta-Pearlman, V.J., P.H. Gunaratne, and A.C. Chinault. 1993. Analysis of a replication initiation sequence from the adenosine deaminase region of the mouse genome. *Mol. Cell. Biol.* **13:** 5931–5942.

Waga, S. and B. Stillman. 1994. Anatomy of a DNA replication fork revealed by reconstitution of SV40 replication in vitro. *Nature* **369:** 207–212.

Zannis-Hadjopoulos, M., T.O. Nielsen, A. Todd, and G.B. Price. 1995. Autonomous replication in vivo and in vitro of clones spanning the region of the DHFR origin of bidirectional replication (oriβ). *Gene* **151:** 273–277.

38

DNA Replication in Plants

Jack Van't Hof
Biology Department
Brookhaven National Laboratory
Upton, New York 11793

The subject of this chapter is chromosomal DNA replication in higher plants. The discussion purposely is limited to results obtained with plants. References to viruses, plasmids, organelles, yeast, and other types of cells are few. Most of the pioneering work and insight on chromosomal DNA replication is traceable to research done with these organisms, but a review of the work need not be repeated here. This chapter has four sections. The first covers the genomic diversity of higher plants; the second covers the replicon, its properties, its hierarchical organization, its relation to the S phase of the cell cycle, and its involvement in chromosomal DNA maturation. The third section discusses the replication of the ribosomal genes. At the end of the chapter is a summary that lists salient points known about chromosomal DNA replication in higher plants.

DIVERSITY: A FEATURE OF THE PLANT GENOME

A striking feature among higher plants is the difference in genome size. Measurements from more than 2000 species show that plant genomes vary at least 2500-fold and cover a range from 0.05 pg of DNA in *Cardamine amara* to 127.5 pg of DNA in *Fritillaria assyricaca* (Bennett and Smith 1976, 1991; Bennett et al. 1982). The chromosome number of different species, likewise, varies widely. For example, *Happlopappus gracilis* has $n = 2$, whereas *Senescio biserratus* has $n = 100$. The plant kingdom also is populated with allopolyploids which are the products of hybridization between diploid species and which, at first glance, appear as normal diploids. The existence of allopolyploids indicates that the replicative machinery of each diploid parent is compatible when mixed in the same nucleus. Two familiar and important allopolyploid plants are commercial wheat and cotton. Given the diversity of genome size and chromosome number among plants, it is not surprising that the average length of duplex DNA per chromatid varies from about a centimeter (*Arabidopsis*) to more than a meter (*Vicia faba*).

DNA Replication in Eukaryotic Cells

THE PLANT REPLICON

Replicon Size and Replication Fork Rate in Seedling Root Tip Cells Are Independent of Genome Size

There are two major groups of higher plants: dicots and monocots. Experiments with cells from the root tip meristem of plants from both groups show that a difference in genome size has little to no effect on replicon size (distance between replication initiation sites) and the rate of movement of a single replication fork. The sizes of replicons and fork rates do differ, however, between the groups. DNA fiber autoradiography shows that replicon size and fork rate of dicots are 66 ± 10.6 kb and 24 ± 4.2 kb per hour, respectively (Van't Hof and Bjerknes 1981). Monocots, on the other hand, have an average replicon size of 47.4 ± 13.2 kb and a fork rate of 5.0 ± 3.1 kb per hour (Francis et al. 1985a). The reason for these differences between dicots and monocots is not understood.

Replicon Families and S-phase Duration in *Arabidopsis thaliana*

A. thaliana has chromosomal DNA replicons, seen by fiber autoradiography, that average 72 kb and a single fork rate of 17 kb per hour. With bidirectional replication, the two forks replicate 72 kb of DNA in a little more than 2 hours (Van't Hof et al. 1978). The genome (0.2 pg of DNA; Bennett and Smith 1976) has about 2600 replicons and two replicon families; i.e., groups of clustered replicons that function simultaneously. The larger family with 1900 members replicates first in S phase, and the smaller one with 700 replicons begins about 35 minutes after the first. S phase ends when the smaller family completes replication. The total time needed to replicate the genome is about 3 hours. Hence, the S-phase duration in *A. thaliana* is determined by the number of replicon families, replication fork rate, and the interval between the onset of replication by one family and the next.

Replicon Size and Replication Fork Rate Are Not Genetically Fixed

Triticale, a man-made allohexaploid produced by a cross of diploid rye (*Secale cereale*, cv. UC 90) with an allotetraploid wheat (*Triticum turgidum* var. *durum*, cv. Cocorit), has replicon properties that differ from those of each parent (Kidd et al. 1992). The genome size of *triticale* is 21.5 pg, and those of its rye and wheat parents, respectively, are 8.3 pg and 13.3 pg. The parents have replicons in the 50- to 60-kb range, whereas the offspring, *triticale*, has replicons of about 22 kb. Thus, previously quiescent replication initiation sites in the parental genomes function in the hybrid. The increased initiation sites (smaller replicons) in *triticale* are accompanied by slower fork movement. In the hybrid, the fork rate is 4.5 kb per hour, whereas the rye and wheat parents, respectively, have rates of 21 kb and 14.1 kb per hour. The hybrid, therefore, has neither the replicon size nor the fork rate of either parent.

Factors Affecting the Number of Replication Initiation Sites (Origins)

The results obtained with *triticale* show that the number of useful replication sites in a genome depends on other factors besides nucleotide sequence. Replicon origins exist in two states, an active or functional state and a quiescent state. The first evidence for two states came from work with *Drosophila* (Blumenthal et al. 1974). Blumenthal et al. discovered that in rapidly dividing nuclei of the zygote there are many more initiation sites than in cultured cells. Thus, replication origins used during the early stages of embryogenesis are not used by mature cells.

The quiescent and active states are reversible, however. When an apical meristem is given a stimulating exposure of light, the vegetative meristem switches to forming cells that will develop as flowers. Responding cells divide more rapidly, their replication forks move faster, and quiescent origins are activated (Ormrod and Francis 1986; Jacqmard and Houssa 1988). A similar response is seen in apical cells treated with certain plant hormones (Houssa et al. 1990, 1994).

Activation of quiescent replicative origins also occurs when replication forks are stalled because the parental templates are cross-linked by psoralen (Francis et al. 1985b). Replication is initiated between stalled forks, thereby reducing measured replicon size by about one-half. A similar phenomenon occurs in soybean cells recovering from treatment with fluorodeoxyuridine (Cress et al. 1978).

Since the frequency of origins is subject to developmental, environmental (light), hormonal, and stressful factors, the nature of the sequences at which replication begins must be questioned. Certainly, more than nucleotide sequence is involved in determining where and when potential replication origins are used by plant cells.

Chromosomal DNA Maturation Occurs Stepwise

Studies using DNA fiber autoradiography show that chromosomal DNA replication is segmental, having a hierarchical organization. The elemental unit is the replicon. Often, tandem replicons replicate DNA simultaneously, forming a cluster of active replicons. These clusters are members of families that, by definition, function simultaneously even if located on different chromosomes. Since the DNA molecule in a metaphase chromatid is one lengthy polynucleotide chain, the replicative segments forming a nascent strand must be joined. Chromosome maturation occurs when the nascent segments are joined. In plant and other cells, maturation is closely tied to progression in the cell cycle (Van't Hof 1980; Schvartzman et al. 1981). The process as seen by alkaline sucrose velocity sedimentation of DNA from nuclei of synchronized pea root cells is diagramed in Figure 1. Replicon-sized nascent molecules formed during S phase persist until cells reach late S or G_2 phase. At this time, stepwise ligation occurs. The replicon-sized molecules are joined, giving rise to cluster-sized molecules, and these molecules are joined to give chromosomal-sized molecules. This stepwise increase in size is dependent on intracellular nucleotide concentration as it is accelerated, becoming independent of cell cycle position, if cells are given exogenous thymidine (Schvartzman et al. 1984).

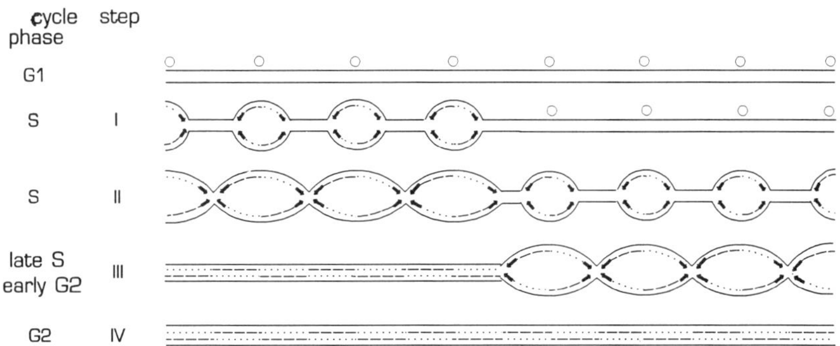

Figure 1 Diagram showing the stepwise replication and maturation of chromosomal DNA during the S and G_2 phases of the cell cycle. (*Top*) Parental chromosomal duplex in the G_1 phase. The circles represent replicon origins. (Step I) Conventional bidirectional replication by four replicons in a cluster. The nascent chains are attached to the arrows. (Step II) Convergence of replication forks of neighboring replicons within the cluster leaving gaps between nascent chains. To the right of the first cluster is a second beginning replication. (Step III) Gaps between nascent chains in the first cluster are sealed giving a cluster-sized molecule. The gaps remain between chains in the second cluster. (Step IV) Gaps are sealed in the second cluster and this cluster is joined to the first, producing chromosomal-sized DNA. Note that time between the replication of the first cluster and that of the second cluster may be greater than shown.

REPLICATION OF REPEATED RIBOSOMAL GENES (rDNA)

rDNA Occurs Mostly in Clusters

Repeated ribosomal genes in plants occur in clusters at a locus known as the nucleolar organizer. The number of tandem repeats in a cluster varies from about 570 in *A. thaliana* (Pruitt and Meyerowitz 1986) up to 7000 in *V. faba* (Rogers and Bendich 1987). The organization of a single repeat is generally conserved among plants, but the length of a repeat within a cluster is species-dependent (Appels and Honeycutt 1986; Hemleben et al. 1988). In pea, for instance, there are two clusters, one with repeats of about 9 kb and another with repeats of about 8.6 kb (Jorgensen et al. 1987). The difference in the length of a repeat is due to the number of subrepeats in the intergenic spacer region located between the 25S and 18S genes. The organization of the spacer region is conserved among species and contains sequences responsible for replicative functions.

Organization of the Intergenic Spacer Region

In Figure 2 is a map of the intergenic spacer region of pea, which contains 3282 bases. The numbering of bases begins at the 3′ end of the 25S gene and ends at the coding region of the 18S gene. Hernández et al. (1993) published the sequence of the spacer region and showed by two-dimensional gel electrophoresis that replication forks are retarded at nucleotides 158–226. They also showed that the DNA is bent at nucleotides 2178–2199 just upstream of ARS consensus sequences at nucleotides 2270–2357. The origin of replication is located at or near nucleotide 1486 (Hernández et al. 1988; Van't Hof and Lamm 1992). The start site for RNA polymerase I at nucleotide 2506 was mapped by Piller et al. (1990).

In the box below the map in Figure 2 is presented a list of six plant species whose spacer regions have a pattern similar to that of pea with respect to the position of the RNA polymerase I start site and ARS consensus sequences. The column of numbers in the box gives the number of bases that separate the start site and consensus sequence in each species. As shown, this distance ranges from 200 to 339 bases. The replication origin in pea (cv. Alaska) is among the subrepeats located 784 bases upstream of the ARS consensus sequences. Replication initiation sites have not been determined for the other species listed, but it is likely that the sites are upstream of the ARS sequence(s) in these plants also.

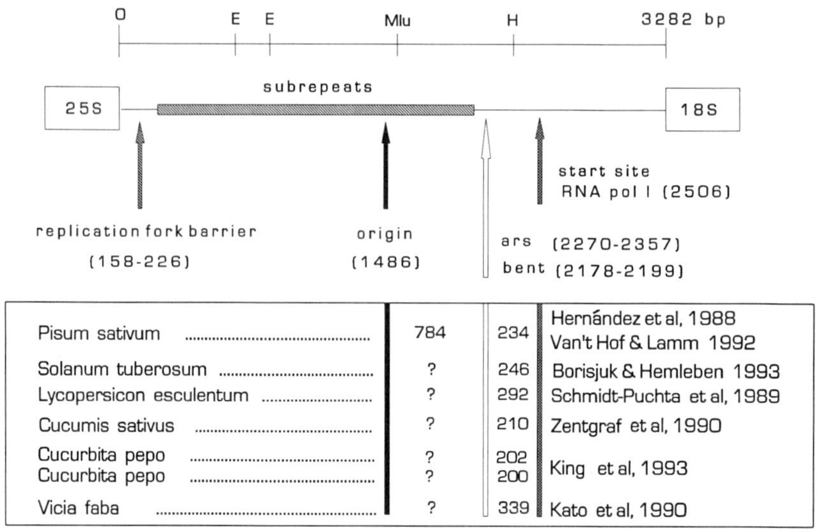

Figure 2 Organization of the ribosomal DNA spacer region of pea *(Pisum sativum* cv. Alaska). At the top, a skeletal map showing the position of restriction enzyme sites, *Eco*RI (E), *Mlu*, and *Hin*dIII (H). The numbers 0 and 3282 refer to nucleotide bases between the 25S and 18S genes seen below the map. The arrowheads indicate positions of known functional sequences, fork barrier, origin, ARS consensus sequences, and RNA pol I start site. The numbers in parentheses are the bases involved in each function. Note: A segment of bent DNA is upstream from the ARS consensus sequences. See text and Hernández et al. (1993) for more details. In the box is a partial list of plant species with known positions of ARS consensus sequences and RNA pol I start sites. Of these, only *P. sativum* has a mapped replication origin. The origin is located at about base number 1486 marked by the vertical solid bar in the box. 784 bases separate the origin from the segment containing ARS consensus sequences marked by the vertical open bar. 234 bases separate the ARS consensus sequences from the RNA pol I start site, whose position is marked by the stippled bar. The numbers between the open and stippled bars are the number of bases separating the ARS consensus sequences and the RNA pol I start site in each species.

Two Mechanisms of rDNA Replication in Pea

Experimental results support two mechanisms of replication for the rDNA in pea (Van't Hof and Lamm 1991, 1992; Hernández et al. 1993). The mechanisms are diagramed in Figure 3. They are labeled conventional and displacement loop. To the right of each diagram is another depicting the expected product produced by branch migration or strand destabilization. In the conventional mechanism, replication begins at an

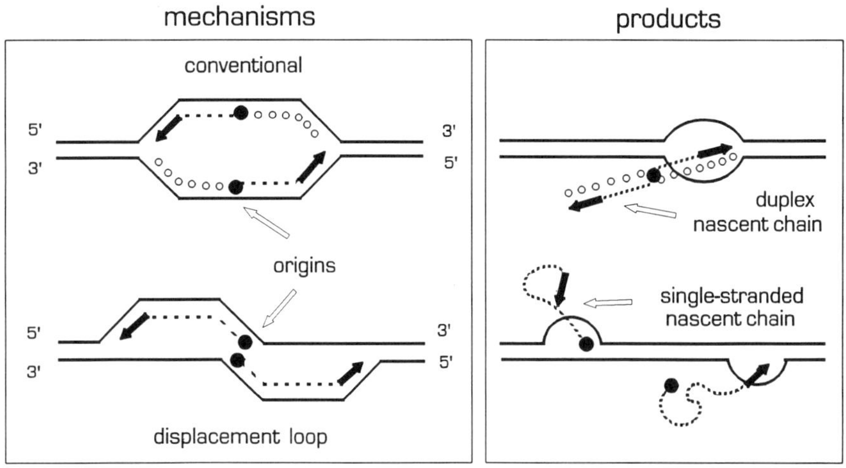

Figure 3 Diagrams showing conventional bidirectional and displacement loop mechanisms of replication of ribosomal DNA and extruded products of each mechanism produced by branch migration. *Solid lines* represent the parental DNA duplex; *dashed lines*, nascent chains elongated 5′→3′; *open circles*, nascent chains elongated 3′→5′; *closed circles*, replication origins. The *arrows* show the direction of chains elongating 5′→3′.

origin and proceeds bidirectionally along the parental templates via leading and lagging forks. Replication also begins at an origin in the displacement mechanism, but strand elongation proceeds only 5′→3′ along the templates. Semi-conservatively replicated DNA is achieved by cooperative converging forks from tandem origins that meet, resolve the supercoiled structure between them, and resume replication moving 5′→3′ in opposite directions along the single-stranded parental templates.

The mechanisms are distinguishable by the nature of the nascent chain produced by either branch migration or destabilization of the parental duplex. The conventional mechanism produces double-stranded nascent molecules, whereas the displacement mechanism produces single-stranded nascent chains.

Hernández et al. (1993) showed, by two-dimensional gel electrophoresis, that the rDNA of pea replicates via the conventional mechanism. To detect replication by the conventional mechanism requires isolation of DNA without organic solvents and about 10 μg of DNA per gel. On the other hand, Van't Hof and Lamm (1991 1992), using the same technique, show that rDNA extracted from nuclei by organic solvents has specific single-stranded products that are 5′→3′ or 3′→5′. About 0.5

μg of DNA per gel is required to see the single-stranded products. In addition, electron microscopy of DNA extracted from pea nuclei without organic solvents shows an abundance of "extrachromosomal" duplex molecules with single-stranded branches (Krimer and Van't Hof 1983) that contain rDNA sequences (Kraszewska et al. 1985). A duplex molecule with single-stranded branches is the conformation expected of displacement loop replication.

Finally, data obtained from a wide variety of plant and animal cells (*Physarum*, *Tetrahymena*, yeast, *Xenopus*, human, and peas) show that rDNA is replicated by the conventional mechanism. Given this fact, the displacement loop mechanism must be viewed as an alternative mechanism of rDNA replication in plants.

SUMMARY

1. Replicon size and replication fork rate in higher plants are independent of genome size when measured in the same tissue under identical conditions.
2. Replicon size and replication fork rate differ between dicotyledonous and monocotyledonous plants.
3. Neither replicon size nor fork rate is genetically determined.
4. Replicon size and fork rate are changed by light stimuli and hormonal treatment in tissue involved in the transition from a vegetative to a floral state.
5. Chromosomal maturation occurs stepwise and is dependent on intracellular thymidine concentration.
6. Functional replicative sequences in ribosomal DNA are located in the intergenic spacer region between the 25S and 18S genes. These sequences include a replication fork barrier, the origin of replication, and ARS consensus sequences.
7. Ribosomal DNA repeats are replicated by two mechanisms, a conventional bidirectional mechanism and a displacement loop mechanism.

ACKNOWLEDGMENT

This research was supported by the Office of Health and Environmental Research, U.S. Department of Energy.

REFERENCES

Appels, R. and R.L. Honeycutt. 1986. rDNA: Evolution over a billion years. In *DNA systematics* (ed. S.K. Dutta), vol. 2, pp. 82–135. CRC Press, Boca Raton, Florida.

Bennett, M.D. and J.B. Smith. 1976. Nuclear DNA amounts in angiosperms. *Philos. Trans. R. Soc. Lond. B* **274:** 227–274.

———. 1991. Nuclear DNA amounts in angiosperms. *Philos. Trans. R. Soc. Lond. B* **334:** 309–345.

Bennett, M.D., J.B. Smith, and J.S. Heslop-Harrison. 1982. Nuclear DNA amounts in angiosperms. *Philos. Trans. R. Soc. Lond. B.* **216:** 179–199.

Blumenthal, A.B., H.J. Kriegstein, and D.S. Hogness. 1974. The units of DNA replication in *Drosophila melanogaster* chromosomes. *Cold Spring Harbor Symp. Quant. Biol.* **38:** 195–203.

Borisjuk, N. and V. Hemleben. 1993. Nucleotide sequence of the potato rDNA intergenic spacer. *Plant Mol. Biol.* **21:** 381–384.

Cress, D.E., P.J. Jackson, A. Kadouri, Y.E. Chu, and K.G. Lark. 1978. DNA replication in soybean protoplasts and suspension-cultured cells: Comparison of exponential and fluorodeoxyuridine synchronized cultures. *Planta* **143:** 241–253.

Francis, D., A.D. Kidd, and M.D. Bennett. 1985a. DNA replication in relation to DNA C values. In *The cell division cycle in plants* (ed. J.A. Bryant and D. Francis), pp. 61–82. Cambridge University Press, Cambridge.

Francis, D., N.D. Davies, J.A. Bryant, S.G. Hughes, D.R. Sibson, and P.N. Fitchett. 1985b. Effects of psoralen on replicon size and mean rate of DNA synthesis in partially synchronized cells of *Pisum sativum* L. *Exp. Cell Res.* **158:** 500–508.

Hemleben, V., M. Ganal, J. Gerstner, K. Schiebel, and R.A. Torres. 1988. Organization and length heterogeneity of plant ribosomal RNA genes. In *Architecture of eukaryotic genes* (ed. G. Kahl), pp. 371–383. VCH Verlagsgesellschaft, Weinheim.

Hernández, P., C.A. Bjerknes, S.S. Lamm, and J. Van't Hof. 1988. Proximity of an ARS consensus sequence to a replication origin of pea *(Pisum sativum)*. *Plant Mol. Biol.* **10:** 413–422.

Hernández, P., L. Martín-Parras, M.L. Martínez-Robles, and J.B. Schvartzman. 1993. Conserved features in the mode of replication of eukaryotic ribosomal RNA genes. *EMBO J.* **12:** 1475–1485.

Houssa, C., A. Jacqmard, and G. Bernier. 1990. Activation of replicon origins as a possible target for cytokinins in shoot meristem of *Sinapis*. *Planta* **181:** 324–326.

Houssa, C., G. Bernier, A. Pieltain, J-M. Kinet, and A. Jacqmard. 1994. Activation of latent DNA-replication origins: A universal effect of cytokinins. *Planta* **193:** 247–250.

Jacqmard, A. and C. Houssa. 1988. DNA fiber replication during a morphogenetic switch in the shoot meristematic cells of a higher plant. *Exp. Cell Res.* **179:** 454–461.

Jorgensen, R.A., R.E. Cuellar, W.F. Thompson, and T.A. Kavanagh. 1987. Structure and variation in ribosomal RNA genes of pea. *Plant Mol. Biol.* **8:** 3–12.

Kato, A., T. Nakajima, J. Yamashita, K. Yakura, and S. Tanifuji. 1990. The structure of the large spacer region of the rDNA in *Vicia faba* and *Pisum sativum*. *Plant Mol. Biol.* **14:** 983–993.

Kidd, A.D., D. Francis, and M.D. Bennett. 1992. Replicon size, rate of DNA replication, and the cell cycle in a primary hexaploid *triticale* and its parents. *Genome* **35:** 126–132.

King, K., R.A. Torres, U. Zentgraf, and V. Hemleben. 1993. Molecular evolution of the intergenic spacer in the nuclear ribosomal RNA genes of cucurbitaceae. *J. Mol. Evol.* **36:** 144–152.

Kraszewska, E.K., C.A. Bjerknes, S.S. Lamm, and J. Van't Hof. 1985. Extrachromosomal DNA of pea-root (*Pisum sativum*) has repeated sequences and ribosomal genes.

Plant Mol. Biol. **5:** 353–361.

Krimer, D.B. and J. Van't Hof. 1983. Extrachromosomal DNA of pea *(Pisum sativum)* root-tip cells replicates by strand displacement. *Proc. Natl. Acad. Sci.* **80:** 1933–1937.

Ormrod, J.C. and D. Francis. 1986. Mean rate of DNA replication and replicon size in the shoot apex of *Silene coeli-rosa* L. during the initial 120 minutes of the first day of floral induction. *Protoplasma* **130:** 206–210.

Piller, K.J., S.R. Baerson, N.O. Polans, and L.S. Kaufman. 1990. Structure analysis of the short length ribosomal DNA variant from *Pisum sativum* L. cv. Alaska. *Nucleic Acids Res.* **18:** 3135–3145.

Pruitt, R.E. and E.M. Meyerowitz. 1986. Characterization of the genome of *Arabidopsis thaliana*. *J. Mol. Biol.* **187:** 169–183.

Rogers, S.O. and A.J. Bendich. 1987. Heritability and variability in ribosomal RNA genes of *Vicia faba*. *Genetics* **117:** 285–295.

Schmidt-Puchta, W., I. Günther, and H.I. Sänger. 1989. Nucleotide sequence of the intergenic spacer (IGS) of the tomato ribosomal DNA. *Plant Mol. Biol.* **13:** 251–253.

Schvartzman, J.B., D.B. Krimer, and J. Van't Hof. 1984. The effects of different thymidine concentrations on DNA replication in pea-root cells synchronized by a protracted 5-fluorodeoxyuridine treatment. *Exp. Cell Res.* **150:** 379–389.

Schvartzman, J.B., B. Chenet, C.A. Bjerknes, and J. Van't Hof. 1981. Nascent replicons are synchronously joined at the end of S phase or during G2 phase in peas. *Biochim. Biophys. Acta* **653:** 185–192.

Van't Hof, J. 1980. Pea (*Pisum sativum*) cells arrested in G2 have nascent DNA with breaks between replicons and replication clusters. *Exp. Cell Res.* **129:** 231–237.

Van't Hof, J. and C.A. Bjerknes. 1981. Similar replicon properties of higher plant cells with different S periods and genome sizes. *Exp. Cell Res.* **136:** 461–465.

Van't Hof, J. and S.S. Lamm. 1991. Single-stranded replication intermediates of ribosomal DNA replicons of pea. *EMBO J.* **10:** 1949–1953.

———. 1992. Site of initiation of replication of the ribosomal genes of pea *(Pisum sativum)* detected by two-dimensional gel electrophoresis. *Plant Mol. Biol.* **20:** 377–382.

Van't Hof, J., A. Kuniyuki, and C.A. Bjerknes. 1978. The size and number of replicon families of chromosomal DNA in *Arabidopsis thaliana*. *Chromosoma* **68:** 269–285.

Zentgraf, U., M. Ganal, and V. Hemleben. 1990. Length heterogeneity of the rRNA precursor in cucumber *(Cucumis sativus)*. *Plant Mol. Biol.* **15:** 465–474.

39
Mitochondrial DNA Replication

David A. Clayton
Department of Developmental Biology
Beckman Center for Molecular and Genetic Medicine
Stanford University School of Medicine
Stanford, California 94305-5427

Eukaryotic mitochondrial organelles contain a genome separate and distinct from that of the nucleus. Although limited in genetic content, expression of mitochondrial DNA (mtDNA) is an essential function in most eukaryotes, and therefore the genome is not dispensable. Strategies designed to define both *cis*-acting elements and *trans*-acting factors required for mtDNA replication have been developed over the last 25 years. Both in vivo and in vitro approaches have been employed to describe the mode of mammalian mtDNA replication. This chapter first reviews the basics of earlier works and then turns to more recent investigations that focus on early steps in the replication process. Finally, comparison is made to yeast mtDNAs. The novel kinetoplast DNA is reviewed by Torri et al. (this volume).

BASIC BACKGROUND

Although mtDNA comprises typically less than 1% of a metazoan cell's DNA population, the cellular copy number is 10^3 to 10^4, given the relatively small size of these genomes (~16 kb). It is usually assumed that mtDNA is present in several copies per mitochondrion, but the organelle population is likely dynamic in the living cell, and the distribution of mtDNA may therefore be variable. Replication is under relaxed control and there is no apparent accounting of DNA origins that ensures that each molecule is replicated once and only once per cell cycle, a requirement assumed to be strictly enforced in the case of chromosomal DNA origins of replication. Consistent with this is the lack of any sharp restriction on mtDNA replication with regard to cell cycle phase (Clayton 1982).

The signature form of mammalian mtDNA is the displacement-loop (D-loop) molecule, which maintains a short piece of nascent strand at the leading-strand origin of replication. The leading strand has historically

DNA Replication in Eukaryotic Cells
 0-87969-459-9/96 $5 + .00

been termed the heavy (H) strand because of its greater intrinsic buoyant density in alkaline cesium chloride gradients. Therefore, the opposite strand of the helix has been termed the light (L) strand. The D loop is thus defined as a three-stranded structure with the nascent leading H strand defining the origin of leading-strand replication (O_H) at its 5′ end. D-loop strands are variable in size in a species-specific manner and are turned over more rapidly than the rate of genomic replication would require. A commitment to replication is defined by a molecule in which DNA synthesis has proceeded past the D-loop region around the molecule; this occurs unidirectionally with continuing displacement of the parental H strand. After the growing H strand has elongated to two-thirds or more of its total length, the origin of lagging L-strand replication (O_L) is exposed on the displaced parental H strand and initiation of daughter L-strand synthesis begins. This asynchrony in the functioning of the two origins of mtDNA replication results in two distinct progeny circles being segregated, one of which requires single-strand gap filling prior to closure. Both newly synthesized circles are initially of zero superhelix density and undergo a stepwise introduction of negative superhelical turns. Synthesis of a new D loop then completes the cycle. The topological nature of replicative intermediates, their occurrence, and their form have been reviewed previously (Clayton 1982).

REPLICATION OF MAMMALIAN mtDNA

Productive mtDNA replication begins by an initiation of leading (H)-strand synthesis that results in strand elongation over the entire length of the genome. Since initiation of lagging (L)-strand synthesis only occurs after O_L is exposed as a single-stranded template, leading-strand synthesis is the dominant event in dictating genomic replication. Although the 5′-end map positions of short D-loop strands and elongated leading strands are the same, there has been no reported experiment that determines whether elongation occurs from preexisting D-loop strands or whether it requires initiation and uninterrupted synthesis through the D loop and then beyond.

Initiation of Heavy-strand Synthesis at the Leading-strand Origin

Experiments performed 10 years ago documented the 5′ and 3′ ends of both DNA and RNA species complementary to the D-loop region of human (Chang and Clayton 1985) and mouse (Chang et al. 1985) mtDNAs.

These data refined the map positions of these species because of improved sizing technologies for nucleic acids and were consistent with previous assignments (Gillum and Clayton 1979). In summary, there are several sites of likely transition from RNA synthesis to DNA synthesis. For both human and mouse species it is possible, in the majority of cases, to align RNA 3′ ends with DNA 5′ ends, consistent with the possibility that the RNAs are involved in priming replication. In the case of mouse, a species was isolated that consisted of RNA at its 5′ end and DNA throughout most of its distal portion (Chang et al. 1985). Importantly, the 5′ end of the RNA portion of a primed nascent leading strand mapped at the transcription promoter of this strand. The organization of the genome is such that all genes served by this promoter lie downstream from the origin. Thus, in principle, one round of transcription could be sufficient for a priming event and transcription of all of the genes on this strand.

Initiation of Light-strand Synthesis at the Lagging-strand Origin

The origin of lagging (L)-strand replication, O_L, located well away from O_H (Fig. 1), is approximately 30 nucleotides in size. It is interesting that this small noncoding region is tightly flanked by tRNA genes. The origin itself has the potential for forming a predictably stable stem-loop structure, and this feature is highly conserved among sequenced vertebrate mtDNAs, with the exception of chicken (Desjardins and Morais 1990).

Our current understanding of how O_L functions required the development of an in vitro replication system capable of initiation of L-strand synthesis that produced nascent strands which were the same as those that had been previously identified from in vivo mitochondrial nucleic acid isolates. The in vitro run-off replication assay (Wong and Clayton 1985a,b) revealed a human mtDNA primase with the ability to recognize O_L and initiate priming and DNA synthesis. Daughter L-strand synthesis is primed by RNA synthesis complementary to the T-rich loop structure of O_L. In turn, the transition from RNA synthesis to DNA synthesis occurs near the base of the stem. The site of transition from RNA synthesis to DNA synthesis is contained within an immediately adjacent tRNA gene. Furthermore, a mutational analysis demonstrated that although the small O_L sequence was necessary and sufficient to support initiation of DNA replication (Hixson et al. 1986), a short pentanucleotide sequence at the base of the O_L stem was required to maintain a correctly functioning origin. This appears to involve the ability of mtDNA polymerase to engage a primer, as the mutations examined by Hixson et al. (1986) were

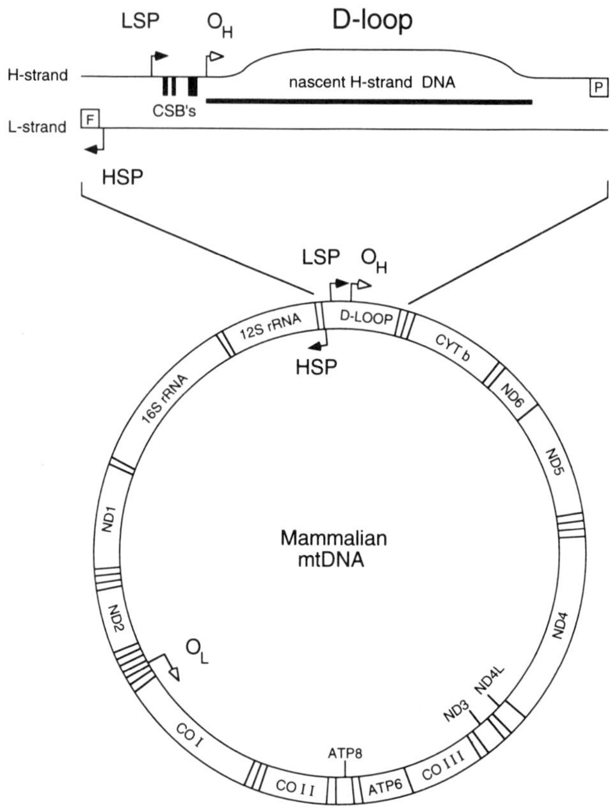

Figure 1 Representation of mammalian mtDNA. The small boxes represent the 22 tRNA genes (F, phenylalanine; P, proline). The 12S and 16S rRNA genes and D-loop region are shown. O_H and O_L are the respective origins of H- and L-strand synthesis. HSP and LSP are the respective promoters for transcripts copied from the H and L template strands. Arrows denote the directions of synthesis. The protein-coding genes are cytochrome *c* oxidase subunits I, II, and III (COI, COII, COIII); ATPase subunits 6 and 8 (ATP6, ATP8); cytochrome *b* (CYTb); and subunits 1, 2, 3, 4, 4L, 5, and 6 of NADH dehydrogenase (ND1, ND2, ND3, ND4, ND4L, ND5, ND6). ND6 and eight tRNA genes are encoded by H-strand sequence (L strand is the template strand); all the other genes are encoded by L-strand sequence. (CSB's) Conserved sequence blocks.

not obviously impaired in priming function, but instead were deficient in the process of elongating primers into DNA strands.

Elongation and Maturation of Progeny Strands

The basic details on the mode of elongation and maturation of both leading and lagging strands of mammalian mtDNA have been determined by isotopic labeling and electron microscopic visualization approaches (Clayton 1982). There has been no extensive further work along these lines, although there have been attempts to utilize isolated organelles as a way to study various steps of mtDNA replication (for examples, see Dunon-Bluteau et al. 1987; Enríquez et al. 1994). Although this approach may lead to insights on regulating replication, it has some limitations, since it is not possible to define the molecules responsible for various events in the process nor to test easily for the efficacy of altered reaction conditions, different templates, and requirements for known, purified entities.

TRANS-ACTING FACTORS INVOLVED IN REPLICATION

The close physical and functional relationships between leading-strand mtDNA replication and transcription have resulted in data from studies of each process that merge at the level of transcription initiation from the strong promoter upstream of O_H. Promoter-specific transcription requires the presence of the transcription factor mtTFA (formerly termed mtTF1) in addition to a mitochondrial RNA (mtRNA) polymerase fraction that alone is incapable of selective transcription (Fisher et al. 1987). Mature human mtTFA (h-mtTFA) is a small (204 amino acids) basic protein that binds immediately upstream of the mitochondrial promoters. There is an excellent correlation between DNA binding by h-mtTFA and the ability of that protein to support correct and efficient transcription initiation (Dairaghi et al. 1995 and references therein), and Fisher et al. (1992) were able to demonstrate that h-mtTFA has the capacity to unwind and bend DNA, thereby suggesting the manner in which at least one part of the transcription initiation process takes place.

Human mtTFA contains two domains characteristic of high mobility group (HMG) proteins, and these two domains (HMG boxes) comprise most of the molecule. The two HMG boxes are interrupted by a 27-amino-acid linker sequence, and the protein contains a 25-residue, carboxy-terminal tail. Mutational analysis of the protein has recently shown that this short tail is critical for DNA binding at the promoter and for transcriptional activation (Dairaghi et al. 1995).

Processing of Mitochondrial Transcripts for Replication

The first isolated mammalian mtRNA processing activity was RNase MRP (for *m*itochondrial *R*NA *p*rocessing), which was isolated from mouse and human mitochondrial fractions and shown to be a site-specific endoribonuclease containing an essential RNA component (Chang and Clayton 1987). RNase MRP is a ribonucleoprotein with its RNA component and very likely all of its protein components encoded by nuclear genes. The sequence of the nuclear gene for the RNA component of RNase MRP has been obtained for vertebrates and fungi (Paluh and Clayton 1995; for sequence comparisons of earlier reported sequences, see Schmitt et al. 1993), and the genes are distinctly similar. Inspection of the coding region and surrounding sequences indicates that it is very likely that these genes function as polymerase II/polymerase III transcription units, which may represent an ancient class of nuclear gene. The genes are single copy in the cases of both mouse and human (Hsieh et al. 1990), and it is interesting to note that there is extensive sequence similarity in the 5′-flanking region that surpasses that typical for other small RNA genes of nuclear origin. There are two protein components identified for *Saccharomyces cerevisiae* RNase MRP: One is shared with RNase P (Lygerou et al. 1994) and the other appears unique to RNase MRP (Schmitt and Clayton 1994).

RNase MRP cleaves mtRNA transcripts that are complementary to the leading-strand origin of mammalian mtDNA, and therefore, it is a candidate for playing a role in primer RNA metabolism. The predominant site of cleavage on single-strand mtRNA is one that may occur in vivo as well, given the fact that transcript termini map at this position (Chang and Clayton 1985; Chang et al. 1985). RNase MRP could produce 3′-OH groups for extension by mtDNA polymerase or could be somehow involved in stabilizing primers at the 5′ end or in 5′-end editing of transcripts (Chang and Clayton 1987); new evidence is consistent with any or all of these roles (see section below on regulation).

Mammalian mtDNA Replication Proteins

mtDNA polymerase, commonly termed DNA polymerase-γ, has been studied sporadically over the years, but only recently has information become available on its actual size and some of its functional features. There are several reports of an associated 3′→5′ exonuclease activity being an intrinsic part of the polymerase; it likely confers a proofreading capacity consistent with studies on the error rate of DNA polymerase-γ (Kunkel and Soni 1988; Insdorf and Bogenhagen 1989b; Kaguni and Olson 1989 [for *Drosophila melanogaster*]). The catalytic subunit of the holoenzyme appears to be of significant size, approximately 140 kD in *Xenopus laevis* (Insdorf and Bogenhagen 1989a; a similar situation may prevail for *D. melanogaster* [Wernette and Kaguni 1986]); in both these cases there appears to be a stoichiometric, smaller subunit present in the most highly purified preparations. Human mtDNA polymerase has been partially purified (Gray and Wong 1992) and has recently been cloned by virtue of its conservation to yeast mtDNA polymerases (W.P. Copeland, pers. comm.). This significant advance should permit a comprehensive study of the structure, functional parameters, and intermitochondrial distribution of this central activity in mtDNA replication.

There are several instances of the identification of mtDNA-binding proteins that display a preference for single-stranded DNA (Pavco and Van Tuyle 1985; Van Tuyle and Pavco 1985; Mignotte et al. 1988). These are likely to be standard single-stranded DNA-binding proteins, because their sequences closely resemble *Escherichia coli* SSB (Tiranti et al. 1993), which has facilitated their cloning. It will be of interest to learn the possible association of mitochondrial SSB with mtDNA polymerase in the replication complex and to determine whether mutations or deficiencies in mitochondrial SSB protein can cause mtDNA deletions in human disease. In the case of *Drosophila, E. coli* SSB is able to stimulate the rate of DNA synthesis in vitro by mtDNA polymerase (Williams and Kaguni 1995). The current working model for mammalian mtDNA replication is depicted in Figure 4 of the chapter by Brush and Kelly (this volume).

REGULATING mtDNA REPLICATION IN THE MAMMALIAN CELL

Regulation of mammalian mtDNA replication most likely involves molecular events surrounding initiation of leading-strand DNA synthesis at the D loop. As outlined earlier, O_L is well separated physically in the genome, and there is no evidence to indicate that a round of productive replication can begin there. Our knowledge of the exact characteristics of the DNA primase activity that recognizes O_L has not advanced beyond work described in an earlier review (Clayton 1991). This is due, at least in part, to the very low abundance of this activity in the cell. However, we can describe some of the fundamental aspects of initiation of leading-strand synthesis that enable, in turn, some predictions on how regulation of replication may occur.

Step one in the process is initiation of transcription at the LSP to provide primers. Initiation requires mtTFA and a partially purified mtRNA polymerase. There are currently no known differences in the requirements for initiation that reflect the eventual fate of the transcripts. That is, the nucleus-encoded transcription proteins, and their interactions with the template and each other, could be identical in producing RNA for priming or for transcripts that represent tRNAs or that encode a protein. Thus, the absence or impairment of function of these proteins should affect both mtDNA replication and gene expression. It is known that h-mtTFA is greatly reduced in some human tissues and cell lines with little or no mtDNA (Larsson et al. 1994; Poulton et al. 1994), but a direct loss of mtDNA due to down-regulation of h-mtTFA remains to be shown.

The second important event in initiation appears to be the formation of stable RNA-DNA hybrids over the origin region (Fig. 2). The principal feature of mtDNA responsible for this is the highly conserved *c*onserved *s*equence *b*lock II (CSBII) element. This short (usually defined as between 15 and 20 nucleotides) sequence is atypical in two obvious ways. It is not only more than 90% G+C-rich, but all of the guanosines are in one strand; RNA transcribed across CSBII thus contains this virtual guanosine homopolymer run. The requirements for stable RNA-DNA hybrid formation are clearest for yeast putative mtDNA origin sequences (Xu and Clayton 1995). The yeast CSBII sequence homolog is the necessary and sufficient element for sponsoring RNA-DNA hybrid formation, and the natural polarity of the sequences (guanosines in the RNA strand, cytosines in the DNA strand) must be maintained. The same basic phenomenon is seen for mammalian mtDNA, and in this case, CSBII is required along with some dependence on the two other conserved sequences at the origin, CSBIII and CSBI, with respect to efficiency of hybrid formation (Xu and Clayton 1996).

Cleavage of D-loop region RNA-DNA hybrids has been achieved (D.Y. Lee and D.A. Clayton, unpubl.) with highly purified preparations of RNase MRP. These sites of cleavage align well with mapped termini of primer RNAs and nascent mtDNA 5′ ends at the leading-strand origin. Thus, a more refined model for initiation of leading-strand synthesis invokes stable RNA-DNA hybrid formation followed by R-loop processing to achieve synthesis by mtDNA polymerase (Fig. 3).

COMPARING MAMMALIAN AND YEAST mtDNA REPLICATION

Although the field of yeast mitochondrial biology and genetics has a longer history than that of mammalian studies, the molecular details of yeast mtDNA replication are largely unknown. This is due in part to the greater complexity of yeast mitochondrial genomes. In the case of wild-type mtDNAs, genome size is variable among strains and species. In cases where yeast mtDNA has been studied physically, it is apparent that the genome can be found in different circular and linear forms. This situation makes it difficult to perform the type of biochemical analyses

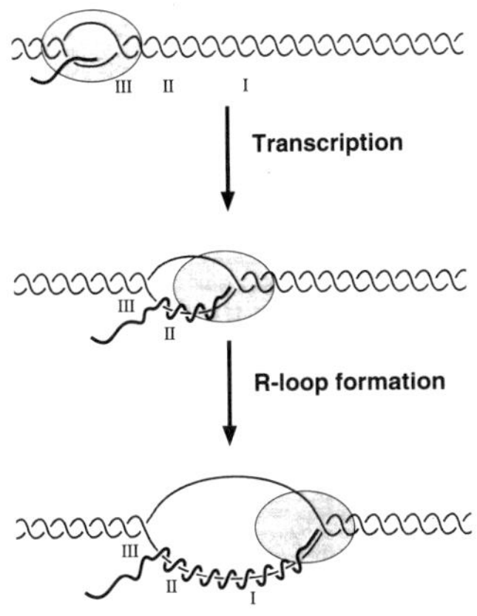

Figure 2 Diagram illustrating formation of a persistent RNA-DNA hybrid via transcription at the leading-strand origin. Roman numerals represent CSBIII, CSBII, and CSBI, respectively. The gray filled oval represents the mtRNA polymerase complex.

that have been employed for the mammalian systems, where the rate of replication is slow and where, in retrospect, the mode of replication (unidirectional synthesis from two widely spaced origins) has advantaged these studies.

Another issue is the aggressive rate of recombination in the fungal mtDNA systems, in contrast to mammals, where mtDNA recombination is at most a rare occurrence. Therefore, the problem of distinguishing between replication and recombination events is acute for yeast and essentially nonexistent for mammals. Finally, the small closed circular form of mammalian mtDNAs and a less aggressive background of nucleases have facilitated isolation of intact genomes.

Some parallels may exist between the yeast and mammalian systems. Very recent cloning efforts point to homologs between transcription, replication, and accessory protein sequences. Putative origin sequences in yeast have some conserved elements in common with mammals, and these possible shared features have been reviewed in detail previously (Schmitt and Clayton 1993). However, it is likely that yeast mtDNA can be propagated by multiple mechanisms, at least one of which is a rolling circle (Maleszka et al. 1991).

CONCLUSIONS AND PROSPECTS

The overall mode of mammalian mtDNA replication has been known for some time. During the last decade, the nature of both *cis*- and *trans*-

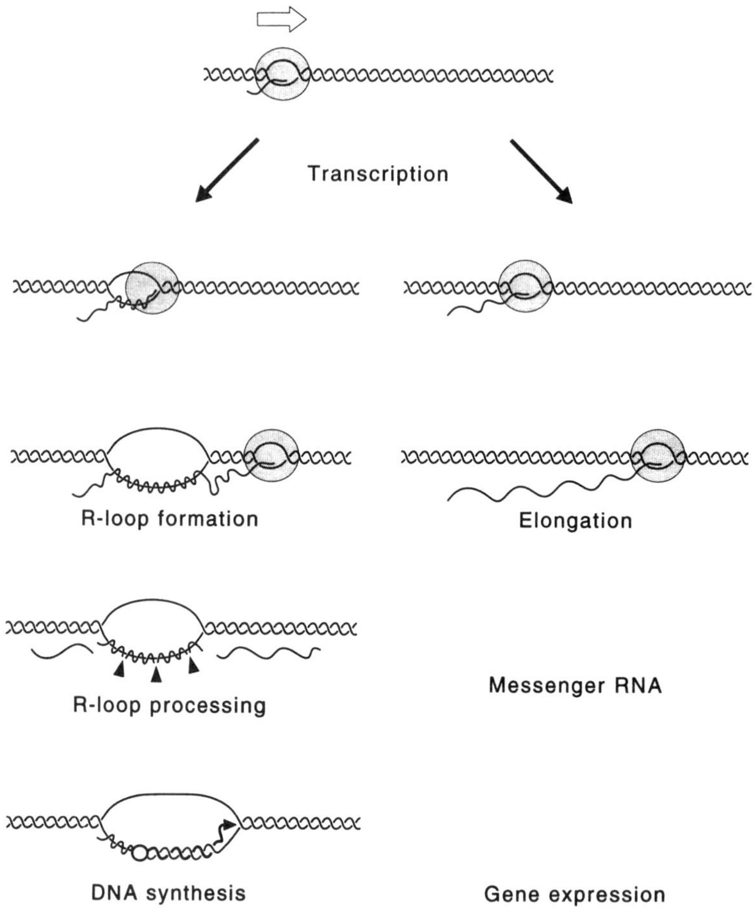

Figure 3 Model for transcript fates at the leading-strand origin. The leftward portion depicts RNA–DNA hybrid formation and subsequent processing of the RNA strand to form primers for replication. The rightward portion represents orthodox transcription, which results in release of the transcript for processing into eight tRNAs and one mRNA. The gray filled circle represents the mtRNA polymerase complex.

acting elements has been revealed for mammalian mtDNA, and the most recent efforts are elucidating the molecular details of nuclear gene-encoded proteins that form the set of activities required for mtDNA replication. Thus, new approaches to understanding mtDNA replication should now be available, and efforts to learn the full spectrum of protein-DNA interactions in the D-loop region should be revealing (see Cantatore et al. 1995 and references therein).

The possibility of developing an in vitro D-loop replication system is now realistic, given that both of the relevant polymerases and key additional activities are now, or soon will be, available as cloned entities. This should facilitate a good understanding of D-loop metabolism and could shed light on possible mechanisms of faulty replication that lead, for example, to deleted forms of mtDNA. Deleted forms of human mtDNA can cause human disease (for review, see Larsson and Clayton 1995), and at present their basis of formation is unknown. A new avenue of research is to study the effect of mutated mtDNA replication proteins in the context of the organism as a whole. Both human and mouse genes for relevant activities will soon be in hand and available for manipulation. It should be possible to introduce variants of mouse mtTFA, mouse polymerases, and other proteins in the animal itself and, thereby, to study altered function with all other components present. Such model systems should also provide the best opportunities to test cellular phenotypic consequences of deficiencies in mtDNA replication and genome maintenance.

ACKNOWLEDGMENT

Research from this laboratory has been supported by grant R37 GM-33088 from the National Institute of General Medical Sciences.

REFERENCES

Cantatore, P., L. Daddabbo, F. Fracasso, and M.N. Gadaleta. 1995. Identification by *in organello* footprinting of protein contact sites and of single-stranded DNA sequences in the regulatory region of rat mitochondrial DNA. Protein binding sites and single-stranded DNA regions in isolated rat liver mitochondria. *J. Biol. Chem.* **270:** 25020–25027.

Chang, D.D. and D.A. Clayton. 1985. Priming of human mitochondrial DNA replication occurs at the light-strand promoter. *Proc. Natl. Acad. Sci.* **82:** 351–355.

———. 1987. A novel endoribonuclease cleaves at a priming site of mouse mitochondrial DNA replication. *EMBO J.* **6:** 409–417.

Chang, D.D., W.W. Hauswirth, and D.A. Clayton. 1985. Replication priming and transcription initiate from precisely the same site in mouse mitochondrial DNA. *EMBO J.* **4:** 1559–1567.

Clayton, D.A. 1982. Replication of animal mitochondrial DNA. *Cell* **28:** 693–705.

———. 1991. Replication and transcription of vertebrate mitochondrial DNA. *Annu. Rev. Cell Biol.* **7:** 453–478.

Dairaghi, D.J., G.S. Shadel, and D.A. Clayton. 1995. Addition of a 29 residue carboxyl-terminal tail converts a simple HMG box-containing protein into a transcriptional activator. *J. Mol. Biol.* **249:** 11–28.

Desjardins, P. and R. Morais. 1990. Sequence and gene organization of the chicken mitochondrial genome. A novel gene order in higher vertebrates. *J. Mol. Biol.* **212:** 599–634.

Dunon-Bluteau, D., A. Cordonnier, and G. Brun. 1987. DNA synthesis in a mitochondrial lysate of *Xenopus laevis* oocytes. H strand replication *in vitro*. *J. Mol. Biol.* **197:** 175–185.

Enríquez, J.A., J. Ramos, A. Pérez-Martos, M.J. López-Pérez, and J. Montoya. 1994. Highly efficient DNA synthesis in isolated mitochondria from rat liver. *Nucleic Acids Res.* **22:** 1861–1865.

Fisher, R.P., J.N. Topper, and D.A. Clayton. 1987. Promoter selection in human mitochondria involves binding of a transcription factor to orientation-independent upstream regulatory elements. *Cell* **50:** 247–258.

Fisher, R.P., T. Lisowsky, M.A. Parisi, and D.A. Clayton. 1992. DNA wrapping and bending by a mitochondrial high mobility group-like transcriptional activator protein. *J. Biol. Chem.* **267:** 3358–3367.

Gillum, A.M. and D.A. Clayton. 1979. Mechanism of mitochondrial DNA replication in mouse L-cells: RNA priming during the initiation of heavy-strand synthesis. *J. Mol. Biol.* **135:** 353–368.

Gray, H. and T.W. Wong. 1992. Purification and identification of subunit structure of the human mitochondrial DNA polymerase. *J. Biol. Chem.* **267:** 5835–5841.

Hixson, J.E., T.W. Wong, and D.A. Clayton. 1986. Both the conserved stem-loop and divergent 5′-flanking sequences are required for initiation at the human mitochondrial origin of light-strand DNA replication. *J. Biol. Chem.* **261:** 2384–2390.

Hsieh, C.-L., T.A. Donlon, B.T. Darras, D.D. Chang, J.N. Topper, D.A. Clayton, and U. Francke. 1990. The gene for the RNA component of the mitochondrial RNA-processing endoribonuclease is located on human chromosome 9p and on mouse chromosome 4. *Genomics* **6:** 540–544.

Insdorf, N.F. and D.F. Bogenhagen. 1989a. DNA polymerase γ from *Xenopus laevis*. I. The identification of a high molecular weight catalytic subunit by a novel DNA polymerase photolabeling procedure. *J. Biol. Chem.* **264:** 21491–21497.

———. 1989b. DNA polymerase γ from *Xenopus laevis*. II. A 3′→5′ exonuclease is tightly associated with the DNA polymerase activity. *J. Biol. Chem.* **264:** 21498–21503.

Kaguni, L.S. and M.W. Olson. 1989. Mismatch-specific 3′→5′ exonuclease associated with the mitochondrial DNA polymerase from *Drosophila* embryos. *Proc. Natl. Acad. Sci.* **86:** 6469–6473.

Kunkel, T.A. and A. Soni. 1988. Exonucleolytic proofreading enhances the fidelity of DNA synthesis by chicken embryo DNA polymerase-γ. *J. Biol. Chem.* **263:** 4450–4459.

Larsson, N.-G. and D.A. Clayton. 1995. Molecular genetic aspects of human mitochondrial disorders. *Annu. Rev. Genet.* **29:** 151–178.

Larsson, N.-G., A. Oldfors, E. Holme, and D.A. Clayton. 1994. Low levels of mitochondrial transcription factor A in mitochondrial DNA depletion. *Biochem. Biophys. Res. Commun.* **200:** 1374–1381.

Lygerou, Z., P. Mitchel, E. Petfalski, B. Séraphin, and D. Tollervey. 1994. The *POP1* gene encodes a protein component common to the RNase MRP and RNase P ribonucleoproteins. *Genes Dev.* **8:** 1423–1433.

Maleszka, R., P.J. Skelly, and G.D. Clark-Walker. 1991. Rolling circle replication of DNA in yeast mitochondria. *EMBO J.* **10:** 3923–3929.

Mignotte, B., J. Marsault, and M. Barat-Gueride. 1988. Effects of the *Xenopus laevis* mitochondrial single-stranded DNA-binding protein on the activity of DNA polymerase γ. *Eur. J. Biochem.* **174:** 479–484.

Paluh, J. and D.A. Clayton. 1995. *Schizosaccharomyces pombe* RNase MRP RNA is homologous to metazoan RNase MRP RNAs and may provide clues to interrelationships between RNase MRP and RNase P. *Yeast* **11:** 1249–1264.

Pavco, P.A. and G.C. Van Tuyle. 1985. Purification and general properties of the DNA-binding protein (P16) from rat liver mitochondria. *J. Cell Biol.* **100:** 258–264.

Poulton, J., K. Morten, C. Freeman-Emmerson, C. Potter, C. Sewry, V. Dubowitz, H. Kidd, J. Stephenson, W. Whitehouse, F.J. Hansen, M. Parisi, and G. Brown. 1994. Deficiency of the human mitochondrial transcription factor h-mtTFA in infantile mitochondrial myopathy is associated with mtDNA depletion. *Hum. Mol. Genet.* **3:** 1763–1769.

Schmitt, M.E. and D.A. Clayton. 1993. Conserved features of yeast and mammalian mitochondrial DNA replication. *Curr. Opin. Genet. Dev.* **3:** 769–774.

———. 1994. Characterization of a unique protein component of yeast RNase MRP: An RNA-binding protein with a zinc-cluster domain. *Genes Dev.* **8:** 2617–2628.

Schmitt, M.E., J.L. Bennett, D.J. Dairaghi, and D.A. Clayton. 1993. Secondary structure of RNase MRP RNA as predicted by phylogenetic comparison. *FASEB J.* **7:** 208–213.

Tiranti, V., M. Rocchi, S. DiDonato, and M. Zeviani. 1993. Cloning of human and rat cDNAs encoding the mitochondrial single-stranded DNA-binding protein (SSB). *Gene* **126:** 219–225.

Van Tuyle, G.C. and P.A. Pavco. 1985. The rat liver mitochondrial DNA-protein complex: Displaced single strands of replicative intermediates are protein coated. *J. Cell Biol.* **100:** 251–257.

Wernette, C.M. and L.S. Kaguni. 1986. A mitochondrial DNA polymerase from embryos of *Drosophila melanogaster*. Purification, subunit structure, and partial characterization. *J. Biol. Chem.* **261:** 14764–14770.

Williams, A.J. and L.S. Kaguni, 1995. Stimulation of *Drosophila* mitochondrial DNA polymerase by single-stranded DNA-binding protein. *J. Biol. Chem.* **270:** 860–865.

Wong, T.W. and D.A. Clayton. 1985a. Isolation and characterization of a DNA primase from human mitochondria. *J. Biol. Chem.* **260:** 11530–11535.

———. 1985b. In vitro replication of human mitochondrial DNA: Accurate initiation at the origin of light-strand synthesis. *Cell* **42:** 951–958.

Xu, B. and D.A. Clayton. 1995. A persistent RNA-DNA hybrid is formed during transcription at a phylogenetically conserved mitochondrial DNA sequence. *Mol. Cell. Biol.* **15:** 580–589.

———. 1996. RNA-DNA hybrid formation at the human mitochondrial heavy-strand origin ceases at replication start sites: An implication for RNA-DNA hybrids serving as primers. *EMBO J.* **15:** (in press).

40

Kinetoplast DNA Replication

Al F. Torri, Laura Rocco Carpenter, and Paul T. Englund
Department of Biological Chemistry
Johns Hopkins University School of Medicine
Baltimore, Maryland 21205

The order Kinetoplastida is composed of several families of flagellated protozoa which are among the most ancient of the eukaryotes. The most prominent members of this order are parasites that are responsible for major tropical diseases. These parasites include *Trypanosoma brucei*, the African trypanosome; *Trypanosoma cruzi*, the South American trypanosome; and several species of *Leishmania*. Other kinetoplastids include *Crithidia fasciculata*, an insect parasite, and species of *Phytomonas*, which are parasites of plants. All of the kinetoplastids possess a distinctive disk-shaped structure, known as the kinetoplast, within their single mitochondrion. The kinetoplast is composed of the cell's mitochondrial DNA, known as kinetoplast DNA (kDNA). kDNA has a structure unlike that of any other DNA in nature. It consists of two types of circular DNA molecules, maxicircles and minicircles, which are topologically interlocked into a single, massive network. Figure 1 shows an electron micrograph of part of a kDNA network from *C. fasciculata* (for reviews on kDNA, see Ray 1987; Simpson 1987; Stuart and Feagin 1992; Shapiro and Englund 1995).

There is substantial information on the genetic function of maxicircles and minicircles. Maxicircles resemble conventional mitochondrial DNA from mammals or yeast, in that they encode rRNAs and proteins involved in mitochondrial energy transduction (such as cytochrome oxidase subunits). A unique feature of maxicircles is that many of their transcripts undergo RNA editing, a process by which uridine residues are inserted or deleted at specific sites (for review, see Hajduk et al. 1993). In some cases, editing occurs on a massive scale, with half of the nucleotides in the coding region of the mRNA being uridines introduced by editing. Minicircles play a crucial role in RNA editing, encoding small guide RNAs that control editing specificity.

In this review, we discuss kDNA structure and its replication mechanism, focusing primarily on the organism *C. fasciculata*. Currently, there is no system to perform kDNA replication in vitro and only a few

DNA Replication in Eukaryotic Cells

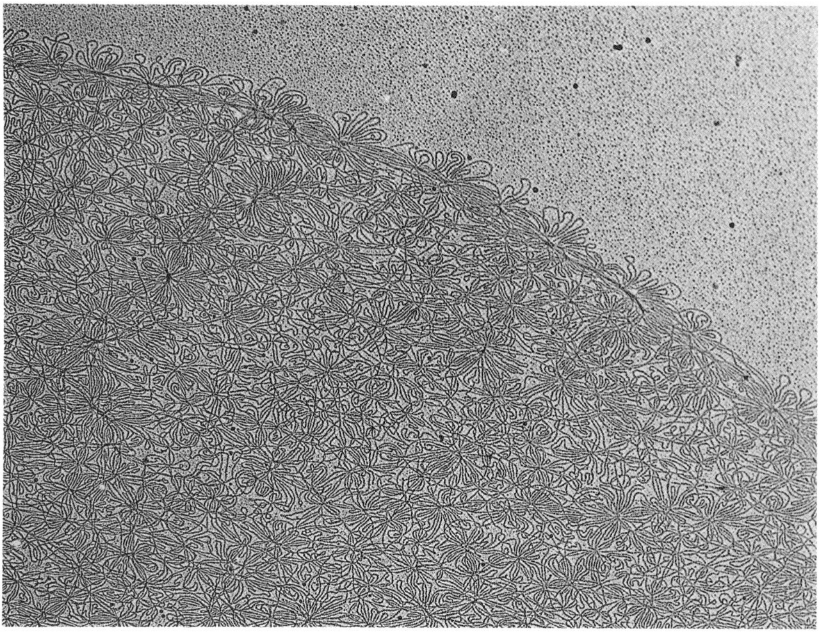

Figure 1 Part of a *C. fasciculata* kDNA network, shown by electron microscopy. Each small loop represents a 2.5-kb minicircle.

proteins involved in replication have been studied. Nevertheless, there is considerable information about the novel mechanism by which a DNA network undergoes replication, and this constitutes the major focus of this chapter (for reviews on kDNA replication, see Ray 1987; Shlomai 1994; Shapiro and Englund 1995).

STRUCTURE OF kDNA

A *C. fasciculata* network has about 25 maxicircles (38 kb) and 5000 minicircles (2.5 kb). Electron microscopy of isolated kDNA reveals that the network is a planar structure, about 10 μm by 15 μm in size, with an elliptical shape (Pérez-Morga and Englund 1993a). The minicircles are arranged in a monolayer, with a structure resembling that of chain mail (see diagram in Fig. 2A). The maxicircles are also catenated to the network, although it is not clear how they are organized. Treatment of the network with a topoisomerase II results in complete decatenation of the minicircles and maxicircles (Marini et al. 1980). In networks not undergoing replication, all of the minicircles are covalently closed (Englund 1978). However, unlike DNA circles in other cells, the mini-

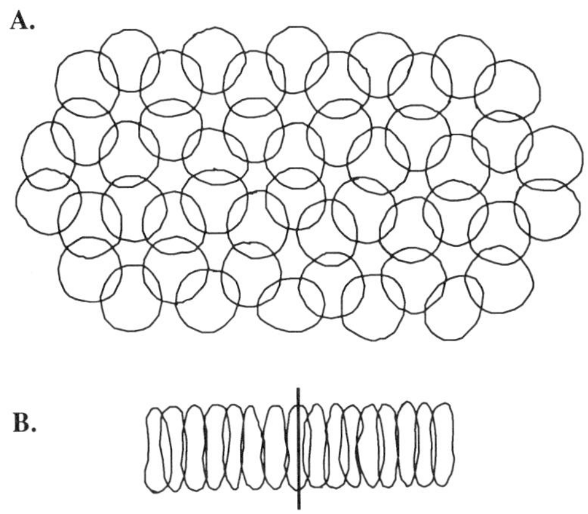

Figure 2 (*A*) Diagram of a segment of an isolated *C. fasciculata* kDNA network, in which each minicircle lies flat in a plane. (*B*) Diagram of a section through the kinetoplast disk in vivo (Delain and Riou 1969). Instead of lying in a plane as in *A*, each minicircle stands perpendicular to the plane. The vertical line shows the disk axis. (Reprinted, with permission, from Shapiro and Englund 1995.)

circles are relaxed rather than negatively supercoiled (Rauch et al. 1993). Minicircles in a *C. fasciculata* network not undergoing replication are each linked to an average of three neighbors (Chen et al. 1995). Unlike some other kinetoplastid parasites, which have a network with many different minicircle sequence classes, *C. fasciculata* networks have a single predominant minicircle class and only several minor classes (Sugisaki and Ray 1987).

Electron micrographs of thin sections and confocal fluorescence microscopy (using an acridine stain) indicate that in vivo the *C. fasciculata* network is condensed into a disk about 1 μm in diameter and 0.4 μm thick (Ferguson et al. 1992). Figure 2B shows a model for the organization of the network inside the mitochondrial matrix. The network in vivo is a monolayer, but in contrast to the isolated networks, in which minicircles lie in the plane of the network, the minicircles in vivo stand erect, perpendicular to the plane of the network. The minicircles stretch out and the strands run parallel to the disk's axis. The thickness of the kinetoplast disk is roughly half the circumference of an individual minicircle.

REPLICATION OF kDNA

During each cell cycle, every minicircle and maxicircle undergoes a single round of replication. In contrast to mitochondrial DNA in mammalian cells, which replicates throughout the entire cell cycle (Clayton 1982), kDNA replicates only during a discrete S phase, approximately coincident with the S phase of nuclear DNA (Cosgrove and Skeen 1970).

Network Replication

Problems in Replicating a DNA Network

One unique problem in replicating kDNA is topological. How do you replicate a minicircle that is interlocked to several neighbors? The kinetoplastid solves this problem by unlinking the minicircles from the network in a topoisomerase-mediated reaction (Englund 1979). The free minicircles can then undergo replication by a standard mechanism (discussed below), unhindered by the constraints of catenation. The network is never completely decatenated; instead there are several hundred free minicircles at any time during the S phase that are undergoing replication. After minicircle replication, the progeny are attached, in another topoisomerase reaction, to the network periphery. Another problem associated with kDNA replication is bookkeeping. How does the kinetoplastid keep track of which minicircle has replicated and which has not? The parasite solves this problem by maintaining nicks or gaps in minicircles after their replication (in this chapter we refer to these interruptions as "nicks") (Englund 1978). These distinguish minicircles that have replicated from those covalently closed molecules that have not. Covalent closure of the minicircles occurs late in the cell cycle, after all have undergone replication.

Complexes of Replication Proteins

Once minicircles are released from the network, presumably from random sites in the network's interior, they migrate to one of two protein complexes that are situated on opposite sides of the kinetoplast disk (see Fig. 3A). These complexes contain a topoisomerase II (Melendy et al. 1988), a DNA polymerase-β, and minicircles that are probably replication intermediates (Ferguson et al. 1992). They probably also contain the other enzymatic machinery needed for minicircle replication. Since the polymerase-β is likely involved in repair of gaps in newly replicated minicircles (see below), a replicative polymerase is probably also present in these structures. After replication, the nicked progeny minicircles are attached to the network adjacent to these complexes, as shown in Figure 3B.

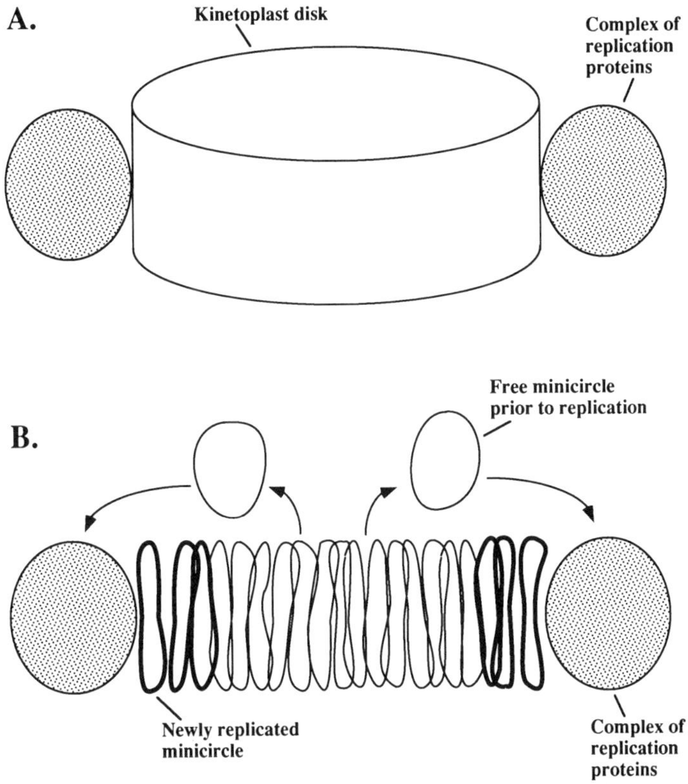

Figure 3 (*A*) Diagram of the kinetoplast disk, showing the two antipodal complexes of replication proteins. (*B*) Model for kDNA replication in vivo. The diagram shows a section through the kinetoplast disk. Minicircles in bold have undergone replication and are nicked. See text for details. (Reprinted, with permission, from Shapiro and Englund 1995.)

Structure of Networks Undergoing Replication

Visualization of partly replicated *C. fasciculata* networks either by electron microscopy after isolation (Pérez-Morga and Englund 1993a) or by fluorescence in situ hybridization (Ferguson et al. 1992) reveals that newly replicated minicircles, containing nicks, are uniformly distributed around the network periphery. Figure 4 shows networks at different stages of replication that had been labeled in vitro at their endogenous nicks with [^{3}H]dTTP and DNA polymerase. The labeled DNA was spread on a grid, coated with photographic emulsion, and, after exposure and development, examined by electron microscope autoradiography (Pérez-Morga and Englund 1993a). The silver grains mark the location

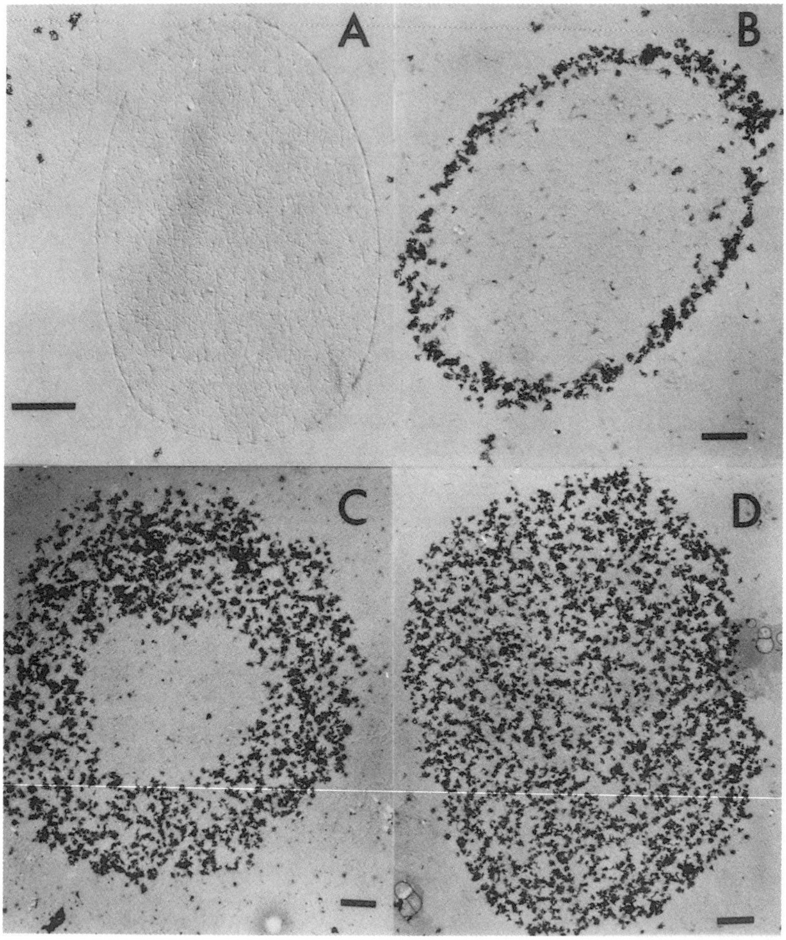

Figure 4 Electron micrographs of isolated kDNA networks at different stages of replication. The networks were radiolabeled with [^{3}H]dTTP at endogenous nicks using DNA polymerase and then subjected to autoradiography. The silver grains indicate the locations of nicked minicircles. (*A*) Prereplication network. (*B*) Network in an early stage of replication. (*C*) Network in a later stage of replication. (*D*) Fully replicated network with all minicircles nicked. Bars, 2 μm. (Reprinted, with permission, from Pérez-Morga and Englund 1993a [copyright Rockefeller University Press].)

of nicked minicircles. Panel A shows a prereplication network containing 5000 minicircles; there are no silver grains because the minicircles are all covalently closed. Panel B shows a network in an early stage of replication. The silver grains are distributed in a narrow and uniform ring around the network's periphery. Panel C shows a later stage, resembling a donut, with a thicker ring and a smaller unlabeled region in the center (the donut hole). Finally, in Panel D, the network is uniformly labeled. At this stage the network contains 10,000 minicircles, double the number that were present in the prereplication network, and all are nicked. The further processing of this double-size network is discussed below.

The Spinning Kinetoplast Model

The uniform distribution of nicked minicircles around the network periphery presented a dilemma. If minicircles are attached to the network adjacent to the two protein complexes, how do they become distributed around the network periphery? This question has been addressed by autoradiography of *C. fasciculata* networks labeled in vivo with pulses of [^{3}H]thymidine (Pérez-Morga and Englund 1993b). In cells labeled for 1–3 minutes, silver grains were concentrated in two peripheral positions, on opposite sides of the network (Simpson and Simpson 1976; Pérez-Morga and Englund 1993b), exactly as expected if minicircle attachment occurred adjacent to the two protein complexes (see example in Fig. 5A). In cells labeled for longer periods (20–60 min), the silver grains were distributed around the network periphery, forming donut-shaped structures (not shown). A clue to the mechanism by which minicircles are distributed around the network came from inspection of networks labeled for 6 minutes (Pérez-Morga and Englund 1993b). As shown in Figure 5B, the silver grains were concentrated in two zones, but together they nearly filled the entire periphery. Surprisingly, the silver grains in both zones form a density gradient. This gradient is due to the rising specific radioactivity of the dTTP during the 6-minute period of the labeling, as the exogenous [^{3}H]thymidine mixes with the internal pool of DNA precursors. The presence of two labeling zones is consistent with the labeled minicircles being attached at two sites. The fact that the silver grains have approximately the same density on opposite sides of the network indicates that the two minicircle attachment sites are always positioned opposite each other. The gradient of silver grains indicates that minicircles are attached sequentially around the network periphery. This result led to the surprising conclusion that there is a relative movement of the kinetoplast disk and the two protein complexes. The kinetoplast disk could be fixed, and the complexes could move around its periphery. Alternatively, the complexes could be fixed, and the kinetoplast disk could rotate between them (see model in Fig. 6). Although much more information is needed before either model can be

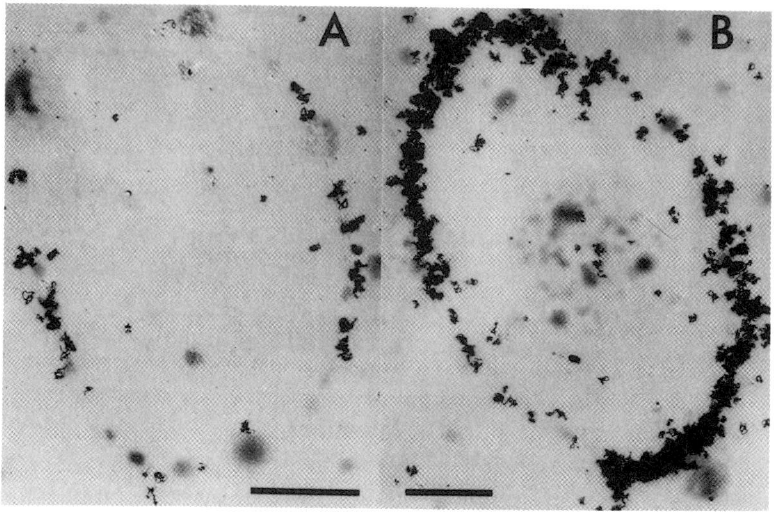

Figure 5 Electron micrographs of kDNA networks labeled in vivo with [^{3}H]thymidine. (*A*) 3-min label. (*B*) 6-min label. After isolation the networks were subjected to autoradiography. Bars, 3 μm. (Reprinted, with permission, from Pérez-Morga and Englund 1993b [copyright Cell Press].)

proven, some preliminary findings favor the latter possibility (Pérez-Morga and Englund 1993b). We assume that the spinning kinetoplast model is correct for the rest of this discussion. Since nearly the entire network periphery is labeled in 6 minutes, the kinetoplast must rotate approximately once every 12 minutes. Presumably, it rotates continuously in the same direction and at the same rate, allowing minicircle attachment in a spiral pattern. It probably takes about seven turns of the kinetoplast disk to replicate the entire network.

The Final Stages of Network Replication

At the end of S phase the network contains 10,000 minicircles, twice the number prior to replication, and all are nicked. At this time the nicks are repaired, and all the minicircles become covalently closed (Pérez-Morga and Englund 1993a). Then the structure splits in two. This scission is probably mediated by a topoisomerase that unlinks neighboring circles between the two nascent progeny networks, but how this reaction is controlled is completely unknown. The progeny networks, each identical to the parent, then segregate into the two daughter cells.

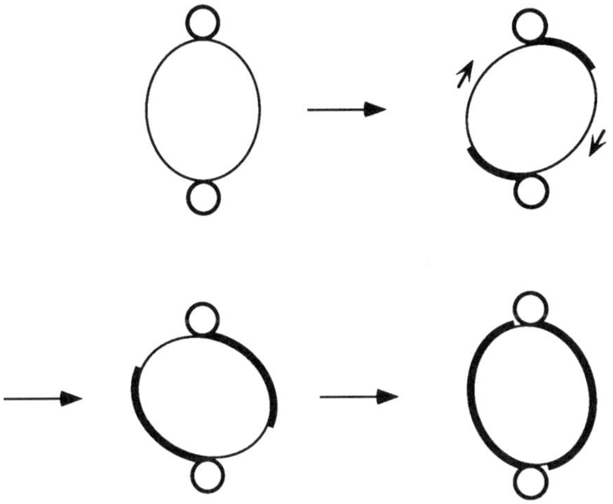

Figure 6 Model showing the rotation of the *C. fasciculata* kinetoplast in vivo. The kinetoplast is depicted as an ellipse and the two complexes of replication proteins as small circles. The bold lines represent a row of newly replicated minicircles. Small arrows indicate the direction of rotation of the kinetoplast. See text for discussion. (Reprinted, with permission, from Pérez-Morga and Englund 1993b [copyright Cell Press].)

Replication Mechanism for Minicircles and Maxicircles

Minicircles

Minicircle replication has been studied extensively in *C. fasciculata* and *Trypanosoma equiperdum*. Replication begins after the covalently closed minicircle is decatenated from the network, and it is thought to occur in one of the two protein complexes (discussed above). Minicircle replication occurs via θ intermediates and is unidirectional (Englund 1979; Ryan and Englund 1989b). Leading-strand synthesis initiates by RNA priming complementary to the sequence GGGGTTGGTGTA, a sequence found with only minor variation in a conserved region of minicircles from all kinetoplastids that have been examined (Ntambi et al. 1986; Birkenmeyer et al. 1987). *C. fasciculata* has two such sequences, positioned 180° apart, and either can serve as the site for the initiation of replication (Birkenmeyer et al. 1987). A second sequence, usually ACGCCC, is located 70–90 nucleotides downstream from the origin; the first Okazaki fragment initiates complementary to this 6-mer (Ryan and Englund 1989a). In both *C. fasciculata* and *T. equiperdum*, the leading strand is synthesized continuously around the molecule and the lagging-strand Okazaki fragments are roughly 100 nucleotides in size (Kitchin et al. 1984; Birkenmeyer and Ray 1986; Birkenmeyer et al. 1987; Ryan and Englund 1989a). Nothing is known about the priming mechanism either for the leading strand or for the Okazaki fragments, and, in fact, RNA primers have not yet been detected on the Okazaki fragments. As mentioned above, covalent closure of the progeny minicircles occurs only after all minicircles have replicated. A unique gap opposite the GGGGTTGGTGTA sequence in the continuously synthesized strand and discontinuities flanking the first Okazaki fragment are the last interruptions to be repaired (Birkenmeyer et al. 1987; Ryan and Englund 1989a,b).

Maxicircles

Recent studies, involving electron microscopy of *C. fasciculata* maxicircles released from networks by restriction enzyme cleavage, revealed branched molecules derived from replication intermediates. Analysis of these molecules indicated that maxicircle replication initiates at a unique site and proceeds unidirectionally as a θ-structure (Rocco Carpenter and Englund 1995). The maxicircle replication origin is within the "variable region," a noncoding segment of the molecule that contains many repetitive sequences. Maxicircles and minicircles replicate simultaneously. However, in contrast to minicircles, the maxicircles replicate while still linked to the network (Hajduk et al. 1984).

Enzymes and Proteins Involved in kDNA Replication

Little is known about the enzymes and proteins involved in the replication of kDNA, and the lack of an in vitro replication system has complicated the search for these proteins. In this section we describe a few well-characterized proteins implicated in kDNA replication.

DNA Polymerase-β

An unusual DNA polymerase has been purified from a *C. fasciculata* mitochondrial fraction (Torri and Englund 1992) and shown by immunofluorescence to localize to the two protein complexes adjacent to the kinetoplast (Ferguson et al. 1992). On the basis of the recovery of activity during purification, this enzyme is present in roughly 60,000 copies per cell. This polymerase is small (42 kD), nonprocessive, deficient in proofreading exonuclease activity, and error-prone (Torri et al. 1994). It prefers a DNA substrate with small gaps. These properties differ from those expected for a conventional mitochondrial DNA polymerase-γ and are instead typical of a DNA polymerase-β (pol-β). Pol-β has previously been found only in the nucleus of eukaryotic cells and is thought to be involved in DNA repair. Analysis of the *C. fasciculata* mitochondrial DNA polymerase sequence reveals similarities to human pol-β; however, the mitochondrial enzyme possesses a cleaved amino-terminal pre-sequence that closely resembles mitochondrial import signals on other *C. fasciculata* proteins (Torri and Englund 1995). Therefore, this *C. fasciculata* enzyme is the first example of a mitochondrial pol-β. This enzyme may have been specially imported into the mitochondrion to assist in repairing the many small gaps between Okazaki fragments in newly replicated minicircles. The fact that this is a repair polymerase implies that the true replicative enzyme, which may resemble a polymerase-γ, remains to be discovered.

Topoisomerases

Two type II topoisomerases have been identified in the mitochondrion of *C. fasciculata*. One of these enzymes, containing four 60-kD subunits (Shlomai and Zadok 1983; Shlomai et al. 1984), appears to be localized within the kinetoplast disk (Shlomai 1994). From its intracellular location, this enzyme could be involved in releasing covalently closed minicircles from the network so that they can be replicated. The other topoisomerase, a dimer of 132-kD subunits (Melendy and Ray 1989), localizes to the two protein complexes (Melendy et al. 1988). This topoisomerase II most likely facilitates minicircle replication, and it could be responsible for attaching the newly replicated nicked minicircles onto the network periphery.

Origin-binding Protein

A *C. fasciculata* single-strand-specific DNA-binding protein interacts specifically with the GGGGTTGGTGTA sequence that is part of the minicircle origin of replication (see above) (Tzfati et al. 1992). It is a homodimeric protein, with 13.5-kD subunits, and its sequence predicts five CCHC zinc fingers (Abeliovich et al. 1993). This protein may be involved in replication initiation, binding to the origin after strand separation. In addition, it could function after the minicircle has completed replication, binding to and preserving the unique gap in the newly synthesized leading strand. In this way it would prevent covalent closure and ensure that the minicircle replicated only once per generation. Removal of this protein after all minicircles have replicated would then allow gap repair and covalent closure. The localization of this protein within the kinetoplast has not yet been reported.

Other kDNA-binding Proteins

Although they are probably not involved directly in kDNA replication, a family of highly basic proteins that could be involved in condensing and organizing kDNA in vivo has recently been reported. These proteins, 15 kD to 21 kD in size, were isolated by reversible cross-linking to *C. fasciculata* kDNA in vivo using formaldehyde (Xu and Ray 1993). One of these proteins, p16, has been shown by immunoelectron microscopy to localize within the kinetoplast disk (D. Ray, pers. comm.).

CONCLUSION

Although many features of the replication of the kDNA network are now understood, there are important issues that are still unexplored. One concerns the spinning kinetoplast. Does the kinetoplast really rotate, and if it does, what mechanisms are responsible? A second issue concerns the nature of the two complexes of replication proteins. Are all proteins required for minicircle replication present in these complexes, and how are they organized? Are the complexes permanent fixtures of the cell or do they assemble only during the S phase? Can we isolate them intact? These questions provide exciting opportunities for future investigation.

ACKNOWLEDGMENTS

Work in the authors' laboratory was supported by grants from the National Institutes of Health (GM-27608) and the MacArthur Foundation.

REFERENCES

Abeliovich, H., Y. Tzfati, and J. Shlomai. 1993. A trypanosomal CCHC-type zinc finger protein which binds the conserved universal sequence of kinetoplast DNA minicircles: Isolation and analysis of the complete cDNA from *Crithidia fasciculata*. *Mol. Cell. Biol.* **13:** 7766–7773.

Birkenmeyer, L. and D.S. Ray. 1986. Replication of kinetoplast DNA in isolated kinetoplasts from *Crithidia fasciculata*. Identification of minicircle DNA replication intermediates. *J. Biol. Chem.* **261:** 2362–2368.

Birkenmeyer, L., H. Sugisaki, and D.S. Ray. 1987. Structural characterization of site-specific discontinuities associated with replication origins of minicircle DNA from *Crithidia fasciculata*. *J. Biol. Chem.* **262:** 2384–2392.

Chen, J., C.A. Rauch, J.H. White, P.T. Englund, and N.R. Cozzarelli. 1995. The topology of the kinetoplast DNA network. *Cell* **80:** 61–69.

Clayton, D.A. 1982. Replication of animal mitochondrial DNA. *Cell* **28:** 693–705.

Cosgrove, W.B. and M.J. Skeen. 1970. The cell cycle in *Crithidia fasciculata*. Temporal relationships between synthesis of deoxyribonucleic acid in the nucleus and in the kinetoplast. *J. Protozool.* **17:** 172–177.

Delain, E. and G. Riou. 1969. Ultrastructure du DNA du kinétoplaste de *Trypanosoma cruzi* cultivé in vitro. *C. R. Acad. Sci.* **268:** 1225–1227.

Englund, P.T. 1978. The replication of kinetoplast DNA networks in *Crithidia fasiculata. Cell* **14:** 157–168.

———. 1979. Free minicircles of kinetoplast DNA in *Crithidia fasciculata. J. Biol. Chem.* **254:** 4895–4900.

Ferguson, M., A.F. Torri, D.C. Ward, and P.T. Englund. 1992. In situ hybridization to the *Crithidia fasciculata* kinetoplast reveals two antipodal sites involved in kinetoplast DNA replication. *Cell* **70:** 621–629.

Hajduk, S.L., M.E. Harris, and V.W. Pollard. 1993. RNA editing in kinetoplastid mitochondria. *FASEB J.* **7:** 54–63.

Hajduk, S.L., V.A. Klein, and P.T. Englund. 1984. Replication of kinetoplast DNA maxicircles. *Cell* **36:** 483–492.

Kitchin, P.A., V.A. Klein, B.I. Fein, and P.T. Englund. 1984. Gapped minicircles. A novel replication intermediate of kinetoplast DNA. *J. Biol. Chem.* **259:** 15532–15539.

Marini, J.C., K.G. Miller, and P.T. Englund. 1980. Decatenation of kinetoplast DNA by topoisomerases. *J. Biol. Chem.* **255:** 4976–4979.

Melendy, T. and D.S. Ray. 1989. Novobiocin affinity purification of a mitochondrial type II topoisomerase from the trypanosomatid *Crithidia fasciculata. J. Biol. Chem.* **264:** 1870–1876.

Melendy, T., C. Sheline, and D.S. Ray. 1988. Localization of a type II DNA topoisomerase to two sites at the periphery of the kinetoplast DNA of *Crithidia fasciculata. Cell* **55:** 1083–1088.

Ntambi, J.M., T.A. Shapiro, K.A. Ryan, and P.T. Englund. 1986. Ribonucleotides associated with a gap in newly replicated kinetoplast DNA minicircles from *Trypanosoma equiperdum. J. Biol. Chem.* **261:** 11890–11895.

Pérez-Morga, D. and P.T. Englund. 1993a. The structure of replicating kinetoplast DNA networks. *J. Cell Biol.* **123:** 1069–1079.

———. 1993b. The attachment of minicircles to kinetoplast DNA networks during replication. *Cell* **74:** 703–711.

Rauch, C.A., D. Pérez-Morga, N.R. Cozzarelli, and P.T. Englund. 1993. The absence of supercoiling in kinetoplast DNA minicircles. *EMBO J.* **12:** 403–411.

Ray, D.S. 1987. Kinetoplast DNA minicircles: High-copy-number mitochondrial plasmids. *Plasmid* **17:** 177–190.

Rocco Carpenter, L. and P.T. Englund. 1995. Kinetoplast maxicircle DNA replication in *Crithidia fasciculata* and *Trypanosoma brucei. Mol. Cell. Biol.* **15:** 6794–6803.

Ryan, K.A. and P.T. Englund. 1989a. Replication of kinetoplast DNA in *Trypanosoma equiperdum.* Minicircle H strand fragments which map at specific locations. *J. Biol. Chem.* **264:** 823–830.

———. 1989b. Synthesis and processing of kinetoplast DNA minicircles in *Trypanosoma equiperdum. Mol. Cell. Biol.* **9:** 3212–3217.

Shapiro, T.A. and P.T. Englund. 1995. The structure and replication of kinetoplast DNA. *Annu. Rev. Microbiol.* **49:** 117–143.

Shlomai, J. 1994. The assembly of kinetoplast DNA. *Parasitol. Today* **10:** 341–346.

Shlomai, J. and A. Zadok. 1983. Reversible decatenation of kinetoplast DNA by a DNA topoisomerase from trypanosomatids. *Nucleic Acids Res.* **11:** 4019–4034.

Shlomai, J., A. Zadok, and D. Frank. 1984. A unique ATP-dependent DNA topoisomerase from trypanosomatids. *Adv. Exp. Med. Biol.* **179:** 409–422.

Simpson, A.M. and L. Simpson. 1976. Pulse-labeling of kinetoplast DNA: Localization of 2 sites of synthesis within the networks and kinetics of labeling of closed minicircles. *J. Protozool.* **23:** 583–587.

Simpson, L. 1987. The mitochondrial genome of kinetoplastid protozoa: Genomic organization, transcription, replication, and evolution. *Annu. Rev. Microbiol.* **41:** 363–382.

Stuart, K. and J.E. Feagin. 1992. Mitochondrial DNA of kinetoplastids. *Int. Rev. Cytol.* **141:** 65–88.

Sugisaki, H. and D.S. Ray. 1987. DNA sequence of *Crithidia fasciculata* kinetoplast minicircles. *Mol. Biochem. Parasitol.* **23:** 253–263.

Torri, A.F. and P.T. Englund. 1992. Purification of a mitochondrial DNA polymerase from *Crithidia fasciculata*. *J. Biol. Chem.* **267:** 4786–4792.

———. 1995. A DNA polymerase β in the mitochondrion of the trypanosomatid *Crithidia fasciculata*. *J. Biol. Chem.* **270:** 3495–3497.

Torri, A.F., T.A. Kunkel, and P.T. Englund. 1994. A β-like DNA polymerase from the mitochondrion of the trypanosomatid *Crithidia fasciculata*. *J. Biol. Chem.* **269:** 8165–8171.

Tzfati, Y., H. Abeliovich, I. Kapeller, and J. Shlomai. 1992. A single-stranded DNA-binding protein from *Crithidia fasciculata* recognizes the nucleotide sequence at the origin of replication of kinetoplast DNA minicircles. *Proc. Natl. Acad. Sci.* **89:** 6891–6895.

Xu, C. and D.S. Ray. 1993. Isolation of proteins associated with kinetoplast DNA networks in vivo. *Proc. Natl. Acad. Sci.* **90:** 1786–1789.

Index